HARCOURT
Math

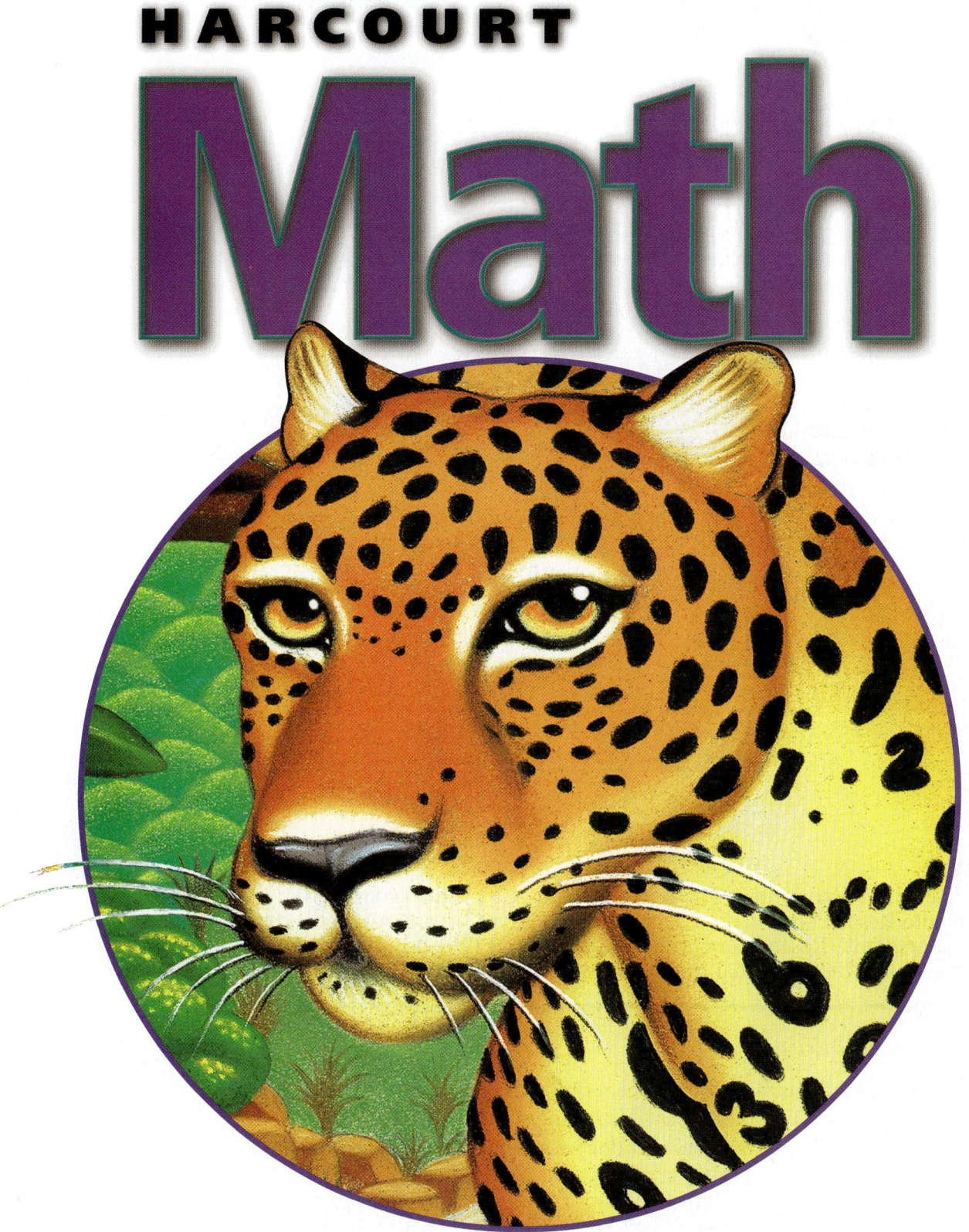

TEACHER'S EDITION

Harcourt School Publishers

Orlando • Boston • Dallas • Chicago • San Diego
www.harcourtschool.com

GRADE **6**

VOLUME 1

Photo Credits:

Unit Overview Pages:

Unit 1 David Parker/Science Photo Library/Photo Researchers; Unit 2 Ed Young/Science Photo Library/Photo Researchers; Unit 4 Sepp Seitz/Woodfin Camp & Associates; Unit 6 Charles D. Winters/Photo Researchers; Unit 7 Robert Frerck/Woodfin Camp & Associates; Unit 8 Harcourt; Unit 9 Michal Heron/Woodfin Camp & Associates.

Copyright © 2002 by Harcourt, Inc.

All rights reserved. No part of this publication may be reproduced or transmitted in any form or by any means, electronic or mechanical, including photocopy, recording, or any information storage and retrieval system, without permission in writing from the publisher.

Requests for permission to make copies of any part of the work should be mailed to the following address:

School Permissions and Copyrights, Harcourt, Inc.
6277 Sea Harbor Drive
Orlando, Florida 32887-6777

HARCOURT and the Harcourt Logo are trademarks of Harcourt, Inc.

Printed in the United States of America

ISBN 0-15-320762-0

3 4 5 6 7 8 9 10   048   10 09 08 07 06 05 04 03 02

Teacher's Edition Contents

Volume 1

About *Harcourt Math*	T5
Authors and Mathematics Advisors	T6
Program Consultants and Reviewers	T8
Harcourt Math: The Solution for Success	T10
A System for Mathematics Instruction	T12
Reaching All Learners	T14
Intervention	T15
Assessment	T16
Family Support	T18
Using Technology	T19
Components Chart	T20

Pupil Edition

UNIT 1 Number Sense and Operations
Chapters 1-4

Chapter 1	Whole Number Applications	14
Chapter 2	Operation Sense	34
Chapter 3	Decimal Concepts	50
Chapter 4	Decimal Operations	64
	Math Detective, Challenge	86
	Study Guide and Review	88
	Performance Assessment, Technology Linkup	90
	Problem Solving on Location	91A

UNIT 2 Statistics and Graphing
Chapters 5-6

Chapter 5	Collect and Organize Data	92
Chapter 6	Graph Data	118
	Math Detective, Challenge	138
	Study Guide and Review	140
	Performance Assessment, Technology Linkup	142
	Problem Solving on Location	143A

UNIT 3 Fraction Concepts and Operations
Chapters 7-10

Chapter 7	Number Theory	144
Chapter 8	Fraction Concepts	158
Chapter 9	Add and Subtract Fractions and Mixed Numbers	174
Chapter 10	Multiply and Divide Fractions and Mixed Numbers	198
	Math Detective, Challenge	220
	Study Guide and Review	222
	Performance Assessment, Technology Linkup	224
	Problem Solving on Location	225A

UNIT 4 Algebra: Integers
Chapters 11-12

Chapter 11	Number Relationships	226
Chapter 12	Operations with Integers	240
	Math Detective, Challenge	262
	Study Guide and Review	264
	Performance Assessment, Technology Linkup	266
	Problem Solving on Location	267A

Student Handbook	H1
Problem of the Day Solutions	PD1
Scope and Sequence	SC1
Correlation to Standardized Tests	SC29
Manipulatives Chart	SC32
Research Articles	RS1
Professional Handbook	PH1
Bibliography for Students	BI1
Computer Software Bibliography	BI3
Professional Bibliography	BI7
Index	BI14
NCTM Standards	NA1

Volume 2

About *Harcourt Math* T5
Authors and Mathematics Advisors T6
Program Consultants and Reviewers T8
Harcourt Math: The Solution for Success T10
A System for Mathematics Instruction T12
Reaching All Learners T14
Intervention T15
Assessment T16
Family Support T18
Using Technology T19
Components Chart T20

Pupil Edition

UNIT 5 Algebra: Expressions and Equations Chapters 13–15

Chapter 13	Expressions 268
Chapter 14	Addition and Subtraction Equations 282
Chapter 15	Multiplication and Division Equations 294
	Math Detective, Challenge 310
	Study Guide and Review 312
	Performance Assessment, Technology Linkup 314
	Problem Solving on Location ... 315A

UNIT 6 Geometry and Spatial Reasoning Chapters 16–19

Chapter 16	Geometric Figures 316
Chapter 17	Plane Figures 330
Chapter 18	Solid Figures 348
Chapter 19	Congruence and Similarity 362
	Math Detective, Challenge 376
	Study Guide and Review 378
	Performance Assessment, Technology Linkup 380
	Problem Solving on Location ... 381A

UNIT 7 Ratio, Proportion, Percent, and Probability Chapters 20–23

Chapter 20	Ratio and Proportion 382
Chapter 21	Percent and Change 404
Chapter 22	Probability of Simple Events ... 426
Chapter 23	Probability of Compound Events . 440
	Math Detective, Challenge 454
	Study Guide and Review 456
	Performance Assessment, Technology Linkup 458
	Problem Solving on Location ... 459A

UNIT 8 Measurement Chapters 24–27

Chapter 24	Units of Measure 460
Chapter 25	Length and Perimeter 476
Chapter 26	Area 492
Chapter 27	Volume 510
	Math Detective, Challenge 526
	Study Guide and Review 528
	Performance Assessment, Technology Linkup 530
	Problem Solving on Location ... 531A

UNIT 9 Algebra: Patterns and Relationships Chapters 28–30

Chapter 28	Algebra: Patterns 532
Chapter 29	Geometry and Motion 548
Chapter 30	Algebra: Graph Relationships ... 566
	Math Detective, Challenge 586
	Study Guide and Review 588
	Performance Assessment, Technology Linkup 590
	Problem Solving on Location ... 591A

Student Handbook H1
Problem of the Day Solutions PD14
Bibliography for Students BI1
Computer Software Bibliography BI3
Professional Bibliography BI7
Index BI14

About Harcourt Math

Dear Educator,

HARCOURT MATH is based on these principles:

▶ All children can experience success in learning mathematics.

▶ A balanced mathematics curriculum promotes conceptual, computational, and problem solving proficiency.

▶ Problem solving is the focus of mathematics instruction and is best developed by learning problem solving strategies and solving a balance of multistep, non-routine, and real-world problems.

▶ Computational and procedural skills practice develops mathematical proficiency.

▶ Concrete materials and pictorial models aid the understanding of mathematical concepts.

▶ Mathematical reasoning must permeate all aspects of mathematics instruction.

▶ Assessing children's prior knowledge identifies the starting point for new learning.

▶ Intervention options ensure student achievement in mathematics.

▶ A strong partnership among home, school, and the community promotes students' success in mathematics.

▶ A variety of tools that are consistent, ongoing, and aligned with instruction can best assess student progress.

▶ Regular mathematics instruction that includes English-language acquisition strategies benefits children with a primary language other than English.

HARCOURT MATH provides you with the resources to develop students' mathematical proficiency and appreciation for the power and usefulness of mathematics in everyday life.

Authors

Senior Author

Evan M. Maletsky
Professor of Mathematics
Montclair State University
Upper Montclair, New Jersey

Angela Giglio Andrews
Math Teacher, Scott School
Naperville District #203
Naperville, Illinois

Grace M. Burton
Chair, Department of Curricular Studies
Professor, School of Education
University of North Carolina at Wilmington
Wilmington, North Carolina

Jennie M. Bennett
Instructional Mathematics Supervisor
Houston Independent School District
Houston, Texas

Howard C. Johnson
Dean of the Graduate School
Associate Vice Chancellor for Academic Affairs
Professor, Mathematics and Mathematics Education
Syracuse University
Syracuse, New York

Lynda Luckie
Administrator/Math Specialist
Gwinnett County Public Schools
Lawrenceville, Georgia

Joyce C. McLeod
Visiting Professor
Rollins College
Winter Park, Florida

Vicki Newman
Classroom Teacher
McGaugh Elementary School
Los Alamitos Unified School District
Seal Beach, California

Janet K. Scheer
Executive Director
Create A Vision
Foster City, California

Karen A. Schultz
College of Education
Georgia State University
Atlanta, Georgia

Mathematics Advisors

The development of **HARCOURT MATH** was guided by prominent, accomplished mathematicians from across the United States. Their guidance helped ensure accurate mathematics and appropriate conceptual development.

David Wright
Professor of Mathematics
Brigham Young University
Provo, Utah
Grades K–2

Richard Askey
Professor of Mathematics
University of Wisconsin
Madison, Wisconsin
Grades 3–4

David Singer
Professor of Mathematics
Case Western Reserve University
Cleveland, Ohio
Grade 5

Jerome Dancis
Associate Professor of Mathematics
University of Maryland
College Park, Maryland
Grade 6

Tom Roby
Assistant Professor of Mathematics
California State University
Hayward, California
Grade 6

Contributions to the Professional Handbook

The Professional Handbook, found behind the fourth tab in Volume 1 of this Teacher's Edition, includes articles and essays written by leading mathematicians and researchers who share their insights into the structure of mathematics and topics important to elementary mathematics instruction.

Roger Howe
Professor of Mathematics
Yale University
New Haven, Connecticut
Marvelous Decimals
Doing Decimal Arithmetic: Addition
Doing Decimal Arithmetic: Subtraction
Estimation and Approximation
Estimation and Arithmetic

Liping Ma
Mathematics Researcher
and Educator
Palo Alto, California
A "Blind Spot" in the Order of Operations

Tom Roby
Models for Fractions

David Wright
Geometry and Measurement

Program Consultants

Program Consultants and Specialists

Janet S. Abbott
Mathematics Consultant
California

Elsie Babcock
Director, Mathematics and Science Center
Mathematics Consultant
Wayne Regional Educational Service Agency
Wayne, Michigan

William J. Driscoll
Professor of Mathematics
Department of Mathematical Sciences
Central Connecticut State University
New Britain, Connecticut

Lois Harrison-Jones
Education and Management Consultant
Dallas, Texas

Arax Miller
Curriculum Coordinator and English Department
  Chairperson
Chamlian School
Glendale, California

Rebecca Valbuena
Language Development Specialist
Stanton Elementary School
Glendora, California

Reviewers and Field-Test Teachers

Daarina Abdus-Samad
Teacher
Norma Coombs Alternative School
Pasadena, California

Britta Abinger
Teacher
Corkery School
Chicago, Illinois

Kathy Albrecht
Teacher
Heritage Oak Elementary
Roseville, California

Ann Allison
Teacher
Woodward Elementary
Lock Haven, Pennsylvania

Terri Battenburg
Teacher
Vencil Brown Elementary
Roseville, California

Caye Baxter
Teacher
Lindbergh Schweitzer Elementary
San Diego, California

Sister Mary Berryman
Teacher
St. Peter Celestine School
Cherry Hill, New Jersey

Hazel Bills
Teacher
Corkery School
Chicago, Illinois

David A. Bond
Gifted and Talented
  Resource Teacher
Bollman Bridge Elementary
Ellicott City, Maryland

Dr. Judith C. Branch-Boyd
City-wide Mathematics Coordinator
Medill Professional Training Center
Glen Ellyn, Illinois

Bonnie Bray
Teacher
Warren Road Elementary
Augusta, Georgia

Gail L. Clark
Teacher
Southhampton School #2
Delran, New Jersey

Mary Couch
Teacher
Georgetown Elementary
Georgetown, California

Alison Cox
Teacher
Malcolm Elementary
Laguna Niguel, California

Erin Cronin
Teacher
San Ysidro Middle School
San Ysidro, California

Deloris Cureton
Teacher
East Lake Elementary
Decatur, Georgia

Carolyn A. Day
Director of Programs: Math/Science
Dayton Public Schools
Dayton, Ohio

Kathleen Duarte
Teacher
Callie Kirkpatrick School
Menifee, California

Vickie Eilenberger
Teacher
Fremont Elementary
Santa Ana, California

Kelly L. Fleming
Teacher
Windy Hill Elementary
Owings, Maryland

Ellen Galdieri
Teacher
Dowell Elementary
Huntingtown, Maryland

Susan Gaspich
Teacher
Chelsea Heights Elementary
Atlantic City, New Jersey

Kathryn George
Teacher
East Park Elementary
Danville, Illinois

Lou Gerbi
Teacher
Westinghouse Elementary
Wilmerding, Pennsylvania

James Giordano
Teacher
W.B. Powell Elementary
Washington, D.C.

Susan Googins
Teacher
Philip Schuyler Elementary
Albany, New York

Jerri A. Hall
Teacher
Miller Middle School
Macon, Georgia

Becky Hamilton
Teacher
Scull Elementary
North Huntingdon, Pennsylvania

Elizabeth Harris
Teacher
Springfield Elementary
Providence, Rhode Island

Lottie Harris
Teacher
Jackson Road Elementary
Griffin, Georgia

Diane Hastings
Teacher
Thoreau Park
Parma, Ohio

Earl Heddle
Teacher
Richland Elementary
Gibsonia, Pennsylvania

and Reviewers

Reviewers and Field-Test Teachers continued

Sarah Hillyer
Teacher
Craddock Elementary
Aurora, Ohio

Tim Horton
Teacher
Quarryville Elementary
Quarryville, Pennsylvania

Heather Hunt
Teacher
Horizon Elementary
Hanover Park, Illinois

Maureen Irvine
Charles Hoffman Elementary
Running Springs, California

Michelle Jaronik
Coordinator, Gifted Math
Lincoln Center
Waukegan, Illinois

Denise Jones
Teacher
Dobbs Elementary
Atlanta, Georgia

Sunyong Kim
Teacher
Wilton Place Elementary
Sacramento, California

Inell Lemon
Math Coordinator
Jensen Scholastic Academy
Chicago, Illinois

Catherine Lewis
Teacher
John Ehrhardt Elementary
Elk Grove, California

Marie S. Massey
Teacher
Reese Road Elementary
Columbus, Georgia

Sarah Meadows
Title I Instructional Liaison
Topeka Public
Topeka, Kansas

Ruth Harbin Miles
Math Coordinator
Unified School District 233
Olathe, Kansas

Elaine Millie
Teacher
Bryant Elementary
Sioux City, Iowa

Susan Milstein
Teacher
Dag Hammerskjold School
Brooklyn, New York

Kathleen Mineau
Teacher
Public School #19
Albany, New York

Marilyn Moore
Teacher
Winterville Elementary
Winterville, Georgia

Penny L. Moore
Teacher
Terrace View Elementary
Grand Terrace, California

Ethel T. Munro
Teacher
Windom Elementary
Orchard Park, New York

Edward H. Nakamura
Teacher
Lakewood School
Lodi, California

Helen Hyun-Jou Park
Teacher
Wilton Place Elementary
Los Angeles, California

Carol G. Parker
Teacher
Valley of Enchantment Elementary
Lake Arrowhead, California

Beth Peery
Learner Support Strategist
Cheatham Hill Elementary
Marietta, Georgia

Maritza Perez
Teacher
Huff Elementary
Elgin, Illinois

Jill E. Perkins
Teacher
Edwardsburg Primary
Cassopolis, Michigan

Eloise Preiss
Assistant Superintendent of Curriculum and Instruction
Turlock School District
Turlock, California

Suzanne Regali
Teacher
C.W. Holmes School
Derry, New Hampshire

Pauline E. Robinson
Teacher
Star Hill Elementary
Dover, Delaware

Hector Ruiz
Teacher
Luther Burbank Elementary
Artesia, California

Telkia Rutherford
Mathematics Support Manager
Department of Instruction
Chicago, Illinois

Theresa Shields
Teacher
Parkside Elementary
Camden, New Jersey

Manwella Smith
Teacher
Prairie Elementary
Sacramento, California

Valerie J. Spindler
Teacher
L.V. Denti Elementary
Ava, New York

Leigh Ann Spitzer
Teacher
Washington Elementary
Kingsburg, California

Sheila Taylor
Teacher
West Wortham Elementary/ Middle School
Saucier, Mississippi

Sylvia Teahan
Teacher
New Albany Elementary
Cinnaminson, New Jersey

Mary Thomas
Math Coordinator
Jersey Shore Area Junior High
Jersey Shore, Pennsylvania

Fran Threewit
Reading and Math Specialist
Kenwood Elementary School
Kenwood, California

Beverly J. Tornberg
Teacher
Valhalla Elementary
Pleasant Hill, California

Peter Tuttle
Teacher
Noble Elementary
Cleveland Heights, Ohio

Michelle Vancheri
Teacher
School #11
Paterson, New Jersey

Shelia R. Wells
Foster Elementary
Compton, California

Claudia West
Teacher
Westlake Hills Elementary
Westlake Village, California

Beverly A. White
Teacher
Cecil Elementary
Cecil, Pennsylvania

Jill Wilke
Teacher
Brea Country Hills School
Brea, California

HARCOURT MATH

The Solution for Success

Students learn mathematics by direct instruction, hands-on experiences, step-by-step models that build conceptual understanding, and ample practice that requires the use of problem solving skills and strategies. **HARCOURT MATH** provides a balance of computational and procedural skills, conceptual understanding, and problem solving.

A System for Mathematics Instruction
Harcourt Math provides a consistent 4-step method of instruction that ensures success for teachers and students.

Reaching All Learners
Harcourt Math offers a wide range of strategies and activities so that all students can have success in mathematics.

Intervention
Harcourt Math includes daily intervention strategies that diagnose students' difficulties with mathematics while providing intervention resources that will bring success to every learner.

Assessment
Harcourt Math assessment tools ensure student success by measuring achievement before, during, and after instruction.

Family Support
Harcourt Math builds a strong partnership among home, school, and the community that will bring students success in mathematics.

Using Technology
Harcourt Math increases student success by offering an array of technology products that promotes mathematics learning.

A System for Mathematics Instruction

1 INTRODUCE

Teachers introduce the topic by assessing prior knowledge, reviewing prerequisite skills, and setting a purpose for learning.

In INTRODUCE, HARCOURT MATH provides:

- **Warm-Up Resources** or **Getting Started Options** on the Lesson Planning page that start each lesson and promote continual review of computation and problem solving skills.
- **Quick Review** of prerequisite skills.
- **Why Learn This?** for helping students understand short- and long-term goals.

2 TEACH

In this step, concrete experiences promote reasoning and provide for conceptual development. Problem solving strategies and computational procedures are taught via step-by-step, direct instruction.

In TEACH, HARCOURT MATH provides:

- Step-by-step **Guided Instruction** to cement understanding, promote **Reasoning**, and monitor achievement for the lesson objective.
- **Modifying Instruction** for adapting the lesson for students with special needs.
- Questioning strategies for assessing student comprehension.
- Concrete experiences, step-by-step pictorial models, and examples that link to abstract processes and procedures.

Intervention and Extension Resources in all four steps provide alternative teaching strategies, ways to meet individual needs, and intervention options when students fall behind.

HARCOURT MATH employs a consistent method of instruction for developing mathematical proficiency. This teaching model is provided in every lesson and includes the following four steps.

3 PRACTICE

Daily practice is essential for conceptual development, computational proficiency, and development of reasoning and problem solving strategies.

In PRACTICE, HARCOURT MATH provides:

- **Guided Practice** and **Check** for identifying students who are having difficulty.
- **Common Error Alert** that identifies and describes common mistakes and offers suggestions for help.
- **Independent Practice** that is conceptual, procedural, reasoning- and problem-based.
- **Scaffolded Instruction** for helping students be successful problem solvers.
- **Mixed Review and Test Prep** that provides cumulative review in a variety of formats.

4 ASSESS

Assessment in every lesson helps students summarize their learning and check their progress. It helps teachers bring closure to the lesson, make critical decisions about future instruction, and provide appropriate interventions for individual students.

In ASSESS, HARCOURT MATH provides:

- Opportunities to assess in different ways, as well as to address different learning styles:

  Discuss

  Write

  Lesson Quiz

- Continuous monitoring of student achievement of the lesson objective. Students are actively involved in summarizing the content of the lesson through oral discussion and through writing.

In addition to lesson assessment, multiple means of assessment are provided for every chapter. See the end-of-chapter pages and the **Assessment Guide.**

Reaching All Learners

UNIVERSAL ACCESS

HARCOURT MATH offers a wide range of strategies and activities so that all students can experience success in mathematics.

Alternative Teaching Strategy

Have students **identify the multiplication property** that each equation represents. Then students can fill in the missing number or symbol.

* × 1 = __ Property of One; *

0 × * = __ Property of Zero; 0

* × ♥ × ◇ = * × ◇ × __ Order Property; ♥

* × ♥ × ◇ × 0 = __ Property of Zero; 0

(* × ♥) × ◇ = * × (♥ × ◇) Grouping Property

VISUAL
LOGICAL/MATHEMATICAL

Alternative Teaching Strategies provide alternative ways to develop lesson concepts, and they address different learning modalities.

Advanced Learners

Challenge students to work with a partner to **write and solve decimal riddles** similar to the following:

- My decimal number is less than 0.5 but greater than 0.35.
- The digit in the tenths place is 1 less than the digit in the hundredths place.
- The sum of the digits is 9. What is the number? 0.45

Check students' work.

AUDITORY
VERBAL/LINGUISTIC

Strategies for **Advanced Learners** provide challenging activities that allow students to go beyond the lesson objectives and move forward in the curriculum.

Early Finishers

Have students write and administer a survey, and **compare data sets** between their class and their school.

Ask students to think of a multiple-choice question they would like to ask their classmates and schoolmates.

When the survey is ready, have the students administer it to their class and then to the student body. Students should display the results of both surveys as circle graphs. Check students' work.

AUDITORY, VISUAL
VISUAL/SPATIAL

Activities for **Early Finishers** help cement the learning and provide greater depth of understanding.

English Language Learners

Help students **understand the terms** *divisor, dividend, and quotient.* Use an example problem and point to the number that represents each term. Then, write the word and give the definition. Give students several other examples and have them label the divisor, dividend, and quotient in each problem. Then, have students write a division problem, labeling the divisor, dividend, and quotient and explaining what each term means. Check students' work.

VISUAL
VERBAL/LINGUISTIC

Activities for **English Language Learners** include graphics and objects that accompany oral and written instruction, to meet the needs of students acquiring English.

Curriculum Connections

Share the following with students to help them **practice different ways to read and write numbers.**

The ancient Roman numeral system goes back to about 500 B.C. It is still used sometimes today, but its flaw is that it lacks a zero. The Romans were not consistent in how they used Roman numerals. They might write a number as XXXXVIIII, XXXXIX, or IL. We now follow certain rules when using Roman numerals. Ask:

- What number do these symbols represent? 49
- What is the correct way to write the number today? Explain. IL; because a numeral cannot be written more than three times in a row

VISUAL; LOGICAL/MATHEMATICAL

Curriculum Connections reinforce or extend mathematics learning by connecting math to other subject areas.

Special Needs

Have students **identify the LCD and subtract mixed numbers** by completing the following activity.

Instruct students to copy these examples and match them with their equivalents. Then students can find the difference by using fraction bars.

$3\frac{1}{4} - 1\frac{1}{8}$ $3\frac{1}{4} - 1\frac{2}{4}$ $1\frac{3}{4}$

$3\frac{1}{4} - 1\frac{2}{3}$ $3\frac{2}{8} - 1\frac{1}{8}$ $2\frac{1}{8}$

$3\frac{1}{4} - 1\frac{1}{2}$ $3\frac{5}{20} - 1\frac{14}{20}$ $1\frac{11}{20}$

$3\frac{1}{4} - 1\frac{7}{10}$ $3\frac{3}{12} - 1\frac{8}{12}$ $1\frac{7}{12}$

VISUAL
LOGICAL/MATHEMATICAL

Strategies for **Special Needs** ensure the inclusion of all students in mathematics learning.

Intervention

HARCOURT MATH includes daily intervention strategies that diagnose students' difficulties with mathematics and provide intervention resources to bring success to every learner.

Before Beginning Each Chapter...

The Assessing Prior Knowledge—Check What You Know section assesses students' knowledge of prerequisite skills. Several intervention options will readily help students:

Troubleshooting
Found in the back of the *Pupil Edition*, targeted mini-lessons present pictorial models and plenty of practice as reteaching aids to help students get back on track. (Gr. 3-6)

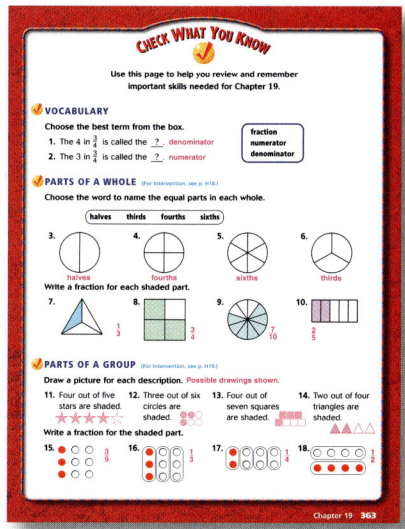

Intervention Strategies and Activities
Several components help students having difficulty with a particular math concept or skill. Activities provide direct instruction, conceptual models, and scaffolded practice. Built-in assessment determines when students can move on to the next level in the chapter concepts. (Gr. 1-6)

While Using the Chapter...

Intervention offered for each lesson ensures success.

Lesson Check
Found within every Grade 3-6 lesson, the Check section helps you ensure that students understand the lesson concept.

Alternative Teaching Strategies
These strategies offer optional ways to develop lesson concepts and to address different learning modalities.

Reteach Workbook
This workbook offers an alternative approach for each lesson concept. It includes a worksheet for every lesson. (Gr. 1-6)

T15

Assessment

HARCOURT MATH assessment tools provide measures of achievement before, during, and after instruction, to help you ensure student success.

▶ Entry-Level Assessment

Inventory Test

This test, provided in the *Assessment Guide*, may be administered at the beginning of the school year to determine a baseline for student mastery of the grade-level objectives. The baseline may also be used to evaluate a student's future growth when compared to subsequent tests.

Assessing Prior Knowledge– Check What You Know

This feature appears at the beginning of every chapter. It may be used before chapter instruction begins in order to determine whether students possess crucial prerequisite skills. Tools for intervention are provided.

Pretests

The Chapter Tests, Form A (multiple choice) or Form B (free response), may be used as pretests to measure what students already may have mastered before instruction begins. The tests are provided in the *Assessment Guide*.

▶ Progress Monitoring

Daily Assessment

These point-of-use strategies allow you to continually adjust instruction so that all students are constantly progressing toward mastery of the grade-level objectives. These strategies, which appear in every lesson of this *Teacher's Edition*, include the **Quick Review**, the **Mixed Review and Test Prep**, and the **Assess** section of the lesson (Discuss, Write, and Lesson Quiz).

Intervention

While monitoring students' progress, you may determine that intervention is needed. The **Intervention and Extensions Resources** page for each lesson suggests several options for meeting individual needs.

Student Self-Assessment

Students evaluate their own work through checklists, portfolios, and journals. Suggestions are provided in the *Assessment Guide*.

Summative Evaluation

Formal Assessment

Several options are provided to help you determine whether students have achieved the goals defined by grade-level objectives. These options are provided at the end of each chapter and unit and at the end of the year.

- **Chapter Review/Test in the *Pupil Edition***
- **Standardized Test Prep, in the *Pupil Edition*, for cumulative review**
- **Chapter Test, Form A and B in the *Assessment Guide***
- **Unit Tests in the *Assessment Guide***

Performance Assessment

Performance Tasks for every unit are provided in the ***Performance Assessment*** component and in Performance Assessment in the ***Pupil Edition***.

The Harcourt Electronic Test System— Math Practice and Assessment

This technology component provides you with the opportunity to make and grade chapter tests electronically. You can customize the tests to meet individual needs or create Standards-based tests from a bank of test items. This component also includes a management system for generating reports.

Test Preparation

Test Prep in the *Pupil Edition*

When faced with the challenge of taking standardized tests, students need to feel confident that the test will be an accurate reflection of what they know or need to learn. Learning to take standardized tests is essentially the same as being a good problem solver. The steps are the same: understand, plan, solve, and check.

To help students prepare for tests, the **Mixed Review and Test Prep** at the end of each lesson provides some items in standardized-test format. In addition, the **Standardized Test Prep** pages at the end of each chapter provide practice in solving problems in a standardized-test format. They include practical test-taking tips that give students ongoing strategies for analyzing problems and finding the best way to solve them. **Write What You Know** questions provide students the opportunity to practice taking standarized tests that include written response items.

Family Support

HARCOURT MATH encourages a strong partnership between home, school, and the community to help students succeed in mathematics.

Family Involvement Activities

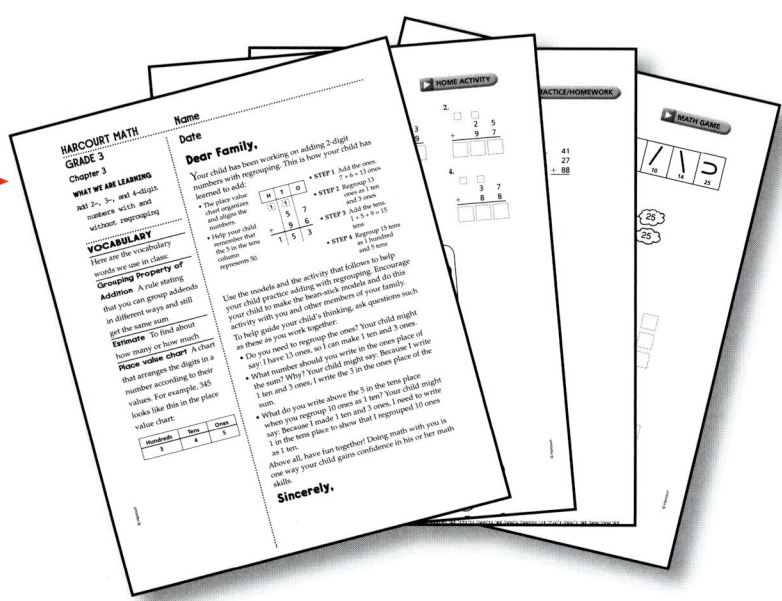

An informative, four-page family letter accompanies every **HARCOURT MATH** chapter. Each letter offers:

- A family-friendly explanation of the expectations for student learning
- Models of chapter skills
- Tips that enable family members to assist students
- Homework practice
- A math game for family fun

Activities and Games for Home or School

Colorful activity cards and gameboards work well in classroom centers and are fun for family members to share at home.

The Learning Site

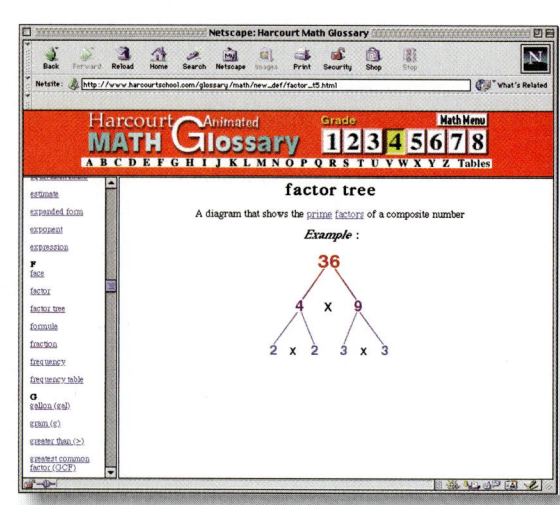

Harcourt's easy-to-use website offers family members interactive learning games, a Multimedia Math Glossary, and a special page with tips and ideas for parents.

Visit the **HARCOURT MATH** website at **www.harcourtschool.com**.

THE LEARNING SITE

Using Technology

HARCOURT MATH includes an array of technology products that can be used to promote mathematics learning.

Harcourt Math Newsroom Videos

 Harcourt School Publishers, CNN, and Turner Learning deliver an exceptional video series to supplement the math curriculum. Through CNN news stories, students explore architecture, the natural world, astronomy, and the many other areas in which mathematics impacts their everyday lives. (Gr. 3-6)

The Harcourt Electronic Test System — Math Practice and Assessment

 The Harcourt Electronic Test System offers computer-administered practice and testing and an instructional management system. It makes organizing classroom information and tracking student progress easy and accurate. (Gr. 1-6)

◀ Intervention Strategies and Activities CD-ROM

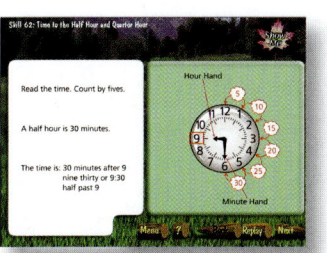

 This CD-ROM offers independent learning, conceptual models, and scaffolded practice to students having difficulty with a particular math concept or skill. Built-in assessment determines when students can move to the next level in the chapter concepts. (Gr. 1-6)

◀ The Harcourt Learning Site

 This user-friendly site offers interactive experiences that reinforce math concepts, provides professional development for teachers, and includes school-home resources for parents. (Gr. K-6)

Stanley's Sticker Stories

 Children can use this program to create animated number books involving number sense, addition, and subtraction. Reading and language skills develop naturally from these experiences. (Gr. K-2)

◀ Mighty Math Software

 Mighty Math software was developed jointly by Harcourt and Edmark. It makes math fun and boosts the skills and concepts taught in **HARCOURT MATH.** Activities increase in difficulty as students progress. (Gr. K-6)

◀ E-Lab

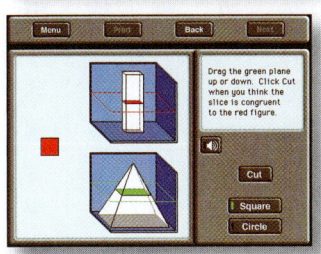

 Located at The Learning Site, E-Lab activities enable students to explore and extend key concepts developed in Hands-On and Lab Activity lessons. (Gr. 3-6)

Components

HARCOURT MATH COMPONENTS	K	1	2	3	4	5	6
Daily Lesson Support							
Pupil Edition		■	■	■	■	■	■
Unit/Chapter Books		■	■				
Big Book	■						
Teacher's Edition	■	■	■	■	■	■	■
Reteach Workbook—Pupil and Teacher's Editions		■	■	■	■	■	■
Practice Workbook—Pupil and Teacher's Editions	■	■	■	■	■	■	■
Problem Solving and Reading Strategies Workbook—Pupil and Teacher's Editions		■	■	■	■	■	■
Challenge Workbook—Pupil and Teacher's Editions		■	■	■	■	■	■
Teacher's Resource Book	■	■	■	■	■	■	■
Lesson Resource Organizer	■						
Daily Transparencies		■	■	■	■	■	■
Problem Solving Teaching Transparencies: Scaffolded Instruction		■	■				
Teaching Transparencies				■	■	■	■
Vocabulary Cards with Teacher's Activity Guide	■	■	■				
Success for English Language Learners		■	■	■	■	■	■
Answer and Solution Key							■
Intervention							
Intervention Strategies and Activities—Kit with Cards		■	■	■	■	■	■
Intervention Strategies and Activities—Teacher's Guide with Copying Masters		■	■	■	■	■	■
Intervention Strategies and Activities—CD-ROM		■	■	■	■	■	■
Check What You Know: Intervention Practice Book—Pupil Edition and Answer Key		■	■	■	■	■	■
Check What You Know: Enrichment Book—Pupil Edition and Answer Key		■	■	■	■	■	■
Intervention Strategies and Activities—Teaching Transparencies		■	■	■	■	■	■
Assessment Resources							
Assessment Guide	■	■	■	■	■	■	■
Performance Assessment		■	■	■	■	■	■
Math Practice and Assessment—CD-ROM		■	■	■	■	■	■
Family Involvement							
Family Involvement Activities—English	■	■	■	■	■	■	■
Family Involvement Activities—in Other Languages	■	■	■	■	■	■	■
Activities and Games for Home or School	■	■	■	■	■	■	■
Literature and Music							
Math Readers	■	■	■				
Literature Big Books and Little Books	■						
Literature Books on Tape	■						
Math Jingles™ and Math Jingles™ Copying Masters	■	■	■	■	■	■	■
Technology Resources							
Harcourt Math Newsroom Videos (CNN)				■	■	■	■
Stanley's Sticker Stories® Software	■						
Mighty Math® Software	■	■	■				
E-Lab® and E-Lab Recording Sheets				■	■	■	■
Calculator Handbook				■	■	■	■
Math Practice and Assessment—CD-ROM		■	■	■	■	■	■
Intervention Strategies and Activities—CD-ROM		■	■	■	■	■	■
Harcourt Math Lesson Planner and Resources—CD-ROM	■	■	■	■	■	■	■
The Learning Site		■	■	■	■	■	■
Manipulative Options							
My Manipulatives and Workmats	■	■	■				
Core Manipulative Kit and Teacher Modeling Kit	■	■	■	■	■	■	■
Build-a-Kit® Manipulatives	■	■	■	■	■	■	■

HARCOURT
Math

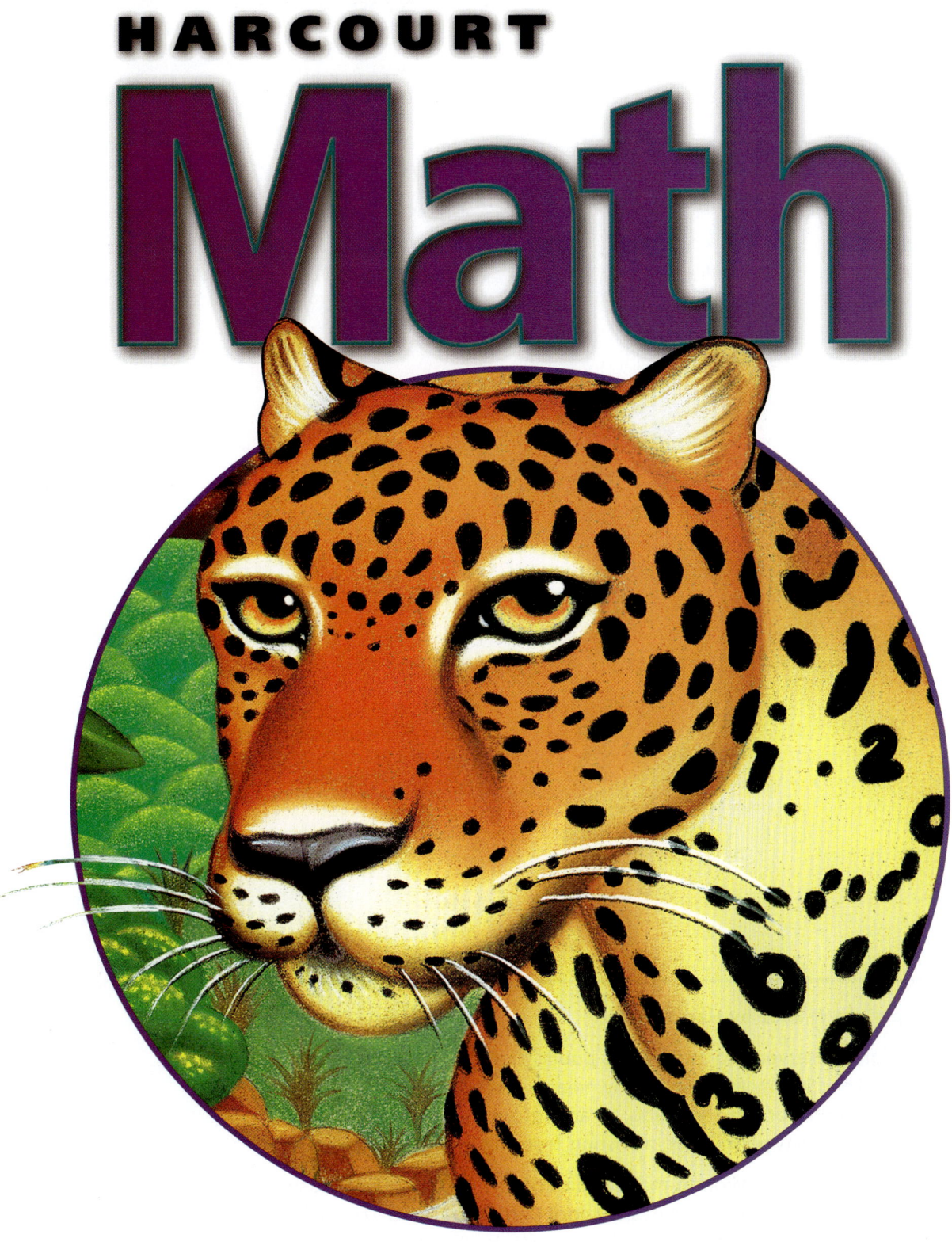

Harcourt School Publishers

Orlando • Boston • Dallas • Chicago • San Diego
www.harcourtschool.com

Copyright © 2002 by Harcourt, Inc.

All rights reserved. No part of this publication may be reproduced or transmitted in any form or by any means, electronic or mechanical, including photocopy, recording, or any information storage and retrieval system, without permission in writing from the publisher.

Requests for permission to make copies of any part of the work should be mailed to the following address: School Permissions and Copyrights, Harcourt, Inc., 6277 Sea Harbor Drive, Orlando, Florida 32887-6777.

HARCOURT and the Harcourt Logo are trademarks of Harcourt, Inc.

Printed in the United States of America

ISBN 0-15-320750-7

2 3 4 5 6 7 8 9 10 032 10 09 08 07 06 05 04 03 02 01

Senior Author

Evan M. Maletsky
Professor of Mathematics
Montclair State University
Upper Montclair, New Jersey

Mathematics Advisors

Jerome Dancis
Associate Professor of Mathematics
University of Maryland
College Park, Maryland

Tom Roby
Assistant Professor of Mathematics
California State University
Hayward, California

Authors

Angela Giglio Andrews
Math Teacher, Scott School
Naperville District #203
Naperville, Illinois

Jennie M. Bennett
Instructional Mathematics Supervisor
Houston Independent School District
Houston, Texas

Grace M. Burton
Chair, Department of Curricular Studies
Professor, School of Education
University of North Carolina
  at Wilmington
Wilmington, North Carolina

Howard C. Johnson
Dean of the Graduate School
Associate Vice Chancellor
  for Academic Affairs
Professor, Mathematics and
  Mathematics Education
Syracuse University
Syracuse, New York

Lynda A. Luckie
Administrator/Math Specialist
Gwinnett County Public Schools
Lawrenceville, Georgia

Joyce C. McLeod
Visiting Professor
Rollins College
Winter Park, Florida

Vicki Newman
Classroom Teacher
McGaugh Elementary School
Los Alamitos Unified School District
Seal Beach, California

Janet K. Scheer
Executive Director
Create A Vision
Foster City, California

Karen A. Schultz
College of Education
Georgia State University
Atlanta, Georgia

Program Consultants and Specialists

Janet S. Abbott
Mathematics Consultant
California

Lois Harrison-Jones
Education and Management
  Consultant
Dallas, Texas

Elsie Babcock
Director, Mathematics and Science
  Center; Mathematics Consultant
Wayne Regional Educational
  Service Agency
Wayne, Michigan

Arax Miller
Curriculum Coordinator and English
  Department Chairperson
Chamlian School
Glendale, California

William J. Driscoll
Professor of Mathematics
Department of Mathematical Sciences
Central Connecticut State University
New Britain, Connecticut

Rebecca Valbuena
Language Development Specialist
Stanton Elementary School
Glendora, California

UNIT 1
CHAPTERS 1–4
Number Sense and Operations

Chapter 1

WHOLE-NUMBER APPLICATIONS **14**
- ✓ Check What You Know 15
- 1 Estimate with Whole Numbers 16
- 2 Use Addition and Subtraction 20
- 3 Use Multiplication and Division 22
  Problem Solving: Linkup to Reading
- 4 **Problem Solving Strategy:** Predict and Test 26
- 5 **ALGEBRA** Use Expressions 28
- 6 **ALGEBRA** Mental Math and Equations 30
  Chapter 1 Review/Test 32
- ⭐ Standardized Test Prep 33
  Intervention: Troubleshooting H3–H4
  Extra Practice H32

Chapter 2

OPERATION SENSE **34**
- ✓ Check What You Know 35
- 1 Mental Math: Use the Properties 36
  Problem Solving: Thinker's Corner
- 2 **ALGEBRA** Exponents 40
- 3 **MATH LAB** Explore Order of Operations 42
- 4 **ALGEBRA** Order of Operations 44
- 5 **Problem Solving Skill:** Sequence and Prioritize Information 46
  Chapter 2 Review/Test 48
- ⭐ Standardized Test Prep 49
  Intervention: Troubleshooting H2, H15
  Extra Practice H33

TECHNOLOGY LINK

Harcourt Math Newsroom Video:
Chapter 1, p. 18

E-Lab:
Chapter 2, p. 43
Chapter 4, p. 75

Mighty Math Calculating Crew:
Chapter 1, p. 24
Chapter 3, p. 54
Chapter 4, p. 67

Multimedia Glossary:
The Learning Site at www.harcourtschool.com/mathglossary

iv

Chapter 3

DECIMAL CONCEPTS **50**

- ✓ Check What You Know 51
- 1 Represent, Compare, and Order Decimals 52
  Problem Solving: Thinker's Corner
- 2 **Problem Solving Strategy:** Make a Table 56
- 3 Estimate with Decimals 58
- 4 Decimals and Percents • **MATH LAB** Activity 60
  Chapter 3 Review/Test 62
- ⭐ Standardized Test Prep 63
  Intervention: Troubleshooting H2–H4
  Extra Practice H34

Chapter 4

DECIMAL OPERATIONS **64**

- ✓ Check What You Know 65
- 1 Add and Subtract Decimals 66
  Problem Solving: Linkup to Science
- 2 Multiply Decimals • **MATH LAB** Activity 70
- 3 • **MATH LAB** Explore Division of Decimals 74
- 4 Divide with Decimals • **MATH LAB** Activity 76
  Problem Solving: Thinker's Corner
- 5 **Problem Solving Skill:** Interpret the Remainder 80
- 6 **ALGEBRA** Decimal Expressions and Equations 82
  Chapter 4 Review/Test 84
- ⭐ Standardized Test Prep 85
  Intervention: Troubleshooting H4–H5, H13
  Extra Practice H35

UNIT WRAPUP

Problem Solving: Math Detective 86
Challenge: Scientific Notation 87
Study Guide and Review 88
Performance Assessment 90
Technology Linkup Data ToolKit: Use Spreadsheet Formulas to Find Sums and Differences 91
🗺 **Problem Solving:** On Location with Energy 91A

v

UNIT 2
CHAPTERS 5–6
Statistics and Graphing

Chapter 5

COLLECT AND ORGANIZE DATA **92**
- ✓ Check What You Know 93
- 1 Samples 94
- 2 Bias in Surveys 98
- 3 **Problem Solving Strategy:** Make a Table 100
- 4 Frequency Tables and Line Plots 102
- 5 Measures of Central Tendency 106
   Problem Solving: Linkup to Social Studies
- 6 Outliers and Additional Data 109
- 7 Data and Conclusions 112
   Chapter 5 Review/Test 116
- ⭐ Standardized Test Prep 117
   Intervention: Troubleshooting H5–H6
   Extra Practice H36

TECHNOLOGY LINK

Harcourt Math Newsroom Video:
Chapter 5, p. 95

E-Lab:
Chapter 6, p. 129

Data Toolkit:
Chapter 5, p. 102
Chapter 6, p. 122

Multimedia Glossary:
The Learning Site at www.harcourtschool.com/mathglossary

vi

Chapter 6
GRAPH DATA 118

- ✓ Check What You Know 119
- 1 Make and Analyze Graphs 120
- 2 Find Unknown Values 124
- 3 Stem-and-Leaf Plots and Histograms 126
- 4 ● **MATH LAB** Explore Box-and-Whisker Graphs 129
- 5 Box-and-Whisker Graphs 130
- 6 Analyze Graphs 132
  - **Problem Solving:** Linkup to Reading
  - Chapter 6 Review/Test 136
- ★ Standardized Test Prep 137
  - **Intervention:** Troubleshooting H6–H7
  - **Extra Practice** H37

UNIT WRAPUP

Problem Solving: Math Detective 138
Challenge: Explore Scatterplots 139
Study Guide and Review 140
Performance Assessment 142
Technology Linkup E-Lab: Exploring
    Box-and-Whisker Graphs 143
    Problem Solving: On Location In Oregon 143A

UNIT 3
CHAPTERS 7–10
Fraction Concepts and Operations

Chapter 7

NUMBER THEORY **144**

    Check What You Know 145
1. Divisibility • **MATH LAB** Activity 146
2. Prime Factorization 148
3. Least Common Multiple and Greatest Common Factor 150
   Problem Solving: Linkup to Careers
4. Problem Solving Strategy: Make an Organized List .. 154
   Chapter 7 Review/Test 156
   ⭐ Standardized Test Prep 157
   Intervention: Troubleshooting H7–H8
   Extra Practice H38

Chapter 8

FRACTION CONCEPTS **158**

    Check What You Know 159
1. Equivalent Fractions and Simplest Form • **MATH LAB** Activity 160
   Problem Solving: Thinker's Corner
2. Mixed Numbers and Fractions 164
3. Compare and Order Fractions 166
4. • **MATH LAB** Explore Fractions and Decimals 168
5. Fractions, Decimals, and Percents 169
   Chapter 8 Review/Test 172
   ⭐ Standardized Test Prep 173
   Intervention: Troubleshooting H3, H9
   Extra Practice H39

Chapter 9

ADD AND SUBTRACT FRACTIONS AND MIXED NUMBERS **174**

    Check What You Know 175
1. Estimate Sums and Differences 176
2. • **MATH LAB** Model Addition and Subtraction 180
3. Add and Subtract Fractions 182
   Problem Solving: Thinker's Corner
4. Add and Subtract Mixed Numbers 186
   Problem Solving: Linkup to Reading

viii

TECHNOLOGY LINK

Harcourt Math Newsroom Video:
Chapter 9, p. 178

E-Lab:
Chapter 8, pp. 162, 168
Chapter 9, pp. 181, 191
Chapter 10, p. 208

Mighty Math Calculating Crew:
Chapter 8, p. 171
Chapter 9, p. 188

Mighty Math Number Heroes:
Chapter 9, p. 183

Multimedia Glossary:
The Learning Site at
www.harcourtschool.com/
mathglossary

Chapter 9 (continued)

5	**MATH LAB** Rename to Subtract	190
6	Subtract Mixed Numbers	192
7	**Problem Solving Strategy:** Draw a Diagram	194
	Chapter 9 Review/Test	196
★	Standardized Test Prep	197
	Intervention: Troubleshooting	H9–H10
	Extra Practice	H40

Chapter 10

MULTIPLY AND DIVIDE FRACTIONS AND MIXED NUMBERS 198

✓	Check What You Know	199
1	Estimate Products and Quotients	200
2	Multiply Fractions	202
3	Multiply Mixed Numbers	206
4	**MATH LAB** Division of Fractions	208
5	Divide Fractions and Mixed Numbers	210
	Problem Solving: Linkup to Careers	
6	**Problem Solving Skill:** Choose the Operation	214
7	**ALGEBRA** Fraction Expressions and Equations	216
	Chapter 10 Review/Test	218
★	Standardized Test Prep	219
	Intervention: Troubleshooting	H10–H11
	Extra Practice	H41

UNIT WRAPUP

Problem Solving: Math Detective	220
Challenge: Mixed Numbers and Time	221
Study Guide and Review	222
Performance Assessment	224
Technology Linkup Mighty Math Astro Algebra: Fractions, Decimals, and Percents	225
Problem Solving: On Location In South Carolina	225A

UNIT 4
CHAPTERS 11–12

Algebra: Integers

Chapter 11
ALGEBRA NUMBER RELATIONSHIPS 226
- Check What You Know 227
1. Understand Integers 228
2. Rational Numbers 230
   Problem Solving: Linkup to Reading
3. Compare and Order Rational Numbers 234
4. **Problem Solving Strategy:** Use Logical Reasoning ... 236
   Chapter 11 Review/Test 238
   ★ Standardized Test Prep 239
   Intervention: Troubleshooting H12–H13
   Extra Practice H42

TECHNOLOGY LINK

Harcourt Math Newsroom Video:
Chapter 11, p. 234

E-Lab:
Chapter 12, p. 249

Mighty Math Astro Algebra:
Chapter 12, pp. 253

Multimedia Glossary:
The Learning Site at
www.harcourtschool.com/mathglossary

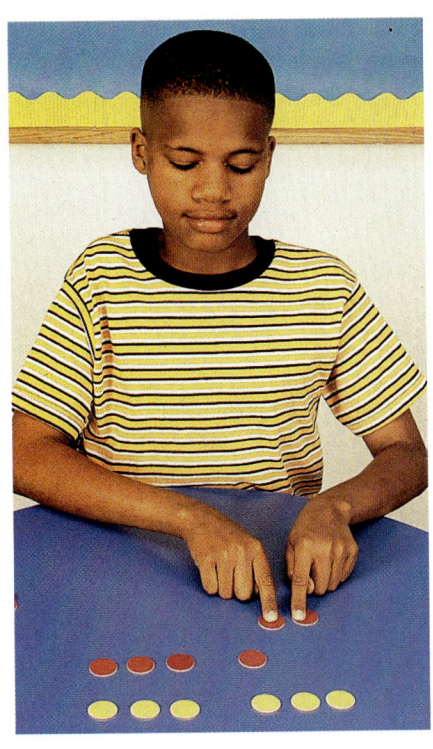

Chapter 12
ALGEBRA OPERATIONS WITH INTEGERS 240

    Check What You Know 241
1 MATH LAB Model Addition of Integers 242
2 Add Integers 244
    Problem Solving: Thinker's Corner
3 MATH LAB Model Subtraction of Integers 248
4 Subtract Integers 250
5 Multiply and Divide Integers • MATH LAB Activity .252
6 Explore Operations with Rational Numbers 256
    Problem Solving: Linkup to Careers
    Chapter 12 Review/Test 260
    Standardized Test Prep 261
    Intervention: Troubleshooting H12–H13
    Extra Practice H43

UNIT WRAPUP

Problem Solving: Math Detective 262
Challenge: Negative Exponents 263
Study Guide and Review 264
Performance Assessment 266
Technology Linkup Mighty Math Calculating Crew:
    Operations with Integers 267
    Problem Solving: On Location In Kentucky 267A

UNIT 5
CHAPTERS 13–15

Algebra: Expressions and Equations

Chapter 13
 EXPRESSIONS **268**
- ✓ Check What You Know 269
- 1 Write Expressions 270
- 2 Evaluate Expressions 272
  Problem Solving: Linkup to Reading
- 3 ● **MATH LAB** Squares and Square Roots 276
- 4 Expressions with Squares and Square Roots 278
  - Chapter 13 Review/Test 280
- ★ Standardized Test Prep 281
  - **Intervention:** Troubleshooting H8, H15
  - **Extra Practice** H44

Chapter 14
 ADDITION AND SUBTRACTION EQUATIONS **282**
- ✓ Check What You Know 283
- 1 Connect Words and Equations 284
- 2 ● **MATH LAB** Model and Solve One-Step Equations 286
- 3 Solve Addition Equations 287
  Problem Solving: Thinker's Corner
- 4 Solve Subtraction Equations 290
  - Chapter 14 Review/Test 292
- ★ Standardized Test Prep 293
  - **Intervention:** Troubleshooting H16–H17
  - **Extra Practice** H45

TECHNOLOGY LINK

Harcourt Math Newsroom Video:
Chapter 13, p. 273

E-Lab:
Chapter 13, p. 277
Chapter 14, p. 286
Chapter 15, pp. 296, 305

Mighty Math Astro Algebra:
Chapter 15, p. 298

Multimedia Glossary:
The Learning Site at www.harcourtschool.com/mathglossary

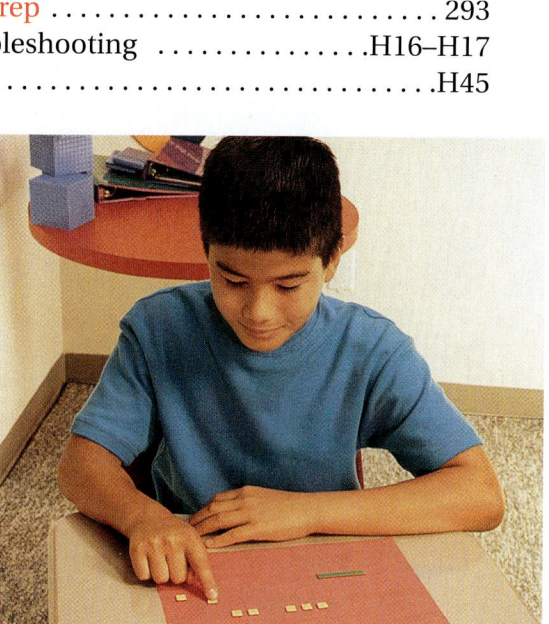

xii

Chapter 15
ALGEBRA MULTIPLICATION AND DIVISION EQUATIONS **294**

	Check What You Know 295
1	**MATH LAB** Model Multiplication Equations 296
2	Solve Multiplication and Division Equations 297
3	Use Formulas 300
	Problem Solving: Linkup to Careers
4	**MATH LAB** Two-Step Equations 304
5	**Problem Solving Strategy:** Work Backward 306
	Chapter 15 Review/Test 308
	Standardized Test Prep 309
	Intervention: Troubleshooting H11, H17–H18
	Extra Practice H46

UNIT WRAPUP

Problem Solving: Math Detective 310
Challenge: Reflexive, Symmetric, and
   Transitive Properties 311
Study Guide and Review 312
Performance Assessment 314
Technology Linkup Data Toolkit: Use a Spreadsheet
   to Complete Function Tables 315
   Problem Solving: On Location
      In Pennsylvania 315A

xiii

UNIT 6
CHAPTERS 16–19
Geometry and Spatial Reasoning

Chapter 16
GEOMETRIC FIGURES 316
- ✓ Check What You Know 317
- 1 Points, Lines, and Planes 318
- 2 **MATH LAB** Angles 320
- 3 Angle Relationships 322
     Problem Solving: Linkup to Art
- 4 Classify Lines 326
     Chapter 16 Review/Test 328
- ★ Standardized Test Prep 329
     Intervention: Troubleshooting H18–H19
     Extra Practice H47

Chapter 17
PLANE FIGURES 330
- ✓ Check What You Know 331
- 1 Triangles **MATH LAB** Activity 332
     Problem Solving: Thinker's Corner
- 2 **Problem Solving Strategy:** Find a Pattern 336
- 3 Quadrilaterals **MATH LAB** Activity 338
     Problem Solving: Linkup to Reading
- 4 Draw Two-Dimensional Figures 342
- 5 Circles 344
     Chapter 17 Review/Test 346
- ★ Standardized Test Prep 347
     Intervention: Troubleshooting H14, H18–H19
     Extra Practice H48

TECHNOLOGY LINK

Harcourt Math Newsroom Video:
Chapter 17, p. 338

E-Lab:
Chapter 18, p. 353
Chapter 19, p. 369

Mighty Math Cosmic Geometry:
Chapter 16, pp. 321, 323
Chapter 17, pp. 334, 345

Multimedia Glossary:
The Learning Site at www.harcourtschool.com/mathglossary

xiv

Chapter 18

SOLID FIGURES **348**

- Check What You Know 349
1. Types of Solid Figures • **MATH LAB** Activity 350
2. Different Views of Solid
   Figures • **MATH LAB** Activity 353
3. **MATH LAB** Models of Solid Figures 356
4. Problem Solving Strategy:
   Solve a Simpler Problem 358
   Chapter 18 Review/Test 360
   Standardized Test Prep 361
   Intervention: Troubleshooting H20
   Extra Practice H49

Chapter 19

CONGRUENCE AND SIMILARITY **362**

- Check What You Know 363
1. Construct Congruent Segments
   and Angles • **MATH LAB** Activity 364
2. Bisect Line Segments and
   Angles • **MATH LAB** Activity 368
3. **MATH LAB** Construct Parallel Lines 371
4. Similar and Congruent Figures 372
   Problem Solving: Thinker's Corner
   Chapter 19 Review/Test 374
   Standardized Test Prep 375
   Intervention: Troubleshooting H14, H22
   Extra Practice H50

UNIT WRAPUP

Problem Solving: Math Detective	376
Challenge: Explore Perspective	377
Study Guide and Review	378
Performance Assessment	380
Technology Linkup Mighty Math Cosmic Geometry: Lines, Angles, and Plane Figures	381
Problem Solving: On Location Geometric Patterns and Structures	381A

xv

UNIT 7
CHAPTERS 20–23

Ratio, Proportion, Percent, and Probability

Chapter 20

RATIO AND PROPORTION **382**
- ✓ Check What You Know 383
- 1 Ratios and Rates 384
- 2 **MATH LAB** **ALGEBRA** Explore Proportions 387
- 3 **Problem Solving Strategy:** Write an Equation 388
- 4 **ALGEBRA** Ratios and Similar Figures • **MATH LAB** Activity 390
   Problem Solving: Linkup to Careers
- 5 **ALGEBRA** Proportions and Similar Figures 394
- 6 **ALGEBRA** Scale Drawings • **MATH LAB** Activity 397
- 7 **ALGEBRA** Maps 400
   Chapter 20 Review/Test 402
- ★ Standardized Test Prep 403
   Intervention: Troubleshooting H21–H22
   Extra Practice H51

Chapter 21

- ✓ **PERCENT AND CHANGE** **404**
   Check What You Know 405
- 1 Percent 406
- 2 Percents, Decimals, and Fractions 408
   Problem Solving: Thinker's Corner
- 3 Estimate and Find Percent of a Number • **MATH LAB** Activity 412
   Problem Solving: Linkup to Reading
- 4 **MATH LAB** Construct Circle Graphs 416
- 5 Discount and Sales Tax 418
- 6 Simple Interest • **MATH LAB** Activity 422
   Chapter 21 Review/Test 424
- ★ Standardized Test Prep 425
   Intervention: Troubleshooting H22–H23
   Extra Practice H52

TECHNOLOGY LINK

Harcourt Math Newsroom Video:
Chapter 20, p. 391

E-Lab:
Chapter 20, pp. 387, 398
Chapter 21, p. 417
Chapter 22, p. 435

Mighty Math Cosmic Geometry:
Chapter 20, p. 397

Mighty Math Number Heroes:
Chapter 23, p. 448

Multimedia Glossary:
The Learning Site at www.harcourtschool.com/mathglossary

xvi

Chapter 22

PROBABILITY OF SIMPLE EVENTS **426**

   ✓ Check What You Know 427
1   Theoretical Probability 428
      Problem Solving: Thinker's Corner
2   **Problem Solving Skill:**
      Too Much or Too Little Information 432
3   **MATH LAB** Simulations 434
4   Experimental Probability 436
      Chapter 22 Review/Test 438
  ⭐ Standardized Test Prep 439
      Intervention: Troubleshooting H9, H23–H24
      Extra Practice H53

Chapter 23

PROBABILITY OF COMPOUND EVENTS **440**

   ✓ Check What You Know 441
1   **Problem Solving Strategy:**
      Make an Organized List 442
2   Compound Events 444
      Problem Solving: Thinker's Corner
3   Independent and Dependent Events 447
4   Make Predictions 450
      Chapter 23 Review/Test 452
  ⭐ Standardized Test Prep 453
      Intervention: Troubleshooting H23–H24
      Extra Practice H54

UNIT WRAPUP

Problem Solving: Math Detective 454
Challenge: Percent of Increase and Decrease 455
Study Guide and Review 456
Performance Assessment 458
Technology Linkup E-Lab: Simulations 459
  Problem Solving: On Location
    In New York City .... .459A

xvii

UNIT 8
CHAPTERS 24–27
Measurement

Chapter 24

UNITS OF MEASURE **460**
- ✓ Check What You Know 461
1. *Algebra* Customary Measurements 462
2. *Algebra* Metric Measurements 464
3. *Algebra* Relate Customary and Metric 466
4. Appropriate Tools and Units ● **MATH LAB** Activity 468
   Problem Solving: Thinker's Corner
5. **Problem Solving Skill:** Estimate or Find Exact Answer 472
   Chapter 24 Review/Test 474
   ⭐ Standardized Test Prep 475
   Intervention: Troubleshooting H25–H26
   Extra Practice H55

Chapter 25

LENGTH AND PERIMETER **476**
- ✓ Check What You Know 477
1. ● **MATH LAB** Estimate Perimeter 478
2. Perimeter ● **MATH LAB** Activity 479
3. **Problem Solving Strategy:** Draw a Diagram 482
4. Circumference ● **MATH LAB** Activity 484
   Problem Solving: Linkup to Social Studies
5. ● **MATH LAB** The Pythagorean Theorem 488
   Chapter 25 Review/Test 490
   ⭐ Standardized Test Prep 491
   Intervention: Troubleshooting H23, H26–H27
   Extra Practice H56

TECHNOLOGY LINK

Harcourt Math Newsroom Video:
Chapter 25, p. 485

E-Lab:
Chapter 26, p. 501
Chapter 27, p. 520

Multimedia Glossary:
The Learning Site at www.harcourtschool.com/mathglossary

xviii

Chapter 26

AREA 492

✓ Check What You Know 493
1 Estimate and Find Area 494
   Problem Solving: Thinker's Corner
2 **Algebra** Areas of Parallelograms
   and Trapezoids • **MATH LAB** Activity 498
3 • **MATH LAB** Area of a Circle 501
4 **Algebra** Areas of Circles 502
5 **Algebra** Surface Areas of Prisms and Pyramids 504
   Problem Solving: Thinker's Corner
   Chapter 26 Review/Test 508
★ Standardized Test Prep 509
   Intervention: Troubleshooting H18, H20, H27
   Extra Practice H57

Chapter 27

VOLUME 510

✓ Check What You Know 511
1 Estimate and Find Volume • **MATH LAB** Activity ... 512
   Problem Solving: Linkup to Reading
2 Problem Solving Strategy: Make a Model 516
3 **Algebra** Volumes of
   Pyramids • **MATH LAB** Activity 518
4 • **MATH LAB** Volume of a Cylinder 520
5 Volumes of Cylinders 521
   Chapter 27 Review/Test 524
★ Standardized Test Prep 525
   Intervention: Troubleshooting H27–H28
   Extra Practice H58

UNIT WRAPUP

Problem Solving: Math Detective 526
Challenge: Networks 527
Study Guide and Review 528
Performance Assessment 530
Technology Linkup Data Toolkit: Use a Spreadsheet to
   Find the Perimeter and Area of Rectangles 531
   Problem Solving: On Location In Illinois 531A

xix

UNIT 9
CHAPTERS 28–30
Algebra: Patterns and Relationships

Chapter 28

ALGEBRA PATTERNS **532**
- Check What You Know 533
1. **Problem Solving Strategy:** Find a Pattern 534
2. Patterns in Sequences 536
3. Number Patterns and Functions 539
   Problem Solving: Thinker's Corner
4. Geometric Patterns 543
   Chapter 28 Review/Test 546
- **Standardized Test Prep** 547
   Intervention: Troubleshooting H18, H29
   Extra Practice H59

Chapter 29

GEOMETRY AND MOTION **548**
- Check What You Know 549
1. Transformations of Plane Figures 550
2. Tessellations ● **MATH LAB** Activity 553
3. **Problem Solving Strategy:** Make a Model 556
4. Transformations of Solid Figures ● **MATH LAB** Activity 558
5. Symmetry ● **MATH LAB** Activity 560
   Problem Solving: Linkup to Reading
   Chapter 29 Review/Test 564
- **Standardized Test Prep** 565
   Intervention: Troubleshooting H30–H31
   Extra Practice H60

TECHNOLOGY LINK

Harcourt Math Newsroom Video:
Chapter 28, p. 539

E-Lab:
Chapter 28, p. 541
Chapter 30, p. 581

Mighty Math Cosmic Geometry:
Chapter 29, p. 554

Mighty Math Astro Algebra:
Chapter 30, p. 569

Multimedia Glossary:
The Learning Site at www.harcourtschool.com/mathglossary

xx

Chapter 30

ALGEBRA GRAPH RELATIONSHIPS **566**
  Check What You Know 567
1 Inequalities on a Number Line 568
2 Graph on the Coordinate Plane 570
  Problem Solving: Linkup to Science
3 Graph Functions 574
4 Problem Solving Skill: Make Generalizations 576
5 ● MATH LAB Explore Linear and
  Nonlinear Relationships 578
6 Graph Transformations ● MATH LAB Activity 580
  Chapter 30 Review/Test 584
  Standardized Test Prep 585
  Intervention: Troubleshooting H29–H31
  Extra Practice H61

UNIT WRAPUP

Problem Solving: Math Detective 586
Challenge: Stretching Figures 587
Study Guide and Review 588
Performance Assessment 590
Technology Linkup Calculator: Explore
     Graphing Equations 591
  Problem Solving: On Location In Maryland 591A

STUDENT HANDBOOK

Table of Contents H1
Troubleshooting H2–H31
Extra Practice H32–H61
Sharpen Your Test-Taking Skills H62–H65
Skills Review H66–H68
Glossary H69–H78
Selected Answers H79–H93
Index H94–H107
Table of Measures .... Back Cover

xxi

WELCOME!

The authors of *Harcourt Math* want you to be a good mathematician, but most of all we want you to enjoy learning math and feel confident that you can do it. We invite you to share your math book with family members. Take them on a guided tour through your book!

THE GUIDED TOUR

Choose a chapter you are interested in. Show your family some of these things in the chapter that will help you learn.

✓ CHECK WHAT YOU KNOW

Do you need to review any skills before you begin the next chapter? If you do, you will find help in the Handbook in the back of your book.

THE MATH LESSONS

- ☑ **Quick Review** to check the skills you need for the lesson.

- ☑ **Learn section** to help you study problems, models, examples, and questions that give you different ways to learn.

- ☑ **Check for Understanding** to make sure you understood the lesson.

- ☑ **Practice and Problem Solving** to help you practice what you have just learned.

- ☑ **Mixed Review and Test Prep** to keep your skills sharp and prepare you for important tests. Look back at the pages shown by each problem to get help if you need it.

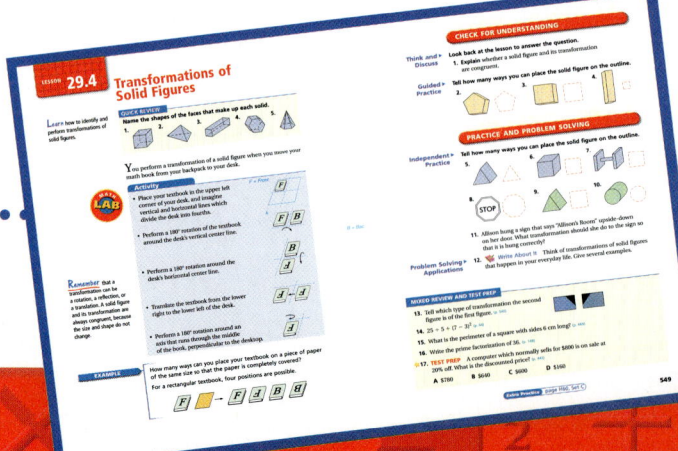

xxii

STUDENT HANDBOOK

Now show your family the **Student Handbook** in the back of your book. The sections will help you in many different ways.

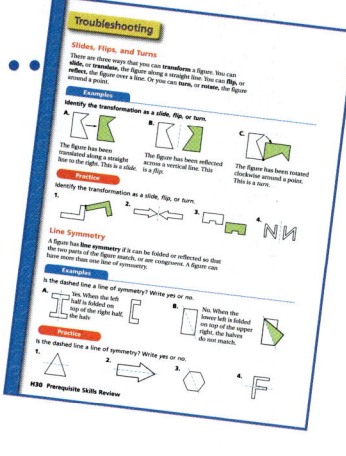

- ✓ **Troubleshooting** will help you review and remember skills from last year.

- ✓ **Extra Practice** will help you make sure that you are ready to move on to the next lesson.

- ✓ **Sharpen Your Test-Taking Skills** will help you feel confident that you can do well on a test.

- ✓ **Skills Review** will help you review addition, subtraction, multiplication, and division of whole numbers and decimals to improve your skills.

- ✓ **Answers to Selected Exercises** will help you check your answers to part of an assignment.

Invite your family members to:

- ▶ talk with you about what you are learning.
- ▶ help you correct errors you have made on completed work.
- ▶ help you set a time and find a quiet place to do math homework.
- ▶ solve problems with you as you play, shop, and do household chores together.
- ▶ visit **The Learning Site** at **www.harcourtschool.com**
- ▶ have ***FUN WITH MATH!***

Have a great year!

The Authors

xxiii

FOCUS ON PROBLEM SOLVING

Good problem solvers need to be good thinkers. They also need to know the strategies listed on page 1. This plan can help you learn how to think through a problem.

Analyze the problem.

What are you asked to find?	Restate the question in your own words.
What information is given?	Look for numbers. Find how they are related.
Is there information you will not use? If so, what?	Decide whether you need all the information you are given.

Choose a strategy to solve.

What strategy will you use?	Think about some problem solving strategies you can use. Then choose one.

Solve the problem.

How can you use the strategy to solve the problem?	Follow your plan. Show your solution.

Check your answer.

Look back at the problem. Does the answer make sense? Explain.	Be sure you answered the question that is asked.
What other strategy could you use?	Solving the problem by another method is a good way to check your work.

Try It

Here's how you can use the problem-solving steps to solve a problem.

Draw a Diagram

During summer vacation, Kayla will visit her cousin and then her grandmother. She will be gone for 5 weeks and 2 days and will spend 9 more days with her cousin than with her grandmother. How long will she stay with each?

PROBLEM SOLVING STRATEGIES
- ▶ Draw a Diagram or Picture
- Make a Model
- Predict and Test
- Work Backward
- Make an Organized List
- Find a Pattern
- Make a Table or Graph
- Solve a Simpler Problem
- Write an Equation
- Use Logical Reasoning

Analyze
Kayla will visit her cousin and grandmother for 5 weeks and 2 days this summer. I need to find out how long she will stay with each if she spends 9 more days with her cousin than with her grandmother.

Choose
I can *draw a diagram* to show how long she will stay. I can use boxes for the length of each stay. The length of the boxes will represent the lengths of the stays.

Solve
5 weeks and 2 days is 37 days in all. So, my diagram looks like this:

Step 1

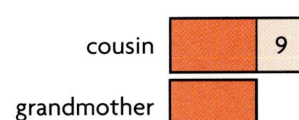

Sum of all parts (days) = 37.
37 − 9 = 28
28 ÷ 2 = 14

Step 2

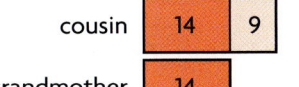

= 23
= 14   So, each part = 14.

So, Kayla will stay with her cousin for 23 days and with her grandmother for 14 days.

Check
23 days is 9 days longer than 14 days. The total of the two stays is 23 + 14, or 37 days. This solution fits the description of Kayla's vacation trip.

Focus on Problem Solving 1

Draw a Diagram

Eight basketball teams will play in a tournament. A team is out after it loses once. How many games does a team have to win in order to win the tournament?

Analyze

There are 8 teams. If a team loses, it is out of the tournament. If a team wins, it keeps playing.

Choose

You can *draw a diagram* to show which teams play in each game of the tournament. Use the diagram to show each pair of teams playing against each other.

Team 1
Team 2

Solve

Use the diagram to pair up the 8 teams. Make up a winner for each game. Then show the next games until the tournament has a winner.

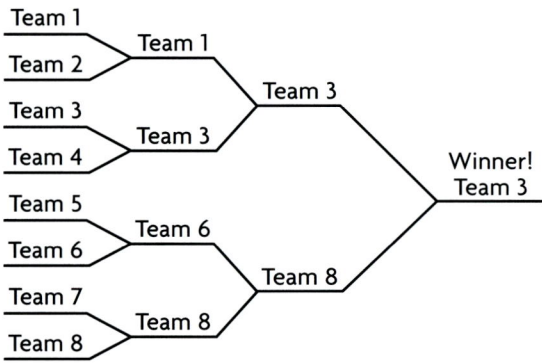

So, a team must win 3 games to win the tournament.

Check

Follow Team 3 through the tournament listing. Count the number of times it wins.

Problem Solving Practice

1. **What if** half as many teams play in the tournament? How many games will have to be played to determine the winner?
   2 games will determine the winner.
2. After the game, the teams have sandwiches at their coaches' houses. You can build your own sandwich with lettuce, tomato, and ham. How many different ways can you stack the 3 fillings to make a sandwich? **6 different ways**

2 Focus on Problem Solving

Make a Model

Alice has three pieces of ribbon. Their lengths are 7 in., 10 in., and 12 in. How can Alice use these ribbons to measure a length of 15 in.?

Analyze — Each of the three ribbons is a different length. No ribbon is 15 in. long, but you can combine the ribbon lengths to measure 15 in.

Choose — You can *make a model* of each piece of ribbon. Then you can organize the ribbons to show new lengths.

Solve — When you put two pieces together, you can form lengths of 17, 19, and 22 in. All of these are too long.

One Way
Alice can place the 7-in. piece above the 10-in. piece to show that the 10-in. piece extends 3 in. beyond it. Then that 3 in. piece and the 12-in. piece together will make 15 in.

Another Way
Alice can place the 7-in. piece above the 12-in. piece to show that the 12-in. piece extends 5 in. beyond it. Then that 5 in. and the 10-in. piece together will make 15 in.

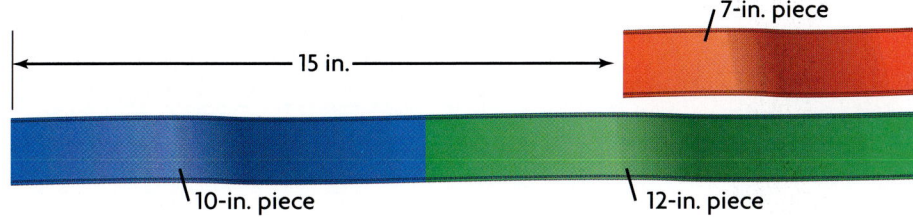

Check — Check that you used the correct lengths for the ribbon pieces. Then add the lengths of the two longer pieces and subtract the length of the short one. The result is 15 in.

Problem Solving Practice

1. Draw pictures of models you could make to measure other lengths with the three pieces of ribbon. **Check students' drawings.**

2. **What if** Andy is building a stand by using four cubes? He stacks the cubes, one on top of the other, and paints the outside of each cube (not the bottom). How many faces of the cubes are painted? **4 faces of bottom 3 cubes, 5 faces of top cube; 17 faces in all**

Focus on Problem Solving 3

Predict and Test

Margaret sells tickets to a pizza dinner at her school. An adult ticket costs $3.50. A student ticket costs $2.50. Margaret sells 10 more adult tickets than student tickets. She collects $155 in ticket sales. A total of 450 people attend the pizza dinner. How many student tickets has Margaret sold?

Analyze Student tickets cost $2.50 and adult tickets cost $3.50. Margaret sells 10 more adult tickets than student tickets. Her sales are $155. The number of people at the dinner is given, but this information is not needed.

Choose You can use *predict and test* to solve. Use number sense and the needed information to predict how many student tickets Margaret has sold. Then test your prediction and revise it if needed.

Solve Make a table to show each prediction and its result. Be sure you have 10 more adult tickets than student tickets each time.

PREDICTION		TEST	
Adult	Student	Sales	
20	10	(20 × $3.50) + (10 × $2.50) = $95	too low, revise
40	30	(40 × $3.50) + (30 × $2.50) = $215	too high, revise
30	20	(30 × $3.50) + (20 × $2.50) = $155	✓ correct

So, Margaret has sold 20 student tickets.

Check Check that each multiplication is correct: 30 × $3.50 = $105, and 20 × $2.50 = $50; $105 + $50 = $155. The answer checks.

Problem Solving Practice

1. The sum of Margaret's and her younger brother's ages is 38. The difference between their ages is 8. How old is Margaret's brother? **15 years old**

2. Tables for the pizza dinner are set up in 3 rooms. One room has 11 tables to seat 98 people. Some tables are for 8 people, and others are for 10 people. How many tables for 8 people are set up in this room? **6 tables**

4 Focus on Problem Solving

Work Backward

Joe and his brother Tim go shopping. At the toy store, they use half of their money to buy a video game. Then they go to the pizza parlor and spend half of the money they have left on a pizza. Then they spend half of the remaining money to rent a video. After these stops, they have $4.50 left. How much money did they have at the start?

Analyze

You need to find how much money they had at the start. You know that they had $4.50 left and that they spent half of their money at each of three stops.

Choose

Start with the amount you know they have left—$4.50—and *work backward*. Knowing how much they have left will help you calculate first how much they spent to rent the video, then how much the pizza cost, then the price of the video game, and finally how much money they had at the start.

Solve

amount they had left → $4.50

$4.50 → twice that amount before a video → 2 × $4.50 = $9

$9 → twice that amount before a pizza → 2 × $9 = $18

$18 → twice that amount before a video game → 2 × $18 = $36

So, the amount they had at the start was $36.

Check

Put $36 in the original problem to check the amount they spent at each stop.

$36 ÷ 2 = $18, $18 ÷ 2 = $9, $9 ÷ 2 = $4.50

The amount they had left matches the amount given in the problem.

Problem Solving Practice

1. The Lauber family has 4 children. Joe is 5 years younger than his brother Mark. Tim is half as old as his brother Joe. Mary, who is 10, is 3 years older than Tim. How old is Mark?
Mark is 19 yr old.
2. If you divide this mystery number by 4, add 8, and multiply by 3, you get 42. What is the mystery number? **24**

Focus on Problem Solving 5

Make an Organized List

At Fun City Amusement Park you can throw 3 darts at a target to score points and win prizes. If each dart lands within the target area, how many different total scores are possible?

Analyze

If all 3 darts hit the center circle, you get 30 points. This is the highest score. If all 3 darts hit the outside circle, you get 6 points. This is the lowest score. If the 3 darts hit a different combination, other scores are possible.

Choose

Make an organized list to determine all possible hits and score totals. List the value of each dart and the total for all three darts.

Solve

First consider all 3 darts hitting the center circle. List the value of each dart and the total score. Then consider 2 darts hitting the center circle and the third dart hitting a different circle. List the value of each dart and the total scores. Do the same for 1 dart hitting the center circle and no darts hitting the center circle.

3 Darts Hit Center	2 Darts Hit Center	1 Dart Hits Center	0 Darts Hit Center
10 + 10 + 10 = 30	10 + 10 + 5 = 25	10 + 5 + 5 = 20	5 + 5 + 5 = 15
	10 + 10 + 2 = 22	10 + 5 + 2 = 17	5 + 5 + 2 = 12
		10 + 2 + 2 = 14	5 + 2 + 2 = 9
			2 + 2 + 2 = 6

So, there are 10 possible scores.

Check

Make sure that all possible combinations of scores are listed and that each set of scores in the list is different.

Problem Solving Practice

1. The Yogurt Store at Fun City sells 3 flavors of yogurt: chocolate, vanilla, and strawberry. You want to get a scoop of each flavor in a waffle cone. How many different ways can the scoops be arranged? **6 ways**

2. How many ways can you make change for a quarter by using dimes, nickels, and pennies? **12 ways**

6 Focus on Problem Solving

Find a Pattern

A contractor can build stairways to a deck or patio with any number of steps. She uses the pattern at the right to build them. How many blocks are needed to build a stairway with 7 steps?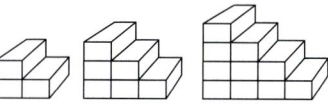

Analyze As the number of steps increases, so does the number of blocks. You must find the number of blocks needed for a stairway with 7 steps.

Choose You can *find a pattern* for the number of blocks needed. For each stairway, count the number of blocks to make each step and then find the total number of blocks. Use the pattern to find the number of blocks for 7 steps.

Solve The pattern shows that the number of the step in the stairway is the same as the number of blocks needed to make it. The 2nd step is made with 2 blocks, the 3rd step is made with 3 blocks, and so on.

Number of Step	Side View	Number of Blocks	Cumulative Total
2		2	2 + 1 = 3
3		3	3 + 2 + 1 = 6
4		4	4 + 3 + 2 + 1 = 10

So, a stairway with 7 steps has 7 blocks in the 7th step, 6 blocks in the 6th step, and so on. The total number of blocks is 7 + 6 + 5 + 4 + 3 + 2 + 1 = 28 blocks.

Check Sketch a stairway with 7 steps. Check the number of blocks needed. The number in the sketch matches the answer.

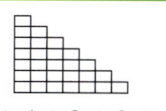

7 + 6 + 5 + 4 + 3 + 2 + 1 = 28

Problem Solving Practice

1. A cereal company adds baseball cards to certain boxes of cereal. Cards are added to the 3rd box, the 6th box, the 11th box, and the 18th box of cereal. If this pattern continues, how many boxes will have baseball cards in them when a case of 40 is ready to be shipped? Explain.

   6 boxes; the pattern is add the next odd number to each sum: 3 + 3 = 6, 6 + 5 = 11, 11 + 7 = 18, 18 + 9 = 27, 27 + 11 = 38.

Describe the pattern and find the missing numbers.

2. 1, 4, 16, 64, 256, ■, ■, 16,384   multiply by 4; 1,024; 4,096

Focus on Problem Solving 7

Make a Table

In math, Mrs. Laurence gave her class a 100-point test. Students who score 80 or above are eligible to join the Math Club. The test scores are in the box below. How many students scored 80 or above? How many students scored below 80?

90	83	80	77	78	91	92	73
62	83	79	88	72	85	93	84
75	68	82	75	94	70	98	82

Analyze You have the test scores for the class. You need to find how many students scored below 80 and how many students scored 80 or above.

Choose You can *make a table* and tally the test scores. This organizes the data and makes it easier to answer questions about the scores.

Solve Make a two-column table for the range of scores. As you read each test score, place a tally mark across from the appropriate range. Be sure you end up with 24 tallies, one for each student.

SCORES	TALLIES
60–69	||
70–79	ЖЖ |||
80–89	ЖЖ |||
90–99	ЖЖ |

Now use the table to answer the questions.

SCORES OF 80 OR ABOVE	SCORES BELOW 80
8 students scored 80–89.	8 students scored 70–79.
6 students scored 90–99.	2 students scored 60–69.

So, 14 students scored 80 or above, and 10 students scored below 80.

Check Recount the tallies in each row and check your addition.

Problem Solving Practice

1. Starting at 3:00, the director will give each student 5 minutes to try out for the talent show. The students, in order, are Ben, Tarek, Jan, Ed, and Frank. At what time does Frank start? **3:20**

2. Kelly reads stories to children at the library. There are three sessions, each lasting 45 minutes, with 30 minutes between sessions. If Kelly starts reading at 10:00 A.M., at what time does she finish? **1:15 P.M.**

8 Focus on Problem Solving

Solve a Simpler Problem

Make a table of the first 2 odd numbers and their sum, the first 3 odd numbers and their sum, and so on. How does the table show a pattern? What is the sum of the first 20 odd numbers?

Analyze You know how to begin finding sums for the first 2 odd numbers and the first 3 odd numbers. You need to find a pattern in the table to help you find the sum of the first 20 odd numbers.

Choose You can begin by *solving a simpler problem*. Start with sums for the first 2 odd numbers, followed by sums for 3 odd numbers, 4 odd numbers, and 5 odd numbers.

Solve Organize the data in a table and look for a pattern in the sums.

ODD NUMBERS	SUM	PATTERN
1 + 3	4	2^2
1 + 3 + 5	9	3^2
1 + 3 + 5 + 7	16	4^2
1 + 3 + 5 + 7 + 9	25	5^2

The table shows a pattern in the sums that relates to square numbers. Extend the pattern in the table to 20 odd numbers.

    6 odd numbers → 36

    7 odd numbers → 49

    8 odd numbers → 64

20 odd numbers → 400

So, the sum of the first 20 odd numbers is 20^2, or 400.

Check Extend the table a few more rows to check that there really is a pattern of square numbers. For example, find sums for the first 9, 10, 11, and 12 odd numbers. The answers check.

Problem Solving Practice

1. Martha has 5 pairs of slacks and 4 blouses for school. How many different outfits can she make with these items? **20 outfits**

2. What is the least 5-digit number that can be divided by 50 with a remainder of 17? **1,017**

Focus on Problem Solving 9

Write an Equation

A school district can afford to build a gym with three basketball courts with an area of 18,700 square feet. The gym needs to be 110 feet wide to leave walking room around the courts. How long will the gym be?

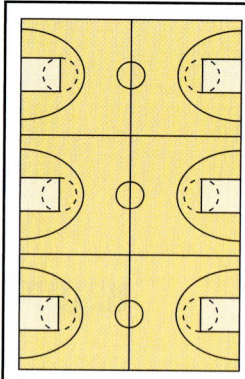

Analyze

The gym is rectangular. To find the area a rectangle covers, you multiply the length by the width. You know the area and the width of the gym and need to find its length.

Choose

You can *write an equation* for the area of the rectangle. Begin with the formula for the area of a rectangle. Use the numbers you know to find the missing number.

Solve

You find the area of a rectangle by multiplying its length by its width.

$A = l \times w$	Write the formula.
$18,700 = l \times 110$	Use the numbers you know.
$l = 18,700 \div 110$	Write a related division equation.
$l = 170$	

So, the length of the gym should be 170 feet.

Check

Place your answer in the original equation.

$A = l \times w$
$A = 170 \times 110$
$A = 18,700$

The product matches the information in the problem.

Problem Solving Practice

1. A rectangular box of crackers has a volume of 128 cubic inches. The area of its base is 16 square inches. What is the height of the box?
   $128 = 16 \times h$, or $128 \div 16 = h$; $h = 8$ in.

2. The perimeter of an isosceles triangle is 70 cm. The sides of equal length each measure 28 cm. What is the length of the unknown side? $70 = 28 + 28 + s$, or $70 - 56 = s$, $s = 14$ cm

10 Focus on Problem Solving

Use Logical Reasoning

Christine, Elizabeth, and Sharon like different sports. One girl likes to snow ski, one likes to run track, and the other likes to swim. Elizabeth is a good friend of the girl who likes track. Sharon is shorter than the girl who likes to ski. Elizabeth does not like cold weather. Which girl likes which sport?

Analyze

You know that there are three girls and that each one likes a different sport. Clues will help you determine which girl likes which sport.

Choose

You can *use logical reasoning* to solve this problem.

Solve

Make a table. List the sports and the names of the friends. Work with the clues one at a time. Write "yes" in a box if the clue applies to that girl. Write "no" in a box if the clue does not apply. Only one box in each row and column can have a "yes" in it.

a. Elizabeth is a good friend of the girl who likes track, so Elizabeth does not run track. Write "no" in the appropriate box.

	ski	track	swim
Christine	yes	no	no
Elizabeth	no	no	yes
Sharon	no	yes	no

b. Sharon is shorter than the girl who likes to ski, so Sharon does not ski. Write "no" in the appropriate box.

c. Elizabeth does not like cold weather. Write "no" in the appropriate box. With the other sports eliminated, Elizabeth must be the person who likes to swim. Write "yes" in the appropriate box. For Christine and Sharon, write "no" in each box under *swim*.

So, Christine likes to snow ski, Sharon likes track, and Elizabeth likes to swim.

Check

Compare your answers to the clues in the problem. Make sure none of your conclusions conflict with the clues.

Problem Solving Practice

Mike: 3-cheese; Brent: sausage; Tim: pepperoni

- Brent, Michael, and Tim always order their favorite pizzas: pepperoni, three-cheese, and sausage. Brent is allergic to pepperoni. Tim is going to see a movie with the boy who loves sausage. Mike does not like meat on his pizza. Which kind of pizza does each boy order?

Compare Strategies

Mrs. Hagen is eager to plant her spring garden. She wants to plant 24 tomato plants in a rectangular array. For each plant, there will be a 1-foot square of space. How many different arrays can she make?

Analyze There are 24 plants to be put into a rectangular array. You need to find how many arrays are possible.

Choose Lexi and Jake use different strategies to solve this problem. Lexi chooses to *draw a diagram*, and Jake chooses to *write an equation*.

Solve

Lexi's Way: Draw a Diagram

Lexi uses graph paper to draw diagrams of rectangles to show all possible arrays. Each rectangle represents a garden that covers 24 square feet. The rectangles are 1×24, 2×12, 3×8, and 4×6.

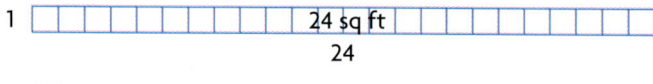
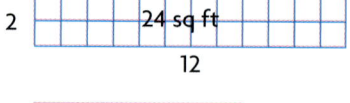
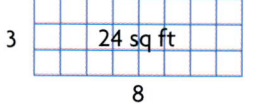
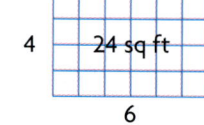

Jake's Way: Write an Equation

Jake uses the equation $A = l \times w$ for finding the area of a rectangle to identify the different rectangles that could be formed. Each length and width needs to be a whole number.

$A = l \times w$
$24 = 1 \times 24$
$24 = 2 \times 12$
$24 = 3 \times 8$
$24 = 4 \times 6$

Think: What are the factors of 24?

So, Lexi and Jake both find that Mrs. Hagen can choose from four rectangular arrays.

Check Lexi can do the problem Jake's way, and Jake can try it Lexi's way.

Problem Solving Practice

1. Mr. Sargent is buying math games for his class. One game costs $2.95, one costs $3.75, and one costs $6.00. He wants 6 of each game for his class. What is the total cost of these games? **$76.20**
2. How many triangles are in the figure at the right? **25**

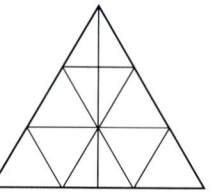

12 Focus on Problem Solving

Multistep Problems

The Enrichment Program Committee at the Franklin School orders 500 T-shirts to sell at the school Field Day. Each T-shirt costs the school $4.20 and sells for $13.50. All 500 T-shirts are sold. After paying for the T-shirts, the school has reached its goal of raising $4,000. How much money has the school made?

Look at the facts in the problem. T-shirts cost $4.20 each and sell for $13.50. The committee orders and sells 500 T-shirts. You need to find how much money has been made after the costs are deducted. You know your answer should be at least $4,000.

One Way

First subtract to find how much the school has made on each T-shirt.
　　money earned on 1 T-shirt: $13.50 − $4.20 = **$9.30**

Then multiply to find the total earned.
　　money earned on 500 T-shirts: $9.30 × 500 = **$4,650**

Another Way

Find the total sales and subtract the total cost.

　　Total sales: 500 × $13.50 = **$6,750**

　　Total cost: 500 × $4.20 = **$2,100**

　　Total earned = $6,750 − $2,100 = **$4,650**

So, the school has made $4,650 on the sale of the T-shirts.

Problem Solving Practice

1. At the school Field Day, frozen yogurt cones sell for $2.25. Lin buys one for herself and one for each of her 4 friends. How much change does she get back from $20? **$8.75**

2. Fran is making a cabinet for her shell collection. She buys 3 boards at $2.75 each and 4 hinges at $0.99 each. What is the total cost of the supplies? **$12.21**

3. Mrs. Smith bought bread for $0.98, eggs for $1.17, ground turkey for $3.18, and 3 bunches of carrots for $0.65 each. She gave the clerk $10.03. How much change should Mrs. Smith receive? **$2.75**

Focus on Problem Solving

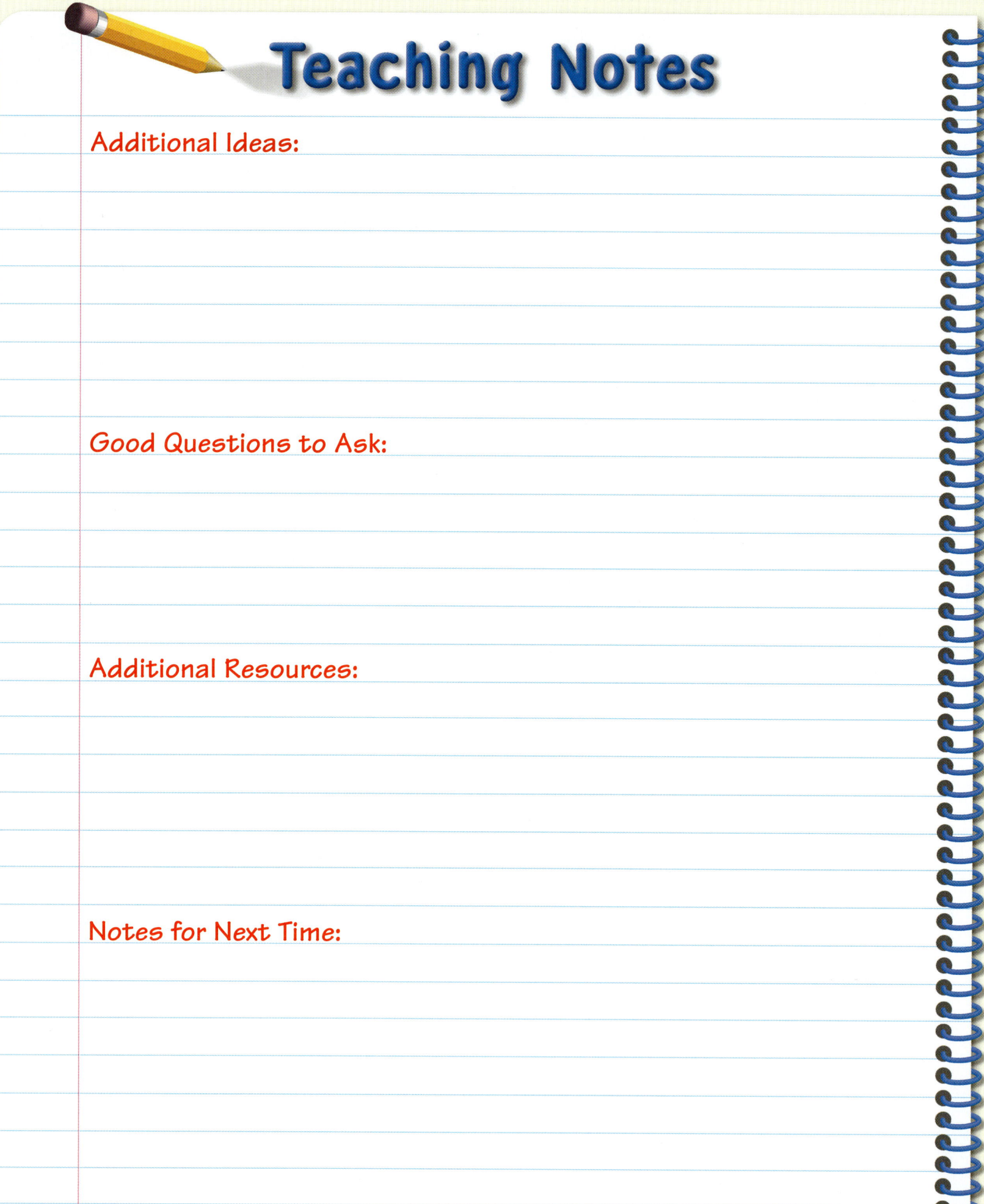

UNIT 1 Number Sense and Operations

UNIT AT A GLANCE

CHAPTER 1
Whole Number Applications 14
- **Lesson 1** — Estimate with Whole Numbers
- **Lesson 2** — Use Addition and Subtraction
- **Lesson 3** — Use Multiplication and Division
- **Lesson 4** — Problem Solving Strategy: *Predict and Test*
- **Lesson 5** — Algebra: Use Expressions
- **Lesson 6** — Algebra: Mental Math and Equations

CHAPTER 2
Operation Sense 34
- **Lesson 1** — Mental Math: Use the Properties
- **Lesson 2** — Algebra: Exponents
- **Lesson 3** — Math Lab: Explore Order of Operations
- **Lesson 4** — Algebra: Order of Operations
- **Lesson 5** — Problem Solving Skill: *Sequence and Prioritize Information*

CHAPTER 3
Decimal Concepts 50
- **Lesson 1** — Represent, Compare, and Order Decimals
- **Lesson 2** — Problem Solving Strategy: *Make a Table*
- **Lesson 3** — Estimate with Decimals
- **Lesson 4** — Decimals and Percents

CHAPTER 4
Decimal Operations 64
- **Lesson 1** — Add and Subtract Decimals
- **Lesson 2** — Multiply Decimals
- **Lesson 3** — Math Lab: Explore Division of Decimals
- **Lesson 4** — Divide with Decimals
- **Lesson 5** — Problem Solving Skill: *Interpret the Remainder*
- **Lesson 6** — Algebra: Decimal Expressions and Equations

UNIT 1: Number Sense and Operations

Assessment Options

Assessing Prior Knowledge
Determine whether students have the required prerequisite concepts and skills.
Check What You Know, PE pp. 15, 35, 51, 65

Test Preparation
Provide review and practice for chapter and standardized tests.
Standardized Test Prep, PE pp. 33, 49, 63, 85
Mixed Review and Test Prep
   See the last page of each PE skill lesson.
Study Guide and Review, PE pp. 88–89

Formal Assessment
Assess students' mastery of chapter concepts and skills.
Chapter Review/Test
   PE pp. 32, 48, 62, 84
Pretest and Posttest Options
   Chapter Test, Form A
      pp. AG9–10, 13–14, 17–18, 21–22
   Chapter Test, Form B
      pp. AG11–12, 15–16, 19–20, 23–24
Unit 1 Test • Chapters 1–4, pp. AG25–32

Daily Assessment
Obtain daily feedback on students' understanding of concepts.
Quick Review
   See the first page of each PE lesson.
Mixed Review and Test Prep
   See the last page of each PE skill lesson.
Number of the Day
   See the first page of each TE skill lesson.
Problem of the Day
   See the first page of each TE skill lesson.
Lesson Quiz
   See the *Assess* section of each TE skill lesson.

Performance Assessment
Assess students' understanding of concepts applied to real-world situations.
Performance Assessment (Tasks A–B)
   PE, p. 90; pp. PA3–4

Student Self-Assessment
Have students evaluate their own work.
How Did I Do?, p. AGxvii
A Guide to My Math Portfolio, p. AGxix
Math Journal
   See *Write* in the *Assess* section of each TE skill lesson and TE pages 16B, 40B, 58B, 74.

Harcourt Electronic Test System Math Practice and Assessment
Make and grade chapter tests electronically.
This software includes:
- multiple-choice items
- free-response items
- customizable tests
- the means to make your own tests

Portfolio
Portfolio opportunities appear throughout the Pupil and Teacher's Editions.
Suggested work samples:
Problem Solving Project, TE pp. 14, 34, 50, 64
Write About It, PE pp. 19, 29, 39, 47, 61, 69
Chapter Review/Test, PE pp. 32, 48, 62, 84

KEY   **AG** Assessment Guide   **TE** Teacher's Edition   **PA** Performance Assessment   **PE** Pupil Edition

Correlation to STANDARDIZED TESTS

LEARNING GOAL		TAUGHT IN LESSONS	CAT	CTBS/ TERRA NOVA	ITBS	MAT	SAT	YOUR STATE TEST
1A	To write whole number estimates	1.1	•	•	•	•	•	
1B	To write whole number sums, differences, products, and quotients	1.2, 1.3	•	•	•	•	•	
1C	To evaluate expressions, and to use mental math to solve equations involving addition, subtraction, multiplication, or division	1.5, 1.6	•	•	•	•	•	
1D	To solve problems by using an appropriate problem solving strategy such as *predict and test*	1.4				•		
2A	To write whole number sums, differences, products, and quotients using number properties and mental math	2.1	•	•	•	•	•	
2B	To evaluate expressions using exponents	2.2	•	•	•	•	•	
2C	To evaluate expressions using order of operations	2.3, 2.4	•	•	•	•	•	
2D	To solve problems by using an appropriate problem solving skill such as *sequence and prioritize information*	2.5	•	•	•	•	•	
3A	To write, compare, and order decimals	3.1, 3.2	•	•	•	•	•	
3B	To write estimates of decimal sums, differences, products, and quotients	3.3	•	•	•	•	•	
3C	To write decimals as percents and percents as decimals	3.4	•	•				
3D	To solve problems by using an appropriate problem solving strategy such as *make a table*	3.2		•				
4A	To write sums, differences, products, and quotients of decimals	4.1, 4.2, 4.3, 4.4	•	•	•	•	•	
4B	To evaluate expressions and use mental math to solve equations involving decimals	4.6		•			•	
4C	To solve problems by using an appropriate skill such as *interpret the remainder*	4.5	•	•	•	•	•	

Number Sense and Operations 14C

UNIT 1 Number Sense and Operations
Technology Links

🌐 The Harcourt Learning Site
Visit The Harcourt Learning Site for related links, activities, and resources:
- Exponents and order of operations activity *(Use with Chapter 2.)*
- *Multimedia Math Glossary*
- E-Lab interactive learning experiences
- professional development and instructional resources

www.harcourtschool.com

Harcourt Math Newsroom Videos
These videos bring exciting news events to your classroom from the leaders in news broadcasting. For each unit, there is a **Harcourt Math Newsroom Video** that helps students see the relevance of math concepts to their lives. You may wish to use the data and concepts shown in the video for real-life problem solving or class projects.

See **Technology Linkup** in the Pupil Edition, p. 91.

TECHNOLOGY CORRELATION

Intervention Strategies and Activities This CD-ROM helps you assess students' knowledge of prerequisite concepts and skills.

Mighty Math CD-ROM Series includes *Calculating Crew, Number Heroes,* and *Astro Algebra.* These provide levels of difficulty that increase from A up to Z.

E-Lab is a collection of electronic learning activities.

The chart below correlates technology activities to specific lessons.

LESSON	ACTIVITY/LEVEL	SKILL
1.1	**Calculating Crew** • *Nautical Number Line,* Levels C, H **Harcourt Math Newsroom Video** • *Weather-Stacked Stadium*	Estimate with whole numbers
1.2	**Number Heroes** • *Quizzo,* Levels L, M	Use addition and subtraction
1.3	**Number Heroes** • *Quizzo,* Levels S, U, Y **Calculating Crew** • *Nick Knack,* Level V	Use multiplication and division
1.5	**Astro Algebra** • *Red,* Level J	Use expressions
1.6	**Astro Algebra** • *Red,* Level E	Use mental math and equations
2.2	**Astro Algebra** • *Red,* Level S *Calculator Handbook, pp. 1–2*	Use exponents
2.3	**E-Lab** • *Order of Operations* **Astro Algebra** • *Red,* Level B	Use order of operations
2.4	**Astro Algebra** • *Red,* Level B	Use order of operations
3.1	**Calculating Crew** • *Nautical Number Line,* Levels O, P	Represent, compare, and order decimals
4.1	**Calculating Crew** • *Nautical Number Line,* Level R	Add and subtract decimals
4.2	**Calculating Crew** • *Nautical Number Line,* Level U *Calculator Handbook, p. 5*	Multiply decimals
4.3	**E-Lab** • *Exploring Division of Decimals*	Divide decimals
4.4	**Number Heroes** • *Quizzo,* Level W *Calculator Handbook, p. 6*	Divide decimals

For the Student

 Intervention Strategies and Activities

Review and practice the prerequisite skills for Chapters 1–4.

E-LAB These interactive learning experiences reinforce and extend the skills taught in Chapters 1–4.
- Skill development
- Practice

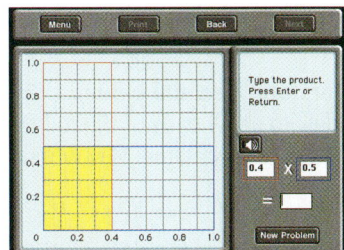

 Mighty Math

The learning activities in this comprehensive math software series complement, enrich, and enhance the Pupil Edition lessons.

 Calculating Crew
- *Nautical Number Line*
- *Nick Knack*

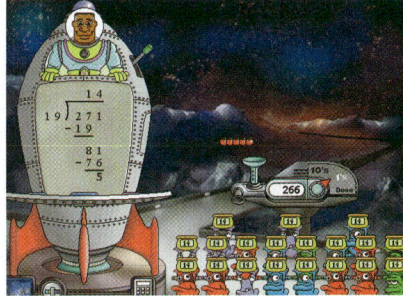

 Number Heroes • *Quizzo*

 Astro Algebra • *Red*

For the Teacher

 Teacher Support Software

- **Intervention Strategies and Activities** Provide instruction, practice, and a check of the prerequisite skills for each chapter.
- **Electronic Lesson Planner** Quickly prepare daily and weekly lessons for all subject areas.
- **Harcourt Electronic Test System Math Practice and Assessment** Edit and customize Chapter Tests or construct unique tests from large item banks.

For the Parent

 The Harcourt Learning Site

- Encourage parents to visit The Harcourt Learning Site to help them reinforce mathematics vocabulary, concepts, and skills with their children.
- Have them click on *Math* for vocabulary, activities, real-life connections, and homework tips for Chapters 1–4.

www.harcourtschool.com

Internet

Teachers can find number sense and operations activities and resources.

Students can learn more about exponents and order of operations and reinforce the critical concepts and skills for Chapters 1–4.

Parents can use The Harcourt Learning Site's resources to help their children with the vocabulary, concepts, and skills needed for Chapters 1–4.

Visit The Harcourt Learning Site
www.harcourtschool.com

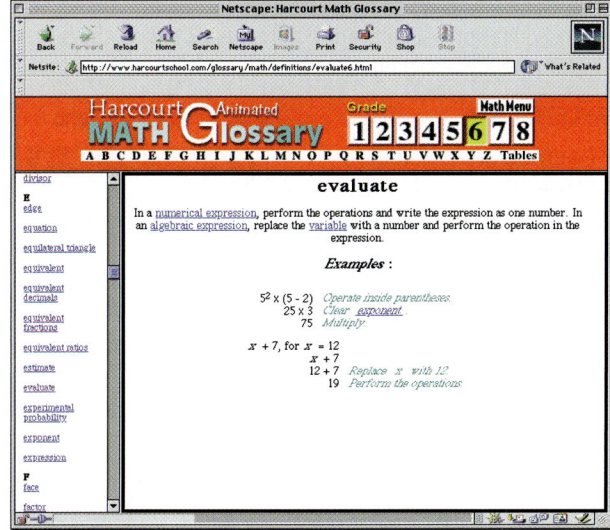

Number Sense and Operations 14E

UNIT 1 Number Sense and Operations

Reaching All Learners

ADVANCED LEARNERS

MATERIALS *For each pair* number lines, p. TR12

Challenge students to **compare and order decimals.** Display the following:

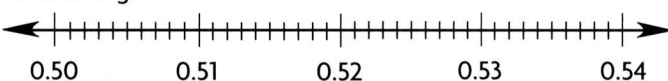

What hundredth is closest to 0.507? 0.51
What hundredth is closest to 0.528? 0.53
What hundredth is closest to 0.513? 0.51

Ask students to show where each decimal is placed on the number line. Then ask them to write the answer to each question. Challenge students to create a similar set of questions for a number line with decimals 0.74, 0.75, 0.76, 0.77. *Use with Lesson 3.1.*

VISUAL; LOGICAL/MATHEMATICAL

SPECIAL NEEDS

MATERIALS *For each pair* decimal models (tenths), p. TR6; decimal models (hundredths), p. TR7; markers or colored pencils

To help students **understand decimal values,** ask them to shade a pair of decimal models to show each of the following statements. Then have them complete each statement by writing < or >.

0.23 ● 0.3 <   0.62 ● 0.08 >   0.5 ● 0.05 >

Have a volunteer read each statement aloud. Ask students to write and illustrate three more comparison statements. *Use with Lessons 3.1 or 3.4.*

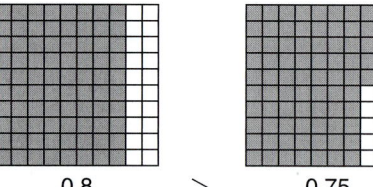

0.8 > 0.75

AUDITORY; VISUAL/SPATIAL

BLOCK SCHEDULING

INTERDISCIPLINARY COURSES

- Astronomy—Relate writing large numbers, such as distance or volume, to scientific notation.

- Social Studies—Round large numbers to a reasonable place value in order to compare population data quickly.

- Physical Fitness—Use historical Olympic data to compare and order Olympic finish times for various events.

- Health Education—Explore the percents noted on the sides of common food packaging.

COMPLETE UNIT

Unit 1 may be presented in
- fourteen 90-minute blocks.
- seventeen 75-minute blocks.

INTERDISCIPLINARY SUGGESTIONS

PURPOSE To connect *Number Sense and Operations* to other subjects with these activities

CHAPTER 1 — Social Studies

Students round numbers to compare data. First students find and round the area of each state in the United States to the nearest thousand. Then they sort the states into three categories by size.

CHAPTER 3 — Earth Science

Students research the largest earthquakes recorded in the twentieth century. They then make a table listing the earthquakes by magnitude from least to greatest based on the Richter Scale (for example: Japan—7.7, Peru—7.8, India—8.7).

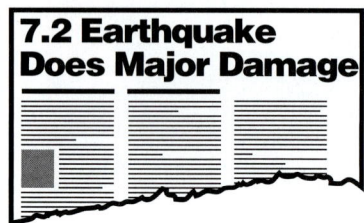

VISUAL; VERBAL/LINGUISTIC

ACTIVITIES AND GAMES FOR HOME OR SCHOOL

You may choose to use this activity and game in the classroom or send it home for students to do with family members.

ACTIVITY • *ZAP Algebra*

PURPOSE To solve algebraic equations *Use after Lesson 1.6.*

VARIATION After students have worked with decimal operations, you can adapt the activity by making 9 new game cards:

$0.8 + 0.2 = z$ $4.7 + 2.3 = a$ $0.3 + 3.7 = p$
$9.9 - 3.9 = z$ $11.6 - 6.6 = a$ $26.1 - 17.1 = p$
$3.2 \times 2.5 = z$ $4 \times 0.5 = a$ $2.5 \times 2 = p$

GAME • *Australian Expedition*

PURPOSE To add and subtract decimals (money)
Use after Lesson 4.1.
KINESTHETIC; BODILY KINESTHETIC

ENGLISH LANGUAGE LEARNERS

VOCABULARY PREVIEW Have students make a three-column chart for the new vocabulary. In the first column, have them write the word and in the second column, what they think the word means. Then in the third column, have them record the definition they learn. *Use with Lessons 1.1–3.4.*

Word	What I Think It Means	Definition

NUMERICAL EXPRESSION Numerical expressions are to equations as phrases are to sentences. Have students give examples of each.

Use with Lesson 1.5.

ORDER OF OPERATIONS Review the meaning of the word *order* by discussing phrases such as "keeping things in order." *Use with Lesson 2.3.*

AUDITORY; VERBAL/LINGUISTIC

LITERATURE CONNECTIONS

These books provide students with additional ways to explore whole number and decimal applications and operations.

FROM THE MIXED-UP FILES OF MRS. BASIL E. FRANKWEILER by E. L. Konigsberg (Atheneum, 1987) tells of Claudia and James, who run away from their "boring" home to the Museum of Natural History.

- Suppose Claudia and James stay away for a week and have $30. Have students devise a reasonable amount to spend per meal per person. *Use with Lesson 1.3.*

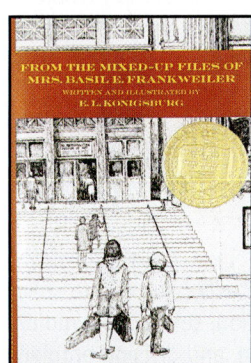

Not for a Billion, Gazillion Dollars by Paula Danziger (Delacorte Press, 1992) tells how Matthew learns about money management and friendship.

- Matthew saved $100 in coins. Have students decide if it is possible for him to have saved a total of 15,000, 1,000, or 800 coins. *Use with Lesson 4.6.*

Number Sense and Operations 14G

CHAPTER 1 Whole Number Applications

CHAPTER PLANNER

PACING OPTIONS
| Compacted | 5 Days |
| Expanded | 10 Days |

Getting Ready for Chapter 1 • Assessing Prior Knowledge and INTERVENTION (See PE and TE page 15.)

LESSON	NCTM STANDARDS	PACING	VOCABULARY*	MATERIALS	RESOURCES AND TECHNOLOGY
1.1 Estimate with Whole Numbers pp. 16–19 **Objective** To use estimation strategies to find sums, differences, products, and quotients of whole numbers	1, 5, 6, 8	2 Days	**clustering** **underestimate** **overestimate** compatible numbers		Reteach, Practice, Challenge, Problem Solving 1.1 Worksheets Extra Practice p. H32, Set A Transparency 1.1 Calculating Crew • *Nautical Number Line* Harcourt Math Newsroom Video • *Weather-Stacked Stadium*
1.2 Use Addition and Subtraction pp. 20–21 **Objective** To use addition and subtraction of whole numbers to solve real-life problems	1, 5, 6, 8	1 Day			Reteach, Practice, Challenge, Problem Solving 1.2 Worksheets Extra Practice p. H32, Set B Transparency 1.2 Number Heroes • *Quizzo*
1.3 Use Multiplication and Division pp. 22–25 **Objective** To use multiplication and division of whole numbers to solve real-life problems	1, 2, 5, 6, 8, 9	2 Days		calculator	Reteach, Practice, Challenge, Problem Solving 1.3 Worksheets, Extra Practice p. H32, Set C Transparency 1.3 Number Heroes • *Quizzo* Calculating Crew • *Nick Knack* Math Jingles™ CD 5–6
1.4 Problem Solving Strategy: *Predict and Test* pp. 26–27 **Objective** To use the strategy *predict and test* to solve addition, subtraction, multiplication, and division problems with whole numbers	1, 6, 7	1 Day		calculator	Reteach, Practice, Challenge, Reading Strategy 1.4 Worksheets Transparency 1.4 Problem Solving Think Along, p. TR1
1.5 Algebra: Use Expressions pp. 28–29 **Objective** To identify, write, and evaluate numerical and algebraic expressions involving whole numbers	1, 2, 6, 7, 8	1 Day	**numerical expression** **variable** **algebraic expression** **evaluate**		Reteach, Practice, Challenge, Problem Solving 1.5 Worksheets Extra Practice p. H32, Set D Transparency 1.5 Astro Algebra • *Red* Math Jingles™ CD 5–6
1.6 Algebra: Mental Math and Equations pp. 30–31 **Objective** To solve equations with whole numbers by using mental math and substitution	1, 2, 6, 8	1 Day	**equation** **solution** variable		Reteach, Practice, Challenge, Problem Solving 1.6 Worksheets Extra Practice p. H32, Set E Transparency 1.6 Astro Algebra • *Red*

Ending Chapter 1 • Chapter 1 Review/Test, p. 32 • Standardized Test Prep, p. 33

***Boldfaced** terms are new vocabulary. Other terms are review vocabulary.

CHAPTER AT A GLANCE

Vocabulary Development

The boldfaced words are the new vocabulary terms in the chapter. Have students record the definitions in their Math Journals.

- **clustering**, p. 16
- **underestimate**, p. 17
- **overestimate**, p. 17
- **numerical expression**, p. 28
- **variable**, p. 28
- **algebraic expression**, p. 28
- **evaluate**, p. 28
- **equation**, p. 30
- **solution**, p. 30

Vocabulary Cards
Have students use the Vocabulary Cards on *Teacher's Resource Book* pp. 117–120 to make graphic organizers or word puzzles. The cards can also be added to a file of mathematics terms.

NCTM Standards

1. **Number and Operations**
   Lessons 1.1, 1.2, 1.3, 1.4, 1.5, 1.6
2. **Algebra**
   Lessons 1.3, 1.5, 1.6
3. **Geometry**
4. **Measurement**
5. **Data Analysis and Probability**
   Lessons 1.1, 1.2, 1.3
6. **Problem Solving**
   Lessons 1.1, 1.2, 1.3, 1.4, 1.5, 1.6
7. **Reasoning and Proof**
   Lessons 1.4, 1.5
8. **Communication**
   Lessons 1.1, 1.2, 1.3, 1.5, 1.6
9. **Connections**
   Lesson 1.3
10. **Representation**

Writing Opportunities

PUPIL EDITION
- Write About It, pp. 19, 29
- Write a Problem, p. 21
- What's the Error?, p. 25
- What's the Question?, pp. 27, 31

TEACHER'S EDITION
- Write—See the *Assess* section of each TE lesson.
- Writing in Mathematics, p. 16B

ASSESSMENT GUIDE
- How Did I Do?, p. AGxvii

Family Involvement Activities

These activities provide:
- Letter to the Family
- Math Vocabulary
- Family Game
- Practice (Homework)

Family Involvement Activities, p. FA1

14I

CHAPTER 1

Whole Number Applications

MATHEMATICS ACROSS THE GRADES

SKILLS TRACE ACROSS THE GRADES

GRADE 5
Estimate sums, differences, products, and quotients; find sums and differences; multiply by 2-digit numbers; divide by 2-digit divisors; evaluate and write whole number expressions

GRADE 6
Write whole number estimates, sums, differences, products, and quotients; evaluate expressions with whole numbers and use mental math to solve 1-step equations

GRADE 7
Write and evaluate numerical and algebraic expressions; solve multiple step equations; solve inequalities

SKILLS TRACE FOR GRADE 6

LESSON	FIRST INTRODUCED	TAUGHT AND PRACTICED	TESTED	REVIEWED
1.1	Grade 4	PE pp. 16–19, H32, p. RW1, p. PW1, p. PS1	PE p. 32, pp. AG9–12	PE pp. 32, 33, 88–89
1.2	Grade 4	PE pp. 20–21, H32, p. RW2, p. PW2, p. PS2	PE p. 32, pp. AG9–12	PE pp. 32, 33, 88–89
1.3	Grade 4	PE pp. 22–25, H32, p. RW3, p. PW3, p. PS3	PE p. 32, pp. AG9–12	PE pp. 32, 33, 88–89
1.4	Grade 4	PE pp. 26–27, p. RW4, p. PW4, p. PS4	PE p. 32, pp. AG9–12	PE pp. 32, 33, 88–89
1.5	Grade 5	PE pp. 28–29, H32, p. RW5, p. PW5, p. PS5	PE p. 32, pp. AG9–12	PE pp. 32, 33, 88–89
1.6	Grade 5	PE pp. 30–31, H32, p. RW6, p. PW6, p. PS6	PE p. 32, pp. AG9–12	PE pp. 32, 33, 88–89

KEY
PE Pupil Edition
PW Practice Workbook
PS Problem Solving Workbook
AG Assessment Guide
RW Reteach Workbook

Looking Back  Prerequisite Skills

To be ready for Chapter 1, students should have the following understandings and skills:

- **Vocabulary**—Factor, *product, divisor, dividend, sum, difference*
- **Place Value of Whole Numbers**—name place value to hundred billions
- **Round Whole Numbers**—round to nearest thousand or ten-thousand

Check What You Know

Use page 15 to determine students' knowledge of prerequisite concepts and skills.

Intervention

Help students prepare for the chapter by using the intervention resources described on TE page 15.

Looking at Chapter 1  Essential Skills

- develop skill and accuracy with whole number operations.
- use the strategy *predict and test* to solve problems.
- understand the function of variables in algebraic expressions.
- make the connection between word expressions and algebraic and numerical expressions.
- **develop skill in solving equations with one variable by using mental math.**

EXAMPLE

Solve. $n + 6 = 13$

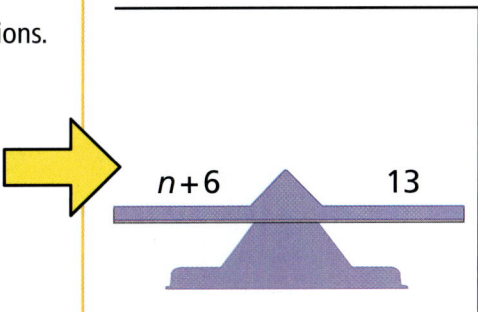

Think: What number plus 6 is 13?

$7 + 6 = 13$

So, $n = 7$.

Looking Ahead  Applications

Students will apply what they learn in Chapter 1 to the following new concepts:

- Operations with Decimals (Chapter 4)
- Operations with Integers (Chapter 12)
- Writing Expressions (Chapter 13)
- Solving Equations (Chapters 14 and 15)

Whole Number Applications 14K

Whole Number Applications

INTRODUCING THE CHAPTER

Tell students that operations with whole numbers and estimation are often used to compare data. Have students formulate a question about the photograph that can be answered by applying what they know about whole numbers. Possible answer: Using estimation, what is the total number of hours for the five shuttle missions shown on the graph? about 1,000 hr

USING DATA

To begin the study of this chapter, have students

- Write a number sentence to tell the total length of the three shortest missions to the nearest day. $1 + 5 + 8 = 14$ days
- Determine which missions had a combined length of time almost equal to the length of the 1996 mission. STS-1, STS-9 and STS-34
- Display the data in the graph in another way. Check students' work.

PROBLEM SOLVING PROJECT

Purpose To use whole numbers to solve a problem

Grouping pairs or small groups

Background The longest space shuttle flight (through 1/1/98) was STS-80, which lasted 423 hr, 53 min, 18 sec from Nov. 19–Dec. 7, 1996.

Analyze, Choose, Solve, and Check

Have students

- Research the five longest space shuttle flights.
- Make a table with the mission dates, the length of the missions rounded to the nearest day, and the total distance each traveled.
- Write a summary of the data by describing how the lengths and distances of the missions compare to each other.

Check students' work.

 Suggest that students place the tables and summaries in their portfolios.

CHAPTER 1 Whole Number Applications

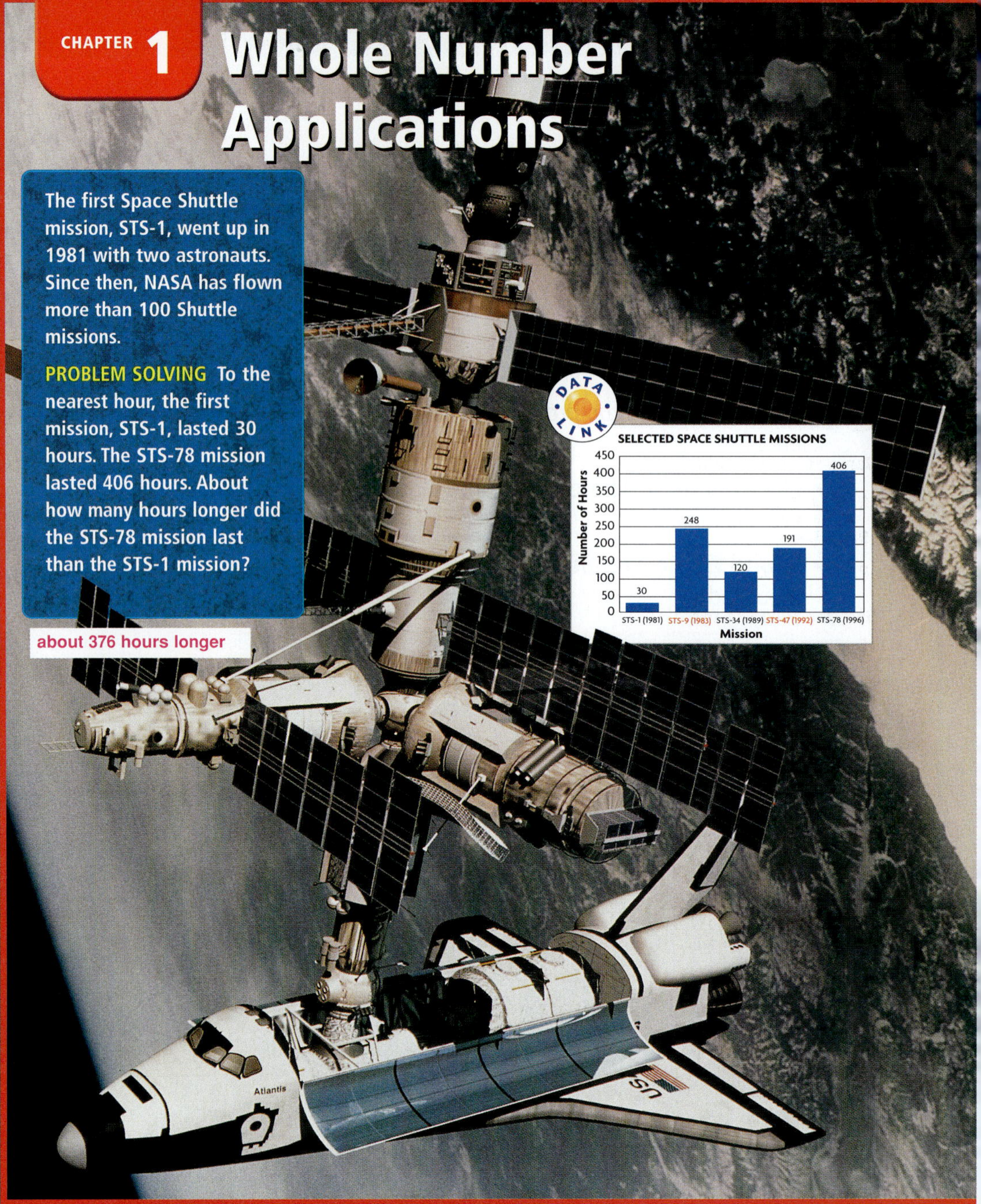

The first Space Shuttle mission, STS-1, went up in 1981 with two astronauts. Since then, NASA has flown more than 100 Shuttle missions.

PROBLEM SOLVING To the nearest hour, the first mission, STS-1, lasted 30 hours. The STS-78 mission lasted 406 hours. About how many hours longer did the STS-78 mission last than the STS-1 mission?

about 376 hours longer

SELECTED SPACE SHUTTLE MISSIONS

Mission	Number of Hours
STS-1 (1981)	30
STS-9 (1983)	248
STS-34 (1989)	120
STS-47 (1992)	191
STS-78 (1996)	406

Why learn math? Explain that astronauts and support staff use whole number operations to determine all kinds of essential information they need to travel successfully through space. For example, they need to know how long their oxygen supply can last, how much the payload weighs, and how much food they will need to bring for specific missions. Ask: What other ways do you think astronauts use math? Possible answer: to determine how much fuel they will need for maneuvering in space; how long it will take to perform mission experiments; landing time adjustments in case of bad weather

Check What You Know

Use this page to help you review and remember important skills needed for Chapter 1.

✓ Vocabulary

Choose the best term from the box.

> factor
> difference
> dividend
> divisor
> product
> sum

1. In 8 × 3 = 24, the ? is 24. **product**
2. In 48 ÷ 12 = 4, the ? is 12. **divisor**
3. In 40 − 30 = 10, the ? is 10. **difference**
4. In 4)‾412, the ? is 412. **dividend**
5. In 3 + 8 + 5 = 16, the ? is 16. **sum**

✓ Place Value of Whole Numbers (For Intervention, see p. H3.)

Name the place value of the digit 8.

6. 56,485,013 — **ten thousands**
7. 2,403,815 — **hundreds**
8. 4,568,137 — **thousands**
9. 423,859,100 — **hundred thousands**
10. 207,108,629 — **thousands**
11. 813,593,005 — **hundred millions**
12. 375,178,955,276 — **millions**
13. 351,480,994,133 — **ten millions**
14. 748,630,123,944 — **billions**

Give the value of the blue digit.

15. 321,497,580 — **90,000 or 9 ten thousands**
16. 321,497,580 — **500 or 5 hundreds**
17. 321,497,580 — **300,000,000 or 3 hundred millions**
18. 628,705,814,956 — **5,000,000 or 5 millions**
19. 628,705,814,956 — **20,000,000,000 or 2 ten billions**
20. 628,705,814,956 — **800,000 or 8 hundred thousands**

✓ Round Whole Numbers (For Intervention, see p. H4.)

Round to the nearest thousand.

21. 2,467 **2,000**
22. 5,609 **6,000**
23. 28,500 **29,000**
24. 299 **0**
25. 34,831 **35,000**
26. 19,089 **19,000**
27. 6,136 **6,000**
28. 43,712 **44,000**
29. 134,612 **135,000**
30. 217,501 **218,000**
31. 59,832 **60,000**
32. 539,513 **540,000**

Round to the nearest ten thousand.

33. 51,677 **50,000**
34. 228,260 **230,000**
35. 12,435 **10,000**
36. 78,562 **80,000**
37. 639,108 **640,000**
38. 40,500 **40,000**
39. 90,499 **90,000**
40. 381,810 **380,000**

LOOK AHEAD

In Chapter 1 you will
- estimate and compute with whole numbers
- evaluate expressions
- use mental math to compute and to solve equations

Assessing Prior Knowledge

Use the **Check What You Know** page to determine whether your students have mastered the prerequisite skills critical for this chapter.

Intervention

- **Diagnose and Prescribe**

  Evaluate your students' performance on this page to determine whether intervention is necessary or if enrichment is appropriate. Options that provide instruction, practice, and a check are listed in the chart below.

✓ CHECK WHAT YOU KNOW RESOURCES

Intervention Card, Copying Master, or CD-ROM

Intervention Strategies and Activities Teaching Transparencies

Intervention Practice Book

Enrichment Book

Were students successful with
✓ CHECK WHAT YOU KNOW?

OPTIONS

NO — INTERVENE

✓ PLACE VALUE OF WHOLE NUMBERS, 6–20
How to Help
Troubleshooting, Pupil Edition p. H3
Intervention Strategies and Activities, Skill 1

✓ ROUND WHOLE NUMBERS, 21–40
How to Help
Troubleshooting, Pupil Edition p. H4
Intervention Strategies and Activities, Skill 2

YES — ENRICH

Check What You Know Enrichment Book, pp. 1–2

LESSON 1.1

Estimate with Whole Numbers

LESSON PLANNING

Objective To use estimation strategies to find sums, differences, products, and quotients of whole numbers

Intervention for Prerequisite Skills
Place Value of Whole Numbers, Round Whole Numbers (For intervention strategies, see page 15.)

NCTM Standards
1. Number and Operations
5. Data Analysis and Probability
6. Problem Solving
8. Communication

Vocabulary

clustering a method used in estimation when all addends have about the same value

underestimate an estimate that is less than the exact answer

overestimate an estimate that is greater than the exact answer

Math Background

Estimation is a powerful tool that students can use to verify that answers are reasonable or to solve problems when exact answers are not needed. Students may choose from the following estimation techniques:

- **rounding**, used in addition, subtraction, and multiplication, where numbers are rounded to a specified place value before computation occurs.
- **clustering**, used when numbers in a group that are being added are near a single value, where numbers are rounded to the common value.
- **compatible numbers**, used in finding a quotient, where numbers are rounded to values that can be easily divided.

WARM-UP RESOURCES

 **NUMBER OF THE DAY**

Take the number of the year you were born. Round it to the nearest 10, 100, and 1,000. 1990, 2000, 2000

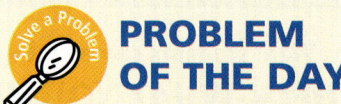 **PROBLEM OF THE DAY**

Find the three-digit number that rounds to 440 and includes a digit that is the quotient of 24 ÷ 3. Is there more than one possible answer? Explain. 438; No. The numbers that round to 440 are 435–444 and 24 ÷ 3 = 8. Only one of those numbers has 8 as a digit.

Solution Problem of the Day tab, p. PD1A

 **DAILY FACTS PRACTICE**

Have students practice addition facts by completing Set A of *Teacher's Resource Book,* p. TR93.

16A   Chapter 1

INTERVENTION AND EXTENSION RESOURCES

ALTERNATIVE TEACHING STRATEGY  ELL

Ask students to **practice estimating with whole numbers.** Have students round to the nearest ten to estimate the total annual energy cost for the use of these appliances. **$400** Then have them estimate the annual cost for all the families in the class. **Estimates will vary.**

Appliance	Average Annual Energy Cost
Washer/Dryer	$77
Color TV	$22
Range/Oven	$42
Refrigerator/Freezer	$239
Microwave oven	$7
Dishwasher	$12

See also page 18.

VISUAL
VISUAL/SPATIAL

MULTISTEP AND STRATEGY PROBLEMS

The following multistep and strategy problems are provided in Lesson 1.1:

Page	Item
19	55, 59, 60

CAREER CONNECTION

Explain that a journalist is a writer or editor for a newspaper or magazine. Have students **write newspaper headlines that include exact numbers and estimates.** Two examples are given.

TEMPERATURES REACH A RECORD-BREAKING HIGH OF 102°F
exact

MORE THAN 2,000 TURN OUT FOR MEMORIAL DAY PARADE
estimate

Check students' work.

VISUAL
VERBAL/LINGUISTIC

WRITING IN MATHEMATICS

 **Reinforce the concept of estimating whole numbers** by having students complete the following activity. Ask students to work in pairs to write problems that use the populations per square mile of these cities. Then have students exchange papers and have their partners estimate the answers.

City	Population
Cairo	97,106
Hong Kong	237,501
New York City	24,089
Bangkok	58,379
Bombay	127,461

TECHNOLOGY LINK

- **Intervention Strategies and Activities CD-ROM** • *Skills 1, 2*
- **Calculating Crew** • *Nautical Number Line,* Levels C, H
- **Harcourt Math Newsroom Video** • *Weather-Stacked Stadium*

LESSON 1.1 ORGANIZER

Objective To use estimation strategies to find sums, differences, products, and quotients of whole numbers

Vocabulary clustering, underestimate, overestimate *Review* compatible numbers

1 Introduce

QUICK REVIEW provides review of prerequisite skills.

Why Learn This? You can use this skill to solve problems that do not need an exact answer, such as estimating the distance a flying disc is thrown. *Share the lesson objective with students.*

2 Teach

Guided Instruction

- Discuss with students the accuracy of the estimate 460 ft at the top of page 16.

  If you rounded the addends to the nearest hundred, what would your estimate be? 400 ft

  Find the exact answer. 465 ft

  Which estimate is closer? 460 ft

- Lead students to draw a conclusion about rounding and the accuracy of an estimate.

  How does the place you round to affect the accuracy of the estimate? Rounding to numbers with a lesser place value results in a more accurate estimate.

ADDITIONAL EXAMPLE

Example 1, p. 16

Use clustering to estimate the sum.

  5,889
  6,019
 +5,999

The sum is about 18,000.

See also page 17.

LESSON 1.1

Estimate with Whole Numbers

Learn how to estimate sums, differences, products, and quotients of whole numbers.

QUICK REVIEW
1. 30 + 40 70
2. 25 + 60 85
3. 70 − 20 50
4. 100 − 45 55
5. 1,200 + 1,200 + 1,200 3,600

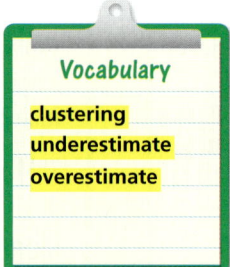

Vocabulary
clustering
underestimate
overestimate

Remember that when rounding, you look at the digit to the right of the place to which you are rounding.
- If that digit is 5 or greater, round up.
- If that digit is less than 5, round down.

The longest throw of a flying disc was 656 ft 2 in., made by Scott Stokely in 1995. Carmen, David, and Leona each threw a disc. Is the total distance of the three throws close to the record distance?

You don't need an exact sum to answer the question, so estimate.

  134 →   130
  148 →   150   *Round each number*
 +183 → +180   *to the nearest ten.*
           460

NAME	DISTANCE
David	134 ft
Leona	148 ft
Carmen	183 ft

The estimate, 460, is not close to the record distance of 656 ft, so the total distance is not close either.

Use **clustering** to estimate a sum when the addends are about the same.

EXAMPLE 1

Estimate 1,802 + 2,182 + 1,999.

  1,802   *The three addends all cluster around 2,000, so*
  2,182   *use 2,000 for each number.*
 +1,999

3 × 2,000 = 6,000   *Multiply.*

So, the sum is about 6,000.

You can use rounding to estimate a difference.

EXAMPLE 2

Estimate 31,928 − 20,915.

Round to the nearest	*Round to the nearest*
ten thousand.	*thousand.*
30,000	32,000
−20,000	−21,000
10,000	11,000

So, both 10,000 and 11,000 are reasonable estimates.

16 Chapter 1

RETEACH 1.1

Estimate with Whole Numbers

You can use compatible numbers to estimate a quotient. **Compatible numbers** are helpful to use because they divide without a remainder, are close to the actual numbers, and are easy to compute mentally.

Oakdale Middle School is collecting recycled cans. The school has set a goal of collecting 2,788 cans. There are 38 homerooms in the school. About how many cans should each homeroom collect for the school to reach its goal?

Because you are asked "about how many," an estimate is appropriate for the answer. Use compatible numbers to estimate.

Step 1: Look at the actual numbers that make up the problem. Think about numbers that are close to the real numbers that will divide without a remainder.

2,788 ÷ 38 → 2,800 ÷ 40

Step 2: Divide. → 2,800 ÷ 40 = 70

So, each homeroom should collect about 70 cans.

Complete to show how compatible numbers are used to estimate the quotient.

1. 2,615 ÷ 47 → 2,500 ÷ _50_ = _50_
2. 3,104 ÷ 62 → 3,000 ÷ _60_ = _50_
3. 3,591 ÷ 88 → _3,600_ ÷ 90 = _40_
4. 4,733 ÷ 74 → 4,800 ÷ _80_ = _60_
5. 7,105 ÷ 77 → _7,200_ ÷ 80 = _90_
6. 5,511 ÷ 62 → 5,400 ÷ _60_ = _90_
7. 15,843 ÷ 381 → 16,000 ÷ _400_ = _40_
8. 20,972 ÷ 287 → 21,000 ÷ _300_ = _70_
9. 29,100 ÷ 307 → _30,000_ ÷ 300 = _100_
10. 95,347 ÷ 795 → 96,000 ÷ _800_ = _120_

Use compatible numbers to estimate the quotient. *Possible estimates are given.*

11. 434 ÷ 68 _6_
12. 394 ÷ 5 _80_
13. 448 ÷ 15 _30_
14. 986 ÷ 102 _10_
15. 627 ÷ 89 _7_
16. 554 ÷ 63 _9_
17. 293 ÷ 31 _10_
18. 705 ÷ 97 _7_
19. 1,246 ÷ 43 _30_
20. 2,779 ÷ 28 _100_
21. 3,896 ÷ 38 _100_
22. 7,164 ÷ 78 _90_

PRACTICE 1.1

Estimate with Whole Numbers

Vocabulary

1. When both factors in a multiplication problem are rounded up to estimate the product, the estimate is an __overestimate__.

2. When all addends are about the same, you can use __clustering__ to estimate their sum.

Estimate the sum or difference. *Possible estimates are given.*

3. 2,489 1,601 +2,109 6,000	4. 398 415 +368 1,200	5. 4,723 +2,198 6,900	6. 7,132 6,594 +7,301 21,000	7. 5,401 +9,188 14,600
8. 478 − 26 450	9. 263 −211 50	10. 5,877 −5,318 600	11. 8,528 −6,491 2,000	12. 8,903 −4,575 4,300

Estimate the product or quotient. *Possible estimates are given.*

13. 53 × 8 400	14. 76 × 9 720	15. 72 ×28 2,100	16. 47 ×53 2,500	17. 660 × 42 28,000
18. 371 × 78 32,000	19. 68 ×37 2,800	20. 480 ×192 100,000	21. 375 ×591 240,000	22. 824 ×693 560,000

23. 331 ÷ 5 _70_
24. 643 ÷ 9 _70_
25. 1,827 ÷ 59 _30_
26. 5,543 ÷ 77 _70_
27. 9,165 ÷ 28 _300_
28. 6,281 ÷ 875 _7_
29. 7,118 ÷ 614 _12_
30. 8,215 ÷ 897 _9_

Mixed Review

Round to the nearest 1,000.

31. 4,571 _5,000_
32. 8,445 _8,000_
33. 1,902 _2,000_
34. 6,679 _7,000_

Find the product.

35. 6 × 6 × 6 _216_
36. 3 × 3 × 3 × 3 _81_
37. 4 × 4 × 4 × 4 _256_

You can show an estimate by using the "approximately equal to" symbol, ≈.

5,125 − 1,920 ≈ 3,000   *Read: 5,125 − 1,920 is approximately equal to 3,000.*

An estimate that is less than the exact answer is an **underestimate**.
An estimate that is greater than the exact answer is an **overestimate**.

```
  366              370
 +198             +200     Round up.
 ----             ----
  564 Exact answer 570     Overestimate

  144              100
 ×123             ×100     Round down.
 ----             ----
17,712 Exact answer 10,000 Underestimate
```

EXAMPLE 3

Students set up 28 rows of 36 seats each for a talent show in the school cafeteria. About how many programs should the school print for the show?

Estimate 28 × 36. To make sure there are enough programs, find an overestimate.

Remember that compatible numbers are numbers that divide without a remainder, are close to the actual numbers, and are easy to compute with mentally.

```
 28      30      Round each factor up to the nearest ten.
×36  →  ×40

         30      Multiply. Because each factor is rounded up,
        ×40      the product is an overestimate.
       ----
       1,200
```

So, the school should print about 1,200 programs.

To estimate a quotient, use rounding or compatible numbers.

EXAMPLE 4

The employees of the Briar Creek office building have collected 1,545 lb of paper to be recycled. The building has 36 offices. About how many pounds of paper, on average, did the employees in each office collect?

Estimate 1,545 ÷ 36.

1,600 ÷ 40   *4 is compatible with 16, so use 1,600 ÷ 40.*
1,600 ÷ 40 = 40   *Divide.*

So, the employees in each office collected an average of about 40 lb.

• Is 1,488 ÷ 36 easier to estimate using rounding or compatible numbers? Explain. **Possible answer: Compatible numbers; 1,600 ÷ 40 is easier to compute than 1,500 ÷ 40.**

Math Idea ▶ Some of the strategies you can use to estimate are rounding, clustering, and compatible numbers.

• Discuss with students that sometimes it is better to overestimate than underestimate an answer. It depends on the situation.

REASONING When would it be better to overestimate than to underestimate? Possible answer: when it is necessary to have enough of something, such as chairs for a performance

ADDITIONAL EXAMPLES

Example 2, p. 16

Use rounding to estimate the difference.

785
−262

The difference is 500 when rounding to the nearest hundred and it is 530 when rounding to the nearest ten.

Example 3, p. 17

Each of the 47 students who enter the Barbour County Math Competition must solve 32 different problems. About how many problems will the sponsors of the competition need to write for the competitors? about 1,500 problems

Example 4, p. 17

Jeff made a total of $1,325 mowing lawns for 63 days this past summer. About how much did he earn each day? Using compatible numbers, 1,200 ÷ 60, shows that Jeff earned about $20 each day.

CHALLENGE 1.1

Estimating Populations

POPULATION OF THE MIDDLE COLONIES: 1670 – 1750

Colony	1670	1690	1710	1730	1750
Delaware	700	1,482	3,645	9,170	28,704
New Jersey	1,000	8,000	19,872	35,510	71,393
New York	5,754	13,909	21,625	48,594	76,696
Pennsylvania	—	11,450	24,450	51,707	119,666

The table shows how the population of the middle colonies changed from 1670 to 1750. Use the table to answer the questions. Estimate to the nearest thousand.

1. About how many people lived in either Delaware or New York in 1690?
   about 15,000 people

2. About how many people lived in either Pennsylvania or New Jersey in 1730?
   about 88,000 people

3. About how many more people lived in New York than in Delaware in 1710?
   about 18,000 more people

4. About how many more people lived in Pennsylvania in 1750 than in 1690?
   about 109,000 more people

5. About how many people lived in the middle colonies in 1670?
   about 8,000 people

6. About how many people lived in the middle colonies in 1750?
   about 297,000 people

7. About how many more people lived in the middle colonies in 1750 than in 1670?
   about 289,000 more people

PROBLEM SOLVING 1.1

Estimate with Whole Numbers   *Analyze Choose Solve Check*

Write the correct answer.

1. Use clustering to estimate the sum.
   7,843
   8,213
   + 8,107
   8,000 × 3 = 24,000

2. Use rounding to estimate the product.
   33 × 21
   30 × 20 = 600

3. The local museum estimates that about 5,475 people visited the museum in the last 9 days. About how many people visited the museum each day?
   Possible estimate: about 600 people

4. Ruby made a quilt using 588 squares. There were 28 rows of squares in the quilt. About how many squares were in each row?
   20 squares

Choose the letter for the best answer.

5. What is the place value of the underlined digit?
   1,345.835
   A hundredths
   B tenths
   C tens
   D hundreds

6. What is 2,768 rounded to the nearest hundred?
   F 3,000
   G 2,800
   H 2,770
   J 2,700

7. The Rockwells traveled 4,476 miles in 11 days. Each day they traveled about the same number of miles. What is a good estimate of how many miles they traveled each day?
   A 200 mi
   B 300 mi
   C 400 mi
   D 500 mi

8. June gets paid about $1,550 each month. What is a reasonable estimate of how much she makes in a year?
   F Less than $10,000
   G Between $10,000 and $15,000
   H Between $15,000 and $20,000
   J More than $20,000

9. **Write About It** Explain how to use clustering to estimate the sum of 385 + 408 + 396 + 411.
   The four addends all cluster around 400, so use 400 for each number: 4 × 400 = 1,600.

LESSON 1.1

3 Practice

Guided Practice

Do Check for Understanding Exercises 1–18 with your students. Identify those having difficulty and use lesson resources to help.

As students work through Exercises 3, 7, 12, and 16, have them rewrite the problems with the numbers they will use to make the estimates.

COMMON ERROR ALERT

When estimating products, students may make a place-value error in the product. Have students add the number of zeros in the rounded factors to any zeros in the basic fact product, and verify that the total number of zeros equals the number of zeros in the product.

Independent Practice

Note that Exercises 55, 59, and 60 are **multi-step or strategy problems**. Assign Exercises 19–55.

Multistep or Strategy Problem To solve Exercise 60, students can convert $1\frac{1}{4}$ hr to minutes and then subtract 50 min from 75 min. Guide students to see that the conversion to minutes makes finding the difference easier.

CHECK FOR UNDERSTANDING

Think and Discuss Look back at the lesson to answer each question.

1. **Explain** why 6,000 + 1,500 is an overestimate or an underestimate of 6,108 + 1,524.
   underestimate since both addends are rounded down
2. **Describe** how to estimate a quotient by using compatible numbers. Give an example to illustrate your answer.
   Answers will vary.

Guided Practice Estimate the sum or difference. Possible estimates are given.

3. 723 +819 **1,500**	4. 2,940 3,140 +2,834 **9,000**	5. 4,480 4,100 +3,967 **12,000**	6. 5,449 4,869 +4,834 **15,000**
7. 667 −133 **600**	8. 8,855 −2,268 **7,000**	9. 34,855 −11,268 **24,000**	10. 67,184 −49,650 **17,000**

Estimate the product or quotient. Possible estimates are given.

11. 36 × 9 **360**	12. 59 ×33 **1,800**	13. 48 ×29 **1,500**	14. 490 × 66 **35,000**
15. 321 ÷ 4 **80**	16. 1,544 ÷ 28 **50**	17. 4,156 ÷ 64 **70**	18. 8,429 ÷ 39 **200**

PRACTICE AND PROBLEM SOLVING

Independent Practice Estimate the sum or difference. Possible estimates are given.

19. 1,700 2,008 +2,324 **6,000**	20. 293 348 +343 **900**	21. 5,765 5,948 +6,324 **18,000**	22. 43,643 +84,211 **128,000**
23. 389 − 43 **350**	24. 3,556 −3,339 **300**	25. 44,123 −29,512 **14,000**	26. 667,184 −249,650 **420,000**

27. 17,809 − 2,145 **16,000** 28. 321,059 + 42,950 **364,000**

TECHNOLOGY LINK
To learn more about estimation with whole numbers, watch the **Harcourt Math Newsroom Video** *Weather Stacked Stadium.*

Estimate the product or quotient. Possible estimates are given.

29. 364 × 12 **4,000**	30. 53 ×41 **2,000**	31. 482 ×299 **150,000**	32. 1,874 × 582 **1,200,000**
33. 1,844 ÷ 22 **90**	34. 3,575 ÷ 56 **60**	35. 6,435 ÷ 529 **12**	36. 21,416 ÷ 521 **40**

37. 4,135 × 784 **3,200,000**   38. 62,217 ÷ 4,889 **12**

Possible estimates are given. Tell whether the estimate is an *overestimate* or an *underestimate*. Then show how the estimate was determined.

39. 352 + 675 ≈ 1,100
    over; 400 + 700
40. 4,134 + 47 ≈ 4,100
    under; 4,100 + 0
41. 96 × 19 ≈ 2,000
    over; 100 × 20
42. 709 + 151 ≈ 850
    under; 700 + 150
43. 291 × 28 ≈ 9,000
    over; 300 × 30
44. 25 × 29 ≈ 900
    over; 30 × 30

18 Chapter 1

ALTERNATIVE TEACHING STRATEGY SCAFFOLDED INSTRUCTION

Purpose Students use base-ten blocks and number lines to learn about estimation.

Materials *For each group* base-ten blocks: 4 hundreds, 16 tens

Display the following problem and ask students to estimate the sum to the nearest ten:

134
148
+183

Then draw a number line and have students locate each of the addends on the number line.

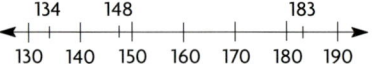

Ask a volunteer to identify the tens between which 134 falls and to identify the ten closer to 134. **130 and 140; 130**

Have each group model the number 130 by using hundreds and tens. Repeat for the remaining two numbers, having groups model 150 and 180 by using hundreds and tens.

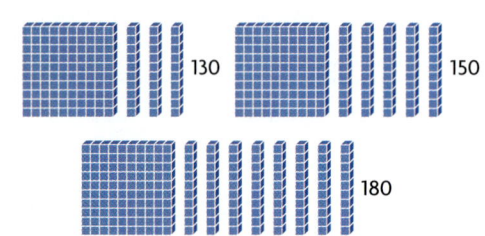

Write the estimates next to the problem:

134 130
148 150
+183 +180

Have groups model the sum of 130, 150, and 180, regrouping to show the final answer as 460.

18 Chapter 1

Use estimation to compare. Write < or > for ●.
Possible answers are given.

45. 614 × 41 ● 21,119 + 1,899 **>**
46. 18,391 ÷ 19 ● 59 × 21 **<**
47. 4,012 − 3,508 ● 3,624 ÷ 6 **<**
48. 12,283 + 19,971 ● 209,910 ÷ 7 **>**
49. 711 × 63 ● 28,520 + 16,990 **<**
50. 513 × 52 ● 29,190 − 1,986 **<**

Problem Solving ▶ Applications

Use Data For 51–52, use the table.
Possible estimates are given.

51. About how many more library books does Harvard have than the University of Illinois at Urbana? **about 4,500,000 books**

52. Estimate the total number of library books at Yale and the University of California at Berkeley. **about 17,000,000 books**

LARGEST UNIVERSITY LIBRARIES IN THE UNITED STATES (1999)

Library	Books
Harvard University	12,877,360
Yale University	9,485,823
University of Illinois–Urbana	8,474,737
University of California–Berkeley	8,078,685
University of Texas	7,019,508

53. The theater of the Natural Science and History Museum was filled to capacity for 276 shows. The theater holds 36 people. How many people attended the shows? **9,936 people**

54. Possible answer: Compatible numbers; 750 ÷ 75 is easier to compute than 760 ÷ 70.

54. **Write About It** Is it easier to use rounding or compatible numbers to estimate 756 ÷ 74? Explain.

55. Two numbers, each rounded to the nearest hundred, have a product of 60,000. What are two possible numbers?
Possible answer: 235 and 345

MIXED REVIEW AND TEST PREP

For 56–58, find the perimeter and area.

56. 12 ft × 12 ft square
48 ft; 144 ft²

57. triangle 12 in., 15 in., 9 in.
36 in.; 54 in.²

58. 10 cm × 3 cm rectangle
26 cm; 30 cm²

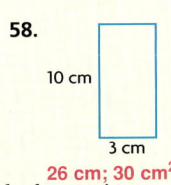

★ 59. **TEST PREP** Ms. Cannon baked a total of 60 apple and blueberry pies. She baked a dozen more apple pies than blueberry pies. How many apple pies did she bake? **C**

A 18   B 24   C 36   D 48

★ 60. **TEST PREP** Allen and Mei began working at the same time. It took Allen 50 minutes to mow the lawn, while Mei took 1¼ hr to paint the fence. How much longer did Mei work than Allen? **J**

F 85 min   G 65 min   H 35 min   J 25 min

EXTRA PRACTICE page H32, Set A

For Exercises 51 and 52, have students round each of the numbers in the table to the nearest million. Then help them identify which parts of the table they will use for each problem.

MIXED REVIEW AND TEST PREP

Exercises 56–60 provide **cumulative review** (Grade 5).

4 Assess

Summarize the lesson by having students:

DISCUSS Would the exact answer to Exercise 30 be greater than or less than the estimated answer? Why? **greater than; because both factors were rounded down**

WRITE A total of 367 students completed a 6-mile walk. Give two ways to estimate the total mileage walked by the students. Possible answer: Method 1: Round 367 to 400 and multiply. 400 × 6 = 2,400 miles; Method 2: Round 367 to 370 and multiply. 370 × 6 = 2,220 miles

Lesson Quiz

Transparency 1.1

Estimate the sum or difference.
Possible estimates are given.

1. 756 + 229 = **1,000**
2. 3,214 − 1,570 = **1,600**
3. 4,893 + 6,275 + 1,025 = **12,000**

Estimate the quotient. Possible estimates are given.

4. 415 ÷ 7 **60**   5. 4,765 ÷ 5 **900**

19

LESSON 1.2 Use Addition and Subtraction

LESSON PLANNING

Objective To use addition and subtraction of whole numbers to solve real-life problems

Intervention for Prerequisite Skills
Place Value of Whole Numbers, Round Whole Numbers (For intervention strategies, see page 15.)

NCTM Standards
1. Number and Operations
5. Data Analysis and Probability
6. Problem Solving
8. Communication

Math Background
Consider the following as you help students work successfully with the addition algorithm.

- Rounding is a useful tool for determining the reasonableness of an answer, but it does not provide a check that the answer is correct.
- Aligning the digits by place value is essential for finding the correct sum.
- Regrouping may occur from column to column; that is, 10 ones are regrouped as 1 ten; 10 tens are regrouped as 1 hundred, and so on.

Students may check addition by adding up instead of down and check subtraction by adding the difference to the number above.

WARM-UP RESOURCES

NUMBER OF THE DAY

If you add 80 to me and subtract 3, you get 102. What number am I? 25

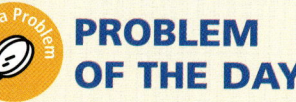 
PROBLEM OF THE DAY

Presidents George Washington, John Adams, and Thomas Jefferson lived a total of 240 years. Adams lived the longest, 23 years longer than Washington. Jefferson lived 16 years longer than Washington. How old was each President when he died? Washington, 67; Adams, 90; Jefferson, 83

Solution Problem of the Day tab, p. PD1A

DAILY FACTS PRACTICE

Have students practice addition and subtraction facts by completing Set B of *Teacher's Resource Book*, p. TR93.

INTERVENTION AND EXTENSION RESOURCES

ALTERNATIVE TEACHING STRATEGY — ELL

Materials For each pair base-ten blocks: 19 hundreds, 19 tens, and 19 ones

Have students **practice addition and subtraction.** Display two addition and two subtraction problems using 2- and 3-digit numbers. Have pairs of students work together to model the sums and differences using the hundreds, tens, and units models. After they have modeled all the problems, let each student use paper and pencil to find the answers.

Check students' work.

KINESTHETIC
INTERPERSONAL/SOCIAL

EARLY FINISHERS

Students can improve their ability to **add and subtract.** Have students use each of the digits 1–6 once in the following addition problem.

$$\begin{array}{r} \square\square\square \\ +\,\square\square\square \\ \hline 1{,}0\,4\,7 \end{array}$$

632 + 415

Challenge students to make up a similar subtraction problem to share with others.

VISUAL
LOGICAL/MATHEMATICAL

MULTISTEP AND STRATEGY PROBLEMS

The following multistep and strategy problems are provided in Lesson 1.2:

Page	Item
21	14, 16

ADVANCED LEARNERS

Challenge students to **add a column of numbers** mentally using a zigzag pattern.

$$\begin{array}{r} 84 \\ 33 \\ +56 \end{array} \quad \text{Think:} \quad 84 + 30 = 114$$

$$114 + 3 = 117$$
$$117 + 50 = 167$$
$$167 + 6 = 173$$

Use this method to find the following sums.

$$\begin{array}{r} 59 \\ 42 \\ +36 \\ \hline 137 \end{array} \quad \begin{array}{r} 46 \\ 29 \\ +65 \\ \hline 140 \end{array} \quad \begin{array}{r} 47 \\ 38 \\ 93 \\ +44 \\ \hline 222 \end{array}$$

VISUAL
VISUAL/SPATIAL

TECHNOLOGY LINK

- **Intervention Strategies and Activities CD-ROM** • *Skills* 1, 2
- **Number Heroes** • *Quizzo,* Levels L, M

LESSON 1.2 ORGANIZER

Objective To use addition and subtraction of whole numbers to solve real-life problems

1 Introduce

QUICK REVIEW provides review of prerequisite skills.

Why Learn This? You can use this skill to help you solve more complex word problems. *Share the lesson objective with students.*

2 Teach

Guided Instruction

- As you discuss Examples 1 and 2, have students describe how to choose the correct operation to solve the problem.

  How did you know whether to add or subtract in Example 1? In Example 2? You are joining amounts of time, so you add; You are comparing two numbers, so you subtract.

- Help students evaluate the role of an estimate in checking their work.

  REASONING Does an estimate show that your answer is correct? Explain. Possible answer: No; it shows that your answer is reasonable. You still need to check to be sure it is correct.

- Present this multi-step problem and guide students through the steps.

  How many more hours were spent in space for the Apollo and Skylab programs than for the Mercury and Gemini programs? 17,864 hr

  What do you need to find out before you can solve this problem? the number of hours in space for both the Apollo and Skylab programs and the Mercury and Gemini programs

ADDITIONAL EXAMPLE

Example 1, p. 20

Captain Juarez is a pilot. He flew 256 miles during week 1 of January. Week 2 he flew 2,753 miles, week 3 he flew 8,809 miles, and week 4 he flew 14,174 miles. How many miles did Captain Juarez fly in January? estimate: 26,000; exact sum: 25,992 mi.

LESSON 1.2 Use Addition and Subtraction

Learn how to add and subtract whole numbers.

QUICK REVIEW
Round to the nearest thousand.
1. 3,841 **4,000**   2. 2,490 **2,000**
3. 7,450 **7,000**   4. 8,500 **9,000**
5. Add. 4,256 + 1,725 **5,981**

Remember that you add when joining groups of different sizes and subtract when taking away or comparing groups.

PROGRAM	YEARS	TOTAL HOURS IN SPACE
Mercury	1961–1963	54
Gemini	1965–1966	1,940
Apollo	1968–1972	7,506
Skylab	1973–1974	12,352

During the early years of space flight, the total time United States astronauts spent in space increased with each program.

EXAMPLE 1

How many hours did U.S. astronauts spend in space?

Find 54 + 1,940 + 7,506 + 12,352.

Estimate to check for reasonableness.

Round to the nearest thousand.

```
    54           0              54
 1,940       2,000           1,940
 7,506   →   8,000           7,506
+12,352     +12,000         +12,352
            22,000           21,852
```

Compare your estimate. 21,852 is close to 22,000, so the sum is reasonable.

So, U.S. astronauts spent 21,852 hours in space.

EXAMPLE 2

How many more hours did U.S. astronauts spend in Skylab than on the Gemini missions?

Find 12,352 − 1,940.

Estimate to check for reasonableness.   Find the difference.

```
 12,352        12,000         12,352
− 1,940   →   − 2,000        − 1,940
              10,000         10,412
```

Compare your estimate. 10,412 is close to 10,000, so the difference is reasonable.

So, U.S. astronauts spent 10,412 more hours in Skylab than on the Gemini missions.

20 Chapter 1

CHECK FOR UNDERSTANDING

Think and Discuss Look back at the lesson to answer each question.

1. **Explain** how you know whether to add or subtract when solving a word problem. *Answers will vary.*

2. **Explain** why it is a good idea to find an estimate before or after you find the exact answer. *to determine if your exact answer is reasonable*

Guided Practice Find the sum or difference. Estimate to check.

3. $835 + 604$ **1,439**
4. $6,901 + 342 + 67$ **7,310**
5. $40,190 - 13,982$ **26,208**

PRACTICE AND PROBLEM SOLVING

Independent Practice Find the sum or difference. Estimate to check.

6. $9,500 - 289$ **9,211**
7. $21,670 + 14,704$ **36,374**
8. $31,227 + 56,995$ **88,222**
9. $999,999 + 111,385$ **1,111,384**
10. $987,654 - 456,789$ **530,865**
11. $50,000,000 - 3,604,381$ **46,395,619**

Solve.

12. $1,485 + 2,019 + 1,310 + 3,665 + 798$ **9,277**
13. $43,875 + 81,420 - 38,288 + 12,108 - 23,990$ **75,125**

Problem Solving Applications **Use Data** For 14–15, use the graph.

International space station

14. How many more calories are in 1 cup of almonds than in 1 cup of carrots mixed with 1 cup of raisins? Estimate to check. **384 calories**

15. **Write a problem** that uses data from the graph and that can be solved by adding or subtracting. *Problems will vary.*

16. When the international space station is completed, it will be 290 ft long, which is 206 ft longer than Skylab. Skylab was 41 ft longer than Mir, the Russian space station. How long is Mir? **43 ft**

MIXED REVIEW AND TEST PREP

17. $6,785 + 4,521$ **11,306**
18. Complete. 36 in. = ■ ft **3**
19. List the factors of 24. **1, 2, 3, 4, 6, 8, 12, 24**
20. List the first six multiples of 9. **9, 18, 27, 36, 45, 54**
21. **TEST PREP** Which is a prime number? **C**
    A 15 B 27 C 31 D 50

EXTRA PRACTICE page H32, Set B

21

---

3 Practice

Guided Practice

Do Check for Understanding Exercises 1–5 with your students. Identify those having difficulty and use lesson resources to help.

COMMON ERROR ALERT

In an addition problem, if students' answers are off by 1 in any place, they may be neglecting to add regrouped ones, tens, hundreds, and so on. Make sure they show the regroupings as they work the problems.

```
   376        11
   224       376
 + 538       224
 -----     + 538
 1,028     -----
           1,138
```

Independent Practice

Note that Exercises 14 and 16 are **multistep or strategy problems**. Assign Exercises 6–16.

Scaffolded Instruction Use the prompts on Transparency 1 to guide instruction for the multistep or strategy problem in Exercise 16.

MIXED REVIEW AND TEST PREP

Exercises 17–21 provide **cumulative review** (Grade 5 and Chapter 1).

4 Assess

Summarize the lesson by having students:

DISCUSS How do you decide if your answer is reasonable? *Possible answer: Compare the answer to an estimate. If it is close to the estimate, it is reasonable.*

 **WRITE** How did you solve Exercise 14? *Possible answer: I added 46 and 419 and then subtracted the sum from 849.*

Lesson Quiz

Estimate. Then find the solution.

1. $32,421 + 14,085$ **46,000; 46,506**
2. $54,273 - 21,509$ **32,000; 32,764**
3. Randall earns $26,855 a year and his wife, Karyn, earns $32,425. How much do they earn together? **$59,000; $59,280**

21

CHALLENGE 1.2

Shopping at the Airport

Airports across the nation are discovering a captive audience for shopping. Millions of people spend time—and money—in airports.

The table shows the number of passengers at five airports in 1998 and the annual sales concessions at the airports.

Airport	Number of Passengers	Annual Sales Concessions
Portland International	12,739,851	$45.2 million
Pittsburgh International	20,500,000	$87.1 million
Denver International	36,831,400	$89.0 million
Baltimore/Washington International	15,000,000	$35.0 million
Los Angeles International	59,730,530	$158.2 million

1. Estimate the total number of passengers who used the Los Angeles and Denver airports during 1998.
   Possible answer: about 100,000,000 passengers

2. Which airport had the fewest passengers? *Portland*

3. What was the total number of passengers who used these five airports during 1998? *144,801,781 passengers*

4. How many more passengers used the Denver airport than used the Portland airport? *24,091,549 more passengers*

5. Portland International had fewer passengers than Baltimore/Washington International. How did their annual sales compare?
   Portland had $10.2 million more in sales.

6. Find the difference between the greatest annual sales and the least annual sales. *$123.2 million (or $123,200,000)*

7. What was the total annual sales concessions for these five airports during 1998? *$414.5 million (or $414,500,000)*

PROBLEM SOLVING 1.2

Use Addition and Subtraction

Analyze • Choose • Solve • Check

Solve.

1. In 1995, there were about 58,000 farms in North Carolina and about 22,000 farms in South Carolina. There were about 100,000 farms in Iowa in 1995. About how many more farms were there in Iowa than in North Carolina and South Carolina combined in 1995?
   about 20,000 more farms

2. Carrie participated in a bird census during three days last week. She counted 435 birds on Monday, 206 birds on Tuesday, and 359 birds on Wednesday. How many birds did she count in all during these three days?
   1,000 birds

3. Give the value represented by the digit 8 in the number 258,034,199.
   8 million

4. Use clustering to estimate the sum.
   $65 + 57 + 62 + 54$
   $4 \times 60 = 240$

Choose the letter for the best answer.

5. In 1999, a world record for the largest gathering of twins was set in Taipei, Taiwan, with 3,961 pairs of twins in attendance. The number of twins shattered the previous record of 2,900 pairs set in Twinsburg, Ohio, in 1998. What is a reasonable estimate of the increase in the number of pairs of twins?
   A 60 pairs
   B 160 pairs
   C 900 pairs
   D 1,100 pairs

6. When a children's museum opened near Roberto's home, he was among 14,756 children who visited it during the first month it was open. The next month, 18,355 children visited, while 27,982 children visited during the third month. What is a reasonable estimate of the number of children who visited the museum during the first three months it was open?
   F 40,000 children H 60,000 children
   G 50,000 children **J 70,000 children**

7. What 2 numbers have a sum of 4,949 and a difference of 1,963?
   A 1,999 and 2,950
   B 1,493 and 3,456
   C 1,358 and 3,591
   D 1,078 and 3,871

8. Which is the greatest number of the four shown below?
   23,887; 32,109; 24,999; 32,190
   F 23,887
   G 32,109
   H 24,999
   J 32,190

9. **Write About It** Which operation would you use to solve a problem in which you are asked to find an amount of increase? Explain.
   Subtraction; an amount of increase or change implies a difference.
   To find the difference between two numbers, you should subtract.

LESSON 1.3

Use Multiplication and Division

LESSON PLANNING

Objective To use multiplication and division of whole numbers to solve real-life problems

Intervention for Prerequisite Skills
Place Value of Whole Numbers, Round Whole Numbers (For intervention strategies, see page 15.)

Materials calculator

NCTM Standards
1. Number and Operations
2. Algebra
5. Data Analysis and Probability
6. Problem Solving
8. Communication
9. Connections

Math Background

These ideas will help students understand the steps in the multiplication and division algorithms:

- When you find the product of multi-digit numbers, the partial products except for the first one must end with one, two, or more zeros, as appropriate, or be moved to the left the appropriate number of places. The sum of all partial products is then taken.
- Estimating helps you choose the place for the first digit in the quotient.
- The steps in the division algorithm—divide, multiply, subtract, and compare—are repeated for each digit in the quotient.

WARM-UP RESOURCES

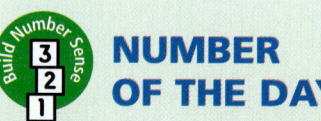
NUMBER OF THE DAY

Begin with your age. Multiply by 4, add 20, divide by 2, and subtract 10. Describe the number you get as an answer. It is 2 times your age.

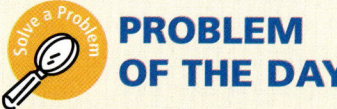
PROBLEM OF THE DAY

Find the product. Compare the product with the first factor. Write a rule for multiplying a 2-digit number by 11 and a rule for multiplying greater numbers by 11.

1. 13 × 11 = __?__ 143
2. 72 × 11 = __?__ 792
3. 326 × 11 = __?__ 3,586
4. 6,045 × 11 __?__ 66,495

Solution Problem of the Day tab, p. PD1A

DAILY FACTS PRACTICE

Have students practice multiplication and division facts by completing Set C of *Teacher's Resource Book,* p. TR93.

22A Chapter 1

INTERVENTION AND EXTENSION RESOURCES

ALTERNATIVE TEACHING STRATEGY — ELL

Have students **practice multiplication and division** by doing the following. Display the multiplication exercises. Have students tell to which place they would round each number to estimate the product and then name the greatest place they will have in the product.

82 × 31 tens, tens, thousands

437 × 129 hundreds, hundreds, ten-thousands

Display these division exercises. Have students tell which will be the greatest place in the quotient.

873 ÷ 9 tens

7,894 ÷ 32 hundreds

See also page 24.

VISUAL
VERBAL/LINGUISTIC

MULTISTEP AND STRATEGY PROBLEMS

The following multistep and strategy problems are provided in Lesson 1.3:

Page	Item
25	42, 43

SCIENCE CONNECTION

Increase students' ability to **apply multiplication and division to real-life problems**. Have students apply their knowledge of multiplication by using the following information.

The average amount of garbage generated by each person in the United States in one year includes 190 lb of plastic, 85 lb of glass, 72 lb of aluminum cans, and 24 lb of plastic containers.

- As a class, students can calculate how much garbage is generated by the entire class. Have students make a pictograph showing the results.
- Have students find the total amount of each type of garbage generated by the number of people in their homes.

Check students' work.

VISUAL
LOGICAL/MATHEMATICAL

ADVANCED LEARNERS

Challenge students to **use multiplication mentally.** Ask students to solve the following multiplication problem. Have them replace the boxes with the numbers 1–6. They can use each number only once.

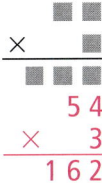

```
   5 4
 ×   3
─────
 1 6 2
```

VISUAL
LOGICAL/MATHEMATICAL

TECHNOLOGY LINK

- **Intervention Strategies and Activities CD-ROM** • *Skills 1, 2*
- **Number Heroes** • *Quizzo,* Levels S, U, Y
- **Mighty Math Calculating Crew** • *Captain Nick Knack,* Level V
- **Math Jingles™ CD 5–6** • *Track 4*

LESSON 1.3 ORGANIZER

Objective To use multiplication and division of whole numbers to solve real-life problems

Materials calculator

1 Introduce

QUICK REVIEW provides review of prerequisite skills.

Why Learn This? You can use this skill to determine the total amount of money earned by selling car wash tickets for a school fundraiser, or to determine average test scores. *Share the lesson objective with students.*

2 Teach

Guided Instruction

- As you work through Example 1, discuss with students how to tell if an answer is reasonable.

  In Example 1, is the estimate close enough to the actual product to be reasonable? Explain. Possible answer: Yes; both numbers are more than 2,000.

- In Example 2, ask students to think about another way of rounding to estimate the product.

  How else could you have rounded the factors? Possible answer: to 10,000 and 100

ADDITIONAL EXAMPLES

Example 1, p. 22

A warehouse has 831 shelves. How many boxes are in the warehouse if there are 32 boxes on each shelf? estimate: $800 \times 30 = 24,000$; exact answer: 26,592 boxes

Example 2, p. 22

Monthly train tickets cost $132. How much will the transit company receive if it sells 13,316 monthly tickets? estimate: $130 \times 13,000 = 1,690,000$; exact answer: $1,757,712

LESSON 1.3

Use Multiplication and Division

Learn how to multiply and divide whole numbers.

QUICK REVIEW
1. 7×3 21   2. $72 \div 9$ 8   3. 12×4 48   4. $300 \div 25$ 12
5. 800×40 32,000

Sometimes you can multiply to solve a word problem.

EXAMPLE 1

Sixth-grade students sold 132 books of carnival ride coupons. How many ride coupons did they sell if there were 18 in each book?

Multiply. 132×18 Estimate. $130 \times 20 = 2,600$

Remember that you multiply when joining equal-sized groups, and you divide when separating into equal-sized groups or when finding how many in each group.

```
   132
 ×  18
 1 056
+1 32
 2,376
```

Compare the exact product to your estimate. Since 2,376 is close to the estimate of 2,600, the exact product is reasonable.

So, the students sold 2,376 coupons.

You can omit the zero placeholders when you multiply. Just be careful to line up the products correctly.

Correct	Incorrect
132	132
× 24	× 24
528	528
+264	+264

EXAMPLE 2

Season tickets to an amusement park are on sale for $125 each. On the first day of the sale, the amusement park sold 12,383 tickets. How much money did the amusement park receive for season tickets that day?

Multiply. $12,383 \times 125$

Estimate. $12,000 \times 130 = 1,560,000$

```
   12,383
 ×    125
   61 915
  247 66
+1 238 3
1,547,875
```

Compare the exact product to your estimate. Since 1,547,875 is close to the estimate of 1,560,000, the exact product is reasonable.

So, the amusement park received $1,547,875.

22 Chapter 1

RETEACH 1.3

Use Multiplication and Division

PRACTICE 1.3

Use Multiplication and Division

Sometimes you have to use division to solve word problems.

EXAMPLE 3

Remember that the procedure for dividing is divide, multiply, subtract, compare, and bring down.

Mrs. Lopez is redesigning the company cafeteria to seat 540 employees. Each table in her design seats 12 employees. How many tables will she need?

Divide. $540 \div 12$

Estimate. $480 \div 12 = 40$

$$\begin{array}{r} 45 \\ 12\overline{)540} \\ -48 \\ \hline 60 \\ -60 \\ \hline 0 \end{array}$$

Compare the exact quotient to your estimate. Since 45 is close to the estimate of 40, the exact quotient is reasonable.

So, Mrs. Lopez will need 45 tables in her design.

- What if each table seats 10 employees? How many tables will Mrs. Lopez need? **54 tables**

Sometimes a division problem has a zero in the quotient.

EXAMPLE 4

A school collected 2,568 newspapers. The newspapers were bundled in packages of 25. How many packages of newspapers did the school bundle?

Divide. $2,568 \div 25$

Estimate. $2,500 \div 25 = 100$

$$\begin{array}{r} 102 \text{ r}18 \\ 25\overline{)2,568} \\ -25 \\ \hline 06 \\ -0 \\ \hline 68 \\ -50 \\ \hline 18 \end{array}$$

Compare the exact answer to your estimate. Since 102 r18 is close to the estimate of 100, the exact quotient is reasonable.

So, the school bundled 102 packages of newspapers.

In Example 4 there is a remainder. Some calculators allow you to show a whole-number remainder.

2,568 ÷R 25 = 102 R18

You can express a remainder with an *r*, or you can express it as a fractional part of the divisor or as a decimal. The quotient and remainder in Example 4 can also be expressed as $102\frac{18}{25}$ or as 102.72.

23

- Discuss with students the choice of compatible numbers with which to estimate.

  In Example 3, what compatible numbers would you use to estimate the product to ensure you had enough tables? 600 ÷ 12 = 50

- Review with students how to interpret and use remainders.

  In Example 4, why do you drop the remainder? The question asks about bundles and each bundle must have 25 newspapers.

REASONING Interpreting remainders helps students develop reasoning skills.

- Have students discuss the remainder in this situation.

  The instructions in a bow-making kit say to use 5 yd of ribbon to make each bow. How many bows could you make with 48 yd of ribbon? 9 bows

  Did you drop the remainder or increase the quotient by 1? Explain. drop the remainder; You don't have enough ribbon to make another bow.

Remind students that some calculators have a ÷R key that can be used to find a whole-number remainder.

ADDITIONAL EXAMPLES

Example 3, p. 23

Each table can seat 15 people. There are 465 people. How many tables are needed? estimate: 450 ÷ 15 = 30; exact answer: 31 tables

Example 4, p. 23

The Craft Club makes wooden houses out of craft sticks. It has 4,356 sticks. Each house takes 43 sticks to build. How many houses can the Craft Club build? estimate: 4,000 ÷ 40 = 100; exact answer: 101 r13; 101 houses

---

CHALLENGE 1.3

Number Crossword

Solve each problem. Complete the puzzle with the answers.

Across
1. 1,685 × 124 = **208,940**
2. 65 × 104 = **6,760**
3. 2,549 × 317 = **808,033**
4. 596 × 240 = **143,040**
5. 99 × 5 = **495**
6. 6,058 × 847 = **5,131,126**
7. 351 × 208 = **73,008**
8. 872 × 234 = **204,048**

Down
1. 6,150 ÷ 3 = **2,050**
4. 3,454 ÷ 22 = **157**
5. 855 ÷ 19 = **45**
9. 25,344 ÷ 36 = **704**
10. 16,740 ÷ 54 = **310**
11. 7,968 ÷ 16 = **498**
12. 46,295 ÷ 47 = **985**
13. 17,568 ÷ 36 = **488**
14. 14,761 ÷ 29 = **509**
15. 32,625 ÷ 87 = **375**
16. 9,672 ÷ 93 = **104**
17. 7,896 ÷ 12 = **658**

PROBLEM SOLVING 1.3

Use Multiplication and Division

Analyze Choose Solve Check

Write the correct answer.

1. Larry washed 58 windows. He charged $4 for every window he washed. How much money did he make washing windows?
   $232

2. Claire had 108 balloons that she wanted to give to her 6 friends. If each person got the same number, how many balloons did each person get?
   18 balloons

3. Write the numbers in order from least to greatest. Use <.
   80,808, 80,080, 80,088
   80,080 < 80,088 < 80,808

4. What is the value of the 2 in 3,927,648?
   20 thousand; 20,000

Choose the letter for the best answer.

5. What is the difference between 2,403,615 and 1,417,528?
   A 1,096,133
   B 1,086,197
   (C) 986,087
   D 985,987

6. What is the product of 1,010 and 100?
   F 1,010,000
   G 110,101
   (H) 101,000
   J 100,110

7. Pauline rides to and from school on her bike every day. Each round-trip is 6 miles. What is a good estimate for the number of miles she rides in 180 school days?
   (A) 1,000 mi
   B 1,500 mi
   C 2,000 mi
   D 2,500 mi

8. Sam has 9 friends in the gardening club. He orders 340 tomato seeds for his friends to share. What is a good estimate of how many seeds each person would get if they share the seeds equally?
   (F) 40 seeds
   G 25 seeds
   H 20 seeds
   J 15 seeds

9. **Write About It** Which operation would you use to solve a problem in which objects are being shared equally? Explain your choice.
   Division; sharing equally means the same number of objects are put in each group. You can divide to find equal groups.

LESSON 1.3

3 Practice

Guided Practice

Do Check for Understanding Exercises 1–13 with your students. Identify those having difficulty and use lesson resources to help.

COMMON ERROR ALERT

If students have difficulty lining up partial products correctly, they may want to continue to use zero as a place holder or do their work on grid paper.

Independent Practice

Note that Exercises 42 and 43 are **multistep or strategy problems**. Assign Exercises 14–45.

Multistep or Strategy Problem To solve Exercise 42, students can find the amount to be paid each month by dividing $1,176 by 14 and then adding $3.50 to the quotient. Guide students to conclude that the division must be completed before the addition.

CHECK FOR UNDERSTANDING

Think and Discuss

1. There were 102 full packages. The 18 newspapers left over were not enough to make a full package.

Look back at the lesson to answer each question.

1. **Explain** why the school bundled 102 packages of newspapers instead of 103 packages of newspapers in Example 4.
2. **Tell** the different ways to express the remainder for the division problem $153 \div 6$. **as a remainder, a decimal, and a fraction; 25 r3, 25.5, and $25\frac{1}{2}$**

Guided Practice

Multiply or divide. Estimate to check.

3. $1,113 \times 712$ **792,456**
4. $2,115 \div 72$ **29.375**
5. $16,225 \times 219$ **3,553,275**

Multiply or divide.

6. 13×14 **182**
7. $8)\overline{432}$ **54**
8. $12)\overline{144}$ **12**
9. 962×40 **38,480**
10. 159×340 **54,060**
11. $7,658 \times 111$ **850,038**
12. $7,044 \div 14$ **503 r2**
13. $1,068 \div 19$ **56 r4**

PRACTICE AND PROBLEM SOLVING

Independent Practice

Multiply or divide. Estimate to check.

14. $2,250 \div 18$ **125**
15. $4,904 \times 196$ **961,184**
16. $193,200 \div 46$ **4,200**
17. $7,021 \times 498$ **3,496,458**
18. $249,900 \div 49$ **5,100**
19. $24,587 \times 71$ **1,745,677**

TECHNOLOGY LINK
More Practice: Use **Mighty Math Calculating Crew**, *Captain Nick Knack*, Level V.

Multiply or divide.

20. $16)\overline{1,664}$ **104**
21. 298×89 **26,522**
22. $5,233 \times 238$ **1,245,454**
23. $52)\overline{728}$ **14**
24. $4)\overline{412}$ **103**
25. 380×55 **20,900**
26. $2,382 \times 12$ **28,584**
27. $24)\overline{626}$ **26 r2**
28. 327×123 **40,221**
29. $26)\overline{2,314}$ **89**
30. $68)\overline{24,820}$ **365**
31. $5,470 \times 240$ **1,312,800**
32. $29)\overline{13,253}$ **457**
33. $6,378 \times 291$ **1,855,998**
34. $2,009 \times 562$ **1,129,058**
35. $120)\overline{10,080}$ **84**

Divide. Write the remainder as a fraction.

36. $5)\overline{49}$ **$9\frac{4}{5}$**
37. $7,349 \div 20$ **$367\frac{9}{20}$**
38. $386 \div 15$ **$25\frac{11}{15}$**
39. $4)\overline{3,385}$ **$846\frac{1}{4}$**

40. **ALGEBRA** What is the least whole number, *n*, for which it is true that $n \div 8 > 542 + 258$? **6,401**

41. **ALGEBRA** What is the least whole number, *n*, for which it is true that $70 \times n > 29,000$? **415**

24 Chapter 1

ALTERNATIVE TEACHING STRATEGY — SCAFFOLDED INSTRUCTION

Purpose Students use an activity to help develop proficiency with the division algorithm.

Materials *For each pair* 1-inch graph paper, p. TR62; 1-inch by 1-inch squares of paper

Give each pair of students a piece of graph paper on which you have written a division problem that will have a zero in the quotient. The problem should be written with one digit to a graph square.

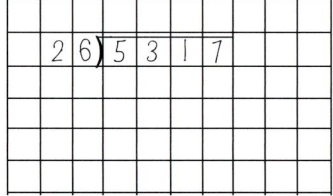

Then give each group 1-inch squares of paper with the digits of the quotient and remainder, written one to a square.

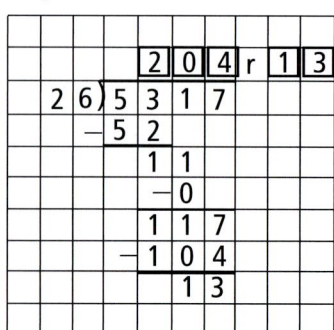

Each pair works together to decide where to place the first digit in the quotient and which digit to use. When they decide, they place the small square with that digit in the proper place on the graph paper and then multiply it by the divisor. The graph paper will help students align their digits properly.

They repeat the process with the second digit in the quotient, until the problem is complete. The square(s) for the remainder can then be placed beside the quotient.

When each group is done, have students show and explain their division problem to the rest of the class.

24 Chapter 1

Problem Solving ▶ Applications

42. Yuji's parents bought an entertainment center for $1,176. They plan to pay for it with 14 equal monthly payments. How much will each payment be if a $3.50 service charge is added every month? **$87.50**

43. Lincoln Middle School had a car wash to raise money. The students charged $3.00 for every car and $5.00 for every van. If they washed 23 cars and 18 vans, how much did they earn? **$159**

44. **? What's the Error?** Describe the error. Then solve the problem correctly. **forgot to write a zero in the quotient; 620 r15**

    $$\begin{array}{r}62\text{ r}15\\49\overline{)30{,}395}\end{array}$$

45. There are 5 children in Brenda's family. Brenda is 16 years old. Brenda's twin sisters are 6 years old, and her brothers are 10 and 12 years old. What is the mean of the 5 children's ages? **10**

MIXED REVIEW AND TEST PREP

46. How much more than 96,784 is 142,981? (p. 20) **46,197**

47. Complete. 4 lb = ■ oz **64**

48. Order 804, 824, 818, and 803 from least to greatest. **803, 804, 818, 824**

49. **TEST PREP** How many meters are in 1,800 millimeters? **B**

    A 0.18 m B 1.8 m C 18 m D 180 m

50. **TEST PREP** Teresa is buying perfume for her mother. The prices are $16.19, $15.89, $15.99, and $17.00. How much will she save by buying the least expensive instead of the most expensive? **J**

    F $0.01 G $0.11 H $0.81 J $1.11

PROBLEM SOLVING  to Reading

Strategy • Use Context Many word problems contain words such as *more than, fewer than, twice as many,* and *total*. Be sure to interpret these words within the context of the problem before you choose an operation.

Use Data For each problem, write the words that help you choose the operation. Then solve the problem.

1. How many more calories will you burn in 1 hr by skiing than by hiking? **words: more than; operation: subtraction; 135 calories**

2. On Saturday, Connie spent an hour in gymnastics class and then walked for 1 hr. How many total calories did she burn during these two activities? **word: total; operation: addition; 240 calories**

3. Clara burned half as many calories while raking the lawn for 1 hr as she did while jogging for 1 hr. How many calories did she burn while raking the lawn? **words: half as many; operation: division; 112 calories**

CALORIES BURNED PER HOUR	
Activity	Calories
Walking (at 2 mi per hr)	112
Gymnastics	128
Hiking	191
Jogging	224
Cross-country skiing	326

EXTRA PRACTICE page H32, Set C

25

MIXED REVIEW AND TEST PREP

Exercises 46–50 provide **cumulative review** (Grade 5 and Chapter 1).

LINKUP to READING

• Have students look at the table. Ask:

How are the activities in the table arranged? from the least amount of calories burned to greatest amount of calories burned

Do you think every person burns the same number of calories listed for each exercise? Explain. No; the number of calories burned would vary with the age and weight of the person.

REASONING If you burned twice as many calories walking 4 mi per hour as walking 2 mi per hour, what exercise would burn the same number of calories as walking 4 mi per hour? jogging

4 Assess

Summarize the lesson by having students:

DISCUSS What is the value of estimating solutions when solving word problems? Comparing estimates to exact answers helps you determine the reasonableness of your solution.

WRITE Show the different ways you can present the remainder in this division problem: 739 ÷ 10. 73 r9, $73\frac{9}{10}$, 73.9

Lesson Quiz

Transparency 1.3

Find the product.

1. 34 × 8 = **272**
2. 420 × 13 = **5,460**
3. 4,216 × 51 = **215,016**

Find the quotient.

4. 5)180 = **36**
5. 22)1,659 = **75 r9**
6. 37)8,475 = **229 r2**

25

LESSON 1.4

Problem Solving Strategy: Predict and Test

LESSON PLANNING

Objective To use the strategy *predict and test* to solve addition, subtraction, multiplication, and division problems with whole numbers

Intervention for Prerequisite Skills
Place Value of Whole Numbers (For intervention strategies, see page 15.)

Lesson Resources Problem Solving Think Along, p. TR1

NCTM Standards
1. Number and Operations
6. Problem Solving
7. Reasoning and Proof

Math Background
The idea behind the *predict and test* strategy is that the student will:

- think about a reasonable or logical solution to the problem and give an answer based on that logic.
- test the predicted solution in the context of the problem to decide if it satisfies all the conditions.
- use the result of the test to change the predicted answer as needed.

The process is continued until the correct answer is obtained.

WARM-UP RESOURCES

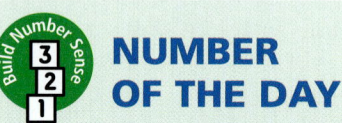 

NUMBER OF THE DAY

The sum of a smaller number and a larger number is 20. Their product is 64. What is the larger number? 16

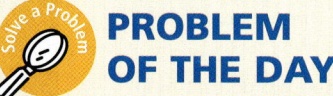 

PROBLEM OF THE DAY

At noon on Monday, Latisha sets her watch to the correct time. If her watch loses one minute each hour and she does not correct it, what time will her watch show at noon on Wednesday? 11:12 At noon on what day will her watch show 10:00? Saturday

Solution Problem of the Day tab, p. PD1A

DAILY FACTS PRACTICE

Have students practice subtraction facts by completing Set D of *Teacher's Resource Book*, p. TR93.

INTERVENTION AND EXTENSION RESOURCES

ALTERNATIVE TEACHING STRATEGY — ELL

Have students work in small groups to **practice the strategy** *predict and test.* Ask: How many students does it take to reach across the classroom with their arms out and fingertips touching?

First, ask students to come up with an estimate they feel is reasonable without using any measuring tools. Have them explain their estimate.

Once the estimates are made, have students determine how many students are needed to span the classroom. Make a table to compare their estimates with the actual number they determined. Check students' work.

KINESTHETIC
INTERPERSONAL/SOCIAL

SOCIAL STUDIES CONNECTION

Students can **apply the strategy** *predict and test* **to everyday problems**. Have students work in small groups to predict the number of students in their class born in each month of the year. Then tabulate the class birthdays by month and compare totals with the predictions.

Regroup the students into 12 groups according to the months in which they were born. Have group members research historical events that occurred on their birthdays or during the month that they were born. Then have them make a poster describing and illustrating the events. Check students' work.

VISUAL
VISUAL/SPATIAL

READING STRATEGY

Compare Have students use the reading strategy *compare* to help them understand the problem on page 26. Have students compare the information about correct answers with the information about incorrect answers. They should recognize that for each correct answer, 4 points are added to the score, and for each incorrect answer, 1 point is subtracted.

Direct students' attention to the Solve section on page 26. Ask: When you check, what are you comparing your prediction to? Tammy's score of 85 points

ADVANCED LEARNERS

Challenge students to **use the strategy** *predict and test* to complete the magic square. The oldest-known magic square is the *lo-shu*, which was discovered about 4,000 years ago in China. Some of the numbers from the *lo-shu* are shown below. Have students complete the square and find the sum. top left = 6; middle = 5; bottom left = 2; sum = 15

	1	8
7		3
	9	4

VISUAL
VISUAL/SPATIAL

MULTISTEP AND STRATEGY PROBLEMS

The following multistep and strategy problems are provided in Lesson 1.4:

Page	Item
27	1–11

TECHNOLOGY LINK
Intervention Strategies and Activities CD-ROM • *Skill 1*

LESSON 1.4 ORGANIZER

Objective To use the strategy *predict and test* to solve addition, subtraction, multiplication, and division problems with whole numbers

Lesson Resources Problem Solving Think Along, p. TR1

1 Introduce

QUICK REVIEW provides review of prerequisite skills.

Why Learn This? You can use this strategy to find the number of specific types of coins needed to make a given amount of money or to solve equations with two variables. *Share the lesson objective with students.*

2 Teach

Guided Instruction

- As you work through the problem with students, be sure they understand the scoring process.

  How many points would be given for 10 right answers? 40 points

  How many points would be subtracted for 3 wrong answers? 3 points

  How would you find the total score for 10 right answers and 3 wrong answers? 40 − 3 = 37

- Encourage students to consider another starting point for their prediction/test table.

REASONING **What would be another logical starting point for your predictions of correct and incorrect answers?** Possible answer: Start just below a perfect score of 25 correct answers at 24 correct and 1 incorrect and work backward.

- Discuss with students the reasoning for this scoring method.

  Why do you think a test might be scored this way? to discourage guessing

Algebraic Thinking The process of making a prediction and then testing it is much like substituting different values for a variable in an algebraic equation.

26 Chapter 1

LESSON 1.4

PROBLEM SOLVING STRATEGY
Predict and Test

Learn how to use the strategy *predict and test* to solve problems with whole numbers.

QUICK REVIEW

1. 22 + 62 = **84**
2. 88 − 47 = **41**
3. 16 × 20 = **320**
4. 9)360 = **40**
5. 150 − 72 = **78**

You can solve some problems by using your number sense to predict a possible answer. You should then test your answer and revise your prediction if necessary.

There were 25 problems on a test. For each correct answer, 4 points were given. For each incorrect answer, 1 point was subtracted. Tania answered all 25 problems. Her score was 85. How many correct answers did she have?

Analyze
What are you asked to find? number of correct answers

What facts are given? total number of problems; number of points for correct and incorrect; Tania's score

Is there any numerical information you will not use? If so, what? no

Choose
What strategy will you use?

You can use the strategy *predict and test*. Use the given information and your number sense to predict about how many correct answers Tania had. Then test your prediction, and revise it if necessary.

Solve
How will you solve the problem?

Make a table to show your prediction, your test of it, and any revisions you need. Be sure that the total of correct and incorrect problems is 25.

PREDICTION		TEST
Correct	Incorrect	SCORE
20	5	(20 × 4) − 5 = 75 *too low, so revise*
21	4	(21 × 4) − 4 = 80 *too low, so revise*
22	3	(22 × 4) − 3 = 85 ← *correct*

So, Tania had 22 correct answers.

Check
How can you check your answer? Answers will vary.

What if Tania's score were 65? How many incorrect answers would she have? 7 incorrect answers

26 Chapter 1

RETEACH 1.4

Problem Solving Strategy
Using Predict and Test

Looking for clues in a problem can help you find its answer. You can use the clues to help you guess and check different answers until you find the right one.

Valley Middle School is holding a canned food drive. Sixth-grade students have collected 150 more cans than seventh-grade students. Together, the students in both grades have collected a total of 530 cans. How many cans did the sixth graders collect? How many cans did the seventh graders collect?

Step 1: Think about what you know.
- You are asked to find the number of cans collected by each grade.
- You know the total number of cans collected and how many more cans the sixth graders collected than the seventh graders.

Step 2: Plan a strategy to solve.
- Use the *predict and test* strategy.
- Use these clues: total cans collected is 530; the difference between amounts collected by sixth and seventh graders is 150.

Step 3: Solve.
- Use a table to record your predictions and tests. Try to predict in an organized way to help you get closer to the exact answer.

PREDICT		TEST	
Sixth Graders	Seventh Graders	Clue 1: The sum is 530.	Clue 2: The difference is 150.
330	200	330 + 200 = 530 ✓	330 − 200 = 130 ✗ Difference is too low.
350	180	350 + 180 = 530 ✓	350 − 180 = 170 ✗ Difference is too high.
340	190	340 + 190 = 530 ✓	340 − 190 = 150 ✓ Both clues are satisfied.

Use the strategy *predict and test* with a table to help you solve.

1. In the problem above, what if the sixth graders had collected 120 more cans than the seventh graders? How many cans would each grade have collected?

   sixth grade: 325 cans; seventh grade: 205 cans

2. Tony collected 85 cans of either soup or fruit. He collected 15 more cans of soup than fruit. How many cans of soup did he collect? How many cans of fruit did he collect?

   50 soup; 35 fruit

PRACTICE 1.4

Problem-Solving Strategy: Predict and Test

Solve by predicting and testing.

1. Ryan bought a total of 40 juice boxes. He bought 8 more boxes of apple juice than of grape juice. How many of each kind did he buy?

   24 apple juice, 16 grape juice

2. The perimeter of a rectangular garden is 56 ft. The length is 4 ft more than the width. What are the dimensions of the garden?

   $l = 16$ ft; $w = 12$ ft

3. The Hawks soccer team played a total of 24 games. They won 6 more games than they lost, and they tied 2 games. How many games did they win?

   14 games

4. Rico collected a total of 47 rocks. He gathered 5 more jagged rocks than smooth rocks. How many of each kind of rock did he collect?

   26 jagged rocks, 21 smooth rocks

5. Matt has earned $75. To buy a bicycle, he needs twice that amount plus $30. How much does the bicycle cost?

   $180

6. The perimeter of a rectangular lot is 190 ft. The width of the lot is 15 ft more than the length. What are the dimensions of the lot?

   $w = 55$ ft; $l = 40$ ft

7. The Wolverines swimming team won a total of 15 first- and second-place medals at their last swim meet. If they won 7 more first-place medals than second-place medals, how many first-place medals did they win?

   11 first-place medals

8. Valley High School's football team played a total of 16 games. They won twice as many games as they lost. If they tied one game, how many games did the team win?

   10 games

Mixed Review

Find the product or quotient. Estimate to check. Possible estimates are given.

9. 306 × 582
   180,000; 178,092

10. 8,246 ÷ 38
    200; 217

11. 21,420 ÷ 51
    400; 420

Tell whether the estimate is an *overestimate* or *underestimate*. Then show how the estimate was determined.

12. 1,872 + 4,774 ≈ 7,000
    overestimate; 2,000 + 5,000

13. 321 × 82 ≈ 24,000
    underestimate; 300 × 80

PROBLEM SOLVING PRACTICE

Solve by predicting and testing.

1. Rodney bought a total of 40 oranges and apples. He bought 14 fewer apples than oranges. How many of each fruit did he buy? **13 apples, 27 oranges**

2. The Mighty Tigers soccer team played a total of 25 games. They won 9 more games than they lost, and 2 games ended in ties. How many games did they win? **16 games**

PROBLEM SOLVING STRATEGIES
- Draw a Diagram or Picture
- Make a Model
- ▶ Predict and Test
- Work Backward
- Make an Organized List
- Find a Pattern
- Make a Table or Graph
- Solve a Simpler Problem
- Write an Equation
- Use Logical Reasoning

3. The perimeter of a rectangular garden is 40 ft. If the length is 6 ft more than the width, what are the length and width of the garden? **B**

   A $l = 12$ ft, $w = 6$ ft
   B $l = 13$ ft, $w = 7$ ft
   C $l = 6$ ft, $w = 12$ ft
   D $l = 7$ ft, $w = 13$ ft

4. The perimeter of a rectangular lawn is 28 yd. If the length is 4 yd more than the width, what are the length and width of the lawn? **G**

   F $l = 10$ yd, $w = 4$ yd
   G $l = 9$ yd, $w = 5$ yd
   H $l = 5$ yd, $w = 9$ yd
   J $l = 4$ yd, $w = 10$ yd

MIXED STRATEGY PRACTICE

5. Rosalia waters her tomato plants every other day. She waters her pepper plants every 3 days. If she waters both on April 20, what are the next three dates on which she will water both? **April 26, May 2, and May 8**

6. Stacy spent a total of $28.45. She bought a ticket for a basketball game for $8.50, food for $7.95, and some T-shirts for $6.00 each. How many T-shirts did she buy? **2 T-shirts**

7. Sam has 98 baseball cards. This is 2 more than twice as many as Paul has. How many cards does Paul have? **48 cards**

8. Use the table below. If the pattern continues, how many miles in all will four runners run on the fifth day? **72 mi**

9. Melina and her two sisters collect stamps. Melina has twice as many as her older sister, who has 33 stamps. Melina has three times as many as her younger sister. How many stamps do they have in all? **121 stamps**

10. ❓ **What's the Question?** The sum of the ages of Jeff, Elijah, and Stefan is 41. Jeff is 14 years old. Stefan is 3 years older than Elijah. The answer is 12 years old. **Possible question: How old is Elijah?**

11. The train that leaves at 11:45 A.M. usually arrives in New York City 34 minutes after that. Today it arrived at 12:24 P.M. How late was the train? **5 min**

27

3 Practice

Guided Practice

Note that Exercises 1–4 are **multistep or strategy problems**. Do Problem Solving Practice Exercises 1–4 with your students. Identify those having difficulty and use lesson resources to help.

Independent Practice

Note that Exercises 5–11 are **multistep or strategy problems**. Assign Exercises 5–11.

4 Assess

Summarize the lesson by having students:

DISCUSS How would you use *predict and test* to solve this problem: A poster costs $4 less than a banner. Together, the poster and banner cost $20. What is the price of the poster? **Possible answer: Choose two numbers with a difference of 4 and check to see if the sum is 20. 10 + 14 = 24, so I would try two lesser numbers, 8 and 12. 8 + 12 = 20. So the poster costs $8 and the banner costs $12.**

📖 **WRITE** Explain how you solved Exercise 2. **Possible answer: I made a table to keep track of win/loss predictions and tests that total 23 due to 2 ties.**

Lesson Quiz

Predict and test to solve.

1. Juan paid $21 for two model airplanes. One airplane cost $7 more than the other. What was the cost of each airplane? **$7, $14**

2. The Lorenzos traveled 300 more miles during the first week of their camping trip than they did the second week. They traveled a total of 2,280 miles. How many miles did they travel each week? **first week, 1,290 mi; second week, 990 mi**

CHALLENGE 1.4

Patterns, Patterns

Draw the next three figures of the pattern. Then describe the rule used to form the pattern.

1. Rule: Triangle, square, then repeat adding one triangle and one square, repeating

2. Rule: Circle, square, diamond, then reverse

3. Rule: Circle, diamond, circle, then replace circles with squares, then back to circle, diamond, circle, repeating

4. Rule: Square 1/4 shaded, square 1/2 shaded, square entirely shaded, then reverse

Give the next three numbers in the pattern. Then describe the rule.

5. 1, 2, 4, 8, 16, 32, **64**, **128**, **256**, ...
   Rule: Multiply by 2.

6. 3, 2, 4, 3, 5, 4, **6**, **5**, **7**, ...
   Rule: Subtract 1, add 2.

7. 3, 6, 9, 15, 24, 39, **63**, **102**, **165**, ...
   Rule: Add the two previous terms.

READING STRATEGY 1.4

Compare   Analyze Choose Solve Check

When you **compare** two or more things, you examine how they are alike. It can be helpful to compare information in a problem. Read the following problem.

VOCABULARY
compare

Ralph has some chickens and some pigs. Together, the animals have 38 legs. They have 15 heads. How many of each kind of animal does he have?

This is a problem for which you might want to use the *predict and test* strategy. When you use this strategy, you think of possible solutions. Then you compare to see whether your solution fits the information given in the problem. You can use a table to compare information.

1. Complete the table. Compare the information about heads and legs in the chart with the information given in the problem.

Predict		Test	
Number of Chickens	Number of Pigs	Number of Legs	Number of Heads
7	8	46	15
9	6	42	15
10	5	40	15
11	4	38	15

2. Solve the problem. **11 chickens and 4 pigs**

Make a table to compare the facts. Solve.

3. The Ping-Pong Paddlers table-tennis team played 15 games. They lost 4 fewer games than they won. They tied 2 more games than they lost. What was the team's record?
   7 wins, 3 losses, 5 ties

4. Janine bought 20 pieces of fruit. Ten can be eaten without peeling. Eight are yellow and 6 are orange. She has 2 more pears than bananas. She bought grapefruit, lemons, bananas, apples, yellow pears, and oranges. How many of each fruit did she buy?
   4 pears, 1 grapefruit, 2 bananas, 1 lemon, 6 apples, 6 oranges

LESSON 1.5

Algebra: Use Expressions

LESSON PLANNING

Objective To identify, write, and evaluate numerical and algebraic expressions involving whole numbers

Intervention for Prerequisite Skills
Place Value of Whole Numbers (For intervention strategies, see page 15.)

NCTM Standards
1. Number and Operations
2. Algebra
6. Problem Solving
7. Reasoning and Proof
8. Communication

Vocabulary

numerical expression a mathematical phrase that includes only numbers and operation symbols

variable a letter or symbol that stands for one or more numbers

algebraic expression an expression that is written using one or more variables

evaluate to find the answer to an expression

Math Background

One of the first steps in using equations to solve problems is to learn to interpret and evaluate expressions. Consider the following as you help students understand how to work with expressions:

- A numerical expression uses numbers and operation symbols to express the ideas given in a word expression.
- An algebraic expression uses numbers, operation symbols, and variables to express a quantitative idea.
- Evaluating an expression means simplifying it or replacing the variable with a number and simplifying it.

It is important for students to understand that an expression is like a phrase, or part of a sentence. An expression does not contain an equal sign or an inequality sign.

WARM-UP RESOURCES

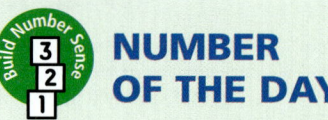 

NUMBER OF THE DAY

I am a number. Add 5 to me and multiply the result by 8. Subtract 40 and you get 48. What number am I? 6

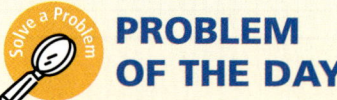 

PROBLEM OF THE DAY

In a number game, when Tina says *three*, Jay says *ten*. When Tina says *five*, Jay says *sixteen*. When Tina says *nine*, Jay says *twenty-eight*. When Tina says *eight*, what does Jay say? twenty-five If Jay says *one*, what number has Tina said? zero

Solution Problem of the Day tab, p. PD1B

DAILY FACTS PRACTICE

Have students practice addition and subtraction facts by completing Set E of *Teacher's Resource Book*, p. TR93.

INTERVENTION AND EXTENSION RESOURCES

ALTERNATIVE TEACHING STRATEGY — ELL

Materials 20 index cards

Have students **identify numerical and algebraic expressions.** Display 10 numerical and 10 algebraic expressions. Write the corresponding word expressions on cards, one to a card. Divide the class into small groups and give each group several cards. Then choose the expressions one at a time. Let the groups decide who has the corresponding word expression. Call on a volunteer from the group to read the word expression. Check students' work.

VISUAL
BODILY/KINESTHETIC

MULTISTEP AND STRATEGY PROBLEMS

The following multistep and strategy problems are provided in Lesson 1.5:

Page	Item
29	23, 28

CAREER CONNECTION

Encourage students to **apply their knowledge of numerical and algebraic expressions to real life.** Read the following to students:

Shari works as a plumber. When she comes to your home to fix a plumbing problem, she charges a basic service fee of $65. Then she charges $35 per hour for every hour or part of an hour she spends at your home. She uses the expression $65 + 35h$ to determine how much you owe her. Find out how much you would owe Shari if she spent the following amounts of time at your home.

a. 45 min $100

b. 1 hr 15 min $135

c. 6 hr 10 min $310

AUDITORY
VISUAL/SPATIAL

EARLY FINISHERS

Materials *For each pair* 6 index cards

Have students work in pairs to **practice using expressions.** First have them write 6 simple algebraic expressions on cards, one to a card. Then one student names a number between 1 and 50. The other student draws a card and evaluates the algebraic expression for that number. They then check the work and trade roles. Answers will vary.

AUDITORY
VERBAL/LINGUISTIC

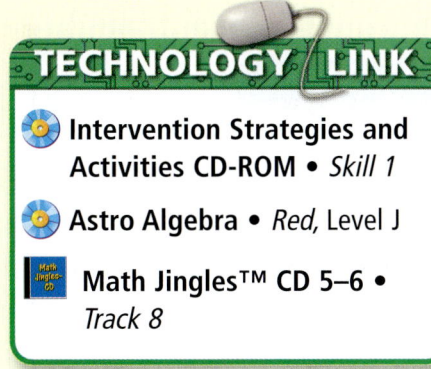

TECHNOLOGY LINK

- Intervention Strategies and Activities CD-ROM • *Skill 1*
- Astro Algebra • *Red,* Level J
- Math Jingles™ CD 5–6 • *Track 8*

LESSON 1.5 ORGANIZER

Objective To identify, write, and evaluate numerical and algebraic expressions involving whole numbers

Vocabulary numerical expression, variable, algebraic expression, evaluate

1 Introduce

QUICK REVIEW provides review of prerequisite skills.

Why Learn This? You can apply this skill to solving equations and more challenging word problems. Share the lesson objective with students.

2 Teach

Guided Instruction

- Help students compare the phrases "numerical expression" and "algebraic expression."
  What does the word *numerical* make you think of? numbers
  How is an algebraic expression different from a numerical expression? The algebraic expression contains one or more variables.
- Direct students' attention to the other examples of algebraic expressions.
  What is another way to write $y \div 2$? $\frac{y}{2}$

REASONING Write the other examples of algebraic expressions in words. five more than n; seven times (or multiplied by) a; three less than k (or k minus 3); y divided by 2; six times five times b

ADDITIONAL EXAMPLES

Example 1, p. 28

Write a numerical or algebraic expression for the word expression.

A. the number of slices in 3 dozen oranges, each of which has 10 slices $3 \times 12 \times 10$, $3(12)(10)$, or $3 \cdot 12 \cdot 10$

B. the height of a 17-step stairway of which each step is y in. high $17 \times y$, $17(y)$ or $17y$, or $17 \cdot y$

Example 2, p. 28

Evaluate each expression.
A. $s - 16$, for $s = 125$ 109
B. $t \div 7 \times 2$, for $t = 140$ 40

28 Chapter 1

LESSON 1.5

ALGEBRA
Use Expressions

Learn how to identify, write, and evaluate expressions involving whole numbers.

QUICK REVIEW
1. $23 + 14$ 37
2. $67 - 40$ 27
3. 15×6 90
4. $180 \div 30$ 6
5. 25×8 200

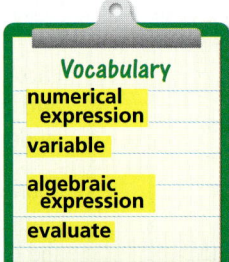

Vocabulary
numerical expression
variable
algebraic expression
evaluate

In a basketball game, a team scored 27 points in the first half and 38 points in the second half. To represent the total points, you could use a numerical expression. A **numerical expression** is a mathematical phrase that includes only numbers and operation symbols.

$$27 + 38 \leftarrow \text{total points}$$

Here are other examples of numerical expressions.

$60 + 25 \qquad 42 \div 7 \qquad 16 - 3 \qquad 51 \times 36 \qquad 30 + 12 + 41$

If you didn't know how many points the team scored in the second half, you could use a variable to represent the points. A **variable** is a letter or symbol that can stand for one or more numbers. An expression that includes a variable is called an **algebraic expression**.

$$27 + p \leftarrow \text{Use } p \text{ to represent points scored in second half.}$$

Here are other examples of algebraic expressions.

$5 + n \qquad 7 \times a \qquad k - 3 \qquad y \div 2 \qquad 6 \times 5 \times b$

In an expression, there are several ways to show multiplication.

$7 \times a$ can be written as $7a$, $7(a)$, or $7 \cdot a$.

Word expressions can be translated into numerical or algebraic expressions.

EXAMPLE 1

Write a numerical or algebraic expression for the word expression.

A. three dollars less than five dollars $5 - 3$

B. two times a distance, d $2 \times d$, $2(d)$, $2d$, or $2 \cdot d$

To **evaluate** a numerical expression, you find its value. To evaluate an algebraic expression, replace the variable with a number and then find the value.

EXAMPLE 2

Evaluate each expression.

A. $a + 150$, for $a = 18$

$a + 150$ *Replace a*
$18 + 150$ *with 18.*
168 *Add.*

B. $b \div 10 \times 3$, for $b = 120$

$b \div 10 \times 3$ *Replace b with*
$120 \div 10 \times 3$ *120.*
12×3 *Divide and*
36 *then multiply.*

28 Chapter 1

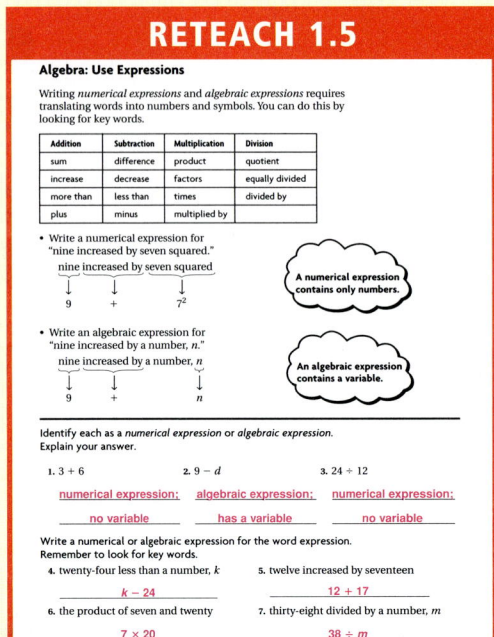

RETEACH 1.5

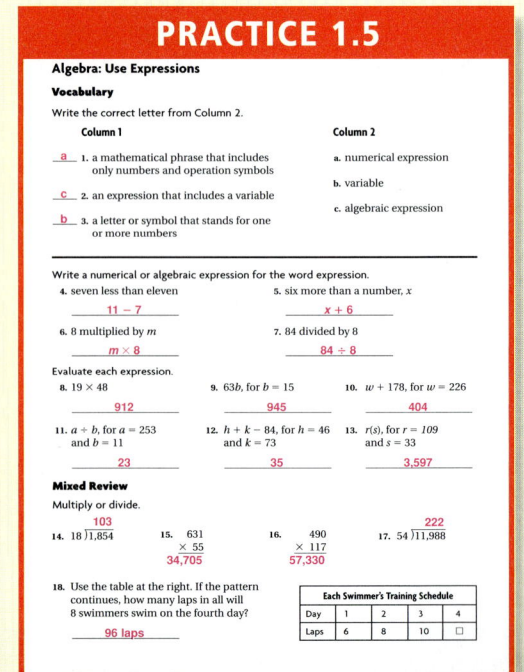

PRACTICE 1.5

CHECK FOR UNDERSTANDING

Think and Discuss Look back at the lesson to answer each question.
1. **Explain** the difference between a numerical expression and an algebraic expression. Give some examples of each. **An algebraic expression has one or more variables; examples will vary.**
2. **Show** four different ways to write an algebraic expression for the product of the number 10 and the variable g. **$10 \times g$, $10g$, $10(g)$, or $10 \cdot g$**

Guided Practice Write a numerical or algebraic expression for the word expression.
3. forty-six less than one hundred twenty-five **$125 - 46$**
4. one hundred seven more than y **$y + 107$**
5. y divided by fifteen **$y \div 15$, or $\frac{y}{15}$**

Evaluate each expression.
6. 21×15 **315**
7. $100 - g$, for $g = 54$ **46**
8. $s \div 8$, for $s = 720$ **90**

PRACTICE AND PROBLEM SOLVING

Independent Practice Write a numerical or algebraic expression for the word expression.
9. twenty-five $20 bills **$25 \times 20$**
10. q more than two hundred fifteen **$215 + q$**
11. seventy-six decreased by k **$76 - k$**
12. x divided by fourteen **$x \div 14$, or $\frac{x}{14}$**

Evaluate each expression.
13. 15×31 **465**
14. $3,021 + 915$ **3,936**
15. $10,340 - 1,340$ **9,000**
16. $k - 65$, for $k = 95$ **30**
17. $\frac{d}{7} \times 2$, for $d = 490$ **140**
18. $100b$, for $b = 54$ **5,400**
19. $m \div n$, for $m = 1,230$ and $n = 410$ **3**
20. cd, for $c = 5$ and $d = 200$ **1,000**

Problem Solving Applications
21. Let n represent the number of free throws Nathan scored. Bryan scored 12 more free throws than Nathan. Write an algebraic expression to show how many free throws Bryan scored. **$n + 12$**
22. ✍ **Write About It** Explain how to evaluate an algebraic expression when you know the value of each variable. Give an example. **Replace each variable with its value and perform each operation. Examples will vary.**
23. **REASONING** Tiffany, Deidre, Luisa, Kendall, and Ann Marie are runners. Kendall can outrun Luisa and Tiffany, but Deidre can outrun Kendall. Luisa can outrun Ann Marie, but Deidre can outrun Luisa. Which one of the girls is the fastest runner? **Deidre**

MIXED REVIEW AND TEST PREP

24. 530×42 (p. 22) **22,260**
25. $3,870 \div 18$ (p. 22) **215**
26. $1,234 + 453$ (p. 20) **1,687**
27. $8,000 - 357$ (p. 20) **7,643**

★28. **TEST PREP** Christina bought 3 pens and 1 notebook for $5.85. A pen cost $1.20. How much did the notebook cost? **D**

A $1.20   B $1.25   C $1.85   D $2.25

EXTRA PRACTICE page H32, Set D

29

---

3 Practice

Guided Practice
Do Check for Understanding Exercises 1–8 with your students. Identify those having difficulty and use lesson resources to help.

Independent Practice
Note that Exercises 23 and 28 are **multistep or strategy problems**. Assign Exercises 9–23.

Algebraic Thinking The ability to translate words into symbols is fundamental to solving word problems and to all future work in algebra.

Before assigning Exercises 9–21 remind students that a variable holds the place of a number, much like the symbol  that they have seen in previous expressions.

MIXED REVIEW AND TEST PREP
Exercises 24–28 provide **cumulative review** (Grade 5 and Chapter 1).

4 Assess

Summarize the lesson by having students:

DISCUSS What can a variable represent? **A variable can represent one or more numbers.**

 **WRITE** Describe three algebraic and three numerical expressions. Then rewrite them as word expressions. **Check students' work.**

Lesson Quiz

Transparency 1.5

Match the word expression with the numerical expression.

1. 7 times s **c**   a. $s - 7$
2. 7 less than s **a**   b. $q + 7$
3. q minus s **e**   c. $7 \times s$
4. twice q plus 7 **f**   d. $\frac{q}{s}$
5. 7 more than q **b**   e. $q - s$
6. q divided by s **d**   f. $2q + 7$

Evaluate the expression.

7. $5r - 6$, for $r = 9$ and $r = 12$ **39; 54**

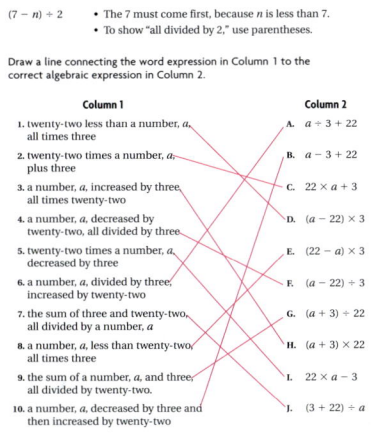

29

LESSON 1.6

Algebra: Mental Math and Equations

LESSON PLANNING

Objective To solve equations with whole numbers by using mental math and substitution

Intervention for Prerequisite Skills
Place Value of Whole Numbers (For intervention strategies, see page 15.)

NCTM Standards
1. Number and Operations
2. Algebra
6. Problem Solving
8. Communication

Vocabulary

equation an algebraic or numerical sentence that shows two quantities are equal

solution a value that, when substituted for the variable, makes an equation true

Math Background

Before students learn to solve problems by using the properties of equality, it is important that students understand what a solution is. Accordingly, in this lesson students:

- test a possible solution by substituting it in the equation to see if the resulting statement is true.
- solve simple equations by using mental math.

To solve by using mental math, students rely on number facts. This process of using related number facts and then testing the solution strengthens the students' understanding of what constitutes a solution of an equation.

WARM-UP RESOURCES

 **NUMBER OF THE DAY**

Take the number that represents the day of the month. Find the value for each of these expressions: $n + 8$, $n - 1$, $n \times 5$, and $n \div 2$. Check students' answers.

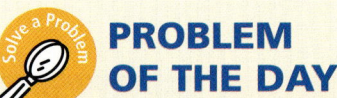 **PROBLEM OF THE DAY**

Martin saves n dollars each week. Kara saves twice as much as Martin. In 15 weeks their combined savings total $450. How much does Martin save each week? How can you use mental math to solve it? $10

Solution Problem of the Day tab, p. PD1B

 **DAILY FACTS PRACTICE**

Have students practice multiplication and division facts by completing Set F of *Teacher's Resource Book*, p. TR93.

INTERVENTION AND EXTENSION RESOURCES

ALTERNATIVE TEACHING STRATEGY

Materials *For each pair* number cube, p. TR75

Have students work in pairs to **write equations for each other to solve.** Display this formula:

variable + low roll = high roll

Have one student in each pair roll the number cube twice and use the formula to write an equation for his or her partner to solve. If the two rolls are equal, then the high roll and the low roll will be the same number. Have students record their equations and solutions. Answers will vary.

KINESTHETIC
INTERPERSONAL/SOCIAL

MULTISTEP AND STRATEGY PROBLEMS

The following multistep and strategy problems are provided in Lesson 1.6:

Page	Item
31	25

ENGLISH LANGUAGE LEARNER

Explore the word *equation* with your students. Display the words *equal, equality,* and *equinox.* Have students suggest meanings of these words. Record their ideas. You may want to explain that the equinox occurs when day and night are of approximately equal length, the beginning of spring and fall. Then ask:

- What do these words have in common? all have *equ* as part of the word
- What does *equation* mean? an algebraic sentence with parts that are equal on either side of the equal sign
- In what ways is the word *equation* similar to the other three words? They all deal with things being equal.

Have students illustrate each word for a class poster and add "equ" words throughout the unit.

VISUAL
VISUAL/SPATIAL

SPECIAL NEEDS

Provide additional **practice with variables and solutions.** Have each student write a number less than 15 on a sheet of paper. Then write this equation and read it to the students: $2n = 20$

Ask a volunteer to come to the board and place his or her number over the variable in the equation. Have other students decide if that student's number is a solution to the equation. Call on another volunteer to explain why it is a solution or why it is not.

Repeat with other equations, such as $a + 7 = 18$ and $b - 9 = 11$. Check students' work.

VISUAL
VISUAL/SPATIAL

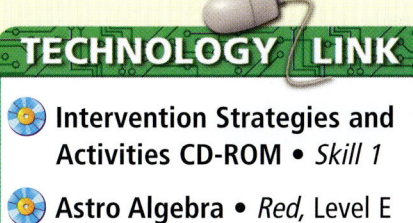

TECHNOLOGY LINK
- Intervention Strategies and Activities CD-ROM • *Skill 1*
- Astro Algebra • *Red,* Level E

LESSON 1.6 ORGANIZER

Objective To solve equations with whole numbers by using mental math and substitution

Vocabulary equation, solution
Review variable

1 Introduce

QUICK REVIEW provides review of prerequisite skills.

Why Learn This? You can apply this skill to future math problems to determine unknown numbers in an equation. *Share the lesson objective with students.*

2 Teach

Guided Instruction

- Ask students to describe an equation.

  What symbol differentiates an equation from an expression? an equal sign

- After working through Example 1, ask:

  REASONING Are there any other numbers that are not solutions of $12p = 108$? yes, any number other than 9

  Are there any other numbers that are solutions? Explain. No; When you replace p with any other number, the statement is no longer true.

- Help students explore ways to use mental math to solve the equations.

  How did you use mental math to solve Example 2? Possible answer: I remembered that $8 + 8 = 16$.

  How could you use subtraction to help you solve Example 2? Think that $16 - 8 = 8$, so $8 + 8 = 16$.

ADDITIONAL EXAMPLES

Example 1, p. 30

Which of the following numbers, 6, 7, or 8, is a solution of the equation $9n = 72$? 8

Example 2, p. 30

Solve the equation $15 = t + 9$ by using mental math. $t = 6$; The solution is 6.

LESSON 1.6

ALGEBRA
Mental Math and Equations

Learn how to use mental math to solve equations.

QUICK REVIEW

Evaluate each expression.
1. 25×8 **200**
2. $80 + d$, for $d = 37$ **117**
3. $g \div 8$, for $g = 72$ **9**
4. $450 - 225$ **225**
5. Find the missing factor. ■ $\times 8 = 24$ **3**

Vocabulary
equation
solution

An **equation** is a statement showing that two quantities are equal. These are equations:

$$6 + 7 = 13 \qquad 24 \div 3 = 8 \qquad k - 3 = 1 \qquad 2d = 18 \qquad a + b = 11$$

If an equation contains a variable, you can solve the equation by finding the value of the variable that makes the equation true. That value is the **solution**.

EXAMPLE 1

Which of the numbers 8, 9, and 10 is a solution of the equation $12p = 108$?

Remember that a variable is a letter or symbol that stands for one or more numbers.

Replace p with 8.	Replace p with 9.	Replace p with 10.
$12(8) \stackrel{?}{=} 108$	$12(9) \stackrel{?}{=} 108$	$12(10) \stackrel{?}{=} 108$
$96 = 108$ false	$108 = 108$ true	$120 = 108$ false

The solution is 9 because $12(9) = 108$.

- Which of the numbers 4, 5, and 6 is a solution of the equation $222 \div n = 37$? **6**

Some equations with variables can be solved by using mental math. Think of the value of the variable that makes the equation true. Then check your answer.

EXAMPLE 2

The Statue of Liberty's hand is about 16 ft long. The index finger is 8 ft long. What is the length of the palm of her hand? Solve the equation $16 = c + 8$ by using mental math.

$16 = c + 8$ *What number added to 8 gives 16?*
$8 = c$ *The solution is 8.*

Check:
$16 = 8 + 8$ *Replace c with 8.*
$16 = 16$ *$8 + 8$ is equal to 16.*

- Solve the equation $m \times 7 = 56$ by using mental math. $m = 8$

30 Chapter 1

RETEACH 1.6

Algebra: Mental Math and Equations

You can use the number facts you know to help solve equations. Remember, you can use fact families to find missing numbers.

- Solve the equation $8 + x = 12$ by using mental math.

  Related fact: $12 - 8 = 4$
  So, $8 + 4 = 12$.
  $8 + x = 12$
  $x = 4$ The solution is 4.
  Check to be sure your answer is correct.
  $8 + 4 = 12$ Replace x with 4.
  $12 = 12$ $8 + 4$ is equal to 12.

- Solve the equation $y \times 7 = 35$ by using mental math.

  Related fact: $35 \div 7 = 5$
  So, $5 \times 7 = 35$.
  $y \times 7 = 35$
  $y = 5$ The solution is 5.
  Check to be sure your answer is correct.
  $5 \times 7 = 35$ Replace y with 5.
  $35 = 35$ 5×7 is equal to 35.

- Solve the equation $m - 9 = 8$ by using mental math.

  Related fact: $9 + 8 = 17$.
  So, $17 - 9 = 8$.
  $m - 9 = 8$
  $m = 17$ The solution is 17.
  Check to be sure your answer is correct.
  $17 - 9 = 8$ Replace m with 17.
  $8 = 8$ $17 - 9$ is equal to 8.

- Solve the equation $d \div 6 = 8$ by using mental math.

  Related fact: $8 \times 6 = 48$
  So, $48 \div 6 = 8$.
  $d \div 6 = 8$
  $d = 48$ The solution is 48.
  Check to be sure your answer is correct.
  $48 \div 6 = 8$ Replace d with 48.
  $8 = 8$ $48 \div 6$ is equal to 8.

Solve each equation by using mental math.

1. $a - 5 = 9$
   Related fact:
   $9 + 5 =$ **14**
   The solution is **14**.

2. $\frac{k}{5} = 5$
   Related fact:
   $5 \times 5 =$ **25**
   The solution is **25**.

3. $3r = 18$
   Related fact:
   $18 \div 3 =$ **6**
   The solution is **6**.

4. $10 + f = 15$
   f = 5

5. $n \div 12 = 3$
   n = 36

6. $8s = 64$
   s = 8

7. $w - 20 = 10$
   w = 30

8. $n \times 9 = 81$
   n = 9

9. $x + 9 = 16$
   x = 7

10. $\frac{m}{10} = 10$
    m = 100

11. $g - 30 = 6$
    g = 36

12. $16 = c + 100$
    c = 1,600

PRACTICE 1.6

Algebra: Mental Math and Equations

Determine which of the given values is a solution of the equation.

1. $4d = 28$;
   $d = 7, 8,$ or 9
   d = 7

2. $50 - t = 28$;
   $t = 20, 21,$ or 22
   t = 22

3. $42 \div n = 6$;
   $n = 5, 6,$ or 7
   n = 7

4. $72 + v = 85$;
   $v = 12, 13,$ or 14
   v = 13

5. $m + 7 = 18$;
   $m = 9, 10,$ or 11
   m = 11

6. $s - 17 = 10$;
   $s = 26, 27,$ or 28
   s = 27

7. $c \div 8 = 3$;
   $c = 22, 23,$ or 24
   c = 24

8. $155 = 5k$;
   $k = 30, 31,$ or 32
   k = 31

9. $8 = 25 - x$;
   $x = 17, 18,$ or 19
   x = 17

Solve each equation by using mental math.

10. $e + 6 = 20$
    e = 14

11. $x \div 2 = 10$
    x = 20

12. $6 \times h = 300$
    h = 50

13. $s - 18 = 40$
    s = 58

14. $92 = b + 7$
    b = 85

15. $90 \div t = 15$
    t = 6

16. $m - 150 = 420$
    m = 570

17. $8 \times n = 72$
    n = 9

18. $f - 6 = 98$
    f = 104

19. $c \times 4 = 40$
    c = 10

20. $63 = d \times 7$
    d = 9

21. $k + 28 = 32$
    k = 4

22. $9x = 180$
    x = 20

23. $6 = v - 58$
    v = 64

24. $w \div 9 = 12$
    w = 108

25. $p + 62 = 100$
    p = 38

Mixed Review

Find the sum or difference. Estimate to check. Possible estimates are given.

26.   390
       + 789
       1,200;
       1,179

27.   9,056
       − 1,732
       7,000;
       7,324

28.   1,978
       + 693
       2,700;
       2,671

29.   47,813
       − 9,507
       40,000;
       38,306

30.   73,681
       + 50,342
       120,000;
       124,023

Evaluate each expression.

31. $n + 701$, for $n = 510$
    1,211

32. $50p$, for $p = 53$
    2,650

33. $r \times s$, for $r = 12$ and $s = 30$
    360

34. $h + g$ for $h = 65$ and $g = 41$
    106

CHECK FOR UNDERSTANDING

Think and Discuss Look back at the lesson to answer each question.
1. **Tell** whether 4 is a solution of the equation $x + 3 = 9$. If it is not, find the solution. **No; 6**
2. **Give an example** of an equation with a solution of 5. **Possible answer: $b \times 2 = 10$**

Guided Practice Determine which of the given values is the solution of the equation.
3. $f \div 7 = 3$; $f = 19, 20,$ or 21 **$f = 21$**
4. $t + 9 = 20$; $t = 10, 11,$ or 12 **$t = 11$**

Solve each equation by using mental math.
5. $7 = x + 3$ **$x = 4$**
6. $\frac{h}{9} = 3$ **$h = 27$**
7. $4 \times k = 16$ **$k = 4$**

PRACTICE AND PROBLEM SOLVING

Independent Practice Determine which of the given values is the solution of the equation.
8. $3h = 39$; $h = 11, 12,$ or 13 **$h = 13$**
9. $17 - x = 12$; $x = 5, 6,$ or 7 **$x = 5$**
10. $48 + s = 57$; $s = 8, 9,$ or 10 **$s = 9$**
11. $3 = 54 \div k$; $k = 16, 17,$ or 18 **$k = 18$**

Solve each equation by using mental math.
12. $p - 7 = 7$ **$p = 14$**
13. $9m = 81$ **$m = 9$**
14. $13 + r = 30$ **$r = 17$**
15. $x - 16 = 4$ **$x = 20$**
16. $h \div 8 = 7$ **$h = 56$**
17. $14 = k - 15$ **$k = 29$**
18. $87 = e \div 10$ **$e = 870$**
19. $12 \times v = 240$ **$v = 20$**
20. $t \div 6 = 125$ **$t = 750$**
21. $12 + 4 + d = 25$ **$d = 9$**
22. $3 \times 4 = c - 8$ **$c = 20$**
23. $p + 14 = 32 - 12$ **$p = 6$**

Problem Solving Applications
24. The equation $w + 12 = 40$ describes the number of men and women riding the bus to a convention. If w is the number of women riding the bus, how many men are riding the bus? **12 men**
25. Mr. Murakami teaches 5 classes of 25 students each. One hundred of his students are sixth graders. How many are not sixth graders? **25 students**
26.  **What's the Question?** A roller coaster has 7 cars. Fifty-six people can ride the roller coaster at one time. The answer is 8. **Possible question: How many people can ride in each car?**

MIXED REVIEW AND TEST PREP

27. Evaluate $a + 14$ for $a = 27$. (p. 28) **41**
28. $525 \div 25$ (p. 22) **21**
29. Find $4,310 - 1,900 + 3,450 - 870$. (p. 20) **4,990**
30. Find the greatest common factor of 15 and 35. **5**
31.  **TEST PREP** Andre left the house at 8:45 A.M. He arrived home $4\frac{1}{2}$ hours later. At what time did Andre arrive home? **D**

A 11:45 A.M.   B 12:45 A.M.   C 12:45 P.M.   D 1:15 P.M.

EXTRA PRACTICE page H32, Set E

31

CHAPTER 1

REVIEW/TEST

Purpose To check understanding of concepts, skills, and problem solving presented in Chapter 1

USING THE PAGE

The Chapter 1 Review/Test can be used as a **review** or a **test**.
- Items 1–3 check understanding of concepts and new vocabulary.
- Items 4–38 check skill proficiency.
- Items 39–40 check students' abilities to choose and apply problem solving strategies to real-life problems involving whole numbers.

 Suggest that students place the completed Chapter 1 Review/Test in their portfolios.

USING THE ASSESSMENT GUIDE

- Multiple-choice format of Chapter 1 Posttest—See *Assessment Guide*, pp. AG9–10.
- Free-response format of Chapter 1 Posttest—See *Assessment Guide*, pp. AG11–12.

USING STUDENT SELF-ASSESSMENT

The How Did I Do? survey helps students assess what they have learned and how they learned it. This survey is available as a copying master in *Assessment Guide*, p. AGxii.

CHAPTER 1 REVIEW/TEST

1. **VOCABULARY** A way to estimate a sum when all of the addends are about the same is ? . (p. 16)
   clustering

2. **VOCABULARY** A letter or symbol that can stand for one or more numbers is a(n) ? . (p. 28)
   variable

3. **VOCABULARY** A statement showing that two quantities are equal is a(n) ? . (p. 30)
   equation

Estimate. (pp. 16–19) Possible estimates are given.

4. 593 + 724 = **1,300**
5. 1,420 + 5,791 = **7,200**
6. 935 − 549 = **400**
7. 2,371 − 1,456 = **900**
8. 43,816 − 39,972 = **4,000**
9. 48 × 6 **300**
10. 308 × 67 **21,000**
11. 374 ÷ 7 **50**
12. 276 ÷ 42 **7**
13. 3,764 ÷ 591 **6**

Find the sum or difference. (pp. 20–21)

14. 4,762 + 39,038 = **43,800**
15. 9,724 − 286 = **9,438**
16. 50,031 − 9,352 = **40,679**
17. 737 + 4,650 + 11,821 = **17,208**
18. 678,040 − 329,193 = **348,847**

Multiply or divide. (pp. 22–25)

19. 526 × 42 **22,092**
20. 123 × 12 **1,476**
21. 2,250 ÷ 18 **125**
22. 189 × 108 **20,412**
23. 40)3,206 **80 r6**

Evaluate each expression for $a = 63$, $b = 150$, and $c = 7$. (pp. 28–29)

24. $a + 305$ **368**
25. $b - 36$ **114**
26. $300 \div b$ **2**
27. $3c$ **21**
28. $a \div 9$ **7**
29. $215 - b$ **65**
30. $112 \div c$ **16**
31. $a \times 5$ **315**
32. $2a + 4$ **130**
33. $a + b$ **213**

Solve each equation by using mental math. (pp. 30–31)

34. $3m = 27$ **$m = 9$**
35. $14 = q + 6$ **$q = 8$**
36. $20 = y - 9$ **$y = 29$**
37. $w \div 50 = 5$ **$w = 250$**
38. $74 + a = 85$ **$a = 11$**

Solve. (pp. 26–27)

39. At school during spirit week, Ming sold a total of 36 red and blue ribbons. She sold 6 more red ribbons than blue ribbons. How many of each color did she sell? **15 blue and 21 red ribbons**

40. Colton worked two days on a project for school. He worked a total of 195 minutes. If he worked 45 minutes longer on the first day, how long did Colton work each day? **120 min, 75 min**

Looking Back  Prerequisite Skills

To be ready for Chapter 2, students should have the following understandings and skills:

- **Repeated Multiplication**—find products of repeated factors
- **Properties**—Commutative and Associative of Addition and Multiplication, Distributive, Identity of Zero and One, and Property of Zero
- **Use of Parentheses**—solve equations; evaluate expressions

Check What You Know

Use page 35 to determine students' knowledge of prerequisite concepts and skills.

Intervention

Help students prepare for the chapter by using the intervention resources described on TE page 35.

Looking at Chapter 2  Essential Skills

Students will

- apply the properties and other mental math strategies with whole number operations.
- find the value of numbers written in exponent form.
- **use the order of operations to evaluate expressions involving whole numbers.**
- use the skill *sequence and prioritize information* to solve problems.

EXAMPLE

$12 + (7 + 1) \div 2^2$

$12 + (8) \div 2^2$	Operate inside parentheses.
$12 + 8 \div 4$	Clear exponents.
$12 + 2$	Divide.
14	Add.

Looking Ahead  Applications

Students will apply what they learn in Chapter 2 to the following new concepts:

- Decimal Expressions (Chapter 4)
- Operations with Integers (Chapters 11 and 12)
- Expressions with Squares (Chapter 13)
- Solve Equations (Chapters 14 and 15)

Operation Sense **34D**

CHAPTER 2

Operation Sense

INTRODUCING THE CHAPTER

Tell students that they can use operation sense to relate the structure of the human body to the body structure of a variety of animals. Have students focus on the photograph. Ask them to write another expression that tells how many cervical vertebrae flamingos have. **Possible answer: $3^2 + 10$**

USING DATA

To begin the study of this chapter, have students
- Write an expression for the total number of cervical vertebrae of 7 people. **7×7**
- Which bird has $(36 - 16) + 5$ vertebrae? **swan**
- Which bird has twice as many cervical vertebrae as a human does? **Owl**
- Formulate and answer three questions about the data. **Check students' work.**

PROBLEM SOLVING PROJECT

Purpose To use operation sense to solve a problem.

Grouping pairs or small groups

Background The human backbone has a total of 24 vertebrae (7 cervical, 12 thoracic, 5 lumbar and two fused bones (sacrum and Coccyx).

Analyze, Choose, Solve, and Check

Have each group
- Research the number of vertebrae in the backbones of three other animals.
- Make number riddles by using exponents and parentheses in expressions for the number of vertebrae each animal has.
- Exchange and evaluate each other's expressions.

Check students' work.

 Suggest that students place the riddles and solutions in their portfolios.

CHAPTER 2 Operation Sense

about 3 times as many

A bird can reach almost any part of its body with its beak because it has an extremely flexible neck. The bones in the neck are called cervical vertebrae. Humans have 7 cervical vertebrae. Most birds have more cervical vertebrae than humans or other mammals. Flamingos have $4^2 + 8 \times 2 - 13$ cervical vertebrae.

PROBLEM SOLVING About how many times as many cervical vertebrae does a flamingo have as a human?

CERVICAL VERTEBRAE

Animal	Number of Cervical Vertebrae
Human	
Owl	
Pigeon	
Swan	
Calif. Condor	

Why learn math? Explain that biologists study all aspects of animal life. They collect and use the data for a variety of purposes. For example, by studying the bones of an animal, biologists can tell the animal's height, weight, and often its diet. Ask: How could you use math to care for animals? **Possible answer: To calculate the weekly cost of pet supplies; to measure appropriate daily food amounts**

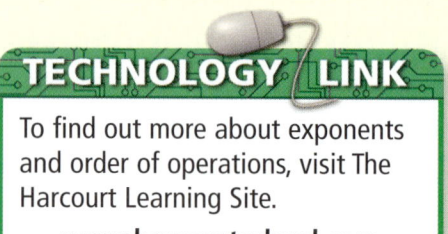

TECHNOLOGY LINK

To find out more about exponents and order of operations, visit The Harcourt Learning Site.

www.harcourtschool.com

Check What You Know

Use this page to help you review and remember important skills needed for Chapter 2.

Repeated Multiplication (For Intervention, see p. H15.)

Find the product.

1. $3 \times 3 \times 3$ **27**
2. $2 \times 2 \times 2 \times 2$ **16**
3. $4 \times 4 \times 4$ **64**
4. $5 \times 5 \times 5 \times 5$ **625**
5. $10 \times 10 \times 10 \times 10$ **10,000**
6. $9 \times 9 \times 9$ **729**
7. $8 \times 8 \times 8$ **512**
8. $6 \times 6 \times 6$ **216**
9. $7 \times 7 \times 7$ **343**
10. $10 \times 10 \times 10$ **1,000**
11. $4 \times 4 \times 4 \times 4$ **256**
12. $5 \times 5 \times 5$ **125**

Properties (For Intervention, see p. H2.)

Name the property illustrated.

13. $48 + 13 + 5 = 13 + 48 + 5$ **Commutative of Addition**
14. $8 \times (3 + 1) = (8 \times 3) + (8 \times 1)$ **Distributive**
15. $0 \times 999 = 0$ **Property of Zero**
16. $15 \times 1 = 15$ **Identity of Multiplication**
17. $(9 + 5) + 10 = 9 + (5 + 10)$ **Associative of Addition**
18. $27 + 36 = 36 + 27$ **Commutative of Addition**
19. $(4 + 2) \times 7 = (4 \times 7) + (2 \times 7)$ **Distributive**
20. $7 \times 9 \times 2 = 2 \times 9 \times 7$ **Commutative of Multiplication**
21. $1 \times 148 = 148$ **Identity of Multiplication**
22. $(2 \times 9) \times 5 = 2 \times (9 \times 5)$ **Associative Property of Multiplication**
23. $8 \times (20 + 8) = (8 \times 20) + (8 \times 8)$ **Distributive Property**
24. $6 \times 3 = 3 \times 6$ **Commutative of Multiplication**

Use of Parentheses (For Intervention, see p. H15.)

Solve the equation.

25. $5 \times (4 + 3) = (a \times 4) + (a \times 3)$ $a = 5$
26. $7 + (3 + 9) = (7 + m) + 9$ $m = 3$
27. $4 + 6 + 3 = 4 + (6 + r)$ $r = 3$
28. $(9 \times 4) + (9 \times 2) = h \times (4 + 2)$ $h = 9$
29. $4 \times (8 - 4) = (4 \times t) - (4 \times 4)$ $t = 8$
30. $(5 + 6) + (4 + 7) = (5 + 4) + (r + 7)$ $r = 6$

Evaluate the expression.

31. $(4 + 7) + 9$ **20**
32. $5 \times (6 + 2)$ **40**
33. $(1 + 7) \times 4$ **32**
34. $(7 + 8) + 3$ **18**
35. $4 + (7 - 3)$ **8**
36. $6 \times (9 - 3)$ **36**
37. $2 \times (3 + 5 + 8)$ **32**
38. $(8 + 5) + (3 + 1)$ **17**
39. $4 + (5 + 7) + 2$ **18**
40. $2 \times (3 \times 1)$ **6**

LOOK AHEAD

In Chapter 2 you will
- use properties and mental math to find sums, differences, products, and quotients
- use exponents
- use order of operations

Assessing Prior Knowledge

Use the **Check What You Know** page to determine whether your students have mastered the prerequisite skills critical for this chapter.

Intervention

- **Diagnose and Prescribe**

  Evaluate your students' performance on this page to determine whether intervention is necessary or if enrichment is appropriate. Options that provide instruction, practice, and a check are listed in the chart below.

CHECK WHAT YOU KNOW RESOURCES

Intervention Card, Copying Master, or CD-ROM

Intervention Strategies and Activities Teaching Transparencies

Intervention Practice Book

Enrichment Book

Were students successful with
CHECK WHAT YOU KNOW?

OPTIONS

NO — INTERVENE

- **REPEATED MULTIPLICATION, 1–12**
  How to Help
  Troubleshooting, Pupil Edition p. H15
  Intervention Strategies and Activities, Skill 12

- **PROPERTIES, 13–24**
  How to Help
  Troubleshooting, Pupil Edition p. H2
  Intervention Strategies and Activities, Skills 13–14

- **USE OF PARENTHESES, 25–40**
  How to Help
  Troubleshooting, Pupil Edition p. H15
  Intervention Strategies and Activities, Skill 40

YES — ENRICH

Check What You Know Enrichment Book, pp. 3–4

LESSON 2.1

Mental Math: Use the Properties

LESSON PLANNING

Objective To use properties and mental math to find sums, differences, products, and quotients

Intervention for Prerequisite Skills

Properties, Use of Parentheses (For intervention strategies, see page 35.)

NCTM Standards
1. Number and Operations
6. Problem Solving
8. Communication

Vocabulary

compensation an estimation strategy in which you change one addend to a multiple of ten and then adjust the other addend to keep the balance

Math Background

The properties of numbers serve as a basis for operating within our number system. Consider these ideas as you help students understand properties:

- Changing the order of numbers, based on the Commutative Property, or the grouping of numbers, based on the Associative Property, may result in numbers that are simpler to compute.
- Compensation is based on the Identity Property of Zero. Adding a number to a sum and then subtracting it is the same as adding zero to it.
- Thinking of one factor in a multiplication exercise as a sum or difference allows the use of the Distributive Property.

WARM-UP RESOURCES

 **NUMBER OF THE DAY**

Calculate your age in number of months. **Possible answer: for 11 years and 8 months—140 months**

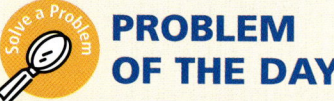 **PROBLEM OF THE DAY**

Replace the ■ with the digits 0–9 to make correct number sentences. Use each digit only once.

■ × ■ = 18 **2 × 9 = 18**

■ × ■ = 24 **3 × 8 = 24**

■ × ■ = 0   **5 × 0 = 0**

■ × ■ = 28 **4 × 7 = 28**

■ × ■ = 6   **6 × 1 = 6**

Solution Problem of the Day tab, p. PD2

 **DAILY FACTS PRACTICE**

Have students practice addition facts by completing Set G of *Teacher's Resource Book,* p. TR93.

36A Chapter 2

INTERVENTION AND EXTENSION RESOURCES

ALTERNATIVE TEACHING STRATEGY ELL

Materials *For each pair* index cards

Have students work in pairs to **practice using properties.**

- They prepare five index cards by writing the name of a multiplication property on each.
- One student draws a card, and both students write one multiplication sentence illustrating the property.
- After all the cards have been drawn, students combine their examples, mix them up, and exchange with another pair.
- They then identify the properties for the new set of problems. Check students' work.

See also page 38.

VISUAL
BODILY/KINESTHETIC

SPECIAL NEEDS

To encourage students to **apply properties to real-life problems**, present the following:

- Suppose you were to take 2 showers, brush your teeth 3 times, and wash dishes 2 times today. How much water would you use? 106 gal

Activity	Water Used
Washing dishes	20 gal
Taking shower	30 gal
Brushing teeth	2 gal

VISUAL
VERBAL/LINGUISTIC

MULTISTEP AND STRATEGY PROBLEMS

The following multistep and strategy problems are provided in Lesson 2.1:

Page	Item
39	53, 54, 59

VOCABULARY STRATEGY

Present the **Associative and Distributive Properties** by discussing the meaning of the terms *associate* and *distribute*.

- To act out the term *associate*, have three students stand up in front of the class spaced evenly apart.
- Then have two of the students move closer together. Describe the action as associating.
- Now ask a volunteer to relate the action to the property.
- To illustrate the term *distribute*, have one student pass out a piece of paper to each student. Describe the action as distributing the paper.
- Now ask a volunteer to relate the action to the property.
- Have volunteers practice using the words *associate* and *distribute*.

KINESTHETIC
BODILY/KINESTHETIC

TECHNOLOGY LINK

Intervention Strategies and Activities CD-ROM • *Skills 13–14, 40*

LESSON 2.1 ORGANIZER

Objective To use properties and mental math to find sums, differences, products, and quotients

Vocabulary compensation *Review* Commutative Property, Associative Property, Distributive Property

1 Introduce

QUICK REVIEW provides review of prerequisite skills.

Why Learn This? You can use this skill to help you determine the number of favors needed for a party. *Share the lesson objective with students.*

2 Teach

Guided Instruction

• *Review the Distributive Property.*

How would you use the Distributive Property to multiply a number by 7? Break the 7 into addends, 5 + 2 or 3 + 4. Multiply the number by each addend and add the products.

REASONING **Ellen scheduled four 45-minute classes and four 10-minute breaks. How could you use the Distributive Property to find the number of minutes she scheduled in all?** Possible answer: (4 × 45) + (4 × 10) = 4 × (45 + 10) = 4 × 55 = (4 × 50) + (4 × 5) = 200 + 20 = 220

• *Refer students to Example 2.*

Modifying Instruction In the Commutative Property part of Example 2, the Associative Property is also used twice:
(8 × 6) × 5 = 8 × (6 × 5) by Associative Property
= 8 × (5 × 6) by Commutative Property
= (8 × 5) × 6 by Associative Property

What is the purpose of the parentheses in the Associative Property? to regroup factors that are easier to multiply

ADDITIONAL EXAMPLE

Example 1, p. 36

Use the Distributive Property to solve 9 × 43.

$9 \times 43 = 9 \times (40 + 3)$
$= (9 \times 40) + (9 \times 3)$
$= (360) + (27)$
$= 387$

36 Chapter 2

LESSON 2.1 — MENTAL MATH
Use the Properties

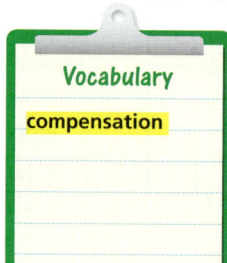

Learn how to use properties and mental math to find sums, differences, products, and quotients.

QUICK REVIEW

Solve by using mental math.
1. $d + 13 = 33$ 2. $25 - g = 17$ $g = 8$
3. $6m = 54$ $m = 9$ 4. $9 = t \div 11$ $t = 99$
5. $340 + 34 + 3$ 377

1. $d = 20$

Vocabulary
compensation

Remember these properties:

Commutative Property
$4 + 8 = 8 + 4$
$6 \times 7 = 7 \times 6$

Associative Property
$(6 + 8) + 5 = 6 + (8 + 5)$
$(2 \times 9) \times 5 = 2 \times (9 \times 5)$

Distributive Property
$12 \times 32 = 12 \times (30 + 2) = (12 \times 30) + (12 \times 2)$

EXAMPLE 1

EXAMPLE 2

Use the table to find how many bones the spine, chest, and shoulders have in all. Mentally find the sum by first reordering using the Commutative Property.

$26 + 25 + 4 = 26 + 4 + 25$ *Commutative Property*
$= 30 + 25$ *Use mental math.*
$= 55$

Mentally find the sum by regrouping using the Associative Property.

$(26 + 25) + 4 = 26 + (25 + 4)$ *Associative Property*
$= 26 + 29$ *Use mental math.*
$= 55$

So, the spine, chest, and shoulders have a total of 55 bones.

You can use the Distributive Property to mentally solve a problem.

How many bones are in 5 models of the human spine?
$5 \times 26 = 5 \times (20 + 6)$ *Break 26 into parts.*
$= (5 \times 20) + (5 \times 6)$ *Use the Distributive Property. Multiply mentally.*
$= 100 + 30$ *Add the products.*
$= 130$

So, there are 130 bones in 5 models.

You can also use the Commutative and Associative Properties. Try to make partial products that end in 0.

James has 8 storage boxes on each of 6 shelves. Each box contains 5 items. How many items are there altogether?

Commutative Property	Associative Property
$(8 \times 6) \times 5 = (8 \times 5) \times 6$	$(8 \times 6) \times 5 = 8 \times (6 \times 5)$
$= 40 \times 6$	$= 8 \times 30$
$= 240$	$= 240$

BONES IN THE HUMAN BODY

Part	Number of Bones
Head	28
Spine	26
Throat	1
Chest	25
Shoulders	4
Arms	6
Hands	54
Legs	10
Feet	52

36 Chapter 2

RETEACH 2.1

Use the Properties

One way to find a sum or product mentally is to use a number property.

Commutative Property
Numbers can be added in any order without changing the sum.
$45 + 29 + 55 = 29 + 45 + 55$
Order has been changed.

Associative Property
Addends can be grouped differently. The sum is always the same.
$(45 + 29) + 55 = 29 + (45 + 55)$
Grouping has been changed.

Numbers can be multiplied in any order without changing the product.
$5 \times 13 \times 8 = 5 \times 8 \times 13$
Order has been changed.

Factors can be grouped differently. The product is always the same.
$(5 \times 13) \times 8 = 5 \times (8 \times 13)$
Grouping has been changed.

Distributive Property
$25 \times 23 = 25 \times (20 + 3) = (25 \times 20) + (25 \times 3)$
Product of a number and a sum   Sum of two products

1. Complete to show how to find the sum. Name the reason for each step.

$19 + 45 + 21 + 5 = 19 + \underline{21} + \underline{45} + 5$ → Commutative Property
$= (19 + 21) + (45 + 5)$ → Associative Property
$= 40 + 50$ → Use mental math.
$= \underline{90}$

Add. Use mental math.
2. $16 + 9 + 24$ 3. $33 + 26 + 17 + 44$ 4. $21 + 14 + 29 + 36$
   $\underline{49}$ $\underline{120}$ $\underline{100}$

Complete to show how to use the Distributive Property to find each product.
5. $8 \times 14 = 8 \times (\underline{10} + 4)$ 6. $9 \times 34 = 9 \times (30 + \underline{4})$
   $= (8 \times \underline{10}) + (8 \times \underline{4})$ $= (9 \times \underline{30}) + (9 \times \underline{4})$
   $= \underline{80} + \underline{32}$ $= \underline{270} + \underline{36}$
   $= \underline{112}$ $= \underline{306}$

PRACTICE 2.1

Use the Properties

Vocabulary

Write the correct letter from Column 2.

Column 1		Column 2
b 1. Associative Property	a.	$58 + 72 = (58 + 2) + (72 - 2)$
c 2. Commutative Property	b.	$3 \times (2 \times 4) = (3 \times 2) \times 4$
a 3. compensation	c.	$10 \times 23 = 23 \times 10$
e 4. Distributive Property	d.	$18x = 18$
d 5. Identity Property of One	e.	$6 \times 24 = 6 \times (20 + 4)$

Use mental math to find the value.

6. $37 + 14$ $\underline{51}$ 7. $65 - 23$ $\underline{42}$ 8. 18×6 $\underline{108}$
9. $258 \div 3$ $\underline{86}$ 10. 18×22 $\underline{396}$ 11. $141 \div 3$ $\underline{47}$
12. $78 - 45$ $\underline{33}$ 13. $49 + 14$ $\underline{63}$ 14. $41 + 18$ $\underline{59}$
15. 19×11 $\underline{209}$ 16. $37 - 11$ $\underline{26}$ 17. $366 \div 6$ $\underline{61}$
18. $320 \div 5$ $\underline{64}$ 19. $59 + 26$ $\underline{85}$ 20. $74 - 23$ $\underline{51}$
21. 15×51 $\underline{765}$ 22. $88 - 54$ $\underline{34}$ 23. 43×21 $\underline{903}$
24. $465 \div 15$ $\underline{31}$ 25. $56 + 15$ $\underline{71}$ 26. 15×48 $\underline{720}$
27. $32 + 35$ $\underline{67}$ 28. $153 \div 9$ $\underline{17}$ 29. $96 - 25$ $\underline{71}$
30. $37 + 14 + 43$ $\underline{94}$ 31. $(7 \times 12) + (7 \times 18)$ $\underline{210}$ 32. $5 \times 33 \times 6$ $\underline{990}$

Mixed Review

Evaluate each expression for $a = 72, b = 28,$ and $c = 8$.
33. $b \times 7$ 34. $a + b + 362$ 35. $a \div c$ 36. $225 - a$
   $\underline{196}$ $\underline{462}$ $\underline{9}$ $\underline{153}$

Solve each equation using mental math.
37. $n \times 8 = 56$ 38. $19 + w = 36$ 39. $h + 20 = 35$ 40. $98 - x = 59$
   $n = 7$ $w = 17$ $h = 700$ $x = 39$

A strategy you can use for some addition and subtraction problems is **compensation**. For addition, change one number to a multiple of 10 and then adjust the other number to keep the balance.

EXAMPLE 3

Mr. Forge and his friends play basketball for an hour on Fridays and Saturdays. On Friday they scored a total of 44 points, and on Saturday they scored 57 points. Use compensation to find the total points scored for both days.

$44 + 57 = (44 + 6) + (57 - 6)$ *Add 6 to 44 and subtract 6 from 57.*
$ = 50 + 51$ *Use mental math to add.*
$ = 101$

So, the total points scored is 101.

When you use compensation to subtract, you have to do the same thing to each number. Since it's easy to subtract numbers ending in zero, try to make the second number a multiple of 10.

EXAMPLE 4

Use compensation to find $128 - 56$.

$128 - 56 = (128 + 4) - (56 + 4)$ *Add 4 to 128 and to 56 before subtracting.*
$ = 132 - 60$
$ = 72$

So, the difference is 72.

You can sometimes divide mentally by breaking a number into smaller parts that are each divisible by the divisor.

EXAMPLE 5

Use mental math to find $396 \div 4$.

$396 = 360 + 36$ *Break 396 into parts.*
$360 \div 4 = 90$ and $36 \div 4 = 9$ *Divide each part by 4 mentally.*
$90 + 9 = 99$ *Add the parts of the quotient.*

So, $396 \div 4 = 99$.

- Tell another way to break 396 into two parts to divide by 4.
  Answers will vary. Possible answers: 320 and 76; 324 and 72; 200 and 196

Math Idea ▶ Using the number properties and other mental math strategies will help you add, subtract, multiply, and divide mentally.

CHECK FOR UNDERSTANDING

Think and Discuss

Look back at the lesson to answer each question.

1. **Tell** how using the Associative Property in Example 2 made the problem easier to solve. **Possible answer: It is easier to find 8×30 than 48×5.**

2. **Explain** two ways to use compensation to find $349 + 138$ mentally.
   Possible answer: Add 1 to 349, subtract 1 from 138, add 350 and 137 to get 487; add 2 to 138, subtract 2 from 349, add 140 and 347 to get 487.

37

CHALLENGE 2.1

Solve It

Use mental math to solve each problem in the Decoder Box. Find the value in the Tip Box. Each time the value appears, write the letter of that problem above it. When you have solved all the problems, you will have discovered the math tip.

Decoder Box

A $63 + 27 =$	90	N $13 \times 6 =$	78
B $112 - 14 =$	98	O $4 \times 11 \times 3 =$	132
C $480 \div 8 =$	60	P $198 \div 9 =$	22
D $6 \times 7 \times 10 =$	420	Q $5 + 34 + 4 =$	43
E $55 \times 3 =$	165	R $15 \times 4 \times 2 =$	120
F $397 - 158 =$	239	S $16 \times 7 =$	112
H $2 \times 13 \times 5 =$	130	T $25 \times 6 =$	150
I $7 \times 21 =$	147	U $440 \div 5 =$	88
K $803 - 571 =$	232	V $25 + 19 + 4 =$	48
L $9 \times 2 \times 6 =$	108	Y $197 + 326 =$	523
M $8 \times 5 \times 4 =$	160	Z $1,135 - 797 =$	338

Tip Box

F	A	C	T	O	R	S		C	A	N		B	E
239	90	60	150	132	120	112		60	90	78		98	165

M	U	L	T	I	P	L	I	E	D		I	N
160	88	108	150	147	22	108	147	165	420		147	78

A	N	Y		O	R	D	E	R
90	78	523		132	120	420	165	120

Now use the Decoder Box to help you find the answer to a riddle. What is useful only when it's used up?

A	N		U	M	B	R	E	L	L	A
90	78		88	160	98	120	165	108	108	90

PROBLEM SOLVING 2.1

Use the Properties   *Analyze Choose Solve Check*

Write the correct answer.

1. Use compensation to add.
   $48 + 35$

   $48 + 35 = (48 + 2) + (35 - 2) =$
   $50 + 33 = 83$

2. Use mental math to find the value of $(13 + 12) + 7$.

   32

3. In the auditorium, there are 32 rows of seats. Each row has 24 chairs. How many students can the auditorium seat?

   768 students

4. Brock sorted his toy cars into five groups. The groups contained 18, 22, 16, 7, and 14 cars. Use mental math to find the total number of cars.

   77 cars

Choose the letter for the best answer.

5. Which expression shows how to use compensation to subtract 22 from 47?

   A $(47 - 2) - (22 - 2)$
   B $(47 + 3) - (22 - 2)$
   C $(47 - 20) - (22 - 2)$
   D $47 - 22$

6. What is the value of the underlined digit in 9,987.6<u>5</u>32?

   F 5 tens
   G 5 ones
   H 5 tenths
   I 5 hundredths

7. If you swim between 35 and 45 minutes a day, what is a reasonable estimate of the number of minutes you swim in 15 days?

   A Less than 300
   B Between 300 and 500
   C Between 500 and 700
   D More than 700

8. Which equation illustrates the Commutative Property?

   F $(2 \times 3) \times 4 = (2 \times 3) \times 4$
   G $(2 \times 3) \times 4 = (2 \times 3) \times 4$
   H $(2 \times 3) \times 4 = (3 \times 2) \times 4$
   J $(2 \times 3) \times 4 = 6 \times 4$

9. **Write About It** Explain how to use the Distributive Property to multiply 48 and 17.

   You rewrite 17 as $10 + 7$ and then multiply the two addends by 48.
   $48 \times 17 = (48 \times 10) + (48 \times 7) = 480 + 336 = 816$

Modifying Instruction In Example 4, show students that neither subtraction nor division is associative: $7 - (5 - 1) = 7 - 4 = 3$ but $(7 - 5) - 1 = 2 - 1 = 1$; $8 \div (4 \div 2) = 8 \div 2 = 4$ but $(8 \div 4) \div 2 = 2 \div 2 = 1$

Modifying Instruction In Example 5, show students that this strategy works with subtraction, as well as with addition: $\frac{396}{4} = \frac{(400 - 4)}{4} = 100 - 1 = 99$

Algebraic Thinking Understanding the properties and the mental math strategies of compensation and breaking dividends into easily divisible parts will help students write, simplify, and evaluate algebraic expressions.

ADDITIONAL EXAMPLES

Example 2, p. 36

The Office Supply Store has 4 boxes of binders on each of 6 shelves. There are 5 binders in each box. How many binders are there altogether?

Commutative Property

$(4 \times 6) \times 5 = (4 \times 5) \times 6$
$ = 20 \times 6$
$ = 120$

Associative Property

$(4 \times 6) \times 5 = 4 \times (6 \times 5)$
$ = 4 \times 30$
$ = 120$

So, there are 120 binders altogether.

Example 3, p. 37

Use compensation to find $46 + 97$.

$46 + 97 = (46 + 4) + (97 - 4)$
$ = 50 + 93$
$ = 143$

So, the sum is 143.

Example 4, p. 37

Use compensation to find $147 - 98$.

$147 - 98 = (147 + 2) - (98 + 2)$
$ = 149 - 100$
$ = 49$

So, the difference is 49.

Example 5, p. 37

Use mental math to find $450 \div 6$.

$450 = 420 + 30$
$420 \div 6 = 70$ and $30 \div 6 = 5$
$70 + 5 = 75$

So, $450 \div 6 = 75$.

LESSON 2.1

3 Practice

> **COMMON ERROR ALERT**
>
> When using the compensation strategy to add and subtract, some students may change one number to a multiple of ten and forget to adjust the other number. Have students explain how they plan to use the strategy before they make mental computations.
>
> *Error*
> $$76 + 37 = (76 + 4) + 37$$
> $$= 80 + 37$$
> $$= 117$$

Guided Practice

Do Check for Understanding Exercises 1–18 with your students. Identify those having difficulty and use lesson resources to help.

Independent Practice

Note that Exercises 53, 54, and 59 are **multi-step or strategy problems**. Assign Exercises 19–55.

Multistep or Strategy Problem To solve Exercise 53, students must first find the number of votes on Tuesday by multiplying Monday's votes, 23, by 3. Students can then add the votes of the 4 days together, then subtract from the necessary 250 votes. Guide students to conclude that the multiplication must be completed before the subtraction.

Guided Practice — Use mental math to find the value.

3. 12×17	4. $45 + 9 + 15$	5. $124 + 17 + 16$	6. 9×36
204	69	157	324
7. $(6 + 37) + 13$	8. $2 \times 9 \times 50$	9. 5×29	10. 11×43
56	900	145	473
11. $39 + 16$	12. $83 + 38$	13. $426 \div 3$	14. $16 + 35$
55	121	142	51
15. $279 \div 3$	16. $137 - 51$	17. $65 - 22$	18. $567 \div 7$
93	86	43	81

PRACTICE AND PROBLEM SOLVING

Independent Practice — Use mental math to find the value.

19. 24×7	20. $73 - 27$	21. 45×11	22. 12×35
168	46	495	420
23. $87 + 98$	24. $(12 + 23) + 8$	25. 4×27	26. $18 + 26$
185	43	108	44
27. 4×53	28. $64 - 29$	29. $24 + 32 + 16$	30. 19×14
212	35	72	266
31. $126 + 118$	32. $293 - 137$	33. $765 \div 9$	
244	156	85	
34. $32 + 36$	35. $19 + 26$	36. $4 \times 6 \times 50$	
68	45	1,200	
37. $25 \times 30 \times 2$	38. $172 \div 4$	39. $1,526 - 498$	
1,500	43	1,028	
40. $40 \times 15 \times 2$	41. $(4 \times 33) + (4 \times 7)$	42. $(6 \times 24) + (6 \times 6)$	
1,200	160	180	
43. $192 \div n$ for $n = 3$		44. $c \times 9 \times 5$ for $c = 8$	
64		360	
45. $p \div 12$ for $p = 624$		46. $a + 19 + 32$ for $a = 18$	
52		69	

Name each missing reason.

47. $80 \times 3 = (8 \times 10) \times 3$ 80 means 8×10.
       $= 8 \times (10 \times 3)$ Associative Property of Multiplication
       $= 8 \times (3 \times 10)$ __?__ Commutative Property of Multiplication
       $= (8 \times 3) \times 10$ __?__ Associative Property of Multiplication
       $= 24 \times 10$ __?__ $8 \times 3 = 24$
       $= 240$ __?__ $24 \times 10 = 240$

48. **What if** the product of three whole numbers is 210? Without using 1 as a factor, what are the possible choices for the numbers? Possible answers: 3, 7, 10; 3, 5, 14; 5, 6, 7; 2, 3, 35; 2, 7, 15; 2, 5, 21

Problem Solving Applications — **Use Data** For 49–51, use the data below.

49. Use mental math to find how many CDs were bought in all. Explain how you got your answer. **72 CDs;** Explanations will vary.

50. If Nick and Selena each gave 12 CDs to Brenda, how many would Brenda have then? **36 CDs**

51. How many CDs would Brenda, Selena, Ricardo, and Nick each have if they shared their CDs equally? **18 CDs each**

CDs Bought: Brenda 12, Selena 17, Nick 25, Ricardo 18

38   Chapter 2

ALTERNATIVE TEACHING STRATEGY   SCAFFOLDED INSTRUCTION

Purpose Students practice using compensation in a game of mental addition.

Materials *For each group of 4* 9-section spinner, p. TR73

Give each of several groups a spinner with 9 equal sections labeled as shown:

One student serves as Recorder. Each group uses the spinner to generate 2 two-digit numbers to add. The recorder writes the numbers on a sheet of paper. For example, 54 and 28.

The first student decides which number to adjust and states the adjusted number; for example, 30.

The second student decides how to compensate with the second number, thinking "2 has been added to 28, so I should subtract 2 from 54." That student should then state the second adjusted number; for example, 52.

The third student finds the sum mentally and states it; for example, 82.

The Recorder records each of the numbers as the students say them. The recording sheet might look like this:

```
   54        30
 + 28      + 52
            82
```

The group members then find the sum of the two original numbers to check their mental math. If they are correct, they get a point.

Have students repeat the activity, with each student assuming a different role. The group with the most points at the end of a certain time is the winner.

38   Chapter 2

52. A toy store has 7 boxes on each of 4 shelves. Each box has 25 items in it. How many items altogether are on the 4 shelves? **700 items**

53. Ann needs 250 signatures on a petition. On Monday she got 23 signatures, on Tuesday she got 3 times as many as on Monday, and on Wednesday and Thursday she got 45 each. How many more signatures does she need? **68 more signatures**

54. Jocelyn has $253.47. Her aunt gives her $87.95 more. Jocelyn buys a pair of shoes for $39.99, three T-shirts for $7.77 each, and two pairs of jeans for $60.22. Use estimation to find about how much money Jocelyn has left. **about $160**

55. **Write About It** Explain how to use compensation to add two numbers. Give an appropriate example to support your explanation. **Change one addend to a multiple of 10, and adjust the other addend to keep the balance. Examples will vary.**

MIXED REVIEW AND TEST PREP

56. Use mental math to solve. $a \div 7 = 21$ (p. 30) **$a = 147$**

57. Multiply. 732×46 (p. 22) **33,672**

58. Divide. $64,270 \div 35$ (p. 22) **1,836 r10**

59. TEST PREP Rob changes 4 quarts of oil in his car every 3,000 miles. How many quarts of oil will Rob have used after driving 9,000 miles? (p. 22) **C**

A 3   B 9   C 12   D 36

60. TEST PREP Joe bought 3 basketballs for $22.99 each and a net for $5.99. Which number sentence can be used to find the total cost of the basketballs and net? (p. 28) **G**

F $3 \times (22.99 + 5.99)$ H $(3 \times 5.99) + 22.99$
G $(3 \times 22.99) + 5.99$ J $(3 + 22.99) + 5.99$

PROBLEM SOLVING — Thinker's Corner

MATH FUN Practice using mental math strategies to solve this puzzle.

1. Copy the diagram. Place the values of the expressions below in the circles so that every sum of three numbers in a line is the same.

$84 \div 2$ **42**
$36 + 8$ **44**
$28 + 16 + 2$ **46**
$3 \times 8 \times 2$ **48** $8 + 14 + 32$ **54**
$28 + 22$ **50** $448 \div 8$ **56**
4×13 **52** $4 + 38 + 16$ **58**

2. Use mental math. What is the sum of each row of three numbers? **150**

EXTRA PRACTICE page H33, Set A

39

LESSON 2.2

Algebra: Exponents

LESSON PLANNING

Objective To represent numbers by using exponents

Intervention for Prerequisite Skills
Repeated Multiplication (For intervention strategies, see page 35.)

Lesson Resources Calculator Handbook p. 1

NCTM Standards
1. Number and Operations
2. Algebra
6. Problem Solving
8. Communication

Vocabulary

exponent a number that tells how many times a base is to be used as a factor

base a number used as a repeated factor

Math Background

Exponential notation is a convenient way to write multiplication of repeated factors in compact form. Exponents are used in many situations in mathematics. In order for students to understand exponential notation, it is important that they understand these ideas:

- The base in an exponential expression is the number used as a factor.
- The exponent tells how many times the base is used as a factor.
- The value of a base raised to an exponent n is called the nth power of the base.

In order for the properties of exponents to be consistent, a nonzero number to the zero power is 1.

WARM-UP RESOURCES

 **NUMBER OF THE DAY**

Use today's calendar date. Square it. Identify the base and the exponent. Check students' answers. The base is the number of the day and the exponent is 2.

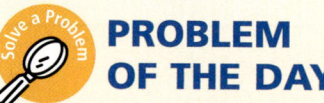 **PROBLEM OF THE DAY**

Replace the letters a, b, and c with the numbers 3, 4, and 5 to make a true sentence.

$2^a + 2^a = b^c$ $2^5 + 2^5 = 4^3$

Solution Problem of the Day tab, p. PD2

 **DAILY FACTS PRACTICE**

Have students practice multiplication facts by completing Set A of *Teacher's Resource Book*, p. TR94.

40A Chapter 2

INTERVENTION AND EXTENSION RESOURCES

ALTERNATIVE TEACHING STRATEGY

Materials *For each pair* 64 square tiles and 64 centimeter cubes

Ask pairs of students to **use area models to represent squares** and find their values.

1. 2^2 4   2. 1^2 1
3. 4^2 16   4. 5^2 25

Then have students use volume models to represent these cubed numbers and find their values.

5. 2^3 8   6. 1^3 1
7. 4^3 64   8. 3^3 27

KINESTHETIC
VISUAL/SPATIAL

MULTISTEP AND STRATEGY PROBLEMS

The following multistep or strategy problem is provided in Lesson 2.2:

Page	Item
41	34

ENGLISH LANGUAGE LEARNERS

Help students acquiring English to **develop their understanding of base.** Have them relate the idea of base to the number on the bottom of an exponential expression, just as the base of some geometric solids is the foundation that the shape stands on. Have the whole class join in finding expressions in which the word *base* means "foundation." Possible answer: the base of a pyramid or basement of a building

Have groups of students illustrate the words they find and present them to the class.

AUDITORY
INTERPERSONAL/SOCIAL

WRITING IN MATHEMATICS

 Reinforce students' understanding of squaring numbers. Read the following question to students: Can you cover a square area with 10 square tiles or 12 square tiles?

Have them write a paragraph to explain their answer.

Possible answer: No; in order to cover a square area, you need the same number of tiles on each side, so you need a number of tiles that forms a square, such as 4 or 9. Neither 10 nor 12 tiles will cover a square area.

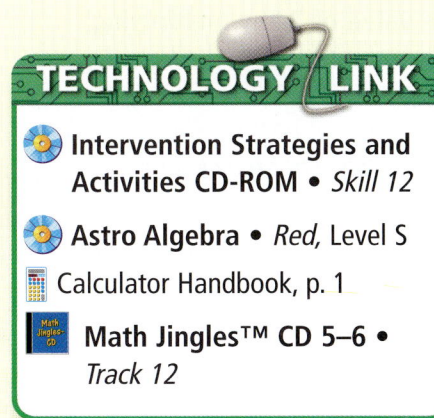

TECHNOLOGY LINK

- Intervention Strategies and Activities CD-ROM • *Skill 12*
- Astro Algebra • *Red,* Level S
- Calculator Handbook, p. 1
- Math Jingles™ CD 5–6 • *Track 12*

LESSON 2.2 ORGANIZER

Objective To represent numbers by using exponents

Vocabulary exponent, base *Review* factor

Lesson Resources Calculator Handbook, pp. 1–2

1 Introduce

QUICK REVIEW provides review of prerequisite skills.

Why Learn This? In science, you can use this skill to write large numbers in a shortened form. *Share the lesson objective with students.*

2 Teach

Guided Instruction

- Help students verbalize the relationship between the exponent and the base.

  What does the exponent tell you? the number of times the base is used as a factor

- Demonstrate to students a simple rule for evaluating base 10 exponents.

  10^n = 1 followed by *n* zeros. **Write the exponent form of 1,000,000.** 10^6

- Call to students' attention a common error in evaluating exponents.

  Juan found the value of 2^3 by multiplying 2 by 3. What mistake did he make? He multiplied the base by the exponent rather than using the base 2 as a factor 3 times.

- Guide students to apply what they know about exponents.

REASONING **Numbers with the exponent 3 are called cubes or cubed numbers. Why do you think the name *cubed* is used?** The volume of a cube is equal to the length of one side of the cube raised to a power of 3.

ADDITIONAL EXAMPLES

Example 1, p. 40

Find the value of 3^3. $3^3 = 3 \times 3 \times 3 = 27$

Example 2, p. 40

Express 16 by using an exponent and the base 2. $16 = 2 \times 2 \times 2 \times 2 = 2^4$

40 Chapter 2

LESSON 2.2

ALGEBRA
Exponents

Learn how to represent numbers by using exponents.

QUICK REVIEW
1. 3×3 9
2. 6×6 36
3. $4 \times 4 \times 4$ 64
4. $2 \times 2 \times 2 \times 2$ 16
5. $9 \times 9 \times 9$ 729

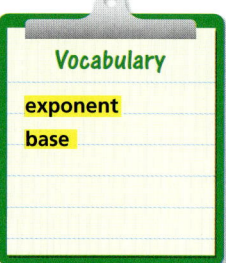

Vocabulary
exponent
base

Remember that when you multiply two or more numbers to get a product, the numbers multiplied are called factors.
$8 \times 3 \times 4 = 96$
The numbers 8, 3, and 4 are factors of 96.

Some football stadiums can seat over 100,000 people. Large numbers can be hard to understand. On the right are four ways to write 100,000 using smaller numbers.

$10 \times 10,000$
$10 \times 10 \times 1,000$
$10 \times 10 \times 10 \times 100$
$10 \times 10 \times 10 \times 10 \times 10$

Another way to write 100,000 is by using exponents. An **exponent** shows how many times a number called the **base** is used as a factor.

$$10^5 = 10 \times 10 \times 10 \times 10 \times 10 = 100,000$$

base — equal factors

EXPONENT FORM	READ	VALUE
10^1	The first power of ten	10
$10^2 = 10 \times 10$	Ten squared, or the second power of ten	100
$10^3 = 10 \times 10 \times 10$	Ten cubed, or the third power of ten	1,000

EXAMPLE 1 Find the values of 2^4, 4^2, and 6^3.

$2^4 = 2 \times 2 \times 2 \times 2$ = 16
2 is a factor four times.

$4^2 = 4 \times 4$ = 16
4 is a factor two times.

$6^3 = 6 \times 6 \times 6$ = 216
6 is a factor three times.

Note: The first power of any number equals that number.
$6^1 = 6 \qquad 9^1 = 9 \qquad 10^1 = 10$

The zero power of any number, except zero, is defined to be 1.
$6^0 = 1 \qquad 19^0 = 1 \qquad 10^0 = 1$

EXAMPLE 2 Write 125 using an exponent and the base 5.

$125 = 5 \times 25 = 5 \times 5 \times 5$ *Find the equal factors.*
$= 5^3$ *Write the base and the exponent.*

So, $125 = 5^3$.

40 Chapter 2

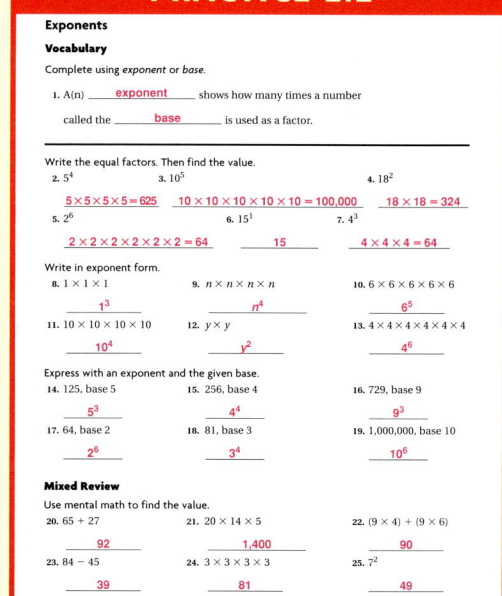

CHECK FOR UNDERSTANDING

Think and Discuss ▶ Look back at the lesson to answer each question.
1. **Tell** how many zeros there are in the standard form of 10^7. **7**
2. **Explain** how to write $6 \times 6 \times 6 \times 6$ using exponents. **6 will be the base and 4 is the exponent; 6^4.**

Guided Practice ▶ Write the equal factors. Then find the value.
3. 2^3 4. 5^2 5. 3^4 6. 9^3 7. 1^4
$2 \times 2 \times 2; 8$ $5 \times 5; 25$ $3 \times 3 \times 3 \times 3;$ $9 \times 9 \times 9;$ $1 \times 1 \times$
                                            81   729 $1 \times 1; 1$

PRACTICE AND PROBLEM SOLVING

For 10, 15, 17, and 20, see left.

Independent Practice ▶ Write the equal factors. Then find the value.

10. $1 \times 1 \times 1 \times 1 \times 1 \times 1 \times 1 \times 1 \times 1 \times 1 \times 1 \times 1; 1$

15. $10 \times 10 \times 10 \times 10 \times 10 \times 10 \times 10 \times 10;$ 100,000,000

17. $2 \times 2 \times 2 \times 2 \times 2 \times 2 \times 2 \times 2 \times 2 \times 2; 1,024$

20. $15 \times 15; 225$

$4 \times 4 \times 4 \times 4 \times 4; 1,024$ $2 \times 2 \times 2 \times 2 \times 2; 32$
8. 4^5 9. 7^3 10. 1^{12} 11. 5^3 12. 2^5
       $7 \times 7 \times 7; 343$ $5 \times 5 \times 5; 125$
13. 34^2 14. 13^2 15. 10^8 16. 20^2 17. 2^{10}
$34 \times 34; 1,156$ $13 \times 13; 169$ $20 \times 20; 400$
18. 10^4 19. 30^1 20. 15^2 21. 25^1 22. 90^2
$10 \times 10 \times 10 \times 10; 10,000$ 25 $90 \times 90; 8,100$

Write in exponent form.

23. $12 \times 12 \times 12$ 12^3 24. $1 \times 1 \times 1 \times 1 \times 1$ 1^5 25. $4 \times 4 \times 4 \times 4$ 4^4

26. $2 \times 2 \times 2 \times 2 \times 2$ 27. $n \times n$ n^2 28. $y \times y \times y \times y$ y^4
     2^5

Express with an exponent and the given base.

29. 64, base 8 8^2 30. 216, base 6 6^3 31. 10,000; base 10 10^4

32. Write 64 using a base of 8, a base of 4, and a base of 2. $8^2; 4^3; 2^6$

Problem Solving Applications

33. **Use Data** When did Oregon first have a population greater than 10^6? Explain. **See left.**

33. 1940; 10^6 = 1 million = 10 hundred thousand; the value is greater than 10 hundred thousand

34. Ben saves movie ticket stubs. He puts them in albums with 2^4 pages. Each page holds 3^2 stubs. How many albums does he need for 720 stubs? **5 albums**

35. Possible question: How many more games does Scott have than Aaron?

35. ❓ **What's the Question?** Scott has 3^2 video games and Aaron has 2^3 games. The answer is 1. **See left.**

Population of Oregon

MIXED REVIEW AND TEST PREP

36. Use mental math to find the value of $279 \div 3$. (p. 30) **93**
37. Round 45,621 to the nearest thousand. (p. 16) **46,000**
38. $943,012 - 57,806$ (p. 20) **885,206**   39. $32,047 \div 43$ (p. 22) **745 r12**
⭐ 40. **TEST PREP** Which expression is equivalent to $(3 + 5) \times 2$? (p. 28) **B**
     A $(7 - 3) + 10$ **B** $(4 \times 2) + 8$ **C** $(2 \times 3) + 5$ **D** $10 + (7 \times 10)$

EXTRA PRACTICE page H33, Set B

41

---

3 Practice

Guided Practice
Do Check for Understanding Exercises 1–7 with your students. Identify those having difficulty and use lesson resources to help.

Independent Practice
Note that Exercise 34 is a **multistep or strategy problem**. Assign Exercises 8–35.

Algebraic Thinking Seeing the relationship between the base and the exponent is an important step in developing relationship thinking. To ensure that students understand this concept, check their answers to Exercises 23–32.

MIXED REVIEW AND TEST PREP
Exercises 36–40 provide **cumulative review** (Chapters 1–2).

4 Assess

Summarize the lesson by having students:

DISCUSS How do you solve Exercise 17? Possible answer: Write 2 as a factor 10 times and multiply.

WRITE Explain how to find the value of a number with an exponent. Possible answer: The exponent tells how many times to write the base as a factor. Then multiply to solve.

Lesson Quiz
Transparency 2.2

Write the equal factors. Then find the value.
1. 10^5 $10 \times 10 \times 10 \times 10 \times 10; 100,000$
2. 11^2 $11 \times 11; 121$
3. 3^4 $3 \times 3 \times 3 \times 3; 81$

Express with an exponent and the given base.
4. 1,331; base 11 11^3 5. 625; base 5 5^4
6. 7,776; base 6 6^5

41

CHALLENGE 2.2
Puzzling Exponents
Complete the puzzle with the values.

Across
1. 11^3
2. 5^3
3. 17^2
4. 7^4
5. 14^3
6. 9^4
7. 5^5
8. 10^4
9. 16^2
10. 2^7
11. 15^3

Down
1. 25^3
2. 6^3
3. 30^2
4. 6^4
5. 2^{10}
6. 7^3
7. 20^2
8. 12^3
9. 24^2
10. 3^7

PROBLEM SOLVING 2.2
Exponents
Write the correct answer.

1. Write in exponent form.
$5 \times 5 \times 5 \times 5 \times 5 \times 5 \times 5$
5^8

2. Compare the fractions $\frac{3}{4}$ and $\frac{7}{8}$. Use < or >.
$\frac{3}{4} < \frac{7}{8}$ or $\frac{7}{8} > \frac{3}{4}$

3. Claire is working on her reading assignment for school. On Monday she read three pages. Then, on each day after the first day, she read triple the amount of the previous day. Using exponent form, write the number of pages she will read on the fifth day.
3^5

4. Bill needs to know the decimal equivalent of $\frac{3}{16}$ to solve a problem in his math homework. He changes the fraction to a decimal by dividing the numerator by the denominator. What decimal does he get?
0.1875

Choose the letter for the best answer.

5. Find the value of 7^3.
A 73
B 343
C 21
D 10

6. Which is the exponent form of $n \times n \times n \times n \times n$?
F n^5 H $5n$
G 5^n J $5n^5$

7. Which group of numbers is listed from greatest to least?
A 3.045, 3.04, 3.05
B 4.2, 4.013, 4.01
C 2.7, 2.86, 2.68
D 5.10, 5.010, 5.02

8. A salesman travels 517 miles a week to cover his territory. Which is a good estimate for the number of miles he travels in 4 weeks?
F 500 mi
G 1,000 mi
H 1,500 mi
J 2,000 mi

9. **Write About It** Explain how you can tell which is greater, 8^6 or 12^6, without finding their values.
The one with the greater base is greater because they have the same exponent. 12^6

LESSON 2.3

ORGANIZER

Objective To explore how to evaluate expressions by using order of operations

Vocabulary order of operations, algebraic operating system

Materials For each student calculator

Lesson Resources E-Lab Recording Sheet • *Order of Operations*

Intervention for Prerequisite Skills
Use of Parentheses (For intervention strategies, see page 35.)

Using the Pages

Point out to students that just as we construct models in a certain order, we must also perform expression operations in a certain order.

Activity 1

Some students will find the value for $23 + 12 \times (6 - 2)^2$ by performing the operations in the order they are presented. After they have reviewed the correct order of operations, have them find the value again.

Reinforce students' understanding that expressions which contain the same numbers and operation signs may have different solutions.

Evaluate the following expressions, and explain why the solutions are different.
$(23 + 12) \times (6 - 2)$ **140** $23 + (12 \times 6) - 2$ **93**
$23 + 12 \times (6 - 2)$ **71** Possible answer: The parentheses dictate which operations are performed first on which numbers.

Think and Discuss

Ask students to evaluate this expression:
$13 + 5 - 3 \times (21 \div 7)$
Was this expression easier to evaluate than those you evaluated in Activity 1? Why?
Possible answer: Yes, because the mnemonic helped me remember the order of operations.

Practice

Ask students why they subtracted before clearing the exponents in Exercises 1, 4, and 5. You must perform the operation in parentheses first, which was subtraction in these 3 exercises.

LESSON 2.3 — Explore Order of Operations

Explore how to evaluate expressions by using order of operations.

You need a calculator.

QUICK REVIEW
1. $16 + 24$ **40**
2. $(24 \div 6) \div 2$ **2**
3. 14×10 **140**
4. $(75 - 21) + 8$ **62**
5. $4 \times (6 + 2)$ **32**

Vocabulary
order of operations
algebraic operating system (AOS)

Erin's older brother, Todd, was working on his homework. At the top of the page, he wrote, "**P**lease **E**xcuse **M**y **D**ear **A**unt **S**ally." Erin didn't understand. She said, "We don't have an aunt named Sally."

Todd said, "You'll see why I wrote this on my paper."

Activity 1

• Use paper and pencil to find the value of $23 + 12 \times (6 - 2)^2$. **215**
• How does your answer compare with the answer of a classmate? Answers may vary.

Math Idea ▶ When you find the value of an expression with more than one operation, you need to use the **order of operations**.

1. Perform operations in parentheses.
2. Clear exponents.
3. Multiply and divide from left to right.
4. Add and subtract from left to right.

• Use the order of operations to evaluate $(41 + 31) \div 2^2 - 8$. **10**

Think and Discuss

• What do the underlined letters in "**P**lease **E**xcuse **M**y **D**ear **A**unt **S**ally" represent? parentheses, exponents, mutiplication, division, addition, and subtraction
• List the order of operations you would use to find the value of $3^3 + 5 - 3 \times (21 \div 7)$. Explain why the order is important. parentheses, exponents, multiplication, addition, subtraction; so that the value of an expression is always the same

Practice

Tell the order in which you would perform the operations in each expression. Then find the value of the expression.

1. $(120 - 14) + 4^2 \times 3$ parentheses, exponent, multiplication, addition; 154
2. $3 + 5^2 \times 2 \div 10 - 4$ exponent, multiplication, division, addition, subtraction; 4
3. $9 \times 1 + 12 \times 2 \div 8 + 4$ multiplication, division, addition; 16
4. $5 + (7 - 4)^2 - 8 \div 2$ parentheses, exponent, division, addition, subtraction; 10
5. $16 - 2^3 - (9 - 7) \times 4$ parentheses, exponent, multiplication, subtraction; 0

42 Chapter 2

MATH CONNECTION: ALGEBRA

Review with students the effect of **parentheses in expressions**. Point out that parentheses do not always change the value of an expression. Have students evaluate this expression with and without parentheses: $(2 \times 3) + (5 \times 6) - 4$ **32**

They should find that, with or without the parentheses, the value is 32.

Have students write one expression in which the parentheses are essential and one in which they are not. Possible answer: $(4 + 5) \times 2 - 8$ and $(3 \times 5) + 8$

VISUAL; VERBAL/LINGUISTIC

Intervention and Extension Resources

WRITING IN MATHEMATICS

Have students compare and contrast using the order of operations and using a calculator that uses an AOS to solve an expression. Tell them to include examples of each. Check students' work.

42 Chapter 2

You can use a calculator to evaluate expressions with more than one operation. Some calculators use an **algebraic operating system (AOS)** that automatically follows the order of operations.

Activity 2

- Use your calculator to find the value of $8 \div 2 + 6 \times 3 - 4$. **Answers may vary.**
- Following the order of operations, use paper and pencil to find the value of $8 \div 2 + 6 \times 3 - 4$. **18**
- Exchange papers with a classmate, and check each other's work.

Think and Discuss

- How does the calculator value for $8 \div 2 + 6 \times 3 - 4$ compare with the value you got by using paper and pencil? Does your calculator use an AOS? **Answers may vary.**

To find the value of an expression with a calculator that does not use an AOS, follow the order of operations or use the memory keys.

Follow the order of operations to find the value of the expression $2 + 6 \times 3^2 - 4$.

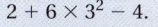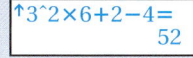

Use the memory keys to find the value of the expression $9^2 + 6 \div 2 \times 4$.

- When you enter values into a calculator that does not have an AOS, how do you know which values to enter first? **Use the order of operations to identify order.**

Practice

1. A calculator shows the display 9 as the value of $12 + 15 \div 3$. Does the calculator use an AOS? Explain. **No; 17 is the correct value, not 9.**
2. How could you use memory keys to evaluate the expression $12 + 15 \div 3$?
3. How could you use the order of operations to evaluate the expression $12 + 15 \div 3$? **$15 \div 3 + 12$**

2. 1 2  1 5 3

Use a calculator to find the value.

4. $12 + 8 \times 4^2$ **140**
5. $9 + (6 - 2) \times 5$ **29**
6. $18 \div (6 - 4) + 5$ **14**

TECHNOLOGY LINK
More Practice: Use E-Lab, *Order of Operations.*
www.harcourtschool.com/elab2002

MIXED REVIEW AND TEST PREP

7. Find the value for 6^3. (p. 40) **216**
8. $34,056 + 2,207$ (p. 20) **36,263**
9. Evaluate $6p - q$ for $p = 7$ and $q = 11$. (p. 28) **31**
10. $807 \div 45$ (p. 22) **17 r42**

★ **11. TEST PREP** A machine makes 2^5 bolts per second. How many minutes does it take to make 9,600 bolts? (p. 40) **A**

A 5   B 12   C 300   D 1,920

43

LESSON 2.4

Algebra: Order of Operations

LESSON PLANNING

Objective To use the order of operations

Intervention for Prerequisite Skills

Use of Parentheses (For intervention strategies, see page 35.)

Lesson Resources Calculator Handbook p. 2

NCTM Standards
1. Number and Operations
2. Algebra
6. Problem Solving
8. Communication

Math Background

Without the order of operations it would be impossible to determine a single value for many expressions. Consider the following ideas as you help students understand order of operations:

- No matter what operations are involved within parentheses, these operations are done first.
- Before basic operations are completed, exponents must be cleared.
- If operations occur in the order divide, multiply, divide, they should be done in that order. In other words, multiplication should not always be done before division. Similarly, addition should not be done before subtraction unless they occur in that order.

WARM-UP RESOURCES

 **NUMBER OF THE DAY**

Find the value of the expression (3 + your age) × 4 − 3^2. Possible answer: (3 + 11) × 4 − 3^2 = 47

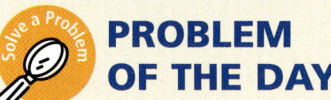 **PROBLEM OF THE DAY**

Complete the expression using the numbers 3, 4, and 5 so that it equals 19.

____ + ____ × ____

4 + 5 × 3

Solution Problem of the Day tab, p. PD2

 **DAILY FACTS PRACTICE**

Have students practice multiplication and division facts by completing Set C of *Teacher's Resource Book*, p. TR94.

INTERVENTION AND EXTENSION RESOURCES

ALTERNATIVE TEACHING STRATEGY — ELL

Some students may still have difficulty **understanding the need for the order of operations.** Ask them to act out the procedure for a common activity, such as making a telephone call, to stress the idea that order is important. Point out that it makes a difference whether you lift the receiver and then touch the numbers or touch the numbers and then lift the receiver. The results are not the same.

KINESTHETIC
BODILY/KINESTHETIC

MULTISTEP AND STRATEGY PROBLEMS

The following multistep and strategy problems are provided in Lesson 2.4:

Page	Item
45	20, 21, 23, 28

SCIENCE CONNECTION

Encourage students to **write two word problems using the order of operations.** Have students use the following data about the five longest snakes in the world. The solution to each problem should require two operations. Have students exchange problems and solve.

AVERAGE LENGTH OF SNAKES	
Reticulated python	35 ft
Anaconda	28 ft
Indian python	25 ft
Diamond python	21 ft
King cobra	19 ft

Possible answer: How much longer are two reticulated pythons than one king cobra? 51 ft

VISUAL
LOGICAL/MATHEMATICAL

ADVANCED LEARNERS

Challenge students to **use the order of operations** to write an expression that equals 100 by using the number 5 five times. The expression may include any number of $+$, $-$, $\times$, and $\div$ signs as well as parentheses.

Possible answers: $(5 \times 5 \times 5) - (5 \times 5)$ or $(5 + 5 + 5 + 5) \times 5$

VISUAL
LOGICAL/MATHEMATICAL

TECHNOLOGY LINK

- Intervention Strategies and Activities CD-ROM • *Skill 40*
- Astro Algebra • *Red,* Level B
- Math Jingles™ CD 5–6 • *Track 11*

LESSON 2.4 ORGANIZER

Objective To use the order of operations

Lesson Resources Calculator Handbook, p. 2

1 Introduce

QUICK REVIEW provides review of prerequisite skills.

Why Learn This? You can use the order of operations to solve problems that include more than one operation. *Share the lesson objective with students.*

2 Teach

Guided Instruction

- Check students' understanding of the examples.

  Why do you start with exponents in Example 1? There are no parentheses.

  In Example 2, what do you do first? Subtract within the parentheses.

REASONING **In Example 2, would you get a different answer if you did not use parentheses? Explain.** Yes; without the parentheses, you would divide 93 by 3 instead of by 1.

ADDITIONAL EXAMPLES

Example 1, p. 44

Tell the operations used to evaluate the expression.
$54 \div 6 + 4 \times 2^3$ clear exponents, divide, multiply, add; 41

Example 2, p. 44

Find the value of the expression.
$186 + 68 \div 2 - 2 \times (5 \times 2^2)$

$186 + 68 \div 2 - 2 \times (5 \times 2^2)$	Operate inside parentheses
$186 + 68 \div 2 - 2 \times (5 \times 4)$	Clear exponents
$186 + 68 \div 2 - 2 \times 20$	Divide
$186 + 34 - 2 \times 20$	Multiply
$186 + 34 - 40$	Add
$220 - 40$	Subtract
180	

LESSON 2.4

ALGEBRA
Order of Operations

Learn how to use the order of operations.

QUICK REVIEW
1. $3 + (9 + 7)$ 19
2. $(7 - 3) + 4$ 8
3. $4 \times (2 + 5)$ 28
4. $(9 - 2) \times 7$ 49
5. $8 + 7 \times 10$ 78

To evaluate an expression that contains more than one operation, you use rules called the order of operations. The first letters of these words help you remember the order of operations.

Please Excuse My Dear Aunt Sally

Parentheses
Exponents
Multiplication or **D**ivision
Addition or **S**ubtraction

EXAMPLE 1

Tell the operations used to evaluate the expression.

$35 \div 7 + 5 \times 3^2$	*Clear exponents.*
$35 \div 7 + 5 \times 9$	*Divide.*
$5 + 5 \times 9$	*Multiply.*
$5 + 45$	*Add.*
50	

EXAMPLE 2

Find the value of the expression $285 + 93 \div (3 - 2) \times 3 \times 4^2$.

$285 + 93 \div (3 - 2) \times 3 \times 4^2$	*Operate inside parentheses.*
$285 + 93 \div 1 \times 3 \times 4^2$	*Clear exponents.*
$285 + 93 \div 1 \times 3 \times 16$	*Divide.*
$285 + 93 \times 3 \times 16$	*Multiply twice.*
$285 + 4,464$	*Add.*
$4,749$	

CHECK FOR UNDERSTANDING

Think and Discuss ▶ Look back at the lesson to answer each question.

1. **Show** where to insert parentheses to make this equation true.
   $420 - 100 \div 40 = 8$ $(420 - 100) \div 40 = 8$

2. **Tell** which operation you would do last to evaluate the expression $7 + 8 - 2^3 \div 4$. subtraction

Guided Practice ▶ Evaluate the expression.

3. $30 - 15 \div 3$ 25
4. $5^2 - (40 \div 4) \div 2$ 20
5. $5^2 + 10^2 \div 25 - 1$ 28

RETEACH 2.4

Order of Operations

When an expression involves more than one operation, you use the order of operations to evaluate it.
1. Operate inside parentheses.
2. Clear exponents.
3. Multiply and divide from left to right.
4. Add and subtract from left to right.

What is the value of $2 \times 3 + (6 - 2) \div 2$?

Step 1 Simplify inside parentheses.
Step 2 There are no exponents.
Step 3 Multiply and divide in order from left to right.
Step 4 Add and subtract in order from left to right.

Evaluate the expression.
1. $5 + 4 + 7 \times 3$
   $5 + 4 + \underline{21}$
   $\underline{9} + \underline{21}$
   $\underline{30}$
2. $28 \div 4 - (4 + 3)$
   $28 \div 4 - \underline{7}$
   $\underline{7} - \underline{7}$
   $\underline{0}$
3. $2^2 + 16 - 8$
   $\underline{4} + 16 - 8$
   $\underline{20} - 8$
   $\underline{12}$
4. $(6 + 4)^2 \times 8$
   $(\underline{10})^2 \times 8$
   $\underline{100} \times 8$
   $\underline{800}$

Find the value of each expression. From the box at the right, choose the letter that corresponds to the value.

A 1	E 16	
B 5	F 18	
C 7	G 20	
D 10	H 32	

5. $(5 \times 6) + (4 + 2)$ B
6. $10 + 15 - 9 \times 2$ C
7. $3^2 + 1 - (18 - 9)$ A
8. $(4 + 8) \times 3 + 6 + 10$ E
9. $2 + 5^2 - (4 + 3)$ G
10. $4 \div 2 \times 6 - 3 + 3^2$ F

PRACTICE 2.4

Order of Operations

Give the correct order of operations.
1. $100 + 6^2 - 9$
   Clear exponents.
   Add. Subtract.
2. $(52 - 49)^2 \div 9$
   Operate inside parentheses.
   Clear exponents. Divide.
3. $(5^2 + 1) \div 2$
   Inside parentheses: Clear exponents and add. Divide.
4. $(9 + 2) \times (16 - 12)^2$
   Operate inside parentheses.
   Clear exponents. Multiply.

Evaluate the expression.
5. $27 \div 3 + 1$ 10
6. $(6 + 8) \times (9 - 8)$ 14
7. $(6 + 7^2) \div 5 \times 2$ 22
8. $(12 + 2)^3 + (2^3 + 1^3)$ 225
9. $(15 - 5)^2 - (4 \times 3)$ 88
10. $(57 + 3) \times 2^4$ 960
11. $(19 + 9) \div (2^3 - 1) + 20$ 24
12. $(3 \times 7^2) - (5^3 - 9^2) + 10^2$ 203
13. $3 \times (10^2 - 65) + (5^2 \times 2)$ 155

Evaluate the expression for $s = 5$ and $t = 12$.
14. $50 \div s + 7$ 17
15. $s^2 + 150$ 175
16. $2 \times t - 18$ 6
17. $t^2 - 3 \times 8$ 120
18. $15 + t \div 6$ 17
19. $27 + 9 \times s$ 72

Mixed Review
Use mental math to find the value.
20. 12×7 84
21. $37 + 62$ 99
22. $434 \div 7$ 62
23. $1{,}731 - 605$ 1,126

Write in exponent form.
24. $8 \times 8 \times 8 \times 8$ 8^4
25. $6 \times 6 \times 6 \times 6 \times 6$ 6^5
26. $n \times n \times n \times n \times n$ n^5

---

44 Chapter 2

PRACTICE AND PROBLEM SOLVING

Independent Practice

Evaluate the expression.

6. $45 \div 15 + 2 \times 3$ **9**
7. $7 \times 2^2 + 6 - 9$ **25**
8. $3 + 4 \times 250$ **1,003**
9. $12 + (36 \div 4)^2 - 25$ **68**
10. $2^6 - (27 - 9) + (5^2 - 21)$ **49**
11. $(43 + 57) \times (9 - 6)^0$ **100**
12. $4^4 - (5^3 - 7^2) + (3^3 - 25)^3$ **188**
13. $(6^2 + 3^2)^2 \div 5 \times 3 + 3$ **1,218**
14. $(24 + 1^8) \times (7 - 5)^3$ **200**
15. $(7 \times 4)^2 - (34 + 1^8) \times 2^3$ **504**

ALGEBRA Evaluate the expression for $a = 4$ and $b = 7$.

16. $21 \div b + 8$ **11**
17. $a \times 31 - 8^2$ **60**
18. $(8 - a) \div 2 + 7$ **9**
19. $b^2 \div 7 \times (6 + 5)$ **77**
20. $(a^2 + b^2) \times 4$ **260**
21. $(b^2 - a^2) \div 3$ **11**

Problem Solving Applications

For 22–23, write and evaluate an expression to solve.

22. Heather mailed 3 packages that cost $2.50 each and 2 packages that cost $1.50 each. How much did she spend on postage?
   $(3 \times \$2.50) + (2 \times \$1.50) = \$10.50$
23. Minh bought a watermelon for $3.25 and 3 cantaloupes for $1.29 each. He gave the clerk $20. How much change did he get in return? $\$20 - (\$3.25 + 3 \times \$1.29) = \$12.88$
24. **What's the Error?** Joe and Brett found the value of $2 + 6 \times 3^2 - 4$. Joe said the answer is 68 and Brett said the answer is 52. Decide who made the error and describe what the error is.
   Joe; he did not follow the order of operations correctly.
25. The Island Theater holds 236 people. It was filled to capacity for each of its 43 shows last week. This week 8,299 people attended shows at the theater. How many people attended shows at the Island Theater during these two weeks? **18,447 people**

MIXED REVIEW AND TEST PREP

26. What are the equal factors and the value for 7^3? (p. 40) $7 \times 7 \times 7$; 343
27. Use mental math to find the value. 34×6 (p. 36) **204**
28. Subtract. $2,500 - 1,646$ (p. 20) **854**
29. Divide. $163 \div 5$ (p. 22) **32 r3**
30. **TEST PREP** Ricky's math class collected a total of 1,364 pennies and quarters. They collected 234 more pennies than quarters. How much money did Ricky's class collect? (p. 26) **A**

   A $149.24
   B $158.34
   C $229.40
   D $237.12

EXTRA PRACTICE page H33, Set C

45

LESSON 2.5

Problem Solving Skill: *Sequence and Prioritize Information*

LESSON PLANNING

Objective To use the skill *sequence and prioritize information* to solve problems

Intervention for Prerequisite Skills
Use of Parentheses (For intervention strategies, see page 35.)

Lesson Resources Problem Solving Think Along, p. TR1

NCTM Standards
1. Number and Operations
6. Problem Solving
8. Communication
9. Connections

Math Background
The following experiences will help students *sequence and prioritize information*:
- Put data in numerical order.
- Put steps in logical order.
- Choose important steps or information and disregard less important steps or information.

This is a skill students will use often in solving real-life problems. Learning to put steps in logical order involves the ability to foresee what would happen if they were done out of order.

WARM-UP RESOURCES

 **NUMBER OF THE DAY**

Start with your favorite number. Write an expression that includes subtracting 6, multiplying by 2, adding 3, and subtracting the starting number to end up with your original number. Possible answer: $(n + 3) \times 2 - 6 - n$

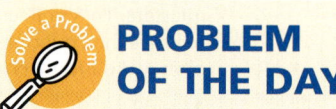 **PROBLEM OF THE DAY**

Look at the following figure. Start at point *A*. Write the sequence that allows you to go around the entire figure, covering each segment only once. Is there only one way? Can you do the same thing if you start at *B*? AFDEACD; no, there is more than one sequence; no.

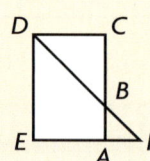

Solution Problem of the Day tab, p. PD2

 **DAILY FACTS PRACTICE**

Have students practice division facts by completing Set D of *Teacher's Resource Book*, p. TR94.

46A Chapter 2

INTERVENTION AND EXTENSION RESOURCES

REACHING ALL LEARNERS

ALTERNATIVE TEACHING STRATEGY — ELL

Materials *For each student* nine $10 bills and two $5 bills of play money

Have students practice the skill *sequence and prioritize information* by giving each student $100 of play money. Then give students the following instructions:

- Save half of what is left.
- Pay bills of $60.
- Buy a bus ticket for $30.

Have them use their money to illustrate the correct order of the steps. Possible answer: Pay bills, buy a bus ticket, save.

KINESTHETIC
LOGICAL/MATHEMATICAL

READING STRATEGY

Sequence Information is not always presented in the order that it is needed. Point out that it was necessary to reorder the information to solve the problem on page 46. Present the following situation and have students suggest an order and explain their reasoning.

Mrs. Brown wants to pay her gardener, withdraw cash from the bank, and shop for groceries. Possible answer: Withdraw cash, pay the gardener, buy groceries; the cash is needed to do the other activities. Pay the gardener first to be sure you don't spend so much on groceries that there is not enough left to pay the gardener.

MULTISTEP AND STRATEGY PROBLEMS

The following multistep and strategy problems are provided in Lesson 2.5:

Page	Item
47	4, 5, 7, 8, 9

TECHNOLOGY

Students can **apply the skill** *sequence and prioritize information* by working in small groups to research an important scientific discovery or technology. They might want to consider personal computers, compact discs, lasers, the Internet, or space travel. Have them research the development of the technology or discovery, highlighting important inventions, discoveries, or earlier technologies that led to it. Then ask them to prepare an oral presentation for the class that includes a poster or other visual representation of the information. Check students' work.

AUDITORY
INTERPERSONAL/SOCIAL

EARLY FINISHERS

Have students make a time line of the following composers' lives to demonstrate the skill *sequence and prioritize information:*

Ludwig van Beethoven, Leonard Bernstein, Frederic Chopin, Nikolay Rimsky-Korsakov, Antonio Vivaldi

Suggest that students use an encyclopedia or other reference source to find the date of birth, death, or first major work for each composer.

Play a recording for the class of a piece of music composed by one or more of the composers.

Dates of birth: Vivaldi (1678), Beethoven (1770), Chopin (1810), Rimsky-Korsakov (1844), Bernstein (1918)

VISUAL
MUSICAL/RHYTHMIC

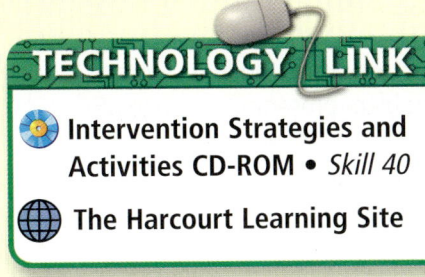

TECHNOLOGY LINK
- Intervention Strategies and Activities CD-ROM • *Skill 40*
- The Harcourt Learning Site

LESSON 2.5 ORGANIZER

Objective To use the skill *sequence and prioritize information* to solve problems

Lesson Resources Problem Solving Think Along, p. TR1

1 Introduce

QUICK REVIEW provides review of prerequisite skills.

Why Learn This? In the future, you can use this skill to plan a trip to an amusement park and visit a maximum number of attractions during the day. *Share the lesson objective with students.*

2 Teach

Guided Instruction

- Point out that there are many activities in which order is important and in which some steps are more important than others.

  What kinds of activities can you think of where you need to follow a series of steps? Students may suggest making lunch, playing games, or getting dressed.

  When you make a *Things to do* list, how do you prioritize the items on your list? Possible answer: by giving higher priority to items with a deadline or parental/teacher requests

- As you study the example, ask:

  How do you know you cannot pay bills as the first step? You cannot pay bills without having money.

Modifying Instruction Have students write the steps of the problem presented on page 46 on index cards, one to a card, and move the cards around to determine the most logical order.

LESSON 2.5

PROBLEM SOLVING SKILL
Sequence and Prioritize Information

Learn how to solve problems by sequencing and prioritizing information.

QUICK REVIEW
1. $3 + (8 \times 2)$ **19**
2. $6^2 + 18$ **54**
3. $40 - (3 + 2)^2$ **15**
4. $4 + 8 \div 2$ **8**
5. $15 - 2 \times 3 + 7$ **16**

Mrs. Rucki gets paid on the first workday of each month. She deposits her check in the bank and uses the money to pay her bills. Of the money that's left she uses part for spending money and puts the rest into savings.

On March 1, the following occurred:
- Mrs. Rucki had bills of $740, $85, $102, $33, and $52 to pay.
- She received her monthly paycheck for $1,570.
- She wanted to save at least $300 from her check.
- She needed at least $250 spending money.

The sequence, or order, in which Mrs. Rucki does these things is important. For example, she cannot use the money until the bank gives her credit for her deposit. That may take three days.

Mrs. Rucki made the list at the left. The list shows the things to do and the order she planned to do them.

- Why does it make sense for Mrs. Rucki to write "Put the rest into my savings account" last? so that she pays all her bills and has enough money to spend

Suppose Mrs. Rucki follows the sequence above. How much will she have left to put into her savings account?

$740 + \$85 + \$102 + \$33 + \$52 = \$1,012$ *Find the total of her bills.*

$\$1,570 - \$1,012 = \$558$ *Find the amount left after she pays her bills.*

$\$558 - \$250 = \$308$ *Find what is left after she withdraws spending money.*

So, Mrs. Rucki could put $308 into her savings account.

Math Idea ▶ The order in which parts of an activity are carried out is often important to success. Some parts of an activity may be more important than others.

46 Chapter 2

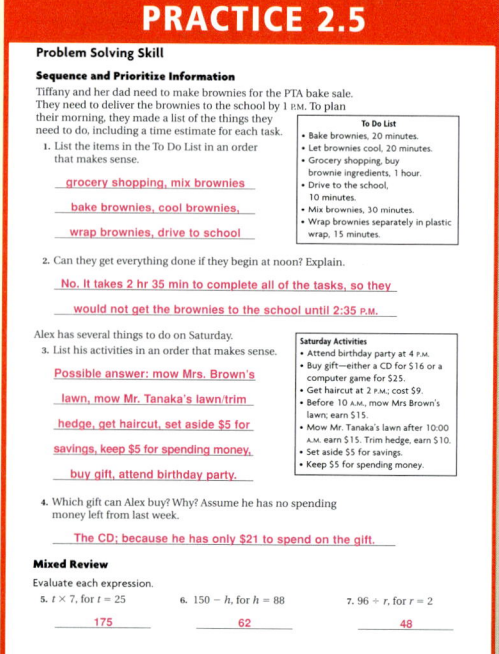

PROBLEM SOLVING PRACTICE

For 1–3, use the schedule at the right. Each show is 50 minutes long.

Ocean World Park Show Times

Time	Show
9:00, 12:00	Underwater Acrobats
9:00, 3:00	Whale Acts
10:00, 2:00	Animal Acts
10:00, 1:00	Water Skiing
11:00, 4:00	Aquarium Tour

1. Jennifer and her family will visit Ocean World Park from 9:30 to 4:00. They want to see the Water Skiing show at 10:00. Name the order in which they can see all the other shows. **Aquarium Tour, Underwater Acrobats, Animal Acts, Whale Acts**

2. The Jackson family wants to see the Underwater Acrobats at 12:00 and then take 45 minutes to eat lunch. What other show could they see before leaving the park at 3:00? **B**
   A Aquarium Tour
   B Animal Acts
   C Whale Acts
   D Water Skiing

3. Michele wants to go on the Aquarium Tour at 11:00. Before she leaves the park, she wants to see the Animal Acts show at 2:00. At which other time could she see a show after the Aquarium Tour? **H**
   F 9:00   H 1:00
   G 10:00 J 2:00

MIXED APPLICATIONS

4. Ed wants to buy 3 boxes of cereal at $2.99 each, 2 melons at $1.29 each, and 5 cans of juice at $0.79 each. Ed has $12.80. How much more money does he need to buy all of the items? **$2.70**

5. At the Discount Book Barn, books cost $5.00 and magazines cost $1.50. José bought some books and magazines for $21. How many of each did he buy? **3 books, 4 magazines**

6. **Use Data** Use the graph below. The highest waterfall in the world is Angel Falls in Venezuela. About how much higher is Angel Falls than Ribbon Falls? **Possible answer: about 1,500 feet**

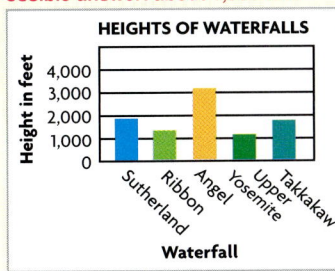

7. Tim has saved $1,475 to make a down payment on a new car. He wants to save a total of three times that amount plus $275. How much money does Tim want to save for a down payment? **$4,700**

8. Ron and his two cousins own videotapes. Ron has twice as many as his older cousin, who has 27 videos. Ron has three times as many as his younger cousin. How many videotapes do they have in all? **99 videotapes**

9. Harry hammered nails into a board to make a circular pegboard. The nails were the same distance apart, and the sixth nail was directly opposite the eighteenth nail. How many nails formed the circle? **24 nails**

10. **Write About It** Make a list of the steps you would follow to make a scrambled egg. **Possible answer: Crack an egg into a bowl. Stir the egg with a fork. Put butter in a pan. Heat the butter. Pour the egg into the pan. Let the egg cook.**

47

3 Practice

Guided Practice

Do Problem Solving Practice Exercises 1–3 with your students. Identify those having difficulty and use lesson resources to help.

When students approach Exercise 1, have them begin by listing the available hours and then fill in show names next to the times.

Independent Practice

Note that Exercises 4, 5, 7, 8, and 9 are **multi-step or strategy problems**. Assign Exercises 4–10.

4 Assess

Summarize the lesson by having students:

DISCUSS Talk about the steps you might follow in crossing the street. Are any steps more important than others? Possible answer: go to a crosswalk, look for traffic, cross; it is very important to look for traffic.

WRITE Explain why it is important to use the skill *sequence information* when evaluating a numerical expression. If you do not use the correct sequence when you evaluate the expression, you may get a wrong answer.

Lesson Quiz

1. Find a sequence of programming that allows Sean to finish each block on the half hour or hour of his radio show.

Programs	Min	Ads	Min
Country Music	28	Carpet Store	3
Golden Oldies	26	Restaurant	2
Rap	27	Grocery Store	4

Possible answer: country music, restaurant ad; rap, carpet store ad; golden oldies, grocery store ad

2. Write a sequence of steps for calling a store to get information. Possible answer: Look up the number, pick up the phone, listen for a dial tone, enter the number, ask for the appropriate department, ask the question.

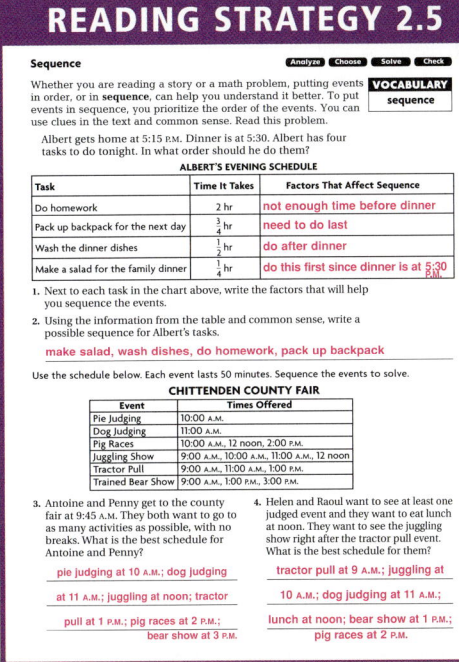

47

CHAPTER 2

REVIEW/TEST

Purpose To check understanding of concepts, skills, and problem solving presented in Chapter 2

USING THE PAGE

The Chapter 2 Review/Test can be used as a **review** or a **test**.

- Items 1–3 check understanding of concepts and new vocabulary.
- Items 4–29 check skill proficiency.
- Items 30–33 check students' abilities to choose and apply problem solving strategies to real-life problems involving operations.

Portfolio Suggest that students place the completed Chapter 2 Review/Test in their portfolios.

USING THE ASSESSMENT GUIDE

- Multiple-choice format of Chapter 2 Posttest—See *Assessment Guide*, pp. AG13–14.
- Free-response format of Chapter 2 Posttest—See *Assessment Guide*, pp. AG15–16.

USING STUDENT SELF-ASSESSMENT

The How Did I Do? survey helps students assess what they have learned and how they learned it. This survey is available as a copying master in *Assessment Guide*, p. AGxvii.

CHAPTER 2 REVIEW/TEST

1. **VOCABULARY** In the expression 2^3, the number 2 is the __?__. (p. 40) **base**
2. **VOCABULARY** To find the value of an expression that has more than one operation, you need to use the __?__. (p. 42) **order of operations**
3. **VOCABULARY** In the expression 8^4, the number 4 is the __?__. (p. 40) **exponent**

Use mental math to find the value. (pp. 36–39)

4. $19 + 43$ **62**
5. $76 - 37$ **39**
6. $32 + (48 + 83)$ **163**
7. 26×12 **312**
8. $5,986 \times 1$ **5,986**
9. $4 \times 6 \times 25$ **600**

Write the equal factors. Then find the value. (pp. 40–41)

10. 6^2 **6 × 6; 36**
11. 9^2 **9 × 9; 81**
12. 3^5 **3 × 3 × 3 × 3 × 3; 243**
13. 5^4 **5 × 5 × 5 × 5; 625**
14. 2^0 **1**
15. 10^6 **10 × 10 × 10 × 10 × 10 × 10; 1,000,000**
16. 7^5 **7 × 7 × 7 × 7 × 7; 16,807**
17. 25^1 **25; 25**
18. 4^3 **4 × 4 × 4; 64**
19. 12^2 **12 × 12; 144**

Evaluate the expression. (pp. 44–45)

20. $4 \times 5 - 6 \times 3$ **2**
21. $(12 \times 7) + 9^2$ **165**
22. $13 + 4 \times (20 + 35)$ **233**
23. $4^3 - 4 \times 12$ **16**
24. $36 \div (24 - 18) + 9$ **15**
25. $45 - (12 \times 3) \div 6$ **39**
26. $16 \times 9 \div 2^3$ **18**
27. $(100 - 28) \div 3^2$ **8**
28. $43 + (6^2 - 3^3)^2$ **124**
29. $(19 + 9^2)^2 \div 5^2 + 94$ **494**

Use Data For 30–31, use the table at the right.

30. Michael wants to surprise his family by having dinner ready when they get home. The table shows the length of time each of the foods needs to cook. If Michael wants to have everything ready to eat at the same time, in what order should he start cooking the food? (pp. 46–47) **chicken, rice, stuffing, dinner rolls, peas**

31. Suppose Michael decides to cook baked potatoes instead of rice. The potatoes take 75 minutes to bake. In what order should he start cooking the food? (pp. 46–47) **potatoes, chicken, stuffing, dinner rolls, peas**

32. Doreen has $40.00. She wants to buy 3 pairs of socks for $2.75 each, gloves for $9.99, and 2 T-shirts for $8.79 each. How much money will she have left? (pp. 42–43) **$4.18**

33. Each of three pyramids of Egypt is made up of about 2.5 million large stones. Is the total number of stones greater than or less than 10^8? (pp. 40–41) **less than**

Cooking Times for Food

Type of Food	Cooking Time
Chicken	55 Minutes
Peas	5 Minutes
Rice	20 Minutes
Stuffing	15 Minutes
Dinner Rolls	12 Minutes

CHAPTER 2 TEST, page 1

Choose the best answer.

For 1–6, use mental math to find the value.

1. $19 + 254$
   - A 263
   - B 265
   - **C 273**
   - D 275

2. $2 \times 7 \times 40$
   - F 280
   - G 360
   - **H 560**
   - J 650

3. $395 - 87$
   - A 318
   - **B 308**
   - C 306
   - D 288

4. $225 \div 5$
   - **F 45**
   - G 41
   - H 35
   - J 31

5. $6,784 \times 1$
   - A 6,785
   - **B 6,784**
   - C 1
   - D 0

6. $20 \times 37 \times 5$
   - F 137
   - G 185
   - H 925
   - **J 3,700**

For 7–10, find the value.

7. 8^5
   - A 390,625
   - B 262,144
   - **C 32,768**
   - D 4,096

8. 5^6
   - **F 15,625**
   - G 5,600
   - H 3,125
   - J 25

9. 7^5
   - A 49
   - B 2,401
   - C 7,500
   - **D 16,807**

10. 6^4
    - F 216
    - **G 1,296**
    - H 7,776
    - J 46,656

For 11–15, evaluate the expression.

11. $3^3 + 4 \times 5$
    - A 29
    - **B 47**
    - C 155
    - D 180

12. $42 - (6 \div 3) \times (5 + 3)$
    - F 320
    - G 312
    - H 168
    - **J 26**

13. $33 \times (4 - 2) - 4^2$
    - **A 50**
    - B 58
    - C 82
    - D 560

14. $(44 + 4) \times (2 + 3^2)$
    - F 39
    - G 55
    - H 110
    - **J 121**

15. $(72 \div 9) + 13^2 - 8$
    - A 22
    - B 42
    - **C 169**
    - D 433

CHAPTER 2 TEST, page 2

16. Ling has written 8 pages each day for the last 30 days. If she has to write a total of 400 pages in 46 days, how many pages will she have to write per day during the remaining time in order to meet her goal?
    - **F 10 pages**
    - G 15 pages
    - H 25 pages
    - J 50 pages

17. Robin baby-sits 5 hours a week for $6 per hour. He mows lawns for $10 each twice a week. How much will he make in 12 weeks?
    - A $6,000
    - **B $600**
    - C $480
    - D $380

For 18–21, evaluate the expression for $a = 8$ and $b = 3$.

18. $9 + a^2 \div (12 - 4)$
    - F 64
    - **G 17**
    - H 11
    - J 3

19. $b \times 5 + 43$
    - **A 58**
    - B 88
    - C 144
    - D 645

20. $80 + a \times (28 - 23)$
    - F 450
    - **G 50**
    - H 15
    - J 2

21. $48 \div b - 7$
    - A 23
    - B 11
    - **C 10**
    - D 9

22. Deborah's train ride takes 8 hours. She reads 30 pages per hour. How many pages will she read if she sleeps for 2 hours and reads the rest of the time?
    - F 300 pages
    - G 240 pages
    - **H 180 pages**
    - J 38 pages

For 23–25, use the following chart.

GUIDED TOUR TIMES	
1 Expressionists	10 A.M., 4 P.M.
2 American Painters	10 A.M., 3 P.M.
3 Dutch Painters	12 P.M., 3 P.M.
4 Sculpture	1 P.M., 4 P.M.
5 Impressionists	2 P.M., 4 P.M.

Each tour lasts 50 minutes.

23. Miranda wants to take the Sculpture tour at 1 P.M. If she arrives at the museum at 10 A.M. and leaves at 4 P.M., in which order could she take all of the tours?
    - A 1, 2, 3, 4, 5
    - **B 1, 3, 4, 5, 2**
    - C 2, 5, 4, 3, 1
    - D 5, 4, 3, 2, 1

24. Ivan wants to take the 12 P.M. Dutch Painters tour. If he takes the tour and then eats for 45 minutes, what other tour could he take before he leaves the museum at 3:00 P.M.?
    - F 1
    - G 3
    - **H 4**
    - J 5

25. Susan wants to take the 10 A.M. Expressionists tour. If she plans to leave the museum by 2 P.M., which other tours could she take?
    - **A 3, 4**
    - B 1, 2
    - C 2, 3
    - D 5, 4

48 Chapter 2

CHAPTERS 1–2 ★ STANDARDIZED TEST PREP

Look for important words.
See item **5**.

An important word is *not*. The words *not true* mean you need to find the one expression among the answer choices that is *not* an equality.

Also see problem **2**, p. H62.

Choose the best answer.

1. Which can be expressed as 3^5? **A**

   A $3 \times 3 \times 3 \times 3 \times 3$ C 5×3

   B $5 \times 5 \times 5$ D 3×3

2. Lori has 79 stickers. This is three more than twice as many as Beth has. How many stickers does Beth have? **H**

   F 34 H 38

   G 36 J 40

3. Calvin spent $43 on a video game and a CD. The video game cost $13 more than the CD. How much did the video game cost? **B**

   A $13 C $30

   B $28 D $56

4. Kim's class has been saving pennies to go on a field trip to the Gem and Mineral Museum. The class has collected 10,000 pennies. How is 10,000 written using an exponent with the base 10? **H**

   F 1^4 H 10^4

   G 10^2 J 10^5

5. Which number sentence is **not** true? **D**

   A $30 \times 12 \times 8 = 12 \times 30 \times 8$

   B $30 + 8 + 12 = 30 + 12 + 8$

   C $30 - 12 + 8 = 30 + 8 - 12$

   D $30 + 8 \times 12 = 30 \times 12 + 8$

6. Which expression is equivalent to $(51 + 32) + 71$? **G**

   F $(51 + 71) + (32 + 71)$

   G $51 + (32 + 71)$

   H $93 + 51$

   J $19 + 71$

7. What is the value of $42 - d$ for $d = 6$? **B**

   A 7 C 48

   B 36 D Not here

8. Which algebraic expression represents the expression "13 less than a number, t"? **J**

   F $13 \times t$ H $13 - t$

   G $t + 13$ J $t - 13$

9. Which is the best estimate for this sum?
   $15{,}192 + 3{,}751 + 1{,}551$ **D**

   A 18,000 C 20,000

   B 19,000 D 21,000

10. Which is another way to write $y \times y \times y$? **H**

    F $3 \times y$ H y^3

    G $y \times 3y$ J y^4

Write What You Know

See below.

11. Explain how you can use compensation to find the sum $48 + 57$ mentally.

12. Evaluate $(3^2 + 7) \div (4 - 2)$. Explain how you used the order of operations to find your answer.

49

Write What You Know • Written Response

11. Start by adding 2 to 48 and subtracting 2 from 57:
    $48 + 57 = (48 + 2) + (57 - 2) = 50 + 55$.
    Now add again: $50 + 55 = 50 + 50 + 5 = 100 + 5 = 105$.

12. First work inside the first set of parentheses, evaluating the exponent first and then the sum: $(3^2 + 7) \div (4 - 2) = (9 + 7) \div (4 - 2) = 16 \div (4 - 2)$. Then work inside the second set of parentheses: $16 \div (4 - 2) = 16 \div 2 = 8$.

STANDARDIZED TEST PREP •
Chapters 1–2

USING THE PAGE

This page may be used to help students get ready for standardized tests. The test items are written in the same style and arranged in the same format as those on many state assessments. The page is cumulative. It covers math objectives and essential skills that have been taught up to this point in the text. Most of the items represent skills from the current chapter, and the remainder represent skills from earlier chapters.

This page can be assigned at the end of the chapter as classwork or as a homework assignment. You may want to have students use individual recording sheets presented in a multiple-choice (standardized) format. A Test Answer Sheet is available as a blackline master in *Assessment Guide* (p. AGxlii).

You may wish to have students describe how they solved each problem and share their solutions.

ITEM ANALYSIS

Item	Learning Goal	Item	Learning Goal
1	2B	7	1C
2	2C	8	1C
3	2C	9	1A
4	2B	10	2B
5	2A	11	2A
6	2A	12	2C

Written response items for the Write What You Know are available as a blackline master in *Performance Assessment*.

SCORING RUBRIC • WRITE WHAT YOU KNOW

2 Demonstrates a complete understanding of the problem and chooses an appropriate strategy to determine the solution

1 Demonstrates a partial understanding of the problem and chooses a strategy that does not lead to a complete and accurate solution

0 Demonstrates little understanding of the problem and shows little evidence of using any strategy to determine a solution

Operation Sense **49**

CHAPTER 3

Decimal Concepts

CHAPTER PLANNER

PACING OPTIONS
| Compacted | 4 Days |
| Expanded | 7 Days |

Getting Ready for Chapter 3 • Assessing Prior Knowledge and **INTERVENTION** (See PE and TE page 51.)

LESSON	NCTM STANDARDS	PACING	VOCABULARY*	MATERIALS	RESOURCES AND TECHNOLOGY
3.1 Represent, Compare, and Order Decimals pp. 52–55 **Objective** To use place value to express, compare, and order decimals	**1, 6, 8**	2 Days		*For Thinker's Corner* 10-section spinner, place-value chart	Reteach, Practice, Challenge, Problem Solving 3.1 Worksheets Extra Practice p. H34, Set A ▢ Transparency 3.1 ◉ Calculating Crew • *Nautical Number Line*
3.2 Problem Solving Strategy: *Make a Table* pp. 56–57 **Objective** To use the strategy *make a table* to solve problems	**1, 6, 7, 10**	1 Day		🖩	Reteach, Practice, Challenge, Reading Strategy 3.2 Worksheets ▢ Transparency 3.2 Problem Solving Think Along, p. TR1
3.3 Estimate with Decimals pp. 58–59 **Objective** To estimate decimal sums, differences, products, and quotients	**1, 6, 8**	1 Day			Reteach, Practice, Challenge, Problem Solving 3.3 Worksheets Extra Practice p. H34, Set B ▢ Transparency 3.3
3.4 Decimals and Percents pp. 60–61 **Objective** To write a decimal as a percent and a percent as a decimal	**1, 6, 8, 10**	1 Day	**percent**	*For each student* two 10 × 10 grids (decimal squares)	Reteach, Practice, Challenge, Problem Solving 3.4 Worksheets Extra Practice p. H34, Set C ▢ Transparency 3.4

Ending Chapter 3 • **Chapter 3 Review/Test,** p. 62 • **Standardized Test Prep,** p. 63

***Boldfaced** terms are new vocabulary. Other terms are review vocabulary.

CHAPTER AT A GLANCE

Vocabulary Development

The boldfaced word is the new vocabulary term in the chapter. Have students record the definition in their Math Journals.

percent, p. 60

Vocabulary Cards
Have students use the Vocabulary Cards on *Teacher's Resource Book* pp. TR121–122 to make graphic organizers or word puzzles. The cards can also be added to a file of mathematics terms.

NCTM Standards

1. **Number and Operations**
   Lessons 3.1, 3.2, 3.3, 3.4
2. **Algebra**
3. **Geometry**
4. **Measurement**
5. **Data Analysis and Probability**
6. **Problem Solving**
   Lessons 3.1, 3.2, 3.3, 3.4
7. **Reasoning and Proof**
   Lesson 3.2
8. **Communication**
   Lessons 3.1, 3.3, 3.4
9. **Connections**
10. **Representation**
    Lessons 3.2, 3.4

Writing Opportunities

PUPIL EDITION
- What's the Error?, p. 55
- What's the Question?, p. 57
- Write a Problem, p. 59
- Write About It, p. 61

TEACHER'S EDITION
- Write—See the *Assess* section of each TE lesson.
- Writing in Mathematics, p. 58B

ASSESSMENT GUIDE
- How Did I Do?, p. AGxvii

Family Involvement Activities

These activities provide:
- Letters to the Family
- Math Vocabulary
- Family Game
- Practice (Homework)

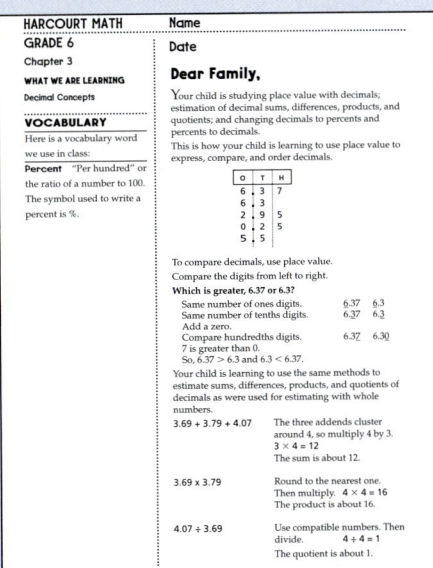

Family Involvement Activities, p. FA9

CHAPTER 3 Decimal Concepts

MATHEMATICS ACROSS THE GRADES

SKILLS TRACE ACROSS THE GRADES

GRADE 5
Round decimals; estimate decimal sums and differences; add, subtract, multiply, and divide decimals; write fractions as decimals

GRADE 6
Write, compare, and order decimals; estimate decimal sums, differences, products, and quotients; write decimals as percents and percents as decimals

GRADE 7
Estimate and find decimal sums, differences, products, and quotients; write decimals with scientific notation; write fractions as repeating and terminating decimals

SKILLS TRACE FOR GRADE 6

LESSON	FIRST INTRODUCED	TAUGHT AND PRACTICED	TESTED	REVIEWED
3.1	Grade 5	PE pp. 52–55, H34, p. RW11, p. PW11, p. PS 11	PE p. 62, pp. AG17–20	PE pp. 62, 63, 88–89
3.2	Grade 4	PE pp. 56–57, p. RW12, p. PW12, p. PS 12	PE p. 62, pp. AG17–20	PE pp. 62, 63, 88–89
3.3	Grade 5	PE pp. 58–59, H34, p. RW13, p. PW13, p. PS 13	PE p. 62, pp. AG17–20	PE pp. 62, 63, 88–89
3.4	Grade 5	PE pp. 60–61, H34, p. RW14, p. PW14, p. PS 14	PE p. 62, pp. AG17–20	PE pp. 62, 63, 88–89

KEY PE Pupil Edition PS Problem Solving Workbook RW Reteach Workbook
PW Practice Workbook AG Assessment Guide

Looking Back  Prerequisite Skills

To be ready for Chapter 3, students should have the following understandings and skills:

- **Represent Decimals**—write decimals for models
- **Write and Read Decimals**—decimals to hundredths
- **Compare Whole Numbers**—use <, > or =
- **Round Decimals**—round to nearest whole number

Check What You Know

Use page 51 to determine students' knowledge of prerequisite concepts and skills.

Intervention

Help students prepare for the chapter by using the Intervention resources described on TE page 51.

Looking at Chapter 3  Essential Skills

Students will

- develop skill in writing, comparing, and ordering decimals.
- use the strategy *make a table* to solve problems.
- develop skill estimating decimal sums, differences, products, and quotients.
- make the connection between graphic representations of decimals and percents to write decimals as percents and percents as decimals.

EXAMPLE
Write 0.2 as a percent.

Model Percents	Write Percents
	0.2 is 20 hundredths. So, $0.2 = 20\%$

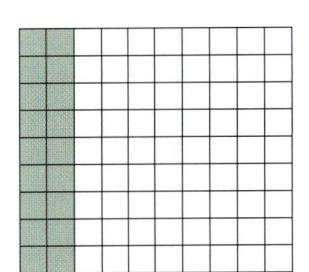

Looking Ahead  Applications

Students will apply what they learn in Chapter 3 to the following new concepts:

- Decimal Operations (Chapter 4)
- Representing Fractions as Decimals (Chapter 8)
- Solving Decimal Equations (Chapter 4)
- Equivalent Forms of Decimals, Fractions, and Percents (Chapter 21)

Decimal Concepts 50D

CHAPTER 3

Decimal Concepts

INTRODUCING THE CHAPTER

Tell students that the set of decimal numbers includes numbers such as 44, 0.38, and 5.762. Have students focus on the Party Mix Ingredients grid (decimal model) and compare decimals. Then have them list the ingredients in order from least to greatest. *wheat chips, puffed corn, puffed wheat, corn chips*

USING DATA

To begin the study of this chapter, have students

- Write the correct decimal number for puffed corn if it were doubled in the party mix. 0.36
- Write the correct decimal number for puffed wheat if it were halved in the party mix. 0.12
- Write the decimal number by which corn chips exceed wheat chips in this recipe. 0.26

PROBLEM SOLVING PROJECT

Purpose To use decimals to solve problems

Grouping pairs or small groups

Background The U.S. Department of Agriculture recommends a diet low in fat and high in complex carbohydrates. When you choose food for a party, you can select food that both tastes good and is good for you nutritionally.

Analyze, Choose, Solve, and Check

Have students

- Choose healthful foods, based on their own research, for their own party mix.
- Decide how much of each food they want to put in their party mix.
- Make a grid (decimal model) showing the different parts of their party mix.

Portfolio Students may want to display their party mix grids in the classroom before placing them in their portfolios.

CHAPTER 3 Decimal Concepts

puffed corn and wheat chips

Wheat and corn are very important crops to farmers in the United States. These crops are grown across vast fields in Kansas and many other states. We eat wheat and corn at almost every meal in our cereals, breads, and pastas. We even eat them at parties.

PROBLEM SOLVING Which ingredients of the party mix occur in almost equal amounts?

Party Mix Ingredients

- Corn Chips 0.42
- Puffed Corn 0.18
- Wheat Chips 0.16
- Puffed Wheat 0.24

Why learn math? Explain that nutritionists use mathematics when they evaluate foods for fat content and customize diets to people's individual needs. For example, using skim milk instead of whole milk may cut a recipe's fat content by 0.30, or 30%. Ask: What other professions might use decimals and percents? *Possible answer: pharmacists, teachers, accountants, bankers*

Check What You Know

Use this page to help you review and remember important skills needed for Chapter 3.

✓ Represent Decimals (For Intervention, see p. H2.)

Write the decimal that is modeled.

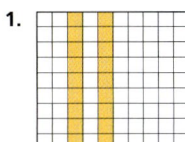 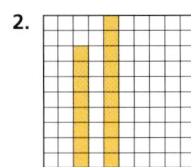 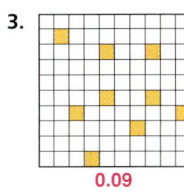 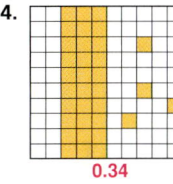

1. 0.2
2. 0.18
3. 0.09
4. 0.34

✓ Write and Read Decimals (For Intervention, see p. H3.)

Write the decimal.

5. 836 and 23 hundredths **836.23**
6. 364 and 2 tenths **364.2**
7. 93 thousand, 450 and 38 hundredths **93,450.38**
8. 595 thousand, 821 and 9 tenths **595,821.9**
9. 306 thousand, 7 and 6 hundredths **306,007.06**

Complete to show how to read the numbers.

10. 1,463.05 1 thousand, 463 and __?__ **5 hundredths**
11. 204.7 204 and __?__ tenths **7**
12. 32,617.45 32 __?__, 617 and 45 __?__ **thousand; hundredths**
13. 4,382.1 4 thousand, 382 and __?__ **1 tenth**

✓ Compare Whole Numbers (For Intervention, see p. H3.)

Compare the numbers. Write <, >, or = for ●.

14. 143 ● 140 **>**
15. 808 ● 880 **<**
16. 716 ● 716 **=**
17. 691 ● 961 **<**
18. 94 ● 49 **>**
19. 405 ● 305 **>**
20. 383 ● 383 **=**
21. 5,937 ● 397 **>**
22. 4,062 ● 4,206 **<**
23. 689 ● 648 **>**
24. 6,098 ● 6,908 **<**
25. 4,801 ● 4,108 **>**

✓ Round Decimals (For Intervention, see p. H4.)

Round to the nearest whole number.

26. 3.64 **4**
27. 1.49 **1**
28. 6.938 **7**
29. 41.8 **42**
30. 18.70 **19**
31. 72.06 **72**

Round to the nearest tenth.

32. 69.64 **69.6**
33. 26.37 **26.4**
34. 52.489 **52.5**
35. 8.630 **8.6**
36. 9.479 **9.5**
37. 14.507 **14.5**
38. 90.63 **90.6**
39. 55.58 **55.6**
40. 26.397 **26.4**

LOOK AHEAD

In Chapter 3 you will
- represent, compare, and order decimals
- estimate with decimals
- use decimals and percents

LESSON 3.1

Represent, Compare, and Order Decimals

LESSON PLANNING

Objective To use place value to express, compare, and order decimals

Intervention for Prerequisite Skills
Represent Decimals, Write and Read Decimals, Compare and Order Whole Numbers (For intervention strategies, see page 51.)

Materials *For Thinker's Corner* 10-section spinner, p. TR73; place-value chart

NCTM Standards
1. Number and Operations
6. Problem Solving
8. Communication

Math Background
Decimals can be compared and ordered using models such as place-value charts and number lines. Consider the following ideas as you help students understand the process of comparing and ordering decimals:

- The decimal with the greatest whole-number part is the greatest.
- Decimals can be compared, place by place, starting with the greatest place. The decimal that has a greater digit in a given place is greater.
- Adding zeros so that decimals have the same number of places helps to compare decimals.
- When comparing several decimals, it is helpful to compare two at a time until they are in order.

WARM-UP RESOURCES

 **NUMBER OF THE DAY**

Subtract your age from your age multiplied by 100. When you divide the result by 11 and then divide the quotient by 9, what number do you get? The answer will be the student's age.

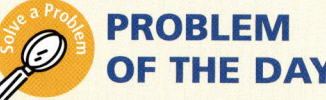 **PROBLEM OF THE DAY**

The money that Mrs. Frey deposited in her bank account was in $10 bills. The sum of the digits in the amount she deposited was 18. If she had deposited $10 more, the sum of the digits would have been 1. How much did Mrs. Frey deposit? $990

Solution Problem of the Day tab, p. PD3

 **DAILY FACTS PRACTICE**

Have students practice multiplication facts by completing Set E of *Teacher's Resource Book,* p. TR94.

52A Chapter 3

INTERVENTION AND EXTENSION RESOURCES

ALTERNATIVE TEACHING STRATEGY

Have students **review ordering whole numbers** by making a time line from A.D. 1000 to 2000 with intervals of 100 years. Have them locate the year in which each of these items was invented and put the information on the time line. Students may wish to illustrate their work.

telescope—1608
printing press—1450
computer—1943
telephone—1876
navigational compass—1086

Check students' work.

See also page 54.

VISUAL
LOGICAL/MATHEMATICAL

SPECIAL NEEDS ELL

Materials *For each student* index card

Engage students physically in **ordering decimals.** Write a different one- or two-place decimal on each index card, and distribute 1 card to each student. Have students form a line ordering the decimals from greatest to least. The first student in line holds the card with the greatest decimal; the last student holds the card with the least decimal.

Then have the students turn the cards over and write a different decimal. Have the students form a new line ordering the decimals they have written from least to greatest. Check students' work.

KINESTHETIC
LOGICAL/MATHEMATICAL

MULTISTEP AND STRATEGY PROBLEMS

The following multistep or strategy problem is provided in Lesson 3.1:

Page	Item
55	47

TECHNOLOGY • *DATA TOOLKIT*

Show students how to **order data** from greatest to least or least to greatest by using a spreadsheet program, such as *Data ToolKit*:

- Insert the following data into the spreadsheet, and then highlight both columns. Bring down the Spreadsheet menu, select *Sort*, and click on the word *Descending* (greatest to least).

Energy Use (in quadrillion BTUs)
   Canada   14.36
   China   30.18
   United States 66.68

Check students' work.

VISUAL
LOGICAL/MATHEMATICAL

TECHNOLOGY LINK

 Intervention Strategies and Activities CD-ROM • *Skills 3, 32, 33*

 Calculating Crew • *Nautical Number Line,* Levels O, P

LESSON 3.1 ORGANIZER

Objective To use place value to express, compare, and order decimals

Materials For Thinker's Corner 10-section spinner, p. TR73; place-value chart

1 Introduce

QUICK REVIEW provides review of prerequisite skills.

Why Learn This? Expressing, comparing, and ordering decimals helps you compare prices, measurements, and batting averages. *Share the lesson objective with students.*

2 Teach

Guided Instruction

- *Discuss with students how the value of the digit 3 changes depending on its position in the place-value chart on page 52.*

  How does the value of the 3 in the second number compare with that of the 3 in the first number? It is 10,000 times as great.

  How does the value of the 3 in the first number compare with that of the 3 in the third number? It is 10 times as great.

- *Draw students' attention to the different ways the numbers are expressed in Example 1.*

  How can you remember the difference between *standard form* and *expanded form*? Possible answer: Standard form is the usual way to express a number. Expanded form is the longer way to express a number as addends.

Algebraic Thinking Understanding that the same number or value can be represented in more than one way is a fundamental concept needed for algebra. A decimal number can be written in standard form, in expanded form, in word form, as a fraction, or as a percent.

LESSON 3.1

Represent, Compare, and Order Decimals

Learn how to use place value to express, compare, and order decimals.

QUICK REVIEW
Compare the numbers. Write <, >, or = for ●.
1. 124 ● 134 <
2. 143 ● 134 >
3. 909 ● 990 <
4. 365 ● 346 >
5. 2,047 ● 2,047 =

Remember that when reading a number with a decimal point, read the decimal point as "and." Read 8.2 as "eight and two tenths."

Genna read that it costs $0.03 to use her hair dryer for 30 minutes, $0.08 to use her clock for a month, and $0.53 to wash and dry a load of laundry.

Place value helps you understand numbers. The digits and the position of each digit determine a number's value. Read each number on the place-value chart. These numbers are part of the decimal system. Notice that 3 has a value of 3 thousandths, 3 tens, or 3 ten-thousandths, depending on its position in the number.

	PLACE VALUE								
	Ten Thousands	Thousands	Hundreds	Tens	Ones	Tenths	Hundredths	Thousandths	Ten Thousandths
0.053					0 .	0	5	3	
32.4				3	2 .	4			
8.0023					8 .	0	0	2	3

When you read and write numbers, you are using place value.

EXAMPLE 1

A. Standard form: 0.053
   Expanded form: 0.05 + 0.003
   Word form: *fifty-three thousandths*

B. Standard form: 32.4
   Expanded form: 30 + 2 + 0.4
   Word form: *thirty-two and four tenths*

C. Standard form: 8.0023
   Expanded form: 8 + 0.002 + 0.0003
   Word form: *eight and twenty-three ten-thousandths*

Math Idea ▶ Knowing the place value of digits will help you read, write, and calculate numbers correctly, including decimal numbers.

52 Chapter 3

RETEACH 3.1
Represent, Compare, and Order Decimals

The numbers you use every day are part of the decimal system. To find the value of a number, look at the digits and the position of each digit. A place-value chart can help you.
Knowing place values is useful when comparing decimal numbers. Mindy is asked to list 18.3, 17.8, and 24.1 in order from greatest to least.

Step 1 Mindy compares the first two numbers.
- She starts at the left. Both numbers have the digit 1 in the tens place. 18.3 ↔ 17.8
- So, Mindy looks at the digits in the ones place. The first number has the digit 8 in the ones place, while the second number has the digit 7. 18.3 ↔ 17.8
- Since 8 > 7, 18.3 > 17.8.

Step 2 Mindy compares the third number to the greatest number so far, the first number. 24.1 ↔ 18.3
- The third number has the digit 2 in the tens place, while the first number has the digit 1 in the tens place.
- Since 2 > 1, 24.1 > 18.3.

Using what she has discovered, Mindy makes the list: 24.1; 18.3; 17.8.

Give the position and the value of each underlined digit.
1. 30.1<u>9</u>4
   a. position __tenths__
   b. value __1 tenth; 0.1__
2. 4,082,113.7<u>2</u>3
   a. position __thousandths__
   b. value __3 thousandths; 0.003__

Compare the numbers. Write <, >, or = in the ●.
3. 9.03 ● 0.93
4. 0.210 ● 0.012
5. 8.241 ● 8.24

Write the numbers in order from least to greatest.
6. 42.05; 45.02; 40.52   __40.52; 42.05; 45.02__
7. 19.7; 19.007; 19.07   __19.007; 19.07; 19.7__
8. 0.59; 0.95; 0.6   __0.59; 0.6; 0.95__

PRACTICE 3.1
Represent, Compare, and Order Decimals

Write the value of the underlined digit.
1. 485.0<u>3</u>6   __6 thousandths__
2. 16,005.8<u>4</u>5   __4 hundredths__
3. 8,492.<u>7</u>792   __7 tenths__

Write the number in expanded form.
4. 5.71   __5 + 0.7 + 0.01__
5. 85.083   __80 + 5 + 0.08 + 0.003__
6. 0.4625   __0.4 + 0.06 + 0.002 + 0.0005__
7. 17.00157   __10 + 7 + 0.001 + 0.0005 + 0.00007__

Compare the numbers. Write <, >, or = for ○.
8. 15.4 ○ 14.5 __>__
9. 5.67 ○ 5.76 __<__
10. 43.90 ○ 43.9 __=__
11. 7.91 ○ 9.17 __<__
12. 765.28 ○ 762.58 __>__
13. 0.234 ○ 2.304 __<__

Write the numbers in order from least to greatest.
14. 3,224; 2,432; 3,422   __2,432; 3,224; 3,422__
15. 88.5; 85.8; 58.8   __58.8; 85.8; 88.5__
16. 6.21; 6.02; 6.12   __6.02; 6.12; 6.21__

Write the numbers in order from greatest to least
17. 0.005; 0.500; 0.050   __0.500; 0.050; 0.005__
18. 317.8; 318.7; 371.8   __371.8; 318.7; 317.8__
19. 16.04; 14.6; 16.4   __16.4; 16.04; 14.6__

Mixed Review
Evaluate each expression.
20. $4 + 3^3 \times 2 - (6 - 1)$ __53__
21. $(11 + 16) \div 3 + (4 - 2)^2$ __13__
22. $45 + (6^2 - 11) \times 2$ __95__

Solve each equation by using mental math.
23. $m - 7 = 36$ $m = 43$
24. $9x = 63$ $x = 7$
25. $a \div 6 = 14$ $a = 84$

Evaluate each expression for $a = 6$, $b = 120$, and $c = 54$.
26. $b + 295$ __415__
27. $93 - c$ __39__
28. $b \div a$ __20__

Mark notices that one jar of cinnamon contains 2.6 oz and another contains 2.3 oz. He wants to buy the jar with the greater amount of cinnamon.

You can use a number line to compare 2.6 and 2.3.

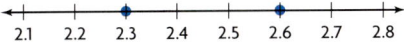

Since 2.6 is to the right of 2.3 on the number line, 2.6 is greater than 2.3.	Since 2.3 is to the left of 2.6 on the number line, 2.3 is less than 2.6.
2.6 > 2.3 ↓ greater than	2.3 < 2.6 ↓ less than

So, the 2.6-oz jar has more cinnamon.

You can also use place value to compare decimal numbers.

TECHNOLOGY LINK
More Practice: Use **Mighty Math Calculating Crew**, *Nautical Number Line*, Levels O and P.

EXAMPLE 2

Compare 7.28 and 7.2. Use < or >.

Start at the left.

7.28	7.2	Compare the ones digits. They are the same.
7.28	7.2	Compare the tenths digits. They are the same.
7.28	7.20	Add a zero so both numbers have the same number of places. Compare the hundredths digits. 8 is greater than 0.

So, 7.28 > 7.2, and 7.2 < 7.28.

Remember that you can add a zero to the right of a decimal without changing its value.
7.2 = 7.20
2 tenths is the same as 20 hundredths.

• Which is greater, 7.2 or 7.08? Explain. **7.2; compare the tenths, 2 > 0.**

You can use place value to order two or more decimal numbers.

EXAMPLE 3

The prices for the same kind of CD player in four different stores are $132.95, $132.50, $130.25, and $135.25. Order the prices of the CD players from least to greatest.

Compare every possible pair of numbers.

$132.95 > $132.50 $132.95 > $130.25 $132.95 < $135.25
$132.50 > $130.25 $132.50 < $135.25 $130.25 < $135.25

So, the list of numbers in order from least to greatest is $130.25, $132.50, $132.95, $135.25.

• List 9.365, 9.271, 9.356, and 9.065 in order from greatest to least. **9.365, 9.356, 9.271, 9.065**

• Engage students in a discussion of the greater than and less than symbols.

Describe a method you use to differentiate between > and <. Possible answer: The symbol always opens to the greater number and narrows toward the lesser number.

ADDITIONAL EXAMPLES

Example 1, p. 52

Use place value to complete the following:

A. Standard form: 72.514
Expanded form: 70 + 2 + 0.5 + 0.01 + 0.004
Word form: seventy-two and five hundred fourteen thousandths

B. Standard form: 64.8
Expanded form: 60 + 4 + 0.8
Word form: sixty-four and eight tenths

C. Standard form: 0.0316
Expanded form: 0.03 + 0.001 + 0.0006
Word form: three hundred sixteen ten-thousandths

Example 2, p. 53

Compare 5.6 and 5.62. Use < or >.

5.6	5.62	Compare the digits. Start at the left. Same number of ones.
5.6	5.62	Same number of tenths.
5.60	5.62	Add a zero so both numbers have the same number of places. Compare the hundredths.

So, 5.6 < 5.62, and 5.62 > 5.6.

Example 3, p. 53

Philip Lewis priced three Internet services. The monthly rates were $19.95, $20.85, and $19.97. Order the rates from least to greatest. Which is the lowest rate? $19.95 < $19.97 < $20.85; $19.95 is the lowest rate.

53

LESSON 3.2
Problem Solving Strategy: *Make a Table*

LESSON PLANNING

Objective To use the strategy *make a table* to solve problems

Intervention for Prerequisite Skills
Compare and Order Whole Numbers (For intervention strategies, see page 51.)

Lesson Resources Problem Solving Think Along, p. TR1

NCTM Standards
1. Number and Operations
6. Problem Solving
7. Reasoning and Proof
10. Representation

Math Background
As you introduce this strategy, remind students of the following:
- It may be helpful to use the strategy *make a table* if there is a large amount of data.
- This strategy can be used to order data, summarize data, organize data, and to help show a pattern.
- When the data are arranged in the table, it is often easier to solve the stated problem.

WARM-UP RESOURCES

 **NUMBER OF THE DAY**

Compare the number representing the month of the year to that representing the day of the month. **Possible answer for Feb. 21: 2 < 21**

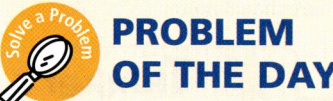 **PROBLEM OF THE DAY**

Jake knows that Uranus is farther from Earth than Saturn but not as far as Neptune. Match each planet with its distance from Earth.

0.744 billion mi

2.7 billion mi

1.6 billion mi

Saturn, 0.744 billion mi
Uranus, 1.6 billion mi
Neptune, 2.7 billion mi

Solution Problem of the Day tab, p. PD3

 **DAILY FACTS PRACTICE**

Have students practice division facts by completing Set F of *Teacher's Resource Book*, p. TR94.

INTERVENTION AND EXTENSION RESOURCES

REACHING ALL LEARNERS

ALTERNATIVE TEACHING STRATEGY

Materials *For each student* 6 index cards

Help students **organize data for a table.** Provide them with 6 index cards. Have them write each skate model and price from page 56 on a card, one to a card.

Then have students compare pairs of cards and order the cards based on price. Once the cards are in order, have them write the information in a table. Check students' work.

KINESTHETIC
VISUAL/SPATIAL

READING STRATEGY

Use Graphic Aids One way to solve the problem on page 56 is simply to list the numbers in order. However, the graphic aid of a table also points out to students the model number of the skates to which each price is attached, thus making it easier to determine which are the preferred models.

In solving Exercises 1 and 2 on page 57, what data will you put in your table? the name of each team and its standing, ordered from least to greatest or greatest to least

MULTISTEP AND STRATEGY PROBLEMS

The following multistep and strategy problems are provided in Lesson 3.2:

Page	Item
57	1–10

ENGLISH LANGUAGE LEARNERS

Reinforce students' understanding of ordering data. Ask them to talk about another situation where different model numbers or names are used to describe items. Students may suggest cars, calculators, or computers.

Have them work in small groups to list several different models. Then have them decide which one would be most expensive, second most expensive, second least expensive, and least expensive. You may wish to assign reasonable whole-number prices to the various cars or computers to help students decide. Have students share their lists with the class by using the words *least* and *most expensive.* Check students' work.

AUDITORY
INTERPERSONAL/SOCIAL

EARLY FINISHERS

Encourage students to **practice making a table.** Ask them to look up the batting averages or other sports data for players or teams of their choice. Once they have collected the data, have them arrange the data in order in a table.

When students have completed their tables, have them write three statements about the people or teams, such as "Smith has the second-highest batting average." Check students' work.

AUDITORY
VISUAL/SPATIAL

Intervention Strategies and Activities CD-ROM • *Skill 3*

LESSON 3.2 ORGANIZER

Objective To use the strategy *make a table* to solve problems

Lesson Resources Problem Solving Think Along, p. TR1

1 Introduce

QUICK REVIEW provides review of prerequisite skills.

Why Learn This? In the future you can use this strategy to solve problems involving patterns. Share the lesson objective with students.

2 Teach

Guided Instruction

- Be sure students understand the difference between second least expensive and second most expensive models.

  If you arrange a list of prices in order with the least amount at the top, where would you place the second most expensive? second from the bottom

- Have students examine the table used to solve the problem.

  Why is *make a table* a good strategy for solving this problem? A table helps you to organize the information in the problem in a visual manner that makes the solution more apparent.

  How do you begin making the table? Possible answer: To order numbers, I must compare them. I see that 78 is less than the other numbers and 78.50 is less than 78.99, so I put 78.50 first.

REASONING **Describe another method of arranging the data in a table.** Arrange the prices in order from greatest to least and again select the price that is second from the top and the one that is second from the bottom.

56 Chapter 3

LESSON 3.2

PROBLEM SOLVING STRATEGY
Make a Table

Learn how to solve problems by organizing data in a table.

1. 276, 265, 263, 224
2. 375, 364, 356, 325
3. 8,903; 8,658; 8,586; 8,459
4. 2,961; 2,906; 2,609; 2,169
5. 3,765; 3,764; 3,456; 3,425

QUICK REVIEW

Order the numbers from greatest to least.

1. 263, 224, 276, 265
2. 364, 356, 375, 325
3. 8,658; 8,586; 8,459; 8,903
4. 2,961; 2,906; 2,609; 2,169
5. 3,764; 3,456; 3,765; 3,425

Andrew chooses new in-line skates from the models below.

RC-204 $99.95; PRX-100 $78.50; D-500 $99.99; OP-1000 $78.99; ZZ-2 $91.50; ZA-45 $99.25

Andrew wants to buy the second most expensive model. His parents want him to buy the second least expensive model. Which models are these? What is the difference in price of the two models?

Analyze What are you asked to find? the difference in price of the second most and second least expensive models of in-line skates
What information is given? the model numbers and prices of different styles of in-line skates

Choose What strategy will you use?

You can use the strategy *make a table* to show the prices of the in-line skates in order from least to greatest.

Solve How will you solve the problem?

Compare the prices and order them in a table. Then find the second most expensive model, the second least expensive model, and the difference in their prices.

$78.50 < $78.99 < $91.50 < $99.25 < $99.95 < $99.99

MODEL	PRICE
PRX-100	$78.50
OP-1000	$78.99	← second least expensive
ZZ-2	$91.50
ZA-45	$99.25
RC-204	$99.95	← second most expensive
D-500	$99.99

Subtract: $99.95 − $78.99 = $20.96

So, the difference in price of models RC-204 and OP-1000 is $20.96.

Check How can you check your answer? Read the prices from top to bottom; make sure that each is greater than the one above it.
What if Andrew chose model ZZ-2? How much more would he spend than if he bought model OP-1000? $12.51

56 Chapter 3

RETEACH 3.2

Problem Solving Strategy: Make a Table

Putting data in numerical order in a table can often help you determine the greatest or the least piece of data.

The areas of some sports fields are given below.
Basketball ....427 square yards   Ice hockey ....2,222 square yards
Football ....6,400 square yards   Tennis (doubles) ......312 square yards

The area of an Olympic swimming pool is 5,135 square yards less than the greatest area above and 953 square yards greater than the least area. What is the area of an Olympic swimming pool?

Step 1 Think about what you know and what you are asked to find.
• You know the sizes of different sports fields.
• You need to find the area of the pool.

Step 2 Plan a strategy to solve.
• Use the strategy *make a table* to order the data.
• Use the table entries to find the area of the pool.

Step 3 Carry out the strategy.
The greatest area is 6,400 square yards and the least area is 312 square yards. Use the greatest area to find the area of an Olympic swimming pool.

Tennis (doubles)	312 square yards
Basketball	427 square yards
Ice hockey	2,222 square yards
Football	6,400 square yards

6,400 − 5,135 = 1,265
So, the area of an Olympic pool is 1,265 square yards.

Now find the difference between the area of an Olympic swimming pool and the least area to check your answer.

1,265 − 312 = 953 The answer checks.

Solve the problem by making a table.

1. The areas of some other sports fields are boxing, 44 yd²; fencing, 33 yd²; judo, 306 yd²; karate, 75 yd²; and kendo, 132 yd². The area used for wrestling is 10 yd² greater than the least area above. How does the area for wrestling compare to that for karate? The area for karate is 32 yd² greater.

2. Baseball fields are not a standard size. However, the bases on the field are the corners of a square whose area is 5,500 yd² less than the area of a football field and 867 yd² greater than the area for a fencing match. What is the area of this square? 900 yd²

PRACTICE 3.2

Problem Solving Strategy: Make a Table

Solve the problem by making a table.

1. Earthquakes are measured using the Richter scale. The greater the number, the greater the magnitude (or strength). Some of the strongest earthquakes during the twentieth century had magnitudes of 7.2, 8.9, 8.4, 8.7, 8.3, 8.6, 7.7, and 8.1. The San Francisco earthquake of 1906 had the fifth highest magnitude of those given above. What was its magnitude on the Richter scale? 8.3

2. Late in 1999, one U.S. dollar was worth the following amounts in five other countries' money.
Australian dollar   1.5798
Brazilian real   1.8780
Canadian dollar   1.4796
German mark   1.9524
Swiss franc   1.5919
In which country could one U.S. dollar be exchanged for the greatest amount of that country's money? Germany

3. Danny is doing library research on animals. He has spent 25 minutes reading about insects. He thinks he will need the same amount of time for each of 5 other types of animals. If he began at 9:45 A.M., at what time would he finish? 12:15 P.M.

4. A theater is showing two films. The starting times for the first film are every even hour, beginning at noon. The starting times for the second film are every odd hour, beginning at 1:00 P.M. If the last show begins at 10:00 P.M., how many times are both films shown? 11 times (1st: 6; 2nd: 5)

Use the table at the right for 5 and 6. The numbers are amounts of energy in quadrillion BTUs.

Country	Energy Produced	Energy Used
United States	66.68	82.19
Great Britain	9.23	9.68
China	30.18	29.22
Canada	14.36	10.97
India	6.94	8.51
Russia	45.66	32.72

5. In which country is the difference between amount of energy produced and amount used the greatest? United States

6. In which country is the difference between amount of energy produced and amount used the least? Great Britain

Mixed Review
Use mental math to find the value.

7. 67 + 83 + 33   183
8. 449 − 398   51
9. 203 + 178 + 22   403

Write which operation you would do first.

10. 8 − 5 + 7   subtraction
11. 16 + 4 ÷ 2   division
12. (10 + 4) × 2   addition

PROBLEM SOLVING PRACTICE

Solve the problem by making a table.

Below are the fractions of games won by 8 baseball teams.

Hawks	0.650	Bulldogs	0.725
Tigers	0.750	Lions	0.490
Angels	0.675	Flames	0.700
Dolphins	0.550	Giants	0.695

1. Which team is in second place? **C**

   A Tigers **B** Flames **C** Bulldogs **D** Angels

2. How many teams are behind the Giants? **H**

   F 2 **G** 3 **H** 4 **J** 5

Use Data For 3–4, reorder the data in the table at the right from greatest to least.

3. Which appliance uses the greatest amount of electricity? **air conditioner**

4. Which appliance uses the least amount of electricity? **color TV**

PROBLEM SOLVING STRATEGIES
- Draw a Diagram or Picture
- Make a Model
- Predict and Test
- Work Backward
- Make an Organized List
- Find a Pattern
- ▶ Make a Table or Graph
- Solve a Simpler Problem
- Write an Equation
- Use Logical Reasoning

Electricity Used By Appliances

Appliance	Electricity (In Kilowatts)
Refrigerator	0.6
Air conditioner	1.5
Color TV	0.33
Iron	1.2
Coffeepot	0.9
Toaster	1.2

MIXED STRATEGY PRACTICE

5. A calculator, pen, and notebook cost $14.00 altogether. The calculator costs $9.00 more than the pen and $8.50 more than the notebook. How much does each item cost? **calculator $10.50, pen $1.50, notebook $2.00**

6. Kelly left the house with $16.00. She had $4.50 left after buying a movie ticket for $6.75, buying two snacks for $1.75 each, and paying for a bus ride. How much did she pay for the bus ride? **$1.25**

7. Carlene makes greeting cards. It costs $0.20 to make each card. She then sells them for $0.75 each. How many cards does she need to sell in order to make a profit of $33.00? **60 cards**

8. Frank walks 5 blocks to school for every 3 blocks Robert walks. They walk a total of 24 blocks. How many blocks from school does Robert live? **9 blocks**

9. In a survey, teens prefer the Boomer portable stereo over the Blaster, but not as much as the Soundmaster. The Tekesound was preferred above all the others. Which portable stereo was least preferred? **Blaster**

10. ❓ **What's the Question?** Glen read 69 pages each day for 6 days. He then read 23 pages each day for 4 days. The answer is about 500.
    Possible question: About how many pages did Glen read altogether?

57

3 Practice

Guided Practice

Note that Exercises 1–4 are **multistep or strategy problems**. Do Problem Solving Practice Exercises 1–4 with your students. Identify those having difficulty and use lesson resources to help.

In the data for Exercises 1–2, point out that expressing fractions here as decimals does not change the fact that they are fractions. Demonstrate this point by showing that 0.725 represents $\frac{725}{1,000}$.

Independent Practice

Note that Exercises 5–10 are **multistep or strategy problems**. Assign Exercises 5–10.

Scaffolded Instruction Use the prompts on Transparency 3 to guide instruction for the multistep or strategy problem in Exercise 7.

Encourage students to use the strategy *make a table* in Exercise 8 to show how many blocks Robert walks if Frank walks 5 blocks, 10 blocks, and so on. They should then record the total of each combination until their total is 24.

4 Assess

Summarize the lesson by having students:

DISCUSS Talk about how you can make a table to find the best buy among six boxes of cereal. Determine the unit price, or price per ounce, for each kind and select the one with the least cost per ounce.

✏️ **WRITE** Compose a problem that can be solved by using a table. Ask another student to solve the problem, and then check your classmate's work.
Check students' work.

Lesson Quiz

1. Clarice spent $3.45, $5.35, $2.55, $4.15, and $4.80 on lunches. Which lunch cost both more than and less than two of the other lunches? **the mid-priced lunch of $4.15**

2. Sean spent $39 on 10 pens and 8 notebooks. The pens cost half as much as the notebooks. How much did he spend on each pen and on each notebook? **$1.50 per pen and $3 per notebook**

CHALLENGE 3.2

Come Fly with Me!

In 1995, about 47,000,000 visitors arrived in the United States by airplane. This was about 4,000,000 more than the number of Americans who traveled by airplane from the United States to other countries. Some of the countries from which the greatest number of visitors came were Canada (7,262,000), France (2,045,000), Germany (3,125,000), Japan (5,676,000), Mexico (4,884,000), and the United Kingdom (6,648,000). Complete the table below, ordering the countries from greatest number of visitors to the United States to the least number. Then use the table to do 2–6.

Airline Passenger Arrivals in the United States

Country	Number of Visitors
Canada	7,262,000
United Kingdom	6,648,000
Japan	5,676,000
Mexico	4,884,000
Germany	3,125,000
France	2,045,000

1. Approximately how many Americans left the United States by airplane to visit other countries in 1995? **about 43,000,000 Americans**

2. From which two countries came the greatest number of visitors? **Canada and the United Kingdom**

3. About how many total visitors were there from these two countries? **about 14,000,000 visitors**

4. About how many visitors flew to the United States from countries other than those included in the table? **Possible answer: about 17,000,000 visitors**

5. In 1995, about 1,580,000 people flew to the United States from the Netherlands. About how many fewer people came from the Netherlands than from the last country listed in the table? **Possible answer: about 500,000 visitors**

6. About three times as many visitors flew to the United States from France in 1995 as flew here in 1990. Approximately how many visitors flew to the United States from France in 1990? **Possible answer: about 700,000 visitors**

READING STRATEGY 3.2

Use Graphic Aids   Analyze Choose Solve Check

You have used **graphic aids** such as tables to find information. You can make a table to organize data with numbers to help you solve problems. Read the following problem.

VOCABULARY graphic aids

Five friends have saved different amounts of money. Bob has $18.94; Dot, $25.37; Carol, $9.59; Ruth, $34.75; and Ann, $12.38. Who has saved the second greatest amount of money? the second least amount?

1. Order the data in the table below to make the problem easier to solve.

Name	Amount Saved
Ruth	$34.75
Dot	$25.37
Bob	$18.94
Ann	$12.38
Carol	$9.59

2. Solve the problem. **Dot has saved the second greatest amount; Ann has saved the second least amount.**

3. Explain the strategy you used to solve the problem. **I ordered the data in a table. Then I could easily see who saved the second greatest and second least amounts of money.**

Reorder the data in the table to solve.

MR. FRENCH'S OFFICE	
Equipment	Price
scanner	$299
copy machine	$1,769
printer	$995
phone system	$488
computer	$2,500
fax machine	$547

GIRLS' BASKETBALL	
Team	Games Won and Lost
Diamonds	1 win, 3 losses
Tigers	0 wins, 4 losses
Hawks	3 wins, 1 loss
Astros	2 wins, 2 losses
Rubies	1 win, 3 losses

4. Mr. French is buying new office equipment. The store requires him to pay for the least and most expensive items in advance. How much does he have to pay for? **$2,799**

5. There are five girls' basketball teams in the district. Which team is in second place? **Astros**

57

LESSON 3.3

Estimate with Decimals

LESSON PLANNING

Objective To estimate decimal sums, differences, products, and quotients

Intervention for Prerequisite Skills
Round Decimals (For intervention strategies, see page 51.)

NCTM Standards
1. Number and Operations
6. Problem Solving
8. Communication

Math Background
Remind students of what they learned about estimating with whole numbers as they begin to estimate with decimals:

- Students may estimate by rounding, clustering, and by using compatible numbers.
- As students gain estimating experience, they will learn to determine the best strategy for different situations.
- Estimation with decimals helps to determine if an answer is a reasonable one. An estimate may also be used when it is not necessary to obtain an exact answer.

WARM-UP RESOURCES

 **NUMBER OF THE DAY**

Estimate the product of your age in months and the number of days in a month. What does your answer represent? **Possible answer: an estimate of your age in days**

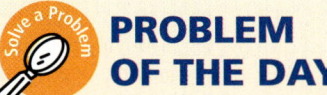 

Randall, Kira, and Sean have 100 baseball cards in all. Randall has twice as many as Kira and 10 more than Sean. How many cards does each person have? **Randall has 44 cards, Kira has 22 cards, and Sean has 34 cards.**

Solution Problem of the Day tab, p. PD3

Have students practice subtraction facts by completing Set G of *Teacher's Resource Book,* p. TR94.

INTERVENTION AND EXTENSION RESOURCES

ALTERNATIVE TEACHING STRATEGY — ELL

Materials For each group 3 $10-bills, 6 $1-bills, 16 dimes, and 25 pennies in play money

Have students **model each decimal** in Example 1 on page 58 by using the play money. By looking at each model, students should be able to see more clearly that each amount is close to $2, and so using clustering is sensible. Then have them model amounts such as $11.35, $10.72, and $11.22 and estimate the sum by clustering. $3 \times \$11 = \$33$

KINESTHETIC
LOGICAL/MATHEMATICAL

MULTISTEP AND STRATEGY PROBLEMS

The following multistep and strategy problems are provided in Lesson 3.3:

Page	Item
59	27, 28

WRITING IN MATHEMATICS

Have each student **apply mental math strategies** by choosing one of the following: rounding, clustering, or compatible numbers. Then ask them to write a description of when they might use the strategy and how they would apply it. Possible answer: clustering; I would apply it when there are several numbers very close in value. I would pick one number close to the given numbers and multiply by the number of values.

EARLY FINISHERS

Materials For each pair spinner, labeled 0–9, p. TR73

To **reinforce estimation,** give each pair of students a spinner. The first student spins it four times, recording the digit spun each time. The second player makes two decimal numbers from the digits. Students take turns estimating the sum, difference, and product of the two numbers formed. Students then switch roles and repeat the activity. Check students' work.

VISUAL
INTERPERSONAL/SOCIAL

TECHNOLOGY LINK
Intervention Strategies and Activities CD-ROM • *Skill 35*

LESSON 3.3 ORGANIZER

Objective To estimate decimal sums, differences, products, and quotients

1 Introduce

QUICK REVIEW provides review of prerequisite skills.

Why Learn This? Decimal estimates will help you determine a car's gas mileage. *Share the lesson objective with students.*

2 Teach

Guided Instruction

• *Review the estimating strategies.*

Why would you use clustering in Example 1? All the numbers are close to 2.

Explain how you could use another estimation strategy in Example 1. Round each charge to the nearest dollar and add.

In Example 2B, why choose 160 as one of the compatible numbers? 8 divides evenly into 16 and 160 is close to 162.

ADDITIONAL EXAMPLES

Example 1, p. 58

Jasmine bought a small salad for $2.85, a chicken wrap for $3.29, and a fruit smoothie for $3.25. About how much did she pay for the meal? about $9.00

Example 2, p. 58

Estimate.

A. 57.6 × 13.21
Round to the nearest ten. 60 × 10 = 600
So, 57.6 × 13.21 ≈ 600.

B. 249.8 ÷ 3.81
Use compatible numbers. 240 ÷ 4 = 60
So, 249.8 ÷ 3.81 ≈ 60.

58 Chapter 3

LESSON 3.3 Estimate with Decimals

Learn how to estimate decimal sums, differences, products, and quotients.

QUICK REVIEW
Round each number to the nearest 100.
1. 346 **300**   2. 2,506 **2,500**   3. 54,067 **54,100**
4. 30,463 **30,500**   5. 156,556 **156,600**

You can estimate sums, differences, products, and quotients of decimals. To estimate with decimal numbers, use the same methods you used to estimate with whole numbers.

EXAMPLE 1

A long-distance phone company charges the rates shown at the right for calls from the United States. DeAnn made one-minute calls to India, Jordan, and Pakistan. About how much did the three calls cost?

Estimate $1.79 + $1.87 + $2.17.

$1.79
$1.87
+$2.17   *The three addends cluster around $2.00. So, multiply $2.00 by 3.*

3 × $2.00 = $6.00 *Multiply.*

So, the three calls cost about $6.00.

WIRED WORLD PHONE COMPANY	
Country	Rate per Minute
Argentina	$0.39
China	$0.49
France	$0.13
Germany	$0.09
India	$1.79
Ireland	$0.15
Jordan	$1.87
Pakistan	$2.17

EXAMPLE 2

Remember that the symbol ≈ means "is approximately equal to."

Estimate.

A. 36.4 × 18.25
Round to the nearest ten.

36.4   40
×18.25 →   ×20
            800

So, 36.4 × 18.25 ≈ 800.

B. 162.8 ÷ 8.16
Use compatible numbers.

                20
8.16)162.8 → 8)160

So, 162.8 ÷ 8.16 ≈ 20.

CHECK FOR UNDERSTANDING

Think and Discuss Look back at the lesson to answer each question.

1. **Tell** how you could estimate the sum of 4.79, 18.99, and 3.09.
   Possible answer: Round each number to the nearest whole number and add.

2. **Explain** how to use compatible numbers to estimate 423.2 ÷ 2.7.
   Possible answer: Use 420 and 3; 420 ÷ 3 = 140.

58 Chapter 3

RETEACH 3.3

Estimate with Decimals

When you estimate with decimals, you want to get an idea of the size of the result. Working with whole numbers can help you estimate quickly.

Estimate 19.7 + 40.13 + 100.4.
Step 1 Round each number to the nearest whole number.
19.7 → 20
40.13 → 40
100.4 → 100
Step 2 Add the rounded numbers.
20 + 40 + 100 = 160
So, a good estimate for the sum is 160.

Estimate $78.31 − $49.47.
Step 1 Round to the nearest ten.
$78.31 → $80
$49.47 → $50
Step 2 Subtract the rounded numbers.
$80 − $50 = $30
So, a good estimate for the difference is $30.

Estimate 62.88 × 28.97.
Step 1 Round to the nearest ten.
62.88 → 60
28.97 → 30
Step 2 Multiply the rounded numbers.
60 × 30 = 1,800
So, a good estimate for the product is 1,800.

Estimate 54.67 ÷ 8.56.
Step 1 Find compatible numbers close to those given.
54.67 → 54
8.56 → 9
Step 2 Divide the compatible numbers.
54 ÷ 9 = 6
So, a good estimate for the quotient is 6.

Estimate. **Possible estimates are given.**
1. 19.82 + 51.5 **70**
2. 149.2 ÷ 23.8 **6**
3. 784.49 − 610.88 **200**
4. 39.66 × 6.75 **280**
5. 1003.2 − 796.1 **200**
6. 7.86 + 10.03 **18**
7. 98.15 ÷ 8.23 **800**
8. 82.88 ÷ 9.31 **9**
9. 38.8 × 9.12 **360**
10. 161.10 ÷ 7.84 **20**
11. 108.46 + 392.54 **500**
12. 80.55 − 67.86 **10**
13. 57.93 × 21.5 **1,200**
14. 119.4 + 42.3 **3**
15. 53.3 + 39.2 **90**
16. 48.28 ÷ 6.82 **8**
17. 28.7 × 61.75 **1,800**
18. 982.3 − 498.7 **500**
19. 411.9 + 298.34 + 128.6 **800**
20. $49.28 + $32.61 + $18.95 **$100**

PRACTICE 3.3

Estimate with Decimals

Estimate. **Possible estimates are given.**
1. 3.8 + 7.9 **12**
2. 7.1 × 6.2 **42**
3. 23.18 − 19.09 **4**
4. 12.2 ÷ 5.9 **2**
5. 4.09 × 6.18 **24**
6. 83.89 + 17.66 **102**
7. 162.3 − 15.7 **10**
8. 31.6 − 8.23 **23**
9. 7.7 + 118.2 **126**
10. 101.2 − 34.9 **66**
11. $35.99 − $6.02 **$30**
12. 19.8 × 21.3 **400**
13. $124.66 × 3 **$375**
14. 10.6 + 19.01 **30**
15. 81.3 × 9.6 **800**
16. 810.1 − 69.9 **740**
17. 602.5 + 87.3 **690**
18. 397.9 × 21 **8,000**
19. 502.03 ÷ 4.9 **100**
20. $88.20 + $79.10 **$170**
21. 1.8 + 2.9 + 11.8 **17**
22. $203.99 ÷ 21 **$10**
23. $199.50 − $53.99 **$145**
24. 8.8 × 7.1 **63**
25. 67.2 + 11.9 + 107.44 **190**
26. 889.52 − 402.68 **490**

Mixed Review
Write in exponent form.
27. 4 × 4 × 4 4^3
28. 2 × 2 × 2 × 2 2^4
29. 6 × 6 6^2
30. 1 × 1 × 1 × 1 × 1 1^5
31. 7 × 7 × 7 × 7 7^4
32. 8 × 8 × 8 8^3
33. 9 × 9 × 9 9^3
34. 3 × 3 × 3 × 3 3^4

Find the value.
35. 5^2 **25**
36. 2^5 **32**
37. 8^2 **64**
38. 1^4 **1**

Guided Practice — Estimate. Possible estimates are given.

3. 18.7 + 23.1
   40
4. 123.76 ÷ 9
   12
5. 185.32 − 101.99
   90
6. 39.83 × 36
   1,600
7. 67.8 + 66.1 + 71.7
   210
8. 817.3 × 11
   8,000

PRACTICE AND PROBLEM SOLVING

Independent Practice — Estimate. Possible estimates are given.

9. 6.7 + 9.4 + 15.82
   32
10. 12.2 × 8.3
    100
11. 82.5 ÷ 9.3
    9
12. $266.08 − $97.30
    $170
13. 9.8 + 38.2
    50
14. 31.5 × 2.8
    90
15. 6.8 × 18.2
    140
16. 103.08 − 45.32
    50
17. 56.20 × 30.7
    1,800
18. 689.89 − 98.5
    590
19. 1,038.54 × 26.12
    26,000
20. 234.91 ÷ 5.79
    40
21. $7,805.90 + 9,158.43
    $17,000
22. 81.5 × 23.1 ÷ 3.9
    400
23. 18.2 × (7.2 − 4.9)
    36

Estimate to compare. Write < or > for each ●.

24. 4.32 × 8.56 ● 40 **<**
25. 25 ● 81.27 ÷ 4.1 **>**
26. 34.6 − 12.4 ● 14 **>**

Problem Solving Applications — Use Data For 27–29, the table shows the types of waste that make up a typical 100 pounds of garbage in the United States.

27. About how many pounds of newspapers and other paper are included in every 300 pounds of garbage thrown away? **about 120 lb**

28. Possible answer: about 20 lb more

28. About how many more pounds of food waste than glass are thrown away for every 500 pounds of garbage?

What We Throw Away

Type of Garbage	Weight (in lb)
Food Waste	8.6
Yard Waste	17.4
Newspapers	7.9
Other Paper	30.8
Metals	9.2
Glass	5.2
Other	20.9

29. ✎ Write a problem involving estimation. Use the data about yard waste, other paper, and metals shown in the table. **Check students' problems.**

MIXED REVIEW AND TEST PREP

30. Write the value of the digit 8 in the number 342.285. (p. 52) **8 hundredths**
31. Write 2.523, 2.325, 2.532, and 2.235 in order from least to greatest. (p. 52)
    2.235, 2.325, 2.523, 2.532
32. 20,817 − 19,805 (p. 20) **1,012**
33. 25,801 ÷ 23 (p. 22) **1,121 r18**
★ 34. **TEST PREP** Evaluate $a \times 32$ for $a = 426$. (p. 28) **D**

    A 472   B 2,130   C 13,522   D 13,632

EXTRA PRACTICE page H34, Set B

59

LESSON 3.4 Decimals and Percents

LESSON PLANNING

Objective To write a decimal as a percent and a percent as a decimal

Intervention for Prerequisite Skills
Represent Decimals, Write and Read Decimals (For intervention strategies, see page 51.)

Materials *For each student* two 10 × 10 grids (decimal squares), p. TR7

NCTM Standards
1. Number and Operations
6. Problem Solving
8. Communication
10. Representation

Vocabulary
percent "per hundred"

Math Background

Most students are familiar with using percents to relay information. Percents are easy to use to compare information because they are based on a common number, 100. The word *percent* has a Latin origin meaning "per hundred." Based on this definition, percents can be:

- shown on a 10 × 10 grid (decimal square).
- written as decimals.
- written as fractions.

Students reinforce their understanding of place value when they write percents as decimals and one- or two-place decimals as percents. They will build on these skills to write percents as fractions and fractions as percents.

WARM-UP RESOURCES

 **NUMBER OF THE DAY**

Think of the school's grading system. At least what percent of a 100-point test must you answer correctly to make an A? Answers will vary.

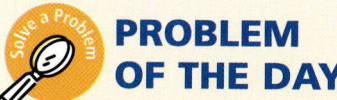 

An estimate of the sum of two decimals is 27 and an estimate of the product is 140. Give two decimals that satisfy these requirements. Possible answer: 6.85 and 19.61

Solution Problem of the Day tab, p. PD3

Have students practice addition facts by completing Set A of *Teacher's Resource Book*, p. TR95.

60A Chapter 3

INTERVENTION AND EXTENSION RESOURCES

ALTERNATIVE TEACHING STRATEGY

Materials *For each student* at least one 10 × 10 grid (decimal square), p. TR7

To **model the concept of decimals,** give each student one or more 10 × 10 grids. Ask students to make designs on the grids, using only whole squares. Suggest students make animals, letters, patterns, or other designs of their choosing. When they have finished, ask each person to determine what percent of the grid is taken up by the design. Ask them to write the number on the back of the design.

Students can then exchange designs, find the percent of the grid covered by the design, and compare his or her answer to the one written on the back. Check students' work.

VISUAL
LOGICAL/MATHEMATICAL

ENGLISH LANGUAGE LEARNERS

Materials *For each pair* 100 pennies and 1 $1-dollar bill of play money

Help students **model the concept of percent.** Give them play money (pennies and dollars) and talk about the fact that there are 100 pennies in a dollar. Then have students count out 18 pennies and write the money amount as a decimal and percent of a dollar. $0.18; 18%

Repeat the activity with other amounts, having students model the amounts and write the money amounts as decimals and as percents of a dollar. Check students' work.

Ask students to describe other monetary systems they may know about (such as peso, centavo, yen, or sen) that also have 100 smaller units in 1 larger unit.

KINESTHETIC
INTERPERSONAL/SOCIAL

MULTISTEP AND STRATEGY PROBLEMS

The following multistep or strategy problem is provided in Lesson 3.4:

Page	Item
61	24

SOCIAL STUDIES CONNECTION

Encourage students to **relate percents and decimals.** According to a recent poll, teens want to be popular, but they really want to be remembered for their achievements. In response to the question *What title would you like to see under your yearbook picture?*, 54% said *most likely to succeed.* Of the rest, 18% would like to be *class valedictorian,* 11% would like to be *best athlete,* 8% would like to be *class clown,* and 5% would like to be *most popular.*

Have students write each percent as a decimal. 0.54, 0.18, 0.11, 0.08, 0.05

VISUAL
LOGICAL/MATHEMATICAL

Intervention Strategies and Activities CD-ROM • *Skills 32, 33*

LESSON 3.4 ORGANIZER

Objective To write a decimal as a percent and a percent as a decimal

Vocabulary percent

Materials For each student two 10 × 10 grids (decimal squares), p. TR7

1 Introduce

QUICK REVIEW provides review of pre-requisite skills.

Why Learn This? Writing a decimal as a percent helps you convert the number of correct answers on a test to a letter grade. *Share the lesson objective with students.*

2 Teach

Guided Instruction

- Discuss the circle graph showing the results of the survey.

  About what fraction of those surveyed eat cold cereal? How do you know? About $\frac{1}{4}$; about one-fourth of the circle is used to represent that category.

  What percent represents about $\frac{1}{3}$ of the circle and therefore about $\frac{1}{3}$ of those surveyed? 34%

- Reinforce students' understanding of the word *percent*.

  What is another example of using *cent* to mean "hundredth"? In our money system, 100 cents equal 1 dollar.

  How are data from surveys usually reported? usually in percents

REASONING **Would you write 4% as 0.4 or 0.04? Explain.** 0.04; both 4% and 0.04 are the same as 4 hundredths, but 0.4 is 4 tenths.

ADDITIONAL EXAMPLES

Example A, p. 60

Write 0.02 as a percent.

0.02 is 2 hundredths. 0.02 = 2%

Example B, p. 60

Write 81% as a decimal.

81% is 81 hundredths. 81% = 0.81

60 Chapter 3

LESSON 3.4 Decimals and Percents

Learn how to write a decimal as a percent and a percent as a decimal.

QUICK REVIEW
1. 2,457 + 4,541 **6,998**
2. 3,470 − 350 **3,120**
3. 6 × 5 **30**
4. 50 ÷ 5 **10**
5. Write the decimal for $\frac{2}{100}$. **0.02**

Vocabulary
percent

The graph at the right shows the responses to a question about what people in the United States like to eat for breakfast.

BREAKFAST FOODS

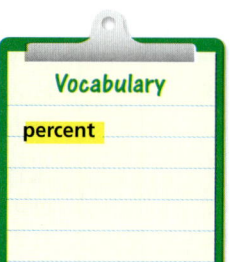

Percent means "per hundred" or "hundredths." The symbol used to write a percent is %.

40 percent: $40\% = \frac{40}{100}$

So, 40 out of 100 have toast or a roll.

26 percent: $26\% = \frac{26}{100}$

So, 26 out of 100 have cold cereal.

Activity

MATH LAB

You need: two 10 × 10 grids (decimal squares)

- On one grid, shade complete squares to make the first letter of your first name. On the other grid, shade complete squares to make the first letter of your last name. Make each letter as large as possible. Some examples are shown below.

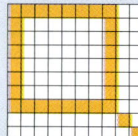 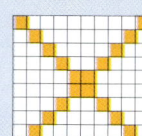 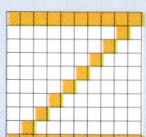

- Since 30 out of 100 squares are shaded for the letter Q, you can write 0.30 or 30%. What decimals and percents can be written for the squares shaded for X and Z? **X: 0.20, 20%; Z: 0.28, 28%**

- Count the number of complete squares you shaded on each of your grids. What percent of the squares are shaded? **Check students' grids.**

You can think about place value when you change decimals to percents or percents to decimals.

Remember that when you read a decimal, you name the place with the least value. For 0.93, the place with the least value is hundredths. The number is read as "93 hundredths."

EXAMPLES

A. Write 0.08 as a percent.
0.08 is 8 hundredths.
So, 0.08 = 8%.

B. Write 32% as a decimal.
32% is 32 hundredths.
So, 32% = 0.32.

60 Chapter 3

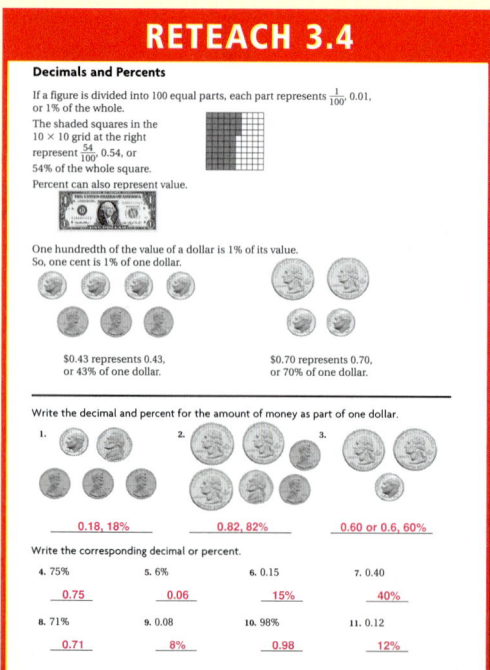

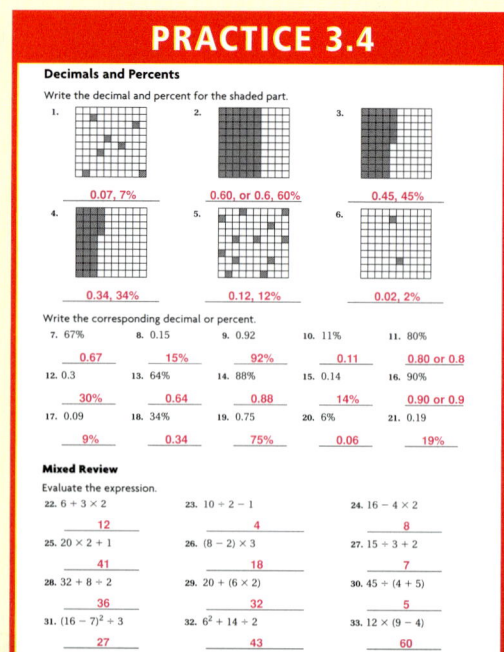

Booksavers of Ephrata
1054 South State St.
Ephrata, PA 17522
UNITED STATES
books@dejazzd.com

To: Jeff Chung
3 Fifth Street
Closter, NJ 07624
UNITED STATES

Marketplace:	Amazon US
Order Number:	1059347
Ship Method:	Standard
Customer Name:	Jeff Chung
Order Date:	3/1/2016
Marketplace Order #:	116-3267059-2875427
Email:	9gbzq1jk5f673s5@marketplace.amazon.com

Items:

Qty	Item	Locator	Condition	Price
1	Te Grade 6 Vol1 Harcourt Math 2002 HSP SKU: mon0000073712 ISBN: 0153207620 - Books	C2	Like New	$9.39

Subtotal:	$9.39
Shipping:	$3.99
Total Tax:	$0.00
Total:	$13.38

Notes:

Thanks for your order!
If you have any questions or concerns regarding this order, please contact us at books@dejazzd.com

Booksavers of Ephrata
1054 South State St.
Ephrata, PA 17522
UNITED STATES

CHECK FOR UNDERSTANDING

Think and Discuss
Look back at the lesson to answer the question.
1. **Discuss** how you know that Sharon has 18 red cars if 18% of her 100 model cars are red. *18% means 18 per hundred, so 18 cars are red.*

Guided Practice
Write the decimal and percent for the shaded part.

2.
0.35, 35%

3.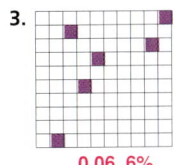
0.06, 6%

4.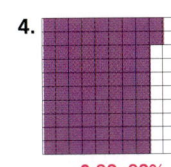
0.82, 82%

Write the corresponding decimal or percent.
5. 70% **0.7, or 0.70**
6. 0.20 **20%**
7. 0.03 **3%**
8. 84% **0.84**
9. 50% **0.5, or 0.50**

PRACTICE AND PROBLEM SOLVING

Independent Practice
Write the decimal and percent for the shaded part.

10.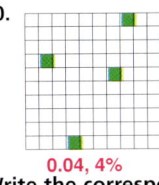
0.04, 4%

11.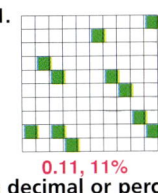
0.11, 11%

12.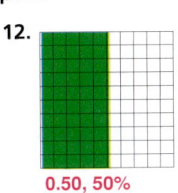
0.50, 50%

Write the corresponding decimal or percent.
13. 62% **0.62**
14. 0.05 **5%**
15. 28% **0.28**
16. 45% **0.45**
17. 53% **0.53**
18. 0.63 **63%**
19. 0.85 **85%**
20. 33% **0.33**
21. 0.4 **40%**
22. 7% **0.07**

Favorite Types of Music
Alternative	21
Rhythm and Blues	14
Rap	21
Rock	20
Other or none	24

Problem Solving Applications
23. **Use Data** The table shows how 100 teens responded to a survey. Write a decimal and a percent to show the number of teens who did not choose Alternative music. **0.79, 79%**

24. What percent shows how many more students chose Rap and Rock music than Rhythm and Blues? **27%**

25. **Write About It** Explain how to write 0.6 as a percent. *Write 0.6 as 0.60 to show hundredths. Then write 60 hundredths as 60%.*

MIXED REVIEW AND TEST PREP

26. Order 27.8, 27.5, 27.82 from least to greatest. (p. 52) **27.5, 27.8, 27.82**
27. Evaluate. $35 + 17 \times 3^2 - 16$ (p. 44) **172**
28. 168×92 (p. 22) **15,456**
29. $3,470 \div 42$ (p. 22) **82 r26**
30. **TEST PREP** Which shows the sum $34,904 + 15,456 + 6,943$? (p. 20) **B**

A 55,920   B 57,303   C 68,870   D 72,780

EXTRA PRACTICE page H34, Set C

61

CHAPTER 3

REVIEW/TEST

Purpose To check understanding of concepts, skills, and problem solving presented in Chapter 3

USING THE PAGE

The Chapter 3 Review/Test can be used as a **review** or a **test**.

- Item 1 checks understanding of concepts and new vocabulary.
- Items 2–28 check skill proficiency.
- Items 29–33 check students' abilities to choose and apply problem solving strategies to real-life decimal problems.

 Suggest that students place the completed Chapter 3 Review/Test in their portfolios.

USING THE ASSESSMENT GUIDE

- Multiple-choice format of Chapter 3 Posttest—See *Assessment Guide*, pp. AG17–18.
- Free-response format of Chapter 3 Posttest—See *Assessment Guide*, pp. AG19–20.

USING STUDENT SELF-ASSESSMENT

The How Did I Do? survey helps students assess what they have learned and how they learned it. This survey is available as a copying master in *Assessment Guide*, p. AGxvii.

CHAPTER 3 REVIEW/TEST

1. **VOCABULARY** A word that means "per hundred" is __?__. (p. 60) **percent**

Write the value of the blue digit. (pp. 52–55)

2. 3.24**9**7 **nine thousandths** 3. 14.**5**805 **five tenths** 4. 0.0**9**003 **nine hundredths**

5. 628.040**2** **two ten-thousandths** 6. 1.817**3**8 **three ten-thousandths** 7. 78.05**1**24 **one thousandth**

Write the numbers in order from least to greatest. (pp. 52–55)

8. 2.365, 2.305, 2.3, 2.35, 2.035
   2.035; 2.3; 2.305; 2.35; 2.365
9. 125.3, 124.32, 125.33, 12.245, 120.4
   12.245; 120.4; 124.32; 125.3; 125.33

Estimate. (pp. 58–59) **Possible estimates are given.**

10. 27.6 + 135.2 **165** 11. 4.8 × 2.3 **10** 12. 30.7 − 6.25 **24** 13. 89.75 ÷ 8 **11**

14. 8.45 + 8.99 + 9.2 **27** 15. 219.48 − 107.43 **100** 16. 416.2 × 31 **12,000** 17. 40.02 ÷ 6.3 **7**

Write the decimal and percent for the shaded part. (pp. 60–61)

18. 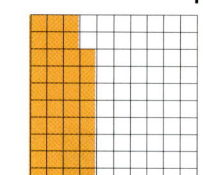 **0.38; 38%** 19. **0.60; 60%** 20. 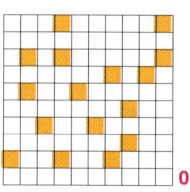 **0.16; 16%**

Write the corresponding decimal or percent. (pp. 60–61)

21. 74% **0.74** 22. 0.07 **7%** 23. 39% **0.39** 24. 0.61 **61%**

25. 0.6 **60%** 26. 3% **0.03** 27. 0.04 **4%** 28. 84% **0.84**

Solve.

29. The county library charges a fine of $0.10 a day for overdue books. The university library charges a fine of $0.50 for the first day and $0.05 for each additional day. On what day would overdue books have the same fine at both libraries? (pp. 56–57) **on the ninth day**

30. Planes leave Washington, D.C., for New York City every 45 min. The first plane leaves at 5:45 A.M. What is the departure time closest to 4:30 P.M.? (pp. 56–57) **4:15 P.M.**

31. Donna is on the decoration committee. She spent $15.90 on streamers, $12.15 on balloons, $6.84 on tape, $19.98 on banner paper, and $13.22 on banner paint. What is a reasonable estimate of the amount she spent? (pp. 58–59) **Possible answer: about $68.00**

32. Frank earns $6.25 per hour. One week he worked 16 hours. About how much did Frank earn that week? (pp. 58–59) **Possible answer: about $96**

33. Kirk has to list 125.3, 124.32, 125.33, 12.345, 120.4 in order from greatest to least. Which number should he list third? (pp. 52–55) **124.32**

CHAPTERS 1–3 ★ STANDARDIZED TEST PREP

 Get the information you need.
See item **7**.
Recall that a percent is the ratio of a number to 100. Write the ratio as a decimal.
Also see problem **3**, p. H63.

Choose the best answer.

1. Which of these numbers rounds to 450 when rounded to the nearest ten and to 500 when rounded to the nearest hundred? **C**

   A 415 C 452
   B 428 D 478

2. Earth's orbit is more than 100,000,000 kilometers from the sun. How is this number written in exponential notation? **G**

   F 10^7 km   H 10^9 km
   G 10^8 km   J 10^{10} km

3. What is the value of $2 + 3 \times 4$? **B**

   A 11   C 20
   B 14   D Not here

4. Each of the 42 members of the band contributed $3.75 toward a gift for the band director. Which is a reasonable estimate of the total amount collected? **J**

   F Between $80 and $90
   G Between $90 and $110
   H About $120
   J About $160

5. Kate has 4 bags of birdseed that have masses of 2.5 kilograms, 1.25 kilograms, 1.9 kilograms, and 2.15 kilograms. Which shows the bags in order from least mass to greatest? **D**

   A 1.9 kg, 2.15 kg, 2.5 kg, 1.25 kg
   B 2.5 kg, 2.15 kg, 1.9 kg, 1.25 kg
   C 2.15 kg, 1.25 kg, 2.5 kg, 1.9 kg
   D 1.25 kg, 1.9 kg, 2.15 kg, 2.5 kg

6. A restaurant manager bought a total of 73 apples and oranges. She bought 11 more oranges than apples. How many of each kind of fruit did she buy? **F**

   F 31 apples and 42 oranges
   G 27 apples and 46 oranges
   H 36 apples and 37 oranges
   J 42 apples and 31 oranges

7. How is 6% written as a decimal? **C**

   A 6.0   B 0.6   C 0.06   D 0.006

8. Which is greater than 16.30? **F**

   F 16.45   H 16.03
   G 16.23   J 1.730

9. Which is the value of $(3 - 2) \times 5 + 6^2$? **C**

   A 29   B 40   C 41   D 122

10. Which of these is in order from least to greatest? **F**

    F 0.0310, 0.301, 0.310
    G 0.310, 0.0310, 0.301
    H 0.301, 0.0310, 0.310
    J 0.0310, 0.310, 0.301

Write What You Know

11. With tax, a CD player costs $77.75. Ken saves $4.85 each week. Find a reasonable estimate of the number of weeks he must save for the new CD player. Explain your method.

12. Explain how you can write a decimal as a percent. Use your method to write 0.36 as a percent.

STANDARDIZED TEST PREP • Chapters 1–3

USING THE PAGE

This page may be used to help students get ready for standardized tests. The test items are written in the same style and arranged in the same format as those on many state assessments. The page is cumulative. It covers math objectives and essential skills that have been taught up to this point in the text. Most of the items represent skills from the current chapter and the remainder represent skills from earlier chapters.

This page can be assigned at the end of the chapter as classwork or as a homework assignment. You may want to have students use individual recording sheets presented in a multiple choice (standardized) format. A Test Answer Sheet is available as a blackline master in *Assessment Guide* (p. AGxlii).

You may wish to have students describe how they solved each problem and share their solutions.

ITEM ANALYSIS

Item	Learning Goal	Item	Learning Goal
1	Grade 5	7	3C
2	2B	8	3A
3	2C	9	2C
4	3B	10	3A
5	3A	11	3D
6	1D	12	3C

Written Response items for Write What You Know are available as a blackline master in *Performance Assessment*.

SCORING RUBRIC • WRITE WHAT YOU KNOW

2 Demonstrates a complete understanding of the problem and chooses an appropriate strategy to determine the solution

1 Demonstrates a partial understanding of the problem and chooses a strategy that does not lead to a complete and accurate solution

0 Demonstrates little understanding of the problem and shows little evidence of using any strategy to determine a solution

---

Write What You Know • Written Response

11. Round the cost to the nearest ten dollars and the amount saved to the nearest dollar: $77.85 ≈ $80 and $4.85 ≈ $5. The numbers 80 and 5 are compatible. Use division to estimate: 80 ÷ 5 = 16. So, it will take Ken about 16 weeks.

12. Possible explanation: Think about place value when you write a decimal as a percent. Since percent is "per hundredths," think of the decimal in hundredths. The number of hundredths is the percent you are looking for. 0.36 is 36 hundredths. So, 0.36 = 36%.

Decimal Concepts

CHAPTER 4 Decimal Operations

CHAPTER PLANNER

PACING OPTIONS
| Compacted | 4 Days |
| Expanded | 9 Days |

Getting Ready for Chapter 4 • Assessing Prior Knowledge and INTERVENTION (See PE and TE page 65.)

LESSON	NCTM STANDARDS	PACING	VOCABULARY*	MATERIALS	RESOURCES AND TECHNOLOGY
4.1 Add and Subtract Decimals pp. 66–69 **Objective** To add and subtract decimals	1, 2, 6, 8, 9	2 Days			Reteach, Practice, Challenge, Problem Solving 4.1 Worksheets Extra Practice p. H35, Set A Transparency 4.1 **Calculating Crew** • *Nautical Number Line*
4.2 Multiply Decimals pp. 70–73 **Objective** To multiply decimals	1, 6, 8, 10	1 Day		For each *student* decimal squares, colored pencils	Reteach, Practice, Challenge, Problem Solving 4.2 Worksheets Extra Practice p. H35, Set B Transparency 4.2 **Calculating Crew** • *Nautical Number Line* Calculator Handbook p. 5
4.3 Math Lab: Explore Division of Decimals pp. 74–75 **Objective** To use a model to divide decimals	1, 8, 10			For each *student* decimal squares, colored pencils, scissors	E-Lab • *Exploring Division of Decimals*; E-Lab Recording Sheet Math Jingles™ CD 5–6
4.4 Divide with Decimals pp. 76–79 **Objective** To divide a decimal by a whole number and a decimal by a decimal	1, 6, 8	2 Days (For Lessons 4.3 and 4.4)			Reteach, Practice, Challenge, Problem Solving 4.4 Worksheets Extra Practice p. H35, Set C Transparency 4.4 **Number Heroes** • *Quizzo* Calculator Handbook p. 6
4.5 Problem Solving Skill: *Interpret the Remainder* pp. 80–81 **Objective** To solve problems by using the skill *interpret the remainder*	1, 6, 8	1 Day			Reteach, Practice, Challenge, Reading Strategy 4.5 Worksheets Transparency 4.5 Problem Solving Think Along, p. TR1 Math Jingles™ CD 5–6
4.6 Algebra: Decimal Expressions and Equations pp. 82–83 **Objective** To evaluate expressions with decimals and to use mental math and substitution to solve equations with decimals	1, 2, 6, 8	1 Day			Reteach, Practice, Challenge, Problem Solving 4.6 Worksheets Extra Practice p. H35, Set D Transparency 4.6 Math Jingles™ CD 5–6

Ending Chapter 4 • Chapter 4 Review/Test, p. 84 • **Standardized Test Prep,** p. 85
Ending Unit 1 • Math Detective, p. 86; **Challenge,** p. 87; **Study Guide and Review,** pp. 88–89; **Performance Assessment,** p. 90; **Technology,** p. 91; **Problem Solving: On Location,** pp. 91A–91B

*****Boldfaced** terms are new vocabulary. Other terms are review vocabulary.

CHAPTER AT A GLANCE

Vocabulary Development

There are no new words introduced in this chapter, but mathematics vocabulary is reinforced visually and verbally. Encourage students to review the mathematics vocabulary in their math journals.

Writing Opportunities

PUPIL EDITION
- Write a Problem, p. 73
- What's the Error?, pp. 79, 83
- Write About It, p. 69
- What's the Question?, p. 81

TEACHER'S EDITION
- Write—See the *Assess* section of each TE lesson.
- Writing in Mathematics, p. 74

ASSESSMENT GUIDE
- How Did I Do?, p. AGxvii

NCTM Standards

1. **Number and Operations**
   Lessons 4.1, 4.2, 4.3, 4.4, 4.5, 4.6
2. **Algebra**
   Lessons 4.1, 4.6
3. **Geometry**
4. **Measurement**
5. **Data Analysis and Probability**
6. **Problem Solving**
   Lessons 4.1, 4.2, 4.4, 4.5, 4.6
7. **Reasoning and Proof**
8. **Communication**
   Lessons 4.1, 4.2, 4.3, 4.4, 4.5, 4.6
9. **Connections**
   Lesson 4.1
10. **Representation**
    Lessons 4.2, 4.3

Family Involvement Activities

These activities provide:
- Letters to the Family
- Math Vocabulary
- Family Game
- Practice (Homework)

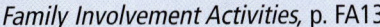

Family Involvement Activities, p. FA13

CHAPTER 4 Decimal Operations

MATHEMATICS ACROSS THE GRADES

SKILLS TRACE ACROSS THE GRADES

GRADE 5	GRADE 6	GRADE 7
Add, subtract, multiply, and divide with decimals	Use addition, subtraction, multiplication, and division to solve problems involving decimals; evaluate expressions and solve one-step equations with decimals	Use addition, subtraction, multiplication, and division to solve problems involving decimals; write and evaluate expressions using decimals; solve equations using decimals

SKILLS TRACE FOR GRADE 6

LESSON	FIRST INTRODUCED	TAUGHT AND PRACTICED	TESTED	REVIEWED
4.1	Grade 4	PE pp. 66–69, H35, p. RW15, p. PW15, p. PS15	PE p. 84, pp. AG21–24	PE pp. 84, 85, 88–89
4.2	Grade 5	PE pp. 70–73, H35, p. RW16, p. PW16, p. PS16	PE p. 84, pp. AG21–24	PE pp. 84, 85, 88–89
4.3	Grade 5	PE pp. 74–75	PE p. 84, pp. AG21–24	PE pp. 84, 85, 88–89
4.4	Grade 5	PE pp. 76–79, H35, p. RW17, p. PW17, p. PS17	PE p. 84, pp. AG21–24	PE pp. 84, 85, 88–89
4.5	Grade 4	PE pp. 80–81, p. RW18, p. PW18, p. PS18	PE p. 84, pp. AG21–24	PE pp. 84, 85, 88–89
4.6	Grade 6	PE pp. 82–83, H35, p. RW19, p. PW19, p. PS19	PE p. 84, pp. AG21–24	PE pp. 84, 85, 88–89

KEY   **PE** Pupil Edition   **PS** Problem Solving Workbook   **RW** Reteach Workbook
       PW Practice Workbook   **AG** Assessment Guide

Looking Back  Prerequisite Skills

To be ready for Chapter 4, students should have the following understandings and skills:

Whole-Number Operations—add, subtract, multiply, and divide whole numbers

Multiply Decimals by 10, 100, and 1,000—multiply one-, two-, and three-place decimals by 10, 100, and 1,000

Remainders—divide whole numbers and write remainders as decimals and as fractions

Check What You Know

Use page 65 to determine students' knowledge of prerequisite concepts and skills.

Intervention

Help students prepare for the chapter by using the intervention resources described on TE page 65.

Looking at Chapter 4  Essential Skills

Students will

- develop skill and accuracy adding, subtracting, multiplying, and dividing decimals.
- interpret remainders meaningfully in real-world division problems.
- **make the connections between models and the division algorithm for dividing decimals.**
- apply what they learned about whole number expressions and equations to decimal expressions and equations.

EXAMPLE

$1.2 \div 0.4$ or $0.4\overline{)1.2}$

Model	Algorithm
(grid model showing 0.4, 0.4, 0.4)	$0.4\overline{)1.2}$ Make the divisor a whole number.
	$\begin{array}{r}3.0\\4\overline{)12.0}\\\underline{12}\\0\end{array}$ Place the decimal point. Divide.

Looking Ahead  Applications

Students will apply what they learn in Chapter 4 to the following new concepts:

- Terminating and Repeating Decimals (Chapter 8)
- Solving Two-Step Equations (Chapter 15)
- Operations with Rational Numbers (Chapter 11)

Decimal Operations 64D

CHAPTER 4

Decimal Operations

INTRODUCING THE CHAPTER

Tell students that they can apply the same operations to decimals as to whole numbers. Have students read the page and examine the graph. Ask them how many gigabytes a megabyte equals. **0.001 gigabytes**

USING DATA

To begin the study of this chapter, have students

- Determine how many gigabytes of storage a computer with one Zip® disk and one Jaz® disk has. **1.1 GB**
- Express the approximate number of gigabytes each type of disk listed on the graph holds. **Zip® disk: 0.1; CD-ROM: 0.7; Jaz® disk: 1; DVD-RAM: 5.2**
- Make a graph showing the storage capacity of each disk in megabytes. **Check students' work.**

PROBLEM SOLVING PROJECT

Purpose To use decimals to solve a problem

Grouping pairs or small groups

Background A 4.75-inch disk can hold a gigabyte of information which is encoded and stored on a spiral track that is more than 4.8 kilometers long.

Analyze, Choose, Solve, and Check

Have students

- Find the total amount of storage capacity possible for 3 Zip® disks, 5 CD-ROM disks, 2 Jaz® disks, and 1 DVD-RAM disk. **10.75 GB**
- Make a graph to display the total storage capacity in gigabytes for the disks described above.

Check students' work.

 Suggest that students place the graphs in their portfolios.

CHAPTER 4 Decimal Operations

1,440,000 bytes

Today's computers use optical laser writing technology to store billions of bytes of information. A Jaz® disk holds 1 gigabyte of information. A gigabyte is about 1,000,000,000 (10^9) bytes. Sometimes capacity is given in megabytes. A megabyte is about 1,000,000 (10^6) bytes.

PROBLEM SOLVING A 3.5-inch floppy disk holds just 1.44 megabytes of information. About how many bytes is that?

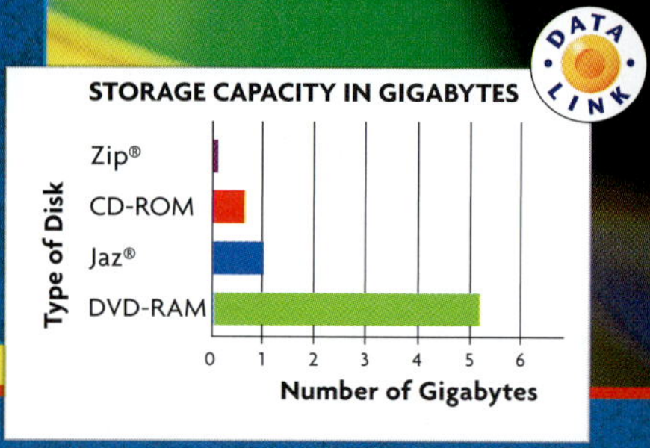

STORAGE CAPACITY IN GIGABYTES

64 Chapter 4

Why learn math? Explain that computer applications programmers write commercial programs used in businesses, schools, and homes. Programmers need to calculate the amount of space their programs will need in order to be stored on both the internal and external memory devices. Ask: How do you use decimal operations when buying computer programs for your computer? **Possible answer: to calculate the total cost of the computer program or supplies and the amount of change due**

Check What You Know

Use this page to help you review and remember important skills needed for Chapter 4.

✓ Whole-Number Operations (For Intervention, see p. H4.)

Add or subtract.

1. $7 + 28 + 12$ **47**
2. $45 - 15$ **30**
3. $63 - 19$ **44**
4. $19 + 41 + 27 + 23$ **110**
5. $34 - 17 - 7$ **10**
6. $27 + 56 + 100$ **183**
7. $143 + 79$ **222**
8. $213 - 88$ **125**

Multiply.

9. 63×4 **252**
10. 49×9 **441**
11. 19×76 **1,444**
12. 88×32 **2,816**
13. 80×50 **4,000**
14. 75×11 **825**
15. 200×15 **3,000**
16. 340×20 **6,800**

Divide.

17. $4\overline{)96}$ **24**
18. $5\overline{)127}$ **25 r2**
19. $9\overline{)423}$ **47**
20. $7\overline{)760}$ **108 r4**
21. $32\overline{)448}$ **14**
22. $20\overline{)3,660}$ **183**
23. $37\overline{)1,073}$ **29**
24. $23\overline{)4,715}$ **205**

✓ Multiply Decimals by 10, 100, and 1,000 (For Intervention, see p. H13.)

Multiply.

25. 4.3×10 **43**
26. $9.61 \times 1,000$ **9,610**
27. 8.4×100 **840**
28. $25.397 \times 1,000$ **25,397**
29. 194.05×100 **19,405**
30. 408.08×10 **4,080.8**

✓ Remainders (For Intervention, see p. H5.)

Divide. Write the remainder as a decimal.

31. $4\overline{)35}$ **8.75**
32. $5\overline{)56}$ **11.2**
33. $8\overline{)100}$ **12.5**
34. $12\overline{)243}$ **20.25**
35. $15\overline{)2,412}$ **160.8**

Divide. Write the remainder as a fraction.

36. $6\overline{)45}$ **$7\frac{1}{2}$**
37. $8\overline{)77}$ **$9\frac{5}{8}$**
38. $12\overline{)134}$ **$11\frac{1}{6}$**
39. $14\overline{)550}$ **$39\frac{2}{7}$**
40. $18\overline{)459}$ **$25\frac{1}{2}$**

LOOK AHEAD

In Chapter 4 you will
- add, subtract, multiply, and divide decimals
- evaluate decimal expressions and solve equations

LESSON 4.1 Add and Subtract Decimals

LESSON PLANNING

Objective To add and subtract decimals

Intervention for Prerequisite Skills

Whole-Number Operations (For intervention strategies, see page 65.)

NCTM Standards
1. Number and Operations
2. Algebra
6. Problem Solving
8. Communication
9. Connections

Math Background

The skills needed to add or subtract decimals are the same as those needed to add or subtract whole numbers.

Consider the following as you help students understand the procedures for adding and subtracting decimals.

- When adding or subtracting decimals in a column, align decimal points so the corresponding digits in each place are lined up properly.
- Before solving an addition or subtraction problem, place the decimal point for the answer below the decimal points in the problem.
- When decimals do not have the same number of places, add zeros as placeholders. These zeros do not change the value of the decimal.

It is important for students to estimate and to check the reasonableness of their answers.

WARM-UP RESOURCES

 **NUMBER OF THE DAY**

Suppose you are saving money and that each day you save d cents, where d is the number of the day of the month. What is the most you will save in one week of any month? **$1.96**

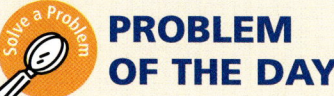 **PROBLEM OF THE DAY**

Replace each [♥] with a different digit from 0–9 to make a true number sentence.

[♥].[♥][♥][♥] + [♥][♥].[♥][♥] + [♥].[♥] = 22.815

Possible answer: 0.725 + 13.69 + 8.4 = 22.815

Solution Problem of the Day tab, p. PD4

 **DAILY FACTS PRACTICE**

Have students practice addition and subtraction facts by completing Set B of *Teacher's Resource Book,* p. TR95.

INTERVENTION AND EXTENSION RESOURCES

ALTERNATIVE TEACHING STRATEGY

Materials *For each student* newspaper ads

Reinforce adding and subtracting decimals. Provide students with newspaper ads. Have them make lists of two or more items that they could buy for less than $20, not including any sales tax. Students should show their work, first estimating, then finding the total price of the items, and finally finding the change they would receive if they paid with a $20 bill.

Answers will vary.

See also page 68.

VISUAL
BODILY/KINESTHETIC

MULTISTEP AND STRATEGY PROBLEMS

The following multistep and strategy problems are provided in Lesson 4.1:

Page	Item
69	40, 43

SOCIAL STUDIES CONNECTION

Have students **practice writing addition and subtraction word problems with decimals** by using the data below.

Form of Advertising	Money Spent
Newspaper	$31.0 billion
Television	$10.6 billion
Radio	$9.5 billion
Direct Mail	$27.3 billion
Yellow Pages	$9.5 billion
Other	$21.7 billion

Check students' work.

VISUAL
LOGICAL/MATHEMATICAL

EARLY FINISHERS

Materials *For each pair* 10 index cards

Encourage students to **estimate decimal sums and differences.** Have each pair write a different decimal on each of the 10 cards. The decimals should be less than 100 and have from 1 to 3 places to the right of the decimal point.

The students then use the cards to generate addition and subtraction problems. One student selects two cards and computes either the exact sum or difference.

The other student rounds the numbers on the cards and estimates the sum or difference. Then they compare the exact answer with the estimate and discuss any adjustments that need to be made to either. Have students reverse roles and repeat the activity.
Check students' work.

VISUAL
BODILY/KINESTHETIC

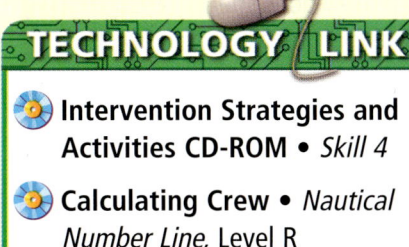

TECHNOLOGY LINK

- **Intervention Strategies and Activities CD-ROM** • *Skill 4*
- **Calculating Crew** • *Nautical Number Line,* Level R

LESSON 4.1

3 Practice

Guided Practice

Do Check for Understanding Exercises 1–12 with your students. Identify those having difficulty and use lesson resources to help.

COMMON ERROR ALERT

If students have difficulty subtracting decimals, emphasize the importance of using zeros to make equivalent decimals so that all the decimals have the same number of places and are aligned properly.

Independent Practice

Note that Exercises 40 and 43 are **multistep or strategy problems**. Assign Exercises 13–43.

Multistep or Strategy Problem To solve Exercise 40, students can add the four amounts of money Karen has saved and then subtract the total from $165.45.

CHECK FOR UNDERSTANDING

Think and Discuss Look back at the lesson to answer each question.

1. **Tell** why it is important to align the decimal points when you add or subtract. **Possible answer: If the decimal points are not aligned, like places will not be aligned correctly to add or subtract.**
2. **Explain** the steps you would use to find 67 − 34.58. **Possible answer: Write 67 as 67.00. Align the decimal points, and subtract.**

Guided Practice Add or subtract. Estimate to check.

3. $6.18 − 5.55 **$0.63**
4. 0.45 + 0.5 + 1.349 **2.299**
5. 6 + 5.43 + 1.4 + 5.755 **18.585**
6. 10.72 − 1.3 **9.42**

7. 3.2 + 2.68 + 15.043 **20.923**
8. 142.108 − 63.8 **78.308**

Copy the problem. Place the decimal point correctly in the answer.

9. 37.5 − 0.19 = 3731 **37.31**
10. 0.431 + 1.549 + 2.017 = 3997 **3.997**
11. 6 + 118.59 + 0.35 = 12494 **124.94**
12. 9.7 − 3.01 = 669 **6.69**

PRACTICE AND PROBLEM SOLVING

Independent Practice Add or subtract. Estimate to check.

13. 50.28 + 37.52 **87.80**
14. 153.95 + 434.16 **588.11**
15. 805.41 + 633.25 **1,438.66**
16. 31.62 − 5.8 **25.82**

17. 3.2 − 2.6 **0.6**
18. 735.1 + 37 + 105.73 **877.83**
19. 370.92 − 83.247 **287.673**
20. 275.2 − 86.05 **189.15**
21. 123.1 + 140 + 225.45 **488.55**
22. $8 + $215.49 + $0.75 **$224.24**
23. 620.87 − 91.386 **529.484**
24. 56.60 − 8.476 **48.124**

Copy the problem. Place the decimal point correctly in the answer.

25. 23.64 + 233.5 = 25714 **257.14**
26. $25.67 + $7.16 + $0.35 = $3318 **$33.18**
27. 11.2 − 1.78 = 942 **9.42**
28. 4.98 − 3.235 = 1745 **1.745**

Estimate to determine if the given sum is reasonable. Write *yes* or *no*.

29. 14.78 + 122.4 = 137.18 **yes**
30. $32.76 + $8.09 + $0.49 = $41.34 **yes**
31. 58.02 − 9.473 = 3.671 **no**
32. 427.7 − 39.27 = 388.43 **yes**

ALGEBRA Evaluate each expression for $d = 4.3$.

33. $d − 3.05$ **1.25**
34. $1 + d + 0.7$ **6**
35. $8 + d$ **12.3**
36. $37.60 − d$ **33.30**
37. $d − 2.084$ **2.216**
38. $(d + 16.05) − 4.5$ **15.85**

68 Chapter 4

ALTERNATIVE TEACHING STRATEGY   SCAFFOLDED INSTRUCTION

Purpose Students use decimal squares to learn about subtracting decimals.

Materials *For each group* 10 decimal squares, p. TR7; colored pencil

Have groups use their decimal squares to model 4.17.

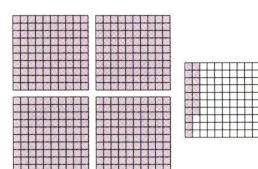

Display the following: 4.17 − 3.25

Discuss with students and demonstrate how to use the decimal squares to show the subtraction of 4.17 − 3.25. Students should realize that each small square within the decimal square stands for 1 hundredth.

Have students cross out decimal squares for 3.25. They can then use their models to copy and complete the subtraction sentence.

4.17 − 3.25 = 0.92

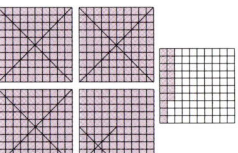

Have students then use decimal squares to find the difference 3.81 − 2.07. **1.74**

Ask volunteers to share their models and solutions.

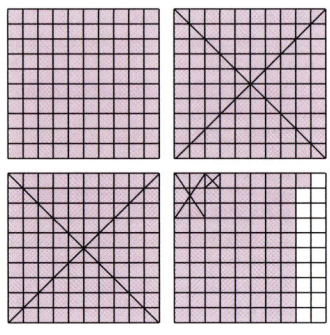

Check students' work.

68 Chapter 4

Problem Solving Applications

39. Jake's batting average is 0.325. Last year it was 0.235. What is the difference between his average last year and this year? **0.090**

40. Karen has saved $15.75, $18.36, $9.07, and $20.37 to buy a camera. How much more does she need to save to buy a camera that costs $165.45? **$101.90**

41. Estimate 30.53 + 95.7 + 75.12. Is the sum more than or less than your estimate? Explain how you know. **Possible estimate: 210; more than; explanations will vary.**

42.  Write About It Why is it important to estimate the answer when you add and subtract decimals?

43. Four clubs collected 3,905; 3,950; 3,590; and 3,509 lbs of paper each. Order the four weights of paper. Then find the difference between the least and the greatest weights of paper collected.

42. You may have put the decimal point in the wrong place, so you want to make sure the answer is reasonable.

MIXED REVIEW AND TEST PREP

44. Write 0.05 as a percent. (p. 60) **5%**   **45.** Write the decimal for 46%. (p. 60) **0.46**

46. Evaluate $a \div c$ for $a = 4{,}602$ and $c = 37$. (p. 28) **124 r14**

★**47. TEST PREP** Which shows the value of 7^4? (p. 40) **D**

   A 283   **B** 343   **C** 2,381   **D** 2,401

★**48. TEST PREP** Tanya bought milk for $2.09, two loaves of bread for $1.05 each, cheese for $4.50, and three bottles of juice for $3.00 each. How much did Tanya pay for the items? (p. 20) **H**

   F $12.04   **G** $12.64   **H** $17.69   **J** $23.04

PROBLEM SOLVING LINKUP to Science

Microbiologist Microbiologists are scientists who specialize in the study of the tiny cells that make up every living thing. While cells vary widely in size, most plant and animal cells are so small that they can be seen only with a microscope. Microbiologists often must use decimals when measuring the sizes of cells and their structures, as in these examples:

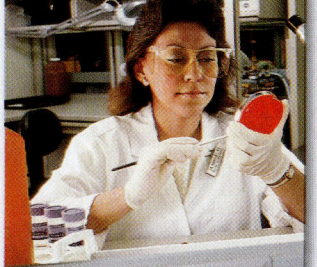

Average plant cell	0.000035 meter
Small bacterium	0.0000002 meter
Cell wall or membrane	0.0000000075 meter

- Use your school library to find the sizes of blood cells, skin cells, and nerve cells in the human body. How much larger or smaller is each of these cell types than the average plant cell?

EXTRA PRACTICE page H35, Set A

69

MIXED REVIEW AND TEST PREP
Exercises 44–48 provide **cumulative review** (Chapters 1–4).

LINKUP to SCIENCE

- As students read the Linkup to Science, have them identify the places in the numbers shown.

  In the number that tells the size of the average plant cell, what place does the 3 name? **hundred-thousandths**

  Read the number that describes the thickness of a cell wall or membrane. **seventy-five ten-billionths**

REASONING A micron is 1 millionth of a meter. How many microns long are the average plant cell and the small bacteria? **35 microns; 0.2 micron**

4 Assess

Summarize the lesson by having students:

DISCUSS Gary says that 24.3 − 17.56 = 6.86. What might he have done wrong? Possible answer: He forgot to write a zero after the 3 before he subtracted and just brought the 6 down into the hundredths place.

WRITE What are some things you should be careful to do when you add and subtract decimals? Possible answer: Be sure decimal points are lined up correctly. If one of the numbers is a whole number, place a decimal point and zeros after the last digit. Estimate the answer so that you will have some way to know whether your computed answer is reasonable.

Lesson Quiz

Transparency **4.1**

1. 4.9 + 18.8 **23.7**
2. 0.7 + 2.045 **2.745**
3. 8.5 − 7.8 **0.7**
4. 51.3 − 5.13 **46.17**
5. 145.89 − 65.036 **80.854**
6. 0.45 + 3.2 + 45.683 **49.333**

READING STRATEGY

K-W-L Chart Before having students read the Linkup, have them look at the photograph. Ask them to predict what the Linkup will be about. Then have students make a three-column chart headed What I Know, What I Want to Know, and What I Learned. Ask them to fill in the first two columns. Have them fill in the third column as they read the paragraph.

K-W-L Chart

What I Know	What I Want to Know	What I Learned

LESSON 4.2
Multiply Decimals

LESSON PLANNING

Objective To multiply decimals

Intervention for Prerequisite Skills
Whole-Number Operations, Multiply by 10, 100, and 1,000 (For intervention strategies, see page 65.)

Materials *For each student* decimal squares, p. TR7; colored pencils

Lesson Resources Calculator Handbook p. 5

NCTM Standards
1. Number and Operations
6. Problem Solving
8. Communication
10. Representation

Math Background

Students need the skills developed with whole number multiplication for multiplying decimals. To extend their skills in multiplying decimals, stress these points:

- Initially disregard the decimal points and multiply as if the numbers were whole numbers.
- The decimal point is placed in the product according to the sum of the number of places in the factors. That is, if you multiply tenths (1 place) by tenths (1 place), the answer will contain hundredths (1 + 1 = 2 places).
- An estimate is especially helpful because it shows if you have placed the decimal point correctly.

WARM-UP RESOURCES

 **NUMBER OF THE DAY**

I am a decimal less than 1. If you multiply a number by me, the number may gain or lose one zero, but all the other digits will still be there. What number am I? 0.1

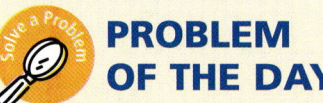 **PROBLEM OF THE DAY**

Rhonda is making an input/output table. When she sees 5, she writes 3.0. When she sees 10, she writes 6.0. When she sees 4, she writes 2.4. What will she write when she sees 12? 7.2 What did she see if she wrote 4.2? 7

Solution Problem of the Day tab, p. PD4

 **DAILY FACTS PRACTICE**

Have students practice multiplication facts by completing Set C of *Teacher's Resource Book*, p. TR95.

70A Chapter 4

INTERVENTION AND EXTENSION RESOURCES

ALTERNATIVE TEACHING STRATEGY — ELL

To help students better **understand the concept of multiplying decimals,** have them try this approach: Find 0.5 × 0.8. What fraction in simplest form is equal to 0.5? $\frac{1}{2}$ What is half of 0.8? 0.4 So, 0.5 × 0.8 = 0.4. Repeat this activity with 0.5 × 0.4 and with 0.5 × 0.6. 0.2; 0.3

See also page 72.

AUDITORY
LOGICAL/MATHEMATICAL

MULTISTEP AND STRATEGY PROBLEMS

The following multistep and strategy problems are provided in Lesson 4.2:

Page	Item
73	43, 44

CAREER CONNECTION

To **reinforce decimal multiplication,** display this ad. Ask students: Which job pays the most per week? How much does it pay? lawn mowing; $72.75

PART-TIME SUMMER JOBS		
Car Washing	Lawn Mowing	Bagging Groceries
$4.95/hr	$4.85/hr	$5.05/hr
14 hr a wk	15 hr a wk	13 hr a wk

VISUAL
VISUAL/SPATIAL

ADVANCED LEARNERS

Materials *For each pair* a set of 8 cards, each with one of the following numbers: 0.2, 0.04, 0.008, 0.0016, 0.5, 0.25, 0.125, 0.0625

Challenge students to **extend their understanding of multiplying decimals.**

- Students should shuffle the cards and stack them face down.
- Students take turns turning over the top card.
- Each time a new card is turned over, both students try to find the least whole number by which the decimal can be multiplied to give a whole number product.
- The first student to find the answer earns a number of points equal to the number of decimal places in the number on the card.

0.2 (5), 0.04 (25), 0.008 (125), 0.0016 (625), 0.5 (2), 0.25 (4), 0.125 (8), 0.0625 (16)

VISUAL
INTERPERSONAL/SOCIAL

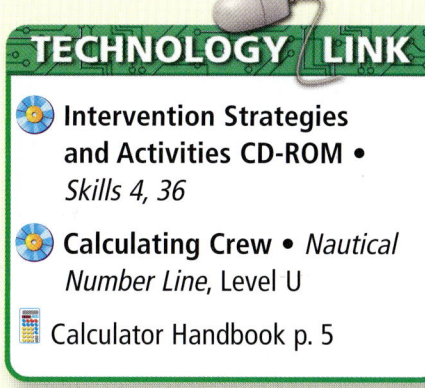

TECHNOLOGY LINK

- **Intervention Strategies and Activities CD-ROM** • *Skills 4, 36*
- **Calculating Crew** • *Nautical Number Line,* Level U
- Calculator Handbook p. 5

LESSON 4.2 ORGANIZER

Objective To multiply decimals

Materials *For each student* decimal squares, p. TR7; colored pencils

Lesson Resources Calculator Handbook p. 5

1 Introduce

QUICK REVIEW provides review of prerequisite skills.

Why Learn This? Multiplying decimals will help you determine the total cost for several of the same item. *Share the lesson objective with students.*

2 Teach

Guided Instruction

- Have students recall the relationship between addition and multiplication.

  How would you write an addition expression for the activity exercise? 0.14 + 0.14 + 0.14

- Remind students of the importance of estimating solutions.

  In Example 1, why do you not place the decimal after the 1 or the second 0? The solution must be close to the estimated product of $9. It could not be $1.01 or $100.80.

ADDITIONAL EXAMPLE

Example 1, p. 70

Rick sold 8 comics for $1.35 each. How much did he get paid for the comics? estimate: $8; answer: $10.80

LESSON 4.2 Multiply Decimals

Learn how to multiply decimals.

You need colored pencils and decimal squares.

QUICK REVIEW

1. 42 × 7	2. 83 × 4	3. 52 × 6	4. 93 × 5	5. 36 × 4
294	332	312	465	144

You can use a model to find the product of a decimal number and a whole number.

Activity

- To find 3 × 0.14, shade 0.14, or 14 small squares, three times. Use a different color and shade a different group of 14 small squares each time.
- Count the number of shaded squares. What is 3 × 0.14? **0.42**

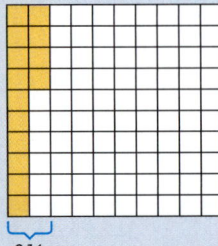
0.14

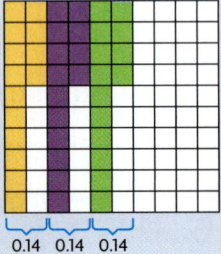
0.14 0.14 0.14

- Use a decimal square to find 5 × 0.18.
- Describe how you shaded your decimal square. **Possible answer: I shaded groups of 18 small squares five times.**
- What is 5 × 0.18? **0.90**

Sometimes, when the factors are greater, as in 8 × 1.52, it is easier to compute the product by not using decimal squares.

EXAMPLE 1

Ed buys 9 sports cards at $1.12 each. How much does he spend?

Multiply. $1.12 × 9

Estimate. $1.12 × 9 → $1 × 9 = $9

Find the answer.

$1.12
× 9

$10.08

Multiply as with whole numbers. Since the estimate is $9, place the decimal point after the 10.

Since the estimate is $9, the answer is reasonable.

So, Ed pays $10.08 for sports cards.

70 Chapter 4

RETEACH 4.2

Multiply Decimals

Marty worked 26.5 hours this week. He earns $6.40 per hour. How much money did Marty earn this week?

To solve, you need to find the product 26.5 × 6.40.

Step 1: Multiply as you would with whole numbers.
$6.40
× 26.5

169600

Step 2: Count the number of decimal places in the factors.
$6.40 → 2 decimal places
× 26.5 → 1 decimal place
169600 → 3 decimal places

Step 3: Starting at the right side of the answer, count over that number of places. This is where the decimal point is placed.
$6.40
× 26.5
$169.600

Marty earned $169.60 this week.

Complete to solve.

1. Sean worked 34 hours this week. He earns $9.25 an hour. How much money did Sean earn this week?
   a. What multiplication problem will you use? __34 × $9.25__
   b. How many decimal places are in the factors? __2__
   c. How much did Sean earn? __$314.50__

2. Denise earns $4.65 an hour. Last week she worked 17 hours. How much did she earn last week?
   a. What multiplication problem will you use? __17 × $4.65__
   b. How many decimal places are in the factors? __2__
   c. How much did Denise earn? __$79.05__

Multiply.

3. 23.1 × 5.7   4. 28.44 × 7   5. 25.4 × 24.55
   __131.67__   __199.08__   __623.57__

6. 16.6 × 0.24   7. 101.01 × 8.8   8. 2.32 × 11.48
   __3.984__   __888.888__   __26.6336__

PRACTICE 4.2

Multiply Decimals

Tell the number of decimal places there will be in the product.

1. 6.3 × 0.75   2. 9.7 × 48.8   3. 5.96 × 62.15   4. 37.6 × 8.3
   __3__   __2__   __4__   __2__

5. 32.08 × 7.3 6. 428.9 × 5.6 7. 897.3 × 5.3   8. 186.472 × 9.6
   __3__   __2__   __2__   __4__

Place the decimal point in the product.

9. 6.17 × 8.2 = 50594   10. 24.01 × 8.51 = 2043251   11. 8.94 × 5.27 = 471138
   __50.594__   __204.3251__   __47.1138__

12. 8.04 × 1.7 = 13668   13. 19.6 × 5.8 = 11368   14. 30.7 × 8.33 = 255731
    __13.668__   __113.68__   __255.731__

Multiply. Estimate to check.

15. 5 × 0.9   16. 9 × 1.2   17. 4 × 3.47   18. $18.93 × 7
    __4.5__   __10.8__   __13.88__   __$132.51__

19. 5.55 × 9   20. 5 × 2.89   21. 31.82 × 4   22. 4.61 × 8
    __49.95__   __14.45__   __127.28__   __36.88__

23. 2.49 × 6   24. 35.98 × 6.3   25. 73.02 × 9.1   26. 8.5 × 16.03
    __14.94__   __226.674__   __664.482__   __136.255__

27. 3.91 × 6.22   28. 164.5 × 0.03   29. 28.14 × 1.52   30. 6.114 × 3.72
    __24.3202__   __4.935__   __42.7728__   __22.74408__

Mixed Review

Write the decimal and the percent for the shaded part.

31. __0.33, 33%__   32. __0.40, or 0.4, 40%__   33. __0.52, 52%__

---

70 Chapter 4

Multiply a Decimal by a Decimal

You can use a decimal square or paper and pencil to find the product of two decimals.

EXAMPLE 2

Multiply. 0.2×0.6

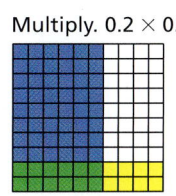

Shade 6 columns blue for 0.6.

Shade 2 rows yellow for 0.2.

The area in which the shading overlaps shows the product, or 0.2 of 0.6.

So, $0.2 \times 0.6 = 0.12$.

Place the decimal point in a product by estimating or by adding the number of decimal places in the factors.

$$\begin{array}{r} 0.2 \\ \times 0.6 \\ \hline 0.12 \end{array} \begin{array}{l} \leftarrow 1 \text{ decimal place} \\ \leftarrow 1 \text{ decimal place} \\ \leftarrow 1 + 1, \text{ or } 2, \text{ decimal places} \end{array}$$

EXAMPLE 3

Mr. Ponti works 37.5 hr per week. He earns $8.70 an hour. How much does he earn in a week?

Multiply. $\$8.70 \times 37.5$

Estimate. $\$8.70 \times 37.5 \rightarrow \$9 \times 38 = \$342$

Find the answer.

$$\begin{array}{r} \$8.70 \\ \times\ 37.5 \\ \hline 4350 \\ 6090 \\ +2610 \\ \hline 326.250 \end{array}$$

← 2 decimal places
← 1 decimal place
Multiply as with whole numbers.
Place the decimal point in the product.
← 3 decimal places

Since the estimate is $342, the answer is reasonable.

So, Mr. Ponti earns $326.25.

When you multiply decimals, you sometimes have to insert zeros in the answer.

EXAMPLE 4

Multiply. 0.037×0.062

$$\begin{array}{r} 0.037 \\ \times 0.062 \\ \hline 74 \\ +222 \\ \hline 0.002294 \end{array}$$

← 3 decimal places
← 3 decimal places
Multiply as with whole numbers.
Place the decimal point in the product.
The answer must have 6 decimal places, so place 2 zeros to the left of 2.

So, $0.037 \times 0.062 = 0.002294$.

- Remind students that Examples 2 and 4 show that the product of two factors less than 1 is always less than each of the factors.

  What part of the model in Example 2 shows the product? the part where the two colored sections overlap

  How does this double shading show that the product is less than each of the factors? The overlapped portion is smaller than each of the individual sections representing the factors.

Modifying Instruction You may want to guide students in discovering for themselves the rules for proper decimal placement in products. Display exercises similar to those in Examples 2–4 and have students solve with calculators. Then ask them to compare the number of decimal places in the factors to the number of decimal places in the products.

ADDITIONAL EXAMPLES

Example 2, p. 71

Multiply. 0.4×0.7

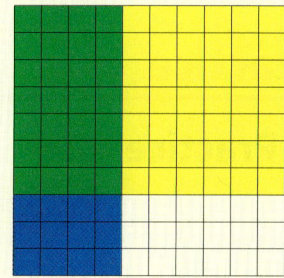

So, $0.4 \times 0.7 = 0.28$

Example 3, p. 71

Marshall bought 22.5 lb of meat from Grocery Warehouse. He paid $5.79 per lb. How much did he pay for the meat? estimate: $120; answer: $130.28

Example 4, p. 71

Multiply. 0.023×0.054

$$\begin{array}{r} 0.023 \\ \times 0.054 \\ \hline 92 \\ +115 \\ \hline 0.001242 \end{array}$$

3 decimal places
3 decimal places
Multiply as with whole numbers.
Place the decimal point. The answer must have 6 decimal places, so place 2 zeros to the left of 1.

So, $0.023 \times 0.054 = 0.001242$.

CHALLENGE 4.2

Puzzling Problems

Find each product. Locate the product in the Tip Box. (Hint: Not all products are in the Tip Box.)
Each time the product appears, write the letter of that exercise above it. When you have solved all the problems, you will discover a math tip for multiplying decimals.

A $5.26 \times 7.5 =$	39.45	N $2.008 \times 1.2 =$	2.4096
B $0.12 \times 1.2 =$	0.144	O $6.1 \times 0.42 =$	2.562
C $3 \times 0.009 =$	0.027	P $0.54 \times 2.9 =$	1.566
D $2.9 \times 2.03 =$	5.887	R $0.3 \times 30 =$	9
E $8 \times 2.5 =$	20	S $1.8 \times 2.3 =$	4.14
G $0.15 \times 0.07 =$	0.0105	T $6 \times 1.7 =$	10.2
H $7.4 \times 6.8 =$	50.32	U $5.9 \times 0.04 =$	0.236
I $0.04 \times 40.5 =$	1.62	V $8.5 \times 6.3 =$	53.55
J $0.25 \times 3.8 =$	0.95	W $46.7 \times 2.3 =$	107.41
M $190 \times 0.03 =$	5.7		

Tip Box

R	E	M	E	M	B	E	R
9	20	5.7	20	5.7	0.144	20	9

T	O		E	S	T	I	M	A	T	E
10.2	2.562		20	4.14	10.2	1.62	5.7	39.45	10.2	20

T	H	E		P	R	O	D	U	C	T
10.2	50.32	20		1.566	9	2.562	5.887	0.236	0.027	10.2

PROBLEM SOLVING 4.2

Multiply Decimals

Analyze • Choose • Solve • Check

Write the correct answer.

1. Which is greater, 0.108 or 0.091? Use > or <.

   0.108 > 0.091 or 0.091 < 0.108

2. Sonia wrote a check for $27.86. What is the number of dollars written in words?

   twenty-seven and eighty-six hundredths

Choose the letter for the best answer.

3. Walter grew a pumpkin that weighed 38.73 pounds. Bill grew a pumpkin that weighed 42.1 pounds. How many more pounds did Bill's pumpkin weigh than Walter's pumpkin?
   A 4.67 more pounds
   B 4.63 more pounds
   C 3.67 more pounds
   (D) 3.37 more pounds

4. Ted wants to use a special wallpaper border in his living room. He has three pieces of border that are 11.7 meters, 6.05 meters, and 24.75 meters long. How many meters of border does he have in all?
   F 24.75 meters
   G 31.97 meters
   (H) 42.5 meters
   J 641.45 meters

5. A pencil costs $0.85 and a pen costs $1.76. Wayne buys 12 pencils and 8 pens. Which expression can be used to find the total cost of Wayne's purchases?
   (A) $(12 \times 0.85) + (8 \times 1.76)$
   B $(12 \times 0.85) + (8 + 1.76)$
   C $(12 \times 0.85) \times (8 \times 1.76)$
   D $(12 + 0.85) \times (8 + 1.76)$

6. A grocery store needs to stock a new cereal on the shelf. There are 8 shelves that can hold 6 boxes in each row. What else do you need to know to find out how many boxes of the cereal the store can put out at once?
   F The height of the box
   (G) How many rows of boxes fit on a shelf
   H How much a box of cereal costs
   J The brand of cereal

7. **Write About It** Explain how you could use a decimal square to model the product 0.3×0.2.

   Use a hundred square. Shade 2 columns one color for 0.2 and shade 3 rows another color for 0.3. The area in which the shading overlaps shows the product, or 0.3 of 0.2.

LESSON 4.2

• After students work Example 5, ask:

REASONING Show how you could use the Distributive Property to multiply 4 and 11.08. $(4 \times 11) + (4 \times 0.08) = 44 + 0.32 = 44.32$

ADDITIONAL EXAMPLE

Example 5, p. 72

Multiply. 8×5.7
Estimate. $8 \times 5.7 \rightarrow 8 \times 6 = 48$

Find the answer.
$8 \times 5.7 = (8 \times 5) + (8 \times 0.7)$ Use the Distributive Property.
$\quad\quad\quad = 40 + 5.6$
$\quad\quad\quad = 45.6$

Since the estimate is 48, place the decimal point after the 45.

So, $8 \times 5.7 = 45.6$.

3 Practice

Guided Practice

Do Check for Understanding Exercises 1–13 with your students. Identify those having difficulty and use lesson resources to help.

Independent Practice

Note that Exercises 43 and 44 are **multistep or strategy problems**. Assign Exercises 14–46.

EXAMPLE 5

You can also use the Distributive Property to multiply with decimals.

Multiply. 6×5.9

Use estimation to place the decimal point in the product.
Estimate. $6 \times 5.9 \rightarrow 6 \times 6 = 36$

Find the answer.
$6 \times 5.9 = (6 \times 5) + (6 \times 0.9)$ Use the Distributive Property.
$\quad\quad\quad = 30 + 5.4$
$\quad\quad\quad = 35.4$ Since the estimate is 36, place the decimal point after the 35.

Since the estimate is 36, the answer is reasonable.
So, $6 \times 5.9 = 35.4$.

CHECK FOR UNDERSTANDING

Think and Discuss Look back at the lesson to answer the question.

1. **Explain** how you would place the decimal point in the product 0.27×0.476. Count the decimal places in the factors. Then count that number of places from the right in the product.

Guided Practice Use the decimal square shown to help you multiply.

2.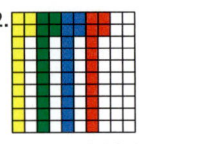
4×0.12 **0.48**

3.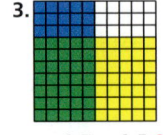
0.7×0.5 **0.35**

4.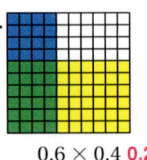
0.6×0.4 **0.24**

Tell the number of decimal places there will be in the product.

5. 3.62×7 **2**
6. 2.15×8.18 **4**
7. 4.04×5.2 **3**

Copy the problem. Place the decimal point in the product.

8. $9 \times 5.4 = 486$ **48.6**
9. $0.7 \times 4.1 = 287$ **2.87**
10. $2.2 \times 0.55 = 1210$ **1.210**

Multiply. Estimate to check.

11. 0.42×2.9 **1.218**
12. 1.25×0.4 **0.500**
13. 3.23×8 **25.84**

PRACTICE AND PROBLEM SOLVING

Independent Practice Use the decimal square shown to help you multiply.

14.
5×0.18 **0.90**

15.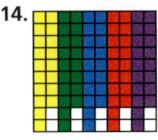
0.3×0.8 **0.24**

16.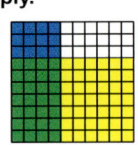
0.7×0.4 **0.28**

72 Chapter 4

ALTERNATIVE TEACHING STRATEGY   SCAFFOLDED INSTRUCTION

Purpose Students analyze errors involving misplaced decimal points in completed problems and correct them.

Display the following problem:

$$\begin{array}{r} 3.07 \\ \times\ 0.7 \\ \hline 0.02149 \end{array}$$

Have students work in pairs to find and correct the error.

Students should notice that the number of digits in the factors was counted to place the decimal point rather than the number of decimal places.

Have a volunteer show the correct solution and explain how to place the decimal point in the answer.

$$\begin{array}{rl} 3.07 & \text{2 places} \\ \times\ 0.7 & \text{+1 place} \\ \hline 2.149 & \text{3 places} \end{array}$$

Next have partners repeat the process with the following problem:

$$\begin{array}{r} 21.4 \\ \times\ 1.6 \\ \hline 1284 \\ +2140 \\ \hline 3.424 \end{array}$$

Guide students to realize that the number of places before the decimal point in each factor was counted instead of the number of places after the decimal point in each factor.

$$\begin{array}{rl} 21.4 & \text{1 place} \\ \times\ 1.6 & \text{+1 place} \\ \hline 1284 & \\ +2140 & \\ \hline 34.24 & \text{2 places} \end{array}$$

To help students avoid such errors, encourage them to circle each digit after the decimal point in each factor and then to count the circles. This total will be the number of decimal places they need in the answer.

Check students' work.

Tell the number of decimal places there will be in the product.

17. 3.79 × 8.2 **3**
18. 0.876 × 0.2 **4**
19. 1.3842 × 0.91 **6**

Copy the problem. Place the decimal point in the product.

20. 6.37 × 2.91 = 185367 **18.5367**
21. 20.4 × 9.52 = 194208 **194.208**
22. 7.32 × 3 = 2196 **21.96**
23. 0.82 × 0.5 = 410 **0.410**
24. 32.5 × 0.06 = 1950 **1.950**

Multiply. Estimate to check.

25. 3 × 4.6 **13.8**
26. 9 × 2.5 **22.5**
27. 7.3 × 5 **36.5**
28. 8.2 × 5 **41.0**
29. 1.2 × 4.1 **4.92**
30. 0.9 × 6.3 **5.67**
31. 0.2 × 0.4 **0.08**
32. 6.3 × 0.9 **5.67**
33. 0.21 × 2.1 **0.441**
34. 6.15 × 2.4 **14.760**
35. 4.08 × 1.35 **5.5080**
36. 6.21 × 0.95 **5.8995**
37. 24.63 × 1.09 **26.8467**
38. 29.147 × 5.61 **163.51467**
39. 0.189 × 2.09 **0.39501**
40. 118.001 × 0.37 **43.66037**
41. 148.9 × 0.006 **0.8934**
42. 1,200.5 × 8.2 **9,844.10**

Problem Solving ▶ Applications

43. The recipe for a pie calls for 2.25 lb of apples and 0.75 lb of walnuts. Apples are $1.60 per pound and walnuts are $4.95 per pound. How much will the apples and walnuts cost in all? **$7.31**

44. Keith has a wall that is 5.2 m wide. He has 3 bookcases that are each 1.9 m wide. Is there enough room for all of the bookcases to be placed against the wall? Explain. **No, 1.9 × 3 = 5.7; 5.7 > 5.2**

45. ✏ Write a problem that uses multiplication of two decimals to find the answer. The product must have four decimal places. **Check students' problems.**

46. Use Data Look at the graph at the right. Rochelle had to earn 134 points in the first four rounds to advance in a competition. Rochelle says that she did advance. Is this a reasonable statement? Explain. **yes; round each score for rounds 1 – 4 to the nearest ten and add; 260 > 134**

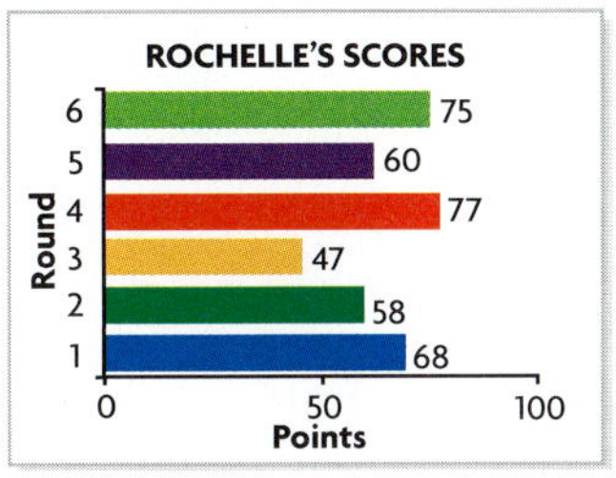

ROCHELLE'S SCORES

Round 6: 75
Round 5: 60
Round 4: 77
Round 3: 47
Round 2: 58
Round 1: 68

Points

MIXED REVIEW AND TEST PREP

47. Add. 46.2 + 3.45 + 16 (p. 66) **65.65**
48. Subtract. 604.5 − 76.38 (p. 66) **528.12**
49. Evaluate the expression $a \div c$ for $a = 14{,}067$ and $c = 47$. (p. 28) **299 r14**

★ **50. TEST PREP** Which shows the value of m when $8 = m \div 9$? (p. 36) **B**
   A 70   **B** 72   **C** 720   **D** 7,200

★ **51. TEST PREP** Which is the best estimate for $21{,}563 \div 43$? (p. 16) **F**
   F 500   **G** 700   **H** 5,000   **J** 7,000

EXTRA PRACTICE page H35, Set B

73

LESSON 4.3

ORGANIZER

Objective To use a model to divide decimals

Materials *For each student* decimal squares, p. TR7; colored pencils; scissors

Lesson Resources E-Lab Recording Sheet • *Exploring Division of Decimals*

Intervention for Prerequisite Skills
Whole-Number Operations (For intervention strategies, see page 65.)

Using the Pages

Discuss situations in which students may need to divide decimals, such as sharing money with other people.

Activity 1

In this activity students use decimal squares to model dividing decimals by whole numbers. This hands-on experience will help students visualize the concept of equal groups.

Think and Discuss

After discussing the questions, ask students to:

Summarize the relationship between the quotients of these identical digit dividends expressed as a whole number and then as hundredths. The quotient of the whole number dividends are 100 times as great as the quotient of the hundredths dividend.

Practice

Observe students as they model each exercise and ask:

In Exercise 2, how many shaded wholes did you put into each group? none **Why?** 3 is not great enough to be placed into 5 groups

How many hundredths did you put into each group? 65 hundredths

---

LESSON 4.3

Explore Division of Decimals

Explore how to use a model to divide decimals.

You need decimal squares, colored pencils, scissors.

QUICK REVIEW
1. 6)372   62
2. 8)427   53 r3
3. 5)524   104 r4
4. 4)123   30 r3
5. 252 ÷ 6   42

You can shade and cut apart decimal squares to divide a decimal by a whole number.

Remember

1 tenth (0.1) = 1 column

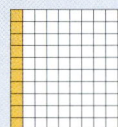

1 hundredth (0.01) = 1 small square

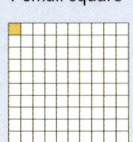

Activity 1

Find 3.66 ÷ 3.

• Shade 3.66 decimal squares.

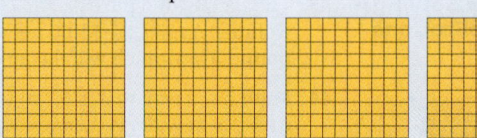

• Divide the shaded wholes into 3 equal groups. Divide the 66 hundredths into 3 equal groups.

 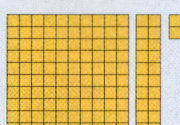

• What decimal names each group? What is the quotient? **1.22; 1.22**

• Use decimal squares to find 1.32 ÷ 4.
  Check students' models; 0.33.

Think and Discuss

• Find 366 ÷ 3. How is the quotient the same as for 3.66 ÷ 3? How is it different?
  122; it has the same digits; it is 100 times as great.

• Find 132 ÷ 4. How is the quotient the same as for 1.32 ÷ 4? How is it different?
  33; it has the same digits; it is 100 times as great.

Practice

Use decimal squares to find the quotient.

1. 4.04 ÷ 4   **1.01**
2. 3.25 ÷ 5   **0.65**
3. 1.35 ÷ 3   **0.45**

74   Chapter 4

SPECIAL NEEDS ELL

Materials *For each student* play money— 10 quarters, 10 dimes, 10 nickels

Display this **model for dividing decimals:**

$1.00 ÷ $0.25 = 4

$0.25 $0.25 $0.25 $0.25

Have students use play money to find quotients for the following: $0.30 ÷ $0.05 and $1.60 ÷ $0.80. **6; 2;** Check students' models.

KINESTHETIC; LOGICAL/MATHEMATICAL

Intervention and Extension Resources

WRITING IN MATHEMATICS

Ask students to write a paragraph explaining the difference in how they would **use models to solve these decimal exercises:** 5.5 ÷ 5 and 5.5 ÷ 1.1. Possible answer: In the first case, you would divide 5.5 into 5 groups and count 1.1 in each group, showing that 5.5 ÷ 5 = 1.1. In the second case, you would divide 5.5 into equal groups of 1.1 and count the groups, showing 5.5 ÷ 1.1 = 5.

You can shade and cut apart decimal squares to divide a decimal by a decimal.

Activity 2

Find 3.6 ÷ 1.2.

- Shade 3.6 decimal squares.

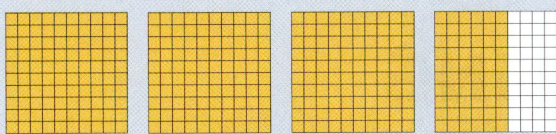

- Cut apart the 6 tenths.
- Divide the shaded squares and shaded tenths into equal groups of 1.2. How many groups of 1.2 are in 3.6? What is the quotient? **3; 3**

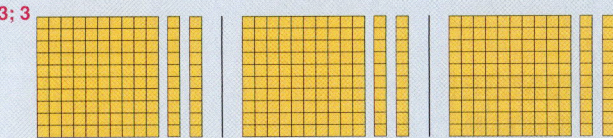

- Use decimal squares to find 3.2 ÷ 1.6.
- How many groups of 1.6 are in 3.2? **2**

Think and Discuss

- Find 36 ÷ 12. How is the quotient the same as for 3.6 ÷ 1.2? How is the problem different from 3.6 ÷ 1.2? **3; it is exactly the same; the divisor and dividend are each 10 times as great.**
- You know that 3.6 ÷ 12 = 0.3 and 3.6 ÷ 1.2 = 3. What do you think 3.6 ÷ 0.12 equals? **30**

Practice

Use decimal squares to find the quotient.

1. 7.8 ÷ 1.3 **6**
2. 5.6 ÷ 0.8 **7**
3. 1.56 ÷ 0.52 **3**
4. 5.5 ÷ 1.1 **5**
5. 3.6 ÷ 0.9 **4**
6. 1.8 ÷ 0.45 **4**
7. 0.42 ÷ 0.14 **3**
8. 10.8 ÷ 2.7 **4**
9. 0.64 ÷ 0.08 **8**

MIXED REVIEW AND TEST PREP

10. Multiply. 64.7 × 3.6 (p. 70) **232.92**
11. Write the decimal for 57%. (p. 60) **0.57**
12. Write the value of 3^4. (p. 40) **81**
13. Divide. 3,759 ÷ 42 (p. 22) **89 r21**
14. **TEST PREP** Evaluate the expression. 12 × 25 ÷ (12 + 18) (p. 44) **B**
    A 8   B 10   C 43   D 300

75

Activity 2

Unlike Activity 1, in which the number of decimal squares that should go into each group was being determined, in this activity the number of groups is being determined.

Think and Discuss

REASONING Explain how you know that 3.6 ÷ 0.12 = 30. Possible answer: by using a pattern: 3.6 ÷ 12 = 0.3, 3.6 ÷ 1.2 = 3, so 3.6 ÷ 0.12 = 30. As the decimal in the divisor moves one place to the left, the decimal in the quotient moves one place to the right.

Practice

Before students begin the exercises, ask:

Which exercise could you solve by using mental math? Explain. Most students will say Exercise 2: 56 ÷ 8 = 7, so 5.6 ÷ 0.8 = 7 and Exercise 5: 36 ÷ 9 = 4, so 3.6 ÷ 0.9 = 4.

Have students model all six exercises even if they were able to find the answers by using mental math. The conceptual understanding they develop by modeling will help them solve more complex problems later.

MIXED REVIEW AND TEST PREP

Exercises 10–14 provide **cumulative review** (Chapters 1–4).

Oral Assessment

Have students explain how they would use decimal squares to model the equations.

1. 0.64 ÷ 2 = 0.32 Possible answer: Shade 64 squares of a decimal square. Divide the 64 squares into 2 equal groups. Each group will have 32 squares, each of which represents a hundredth.

2. 0.9 ÷ 0.05 = 18 Possible answer: Shade 9 tenths of a decimal square. Count the number of groups of 5 hundredths. There are 18 groups of 5 hundredths.

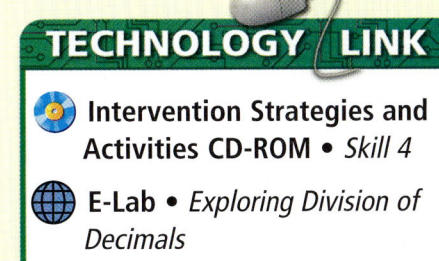

USING E-LAB

Students explore division of decimals by using computer models for decimal multiplication.

The E-Lab Recording Sheets and activities are available on the E-Lab website.

www.harcourtschool.com/elab2002

TECHNOLOGY LINK

Intervention Strategies and Activities CD-ROM • *Skill 4*

E-Lab • *Exploring Division of Decimals*

75

LESSON 4.4 Divide with Decimals

LESSON PLANNING

Objective To divide a decimal by a whole number and a decimal by a decimal

Intervention for Prerequisite Skills
Whole-Number Operations, Multiply Decimals by 10, 100 and 1,000; Remainders (For intervention strategies, see page 65.)

Materials *For each student* calculator

Lesson Resources Calculator Handbook p. 6

NCTM Standards
1. Number and Operations
6. Problem Solving
8. Communication

Math Background

Emphasize these steps as you teach the division algorithm as applied to decimals.

- The division algorithm is defined for a whole number divisor. To divide one decimal by another, first multiply the divisor by a power of ten to make it a whole number. The dividend must be multiplied by the same power of ten. To do this, "move the decimal point to the right" in both divisor and dividend the same number of places. Add zeros to the dividend if needed.
- The decimal point in the quotient is then placed over the decimal point in the dividend. If the quotient is less than 1, insert zeros between the decimal point and first nonzero digit, if needed.
- Estimating the quotient helps locate the position of the first nonzero digit.
- Continue using the steps of whole number division.

WARM-UP RESOURCES

NUMBER OF THE DAY

Write and solve 5 division problems using the number of inches in a yard. Possible answers: 36 ÷ 12 = 3, 72 ÷ 36 = 2, 108 ÷ 3 = 36, 36 ÷ 2 = 18, 36 ÷ 9 = 4

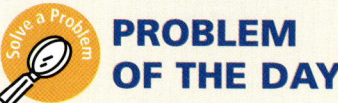 

The sum of two decimal numbers is 9.3. Their difference is 4.3, and their product is 17.00. What are the numbers? 2.5 and 6.8

Solution Problem of the Day tab, p. PD4

Have students practice division facts by completing Set E of *Teacher's Resource Book*, p. TR95.

INTERVENTION AND EXTENSION RESOURCES

ALTERNATIVE TEACHING STRATEGY — ELL

Remind students that they can **use repeated subtraction to find answers to division problems** that have whole number quotients. Have students use repeated subtraction to find each of the following quotients:

$0.18 \div 0.03$ **6**

$2.4 \div 0.6$ **4**

$6 \div 1.5$ **4**

$0.72 \div 0.09$ **8**

See also page 78.

VISUAL
LOGICAL/MATHEMATICAL

MULTISTEP AND STRATEGY PROBLEMS

The following multistep and strategy problems are provided in Lesson 4.4:

Page	Item
79	39, 41

EARLY FINISHERS

Remind students how to **check decimal division**. If a division calculation has been done correctly, the product of the quotient and the divisor should equal the dividend. Have students use this method to check their answers for six of the division exercises on page 79. Answers will vary.

VISUAL
LOGICAL/MATHEMATICAL

ADVANCED LEARNERS

Challenge students to solve this **decimal division puzzle** by finding three different digits A, B, C such that:

$$C.BC \div A.A = C.C$$

There are three possibilities: $A = 1, B = 4, C = 2$; $A = 1, B = 6, C = 3$; $A = 1, B = 8, C = 4$.

If students need encouragement to get started, you may wish to offer one or both of these hints:

- In one possibility, all digits are less than 5.
- Let $A = 1$.

VISUAL
LOGICAL/MATHEMATICAL

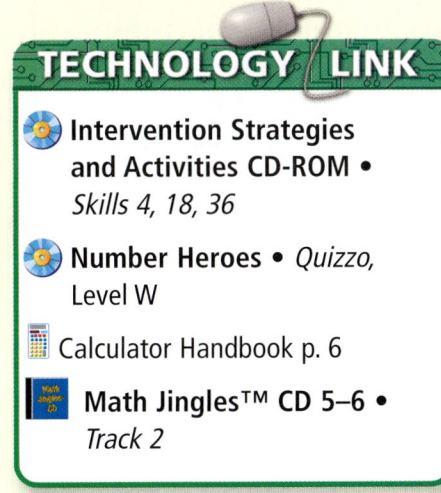

TECHNOLOGY LINK

Intervention Strategies and Activities CD-ROM • *Skills 4, 18, 36*

Number Heroes • *Quizzo, Level W*

Calculator Handbook p. 6

Math Jingles™ CD 5–6 • *Track 2*

LESSON 4.4 ORGANIZER

Objective To divide a decimal by a whole number and a decimal by a decimal

Materials For each student calculator

Lesson Resources Calculator Handbook p. 6

1 Introduce

QUICK REVIEW provides review of pre-requisite skills.

Why Learn This? Dividing a decimal by a decimal will enable you to determine the number of unit-priced items you can purchase with a given amount of money. *Share the lesson objective with students.*

2 Teach

Guided Instruction

- Remind students of the importance of placing the decimal correctly in the quotient before applying the division algorithm.

REASONING In Example 1, why would estimating using compatible numbers not be enough to help you determine the reasonableness of your quotient? **Possible answer: The estimate would tell you that the digits in the quotient would be close to 90, but the decimal placement would depend on where the decimal is in the dividend.**

Why would 0.90 be an incorrect quotient? **The first non-zero digit in the decimal dividend is less than the divisor, making a zero to the right of the decimal in the quotient necessary.**

ADDITIONAL EXAMPLE

Example 1, p. 76

Divide. $0.425 \div 5$ **0.085**

LESSON 4.4

Divide with Decimals

Learn how to divide a decimal by a whole number and how to divide a decimal by a decimal.

QUICK REVIEW
1. $8)\overline{373}$ **46 r5**
2. $6)\overline{205}$ **34 r1**
3. $19)\overline{836}$ **44**
4. $27)\overline{434}$ **16 r2**
5. $294 \div 14$ **21**

Remember that compatible numbers are numbers that divide without a remainder, are close to the actual numbers, and are easy to compute mentally.

Jack wants to transfer photographs onto a CD to use with his computer. He bought a box of 5 blank CDs for $26.45. What was the cost of each CD?

Dividing a decimal by a whole number is like dividing whole numbers.

Divide. $\$26.45 \div 5$

Use compatible numbers to estimate.

$\$26.45 \div 5 \rightarrow \$25 \div 5 = \$5$

Find the answer.

$$\begin{array}{r} 5.29 \\ 5)\overline{26.45} \\ -25 \\ \hline 14 \\ -10 \\ \hline 45 \\ -45 \\ \hline 0 \end{array}$$

Place a decimal point above the decimal point in the dividend.

Divide.

Since the estimate is $5, the answer is reasonable. So, each CD cost $5.29.

Sometimes you have to place a zero in the quotient when the dividend is less than the divisor.

EXAMPLE 1

Divide. $0.285 \div 3$

$$\begin{array}{r} 0.095 \\ 3)\overline{0.285} \\ -0 \\ \hline 28 \\ -27 \\ \hline 15 \\ -15 \\ \hline 0 \end{array}$$

Place a decimal point above the decimal point in the dividend.

Divide. Since 2 tenths is less than 3 ones, place a zero in the tenths place in the quotient.

So, $0.285 \div 3 = 0.095$.

76 Chapter 4

RETEACH 4.4

Divide with Decimals

Mary has saved $0.35 each day. She now has a total of $9.10. How long has Mary been saving?

To solve, you need to divide 9.10 by 0.35.

Step 1: Make the divisor a whole number by multiplying the divisor by a multiple of 10. Multiply the dividend by the same multiple of 10.

$0.35)\overline{9.10} \rightarrow 35)\overline{910}$
$\times 100 \quad \times 100$

Step 2: Place the decimal point in the quotient directly above the decimal point in the dividend. Divide as you would with whole numbers.

$$\begin{array}{r} 26. \\ 35)\overline{910.} \\ -70 \\ \hline 210 \\ -210 \\ \hline 0 \end{array}$$

Since the remainder is 0, the answer is a whole number. You do not need to show the decimal point.

Mary has been saving for 26 days.

Complete to solve.

1. Dusty bought 7 movie tickets for his friends. The tickets cost a total of $34.65. How much does each friend owe him for the cost of one ticket?
   a. What division problem will you use to solve? **$34.65 ÷ 7**
   b. Do you need to multiply to make the divisor a whole number? **no**
   c. How do you decide where to place the decimal point in the quotient? **Place it directly above the decimal point in the dividend.**
   d. How much does each friend owe? **$4.95**

Place the decimal point in the quotient.
2. $131.52 \div 6.4 = 2055$ **20.55**
3. $50.085 \div 10.6 = 4725$ **4.725**
4. $1936.95 \div 2.22 = 8725$ **872.5**

Find the quotient.
5. $16.88 \div 5$ **3.376**
6. $81.9 \div 18$ **4.55**
7. $332.8 \div 40$ **8.32**
8. $118 \div 12.5$ **9.44**
9. $203.205 \div 5.7$ **35.65**
10. $421.155 \div 14.7$ **28.65**

PRACTICE 4.4

Divide with Decimals

Rewrite the problem so that the divisor is a whole number.
1. $8.5 \div 2.3$ **85 ÷ 23**
2. $6.4 \div 1.3$ **64 ÷ 13**
3. $9.1 \div 0.15$ **910 ÷ 15**
4. $33.17 \div 6.8$ **331.7 ÷ 68**

Place the decimal point in the quotient.
5. $7.48 \div 0.25 = 2992$ **29.92**
6. $116.13 \div 4.2 = 2765$ **27.65**
7. $56.68 \div 0.08 = 7085$ **708.5**

Divide. Estimate to check.
8. $36.9 \div 3$ **12.3**
9. $22.4 \div 7$ **3.2**
10. $37.5 \div 5$ **7.5**
11. $89.6 \div 8$ **11.2**
12. $14)\overline{78.4}$ **5.6**
13. $40)\overline{6.8}$ **0.17**
14. $13)\overline{150.8}$ **11.6**
15. $70)\overline{23.8}$ **0.34**
16. $5.32 \div 0.7$ **7.6**
17. $1.88 \div 0.4$ **4.7**
18. $2.12 \div 0.2$ **10.6**
19. $5.4 \div 0.08$ **67.5**
20. $7.54)\overline{24.882}$ **3.3**
21. $12.6)\overline{806.4}$ **64**
22. $0.91)\overline{6.734}$ **7.4**
23. $10.9)\overline{81.75}$ **7.5**
24. $2.9)\overline{0.3335}$ **0.115**
25. $0.18)\overline{64.296}$ **357.2**
26. $12.3)\overline{84.87}$ **6.9**
27. $8.7)\overline{53.244}$ **6.12**

Mixed Review

Add, subtract, or multiply.
28. $78.94 + 9.66 + 103.71$ **192.31**
29. $1,083.75 - 706.9$ **376.85**
30. 0.072×0.48 **0.03456**
31. $215.6 + 49.87 + 8.351$ **273.821**
32. 42.83×1.91 **81.8053**
33. $65.85 - 39.478$ **26.372**
34. $430.62 - 288.74$ **141.88**
35. $192.6 + 847.56$ **1,040.16**
36. 17.335×8.26 **143.1871**

76 Chapter 4

Activity

- Use a calculator to find the first three quotients in each set.
- Look for a pattern. Try to predict the last quotient in each set.

Set A	Set B
0.48 ÷ 0.03 = ■ 16	0.621 ÷ 0.023 = ■ 27
4.8 ÷ 0.3 = ■ 16	6.21 ÷ 0.23 = ■ 27
48 ÷ 3 = ■ 16	62.1 ÷ 2.3 = ■ 27
480 ÷ 30 = ■ 16	621 ÷ 23 = ■ 27

When the decimal point is moved the same number of places in the dividend and the divisor, the quotient remains unchanged.

- Describe the pattern that helped you predict the last quotient in each set.
- Look at 4.8 ÷ 0.3. Multiply both numbers by 10. How do the quotients for 4.8 ÷ 0.3 and 48 ÷ 3 compare? How does multiplying the divisor and the dividend by 10 affect the quotient? **4.8 × 10 = 48, 0.3 × 10 = 3; the same; does not affect it**

To divide a decimal by a decimal, first multiply the divisor and the dividend by a power of 10 to change the divisor to a whole number.

$$0.7\overline{)62.44} \rightarrow 7\overline{)624.4}$$

THINK: $0.7 \times 10 = 7$

$62.44 \times 10 = 624.4$

EXAMPLE 2

Divide. 22.8 ÷ 0.8

$0.8\overline{)22.8}$ Make the divisor a whole number by multiplying the divisor and dividend by 10.

$0.8 \times 10 = 8$ $22.8 \times 10 = 228$

```
      28.5
   8)228.0
    −16
     68
    −64
      40
     −40
       0
```
Place the decimal point in the quotient. Divide.

Since there is a remainder, place a zero in the tenths place in the dividend, and continue to divide.

So, 22.8 ÷ 0.8 = 28.5.

- Think about 63.7 ÷ 0.24. To change the divisor to a whole number, you multiply by 100. What does the dividend become? **6,370**

- After completing the Activity, emphasize that multiplying both the divisor and the dividend of a division problem by the same number does not change the value of the quotient.

What is the quotient for 280 ÷ 70? 4

When you multiply both divisor and dividend by 100 to get 2,800 ÷ 700, does the quotient change? No, it is still 4.

- Discuss the mathematical meaning of moving the decimal point in the divisor and dividend of a division problem.

A quick way to multiply a decimal by 10 is to move the decimal point one place to the right. If you move the decimal point in 4.58 two places to the right, by what number are you multiplying? 100

If you move the decimal point in 0.674 three places to the right, by what number are you multiplying? 1,000

ADDITIONAL EXAMPLE

Example 2, p. 77

Divide. 39.8 ÷ 0.4 **99.5**

CHALLENGE 4.4

Decimal Solutions

Find each quotient. Locate the quotient in the Tip Box. (Hint: Not all quotients are in the Tip Box.)
Each time the quotient appears, write the letter of that exercise above it. When you have solved all the problems, you will discover a math tip for dividing decimals.

A	6.3 ÷ 0.05 =	126	N	9.3 ÷ 0.6 =	15.5
B	50.2 ÷ 0.01 =	5,020	O	0.0024 ÷ 0.3 =	0.008
C	33.6 ÷ 8 =	4.2	P	8.7 ÷ 17.4 =	0.5
E	5.4 ÷ 0.02 =	270	Q	28.7 ÷ 8.2 =	3.5
F	107.91 ÷ 5.5 =	19.62	R	400.98 ÷ 24.6 =	16.3
G	4.077 ÷ 0.18 =	22.65	S	21.54 ÷ 0.6 =	35.9
H	9 ÷ 0.3 =	30	T	0.4168 ÷ 8 =	0.0521
I	1.6 ÷ 0.4 =	4	U	25 ÷ 0.005 =	5,000
K	5.44 ÷ 1.7 =	3.2	W	0.568 ÷ 0.4 =	1.42
L	0.192 ÷ 0.3 =	0.64	Y	4.48 ÷ 0.08 =	56
M	8.05 ÷ 0.7 =	11.5	Z	6.3 ÷ 0.18 =	35

Tip Box

M	U	L	T	I	P	L	Y	T	O
11.5	5,000	0.64	0.0521	4	0.5	0.64	56	0.0521	0.008

C	H	E	C	K		T	H	E
4.2	30	270	4.2	3.2		0.0521	30	270

A	N	S	W	E	R
126	15.5	35.9	1.42	270	16.3

PROBLEM SOLVING 4.4

Divide with Decimals

Analyze Choose Solve Check

Write the correct answer.

1. Find the quotient.

   $7.4\overline{)153.92}$

   20.8

2. Place the decimal point in the quotient.

   235.468 ÷ 8.6 = 2738

   27.38

3. Jacob bought a new computer for $2,124.00. He is paying $88.50 a month for the computer. For how many months will he have to make payments?

   24 months

4. Selma needs a new notebook that costs $18.75 and a calculator that costs $23.64. How much money does she need to make the purchases?

   $42.39

Choose the letter for the best answer.

5. Which expression is 211.68 ÷ 12.6 rewritten so that the divisor is a whole number?
   - Ⓐ 2116.8 ÷ 126
   - B 21168 ÷ 126
   - C 211.68 ÷ 126
   - D 21168 ÷ 12.6

6. Which is the exponent form of the expression?

   24 × 24 × 24
   - F 3 × 24
   - G 3³
   - H 3²⁴
   - Ⓙ 24³

7. Loraine sleeps between 6 and 8 hours each night. What is a reasonable estimate of the number of minutes she sleeps in a week?
   - A Less than 1,500
   - B Between 1,500 and 2,500
   - Ⓒ Between 2,500 and 3,500
   - D Between 3,500 and 4,500

8. Hunter saves $3.50 each week to buy a CD boxed set that sells for $52.50. He has already saved $10.50. How many more weeks does he need to save money?
   - F 11 weeks
   - Ⓖ 12 weeks
   - H 14 weeks
   - J 15 weeks

9. **Write About It** Describe a pattern you see in the set of problems at the right.

   600 ÷ 10 = 60   6 ÷ 10 = 0.6
   60 ÷ 10 = 6   0.6 ÷ 10 = 0.06

   Possible answer: When dividing by 10, you move the decimal point one place to the left.

LESSON 4.4

- Demonstrate the necessity of sometimes writing a zero in the dividend.

In Example 3, what would the quotient have been if you had not written the zero in the dividend? **132**

ADDITIONAL EXAMPLE

Example 3, p. 77

Divide. 663.5 ÷ 0.25 **2,654**

3 Practice

Guided Practice

Do Check for Understanding Exercises 1–11 with your students. Identify those having difficulty and use lesson resources to help.

COMMON ERROR ALERT

Students may mistakenly move the decimal point all the way to the right in both the divisor and the dividend. Remind them that they must move the decimal point the same number of places in both the divisor and dividend.

Independent Practice

Note that Exercises 39 and 41 are **multistep or strategy problems**. Assign Exercises 12–41.

EXAMPLE 3

Sometimes there aren't enough places in the dividend to move the decimal to the right.

Divide. 158.4 ÷ 0.12

```
0.12)158.40
```
Multiply the divisor and dividend by 100.
0.12 × 100 = 12   158.4 × 100 = 15,840
Write a zero in the dividend.

```
      1320
12)15840
    -12
     38
    -36
     24
    -24
     00
```
Divide.

Since the remainder is zero, the quotient is a whole number. You do not need to put the decimal point in the answer.

So, 158.4 ÷ 0.12 = 1,320.

CHECK FOR UNDERSTANDING

Think and Discuss ▶ Look back at the lesson to answer each question.

1. **Explain** how to change the divisor and the dividend before you solve the problem 55.8 ÷ 0.18. **Multiply both by 100.**
2. **Compare** the quotients 4.5 ÷ 1.5 and 45 ÷ 15. **The quotients are the same.**

Guided Practice ▶ Rewrite the problem so that the divisor is a whole number.

3. 9.6 ÷ 1.6   4. 73.6 ÷ 0.5   5. 48.24 ÷ 2.4
   96 ÷ 16   736 ÷ 5   482.4 ÷ 24

Copy the problem. Place the decimal point in the quotient.

6. 28.50 ÷ 2.50 = 1140   7. 34.178 ÷ 2.3 = 1486   8. 62.44 ÷ 7 = 892
   11.40   14.86   8.92

Divide. Estimate to check.

9. 7.88 ÷ 4 **1.97**   10. 33.66 ÷ 11 **3.06**   11. 0.55)2.42 **4.4**

PRACTICE AND PROBLEM SOLVING

Independent Practice ▶ Rewrite the problem so that the divisor is a whole number.

12. 48.4 ÷ 0.4   13. 8.19 ÷ 0.09   14. 3.7 ÷ 2.1
    484 ÷ 4   819 ÷ 9   37 ÷ 21
15. 2.39 ÷ 0.05   16. 45.218 ÷ 0.23   17. 233.58 ÷ 10.2
    239 ÷ 5   4,521.8 ÷ 23   2,335.8 ÷ 102

Copy the problem. Place the decimal point in the quotient.

18. 325.8 ÷ 3 = 1086   19. 53.07 ÷ 8.7 = 61   20. 10.2 ÷ 2 = 51
    108.6   6.1   5.1
21. 84.87 ÷ 12.3 = 69   22. 274.89 ÷ 1.5 = 18326
    6.9   183.26

78 Chapter 4

ALTERNATIVE TEACHING STRATEGY — SCAFFOLDED INSTRUCTION

Purpose Students use play money to model dividing with decimals.

Materials *For each pair* 1 index card; play money—40 pennies

On index cards, write division exercises like the following, one to a card:

Display the following: 0.05)0.35

Discuss the meaning of the division with the students. They should realize that they are asked to find how many groups of 5 hundredths there are in 35 hundredths.

Direct the students to model the division using the pennies.

- Show 35 pennies.

- Separate pennies into groups of 5.

Review how each part of the division was modeled: the dividend was the total number of pennies, the divisor was the number of pennies in each group, and the quotient was the number of groups.

```
       7
0.05)0.35
```

Give each pair one of the index cards with a division problem. Have them model the division and record their work. Have the pairs exchange index cards and repeat. **Check students' work.**

Divide. Estimate to check.

23. $11\overline{)109.01}$ 9.91
24. $90\overline{)10.8}$ 0.12
25. $60\overline{)12.6}$ 0.21
26. $0.38\overline{)13.3}$ 35
27. $6.41\overline{)135.892}$ 21.2
28. $38.2\overline{)469.86}$ 12.3
29. $44.28 \div 5.4$ 8.2
30. $80.1 \div 9$ 8.9
31. $90.3 \div 6$ 15.05
32. $1.26 \div 0.2$ 6.3
33. $13.2 \div 0.06$ 220
34. $42.5 \div 0.05$ 850

Evaluate each expression.

35. $3 - 1.6 \times 0.4$ 2.36
36. $5 - 2.4 \div 8$ 4.7
37. $0.09 \div 3 + 3$ 3.03

Problem Solving ▶ Applications

38. Emily rents 5 DVD movies for $28.75. Is the price for one movie closer to $5 or to $6? Explain. **$6; $28.75 ÷ 5 = $5.75**

39. Jonelle saves $4.95 every week to buy a video that costs $29.70. She has already saved $9. For how many more weeks does she need to save money to have enough to buy the video? **5 weeks**

40. **What's the Error?** Michael divided 4.25 by 0.25 and got a quotient of 0.17. Explain the error. What is the correct quotient? **He did not multiply both the divisor and dividend by 100; 17.**

41. Andrew reads 22 pages the first day. He plans to increase the number of pages he reads by 4 a day until he finishes the book. How many pages will he be reading at the end of 3 days? **30 pages**

MIXED REVIEW AND TEST PREP

42. Multiply. 2.051×8.6 (p. 70) **17.6386**
43. Is 208.605 greater than or less than 206.605? (p. 52) **greater than**
44. Use mental math to find $210 \div 42$. (p. 36) **5**

★ 45. **TEST PREP** Which shows the value of r when $18 + r = 52$? (p. 30) **A**

A 34   B 30   C 24   D 14

★ 46. **TEST PREP** Jocelyn works 12 hours per week. She earns $7.25 an hour. How much does she earn in 4 weeks? (p. 70) **J**

F $87   G $174   H $290   J $348

PROBLEM SOLVING Thinker's Corner

REASONING The ancient Greeks discovered that there is one particular number that can be used to relate the dimensions of any circle. They called this particular number π (pi) and found that it is the same number regardless of the size of the circle.

- Determine the approximate value of π by dividing the distance around each circle described below (the circumference) by the distance across that circle (the diameter). Is the answer the same for each? **yes**

	Circumference	Diameter	
1. Circle 1	6.28 in.	2 in.	**3.14**
2. Circle 2	9.42 in.	3 in.	**3.14**

EXTRA PRACTICE page H35, Set C

79

LESSON 4.5

Problem Solving Skill: Interpret the Remainder

LESSON PLANNING

Objective To solve problems by using the skill *interpret the remainder*

Intervention for Prerequisite Skills
Whole-Number Operations, Remainders (For intervention strategies, see page 65.)

Lesson Resources Problem Solving Think Along, p. TR1

NCTM Standards
1. Number and Operations
6. Problem Solving
8. Communication

Math Background

The remainder can play different roles in division problems.

- Sometimes the remainder is disregarded, such as when you divide 42 pencils evenly among 12 students.
- Sometimes the remainder indicates that you must round up to the next whole number, such as if you want to know how many cars are needed to transport 22 children if 4 children can go in each car and all children must go.
- Sometimes the remainder is the answer, such as if you want to know how many apples will not be shipped if 203 apples are packed 48 to a box for shipping.

It may be helpful for students to draw a diagram of the division problem to help determine the role played by the remainder in each case.

WARM-UP RESOURCES

 **NUMBER OF THE DAY**

Divide your age in months by 12. What does the solution tell you? *my age in years and months*

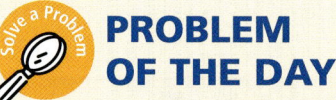 **PROBLEM OF THE DAY**

Marge had five 32-oz bottles of milk. Monday she drank 6 oz from the first bottle, Tuesday she drank 12 oz from the second bottle, Wednesday she drank 18 oz from the third bottle, and so on. Each day she divided the remaining milk among her five cats so that each got a whole number of ounces. She saved the remainder. On the sixth day, she drank what was left. How much did she drink on Day 6? *10 oz*

Solution Problem of the Day tab, p. PD4

 **DAILY FACTS PRACTICE**

Have students practice addition and subtraction facts by completing Set F of *Teacher's Resource Book*, p. TR95.

80A Chapter 4

INTERVENTION AND EXTENSION RESOURCES

ALTERNATIVE TEACHING STRATEGY

Materials 6 index cards

Reinforce students' ability to **interpret remainders**. On each of 6 index cards, write a word problem such as this:

Mitch has $15. Batteries cost $2. How many can Mitch buy? 7 batteries

Call on a volunteer to read the problem and then ask students to solve. Ask: What is the remainder? What does the remainder represent? $1; the amount of money Mitch has left over

Repeat with other word problems in which the remainder might be the answer or might tell students that another bus or box, for instance, is needed. Check students' work.

AUDITORY
VERBAL/LINGUISTIC

ENGLISH LANGUAGE LEARNERS

To help students learn the **vocabulary of division,** bring a group of 10 students to the front of the class. Ask them to group themselves into 3 equal groups. Use word cards or sentence strips and have students identify the terms *dividend, divisor, quotient,* and *remainder*. 10 students, 3 groups, 3 students, 1 student

Ask:
- If you are riding a roller coaster in cars that hold 3 people, how many cars would you need? 4 cars
- How many would be in the fourth car? 1 student
- If the park attendant must fill each car, what does the remainder tell you? One person must ride with two other people to form a new group of three.

KINESTHETIC
INTERPERSONAL/SOCIAL

READING STRATEGY

Use Context Tell students to look for words and phrases in a problem solving situation to help them decide how to interpret the remainder. Have students identify the words or phrases in each of the three situations given on page 80 that give them information about how to interpret the remainder. How many packages will Ms. Gordon need to buy?; How many lengths will Ms. Gordon have?; How many (left-over) prizes were given?

ADVANCED LEARNERS

Challenge students to apply what they've learned about **interpreting remainders.** Have them work in pairs to write a division exercise with a remainder.

- Direct one of the students to select a number between 10 and 50 to use as a dividend and the other student to select a number between 1 and 10 to use as a divisor.
- Then, have students write two word problems based on their division exercise in which the remainder in each problem is interpreted differently. Check students' work.

VISUAL
INTERPERSONAL/SOCIAL

MULTISTEP AND STRATEGY PROBLEMS

The following multistep and strategy problems are provided in Lesson 4.5:

Page	Item
81	5, 7, 8, 9

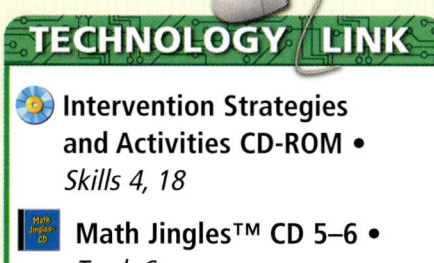

TECHNOLOGY LINK

Intervention Strategies and Activities CD-ROM •
Skills 4, 18

Math Jingles™ CD 5–6 •
Track 6

LESSON 4.5 ORGANIZER

Objective To solve problems by using the skill *interpret the remainder*

Lesson Resources Problem Solving Think Along, p. TR1

1 Introduce

QUICK REVIEW provides review of pre-requisite skills.

Why Learn This? Interpreting remainders will help you know that you have a sufficient number of packaged party favors for a given number of guests. *Share the lesson objective with students.*

2 Teach

Guided Instruction

- Call on a volunteer to read each problem on page 80, and discuss the solutions.

  How many drinks would Ms. Gordon have if she bought 27 packages? Is that enough? 162 drinks; no

  How do you use the remainder 1? The remainder tells you that you will need to buy another package.

  REASONING **How are the remainders different in the second and third problems?** In the second problem you do not need the remainder, but in the third problem it is the answer.

  Does the remainder in the second problem refer to 6 ribbons or 6 feet of ribbon? 6 feet
  What portion of an 8-foot length of ribbon would Ms. Gordon have left over? $\frac{3}{4}$

LESSON 4.5

Learn how to interpret the remainder in a division problem.

PROBLEM SOLVING SKILL
Interpret the Remainder

QUICK REVIEW

1. $6\overline{)480}$ 80
2. $8\overline{)656}$ 82
3. $7\overline{)1,047}$ 149 r4
4. $3\overline{)1,305}$ 435
5. 142 boxes ÷ 14 shelves = 10 shelves with ■ boxes left 2

The sixth-grade class at Hightown Middle School has an annual class picnic. With the help of some students, Ms. Gordon is planning for the picnic this year.

| Ms. Gordon needs 163 boxed drinks for lunch. A package holds 6 boxes. How many packages of drinks will Ms. Gordon need to buy? | $\begin{array}{r} 27\text{ r}1 \\ 6\overline{)163} \\ -12 \\ \hline 43 \\ -42 \\ \hline 1 \end{array}$ | Since 27 packages hold only 162 total boxes, increase the quotient by 1. So, Ms. Gordon needs to buy 28 packages of drinks. |

| Ms. Gordon has 158 ft of ribbon to use for games. She will cut the ribbon into 8-ft lengths. How many 8-ft lengths of ribbon will Ms. Gordon have? | $\begin{array}{r} 19\text{ r}6 \\ 8\overline{)158} \\ -8 \\ \hline 78 \\ -72 \\ \hline 6 \end{array}$ | The remainder is not enough for another 8-ft length of ribbon. Drop the remainder. So, Ms. Gordon will have 19 pieces of ribbon. |

| Students collected 158 prizes. The prizes were put in packages of 5 each. The remaining prizes were given to another class in the school. How many prizes were given to the other class? | $\begin{array}{r} 31\text{ r}3 \\ 5\overline{)158} \\ -15 \\ \hline 8 \\ -5 \\ \hline 3 \end{array}$ | The remainder is your answer. So, 3 prizes were given to the other class. |

Talk About It ▶
- What does the remainder in the first problem mean?
  1 more boxed drink will be needed; 27 packages are not enough.
- **What if** Ms. Gordon buys 28 packages of boxed drinks? How many extra boxed drinks will Ms. Gordon buy?
  5 extra boxed drinks
- **What if** Ms. Gordon has a total of 165 ft of ribbon? How many 8-ft lengths of ribbon will she have? How long is the ribbon that's too short? 20 lengths; 5 ft

80 Chapter 4

RETEACH 4.5

Problem Solving Skill: Interpret the Remainder

When you solve a problem using division, there is often a remainder. When there is a remainder, you must decide what it means for the problem. You need to determine whether the solution to the problem is
- the quotient without the remainder,
- the next whole number greater than the quotient,
- the next whole number less than the quotient, or
- the remainder.

A real-estate agent ordered 190 doughnuts for an open house. The Donut Shoppe packed the order in boxes of 12 doughnuts. How many boxes were needed?

Step 1: Think about what you know and what you are asked to find.
- You know the number of doughnuts ordered and the number that were put in each box.
- You are asked to find the number of boxes that were needed.

Step 2: Decide on a plan to solve.
- Since the same number of doughnuts were put into each box, use division to solve the problem.
- Examine the quotient and remainder and decide how they relate to the problem.

Step 3: Carry out the plan.

$12\overline{)190}$ 15 r10

The quotient 15 means that 15 boxes were filled. The remainder 10 means that there were 10 doughnuts left after the last full box was packed.

So, the Donut Shoppe needed 15 + 1, or 16 boxes in all, to pack the agent's 190 doughnuts.

Interpret the remainder to solve.

1. A school received a shipment of 166 new social studies textbooks. The textbooks had been packed into cartons with 12 books in each full carton. How many cartons were delivered to the school?
   14 cartons (13 full cartons, 1 with 10 books)

2. In the school science laboratory, students generally work in groups of 4. Extra students are divided among the groups to make some groups of 5. If there are 27 students in a class, how many groups will have 5 students?
   3 groups (out of 6)

PRACTICE 4.5

Problem Solving Skill: Interpret the Remainder

Solve the problem by interpreting the remainder.

1. Thirty-seven people are attending a party at a restaurant. In the banquet room, the restaurant staff has set up tables that can each seat 8 people. What is the least number of tables that the group will use?
   5 tables

2. There are 23 pancakes on the griddle at a restaurant. The chef places 4 pancakes on each order. How many orders can the chef fill, and how many pancakes must be added to those remaining to make another order?
   5 orders; 1 pancake

3. A library reading room contains a number of tables that can seat 4 people. What is the least number of tables needed to seat 54 people?
   14 tables

4. A group of 5 friends wants to buy snacks. If each snack costs $0.75 and they have a total of $4.80 to spend, how many snacks can they buy?
   6 snacks

5. The chef at a restaurant uses 3 eggs to make each omelet. If the chef has 200 eggs, how many 3-egg omelets can he make?
   66 omelets

6. A total of 125 hamburgers were sold at a fund-raiser at the last football game. If the hamburger patties came in packages of 8, how many packages were opened?
   16 packages

Mixed Review

Estimate the sum or difference. **Possible estimates are given.**

7. 671 + 902 = 1,600
8. 478 − 310 = 200
9. 831 − 289 = 500
10. 1,226 + 533 = 1,700
11. 661 + 2,403 = 3,100
12. 1,729 − 494 = 1,200
13. 488 − 391 = 100
14. 2,994 + 1,258 = 4,000

Solve each equation by using mental math.

15. $m + 12 = 15$ $m = 3$
16. $5w = 20$ $w = 4$
17. $x - 7 = 8$ $x = 15$
18. $q + 4 = 10 + 6$ $q = 12$
19. $6r = 24$ $r = 4$
20. $y - 9 = 10$ $y = 19$
21. $a - 2 = 8 + 6$ $a = 16$
22. $d + 3 = 21 - 7$ $d = 11$

PROBLEM SOLVING PRACTICE

Solve the problem by interpreting the remainder.
A total of 39 students and adults tour the science museum to see an exhibit about the shaping of the Earth's surface. The tour director can take groups of up to 5 on each tour. All 39 students and adults need to see the exhibit.

1. How many complete groups of 5 people can the tour director take? **C**

   A 5 groups **B** 6 groups **C** 7 groups **D** 8 groups

2. What is the least number of tours the director will have to give to accommodate all 39 students? **H**

   F 6 tours **G** 7 tours **H** 8 tours **J** 9 tours

3. James has $4.39 to buy magnets at the science museum. Each magnet costs $0.95. He wants to buy as many magnets as he can. How many magnets can James buy? **4 magnets**

4. Sharon bought a package of 15 postcards at the museum. She gave the same number of postcards to each of her 4 teachers and kept the ones left over. How many postcards did Sharon keep? **Possible answers: 3, 7 or 11**

MIXED APPLICATIONS

5. A train that is scheduled to arrive at 5:15 P.M. arrives 20 minutes late. If the train left at 9:30 A.M., how long was the trip? **8 hr 5 min**

6. A total of 51 students and teachers are using cars to go on a field trip. Six people ride in each car. How many cars are needed for the field trip? **9 cars**

7. Mark estimates that he needs 1 minute to solve a short homework problem and 5 minutes for each long problem. If he has 25 short and 8 long problems, how long should his homework take? **65 min, or 1 hr 5 min**

8. Tina makes bracelets for her friends. She uses 3 red beads for every 7 yellow beads to make a pattern. For one bracelet, she uses a total of 50 beads. How many of each color does she use? **15 red, 35 yellow**

9. A board game has 3 times as many red playing pieces as blue. It has 5 times as many green pieces as blue. There are 12 blue pieces. How many playing pieces are there in all? **108 pieces**

10. Edgar has twice as many library books as his brother. If Edgar has 10 books in all, how many library books must he return to have the same number as his brother? **5 library books**

11. **What's the Question?** Use the table at the right. Each Presidential term in office is 4 years. The answer is that he served more than one term but fewer than two.
    Possible question: How many terms did Richard Nixon serve?

United States Presidents	
President	Years in Office
Richard Nixon	1969–1974
Gerald Ford	1974–1977
Jimmy Carter	1977–1981
Ronald Reagan	1981–1989
George Bush	1989–1993

81

---

3 Practice

Guided Practice
Do Problem Solving Practice Exercises 1–4 with your students. Identify those having difficulty and use lesson resources to help.

Independent Practice
Note that Exercises 5, 7, 8, and 9 are **multistep or strategy problems**. Assign Exercises 5–11.

Scaffolded Instruction Use the prompts on Transparency 4 to guide instruction for the multistep or strategy problem in Exercise 9.

For students who have limited reading skills, it may be helpful to have them draw pictures that show some of the information given in the problems as you read them aloud. For example, for Exercise 5 have students draw clocks showing the start and end times for the train. For Exercise 8 have them draw the beads.

4 Assess

Summarize the lesson by having students:

DISCUSS Describe how Exercises 3 and 4 are alike and how they are different.
Possible answer: They are alike because you divide to solve each; they are different in the role the remainder plays in the answer.

WRITE Describe three ways you can use a remainder. Possible answer: It can be the answer, it can be ignored, or it can help decide what the answer is.

Lesson Quiz

1. Trent has invited 26 friends to a cookout. He wants 2 hot dogs for each friend, and hot dogs come in packages of 8. How many packages should he buy? **7 packages**

2. Sharlene is making ribbon decorations for the party. Each one takes 18 inches of ribbon. How many decorations can she make from 100 inches of ribbon? **5 decorations**

CHALLENGE 4.5

Interpret the Remainder

In each problem, explain the mistake in reasoning that was made. Describe how the mistake might have been avoided. Possible answers given.

1. All 159 sixth grade students at McKinley Middle School and their 7 teachers are going on a trip. The principal ordered 5 buses that could each carry 31 passengers. He planned to use a school van that could carry 8 passengers as transportation for those people who would not have seats on the bus.
   The 5 buses could carry 155 of the 166 people, leaving 11 people, not 8. The principal could have planned for 6 buses instead.

2. A farmer is putting up a new fence that will measure 64 feet long. He wants to place posts 8 feet apart, beginning at one end, to support the fencing. He orders 8 fence posts.
   The farmer needs 9, not 8, posts. Instead of dividing 64 by 8, he could have drawn a diagram to find the number of posts needed.

3. Margo and her sister have $10.00 to spend on cards. The cards Margo has chosen cost $0.95 each. She tells her sister they can buy 11 cards with the money they have. They stand in line to pay for the 11 cards.
   Margo and her sister do not have enough money. $10.00 divided by $0.95 is 10 r50. The remainder means they only have $0.50 left. This is not enough to buy another card.

4. Eighty-five teenagers have come to register for a new baseball league. Each team will have 9 members. The organizers of the league tell the teens that as soon as 4 more people arrive, they will have exactly enough players for everyone to be assigned to a team.
   85 ÷ 9 = 9 r4; The organizers thought the remainder meant 4 more players were needed, but the remainder meant those not on a team, so, 5 more players are needed.

READING STRATEGY 4.5

Use Context

If there is a word, phrase, or paragraph you do not understand, context can help you. **Context** means the words, phrases, pictures, or graphic aids that go along with what you are reading. Context can help you decide how to interpret the remainder.

VOCABULARY context

Read the following problem.

Thirty-eight sixth graders are going to see a band from Puerto Rico that specializes in Caribbean music. Each driver can take 4 students. How many drivers are needed?

1. Use context to help you decide how to treat the remainder, if there is one. If there is a remainder, should you add 1 to the quotient, drop the remainder, or use it as the answer? Why?
   Whether or not the students fill every car, they still must be transported. Add 1 to the quotient.

2. Solve the problem.
   38 ÷ 4 = 9 R2; 10 drivers

Solve the problem. Use context to help you decide how to interpret the remainder.

3. The band needs 40 minutes of music to make 1 CD. The songs they know last for 2 hours and 42 minutes. How many CDs could they cut now?
   4 CDs

4. Alexis Rivera wants to take some friends to the concert. She has $135 and each ticket costs $30. How many tickets can she buy?
   4 tickets

5. The concert was attended by 1,000 people. If there were 36 seats in a row, how many rows could have been filled?
   27 rows

6. The band has 5,000 copies of their new CD. If 73 music stores each get the same number of copies of the CD, how many CDs will be left over?
   36 CDs

7. How many containers for 1 dozen eggs are needed for 2,000 eggs?
   167 containers

8. If 25 books fit on a shelf, how many shelves are needed for 465 books?
   19 shelves

81

LESSON 4.6

Algebra: Decimal Expressions and Equations

LESSON PLANNING

Objective To evaluate expressions with decimals and to use mental math and substitution to solve equations with decimals

Intervention for Prerequisite Skills
Whole-Number Operations (For intervention strategies, see page 65.)

NCTM Standards
1. Number and Operations
2. Algebra
6. Problem Solving
8. Communication

Math Background
Students use the same skills to evaluate expressions involving decimals and solve decimal equations as they used with whole numbers. As you help students solve problems involving decimal expressions and equations, remind them of the following:

- An algebraic expression may have different values, depending on the number substituted for the variable.

- To evaluate an algebraic expression, replace the variable with a number and perform the indicated operation or operations.

- An equation has one solution. Determine if a number is a solution to an equation by substituting it into the equation. Then simplify. It is a solution if the resulting statement is true.

WARM-UP RESOURCES

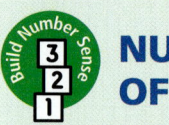 **NUMBER OF THE DAY**  Transparency 4.6

Write an expression that will give you the part of an hour that has passed so far this hour. $\frac{n}{60}$, where n is the number of minutes past the hour

PROBLEM OF THE DAY  Transparency 4.6

Lashonda and Mark each have the same number of coins. Lashonda has $8.25 in quarters. Mark has all dimes. How much more money does Lashonda have than Mark? Lashonda has $4.95 more than Mark.

Solution Problem of the Day tab, p. PD4

 **DAILY FACTS PRACTICE**

Have students practice multiplication and division facts by completing Set G of *Teacher's Resource Book*, p. TR95.

82A Chapter 4

INTERVENTION AND EXTENSION RESOURCES

ALTERNATIVE TEACHING STRATEGY

Display a list of equations on the left and a list of number sentences students can use to **solve the equations** on the right. For example:

$3 \times a = 0.9$ **0.3**	$9.26 - 5.2 = 4.06$
$5.2 + c = 9.26$ **4.06**	$1.5 \times 3 = 4.5$
$d \div 3 = 1.5$ **4.5**	$10.4 + 5.2 = 15.6$
$l - 5.2 = 10.4$ **15.6**	$0.9 \div 3 = 0.3$

Have students copy the equations on paper and then match the equation on the left with the number sentence on the right that helps solve the equation. Have volunteers explain their thinking and show that solutions check.

VISUAL
LOGICAL/MATHEMATICAL

ENGLISH LANGUAGE LEARNERS

Materials 9 index cards

Help students **understand how to evaluate an expression.** Write each number, operation, and variable on an index card: 3.2, +, x, 1.6, 4.8, 4.1, 7.3, 10.5, 13.7

Give three students the cards for +, 3.2, and x. Place them in the front of the room to form the expression $x + 3.2$. Then distribute the remaining cards to other students.

Ask students to evaluate the expression for $x = 1.6$. The student holding the card 1.6 should come to the front of the room and replace the student holding the x. Have students evaluate the expressions and then have the student with the 4.8 card show it.

Repeat the activity for these values of the variable 4.1 and 10.5. **7.3 and 13.7**

KINESTHETIC; BODILY/KINESTHETIC

MULTISTEP AND STRATEGY PROBLEMS

The following multistep or strategy problem is provided in Lesson 4.6:

Page	Item
83	23

SCIENCE CONNECTION

Encourage students to **practice evaluating expressions.** Tell them that the distance a glacier flows in 52 weeks can be determined by substituting the rate of flow into the expression $52r$. Have students find the distance each of the following glaciers would flow in a year.

Alpine glacier

    17.7 yd/wk   **920.4 yd**

Greenland glacier

    54.9 yd/wk   **2,854.8 yd**

Antarctic glacier

    84.6 yd/wk   **4,399.2 yd**

Extend the activity by having students write an expression to show each distance d as miles. HINT: 1 mi = 1,760 yd. To find the distance in miles, use the expression $\frac{d}{1,760}$.

VISUAL; LOGICAL/MATHEMATICAL

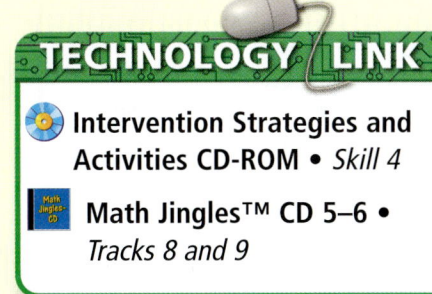

TECHNOLOGY LINK

- Intervention Strategies and Activities CD-ROM • *Skill 4*
- Math Jingles™ CD 5–6 • *Tracks 8 and 9*

LESSON 4.6 ORGANIZER

Objective To evaluate expressions with decimals and to use mental math and substitution to solve equations with decimals

1 Introduce

QUICK REVIEW provides review of prerequisite skills.

Why Learn This? Solving decimal expressions and equations using mental math will prepare you for algebra and more complex problems. Share the lesson objective with students.

2 Teach

Guided Instruction

- Review the meaning of the terms numerical expression and algebraic expression.

  Give an example of a numerical expression involving two decimals. Possible answer: $3.2 + 5.6$

  Give an example of an algebraic expression containing a decimal. Possible answer: $4.5 \times s$

- As students look at the first expression on page 82, ask:

  Why is the a replaced by the 7? 7 people ate lunch and we need to evaluate the expression for $a = 7$.

- As students finish Example 1, ask:

  What operation will you use to find t? Explain. subtraction; you think what number added to $4.24 = 9.48$.

REASONING What do you know about the relationship of n to 15.3 in the equation $n - 10.4 = 15.3$? **Explain.** Possible answer: The variable n is greater than 15.3 because you subtract 10.4 to get 15.3.

Algebraic Thinking Substituting values for variables in expressions helps students become comfortable with the idea that variables stand for numbers. This will make the transition to solving for variables easier.

ADDITIONAL EXAMPLE

Example, p. 82

Solve the equation $n \div 5 = 0.3$ by using mental math. $n = 1.5$

LESSON 4.6

ALGEBRA
Decimal Expressions and Equations

Learn how to evaluate expressions and solve equations with decimals.

QUICK REVIEW

Solve by using mental math.
1. $s + 14 = 36$ 2. $18 \div p = 9$
   $s = 22$ $p = 2$
3. $57 - d = 42$ 4. $w \div 7 = 7$
   $d = 15$ $w = 49$
5. $a \times 8 = 24$
   $a = 3$

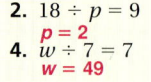

Different groups of friends eat lunch together at a cafe. The cost of the lunch is $6.45 per person. Write an expression to find the total cost of a lunch for a group of friends.

Let a be the number of friends buying lunch.

$a \times 6.45$ or $6.45a$ *Write the expression.*

The number of friends varies. What is the total cost for 7 friends?

$a \times 6.45$ *Write the expression.*
7×6.45 *Replace a with 7.*
45.15 *Multiply.*

So, the total cost is $45.15.

- Evaluate the expression $w \div 3 + 9.3$ for $w = 4.8$. **10.9**

You solved equations with whole numbers by using mental math. You can use the same methods to solve some equations with decimals.

EXAMPLE

Solve the equation $h \div 6 = 0.6$ by using mental math.

$h \div 6 = 0.6$ *What number divided by 6 = 0.6?*
$h = 3.6$ THINK: $6 \times 0.6 = 3.6$.

$h \div 6 = 0.6$ *Check your answer. Replace h with 3.6.*
$3.6 \div 6 = 0.6$
$0.6 = 0.6$

So, $h = 3.6$.

- Solve. $t + 4.24 = 9.48$ $t = 5.24$

RETEACH 4.6

Algebra: Decimal Expressions and Equations

You speak and write using words, for example
 the number of days in a week times the number of weeks
Some words can be expressed as numbers or numerical expressions.
 days × weeks
 7 × 5

An algebraic expression uses a letter to represent a number. For example, if the number of weeks is written as the letter w, the algebraic expression is $7 \times w$.
It can also be written as $7w$.
To evaluate the expression $7w$ to find the number of days in 8 weeks, write $7 \times 8 = 56$.

Match the algebraic expression with the words.
1. number of inches in a foot times the number of feet   **C**
2. number of items in a dozen plus some more items   **A**
3. sum of money divided among several people   **D**
4. total amount of money less the amount spent   **B**

A $12 + s$
B $25 - d$
C $12n$
D $25 \div p$

Solve using mental math.
5. How much is $6y$ if $y = 3$? **18**
6. How much is $8 + r$ if $r = 9$? **17**
7. How much is $24 \div t$ if $t = 3$? **8**
8. How much is $35 - a$ if $a = 15$? **20**
9. How much is $4.5 - q$ if $q = 1.9$? **2.6**
10. How much is $5b$ if $b = 1.1$? **5.5**
11. How much is $3.6 \div z$ if $z = 3$? **1.2**
12. How much is $2.4 + c$ if $c = 4.2$? **6.6**

Evaluate each expression.
13. $5m$ for $m = 3.3$ **16.5**
14. $w + 2$ for $w = 8.4$ **4.2** [sic]
15. $y + 12.4$ for $y = 5.2$ **17.6**
16. $7.4 - b$ for $b = 1.3$ **6.1**
17. $6x$ for $x = 2.5$ **15**
18. $m - 4.6$ for $m = 8.9$ **4.3**

PRACTICE 4.6

Algebra: Decimal Expressions and Equations

Evaluate each expression.
1. $t - 1.2$ for $t = 3$ **1.8**
2. $y + 4.6$ for $y = 2.4$ **7**
3. $8.2 - m$ for $m = 1.1$ **7.1**
4. $2.4 \div a$ for $a = 6$ **0.4**
5. $6g$ for $g = 1.5$ **9**
6. $j - 6.3$ for $j = 9.6$ **3.3**
7. $12.6 + r$ for $r = 4.4$ **17**
8. $4.5 \div p$ for $p = 9$ **0.5**
9. $7.24 - q$ for $q = 1.04$ **6.20 or 6.2**
10. $6.18 \div y$ for $y = 3$ **2.06**
11. $t + 4.66$ for $t = 2.1$ **6.76**
12. $5h$ for $h = 2.4$ **12**

Solve each equation by using mental math.
13. $w + 4.5 = 8$ $w = 3.5$
14. $\frac{k}{3} = 2.5$ $k = 7.5$
15. $1.4 = \frac{t}{2}$ $t = 2.8$
16. $m - 7.6 = 2.4$ $m = 10$
17. $3a = 6.9$ $a = 2.3$
18. $9c = 22.5$ $c = 2.5$
19. $3b = 6.4 + 2.6$ $b = 3$
20. $w + 10.3 = 21.7$ $w = 11.4$
21. $13.7 = d - 3.4$ $d = 17.1$
22. $4.8 = \frac{n}{4}$ $n = 19.2$
23. $\frac{x}{5} = 19.5$ $x = 97.5$
24. $7h = 15.4$ $h = 2.2$

Mixed Review
Estimate. Possible estimates are given.
25. $6.9 + 7.8$ **15**
26. 31.77×6 **180**
27. $63.85 \div 8$ **8**
28. $17.04 - 9.8$ **7**
29. $18.58 + 21.44$ **40**
30. 91.92×4 **360**
31. $54.3 - 19.7$ **34**
32. $80.8 \div 9.2$ **9**

Find the quotient.
33. $88.8 \div 6$ **14.8**
34. $59.4 \div 36$ **1.65**
35. $38.88 \div 7.2$ **5.4**
36. $31.108 \div 2.2$ **14.14**

CHECK FOR UNDERSTANDING

Think and Discuss

Look back at the lesson to answer each question.

1. What if 12 friends went to lunch at the cafe? Show how you would find the total cost of the lunch. $a \times 6.45$; 12×6.45; $77.40
2. Explain how you would solve the equation $3.2 = a \div 2$.
   Possible answer: THINK: $6.4 \div 2 = 3.2$, so $a = 6.4$

Guided Practice

Evaluate each expression.

3. $a + 3.4$ **11.7**
   for $a = 8.3$
4. $1.6 \div b$ **4**
   for $b = 0.4$
5. $9.16 - a$ **5.08**
   for $a = 4.08$

Solve each equation by using mental math.

6. $\frac{4.8}{k} = 8$
   $k = 0.6$
7. $m - 12.7 = 6.3$
   $m = 19$
8. $3t = 21.9$
   $t = 7.3$

PRACTICE AND PROBLEM SOLVING

Independent Practice

Evaluate each expression.

9. $2h$ **4.6**
   for $h = 2.3$
10. $9.6 \div a$ **3.2**
    for $a = 3$
11. $j + 7.1$ **14**
    for $j = 6.9$

12. $4.17 - c$ **3.08**
    for $c = 1.09$
13. $m \div 6 + 3.6$ **3.9**
    for $m = 1.8$
14. $g + h - 3.2$
    for $g = 4.1$ and
    $h = 2.3$ **3.2**

Solve each equation by using mental math.

15. $r + 8.1 = 15.8$
    $r = 7.7$
16. $1.7 = \frac{d}{4}$
    $d = 6.8$
17. $4a = 32.8$
    $a = 8.2$
18. $x - 2.4 = 8.6$
    $x = 11$
19. $p + 11.1 = 28.7$
    $p = 17.6$
20. $3.2r = 7.1 + 5.7$
    $r = 4$

Problem Solving Applications

21. Let n represent the number of miles Jeremy rides his bicycle to attend 6 baseball practices. Write an expression to show how far he travels for one practice. Evaluate the expression for $n = 28.8$ mi.
    $n \div 6$, 4.8

22. **What's the Error?** Explain the error at the right. Give the correct solution. 24.8 should not be added to 30 to find the value of x; $x = 5.2$

    $24.8 + x = 30$
    $x = 54.8$

23. David jogs 3 mi each weekday and 7 mi each Saturday. Ray jogs 18 mi each week. How much farther does David jog in a week? **4 mi**

MIXED REVIEW AND TEST PREP

24. $4.38 \div 7.5$ (p. 76)
    0.584
25. $18 \div 6 + (16 \times 3) - 14$ (p. 44)
    37
26. $6,045 - 973$ (p. 20)
    5,072
27. Order from least to greatest. 3.58, 3.08, 3.85, 3.508 (p. 52) **3.08, 3.508, 3.58, 3.85**

⭐ 28. **TEST PREP** The sum of two numbers is 35. Their difference is less than 10. Which is not a possible pair? (p. 26) **C**

    A 15, 20   **B** 17, 18   **C** 10, 25   **D** 22, 13

EXTRA PRACTICE page H35, Set D

83

---

3 Practice

Guided Practice

Do Check for Understanding Exercises 1–8 with your students. Identify those having difficulty and use lesson resources to help.

COMMON ERROR ALERT

When using mental math to solve equations, students may use an incorrect operation to solve. That is, in solving $t + 4.24 = 9.48$, they may add 4.24 to 9.48 instead of subtracting it. The best way to catch these errors is to check. So, constantly encourage students to substitute each answer in its equation to verify the solution.

Independent Practice

Note that Exercise 23 is a **multistep or strategy problem**. Assign Exercises 9–23.

As students evaluate expressions, have them write the number for the variable on a small piece of paper and place it over the variable in the expression to see it as a numerical expression.

MIXED REVIEW AND TEST PREP

Exercises 24–28 provide **cumulative review** (Chapters 1–4).

4 Assess

Summarize the lesson by having students:

DISCUSS What operation did you use to solve Exercise 21? Explain. division; n is the total miles traveled for 6 practices.

WRITE Explain how to solve an equation such as $5 \times f = 30.5$.
Possible answer: I want a number that gives 30.5 when multiplied by 5, so I can use the related sentence $30.5 \div 5 = 6.1$. Then I check by writing 6.1 in the place of f and multiplying.

Lesson Quiz

Transparency **4.6**

Evaluate each expression for $n = 0.5$.

1. $n + 0.88$ **1.38**
2. $n - 0.02$ **0.48**
3. $3 \times n$ **1.5**
4. $n \div 0.25$ **2**

Solve for the variable.

5. $3 \times a = 4.8$ $a = 1.6$
6. $b - 1.1 = 12$ $b = 13.1$
7. $y + 4.2 = 5.8$ $y = 1.6$

83

CHALLENGE 4.6

Match Up

Match each description with its expression.

1. The cost of 3 shirts and 2 sweaters if each shirt costs s dollars and each sweater costs w dollars — **f**
2. The difference between the distances traveled on 3 days of driving and 2 days of walking if the distance driven each day is s miles, the distance walked each day is w miles, and s is greater than w — **d**
3. The total number of hours studied if you study s hours for 2 weeks and w hours for 3 weeks — **e**
4. The total distance run in training for a race if you run w miles per week for 3 weeks — **c**
5. The price paid for one apple if you pay w dollars for 3 apples — **a**
6. The difference between the greater amount you will earn in May if you work 3 weeks earning w dollars per week and the amount you will earn in June if you work 2 weeks earning s dollars per week — **j**
7. The total number of pages you study if you study 3 pages per hour for w hours and one page per hour for s hours — **h**
8. The price you pay for one container of juice and one container of milk if juice costs w dollars for 3 containers and milk costs s dollars per container — **b**
9. The number of miles you can drive on 3 tanks of gas if your car averages s miles per tankful — **g**
10. The cost of 2 entrees and 1 salad at a restaurant if each entree costs w dollars and salads cost s dollars for 3 salads — **k**

a. $\frac{w}{3}$
b. $\frac{w}{3} + s$
c. $3w$
d. $3s - 2w$
e. $2s + 3w$
f. $3s + 2w$
g. $3s$
h. $3w + s$
j. $3w - 2s$
k. $2w + \frac{s}{3}$

PROBLEM SOLVING 4.6

Algebra: Decimal Expressions and Equations

Analyze • Choose • Solve • Check

Write the correct answer.

1. Each child's meal at a fast-food restaurant costs $2.79. What is the greatest number of these meals that can be bought with $20.00?
   7 meals

2. Felipe is a teenager who is 10 years older than his sister Irene. In 6 years, Felipe will be twice as old as his sister. How old is Felipe now?
   14 years old

3. The winning car in a race had an average speed of 203.7 miles per hour. This was b miles per hour faster than the second-place car. Write an expression for the average speed of the second-place car.
   203.7 − b

4. The round-trip distance between Kaitlin's house and her school is 3.2 miles. Kaitlin rides her bike to school 3 days per week. Write an expression that can be used to find the number of miles Kaitlin rides in w weeks.
   9.6w

Choose the letter for the best answer.

5. At a self-service copy center, the cost of making copies is $0.08 per copy for the first 100 copies, $0.06 per copy for copies 101–200, and $0.05 per copy for any above 200. Stan needs to make 7 copies of a 30-page report. How much should he expect to pay?
   A $16.80   C $12.60
   B $14.50   D $10.50

6. Marla poured out g glasses of juice for a party she is hosting. If each glass contained 0.2 liter of juice, which expression describes the total amount of juice she poured?
   F $0.2g$ H $\frac{g}{0.2}$
   G $\frac{0.2}{g}$ J $0.2 + g$

7. After driving 159.7 miles, Rasheed had r miles left to travel. If the total distance he needed to travel was 201.3 miles, which equation can you use to find the value of r?
   A $r = 159.7 + 201.3$
   B $r + 201.3 = 159.7$
   C $159.7 = 201.3$
   D $201.3 − r = 159.7$

8. At a school cafeteria, 6 carrot sticks are served with each lunch order. Carrot sticks are purchased in bags of 120. If 310 lunches were served today, how many bags of carrots were opened?
   F 13 bags   **G 14 bags**
   H 15 bags   J 16 bags

9. **Write About It** Describe how you decided which operation was needed to find the total distance Kaitlin rides in w weeks in Problem 4.
   Possible answer: Since she rides the same distance each day, multiply to find the number of miles per week. Since she rides the same distance each week, multiply to find the total distance in w weeks.

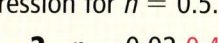

CHAPTER 4

REVIEW/TEST

Purpose To check understanding of concepts, skills, and problem solving presented in Chapter 4

USING THE PAGE

The Chapter 4 Review/Test can be used as a **review** or a **test**.

- Items 1–38 check skill proficiency.
- Items 39–40 check students' abilities to choose and apply problem solving strategies to real-life problems involving operations with decimals.

 Suggest that students place the completed Chapter 4 Review/Test in their portfolios.

USING THE ASSESSMENT GUIDE

- Multiple-choice format of Chapter 4 Posttest—See *Assessment Guide*, pp. AG21–22.
- Free-response format of Chapter 4 Posttest—See *Assessment Guide*, pp. AG23–24.

USING STUDENT SELF-ASSESSMENT

The How Did I Do? survey helps students assess what they have learned and how they learned it. This survey is available as a copying master in *Assessment Guide*, p. AGxvii.

CHAPTER 4 REVIEW/TEST

Add or subtract. Estimate to check. (pp. 66–69)

1. $3.9 + 4 + 5.91 = $ **13.81**
2. $7.6 - 0.95 = $ **6.65**
3. $3.02 + 0.17 + 4.338 = $ **7.528**
4. $19.3 - 2.56$ **16.74**
5. $0.126 + 5.3 + 3.04$ **8.466**
6. $245 - 39.05$ **205.95**

Tell the number of decimal places there will be in the product. Then multiply. (pp. 70–73)

7. 8.3×12.9 **2; 107.07**
8. 7.82×4.5 **3; 35.190**
9. 0.13×2.07 **4; 0.2691**
10. 53.6×1.23 **3; 65.928**
11. 20.01×8.2 **3; 164.082**
12. 6.9×17.4 **2; 120.06**
13. 4.91×6.2 **3; 30.442**
14. 7.02×5.5 **3; 38.610**

Divide. Estimate to check. (pp. 76–79)

15. $1.4\overline{)35}$ **25**
16. $3.7\overline{)5.92}$ **1.6**
17. $0.45\overline{)1.08}$ **2.4**
18. $0.25\overline{)85}$ **340**
19. $22.8 \div 3$ **7.6**
20. $9.72 \div 2.7$ **3.6**
21. $33.33 \div 1.1$ **30.3**
22. $6.9 \div 0.3$ **23**

Evaluate each expression. (pp. 82–83)

23. $(2.3 + c) + 1.7$ for $c = 8$ **12**
24. $3 \times d + b$ for $d = 1.7$ and $b = 5.4$ **10.5**
25. $c \div 2 - a$ for $c = 8$ and $a = 2.3$ **1.7**
26. $(5.4 - a) + d$ for $a = 2.3$ and $d = 1.7$ **4.8**
27. $4 \times a - c$ for $a = 2.3$ and $c = 8$ **1.2**
28. $(c - 5.4) \times c$ for $c = 8$ **20.8**
29. $(d + c) - 5$ for $d = 1.7$ and $c = 8$ **4.7**
30. $(a + d) \div 4$ for $a = 2.3$ and $d = 1.7$ **1**

Solve each equation by using mental math. (pp. 82–83)

31. $c + 14.07 = 32.97$ $c = $ **18.9**
32. $6a = 24.78$ $a = $ **4.13**
33. $8d = 6.4$ $d = $ **0.8**
34. $7.14 - g = 3.24$ $g = $ **3.9**
35. $2.4 + f = 4.76$ $f = $ **2.36**
36. $4y = 29.6$ $y = $ **7.4**
37. $1.86 \div r = 6.2$ $r = $ **0.3**
38. $t + 3.6 = 4.5 + 3.3$ $t = $ **4.2**

Solve. (pp. 80–81)

39. There are 9 golf balls in a box. How many boxes does Hector need if he wants to give away 500 golf balls as souvenirs? **56 boxes**

40. On Tuesday, Sabrina gets an invitation to a party that is in 20 days. She can go to the party if it does not fall on the weekend. Can she go to the party? Explain.
Yes; 20 ÷ 7 = 2 r6, so the party will be on Monday.

84 Chapter 4

CHAPTER 4 TEST, page 1

Choose the best answer.

1. $75.9 + 48.39$
   - A 27.51
   - B 123.29
   - C 124.29
   - D 1,243.2

2. $102.4 - 89.72$
   - F 12.68
   - G 12.72
   - H 13.38
   - J 192.12

3. $18.2 - 5.68$
   - A 12.52
   - B 12.68
   - C 13.48
   - D 23.88

4. $4.7 + 0.25 + 6.09$
   - F 35.79
   - G 13.29
   - H 11.04
   - J 9.14

5. 5.9×4.2
   - A 3.54
   - B 24.78
   - C 247.8
   - D 2,478

6. 8.03×3.22
   - F 2.6726
   - G 24.8566
   - H 25.8566
   - J 258.566

7. 77.32×6.8
   - A 52.5776
   - B 70.52
   - C 84.12
   - D 525.776

8. 34.52×4.8
   - F 165.696
   - G 65.696
   - H 39.32
   - J 29.72

9. $9.03 \div 3$
   - A 30.1
   - B 6.03
   - C 3.1
   - D 3.01

10. $15.75 \div 4.5$
    - F 35
    - G 11.25
    - H 5.3
    - J 3.5

11. $45.9 \div 7.5$
    - A 61.2
    - B 38.4
    - C 6.12
    - D 6.02

12. $59.52 \div 0.96$
    - F 62
    - G 49.2
    - H 9.2
    - J 6.2

CHAPTER 4 TEST, page 2

13. Gordy has 187 CDs. He wants to put them on shelves. Each shelf holds 42 CDs. How many shelves will he be able to completely fill with CDs?
    - A 2
    - B 3
    - C 4
    - D 5

14. Sasha is making craft projects to sell at the fair. Each project will take 3 days to finish. If she can spend 35 days working on the projects, how many projects can she complete?
    - F 32
    - G 12
    - H 11
    - J 5

15. Lee is buying cupcakes for his class picnic. If each cupcake costs $0.50 and he has $12.35, how many cupcakes can he buy?
    - A 25
    - B 24
    - C 12
    - D 5

16. Carole is making flower baskets. Each basket takes 35 minutes to make. She works from 9:00 AM to 5:00 PM. How many flower baskets can she finish?
    - F 14
    - G 13
    - H 12
    - J 3

For 17–20, evaluate each expression for the given value.

17. $e \times 7$ for $e = 5.6$
    - A 0.8
    - B 12.6
    - C 35.2
    - D 39.2

18. $g \div 4.3 \times 7$ for $g = 13.76$
    - F 30.1
    - G 22.4
    - H 13.76
    - J 10.2

19. $6.8 + c - 5.2$ for $c = 7.4$
    - A 25
    - B 19.4
    - C 9.4
    - D 9

20. $5.9 \times f + r$ for $f = 4.2$ and $r = 7.8$
    - F 32.58
    - G 32.08
    - H 16.98
    - J 3.258

For 21–25, solve each equation using mental math.

21. $5.7t = 17.1$
    - A 12.2
    - B 9.2
    - C 9
    - D 3

22. $2.8 \div c = 1.4$
    - F 38
    - G 18
    - H 2
    - J 1.876

23. $44.5 - d = 16.9$
    - A 2.76
    - B 16.9
    - C 27.6
    - D 61.4

24. $58.47 + y = 63.81$
    - F 122.28
    - G 55.34
    - H 5.34
    - J 0.534

25. $8.62 + k = 8.95$
    - A 7.99
    - B 8.62
    - C 16.61
    - D 25.13

CHAPTERS 1–4 ★ STANDARDIZED TEST PREP

Choose the answer.
See item **4**.
If your answer doesn't match one of the choices, check your computation and the placement of the decimal point. If your computation is correct, mark the letter for Not here.

Also see problem **6**, p. H64.

Choose the best answer.

1. Elise bought 4 balloons. Each balloon cost $0.98, including tax. Which is the total cost of the 4 balloons? **B**

   A $3.62 C $3.96
   B $3.92 D $4.00

2. At the end of a trip, the odometer of a car read 4,572.1 miles. The odometer read 2,998.7 miles at the beginning of the trip. How far did the car go? **J**

   F 7,570.8 mi H 2,684.4 mi
   G 7,460.8 mi J 1,573.4 mi

3. $0.27 + 1.098$ **A**

   A 1.368 C 1.125
   B 1.255 D Not here

4. $1.65 \div 0.5$ **G**

   F 0.33 H 33
   G 3.3   J Not here

5. Stu wants to buy a poster that costs $5.99, a desk lamp that costs $22.18, and bookends that cost $18.98. What is the cost of the 3 items he wants to buy, before tax is added? **C**

   A $46.15 C $47.15
   B $47.00 D $48.15

6. What is the value of $x + 16.714$ for $x = 20.3$? **F**

   F 37.014 H 3.614
   G 36.014 J 0.3614

7. A sweatshirt costs $9 more than a T-shirt. Together, the shirts cost $23. What is the cost of each type of shirt? **D**

   A T-shirt: $4, sweatshirt: $19
   B T-shirt: $16, sweatshirt: $7
   C T-shirt: $12, sweatshirt: $11
   D T-shirt: $7, sweatshirt: $16

8. In which pair of numbers is there a 5 in the hundredths place of both numbers? **H**

   F 553.621; 8,516.2
   G 87.35; 62,531.9
   H 36.157; 4,062.058
   J 513.26; 792.56

9. 7.8×0.3 **A**

   A 2.34 C 8.1
   B 3.24 D Not here

Write What You Know See below.

10. Baseballs are sold in boxes of 12. During a typical game, 5 baseballs are used. A team has 20 games to play this season. Tell the steps you would use to find the number of baseballs the coach should buy.

11. What does the equation below mean? Explain the steps you can use to solve it. Then solve.

    $$s + 1.1 = 5.9$$

Write What You Know • Written Response

10. Since the team plays 20 games and needs 5 baseballs per game, I can use multiplication to find the number of baseballs needed: $20 \times 5 = 100$ baseballs. Since a box holds 12 baseballs, I can use division to find the number of boxes the team needs: $100 \div 12 = 8$ r4. To make sure they have enough baseballs, the coach should buy 9 boxes.

11. The equation tells me the sum of some number, s, and 1.1 is 5.9. To solve the equation, find the number which when added to 1.1 has the sum of 5.9. I can use mental math to solve it: $4.8 + 1.1 = 5.9$. So, $s = 4.8$.

UNIT 1

MATH DETECTIVE
Who Am I?

Purpose To use deductive reasoning to solve problems involving the division of whole numbers

USING THE PAGE

- Direct students' attention to the Reasoning section and to Mystery Number 1.

  How did you use the clues in Mystery Number 1 to find the mystery number? Possible answer: by using the inverse operation of multiplication: $9 \times 5 = 45$ and $45 + 4 = 49$

- Have students read the clues for Mystery Number 2.

  What is the first whole number that gives a remainder of 5 when divided by 9? the next whole number that gives a remainder of 5 when divided by 9? Explain. 14; 23; $23 \div 9 = 2$ R5

  How would you find Mystery Number 2? Explain. Possible answer: Find the first multiple of 9 that is greater than 240 and then add 5 to that number: $27 \times 9 = 243$ and $243 + 5 = 248$.

- After students find Mystery Number 4, have them explain their thinking.

  Reasoning Find the least whole number which when divided by 3, 6, or 8 gives a remainder of 3. Explain how finding this mystery number is different from finding Mystery Number 4. 50; Only multiply $6 \times 8 = 48$ and then add 2 because 6 is a multiple of 3. 4, 5, and 7 are not multiples of one another. So all three must be multiplied together to find mystery number 4.

  Think It Over! After students complete the Write About It, have them work with a partner, exchange papers, and compare their methods for solving the cases.

  Encourage a discussion of how to find the mystery number in Stretch Your Thinking. Have students explain how they can use the first mystery number to find the second mystery number in the problem.

86 Unit 1 • Chapters 1–4

PROBLEM SOLVING
MATH DETECTIVE

Who Am I?

REASONING Use the given information plus your knowledge of division to find the mystery number. Be prepared to explain how you solved the mystery.

Mystery Number 1
I'm the least whole number that gives a remainder of 4 when divided by 5 and when divided by 9. Who am I? **49**

Mystery Number 2
I'm the first whole number greater than 240 that gives a remainder of 5 when divided by 9. Who am I? **248**

Mystery Number 3
I'm the first whole number greater than 100 that gives a remainder of 3 when divided by 8 and when divided by 7. Who am I? **115**

Mystery Number 4
I'm the least whole number that when divided by 4, by 5, and by 7 gives a remainder of 3. Who am I? **143**

Think It Over! Possible answers: #1: $9 \times 5 + 4 = 49$; #2: 243 is divisible by 9 and $243 + 5 = 248$; #3: 56 is divisible by 8 and 7, so $2 \times 56 = 112$, and $112 + 3 = 115$; #4: $4 \times 5 \times 7 + 3 = 143$

- **Write About It** Explain how you found each mystery number. See above.

- **Stretch Your Thinking** I'm the least number that when divided by 5 and when divided by 7 produces a remainder of 3. How much do you have to add to me to get a remainder of 0 when you divide by 5 and when you divide by 7?

86 Unit 1

Intervention and Extension Resources

LITERATURE CONNECTION

Have students **use reasoning skills with whole number division.** In the book *From the Mixed Up Files of Mrs. Basil E. Frankweiler* by E. L. Konigsberg, Claudia and James run away to live in the Metropolitan Museum of Art. They stay for several days and while wandering around the museum they become interested in a beautiful statue of an angel and decide to find the name of the mysterious sculptor of the statue. The book describes their adventures as they attempt to solve the mystery.

Claudia and James must use their reasoning to solve the mystery of the sculptor's identity. Ask students to write clues that can be used to solve a mystery involving the division of whole numbers. Have them exchange with a partner and solve. Check students' work.

VISUAL; LOGICAL/MATHEMATICAL

Challenge: Scientific Notation

Learn how to write numbers by using scientific notation.

820,000 pounds lifting off!

The weight of one type of passenger jet is about 820,000 lb. This number, 820,000, is written using *standard notation*.

You can also write the weight using **scientific notation**. Scientific notation is a shorthand method for writing large numbers.

A number written in scientific notation has two parts separated by a multiplication symbol.

$$8.2 \times 10^5$$

The first part is a number that is at least 1 but less than 10. The second part is a power of 10.

To write the number 820,000 in scientific notation:

Count the number of places the decimal point must be moved to the left to form a number that is at least 1 but less than 10.

$$820{,}000 \rightarrow 8.2$$
5 places

Since the decimal point moved 5 places, the power of 10 is 5.

$$820{,}000 = 8.2 \times 10^5$$

The airplane's weight written in scientific notation is 8.2×10^5 lb.

EXAMPLE

About 32,500,000 passengers used Newark International Airport in 1998. Write the number of passengers in scientific notation.

$$32{,}500{,}000 \rightarrow 3.25 \times 10^7$$
7 places

The number of passengers was about 3.25×10^7.

TALK ABOUT IT

- Is 1.2×10^9 greater than 9.98×10^8? Explain. **Yes. The exponent is greater.**
- Show how to correctly write 782.5×10^8 in scientific notation. **7.825×10^{10}**

TRY IT

Write the number in scientific notation.

1. 602,000 **6.02×10^5**
2. 199 **1.99×10^2**
3. 3,400,000 **3.4×10^6**
4. 8,540 **8.54×10^3**
5. 5,010,000 **5.01×10^6**
6. 113,000 **1.13×10^5**
7. 72 **7.2×10^1**
8. 48,900 **4.89×10^4**
9. 26,200,000 **2.62×10^7**

Chapters 1–4 **87**

CHALLENGE
Scientific Notation

Objective To extend the concepts and skills of Chapters 1–4

USING THE PAGE

- Have a volunteer read the caption under the photo. Have the student explain how to read the number in standard notation. Then direct students' attention to the term scientific notation.

Is 68×10^6 written in scientific notation? Explain. No; 68 is greater than 10.

Is 0.3×10^4 written in scientific notation? Explain. No; 0.3 is less than 1.

Reasoning Is 10^5 written in scientific notation? Explain. Yes; $10^5 = 1 \times 10^5$.

How can you check that you have written a number correctly in scientific notation? Convert the number in scientific notation back into standard form.

- Have students complete the Talk About It and then extend their thinking.

Reasoning How would you compare two numbers written in scientific notation? Explain. If the exponents are different, the number with the greater exponent is the greater number. If the exponents are the same, then compare the decimal numbers to find the greater number.

Try It Before assigning Try It Exercises 1–9, have students describe how to write a number in scientific notation.

Intervention and Extension Resources

ALTERNATIVE TEACHING STRATEGY • SCAFFOLDED INSTRUCTION

Reinforce students' understanding of scientific notation. Explain that a number in scientific notation is a product of two factors: (a decimal that is at least 1 but less than 10) × (a power of 10).

Introduce scientific notation with 6,000:

$$6{,}000 = 6 \times 10^3$$

| Move the decimal 3 places left. | 6 is greater than 1 and less than 10. | Use 3 as the exponent, since the decimal moved 3 places. |

Remind students that a number in scientific notation has the same value as the number in standard form. Have students write 90,000,000 in scientific notation. **9×10^7**

Then ask students to write 5.05×10^6 in standard form. **5,050,000**

VISUAL; VERBAL/LINGUISTIC

Number Sense and Operations **87**

UNIT 1 CHAPTERS 1–4

STUDY GUIDE AND REVIEW

Purpose To help students review concepts and skills presented in Chapters 1–4

USING THE PAGES

✓ Assessment Checkpoint

The Study Guide and Review includes content from Chapters 1–4.

Chapter 1
1.1 Estimate with Whole Numbers
1.2 Use Addition and Subtraction
1.3 Use Multiplication and Division
1.4 Problem Solving Strategy: *Predict and Test*
1.5 Algebra: Use Expressions
1.6 Algebra: Mental Math and Equations

Chapter 2
2.1 Mental Math: Use the Properties
2.2 Algebra: Exponents
2.3 Math Lab: Explore Order of Operations
2.4 Algebra: Order of Operations
2.5 Problem Solving Skill: *Sequence and Prioritize Information*

Chapter 3
3.1 Represent, Compare, and Order Decimals
3.2 Problem Solving Strategy: *Make a Table*
3.3 Estimate with Decimals
3.4 Decimals and Percents

Chapter 4
4.1 Add and Subtract Decimals
4.2 Multiply Decimals
4.3 Math Lab: Explore Division of Decimals
4.4 Divide with Decimals
4.5 Problem Solving Skill: *Interpret the Remainder*
4.6 Algebra: Decimal Expressions and Equations

The blue page numbers in parentheses provided with each group of exercises indicate the pages on which the concept or skill was presented. The red number given with each group of exercises identifies the Learning Goal for the concept or skill.

88 Unit 1 • Chapters 1–4

UNIT 1 Study Guide and Review

VOCABULARY

1. An expression that includes a variable is a(n) __?__ expression. (p. 28) **algebraic**
2. When you __?__ a decimal by a power of 10, the decimal point moves one place to the right for each factor of 10. (p. 87) **multiply**

EXAMPLES

EXERCISES

Chapter 1

• **Add and subtract whole numbers.**
(pp. 20–21) **1B**

```
  3,921        3,000
     68       −1,650
  + 205        1,350
  -----
  4,194
```

Find the sum or difference.

3. $756 + 902$ **1,658**
4. $4,293 + 256 + 19$ **4,568**
5. $3,511 − 1,345$ **2,166**
6. $729 + 8 + 3,996$ **4,733**
7. $16,092 − 5,618$ **10,474**
8. $25,080 − 19,387$ **5,693**

• **Multiply and divide whole numbers.**
(pp. 22–25) **1B**

```
    273            203 r16
  ×  86         41)8,339
  -----            −8 2
  1 638              13
 +21 84             − 0
  ------            139
  23,478           −123
                    16
```

Find the product or quotient.

9. 57×38 **2,166**
10. 430×17 **7,310**
11. 276×809 **223,284**
12. $489 \div 26$ **18 r21**
13. $9,671 \div 42$ **230 r11**
14. $9,538 \div 19$ **502**

• **Use mental math to solve equations.**
(pp. 30–31) **1C**

Solve. $w \div 6 = 7$ THINK: What number divided by 6 equals 7?

$w = 42$ $42 \div 6 = 7$

Solve each equation by using mental math.

15. $b + 5 = 12$ **b = 7**
16. $a − 3 = 5$ **a = 8**
17. $3y = 18$ **y = 6**
18. $k \div 5 = 4$ **k = 20**

Chapter 2

• **Use the order of operations to evaluate expressions.** (pp. 44–45) **2C**

Evaluate. $25 \div 5 + (8 − 4)^2 \times 3$

$25 \div 5 + (4)^2 \times 3$ *Operate in parentheses.*
$25 \div 5 + 16 \times 3$ *Clear exponents.*
$5 + 16 \times 3$ *Divide.*
$5 + 48$ *Multiply.*
53 *Add.*

Evaluate the expression.

19. $3 \times 6 + 7^2 − 9$ **58**
20. $(8 + 7) \div 3 + (5 − 3)^3$ **13**
21. $(15 − 3 \times 4) + 8 \div 2$ **7**
22. $9^2 − 20 \times 2 + 5$ **46**
23. $32 + (8^2 − 50) \times 2$ **60**
24. $16 \div 2^3 + 4 \times 3$ **14**

88 Unit 1

Chapter 3

- **Compare and order decimals.** (pp. 52–55) **3A**

Compare 2.8 and 2.83.

2.8	2.83	*Start at the left.*
2.8	2.83	← *same number of ones*
2.8	2.83	← *same number of tenths*
2.80	2.83	*Add a zero to compare.*

2.8 < 2.83, or 2.83 > 2.8

Compare the numbers. Write <, > or = for ●.

25. 3.72 ● 3.7 **>**
26. 5.02 ● 5.021 **<**

Write the numbers in order from least to greatest. Use <.

27. 1.67, 1.76, 1.607, 1.706, 1.076
1.076 < 1.607 < 1.67 < 1.706 < 1.76
28. 0.0014, 0.4001, 0.0401, 0.0041, 0.014
0.0014 < 0.0041 < 0.014 < 0.0401 < 0.4001

- **Write decimals as percents and percents as decimals.** (pp. 60–61) **3C**

Write 0.05 as a percent.
0.05 is 5 hundredths.
So, 0.05 is 5%.

Write the percent or decimal.

29. 0.28 **28%**
30. 7% **0.07**
31. 47% **0.47**
32. 0.6 **60%**

Chapter 4

- **Add and subtract decimals.** (pp. 66–69) **4A**

Find the difference. 8.2 − 6.391

  8.200 *Align the decimal points.*
−6.391 *Use zeros as placeholders.*
  1.809 *Place the decimal point. Subtract.*

Find the sum or difference.

33. 29.6 + 0.935   **30.535**
34. 26.53 + 5.238   **31.768**
35. 76.03 − 58.94   **17.09**
36. 347.31 − 48.896   **298.414**

- **Multiply and divide decimals.** (pp. 70–79) **4A**

Find the quotient. 2.46 ÷ 0.6

```
    4.1
0.6)2.46    Make the divisor a
   −2 4     whole number.
     06     0.6 × 10 = 6
     −6     2.46 × 10 = 24.6
      0     Divide.
```

Find the product or quotient.

37. 75.8 × 6   **454.8**
38. 4.83 × 0.9   **4.347**
39. 3.92 × 0.58   **2.2736**
40. 68.49 × 3.6   **246.564**
41. 36.48 ÷ 12   **3.04**
42. 43.5 ÷ 0.3   **145**
43. 59.04 ÷ 4.8   **12.3**
44. 8.094 ÷ 0.95   **8.52**

- **Evaluate expressions with decimals.** (pp. 82–83) **4B**

Evaluate $r + s + 3$ for $r = 6.9$ and $s = 4.8$.
$r + s + 3$ *Replace r with 6.9 and s with 4.8.*
$6.9 + 4.8 + 3 = 14.7$

Evaluate each expression for $c = 2.5$, $d = 3.2$, and $f = 4.1$.

45. $(c \times 3) + d$ **10.7**
46. $(10 − d) + f$ **10.9**
47. $\dfrac{d}{8}$ **0.4**
48. $f − c + d$ **4.8**

PROBLEM SOLVING APPLICATIONS

49. Jake has half as many pennies as his sister has. Together their pennies are worth 72 cents. How many pennies does Jake have? (p. 26) **1D 24 pennies**

50. Movie tickets cost $3.75 each for the matinee. How many tickets can Ms. Hamil buy with $20? How much money will she have left over? (p. 76) **4C**
5 tickets; $1.25

Chapters 1–4 **89**

Assessment Checkpoint

Portfolio Suggestions The portfolio represents the growth, talents, achievements, and reflections of the mathematics learner. Students might spend a short time selecting work samples for their portfolios and completing A Guide to My Math Portfolio from *Assessment Guide*, page AGxix.

You may want to have students respond to the following questions:

- What new understanding of math have I developed in the past several weeks?
- What growth in understanding or skills can I see in my work?
- What can I do to improve my understanding of math ideas?
- What would I like to learn more about?

For information about how to organize, share, and evaluate portfolios, see *Assessment Guide*, page AGxviii.

Use the item analysis in the **Intervention** chart to diagnose students' errors. You may wish to reinforce content or remediate misunderstandings by using the text pages or lesson resources.

STUDY GUIDE AND REVIEW INTERVENTION

How to Help Options

Learning Goal	Items	Text Pages	Reteach and Practice Resources
1B See page 14C for Chapter 1 learning goals	3–8, 9–14	20–21, 22–25	Worksheets for Lessons 1.2, 1.3
1C See page 14C for Chapter 1 learning goals	15–18	30–31	Worksheets for Lesson 1.6
1D See page 14C for Chapter 1 learning goals	49	26	Worksheets for Lesson 1.4
2C See page 14C for Chapter 2 learning goals	19–24	44–45	Worksheets for Lesson 2.4
3A See page 14C for Chapter 3 learning goals	25–28	52–55	Worksheets for Lesson 3.1
3C See page 14C for Chapter 3 learning goals	29–32	60–61	Worksheets for Lesson 3.4
4A See page 14C for Chapter 4 learning goals	33–36, 37–44	66–69, 70–73, 74–75, 76–79	Worksheets for Lessons 4.1, 4.2, 4.3, 4.4
4B See page 14C for Chapter 4 learning goals	45–48	82–83	Worksheets for Lesson 4.6
4C See page 14C for Chapter 4 learning goals	50	76	Worksheets for Lesson 4.4

Number Sense and Operations **89**

UNIT 1

PERFORMANCE ASSESSMENT

Purpose To provide performance assessment tasks for Chapters 1–4

USING THE PAGE

- Have students work individually or in pairs as an alternative to formal assessment.
- Use the performance indicators and work samples below to evaluate Tasks A–B.

See *Performance Assessment* for

- a complete scoring rubric, p. PA2, for this unit.
- additional student work samples for this unit.
- copying masters for this unit.

 You may suggest that students place completed Performance Assessment tasks in their portfolios.

Performance Indicators

Task A

A student with a Level 3 paper should

____ Determine the steps a computer must go through to evaluate any expression put into it.

____ Make up a numerical expression that includes three different operations, and explain how the computer will evaluate it.

____ Explain how a computer will evaluate a particular expression.

____ Show work and explain how the answers were determined.

Task B

A student with a Level 3 paper should

____ Show understanding of possible ways to spend money on pizza by making a list or table of the information.

____ Use information about pizza purchase to determine how many people to invite to a party.

____ Determine how much pizza to buy, staying within an established budget.

____ Show work and explain how the answers were determined.

Performance Assessment

TASK A • Programming Problem

As part of a larger computer program you are building, you need to make a module that will evaluate any numerical expression the user puts into it.

Evaluate within parentheses; clear exponents; multiply and divide
- Write down the steps the computer must go through to evaluate any expression that is put into it.
left to right; add and subtract left to right.
- Make up your own numerical expression that includes at least three different operations and has part of the expression in parentheses. Explain the steps the computer will go through to evaluate it. *Check students' work.*

- Tell how the computer will evaluate the following expression:
$3 \times 0.2 + 9 \div 0.3 - (4^2 - 2^3)$
$3 \times 0.2 + 9 \div 0.3 - (4^2 - 2^3) = 0.3 \times 0.2 + 9 \div 0.3 - (16 - 8) =$
$3 \times 0.2 + 9 \div 0.3 - (8) = 0.6 + 30 - 8 = 22.6$

TASK B • Pizza Party

Peggy is giving a pizza party to celebrate her birthday. To plan it she needs some information about the pizzas she can order.

Pizza Diameter	Number of Slices	Cost per Pizza
12 inches	4	$7
14 inches	6	$9
16 inches	8	$11

- Suppose Peggy has $30 to spend on the pizzas. What are some different ways to spend the money? Make a list. Which way gets her closest to spending all of the $30? What combination of pizzas should she buy to get the greatest number of slices? *See additional answers p. 91*

Next, she'll have to decide how many friends she can invite.

- How many slices do you think a typical friend will eat? Find a way for Peggy to order exactly enough pizzas so that each friend gets the same number of slices and none are left over. Describe what you would do. Remember to stay within the budget. *See additional answers p. 91*

90 Unit 1

Work Samples for Task A and Task B

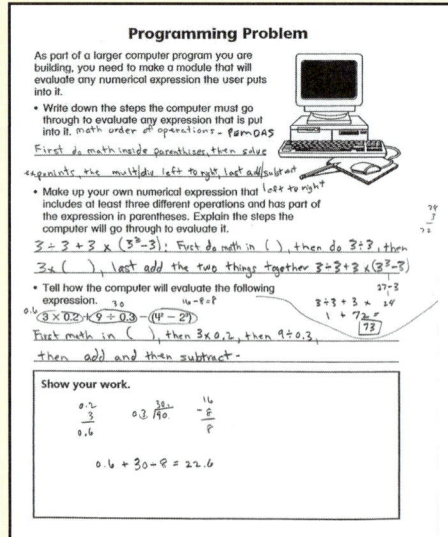

Level 3 Student's work is complete and accurate. Student displays good understanding of the task. Work shows understanding of order of operations.

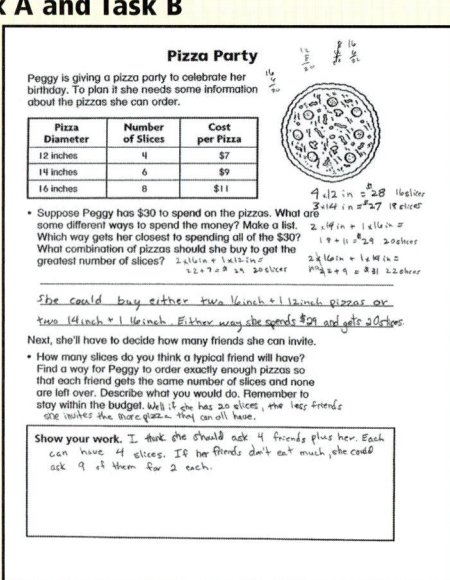

Level 3 The students shows clear understanding of the task. All steps are addressed. Answers are correct, and logical reasoning is evident.

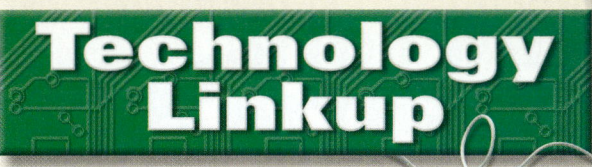

Data ToolKit
Use Spreadsheet Formulas to Find Sums and Differences

The table below shows money Dave earned by helping in his parents' store last year. Dave uses a spreadsheet to find the total amount he earned.

Month	Amount	Month	Amount	Month	Amount
January	$25.15	May	$45.50	September	$47.77
February	$20.95	June	$56.85	October	$32.95
March	$31.25	July	$54.60	November	$68.25
April	$38.75	August	$59.00	December	$75.35

- Enter the months into Column A and the amounts into Column B.
- Click in cell B13. Click f_x.
- Highlight *SUM* and click *OK*.
- Type B1:B12. This tells the spreadsheet to add the cells from B1 to B12.
- Press *Enter*.

So, Dave earned a total of $556.37 last year.

How much more did Dave earn in December than in January?

- Click in an empty cell. Type =.
- Click in cell B12. Type −. Click in cell B1. Type *Enter*.

So, Dave earned $50.20 more in December.

Practice and Problem Solving

Make a spreadsheet to find the sum and difference. Check students' spreadsheets.

1. Kim earned $38.19 in January, $24.35 in February, $68.90 in March, and $47.60 in April. What is the total amount she earned? How much more did she earn in March than in April? **$179.04; $21.30**

2. Chris earned $125.14 in May, $108.75 in June, $98.50 in July, $116.90 in August, and $130.25 in September. What is the total amount he earned? How much more did he earn in September than in July? **$579.54; $31.75**

3. **REASONING** Explain the steps you follow to add and subtract numbers on a spreadsheet. Check that students outline the steps above.

Multimedia Math Glossary www.harcourtschool.com/mathglossary

4. **Vocabulary** Look up *clustering* in the Multimedia Math Glossary. Write a problem that can be answered using the example shown in the glossary.

Chapters 1–4 **91**

Additional Answers for Performance Task B, PE p. 90

- Check students' work. Students should conclude that the maximum number of pieces she can get is 20 if she stays within her budget.
- Students may answer that they would use the factors of 20 to determine the number of friends to invite: 10 would get 2 pieces, 5 would get 4 pieces, and so on. Be sure students show their work and justify their answers.

TECHNOLOGY LINKUP

Objective To use a spreadsheet to calculate sums and differences

USING THE PAGE

- State that a spreadsheet sometimes has advantages over a calculator. Then ask this question.

  Why might it be better to use a spreadsheet than to use a calculator when you are adding several numbers? Possible answer: You can review the numbers you enter, so you're less likely to make a mistake. You can easily insert or delete numbers. And you can add many sets of several numbers almost as easily as you can add one set.

Using Data ToolKit

Students may use Data ToolKit or other spreadsheet software to complete the exercises. Point out that to format the column with money amounts, students should highlight the column and select *Spreadsheet*, then *Number Format*.

In the second part of the Example, explain that clicking in cell B12, pressing −, clicking in cell B1, and clicking *enter*, tells the spreadsheet to subtract the contents of cell B1 from the contents of cell B12.

Practice and Problem Solving

Check to see that students are using the correct spreadsheet formulas to add and subtract the amounts of money. You may want to list the steps on a poster for students to refer to.

Multimedia Math Glossary

Clustering and all other vocabulary words in this unit can be found in the Harcourt Multimedia Math Glossary.
www.harcourtschool.com/mathglossary

Number Sense and Operations **91**

UNIT 1

PROBLEM SOLVING
On Location With Energy

Purpose To provide additional practice for concepts and skills in Chapters 1–4

USING THE PAGE
Wind Energy

- After Exercise 4, have students continue their comparisons of numbers of turbines at different locations.

About how many times as many turbines are there at Lake Benton II as there are at Lakota Ridge? about 9 times as many

There are about 50,000 wind turbines in the world. About what percent of the total number are in California? about 30%

- After Exercise 8, ask these questions.

The turbines at Lake Benton I can serve 43,000 households. About how many households can one turbine serve? about 300

About how many people could obtain electricity from one turbine at Lake Benton I? Explain your thinking. Possible answer: The last problem showed that 1 turbine can serve 300 households. If the average household has 3 people, that means 3 x 300 = 900 people can be served by 1 turbine.

The population of Minnesota is about 4,800,000. How many turbines would be needed to meet the electricity needs of every household? Explain your thinking. Possible answer: The last problem showed that 1 turbine can serve about 900 people. So, about 4,800,000 ÷ 900, or about 5,300 turbines would be needed.

Extension Households usually measure electrical use in kWh, but utility companies measure electrical use in megawatt-hours (MWh). One megawatt-hour is the same as 1,000 kilowatt-hours. Have students find the annual electricity generation at Lakota Ridge and Lake Benton II in megawatt-hours. Lakota Ridge: 30,000 MWh; Lake Benton II: 355,000 MWh

---

PROBLEM SOLVING ON LOCATION
With Energy

Wind Energy

Wind energy has been used for centuries to pump water, grind grain, and power sailboats. Today, many communities use windmills called wind turbines to generate electricity. Several important wind turbine projects are located in Minnesota.

Use Data For 1–8, use the table.

Wind Turbines in Minnesota	
Location	Number of Turbines
Lakota Ridge	15
Lake Benton I	143
Lake Benton II	138

1. Order the locations from greatest number of turbines to least.
Lake Benton I, Lake Benton II, Lakota Ridge

2. How many more turbines are there at Lake Benton I than at Lake Benton II? **5**

3. There are about 50 times as many wind turbines in California as there are at the three sites in Minnesota combined. About how many wind turbines are there in California? **about 15,000**

4. How many times as many turbines are there at Lake Benton II as there are at Lakota Ridge? Write your answer as a fraction in simplest form and as a mixed number. $\frac{46}{5}$; $9\frac{1}{5}$

Electricity is measured in kilowatt-hours (kWh). The turbines at Lakota Ridge generate 30 million kWh per year. The turbines at Lake Benton II generate 355 million kWh per year.

5. About how many kWh are generated each day at Lake Benton II? **about 1 million kWh**

6. About how many kWh are generated each year by each turbine at Lake Benton II? **about 3 million kWh**

7. About how many kWh are generated each year by each turbine at Lakota Ridge? **about 2 million kWh**

8. About how many kWh are generated each day at Lakota Ridge? **about 100,000 kWh**

Solar Energy

Each day, huge amounts of free energy strike the Earth in the form of radiation from the sun. This *solar* energy can be converted to heat and electricity for human use. Solar energy, like wind power, is a *renewable* resource.

The Reilly Elementary School is one of several schools in Chicago that use solar energy. These schools are part of the Million Solar Roofs program. By using the sun's energy, the schools can provide for some or all of their heating needs. This reduces their dependence on oil, coal, and other *nonrenewable* resources.

1. The solar project at Reilly Elementary School installed 120 solar power units at a total cost of $500,000. What was the average cost of a solar power unit, rounded to the nearest hundred dollars? **$4,200**

2. Out of every dollar spent for the Reilly solar units, the Chicago school system paid 10 cents. If the total cost was $500,000, how much money did the Chicago school system provide? **$50,000**

Use Data For 3–5, use the table.

3. A quadrillion is a million billion. How is the amount for fossil fuels written in standard form?
80,000,000,000,000,000 btu

U.S. Energy Consumption by Energy Source	
Source	Amount (in quadrillion btu)
Fossil Fuels	80
Nuclear Power	7
Renewable Sources	7

4. Write an equation you can use to find how much more energy is consumed from fossil fuels than from renewable sources. Then solve the equation.
$80 - 7 = n$; $n =$ 73 quadrillion btu

5. Suppose the United States set a goal of using renewable energy sources for half of its energy use. By how much would the nation's current renewable sources have to increase?
73 quadrillion btu

These wind turbines are located at Lake Benton, Minnesota. When the blades spin, they drive generators, which in turn produce electricity.

USING THE PAGE

Solar Energy

- After Exercise 2, ask these questions about Reilly Elementary School.

  On a sunny day, solar energy can supply 0.15 of the energy needs in a school. If energy costs $120 each day, how much money can be saved at the Reilly Elementary School each day? $18

  What energy savings would you expect to obtain on a cloudy day? Explain. You would expect to save less than $18. On a very cloudy day, you may have very little savings.

- Have students look at the table.

  Do you agree with the statement, "The United States is using renewable resources extensively?" Possible answer: No. Out of 94 quadrillion BTUs, only 7 quadrillion BTUs come from renewable resources.

  Name some common fossil fuels. Possible answers: coal, oil

  Extension One gallon of fuel oil is equivalent to about 138,000 btu. Find the total U.S. energy consumption in gallons of fuel oil. Explain your method. Total consumption is 94 quadrillion btu. Divide to convert to gallons of fuel oil: 94,000,000,000,000,000 ÷ 138,000 is about 680 billion gallons.

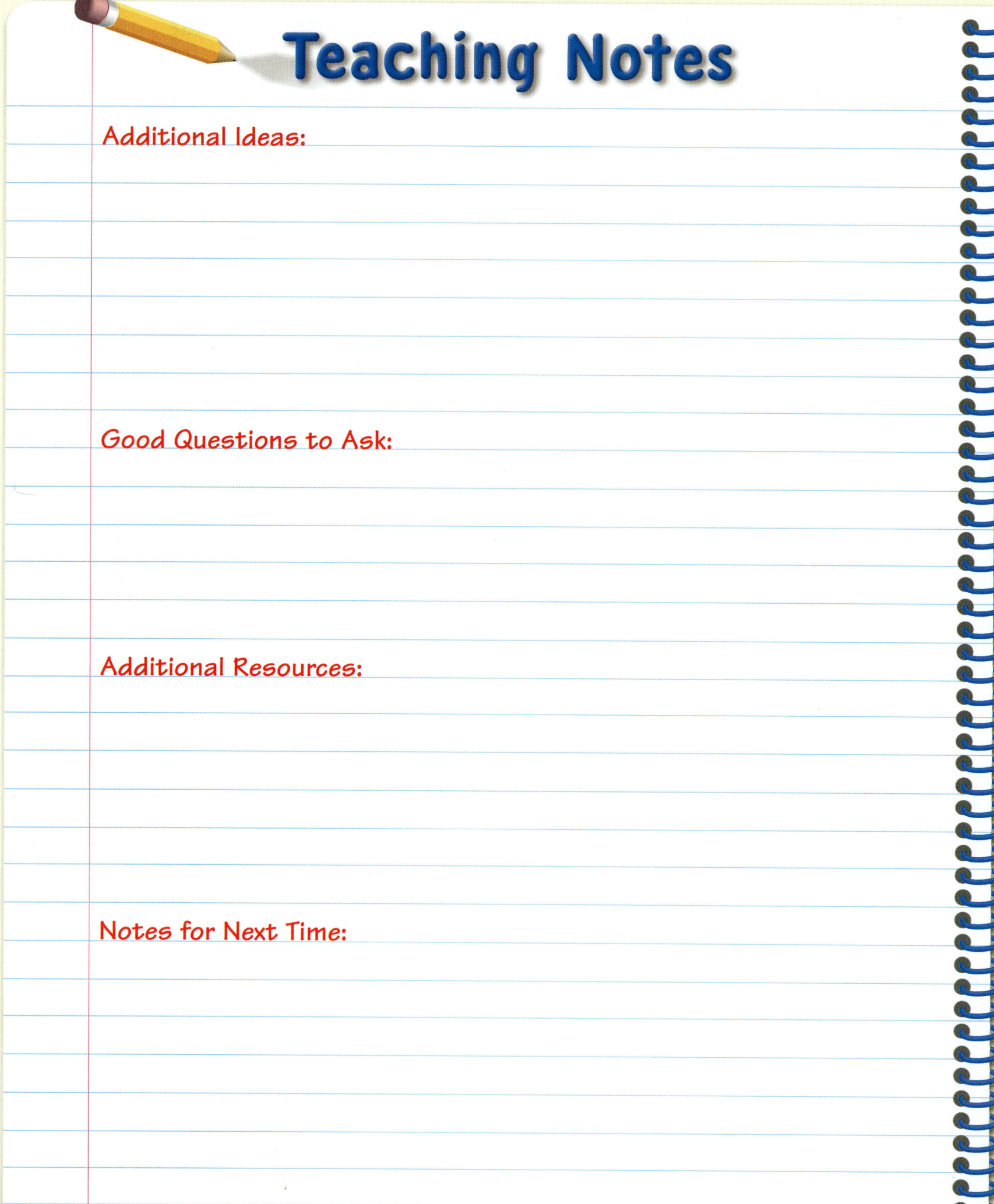

UNIT 2 Statistics and Graphing

UNIT AT A GLANCE

CHAPTER 5
Collect and Organize Data 92

Lesson 1 — Samples
Lesson 2 — Bias in Surveys
Lesson 3 — Problem Solving Strategy: *Make a Table*
Lesson 4 — Frequency Tables and Line Plots
Lesson 5 — Measures of Central Tendency
Lesson 6 — Outliers and Additional Data
Lesson 7 — Data and Conclusions

CHAPTER 6
Graph Data 118

Lesson 1 — Make and Analyze Graphs
Lesson 2 — Find Unknown Values
Lesson 3 — Stem-and-Leaf Plots and Histograms
Lesson 4 — Math Lab: Explore Box-and-Whisker Graphs
Lesson 5 — Box-and-Whisker Graphs
Lesson 6 — Analyze Graphs

Statistics and Graphing **92A**

UNIT 2: Statistics and Graphing

Assessment Options

Assessing Prior Knowledge
Determine whether students have the required prerequisite concepts and skills.
Check What You Know PE pp. 93, 119

Test Preparation
Provide review and practice for chapter and standardized tests.
Standardized Test Prep, PE pp. 117, 137
Mixed Review and Test Prep
  See the last page of each PE skill lesson.
Study Guide and Review, PE pp. 140–141

Formal Assessment
Assess students' mastery of chapter concepts and skills.
Chapter Review/Test
  PE pp. 116, 136
Pretest and Posttest Options
  Chapter Test, Form A
    pp. AG33–34, 37–38
  Chapter Test, Form B
    pp. AG35–36, 39–40
Unit 2 Test • Chapters 5–6, pp. AG41–48

Daily Assessment
Obtain daily feedback on students' understanding of concepts.
Quick Review
  See the first page of each PE lesson.
Mixed Review and Test Prep
  See the last page of each PE skill lesson.
Number of the Day
  See the first page of each TE skill lesson.
Problem of the Day
  See the first page of each TE skill lesson.
Lesson Quiz
  See the Assess section of each TE skill lesson.

Performance Assessment
Assess students' understanding of concepts applied to real-world situations.
Performance Assessment (Tasks A–B)
  PE, p. 142; pp. PA12–13

Student Self-Assessment
Have students evaluate their own work.
How Did I Do?, p. AGxvii
A Guide to My Math Portfolio, p. AGxix
Math Journal
  See Write in the Assess section of each TE lesson and TE pages 106B, 112B, 124B, 126B, 132B.

Harcourt Electronic Test System Math Practice and Assessment
Make and grade chapter tests electronically.
This software includes:
- multiple-choice items
- free-response items
- customizable tests
- the means to make your own tests

Portfolio
Portfolio opportunities appear throughout the Pupil and Teacher's Editions.
Suggested work samples:
Problem Solving Project, TE pp. 92, 118
Write About It, PE pp. 97, 99, 128
Chapter Review/Test, PE pp. 116, 136

KEY **AG** Assessment Guide   **TE** Teacher's Edition   **PA** Performance Assessment   **PE** Pupil Edition

Correlation to STANDARDIZED TESTS

	LEARNING GOAL	TAUGHT IN LESSONS	CAT	CTBS/ TERRA NOVA	ITBS	MAT	SAT	YOUR STATE TEST
5A	To identify types of samples and to determine if they are representative of a given population or biased, and to draw conclusions about a set of data	5.1, 5.2, 5.7						
5B	To organize, read, interpret, and analyze data in frequency tables and line plots	5.4		•	•	•		
5C	To calculate and compare measures of central tendency with and without outliers	5.5, 5.6	•			•		
5D	To solve problems by using appropriate strategies such as *make a table*	5.3		•				
6A	To make and analyze different kinds of graphs and visual displays including circle graphs, bar graphs, stem-and-leaf plots, histograms, and box-and-whisker graphs	6.1, 6.3, 6.4, 6.5	•	•	•	•	•	
6B	To estimate and solve for unknown values by using a graph, arithmetic, logical reasoning, and algebraic techniques	6.2	•	•	•	•	•	
6C	To compare different types of graphs to determine if they are appropriate or misleading	6.6		•		•		

Statistics and Graphing **92C**

UNIT 2 Statistics and Graphing
Technology Links

🌐 The Harcourt Learning Site
Visit The Harcourt Learning Site for related links, activities, and resources:
- Graphs and Graphing activity *(Use with Chapter 6.)*
- *Multimedia Math Glossary*
- E-Lab interactive learning experiences
- professional development and instructional resources

www.harcourtschool.com

Harcourt Math Newsroom Videos
These videos bring exciting news events to your classroom from the leaders in news broadcasting. For each unit, there is a **Harcourt Math Newsroom Video** that helps students see the relevance of math concepts to their lives. You may wish to use the data and concepts shown in the video for real-life problem solving or class projects.

See **Technology Linkup** in the Pupil Edition, p. 143.

TECHNOLOGY CORRELATION

Intervention Strategies and Activities This CD-ROM helps you assess students' knowledge of prerequisite concepts and skills.

Data ToolKit allows students to enter data into a spreadsheet and display it on as many as four different graphs.

E-Lab is a collection of electronic learning activities.

The chart below correlates technology activities to specific lessons.

LESSON	ACTIVITY/LEVEL	SKILL
5.1	**Harcourt Math Newsroom Video** • *What is in a Poll?*	Explore random sampling
5.3	**Data ToolKit** • *Make a Table*	Organize survey results
5.4	**Data ToolKit** • *Make a Frequency Table* **Data ToolKit** • *Make a Line Plot*	Make a frequency table; make a line plot
6.1	**Data ToolKit** • *Make a Table* **Data ToolKit** • *Make a Bar Graph* **Data ToolKit** • *Make a Line Graph* **Data ToolKit** • *Make a Circle Graph*	Make a table; make a bar graph; make a line graph; make a circle graph
6.3	**Data ToolKit** • *Make a Stem-and-Leaf Plot* **Data ToolKit** • *Make a Histogram* **Calculator Handbook,** *pp. 28–29*	Make a stem-and-leaf plot; make a histogram
6.4	**E-Lab** • *Exploring Box-and-Whisker Graphs* **Calculator Handbook,** *pp. 26–27*	Explore box-and-whisker graphs
6.5	**Data ToolKit** • *Make a Table* **Data ToolKit** • *Make a Box-and-Whisker Graph*	Make a table; make a box-and-whisker graph

For the Student

 Intervention Strategies and Activities

Review and practice the prerequisite skills for Chapters 5–6.

E-LAB These interactive learning experiences reinforce and extend the skills taught in Chapters 5–6.

- Skill development
- Practice

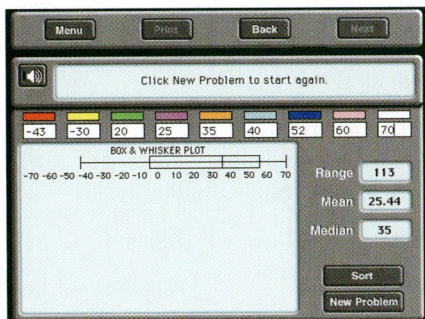

 Data ToolKit

This is a full-functioning graphing and spreadsheet program.

- Make a Table
- Make a Frequency Table
- Make a Line Plot
- Make a Bar Graph
- Make a Line Graph
- Make a Circle Graph
- Make a Stem-and-Leaf Plot
- Make a Histogram
- Make a Box-and-Whisker Graph

For the Teacher

 Teacher Support Software

- **Intervention Strategies and Activities**
  Provide instruction, practice, and a check of the prerequisite skills for each chapter.
- **Electronic Lesson Planner**
  Quickly prepare daily and weekly lessons for all subject areas.
- **Harcourt Electronic Test System Math Practice and Assessment**
  Edit and customize Chapter Tests or construct unique tests from large item banks.

For the Parent

 The Harcourt Learning Site

- Encourage parents to visit The Harcourt Learning Site to help them reinforce mathematics vocabulary, concepts, and skills with their children.
- Have them click on *Math* for vocabulary, activities, real-life connections, and homework tips for Chapters 5–6.

www.harcourtschool.com

Internet

Teachers can find statistics and graphing activities and resources.

Students can learn more about graphs and graphing and reinforce the critical concepts and skills for Chapters 5–6.

Parents can use The Harcourt Learning Site's resources to help their children with the vocabulary, concepts, and skills needed for Chapters 5–6.

Visit The Harcourt Learning Site
www.harcourtschool.com

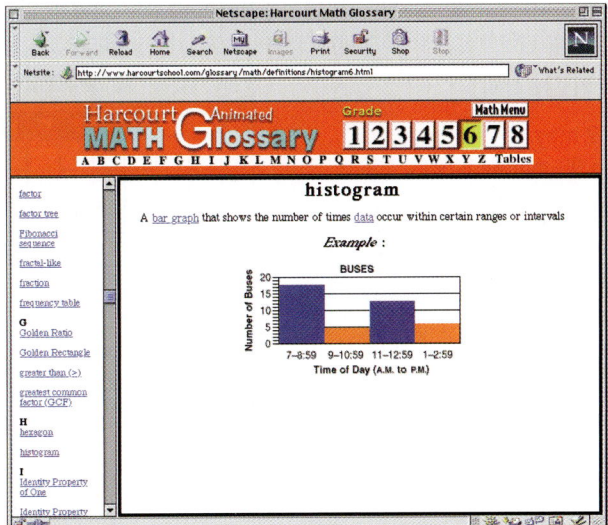

Statistics and Graphing 92E

UNIT 2 Statistics and Graphing

Reaching All Learners

ADVANCED LEARNERS

MATERIALS *For each pair* graph paper, p. TR64; watch with a second hand

Challenge pairs of students to perform this experiment and then **make a box-and-whisker graph using the group data:**

Have one student in each pair estimate when one minute of time has passed, while the other student watches a clock and records the actual time in seconds. Ask each pair to record and display their results.

Direct each pair to use the group data to make a box-and-whisker graph.

After students make their graphs, ask them to write a paragraph analyzing the data. *Use with Lessons 6.4–6.5.*

VISUAL
INTERPERSONAL/SOCIAL

SPECIAL NEEDS

MATERIALS *For each group* 15 counters, a sheet of paper with four circles drawn on it

Have students **find the mean of a data set**. Tell them that a survey was taken in which four people were asked "How many pets do you own?" Display the results of the survey.

Person A – 0 pets   Person B – 2 pets

Person C – 5 pets   Person D – 1 pet

Ask students to place counters in each circle to represent each person's response to the survey question.

To find the mean, have students evenly distribute the counters in the circles. Then have them count to find the average. Connect this activity to the arithmetic method. *Use with Lesson 5.5.*

KINESTHETIC
LOGICAL/MATHEMATICAL

BLOCK SCHEDULING

INTERDISCIPLINARY COURSES
- Social Studies—Use a line plot to record the number of representatives per state in the U.S. House of Representatives.
- Statistics—Look at data trends and patterns in graphs.
- Physical Fitness—Use a stem-and-leaf plot to order the finish times for a 5-kilometer race.
- Geography—Compare weather data in a double-line graph of two cities that are equidistant from the equator—one in the Northern Hemisphere, one in the Southern Hemisphere.

COMPLETE UNIT

Unit 2 may be presented in
- ten 90-minute blocks.
- twelve 75-minute blocks.

INTERDISCIPLINARY SUGGESTIONS

PURPOSE To connect *Statistics and Graphing* to other subjects with these activities

CHAPTER 5—Social Studies

Students research changes in the United States workplace. Ask students to make a double-bar graph to compare the number of women who worked outside the home in 1965 to those who worked outside the home in 1995.

CHAPTER 6—Biology

Students research the approximate pH values for the following items: vinegar, apple juice, carrot juice, drinking water, milk, sea water, milk of magnesia, and household ammonia. Then they make a bar graph comparing the items' levels of acidity.

VISUAL
LOGICAL/MATHEMATICAL

ACTIVITIES AND GAMES FOR HOME OR SCHOOL

You may choose to use this activity and game in the classroom or send it home for students to do with family members.

ACTIVITY • Hot or Not?

PURPOSE To use *range*, *mean*, *median*, and *mode* to describe temperature data *Use after Lesson 5.5.*

VARIATION Have students collect newspaper weather sections for several consecutive days. Remind students that "today's forecast" is a prediction. Invite students to make a double-bar or double-line graph to compare the predictions with the actual temperatures.

GAME • The Dive

PURPOSE To find unknown values using graphs and tables *Use after Lesson 6.2.*

KINESTHETIC; BODILY/KINESTHETIC

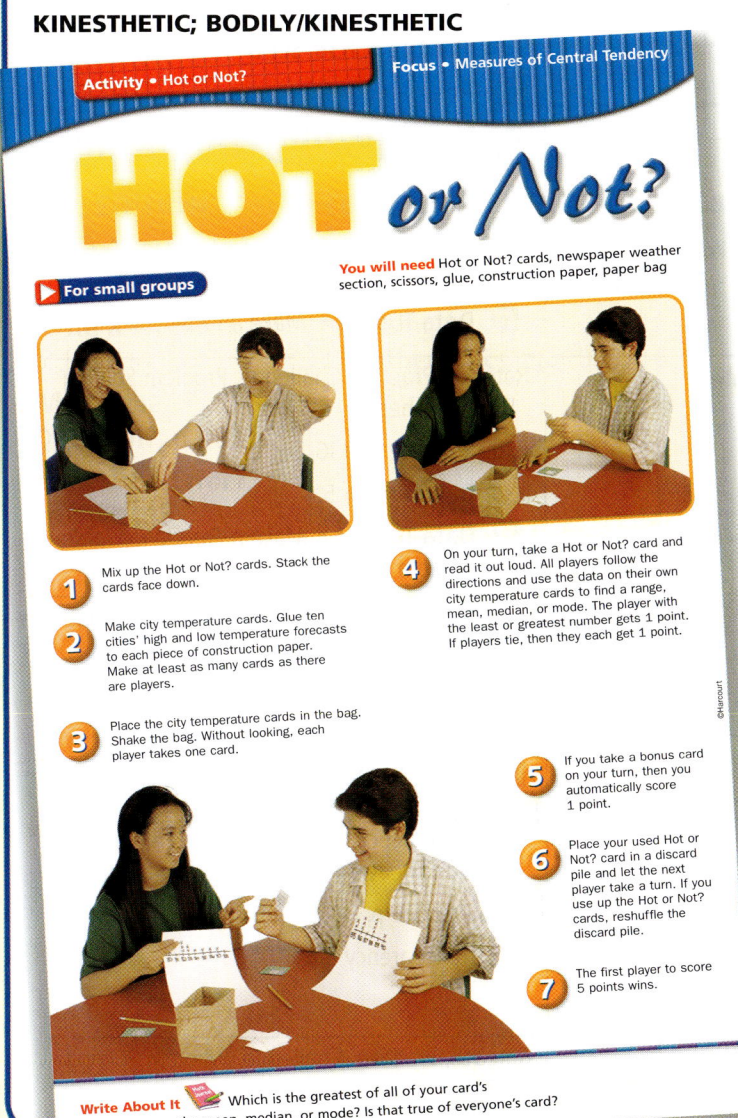

ENGLISH LANGUAGE LEARNERS

Vocabulary Preview Have students make a math dictionary for the vocabulary in the unit. The dictionary will provide students with an easy review tool for tests and will help them build toward independence when answering word problems. Have students make a three-column chart and label the columns. As students learn about each new term, they can fill in their dictionary. *Use with Lessons 5.1–6.4.*

Term	Example	Explanation

Survey Explain that the word *survey* is both a verb and a noun. Have students practice using the word both ways. *Use with Lesson 5.1.*

Outlier To illustrate the term *outlier*, show objects in a cluster and one object set apart. Have students draw an example to illustrate the term. *Use with Lesson 5.6.*

VISUAL; BODILY/KINESTHETIC

LITERATURE CONNECTIONS

These books provide students with additional ways to explore the gathering and reporting of data.

***Exploring the* Titanic** by Robert D. Ballard (Scholastic, Inc., 1988) gives a true account of the author's expedition to find the wreck of the *Titanic*.

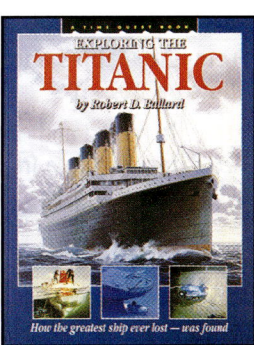

- The depth of the ocean at the *Titanic* wreck site is 12,460 ft. Have students research other shipwrecks or oceanic features to compare their depths to that of *Titanic*. Then have students graph their findings. *Use with Lesson 6.1.*

What Hearts by Bruce Brooks (HarperCollins, 1992) traces Asa's life from first through seventh grade.

- Asa often had to ride his bike 5 to 6 miles to reach his friends' houses. Have students survey their classmates about the longest bike ride each has ever taken and make a box-and-whisker graph to show the results. *Use with Lessons 6.4–6.5.*

CHAPTER 5 Collect and Organize Data

MATHEMATICS ACROSS THE GRADES

SKILLS TRACE ACROSS THE GRADES

GRADE 5
Organize data in tables and line plots; interpret data using measures of central tendency; analyze data in graphs

GRADE 6
Determine if samples are representative of the population or biased and draw conclusions about a set of data; organize and analyze data in frequency tables and line plots; calculate and compare measures of central tendency

GRADE 7
Analyze samples and surveys for bias; collect, organize, and display data in graphs, frequency distributions, line plots, and histograms; draw conclusions about a set of data; determine which measure of central tendency best represents a set of data

SKILLS TRACE FOR GRADE 6

LESSON	FIRST INTRODUCED	TAUGHT AND PRACTICED	TESTED	REVIEWED
5.1	Grade 6	PE pp. 94–97, H36, p. RW20, p PW20, p. PS20	PE p. 116, pp. AG33–36	PE pp. 116, 117, 140–141
5.2	Grade 6	PE pp. 98–99, H36, p. RW21, p. PW21, p. PS21	PE p. 116, pp. AG33–36	PE pp. 116, 117, 140–141
5.3	Grade 4	PE pp. 100–101, p. RW22, p. PW22, p. PS22	PE p. 116, pp. AG33–36	PE pp. 116, 117, 140–141
5.4	Grade 4	PE pp. 102–105, H36, p. RW23, p. PW23, p. PS23	PE p. 116, pp. AG33–36	PE pp. 116, 117, 140–141
5.5	Grade 5	PE pp. 106–108, p. H36, p. RW24, p. PW24, p. PS24	PE p. 116, pp. AG33–36	PE pp. 116, 117, 140–141
5.6	Grade 6	PE pp. 109–111, H36, p. RW25, p. PW25, p. PS25	PE p. 116, pp. AG33–36	PE pp. 116, 117, 140–141
5.7	Grade 6	PE pp. 112–115, H36, p. RW26, p. PW26, p. PS26	PE p. 116, pp. AG33–36	PE pp. 116, 117, 140–141

KEY
PE Pupil Edition PS Problem Solving Workbook RW Reteach Workbook
PW Practice Workbook AG Assessment Guide

Looking Back  Prerequisite Skills

To be ready for Chapter 5, students should have the following understandings and skills:

- **Vocabulary**—*mean, median, mode, range*
- **Reading a Table**—use data in a table to answer questions
- **Mean, Median, and Mode**—find mean, median, and mode for a set of data
- **Range**—find range for a set of data

Check What You Know

Use page 93 to determine students' knowledge of prerequisite concepts and skills.

Intervention

Help students prepare for the chapter by using the intervention resources described on TE page 93.

Looking at Chapter 5  Essential Skills

Students will

- identify different types of samples and determine whether a sample is representative of the population or biased and draw conclusions about a set of data.
- **understand the process of organizing and analyzing data in a frequency table or a line plot.**
- develop skill in calculating the mean, median, and mode and determining how outliers affect them.

EXAMPLE

Use the data to make a line plot.

Number of Migrating Whales Observed				
3	7	4	9	5
7	2	5	7	3
1	5	2	3	7

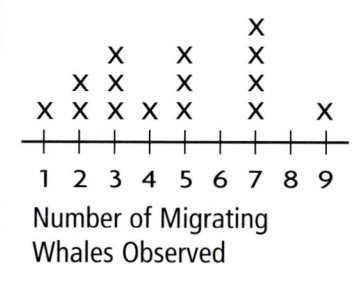

Looking Ahead  Applications

Students will apply what they learn in Chapter 5 to the following new concepts:

- Make, Analyze, and Compare Graphs (Chapter 6)
- Circle Graphs (Chapter 21)
- Find Unknown Values (Chapter 6)
- Make Predictions (Chapter 23)

Collect and Organize Data 92K

CHAPTER 5

Collect and Organize Data

INTRODUCING THE CHAPTER

Tell students that people collect and organize numerical information, or data, to help them understand the world, identify patterns and trends, and make predictions. After students read the paragraph, explain that the 100 pails of water represent all of the water on Earth. Ask them to determine how many of the 100 pails are salt water. **97** Ask students to write 97 parts out of 100 as a decimal. **0.97**

USING DATA

To begin the study of this chapter, have students

- Write a decimal for the part of Earth's total water that is liquid fresh water. **0.01**
- Write decimals to compare the amount of salt water to the amount of fresh water. **0.97 > 0.03**
- Write a decimal to show the part of Earth's fresh water that is ice. **0.67**

PROBLEM SOLVING PROJECT

Purpose To collect and organize data

Grouping pairs

Background Malaspina Glacier in Alaska is the largest glacier in the United States. Yet, at 2,176 km^2, it is significantly smaller than Antarctica's glaciers, which account for about 70% of Earth's fresh water.

Analyze, Choose, Solve, and Check

Have students

- Research the areas of five large glaciers on Earth.
- Organize the data in a logical order and then sort the data using two different sets of criteria, such as area or location.
- Analyze each other's data and determine the criteria used for sorting.

Check students' work.

 Suggest that students place their data lists in their portfolios.

92 Chapter 5

CHAPTER 5 Collect and Organize Data

Hubbard Glacier in Alaska

If you could look at Earth from space, you would see that most of its surface is covered with water. However, most of that water is salt water, which is unusable for drinking, watering crops, or manufacturing.

PROBLEM SOLVING What information about Earth's water supply can you find in the graph?

Possible answer: there is 97 times as much salt water as fresh water; there is twice as much ice as fresh water.

WATER ON THE EARTH

Fresh Water
Ice
Salt Water

If you could put all the water on Earth into 100 buckets, 97 buckets would hold the salt water of the oceans and seas, and 2 buckets would hold the frozen fresh water of glaciers and icecaps. Only 1 bucket would hold liquid fresh water.

92 Chapter 5

Why learn math? Explain that geologists collect data from core samples of ice taken from deep within a glacier. When these data from long ago are compared with more recent data, patterns emerge and predictions can be made. For example, one computer model shows a trend of glacial shrinkage that some scientists predict will cause a continued rise in sea level. Ask: What other jobs might require the collection and organization of data?
Possible answer: book publishers, newspaper journalists, sports statisticians

Check What You Know

Use this page to help you review and remember important skills needed for Chapter 5.

✓ Vocabulary

Choose the best term from the box.

> mean
> median
> mode
> range

1. The middle number in a group of numbers arranged in numerical order is the __?__. **median**
2. The sum of a group of numbers divided by the number of addends is the __?__. **mean**
3. The difference between the greatest number and the least number in a set of data is the __?__. **range**

✓ Reading a Table (For Intervention, see p. H5.)

STUDENTS' FAVORITE SPORTS				
	Gymnastics	Basketball	Baseball	Football
Boys	4	29	16	13
Girls	17	14	9	12

Use the data in the table above to answer the questions.

4. How many boys like baseball the best? **16 boys**
5. How many more girls than boys like gymnastics the best? **13 more girls**
6. Which sport was selected by more boys than any other? **basketball**
7. How many girls like gymnastics or football the best? **29 girls**
8. How many students were surveyed? **114 students**

✓ Mean, Median, and Mode (For Intervention, see p. H6.)

Find the mean for each set of data.

9. 7, 6, 11, 7, 9 **8**
10. 84, 73, 92, 77, 90 **83.2**
11. 8.9, 8.8, 7.9, 8.3 **8.475**
12. 4.6, 5.1, 6.2, 5.4, 4.8, 5.7 **5.3**

Find the median and the mode for each set of data.

13. 16, 32, 24, 10, 48, 32, 28 **28; 32**
14. 10, 15, 7, 8, 12, 13, 13, 19, 21 **13; 13**
15. 12.1, 10.8, 11.6, 10.8, 11 **11; 10.8**
16. 1.4, 0.9, 3.6, 2.5, 0.9, 6.2, 2.4 **2.4; 0.9**

✓ Range (For Intervention, see p. H6.)

Find the range for each set of data.

17. 6, 3, 7, 5, 9 **6**
18. 105, 102, 116, 103, 96 **20**

LOOK AHEAD

In Chapter 5 you will
- identify different kinds of samples and determine if surveys are biased
- organize data in frequency tables and line plots
- work with and analyze measures of central tendency

LESSON 5.1 Samples

LESSON PLANNING

Objective To identify different types of samples and to determine if a sample is representative of the population

Intervention for Prerequisite Skills
Read a Table (For intervention strategies, see page 93.)

NCTM Standards
5. Data Analysis and Probability
6. Problem Solving
8. Communication
9. Connections

Vocabulary

survey a method of gathering information about a group

population the entire group of individuals or objects

sample a group of people or objects chosen from a larger group to provide data to make predictions about a larger group

convenience sample a sample for which the most available individuals or objects in the population were selected

random sample a sample for which every individual or item in the population had an equal chance of being selected

systematic sample a sample for which a pattern was used to select the individuals or items except for the randomly-selected first one

Math Background

Surveys are a common part of our culture. Since it is often difficult to survey an entire population, a sample can be chosen and queried. These ideas will help students understand the process of choosing a sample.

- A random sample is a sample whose individuals were selected from a population in such a way that any member of the population had an equal chance of being chosen.
- A systematic sample is a sample for which individuals were chosen by using a pattern.
- A convenience sample contains people or objects that are at hand or convenient.

It is essential for students to understand that the type of sample used can influence the validity of the results.

WARM-UP RESOURCES

 **NUMBER OF THE DAY**

In every year of 365 days, there are just as many days before me as after me. What date am I? July 2

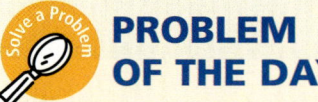 **PROBLEM OF THE DAY**

Dana's survey showed that 3 out of 8 students preferred pepperoni pizza and 1 out of 8 students preferred cheese pizza. How many more of the 72 students surveyed by Dana liked pepperoni pizza than liked cheese pizza? 18 more students like pepperoni pizza.

Solution Problem of the Day tab, p. PD5

 **DAILY FACTS PRACTICE**

Have students practice addition facts by completing Set A of *Teacher's Resource Book*, p. TR96.

INTERVENTION AND EXTENSION RESOURCES

ALTERNATIVE TEACHING STRATEGY

Materials calculator

Discuss with students the words *population* and *sample* and relate them to the class. Ask for a show of hands of all left-handed students. Record that number and the total number of students in the class. Use a calculator to find the percent who are left-handed and record the percent.

Have students consider the class as a sample of a larger population, such as all sixth graders in the school, all people in the school including adults, all people in the town or city, and so on.

Discuss problems that arise when one is collecting data for very large populations and the difficulty of getting a representative sample.

See also page 96.

AUDITORY
VERBAL/LINGUISTIC

MULTISTEP AND STRATEGY PROBLEMS

The following multistep and strategy problems are provided in Lesson 5.1:

Page	Item
97	20, 25

VOCABULARY STRATEGY

Before starting the lesson, **emphasize the vocabulary.** Display all terms on a three-column chart. Have volunteers suggest definitions and record one or two possible definitions for each word or phrase. As each word or phrase is covered in the lesson, ask volunteers to modify the definitions as needed.

VISUAL
VERBAL/LINGUISTIC

ADVANCED LEARNERS

Challenge students to **apply their knowledge of samples.** Ask students to mention as many reasons as they can why the following survey uses a sample that is not representative.

A company wants to survey people about their favorite radio station. The company picks 200 phone numbers at random from the phone book, calls at about 10 A.M., and asks the person who answers what radio station they like the most.

Possible answer: The sample does not include people without phones, people with unlisted numbers, working people who are not at home at 10 A.M., or students who may be in school.

VISUAL
VISUAL/SPATIAL

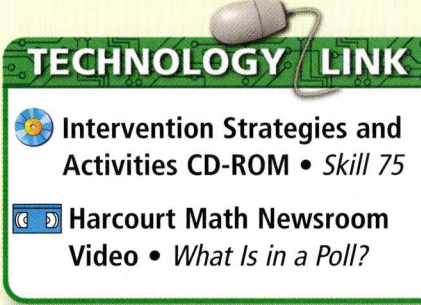

TECHNOLOGY LINK
- Intervention Strategies and Activities CD-ROM • *Skill 75*
- Harcourt Math Newsroom Video • *What Is in a Poll?*

94B

LESSON 5.1 ORGANIZER

Objective To identify different types of samples and to determine if a sample is representative of the population

Vocabulary survey, population, sample, convenience sample, random sample, systematic sample

1 Introduce

QUICK REVIEW provides review of prerequisite skills.

Why Learn This? You will know how to choose a sample to conduct a survey of your classmates. *Share the lesson objective with students.*

2 Teach

Guided Instruction

- *Discuss the meaning of the word* population *in the context of surveys.*

   If the population for a survey about favorite breakfast foods is all sixth graders in your school, would you want to include answers from teachers? from a seventh grader? from a sixth grader who does not like sports? no; no; yes

REASONING **What problems are presented by surveys such as the U.S. census or surveying people in your neighborhood?** Possible answer: It's impossible to reach the entire population or even to find everyone at home during a certain time.

Modifying Instruction If students have difficulty understanding a random sample, show them a spinner with 10 equal sections numbered 0–9. Explain that when you spin the spinner, there is an equal chance of landing on each number. Relate that to each member of a population having the same chance of being chosen for a sample.

ADDITIONAL EXAMPLE

Example 1, p. 94

Suppose Walter wants to find out the favorite sports team of the people in his hometown. Describe the population. Should Walter survey the population or use a sample? Explain. The population is all of the people in the town. He should survey a sample because it may not be possible to survey every person in town.

See also page 95.

94 Chapter 5

LESSON 5.1 Samples

Learn how to identify different types of samples and to determine if a sample is representative of the population.

QUICK REVIEW

1. $240 + 360$ = 600
2. $94 - 60$ = 34
3. 60×20 = 1,200
4. $30\overline{)270}$ = 9
5. $10\overline{)850}$ = 85

Vocabulary
survey
population
sample
convenience sample
random sample
systematic sample

A **survey** is a method of gathering information about a group. Surveys are usually made up of questions or other items that require responses.

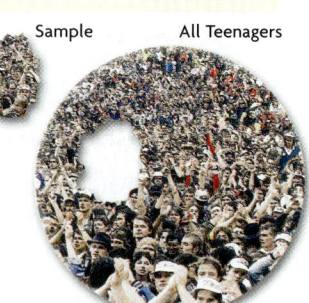
Sample   All Teenagers

You can survey the **population**, the entire group of individuals or objects, such as all teenagers. Or, if the population is large, you can survey a part of the group, called a **sample**.

EXAMPLE 1 Suppose Jenna wants to find out the favorite game among students in her math class. Describe the population. Should Jenna survey the population or use a sample? Explain.

The population consists of all the students in Jenna's math class. Jenna should survey the population since it is small.

- **What if** the population is changed to all of the 1,800 students in Jenna's school? Jenna could survey a sample.

Surveying the population is not always possible, so samples are used.

EXAMPLE 2 Renee asked students at Roosevelt Middle School to indicate their favorite sport. She surveyed two samples and then the entire population of 600 students. How do the results from the two samples compare with the results from the population?

SAMPLE SIZE	NUMBER CHOOSING BASKETBALL	PERCENT OF SAMPLE
50	18	36%
100	33	33%

POPULATION	NUMBER CHOOSING BASKETBALL	PERCENT OF POPULATION
600	192	32%

The results from the samples are different from, but close to, the results from the population.

94 Chapter 5

RETEACH 5.1

Samples

A sample is a part of a population. Samples are used when it would be too time-consuming or too expensive to survey an entire population.
A population may be people, such as all the people in a city, or it may be objects, such as all the books in a bookstore.
A sample is chosen using one of several methods. Three such methods are:
- **Random sample:** Each person or object in a population has the same chance of being chosen.
- **Convenience sample:** Those people or objects that are most readily available are chosen.
- **Systematic sample:** A person or object is chosen randomly and then a pattern is used, such as every fifth person, to choose others.
In general, it is best to use a random sample, since it is most representative of an entire population. It is also important to use the largest sample possible, since the larger the sample, the closer the results will represent the entire population.

1. Carole is surveying members of her community asking them to name their favorite restaurant. Determine the type of sample. Write *convenience, random,* or *systematic*.
   a. She stands outside a local mall and asks people leaving the mall to name their favorite restaurant.   convenience
   b. She opens the telephone book, randomly chooses a name, and then chooses every tenth name that begins with the same letter.   systematic
   c. She knows that every family in her community is listed in the telephone book. She calls 5 people randomly chosen from each page of the book.   random

2. Evan is on a committee to determine if the players in the Tri-Town Soccer League should survey *all the players* or use *a sample* in each situation.
   a. There are 40 players in the league.   all the players
   b. The players are difficult to reach since they live in many areas.   a sample
   c. There are 650 players in the league.   a sample

PRACTICE 5.1

Samples

Determine the type of sample. Write *convenience, random,* or *systematic*.

1. An assembly-line worker randomly selected one microwave oven and then checked every fifteenth oven to see whether it worked.
   systematic

2. Carl selected students to complete a survey by assigning each student's name a number from 1 to 6, rolling a cube numbered 1 to 6, and choosing each student whose name had the number he rolled.
   random

3. A store manager asked the first 50 shoppers to enter her store on Saturday to complete a survey about changes they would like to see made at the store.
   convenience

Tell whether you would survey the population or use a sample. Explain.

4. You want to know the type of computer, if any, that each student in your class has at home.
   population; There are not too many class members to survey them all.

5. You want to know the average number of siblings of all sixth grade students in your school district.
   Possible answer: sample; There are too many sixth graders to survey them all.

6. You want to know your friends' favorite television program.
   Possible answer: population; Your group of friends is not too large to survey them all.

Mixed Review

Evaluate each expression.

7. $9.03 \div x$ for $x = 3$ 3.01
8. $7m$ for $m = 2.2$ 15.4
9. $4.5 - w$ for $w = 1.9$ 2.6
10. $17.4 \div h$ for $h = 5.9$ 23.3
11. $k \div 2$ for $k = 6.4$ 3.2
12. $6.58 + a$ for $a = 0.45$ 7.03

Types of Samples

There are many types of sampling methods. The table below shows three different sampling methods.

TECHNOLOGY LINK
To learn more about samples, watch the **Harcourt Math Newsroom Video** *What is in a Poll?*

TYPE	DEFINITION	EXAMPLE
convenience sample	The most available individuals or objects in the population are selected to obtain results quickly.	Choose a specific location, such as the cafeteria or the library, and survey students as they walk by you.
random sample	Every individual or object in the population has an equal chance of being selected. This produces the sample that is most representative of the population.	Assign a number to each student and then choose students by randomly selecting numbers with a computer.
systematic sample	An individual or object is randomly selected and then others are selected using a pattern.	Randomly choose a student from a list of students, and then choose every fourth student.

EXAMPLE 3

Identify each type of sample.

A. A bicycle shop owner randomly surveys a customer and then surveys every tenth customer after that to determine when customers expect to buy their next new bike.

This is a systematic sample since individuals are selected according to a pattern.

B. A sixth-grade middle school math teacher randomly selects 50 students in her math classes and asks them if they have a home computer.

This is a random sample since everyone has an equal chance of being selected.

C. The owner of a movie rental shop asks customers in the shop on a given day to fill out a card naming their favorite movie.

This is a convenience sample since the most available individuals are selected.

Math Idea ▶ The results from a sample can depend on how the sample is chosen. It is important to select a sample that is representative of the population. For example, if the population includes men and women, then the sample must include both men and women.

95

- As students consider the three sampling methods, discuss difficulties that might arise with obtaining different kinds of samples.

Why might it be hard to obtain a random sample from a large population? Possible answer: You would need to insure that all members of the population have an equal chance of being included in the sample.

Which sampling method would you not use if you wanted to be sure of getting a representative sample? Why? Convenience sampling; it is likely that not every member of the population has an equal chance of being selected due to the time or place used to select the sample.

REASONING Is a systematic sample always representative? Explain. Possible answer: No; suppose you had a list of student names that alternate boy, girl, boy, girl. If you chose every fourth name on the list, the sample would be all girls, which would not represent the population.

ADDITIONAL EXAMPLES

Example 2, p. 94

The audience at the last school concert filled out surveys on their favorite concert song. First, two samples of the surveys were checked. Then the chorus instructor checked all 120 of them. How do the results from the two samples compare with the results from the population?

Sample Size	Number Choosing the National Anthem	Percent of Sample
30	10	33%
50	20	40%

Population	Number Choosing the National Anthem	Percent of Population
120	42	35%

The results from the samples are different from, but close to, the results from the population.

Example 3, p. 95

Identify the type of sample.

The cafeteria aide asks students to name their favorite food for lunch as they walk by. This is a convenience sample, since the most available individuals are selected.

CHALLENGE 5.1

My, How You've Aged!

The graph shows changes in the median age of the United States population from 1800 to 1990. Use the graph to answer the questions.

1. What was the median age of the population in 1900?
   about 23 years old

2. What was the median age of the population in 1840?
   about 18 years old

3. For which two years was the median age the same?
   1950 and 1980

4. By about how many years did the median age increase between the first and last dates shown on the graph?
   by about 17 years

5. What does the graph indicate about median age of the population between 1800 and 1920?
   It increased at a steady rate.

6. What does the graph indicate about the median age of the population between 1950 and 1970?
   It declined at a steady rate.

7. During which 10-year span did the median age remain almost constant?
   between 1800 and 1810

8. During which two 10-year spans did the median age change the most?
   from 1930 to 1940 and from 1980 to 1990

9. What might have caused the change that occurred between 1950 and 1970?
   a disproportionate number of births

10. Based on the data from 1970 to 1990, predict the median age for 2000.
    It is likely the median age will be greater than 35.

PROBLEM SOLVING 5.1

Samples
Write the correct answer.

1. Find the product.
   65.35 × 80.6
   5,267.21

2. Evaluate the expression.
   $(7^2 \times (5 - 3) + 22) \div 40$
   3

3. Fred wanted to find out the favorite color of all the students in his middle school. He surveyed all the students in his class. Is this a random sample? Explain.
   Yes. Students are not assigned on the basis of favorite color.

4. Cecily is ordering sodas for the class party. She asks a student in the lunch line for her favorite soda and then asks every tenth student. What kind of sample is she using?
   a systematic sample

Choose the letter for the best answer.

5. Thad conducted a survey on hair color at his school. His results were 23 students had blonde hair, 38 students had black hair, 7 students had red hair, and 19 students had brown hair. If he sampled 1 out of every 10 students at his school, how many people attend the school?
   A 900 people
   B 870 people
   C 820 people
   D 750 people

6. Jill surveyed students about their choice for a new school color. The results were that 45 people liked red, 33 liked green, 16 liked orange, and 8 liked blue. If she chose a student at random from the school's enrollment list and then asked every tenth student on the list, which describes the school's enrollment and her sample?
   F 1,020 students; random sample
   G 1,002 students; systematic sample
   H 1,020 students; systematic sample
   J 1,002 students; convenience sample

7. The owner of a grocery store ordered 56 cases of cups. Each case holds 16 packages. How many packages of cups did the store owner order?
   A 896 packages   C 1,026 packages
   B 1,006 packages   D 1,128 packages

8. Paul has 1,716 eggs to put into cartons. Each carton holds one dozen eggs. How many cartons does Paul need to store all the eggs?
   F 163 cartons   **H 143 cartons**
   G 153 cartons   J 133 cartons

9. **Write About It** Why does a large sample generally give better results than a small sample?
   Possible answer: The larger the sample, the closer you come to surveying the entire population and the more representative the sample becomes.

LESSON 5.1

3 Practice

Guided Practice

Do Check for Understanding Exercises 1–9 with your students. Identify those having difficulty and use lesson resources to help.

COMMON ERROR ALERT

Some students may have difficulty understanding the specialized use of the words *population* and *sample*. Lead students to see the relation between these words as the same relationship between the words *whole* and *part*.

Independent Practice

Note that Exercises 20 and 25 are **multistep or strategy problems**. Assign Exercises 10–20.

Scaffolded Instruction Use the prompts on Transparency 5 to guide instruction for the multistep or strategy problem in Exercise 20.

Transparency 5

CHECK FOR UNDERSTANDING

Think and Discuss — Look back at the lesson to answer each question.

1. **Explain** why a sample is often used rather than a population when conducting a survey. Possible answer: There may be too many people in the population so it could take too much time and money.
2. **Tell** when you might survey a population rather than a sample. when the population is small
3. **Discuss** why a random sample is more likely to represent a population than a convenience sample. Possible answer: A random sample gives everyone an equal chance of being selected.

Guided Practice — Describe the population.

4. Suppose a random sample of 1,250 women in a city was taken in order to determine the favorite type of music of all women in the city. all women in the city

Tell whether you would survey the population or use a sample. Explain.

5. You want to find out where the students in your class want to go for a field trip. population, because there are not too many students and they are easily available
6. You want to know the favorite mountain bike of teenagers in your state. sample; because there are too many teenagers to survey all of them

For 7–9 determine the type of sample. Write *convenience*, *random*, or *systematic*.

7. A camp director was asked to conduct a survey to find which team game the campers liked best. She randomly selected one camper and then surveyed every fifth person on her list of campers. systematic
8. The names of one hundred high school students are drawn from a box containing all student names. random
9. The owners of a sports store ask shoppers in the store a question about the kind of running shoes they like. convenience

PRACTICE AND PROBLEM SOLVING

Independent Practice — Describe the population.

10. To find out the average salary of female architects, a researcher randomly selected 100 female architects. all female architects

Tell whether you would survey the population or use a sample. Explain.

11. You want to find out the daily exercise habits of the varsity basketball team. population; There are not too many basketball team members.
12. You want to know the favorite kinds of food of all teens in a large city. sample; There are too many teenagers to survey all of them.

96 Chapter 5

ALTERNATIVE TEACHING STRATEGY | SCAFFOLDED INSTRUCTION

Purpose Students do an experiment to compare results of random sampling to those of the whole population.

Materials *For each group* a bag containing 25 crayons—12 red, 5 blue, 4 black, 4 yellow

Distribute a bag of crayons to each group with instructions not to look inside. Present the following information:

An artist asked 25 people to identify their favorite color. Each person put a crayon of his or her favorite color in the bag.

Have groups use a random sample to identify the color most frequently chosen as the favorite.

First students should decide how many crayons to pull from the bag for their random sample.

Have groups record the results of their random sample, then replace the crayons and repeat the experiment. The results of 2 random samples might be as follows:

Random Sample 1

5 crayons

Red—3
Black—1
Yellow—1

Random Sample 2

5 crayons

Red—4
Blue—1

Have groups share their results and predictions about which color was voted as the favorite by the whole population of 25.

Groups can then count all the crayons and compare their random sampling findings with the whole population.

If time permits, have each group put a different combination of crayons in the bag, then exchange bags with another group and repeat the experiment. Check students' work.

For 13–15, determine the type of sample. Write *convenience*, *random*, or *systematic*.

13. A cafeteria worker randomly selected a student and then asked every tenth student who entered the cafeteria a question. systematic
14. When conducting a survey, Rachel selected 50 people by drawing names out of a hat without looking. random
15. To find out the favorite movie of students at her school, a teacher asked students in her math class to complete a survey. convenience

Problem Solving ▶ Applications

16. Jim asked Lake Middle School students to indicate their favorite pet. Some results are in the tables at the right. He surveyed two samples and then the population of 320 students. How do the results from the two samples compare with the results from the population?

SAMPLE SIZE	NUMBER CHOOSING DOGS	PERCENT OF SAMPLE
10	2	20%
50	24	48%

POPULATION	NUMBER CHOOSING DOGS	PERCENT OF POPULATION
320	160	50%

16. The results from the smaller sample are different while the results from the larger sample are about the same.

Dalton wants to conduct a survey to find out which computer game is most popular among the students in his school.

17. How can Dalton get a random sample? Possible answer: He can randomly select names from a list of students in his school.
18. How can Dalton get a convenience sample? Possible answer: He can survey 10 friends he eats lunch with.
19. 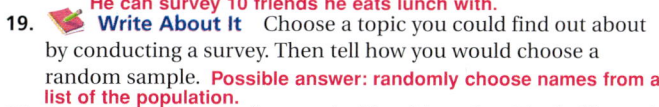 **Write About It** Choose a topic you could find out about by conducting a survey. Then tell how you would choose a random sample. Possible answer: randomly choose names from a list of the population.
20. A target has three circles, one inside of the other. The bull's-eye is 10 points, the middle circle is 6 points, and the outer circle is 4 points. If 3 darts are thrown and only 2 hit the target, what are the possible scores? 20 pts, 16 pts, 14 pts, 12 pts, 10 pts, or 8 pts

MIXED REVIEW AND TEST PREP

Evaluate for $a = 3.1$ and $b = 6.2$. (p. 82)

21. $a + b$  9.3
22. $a \times b$ 19.22
23. $b \div a$ 2

★ 24. **TEST PREP** Which is the product of 5.25 and 1.9? (p. 70) **A**

   A 9.975   **B** 99.75   **C** 997.5   **D** 9,975

★ 25. **TEST PREP** Martha saved $0.50 one week. Then she saved $0.50 more each week than she had the week before. How many weeks did it take to save a total of $10.50? (p. 66) **G**

   F 5 weeks   **G** 6 weeks   **H** 7 weeks   **J** 8 weeks

EXTRA PRACTICE page H36, Set A

MIXED REVIEW AND TEST PREP

Exercises 21–25 provide **cumulative review** (Chapters 1–5).

Multistep or Strategy Problem To solve Exercise 25, students can make a table to keep track of the amount Martha saved. They can keep a running total until the total is $10.50.

4 Assess

Summarize the lesson by having students:

DISCUSS **Give an example of each type of sample.** Possible answer: convenience sample—the first 30 people who walk into a store; systematic sample—every tenth person who walks into the store all day long; random sample—30 people chosen at random from a list of the store's customers

WRITE **Explain the difference between the three sampling methods presented in this lesson.** For a convenience sample, individuals are selected based on their availability. For a random sample, every individual in a given population has an equal chance of being selected. For a systematic sample, an individual is randomly selected, and then other individuals are selected according to a pattern.

Lesson Quiz

Determine the type of sample. Write *convenience, random,* or *systematic*.

1. Every tenth person to enter the movie theater was asked how many movies he or she attends in a month. systematic
2. The owner of the Ice Parlor asked customers in the store today about their favorite yogurt flavors. convenience
3. When conducting a survey of the student population, the teacher selected 100 students by assigning a number to each student in the school and then having a computer randomly select 100 of the numbers. random
4. Every fifth person leaving the hospital is asked about health care. systematic

LESSON 5.2

Bias in Surveys

LESSON PLANNING

Objective To determine whether a sample or a question in a survey is biased

Intervention for Prerequisite Skills
Read a Table (For intervention strategies, see page 93.)

NCTM Standards
5. Data Analysis and Probability
6. Problem Solving
7. Reasoning and Proof
8. Communication

Vocabulary

unbiased sample a sample that is representative of the population

biased sample a sample that is not representative of the population

biased question a question that leads to a specific response or excludes a certain group

Math Background

Statistics are used to inform us or to persuade us to change our opinions. Since governments use statistics in making decisions that affect our lives, it is important that data are gathered in an unbiased manner.

Consider the following as you help students understand how to gather data in an unbiased manner:

- For an unbiased sample, everyone in the population has an equal chance of being chosen. Thus it is important not to choose just one subgroup of the population to interview.
- The questions asked must not lead the respondent to reply in any particular way.

WARM-UP RESOURCES

 **NUMBER OF THE DAY**

Draw squares around the numbers of 4 adjacent days on the calendar for this month. Add all the numbers in the squares, and subtract 4 times the first number. What number do you get? 6

 **PROBLEM OF THE DAY**

In a survey of students about a field trip to a nearby factory, 12 students were undecided, 5 times that many were in favor of the field trip, and half as many were against it as were in favor of it. How many students participated in the survey? 102 students

Solution Problem of the Day tab, p. PD5

 **DAILY FACTS PRACTICE**

Have students practice subtraction facts by completing Set B of *Teacher's Resource Book*, p. TR96.

98A Chapter 5

INTERVENTION AND EXTENSION RESOURCES

ALTERNATIVE TEACHING STRATEGY

Discuss with students the idea that a **question is biased** if it suggests a preferred answer. Display the following biased questions:

- Do you agree with me that dogs are the best pet?
- Do you like hot dogs the best?
- Do you agree that it is more fun to go out to dinner than to eat at home?

Have students rewrite the questions so that they are not biased. Possible answers: What animal would you choose for a pet?; What is your favorite food?; Where do you prefer to eat, at home or in a restaurant?

VISUAL
INTRAPERSONAL/INTROSPECTIVE

ENGLISH LANGUAGE LEARNERS

To help students **understand the meaning of** *biased* and *unbiased*, display the two words. Then write a behavior under *biased*, such as: *I did not choose Maria to be on my team because she is a girl.*

Have students work with partners and talk about why the behavior is biased. Possible answer: It assumes that a girl would not be as good a player as a boy.

Then ask each group to discuss what kind of action would not be biased and record that under *unbiased* on the chart. Possible answer: drawing numbers to determine teams

Finally suggest that students draw a visual for each term.

AUDITORY
INTERPERSONAL/SOCIAL

MULTISTEP AND STRATEGY PROBLEMS

The following multistep or strategy problem is provided in Lesson 5.2:

Page	Item
99	18

SOCIAL STUDIES CONNECTION

Have students **demonstrate their knowledge of biased surveys.** Polls can play a very large part in influencing our decisions. Have students find examples of polls and bring them to class to share.

Give these and similar questions to small groups of students to discuss:

- How do you think it might affect you if you heard a poll on Election Day stating that Candidate A is winning in several states? Possible answer: You might not vote if you thought it would not make a difference.
- How do you think it might affect you if you saw a poll stating that 7 out of 10 college graduates favor a constitutional amendment? Possible answer: You might assume that college educated voters better understand the issue and adjust your vote to theirs.

AUDITORY
VERBAL/LINGUISTIC

Intervention Strategies and Activities CD-ROM • *Skill 75*

LESSON 5.2 ORGANIZER

Objective To determine whether a sample or a question in a survey is biased

Vocabulary unbiased sample, biased sample, biased question

1 Introduce

QUICK REVIEW provides review of pre-requisite skills.

Why Learn This? You can use this skill to help you detect biased samples and survey questions in consumer product studies. *Share the lesson objective with students.*

2 Teach

Guided Instruction

- *Discuss the two main sources of bias in surveys and what to look for to detect bias.*

  What do you need to look for to detect bias in a survey? how the sample is selected and how the questions are worded

  REASONING In a survey of 500 Orange County teachers, would a sample of 50 middle school teachers be biased or unbiased? Explain. biased; the sampling method does not include elementary and high school teachers

ADDITIONAL EXAMPLE

Example 1, p. 98

Keisha surveys students who take karate lessons at a local gym about the number of days per week they jog. There are 150 students in the school. Which samples will be biased? Explain.

A. Randomly survey 30 students over the age of 18.

B. Randomly survey 30 students.

C. Randomly survey 30 male students.

D. Randomly survey 30 new students.

Choices A, C, and D are biased. Choice A excludes students younger than 18, choice C excludes females, and choice D includes only new students.

LESSON 5.2
Bias in Surveys

Learn how to determine whether a sample or question in a survey is unbiased.

QUICK REVIEW
1. $0.25 + 0.75$ **1**
2. $19.5 - 18.0$ **1.5**
3. 30×0.5 **15**
4. $0.35 \div 0.5$ **0.7**
5. $2.5 + 1.5 - 1.75$ **2.25**

Vocabulary
unbiased sample
biased sample
biased question

Does the earth revolve around the sun or does the sun revolve around the earth? In a recent survey of over 1,000 adults, 79% knew the correct answer.

When you collect data from a survey, your sample should represent the whole population. Every individual in that population should have an equal chance of being selected. Then the sample is an **unbiased sample**.

If individuals or groups from the population are not represented in the sample, then the sample is a **biased sample**. If the sample for the survey above included males only, it would be biased since the survey was for all adults.

EXAMPLE 1

Latosha wants to find out how many hours Polk Middle School students spend on the Internet. If she surveys students from this school, which samples below will be biased? Explain.

A. 200 girls
B. 200 athletes
C. 200 randomly selected students
D. Students who ride their bikes to school

Choices A, B, and D are biased. Choice A excludes boys, choice B only includes athletes, and choice D excludes students who don't ride their bikes to school.

Sometimes, questions are biased. A **biased question** leads to a specific response or excludes a certain group.

EXAMPLE 2

Is the following question biased?

Do you agree with a well-known movie critic that movies longer than two hours are boring?

This question is biased since it leads you to agree with the movie critic.

98 Chapter 5

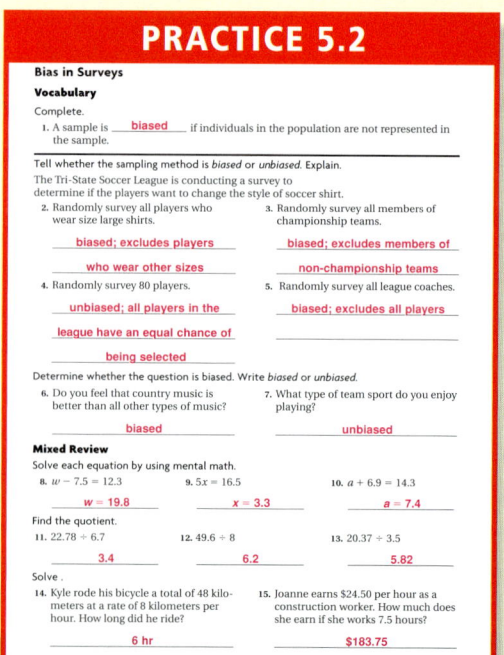

RETEACH 5.2
Bias in Surveys

A sample is **biased** if individuals or groups from the population are not represented in the sample.
Linda wanted to survey the 125 sixth-grade students at her school to find out the number of hours they spent reading each week. She made a list of four sampling methods she could use.
- randomly survey 5 students
- randomly survey 15 sixth-grade boys
- randomly survey 13 students at her school
- randomly survey 15 sixth-grade students

She decided that the first method would not include a large-enough sample. The second method would not include sixth-grade girls. The third method was biased because it would not be exclusively for sixth graders. Linda decided to use the last method. It would contain a large-enough sample, and every member of the sixth grade would have an equal chance of being selected.

1. Marian wants to survey the 58 sixth-grade teachers at Oak Park School to find out the average number of hours they spend grading papers. She makes a list of four sampling methods she could use. Circle the sampling method that is not biased. Then explain how the other three methods are biased.
   a. randomly survey all teachers who have taught for more than 10 years
   b. randomly survey 10 male teachers at the school
   c. randomly survey 8 sixth-grade teachers
   d. randomly survey 8 math teachers at the school
   Method a excludes teachers who have taught less than 10 years. Method b excludes female teachers. Method d excludes teachers who teach subjects other than math.

2. Dan wants to survey the 230 members of a golf club to find out the average number of hours they play golf each week. He makes a list of four sampling methods he could use. Circle the sampling method that is not biased. Then explain how the other three methods are biased.
   a. randomly survey all club members who have won tournaments
   b. randomly survey 30 club members
   c. randomly survey all members who have ever shot a hole-in-one
   d. randomly survey 25 members under the age of 30
   Method a excludes members who have not won tournaments. Method c excludes members who never shot a hole-in-one. Method d excludes members over the age of 30.

PRACTICE 5.2
Bias in Surveys
Vocabulary
Complete.
1. A sample is ___biased___ if individuals in the population are not represented in the sample.

Tell whether the sampling method is *biased* or *unbiased*. Explain.
The Tri-State Soccer League is conducting a survey to determine if the players want to change the style of soccer shirt.

2. Randomly survey all players who wear size large shirts.
   biased; excludes players who wear other sizes

3. Randomly survey all members of championship teams.
   biased; excludes members of non-championship teams

4. Randomly survey 80 players.
   unbiased; all players in the league have an equal chance of being selected

5. Randomly survey all league coaches.
   biased; excludes all players

Determine whether the question is biased. Write *biased* or *unbiased*.

6. Do you feel that country music is better than all other types of music?
   biased

7. What type of team sport do you enjoy playing?
   unbiased

Mixed Review
Solve each equation by using mental math.
8. $w - 7.5 = 12.3$ $w = 19.8$
9. $5x = 16.5$ $x = 3.3$
10. $a + 6.9 = 14.3$ $a = 7.4$

Find the quotient.
11. $22.78 \div 6.7$ 3.4
12. $49.6 \div 8$ 6.2
13. $20.37 \div 3.5$ 5.82

Solve.
14. Kyle rode his bicycle a total of 48 kilometers at an average speed of 8 kilometers per hour. How long did he ride?
    6 hr
15. Joanne earns $24.50 per hour as a construction worker. How much does she earn if she works 7.5 hours?
    $183.75

98 Chapter 5

CHECK FOR UNDERSTANDING

Think and Discuss — Look back at the lesson to answer the question.
1. Write the question in Example 2 so that it is not biased. **Possible question: Do you enjoy movies that are longer than two hours?**

Guided Practice — Tell whether the sample is *biased* or *unbiased*. Explain.

The local gym wants to find out how its members feel about the new exercise equipment.

2. Female members **biased; excludes males**
3. Members under age 20 **biased; excludes members 20 and older**
4. Members who used the gym in August **biased; excludes members who did not use the gym in August**
5. 50 randomly selected members **unbiased; all members have equal chance**

PRACTICE AND PROBLEM SOLVING

Independent Practice — Tell whether the sample is *biased* or *unbiased*. Explain.

The Middletown Mall is conducting a survey of its shoppers to find out which days they prefer to shop.

6. 400 teenage shoppers **biased; excludes people who are not teenagers**
7. 400 randomly selected shoppers **unbiased; all have equal chance**
8. 400 female shoppers **biased; excludes males**
9. Shoppers who enter the record store **biased; does not include people who shop in other stores**

Determine whether the question is biased. Write *biased* or *unbiased*.

10. Is basketball your favorite sport? **unbiased**
11. Do you agree with the fruit industry president that green apples taste better than red apples? **biased**

Problem Solving Applications

12. **REASONING** Students at Memorial Middle School are being surveyed to find out their choice of a new school mascot. If 300 students are randomly selected, do equal numbers of boys and girls have to be chosen? Explain. **No; boys and girls are randomly selected.**

13. 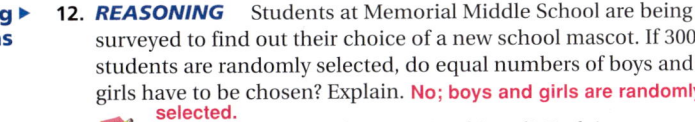 **Write About It** Is this question biased? Explain.
    I think pizza is the best choice for lunch, don't you? **Yes; the question leads you to agree with the person asking the question.**

MIXED REVIEW AND TEST PREP

14. A store owner randomly selects a customer and then surveys every tenth customer. What type of sample is this? (p. 94) **systematic**

Find the quotient. (p. 76)

15. $3.6 \div 0.9$ **4**
16. $1.44 \div 0.12$ **12**
17. $270 \div 0.03$ **9,000**

18. **TEST PREP** Replace ● with the missing operation $3^2 \times (6 \bullet 4) = 90$. (p. 44) **D**

A ÷   B ×   C −   D +

EXTRA PRACTICE page H36, Set B

99

---

3 Practice

Guided Practice
Do Check for Understanding Exercises 1–5 with your students. Identify those having difficulty and use lesson resources to help.

Independent Practice
Note that Exercise 18 is a **multistep or strategy problem**. Assign Exercises 6–13.

MIXED REVIEW AND TEST PREP
Exercises 14–18 provide **cumulative review** (Chapters 1–5).

4 Assess

Summarize the lesson by having students:

DISCUSS Why would you probably not use a convenience sample to get an unbiased sample for a survey? **Not everyone in the population would have an equal chance of being selected.**

WRITE What is a biased question? **Possible answer: a question that leads to a specific response**

Lesson Quiz

Tell whether the sample is biased or unbiased. Explain.

Sarah is doing a survey of 75 chess players to find out their favorite game strategy.

1. Randomly survey 8 members of the Chess Club. **Biased; it excludes chess players who are not members of the Chess Club.**
2. Randomly survey 8 chess players. **Not biased; all players had an equal chance of being selected.**

CHALLENGE 5.2
Bias in Advertising

PROBLEM SOLVING 5.2
Bias in Surveys

LESSON 5.3

Problem Solving Strategy: *Make a Table*

LESSON PLANNING

Objective To use the strategy *make a table* to organize data

Intervention for Prerequisite Skills
Read a Table (For intervention strategies, see page 93.)

Lesson Resources Problem Solving Think Along, p. TR1

NCTM Standards
- 5. Data Analysis and Probability
- 6. Problem Solving
- 10. Representation

Math Background

The strategy *make a table* can be used to:
- Order data
- Summarize data
- Organize data
- Help show a pattern

In statistics, the data collected must be organized or displayed in an enlightening fashion. Organizing data in a table allows students to more quickly answer questions about the data or to use it to make graphs or plots that illustrate the data.

WARM-UP RESOURCES

 **NUMBER OF THE DAY**  Transparency 5.3

How many half-dollars are needed to equal $5.00? How many of each of the other kinds of coins? **10 half-dollars, 20 quarters, 50 dimes, 100 nickels, 500 pennies**

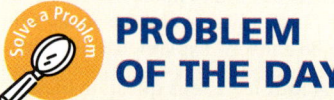 **PROBLEM OF THE DAY**  Transparency 5.3

It takes 4 yd of material and 3 yd of trim to make 2 pumpkin decorations. Rhonda needs to make 10 decorations. How many yards of material and trim will she need? **20 yd of material and 15 yd of trim**

Solution Problem of the Day tab, p. PD5

 **DAILY FACTS PRACTICE**

Have students practice subtraction facts by completing Set C of *Teacher's Resource Book,* p. TR96.

100A Chapter 5

INTERVENTION AND EXTENSION RESOURCES

ALTERNATIVE TEACHING STRATEGY

Materials *For each student* 1-inch paper square

Have students **practice the strategy** *make a table* by displaying a vertical number line labeled *0, 1, 2, 3, 4, 5,* and *more than 5.* Above the number line write *Number of Brothers.*

- Give each student a small paper square.
- Ask them to tape their square to the right of the number naming the number of brothers they have.
- Then ask students how many have no brothers, 1 brother, and so on.
- Use tally marks to show the data, making tallies as you point to the squares.
- Point out how the slash indicates a total of 5 and thus helps organize the tallies.

KINESTHETIC
BODILY/KINESTHETIC

ENGLISH LANGUAGE LEARNERS

Materials 1 small twig

Reinforce the vocabulary associated with surveys and tables by pointing out to students that the word *tally* comes from a Latin word meaning *twig.* Show a twig and ask students to give the word in their language that has the same meaning. Discuss with students how the twig visually resembles a tally mark.

Have each student choose three colors and write them down in a list. Have students survey each other as to which of the three colors they like best. Ask them to record the responses with tally marks and display their tables for the class. Check students' work.

VISUAL
VISUAL/SPATIAL

READING STRATEGY

Use Graphic Aids Remind students that making a table is not the only method to visually convey information. Extend your lesson discussion of other ways to solve the problem on page 100.

- Ask students to make a list of advantages of using a table, such as making the data more readable and usable, helping to avoid mistakes, and so on.
- Ask students how their strategy would change if 100 students had been interviewed. Possible answer: I would use a bar graph to display a greater number of responses rather than count so many tally marks.

EARLY FINISHERS

Encourage students to **apply the strategy** *make a table.* Divide the class into small groups.

- Ask each group to decide on a question they want to have answered by their classmates, such as *What is your favorite ice cream?* or *How many pets do you have?*
- Students then collect data either by interviewing classmates or asking the class to write their answers on slips of paper.
- Then have each group make a table and tally the results.
- Finally, have groups write at least two summarizing statements about the data, such as *14 people like chocolate best*, or *more people like vanilla than the other two flavors combined.* Check students' work.

AUDITORY
INTERPERSONAL/SOCIAL

MULTISTEP AND STRATEGY PROBLEMS

The following multistep and strategy problems are provided in Lesson 5.3:

Page	Item
101	1–10

- Intervention Strategies and Activities CD-ROM • *Skill 75*
- Data ToolKit • *Make a Table*

LESSON 5.3 ORGANIZER

Objective To use the strategy *make a table* to organize data

Lesson Resources Problem Solving Think Along, p. TR1

1 Introduce

QUICK REVIEW provides review of pre-requisite skills.

Why Learn This? After conducting a survey, you can organize your data into a table to solve a problem. *Share the lesson objective with students.*

2 Teach

Guided Instruction

- *Begin by discussing the data shown at the top of the page.*

  Are the numbers recorded in a meaningful list? Explain. No; the numbers are listed in the order students responded.

  What does the number 2 mean? It means that one student saw 2 movies.

- *Consider the table shown in the Solve section.*

  Why does the table make it easier to determine the number of students who saw either 3 or 4 movies? It's quicker to total tally marks grouped by 5 for a given category than to scan an entire list while counting specific numbers.

  Would students who saw 3 movies be counted in both groups described in the problem? Explain. No; students who saw 3 movies would not be counted in the group that saw fewer than 3 movies.

REASONING **Describe another way to find the answers to the questions that does not involve making a table.** Possible answer: You could count all the numbers 3 and greater. That total would tell you the number of students that saw at least 3 movies. Then subtract that number from the total number of students in the class to find the number that saw fewer than 3 movies.

100 Chapter 5

LESSON 5.3

PROBLEM SOLVING STRATEGY
Make a Table

Learn how to solve problems by displaying information in a table.

QUICK REVIEW

Find the value.
1. 6^2 = 36 2. 3^3 = 27 3. 2^3 = 8 4. 2^4 = 16 5. 2^5 = 32

Mrs. Donovan asked the students in her class to tell the number of movies they had seen since the start of the school year. The results are shown at the right.

Number of Movies Seen
1 5 3 2 1 6 1 2 3 4 5
1 1 6 3 4 2 2 2 4 3 2 1
2 2 4 5 1 3 2 1

How many students saw at least 3 movies? How many students saw fewer than 3 movies?

Analyze What are you asked to find? how many students saw at least 3 movies and how many saw fewer than 3 movies
What information is given? the number of movies each student has seen
Is there information you will not use? If so, what? No.

Choose What strategy will you use?
You can make a tally table to organize the data. Then you can use the table to answer the questions.

Solve How can you organize the tally table?
Make a row for each of the numbers of movies the students have seen. Then read the data and make a tally mark for each piece of information next to the appropriate number.

Next, use the table to answer the questions.

There are 6 + 4 + 3 + 2, or 15, students who saw at least 3 movies and 8 + 9, or 17, students who saw fewer than 3 movies.

NUMBER OF MOVIES SEEN	
1								
2									
3						
4					
5				
6			

Check How can you check your answer?
Recount the totals in the tally table.

What if two students were absent that day and they each had seen 5 movies? How many students would have seen at least 3 movies? fewer than 3 movies? 17 students; 17 students

100 Chapter 5

RETEACH 5.3

Problem Solving Strategy: Make a Table

Sometimes, when you have a lot of data, you do not need to know about each value. It may be enough to know how many values fit into a particular category. When this is the case, a tally table can help you solve problems.
Mr. Franks corrected the test papers of the 29 students in his 3rd period class. The scores (out of 100) are given below. How many more students had a score below 90 than had a score of 90 or above?

73, 83, 85, 92, 93, 85, 89, 91, 99, 80, 80, 84, 76, 78, 80, 93, 90, 88, 98, 82, 100, 78, 67, 88, 98, 94, 90, 76, 73

Step 1: Think about what you know and what you are asked to find.
- You know all the grades that the students in the 3rd period class received. You do not know who received each grade, but that information is not needed in this situation.
- You are asked to find the difference between the number of students who scored less than 90 and the number who scored 90 or above.

Step 2: Plan a strategy to solve.
- Use the strategy *make a table*
- Organize the data into a tally table. Make only as many rows as you need to solve the problem. For this problem, you need 2 rows: *less than 90* and *90 or above*.

Step 3: Solve.
- Carry out the strategy.

Test Scores	
Less than 90	|||| |||| |||| |||
90 or above	|||| ||||	

Count the tallies and find the difference. 18 − 11 = 7
So, 7 more students had a score below 90 than had a score of 90 or above.

Use the data below and the strategy *make a table* to help you solve 1–2.
The students in a sixth-grade class were surveyed about the number of minutes they spend on the Internet on a typical day. The results of the survey are given below.
15, 0, 20, 15, 30, 0, 0, 0, 15, 20, 60, 45, 20, 10, 0, 30, 0, 0, 90, 45, 0, 10, 0, 20

1. How many more students said they spend some time on the Internet each day than said they do not spend any time on the Internet?

   6 more students

2. Of those students who use the Internet, how many more of them spend less than 30 minutes online than spend at least 30 minutes online?

   3 more students

PRACTICE 5.3

Problem Solving Strategy: Make a Table

For 1–6, use the data below. Display the data in the table at the right using intervals of 31–40, 41–50, 51–60, and 61–70.

During the basketball season, the Falcons scored the following numbers of points in their games: 63, 52, 47, 51, 60, 49, 48, 54, 61, 52, 40, 38, 57, 46, 44, 63, 70

Number of Points Scored	
31–40	||
41–50	||||
51–60	||||	
61–70	||||

1. How many rows of data are in your table?

   4 rows

2. How many scores are greater than 40 but less than 61?

   11 scores

3. The Falcons won every game in which they scored more than 60 points. How many games did they play in which they scored more than 60 points?

   4 games

4. The team lost every game in which they did not score more than 40 points. How many games did they play in which they did not score more than 40 points?

   2 games

5. The Falcons' record for the season was 10 wins and 7 losses. How many games did they win in which they scored 60 points or fewer?

   6 games

6. With a record of 10 wins and 7 losses, how many games did the Falcons lose when they scored more than 40 points?

   5 games

Solve.

7. Dennis has 5 friends and wants to invite 2 of them to go to a baseball game with him and his family. How many different choices of 2 friends can Dennis make?

   10 different choices

8. Latifah has a project that is due on May 15. She expects the project to take her 3 weeks to complete. What is the latest date on which she could begin her project in order to be done on time?

   April 24

Mixed Review

Write the percent or decimal.

9. 16%   10. 5%   11. 0.55   12. 0.83   13. 0.07

   0.16   **0.05**   **55%**   **83%**   **7%**

PROBLEM SOLVING PRACTICE

Solve the problem by displaying the data in a table.

Members of the math club took a survey to find out how each person gets to school. These are the results.

skateboard	bike	in-line skates	bike
walk	bus	bike	walk
skateboard	bike	bike	bus
walk	skateboard	in-line skates	bike
bus	walk	bus	walk
skateboard	bike	bus	walk

PROBLEM SOLVING STRATEGIES
- Draw a Diagram or Picture
- Make a Model
- Predict and Test
- Work Backward
- Make an Organized List
- Find a Pattern
- ▶ Make a Table or Graph
- Solve a Simpler Problem
- Write an Equation
- Use Logical Reasoning

1. Make a table using the data above. How many categories of data are in your table? **D**
   A 2   B 3   C 4   D 5

2. How many of the math club members walk or ride a skateboard to school? **H**
   F 8   G 9   H 10   J 11

3. How many of the math club members ride a bike or the bus to school?
   12 members

4. How many more math club members walk to school than ride in-line skates to school? **4 more**

MIXED STRATEGY PRACTICE

5. Julia receives a $50.00 gift certificate to a music store. If she wants to buy three $13.99 CDs and two $7.99 cassettes, how much of her own money will Julia have to add to the $50.00 gift certificate? **$7.95**

6. A basketball team scores 23 points, including six 1-point foul shots. How many 2-point and 3-point baskets could they have made? Make a list of all the possibilities. **1 3-point, 7 2-point; 3 3-point, 4 2-point; 5 3-point, 1 2-point**

7. An automobile club sells books of 10 movie tickets for $45.00. If tickets usually cost $8.75 each, how much do you save on 10 tickets by buying the book? **$42.50**

8. A fence separating two gardens is 24 ft long. If there is a post in the ground every 3 ft, how many posts are in the fence? **9 posts**

9. On a wildlife outing, Enrique spots 3 more seagulls than egrets and 5 more egrets than geese. If he spots 8 geese, how many birds does he spot in all? **37 birds**

10. If today is Tuesday, what day of the week will it be 200 days from today? **Saturday**

11. **Write a problem** using the data from the math club survey at the top of the page. **Check students' problems.**

101

3 Practice

Guided Practice

Note that Exercises 1–4 are **multistep or strategy problems**. Do Problem Solving Practice Exercises 1–4 with your students. Identify those having difficulty and use lesson resources to help.

Independent Practice

Note that Exercises 5–11 are **multistep or strategy problems**. Assign Exercises 5–11.

In Exercise 6, you may want to suggest students use the strategy *work backward* so that their 2-point and 3-point shot combinations don't incorrectly total 23 points.

4 Assess

Summarize the lesson by having students:

DISCUSS How did you make the table you used to solve Problem 1 on page 101? Possible answer: I listed each way to get to school. Then I placed a tally for each time the way appeared in the survey results.

 **WRITE** Explain why *make a table* is a good strategy to use to solve Problem 1 on page 101. Possible answer: It organizes the information by substituting grouped symbols for words. These symbols are easier to total and help you answer more quickly.

Lesson Quiz

Solve the problem by displaying the data in a table.

Kara counted the number of birds at her feeder every 5 minutes for an hour. She wrote down 3, 5, 1, 3, 7, 3, 5, 2, 7, 6, 4, 4.

1. How many rows of data are in your table? **7 rows**

2. How many times did she see 2 or fewer birds? **2 times**

3. How many times did she see 4 or more birds? **7 times**

101

LESSON 5.4

Frequency Tables and Line Plots

LESSON PLANNING

Objective To record and organize data by using a frequency table or a line plot

Intervention for Prerequisite Skills
Read a Table, Range (For intervention strategies, see page 93.)

NCTM Standards
5. Data Analysis and Probability
6. Problem Solving
7. Reasoning and Proof
8. Communication
9. Connections
10. Representation

Vocabulary

frequency table a table that organizes the total for each category or group

cumulative frequency a running total of frequencies, shown in a column of a frequency table

Math Background

Unorganized data do not provide much information. They are difficult to interpret. Data can be organized in line plots or frequency tables. The following ideas will help students understand and use frequency tables and line plots:

- In a line plot, the categories are placed horizontally along a line and X marks are placed above the numbers.
- The first step in making a frequency table is to tally the data. Once tallied, the total number for each category is entered in the frequency table.
- The running totals, or cumulative frequencies, give information about two or more categories.

WARM-UP RESOURCES

 **NUMBER OF THE DAY**  Transparency 5.4

Looking at the calendar for one full year, you will see one digit for days more often than any other. What is it and how many times does it appear? 1; 163

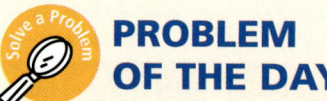 **PROBLEM OF THE DAY**  Transparency 5.4

Anne's line plot of ages of students has a range of 4. Each age has twice as many x's as the previous one. If the last age has 16 x's, how many students did Anne include in her data? 31 students

Solution Problem of the Day tab, p. PD5

 **DAILY FACTS PRACTICE**

Have students practice addition facts by completing Set D of *Teacher's Resource Book,* p. TR96.

102A Chapter 5

INTERVENTION AND EXTENSION RESOURCES

ALTERNATIVE TEACHING STRATEGY

Ask students to **make a tally table for the data** in the table for Example 1 on page 102. Discuss how the tally table and line plot show the same information in different forms. Then add a column labeled *Frequency* to the tally table. Explain that the frequency of a data item tells how many times the data item occurs. Show how to fill in the frequency column of the table.

During discussion, point out that the numbers in the frequency column provide the same information as the tally table and line plot. Use the discussion as a transition to the situation in Example 2 on page 103.

See also page 104.

VISUAL
VISUAL/SPATIAL

ENGLISH LANGUAGE LEARNERS

Materials dictionary

Have students talk about the **meaning of the words** *frequency* and *frequently*.

- Model the words by showing real-life connections such as a menu with multiple pizza entries or a monthly rainfall table for several cities.
- Ask them to look up the words in a dictionary, read the definitions aloud, and identify the meanings that match the way the words are used in this lesson.
- Encourage students to use the words in their own sentences. Possible answer: Pizza is frequently on the menu. The frequency of rain increases in the spring.

VISUAL
VERBAL/LINGUISTIC

MULTISTEP AND STRATEGY PROBLEMS

The following multistep or strategy problem is provided in Lesson 5.4:

Page	Item
105	22

EARLY FINISHERS

Materials *For each group* 2 number cubes numbered 1–6, p. TR75

Encourage students to **practice making frequency tables.** Ask them to work in small groups to see what sums are rolled most often when the 2 cubes are rolled 30 times. Have groups prepare a cumulative frequency table that lists the possible sums in increasing order.

After they have completed their tables, have groups compare and discuss their results. Check students' work.

VISUAL
INTERPERSONAL/SOCIAL

TECHNOLOGY LINK

- Intervention Strategies and Activities CD-ROM • *Skills 75, 76*
- Data ToolKit • *Make a Frequency Table, Make a Line Plot*

LESSON 5.4 ORGANIZER

Objective To record and organize data by using a frequency table or a line plot

Vocabulary frequency table, cumulative frequency *Review* range

1 Introduce

QUICK REVIEW provides review of prerequisite skills.

Why Learn This? Making a frequency table or line plot allows you to quickly analyze and draw conclusions from raw data. *Share the lesson objective with students.*

2 Teach

Guided Instruction

• Discuss with students the meaning of the line plot in Example 1.

Which numbers on the line plot represent data from the original table? the numbers with X's above them

If you made a tally table for the data, how would the number of tally marks after an entry compare with the number of X's for that same number? The number of tally marks would equal the number of X's.

ADDITIONAL EXAMPLE

Example 1, p. 102

TEST SCORES							
100	98	97	96	98	97	99	99
97	96	98	100	100	100	98	98

Use the data above to make a line plot.

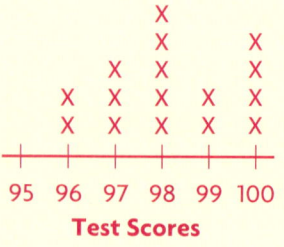

LESSON 5.4

Frequency Tables and Line Plots

Learn how to record and organize data collected in a survey.

QUICK REVIEW

1. 70
   +20   90
2. 35
   +40   75
3. 67
   −30   37
4. 50
   −15   35
5. 240 ÷ 6   40

Vocabulary
frequency table
cumulative frequency

In 2000, the largest iceberg on record contained enough ice to provide everyone on earth with 4 gallons of water per day for 40 years! Most of an iceberg is below the surface of the water. The data in the table below show the number of icebergs a scientist observed every day for 20 days.

NUMBER OF ICEBERGS OBSERVED				
12	15	11	16	15
16	16	15	13	16
16	14	12	14	16
15	13	15	14	15

You can use a line plot to record data.

EXAMPLE 1

Use the data above to make a line plot.

Step 1: Draw a horizontal line.

Step 2: On your line, write the numerical values for the number of icebergs, using vertical tick marks.

Step 3: Plot the data.

TECHNOLOGY LINK
More Practice: Use *Data ToolKit* to make frequency tables and line plots.

Each X represents the number of icebergs the scientist observed during one day.

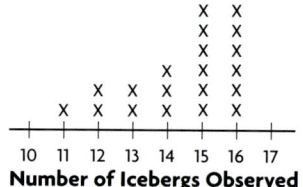

A line plot helps you see if there are any gaps or extremes in the data. You can also see where data cluster. The line plot in Example 1 shows that the scientist observed 15–16 icebergs per day more often than she observed 11–14 icebergs per day.

102 Chapter 5

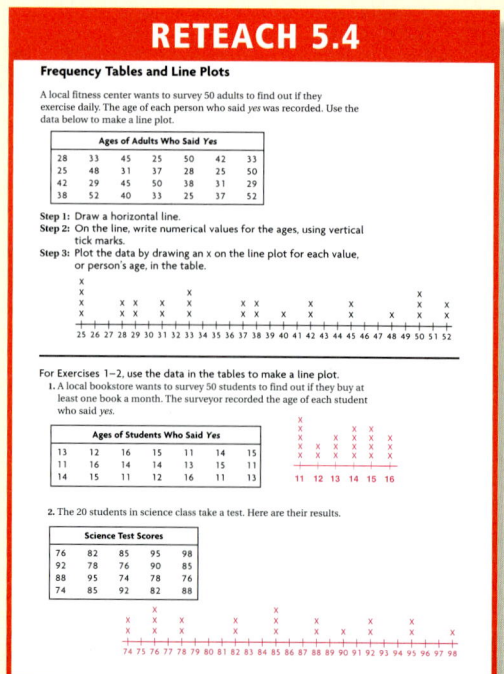

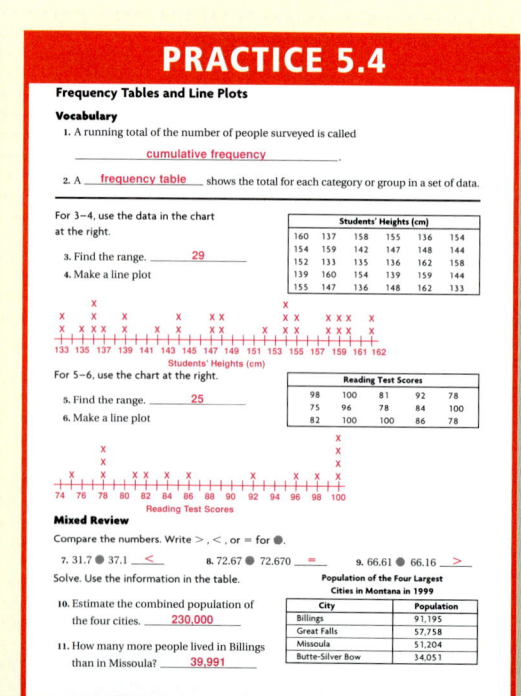

Frequency Tables

A **frequency table** shows the total for each category or group. You can add a cumulative frequency column to the table. **Cumulative frequency** is a running total of the frequencies.

EXAMPLE 2

The local mall surveyed 75 teens regarding the overall quality of food at the mall's food court. Use the data from the survey to make a frequency table. Do you think most teens like the quality of food at the food court?

Quality of Food at Food Court	
Quality	Tally
Very Good	|||| |||| ||||
Good	|||| |||| |||| |||| |||| |||
Fair	|||| |||| ||||
Poor	|||| ||||

List the categories of quality in one column. Record the total for each category in the frequency column. Record the running total in the cumulative frequency column.

QUALITY OF FOOD AT FOOD COURT		
Quality	Frequency	Cumulative Frequency
Very Good	20	20
Good	30	50	← 20 + 30 = 50
Fair	15	65	← 50 + 15 = 65
Poor	10	75	← 65 + 10 = 75

So, most teens liked the quality of food at the food court: 50 out of 75 teens surveyed rated the food as good or very good.

A local mall took a survey to find the number of times people go to the mall in one year. The survey results are shown below.

NUMBER OF TRIPS TO THE MALL									
12	20	5	7	13	11	3	9	19	16
8	17	17	16	1	6	8	5	14	9

EXAMPLE 3

Use the data about the number of trips to the mall to make a cumulative frequency table with intervals.

20 − 1 = 19   *Find the range.*

Since 4 × 5 = 20, which is close to 19, make 4 intervals that include 5 consecutive numbers.   *Use the range to determine intervals.*

Remember that the range of a set of data is the difference between the greatest number and the least number in the set of data.

TRIPS TO THE MALL		
Number of Trips	Frequency	Cumulative Frequency
1-5	4	4
6-10	6	10
11-15	4	14
16-20	6	20

Complete the cumulative frequency table.

103

LESSON 5.4

REASONING In Example 4, there are 30 responses. Why would 3 intervals of 10 consecutive numbers not be enough? The responses are not consecutive numbers but are spread over a range of 37, requiring more intervals.

ADDITIONAL EXAMPLE

Example 4, p. 104

AGES OF BICYCLISTS						
24	15	32	36	44	41	18
17	25	28	31	23	29	32

Use the data above to make a cumulative frequency table with intervals.

Find the range: 44 − 15 = 29

Since 3 × 10 = 30, make 3 intervals that include 10 consecutive numbers. Check students' work.

3 Practice

Guided Practice

Do Check for Understanding Exercises 1–5 with your students. Identify those having difficulty and use lesson resources to help.

Independent Practice

Note that Exercise 22 is a **multistep or strategy problem**. Assign Exercises 6–17.

Cumulative frequency tables with intervals can also be used to organize data about different age groups. The data in the table below show the ages of joggers at the Get Fit exercise trail.

AGES OF JOGGERS									
27	31	11	22	25	42	44	36	33	24
19	34	48	17	28	33	32	30	40	31
38	18	22	23	24	27	36	19	26	32

EXAMPLE 4 Use the data about the ages of joggers at the Get Fit exercise trail to make a cumulative frequency table with intervals.

48 − 11 = 37 *Find the range.*

Since 4 × 10 = 40, which is close to 37, make 4 intervals that include 10 consecutive numbers. *Use the range to determine intervals.*

Complete the cumulative frequency table.

AGES OF JOGGERS		
Age	Frequency	Cumulative Frequency
10–19	5	5
20–29	10	15
30–39	11	26
40–49	4	30

CHECK FOR UNDERSTANDING

Think and Discuss — Look back at the lesson to answer each question.

1. **REASONING** Name the intervals you could have in Example 3 if you made 2 intervals. Explain. 1–10, 11–20; Since 2 × 10 = 20, you would have 2 intervals that each include 10 numbers.
2. **Explain** how to find the size of the survey sample by using the frequency table in Example 4. Look at the last number in the cumulative frequency column.

Guided Practice — For 3–5, use the data in the table.

3. Make a line plot. See Additional Answers, p. 117A.
4. Make a cumulative frequency table. See Additional Answers, p. 117A.
5. Find the range. 16

AGES OF POP MUSIC LISTENERS						
12	13	17	16	15	13	25
28	27	18	15	12	14	23
13	15	17	13	22	18	15

PRACTICE AND PROBLEM SOLVING

Independent Practice — For 6–8, use the data in the table.

6. Make a line plot. See Additional Answers, p. 117A.
7. Make a cumulative frequency table. See Additional Answers, p. 117A.
8. Find the range. 12

NUMBER OF KILOMETERS BIKED						
6	11	7	8	8	5	15
14	12	10	11	6	7	9
15	15	8	9	12	17	10

104 Chapter 5

ALTERNATIVE TEACHING STRATEGY | SCAFFOLDED INSTRUCTION

Purpose Students collect data, use data in a frequency table, and draw conclusions based on the data.

As a class, brainstorm a list of possible survey questions students can ask 50 students throughout the school. For example:

- How many brothers and sisters do you have?
- Do you rate the school lunch as Very Good, Good, Fair, or Poor?
- How many minutes of television do you watch each day?

Have students work in small groups to choose one of the questions to use for their survey. Then have each group survey 50 students from other classes in the school and record the results in a tally chart. A sample of a tally table is shown.

NUMBER OF BROTHERS AND SISTERS	
Number	Tally
0–2	‖‖‖ ‖‖‖ ‖‖‖ ‖‖‖ ‖‖‖ ‖‖‖
3–5	‖‖‖ ‖‖‖ ‖‖‖
More than 5	‖‖‖

The groups can then make a frequency table with appropriate headings. Have groups share their tally tables and frequency tables with the class. A sample of a frequency table is shown.

NUMBER OF BROTHERS AND SISTERS		
Number	Frequency	Cumulative Frequency
0–2	30	30
3–5	15	45
More than 5	5	50

Check students' work.

Kim asked people in her neighborhood about their favorite type of movie. Use the data for 9–10.

FAVORITE MOVIES	
Movies	Tally
Drama	llll llll
Action	llll lll
Comedy	llll llll lll
Mystery	llll llll

9. Organize the data into a frequency table. Include a cumulative frequency column.
Check students' tables.

10. What is the size of the sample?
40 people

For 11–14, use the data in the box at the right.

11. Make a line plot.
See Additional Answers, p. 117A.
12. Find the range. **36**
13. How many numbers would be in each interval if you used the range to make 4 intervals? **9 weights**

WEIGHTS OF STUDENTS' PET DOGS (LB)					
40	10	22	33	44	40
41	42	12	35	20	10
30	25	20	18	46	35
34	35	40	40	35	38

14. Make a cumulative frequency table using 4 intervals for the weights.
See Additional Answers, p. 117A.

Problem Solving ▶ Applications

15. **REASONING** Survey your classmates about their favorite hobbies. Organize the data in a tally table and a frequency table. Can you make a line plot of the data? Explain. **Tables will vary. No, the data are categories and are not numerical.**

16. Decide on appropriate intervals and make a cumulative frequency table using intervals for the data in the table below.
See Additional Answers, p. 117A.

POUNDS OF VEGETABLES CONSUMED YEARLY											
12	16	18	12	9	21	14	17	18	5	23	25	32
17	16	8	9	22	21	14	18	7	23	24	19	13

17. ❓ **What's the Question?** In a set of data, the greatest value is 54 and the least value is 30. The answer is 24. **What is the range?**

MIXED REVIEW AND TEST PREP

Tell whether the question is *biased* or *unbiased*. (p. 98)

18. Do you feel that sixth graders should do two hours of homework each night instead of watching TV? **biased**

19. Whom do you intend to vote for in the upcoming Student Government election? **unbiased**

20. Tell if the expression is numerical or algebraic. $a + 3.9$ (p. 28) **algebraic**

21. **TEST PREP** Which is the value of 4^5? (p. 40) **D**

    A 9 B 20 C 625 D 1,024

22. **TEST PREP** Twenty-seven sixth graders read a total of 135 books. If each student reads the same number of books, how many more books must each student read to reach a total of 243 books read? (p. 22) **G**

    F 2 books G 4 books H 50 books J 108 books

EXTRA PRACTICE page H36, Set C

MIXED REVIEW AND TEST PREP

Exercises 18–22 provide **cumulative review** (Chapters 1–5).

Multistep or Strategy Problem To solve Exercise 20, students can subtract the number of books read so far from 243, and then divide by 27.

4 Assess

Summarize the lesson by having students:

DISCUSS What does the greatest number in the cumulative frequency column of a table tell you about the original set of data? the number of responses in the original set of data

WRITE Which is closer to the original data for a survey, a tally table or a frequency table with intervals? Why? Possible answer: A tally table is more like the original data because it gives information about how often each data item occurs. A frequency table with intervals reports only on groups of data items.

Lesson Quiz

Transparency 5.4

A survey asked people how many pieces of fruit they eat in a week.

Pieces of Fruit Eaten in a Week						
3	1	5	5	2	4	5
9	4	1	2	6	7	4
9	2	7	5	4	3	5

1. Make a line plot for the data.

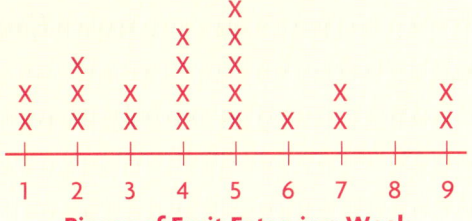

2. Use the data to make a cumulative frequency table with 3 intervals.

PIECES OF FRUIT EATEN IN A WEEK		
Pieces of Fruit	Frequency	Cumulative Frequency
1–3	7	7
4–6	10	17
7–9	4	21

LESSON 5.5 Measures of Central Tendency

LESSON PLANNING

Objective To calculate the mean, median, and mode and to determine their meanings for a set of data

Intervention for Prerequisite Skills
Mean, Median, Mode (For intervention strategies, see page 93.)

Materials *For Thinker's Corner* 10-section spinner, p. TR73

NCTM Standards
2. Algebra
5. Data Analysis and Probability
6. Problem Solving
8. Communication
10. Representation

Math Background

One way to characterize a set of data is to use measures of central tendency, which are single numbers that help describe the set as a whole and are used to compare different sets of data. Consider these ideas when helping students understand the mean, median, and mode.

- The mean of the numbers in a set is the sum of the numbers divided by the number of members in the set.
- The median is the middle number in a set of ordered numbers.
- The mode is the most frequently occurring member of the set.

Of the three measures, only the mode is guaranteed to be a member of the data set.

WARM-UP RESOURCES

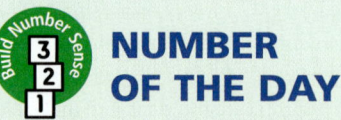 
NUMBER OF THE DAY
Find the range for the number of days of this month.
Answers will vary.

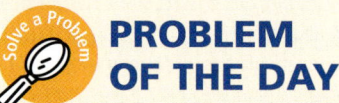 
PROBLEM OF THE DAY
A set of 9 numbers has a mean of 7. When one more number is added to the set, the mean is doubled. What number was added to the data set? 77

Solution Problem of the Day tab, p. PD5

DAILY FACTS PRACTICE
Have students practice addition and subtraction facts by completing Set E of *Teacher's Resource Book,* p. TR96.

INTERVENTION AND EXTENSION RESOURCES

ALTERNATIVE TEACHING STRATEGY ELL

Materials *For each student* 1 index card

Engage students in **finding the mean, median, and mode.** Provide each student with an index card and ask each to write a number from 1 to 10 on it.

Draw a large line plot on a piece of paper and place it on the floor or hang it on a wall.

- Have each student attach his or her number over the appropriate number on the line plot.
- Summarize by stating, for example, "Five people chose 10."
- Then have students use the line plot to find the median of the numbers chosen.
- Also have them find the mode and mean of the numbers. Check students' work.

KINESTHETIC
BODILY/KINESTHETIC

MULTISTEP AND STRATEGY PROBLEMS

The following multistep or strategy problem is provided in Lesson 5.5:

Page	Item
108	14

WRITING IN MATHEMATICS

Have students **describe situations that represent mean, mode, or median** and have their classmates determine which measure of central tendency each situation represents.

Provide this example:
- Most of the cereal brands tested contained 6 grams of sugar per serving. mode

ADVANCED LEARNERS

Challenge students to **use the measures of central tendency** shown to find possible values for a data set containing the given number of data items.

	Mean	Mode	Median	# of items
1.	4	4	4	4
2.	31	37	37	5
3.	47	42	42	5
4.	79	77	78	6

1. Possible answer: 3, 4, 4, 5
2. Possible answer: 13, 25, 37, 37, 43
3. Possible answer: 38, 42, 42, 56, 57
4. Possible answer: 77, 77, 77, 79, 81, 83

VISUAL
INTRAPERSONAL/INTROSPECTIVE

TECHNOLOGY LINK

Intervention Strategies and Activities CD-ROM • *Skills 77, 78*

Math Jingles™ CD 5–6 • *Track 10*

LESSON 5.5 ORGANIZER

Objective To calculate the mean, median, and mode and to determine their meanings for a set of data

Vocabulary *Review* mean, median, mode

Materials 10-section spinner, p. TR73

1 Introduce

QUICK REVIEW provides review of prerequisite skills.

Why Learn This? You can use this skill to summarize data, such as test scores or sports statistics. *Share the lesson objective with students.*

2 Teach

Guided Instruction

• As you discuss the three measures of central tendency, ask:

To find the median of a set of data, when do you need to find the average of the two middle numbers? when there is an even number of data items

REASONING **How can you visually locate the mean, median, and mode on a line plot without looking at the numbers?** The tallest point represents the mode. The point where half the X's represent a lesser value and half the X's represent a greater value would be the median. If the plot were thought of as a seesaw or scale, the balance point represents the mean.

• As students look at Example 2, ask:

Would it be helpful to make a line plot? Explain. Possible answer: Probably not; there are only 5 data items, so a line plot is not necessary to find the mode or median.

ADDITIONAL EXAMPLE

Example 1, p. 106

Use the data to make a line plot. Find the mode and the median.

NUMBER OF BOOKS READ							
3	7	8	1	6	4	3	5
8	7	2	6	1	2	2	3	4

Check students' plots. mode: 3, median: 3.5

106 Chapter 5

LESSON 5.5

Measures of Central Tendency

Learn how to find the mean, median, and mode of a set of data and which best describes the set of data.

QUICK REVIEW
1. $420 ÷ 5$ 84   2. $392 ÷ 7$ 56   3. $320 ÷ 8$ 40   4. $540 ÷ 60$ 9   5. $720 ÷ 9$ 80

Three measures of central tendency are mean, median, and mode. Measures of central tendency can help you to describe a set of data.

In April 1998, amateur rocket builder Gib Reynolds set an American record for 7- to 13-year-olds. His model rocket soared to a height of 47 meters. Data for heights some model rockets traveled, including Gib's, are shown at the right. Find the mean, median, and mode for the data.

ROCKET HEIGHTS (M)				
47	28	33	35	28

Mean: $(47 + 28 + 33 + 35 + 28) ÷ 5 = 171 ÷ 5 = 34.2$ m

Median: 28 28 33 35 47; 33 m   **Mode:** 28 m

Sometimes a line plot can help you to find the mode and the median.

EXAMPLE 1

Use the data to make a line plot. Find the mode and the median.

Remember that the mean is the sum of a group of numbers divided by the number of addends.

The median is the middle number in a group of numbers arranged in numerical order.

The mode is the number that occurs most often.

DAILY TEMPERATURES (°F)										
72	82	83	78	81	78	73	74	75	73	76
71	75	80	83	72	72	78	81	79	82	76

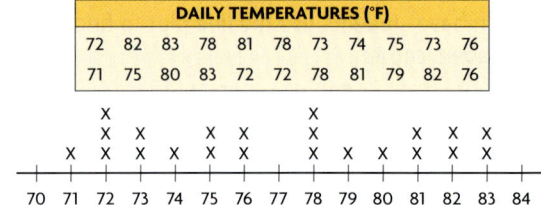

Mode: The modes are 72 and 78 since each occurs three times.

Median: Since there are 22 temperatures, the median is the mean of the 11th and 12th temperatures. $(76 + 78) ÷ 2 = 77$

Math Idea ▶ When you want to summarize a set of data as one value, you can use one of the three measures of central tendency.

EXAMPLE 2

Pedro jogged 6 mi, 5 mi, 2 mi, 2 mi, and 4 mi over 5 days. Which measure of central tendency is most useful to describe the data?

Mean: $(6 + 5 + 2 + 2 + 4) ÷ 5 = 19 ÷ 5 = 3.8$

Median: 2 2 4 5 6; 4   **Mode:** 2

The mean, or 3.8, and the median, 4, are close to most of the data while the mode, or 2, is closer to the lower end of the data. So, the mean or median is most useful to describe the data.

106 Chapter 5

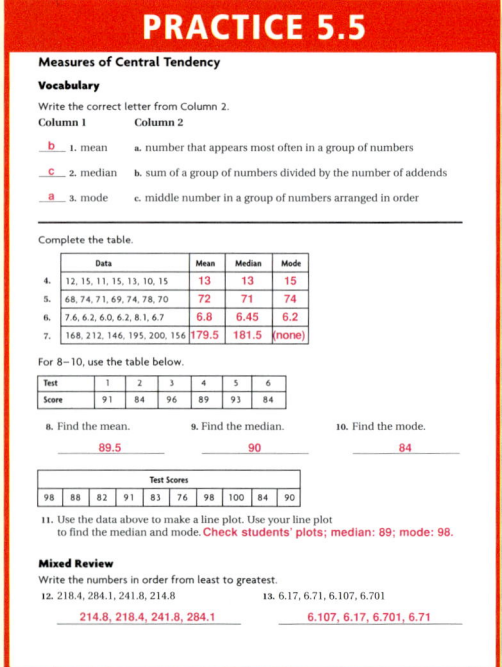

EXAMPLE 3

Sometimes the mean is not the best measure of central tendency to describe a set of data.

The bald eagle, America's national bird, was once a greatly endangered species. In recent decades, the species has made a dramatic recovery. The map at the right shows the number of nests in five zones of Montana. Which measure of central tendency is most useful to describe the data?

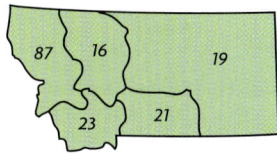

Numbers of nests: 87, 16, 19, 21, 23

Mean: (87 + 16 + 19 + 21 + 23) ÷ 5 = 166 ÷ 5 = 33.2

Median: 16 19 21 23 87 → 21

Mode: No value occurs more than any other, so there is no mode.

The median, 21, is close to most of the data. The high value of 87 makes the mean greater than any of the four other values. So, the median is most useful to describe the data.

CHECK FOR UNDERSTANDING

Think and Discuss — Look back at the lesson to answer the question.

1. **What if,** in Example 3, the region with 87 nests had only 27 nests? Which of the measures would change? Which measure would be most useful to describe the data? The mean would become 21.2. The mean or median would be most useful to describe the data.

Guided Practice — For 2–4, use the table below.

Day	Sun	Mon	Tue	Wed	Thu	Fri	Sat
Hours of Sleep	8	6	7	7	4	10	7

2. Find the mean. **7 hours**
3. Find the median. **7 hours**
4. Find the mode. **7 hours**

Find the mean, median, and mode.

5. 22, 24, 22, 29, 33, 14
   24; 23; 22
6. 124, 120, 132, 133, 119, 90, 87
   115; 120; no mode

PRACTICE AND PROBLEM SOLVING

Independent Practice — For 7–9, use the table below.

Game	1	2	3	4	5	6	7	8
Points Scored	10	12	10	17	18	12	20	24

7. Find the mean. **15.375 points**
8. Find the median. **14.5 points**
9. Find the mode. **10 and 12**

Find the mean, median, and mode.

10. 76, 63, 40, 52, 52, 40, 6, 15
    43; 46; 40 and 52
11. 365, 180, 360, 720, 59
    336.8; 360; no mode

107

REASONING What is a quick way to tell whether the mean will be a useful measure of central tendency? Examine the data for items that are considerably greater or less than the others.

ADDITIONAL EXAMPLES

Example 2, p. 106

A CD producer recorded the following number of songs on her most recent CDs: 15, 10, 18, 17, 10, 16, 12. Find the mean, median, and mode of the data. Then tell which measure of central tendency best represents the data. Mean—14; Median—15; Mode—10; either the mean or the median represents the data best.

Example 3, p. 107

Which measure of central tendency is most useful to describe the data?

Math test scores: 34, 99, 94, 100, 93, 96

The median, or 95, is close to most of the data. The low value of 34 makes the mean less than any of the five other test scores. So, the median is most useful to describe this data.

3 Practice

Guided Practice

Do Check for Understanding Exercises 1–6 with your students. Identify students having difficulty and use lesson resources to help.

COMMON ERROR ALERT

Some students may try to find the median by finding the middle number in an unordered set of data. Remind them to arrange the data from least to greatest before finding the measures of central tendency. This will also help them find the mode.

Error	Correction
9 3 ⑤ 7 7	3 5 ⑦ 7 9
Median = 5	Median = 7

Independent Practice

Note that Exercise 14 is a **multistep or strategy problem**. Assign Exercises 7–15.

CHALLENGE 5.5

Number Puzzles

Use the information to find the numbers in the group. The first one is done for you. Possible answers are given.

1. There are 7 whole numbers in a group. The least number in the group is 6. The greatest number in the group is 16. The mode of the group is 15. The median is 10 and the mean is 11.

   6, 7, 8, 10, 15, 15, 16

2. There are 5 whole numbers in a group. The least number is 7 and the greatest is 14. The mode is 9 and the median is 9. The mean is 10.

   7, 9, 9, 11, 14

3. There are 7 whole numbers in a group. The greatest number is 20 and the least is 8. The median is 12 and the mode is 12. The mean is 13.

   8, 10, 12, 12, 14, 15, 20

4. There are 7 whole numbers in a group. The least number is 5 and the greatest is 15. The mean and the median are 11. The mode is 15.

   5, 7, 9, 11, 15, 15, 15

5. There are 7 whole numbers in a group. The least number is 11 and the greatest is 17. The mean and the median are 14. There is no mode.

   11, 12, 13, 14, 15, 16, 17

6. There are 7 whole numbers in a group. The least number is 15 and the greatest is 33. The mean is 23. The median is 22. The mode is 19.

   15, 19, 19, 22, 25, 28, 33

7. There are 7 whole numbers in a group. The greatest number is 37 and the least is 21. The median is 29. The mean is 28. The mode is 31.

   21, 23, 24, 29, 31, 31, 37

PROBLEM SOLVING 5.5

Measures of Central Tendency

Write the correct answer. Analyze Choose Solve Check

1. Find the mean of the numbers.
   23, 86, 97, 45, 12
   52.6

2. Evaluate the expression below.
   a + b − 12.7 for a = 4.9 and b = 28.6
   20.8

3. Find the median of the numbers.
   13, 8, 9, 16, 18
   13

4. If you survey 1 out of every 10 people, how many would you survey out of a group of 23,800 people?
   2,380 people

Choose the letter for the best answer.

5. Yolanda has received scores of 98, 76, 87, 98, and 80 so far this year on her math tests. What is the mean of her test scores?
   A 98
   Ⓑ 87.8
   C 87.5
   D 87

6. Fred conducted a survey regarding hair color. Which measure of central tendency should he use to report the hair color that occurs most often?
   F range
   G mean
   H median
   Ⓙ mode

7. A pilot logged 87,984 miles of flight time in one month. If he flew the same route every day for 20 days, what is a good estimate for the length of his route?
   A 3,500 mi
   B 4,000 mi
   Ⓒ 4,500 mi
   D 5,000 mi

8. Mr. Jacob works between 9 and 12 hours each day, 5 days a week. What is a reasonable estimate of the number of hours he works in 50 weeks?
   F Less than 400 hr
   G Between 400 and 1,000 hr
   H Between 1,000 and 2,000 hr
   Ⓙ More than 2,000 hr

9. **Write About It** Explain why the mean of a set of data is sometimes a number that is not in the set of data.
   The mean is the sum of a group of numbers divided by the number of addends. This average may or may not be one of the numbers in the group.

107

LESSON 5.5

MIXED REVIEW AND TEST PREP

Exercises 16–20 provide **cumulative review** (Chapters 1–5).

 to SOCIAL STUDIES

- After students understand the table data, have them focus on the Latin America and Caribbean column. Ask:

Do you think the mean age is higher than the median age in Latin America? Explain your thinking. Possible answer: It is probably higher. Since the upper limit for age is much greater than 24.2 years, the mean age is probably pulled upward by the half of the population older than 24.2 years.

REASONING What does it mean to have an age of 35.6 years? How would you estimate that age in years and months? 35.6 years is the same as 35 and $\frac{6}{10}$ of a year. Since there are 365 days in a year, $\frac{6}{10}$ of a year is about 220 days, or a little more than 7 months.

4 Assess

Summarize the lesson by having students:

DISCUSS Can the mean, median, and mode of a set of data all be the same number? Give an example of such a set. Yes; Possible answer: 10, 11, 11, 12, 12, 12, 13, 13, 14; the mean, median, and mode are 12.

WRITE Describe how to find the mode of a set of data using a line plot. Possible answer: It is the number with the greatest number of X's above it.

Lesson Quiz

Transparency 5.5

The table below shows the number of hours Jenny's cat, Sylvester, slept every day for two weeks. Use the table to complete Exercises 1–3.

NUMBER OF HOURS SYLVESTER SLEPT EACH DAY							
Week 1	17	18	20	17	18	19	20
Week 2	12	12	22	18	21	18	19

1. What is the mode? 18
2. What is the mean? about 17.9 hr
3. What is the median? 18 hr

108 Chapter 5

Problem Solving ▶ Applications

12. Check students' line plots. median: 14.5; mode: 13; mean: 17.6. The median is most useful to describe the data since it is near most of the data.

13. The mean, $8.75, or the median, $8.50, are most useful since they are near most of the data.

12. Use the data below to make a line plot. Use your line plot to find the median and the mode. Then use the data to find the mean. Which measure of central tendency would be most useful to describe the data? See answer at left.

YEARLY RAINFALL (IN.)									
47	11	8	14	15	16	13	13	17	22

13. Over the past 8 days, Fernando spent $5, $6, $9, $8, $10, $15, $15, and $2. Which measure of central tendency would be most useful to describe the data? See answer at left.

14. **ALGEBRA** Reggie's average score on five math tests is 92. On his first four tests, Reggie's scores were 88, 97, 93, and 82. What was Reggie's score on his fifth test? 100

15. **What's the Error?** Tim wrote $(7 + 3 + 10 + 4) \div 3 = 8$ to find the mean of 7, 3, 10, and 4. What is his error? What is the correct mean? Tim should divide by 4, not 3; mean is 6.

MIXED REVIEW AND TEST PREP

Use the table for 16–17. (p. 102)

16. How many named blue or red as their favorite color? 49

17. What is the size of the sample? 71 people

Write the decimal as a percent. (p. 60)

18. 0.65 65%
19. 0.9 90%

★ 20. **TEST PREP** Which is the product of 0.56 and 2.5? (p. 70) B

   A 0.14   B 1.4   C 14   D 140

FAVORITE COLOR	FREQUENCY	CUMULATIVE FREQUENCY
Blue	32	32
Red	17	49
Green	22	71

PROBLEM SOLVING  to Social Studies

World Population Population is growing at different rates around the world. In North America, population is growing fairly slowly. In Latin America and Africa, population is growing much more rapidly. The table shows how median ages can differ.

Median Ages				
Africa	Asia	Europe	Latin America and Caribbean	North America
18.3 years	26.0 years	37.4 years	24.2 years	35.6 years

1. Explain what it means to say the median age in Asia is 26.0 years.
   Half of the population is less than 26.0 years old and half is greater than 26.0 years old.

2. Use < to order the median ages from least to greatest. 18.3 < 24.2 < 26.0 < 35.6 < 37.4

108 Chapter 5

EXTRA PRACTICE page H36, Set D

Outliers and Additional Data

LESSON **5.6**

LESSON PLANNING

Objective To determine how outliers and additional data affect the mean, median, and mode

Intervention for Prerequisite Skills

Mean, Median, Mode (For intervention strategies, see page 93.)

NCTM Standards

5. Data Analysis and Probability
6. Problem Solving
7. Reasoning and Proof
8. Communication

Vocabulary

outlier a data value that stands out from other data values in a set and can significantly affect measures of central tendency

Math Background

The following points will help students understand how additional data—especially outliers—affect measures of central tendency:

- If a large outlier is added to a data set, the mode may not be affected. The median may change slightly, and the mean will increase.
- If a very small outlier is added to a data set, the mode may not be affected. The median may change slightly, and the mean will decrease.

A line plot showing the new data or the outliers is helpful in illustrating these concepts.

WARM-UP RESOURCES

 **NUMBER OF THE DAY**  Transparency 5.6

In a class of 23 sixth graders, all but one are 11 years old. Which two measures of central tendency will be the same, and what are they? **the median and the mode; 11 years**

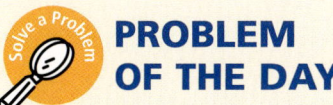 **PROBLEM OF THE DAY**  Transparency 5.6

Rita has 1 brother and 3 sisters. If the mean age of all the children is 5 years, what will their mean age be 7 years from now? **12 years**

Solution Problem of the Day tab, p. PD5

 **DAILY FACTS PRACTICE**

Have students practice multiplication facts by completing Set F of *Teacher's Resource Book*, p. TR96.

109A

INTERVENTION AND EXTENSION RESOURCES

ALTERNATIVE TEACHING STRATEGY

Materials *For each pair* 2 number cubes labeled 1–6, p. TR75

Ask students to work in pairs to **demonstrate how additional data affect measures of central tendency.**

- One student tosses 2 number cubes and the partner records the sum.
- They do this 8 times and find the mean, median, and mode for the 8 sums.
- Have them toss the cubes 2 more times and find the mean, median, and mode for all 10 sums.
- Have them explain how the 2 additional sums affected the mean, median, and mode for the 8 original sums. **Check students' work.**

KINESTHETIC
INTERPERSONAL/SOCIAL

MULTISTEP AND STRATEGY PROBLEMS

The following multistep or strategy problem is provided in Lesson 5.6:

Page	Item
111	7

SPECIAL NEEDS

Materials *For each group* 1 index card

To further **explore the effect of outliers on the mean,** have students work in small groups of 4 or 5. For each group, prepare an index card with a different number. On most of the index cards, the numbers should be greater than 30 but less than 100. On the rest of the cards, write a number less than 5. Give each group one of the cards.

- Have students find the mean age of the students in the group.
- Then have them include the age written on the card and find the mean again.
- Have students compare the two means.
- Ask groups to present the two different means that they found and discuss the effect of including the outlier. **Check students' work.**

VISUAL
VISUAL/SPATIAL

ADVANCED LEARNERS

Challenge students to **describe data sets that have specific measures of central tendency.** Display the following directions. Have students compare their answers.

- List six different numbers with a median of 30. **Possible answer: 10, 20, 25, 35, 40, 50**
- List six different numbers with a mean of 20. **Possible answer: 5, 10, 22, 24, 26, 33**
- List six numbers with a mode and median of 40. **Possible answer: 10, 20, 40, 40, 60, 70**

Have students formulate similar directions and exchange them with classmates to solve. **Check students' work.**

VISUAL
LOGICAL/MATHEMATICAL

Intervention Strategies and Activities CD-ROM • *Skills 77, 78*

LESSON 5.6

Outliers and Additional Data

Learn how outliers and additional data affect the mean, median, and mode.

QUICK REVIEW

Find the mean.
1. 6, 8, 10 **8**
2. 20, 32, 41 **31**
3. 15, 17, 19, 41 **23**
4. 25, 25, 25, 25 **25**
5. 100, 200, 300, 400 **250**

Vocabulary

outlier

When you add data to a data set, some measures of central tendency for the set change. How they change depends on how the new data are related to the original data.

Gina saved dimes for 8 weeks. She recorded the number of dimes in a table. In the ninth week, her parents gave her 25 dimes. In the tenth week, her grandparents gave her 52 dimes. Find the mean, median, and mode for Gina's data, using the values for the 8 weeks of her savings. Then find the measures for all 10 weeks.

EXAMPLE 1

DIMES GINA SAVED

Week	1	2	3	4	5	6	7	8
Dimes	17	18	30	15	1	6	19	6

Data for 8 Weeks
Mean: $(17 + 18 + \ldots + 19 + 6) \div 8 =$ **14**   Mode: **6**
Median: 1 6 6 15 17 18 19 30; $(15 + 17) \div 2 =$ **16**

Data for 10 Weeks
Mean: $(17 + 18 + \ldots + 25 + 52) \div 10 =$ **18.9**   Mode: **6**
Median: 1 6 6 15 17 18 19 25 30 52; $(17 + 18) \div 2 =$ **17.5**

EXAMPLE 2

At age 61, Lev Sarkisov climbed Mount Everest. Suppose he climbed with a group whose ages were 30, 31, 33, 30, 29, 35, 33, 30, and 28. Which measure of central tendency is most affected by adding Sarkisov's age to the group?

Data without Sarkisov's age
Mean: $(30 + 31 + \ldots + 30 + 28) \div 9 =$ **31**   Mode: **30**
Median: 28 29 30 30 **30** 31 33 33 35; **30**

Data with Sarkisov's age
Mean: $(30 + 31 + \ldots + 61) \div 10 =$ **34**   Mode: **30**
Median: 28 29 30 30 **30** 31 33 33 35 61; $(30 + 31) \div 2 =$ **30.5**

The mean increases by 3, the median increases by 0.5, and the mode remains the same. So, the mean is the measure most affected by adding Sarkisov's age.

On May 12, 1999, Armenian Lev Sarkisov became the oldest person to climb Mount Everest.

109

LESSON 5.6 ORGANIZER

Objective To determine how outliers and additional data affect the mean, median, and mode

Vocabulary outlier

1 Introduce

QUICK REVIEW provides review of prerequisite skills.

Why Learn This? This skill is useful in analyzing data and mentally adjusting measures of central tendency. *Share the lesson objective with students.*

2 Teach

Guided Instruction

• As you discuss Example 1, ask:

How do the methods of finding the mean, median, and mode of the data for 10 weeks differ from those for 8 weeks?
Possible answer: To find the mean of the 8-week data, divide by 8; for the 10-week data, divide by 10. To find the median of the 10-week data, insert the additional data items into the original order of the 8-week data and recalculate. The method of finding the mode for each set of data is the same.

REASONING What number can be added to the following data so that all the measures of central tendency remain the same: 21, 22, 23, 25, 25, 29, 30? **25**

ADDITIONAL EXAMPLE

Example 1, p. 109

Tim recorded in a table the number of dimes he saved each week for 5 weeks. He saved 3 dimes in the sixth week and 1 dime in the seventh week. Find the mean, median, and mode for Tim's data by using the values for the 5 weeks and then for all 7 weeks.

DIMES TIM SAVED

Week	1	2	3	4	5
Dimes	6	16	24	17	17

16, 17, 17; 12, 16, 17

See also page 110.

109

- Remind students that the line plot can be viewed as a balance scale and the mean as its balance point. Ask:

REASONING Ignoring the data values, where would you place this balance point on the line plot? *at or to the right of the cluster of 3 X's*

How are the values of the outliers related to the other data if the mean increases when the outliers are added to the other data? *The values of the outliers are greater than the values of the other data.*

ADDITIONAL EXAMPLES

Example 2, p. 109

Melanie has 3 brothers and 3 sisters. Her youngest sister, Tara, is 4 years old. Which measure of central tendency is most affected by adding Tara's age to the group?

Ages of Melanie and Her Siblings
18   21   16   15   19   19   4

The mean decreases by 2; the median decreases by 0.5; the mode remains the same. The mean is most affected.

Example 3, p. 110

Kelly recorded on this line plot the number of magazine subscriptions she and her classmates sold each day during the first 10 days of the school fundraiser.

```
                    X     X
                    X     X
  X                 X     X
  X        X  X  X  X
  +--+--+--+--+--+--+--+--+--+--+--+--+--+
  25 30 35 40 45 50 55 60 65 70 75 80 85
```

A. Find the measures of central tendency for the 8 data values in the line plot if the outliers are not included. *mean, 53.75; median, 52.5; modes, 50 and 60*

B. Find the measures of central tendency for all 10 data values with the outliers included. *mean, 48; median, 50; modes, 50 and 60*

3 Practice

Guided Practice

Do Check for Understanding Exercises 1–4 with your students. Identify those having difficulty and use lesson resources to help.

After Exercise 4, point out that some mathematicians initially consider outliers to be "suspect" and could be possible errors.

Bethany used a line plot to record the number of concert tickets she sold on 10 different days. Most of the data are between 11 and 16. The data values at 32 and 34 are outliers, or extreme values. An **outlier** is a data value that stands out from other data values in a set. Outliers can significantly affect measures of central tendency.

```
              X
  X  X        X
  X  X  X  X  X X                                    X     X
  +--+--+--+--+--+--+--+--+--+--+--+--+--+--+--+--+--+--+--+--+--+--+--+--+--+
  10 11 12 13 14 15 16 17 18 19 20 21 22 23 24 25 26 27 28 29 30 31 32 33 34 35
```

EXAMPLE 3

A. Find the measures of central tendency for the 8 data values in the line plot if the outliers are not included.

Mean: $(11 + 11 + 12 + 12 + 12 + 15 + 15 + 16) \div 8 = 104 \div 8 = $ **13**
Median: 11 11 12 **12 12** 15 15 16; **12**   Mode: **12**

B. Find the measures of central tendency for all the data values with the outliers included.

Mean: $(104 + 32 + 34) \div 10 = 170 \div 10 = $ **17**
Median: 11 11 12 12 **12 15** 15 16 32 34; $(12 + 15) \div 2 = $ **13.5**
Mode: **12**

C. Suppose Bethany sold 1 ticket the next day and 0 tickets the final day. Include these data values with the data in the line plot and find the measures of central tendency for all 12 data values.

Mean: $(170 + 1 + 0) \div 12 = 171 \div 12 = $ **14.25**
Median: 0 1 11 11 12 **12 12** 15 15 16 32 34; **12**   Mode: **12**

- How do the outliers in part B affect the mean, median, and mode? *The mean and median increase; the mode does not change.*

CHECK FOR UNDERSTANDING

Think and Discuss

1. *The median and mode would not change. The mean would not increase as much.*

Guided Practice

Look back at the lesson to answer the question.

1. **What if** Gina's grandparents had given her 32 dimes in Example 1? How would the measures of central tendency be affected? *See answers at left.*

2. Find the mean, median, and mode for the ages in the table. **14.5; 13.5; 12**

AGES OF MEMBERS OF MODEL AIRPLANE CLUB			
15	12	9	13
22	19	12	14

3. Find the mean, median, and mode if a 28-year-old joins the club. **16; 14; 12**

Jeff received these scores on his science quizzes: 75, 70, 70, 45, 100, 70, 70, 80, 75, 80.

4. a. Use the scores to make a line plot. Circle the outliers. *See Additional Answers, p. 117A.*
   b. Find the mean, median, and mode with and without the outliers. *with: 73.5; 72.5; 70; without: 73.75; 72.5; 70*
   c. Why did these outliers have so little effect on the mean, median, and mode? *Possible answer: They were both about 30 points from the mean, median, and mode.*

CHALLENGE 5.6

Describe the Data

For each set of data, tell whether you think the mean, median, or mode gives the best description of the data. Give a reason for your choice. For some sets, you may wish to choose more than one of the measures of central tendency. *Possible answers are given.*

1. Salaries: $50,000; $52,000; $51,000; $400,000; $46,000.
   median; Most of the salaries are clustered around $50,000, while the mean is much higher than most of the salaries and there is no mode

2. Ages of club members in years: 17, 16, 18, 17, 16, 19, 16, 19, 18, 20
   median or mean; The data values are all close together but the mode (16) is the least age and could give the impression that the club members are younger than is the case.

3. Weights of packages in pounds: 3, 5, 3, 3, 3, 3, 3
   mode or median; Only one item is different, but close enough not to have much of an effect.

4. Test scores: 82, 81, 85, 86, 90, 40, 40
   median; Most of the scores are in the range of 80—90, but the mean (72) is too low to be truly representative and the mode gives a completely false impression of the scores

5. Points scored: 23, 24, 26, 25, 24, 22, 21
   mean, median, or mode: All of the data values are close to one another, so the mean, median, and mode are all nearly the same.

6. TV sizes (inches) sold at a store this week: 27, 27, 32, 35, 20, 20
   mode; The mode shows the most-popular TV size, while the mean is not the size of an actual television. (In some situations, the median might not be a TV size either.)

PROBLEM SOLVING 5.6

Outliers and Additional Data   Analyze Choose Solve Check

Write the correct answer.

1. Brittany had test scores of 80, 85, 85, 92, and 90. If her score on the next test is 65, which measures of central tendency change?
   Only the mean changes.

2. While shopping, Debra estimated the sum of $48.99 and $78.85 as $130. How did she know that the result was an overestimate?
   She rounded both numbers up.

3. Robert's scores on six math tests are 90, 80, 80, 85, 88, and 45. How much higher is the mean of his scores without the outlier than when the outlier is included?
   6.6 points higher (84.6 − 78)

4. In Grades 6 through 8 at Adams Middle School, 45% of the members of the computer club are eighth graders and 19% are seventh graders. What percent are sixth graders?
   36%

Choose the letter for the best answer.

5. John wants to pay for a book that costs $28. He has 3 ten-dollar bills, 4 five-dollar bills, and 5 one-dollar bills. In how many different ways can John pay exactly $28 for the book using his money?
   A 1 way   C 3 ways
   B 2 ways D 4 ways

6. A custodian is changing all the lightbulbs in an auditorium. The bulbs come in packages of 4. There are 17 light fixtures in the auditorium and each has 5 bulbs in it. How many packages of bulbs must the custodian open?
   F 20 packages   **H 22 packages**
   G 21 packages   J 23 packages

7. Danielle's first 3 test scores were 86, 87, and 91. If a perfect score is 100, what is the highest mean score she can have after 4 tests?
   A 89
   B 90
   C 91
   D 92

8. The five linemen on the football team weigh 240 pounds, 228 pounds, 230 pounds, 256 pounds, and 266 pounds. The quarterback weighs 172 pounds. How much greater is the mean weight of the five linemen than the mean weight of the six players?
   F 244 pounds   H 20 pounds
   G 232 pounds   **J 12 pounds**

9. **Write About It** In Exercise 3, what is the effect on the median and the mode of Robert's scores if the outlier is removed from his scores?
   The median becomes 85 rather than 82.5.
   The mode remains the same, 80.

PRACTICE AND PROBLEM SOLVING

Independent Practice

a. Abe: mean 7.5, median 6, mode 7; Bart: mean 7.5, median 8.5, mode 0

b. Abe: mean 9.3, median 7, modes 7 and 20; Bart: mean 6.4, median 5, mode 0

c. Abe's: all increase; Bart's: mean/median decrease; mode doesn't change

The table shows the points scored by two basketball players in the first six games of the season.

5. **a.** Find the mean, median, and mode of each player's points for six games.

   b. In the seventh game, Abe scored 20 points and Bart scored 0 points. Find the mean, median, and mode of each player's points for seven games.

   c. How are the measures of central tendency for Abe's and Bart's scores affected by the points scored in the seventh game?

POINTS SCORED		
Game	Abe	Bart
1	7	15
2	5	0
3	4	0
4	20	12
5	7	13
6	2	5

The table shows the number of pets owned by children in a pet club.

6. **a.** Draw a line plot for the data about the number of pets children own. Circle the outliers.
   See Additional Answers, p. 117A.

   b. Find the mean, the median, and the mode for the data with and without the outliers. with: 3.75; 3; 1; without: 2.3; 2.5; 1

   c. How does including the outliers affect the mean? the median? the mode? It increases the mean, increases the median a little, and does not change the mode.

NUMBER OF PETS OWNED			
1	1	2	1
3	12	4	4
3	3	10	1

Problem Solving Applications

7. **a.** Emily scored 75, 85, 35, 85, 70, 10 on her first 6 math quizzes. What score does Emily need on her seventh quiz so that after 7 quizzes she will have a median score of 75? 75

   b. **REASONING** The greatest score Emily can make on a math quiz is 100. Explain whether it is possible for her to have a mean score of 70 after taking the seventh quiz. No. With a perfect score of 100, the mean would be 65.7.

8. **REASONING** The high temperatures in °F over 7 days in town were 72°, 73°, 70°, 68°, 70°, 71°, and 39°. Explain why the mean would not be a good measure to describe the temperatures.
   The 39° temperature is an outlier and significantly decreases the mean.

MIXED REVIEW AND TEST PREP

Find the mean, median, and mode of the data. (p. 106)

9. 5, 6, 8, 6, 5, 7, 5, 4
   5.75, 5.5, 5

10. 42, 56, 44, 38, 10
    38, 42, no mode

11. 10, 12, 12, 10, 8, 5
    9.5, 10, 10 and 12

12. A company claims that its 8-year warranty lasts 4 times as long as any other warranty. What is the longest of the other warranties? (p. 22) 2 years

★ 13. **TEST PREP** There are 180 players in a baseball league. Each team has 12 players. How many teams are in the league? (p. 22) B

   A 12 teams **B** 15 teams **C** 16 teams **D** 18 teams

EXTRA PRACTICE page H36, Set E

111

LESSON 5.6

Independent Practice

Note that Exercise 7 is a **multistep or strategy problem**. Assign Exercises 5–8.

MIXED REVIEW AND TEST PREP
Exercises 9–13 provide **cumulative review** (Chapters 1–5).

4 Assess

Summarize the lesson by having students:

DISCUSS Can data added to a data set affect the mean, median, and mode? Explain. Possible answer: Yes; if most of the data added have a greater value than the original data, the mean and the median will be greater. If most of the data added have a lesser value than the data in the set, the mean and the median will be less. The mode will change only if a new value occurs more often than the original mode.

WRITE When will the mean, median, and mode of a data set remain about the same after additional data are added to the data set? Possible answer: when the added data values do not differ much from the values in the original set or when outliers added to both sides of the data set "balance out" each other

Lesson Quiz

Transparency 5.6

The ages of adults attending a community planning meeting are shown below.

48, 32, 27, 89, 48, 77, 23, 56

1. Find the mean, median, and mode for the data. 50; 48; 48

2. One 5-year-old child also attended the meeting. Find the mean, median, and mode for the ages if the child's age is also included. 45; 48; 48

3. How are the measures of central tendency affected by including this outlier? The mean is decreased by 5. The median and mode are not affected.

111

LESSON 5.7

Data and Conclusions

LESSON PLANNING

Objective To draw conclusions about a set of data

Intervention for Prerequisite Skills

Read a Table (For intervention strategies, see page 93.)

NCTM Standards

5. Data Analysis and Probability
6. Problem Solving
8. Communication

Math Background

This lesson ties together the ideas presented in the chapter by showing students how to draw reasonable conclusions from data collected through samples. Remind students of the following points:

- The population to be surveyed must be defined accurately and clearly.
- To avoid a biased sample, everyone in the population must have an equal chance of being chosen. It is important to sample all segments of the population.
- The questions asked must not lead the respondent to reply in any particular way.

WARM-UP RESOURCES

 **NUMBER OF THE DAY**

A task takes $3\frac{1}{2}$ hr. If you start now, will you finish by 4:30 P.M.? If the current time is 1 P.M. or earlier, students should answer yes.

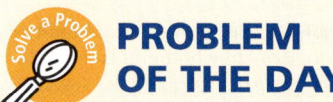 **PROBLEM OF THE DAY**

The mean of these numbers is 14. The greatest number is 21 more than the least. The mode is 18. What are the missing numbers? 3 6 9 ◆ ◆ ◆ ◆

18, 18, 20, 24

Solution Problem of the Day tab, p. PD5

 **DAILY FACTS PRACTICE**

Have students practice division facts by completing Set G of *Teacher's Resource Book*, p. TR96.

INTERVENTION AND EXTENSION RESOURCES

ALTERNATIVE TEACHING STRATEGY ELL

Ask students to apply their knowledge of **collecting and analyzing data.**

- Divide the class into small groups and ask each group to pick a topic on which they would like to gather information, write questions about the topics, and then collect data from a random sample.
- Refer students to the questions on page 112 as they prepare their project to remind them of the importance of a random sample and unbiased questions.
- Then have each group write several conclusions based on their data.
- Ask groups to exchange questions and conclusions and evaluate each other's projects. Check students' work.

See also page 114.

AUDITORY
INTERPERSONAL/SOCIAL

MULTISTEP AND STRATEGY PROBLEMS

The following multistep or strategy problem is provided in Lesson 5.7:

Page	Item
115	18

WRITING IN MATHEMATICS

Materials *For each student* newspaper or magazine

Ask students to **demonstrate what they've learned about collecting data and drawing conclusions.** Find the results of a survey in a newspaper or magazine. Have students write a paragraph discussing what the population is or might be, what kind of a sample might have been used, what questions might have been asked, and what conclusions you can draw from the results. Answers will vary.

CAREER CONNECTION

Have students apply their knowledge of **collecting data** to real life by sharing the following information.

Every 10 years the United States conducts a census of its population. Although questionnaires are mailed to all households, some households get more detailed questionnaires than others. Thus, some of the data collected by census workers is based on samples of the population. The answers to the questions are tabulated and reports are written by Census Bureau staff.

If an area's population is less than 2,500, one out of every two people will be given the detailed questionnaire. Colfax, California, has a population of 1,306. Ask: How many people in Colfax will be sent the detailed questionnaire? 653 people

AUDITORY
VERBAL/LINGUISTIC

Intervention Strategies and Activities CD-ROM • *Skill 75*

LESSON 5.7

3 Practice

Guided Practice

Do Check for Understanding Exercises 1–5 with your students. Identify those having difficulty and use lesson resources to help.

You may wish to do Exercises 3–5 orally with your students to generate discussion. In Exercise 3, encourage students to describe what would be a representative sample.

Independent Practice

Note that Exercise 18 is a **multistep or strategy problem**. Assign Exercises 6–18.

Before assigning Exercise 9, you may want to discuss the *Other* category on the graph. Students may conclude that broccoli is the least favorite vegetable, but each vegetable included in *Other* might have received fewer votes than broccoli.

Multistep or Strategy Problem To solve Exercise 18, students can subtract the price of the glove from the total amount he spent and divide the result by 5.

Guided Practice

For 3, write *yes* or *no* to tell whether the conclusion is valid. Explain your answer.

3. Yes. The sample is random and from the correct population and the question is unbiased.

3. Five hundred randomly selected teenagers from Fresno are asked to name their favorite subject. Eighty percent of these teenagers respond that their favorite subject is math. You conclude that math is the most popular subject among teenagers in Fresno.

Use Data For 4–5, use the data in the graph.

4. The graph at the right shows the results of a survey of 600 randomly selected middle school students. Jen concludes that more middle school students are interested in football than in any other sport. Is her conclusion valid? Explain.

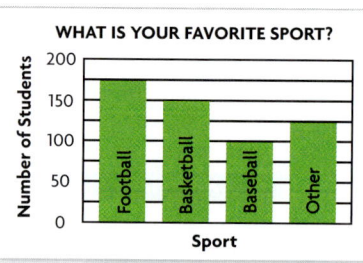

4. Yes. The sample is random and representative; the question is unbiased.

5. Robert concludes that more high school students are interested in football than in any other sport. Is his conclusion valid? Explain.
No. The data is for middle school students.

PRACTICE AND PROBLEM SOLVING

Independent Practice

For 6–7, write *yes* or *no* to tell whether the conclusion is valid. Explain your answer.

6. All the members of the science club were asked if they wanted to take more science classes. Each of them answered yes. You decide that science is becoming more popular among all students. No. The sample is not representative of the student population.

7. No. The sample is not representative of the population.

7. A sample of your friends shows that their favorite color is green. You decide that the same is probably true for the entire school.

Use Data For 8–9, use the data in the graph.

8. The graph at the right shows the results of a survey of 1,000 randomly selected teenagers from Kentucky. José concludes that corn is the favorite vegetable of teenagers from New York. Is his conclusion valid? Explain.

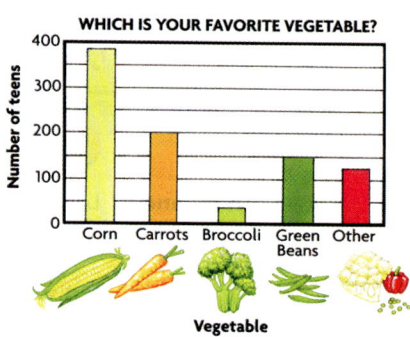

8. No. The data is for teenagers from Kentucky, not New York.

9. Yes. An unbiased question was asked of a random sample from the correct population.

9. Jerome concludes that carrots are the second favorite vegetable among Kentucky teenagers. Is his conclusion valid? Explain.

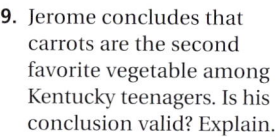

ALTERNATIVE TEACHING STRATEGY | SCAFFOLDED INSTRUCTION

Purpose Students survey two diverse groups to collect data and draw conclusions.

Have students work in small groups. Instruct each group to survey 2 diverse groups such as 20 first graders and 20 teenagers or 20 boys and 20 girls. Surveys should use questions with choices, such as these:

Of the following, which is your favorite after-school activity?
- Bike riding
- Watching television
- Reading
- Using the computer
- Going to the playground

Have each group record the results in a table such as the one at the right.

FAVORITE AFTER-SCHOOL ACTIVITIES		
Activity	First Graders	Teenagers
Bike riding	6	2
Watching TV	4	5
Reading	1	7
Using the computer	3	6
Going to the playground	6	0

Direct groups to make a list of conclusions about each activity, based on the results of their survey. Ask them to share their conclusions with the class.

Sample list of conclusions:

Activity	Conclusions
Bike riding	First graders like to ride bikes more than teenagers do.
Watching TV	Teenagers and children both enjoy watching TV.
Reading	Teenagers like reading after school much more than first graders do.
Computer	Teenagers like to use the computer about twice as much as first graders do.
Playground	Teenagers do not like going to the playground after school, but first graders do.

Check students' work.

Problem Solving ▶ Applications

For a survey completed five years ago, middle school students from Minneapolis were asked, "What is your favorite hobby?"

A research company randomly selected 750 current middle school students from Minneapolis and asked them the same question. The results are in the table.

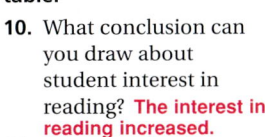

FAVORITE HOBBIES	
Reading	36%
Crafts	7%
Coin Collecting	2%
Stamp Collecting	2%
Building Models	12%
Trading Cards	39%
Other	2%

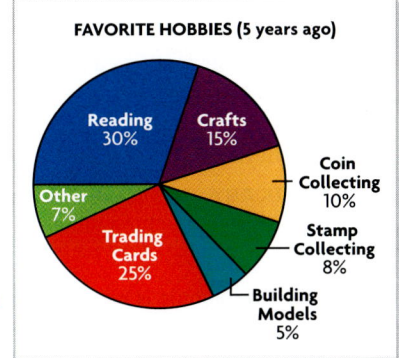

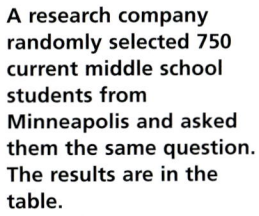

10. What conclusion can you draw about student interest in reading? **The interest in reading increased.**
11. What conclusion can you draw about student interest in crafts? **The interest in crafts decreased.**
12. What conclusion can you draw about student interest in coin collecting? **The interest in coin collecting decreased.**
13. What conclusion can you draw about student interest in trading cards? **The interest in trading cards increased.**
14. What conclusion can you draw about student interest in models? **The interest in building models increased.**
15. How can you tell that your conclusions are valid? **A random sample was selected from the correct population and the question was unbiased.**
16. For which of the hobbies did interest increase the most? decrease the most? **increase: trading cards; decrease: coin collecting**
17.  **What's the Error?** A survey of middle school athletes showed that their favorite exercise is jogging. Cathy looked at the survey results and concluded that all students prefer jogging. Explain Cathy's error. **The sample is biased since it only includes middle school athletes.**
18. Robert spent $60 at the sports equipment store. He bought a glove for $35 and 5 baseballs. What was the price for each baseball? **$5**

MIXED REVIEW AND TEST PREP

Find the mean, median, and mode of the data. (p. 106)

19. 2, 4, 4, 3, 6, 8, 1, 6
4.25, 4, 4 and 6
20. 22, 44, 31, 46, 22
33, 31, 22
21. 6, 8, 11, 10, 2, 5
7, 7, no mode

★ 22. **TEST PREP** Which is the solution of $x + 3 = 10$? (p. 30) **B**

 A $x = 3$ **B** $x = 7$ **C** $x = 10$ **D** $x = 13$

★ 23. **TEST PREP** Which is the solution of $x \div 3 = 6$? (p. 30) **J**

 F $x = 2$ **G** $x = 3$ **H** $x = 9$ **J** $x = 18$

EXTRA PRACTICE page H36, Set F

115

MIXED REVIEW AND TEST PREP

Exercises 19–23 provide **cumulative review** (Chapters 1–5).

4 Assess

Summarize the lesson by having students:

DISCUSS Why do you think someone might use a sample that was not randomly selected or a biased question in a survey? Possible answer: That person might have a personal stake or interest in the results or might want results that will support a certain position.

WRITE What are two conclusions you can draw from the graph in Exercise 5 on page 114? Justify your conclusions. Possible answer: Football is more popular than basketball with students since more chose football than basketball. Baseball is less popular than basketball with students.

Lesson Quiz

Transparency 5.7

1. Forty people shopping at Tabco Grocers stated that Tabco is their favorite grocery store. Is it valid to conclude that most people prefer to shop at Tabco? Explain. No; the sample is not representative.

2. Helena asks a random sample of 45 students at her school of 400 students to name their favorite food. Forty students name pizza. She concludes that most students in the school like pizza best. Is her conclusion justified? Explain. Yes; the sample is random, and 40 out of 45 is most of the students.

3. A random sample of people in a small city shows that 90 people think City Park should be left as it is, 240 people think it should be improved, and 30 people have no opinion. What conclusion can you draw from the survey? Possible answer: Most people feel the park should be improved.

CHAPTER 5

REVIEW/TEST

Purpose To check understanding of concepts, skills, and problem solving presented in Chapter 5

USING THE PAGE

The Chapter 5 Review/Test can be used as a **review** or a **test**.

- Items 1–3 check understanding of concepts and new vocabulary.
- Items 4–12 check skill proficiency.
- Items 13–15 check students' abilities to choose and apply problem solving strategies to problems involving data.

 Suggest that students place the completed Chapter 5 Review/Test in their portfolios.

USING THE ASSESSMENT GUIDE

- Multiple-choice format of Chapter 5 Posttest—See Assessment Guide, pp. AG33–34.
- Free-response format of Chapter 5 Posttest—See Assessment Guide, pp. AG35–36.

USING STUDENT SELF-ASSESSMENT

The How Did I Do? survey helps students assess what they have learned and how they learned it. This survey is available as a copying master in *Assessment Guide*, p. AGxvii.

CHAPTER 5 REVIEW/TEST

For 1–2, determine the type of sample. Write *convenience*, *random*, or *systematic*. (pp. 94-97)

1. A cereal company surveys a person in every third apartment in an apartment building. **systematic**

2. Alicia wants to find out what snack foods students in her school prefer. She randomly surveys 100 students. **random**

3. **VOCABULARY** In a survey, if individuals or groups in the population are not represented by the sample, the sample is __?__ . (p. 98) **biased**

Tell whether the sample is biased or unbiased. (pp. 98-99)

4. A store randomly surveys 10 out of every 100 customers about the quality of its service. **unbiased**

5. A teacher surveys girls about the best day to give a test. **biased**

For 6–9, use the table at the right. (pp. 102-105)

6. What is the sample size? **20**

7. Find the range. **12**

8. Make a line plot. See Additional Answers, p. 117A.

9. Make a frequency table. See Additional Answers, p. 117A.

STUDENTS' HEIGHTS (in inches)									
62	72	63	62	69	70	60	64	66	63
71	62	65	68	63	70	64	62	67	70

Find the mean, median, and mode. (pp. 106-109)

10. 17, 12, 23, 19, 23
    18.8; 19; 23

11. 6.2, 5.5, 8.4, 5.5
    6.4; 5.85; 5.5

12. 265, 235, 171, 253
    231; 244; no mode

13. Ten people in Ana's neighborhood formed a bike club. The table shows the ages of the members. (pp. 106-109)

    a. Find the mean, median, and mode for the ages of the ten members.
       11.6, 11.5, 13
    b. Find the mean, median, and mode if the adult advisor for the club is 27 years old and is a club member. Compare with part *a*.
       13, 12, 13; mean and median increase, mode is unchanged

AGES OF MEMBERS OF BIKE CLUB				
13	14	10	11	13
12	13	9	11	10

14. Janyce wanted to find out where visitors to the local mall come from. She surveyed visitors, tallied her results, and came up with the following percents: California—34%; Idaho—14%; Nevada—11%; Oregon—16%; Montana—8%; Washington—11%; other states—6%. Make a table to organize the information from greatest to least percent. (pp.100-101)

    14. Check students' tables. Order should be California, Oregon, Idaho, Washington or Nevada, Montana, other states.

15. If 50% of the people who visit the mall in Exercise 14 come from within 20 miles of the mall, which is a reasonable conclusion? Why? (pp.100-101)

    A The mall where Janyce took her survey is in Colorado.
    B The mall where Janyce took her survey is in eastern Nevada.
    C The mall where Janyce took her survey is in southern Washington.
    D The mall where Janyce took her survey is in northern California.

    15. D; 50% of the people come from California and Oregon—that suggests a northern California location.

116 Chapter 5

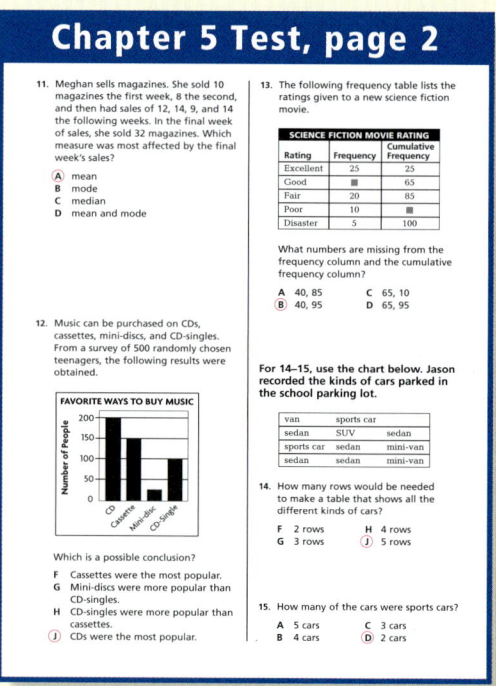

CHAPTERS 1–5 ★ STANDARDIZED TEST PREP

 **Eliminate choices.**
See item **4.**
Keep the question in mind as you read each answer choice. Look for the most important characteristic of a survey sample to decide what was wrong with the survey.
Also see problem **5**, p. H64.

Choose the best answer.

1. Which symbol makes this true? **A**

    8.34 ● 8.342

    A <   C =
    B >   D ÷

2. Which of the following is equivalent to 27 + 43? **G**

    F 25 + 50
    G 30 + 40
    H 27 + 40 + 12
    J 20 + 60

3. Which of the following numbers is between 8.07 and 8.3? **C**

    A 8.06
    B 8.017
    C 8.29
    D 8.312

4. Before Warren ran for class president, he surveyed 20 of his friends, asking "Would you vote for me for class president?" Of those surveyed, 100% answered "yes." Warren was very surprised when he lost the election. What was wrong with his survey? **G**

    F Nothing; his friends changed their minds and didn't vote for him.
    G His survey sample did not represent the student body.
    H He should have included teachers in his survey.
    J His survey did not ask the right question.

5. The ages of students in the Hiking Club are 13, 11, 15, 14, 12, 13, 13. What is the mode of this set of data? **C**

    A 11   B 12   C 13   D 14

6. What is $2 \times 2 \times 2 \times 2 \times 2 \times 2 \times 2$ expressed in exponential notation? **H**

    F 2^2 G 7^2 H 2^7 J 2^6

7. 883.03 ÷ 22.7 **B**

    A 389   C 3.89
    B 38.9   D Not here

8. Jon had the following scores for 5 rounds of golf: 78, 80, 69, 75, 73. What is the mean of the scores? **J**

    F 50   H 74
    G 71   J 75

Write What You Know

See below.

9. Look at the data in Problem 8. Jon thought the range of his golf scores was 5. Explain the mistake Jon made. Then explain the correct way to compute range.

10. The student council wants to conduct a survey to get an idea of what students think about changing the school colors. One council member suggested asking members of the football team. Explain why this might not be a representative sample.

117

Write What You Know • Written Response

9. Jon found the difference of his first and last scores: 78 − 73 = 5. But his scores are not ordered. To find the range correctly, first order the scores from greatest to least: 80, 78, 75, 73, 69. Then find the difference of the first and last scores: 80 − 69 = 11. So, Jon's range is 11.

10. Members of the football team may not be representative of the school as a whole. For example, there may be no members of the lower grades or no girls in that sample. A better sample would be one drawn randomly from the entire student population.

STANDARDIZED TEST PREP • Chapters 1–5

USING THE PAGE

This page may be used to help students get ready for standardized tests. The test items are written in the same style and arranged in the same format as those on many state assessments. The page is cumulative. It covers math objectives and essential skills that have been taught up to this point in the text. Most of the items represent skills from the current chapter, and the remainder represent skills from earlier chapters.

This page can be assigned at the end of the chapter as classwork or as a homework assignment. You may want to have students use individual recording sheets presented in a multiple-choice (standardized) format. A Test Answer Sheet is available as a blackline master in *Assessment Guide* (p. AGxlii).

You may wish to have students describe how they solved each problem and share their solutions.

ITEM ANALYSIS

Item	Learning Goal	Item	Learning Goal
1	3A	6	2B
2	2A	7	4A
3	3A	8	5C
4	5A	9	5C
5	5C	10	5A

Written Response items for **Write What You Know** are available as a blackline master in *Performance Assessment*.

SCORING RUBRIC • WRITE WHAT YOU KNOW

2 Demonstrates a complete understanding of the problem and chooses an appropriate strategy to determine the solution

1 Demonstrates a partial understanding of the problem and chooses a strategy that does not lead to a complete and accurate solution

0 Demonstrates little understanding of the problem and shows little evidence of using any strategy to determine a solution

Collect and Organize Data 117

Lesson 5.4, page 104

3.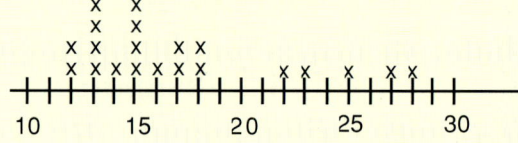

4. Possible table is given.

Age	Frequency	Cumulative Frequency
11-15	11	11
16-20	5	16
21-25	3	19
26-30	2	21

Lesson 5.4, page 104

6.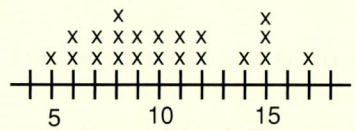

7. Possible table is given.

Number of km	Frequency	Cumulative Frequency
1-5	1	1
6-10	11	12
11-15	8	20
16-20	1	21

Lesson 5.4, page 105

11.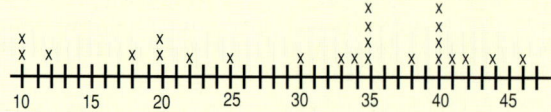

14. Possible table is given.

Weights	Frequency	Cumulative Frequency
10-19	4	4
20-29	4	8
30-39	8	16
40-49	8	24

16. Possible table is given.

Weights	Frequency	Cumulative Frequency
0-9	5	5
10-19	13	18
20-29	7	25
30-39	1	26

PUPIL EDITION
ADDITIONAL ANSWERS

Lesson 5.6, page 110

4.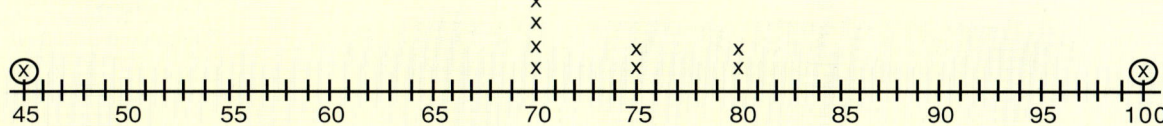

Lesson 5.6, page 111

6.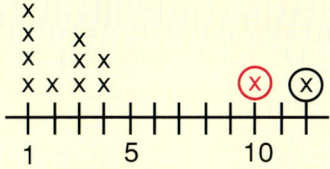

Chapter 5 Review/Test, page 116

8.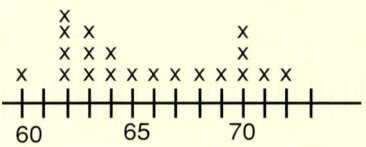

9. Possible table is given.

Heights (in inches)	Frequency	Cumulative Frequency
60-64	10	10
65-69	5	15
70-74	5	20

117B

CHAPTER 6 Graph Data

CHAPTER PLANNER

PACING OPTIONS
Compacted 5 Days
Expanded 10 Days

Getting Ready for Chapter 6 • Assessing Prior Knowledge and INTERVENTION (See PE and TE page 119.)

LESSON	NCTM STANDARDS	PACING	VOCABULARY*	MATERIALS	RESOURCES AND TECHNOLOGY
6.1 Make and Analyze Graphs pp. 120–123 **Objective** To display and analyze data in bar graphs, line graphs, and circle graphs	5, 6, 7, 8, 10	1 Day	multiple-bar graph multiple-line graph		Reteach, Practice, Challenge, Problem Solving 6.1 Worksheets, Extra Practice p. H37, Set A Transparency 6.1 Data ToolKit • *Make a Bar Graph, Make a Line Graph, Make a Circle Graph*
6.2 Find Unknown Values pp. 124–125 **Objective** To estimate unknown values from a graph and to solve for the values by using arithmetic, logic, and algebra	2, 5, 6, 8	1 Day			Reteach, Practice, Challenge, Problem Solving 6.2 Worksheets Extra Practice p. H37, Set B Transparency 6.2
6.3 Stem-and-Leaf Plots and Histograms pp. 126–128 **Objective** To display and analyze data in stem-and-leaf plots and histograms	5, 6, 8, 10	1 Day	stem-and-leaf plot histogram		Reteach, Practice, Challenge, Problem Solving 6.3 Worksheets, Extra Practice p. H37, Set C Transparency 6.3 Data ToolKit • *Make a Stem-and-Leaf Plot* Calculator Handbook, pp. 28–29
6.4 Math Lab: Explore Box-and-Whisker Graphs p. 129 **Objective** To make a box-and-whisker graph and understand its parts	5, 8		box-and-whisker graph lower extreme upper extreme lower quartile upper quartile	*For each pair at least eleven 3- × 5-in. index cards, marker*	E-Lab • *Exploring Box-and-Whisker Graphs*; E-Lab Recording Sheet Calculator Handbook, pp. 26–27
6.5 Box-and-Whisker Graphs pp. 130–131 **Objective** To analyze a box-and-whisker graph	5, 6, 8	1 Day (For Lessons 6.4 and 6.5)			Reteach, Practice, Challenge, Problem Solving 6.5 Worksheets Extra Practice p. H37, Set D Transparency 6.5 Data ToolKit • *Make a Table, Make a Box-and-Whisker Graph*
6.6 Analyze Graphs pp. 132–135 **Objective** To analyze data displays and determine how results and conclusions may have been influenced	5, 6, 7, 8, 10	2 Days			Reteach, Practice, Challenge, Problem Solving 6.6 Worksheets Extra Practice p. H37, Set E Transparency 6.6

Ending Chapter 6 • Chapter 6 Review/Test, p. 136 • **Standardized Test Prep**, p. 137
Ending Unit 2 • Math Detective, p. 138; **Challenge**, p. 139 • **Study Guide and Review**, pp. 140–141; **Performance Assessment**, p. 142; **Technology**, p. 143; **Problem Solving: On Location**, pp. 143A–143B

***Boldfaced** terms are new vocabulary. Other terms are review vocabulary.

CHAPTER AT A GLANCE

Vocabulary Development

The boldfaced words are the new vocabulary terms in the chapter. Have students record the definitions in their Math Journals.

- **multiple-bar graph**, p. 120
- **multiple-line graph**, p. 121
- **stem-and-leaf plot**, p. 126
- **histogram**, p. 127
- **box-and-whisker graph**, p. 129
- **lower extreme**, p. 129
- **upper extreme**, p. 129
- **lower quartile**, p. 129
- **upper quartile**, p. 129

> multiple-bar graph

Vocabulary Cards
Have students use the Vocabulary Cards on *Teacher's Resource Book* pp. TR123–126 to make graphic organizers or word puzzles. The cards can also be added to a file of mathematics terms.

NCTM Standards

1. **Number and Operations**
2. **Algebra**
   Lesson 6.2
3. **Geometry**
4. **Measurement**
5. **Data Analysis and Probability**
   Lessons 6.1, 6.2, 6.3, 6.4, 6.5, 6.6
6. **Problem Solving**
   Lessons 6.1, 6.2, 6.3, 6.5, 6.6
7. **Reasoning and Proof**
   Lessons 6.1, 6.6
8. **Communication**
   Lessons 6.1, 6.2, 6.3, 6.4, 6.5, 6.6
9. **Connections**
10. **Representation**
    Lessons 6.1, 6.3, 6.6

Writing Opportunities

PUPIL EDITION
- Write a Problem, p. 125
- What's the Error?, p. 123
- Write About It, p. 128
- What's the Question?, p. 131

TEACHER'S EDITION
- Write—See the *Assess* section of each TE lesson.
- Writing in Mathematics, pp. 124B, 126B, 132B

ASSESSMENT GUIDE
How Did I Do?, p. AGxvii

Family Involvement Activities

These activities provide:
- Letter to the Family
- Math Vocabulary
- Family Game
- Practice (Homework)

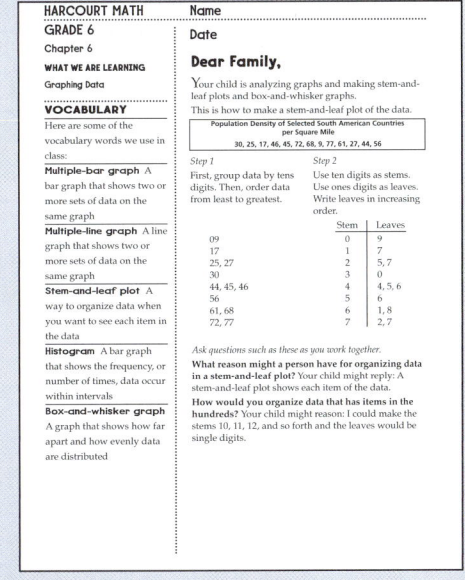

Family Involvement Activities, p. FA21

118B

CHAPTER 6

Graph Data

MATHEMATICS ACROSS THE GRADES

SKILLS TRACE ACROSS THE GRADES

GRADE 5

Display, read, interpret, and analyze data in tables, graphs, and histograms

GRADE 6

Make, analyze, and compare different kinds of graphs and visual displays and determine if they are misleading; estimate and solve for unknown values by using a graph, arithmetic, logical reasoning, and algebraic techniques

GRADE 7

Make, analyze, and compare different kinds of graphs and visual displays, including scatterplots, and determine if they are misleading

SKILLS TRACE FOR GRADE 6

LESSON	FIRST INTRODUCED	TAUGHT AND PRACTICED	TESTED	REVIEWED
6.1	Grade 4	PE pp. 120–123, H37, p. RW27, p. PW27, p. PS27	PE p. 136, pp. AG37–40	PE pp. 136, 137, 140–141
6.2	Grade 6	PE pp. 124–125, H37, p. RW28, p. PW28, p. PS28	PE p. 136, pp. AG37–40	PE pp. 136, 137, 140–141
6.3	Grade 5	PE pp. 126–128, H37, p. RW29, p. PW29, p. PS29	PE p. 136, pp. AG37–40	PE pp. 136, 137, 140–141
6.4	Grade 6	PE p. 129	PE p. 136, pp. AG37–40	PE pp. 136, 137, 140–141
6.5	Grade 6	PE pp. 130–131, H37, p. RW30, p. PW30, p. PS30	PE p. 136, pp. AG37–40	PE pp. 136, 137, 140–141
6.6	Grade 4	PE pp. 132–135, H37, p. RW31, p. PW31, p. PS31	PE p. 136, pp. AG37–40	PE pp. 136, 137, 140–141

KEY   **PE**   Pupil Edition   **PS**   Problem Solving Workbook   **RW**   Reteach Workbook
        PW   Practice Workbook   **AG**   Assessment Guide

Looking Back  Prerequisite Skills

To be ready for Chapter 6, students should have the following understandings and skills:

- **Read Bar Graphs**—use bar graphs to answer questions
- **Read Stem-and-Leaf Plots**—use stem-and-leaf plots to answer questions

Check What You Know

Use page 119 to determine students' knowledge of prerequisite concepts and skills.

Intervention

Help students prepare for the chapter by using the intervention resources described on TE page 119.

Looking at Chapter 6  Essential Skills

Students will

- **display and analyze data in bar graphs, line graphs, circle graphs, stem-and-leaf plots, histograms, and box-and-whisker graphs.**
- use graphs, logical reasoning, and arithmetic to estimate and solve for unknown values.
- compare different types of graphs and identify misleading graphs.

EXAMPLE

Make a stem-and-leaf plot of the data.

Ages of Piano Students

| 18 | 25 | 13 | 46 | 32 |
| 11 | 12 | 41 | 19 | 36 |

Ages of Piano Students

Stem	Leaves
1	1 2 3 8 9
2	5
3	2 6
4	1 6

Looking Ahead Applications

Students will apply what they learn in Chapter 6 to the following new concepts:

- Make a Circle Graph (Chapter 21)
- Display and Analyze Data (Grade 7)
- Collect and Summarize Data (Grade 7)

Graph Data **118D**

CHAPTER 6

Graph Data

INTRODUCING THE CHAPTER

Tell students they can use graphs to represent data visually. Graphs make comparisons, trends, and patterns easier to see. After students read the paragraph, explain that graphs need scales to standardize the data. Ask students to explain how to set up a scale to represent the data in the table. Possible answer: One way is have 1 cm of the graph paper represent 200 km.

USING DATA

To begin the study of this chapter, have students

- Write the diameter of a model of the Popigri crater if 1 cm = 100 km. 0.97 cm
- Write the diameter of a model of the Manicouagan crater if the model of Aitken basin is 10 cm long. 0.4 cm
- Make a bar graph of the data in the chart. Check students' work.

PROBLEM SOLVING PROJECT

Purpose To represent data graphically

Materials *For each pair* centimeter rulers, p. TR22; compass

Background When the Shoemaker-Levy Comet hit Jupiter, one 3-km fragment left an impact site about 13,000 km in diameter. Only about 160 impact craters have been discovered so far on Earth.

Analyze, Choose, Solve, and Check

Have students

- Represent the relative sizes of the craters listed in the table with a graph that shows the craters as concentric circles.
- Determine a scale that will allow a model of the largest crater to fit on a single sheet of paper.
- Use the same scale and calculate the diameter of the model needed to show the 13,000-km impact site on Jupiter.

Check students' work.

 Suggest that students place their calculations and graphs in their portfolios.

CHAPTER 6 Graph Data

about 150 times as long

In July 1994, fragments of Comet Shoemaker-Levy bombarded Jupiter. One impact left a crater about the same diameter as Earth, which is approximately 12,756 km wide. Evidence of past impacts of meteors, asteroids, or comets also appears on Earth and the Moon. One newly discovered crater on the floor of Chesapeake Bay has a diameter of about 80 km.

PROBLEM SOLVING On a bar graph, about how many times as long would the bar for the Jupiter crater be than the bar for the Chesapeake Bay crater?

EARTH'S AND MOON'S LARGEST CRATERS

Location	Approximate Diameter (km)
Popigri, Siberia	97
Manicouagan, Canada	100
Chicxulub, Mexico	180
Orientale basin, Moon	1,300
Ibrium basin, Moon	1,800
Aitken basin, Moon	2,500

Why learn math? Explain that newspaper reporters often use graphs to make their stories more informative and visually appealing. Graphs concisely show data that would take many paragraphs to explain. Ask: In what other jobs might people use graphs to make data easier to understand? Possible answer: bankers, scientists, sales managers

To find out more about graphs and graphing, visit The Harcourt Learning Site.

www.harcourtschool.com

Check What You Know

Use this page to help you review and remember important skills needed for Chapter 6.

✓ Read Bar Graphs (For Intervention, see p. H6.)

For 1–4, use the bar graph at the right.

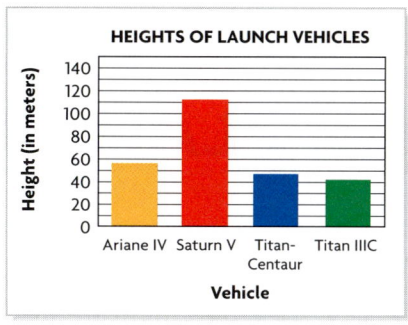

1. List the four launch vehicles in order of size from tallest to shortest. **Saturn V, Ariane IV, Titan-Centaur, Titan IIIC**
2. About how tall is the Saturn V? **about 110 m**
3. About how much taller is the Saturn V than the Ariane IV? **about 55 m**
4. About how much taller is the tallest launch vehicle than the shortest? **about 70 m**

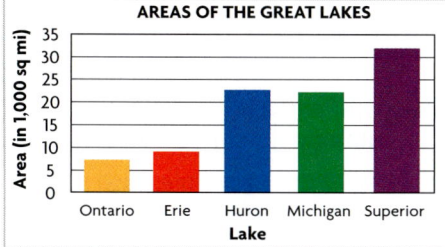

For 5–8, use the bar graph at the left.

5. Name two Great Lakes whose combined area is less than the area of Lake Superior. **Possible answer: Ontario and Erie**
6. What is the area of the third-largest Great Lake? **about 22,000 sq mi**
7. How much larger is the area of Lake Michigan than the area of Lake Ontario? **about 15,000 sq mi**
8. Which Great Lake has an area of about 22,000 sq mi more than the area of Lake Erie? **Lake Superior**

✓ Read Stem-and-Leaf Plots (For Intervention, see p. H7.)

For 9–15, use the stem-and-leaf plot at the right.

MICHAEL'S GOLF SCORES

Stems	Leaves
7	6 7 8 9 9
8	0 2 3 5 5 6 8 9
9	0 0 1 2 3 8

9. What is the score shown by the third stem and fourth leaf? **92**
10. What is the median of Michael's golf scores? **85**
11. What is the range of Michael's golf scores? **22**
12. What is the mode of Michael's golf scores? **79, 85, and 90**
13. How many scores of 90 will Michael need in order to have a mode of 90 for all of his rounds? **3 or more**
14. How many rounds of golf did Michael play? **19**
15. If Michael plays two more rounds and has scores of 86 and 81, what would be the median of the scores? **85**

LOOK AHEAD

In Chapter 6 you will
- make and analyze stem-and-leaf plots, box-and-whisker graphs, histograms, and line plots
- find unknown values
- analyze graphs

119

LESSON 6.1 — Make and Analyze Graphs

LESSON PLANNING

Objective To display and analyze data in bar graphs, line graphs, and circle graphs

Intervention for Prerequisite Skills
Read Bar Graphs (For intervention strategies, see page 119.)

Lesson Resources Calculator Handbook, p. 26

NCTM Standards
5. Data Analysis and Probability
6. Problem Solving
7. Reasoning and Proof
8. Communication
10. Representation

Math Background
Graphs are a highly visual way of presenting data.
- Bar graphs are used to show and compare data in discrete categories that often are not ordered. For example, a bar graph may be used to show speeds or weights of different animals.
- Line graphs are used to show patterns of ordered data, such as changes in temperature or population over time.
- Circle graphs are used to show parts of a whole, such as the amounts for different items in a budget.

Vocabulary
multiple-bar graph a bar graph showing two or more sets of data at once

multiple-line graph a line graph showing two or more sets of data at once

WARM-UP RESOURCES

NUMBER OF THE DAY

If you triple this number and add 8, you get 23. What number is it? 5

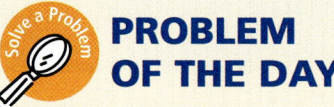 

PROBLEM OF THE DAY

Maurie is the second-youngest of 4 teenagers, all 2 years apart in age. His mother is 3 times as old as he is and 24 years younger than her father. How old is Maurie's grandfather? His grandfather is 69 years old.

Solution Problem of the Day tab, p. PD6

DAILY FACTS PRACTICE

Have students practice multiplication and division facts by completing Set A of *Teacher's Resource Book*, p. TR97.

120A Chapter 6

INTERVENTION AND EXTENSION RESOURCES

ALTERNATIVE TEACHING STRATEGY

Ask students to **practice using bar graphs and line graphs.** Have groups of students gather data on one of the following topics:

- city or state populations
- sizes of the Great Lakes
- heights of the six tallest mountains in the world
- lengths of the five longest tunnels or bridges in the world

Then have them decide what scale is appropriate and round the numbers as needed. They should then make a bar graph or line graph, as appropriate, to show their data. Check students' work.

See also page 122.

VISUAL
VISUAL/SPATIAL

ENGLISH LANGUAGE LEARNERS

Reinforce the **concepts of the words *single*, *double*, and *multiple*** with students. Show pictures of a single bed, double numbers on 2 number cubes, and twins or triplets, and let students tell words in their native languages that have those meanings. Ask them to give examples, such as *double-dip ice cream cones* or *multiple events at a track meet*. Then have students make drawings to illustrate each of the words. Check students' work.

VISUAL
VISUAL/SPATIAL

MULTISTEP AND STRATEGY PROBLEMS

The following multistep or strategy problem is provided in Lesson 6.1:

Page	Item
123	12

SPECIAL NEEDS

Materials magazines, newspapers

Have students work in small groups to **find an example of a line or bar graph** in current magazines and newspapers. Have each group discuss the graph and prepare answers to these questions:

- What does the graph show?
- What does the horizontal scale show?
- What does the vertical scale show?

Then have each group tell the rest of the class about the graph. Check students' work.

AUDITORY
INTERPERSONAL/SOCIAL

TECHNOLOGY LINK

 Intervention Strategies and Activities CD-ROM • *Skill 79*

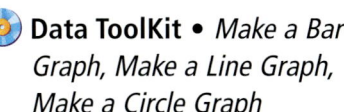 Data ToolKit • *Make a Bar Graph, Make a Line Graph, Make a Circle Graph*

LESSON 6.1 ORGANIZER

Objective To display and analyze data in bar graphs, line graphs, and circle graphs

Vocabulary multiple-bar graph, multiple-line graph

1 Introduce

QUICK REVIEW provides review of mental math skills.

Why Learn This? You can use graphing skills to show the results of a science project, such as how rapidly plants grow under different conditions. *Share the lesson objective with students.*

2 Teach

Guided Instruction

• Conduct a discussion of the opening bar graph.

Which whale weighs the most? How can you tell? Blue whale; it has the tallest bar.

How does the weight of the blue whale compare with that of the fin whale? Its weight is more than 2 times as great.

• For Example 1, help students see that the data contrast in the graph is more quickly discerned than in the table.

Which mammal has the largest weight difference between male and female? gorilla

Did you use the table or the graph to answer that question? Why? Possible answer: the bar graph; It was easier to see the difference.

Why does this graph need a key? You need to be able to tell the difference in the 2 sets of data.

Algebraic Thinking Comparing data in graphs helps students develop their understanding of equivalent and nonequivalent relationships. Have students look for other greater-than and less-than relationships in both graphs.

120 Chapter 6

LESSON 6.1

Make and Analyze Graphs

Learn how to display and analyze data in bar graphs, line graphs, and circle graphs.

QUICK REVIEW

Compare. Use < or >.
1. 5 ● 9 <   2. 25 ● 20 > 3. 87 ● 88 <
4. 19 ● 23   5. 309 ● 304 >
   <

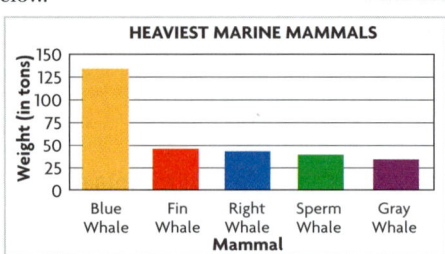

Vocabulary
multiple-bar graph
multiple-line graph

When data are grouped in categories, a bar graph is a good way to display the data. Look at the bar graph below.

HEAVIEST MARINE MAMMALS

The data in the table below show the weights of other mammals.

WEIGHTS OF SOME MAMMALS				
Mammal	Lion	Gorilla	Tiger	Grizzly Bear
Male (lb)	400	450	420	500
Female (lb)	300	200	300	400

A **multiple-bar graph** shows two or more sets of data.

EXAMPLE 1

Use the data in the table above to make a double-bar graph. How do the weights of males compare to the weights of females?

WEIGHTS OF SOME MAMMALS

Determine an appropriate scale.

Use bars of equal width. Use the data to determine the heights of the bars.

Title the graph and both axes. Include a key.

— This means there is a break in the scale.

For each of the mammals in the graph above, the males weigh more than the females.

120 Chapter 6

RETEACH 6.1

Make and Analyze Graphs

At Fox Run School, the sixth-grade and seventh-grade classes are competing to see which class can collect more aluminum cans. The bar graphs show the results for the first three weeks.

To compare the two classes' results more easily, you can show all the data in one graph. The following graph is a double-bar graph.

Weekly Totals

The key tells which bar represents the sixth grade and which represents the seventh grade. A bar graph that shows two or more sets of data is called a multiple-bar graph.

For Exercises 1–3, use the double-bar graph above.

1. How do you know which bar refers to which grade? Read the key.
2. Which grade collected more cans in week 1? in week 2? in week 3? Grade 7; Grade 7; Grade 6
3. Which grade collected more cans altogether? Grade 6

PRACTICE 6.1

Make and Analyze Graphs

Tell if you would use a bar, line, or circle graph to display the data.

1. The amounts of time you spend in your classes in one day. circle or bar graph
2. The amounts of money you spend every day for two weeks. line or bar graph
3. The number of students who play different musical instruments. circle graph
4. The weights of 8 different pets. bar graph
5. Make a multiple-bar graph of the homework data below.

Hours Spent Doing Homework

Name	Science	Math
Nigel	2.5 hr	0.5 hr
Marty	1 hr	1.5 hr
Julie	0.75 hr	2 hr
Luis	1.25 hr	1 hr

Check students' graphs.

6. Make a multiple-line graph of the temperature data below.

Average Low Temperature

Year	Jan	Feb	Mar
1998	5°F	5°F	8°F
1999	6°F	3°F	12°F
2000	10°F	9°F	16°F
2001	12°F	5°F	15°F

7. Gretchen researched the number of new students who came to her school during 5 months of the school years 2000 and 2001. Her data are shown at the right. What kind of graph would you use to display the data? Explain.

Number of New Students

	Sept	Oct	Jan	Feb	Mar
2000	35	12	10	6	9
2001	5	23	14	0	12

multiple-bar graph; to compare two sets of data

Mixed Review

Estimate. Possible estimates are given.

8. 71.3 + 68.6 + 69.7   9. 284.17 ÷ 7.24   10. 979.88 × 31.05
   210   40   30,000

Line Graphs

A bar graph and a line graph are often used to display the same data. However, a line graph is the better choice when the data show change over time. The line graph below shows the change in the price of a stock every five years from 1970 to 2000.

Like a multiple-bar graph, a **multiple-line graph** can show two or more different sets of data on one graph.

MONEY IN SAVINGS ACCOUNTS					
	March	April	May	June	July
Bob	$163	$172	$151	$138	$102
Jan	$43	$55	$76	$79	$96

EXAMPLE 2 Use the data in the table above to make a double-line graph. If the trends continue, how would you describe the amount of money that Bob is saving? that Jan is saving?

Determine an appropriate scale.

Mark a point for each amount saved for Bob and connect the points.

Mark a point for each amount saved for Jan and connect the points.

Title the graph and both axes. Include a key.

The money in Bob's savings account is decreasing. The money in Jan's savings account is increasing.

• Use the graph to predict what will happen to each savings account in August. **Possible answer: If the trends continue, Jan will have more money in her savings account than Bob will have in his savings account.**

• Draw students' attention to the single line graph.

Why doesn't this graph have a key? There is only 1 set of data.

• After students work through Example 2 and predict what will happen to each savings account, ask:

REASONING What amount do you predict will be in each savings account in August? Possible answer: Jan's account: about $110; Bob's account: about $70

ADDITIONAL EXAMPLES

Example 1, p. 120

Use the data in the table to make a multiple-bar graph. Compare the amounts raised by the sixth and seventh grades.

MONEY RAISED IN PTA FUND-RAISERS				
Grade	Holiday Treats	Pizza Party	Family Fun Night	Wrapping Paper
Sixth	$125	$225	$475	$360
Seventh	$75	$190	$290	$225

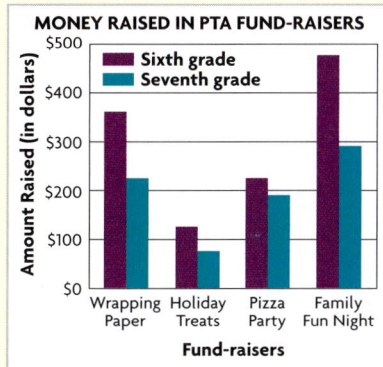

The sixth grade raised more money at each of the fund-raisers.

Example 2, p. 121

Use the data in the table below to make a double-line graph. How would you describe the sales of the two products?

NUMBER OF ITEMS SOLD					
Product	January	February	March	April	May
A	27	53	75	109	158
B	127	103	95	83	71

The sales of Product A are increasing. The sales of Product B are decreasing.

LESSON 6.1

- Tell students that sometimes more than one type of graph is appropriate.

  What other type of graph could you use to display the data in Example 3? a bar graph

ADDITIONAL EXAMPLE

Example 3, p. 122

The graph shows the sales of school apparel at Pace Middle School. About how many T-shirts were sold for each cap sold? **about 4 T-shirts**

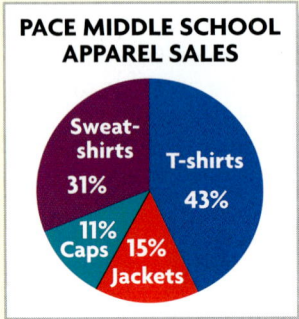

3 Practice

Guided Practice

Do Check for Understanding Exercises 1–4 with your students. Identify those who are having difficulty and use lesson resources to help.

Independent Practice

Note that Exercise 12 is a **multistep or strategy problem**. Assign Exercises 5–16.

Graphs show pictures of data. Bar graphs and line graphs use axes to help you analyze data. A circle graph helps you compare parts to the whole or to other parts.

EXAMPLE 3

In 1987, compact discs (CDs) were new. The circle graph shows the way recorded music was sold in the United States in 1987. About how many cassettes were sold for every CD sold that year?

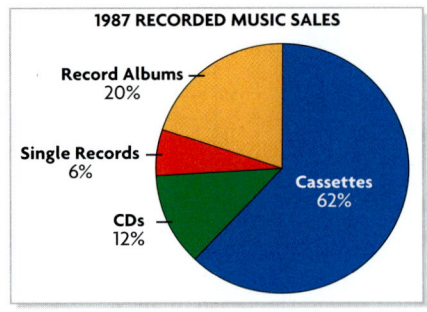

Find the parts that represent cassettes and CDs.

Compare the percent of recorded music sold on cassettes to the percent sold on CDs.

TECHNOLOGY LINK
More Practice: Use *Data Toolkit* to make bar graphs, line graphs, and circle graphs.

Cassettes made up 62% of recorded music sales, while CDs made up 12%. 62 ÷ 12 ≈ 60 ÷ 12, or 5.

So, about 5 cassettes were sold for every CD sold.

- By 1997, the percent of recorded music sold in the form of CDs was about 6 times as great as it had been in 1987. About what percent of recorded music was sold on CDs in 1997? **about 72%**

2. bar graph: data grouped in categories; line graph: data that show change over time; circle graph: data that compare parts to the whole or to other parts

CHECK FOR UNDERSTANDING

Think and Discuss ► Look back at the lesson to answer each question.
1. **Explain** why the graphs in Example 1 and Example 2 have a key. **so you can tell the difference between the sets of data**
2. **Describe** the kind of data you would display in a bar graph, a line graph, and a circle graph. **See above left.**

Guided Practice ► Tell if you would use a bar, line, or circle graph to display the data.

3. The average monthly rainfall in your city over two years **line graph**
4. A family budget divided into types of expenses **circle graph**

PRACTICE AND PROBLEM SOLVING

Independent Practice ► Tell if you would use a bar, line, or circle graph to display the data.

5. The heights of five different students **bar graph**
6. The price of a stock over a period of several months **line graph**
7. The way you spend your weekly allowance **circle graph**
8. The population of six different cities in your state **bar graph**

122 Chapter 6

ALTERNATIVE TEACHING STRATEGY | SCAFFOLDED INSTRUCTION

Purpose Students make a double-bar graph and a double-line graph for temperature data to learn about multiple-bar and multiple-line graphs.

Materials bar graph paper, p. TR61 and 1-cm graph paper, p. TR64 or two colors of construction paper, stapler or glue stick, pushpins, string

Set up two graphs on a bulletin board, a bar graph and a line graph, for displaying temperature data. As a class, record the local daily high and low temperatures in your area for a week.

Have volunteers plot points or glue or staple construction paper bars to represent the temperatures in the bar graph.

For the line graph, ask volunteers to plot points or place pushpins for each temperature. Use string to connect the pushpins and make the double-line graph.

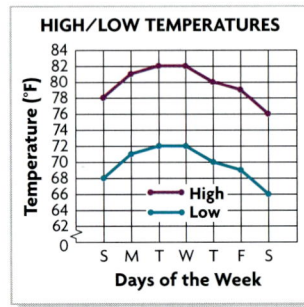

When the graphs are completed discuss the difference between the two graphs. Ask students which graph shows the changes in temperature more clearly.

Encourage students to discuss any trends they notice in the double-line graph. For example, does there seem to be a warming trend? **Check students' work.**

---

122 Chapter 6

9. Make a multiple-bar graph using the data in the table at the right.
   See Additional Answers, p. 137A.

NUMBER OF RAINY DAYS

	April	May	June	July
1999	12	3	13	5
2000	6	7	9	11

10. Make a multiple-line graph using the data in the table at the right.
    See Additional Answers, p. 137A.

AVERAGE STOCK PRICES

	Sep	Oct	Nov	Dec
Stock A	$26	$29	$32	$27
Stock B	$14	$10	$8	$19

Problem Solving Applications

Use Data For 11–13, use the double-line graph below.

11. How do the comedy video rentals compare to the action video rentals?
    11. Each month, more comedies were rented.

12. For the months of March and April combined, about how many more comedy rentals were there than action rentals?
    about 95 more

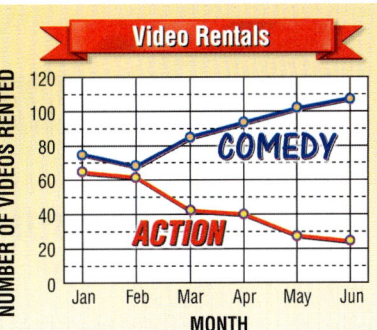

13. **REASONING** If the trends continue, what will happen to the number of comedy and action video rentals in July?
    Comedy video rentals will increase and action video rentals will decrease.

Use Data For 14–16, use the circle graph at left.

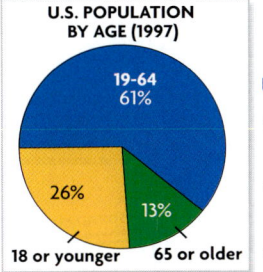

U.S. POPULATION BY AGE (1997)
19–64 61%
26% 13%
18 or younger 65 or older

14. The number of people aged 19–64 is about how many times the number who are 65 or older? **about 4 to 5 times as many**

15. **REASONING** Which of the three age groups do you think spends the most money? Explain. **Possible answer: 19–64, since the group has the most people and is of working age.**

16. **What's the Error?** Jay looked at the circle graph and said that 61% of the U.S. population is aged 19 or older. What is his error?
    He forgot the 13% that are 65 or older.

MIXED REVIEW AND TEST PREP

17. Fifteen students in the gym are asked what their favorite sport is. All of them say "basketball." Ron concludes that the favorite sport of all students is basketball. Is Ron's conclusion valid? Explain. (p. 112)
    No. The sample is not representative of the student population.

Find the product. (p. 70)

18. 6×0.4 **2.4**

19. 0.8×0.7 **0.56**

⭐ 20. **TEST PREP** Which is 0.27×0.4? (p. 70) **A**
   A 0.108   B 0.675   C 1.08   D 10.8

⭐ 21. **TEST PREP** Find the value of 21^2. (p. 40)
   J
   F 21   G 2^3 H 42   J 441

EXTRA PRACTICE page H37, Set A

123

Multistep or Strategy Problem To solve Exercise 12, students can estimate the totals for March and April, and then compare them.

Scaffolded Instruction Use the prompts on Transparency 6 to guide the instruction for the multistep or strategy problem in Exercise 12.

Transparency 6

COMMON ERROR ALERT

Some students may have difficulty determining an appropriate scale with equal intervals for the data. To help them keep the intervals equal, have them work on graph paper.

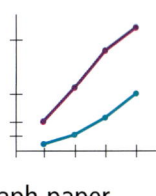

MIXED REVIEW AND TEST PREP

Exercises 17–21 provide **cumulative review** (Chapters 1–6).

4 Assess

Summarize the lesson by having students:

DISCUSS Why does a multiple-bar graph need a key? Possible answer: to indicate which bar shows which set of data

WRITE Make up a word problem using the graph you made for Exercise 9. Possible answer: Which months had more rainy days in 2000 than in 1999?

Lesson Quiz

Transparency 6.1

What type of graph would you use to show the data?

1. the number of VCRs in American and Japanese homes over 5 yr **multiple-line graph**

2. the teachers' and students' favorite holidays **multiple-bar graph**

For 3–4, use the circle graph at the left.

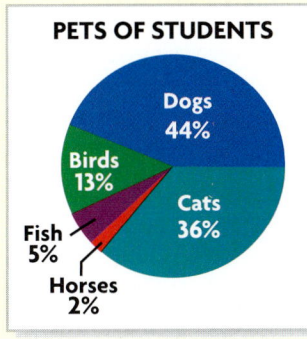

PETS OF STUDENTS
Dogs 44%
Birds 13%
Cats 36%
Fish 5%
Horses 2%

3. About how many students named cats as pets for every student who named fish? **about 6 or 7 students**

4. For every student who has birds for pets, about how many students have dogs? **about 3 or 4 students**

123

LESSON 6.2

Find Unknown Values

LESSON PLANNING

Objective To estimate unknown values from a graph and to solve for the values by using arithmetic, logic, and algebra

Intervention for Prerequisite Skills

Read Bar Graphs (For intervention strategies, see page 119.)

NCTM Standards

2. Algebra
5. Data Analysis and Probability
6. Problem Solving
8. Communication

Math Background

These ideas will help students understand methods of estimating and solving for unknown values.

- On a line graph, if the data continue to change with the same pattern, you can use that idea to predict future values.

- You can use a formula to find an unknown value when given the other values in the formula. For example, given any two values of distance, rate, and time, you can solve for the third value.

WARM-UP RESOURCES

 **NUMBER OF THE DAY**

The number of the day is the number representing the day of the month. Use this number to find how many hours you have practiced this month if you practice from 4 P.M. to 6 P.M. each day. **Possible answer for 21: $21 \times 2 = 42$ hr**

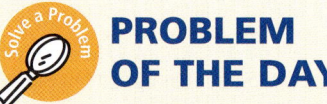 

PROBLEM OF THE DAY

A line graph shows that the temperature at 6 A.M. was 4° warmer than at 4 A.M. In the hours between midnight and 4 A.M., the temperature had fallen an average of 3° per hour. If the temperature at 6 A.M. was 10°F, what was the temperature at midnight? **18°F**

Solution Problem of the Day tab, p. PD6

 **DAILY FACTS PRACTICE**

Have students practice addition facts by completing Set B of *Teacher's Resource Book*, p. TR97.

INTERVENTION AND EXTENSION RESOURCES

ALTERNATIVE TEACHING STRATEGY — ELL

Materials *For each student* 1-cm graph paper, p. TR64

Have students **use graphing to find unknown values.** Display a large graph on a grid with the numbers 1 through 24 vertically (for miles) and 1 through 6 across the bottom (for hours).

Call on volunteers to fill in the appropriate number of blocks over each hour to show the total number of miles Karen walks in 1 hr, 2 hr, and so on, up to 5 hr.

As you work, ask students to copy the pattern on graph paper at their desks. Ask students to describe the pattern and then extend it to find the number of hours it takes Karen to walk 24 mi. Add 4 blocks each time; 6 hr

VISUAL
LOGICAL/MATHEMATICAL

MULTISTEP AND STRATEGY PROBLEMS

The following multistep or strategy problem is provided in Lesson 6.2:

Page	Item
125	15

WRITING IN MATHEMATICS

Have students **practice finding unknown values.** Prepare a graph showing the following data:

Pounds	1	2	3	4	5
Cost	$3	$6	$9	$12	$15

Ask students to write a paragraph describing which of the methods introduced in Examples 1–3 on page 124 they would use to answer the following question and how they would go about using the method:

How much will 6 lb cost?

Possible answers: For the graph: the extended line appears to intersect the 6 lb vertical line at a point across from the $18 cost on the vertical scale. For logical reasoning: the cost increases by $3 when the weight increases by 1 lb.

SOCIAL STUDIES CONNECTION

Have students **practice graphing.** Share this information with students: According to a recent estimate, the world population reached 1 billion in 1804, 2 billion in 1927, 3 billion in 1960, 4 billion in 1974, 5 billion in 1987, and 6 billion on October 12, 1999.

Have students make a line graph to show the increase in population over the years and then predict what the population will be in 2050. Have volunteers give their predictions and explain their reasoning. Check students' work.

VISUAL
VERBAL/LINGUISTIC

Intervention Strategies and Activities CD-ROM • *Skill 79*

124B

LESSON 6.2 ORGANIZER

Objective To estimate unknown values from a graph and to solve for the values by using arithmetic, logic, and algebra

1 Introduce

QUICK REVIEW provides review of mental math skills.

Why Learn This? You can use this skill to predict how long it will take you to drive between two cities. *Share the lesson objective with students.*

2 Teach

Guided Instruction

- Before exploring each example further, point out to students that the times listed are exclusive of any stops or breaks Karen takes while training.

  In Example 1, what pattern do you see in the graph? The graph goes up the same number of units for each unit it goes across.

  Using the same logic and arithmetic used in Example 2, find how long it will take Karen to walk 22 miles. $5\frac{1}{2}$ hr

  REASONING Rewrite the formula in Example 3, with only *t* on the left side of the equal sign. $t = \frac{d}{r}$

ADDITIONAL EXAMPLES

Example 1, p. 124

A bicyclist travels an average of 9 mi per hr. Make a line graph with the data in the table. Estimate how far the bicyclist will travel in 5 hr.

TIME (hr)	1	2	3	4
DISTANCE (mi)	9	18	27	36

Check students' graphs. about 45 mi in 5 hr

Example 3, p. 124

Use the distance formula to find how far the bicyclist will travel in 8 hr if he or she travels at a rate of 9 mi per hr. 72 mi in 8 hr

---

LESSON 6.2

Find Unknown Values

Learn how to estimate unknown values from a graph and solve for the values by using logic, arithmetic, and algebra.

QUICK REVIEW
1. 90×4 — 360
2. 18×3 — 54
3. 20×5 — 100
4. 25×5 — 125
5. 50×4 — 200

For fitness training this week, Karen wants to walk a total of 24 mi. The table above shows how long it will take her to walk certain distances.

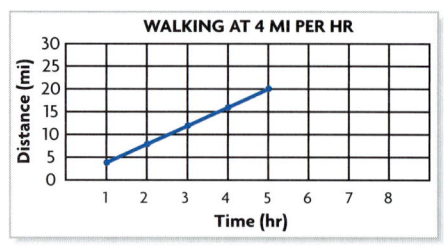

EXAMPLE 1

Make a line graph with the data in the table and use it to estimate how long it will take Karen to walk 24 mi.

Look at the graph. If the pattern continues, it looks as if it will take Karen 6 hr to walk 24 mi.

EXAMPLE 2

Use logical reasoning and arithmetic to find how long it will take Karen to walk 24 mi.

Look at the data in the table. Notice that whenever the time increases by 1 hr, the distance increases by 4 mi.

$5 + 1 = 6$, or 6 hr   *Add 1 to the last time in the table.*

$20 + 4 = 24$, or 24 mi   *Add 4 to the last distance in the table.*

So, it will take Karen 6 hr to walk 24 mi.

You can also use the formula $d = rt$ to solve problems about distance (*d*), rate (*r*), and time (*t*). The distance traveled is a product of the rate of speed and the amount of time.

EXAMPLE 3

Use the formula $d = rt$ to find how long it will take Karen to walk 32 mi if she walks at a rate of 4 mi per hr.

$d = rt$ *Write the formula.*

$32 = 4 \times t$ *Replace d with 32 and r with 4. What number multiplied by 4 gives 32?*

$8 = t$ *The solution is t = 8.*

So, it will take Karen 8 hr to walk 32 mi.

124   Chapter 6

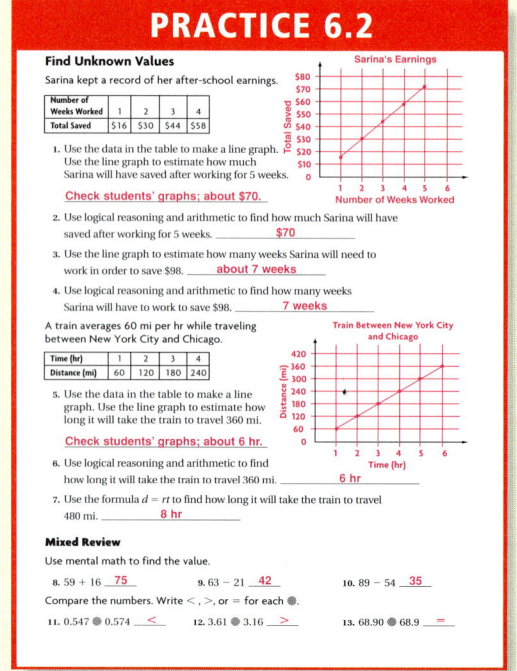

RETEACH 6.2
Find Unknown Values

Line graphs often show a pattern. You can use a line graph to estimate unknown values by extending the graph.

Shauna swims laps in a pool to train for a race. She swims at a rate of 25 m per min. The table below shows how far she swims.

Time (min)	1	2	3	4	5
Distance (m)	25	50	75	100	125

Make a line graph from the data in the table. Use the graph to estimate how many minutes it will take Shauna to swim 150 m.

Step 1 Use the data to draw the graph.
Step 2 Extend the graph until it crosses the 150 m line. Look directly down to the horizontal axis and estimate the value along that axis. If Shauna continues to swim at this rate, it will take her about 6 min to swim 150 m.

You can also use logical reasoning and arithmetic. Since 150 m = 100 m + 50 m, add the number of minutes she takes to swim those two distances: 4 min + 2 min = 6 min.

The formula $d = rt$, where *d* is the distance, *r* is the rate, and *t* is the time, can also be used to find the time Shauna needs to swim 150 m.

$d = rt$ Write the formula.
$150 = 25t$ Substitute the values you know into the formula.
$6 = t$ Solve to find the value of *t*.

So, it will take Shauna 6 min to swim 150 m.

A bicycle racer has averaged 20 mi per hr during the first 4 hr of a race.

Time (hr)	1	2	3	4
Distance (mi)	20	40	60	80

1. Make a line graph. Use the graph to estimate how long it will take the racer to ride 100 mi. Check students' graphs; about 5 hr.
2. Use logical reasoning and arithmetic to find how long it will take the racer to ride 100 mi. _____ 5 hr
3. Use the formula $d = rt$ to find how long it will take the racer to ride 140 mi. _____ 7 hr

PRACTICE 6.2
Find Unknown Values

Sarina kept a record of her after-school earnings.

Number of Weeks Worked	1	2	3	4
Total Saved	$16	$30	$44	$58

1. Use the data in the table to make a line graph. Use the line graph to estimate how much Sarina will have saved after working for 5 weeks. Check students' graphs; about $70.
2. Use logical reasoning and arithmetic to find how much Sarina will have saved after working for 5 weeks. _____ $70
3. Use the line graph to estimate how many weeks Sarina will need to work in order to save $98. _____ about 7 weeks
4. Use logical reasoning and arithmetic to find how many weeks Sarina will have to work to save $98. _____ 7 weeks

A train averages 60 mi per hr while traveling between New York City and Chicago.

Time (hr)	1	2	3	4
Distance (mi)	60	120	180	240

5. Use the data in the table to make a line graph. Use the line graph to estimate how long it will take the train to travel 360 mi. Check students' graphs; about 6 hr.
6. Use logical reasoning and arithmetic to find how long it will take the train to travel 360 mi. _____ 6 hr
7. Use the formula $d = rt$ to find how long it will take the train to travel 480 mi. _____ 8 hr

Mixed Review

Use mental math to find the value.

8. $59 + 16$ 75   9. $63 - 21$ 42   10. $89 - 54$ 35

Compare the numbers. Write <, >, or = for each ●.

11. 0.547 ● 0.574 <   12. 3.61 ● 3.16 >   13. 68.90 ● 68.9 =

CHECK FOR UNDERSTANDING

Think and Discuss Look back at the lesson to answer the question.

1. **Explain** how you used the graph in Example 1 to predict how long it would take Karen to walk 24 mi. **extended the graph until it intersected with 24 mi and found the corresponding time**

Guided Practice Ty averages 50 mi per hr on long car trips. Use the table for 2–4.

Time (hr)	1	2	3	4
Distance (mi)	50	100	150	200

2. Make a line graph. Use the line graph to find how long it will take Ty to drive 250 mi. **See Additional Answers, p. 137A. About 5 hr.**

3. Use logical reasoning and arithmetic to find how long it will take Ty to drive 250 mi. **5 hr**

4. Use the formula $d = rt$ to find how long it will take Ty to drive 350 mi. **7 hr**

PRACTICE AND PROBLEM SOLVING

Independent Practice Juan averages 8 mi per hr while biking. Use the table for 5–7.

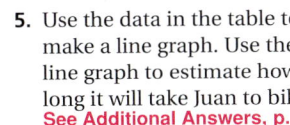

Time (hr)	1	2	3	4
Distance (mi)	8	16	24	32

5. Use the data in the table to make a line graph. Use the line graph to estimate how long it will take Juan to bike 48 mi. **See Additional Answers, p. 137A. About 6 hr.**

6. Use logical reasoning and arithmetic to find how long it will take Juan to bike 48 mi. **6 hr**

7. Use the formula $d = rt$ to find how long it will take Juan to bike 64 mi. **8 hr**

Problem Solving Applications Calvin is training for a marathon. The total distance he has run each week is shown in the table.

8. Make a line graph using the data in the table. **See Additional Answers, p. 137A.**

9. If the trend in Calvin's training continues, about how far will he run in the sixth week of training? **about 45 mi**

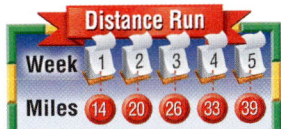

Distance Run

Week	1	2	3	4	5
Miles	14	20	26	33	39

10.  **Write a problem** using Calvin's training data. Explain your solution. **Check students' problems and solutions.**

MIXED REVIEW AND TEST PREP

Find the mean, median, and mode. (p. 106)

11. 9, 6, 4, 4, 5 **5.6, 5, 4**
12. 12, 14, 32, 28 **21.5, 21, none**
13. 2, 9, 9, 9, 6, 5, 8, 4 **6.5, 7, 9**
14. Order 0.76, 0.765, and 0.076 from least to greatest. (p. 52) **0.076, 0.76, 0.765**
15. **TEST PREP** Replace ● with the missing operation. 56 ● 32 ÷ (4 × 2) = 52 (p. 44) **C**

  A ÷   B ×   C −   D +

EXTRA PRACTICE page H37, Set B

125

---

3 Practice

Guided Practice

Do Check for Understanding Exercises 1–4 with your students. Identify those having difficulty and use lesson resources to help.

COMMON ERROR ALERT

When interpreting a graph, students may confuse the horizontal change with the vertical change. To help avoid this problem, have them mark the horizontal and vertical changes on the graph and label each with the size of the change.

Independent Practice

Note that Exercise 15 is a **multistep or strategy problem**. Assign Exercises 5–10.

MIXED REVIEW AND TEST PREP

Exercises 11–15 provide **cumulative review** (Chapters 1–6).

4 Assess

Summarize the lesson by having students:

DISCUSS When you use the distance formula and the rate is given in feet traveled per second, in what unit should time be given? **seconds**

 **WRITE** Describe how you solved Exercise 9. **Possible answer: I extended the line to intersect with the vertical line extending from 6 on the horizontal axis.**

Lesson Quiz

Transparency 6.2

The following table shows how far a giant tortoise can travel.

TIME (min)	1	2	3	4	5
DISTANCE (m)	5	10	15	20	25

1. Make a line graph using the data in the table. **Check students' graphs.**

2. Estimate how far the tortoise will travel in 6 min. **about 30 m**

3. Use the distance formula to find how long it will take the tortoise to travel 100 m. **20 min**

CHALLENGE 6.2

Decisions, Decisions

For each situation, pick the choice that would reach the goal first. Then explain your reasoning.

1. Two planes are heading to the same airport. Plane A is 2,000 mi from the airport, and will average 400 mi per hr for the rest of the trip. Plane B is 1,600 mi from the airport and will average 350 mi per hr for the rest of its trip. Which plane will arrive at the airport first?

   Plane B will reach the airport first. Possible answer: Plane B will arrive in about 4.5 hr, while plane A will arrive in 5 hr.

2. Michael and Wade are reading the same chapter of a book for homework. The chapter begins on page 216 and ends on page 230. Michael has just finished page 222 and is reading at a rate of 1 page every 5 min. Wade is about to begin page 224 and is reading at the rate of 1 page every 6 min. Who will finish reading first?

   Michael will finish first. Possible answer: Michael has 8 pages to read, and reading at 1 page every 5 min, he will finish in 40 min; Wade has 7 pages to read, and reading at 1 page every 6 min, he will finish in 42 min.

3. Two trains leave from the same station. The Express leaves at 1:00 P.M., heading for a city that is 180 mi away. The Express averages 60 mi per hr. The Clipper leaves at 1:30 A.M., averaging 70 mi per hr, heading for a city that is 120 mi away. Which train reaches its destination first?

   The Clipper arrives first. Possible answer: The Express needs 3 hr to make the trip and will arrive at 4:00 P.M. The Clipper needs less than 2 hr and will arrive before 3:30 P.M.

4. Breanna and Luz work in the school library, where a major remodeling project has just been completed. All of the books that were packed away must now be unpacked. They begin unpacking books from cartons at the same time. Breanna unpacks 12 books per min and has 200 books to unpack. Luz unpacks 10 books per min and has 180 books to unpack. Who will finish unpacking first?

   Breanna will finish first. Possible answer: Breanna needs less than 17 min to unpack the cartons (200 ÷ 12 ≈ 16.7 min), while Luz needs 18 min (180 ÷ 10 = 18 min).

PROBLEM SOLVING 6.2

Find Unknown Values

Analyze Choose Solve Check

Write the correct answer.

1. While driving from Cincinnati to Toledo, Ohio, a distance of 200 mi, Jamie averages 40 mi per hr. If she left at 10:30 A.M., at what time should she expect to arrive in Toledo?

   3:30 P.M.

2. In the 1990 Census, the population of Los Angeles was 3,485,557. About how many more people would it take for the population to reach 4,000,000?

   about 500,000 people

3. Kiona kept a record of how much she saved by using an on-line grocery shopping service. Over the first 6 weeks she has used this service, she saved $102. If her savings continue at the same rate, about how much can she expect to save during the twelfth week she uses the shopping service?

   about $17

4. Carmen pays $5.95 per month for a long-distance calling plan that charges her $0.07 per min for her long-distance calls. She averages about 2 hr of long distance calls per month. About how much does she save each month over a plan that charges $0.15 per min for her calls, with no monthly fee?

   about $3.65

Write the letter of the best answer.

5. Ty mows lawns after school for $45 per week. He wants to use the money to pay for a trip that will cost $350. If he spends $20 each week and saves the rest, what is the least number of weeks he must work to pay for the trip?

   A 12 weeks   **C 14 weeks**
   B 13 weeks   D 15 weeks

6. For an art project, you need to cut squares that measure 4 in. on each side from a rectangular sheet of paper that measures 8 in. by 12 in. What is the greatest number of squares that you can cut?

   F 4 squares   H 8 squares
   G 6 squares   J 10 squares

7. An airplane is climbing at a steady rate of 600 ft per min. From the time it reaches an altitude of 3,600 ft, how many more minutes will it take to reach an altitude of 9,000 ft?

   A 6 min   C 8 min
   B 7 min   **D 9 min**

8. Brady counted 240 words on the first page of a reading assignment. If the reading assignment is 6 pages long, about how many words should he expect to read?

   F 1,200 words   **H 1,440 words**
   G 1,340 words   J 1,500 words

9. **Write About It** In Exercise 1, what formula could you use to find the time Jamie would arrive in Toledo?

   Possible answer: Use $d = rt$ to find the number of hours needed; then add the number of hours to 10:30 A.M.

125

LESSON 6.3

Stem-and-Leaf Plots and Histograms

LESSON PLANNING

Objective To display and analyze data in stem-and-leaf plots and histograms

Intervention for Prerequisite Skills
Read Stem-and-Leaf Plots (For intervention strategies, see page 119.)

Lesson Resources Calculator Handbook, p. 25

NCTM Standards
5. Data Analysis and Probability
6. Problem Solving
8. Communication
10. Representation

Math Background

Stem-and-leaf plots and histograms present a more graphic picture of organized data than does a table.

- Stem-and-leaf plots show individual data items organized by stem, which may be the first one, two, or more digits of the number. The leaves display the items' ones digits.
- If a stem-and-leaf plot is turned counterclockwise one quarter turn, its information resembles that of a histogram.
- Histograms show the frequency of data in bars for a range or interval. The vertical axis always shows the frequency and the horizontal axis shows the intervals.

WARM-UP RESOURCES

 **NUMBER OF THE DAY**

Multiply the number of minutes in this class period by 60. Give the product and tell its significance. *Possible answer for 48 minutes: 2,880; the number of seconds in this class period*

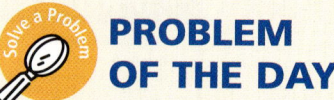 

What is the least number that can be divided evenly by each of the numbers 1 through 12? *27,720*

Solution Problem of the Day tab, p. PD6

 **DAILY FACTS PRACTICE**

Have students practice addition facts by completing Set C of *Teacher's Resource Book*, p. TR97.

Vocabulary

stem-and-leaf plot a method of organizing data that displays each data item's digits in a stem (vertical) and leaf (horizontal) design

histogram a bar graph that shows the frequency, or the number of times, data occur within intervals

INTERVENTION AND EXTENSION RESOURCES

ALTERNATIVE TEACHING STRATEGY — ELL

Display a place-value chart and the **outline for a stem-and-leaf plot.**

Stem	Leaves

Ask volunteers to write the digits for each number from the Card-Tower Competition on page 126 in the place-value chart.

Then have students write them one by one in the stem-and-leaf plot, discussing where to place the tens digit and the ones digit for each number. Check students' work.

VISUAL
VERBAL/LINGUISTIC

MULTISTEP AND STRATEGY PROBLEMS

The following multistep or strategy problem is provided in Lesson 6.3:

Page	Item
128	16

WRITING IN MATHEMATICS

Ask students to **write a paragraph that describes how to make a stem-and-leaf plot** from a given set of data.

Students should include information on how to choose the stem and the leaves and then how to make the plot itself. Possible answer: Examine the numbers. If they are two-digit numbers, choose the tens digits as the stems and the ones digits as leaves. Draw a T and list the stems along the left. Then write each ones digit in the same row as its stem. Finally, arrange the leaves in numerical order.

ADVANCED LEARNERS

Materials reference books, almanacs

Challenge groups of students to make **histograms.** Have each group make a histogram of the populations of all 50 states. Have groups present their graphs and explain how they chose their population intervals.

Check students' work.

VISUAL
INTERPERSONAL/SOCIAL

TECHNOLOGY LINK

- Intervention Strategies and Activities CD-ROM • *Skill 80*
- Data ToolKit • *Make a Stem-and-Leaf Plot*
- Calculator Handbook, pp. 28–29

LESSON 6.3 ORGANIZER

Objective To display and analyze data in stem-and-leaf plots and histograms

Vocabulary stem-and-leaf plot, histogram

Lesson Resources Calculator Handbook, pp. 28–29

1 Introduce

QUICK REVIEW provides review of mental math skills.

Why Learn This? You can use a stem-and-leaf plot to organize test scores for your class. *Share the lesson objective with students.*

2 Teach

Guided Instruction

- Review the parts of a stem-and-leaf plot.

  If you have 10 pieces of data to put in a stem-and-leaf plot, how many leaves will you need? 10

  How is a leaf related to its stem? Possible answer: The stem is the tens digit which might be 0, and the leaf is the ones digit of a piece of data.

- Ask the students to look at Example 1.

  Why are there 3 ones next to the 2 stem? There are three 21's in the data.

ADDITIONAL EXAMPLE

Example 2, p. 126

Use the data for the number of prize tickets collected by students at the Fun Center to make a stem-and-leaf plot. Use the stem-and-leaf plot to help find the mode and the median.

NUMBER OF PRIZE TICKETS COLLECTED						
30	65	53	42	45	66	35
35	49	43	50	72	65	42
66	55	37	47	42	71	34

Number of Prize Tickets Collected

Stem	Leaves
3	0 4 5 5 7
4	2 2 2 3 5 7 9
5	0 3 5
6	5 5 6 6
7	1 2

Mode: 42; Median: 47

126 Chapter 6

LESSON 6.3

Stem-and-Leaf Plots and Histograms

Learn how to display and analyze data in stem-and-leaf plots and histograms.

QUICK REVIEW

1. 8.5 + 4.2
   12.7
2. 125.80 + 11.20
   137.0
3. 10.8 − 8.6
   2.2
4. 225.65 − 5.60
   220.05
5. 10.2 + 2.4 + 3.1
   15.7

Vocabulary
stem-and-leaf plot
histogram

You can use a **stem-and-leaf plot** to organize data when you want to see each item in the data. For a stem-and-leaf plot, choose the stems first and then write the leaves.

EXAMPLE 1

The table shows the number of levels reached, without cards falling, at a card-tower building competition. Use the data to make a stem-and-leaf plot.

CARD-TOWER COMPETITION					
21	18	32	47	50	33
19	21	11	54	31	18
33	42	21	29	16	12

11 12 16 18 18 19
21 21 21 29
31 32 33 33
42 47
50 54

First, group the data by tens digits. Then, order the data from least to greatest.

Card Tower Competition

Stems	Leaves
1	1 2 6 8 8 9
2	1 1 1 9
3	1 2 3 3
4	2 7
5	0 4

Use the tens digits as stems. Use the ones digits as leaves. Write the leaves in increasing order.

The line 4 | 2 7 means 42 and 47.

Bryan Berg built a 19 ft 16½ in. high card tower with 102 levels.

EXAMPLE 2

Use the data from a domino stacking competition to make a stem-and-leaf plot. Then use the stem-and-leaf plot to help find the mode and the median.

NUMBER OF DOMINOES STACKED									
97	88	74	96	98	58	68	90	80	90
72	86	69	78	93	84	99	92	85	

Stems	Leaves
5	8
6	8 9
7	2 4 8
8	0 4 5 6 8
9	0 0 2 3 6 7 8 9

90 occurs more than any other number.

There are 19 scores. The median is the 10th score.

Mode: 90   Median: 86

126 Chapter 6

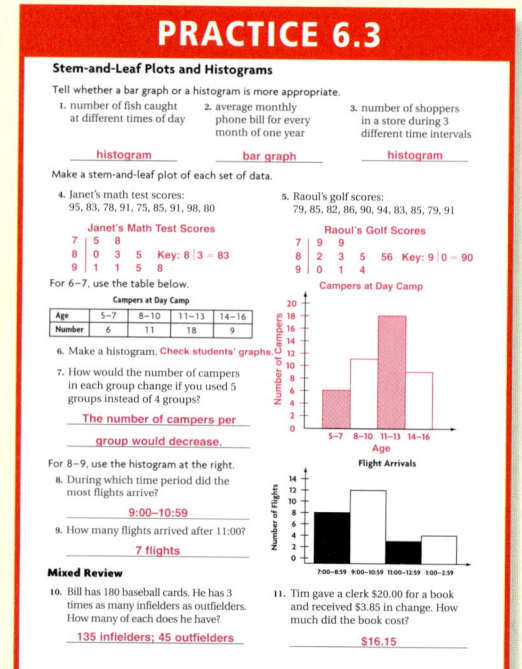

A **histogram** is a bar graph that shows the frequency, or the number of times, data occur within intervals. The bars in a histogram are connected, rather than separated.

BAR GRAPH

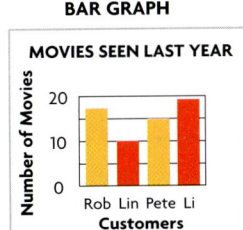

HISTOGRAM

Remember that you can use the range of a set of data to help determine intervals.

This bar graph is used to show information about individual movie customers. The histogram is used to show information about groups of movie customers.

EXAMPLE 3

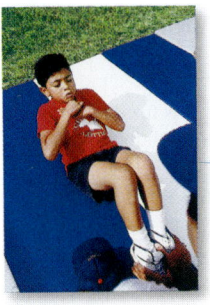

The table below shows the number of sit-ups students in gym class did in one minute. Make a histogram for the data.

NUMBER OF SIT-UPS									
28	19	32	45	44	12	24	32	35	47
55	59	24	25	37	36	38	36	42	41

First, make a frequency table with intervals of 10. Start with 10.

Interval	10-19	20-29	30-39	40-49	50-59
Frequency	2	4	7	5	2

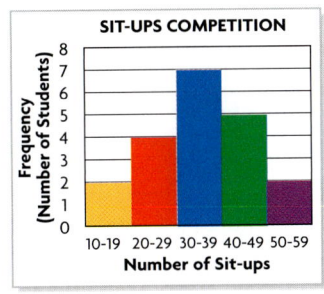

Title the graph and label the scales and axes.

Graph the number of students who did sit-ups within each interval.

1. Possible answer: It helps you compare how many towers were of different heights, such as 10–19 and 40–49.

CHECK FOR UNDERSTANDING

Think and Discuss

Look back at the lesson to answer each question.
1. **Tell** how the stem-and-leaf plot in Example 1 is useful in showing how well people did in the competition. **See above left.**
2. **Explain** how data displayed in a histogram are different from data displayed in a bar graph. **The data in a histogram occur in intervals while data in a bar graph occur in categories.**

127

- Draw students' attention to the bar graph and the histogram.

  How is the bar graph similar to the histogram? Possible answer: Both give customer information on the *x*-axis and the number of movies on the *y*-axis.

  How is the bar graph different from the histogram? Possible answer: The bar graph shows information for only 4 customers. The histogram shows data for more customers and the bars are connected.

ADDITIONAL EXAMPLE

Example 3, p. 127

Make a histogram for the data.

BOOKS READ BY CLUB MEMBERS									
13	53	59	36	75	50	44	64	12	57
19	39	8	61	66	17	48	42	79	

Possible answer:

3 Practice

Guided Practice

Do Check for Understanding Exercises 1–5 with your students. Identify those having difficulty and use lesson resources to help.

In discussing Exercise 2, point out that the horizontal axis of a histogram is always numerical while that of a bar graph may be categorical.

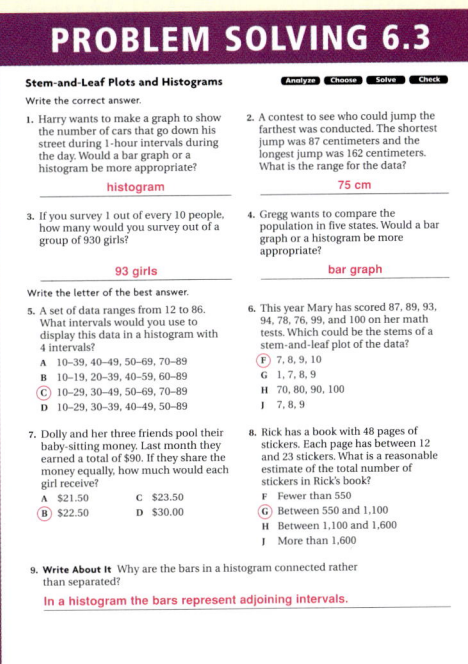

127

Independent Practice

Note that Exercise 16 is a **multistep or strategy problem**. Assign Exercises 6–11.

MIXED REVIEW AND TEST PREP
Exercises 12–16 provide **cumulative review** (Chapters 1–6).

4 Assess

Summarize the lesson by having students:

DISCUSS How would you label the axes of the histogram for Exercise 8? *x*-axis = Ice-Skaters' Ages; *y*-axis = Number of Ice-Skaters

WRITE Why is a histogram more appropriate than a bar graph for displaying the data in Exercise 10? The data can be easily shown in intervals but cannot be easily shown as individual ages.

Lesson Quiz

Transparency **6.3**

1. Make a stem-and-leaf plot of the data: 33, 47, 68, 36, 34, 59, 37, 52, 65, 75, 78, 78

Stem	Leaves
3	3 4 6 7
4	7
5	2 9
6	5 8
7	5 8 8

Tell whether a bar graph or a histogram is more appropriate.

2. the number of teenagers in each of the four largest cities in Japan bar graph
3. the 100 distances in a long-jump competition histogram
4. the number of candy bars sold by students in a school fund-raiser histogram

128   Chapter 6

Guided Practice

3. Make a stem-and-leaf plot of the data 32, 24, 44, 57, 31, 25, 41, 26. See at left.

3. Stem	Leaves
2	4 5 6
3	1 2
4	1 4
5	7

Tell whether a bar graph or a histogram is more appropriate.

4. number of customers at different intervals of time histogram
5. populations of five different states bar graph

PRACTICE AND PROBLEM SOLVING

Independent Practice

6. Make a stem-and-leaf plot of the data 89, 74, 63, 65, 68, 74, 71, 80. See at left.

6. Stem	Leaves
6	3 5 8
7	1 4 4
8	0 9

Tell whether a bar graph or a histogram is more appropriate.

7. heights of the tallest mountains in the United States bar graph
8. ages of 75 ice-skating competitors at a competition histogram

Problem Solving Applications

Use the data in the table for 9.

9. a. Make a stem-and-leaf plot of the data.
   b. Use the stem-and-leaf plot to help find the median and mode. 24.5 in., 28 in.
   c. **REASONING** Can you tell what the mean is for this set of data by looking at the stem-and-leaf plot? Explain. No. You must still compute the mean.

PLANT HEIGHTS IN INCHES					
28	36	25	24	20	32
15	18	28	12	19	26

9. Stem	Leaves
1	2 5 8 9
2	0 4 5 6 8 8
3	2 6

AGES OF MARIA'S MUSIC CUSTOMERS								
10	25	33	14	54	62	29	44	41	40
11	31	41	24	65	16	39	50	51	55
19	22	17	26	31	42	17	18	42	37

10. Use the data in the table above to make a histogram. See Additional Answers, p. 137A.
11. **Write About It** Write a question that can be answered using the data from Exercise 10. Explain your answer. Check students' questions and answers.

MIXED REVIEW AND TEST PREP

12. Thirty students in the gym are asked their favorite sport. They all respond by saying "basketball." Is this valid for all students? (p. 112) no
13. What type of sample does Ron get if he randomly surveys 175 people? (p. 94) random
14. Evaluate 3*k* for *k* = 6.5. (p. 82) 19.5
15. Write 0.73 as a percent. (p. 60) 73%
16. **TEST PREP** Mt. McKinley has a height of 20,320 ft. Mt. Whitney has a height of 14,494 ft. If Brian climbs both mountains and wants to climb a total of 40,000 ft, how many more feet does he need to climb? (p. 20) D

   A 34,814 ft   B 25,506 ft   C 19,680 ft   D 5,186 ft

EXTRA PRACTICE page H37, Set C

LESSON 6.4

Explore Box-and-Whisker Graphs

Explore how to make a box-and-whisker graph and understand its parts.

You need at least eleven 3 in. × 5 in. cards, marker.

Vocabulary
box-and-whisker graph
lower extreme
upper extreme
lower quartile
upper quartile

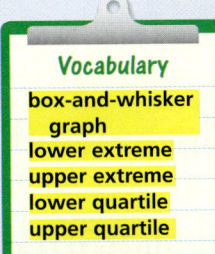

TECHNOLOGY LINK
More Practice: Use E-Lab, *Exploring Box-and-Whisker Graphs*.
www.harcourtschool.com/elab2002

A **box-and-whisker graph** shows how far apart and how evenly data are distributed.

Activity

- Write each of the data in the table on a separate card.

NUMBER OF TROPHIES WON
35 21 24 32 36 20 24 29 27 30

- Order the data from least to greatest. Draw a star on the card with the least value, or **lower extreme**, and the card with the greatest value, or **upper extreme**. 20; 36

- Find the median of the data. If the median is not one of the numbers already written, write it on a card, put it in the middle of the data, and circle it. 28

- Find the median of the lower half of the data. This median is called the **lower quartile**. Circle it. Separate the data to the left of the lower quartile from the rest of the data. 24

- Find the median of the upper half of the data. This median is the **upper quartile**. Circle it. Separate the data to the right of the upper quartile from the rest of the data. 32

Think and Discuss

- Look at your cards. Into how many parts do the lower quartile, the median, and the upper quartile separate the data? 4 parts

- What fraction of the data are to the left of the lower quartile? to the right of the upper quartile? What fraction of the data are between the lower quartile and the upper quartile? $\frac{1}{4}, \frac{1}{4}, \frac{1}{2}$

You have found all you need to make this box-and-whisker graph.

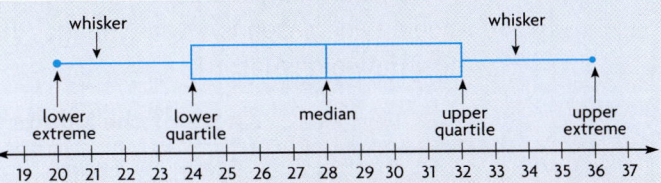

Practice

- Make a box-and-whisker graph of the data.

POINTS SCORED
10 12 9 22 17 7 14 8 11 19

See Additional Answers, p. 137A.

129

LESSON 6.4

ORGANIZER

Objective To make a box-and-whisker graph and understand its parts

Vocabulary box-and-whisker graph, lower extreme, upper extreme, lower quartile, upper quartile

Materials *For each pair* at least eleven 3 in. × 5 in. index cards, marker

Lesson Resources E-Lab Recording Sheet • *Exploring Box-and-Whisker Graphs*, Calculator Handbook, pp. 26–27

Intervention for Prerequisite Skills Read Bar Graphs (For intervention strategies, see page 119.)

Activity

If necessary, assist students as they organize data into the parts that make up a box-and-whisker graph. Record each step as students work with their partners.

Think and Discuss

Have students check to be sure their data are separated into the parts described in the first question and associate them with the graph numbers.

Practice

Have students share their graphs with the class. Ask:

What are the names for certain numbers on the graph? lower extreme, lower quartile, median, upper quartile, upper extreme

Oral Assessment

What are the lower and upper quartiles? lower quartile: the median of the data to the left of the overall median; upper quartile: the median of the data to the right of the overall median

What are the lower and upper extremes? the least number and the greatest number, respectively, in a data set

---

USING E-LAB

Students use a computer tool to make box-and-whisker graphs that they then interpret.

The E-Lab Recording Sheets and activities are available on the E-Lab website.

www.harcourtschool.com/elab2002

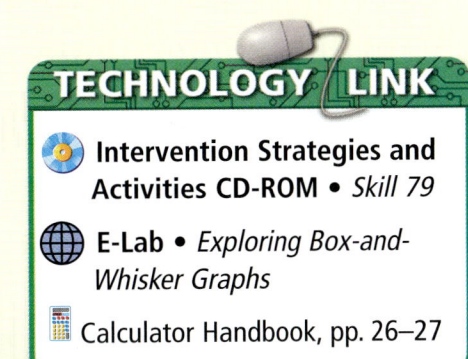

TECHNOLOGY LINK

Intervention Strategies and Activities CD-ROM • *Skill 79*

E-Lab • *Exploring Box-and-Whisker Graphs*

Calculator Handbook, pp. 26–27

129

LESSON 6.5

Box-and-Whisker Graphs

LESSON PLANNING

Objective To analyze a box-and-whisker graph

Intervention for Prerequisite Skills
Read Bar Graphs (For intervention strategies, see page 119.)

NCTM Standards
5. Data Analysis and Probability
6. Problem Solving
8. Communication

Math Background

A box-and-whisker graph provides different information from a histogram or a stem-and-leaf plot about the center, spread, and symmetry of the data.

- The box-and-whisker graph uses the median because it is not as affected by outliers as the mean is.
- You can read the median of the data from the box-and-whisker graph as well as the middle points of the upper and lower halves of the data, called the quartiles.
- Because the median may not be in the data set, the only values you can read that you know must be in the set are the extremes.

WARM-UP RESOURCES

 **NUMBER OF THE DAY**

How much money would you have if you had the same number of quarters as the number for the last day of the month? Possible answer: 31 days—$7.75

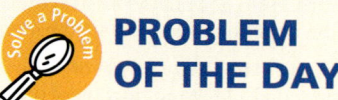 **PROBLEM OF THE DAY**

Jason made a stem-and-leaf plot of the ages of all the adults at a family reunion. His father's age was the median age. Jason is 1 year younger than $\frac{1}{3}$ his father's age. How old is Jason?

Ages of the Adults

Stem	Leaves
2	2 6
3	0 3 7
4	2 3 3
5	5 8
6	8

13 years old

Solution Problem of the Day tab, p. PD6

 **DAILY FACTS PRACTICE**

Have students practice subtraction facts by completing Set E of *Teacher's Resource Book,* p. TR97.

130A Chapter 6

REACHING ALL LEARNERS
INTERVENTION AND EXTENSION RESOURCES

ALTERNATIVE TEACHING STRATEGY — ELL

Help students **make a box-and-whisker graph.** They should draw a number line and plot these data points: 20, 37, 42, 30, 31, 33, 50, 22, 25, 35. Then from the number line, have them draw dotted vertical lines to points above the least and greatest values, the median, and the medians of the upper and lower halves, or quartiles. Direct students to complete the box-and-whisker graph by connecting the extremes to the quartiles with a line and by making a box from the quartiles to include a median line.

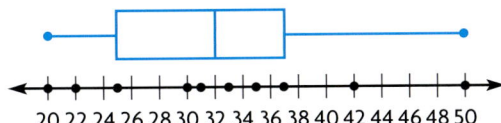

Ask volunteers to relate the whiskers and the box to the data distribution. Check students' work.

VISUAL
INTRAPERSONAL/INTROSPECTIVE

MULTISTEP AND STRATEGY PROBLEMS

The following multistep or strategy problem is provided in Lesson 6.5:

Page	Item
131	13

CAREER CONNECTION

Have students **research the career of statistician.** Tell them that the box-and-whisker graph was invented by statistician John Tukey in the 1960s. Another of his creations is the stem-and-leaf plot, a method of organizing data in order to make comparisons.

Have students find the different types of jobs statisticians might have, and what the employment prospects are for statisticians in government, industry, and education. Then have them make a poster advertising the career of statistician. Check students' work.

VISUAL
VISUAL/SPATIAL

ADVANCED LEARNERS

Challenge students to **make a box-and-whisker graph.** Have them work in pairs to research the high temperatures for several cities in your state over a period of 10 days.

Ask the pairs to make box-and-whisker graphs showing the data for one of the cities researched. Then ask them to exchange graphs with another pair.

Have students write several statements similar to those in Example 2 on page 130 about the graph they have. Then have them compare their conclusions with the actual data. Check students' work.

VISUAL
INTERPERSONAL/SOCIAL

TECHNOLOGY LINK

- Intervention Strategies and Activities CD-ROM • *Skill 79*
- Data ToolKit • *Make a Table, Make a Box-and-Whisker Graph*

LESSON 6.5 ORGANIZER

Objective To analyze a box-and-whisker graph

1 Introduce

QUICK REVIEW provides review of mental math skills.

Why Learn This? You can use this skill to compare how your class does on two different tests. Share the lesson objective with students.

2 Teach

Guided Instruction

- As students look at Example 1, have them review the parts of the graph.

  What value is the lower quartile? the upper quartile? 5; 9

  What does the vertical line in the box indicate? the median

- Direct students' attention to Example 2.

  Why is the box part of the graph in Example 2 divided in half, but the box in Example 1 is not? Possible answer: The median lines fall in different places because the data are distributed differently.

ADDITIONAL EXAMPLES

Example 1, p. 130

The data in the graph show the results of a survey of 20 restaurants for the price of a pizza with 2 toppings. What are the most expensive and the least expensive pizzas? $15; $8

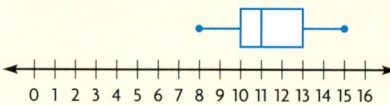

Example 2, p. 130

The graph shows data about the number of CDs sold each day during a sale at The Music Box. What does the graph show about how the data are distributed?

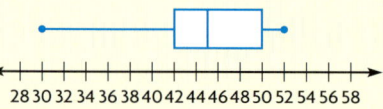

Possible answer: The data in the lowest $\frac{1}{4}$ are spread out. The data in the middle $\frac{1}{2}$ are closer together. On at least one day 30 CDs were sold. On at least one day 52 CDs were sold.

130 Chapter 6

LESSON 6.5 Box-and-Whisker Graphs

Learn how to analyze a box-and-whisker graph.

QUICK REVIEW
Find each for the data 4, 0, 4, 6, 5, 3, 3, 3, 5, and 9.
1. mean 4.2 2. median 4 3. mode 3 4. range 9 5. outliers 0 and 9

The only actual data you can identify from the data set in a box-and-whisker graph are the extremes.

EXAMPLE 1

The data in the box-and-whisker graph represent the diameters in kilometers of some asteroids observed by a scientist. What was the diameter of the largest asteroid? of the smallest asteroid?

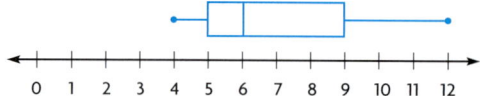

The upper extreme is 12 and the lower extreme is 4. So, the diameter of the largest asteroid was 12 km, and the diameter of the smallest asteroid was 4 km.

- Look again at the graph. Which of the following can you determine—mean, median, mode, range? **median and range**

EXAMPLE 2

Data about the number of meteors seen during a particular hour at different locations during the 1999 Leonids meteor shower are shown in the box-and-whisker graph. What does the graph show about how the data are distributed?

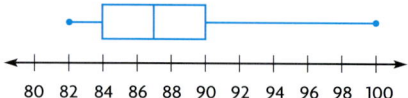

The data in the lowest $\frac{1}{4}$ of the data set are very close together. The data in each part of the middle $\frac{1}{2}$ are farther apart. They are closer to the lower extreme than to the upper extreme. The data in the highest $\frac{1}{4}$ are farther apart than those in the middle. At least one person saw 82 meteors, and at least one person saw 100 meteors.

CHECK FOR UNDERSTANDING

Think and Discuss Look back at the lesson to answer the question.

1. **Explain** why you can't find the mean and the mode when looking at a box-and-whisker graph. **A box-and-whisker graph shows the distribution of data, not all of the data.**

130 Chapter 6

RETEACH 6.5

Box-and-Whisker Graphs

The owner of a gift shop made a box-and-whisker graph to represent the number of customers who came into the shop each day.

The least number in a box-and-whisker graph is called the **lower extreme**. It is the black dot at the end of the left whisker. In this graph, the lower extreme represents the least number of customers to enter the shop in a single day. The lower extreme is 17.

The greatest number in a box-and-whisker graph is called the **upper extreme**. It is the black dot at the end of the right whisker. In this graph, the upper extreme represents the greatest number of customers to enter the shop in one day. The upper extreme is 28.

The range is the difference between the upper and lower extremes. The range of this set of data is 11.

The median of a set of data is the middle number when all numbers are arranged in numerical order. The median of this set is 21.

Another store owner also made a box-and-whisker graph of the customers who entered her store daily.

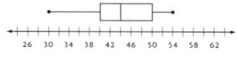

1. What is the lower extreme? **30**
2. What is the upper extreme? **54**
3. What is the lower quartile? **40**
4. What is the upper quartile? **50**
5. What is the range of this set of data? **24**
6. What is the median for this set of data? **44**

PRACTICE 6.5

Box-and-Whisker Graphs

For 1–3, use the box-and-whisker graph below.

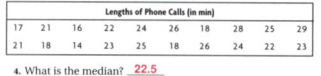

1. What is the median? **23**
2. What are the lower and upper quartiles? **21; 29**
3. What are the lower and upper extremes and the range? **18; 30; 12**

For 4–8, use the data in the chart below.

Lengths of Phone Calls (in min)							
17	21	16	22	24	26	18	28	25	29
21	18	14	23	25	18	26	24	22	23

4. What is the median? **22.5**
5. What are the lower and upper quartiles? **18; 25**
6. What are the lower and upper extremes and the range? **14; 29; 15**
7. Make a box-and-whisker graph. **Check students' graphs.**
8. What fractional part of the data is less than 25 minutes? **$\frac{7}{10}$**

Mixed Review

For 9–10, use the data in the chart above for 4–8.

9. Complete the cumulative frequency table below for the data.

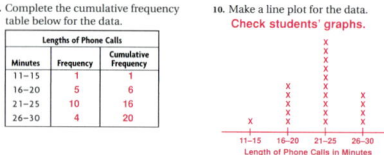

10. Make a line plot for the data. **Check students' graphs.**

Guided Practice For 2–3, use the box-and-whisker graph.

2. What are the median, lower quartile, and upper quartile? **18; 16; 21**

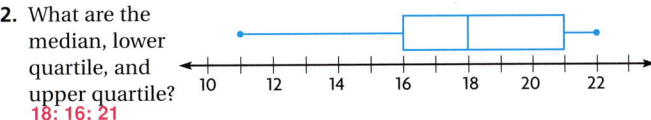

3. What are the lower and upper extremes and the range? **11; 22; 11**

PRACTICE AND PROBLEM SOLVING

Independent Practice For 4–5, use the box-and-whisker graph.

4. What are the median, lower quartile, and upper quartile? **61; 60; 63**

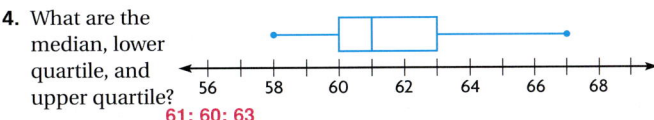

5. What are the lower and upper extremes and the range? **58; 67; 9**

For 6–8, use the data in the table below.

6. What are the median, lower quartile, and upper quartile? **95; 90; 101**

WEIGHTS OF PANTHERS FOOTBALL PLAYERS					
85	102	89	86	104	92
103	97	91	100	100	93

7. What are the lower and upper extremes and the range? **85; 104; 19**

8. Make a box-and-whisker graph and a histogram. Compare them. **See Additional Answers, p. 137A.**

Problem Solving Applications The box-and-whisker graph shows data about the numbers of points scored by the basketball team each game.

9. What are the least and greatest numbers of points? **72 points; 97 points**

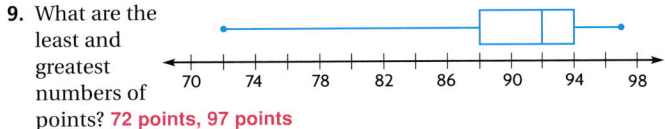

10. What does the graph show about how the data are distributed? **The data in the middle $\frac{1}{2}$ and upper $\frac{1}{4}$ are close together; the data in the lower $\frac{1}{4}$ are more spread out.**

11. ❓ **What's the Question?** Use the box-and-whisker graph for 9–10. The answers are 88 and 94. **What are the lower and upper quartiles of the data?**

MIXED REVIEW AND TEST PREP

12. If you are displaying data about the number of cars that cross an intersection during different intervals of time, is a bar graph or a histogram more appropriate? (p. 126) **histogram**

13. Sue purchases three items that cost $3.73, $9.21, and $2.99. If she pays with $20.00, how much change will she receive? (p. 66) **$4.07**

Write the value of the blue digit. (p. 52)

14. 9.6**4** **0.6, or six tenths**

15. 124.0**2**4 **0.02, or two hundredths**

★ 16. **TEST PREP** Which is $3 \times 3 \times 3 \times 3$ in exponent form? (p. 40) **D**

A 3×4 B 3^2 C 3^3 D 3^4

EXTRA PRACTICE page H37, Set D

131

---

3 Practice

Guided Practice

Do Check for Understanding Exercises 1–3 with your students. Identify those having difficulty and use lesson resources to help.

Independent Practice

Note that Exercise 13 is a **multistep or strategy problem**. Assign Exercises 4–11.

Students may need to be reminded that a box-and-whisker graph divides a set of data into four parts. If necessary, help them identify the numbers that do this.

MIXED REVIEW AND TEST PREP

Exercises 12–16 provide **cumulative review** (Chapters 1–6).

4 Assess

Summarize the lesson by having students:

DISCUSS What three medians divide the box-and-whisker graph at the top of page 131? **16; 18; 21**

WRITE Why can't you use a box-and-whisker graph to find the mean? **Possible answer: A box-and-whisker graph does not show every data item.**

Lesson Quiz

Transparency 6.5

For each exercise, use the box-and-whisker graph below.

1. What is the median? **2.5**

2. What are the lower and upper quartiles? **2; 4**

3. What are the lower and upper extremes? **1; 7**

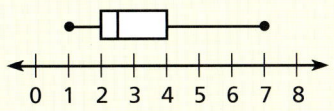

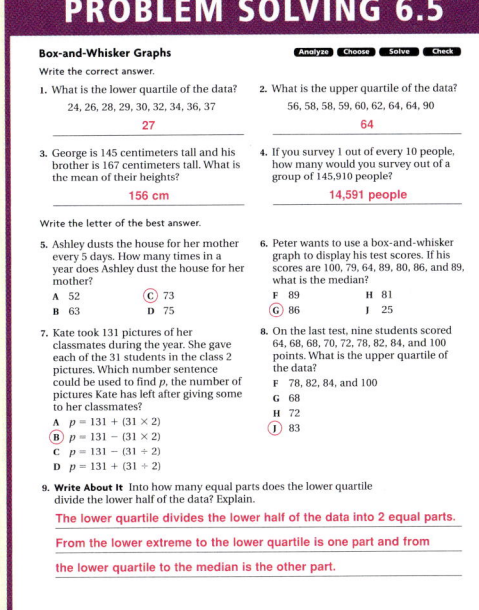

131

LESSON 6.6

Analyze Graphs

LESSON PLANNING

Objective To analyze data displays and determine how results and conclusions may have been influenced

Intervention for Prerequisite Skills
Read Bar Graphs (For intervention strategies, see page 119.)

NCTM Standards
5. Data Analysis and Probability
6. Problem Solving
7. Reasoning and Proof
8. Communication
10. Representation

Math Background
Graphs sometimes show data based on biased questions. They also may show the data in such a way as to give an erroneous impression that is not warranted.

- Students probably cannot tell from a graph if a question was fairly stated. However, they should be aware of this problem as they collect data of their own.
- When looking at a graph, students should learn to examine the scales on the graph to see if they are distorted or abbreviated, which could result in a biased impression of the data.

Because graphs may be distorted intentionally and can influence feelings and opinions, it is important for students to read all graphs very carefully.

WARM-UP RESOURCES

 **NUMBER OF THE DAY**  Transparency 6.6

Start with your age. Multiply by 5 and add any number on a number cube, 1 to 6. Multiply by 2 and subtract the same number on the number cube. What do the digits in your answer tell you? **your age and the number on the number cube**

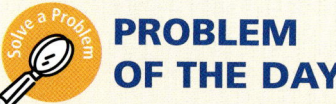  Transparency 6.6

A circle graph shows that half of the 120 ancestors of the students surveyed came to the U.S. from Europe or South America, a quarter came from Africa, and the rest from Asia and Australia. Five times as many came from Europe as from South America. How many of the students' ancestors came from South America? **10 ancestors**

Solution Problem of the Day tab, p. PD6

 **DAILY FACTS PRACTICE**

Have students practice addition and subtraction facts by completing Set F of *Teacher's Resource Book,* p. TR97.

132A Chapter 6

INTERVENTION AND EXTENSION RESOURCES

ALTERNATIVE TEACHING STRATEGY — ELL

Materials index cards

Strengthen students' understanding of biased questions. Write several biased survey questions, one to a card. Then, on other cards, write unbiased questions that can be paired with each of the biased questions. Examples:

Do you prefer the yummy taste of chocolate ice cream, or do you like vanilla, strawberry, or some other flavor?

What is your favorite ice cream: vanilla, chocolate, strawberry, or some other flavor?

Call on volunteers to read each pair of questions and to explain why one of each pair should elicit fairer responses than the other. Have students highlight the phrases that lead to a biased response, and discuss why they created bias. Check students' answers.

See also page 134.

AUDITORY
VERBAL/LINGUISTIC

MULTISTEP AND STRATEGY PROBLEMS

The following multistep or strategy problem is provided in Lesson 6.6:

Page	Item
135	18

WRITING IN MATHEMATICS

Materials *For each student* graph from a magazine or a newspaper

Find several bar graphs and line graphs in newspapers or magazines. Ask each student to choose a graph and write a few lines to describe what the graph shows.

Then have the student write a paragraph about the graph, describing why and how someone might want to change the graph so the appearance would be distorted. For example, break the scale to make it appear as though sales figures increased more than they did, or that the fat content in Product A is greater than in Product B.

Check students' answers.

EARLY FINISHERS

Combine pairs of students into groups of 4 to **explore the effect of biased questions on graphs.** Give each group a topic, such as their classmates' favorite classes. Assign one pair in each group to write a fair question and the other pair to write a question they feel is biased.

Each pair is assigned half the class to poll with their question.

Ask each group of students to display their data in two graphs. Post the graphs. Then have students compare and contrast the graphs to conclude whether the type of question asked affected the outcome. Check students' work.

VISUAL
INTERPERSONAL/SOCIAL

TECHNOLOGY LINK

- Intervention Strategies and Activities CD-ROM • *Skill 79*
- The Harcourt Learning Site
  www.harcourtschool.com

132B

LESSON 6.6 ORGANIZER

Objective To analyze data displays and determine how results and conclusions may have been influenced

1 Introduce

QUICK REVIEW provides review of mental math skills.

Why Learn This? You can examine a graph in a newspaper and decide whether it presents data fairly. *Share the lesson objective with students.*

2 Teach

Guided Instruction

- Examine the graph in Example 1.

  About how many people did Rita interview? about 120 people

  How could you reword the question so it would be less biased? Possible answer: Whom would you choose as the greatest U.S. President—Washington, Lincoln, or Jefferson?

- As you discuss the second example, ask:

  REASONING Can you be positive that Graph A goes with Question 2? Explain. No; it is possible that the biased question results in Graph B.

ADDITIONAL EXAMPLE

Example 2, p. 132

Match each graph with one of the questions. Explain your reasoning.

Question 1: After school would you rather do something fun like skateboard or would you rather bike or read?

Question 2: After school would you rather skateboard, bike, or read?

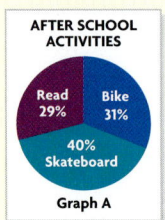
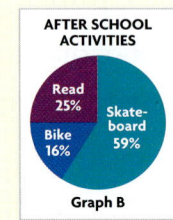

The more evenly divided Graph A probably goes with Question 2. The results in Graph B probably reflect the biased Question 1.

132 Chapter 6

LESSON 6.6

Analyze Graphs

Learn how to analyze data displays and determine how results and conclusions may have been influenced.

QUICK REVIEW

Write the corresponding decimal or percent.
1. 75%   2. 3%   3. 4.6%   4. 0.36   5. 0.07
   0.75   0.03   0.046   36%   7%

Data can be displayed in many different ways. Sometimes, the way a question is asked can influence the results that are displayed.

Rita took a survey asking the following question: Do you agree with me that George Washington was the greatest U.S. President, or would you choose Thomas Jefferson or Abraham Lincoln?

EXAMPLE 1

The results of Rita's survey are displayed in the bar graph shown at the right. Could the way the question was asked have influenced the results? Explain.

Yes. Rita's question is biased, since it leads people to agree with her that George Washington was the greatest U.S. President. As a result, the graph is misleading.

EXAMPLE 2

The results of two other surveys are shown below. Which graph is more likely to come from which question? Explain your reasoning.

Question 1: Would you rather visit the Grand Canyon, Mount Rushmore, or the Statue of Liberty?

Question 2: Would you rather visit the spectacular Grand Canyon, or would you rather visit Mount Rushmore or the Statue of Liberty?

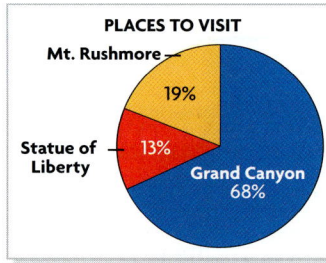
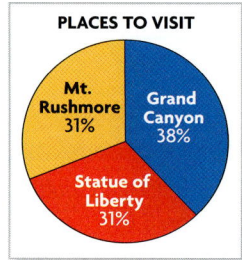

Graph A probably goes with Question 2, since the question is biased and leads people to choose the Grand Canyon. Graph B probably goes with Question 1, since that question is not biased.

132 Chapter 6

RETEACH 6.6

Analyze Graphs

Graphs are sometimes drawn in order to mislead the reader. In order for a graph to provide information honestly, it must meet several requirements. One requirement is that the scale must be accurate.

Stereo City placed an advertisement for a stereo system in a newspaper. The advertisement included the graph at the right, which compares its price for the system to the price of the system at Speaker Town.

In the graph, the bar for Speaker Town appears to be twice the height of the bar for Stereo City. Some readers might think this means Speaker Town's price is twice that of Stereo City. The scale, however, shows that Speaker Town's price for the stereo system is not twice the price at Stereo City. The actual difference between the two prices is only $380 − $340, or $40.

In order to give a true representation of a set of data, any scale used on a graph should follow these rules:
- The scale should begin with zero.
- An interval that makes sense for the data that is shown in the graph should be chosen.
- Every interval on the scale must be the same size.

A consumer research company conducted a survey at both Stereo City and Speaker Town. At each store, customers who had just purchased an item were asked if they would return to the store for their next electronics purchase. The results of the survey are shown in the graph.

1. What percent of Stereo City's customers said they would make their next electronics purchase there? **95%**

2. What percent of Speaker Town's customers said they would make their next electronics purchase there? **80%**

3. About how many times as high is the bar for Stereo City as the bar for Speaker Town? **about 4 times as high**

4. How can you change the graph so that it is not misleading? **Start the scale at zero and have equal intervals.**

PRACTICE 6.6

Analyze Graphs

Renee asked each student in her math class the following question: "Would you rather have some great vanilla ice cream or would you prefer chocolate or strawberry?"

For 1–2, use the graph at the right, which shows the results of her survey.

1. Could the way Renee asked the question have influenced her classmates' answers? Explain. **Yes. The question is biased and could lead people to choose vanilla ice cream.**

2. Tell how you could rewrite the question so it would not influence the results of the survey. **Possible answer: "Which ice cream flavor do you prefer, chocolate, strawberry, or vanilla?"**

A television network used the graph at the right. The network wanted to convince viewers that one of its shows, Show A, was far more popular than one of its competitors' shows, Show B, which airs at the same time.

3. The bar for Show A is about how many times as high as the bar for Show B? **about twice as high**

4. Does twice the percent of the viewing audience watch Show A as watches Show B? **no**

5. How can you change the graph so that it is not misleading? **Adjust the scale to start at zero and have equal intervals.**

Mixed Review

During one day at an airport, an airline experienced flight delays of the following numbers of minutes: 5, 7, 5, 10, 15, 15, 20, 91.

6. Find the mean length of all the flight delays. **21 min**

7. Find the mean of the delays if the outlier is not included. **11 min**

Evaluate each expression.

8. $g + 1.7$ for $g = 3.3$ **5**
9. $5y$ for $y = 1.8$ **9**
10. $p − 4.9$ for $p = 11$ **6.1**

Graphs can communicate information quickly. That's why they are used by advertisers on television and in magazines and newspapers. Some graphs can be misleading and influence conclusions that are drawn.

EXAMPLE 3

Jon looked at the bar graph below and concluded that the Mississippi River is twice as long as the Missouri River. Explain Jon's mistake and tell why his conclusion is wrong.

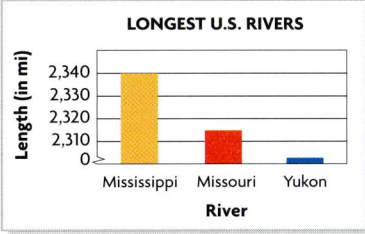

The bar for the Mississippi is twice as long as the bar for the Missouri. However, if you look at the scale, you see that the rivers are about the same length. Because the lower part of the scale is missing, the differences are exaggerated.

When two graphs with different scales show two similar sets of data, comparing the graphs can sometimes be misleading.

EXAMPLE 4

The weekly ticket sales for the Mississippi River tour boat cruise are shown in the graphs below. Deb looked at the graphs and concluded that more tickets were sold in April than in March. Explain Deb's mistake.

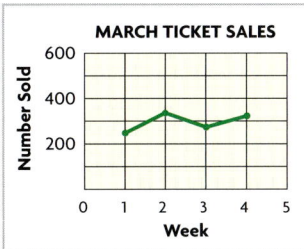 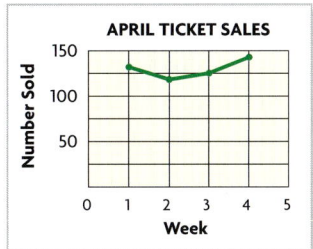

It appears that the April ticket sales were greater than in March since the line for April is higher than the line for March. However, if you look at the scale for each graph, you can see that ticket sales were much greater in March than in April.

1. Possible answer: Do you think George Washington, Thomas Jefferson, or Abraham Lincoln was the greatest U.S. President?

CHECK FOR UNDERSTANDING

Think and Discuss

Look back at the lesson to answer each question.
1. **Tell** how you could rewrite the question in Example 1 so it would not influence the results of the survey. *See above left.*

133

- The visual picture of the data can be misleading with a break (zigzag) in the scale, as in Example 3, or to vary the scales, as in Example 4. Ask:

  What is the actual difference in length of the Mississippi and Missouri Rivers? a little more than 20 mi out of 2,300 mi

  Where would the April ticket sales appear if graphed with the March sales? All April values would be below the March values.

ADDITIONAL EXAMPLES

Example 3, p. 133

Emily looked at the bar graph and concluded that only half as many seventh-grade students participated in the book fair as sixth-grade students. What part of the graph's construction influenced Emily's conclusion?

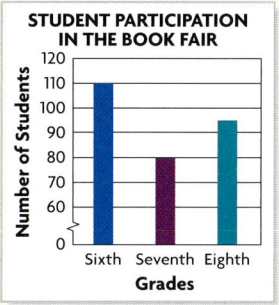

The scale in the number of students exaggerates the difference in student participation.

Example 4, p. 133

These bar graphs show the prices of bicycles sold at two different stores. How are these graphs misleading?

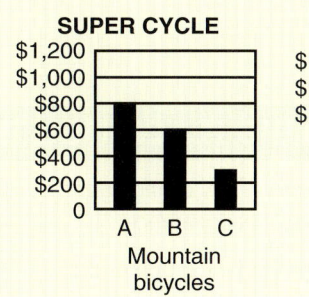 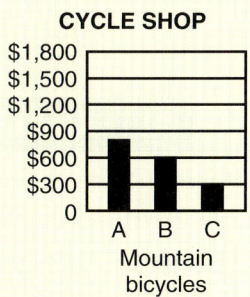

The graphs make it appear that the bicycle prices at Super Cycle are greater than those at Cycle Shop. However, the bicycle prices are actually the same at both stores.

CHALLENGE 6.6

Scale the Heights

For each set of data, assume that you are going to draw a bar graph. Describe the scale you would use for the vertical axis. Include the minimum value, the maximum value, and the size of each interval. *Possible answers are given.*

1. Maximum animal speeds: zebra, 40 mi per hr; lion, 50 mi per hr; grizzly bear, 30 mi per hr; elephant, 25 mi per hr
   minimum: 0; maximum: 50; interval size: 10

2. Average animal life spans: horse, 20 years; leopard, 12 years; Asian elephant, 40 years; rabbit, 5 years
   minimum: 0; maximum: 50; interval size: 5

3. Countries with the fewest people: Nauru, 10,605; Palau, 18,467; San Marino, 25,061; Tuvalu, 10,588
   minimum: 0; maximum: 30,000; interval size: 5,000

4. Professional basketball games won during the 1998–1999 season: Miami Heat, 33; New Jersey Nets, 16; Orlando Magic, 33; Philadelphia 76ers, 28; Washington Wizards, 18
   minimum: 0; maximum: 35; interval size: 5

5. Miles of border shared with the United States: Mexico, 1,933 mi; Pacific Ocean, 7,623 mi; Atlantic Ocean, 2,069 mi; Gulf of Mexico, 1,631 mi
   minimum: 0; maximum: 8,000; interval size: 1,000

6. Number of members of the U.S. House of Representatives: New York, 31; California, 52; Rhode Island, 2; Florida, 23
   minimum: 0; maximum: 60; interval size: 4

7. Warmest temperatures ever recorded, by continent: Africa, 136°F; Asia, 129°F; North America, 134°F; Antarctica, 59°F
   minimum: 0; maximum: 140; interval size: 10

8. Highest point on the continent: North America, 20,320 ft; Australia, 7,310 ft; Asia, 29,035 ft; Europe, 18,510 ft; South America, 22,834 ft
   minimum: 0; maximum: 30,000; interval size: 2,000

9. Most widely used languages in the world (in millions of speakers): Mandarin, 1,075; English, 514; Hindi, 496
   minimum: 0; maximum: 1,200; interval size: 100

PROBLEM SOLVING 6.6

Analyze Graphs

Write the correct answer. Use the graph below for 1–3.

LAND AREAS

Analyze Choose Solve Check

1. After looking at the graph, Casey decided that the area of Argentina was about three times the area of Mexico. Explain why Casey's conclusion is wrong.
   The scale does not begin at zero.

2. Explain how the graph could be fixed so that Casey would not have made the mistake he did.
   Start the vertical scale at zero and make every interval the same size.

3. Use the graph to estimate the total combined area of Mexico and Argentina.
   Possible answer: approximately 1.8 million square miles

Write the letter of the best answer. Use the graph below for 4–6.

DOMINIQUE'S DOG

4. During which of these times did Dominique's dog gain the least amount of weight?
   F from January to February
   G from February to March
   H from March to April
   J from April to May

5. If the scale started at 0 and ended at 60, with intervals of 2, how would the appearance of the graph change?
   A The line would be steeper.
   B The line would be flatter.
   C The line would look the same as it does now.
   D The line would be a straight line.

6. If the scale began at 0, which interval would make Dominique's dog's weight gain seem the greatest?
   F an interval of 2 lb
   G an interval of 5 lb
   H an interval of 10 lb
   J an interval of 15 lb

7. **Write About It** Why does increasing the size of the interval used in the vertical scale of a line graph make the line seem flatter?
   Possible answer: As the interval increases, the range appears to be more compact.

3 Practice

Guided Practice

Do Check for Understanding Exercises 1–6 with your students. Identify those having difficulty and use lesson resources to help.

Independent Practice

Note that Exercise 18 is a **multistep or strategy problem**. Assign Exercises 7–13.

MIXED REVIEW AND TEST PREP
Exercises 14–18 provide **cumulative review** (Chapters 1–6).

Multistep or Strategy Problem To solve Exercise 18, students can work operations in parentheses and then try the different operation signs in the gray circle.

---

2. **Explain** how you could change the graph in Example 3 so it is not misleading. Adjust the scale to include numbers between 0 and 2,300.

Guided Practice

Rosa took a survey, asking the following question: Don't you think that the Mustangs are the best baseball team, or would you choose the Wildcats or Cougars? The results of her survey are displayed in the bar graph at the right.

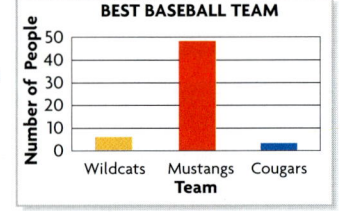

3. Yes. The question is biased and could lead people to choose the Mustangs as the best baseball team.

3. Could the way the question was asked have influenced the results? Explain. See above left.

For 4–6, use the graph at the right. The graph is misleading.

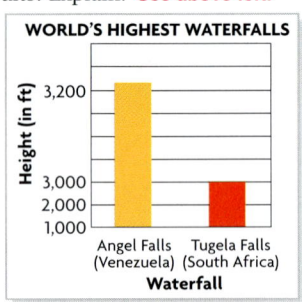

4. About how many times as high is the bar for Angel Falls than for Tugela Falls? about 3 times as high

5. Is Angel Falls 3 times as high as Tugela Falls? Explain. See above left.

5. No. Angel Falls is about 3,200 ft high, while Tugela Falls is about 3,000 ft high.

6. How could you change the graph so it is not misleading? See below left.

6. Adjust the scale to start at zero and have equal intervals.

PRACTICE AND PROBLEM SOLVING

Independent Practice

Miguel took a survey, asking the following question: What is your favorite fruit—apples, bananas, or delicious, juicy Florida navel oranges? The results of Miguel's survey are displayed in the circle graph.

7. Yes. The question is biased and could lead people to choose oranges.

7. Could the way the question was asked have influenced the results? Explain.

For 8–10, use the graph at the right. The graph is misleading.

8. About how many times as high is the bar for Brand B than for Brand A? about 2 times as high

9. Does Brand B cost twice as much as Brand A? Explain. No. Brand B costs $30 and Brand A costs $20.

10. How can you change the graph so it is not misleading? Adjust the scale so the intervals are equal.

134 Chapter 6

ALTERNATIVE TEACHING STRATEGY | SCAFFOLDED INSTRUCTION

Purpose Students explore how changing the scale on a graph affects its appearance.

Materials For each group 1-in. graph paper, p. TR62

Present these data for lengths of the rivers given in Example 3: Mississippi—2,340 mi; Missouri—2,315 mi; Yukon—1,979 mi.

Divide the class into small groups. Have each group draw two different graphs presenting the same information. The first graph uses a scale with intervals of 500 from zero to 2,500.

For the second graph, have them use a scale with intervals of 100 from 1,900 to 2,400, so that part of the scale is missing.

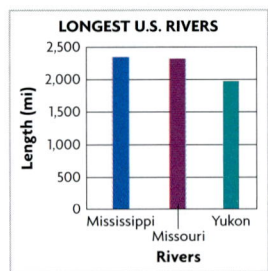

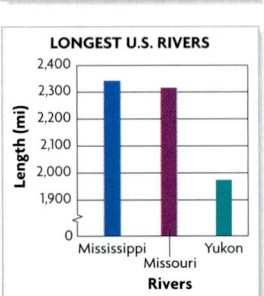

Have each group write a statement comparing the two graphs. Then as a class discuss the different impression each graph gives about the lengths of the rivers. Possible answer: The 1st graph with the scale beginning at zero gives a truer picture of the actual data than the 2nd graph with the scale break. The 1st graph shows that the Mississippi and Missouri Rivers are much closer in length than they appear in the 2nd graph. In the 2nd graph, the Yukon River appears to be much shorter in comparison to the other two rivers than it does in the 1st graph.

134 Chapter 6

Problem Solving Applications

11. Jeff compared the bars but did not look at the scale.

13. Lin looked only at the lines and did not look at the scales.

Boston, Massachusetts

11. Jeff looked at the bar graph at the right and concluded that Los Angeles has three times the population of Chicago. Explain Jeff's mistake and tell why his conclusion is wrong.

12. How could you fix the graph so Jeff would not make a mistake? Adjust the scale to include zero and have equal intervals.

13. Lin looked at the graphs below and concluded that from June to September, the temperatures in Seattle are about the same as the temperatures in Boston. Explain Lin's mistake. See above left.

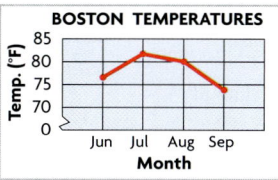
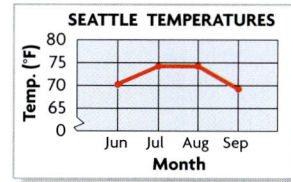

MIXED REVIEW AND TEST PREP

For 14–15, use the box-and-whisker graph. (p. 129)

14. What is the median? 50

15. What are the least and greatest values? 43 and 55

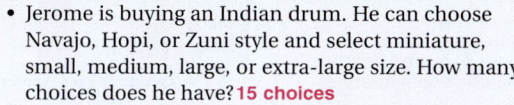

16. Find the mean of the data set 18, 12, 10, 8, 9, 14, and 6. (p. 106) 11

17. **TEST PREP** Which is the value of $2n + 3.7$ for $n = 2.9$? (p. 82) B
   A 8.6   B 9.5   C 10.4   D 21.46

18. **TEST PREP** Replace ● with the missing operation. $(3 + 24) ● 3 \times 2 - 1 = 17$ (p. 44) G
   F ×   G ÷   H +   J −

PROBLEM SOLVING LINKUP to Reading

Strategy • Classify and Categorize
To classify information means to group together similar information. To categorize the information, label the groups. Paige wants to buy a piece of Indian pottery. She can choose Navajo, Hopi, or Zuni style and select small, medium, or large size. By classifying and categorizing the data, you can see that she has 9 different choices.

Style	Size
Navajo	small, medium, large
Hopi	small, medium, large
Zuni	small, medium, large

- Jerome is buying an Indian drum. He can choose Navajo, Hopi, or Zuni style and select miniature, small, medium, large, or extra-large size. How many choices does he have? 15 choices

EXTRA PRACTICE page H37, Set E

135

LINKUP to READING

- After students examine the photograph, encourage them to discuss ways that they could categorize the clothing they wear. Then ask:

How do department stores categorize clothing? Possible answer: by type of garment, for example, shirt, pants, shoes; by style of garment, for example, casual, formal, work.

REASONING What are some ways that students are categorized in your school?
Possible answer: By grade, by boys and girls, by lunch periods, by homeroom teacher.

4 Assess

Summarize the lesson by having students:

DISCUSS Why might someone use some of the techniques illustrated in this lesson to display data? Possible answer: in order to sell a product, sway an audience, or make a point

WRITE Explain how you would examine a graph for bias. Possible answer: Check the scale to see if part is missing or if the intervals are uneven; if available, see if the survey question is biased.

Lesson Quiz
Transparency 6.6

1. Martin asked, "What is your favorite 60's group, the Beatles, the Drifters, or the fabulous Beach Boys?" Do you think his graph will be fair or biased? Explain. Possibly biased; his question favors one group.

Use the graph at the right for 2–3.

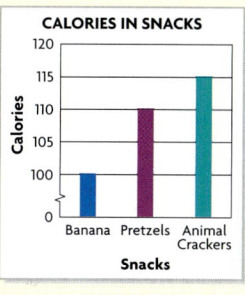

2. Sheri looked at the graph and concluded that a serving of animal crackers has four times as many calories as a banana. Explain her mistake.
The break in the scale exaggerates the difference in calories per serving.

3. How many more calories are in a serving of animal crackers than in a banana? only 15

135

CHAPTER 6

REVIEW/TEST

Purpose To check understanding of concepts, skills, and problem solving presented in Chapter 6

USING THE PAGE

The Chapter 6 Review/Test can be used as a **review** or a **test**.

- Items 1–2 check understanding of concepts and new vocabulary.
- Items 3–20 check skill proficiency.

 Suggest that students place the completed Chapter 6 Review/Test in their portfolios.

USING THE ASSESSMENT GUIDE

- Multiple-choice format of Chapter 6 Posttest—See *Assessment Guide*, pp. AG37–38.
- Free-response format of Chapter 6 Posttest—See *Assessment Guide*, pp. AG39–40.

USING STUDENT SELF-ASSESSMENT

The How Did I Do? survey helps students assess what they have learned and how they learned it. This survey is available as a copying master in *Assessment Guide*, p. AGxvii.

CHAPTER 6 REVIEW/TEST

1. **VOCABULARY** A bar graph that shows frequencies within intervals is a(n) ? . (p. 127) **histogram**

2. **VOCABULARY** A graph that shows how far apart and how evenly data are distributed is a(n) ? . (p. 129) **box-and-whisker graph**

3. What type of graph would best show the highest and lowest temperatures for each of the last four years? (pp. 120–123) **multiple-bar graph**

4. What type of graph would best show high and low temperatures for a week? (pp. 120–123) **multiple-line graph**

5. Make a double-line graph with the data at the right. (pp. 120–123)
   See Additional Answers, p. 137A.

6. Make a double-bar graph with the data at the right. (pp. 120–123)
   See Additional Answers, p. 137A.

END-OF-MONTH STOCK PRICES

	Sep	Oct	Nov	Dec	Jan
Stock A	$80	$74	$45	$50	$52
Stock B	$50	$52	$52	$50	$45

For 7–9, use the graph at the right. (pp. 124–125)

7. About how much did the stock price decrease from Monday to Tuesday? **about $2.50**

8. Describe the pattern in the graph. **Stock prices are falling.**

9. If the trend continues, what do you think the stock price will be on Friday? **Possible answer: about $10**

10. Make a stem-and-leaf plot for the data. (pp. 126–128) **See Additional Answers, p. 137A.**

11. Make a histogram for the data. (pp. 126–128) **See Additional Answers, p. 137A.**

POINTS SCORED

| 33 | 52 | 45 | 47 | 34 | 52 |
| 34 | 58 | 48 | 52 | 46 | 59 |

HEIGHTS OF BUILDINGS (IN FT)

| 20 | 50 | 80 | 20 | 40 | 45 | 85 |
| 25 | 30 | 80 | 60 | 70 | 75 | 55 |

For 12–17, use the following data: 14, 16, 9, 21, 35, 2, 26, 8, 17. (pp. 129–131)

12. Find the upper extreme. **35**
13. Find the lower extreme. **2**
14. Find the upper quartile. **23.5**
15. Find the lower quartile. **8.5**
16. Find the median. **16**
17. Make a box-and-whisker graph. **See Additional Answers, p. 137A.**

For 18–20, use the graph at the right. (pp. 132–135)

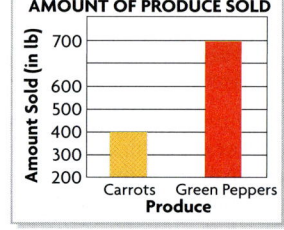

18. About how many times as high is the bar for green peppers as the bar for carrots? **about 3 times as high**

19. Were three times as many pounds of carrots sold as pounds of green peppers? Explain. **No. 700 pounds of carrots and 400 pounds of green peppers were sold.**

20. How could you change the graph so that it is not misleading? **Adjust the scale to start at zero and to include equal intervals.**

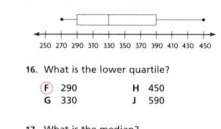
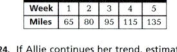

CHAPTERS 1–6 ★ STANDARDIZED TEST PREP

Check your work.
See item **5**.
Think about the different kinds of graphs and the most appropriate data for them. Check that your answer choice could be made into a bar graph.

Also see problem **7**, p. H65.

Choose the best answer.

1. The owner of a music shop made a line graph to show the number of CDs sold during a 4-week period. Which trend does the graph show? **C**

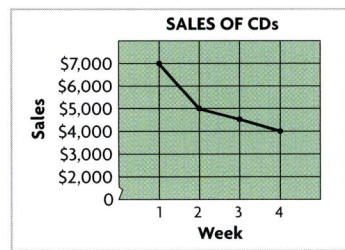

- **A** Sales are increasing.
- **B** Sales are even.
- **C** Sales are decreasing.
- **D** No trend is shown.

2. $22.8 - 3.11$ **G**

- **F** 18.79
- **G** 19.69
- **H** 19.79
- **J** Not here

3. Which of the following is best suited for display in a multiple-bar graph? **C**

- **A** Average temperatures recorded in a town during a 1-year period
- **B** Number of CDs owned by sixth-grade students
- **C** Number of hours sixth- and seventh-grade students spend reading each week
- **D** Changes in water temperature over a 24-hour period

4. Which kind of graph is shown? **G**

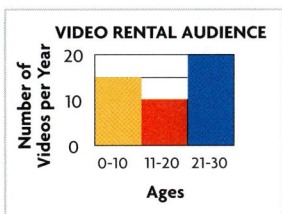

- **F** Circle graph
- **G** Histogram
- **H** Line graph
- **J** Stem-and-leaf plot

5. Which of the following is best suited for display in a bar graph? **A**

- **A** Heights of the tallest buildings in a city
- **B** Frequency of cars stopping at a tollbooth
- **C** Changes in a person's weight over a year
- **D** Part of a day a person spends reading

Write What You Know

See below.

6. Suppose you wanted to show the favorite sports of a sample of teenagers. Would you use a line graph, bar graph, histogram, or stem-and-leaf plot? Justify your choice.

7. Explain how you would find the value of $\frac{12 - 2^3}{4}$. Then find the value.

Write What You Know • Written Response

6. Possible answer: bar graph. This is a good choice because the data will be grouped into categories (different sports), each of which can be assigned its own bar. If you want to compare parts to each other or to the whole, a circle graph is also a good choice.

7. Using the order of operations, find the value of 23, subtract the result from 12, and then divide the result by 4; 1.

STANDARDIZED TEST PREP •
Chapters 1–6

USING THE PAGE

This page may be used to help students get ready for standardized tests. The test items are written in the same style and arranged in the same format as those on many state assessments. The page is cumulative. It covers math objectives and essential skills that have been taught up to this point in the text. Most of the items represent skills from the current chapter, and the remainder represent skills from earlier chapters.

This page can be assigned at the end of the chapter as classwork or as a homework assignment. You may want to have students use individual recording sheets presented in a multiple-choice (standardized) format. A Test Answer Sheet is available as a blackline master in *Assessment Guide* (p. AGxlii).

You may wish to have students describe how they solved each problem and share their solutions.

ITEM ANALYSIS

Item	Learning Goal	Item	Learning Goal
1	6A	5	4A
2	6A	6	6A
3	6A	7	2C
4	6A		

Written response items for the **Write What You Know** are available as a blackline master in *Performance Assessment*.

SCORING RUBRIC • WRITE WHAT YOU KNOW

2 Demonstrates a complete understanding of the problem and chooses an appropriate strategy to determine the solution

1 Demonstrates a partial understanding of the problem and chooses a strategy that does not lead to a complete and accurate solution

0 Demonstrates little understanding of the problem and shows little evidence of using any strategy to determine a solution

Graph Data

PUPIL EDITION
ADDITIONAL ANSWERS

Lesson 6.1, page 123

9.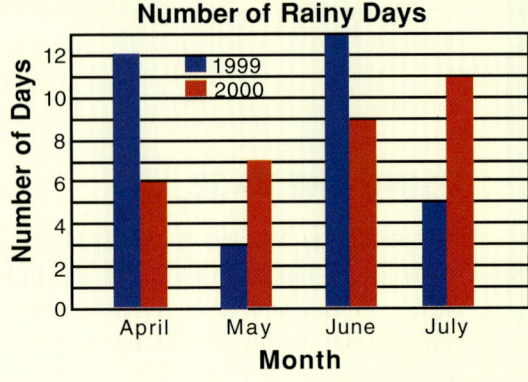

Lesson 6.2, page 125

8.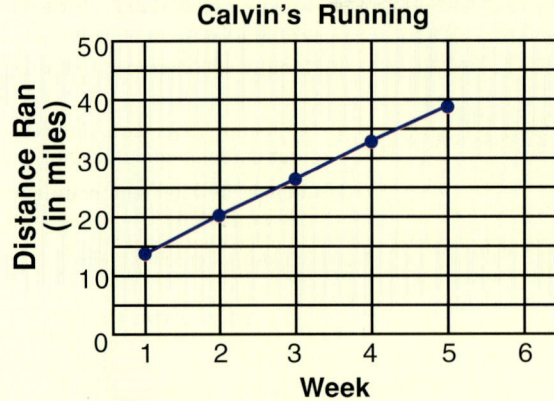

10.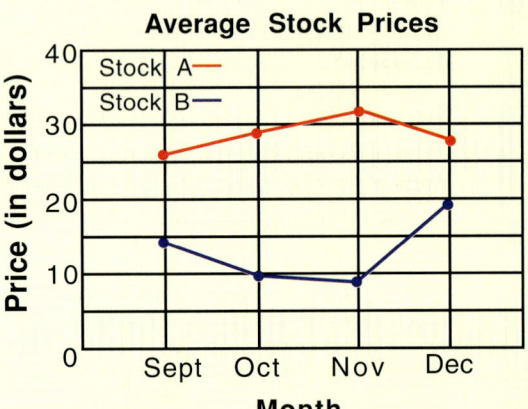

Lesson 6.3, page 128

10.

Lesson 6.2, page 125

2.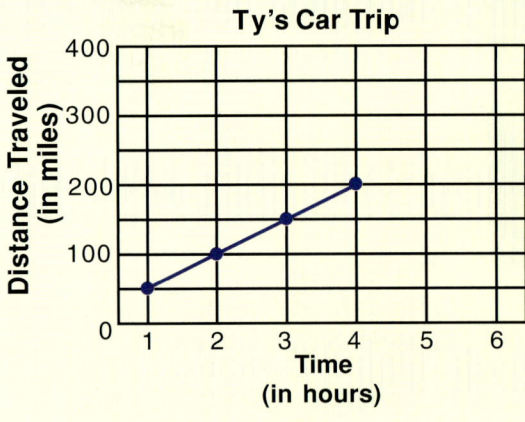

Lesson 6.4, page 129

1.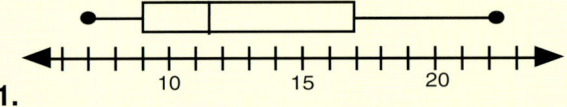

Lesson 6.2, page 125

5.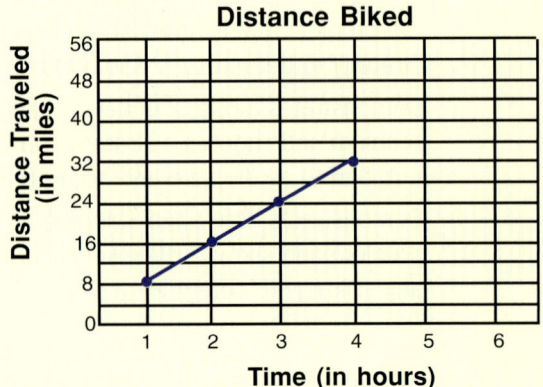

137A   Chapter 6

PUPIL EDITION
ADDITIONAL ANSWERS

Lesson 6.5, page 131

8.

Possible Histogram:

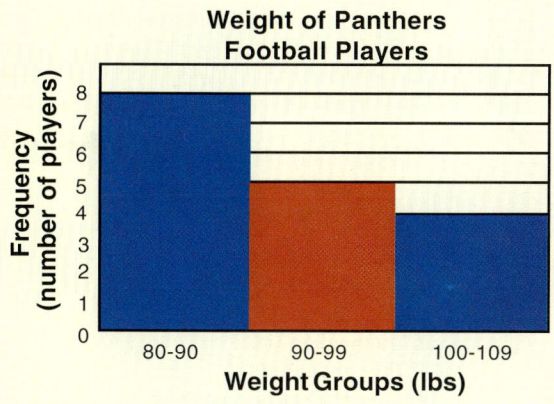

11.

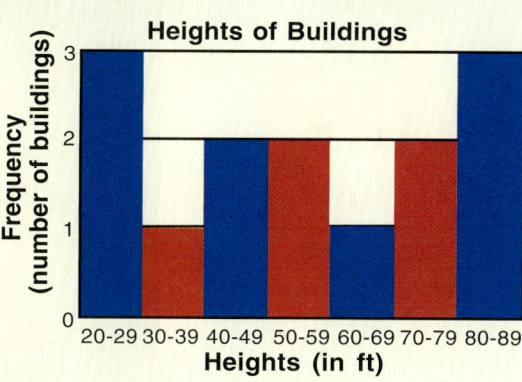

Chapter 6 Review/Test, page 136

5.

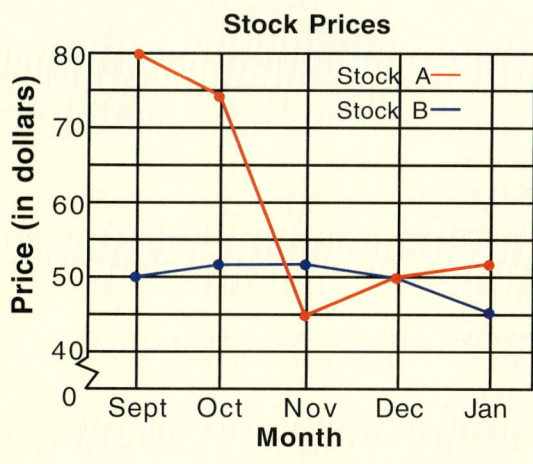

17.

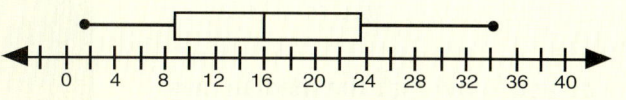

6.

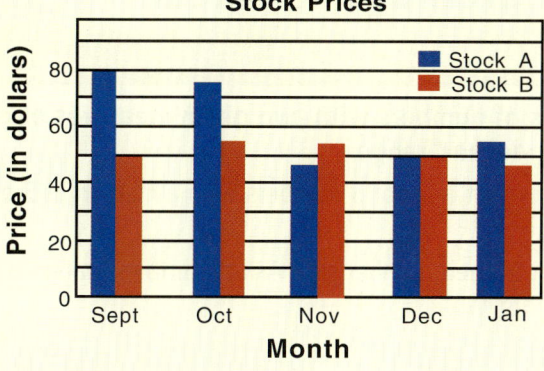

10.
Stem	Leaves
3	3 4 4
4	5 6 7 8
5	2 2 2 8 9

137B

UNIT 2

MATH DETECTIVE
Measure by Measure

Purpose To use deductive reasoning to solve problems involving measures of central tendency

USING THE PAGE

- *Direct students' attention to the Reasoning section and Mystery Number 1.*

  What two numbers can you immediately identify from the clues? Explain. Since the mode is 8 and there are only three numbers in the set with a mean that is not 8, two of the numbers must be 8.

  How would you find the third missing number in the set? Possible answer: Write and solve an equation or predict and test until you find a third number in the set that will yield a mean of 10.

- *Have students read the clues for Mystery Number 2.*

  How do you know that the median must be one of the numbers in the set? There is an odd number (3) of numbers in the set.

  How can you use the range to find the third number in the set? Explain. Add, 16 + 5 = 21. Five must be the least number in the set since the median is 9 and there are only three numbers.

- *After students solve Mystery Number 3, have them explain their thinking.*

  How do you know that the median is not one of the numbers in the set? There are an even number (4) of numbers in the set, so the median must be the average of the middle two numbers in the set.

  Think It Over! After students complete the Write About It, have them explain how they used the clues to find each of the Mystery Number sets. Encourage students to compare and contrast their solution methods.

138 Unit 2 • Chapters 5–6

PROBLEM SOLVING
MATH DETECTIVE

Measure by Measure

REASONING Use the clues and your knowledge of the measures of central tendency to find the set of numbers described. Be prepared to explain how you solved the mystery.

Mystery Number 1
Clues:
1. There are three numbers in the set.
2. The mean is 10.
3. The mode is 8.
What is the set of numbers? 8, 8, 14

Mystery Number 2
Clues:
1. There are three numbers in the set.
2. One number in the set is 5.
3. The median is 9.
4. The range is 16.
What is the set of numbers? 5, 9, 21

Mystery Number 3
Clues:
1. There are four numbers in the set.
2. The median is 12.
3. Two numbers in the set are 3 and 10.
4. The range is 15.
What is the set of numbers? 3, 10, 14, 18

Think It Over!

-  **Write About It** If you know that the sum of a set of six numbers is 150, which of the following measures could you find: mean, median, mode, range? Explain. See above.

  You can find the mean by dividing 150 ÷ 6 = 25. To find the median, mode, or range, you need more information.

- **Stretch Your Thinking** A set of numbers forms the pattern 1, 3, 5, 7, ... If the median of the set is 15, what is the range? 28

138 Unit 2

Intervention and Extension Resources

LANGUAGE ARTS CONNECTION

Ask students to **construct sets of numbers with two given measures of central tendency: mean, median, mode, and range.**

For example, have students construct a set of 5 numbers with a median of 5 and a mean of 6. Possible answer: 2, 4, 5, 9, 10

Then have them write clues about their number sets. Ask them to exchange with a partner and solve. Check students' work.

VISUAL
LOGICAL/MATHEMATICAL

Challenge: Explore Scatterplots

Learn how to read and interpret a scatterplot.

Do you think there is any relationship between the number of people who paint a large building and the number of hours it takes them to finish the job?

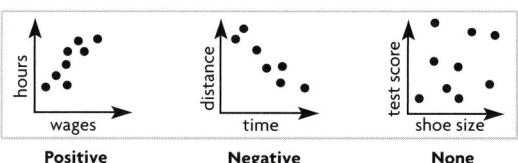

The *scatterplot* displays data for 11 large buildings that were painted. It shows that as the number of painters increased, the time it took them to finish the job tended to decrease.

A scatterplot shows the relationship between two variables.

Positive	Negative	None

When the values of the two variables increase or decrease together, there is a **positive correlation**.

When the values of one variable increase while the others decrease, there is a **negative correlation**.

When the data points show no pattern of increase or decrease, there is **no correlation**.

TALK ABOUT IT

- Tell if the relationship between the speed of a car and the number of hours needed to drive 500 miles has a positive correlation, a negative correlation, or no correlation. **negative correlation**

TRY IT

Sketch a scatterplot that could represent the situation. Then identify the type of correlation between the variables.
Check students' scatterplots.

1. positive
2. none

1. amount of time walking *and* total distance that you walk
2. number of rooms in house *and* street address of house

Write *positive correlation*, *negative correlation*, or *no correlation* to describe the relationship shown in the scatterplot.

3. 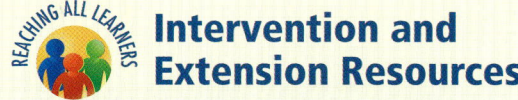 **negative correlation**   4. **positive correlation**

Chapters 5–6 **139**

UNIT 2 CHAPTERS 5–6

STUDY GUIDE AND REVIEW

Purpose To help students review concepts and skills presented in Chapters 5–6

USING THE PAGES

✓ Assessment Checkpoint

The Study Guide and Review includes content from Chapters 5–6.

Chapter 5
- 5.1 Samples
- 5.2 Bias in Surveys
- 5.3 Problem Solving Strategy: *Make a Table*
- 5.4 Frequency Tables and Line Plots
- 5.5 Measures of Central Tendency
- 5.6 Outliers and Additional Data
- 5.7 Data and Conclusions

Chapter 6
- 6.1 Make and Analyze Graphs
- 6.2 Find Unknown Values
- 6.3 Stem-and-Leaf Plots and Histograms
- 6.4 Math Lab: Explore Box-and-Whisker Graphs
- 6.5 Box-and-Whisker Graphs
- 6.6 Analyze Graphs

The blue page numbers in parentheses provided with each group of exercises indicate the pages on which the concept or skill was presented. The red number given with each group of exercises identifies the Learning Goal for the concept or skill.

140 Unit 2 • Chapters 5–6

UNIT 2 Study Guide and Review

VOCABULARY

1. Everyone in the population has the same chance of being selected in a(n) __?__. (p. 95) **random sample**
2. A bar graph that shows the frequency at which data occur within intervals is a(n) __?__. (p. 127) **histogram**
3. Individuals in the population are not represented in the sample if the sample is __?__. (p. 98) **biased**

EXAMPLES

Chapter 5

- **Identify the type of sample.** (pp. 94–97) **5A**

  Determine the sampling method used if workers on an assembly line check every tenth tire.

  This is a systematic sample.

- **Record and organize data.** (pp. 102–105) **5B**

  What is the size of the sample?

  SCORES ON MATH TEST

Score Interval	Frequency	Cumulative Frequency
91–100	5	5
81–90	8	13
71–80	6	19
Below 71	5	24

  There are 24 people in the sample.

- **Find the mean, median, and mode.** (pp. 106–111) **5C**

  24, 20, 24, 21, 26
  Mean: 24 + 20 + 24 + 21 + 26 = 115; 115 ÷ 5 = **23**

  Median: **24**

  Mode: **24**

EXERCISES

Determine the type of sample. Write *convenience*, *random*, or *systematic*.

4. From a computer list of students, each with an equal chance of being selected, 100 students are chosen. **random**
5. Every tenth person walking down the street is surveyed about the President. **systematic**

For 6–7, use the following data.

NUMBER OF MOVIES SEEN IN ONE YEAR

8	12	16	12
9	10	9	8
9	15	20	14
12	7	9	15

6. Make a line plot. **Check students' plots.**
7. Make a cumulative frequency table with intervals. **Check students' tables.**

Find the mean, median, and mode.

8. 2, 7, 9, 4, 6, 8, 6 **6; 6; 6**
9. 5.3, 8.8, 4.7, 6.5, 4.7 **6; 5.3; 4.7**
10. 79, 87, 90, 100, 96, 89, 92, 87 **90; 89.5; 87**
11. Suppose you added the number 10.6 to the data in Exercise 9. Which measure(s) of central tendency would change? **mean and median**

140 Unit 2

Additional Answers, Pupil Edition page 141

15. **Ages**

Stems	Leaves
1	0 0 1 3 4 4 5 5 8 8 9
2	0 1 2

20.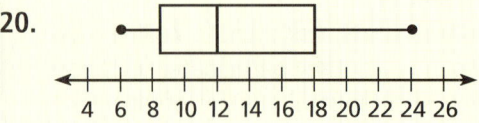

Chapter 6

- **Display data in graphs.** (pp. 120–123) **6A**

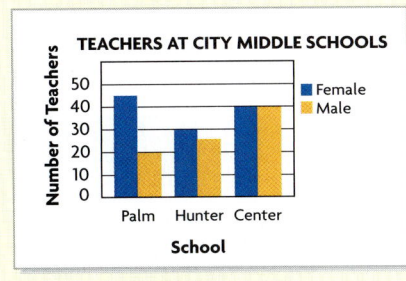

For 12, use the graph at the left.

12. Which schools have more female teachers than male teachers? **Palm and Hunter**
13. Would you use a bar graph, line graph, or circle graph to display a city's temperature readings for 1 month? Explain. **line; shows change over time**
14. Make a multiple-line graph with the data below. **Check students' graphs.**

HIGH AND LOW TEMPERATURES

	Mon	Tues	Wed	Thurs	Fri
Highs	45°	53°	41°	48°	50°
Lows	34°	39°	35°	40°	41°

- **Use stem-and-leaf plots and histograms.** (pp. 126–128) **6A**

In a stem-and-leaf plot, the tens digits of the data are stems, and the ones digits are leaves.

Ages

Stem	Leaves
1	1 3 3
2	0 3 5 5
3	7 7 8

Some students are participating in a jump-rope benefit. Their ages are 10, 15, 18, 20, 15, 11, 13, 10, 14, 14, 18, 19, 22, and 21.

15. Use the data to make a stem-and-leaf plot. **See Additional Answers, page 140.**
16. Use the data to make a histogram. **Check students' graphs.**

- **Make a box-and-whisker graph.** (pp. 129–131) **6A**

Heights of Students (in cm)
135 168 148 160 159 148 163 165 167

Draw a box-and-whisker graph.

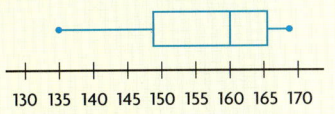

HEIGHTS OF TREES (in ft)

| 8 | 12 | 6 | 9 | 15 | 9 | 16 | 20 | 24 |

For 17–20, use the table below.

17. What is the median? **12 ft**
18. What are the lower and upper quartiles? **8.5 ft, 18 ft**
19. What are the lower and upper extremes? **6 ft, 24 ft**
20. Make a box-and-whisker graph. **See Additional Answers, page 140.**

PROBLEM SOLVING APPLICATIONS

21. Roberto has to pay $27 for his CDs at the record store. In how many ways can he pay, using only bills of $10, $5, and $1? (pp. 100–101) **12 ways**
22. The sixth-grade students are having a car wash. They charge $4 for cars and $6 for SUVs. In how many ways can they earn exactly $100? (pp. 100–101) **9 ways**
23. The machines in the exact change lane of a tollway accept any combination of coins that total exactly 75¢, but they do not accept pennies or half dollars. In how many different ways can a driver pay the toll in an exact change lane? (pp. 100–101) **18 ways**
24. The debate club has 10 members. Each member will debate each of the other members only once. How many debates will they have? (pp. 100–101) **45 debates**

Assessment Checkpoint

Portfolio Suggestions The portfolio represents the growth, talents, achievements, and reflections of the mathematics learner. Students might spend a short time selecting work samples for their portfolios and completing A Guide to My Math Portfolio from *Assessment Guide*, page AGxix.

You may want to have students respond to the following questions:

- What new understanding of math have I developed in the past several weeks?
- What growth in understanding or skills can I see in my work?
- What can I do to improve my understanding of math ideas?
- What would I like to learn more about?

For information about how to organize, share, and evaluate portfolios, see *Assessment Guide*, page AGxviii.

Use the item analysis in the **Intervention** chart to diagnose students' errors. You may wish to reinforce content or remediate misunderstandings by using the text pages or lesson resources.

STUDY GUIDE AND REVIEW INTERVENTION

How to Help Options

Learning Goal		Items	Text Pages	Reteach and Practice Resources
5A	See page 92C for Chapter 5 learning goals	4–5	98–99	Worksheets for Lesson 5.2
5B	See page 92C for Chapter 5 learning goals	6–7	102–105	Worksheets for Lesson 5.4
5C	See page 92C for Chapter 5 learning goals	8–11	106–108, 109–111	Worksheets for Lessons 5.5, 5.6
5D	See page 92C for Chapter 5 learning goals	21–24	100–101	Worksheets for Lesson 5.3
6A	See page 92C for Chapter 6 learning goals	12–14, 15–16, 17–20	120–123, 126–128, 129–131	Worksheets for Lessons 6.1, 6.3, 6.4, 6.5

UNIT 2

PERFORMANCE ASSESSMENT

Purpose To provide performance assessment tasks for Chapters 5–6

USING THE PAGE

- Have students work individually or in pairs as an alternative to formal assessment.
- Use the performance indicators and work samples below to evaluate Tasks A–B.

See *Performance Assessment* for

- a complete scoring rubric, p. PAx, for this unit.
- additional student work samples for this unit.
- copying masters for this unit.

 You may suggest that students place completed Performance Assessment tasks in their portfolios.

Performance Indicators

Task A

A student with a Level 3 paper

____ Determines the mean, median, and mode of a given group of numbers.

____ Interprets the mean, median, and mode for a specific problem and determines which best describes the data for the given purposes.

____ Explains how to choose whether to use mean, median, or mode in a given problem.

____ Shows work and explains how the answers were determined.

Task B

A student with a Level 3 paper

____ Constructs and describes a graph.

____ Uses data to make a table.

____ Writes a question that can be answered by looking at a specific graph.

Performance Assessment

TASK A • Too Much Homework?

For homework, Scott has 3 social studies questions, 3 English questions, 2 science problems, 5 Spanish questions, and 20 math exercises. Scott's friend Beth asks him about how much homework he has in each subject. Would he give the mean, median, or mode of the number of items to make his homework assignment seem as long as possible?

a. Explain your choice. Write down your thinking as you make your decision. **See Additional Answers, p.143**

b. Suppose your friend asks you the same question about the homework you really have. How would you respond?

TASK B • Graph Analysis

Elements in the Earth's Crust: Aluminum 8.1%, Calcium 3.6%, Iron 5.0%, Oxygen 46.6%, Silicon 27.7%, All Others 9.0% **Answers will vary. See** *Performance Assessment* **page 14 for a possible answer.**

a. Organize the data in a table. Use the data to make a graph. Then write a short description of the graph. Include the following:

- What the title and axes represent
- What the parts represent, if it is a circle graph
- What stands out as important or obvious

b. Describe how the data can be used to make a different kind of graph.

c. Write a question that can be answered from either graph.

Work Samples for Task A and Task B

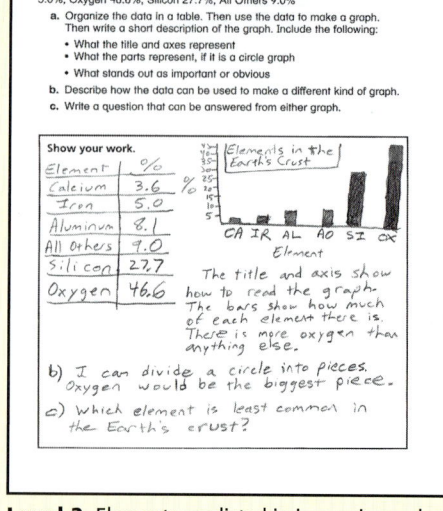

Level 3 The student shows excellent understanding of mean, median, and mode. Answers are accurate and complete.

Level 3 Elements are listed in increasing order. Bar graph is a good choice and size of bars is accurate. Student also chooses a circle graph and asks a good question.

E-Lab • Exploring Box-and-Whisker Graphs

Janie kept track of the lengths of her workouts. She made a table like the one below. Then she made a box-and-whisker graph.

LENGTHS OF WORKOUTS (min)								
45	60	35	46	70	42	63	38	41

You can use E-Lab to make box-and-whisker graphs.

- Click on *Exploring Box-and-Whisker Graphs*.
- Click on *New Problem*.
- Type the length of each of Janie's workouts.
- Press *Enter* after each one.
- Click *Sort* to order the data from least to greatest. Check students' graphs.
- Copy the box-and-whisker graph.

What are the median and lower and upper extremes? 45; 35; 70

What are the lower and upper quartiles? 41; 60

Practice and Problem Solving

Use E-Lab to make a box-and-whisker graph for each table. Copy each graph and answer the questions. Check students' graphs.

1.

NUMBER OF PEOPLE AT DIFFERENT PICNIC PAVILIONS							
15	30	48	24	40	26	55	52

2. What are the median, lower quartile, and upper quartile? 35; 25; 50

3. What are the lower and upper extremes and the range? 15; 55; 40

4.

NUMBER OF STUDENTS IN DIFFERENT AFTER-SCHOOL CLUBS							
10	24	16	30	21	14	26	19

5. What are the median, lower quartile, and upper quartile? 20; 15; 25

6. What are the lower and upper extremes and the range? 10; 30; 20

> **Multimedia Math Glossary** www.harcourtschool.com/mathglossary
>
> 7. **Vocabulary** Locate *frequency table* and *line plot* in the Multimedia Math Glossary. Use the frequency table shown to make a line plot. Use the line plot shown to make a cumulative frequency table.

Chapters 5–6 **143**

TECHNOLOGY LINKUP

Objective To use E-Lab to make box-and-whisker graphs

USING THE PAGE

Review with students that a box-and-whisker graph shows the distribution of the data and the lower and upper extremes, lower and upper quartiles, and median. Ask: **Can you tell the mode by looking at a box-and-whisker graph? the range?** no; yes

Explain to students that E-Lab will only accept 9 values from ⁻70 to 70, so it may not work for all box-and-whisker graphs. (It will work for all box-and-whisker graphs on this page.)

Practice and Problem Solving

As an alternative, you may want to have students first make their own box-and-whisker graphs and then use E-Lab to check them.

To challenge students, have them make a box-and-whisker graph with a lower extreme of 5, an upper extreme of 50, a lower quartile of 22, and an upper quartile of 38. Check students' graphs.

Multimedia Math Glossary

Frequency table, and all other vocabulary words in this unit, can be found in the Harcourt Multimedia Math Glossary.
www.harcourtschool.com/mathglossary

Additional Answers for Performance Task A, PE p. 142

a. Students should begin by determining the mean, median, and mode of the given data: mean—6.6; median—3; mode—3.

Students may select the mean because it is the greatest number and probably best takes into account the larger number of math problems. However, some may select the median, arguing that it is the middle number and therefore the best. Students who choose the mode should reason that it is the most common number of items given.

b. Students' answers may vary. They may argue that they would prefer to find the mean or median number of hours they need to spend on their homework rather than the mean or median of the number of items. For example, 2 science problems could take longer than the math problems if the math requires simple computation. Students' answers should reflect understanding of what the mean, median, and mode tell us about the data.

UNIT 2

PROBLEM SOLVING
On Location in Oregon

Purpose To provide additional practice for concepts and skills in Chapters 5–6

USING THE PAGE
Water Resources and Uses

- *Direct students' attention to the first circle graph.*

  Name two categories that account for about the same amount of water use. Thermoelectric and Agricultural

  Suppose 1,000,000 gallons of freshwater were used for thermoelectric purposes. About how many gallons would be used for industrial/mining purposes? about 200,000

- *Have students look at the second circle graph.*

  Can you tell how many gallons of freshwater are provided by surface water? Explain. No. The graph shows only percents. You can compare ground water to surface water but cannot tell the actual number of gallons used for either.

  Could surface water supply enough freshwater for any three of the categories in the first graph? Explain. It could if thermoelectric and agricultural are not both part of the 3 combined categories. There is not enough surface water to supply both thermoelectric and agricultural needs.

  Extension Have students suppose that the total supply of freshwater amounts to 1,000,000,000 gallons. Then have them use that data to convert each circle graph to a bar graph. Have students work together to discuss methods. Check students' work.

PROBLEM SOLVING ON LOCATION
In Oregon

Water Sources and Uses

Water is a critical resource everywhere in the world. In the United States, we have some very dry states, as well as states that receive an abundance of rainfall. Oregon has both wet and arid regions. On its Pacific coast, there can be 120 inches of rain in a year. But in parts of its eastern region, there might be only 8 inches of rain in a year.

Use Data For 1, use the information above.

1. About how many times as much rainfall can fall on Oregon's coast each year as in parts of its eastern region? **about 15 times**

Use Data For 2–4, use the first circle graph.

2. Which category uses the most freshwater? **Agricultural**

3. Which category uses the least freshwater? **Industrial/Mining**

4. Would you say that thermoelectric and home/commercial uses account for a little less than half or more than half of the freshwater? **a little more than half**

Use Data For 5–6, use both circle graphs.

5. About how many times as much freshwater is supplied by surface water as is supplied by ground water? **about 3–4 times**

6. Could ground water meet the freshwater needs of agriculture? Explain your thinking. **No; ground water supplies only 22% of the freshwater, and agriculture demands 41%.**

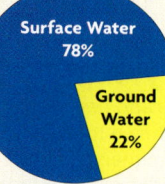

Uses of Freshwater: Agricultural 41%, Thermoelectric 39%, Home/Commercial 12%, Industrial/Mining 8%

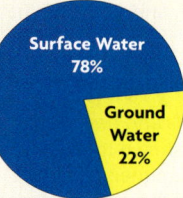

Sources of Freshwater: Surface Water 78%, Ground Water 22%

Dams

Suppose you live in the country. To get water, you would probably dig a well or use a nearby stream. But if you live in a city, you would use other means. Dams, reservoirs, and aqueducts help to collect, store, and transport water to city homes and businesses. In Oregon, more than 20 dams help meet the people's water needs.

Oregon's Bonneville Dam, located 40 miles east of Portland on the Columbia River, allows boats to travel upriver 188 miles.

Use Data For 1–7, use the table.

Oregon Dams		
Name	Storage Capacity	Crest Elevation
Gerber Dam	94,270 acre-feet	4,842 feet
Owyhee Dam	1,183,300 acre-feet	2,675 feet
Thief Valley Dam	26,000 acre-feet	3,143 feet
Warm Springs Dam	192,400 acre-feet	3,409 feet

1. Find the range of storage capacities. **1,157,300 acre-feet**

2. Find the mean storage capacity. **373,992.5 acre-feet**

3. Find the median storage capacity. **143,335 acre-feet**

4. **REASONING** Why is the mean so much greater than the median? **See above.**

5. Is the mean or median a better measure of central tendency for the storage capacities? Explain your thinking. **See above.**

6. **Mental Math** One acre-foot is equivalent to 43,560 cubic feet. Use mental math to estimate the storage capacity of Gerber Dam in cubic feet. **Possible estimate: about 4,000,000,000 cubic feet**

7. Make and label a bar graph showing the crest elevations of the four dams. Explain how you used rounding or estimation in making your graph. **Check students' graphs. Possible answer: I used intervals of 100 feet and rounded each crest elevation to the nearest 100 feet.**

4. Since the figure for Owyhee Dam is so much greater than the three other figures, it acts as an outlier to skew the mean upward.

5. median; the mean is more than twice as great as three of the four storage capacities. The mean distorts the capacities of these other three.

Chapters 5–6 **143B**

USING THE PAGE

Dams

- After Exercise 5, have students focus on the table.

Look at the column labeled *Crest Elevations*. Do you think the difference between the mean and median will be as great as it is between the mean and median for storage capacities? Explain. No; the range for crest elevations is not as great as that for storage capacities.

- Have students look at the graphs they made for Exercise 7.

How does a bar graph make it easier to compare the data? Possible answer: It is easier to see differences in elevations.

Why would you not label the elevation axis 0, 10, 20, . . .? Possible answer: The large elevations would result in a graph too large to draw and one difficult to read.

Would a line graph work as well as a bar graph? Explain. Possible answer: No; a line graph is better for showing change over time. Because you need to compare different dams, a bar graph is a better choice.

Reasoning Why do you think the dam with the greatest elevation does not necessarily have the greatest storage capacity? Possible answer: Capacity depends on surface area of the water, as well as elevation (or height).

Extension Have students investigate the meaning of an acre-foot by finding out what an acre is and how to *compute* an acre-foot. Also have students describe how to picture an acre-foot of water. An acre is a measure of area equal to a square that is approximately 209 feet on each side, or 43,560 square feet. An acre-foot is a measure of volume, or capacity. Picture a 1 acre square field, flooded with water to a height of 1 foot. The volume of the water would be 1 acre-foot.

Statistics and Graphing **143B**

Teaching Notes

Additional Ideas:

Good Questions to Ask:

Additional Resources:

Notes for Next Time:

UNIT 3 Fraction Concepts and Operations

UNIT AT A GLANCE

CHAPTER 7
Number Theory 144

- **Lesson 1** — Divisibility
- **Lesson 2** — Prime Factorization
- **Lesson 3** — Least Common Multiple and Greatest Common Factor
- **Lesson 4** — Problem Solving Strategy: *Make an Organized List*

CHAPTER 8
Fraction Concepts 158

- **Lesson 1** — Equivalent Fractions and Simplest Form
- **Lesson 2** — Mixed Numbers and Fractions
- **Lesson 3** — Compare and Order Fractions
- **Lesson 4** — Math Lab: Explore Fractions and Decimals
- **Lesson 5** — Fractions, Decimals, and Percents

CHAPTER 9
Add and Subtract Fractions and Mixed Numbers 174

- **Lesson 1** — Estimate Sums and Differences
- **Lesson 2** — Math Lab: Model Addition and Subtraction
- **Lesson 3** — Add and Subtract Fractions
- **Lesson 4** — Add and Subtract Mixed Numbers
- **Lesson 5** — Math Lab: Rename to Subtract
- **Lesson 6** — Subtract Mixed Numbers
- **Lesson 7** — Problem Solving Strategy: *Draw a Diagram*

CHAPTER 10
Multiply and Divide Fractions and Mixed Numbers 198

- **Lesson 1** — Estimate Products and Quotients
- **Lesson 2** — Multiply Fractions
- **Lesson 3** — Multiply Mixed Numbers
- **Lesson 4** — Math Lab: Division of Fractions
- **Lesson 5** — Divide Fractions and Mixed Numbers
- **Lesson 6** — Problem Solving Skill: *Choose the Operation*
- **Lesson 7** — Algebra: Fraction Expressions and Equations

UNIT 3: Fraction Concepts and Operations

Assessment Options

Assessing Prior Knowledge
Determine whether students have the required prerequisite concepts and skills.
Check What You Know, PE pp. 145, 159, 175, 199

Test Preparation
Provide review and practice for chapter and standardized tests.
Standardized Test Prep, PE pp. 157, 173, 197, 219
Mixed Review and Test Prep
   See the last page of each PE skill lesson.
Study Guide and Review, PE pp. 222–223

Formal Assessment
Assess students' mastery of chapter concepts and skills.
Chapter Review/Test, PE pp. 156, 172, 196, 218
Pretest and Posttest Options
   Chapter Test, Form A
      pp. AG49–50, 53–54, 57–58, 61–62
   Chapter Test, Form B
      pp. AG51–52, 55–56, 59–60, 63–64
Unit 3 Test • Chapters 7–10, pp. AG65–72

Daily Assessment
Obtain daily feedback on students' understanding of concepts.
Quick Review
   See the first page of each PE lesson.
Mixed Review and Test Prep
   See the last page of each PE skill lesson.
Number of the Day
   See the first page of each TE skill lesson.
Problem of the Day
   See the first page of each TE skill lesson.
Lesson Quiz
   See the *Assess* section of each TE skill lesson.

Performance Assessment
Assess students' understanding of concepts applied to real-world situations.
Performance Assessment (Tasks A–B)
   PE, p. 224; pp. PA21–22

Student Self-Assessment
Have students evaluate their own work.
How Did I Do?, p. AGxvii
A Guide to My Math Portfolio, p. AGxix
Math Journal
   See *Write* in the *Assess* section of each TE skill lesson and TE pages 148B, 164B, 176B, 180, 186B, 202B, 208, 216B.

 Harcourt Electronic Test System Math Practice and Assessment
Make and grade chapter tests electronically.
This software includes:
- multiple-choice items
- free-response items
- customizable tests
- the means to make your own tests

 **Portfolio**
Portfolio opportunities appear throughout the Pupil and Teacher's Editions.
Suggested work samples:
Problem Solving Project, TE pp. 144, 158, 174, 198
Write About It, PE pp. 149, 165, 171, 193, 201, 207, 215
Chapter Review/Test, PE pp. 156, 172, 196, 218

KEY   **AG** Assessment Guide   **TE** Teacher's Edition   **PA** Performance Assessment   **PE** Pupil Edition

For the Student

 Intervention Strategies and Activities

Review and practice the prerequisite skills for Chapters 7–10.

E-LAB These interactive learning experiences reinforce and extend the skills taught in Chapters 7–10.

- Skill development
- Practice

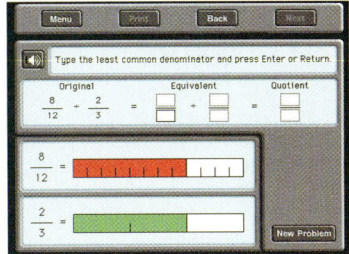

Mighty Math

The learning activities in this comprehensive math software series complement, enrich, and enhance the Pupil Edition lessons.

Calculating Crew • *Nautical Number Line*

Number Heroes • *Fraction Fireworks*

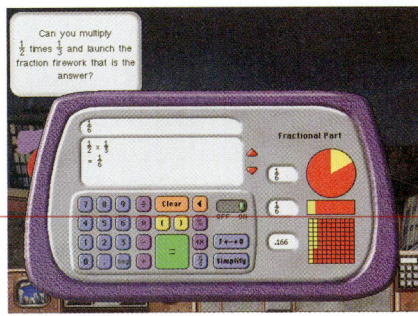

Astro Algebra • *Red*

For the Teacher

 Teacher Support Software

- **Intervention Strategies and Activities**
  Provide instruction, practice, and a check of the prerequisite skills for each chapter.
- **Electronic Lesson Planner**
  Quickly prepare daily and weekly lessons for all subject areas.
- **Harcourt Electronic Test System
  Math Practice and Assessment**
  Edit and customize Chapter Tests or construct unique tests from large item banks.

For the Parent

 The Harcourt Learning Site

- Encourage parents to visit The Harcourt Learning Site to help them reinforce mathematics vocabulary, concepts, and skills with their children.
- Have them click on *Math* for vocabulary, activities, real-life connections, and homework tips for Chapters 7–10.

www.harcourtschool.com

Internet

Teachers can find fraction concepts and operations activities and resources.

Students can learn more about fraction concepts and operations and reinforce the critical concepts and skills for Chapters 7–10.

Parents can use The Harcourt Learning Site's resources to help their children with the vocabulary, concepts, and skills needed for Chapters 7–10.

Visit The Harcourt Learning Site

www.harcourtschool.com

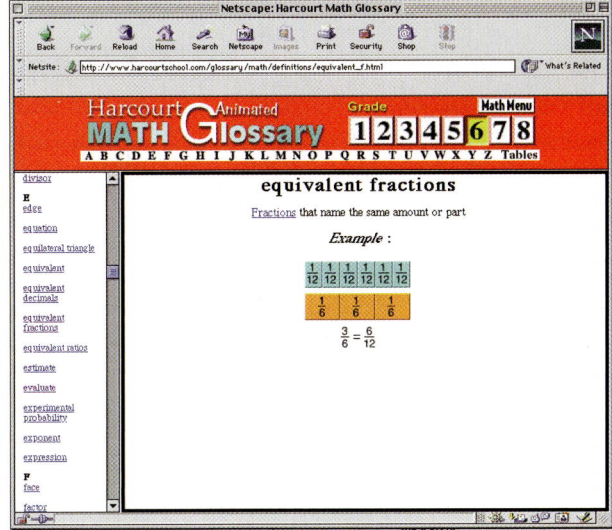

Fraction Concepts and Operations

UNIT 3 Fraction Concepts and Operations

Universal Access: Reaching All Learners

ADVANCED LEARNERS

MATERIALS *For each pair* graph paper, p. TR62; calculator

Challenge students to **explore patterns in equivalent decimals.** Display the following table.

Thirds	$\frac{1}{3}=$ 0.3333	$\frac{2}{3}=$ 0.6666	$\frac{3}{3}=$ 1.0		
Fourths	$\frac{1}{4}=?$	$\frac{2}{4}=?$	$\frac{3}{4}=?$	$\frac{4}{4}=?$	
Fifths	$\frac{1}{5}=?$	$\frac{2}{5}=?$	$\frac{3}{5}=?$	$\frac{4}{5}=?$	$\frac{5}{5}=?$

Ask students to copy and extend the chart to include sixths, sevenths, eighths, ninths, tenths, and elevenths. Then ask them to find decimal equivalents by dividing the numerator by the denominator. Have students discuss the patterns they see and offer suggestions for remembering the decimal equivalents for commonly used fractions. *Use with Lessons 8.5 and 10.1.*

VISUAL, AUDITORY

SPECIAL NEEDS

MATERIALS *For each pair* 3 decimal squares (hundredths), p. TR7; scissors; tape

Have students **convert among decimals, fractions, and percents.** Display the following equivalent expressions statements:

"Eighteen hundredths written in simplest form is nine-fiftieths. Eighteen hundredths also equals eighteen percent. Eighteen percent equals the decimal eighteen hundredths." Have volunteers read the statements aloud.

For each decimal, percent, and fraction shown below, ask students to make a decimal square.

0.15   79% $\frac{3}{5}$

Have them cut out each square and tape it to a sheet of paper. Tell them to write on the bottom of the paper the equivalent fraction, decimal, or percent in numbers and in words. Ask them to read the equivalent expressions statements aloud. *Use with Lesson 8.5.*

AUDITORY; VISUAL/SPATIAL

BLOCK SCHEDULING

INTERDISCIPLINARY COURSES

- Art—Describe geometric designs and fractional parts used in works of art.
- Life Science — Make a chart of the average weight of various animals at birth, written to the nearest $\frac{1}{4}$ pound.
- Ecology — Make a chart with diagrams that illustrate how the fractional width requirements for birdhouse doors vary depending upon species.

COMPLETE UNIT

Unit 3 may be presented in
- fourteen 90-minute blocks.
- seventeen 75-minute blocks.

INTERDISCIPLINARY SUGGESTIONS

PURPOSE To connect *Fraction Concepts and Operations* to other subjects with these activities

Chapter 8 — Social Studies

Students research why and how the metal composition of the penny was changed by the 1981 Coinage Act.

Chapter 10 — Music

Students explore *time* in music as it relates to fractions by comparing the rhythms of two unlike compositions—one traditional and one contemporary.

KINESTHETIC; VISUAL/SPATIAL

ACTIVITIES AND GAMES FOR HOME OR SCHOOL

You may choose to use this activity and game in the classroom or send it home for students to do with family members.

ACTIVITY • Sneak the Straw

PURPOSE To match equivalent fractions and decimals
Use after Lesson 8.5.

VARIATIONS For a more challenging game, replace card pairs with more difficult fraction/decimal equivalents. For example, use $\frac{3}{5}$ instead of $\frac{6}{10}$.

GAME • Winning Ones

PURPOSE To add, subtract, and compare fractions
Use after Lesson 9.3.

KINESTHETIC; BODILY/KINESTHETIC

ENGLISH LANGUAGE LEARNERS — ELL

Vocabulary Preview Ask students to make a math dictionary for the vocabulary in the unit. Have them make a three-column chart and label the columns. As students learn about each new term, they can fill their dictionary. *Use with Lessons 7.2–10.4.*

Term	Example	Explanation

Mixed Number The term *mixed* is an adjective describing the word *number*. Just as a can of mixed nuts has several kinds of nuts, a mixed number is made up of two different types of numbers, a fraction and a whole number. *Use with Lesson 8.2.*

Terminating decimal Help students understand the word *terminating* to mean "having an end." Relate it to the phrase *bus terminal* which is the end of the bus line. *Use with Lesson 8.5.*

VISUAL; VERBAL/LINGUISTIC

LITERATURE CONNECTIONS

These books provide students with additional ways to explore fractions.

Funny and Fabulous Fraction Stories by Dan Greenberg (Scholastic, 1996) consists of 30 fun math tales and problems that reinforce important concepts related to fractions.

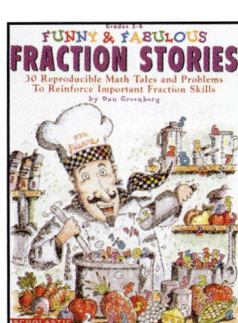

- Read "Enid the Magnificent, Part 1 and Part 2" for additional help with equivalent fractions and decimals. *Use with Lessons 8.4–8.5.*

Island of the Blue Dolphins by Scott O'Dell (Houghton Mifflin, 1990) tells of an American Indian girl who is left behind when her tribe leaves for a better place.

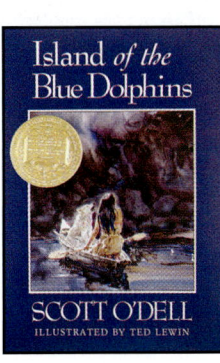

- The island that Karana lives on measures 1 league by 2 leagues. One league is about $4\frac{4}{5}$ kilometers long. Assuming the island is roughly rectangular, have students estimate the perimeter of the island. *Use with Lessons 9.4 and 10.3.*

Fraction Concepts and Operations 144G

CHAPTER 7 Number Theory

CHAPTER PLANNER

PACING OPTIONS
| Compacted | 4 Days |
| Expanded | 7 Days |

Getting Ready for Chapter 7 • Assessing Prior Knowledge and INTERVENTION (See PE and TE page 145.)

LESSON	NCTM STANDARDS	PACING	VOCABULARY*	MATERIALS	RESOURCES AND TECHNOLOGY
7.1 Divisibility pp. 146–147 **Objective** To use divisibility rules	1, 6, 7, 8	1 Day	divisible	hundred chart	Reteach, Practice, Challenge, Problem Solving 7.1 Worksheets Extra Practice p. H38, Set A Transparency 7.1 Math Jingles™ CD 5–6
7.2 Prime Factorization pp. 148–149 **Objective** To write a composite number as the product of prime factors	1, 6, 7, 8	1 Day	**prime factorization**		Reteach, Practice, Challenge, Problem Solving 7.2 Worksheets Extra Practice p. H38, Set B Transparency 7.2
7.3 Least Common Multiple and Greatest Common Factor pp. 150–153 **Objective** To find the least common multiple and greatest common factor of numbers and use them to solve problems	1, 6, 8, 9	2 Days	**least common multiple (LCM)** **greatest common factor (GCF)**		Reteach, Practice, Challenge, Problem Solving 7.3 Worksheets Extra Practice p. H38, Set C Transparency 7.3
7.4 Problem Solving Strategy: *Make an Organized List* pp. 154–155 **Objective** To solve problems by using the strategy *make an organized list*	1, 6, 10	1 Day			Reteach, Practice, Challenge, Reading Strategy 7.4 Worksheets Transparency 7.4 Problem Solving Think Along, p. TR1

Ending Chapter 7 • Chapter 7 Review/Test, p. 156 • **Standardized Test Prep**, p. 157

*****Boldfaced** terms are new vocabulary. Other terms are review vocabulary.

CHAPTER AT A GLANCE

Vocabulary Development

The boldfaced words are the new vocabulary terms in the chapter. Have students record the definitions in their Math Journals.

- **prime factorization**, p. 148
- **least common multiple (LCM)**, p. 150
- **greatest common factor (GCF)**, p. 151

Vocabulary Cards
Have students use the Vocabulary Cards on *Teacher's Resource Book* pp. TR125–128 to make graphic organizers or word puzzles. The cards can also be added to a file of mathematics terms.

NCTM Standards

1. **Number and Operations**
   Lessons 7.1, 7.2, 7.3, 7.4
2. **Algebra**
3. **Geometry**
4. **Measurement**
5. **Data Analysis and Probability**
6. **Problem Solving**
   Lessons 7.1, 7.2, 7.3, 7.4
7. **Reasoning and Proof**
   Lessons 7.1, 7.2
8. **Communication**
   Lessons 7.1, 7.2, 7.3
9. **Connections**
   Lesson 7.3
10. **Representation**
    Lesson 7.4

Writing Opportunities

PUPIL EDITION
- Write About It, p. 149
- What's the Question?, p. 155
- What's the Error?, p. 153

TEACHER'S EDITION
- Write—See the *Assess* section of each TE lesson.
- Writing in Mathematics, p. 148B

ASSESSMENT GUIDE
- How Did I Do?, p. AGxvii

Family Involvement Activities

These activities provide:
- Letter to the Family
- Math Vocabulary
- Family Game
- Practice (Homework)

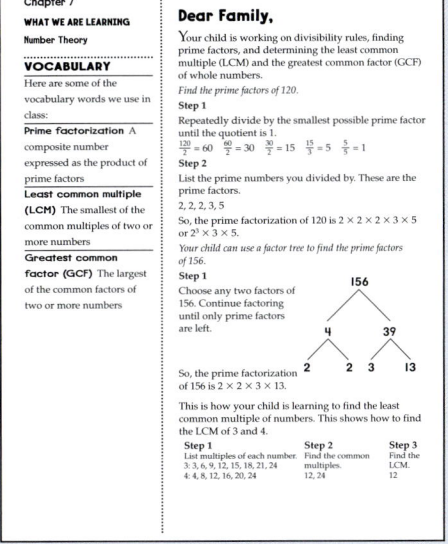

Family Involvement Activities, p. FA25

144I

CHAPTER 7 Number Theory

MATHEMATICS ACROSS THE GRADES

SKILLS TRACE ACROSS THE GRADES

GRADE 5	GRADE 6	GRADE 7
Find least common multiple and greatest common factor of two whole numbers; determine if a number is prime or composite	Use divisibility rules; write prime factorization in exponent form; write and apply greatest common factors and least common multiples	Use and write divisibility rules; write prime factorization in exponent form; write and apply greatest common factors and least common multiples

SKILLS TRACE FOR GRADE 6

LESSON	FIRST INTRODUCED	TAUGHT AND PRACTICED	TESTED	REVIEWED
7.1	Grade 5	PE pp. 146–147, H38, p. RW32, p. PW32, p. PS32	PE p. 156, pp. AG49–52	PE pp. 156, 157, 222–223
7.2	Grade 6	PE pp. 148–149, H38, p. RW33, p. PW33, p. PS33	PE p. 156, pp. AG49–52	PE pp. 156, 157, 222–223
7.3	Grade 5	PE pp. 150–153, H38, p. RW34, p. PW34, p. PS34	PE p. 156, pp. AG49–52	PE pp. 156, 157, 222–223
7.4	Grade 4	PE pp. 154–155, p. RW35, p. PW35, p. PS35	PE p. 156, pp. AG49–52	PE pp. 156, 157, 222–223

KEY PE Pupil Edition PS Problem Solving Workbook RW Reteach Workbook
      PW Practice Workbook AG Assessment Guide

Looking Back  Prerequisite Skills

To be ready for Chapter 7, students should have the following understandings and skills:

- **Vocabulary**—*composite number, multiple, prime number*
- **Prime Numbers**—decide whether a number is prime
- **Composite Numbers**—decide whether a number is composite
- **Multiples**—write multiples of a number
- **Factors**—write all factors of a number

Check What You Know

Use page 145 to determine students' knowledge of prerequisite concepts and skills.

Intervention

Help students prepare for the chapter by using the intervention resources described on TE page 145.

Looking at Chapter 7  Essential Skills

Students will

- understand how to apply and write divisibility rules.
- **develop skill in writing a composite number as the product of prime factors in exponent form.**
- develop skill writing and applying the least common multiple and greatest common factor of two whole numbers.

EXAMPLE

Write the prime factorization of 225.

Factor Tree	Exponent Form

225
5 × 45   45 is not prime.
  5 × 9   9 is not prime.
    3 × 3
So, the prime factorization of 225 is 5 × 5 × 3 × 3.
	225 = 3 × 3 × 5 × 5
	
	225 = $3^2 \times 5^2$
	
	So, the prime factorization of 225 expressed in exponent form is $3^2 \times 5^2$.

Looking Ahead  Applications

Students will apply what they learn in Chapter 7 to the following new concepts:

- Equivalent Fractions and Simplest Form (Chapter 8)
- Adding and Subtracting Mixed Numbers (Chapter 9)
- Adding and Subtracting Fractions (Chapter 9)

Number Theory 144K

CHAPTER 7

Number Theory

INTRODUCING THE CHAPTER

Tell students that number theory is the study of relationships between numbers. Have students focus on the chart to see that relationships between numbers affect how tulip bulbs can be packaged. Ask them to explain whether or not the tulip bulbs can be packaged 21 to a pack with no leftovers. No; $1{,}200 \div 21 = 57$ r3.

USING DATA

To begin the study of this chapter, have students

- Determine if there will be leftover tulip bulbs when 1,200 tulip bulbs are divided into packages of 15. No; $1{,}200 \div 15 = 80$.
- Determine if 60 tulip bulbs could be packaged with a number of tulip bulbs in an equivalent number of packages and no leftovers. No; 60 is not the square of a whole number: $7 \times 7 = 49$ and $8 \times 8 = 64$.
- Use the data in the chart to determine if 1,200 tulip bulbs can be divided into packages of 24 with no leftovers. Yes, since $1{,}200 \div 5 = 240$, then $1{,}200 \div 24 = 50$.

PROBLEM SOLVING PROJECT

Purpose To use number theory to solve problems

Grouping pairs or small groups

Background Bakeries package baked goods in varied amounts. For example, a baker's dozen of cookies is a package of 13 cookies instead of 12.

Analyze, Choose, Solve, and Check

Have students

- Determine if there will be leftovers when packages of 13 are made from 144 baked cookies. Yes; $144 \div 13 = 11$ r1.
- Find out what size packages can be made, without leftovers, from 144 cookies. packages with 2, 3, 4, 6, 8, 9, 12, 16, 18, 24, 36, 48, or 72 cookies
- Make a chart to show their findings.

Check students' work.

 Suggest that students display their charts in the classroom and later add them to their portfolios.

144 Chapter 7

CHAPTER 7 Number Theory

60 packages; no

Tulips are grown commercially in Woodburn, Oregon. As far as the eye can see, bright colors cover the surrounding fields each spring. Bulbs from these plants are collected, packaged, and sold to home gardeners who start tulip gardens of their own on a smaller scale.

PROBLEM SOLVING Suppose 1,200 tulip bulbs are put into packages of 20. How many packages could be made? Would there be any bulbs left over?

PACKAGES OF TULIP BULBS

Tulip Bulbs per Package	Complete Package— Yes or No?
25	Yes
30	Yes
35	No
40	Yes
45	No
50	Yes

144 Chapter 7

Why learn math? Explain that a person who manages a candy factory would use number theory to determine how many packages of candy could be packed each hour or how many boxes of packaged candy would be ready to ship each day. Ask: How do people in other jobs use number theory? Possible answer: A restaurant owner uses number theory to determine how many people could be seated at tables each hour.

TECHNOLOGY LINK

To find out more about number theory, visit The Harcourt Learning Site.

www.harcourtschool.com

Check What You Know

Use this page to help you review and remember important skills needed for Chapter 7.

✓ Vocabulary

Choose the best term from the box.

> composite number
> multiple
> prime number

1. A number whose only factors are 1 and itself is a __prime number__.
2. A number that has more than two factors is a __composite number__.

✓ Prime Numbers (For Intervention, see p. H7.)

Decide whether the number is a prime number. Write *yes* or *no*.

3. 2 yes	4. 5 yes	5. 4 no	6. 9 no
7. 11 yes	8. 21 no	9. 37 yes	10. 26 no
11. 13 yes	12. 7 yes	13. 45 no	14. 70 no

✓ Composite Numbers (For Intervention, see p. H7.)

Decide whether the number is a composite number. Write *yes* or *no*.

| 15. 6 yes | 16. 15 yes | 17. 19 no | 18. 81 yes |
| 19. 24 yes | 20. 53 no | 21. 3 no | 22. 25 yes |

✓ Multiples (For Intervention, see p. H8.)

Write the next three multiples.

23. 4   4, 8, 12, ▪, ▪, ▪   16, 20, 24
24. 10   10, 20, 30, ▪, ▪, ▪   40, 50, 60
25. 12   12, 24, 36, ▪, ▪, ▪   48, 60, 72
26. 8   8, 16, 24, ▪, ▪, ▪   32, 40, 48
27. 5   5, 10, 15, ▪, ▪, ▪   20, 25, 30
28. 11   11, 22, 33, ▪, ▪, ▪   44, 55, 66

Write the first five multiples of each number.

29. 6   6, 12, 18, 24, 30
30. 22   22, 44, 66, 88, 110
31. 30   30, 60, 90, 120, 150
32. 7   7, 14, 21, 28, 35
33. 9   9, 18, 27, 36, 45
34. 13   13, 26, 39, 52, 65

✓ Factors (For Intervention, see p. H8.)

Write all of the factors of each number.

35. 8   1, 2, 4, 8
36. 9   1, 3, 9
37. 11   1, 11
38. 18   1, 2, 3, 6, 9, 18
39. 54   1, 2, 3, 6, 9, 18, 27, 54
40. 32   1, 2, 4, 8, 16, 32

LOOK AHEAD

In Chapter 7 you will
- use divisibility rules
- find prime factors
- determine the LCM and GCF of whole numbers

145

LESSON 7.1 Divisibility

LESSON PLANNING

Objective To use divisibility rules

Intervention for Prerequisite Skills

Multiples, Factors (For intervention strategies, see page 145.)

Materials *For each student* hundred chart

NCTM Standards
1. Number and Operations
6. Problem Solving
7. Reasoning and Proof
8. Communication

Math Background

One topic of number theory is divisibility.

- Students will use the concept of divisibility as they simplify fractions and find common denominators.
- Knowing if a number is divisible by 2, 3, 4, 5, 6, 8, 9, or 10 will help students as they find factors.
- Different observations must be made for 3 and 9, where the sum of the number's digits must be considered.
- Divisibility by 6 is determined by two considerations: divisibility by 2 and divisibility by 3.

WARM-UP RESOURCES

 **NUMBER OF THE DAY**

The number of the day is a common number of pizza servings. It is the smallest number divisible by 2, 3, and 4. What number is it? 12

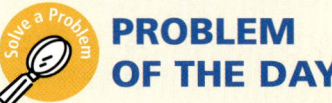 

Corrine is having a party. She has 45 different party favors and wants to give each guest the same number of favors. How many guests could she invite and how many favors would they get? 1, 45; 3, 15; 5, 9; 9, 5; 15, 3; 45, 1

Solution Problem of the Day tab, p. PD7

 **DAILY FACTS PRACTICE**

Have students practice division facts by completing Set G of *Teacher's Resource Book,* p. TR97.

146A Chapter 7

INTERVENTION AND EXTENSION RESOURCES

ALTERNATIVE TEACHING STRATEGY — ELL

Materials *For each group* centimeter cubes

Have students **use centimeter cubes to model division** of a number such as 24 by 2, 3, 4, 5, 6, 8, 9, and 10.

- Divide the class into small groups and give each group several numbers, such as 50, 35, 12, 20, 32, 27, 45. Ask them to model division of their numbers by 2, 3, 4, 5, 6, 8, 9, and 10 and record the results in a table.
- Have the groups of students compare the results.
- Use their results to develop the rules for divisibility.

50 divisible by 2, 5, 10; 35 divisible by 5; 12 divisible by 2, 3, 4, 6; 20 divisible by 2, 4, 5, 10; 32 divisible by 2, 4, 8; 27 divisible by 3, 9; 45 divisible by 3, 5, 9

KINESTHETIC
INTERPERSONAL/SOCIAL

MULTISTEP AND STRATEGY PROBLEMS

The following multistep and strategy problems are provided in Lesson 7.1:

Page	Item
147	28, 29

SPECIAL NEEDS

Divide the class into small groups and have each group **write the rules for divisibility** for 2, 3, 4, 5, and 6.

- Direct each student to say the last two digits of his or her phone number. Then ask the group to work together to decide if the number is divisible by either 2, 3, 4, 5, or 6 by using the rules they have written.
- Suggest that students make a table of their results to share with the class.

Check students' work.

AUDITORY
VISUAL/SPATIAL

ADVANCED LEARNERS

Challenge students to **apply what they have learned about division**. Have students answer and give two examples for each of the following questions: Possible answers are given.

- If a number is divisible by 2 and 3, is it divisible by 6? yes; 18 and 30 are divisible by 2, 3, and 6.
- If a number is divisible by 2 and 4, is it divisible by 8? not necessarily; 12 and 20 are divisible by 2 and 4, but not by 8.
- If a number is divisible by 3 and 5, is it divisible by 15? yes; 30 and 45 are divisible by 3, 5, and 15.
- If a number is divisible by 3 and 6, is it divisible by 18? not necessarily; 30 and 42 are divisible by 3 and 6, but not by 18.

AUDITORY
LOGICAL/MATHEMATICAL

TECHNOLOGY LINK

Intervention Strategies and Activities CD-ROM •
Skills 15, 16

Math Jingles™ CD 5–6 •
Track 7

LESSON 7.1 ORGANIZER

Objective To use divisibility rules
Vocabulary Review divisible
Materials For each student hundred chart

1 Introduce

QUICK REVIEW provides review of prerequisite skills.

Why Learn This? You can use this skill to determine how to divide items, such as party favors, equally among your friends. *Share the lesson objective with students.*

2 Teach

Guided Instruction

- *Review the divisibility rules for 2 and 3.*

  State the divisibility rule for 6 in another way. A number is divisible by 6 if it is even and the sum of the digits is divisible by 3.

- *Discuss the remaining divisibility rules.*

  What is another way of stating the divisibility rule for 10? A number is divisible by 10 if it is divisible by 2 and 5.

- *As students work through the examples, ask:*

  How did you decide 610 was not divisible by 3? The sum of the digits, 6 + 1 + 0, is not divisible by 3.

REASONING If a number is not divisible by 3, will it be divisible by 9? Explain. No; if it does not have a factor of 3, it cannot have a factor of 9.

ADDITIONAL EXAMPLE

Example, p. 146

Determine whether 720 is divisible by 2, 3, 4, 5, 6, 8, 9, or 10.

2: the last digit is zero
3: the sum of the digits is divisible by 3
4: the last two digits form a number divisible by 4
5: the last digit is 0
6: the number is divisible by 2 and 3
8: the three digits form a number divisible by 8
9: the sum of the digits is divisible by 9
10: the last digit is 0

146 Chapter 7

LESSON 7.1

Divisibility

Learn how to tell if a number is divisible by 2, 3, 4, 5, 6, 8, 9, or 10.

QUICK REVIEW
Write *even* or *odd* for each.
1. 34,526 even
2. 5,437 odd
3. 6,230 even
4. 27,343 odd
5. 1,468 even

Remember that a number is divisible by another number if the remainder is zero.

Chris knows that a number is divisible by 2 if the last digit is 0, 2, 4, 6, or 8. He also knows that a number is divisible by 3 if the sum of the digits is divisible by 3. He wonders if there is a rule for numbers that are divisible by 6.

Activity

MATH LAB

You need: hundred chart

- Shade all the numbers divisible by 2.
- Circle all the numbers that are divisible by 3.
- Look at the numbers that are both shaded and circled. Divide these numbers by 6. What do you notice? **All these numbers are divisible by 6.**
- What rule can Chris write about numbers divisible by 6? **Possible answer: a number is divisible by 6 if the number is divisible by 2 and 3.**

Math Idea ▶ You can use divisibility rules to help you decide if a number is divisible by another number.

A number is divisible by	Divisible	Not Divisible
2 if the last digit is even (0, 2, 4, 6, or 8).	11,994	2,175
3 if the sum of the digits is divisible by 3.	216	79
4 if the last two digits form a number divisible by 4.	1,024	621
5 if the last digit is 0 or 5.	15,195	10,007
6 if the number is divisible by 2 and 3.	1,332	44
8 if the last three digits form a number divisible by 8.	5,336	3,180
9 if the sum of the digits is divisible by 9.	144	33
10 if the last digit is 0.	2,790	9,325

EXAMPLES Tell whether each number is divisible by 2, 3, 4, 5, 6, 8, 9, or 10.

A. 610 is divisible by
2; the last digit is even.
5; the last digit is 0 or 5.
10; the last digit is 0.

B. 459 is divisible by
3; the sum of the three digits is divisible by 3.
9; the sum of the digits is divisible by 9.

146 Chapter 7

RETEACH 7.1

Divisibility
The rules for divisibility are:

A number is divisible by:	If:
2	the last digit is 0, 2, 4, 6, or 8.
3	the sum of its digits is divisible by 3.
4	the number formed by the last two digits is divisible by 4.
5	the last digit is 0 or 5.
6	it is divisible by 2 and by 3.
8	the number formed by the last three digits is divisible by 8.
9	the sum of its digits is divisible by 9.
10	the last digit is 0.

To determine whether 3,882 is divisible by 3, follow these steps:
Step 1 Find the sum of the digits of 3,882:
3 + 8 + 8 + 2 = 21
Step 2 Decide whether the sum, 21, is divisible by 3. Since 3 divides 21 evenly (with no remainder), 21 is divisible by 3. So, 3,882 is *divisible* by 3.
To determine whether 3,882 is divisible by 9, decide whether 21 (the sum of the digits, can be divided evenly by 9. Since 21 ÷ 9 = 2 r3, 3,882 is *not divisible* by 9.

To determine whether 7,032 is divisible by 4, follow these steps:
Step 1 Identify the number formed by the last two digits: 32.
Step 2 Decide whether 32 is divisible by 4. Since 4 divides 32 evenly (with no remainder), 32 is *divisible* by 4. So, 7,032 is divisible by 4.
To determine whether 7,032 is divisible by 8, decide whether 032 (the last three digits) is a number divisible by 8. Since 032 ÷ 8 = 4, the number 7,032 is *divisible* by 8.

Determine whether each number is divisible by 2, 3, 4, 5, 6, 8, 9, or 10.

1. 146 — **2**
2. 369 — **3, 9**
3. 195 — **3, 5**
4. 284 — **2, 4**
5. 444 — **2, 3, 4, 6**
6. 512 — **2, 4, 8**
7. 788 — **2, 4**
8. 612 — **2, 3, 4, 6, 9**
9. 2,865 — **3, 5**
10. 4,470 — **2, 3, 5, 6, 10**
11. 6,048 — **2, 3, 4, 6, 8, 9**
12. 3,240 — **2, 3, 4, 5, 6, 8, 9, 10**

PRACTICE 7.1

Divisibility
Tell whether each number is divisible by 2, 3, 4, 5, 6, 8, 9, or 10.

1. 30 — **2, 3, 5, 6, 10**
2. 24 — **2, 3, 4, 6, 8**
3. 115 — **5**
4. 240 — **2, 3, 4, 5, 6, 8, 10**
5. 486 — **2, 3, 6, 9**
6. 235 — **5**
7. 279 — **3, 9**
8. 801 — **3, 9**
9. 145 — **5**
10. 650 — **2, 5, 10**
11. 736 — **2, 4, 8**
12. 1,200 — **2, 3, 4, 5, 6, 8, 10**
13. 207 — **3, 9**
14. 723 — **3**
15. 2,344 — **2, 4, 8**
16. 868 — **2, 4**
17. 694 — **2**
18. 4,464 — **2, 3, 4, 6, 8, 9**
19. 3,894 — **2, 3, 6**
20. 306 — **2, 3, 6, 9**
21. 836 — **2, 4**
22. 5,962 — **2**
23. 2,388 — **2, 3, 4, 6**
24. 792 — **2, 3, 4, 6, 8, 9**
25. 14,730 — **2, 3, 5, 6, 10**
26. 24,456 — **2, 3, 4, 6, 8**
27. 7,677 — **3, 9**
28. 34,248 — **2, 3, 4, 6, 8**

For 29–31 write *T* or *F* to tell whether each statement is true or false. If it is false, give an example that shows it is false.

29. No odd number is divisible by 2. **T**
30. All numbers that are divisible by 4 are also divisible by 2. **T**
31. All numbers that are divisible by 3 are also divisible by 6. **F; 9**
32. A number is between 40 and 50 and is divisible by both 3 and 4. What is the number? **48**

Mixed Review
Add or subtract mentally.

33. 451 − 71 = **380**
34. 898 − 196 = **702**
35. 109 + 46 + 54 = **209**

CHECK FOR UNDERSTANDING

Think and Discuss

sum must be divisible by 3, and for 9 the sum must be divisible by 9.

Look back at the lesson to answer each question.

1. **Compare** the divisibility rules for 3 and 9.
   1. They both involve finding the sum of the digits. For 3 the (see left).
2. **Tell** the advantages of knowing the divisibility rules. See below left.

Guided Practice

Tell whether each number is divisible by 2, 3, 4, 5, 6, 8, 9, or 10.

3. 56 **2, 4, 8**
4. 200 **2, 4, 5, 8, 10**
5. 784 **2, 4, 8**
6. 2,345 **5**
7. 3,009 **3**

PRACTICE AND PROBLEM SOLVING

Independent Practice

2. Possible answer: It saves time because you don't have to do the division to find out if a number is divisible by another number; it helps you find factors of a number.

Tell whether each number is divisible by 2, 3, 4, 5, 6, 8, 9, or 10.

8. 75 **3, 5**
9. 324 **2, 3, 4, 6, 9**
10. 45 **3, 5, 9**
11. 812 **2, 4**
12. 501 **3**
13. 615 **3, 5**
14. 936 **2, 3, 4, 6, 8, 9**
15. 744 **2, 3, 4, 6, 8**
16. 5,188 **2, 4**
17. 4,335 **3, 5**
18. 1,407 **3**
19. 48,006 **2, 3, 6, 9**
20. 7,064 **2, 4, 8**
21. 12,111 **3**
22. 1,044 **2, 3, 4, 6, 9**

For 23–25 write *T* or *F* to tell whether each statement is true or false. If it is false, give an example that shows it is false.

23. All even numbers are divisible by 2. **T**
24. All odd numbers are divisible by 3. **F; Possible example: 617**
25. Some even numbers are divisible by 5. **T**

Problem Solving Applications

26. A number is between 80 and 100 and is divisible by both 5 and 6. What is the number? **90**

27. **REASONING** Use the divisibility rules you know to write a divisibility rule for 15. *if it ends in 0 or 5 and the sum of the digits is divisible by 3*

28. Kirk has 20 trading cards. He has 10 more hockey cards than baseball cards. How many does Kirk have of each type of card? **15 hockey cards, 5 baseball cards**

29. Amber charges $8.25 per hour to baby-sit. One Saturday she baby-sat for 3.5 hr in the morning and 2.25 hr in the evening. How much did she earn baby-sitting on Saturday? Round your answer to the nearest cent. **$47.44**

MIXED REVIEW AND TEST PREP

30. What is the lower quartile of the data 24, 26, 28, 29, 30, 32, 34, 36, 37? (p. 129) **27**
31. Write the percent for 0.05. (p. 60) **5%**
32. 36.4 ÷ 0.28 (p. 76) **130**
33. 858 × 19 (p. 22) **16,302**
34. **TEST PREP** Which is the value of $3^2 + 2 \times 5 - 6$? (p. 44) **A**

   A 13   B 24   C 34   D 49

EXTRA PRACTICE page H38, Set A

147

3 Practice

Guided Practice

Do Check for Understanding Exercises 1–7 with your students. Identify those having difficulty and use lesson resources to help.

COMMON ERROR ALERT

Students may incorrectly decide that a number is divisible by 3 if the last digit is 3. Have students test several numbers with a last digit of 3, such as 43 and 203, for divisibility by 3. Review the divisibility rule. Point out that $4 + 3 = 7$ and $2 + 0 + 3 = 5$, and neither sum is divisible by 3.

Independent Practice

Note that Exercises 28 and 29 are **multistep or strategy problems**. Assign Exercises 8–29.

As students work Exercises 8–22, have them test the numbers for divisibility using one test at a time. Have students write each exercise on a separate line, by recording the number after the exercise number. Then test each one for divisibility by 2, then repeat for 3, and so on.

MIXED REVIEW AND TEST PREP

Exercises 30–34 provide **cumulative review** (Chapters 1–7).

4 Assess

Summarize the lesson by having students:

DISCUSS How could you use a calculator to determine if one number is divisible by another? Divide, and if the answer is a whole number, the first number is divisible by the second.

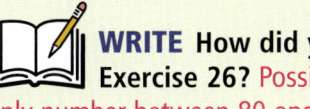

WRITE How did you solve Exercise 26? Possible answer: The only number between 80 and 100 that is divisible by 30 is 90.

Lesson Quiz

Determine whether each number is divisible by 2, 3, 4, 5, 6, 8, 9, or 10.

1. 322 **2**
2. 900 **2, 3, 4, 5, 6, 9, 10**
3. 426 **2, 3, 6**
4. 511 **none**
5. 1,888 **2, 4, 8**

CHALLENGE 7.1

Divisible or Not

Each statement below about divisibility is given as a rule. If you think the rule is correct, write Correct and give an explanation of why it is true. If you think the rule is incorrect, write Incorrect and give a counterexample. A counterexample is an example that shows the rule is not true for every possibility. **Possible explanations and counterexamples are given.**

1. Every number that is divisible by 4 is also divisible by 2.
   Correct; if a number is divisible by 4, it must be an even number, and all even numbers are divisible by 2.

2. If a number is divisible by 6, then it is also divisible by 3.
   Correct; part of the rule for divisibility by 6 is that a number divisible by 3.

3. All numbers that are divisible by 2 are also divisible by 4.
   Incorrect; counterexamples include 6, 10, 14, 18, ...

4. All numbers divisible by 5 are divisible by 2.
   Incorrect; counterexamples include 5, 15, 25, 35, ...

5. If a number is divisible by 9, then it is also divisible by 3.
   Correct; the sum of the digits of any number divisible by 9 will be a number that is also divisible by 3.

6. All numbers divisible by 2 and 4 are also divisible by 8.
   Incorrect; counterexamples include 4, 12, 20, 28, ...

7. If the last digit of a number is 0, the number is not divisible by 9.
   Incorrect; counterexamples include 90, 180, 270, 360, ...

PROBLEM SOLVING 7.1

Divisibility

Analyze Choose Solve Check

Write the correct answer.

1. A bolt manufacturing company has 12,885 bolts to be put into bags. The packing machine can be set to seal either 3, 5, or 6 bolts into each bag. Can the machine be set for any of the three numbers without any bolts being left over? If so, which setting or settings can be used?
   yes; 3 and 5

2. Scott earned $35, $40, $40, $25, and $45 for 5 weeks of part-time work. During a school break, he worked full-time for one week and earned $187. How much greater were his mean weekly earnings with the full-time week included than without it?
   $25 greater

3. What is the least number that is divisible by 2, 3, 4, 5, 6, 8, 9 and 10? What is the least number if 7 is included?
   720; 5,040

4. A supermarket manager wants to make a pyramid of 110 cereal boxes for display. If cereal boxes are packed in cartons of 12, what is the least number of cartons she needs to open?
   10 cartons

Choose the letter for the best answer.

5. A warehouse received 1,448 copies of a book. The manager wants to place them on shelves. Which number of shelves can he use if he wants the same number of books on each shelf?
   A 3 shelves   C 5 shelves
   B 4 shelves   D 6 shelves

6. Amy had a total of $81.60 to spend on 4 gifts. She bought 3 copies of the same book and then had $27 left to spend on a sweater. How much did she pay for each copy of the book?
   F $9.00   H $27.00
   G $18.20   J $54.60

7. Max sells popcorn and potato chips at the ballpark. During one game, he sold a total of 136 bags. He sold 12 fewer bags of chips than popcorn. How many bags of chips did he sell?
   A 124 bags   C 74 bags
   B 84 bags   **D 62 bags**

8. A commercial jet made 3 trips during one 24-hour period, each time carrying the same number of passengers. How many passengers might the plane have carried that day?
   F 516 passengers   H 620 passengers
   G 586 passengers   J 634 passengers

9. **Write About It** Explain how you solved Problem 8.
   Possible answer: I determined which of the possible numbers is divisible by 3, the number of flights.

147

LESSON 7.2

Prime Factorization

LESSON PLANNING

Objective To write a composite number as the product of prime factors

Intervention for Prerequisite Skills
Prime Numbers, Composite Numbers, Factors (For intervention strategies, see page 145.)

NCTM Standards
1. Number and Operations
6. Problem Solving
7. Reasoning and Proof
8. Communication

Math Background

Understanding how to write the prime factorization of a number will help students simplify fractions and find common denominators.

These ideas will help students understand prime factorization.
- When using a factor tree, students may start with any two factors. The final results will be the same.
- When finding factors by dividing, students should use the test of divisibility to decide whether 2 is a factor, then 3, and so on.

Each composite number has only one prime factorization. Changing the order of the factors does not change the factorization. However, it is easier to write a factorization in exponent form when the factors are in order from least to greatest.

WARM-UP RESOURCES

 **NUMBER OF THE DAY**

Write the day of the month as a product of at least two factors. Possible answer: 12th of the month, 6 × 2

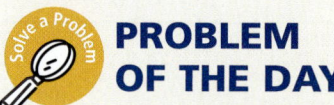 **PROBLEM OF THE DAY**

The sum of the ages of Mr. and Mrs. Olsen and their two children is 108. Their ages are sets of twin prime numbers. What are their ages? NOTE: Twin primes are two prime numbers whose difference is 2.

11 + 13 + 41 + 43 = 108

Solution Problem of the Day tab, p. PD7

 **DAILY FACTS PRACTICE**

Have students practice multiplication facts by completing Set A of *Teacher's Resource Book,* p. TR98.

Vocabulary

prime factorization a number written as the product of all its prime factors

148A Chapter 7

INTERVENTION AND EXTENSION RESOURCES

ALTERNATIVE TEACHING STRATEGY

Materials *For each group* 8 index cards

Have students **make factor trees.** At the top of each card, write a number such as 148, 53, 290, 107, 120, 275, 550, or 333.

1. Give each team of 3 or 4 students a set of cards. One student draws a card and begins a factor tree. Another team member continues the tree, and so on until it is finished.
2. The student who receives a card with a complete prime factorization writes the factorization in exponent form at the bottom of the card.
3. The next team member draws a new card and starts the process over. The team to finish the most correct factorizations in the shortest time wins.

$148 = 2^2 \times 37$; $53 = 1 \times 53$; $290 = 2 \times 5 \times 29$; $107 = 1 \times 107$; $120 = 2^3 \times 3 \times 5$; $275 = 5^2 \times 11$; $550 = 2 \times 5^2 \times 11$; $333 = 3^2 \times 37$

VISUAL
BODILY/KINESTHETIC

ENGLISH LANGUAGE LEARNERS ELL

To **reinforce the meaning of the word** *prime*, display the numbers 1 to 30. Have students circle the prime numbers and discuss with them why each number is or is not prime. Prime numbers between 1 and 30: 2, 3, 5, 7, 11, 13, 17, 19, 23, 29

- After students have identified the primes less than 30, have students write multiplication expressions using the primes as factors. Possible answer: $2 \times 3 \times 5 \times 11$
- Point out that because the expressions use only prime numbers and multiplication, they are also prime factorizations. Ask students to calculate the numbers that are represented by the factorizations.

Possible answer: $2 \times 3 \times 5 \times 11 = 330$

VISUAL
LOGICAL/MATHEMATICAL

MULTISTEP AND STRATEGY PROBLEMS

The following multistep or strategy problem is provided in Lesson 7.2:

Page	Item
149	28

WRITING IN MATHEMATICS

Provide students with additional **practice using prime factorization.** Give students a number, such as 315. Ask them to write a paragraph describing how they would find the prime factors of the number.

Encourage students to use divisibility tests in their process. Point out that making a factor tree or showing the division steps will help make their process clear and logical. Possible answer: Because 315 is divisible by 5, I begin by writing 5×63. Then I can see that 63 is divisible by 3, so I write $5 \times 3 \times 21$. I can factor 21 as 3×7, so $315 = 3 \times 3 \times 5 \times 7 = 3^2 \times 5 \times 7$.

TECHNOLOGY LINK

Intervention Strategies and Activities CD-ROM •
Skills 5, 6, 16

LESSON 7.2 ORGANIZER

Objective To write a composite number as the product of prime factors

Vocabulary prime factorization

1 Introduce

QUICK REVIEW provides review of prerequisite skills.

Why Learn This? Prime factorization can help you find the GCF and LCM of numbers. *Share the lesson objective with students.*

2 Teach

Guided Instruction

- Have students read the first two paragraphs.

  How is a composite number different from a prime number? A composite number has more than two factors; a prime number has exactly two factors.

- Point out that the lesson shows two different methods of finding prime factorization.

  In Example 1, why do you divide first by 2? 2 is the smallest prime factor.

  How can you find factors to use in a factor tree? Possible answer: Use divisibility rules to find a number that divides evenly into a number; that divisor and the resulting quotient are factors of the dividend.

  How does a factor tree help you find a prime factorization? It provides an organized vertical list of the prime factors.

ADDITIONAL EXAMPLES

Example 1, p. 148

Divide to find the prime factorization of 56.

$2 \times 2 \times 2 \times 7$, or $2^3 \times 7$

Example 2, p. 148

Use a factor tree for the prime factorization of 70.

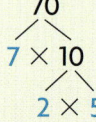

The prime factorization of 70 is $2 \times 5 \times 7$.

LESSON 7.2 Prime Factorization

Learn how to write a composite number as the product of prime numbers.

QUICK REVIEW
Write the equal factors for each.
1. 3^2 2. 2^4 3. 4^3 4. 9^2 5. 5^4
3×3 $2 \times 2 \times 2 \times 2$ $4 \times 4 \times 4$ 9×9 $5 \times 5 \times 5 \times 5$

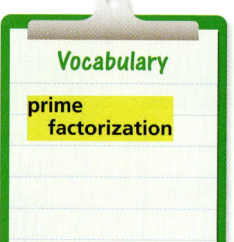

A prime number is a whole number greater than 1 whose only factors are itself and 1. Here are the prime numbers less than 50.

2, 3, 5, 7, 11, 13, 17, 19, 23, 29, 31, 37, 41, 43, 47

A composite number, like 104, has more than two factors. You can write a composite number as the product of prime factors. This is called the **prime factorization** of the number.

You can divide to find the prime factors of a composite number.

EXAMPLE 1 Find the prime factorization of 104.

$$\begin{array}{r}2\,|\underline{104}\\2\,|\underline{\;52}\\2\,|\underline{\;26}\\13\,|\underline{\;13}\\1\end{array}$$

Repeatedly divide by the smallest possible prime factor until the quotient is 1.

$2 \times 2 \times 2 \times 13$ *List the prime numbers you divided by. These are the prime factors.*

So, the prime factorization of 104 is $2 \times 2 \times 2 \times 13$, or $2^3 \times 13$.

Use a factor tree to find the prime factors of a composite number.

EXAMPLE 2 Find the prime factorization of 156.

Choose any two factors of 156. Continue until only prime factors are left.

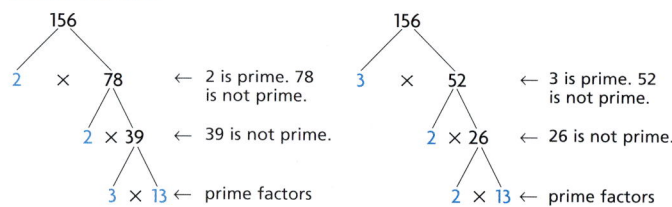

So, the prime factorization of 156 is $2 \times 2 \times 3 \times 13$, or $2^2 \times 3 \times 13$.

Math Idea ▶ Every composite number can be written as a product of two or more prime factors. No matter how you find the prime factors, you will get the same factors, but maybe in a different order.

CHECK FOR UNDERSTANDING

Think and Discuss Look back at the lesson to answer each question.

1. Tell what the prime factorization would be in Example 2 if you started with 4 and 39. $2^2 \times 3 \times 13$

2. Tell how you know when you have finished the prime factorization of a number. All factors are prime numbers.

Guided Practice Use division or a factor tree to find the prime factorization.

3. 12 — $2 \times 2 \times 3$
4. 65 — 5×13
5. 16 — $2 \times 2 \times 2 \times 2$
6. 42 — $2 \times 3 \times 7$

Write the prime factorization in exponent form.

7. 21 — 3×7
8. 28 — $2 \times 2 \times 7$; $2^2 \times 7$
9. 254 — 2×127
10. 908 — $2 \times 2 \times 227$; $2^2 \times 227$

PRACTICE AND PROBLEM SOLVING

Independent Practice Use division or a factor tree to find the prime factorization.

11. 128 — $2 \times 2 \times 2 \times 2 \times 2 \times 2 \times 2$
12. 50 — $2 \times 5 \times 5$
13. 76 — $2 \times 2 \times 19$
14. 108 — $2 \times 2 \times 3 \times 3 \times 3$

Write the prime factorization in exponent form.

15. 18 — $2 \times 3 \times 3$; 2×3^2
16. 302 — 2×151
17. 49 — 7×7; 7^2
18. 217 — 7×31
19. 532 — $2 \times 2 \times 7 \times 19$; $2^2 \times 7 \times 19$
20. 45 — $3 \times 3 \times 5$; $3^2 \times 5$
21. 746 — 2×373
22. 99 — $3 \times 3 \times 11$; $3^2 \times 11$

Solve for *n* to complete the prime factorization.

23. $2 \times n \times 5 = 20$ — $n = 2$
24. $44 = 2^2 \times n$ — $n = 11$
25. $75 = 3 \times 5 \times n$ — $n = 5$

Problem Solving Applications

26. **REASONING** The prime factorization of 50 is 2×5^2. Without dividing or using a factor tree, tell the prime factorization of 100. Possible answer: $2^2 \times 5^2$

27. **REASONING** A number, *c*, is a prime factor of both 12 and 60. What is *c*? $c = 2$ or 3

28. Chris bathed his pet every eighth day in June, beginning on June 8. On what other dates did he bathe his pet in June? 16 and 24

29. **Write About It** Do the prime factors of a number differ depending on which factors you choose first? Explain. No, the order of factors does not affect the value of their product.

MIXED REVIEW AND TEST PREP

30. Is 3,543 divisible by 2, 3, or 9? (p. 146) 3
31. Write the percent for 0.6. (p. 60) 60%
32. $18.3 + 22.6 + 17.03 + 21.99$ (p. 66) 79.92
33. Find the median for the data 13, 8, 9, 16, 18. (p. 106) 13

34. **TEST PREP** A sign is placed in a toy store, asking shoppers to complete a survey about their favorite video game. Which type of sampling is this? (p. 94) A
  A convenience
  B random
  C systematic
  D compatible

EXTRA PRACTICE page H38, Set B

149

3 Practice

Guided Practice

Do Check for Understanding Exercises 1–10 with your students. Identify those having difficulty and use lesson resources to help.

COMMON ERROR ALERT

Simple mistakes can be avoided if students check their work. Remind students that if they multiply the prime factors, the product will be the original number.

Independent Practice

Note that Exercise 28 is a **multistep or strategy problem**. Assign Exercises 11–29.

Algebraic Thinking Writing the factorization in exponent form will help students gain facility working with exponents.

MIXED REVIEW AND TEST PREP

Exercises 30–34 provide **cumulative review** (Chapters 1–7).

4 Assess

Summarize the lesson by having students:

DISCUSS Is $3^2 \times 5 \times 7$ the prime factorization for 305? Explain. No; all the factors are prime, but their product is 315, not 305.

WRITE To find the prime factorization of a number, do you prefer the method shown in Example 1 or the method shown in Example 2? Why? Possible answer: Example 2; drawing a diagram helps me understand the problem.

Lesson Quiz

Transparency 7.2

Write the prime factorization in exponent form.

1. 36 — $2^2 \times 3^2$
2. 54 — 2×3^3
3. 120 — $2^3 \times 3 \times 5$
4. 63 — $3^2 \times 7$

Solve for *y*.

5. $90 = 2 \times 3^2 \times y$ — $y = 5$
6. $84 = 2^2 \times y \times 7$ — $y = 3$

CHALLENGE 7.2

Figure the Prime Factors

Shade one box in each row to show the prime factors of the given number.

PROBLEM SOLVING 7.2

149

LESSON 7.3
Least Common Multiple and Greatest Common Factor

LESSON PLANNING

Objective To find the least common multiple and greatest common factor of numbers and use them to solve problems

Intervention for Prerequisite Skills
Multiples, Factors (For intervention strategies, see page 145.)

NCTM Standards
1. Number and Operations
6. Problem Solving
8. Communication
9. Connections

Vocabulary
least common multiple (LCM) the smallest of common multiples

greatest common factor (GCF) the largest of common factors

Math Background
Prime factorization can be used to find both the least common multiple and the greatest common factor of two or more numbers.

- One way to find a common multiple of two or more numbers is to find their product. The multiple will not necessarily be the LCM. The LCM will contain enough factors so that all the factors of each number are present.
- Some numbers have only one common factor, 1. Others have several. The GCF is the largest of these common factors. One way to find the GCF is to list all the factors of each of the numbers.

In each case, prime factorization provides the most straightforward way to solve LCM and GCF problems.

WARM-UP RESOURCES

 **NUMBER OF THE DAY**

The day of the month is today's number. Find the sum of the first five multiples of the day of the month. **Possible answers: 6th: 6 + 12 + 18 + 24 + 30 = 90; 30th: 30 + 60 + 90 + 120 + 150 = 450**

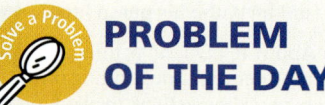 

Find the least values for n and m such that the value of the first expression is twice that of the second. HINT: n and m are both less than 6.

$114 \times n$ and $95 \times m$
$n = 5$ and $m = 3$

Solution Problem of the Day tab, p. PD7

Have students practice division facts by completing Set B of *Teacher's Resource Book*, p. TR98.

INTERVENTION AND EXTENSION RESOURCES

ALTERNATIVE TEACHING STRATEGY

Ask students to **find the GCF and LCM by using prime factorization**. Have them find the prime factors of 45 and 75.

$$45 = 3 \times 3 \times 5 \qquad 75 = 3 \times 5 \times 5$$

Have them identify factors that are shared by the two numbers and underline them.

$$45 = 3 \times \underline{3 \times 5} \qquad 75 = \underline{3 \times 5} \times 5$$

Explain that 3×5, or 15, is the GCF, or greatest common factor, for 45 and 75.

Have students write the prime factorization for the LCM of 45 and 75 by writing all the factors that are common and all the factors that are not common.

$$3 \times \underline{3 \times 5} \times 5 = 225$$

So, the LCM of 45 and 75 is 225.

See also page 152.

VISUAL
LOGICAL/MATHEMATICAL

MULTISTEP AND STRATEGY PROBLEMS

The following multistep and strategy problems are provided in Lesson 7.3:

Page	Item
153	41, 42, Link Up

MATH CONNECTION

Materials reference books

Introduce students to Euclid's ideas about geometry and **ways to find the greatest common factor** by sharing the following:

A famous math book titled *Elements* is more than 2,300 years old. It was written in about 300 B.C. by a Greek mathematician called Euclid. He wrote about geometry and different ways to find the greatest common factor.

Have students find out more about Euclid and his book and share with the class what they learn. Check students' work. The Euclidean algorithm is explained in the Alternative Teaching Strategy, page 152.

VISUAL
BODILY/KINESTHETIC

EARLY FINISHERS

Materials *For each pair* two number cubes labeled 1 to 6

Have students **practice finding the least common multiple**. Give each pair two number cubes. Have them play the following game.

- Roll the number cubes.
- Find and record the LCM of the numbers rolled.
- After five turns each, the winner is the player with the greater number of least common multiples that are 10 or less.

Check students' work.

KINESTHETIC
INTERPERSONAL/SOCIAL

Intervention Strategies and Activities CD-ROM • *Skills 15, 16*

LESSON 7.3 ORGANIZER

Objective To find the least common multiple and greatest common factor of numbers and use them to solve problems

Vocabulary least common multiple (LCM), greatest common factor (GCF)

1 Introduce

QUICK REVIEW provides review of prerequisite skills.

Why Learn This? You can use these skills to add, subtract, multiply, and divide fractions and to write them in simplest form. *Share the lesson objective with students.*

2 Teach

Guided Instruction

- As students look at Examples 1 and 2, help them think about the difference between a common multiple and the least common multiple. Ask:

In the opening problem about Kirk and Amber, the least common multiple is the same as the product of the two numbers. Would that method work for Examples 1 and 2? Explain. No; the resulting multiple would be much greater than the least common multiple.

Why can you find the least common multiple in the opening problem by multiplying the two numbers? 3 and 4 don't have any common factors.

ADDITIONAL EXAMPLE

Example 1, p. 150

Evelyn is making pumpkin pies for the school bake sale. Pie crusts are sold in packages of three. Pie filling is sold in 4-can packages. What is the least number of pie crusts and cans of pie filling Evelyn can buy to have the same number of each? 12 pie crusts and 12 cans of pie filling How many packages of each should she buy? 4 of pie crusts and 3 of filling

See also page 151.

150 Chapter 7

LESSON 7.3

Least Common Multiple and Greatest Common Factor

Learn how to find the LCM and GCF of numbers and use them to solve problems.

QUICK REVIEW

List the first five multiples of each number. See left.
1. 4   2. 2   3. 3
4. 9   5. 6

1. 4, 8, 12, 16, 20
2. 2, 4, 6, 8, 10
3. 3, 6, 9, 12, 15
4. 9, 18, 27, 36, 45
5. 6, 12, 18, 24, 30

Vocabulary
least common multiple (LCM)
greatest common factor (GCF)

Kirk and Amber volunteer during December. Kirk volunteers every fourth day beginning December 4. Amber volunteers every third day beginning December 3. Find the first day they will volunteer together by listing the multiples of 4 and 3.

multiples of 4: 4, 8, **12**, 16, 20, **24**, 28
multiples of 3: 3, 6, 9, **12**, 15, 18, 21, **24**, 27, 30

The multiples that appear in blue are called common multiples. The smallest of the common multiples is called the **least common multiple**, or **LCM**. The LCM of 4 and 3 is 12, the product of the two numbers. So, the first day they volunteer together is December 12.

Examples 1 and 2 show two ways to find the LCM.

EXAMPLE 1

One Way Find the LCM of 12 and 8.
12: 12, **24**, 36, **48**, 60, 72, 84, 96   *List the first eight multiples.*
8: 8, 16, **24**, 32, 40, **48**, 56, 64   *Find the common multiples.*
So, the LCM of 12 and 8 is 24.   *Find the LCM.*

EXAMPLE 2

Another Way Find the LCM of 6, 9, and 18.

Write the prime factorizations.
$6 = 2 \times 3$ $9 = 3 \times 3 = 3^2$ $18 = 2 \times 3 \times 3 = 2 \times 3^2$

Write a product using each prime factor only once.
2×3

For each factor, write the greatest exponent used with that factor in any of the prime factorizations. Multiply.
$2 \times 3^2 = 18$

So, the LCM of 6, 9, and 18 is 18.

150 Chapter 7

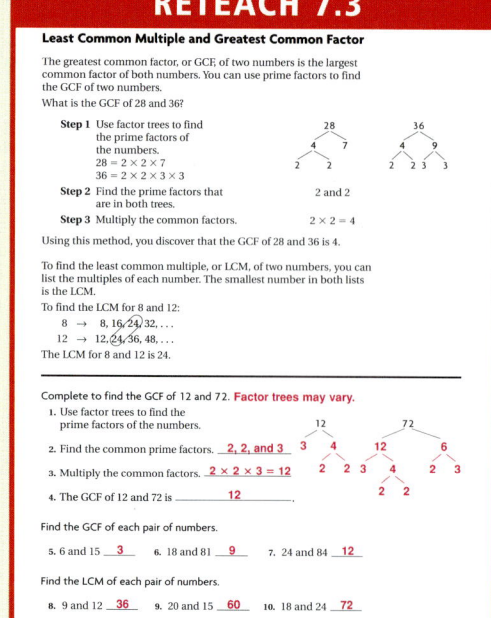

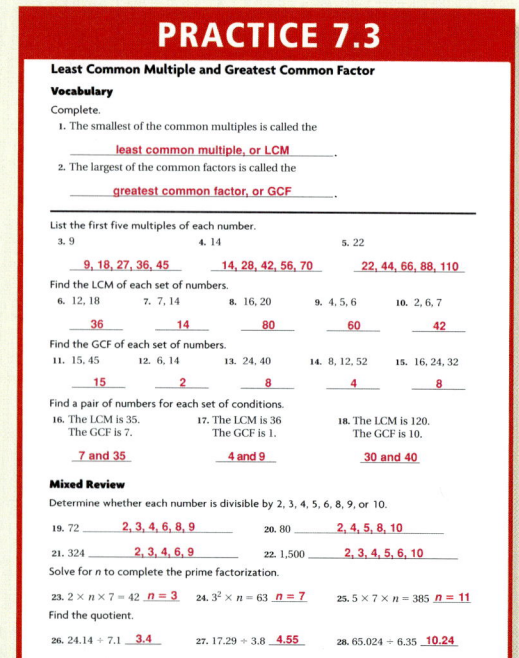

Greatest Common Factor

Factors shared by two or more numbers are called common factors. The largest of the common factors is called the **greatest common factor**, or **GCF**.

To find the GCF of two or more numbers, list all the factors of each number, find the common factors, and then find the greatest common factor.

45: 1, 3, 5, 9, 15, 45   *The common factors are 1, 3, and 9.*
27: 1, 3, 9, 27   *The GCF of 45 and 27 is 9.*

The GCF can be used to solve problems.

EXAMPLE 3

Carlyn has 12 pens and 36 pencils. She is making packages with the same number of each item. What is the greatest number of packages she can make without any items left over? How many of each item will be in each package?

You can find the greatest number of packages by finding the GCF of 12 and 36.

12: 1, 2, 3, 4, 6, 12   *List the factors.*
36: 1, 2, 3, 4, 6, 9, 12, 18, 36   *Find the common factors.*
The GCF of 12 and 36 is 12.   *Find the GCF.*

So, Carlyn can make 12 packages without any items left over.

To find the number of each item, divide the number of pens and the number of pencils by the number of packages.

Pens: $12 \div 12 = 1$ Pencils: $36 \div 12 = 3$

So, there will be 1 pen and 3 pencils in each package.

To find the GCF of two numbers, you can also use their prime factors. List the prime factors, find the common prime factors, and then find their product.

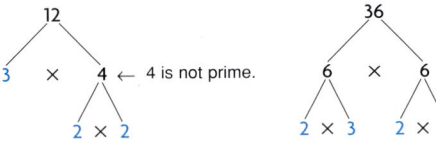

3 × 4 ← 4 is not prime.   6 × 6 ← 6 is not prime.
    2 × 2   2 × 3   2 × 3

The prime factors of 12 are $2 \times 2 \times 3$.

The prime factors of 36 are $2 \times 2 \times 3 \times 3$.

The common prime factors are 2, 2, and 3.

Find the product of the common factors: $2 \times 2 \times 3$, or $2^2 \times 3 = 12$.

The GCF of 12 and 36 is 12.

151

LESSON 7.3

- Using the prime factorization in Example 4, demonstrate the relationship between the product of two numbers and the product of their GCF and LCM.

Compare the product of 48 and 72 to the product of their GCF and LCM.

GCF of 48 and 72 = 24
LCM of 48 and 72: $2 \times 2 \times 2 \times 2 \times 3 \times 3$
$2^4 \times 3^2 = 16 \times 9 = 144$
$48 \times 72 = 3{,}456 \quad 24 \times 144 = 3{,}456$
The product of two numbers and the product of their GCF and LCM are equal.

ADDITIONAL EXAMPLE

Example 4, p. 152

Use prime factors to find the GCF of 30 and 12.
$30 = 2 \times 3 \times 5$
$12 = 2 \times 2 \times 3$
The prime factors are 2, 3, and 5. The common prime factors are 2 and 3. $2 \times 3 = 6$.
So, the GCF of 30 and 12 is 6.

3 Practice

Guided Practice

Do Check for Understanding Exercises 1–14 with your students. Identify those having difficulty and use lesson resources to help.

Independent Practice

Note that Exercises 41 and 42, and the Link Up question are **multistep or strategy problems**. Assign Exercises 15–43.

EXAMPLE 4

Use prime factors to find the GCF of 48 and 72.

48:	$2 \times 2 \times 2 \times 2 \times 3$	Find the prime factors.
72:	$2 \times 2 \times 2 \times 3 \times 3$	
2, 2, 2, and 3		Find the common prime factors.
$2 \times 2 \times 2 \times 3 = 24$		Multiply the common factors.

So, the GCF of 48 and 72 is 24.

CHECK FOR UNDERSTANDING

Think and Discuss — Look back at the lesson to answer each question.

1. **Explain** why you are able to find a greatest common factor but not a greatest common multiple of 2 or more numbers.
   1. Possible answer: A number has a limited number of factors, but its multiples are unlimited.

2. **Tell** the number of pens and pencils there would be in each package in Example 3 if Carlyn made 6 packages.
   2 pens and 6 pencils

Guided Practice — List the first five multiples of each number.

3. 3 *3, 6, 9, 12, 15*
4. 7 *7, 14, 21, 28, 35*
5. 11 *11, 22, 33, 44, 55*
6. 15 *15, 30, 45, 60, 75*

Find the LCM of each set of numbers.

7. 3, 7 *21*
8. 2, 3 *6*
9. 6, 9 *18*
10. 5, 8, 20 *40*

Find the GCF of each set of numbers.

11. 6, 9 *3*
12. 4, 20 *4*
13. 9, 24 *3*
14. 12, 16, 20 *4*

PRACTICE AND PROBLEM SOLVING

Independent Practice — List the first five multiples of each number.

15. 4 *4, 8, 12, 16, 20*
16. 8 *8, 16, 24, 32, 40*
17. 16 *16, 32, 48, 64, 80*
18. 55 *55, 110, 165, 220, 275*
19. 10 *10, 20, 30, 40, 50*
20. 27 *27, 54, 81, 108, 135*
21. 14 *14, 28, 42, 56, 70*
22. 39 *39, 78, 117, 156, 195*

Find the LCM of each set of numbers.

23. 3, 8, 24 *24*
24. 32, 128 *128*
25. 12, 20 *60*
26. 40, 105 *840*
27. 24, 30 *120*
28. 18, 21, 36 *252*
29. 12, 27 *108*
30. 48, 116 *1,392*

Find the GCF of each set of numbers.

31. 16, 18 *2*
32. 15, 18 *3*
33. 21, 306 *3*
34. 16, 24, 40 *8*
35. 25, 33 *1*
36. 200, 215 *5*
37. 24, 32, 40 *8*
38. 630, 712 *2*

Find a pair of numbers for each set of conditions.

39. The LCM is 36. The GCF is 3. *9 and 12*
40. The LCM is 24. The GCF is 2. *6 and 8*

152 Chapter 7

ALTERNATIVE TEACHING STRATEGY SCAFFOLDED INSTRUCTION

Purpose Students use the Euclidean algorithm as an alternate method for finding GCF for greater numbers.

Explain to students that some numbers are too large to use methods such as listing all factors or prime factors in order to find their GCF. Tell them there is another method they can use to find the GCF of greater whole numbers.

Display the following numbers and rules:

791 and 2,034

1. Divide the larger number by the smaller number.

2. If the remainder is not 0, divide the divisor by the remainder.

3. Continue dividing the divisor by the remainder until the remainder is 0.

4. The final divisor is the GCF.

Demonstrate this method of finding the GCF.

```
    2           1           1           3
791)2034   452)791     339)452    113)339
    1582        452         339         339
    452         339         113           0
```

The final divisor is 113, so the GCF for 791 and 2,034 is 113.

Have students work in pairs to use the Euclidean algorithm to find the GCF of 388 and 1,067.

When finished, have a volunteer show his or her solution:

```
       2           1           3
388)1067    291)388     97)291
    776         291         291
    291          97           0
```

The final divisor is 97, so the GCF for 388 and 1,067 is 97.

152 Chapter 7

Problem Solving Applications

41. Peter will distribute cereal samples with pamphlets about good nutrition. The samples come in packages of 15. The pamphlets come in packages of 20.
    a. What is the least number of cereal samples and pamphlets needed to have equal amounts? **60**
    b. How many packages of each does he need? **4 packages of cereal samples, 3 packages of pamphlets**
42. Ruth has 36 markers and 48 erasers. She will put them in bags with the same number of each item. What is the greatest number of bags she can make? **12 bags**
43. **What's the Error?** Jan says the LCM of 10 and 15 is 5. Find her error and the correct answer. **She found the GCF instead of LCM. The LCM is 30.**

MIXED REVIEW AND TEST PREP

44. Write the prime factorization for 45 in exponent form. (p. 148) **5×3^2**
45. Evaluate $a \div c$, for $a = 4{,}602$ and $c = 37$. (p. 28) **124 r14**
46. Find the value of 5^4. (p. 40) **625**
47. **TEST PREP** Which is the exponential notation for $3 \times 3 \times 3 \times 4 \times 4 \times 5 \times 5$? (p. 40) **C**
    A $3 \times 4 \times 5^7$ B $3^3 \times 2^4 \times 2^5$ C $3^3 \times 4^2 \times 5^2$ D $3^3 \times 4^4 \times 5^5$
48. **TEST PREP** Which shows the decimal for 46%? (p. 60) **G**
    F 0.046 G 0.46 H 4.6 J 46

PROBLEM SOLVING LINKUP to Careers

NASA Scientist When astronauts service and repair the Hubble Space Telescope, NASA scientists must set the orbit of the space shuttle so it will meet the Hubble Space Telescope in its orbit. The Hubble Space Telescope takes about 95 minutes to make one orbit around Earth. The space shuttle takes about 90 minutes to go around Earth.

One way to find when the two objects will meet is to find the LCM of the orbit times.

90: 90, 180, 270, 360, 450, 540, … 1,350; 1,440; 1,530; 1,620; 1,710

95: 95, 190, 285, 380, 475, 570, … 1,330; 1,425; 1,520; 1,615; 1,710

So, the space shuttle will meet the Hubble Space Telescope about every 1,710 minutes, or $28\frac{1}{2}$ hours.

- The Russian Mir space station orbits Earth about every 93 minutes. How often would Mir and the space shuttle meet? **about every 2,790 min, or $46\frac{1}{2}$ hrs**

EXTRA PRACTICE page H38, Set C

153

Multistep or Strategy Problem To solve Exercise 42, students can find the GCF for 36 and 48. Guide students to conclude that the GCF is the maximum number of bags.

COMMON ERROR ALERT

Students often confuse the meanings of the terms *LCM* and *GCF*. Have students write the words for each abbreviation. Have them underline the last word and write an expression to show a multiple and to show a factor.

Greatest Common Factor
$5 \times 12 = 60$ 5 is a factor of 60.

Least Common Multiple
$3 \times 8 = 24$ 24 is a multiple of 3.

MIXED REVIEW AND TEST PREP
Exercises 44–48 provide **cumulative review** (Chapters 1–7).

LINKUP to CAREERS

- Have students read the Link Up. Encourage students to share what they know about NASA history.

REASONING The Hubble Telescope passes overhead at noon. After 8:00 P.M., when will it pass overhead again? **9:30 P.M.**

4 Assess

Summarize the lesson by having students:

DISCUSS David has 14 cookies and 35 sticks of gum. Does he have enough to give the same number of each type of treat to each of 7 guests at his party? Explain. **Yes; the GCF of 14 and 35 is 7. Each guest will get 2 cookies and 5 sticks of gum.**

WRITE Explain the difference between the LCM and the GCF. **The LCM is the least number that is a multiple of two or more numbers. The GCF is the greatest number that is a factor of two or more numbers.**

Lesson Quiz

Transparency 7.3

Find the LCM of each set of numbers.
1. 12, 16 **48**   2. 12, 20 **60**
3. 4, 12 **12**   4. 10, 12 **60**

Find the GCF of each set of numbers.
5. 18, 30 **6**   6. 8, 24 **8**
7. 4, 26 **2**   8. 20, 50 **10**

153

READING STRATEGY

K-W-L Chart Before having students read the Link Up, have them look at the title and scan the text for proper nouns. Ask them to predict what the Link Up will be about. Then have students make a three-column chart headed What I Know, What I Want to Know, and What I Learned. Ask them to fill in the first two columns. Have them fill in the third column as they read the paragraph.

K-W-L Chart

What I Know	What I Want to Know	What I Learned

LESSON 7.4
Problem Solving Strategy: Make an Organized List

LESSON PLANNING

Objective To solve problems by using the strategy *make an organized list*

Intervention for Prerequisite Skills
Multiples (For intervention strategies, see page 145.)

Lesson Resources Problem Solving Think Along, p. TR1

NCTM Standards
1. Number and Operations
6. Problem Solving
10. Representation

Math Background
The strategy *make an organized list* is useful when there are several possibilities that can occur in a problem. These ideas may help students to use this strategy:
- Combine this strategy with that of *make a table* to keep track of all possibilities.
- Organize information in a list to avoid omitting possibilities.

WARM-UP RESOURCES

 **NUMBER OF THE DAY**

Today's number is the current time. Write the digits of the time in order. Then find the prime factorization of the number. **Possible answer for 11:12 A.M.: 1,112 = 2 × 2 × 2 × 139**

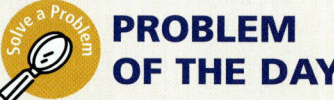 

Jon is making birdhouses. He is cutting 1-in. by 6-in. boards into pieces that are 11 in. long to make the sides of the birdhouses. Which board length would produce the least waste—4 ft, 6 ft, 8 ft, or 10 ft? Explain. **4 ft; possible multiples of 11 are 44, 55, 66, 77, 88, 99, and 110; board lengths are 48 in., 72 in., 96 in., and 120 in.; subtract multiples of 11 from next largest board length to find waste; 48 − 44 = 4; 72 − 66 = 6; 96 − 88 = 8; 120 − 110 = 10.**

Solution Problem of the Day tab, p. PD7

 **DAILY FACTS PRACTICE**

Have students practice multiplication facts by completing Set C of *Teacher's Resource Book*, p. TR98.

INTERVENTION AND EXTENSION RESOURCES

REACHING ALL LEARNERS

ALTERNATIVE TEACHING STRATEGY — ELL

Materials 2 different colors of chalk

Students can **use a number line model to solve the problem** on page 154. Put a large number line on the classroom floor numbered from 0 to 70.

- Ask two volunteers to play the role of wheels with 4-ft and 18-ft circumferences. Give each volunteer a piece of colored chalk.
- Have students call out multiples of four as the person representing the 4-ft wheel marks them on one side of the number line. Repeat the process as the 18-ft wheel volunteer marks multiples of 18 on the opposite side of the number line. Ask:

How many turns each did it take for the wheels to line up again? 9 for the 4-ft wheel; 2 for the 18-ft wheel

How is this way of solving the problem like making an organized list? The number line is like a table or list. The multiples of 4 and 18 are recorded, in order, on the two sides of the number line.

KINESTHETIC
VERBAL/LINGUISTIC

SPECIAL NEEDS

Materials *For each pair* 1 cardboard circle with diameter $3\frac{3}{16}$ in., 1 cardboard circle with diameter $3\frac{13}{16}$ in., 10 ft of butcher paper

Direct students to **build models of wheels to solve a problem.** Ask students to label the smaller cardboard circle with a circumference of 10 and the larger circle with a circumference of 12. Then, ask them to mark a small line at the edge of each circle and a starting line on the butcher paper.

- Have volunteers roll the circles along the sides of the paper starting with the mark on each circle touching the starting line. As one volunteer rolls a circle, another marks each spot where the mark on the circle touches the paper.
- After students have rolled each circle several times and marked the paper, ask:

What does each mark represent for the smaller circle? for the larger circle? 10, 20, 30, and so on; 12, 24, 36, and so on

- Direct students to use their drawings to find the LCM of 10 and 12. 60

KINESTHETIC
INTERPERSONAL/SOCIAL

READING STRATEGY

Synthesize Information Have the students use the reading strategy *synthesize information* to bring together the parts of the problem on page 154. They should recognize that they need to know that if the circumference of the smaller wheel is 4 ft, it will travel 4 ft with each revolution. Ask:

What do the multiples of 4 and 18 represent in this problem? the distance each wheel has rolled after 1 turn, 2 turns, and so on

CAREER CONNECTION

Students can **apply problem solving skills to a career.** Share the following:

- A track and field coach trains athletes for various track and field events. Track events involve running—for example, sprints and relays. Field events involve jumping and throwing—for example, high jump and shot put.
- Marie trains runners for a competition. She wants both runners to run the same distance each day. One runner trains on a 400 yd track, and the other trains on a 300 yd one. What is the least number of laps each should run so that they run the same distance? first runner: 3 laps; second runner: 4 laps

VISUAL
VISUAL/SPATIAL

MULTISTEP AND STRATEGY PROBLEMS

The following multistep and strategy problems are provided in Lesson 7.4:

Page	Item
155	1–9

TECHNOLOGY LINK

- Intervention Strategies and Activities CD-ROM • *Skill 15*
- The Harcourt Learning Site

LESSON 7.4 ORGANIZER

Objective To solve problems by using the strategy *make an organized list*

Lesson Resources Problem Solving Think Along, p. TR1

1 Introduce

QUICK REVIEW provides review of prerequisite skills.

Why Learn This? Making organized lists is helpful when you need to sort, rename, and rearrange files or folders on a computer. *Share the lesson objective with students.*

2 Teach

Guided Instruction

- As you discuss the Boneshaker problem with the class, ask:

REASONING Do you think your feet would go around more often if the larger wheel were in the front or if the smaller wheel were in front? *Possible answer: more times with the smaller wheel in the front because it has to go around more times to cover the same distance as the larger wheel.*

- Discuss the remaining steps in the solution.

What two things do 4, 8, ... represent? *multiples of 4 and the number of feet the small wheel has rolled as it completes a revolution*

- After students read the What If question, ask:

What multiples will you use to solve the What if question? *8, 16, 24, 32, 40, ... ; 18, 36, 54, ...*

Modifying Instruction Once students realize this type of problem calls for the LCM, they may choose to solve by using prime factorization instead of the organized list strategy.

154 Chapter 7

LESSON 7.4

Learn how to solve problems by making an organized list.

PROBLEM SOLVING STRATEGY
Make an Organized List

QUICK REVIEW
List the first four multiples of each number.
1. 3
   3, 6, 9, 12
2. 5
   5, 10, 15, 20
3. 9
   9, 18, 27, 36
4. 11
   11, 22, 33, 44
5. 20
   20, 40, 60, 80

The Boneshaker was invented in 1865. The wooden wheels made for an uncomfortable ride.

Like many early bicycles, the Boneshaker had different sized wheels. Suppose the circumferences of the wheels were 4 ft and 18 ft. How many revolutions would each wheel make before marks on their rims would both be in the same positions again?

Analyze What are you asked to find? *The number of revolutions each wheel will make before the marks would both be in the same positions again.*

What facts are given? *The distance around each wheel.*

Choose What strategy will you use?

You can use the strategy *make an organized list*. List the total distance traveled by each wheel for every complete turn.

Solve How will you solve the problem?

Make a list of multiples of 4 and 18.
 multiples of 4: 4, 8, 12, 16, 20, 24, 28, 32, 36, 40
 multiples of 18: 18, 36, 54

The least common multiple is 36. When the wheels have traveled 36 ft, the marks will both be in the same positions again.

36 is the ninth multiple of 4, so the small wheel makes 9 revolutions.

36 is the second multiple of 18, so the large wheel makes 2 revolutions.

Check How can you check your answer? *Possible answer: draw a diagram*

What if the smaller wheel had a circumference of 8 ft? How many revolutions would each wheel make before the marks were both in the same positions again? *smaller, 9 revolutions; larger, 4 revolutions*

154 Chapter 7

RETEACH 7.4
Problem Solving Strategy: Make an Organized List

Organizing data into a list is one way of making sure that you have considered every possibility for a given situation.

Two trains that ran in Great Britain in the 1930s were the Night Scotsman and the Bournemouth Belle. They both ran between London and other cities. The Night Scotsman left London every 4 days and the Bournemouth Belle left every 6 days. If they both left London on Thursday, March 1, on which other dates in March did they leave London on the same day?

Step 1 Think about what you know and what you are asked to find.
- You know how often each train left from London:
  The Night Scotsman left every 4 days.
  The Bournemouth Belle left every 6 days.
  You know that they both left London on Thursday, March 1.
- You are asked to find the other dates in March on which both trains left London on the same day.

Step 2 Plan a strategy to solve.
- Use the strategy *make an organized list*.
- Make a list of the multiples of 4 and the multiples of 6.
- Use the common multiples to determine the other dates in March on which both trains left London on the same day.

Step 3 Solve.
- Carry out the strategy. Make lists of the multiples.
  multiples of 4: 4, 8, **12**, 16, 20, **24**, 28, 32, ...
  multiples of 6: 6, **12**, 18, **24**, 30, 36, 42, ...
  The common multiples of 4 and 6 that are less than 31 (the number of days in March) are 12 and 24.

So, the trains both left London together 12 days after March 1 and again 24 days after March 1. So, the dates they left on the same day were:
 1 + 12, or March 13 and 1 + 24, or March 25.

Solve the problem by making an organized list.

1. A bus company has routes between Chicago and several other cities. Every 5 hours a bus leaves from Chicago for Detroit and every 3 hours another bus leaves for Cleveland. If buses leave for both cities at 6:00 A.M., when is the next time two buses will leave together for these two cities?
   9:00 P.M. that same day

2. Some buses have seats for 40 people. Other buses can seat only 36. One of each type of bus is filled with families. There are no empty seats and all the families are the same size. What is the greatest number of members that each family can possibly have?
   4 members

PRACTICE 7.4
Problem Solving Strategy: Make an Organized List

Solve the problem by making an organized list.

1. Jack and Ashley begin jogging around a quarter-mile track at the same time. Ashley takes 2 minutes to complete each lap and Jack takes 3 minutes. How many laps will each have run the first time they are side-by-side again at the point where they began?
   Ashley: 3 laps; Jack: 2 laps

2. Terrence is taking two medications for his flu. He begins taking them both at 10:00 P.M. on Tuesday. If he takes one every 8 hours and the other every 10 hours, on what day and at what time will he take the two medications together again?
   2:00 P.M. on Thursday

3. A large high school has a marching band with 64 woodwind players and 72 brass players. All members of the band line up in rows of equal size. Only musicians playing the same instruments are in each row. What is the greatest number of musicians who can be in one row?
   8 musicians

4. Brice plays in a basketball league. In his last game, he scored more than 20 but fewer than 30 points by making a combination of 2- and 3-point shots. If he made 5 more 2-point shots than 3-point shots, how many of each type did he make?
   8 2-point shots, 3 3-point shots

5. Aki is buying franks and buns for a field trip. She sees franks in packages of 6 and buns in packages of 8. There are 70 people going on the trip. What is the least number of each she can buy so there are franks and buns for everyone, with no extra packages?
   9 packages of buns, 12 packages of franks

6. Kiona has 235 CDs. She is buying CD holders for her collection. The two types that she likes hold 20 CDs and 12 CDs each. She wants to buy the same number of each type. What is the least number of each type of CD holder that Kiona will have to buy to hold her entire CD collection?
   8 of each type

Mixed Review
Estimate the sum or difference. *Possible estimates are given.*

7. 80 + 31 + 87
   200
8. 710 − 189
   500
9. 1,208 + 877 + 439
   2,500
10. 7,151 − 2,993
    4,000
11. 67 + 123 + 804
    1,000
12. 920 − 592
    300

PROBLEM SOLVING PRACTICE

Solve the problem by making an organized list.

1. Justin and Amy help with the shopping for their individual families. Justin goes to the store every 3 days and Amy goes every 5 days. They see each other at the store on September 30. On what date will Justin and Amy see each other at the store again? **October 15**

PROBLEM SOLVING STRATEGIES
- Draw a Diagram or Picture
- Make a Model
- Predict and Test
- Work Backward
- ▶ Make an Organized List
- Find a Pattern
- Make a Table or Graph
- Solve a Simpler Problem
- Write an Equation
- Use Logical Reasoning

For 2–3, use the information below.
Susanna buys one bag with 40 snacks and another with 32. She wants to make up small snack packs for a party. All snack packs must have the same number of each snack.

2. If you make lists to find the greatest number of snack packs Susanna can make, what will you include in the lists? **B**

   A addends   **B** factors   **C** multiples   **D** fractions

3. What is the greatest number of snack packs Susanna can make? **H**

   F 2 packs   **G** 4 packs   **H** 8 packs   **J** 15 packs

MIXED STRATEGY PRACTICE

4. Ray begins an eight-week exercise program. He exercises 30 min during the first week, 45 min the second week, and 60 min the third week. If this pattern continues, how many hours and minutes will Ray exercise the sixth week? **1 hr 45 min**

5. DeAnn has a total of 74 in. of yarn to make gifts. She uses 13 in. of yarn for each of the 5 gifts she is making. Find how many inches of yarn DeAnn will have left after making the 5 gifts. **9 in.**

6. A total of 316 middle school teachers are going to a meeting. There are 30 more male teachers than female. How many teachers are male? **173 are male**

7. Christi rides her bicycle 7 blocks south, 3 blocks east, 5 blocks north, and 8 blocks west. How many blocks has she ridden when she crosses her own path? **18 blocks**

8. Use the graph at the right. Look at the area in square miles for each lake. How much larger is Lake Michigan than Lake Erie and Lake Ontario together? **4,920 sq mi**

9. ❓ **What's the Question?** Laura wants to buy the same number of plums and kiwis. Plums are sold in bags of 4 and kiwis are sold in bags of 7. The answer is 28 plums. **Possible question: What is the least number of plums Laura can buy?**

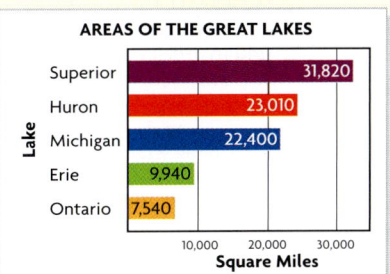

AREAS OF THE GREAT LAKES
- Superior: 31,820
- Huron: 23,010
- Michigan: 22,400
- Erie: 9,940
- Ontario: 7,540

155

---

3 Practice

Guided Practice

Note that Exercises 1–3 are **multistep or strategy problems**. Do Problem Solving Practice Exercises 1–3 with your students. Identify those having difficulty and use lesson resources to help.

Independent Practice

Note that Exercises 4–9 are **multistep or strategy problems**. Assign Exercises 4–9.

Before students begin the exercises, refer them to the list of strategies on the PE page that they might use to solve the problems.

Scaffolded Instruction Use the prompts on Transparency 7 to guide instruction for the multistep or strategy problem in Exercise 7.

4 Assess

Summarize the lesson by having students:

DISCUSS Describe how to find the first ten multiples of a number such as 18. Possible answer: Multiply 18 by 1, by 2, by 3, and so on.

WRITE Describe how to find the LCM of two numbers such as 8 and 14. Possible answer: First, list some multiples of 8. Then list some multiples of 14. Find the smallest number that appears on both lists.

Lesson Quiz

1. Karena shops for groceries every 6 days and for gas every 5 days. How many times a month does she shop for both on the same day? **for 28- to 30-day months—0 times; for 31-day months—1 time**

2. Tim is supposed to give an equal number of red cubes and blue cubes to students in his class. He has 28 red cubes and 35 blue cubes. What is the greatest number of students to whom he can give cubes? **7 students**

CHALLENGE 7.4

Organization is the Key!

Solve each problem by making an organized list of the data.

1. Ana and Juan are going shopping for gifts. They need to go to a toy store (T), a stationery store (S), and a jewelry store (J). Make a list of all the possible orders in which they can visit the stores. How many choices do they have for the order in which they visit the stores?
   S-T-J; S-J-T; J-S-T; J-T-S; T-J-S; T-S-J; 6 choices

2. Suppose Ana needs to stop at her bank before going to any stores. In how many different orders can they make their four stops?
   6 different orders

3. Assume it does not matter whether Ana goes to the bank before shopping or after shopping. In how many different orders can they then make their four stops?
   12 different orders

4. Juan and his three friends, Amy, Vijay, and Yoko, will each talk to one another on the phone today before agreeing where to meet after he finishes shopping. If each friend speaks to the others once, how many phone calls will there be in all? (Keep in mind that once Juan calls Amy, for example, there is no need for Amy to call Juan.)
   6 phone calls

5. When wrapping the three gifts that she has bought, Ana has a choice of three different wrapping papers to use: one with stripes, one with balloons, and one with elephants. She cannot decide whether to wrap all the gifts in the same paper or to use a different paper for each gift. In how many different ways can she wrap the three gifts? (Count each combination of wrapping papers only once.)
   10 ways
   SSS, SSB, SSE, BBB, BBS, BBE, EEE, EES, EEB, SEB

6. If Ana decides to use a different wrapping paper for each gift, in how many different ways can she wrap her three gifts?
   6 ways

READING STRATEGY 7.4

Synthesize Information

Analyze Choose Solve Check

To **synthesize** means to form a whole by combining parts. You can combine new information to make something from the separate parts. One way to do this is to make an organized list. Read the following problem.

VOCABULARY synthesize

Al, Jo-Jo, and Tom are standing on the first step of a staircase. There are 16 steps in all. Al goes up the staircase one step at a time. Jo-Jo skips one step each time. Tom skips two steps each time. On which steps will they all place a foot?

1. Make a list to show on which step each person steps.
   Al: 2, 3, 4, 5, 6, 7, 8, 9, 10, 11, 12, 13, 14, 15, 16
   Jo-Jo: **3, 5, 7, 9, 11, 13, 15**
   Tom: **4, 7, 10, 13, 16**

2. Synthesize the information by finding the number that appears in all three lists. Solve the problem.
   seventh and thirteenth steps

Synthesize the information by making an organized list. Solve.

3. Al, Jo-Jo, and Tom are climbing the staircase, as described above. All three boys start on their left foot. Which is the next step on which they will all place their left foot?
   thirteenth step

4. Al walks 1 mi in 14 min. Jo-Jo bikes 1 mi in 6 min. Tom runs 1 mi in 7 min. If they all start from the same place on a 1-mi track, how many miles will each boy have traveled when they are all on that spot again?
   Al: 3 mi; Jo-Jo: 7 mi; Tom: 6 mi

5. Al, Jo-Jo, and Tom are climbing the staircase. They all start on their left foot. Will all three boys ever step on the same stair with their right foot? Explain.
   No. They all step on the seventh stair, but only Jo-Jo steps with his right foot. They all step on the thirteenth stair, but with their left feet.

6. Jo-Jo jogs every fourth day of each month. Al jogs every sixth day of each month. On which days of each month can they jog together?
   the twelfth and the twenty-fourth

155

CHAPTER 7

REVIEW/TEST

Purpose To check understanding of concepts, skills, and problem solving presented in Chapter 7

USING THE PAGE

The Chapter 7 Review/Test can be used as a **review** or a **test**.
- Items 1–3 check understanding of concepts and new vocabulary.
- Items 4–36 check skill proficiency.
- Items 37–40 check students' abilities to choose and apply problem solving strategies to real-life problems involving number theory.

 Suggest that students place the completed Chapter 7 Review/Test in their portfolios.

USING THE ASSESSMENT GUIDE

- Multiple-choice format of Chapter 7 Posttest—See *Assessment Guide*, pp. AG49–50.
- Free-response format of Chapter 7 Posttest—See *Assessment Guide*, pp. AG51–52.

USING STUDENT SELF-ASSESSMENT

The How Did I Do? survey helps students assess what they have learned and how they learned it. This survey is available as a copying master in *Assessment Guide*, p. AGxvii.

CHAPTER 7 REVIEW/TEST

1. **VOCABULARY** Writing a composite number as the product of prime factors is called ___?___. (p. 148) **prime factorization**
2. **VOCABULARY** The largest of the common factors of two or more numbers is the ___?___. (p. 150) **GCF**
3. **VOCABULARY** The smallest of the multiples of two or more numbers is the ___?___. (p. 150) **LCM**

Tell whether each number is divisible by 2, 3, 4, 5, 6, 8, 9, or 10. (pp. 146–147)

4. 42 **2, 3, 6**
5. 64 **2, 4, 8**
6. 96 **2, 3, 4, 6, 8**
7. 225 **3, 5, 9**
8. 330 **2, 3, 5, 6, 10**
9. 963 **3, 9**
10. 450 **2, 3, 5, 6, 9, 10**
11. 2,385 **3, 5, 9**

Use division or a factor tree to find the prime factorization. Write the prime factorization in exponent form. (pp. 148–149)

12. 9 3^2
13. 8 2^3
14. 14 2×7
15. 18 2×3^2
16. 80 $2^4 \times 5$
17. 12 $2^2 \times 3$
18. 33 3×11
19. 50 2×5^2
20. 49 7^2
21. 98 2×7^2
22. 504 $2^3 \times 3^2 \times 7$
23. 891 $3^4 \times 11$

24. List the first twenty multiples of 4 and the first twenty multiples of 10. What is the least common multiple? What multiples greater than 80 do they have in common? (pp. 150–153)
LCM: 20; all numbers that are multiples of 20

Find the LCM and the GCF of each set of numbers. (pp. 150–153)

25. 3, 9 **9; 3**
26. 2, 6 **6; 2**
27. 6, 4 **12; 2**
28. 10, 15 **30; 5**
29. 8, 12 **24; 4**
30. 9, 27 **27; 9**
31. 15, 25 **75; 5**
32. 25, 115 **575; 5**
33. 27, 189 **189; 27**
34. 6, 8, 12 **24; 2**
35. 6, 9, 12 **36; 3**
36. 8, 16, 20 **80; 4**

Solve.

37. A number is between 60 and 70. It is divisible by 3 and 9. What is the number? (pp. 146–147) **63**
38. Meat patties are sold in packages of 12. Buns are sold in packages of 8. What is the least number of meat patties and buns needed to have an equal number of each? (pp. 150–153) **24 patties and 24 buns**
39. Cashews are sold in 8-oz jars, almonds in 12-oz jars, and peanuts in 16-oz jars. What is the least number of ounces of each type of nuts you can buy to make mixed nuts with equal amounts of each? How many jars of each would you need? (pp. 150–153)
48 oz; 6 jars of cashews, 4 jars of almonds, 3 jars of peanuts
40. Alissa jogs in the park every 3 days. Erin jogs in the park once a week on Saturday. If they met in the park on Saturday, April 30, when will they meet in the park again? (pp. 154–155) **Saturday, May 21**

156 Chapter 7

CHAPTERS 1–7 ★ STANDARDIZED TEST PREP

 **Get the information you need.**
See item **1**.
You know that juice boxes are sold in packages of 8 and rice snacks are sold in packages of 10. Use the least common multiple to find an equal number of treats before you determine how many packages to buy.

Also see problem **3**, p. H63.

1. Hu is buying treats for a party. Juice boxes are sold in packs of 8. Rice snacks are sold in packs of 10. What is the least number of packs of juice and rice snacks he should buy to have an equal number of juice packs and rice snacks? **B**

   A 4 juice packs, 5 rice snack packs
   B 5 juice packs, 4 rice snack packs
   C 9 juice packs, 9 rice snack packs
   D 10 juice packs, 8 rice snack packs

2. What is the least common multiple of 18 and 27? **H**

   F 3   H 54
   G 9   J 486

3. Which number is divisible by both 6 and 9? **B**

   A 45   C 243
   B 216   D 768

4. Carni earned the following amounts baby-sitting: $17, $15, $12, $17, $14. What was the mean amount that Carni earned? **G**

   F $14   H $16
   G $15   J Not here

5. What is the greatest common factor of 16 and 24? **B**

   A 4   B 8   C 48   D 384

6. Marcus has a baseball game every fifth day in April. His first game is on April 5. How many games will there be in April? **J**

   F 3   G 4   H 5   J 6

7. A theater holds 381 people. All the available tickets for 18 shows were sold. Which is a good estimate for the number of people who attended the shows? **C**

   A Less than 4,000 people
   B Between 4,000 and 6,000 people
   C Between 6,000 and 8,000 people
   D More than 8,000 people

8. Eleni bakes brownies for a bake sale at school. She wants to make 4 batches of brownies and package the brownies in boxes of 8. How many boxes will she need if there are 32 brownies in each batch? **G**

   F 12   H 20
   G 16   J 32

Write What You Know

9. Explain how to use a factor tree to find the prime factorization of a number. Then use your method to find the prime factorization of 120.

10. See below. If a number is divisible by both 3 and 4, then it is divisible by 12. Use this fact to write a divisibility rule for 12. Explain your reasoning.

157

Write What You Know • Written Response

9. Start with any two factors of the number that are greater than 1. Draw 2 lines and write one factor at the end of each line. If a factor is prime, stop that branch of the factor tree. If the factor is not prime, continue the process. When all branches have prime numbers at their ends, the factor tree is complete. The prime factorization is equal to the product of the prime factors at the ends of the branches. For 120, the prime factorization is $2^3 \times 3 \times 5$. Check students' factor trees.

10. To be divisible by 12, the number must be divisible by both 3 and 4. So, a number is divisible by 12 if the sum of the digits of the number is divisible by 3 and the last two digits of the number form a number divisible by 4.

STANDARDIZED TEST PREP •
Chapters 1–7

USING THE PAGE

This page may be used to help students get ready for standardized tests. The test items are written in the same style and arranged in the same format as those on many state assessments. The page is cumulative. It covers math objectives and essential skills that have been taught up to this point in the text. Most of the items represent skills from the current chapter, and the remainder represent skills from earlier chapters.

This page can be assigned at the end of the chapter as classwork or as a homework assignment. You may want to have students use individual recording sheets presented in a multiple-choice (standardized) format. A Test Answer Sheet is available as a blackline master in *Assessment Guide* (p. AGxlii).

You may wish to have students describe how they solved each problem and share their solutions.

ITEM ANALYSIS

Item	Learning Goal	Item	Learning Goal
1	7D	6	7D
2	7C	7	1B
3	7A	8	7D
4	5C	9	7B
5	7C	10	7A

Written response items for the **Write What You Know** are available as a blackline master in *Performance Assessment*.

SCORING RUBRIC • WRITE WHAT YOU KNOW

2 Demonstrates a complete understanding of the problem and chooses an appropriate strategy to determine the solution

1 Demonstrates a partial understanding of the problem and chooses a strategy that does not lead to a complete and accurate solution

0 Demonstrates little understanding of the problem and shows little evidence of using any strategy to determine a solution

Number Theory 157

CHAPTER 8 Fraction Concepts

CHAPTER PLANNER

PACING OPTIONS
| Compacted | 4 Days |
| Expanded | 8 Days |

Getting Ready for Chapter 8 • Assessing Prior Knowledge and INTERVENTION (See PE and TE page 159.)

LESSON	NCTM STANDARDS	PACING	VOCABULARY*	MATERIALS	RESOURCES AND TECHNOLOGY
8.1 Equivalent Fractions and Simplest Form pp. 160–163 **Objective** To identify and write equivalent fractions and to write fractions in simplest form	1, 5, 6, 8, 10	2 Days	**equivalent fractions** **simplest form**	*For each group* fraction bars or fraction strips	Reteach, Practice, Challenge, Problem Solving 8.1 Worksheets Extra Practice p. H39, Set A Transparency 8.1 E-Lab • *Equivalent Fractions*; E-Lab Recording Sheet Number Heroes • *Fraction Fireworks*
8.2 Mixed Numbers and Fractions pp. 164–165 **Objective** To write fractions as mixed numbers and mixed numbers as fractions	1, 5, 6, 8, 9	1 Day	**mixed number**		Reteach, Practice, Challenge, Problem Solving 8.2 Worksheets Extra Practice p. H39, Set B Transparency 8.2 Calculating Crew • *Nautical Number Line*
8.3 Compare and Order Fractions pp. 166–167 **Objective** To compare and order fractions	1, 6, 7, 8	1 Day			Reteach, Practice, Challenge, Problem Solving 8.3 Worksheets Extra Practice p. H39, Set C Transparency 8.3
8.4 Math Lab: Explore Fractions and Decimals p. 168 **Objective** To convert fractions to decimals	1, 8			*For each student* graph paper, scissors, colored pencils	E-Lab • *Equivalent Fractions, Decimals, and Mixed Numbers*; E-Lab Recording Sheet Astro Algebra • *Red* Number Heroes • *Fraction Fireworks*
8.5 Fractions, Decimals, and Percents pp. 169–171 **Objective** To convert fractions to decimals, decimals to fractions, and fractions to percents	1, 5, 6, 8	1 Day (For Lessons 8.4 and 8.5)	**terminating decimal** **repeating decimal** percent		Reteach, Practice, Challenge, Problem Solving 8.5 Worksheets Extra Practice p. H39, Set D Transparency 8.5 Astro Algebra • *Red* Calculating Crew • *Nautical Number Line* Calculator Handbook pp. 7–8

Ending Chapter 8 • Chapter 8 Review/Test, p. 172 • **Standardized Test Prep**, p. 173

***Boldfaced** terms are new vocabulary. Other terms are review vocabulary.

CHAPTER AT A GLANCE

Vocabulary Development

The boldfaced words are the new vocabulary terms in the chapter. Have students record the definitions in their Math Journals.

equivalent fractions, p. 160
simplest form, p. 161
mixed number, p. 164
terminating decimal, p. 169
repeating decimal, p.169

equivalent fractions

Vocabulary Cards
Have students use the Vocabulary Cards on *Teacher's Resource Book* pp. TR127–128 to make graphic organizers or word puzzles. The cards can also be added to a file of mathematics terms.

NCTM Standards

1. **Number and Operations**
   Lessons 8.1, 8.2, 8.3, 8.4, 8.5
2. **Algebra**
3. **Geometry**
4. **Measurement**
5. **Data Analysis and Probability**
   Lessons 8.1, 8.2, 8.5
6. **Problem Solving**
   Lessons 8.1, 8.2, 8.3, 8.5
7. **Reasoning and Proof**
   Lesson 8.3
8. **Communication**
   Lessons 8.1, 8.2, 8.3, 8.4, 8.5
9. **Connections**
   Lesson 8.2
10. **Representation**
    Lesson 8.1

Writing Opportunities

PUPIL EDITION
- What's the Question?, p. 163
- What's the Error?, pp. 165, 167
- Write About It, pp. 165, 171

- Write a Problem, p. 163

TEACHER'S EDITION
- Write—See the *Assess* section of each TE lesson.
- Writing in Mathematics, p. 164B

ASSESSMENT GUIDE
- How Did I Do?, p. xvii

Family Involvement Activities

These activities provide:
- Letter to the Family
- Model of Essential Skills
- Math Vocabulary
- Family Game

Family Involvement Activities, p. FA29

CHAPTER 8 Fraction Concepts

MATHEMATICS ACROSS THE GRADES

SKILLS TRACE ACROSS THE GRADES

GRADE 5
Write equivalent fractions; write fractions in simplest form; compare and order fractions

GRADE 6
Write fractions in equivalent and simplest form; convert between, compare, and order fractions and mixed numbers; represent and use equivalent representations for fractions, decimals, and percents

GRADE 7
Write fractions and mixed numbers in equivalent and simplest form; represent fractions as terminating or repeating decimals

SKILLS TRACE FOR GRADE 6

LESSON	FIRST INTRODUCED	TAUGHT AND PRACTICED	TESTED	REVIEWED
8.1	Grade 4	PE pp. 160–163, H39, p. RW36, p. PW36, p. PS36	PE p. 172, pp. AG53–56	PE pp. 172, 173, 222–223
8.2	Grade 4	PE pp. 164–165, H39, p. RW37, p. PW37, p. PS37	PE p. 172, pp. AG53–56	PE pp. 172, 173, 222–223
8.3	Grade 4	PE pp. 166–167, H39, p. RW38, p. PW38, p. PS38	PE p. 172, pp. AG53–56	PE pp. 172, 173, 222–223
8.4	Grade 4	PE p. 168	PE p. 172, pp. AG53–56	PE pp. 172, 173, 222–223
8.5	Grade 6	PE pp. 169–171, H39, p. RW39, p. PW39, p. PS39	PE p. 172, pp. AG53–56	PE pp. 172, 173, 222–223

KEY PE Pupil Edition PS Problem Solving Workbook RW Reteach Workbook
PW Practice Workbook AG Assessment Guide

Looking Back  Prerequisite Skills

To be ready for Chapter 8, students should have the following understandings and skills:

- **Vocabulary**—*denominator, numerator*
- **Compare and Order Whole Numbers**—use <, >, and = symbols to compare whole numbers; order whole numbers from greatest to least
- **Model Fractions**—write the fraction that names the shaded part of a model
- **Model Percents**—write the percent that names the shaded part of a model

Check What You Know

Use page 159 to determine students' knowledge of prerequisite concepts and skills.

Intervention

Help students prepare for the chapter by using the intervention resources described on TE page 159.

Looking at Chapter 8  Essential Skills

Students will

- develop skill in writing equivalent fractions and fractions in simplest form.
- write fractions as mixed numbers and mixed numbers as fractions.
- compare and order fractions.
- **make the connection between graphic representations for fractions, decimals, and percents and converting among fractions, decimals, and percents.**

EXAMPLE

Write the fraction as a decimal.

$$\frac{65}{100}$$

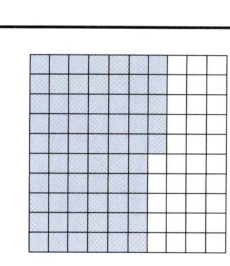

$\frac{65}{100} = 0.65$

So, $\frac{65}{100}$ can be written as 0.65.

Looking Ahead  Applications

Students will apply what they learn in Chapter 8 to the following new concepts:

- Add and Subtract Fractions and Mixed Numbers (Chapter 9)
- Multiply and Divide Fractions and Mixed Numbers (Chapter 10)
- Number Relationships (Chapter 11)
- Rate, Ratio, Percent, and Proportion (Chapter 20)

Fraction Concepts 158D

CHAPTER 8

Fraction Concepts

INTRODUCING THE CHAPTER

Tell students that fractions are used to name equal parts of a whole or of a group. Have students focus on the photo and write a fraction to describe the number of animals. Possible answer: $\frac{2}{3}$ (2 babies to 3 animals)

USING DATA

To begin the study of this chapter, have students

- Find which cat is about half the length of the jaguar. bobcat
- Write a fraction to compare the length of the bobcat to that of the tiger. about $\frac{1}{3}$
- Write three more fractions that could be used to express data relationships on the graph. Answers will vary.

PROBLEM SOLVING PROJECT

Purpose To use fractions to solve a problem

Grouping pairs or small groups

Background The largest member of the cat family is the Siberian tiger. The average length, tip of nose to tail, is 10 ft 4 in.

Analyze, Choose, Solve, and Check

Have students

- Research the average lengths of cats that live in each of five different areas of the world, such as Canada, India, and Brazil.
- Make a table or graph of their results.
- Write fractions that describe the data they have collected.

Check students' work.

 Suggest that students place the tables and graphs in their portfolios.

CHAPTER 8
Fraction Concepts

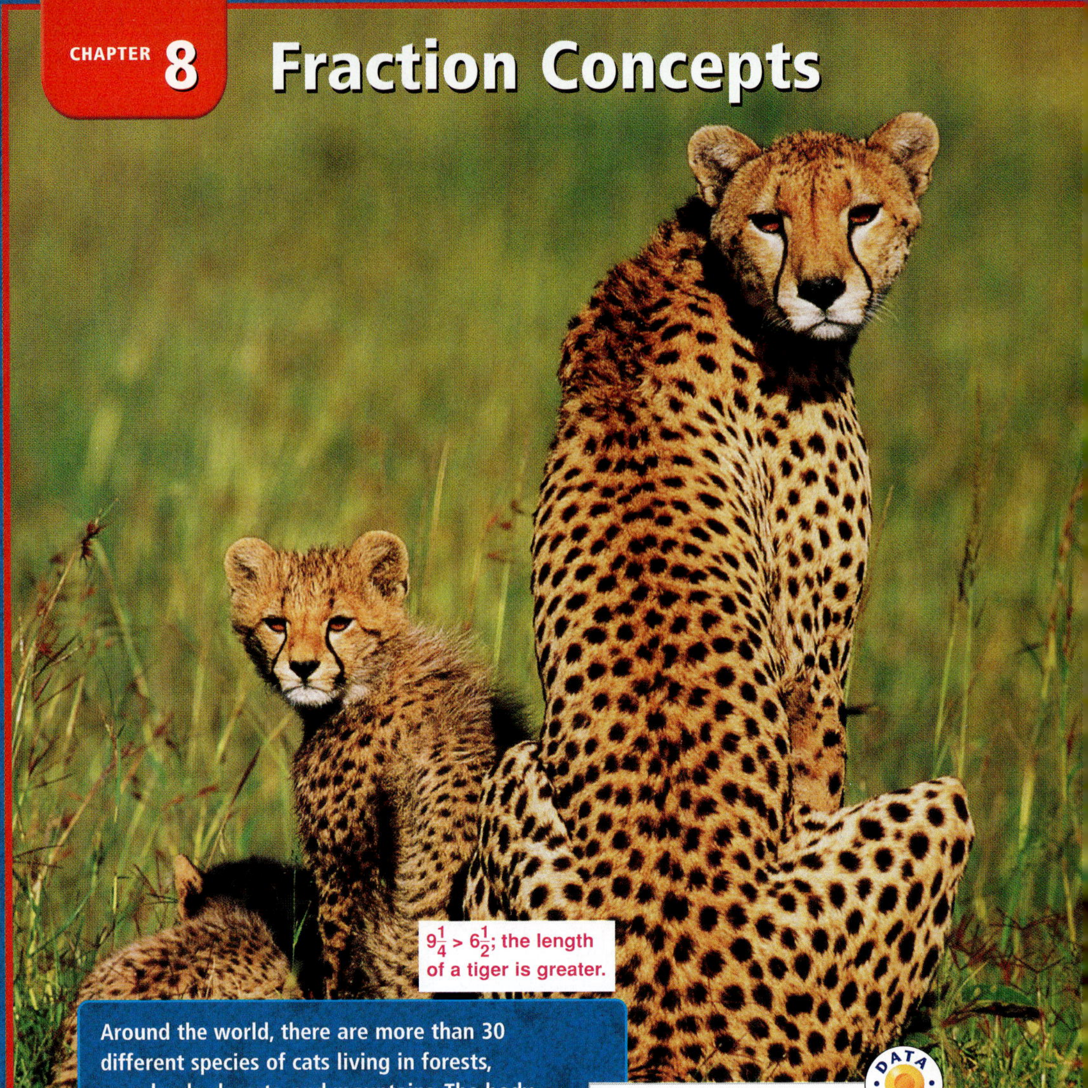

$9\frac{1}{4} > 6\frac{1}{2}$; the length of a tiger is greater.

Around the world, there are more than 30 different species of cats living in forests, grasslands, deserts, and mountains. The body lengths of cats range from 2 feet to 10 feet. The average tiger is $9\frac{1}{4}$ feet long, and the average lion is $6\frac{1}{2}$ feet long. The average bobcat is $3\frac{1}{4}$ feet long.

PROBLEM SOLVING Compare the average length of a tiger to that of a lion.

AVERAGE BODY LENGTHS OF CATS

(Cheetah $4\frac{1}{2}$ ft, Snow Leopard 5 ft, Tiger $9\frac{1}{4}$ ft, Jaguar 6 ft, Lion $6\frac{1}{2}$ ft, Bobcat $3\frac{1}{4}$ ft)

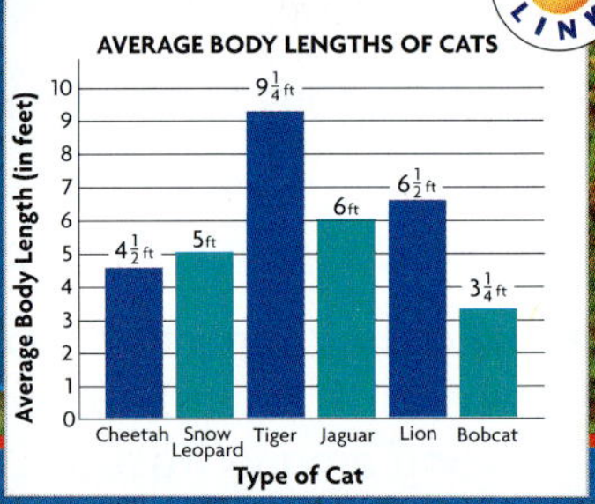

158 Chapter 8

Why learn math? Explain that wildlife biologists collect data about wild animals, including their height, length, and weight. They use these data to study the health and viability of these animals. Ask: What other professions might involve data that contain mixed numbers? Possible answer: architects, chefs, web-page designers

Check What You Know

Use this page to help you review and remember important skills needed for Chapter 8.

✓ Vocabulary

Choose the best term or symbol from the box.

> denominator
> <
> numerator
> >

1. The symbol for "is greater than" is __?__. >
2. The top number of a fraction is called the __?__. **numerator**
3. The bottom number of a fraction is called the __?__. **denominator**

✓ Compare and Order Whole Numbers
(For Intervention, see p. H3.)

Write <, >, or = for each ●.

4. 408 ● 480   **<**
5. 4,279 ● 4,277   **>**
6. 30 tens ● 3 hundreds   **=**
7. 9,315 ● 9,351   **<**
8. 18,808 ● 18,880   **<**
9. 356,782 ● 356,482   **>**

Order the numbers from greatest to least.

10. 3,400; 3,439; 3,399
    3,439; 3,400; 3,399
11. 61,060; 62,000; 61,600
    62,000; 61,600; 61,060
12. 98,450; 98,405; 98,540
    98,540; 98,450; 98,405

✓ Model Fractions (For Intervention, see p. H8.)

Write the fraction for the shaded part.   13. $\frac{5}{6}$ 14. $\frac{2}{5}$ 15. $\frac{1}{3}$ 16. $\frac{4}{8}$

13. 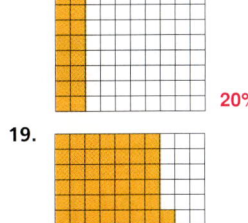   14.   15.   16.

✓ Model Percents (For Intervention, see p. H9.)

Write the percent for the shaded part.

17.   **20%**
18.   **13%**
19.   **75%**
20.   **98%**

LOOK AHEAD

In Chapter 8 you will
- find equivalent fractions
- write fractions in simplest form
- write fractions as mixed numbers and mixed numbers as fractions
- compare and order fractions
- find relationships among fractions, decimals, and percents

159

Assessing Prior Knowledge

Use the **Check What You Know** page to determine whether your students have mastered the prerequisite skills critical for this chapter.

Intervention

- **Diagnose and Prescribe**
  Evaluate your students' performance on this page to determine whether intervention is necessary or if enrichment is appropriate. Options that provide instruction, practice, and a check are listed in the chart below.

✓ CHECK WHAT YOU KNOW RESOURCES

Intervention Card, Copying Master, or CD-ROM

Intervention Strategies and Activities Teaching Transparencies

Intervention Practice Book

Enrichment Book

Were students successful with ✓ CHECK WHAT YOU KNOW?

OPTIONS

NO — INTERVENE

✓ **COMPARE AND ORDER WHOLE NUMBERS, 4–12**
How to Help
Troubleshooting, Pupil Edition p. H3
Intervention Strategies and Activities, Skills 3, 34

✓ **MODEL FRACTIONS, 13–16**
How to Help
Troubleshooting, Pupil Edition p. H8
Intervention Strategies and Activities, Skill 20

✓ **MODEL PERCENTS, 17–20**
How to Help
Troubleshooting, Pupil Edition p. H9
Intervention Strategies and Activities, Skill 21

YES — ENRICH

Check What You Know Enrichment Book, pp. 15–16

Fraction Concepts **159**

LESSON 8.1

Equivalent Fractions and Simplest Form

LESSON PLANNING

Objective To identify and write equivalent fractions and to write fractions in simplest form

Intervention for Prerequisite Skills
Model Fractions (For intervention strategies, see page 159.)

Materials *For each group* fraction bars or fraction strips, p. TR18

Lesson Resources E-Lab Recording Sheet • *Equivalent Fractions*

NCTM Standards
1. Number and Operations
5. Data Analysis and Probability
6. Problem Solving
8. Communication
10. Representation

Vocabulary

equivalent fractions fractions that name the same amount or the same part of a whole

simplest form a fraction in which the numerator and denominator have no common factors other than 1

Math Background

The following experiences will help students understand the concepts of equivalent fractions and fractions in simplest form:

- Using fraction bars to show equivalent fractions.
- Finding equivalent fractions by multiplying or dividing the numerator and denominator of a fraction by the same number.
- Finding the simplest form of a fraction by dividing the numerator and denominator by the greatest common factor.

WARM-UP RESOURCES

 **NUMBER OF THE DAY**

4 and 24 are multiples of this number; 2 and 3 are not. What number is it? 4

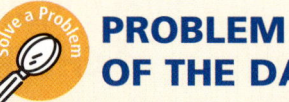 **PROBLEM OF THE DAY**

John has 3 coins, 2 of which are the same. Ellen has 1 fewer coin than John, and Anna has 2 more coins than John. Each girl has only 1 kind of coin. Who has coins that could equal the value of a half-dollar? Ellen and Anna

Solution Problem of the Day tab, p. PD8

 **DAILY FACTS PRACTICE**

Have students practice addition facts by completing Set D of *Teacher's Resource Book*, p. TR98.

INTERVENTION AND EXTENSION RESOURCES

ALTERNATIVE TEACHING STRATEGY

Materials *For each pair* 2 square sheets of paper

Help students **visualize equivalent fractions.** Ask each student to fold one square lengthwise, not diagonally, to model halves, and to label each part $\frac{1}{2}$. Next, have pairs fold and label their squares to solve $\frac{1}{2} = \frac{\blacksquare}{4}$, $\frac{1}{2} = \frac{\blacksquare}{8}$, and $\frac{3}{4} = \frac{\blacksquare}{8}$. **2, 4, 6**

See also page 162.

KINESTHETIC
VISUAL/SPATIAL

ENGLISH LANGUAGE LEARNERS

To help students **understand the concept of fractions,** discuss these part-to-whole relationships. Have them write each relationship in fraction form.

- A room is part of a house. $\frac{room}{house}$
- A flower is part of a garden. $\frac{flower}{garden}$
- A tree is part of a forest. $\frac{tree}{forest}$

Ask volunteers to give other examples of part-to-whole relationships. **Possible answer: An egg is part of a dozen eggs.**

AUDITORY
INTERPERSONAL/SOCIAL

MULTISTEP AND STRATEGY PROBLEMS

The following multistep and strategy problems are provided in Lesson 8.1:

Page	Item
163	63, 65, 71; Thinker's Corner 1–6

EARLY FINISHERS

Have students **demonstrate their understanding of equivalent fractions and simplest form.**

Ask students to write two additional equivalent fractions for the fractions in Exercises 23–26 on page 162. Circle the simplest form fraction. **Possible answers: 23: $\frac{4}{8}$, $\frac{10}{20}$; 24: $\frac{20}{30}$, $\frac{6}{9}$; 25: $\frac{32}{40}$, $\frac{8}{10}$; 26: $\frac{18}{54}$, $\frac{3}{9}$**

Simplest form: **23: $\frac{1}{2}$; 24: $\frac{2}{3}$; 25: $\frac{4}{5}$; 26: $\frac{1}{3}$**

VISUAL
VERBAL/LINGUISTIC

The E-Lab Recording Sheets and activities are available on the E-Lab website. www.harcourtschool.com/elab2002

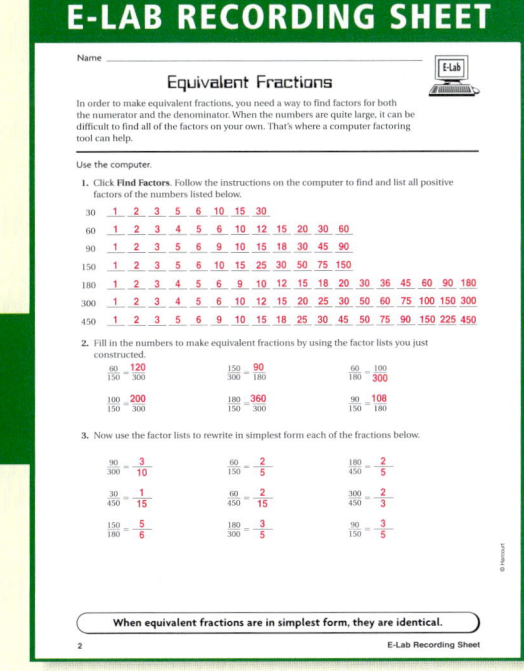

TECHNOLOGY LINK

- **Intervention Strategies and Activities CD-ROM** • *Skill 20*
- **E-Lab** • *Equivalent Fractions*
- **Number Heroes** • *Fraction Fireworks,* Levels D, K, P

LESSON 8.1 ORGANIZER

Objective To identify and write equivalent fractions and to write fractions in simplest form

Vocabulary equivalent fractions, simplest form

Materials For each group fraction bars or fraction strips, p. TR18

Lesson Resources E-Lab Recording Sheet • Equivalent Fractions

1 Introduce

QUICK REVIEW provides review of prerequisite skills.

Why Learn This? You will use equivalent fractions to set up and solve proportions and use simplest form to express solutions to problems. Share the lesson objective with students.

2 Teach

Guided Instruction

- Check students' understanding of the opening model.

  How could you use fraction bars to find how many eighths are equivalent to $\frac{1}{4}$? Place $\frac{1}{8}$ fraction bars along the $\frac{1}{4}$ bar until the lengths are equal.

  How many eighths are equivalent to $\frac{1}{4}$? 2

- Explain that in Example 1, students are being asked to identify a fraction that is equivalent to $\frac{2}{4}$ and has a denominator of 12.

  How do you know which number to multiply the numerator and denominator by? by finding the factor you multiply 4 by to get a product of 12

  What is that factor? 3

ADDITIONAL EXAMPLE

Example 1, p. 160

Complete: $\frac{3}{5} = \frac{\square}{15}$

$\frac{3 \times \boxed{3}}{5 \times \boxed{3}} = \frac{9}{15}$

So, $\frac{3}{5} = \frac{9}{15}$.

160 Chapter 8

LESSON 8.1

Equivalent Fractions and Simplest Form

Learn how to identify and write equivalent fractions, and how to write fractions in simplest form.

QUICK REVIEW

Write all the factors.

1. 12 — 1, 2, 3, 4, 6, 12
2. 10 — 1, 2, 5, 10
3. 16 — 1, 2, 4, 8, 16
4. 25 — 1, 5, 25
5. 20 — 1, 2, 4, 5, 10, 20

Vocabulary
equivalent fractions
simplest form

Fractions that name the same amount or the same part of a whole are called **equivalent fractions**. The figures show that the fractions $\frac{1}{4}$ and $\frac{2}{8}$ are equivalent, because they name the same part of a whole circle.

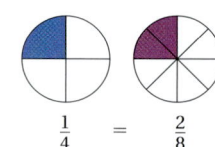

$\frac{1}{4} = \frac{2}{8}$

There are several ways to find equivalent fractions.

Activity

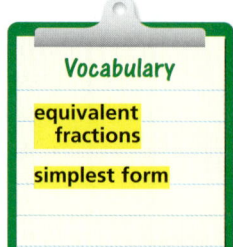

You need: fraction bars

Find how many eighths are equivalent to $\frac{1}{2}$.

- Place $\frac{1}{8}$ fraction bars along the $\frac{1}{2}$ bar until the lengths are equal. How many $\frac{1}{8}$ bars are there? 4
- Complete: $\frac{1}{2} = \frac{\blacksquare}{8}$ 4
- Use fraction bars. How many fourths are equivalent to $\frac{1}{2}$? 2
- Complete: $\frac{1}{2} = \frac{\blacksquare}{4}$ 2

Another way to find an equivalent fraction is to multiply or divide the numerator and denominator of a fraction by the same number, except 0. Doing this does not change the fraction's value, because this is the same as multiplying or dividing by 1.

EXAMPLE 1

Complete: $\frac{2}{4} = \frac{\blacksquare}{12}$

THINK: To get the denominator 12, I need to multiply the denominator 4 by 3. So, to get the missing numerator, I should multiply the numerator 2 by 3.

$\frac{2 \times \boxed{3}}{4 \times \boxed{3}} = \frac{6}{12}$

$\frac{3}{3} = 1$, so the product is still equal to $\frac{2}{4}$.

$\frac{2}{4} = \frac{6}{12}$

160 Chapter 8

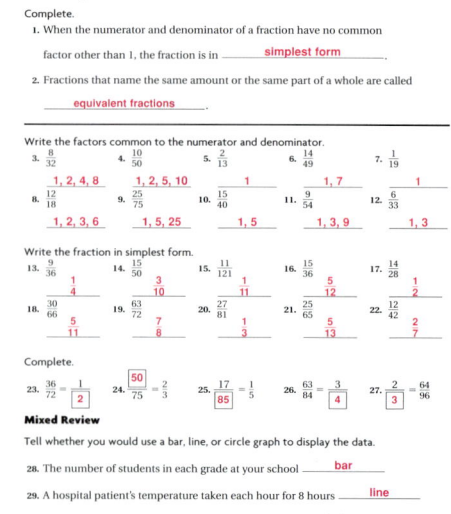

EXAMPLE 2

Complete: $\frac{6}{10} = \frac{\square}{5}$

THINK: I can get the denominator 5 by dividing the denominator 10 by 2. So, to get the missing numerator, I should divide the numerator 6 by 2.

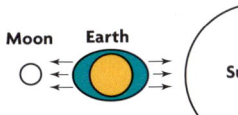

$\frac{6}{10} = \frac{3}{5}$

$\frac{6 \div 2}{10 \div 2} = \frac{3}{5}$ $\frac{2}{2} = 1$, so the quotient is still equal to $\frac{6}{10}$.

Math Idea ▶ When the numerator and denominator of a fraction have no common factors other than 1, the fraction is in **simplest form**.

$\frac{9}{16}$ **is** in simplest form because 9 and 16 have no common factors other than 1.

$\frac{9}{15}$ **is not** in simplest form because 9 and 15 have the common factor 3.

EXAMPLE 3

The sun is much farther from the Earth than the moon is. So, even though the sun is much larger than the moon, the sun's tide-raising force is only $\frac{12}{30}$ of the moon's force. Write $\frac{12}{30}$ in simplest form.

Tides are caused by the gravitational pull of the sun and the moon on Earth's oceans.

12: 1, 2, 3, 4, 6, 12
30: 1, 2, 3, 5, 6, 10, 15, 30   *Find the common factors of 12 and 30.*

$\frac{12}{30} = \frac{12 \div 3}{30 \div 3} = \frac{4}{10}$ *Divide the numerator and denominator by a common factor.*

$\frac{4}{10} = \frac{4 \div 2}{10 \div 2} = \frac{2}{5}$ *Repeat until the fraction is in simplest form.*

So, $\frac{2}{5}$ is the simplest form of $\frac{12}{30}$.

In Example 3, the numerator and denominator were divided by common factors twice to find the simplest form. You can find it in just one step if you divide by the greatest common factor.

Bay of Fundy during low tide

EXAMPLE 4

The world's highest tides occur in Canada's Bay of Fundy. Tides in Seattle, Washington, average only about $\frac{12}{42}$ of tidal heights in the Bay of Fundy. Write $\frac{12}{42}$ in simplest form.

12: 1, 2, 3, 4, 6, 12
42: 1, 2, 3, 6, 7, 14, 21, 42   *Find the GCF of 12 and 42.*
GCF = 6

$\frac{12}{42} = \frac{12 \div 6}{42 \div 6} = \frac{2}{7}$ *Divide the numerator and denominator by the GCF.*

So, $\frac{2}{7}$ is the simplest form of $\frac{12}{42}$.

Bay of Fundy during high tide

161

- Focus on how Examples 1 and 2 are related.

REASONING How are the solutions to Example 1 and Example 2 different? In Example 1, you multiply to change to a greater denominator. In Example 2, you divide to change to a lesser denominator.

- Point out that students can apply what they learned in Example 4 to Example 3.

How could you have used the GCF to write Example 3 in simplest form? I could have divided 12 and 30 by the GCF 6.

ADDITIONAL EXAMPLES

Example 2, p. 161

Complete: $\frac{2}{16} = \frac{\square}{8}$

$\frac{2 \div 2}{16 \div 2} = \frac{1}{8}$

So, $\frac{2}{16} = \frac{1}{8}$.

Example 3, p. 161

In the sixth-grade class, $\frac{12}{18}$ of the students ride the bus to school. Write $\frac{12}{18}$ in simplest form.

12: 1, 2, 3, 4, 6, 12

18: 1, 2, 3, 6, 9, 18

$\frac{12}{18} = \frac{12 \div 3}{18 \div 3} = \frac{4}{6}$

$\frac{4 \div 2}{6 \div 2} = \frac{2}{3}$

So, $\frac{2}{3}$ is the simplest form of $\frac{12}{18}$.

Example 4, p. 161

The weather in central Florida is very warm for more than $\frac{8}{12}$ of the year. Write $\frac{8}{12}$ in simplest form.

8: 1, 2, 4, 8

12: 1, 2, 3, 4, 6, 12

GCF = 4

$\frac{8 \div 4}{12 \div 4} = \frac{2}{3}$

So, $\frac{2}{3}$ is the simplest form of $\frac{8}{12}$.

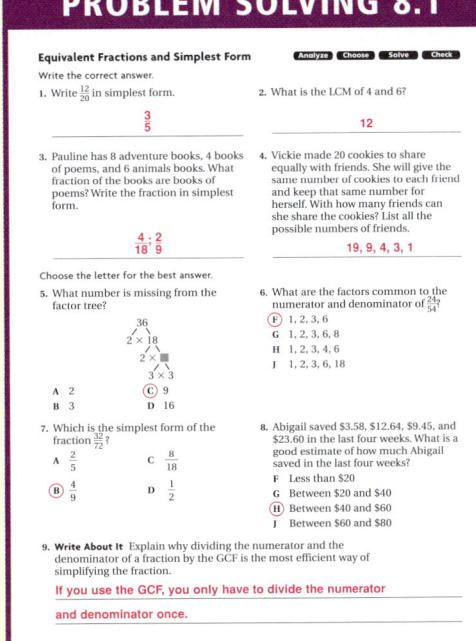

LESSON 8.1

3 Practice

Guided Practice

Do Check for Understanding Exercises 1–22 with your students. Identify those having difficulty and use lesson resources to help.

Independent Practice

Note that Exercises 63, 65, 71, and Thinker's Corner 1–6 are **multistep or strategy problems**. Assign Exercises 23–66.

To help students with Exercises 23–34, have them decide if they need to multiply or divide because of the given numerators or denominators. Then display the following pattern of boxes and have students copy and complete the appropriate pattern for each exercise:

$\frac{\square \times \square}{\square \times \square} = \frac{\square}{\square}$ or $\frac{\square \div \square}{\square \div \square} = \frac{\square}{\square}$

Multistep or Strategy Problem To solve Exercise 63, students can add up the total number of fireballs for the first 6 months and then put March's reported number, 6, over the total number as a fraction. Guide students to find the total and to write and simplify the fraction.

Scaffolded Instruction Use the prompts on Transparency 8 to guide instruction for the multistep problem in Exercise 63.

Transparency 8

CHECK FOR UNDERSTANDING

Think and Discuss
Look back at the lesson to answer each question.

1. Also multiply the denominator by 5; $\frac{10}{15}$

1. **Describe** what you must do to find a fraction equivalent to $\frac{2}{3}$ if you first multiply the numerator by 5. What is the equivalent fraction?

2. **Explain** how you know that $\frac{5}{12}$ is in simplest form.
   5 and 12 have no common factors other than 1.

Guided Practice
Complete.

3. $\frac{3}{5} = \frac{\blacksquare}{20}$ 12
4. $\frac{3}{24} = \frac{\blacksquare}{8}$ 1
5. $\frac{8}{12} = \frac{\blacksquare}{6}$ 4
6. $\frac{3}{4} = \frac{\blacksquare}{24}$ 18
7. $\frac{4}{8} = \frac{\blacksquare}{2}$ 1
8. $\frac{9}{24} = \frac{\blacksquare}{8}$ 3
9. $\frac{8}{18} = \frac{\blacksquare}{36}$ 16
10. $\frac{12}{54} = \frac{\blacksquare}{9}$ 2

Write the factors common to the numerator and denominator.

11. $\frac{4}{8}$ 1, 2, 4
12. $\frac{9}{24}$ 1, 3
13. $\frac{8}{18}$ 1, 2
14. $\frac{12}{54}$ 1, 2, 3, 6

Write the fraction in simplest form.

15. $\frac{4}{32}$ $\frac{1}{8}$
16. $\frac{14}{21}$ $\frac{2}{3}$
17. $\frac{9}{54}$ $\frac{1}{6}$
18. $\frac{48}{54}$ $\frac{8}{9}$
19. $\frac{22}{8}$ $\frac{11}{4}$
20. $\frac{18}{5}$ $\frac{18}{5}$
21. $\frac{9}{30}$ $\frac{3}{10}$
22. $\frac{48}{32}$ $\frac{3}{2}$

PRACTICE AND PROBLEM SOLVING

Independent Practice
Complete.

23. $\frac{1}{2} = \frac{\blacksquare}{10}$ 5
24. $\frac{10}{15} = \frac{\blacksquare}{3}$ 2
25. $\frac{16}{20} = \frac{\blacksquare}{5}$ 4
26. $\frac{9}{27} = \frac{\blacksquare}{3}$ 1
27. $\frac{2}{12} = \frac{1}{\blacksquare}$ 6
28. $\frac{\blacksquare}{36} = \frac{2}{9}$ 8
29. $\frac{21}{24} = \frac{7}{\blacksquare}$ 8
30. $\frac{40}{\blacksquare} = \frac{5}{8}$ 64
31. $\frac{9}{\blacksquare} = \frac{3}{4}$ 12
32. $\frac{3}{16} = \frac{12}{\blacksquare}$ 4
33. $\frac{16}{28} = \frac{4}{\blacksquare}$ 7
34. $\frac{25}{40} = \frac{\blacksquare}{8}$ 5

TECHNOLOGY LINK
More Practice: Use E-Lab, Equivalent Fractions.
www.harcourtschool.com/elab2002

Write the factors common to the numerator and denominator.

35. $\frac{1}{7}$ 1
36. $\frac{9}{30}$ 1, 3
37. $\frac{6}{27}$ 1, 3
38. $\frac{9}{63}$ 1, 3, 9
39. $\frac{10}{35}$ 1, 5
40. $\frac{16}{40}$ 1, 2, 4, 8
41. $\frac{3}{5}$ 1
42. $\frac{8}{10}$ 1, 2

Write the fraction in simplest form.

43. $\frac{4}{24}$ $\frac{1}{6}$
44. $\frac{9}{12}$ $\frac{3}{4}$
45. $\frac{6}{48}$ $\frac{1}{8}$
46. $\frac{10}{15}$ $\frac{2}{3}$
47. $\frac{10}{18}$ $\frac{5}{9}$
48. $\frac{20}{15}$ $\frac{4}{3}$
49. $\frac{18}{90}$ $\frac{1}{5}$
50. $\frac{28}{42}$ $\frac{2}{3}$
51. $\frac{22}{33}$ $\frac{2}{3}$
52. $\frac{24}{30}$ $\frac{4}{5}$
53. $\frac{60}{42}$ $\frac{10}{7}$
54. $\frac{24}{28}$ $\frac{6}{7}$
55. $\frac{16}{64}$ $\frac{1}{4}$
56. $\frac{8}{12}$ $\frac{2}{3}$
57. $\frac{18}{90}$ $\frac{1}{5}$
58. $\frac{48}{18}$ $\frac{8}{3}$
59. $\frac{4^2}{32}$ $\frac{1}{2}$
60. $\frac{3^2}{12}$ $\frac{3}{4}$
61. $\frac{2^3}{12}$ $\frac{2}{3}$
62. $\frac{3^2}{6^2}$ $\frac{1}{4}$

162 Chapter 8

ALTERNATIVE TEACHING STRATEGY | SCAFFOLDED INSTRUCTION

Purpose Students use divisibility rules to write greater fractions in simplest form.

Explain that divisibility rules can be used to find factors. Remind students that a fraction is in simplest form when the numerator and denominator have no common factors other than 1. Model this process for $\frac{84}{93}$.

1. Write 1, 2, 3, 4, 5, 6, 7, 8, and 9. Underline the numerator's factors.
   84: <u>1</u> <u>2</u> <u>3</u> <u>4</u> 5 <u>6</u> <u>7</u> 8 9

2. Write 1, 2, 3, 4, 5, 6, 7, 8, and 9. Underline the denominator's factors.
   93: <u>1</u> 2 <u>3</u> 4 5 6 7 8 9

3. Circle the greatest common factor (GCF) and use it to divide the fraction.
   84: 1 2 ③ 4 5 6 7 8 9
   93: 1 2 ③ 4 5 6 7 8 9
   $\frac{84 \div 3}{93 \div 3} = \frac{28}{31}$

4. Check the new fraction for common factors.
   28: <u>1</u> <u>2</u> 3 <u>4</u> 5 6 <u>7</u> 8 9
   31: <u>1</u> 2 3 4 5 6 7 8 9

   The simplest form for $\frac{84}{93}$ is $\frac{28}{31}$.

5. If there are common factors greater than 1, repeat steps 3 and 4 until 1 is the only common factor.

   Have pairs of students write these fractions in simplest form: $\frac{36}{84}$ and $\frac{125}{200}$. $\frac{3}{7}, \frac{5}{8}$

   Then have each pair write two fractions and challenge other students to write them in simplest form.

   Note: This method will work for numerators and denominators less than 221. It may miss common factors for greater numbers.

   Check students' work.

162 Chapter 8

Problem Solving Applications

AMERICAN METEOR SOCIETY FIREBALL REPORT, 1998	
Month	Number
January	13
February	8
March	20
April	18
May	8
June	13

65. What fraction of the muffins are bran?

63. Use Data Look at the table at the left. A "fireball" is an extremely bright meteor occasionally seen streaking across the night sky. What fraction of the fireballs reported to the American Meteor Society during the first 6 months of 1998 occurred in March? $\frac{1}{4}$

64. Technology Some calculators have a [Simp] key that can be used to simplify fractions. What fraction would this key sequence give? 10 [n] 15 [d] [Simp] [Enter] $\frac{2}{3}$

65. **What's the Question?** Esteban has 6 apple muffins, 2 corn muffins, and 4 bran muffins. The answer is $\frac{1}{3}$ of the muffins.

66. Write a problem about everyday life that involves finding the simplest form of a fraction. Answers will vary.

MIXED REVIEW AND TEST PREP

Evaluate each expression.

67. $(6 \div 3)^3 + 2^4$ (p. 44) **24** **68.** $(6^2 \div 3^2) + 1$ (p. 44) **5** **69.** 3.5×0.01 (p. 70) **0.035**

⭐ **70. TEST PREP** 0.5×1.2 (p. 70) **B**

  A 0.06   B 0.6   C 6   D 6.2

⭐ **71. TEST PREP** An adult's movie ticket costs $7.50, and a child's ticket costs $5.50. Find the total cost for 2 adults and 3 children. (p. 70) **J**

  F $13.00   G $20.50   H $24.00   J $31.50

PROBLEM SOLVING Thinker's Corner

ALGEBRA You can use what you know about equivalent fractions to solve some types of algebraic equations.

A. $\frac{2}{3} = \frac{x-5}{27}$

THINK: I need to find a fraction equivalent to $\frac{2}{3}$ that has 27 as its denominator. I'll multiply by 9.

$\frac{2}{3} = \frac{\square}{27}$ $\frac{2 \times 9}{3 \times 9} = \frac{18}{27}$

So, $x - 5 = 18$. If 5 less than some number is 18, that number must be 23. So, $x = 23$.

B. $\frac{3}{s+2} = \frac{18}{24}$

THINK: I need to find a fraction equivalent to $\frac{18}{24}$ that has 3 as its numerator. I'll divide by 6.

$\frac{3}{\square} = \frac{18}{24}$ $\frac{3}{4} = \frac{18 \div 6}{24 \div 6}$

So, $s + 2 = 4$. If some number plus 2 equals 4, that number must be 2. So, $s = 2$.

Solve the equation.

1. $\frac{x+3}{10} = \frac{1}{2}$ $x = 2$ **2.** $\frac{5}{8} = \frac{c-5}{40}$ $c = 30$ **3.** $\frac{3}{b-4} = \frac{9}{15}$ $b = 9$

4. $\frac{1}{(2+y)} = \frac{6}{24}$ $y = 2$ **5.** $\frac{4}{9} = \frac{(w+4)}{18}$ $w = 4$ **6.** $\frac{(k-5)}{8} = \frac{12}{32}$ $k = 8$

EXTRA PRACTICE page H39, Set A

MIXED REVIEW AND TEST PREP

Exercises 67–71 provide **cumulative review** (Chapters 1–8).

Thinker's Corner

• Focus on the order of the steps students must follow to solve the equations.

In Example A, after you know the factor for 3 and 27, what must you do before you can solve the equation? Explain. Find the numerator; you need to know what $x - 5$ equals before you can solve for x.

In Example B, after you know the factor for 3 and 18, what do you need to do before you can solve the equation? Explain. Find the denominator; you need to know what $s + 2$ equals before you can solve for s.

REASONING Find an equivalent fraction for $\frac{5}{6}$, then use the equivalent fraction to write and solve an addition equation for the numerator. Possible answer: $\frac{15}{18}$; $\frac{5}{6} = \frac{x+4}{18}$; $x = 11$

4 Assess

Summarize the lesson by having students:

DISCUSS Can you multiply the numerator and denominator of a fraction by a factor other than 1 and write the fraction in simplest form? Explain. No; the simplest form means the numerator and denominator have only 1 as a common factor.

WRITE Which of the methods shown in Examples 3 and 4 on page 161 is the more efficient way to write a fraction in simplest form? Explain. Example 4; Example 4's method is more efficient because you divide only once.

Lesson Quiz   Transparency 8.1

Write the fraction in simplest form.

1. $\frac{21}{24}$ $\frac{7}{8}$ **2.** $\frac{12}{16}$ $\frac{3}{4}$ **3.** $\frac{24}{60}$ $\frac{2}{5}$

4. $\frac{12}{18}$ $\frac{2}{3}$ **5.** $\frac{55}{60}$ $\frac{11}{12}$ **6.** $\frac{40}{50}$ $\frac{4}{5}$

7. There are 30 students in the sixth grade. 15 are girls. What fraction of the students are girls? $\frac{1}{2}$

8. The school band has 24 members. $\frac{1}{4}$ of them play the flute. How can you use equivalent fractions to find the number of flute players in the band? $\frac{1}{4} = \frac{6}{24}$; 6 flute players

LESSON 8.2: Mixed Numbers and Fractions

LESSON PLANNING

Objective To write fractions as mixed numbers and mixed numbers as fractions

Intervention for Prerequisite Skills
Model Fractions (For intervention strategies, see page 159.)

NCTM Standards
1. Number and Operations
5. Data Analysis and Probability
6. Problem Solving
8. Communication
9. Connections

Math Background
Consider the following ideas to help students understand how mixed numbers are related to whole numbers and fractions:
- A mixed number can be written as an equivalent fraction whose numerator is greater than its denominator.
- Division can be used to change a fraction greater than 1 to a mixed number.
- Both mathematically and in real-life, fractions greater than 1 are extremely common and useful. So, there is nothing "wrong" with "improper" fractions.

Vocabulary
mixed number a number that includes a whole-number part that is not 0 and a fraction part

WARM-UP RESOURCES

NUMBER OF THE DAY
What fraction of a dollar is a quarter? What fraction of a dollar are 5 dimes? $\frac{1}{4}, \frac{1}{2}$

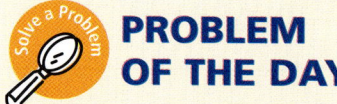
PROBLEM OF THE DAY
At the car show there are 20 vehicles on display. Some are motorcycles and some are cars. All 56 wheels on the vehicles need to be polished. What fraction of the vehicles are motorcycles? $\frac{3}{5}$

Solution Problem of the Day tab, p. PD8

DAILY FACTS PRACTICE
Have students practice subtraction facts by completing Set E of *Teacher's Resource Book,* p. TR98.

INTERVENTION AND EXTENSION RESOURCES

ALTERNATIVE TEACHING STRATEGY — ELL

Materials *For each group* play money—3 one-dollar bills, 20 quarters

Model mixed numbers and fractions greater than 1 with play money. Present 2 one-dollar bills and 1 quarter. Have a volunteer display an equivalent amount using quarters only.

Display the mixed number and the fraction: $2\frac{1}{4}$ and $\frac{9}{4}$.

Have students repeat the activity by using other amounts. **Check students' work.**

KINESTHETIC
VISUAL/SPATIAL

MULTISTEP AND STRATEGY PROBLEMS

The following multistep or strategy problem is provided in Lesson 8.2:

Page	Item
165	41

WRITING IN MATHEMATICS

Have students demonstrate their understanding of how to **write mixed numbers as fractions and fractions as mixed numbers.** Ask them to describe the procedure they would use to write $3\frac{1}{2}$ as a fraction. Then have them write the procedure they would use to express $\frac{15}{6}$ as a mixed number. **Possible answers: multiply 3 × 2 and add 1 to get 7 for the numerator: $\frac{7}{2}$; divide 15 by 6 to get 2 and a remainder of 3 to make the mixed number $2\frac{3}{6}$**

VISUAL
VERBAL/LINGUISTIC

ADVANCED LEARNERS

Materials *For each student* index card

Challenge students to **write mixed numbers as fractions.** Working in small groups, each student should write his or her name and age in years and months on an index card.

Within their groups, have students discuss how to write their ages as mixed numbers and as fractions and then write them on the cards. Have each group order the numbers on the cards from least to greatest. **Check students' work.**

VISUAL
INTERPERSONAL/SOCIAL

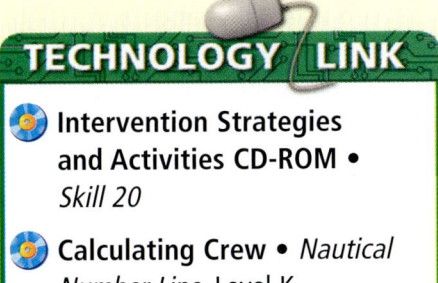

TECHNOLOGY LINK

- **Intervention Strategies and Activities CD-ROM** • *Skill 20*
- **Calculating Crew** • *Nautical Number Line*, Level K

164B

LESSON 8.2 ORGANIZER

Objective To write fractions as mixed numbers and mixed numbers as fractions

Vocabulary mixed number

1 Introduce

QUICK REVIEW provides review of pre-requisite skills.

Why Learn This? You can use this skill to help you express units of measurement, such as feet and yards, as a fraction or a mixed number. Share the lesson objective with students.

2 Teach

Guided Instruction

- Discuss the symbols in the graph.

  How many quarter sections represent 1 minute? 2 minutes? 4; 8

- In Example 1, emphasize the usefulness of changing a fraction to a mixed number.

  How can changing $\frac{11}{4}$ and $\frac{15}{2}$ to mixed numbers help you compare the lengths of the eclipses in 2017 and 2186? Possible answer: It is easier to compare the whole numbers 2 and 7 than fractions with different denominators.

- Have students compare Examples 1 and 2.

  What operations are used to change a fraction to a mixed number? a mixed number to a fraction? division, multiplication, and subtraction; multiplication and addition

Modifying Instruction Demonstrate for students another way to solve Example 2:
$3\frac{2}{5} = 3 + \frac{2}{5} = \frac{3 \times 5}{1 \times 5} + \frac{2}{5}$
$= \frac{15}{5} + \frac{2}{5}$
$= \frac{17}{5}$

ADDITIONAL EXAMPLES

Example 1, p. 164

Harrison studied $\frac{43}{2}$ minutes for his math test. Write $\frac{43}{2}$ as a mixed number. $21\frac{1}{2}$ minutes

Example 2, p. 164

Write $4\frac{5}{6}$ as a fraction. $\frac{29}{6}$

LESSON 8.2

Mixed Numbers and Fractions

Learn how to write fractions as mixed numbers and mixed numbers as fractions.

QUICK REVIEW
1. $4 \times 7 + 2$ 30
2. $19 + 11$ 30
3. $2 \times 7 + 5$ 19
4. $35 + 9$ 44
5. $8 \times 4 + 6$ 38

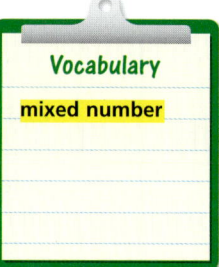

Vocabulary
mixed number

Total solar eclipses are rare, with only three visible in most of the U.S. since 1963. Eclipses last different times. In the graph, each quarter-section of a circle represents $\frac{1}{4}$ minute. How long will the 2017 eclipse last? It is represented by 11 sections, or $\frac{11}{4}$ minutes.

The graph shows that 11 sections equal 2 whole minutes plus $\frac{3}{4}$ of another minute. So, $\frac{11}{4} = 2\frac{3}{4}$. The 2017 eclipse will last $2\frac{3}{4}$ minutes.

The fraction $\frac{11}{4}$ has a value greater than 1 because the numerator is greater than the denominator. Sometimes a fraction such as $\frac{11}{4}$ is called an "improper fraction." Any such fraction can be written as a **mixed number**, like $2\frac{3}{4}$.

A total solar eclipse occurs when the moon passes between the sun and the Earth, "eclipsing" the sun's light.

APPROXIMATE LENGTH OF U.S. TOTAL SOLAR ECLIPSES

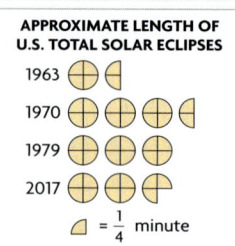

◔ = $\frac{1}{4}$ minute

Math Idea ▶ A mixed number has a whole-number part that is not 0 and a fraction part.

EXAMPLE 1

The longest total solar eclipse in the next 200 years will take place in 2186. It will last about $\frac{15}{2}$ minutes. Write $\frac{15}{2}$ as a mixed number.

$\frac{15}{2} \rightarrow 2\overline{)15}^{\,7\frac{1}{2}}$
$\phantom{\frac{15}{2} \rightarrow 2)}\underline{-14}$
$\phantom{\frac{15}{2} \rightarrow 2)00}1$

Divide the numerator by the denominator. For the fraction part of the quotient, use the remainder as the numerator and the divisor as the denominator. Write the fraction in simplest form.

So, $\frac{15}{2} = 7\frac{1}{2}$. The 2186 eclipse will last $7\frac{1}{2}$ minutes.

You can also write a mixed number as a fraction.

EXAMPLE 2

Write the mixed number $3\frac{2}{5}$ as a fraction.

$3\frac{2}{5} = \frac{3 \times 5}{5} + \frac{2}{5} = \frac{(3 \times 5) + 2}{5} = \frac{17}{5}$

So, $3\frac{2}{5} = \frac{17}{5}$.

Multiply the whole number by the denominator. Add the numerator. Use the same denominator.

164 Chapter 8

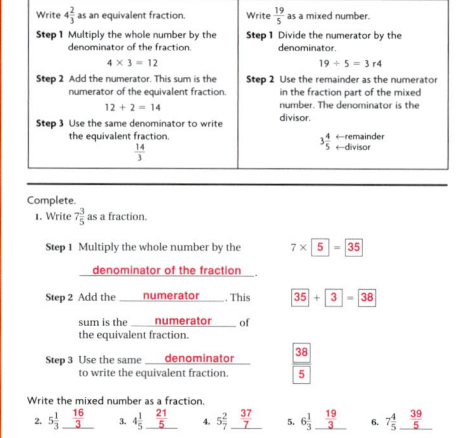

CHECK FOR UNDERSTANDING

Think and Discuss Look back at the lesson to answer the question.

1. **Tell** how you know that a given number is a mixed number.
   It has a whole-number part and a fraction part.

Write the fraction as a mixed number or a whole number.

2. $\frac{5}{3}$ $1\frac{2}{3}$
3. $\frac{7}{2}$ $3\frac{1}{2}$
4. $\frac{15}{5}$ 3
5. $\frac{11}{3}$ $3\frac{2}{3}$
6. $\frac{13}{4}$ $3\frac{1}{4}$

Guided Practice Write the mixed number as a fraction.

7. $1\frac{1}{4}$ $\frac{5}{4}$
8. $1\frac{3}{5}$ $\frac{8}{5}$
9. $2\frac{2}{3}$ $\frac{8}{3}$
10. $3\frac{4}{5}$ $\frac{19}{5}$
11. $5\frac{2}{7}$ $\frac{37}{7}$

PRACTICE AND PROBLEM SOLVING

Independent Practice Write the fraction as a mixed number or a whole number.

12. $\frac{7}{4}$ $1\frac{3}{4}$
13. $\frac{9}{2}$ $4\frac{1}{2}$
14. $\frac{11}{2}$ $5\frac{1}{2}$
15. $\frac{23}{4}$ $5\frac{3}{4}$
16. $\frac{27}{3}$ 9

17. $\frac{31}{6}$ $5\frac{1}{6}$
18. $\frac{18}{11}$ $1\frac{7}{11}$
19. $\frac{90}{7}$ $12\frac{6}{7}$
20. $\frac{104}{13}$ 8
21. $\frac{150}{9}$ $16\frac{2}{3}$

22. $\frac{x}{y}$ for $x = 18$ and $y = 12$ $1\frac{1}{2}$
23. $\frac{a}{b}$ for $a = 55$ and $b = 15$ $3\frac{2}{3}$

Write the mixed number as a fraction.

24. $3\frac{2}{3}$ $\frac{11}{3}$
25. $6\frac{1}{2}$ $\frac{13}{2}$
26. $5\frac{1}{3}$ $\frac{16}{3}$
27. $1\frac{9}{10}$ $\frac{19}{10}$
28. $4\frac{1}{9}$ $\frac{37}{9}$

29. $9\frac{1}{4}$ $\frac{37}{4}$
30. $2\frac{3}{8}$ $\frac{19}{8}$
31. $4\frac{9}{11}$ $\frac{53}{11}$
32. $11\frac{4}{9}$ $\frac{103}{9}$
33. $18\frac{3}{5}$ $\frac{93}{5}$

Problem Solving

34.  **What's the Error?** Marti changed $3\frac{1}{4}$ to $12\frac{1}{4}$. What mistake did she make? What is the correct answer? *See above left.*

   34. She forgot the denominator when she wrote the number of fourths in 3;
   $3\frac{1}{4} = \frac{(3 \times 4) + 1}{4} = \frac{13}{4}$

35. **Write About It** Can any fraction be written as a mixed number? Explain. *No, only a fraction whose numerator is greater than its denominator*

36. **Astronomy** On June 20, 1955, a total solar eclipse lasted 7 min 7 sec. On June 20, 1974, a total solar eclipse lasted 5 min 8 sec. Which lasted longer? How much longer? *June 20, 1955; 1 min 59 sec*

MIXED REVIEW AND TEST PREP

Write the prime factorization of each number. (p.148)

37. 36 $2^2 \times 3^2$
38. 42 $2 \times 3 \times 7$
39. 23 23×1

40. Is the following question biased? If so, rewrite it so that it is unbiased: The film *Time Warp* is great, isn't it? (p. 98) *yes; possible answer: Did you enjoy the film Time Warp?*

⭐ 41. **TEST PREP** Noella is hiking a 25-km trail. She has hiked 3.8 km to the first overlook and another 6.5 km to the second overlook. How many kilometers does she have left to hike? (p. 66) **C**

   A 9.3 km   **B** 10.3 km   **C** 14.7 km   **D** 15.7 km

EXTRA PRACTICE page H39, Set B

165

---

3 Practice

Guided Practice

Do Check For Understanding Exercises 1–11 with your students. Identify those having difficulty and use lesson resources to help.

COMMON ERROR ALERT

When changing mixed numbers to fractions, students often make careless errors in multiplying and/or adding. To check for errors, have students change the fractions back to mixed numbers and compare.

Independent Practice

Note that Exercise 41 is a **multistep or strategy problem**. Assign Exercises 12–36.

MIXED REVIEW AND TEST PREP

Exercises 37–41 provide **cumulative review** (Chapters 1–8).

4 Assess

Summarize the lesson by having students:

DISCUSS Jo has $\frac{11}{4}$ cups of milk. She needs $3\frac{1}{4}$ cups of milk for a recipe. How can Jo tell if she has enough? *Change $\frac{11}{4}$ to a mixed number. Compare it to $3\frac{1}{4}$.*

WRITE Compare the numerator with the denominator in a fraction less than one and in a fraction greater than 1. *The numerator is less than the denominator in a fraction less than one and the numerator is greater than the denominator in a fraction greater than one.*

Lesson Quiz

Write the fraction as a mixed number or a whole number.

1. $\frac{12}{5}$ $2\frac{2}{5}$
2. $\frac{10}{7}$ $1\frac{3}{7}$
3. $\frac{49}{7}$ 7

Write the mixed number as a fraction.

4. $2\frac{3}{5}$ $\frac{13}{5}$
5. $1\frac{7}{8}$ $\frac{15}{8}$
6. $9\frac{1}{3}$ $\frac{28}{3}$

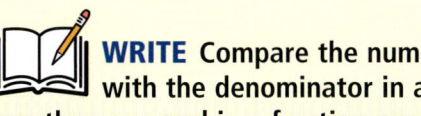

165

LESSON 8.3

Compare and Order Fractions

LESSON PLANNING

Objective To compare and order fractions

Intervention for Prerequisite Skills

Compare and Order Whole Numbers (For intervention strategies, see page 159.)

NCTM Standards
1. Number and Operations
6. Problem Solving
7. Reasoning and Proof
8. Communication

Math Background

Share these ideas with students as they learn how to compare and order fractions:

- Use mental math to compare fractions with the same denominators.
- When comparing fractions with unlike denominators, a number line can be used to find the greater fraction.
- Another way to compare fractions with unlike denominators is to find equivalent fractions with the same denominators.

WARM-UP RESOURCES

 **NUMBER OF THE DAY**

Marcus is in the sixth grade. He says he is 105,120 _____ old. Which word belongs in the blank: *months*, *days*, or *hours*? Explain. hours; Possible answer: Marcus is about 12 years old. 10 years × 12 months = 120; 10 years × 365 days = 3,650; 4,000 days × 25 = 100,000 hours.

 **PROBLEM OF THE DAY**

From 4:00 to 5:30, Jacob, Lisa, and Chelsea took turns playing the same computer game. Jacob played for $\frac{1}{2}$ hour and Lisa played for $\frac{3}{4}$ hour. For how many minutes did Chelsea play the game? 15 minutes

Solution Problem of the Day tab, p. PD8

 **DAILY FACTS PRACTICE**

Have students practice addition and subtraction facts by completing Set F of *Teacher's Resource Book*, page TR98.

166A Chapter 8

INTERVENTION AND EXTENSION RESOURCES

ALTERNATIVE TEACHING STRATEGY — ELL

Materials *For each student* fraction strips, p. TR18

Have students demonstrate how to **use fraction strips to compare fractions with unlike denominators.**

Ask: Which is greater: $\frac{2}{3}$ or $\frac{5}{6}$? $\frac{5}{6}$

Have volunteers model comparisons of other fraction pairs. Display the comparisons. Check students' work.

VISUAL
BODILY/KINESTHETIC

MULTISTEP AND STRATEGY PROBLEMS

The following multistep or strategy problem is provided in Lesson 8.3:

Page	Item
167	26

CAREER CONNECTION

Materials *For each group* reference books

Ask students to **determine how fractions are used** in various careers. Have them work in small groups to prepare an oral report about one of the following careers: stockbroker, engineer, architect, chef. Direct students to describe how fractions relate to each career.

Encourage them to use reference materials and, if possible, to interview a person working in that career. Check students' work.

AUDITORY
INTERPERSONAL/SOCIAL

PHYSICAL FITNESS/HEALTH CONNECTION

Have students **order fractions with unlike denominators** from the survey results below. The data show the fraction of those surveyed who ranked each factor as most important for a good night's sleep.

Good mattress: $\frac{8}{25}$; Healthy diet: $\frac{1}{10}$; Daily exercise: $\frac{1}{5}$; Good pillows: $\frac{2}{25}$; Other factors: $\frac{3}{10}$

Ask students to list the factors in order of importance from greatest to least. good mattress, other factors, daily exercise, healthy diet, good pillows

VISUAL
LOGICAL/MATHEMATICAL

TECHNOLOGY LINK
Intervention Strategies and Activities CD-ROM • *Skill 3*

LESSON 8.3 ORGANIZER

Objective To compare and order fractions

1 Introduce

QUICK REVIEW provides review of prerequisite skills.

Why Learn This? You can use this skill to help compare customary units of measurement that are expressed as fractions. *Share the lesson objective with students.*

2 Teach

Guided Instruction

- Work through the Example, and discuss how finding the LCM helps you compare fractions.

   How did you use the LCM to compare fractions? I wrote an equivalent fraction with the LCM 20 as the denominator for each fraction. The fraction with the greater numerator was the greater number.

REASONING Another way to solve the Example is to compare each fraction to $\frac{1}{2}$: $\frac{7}{10} > \frac{1}{2}$ and $\frac{1}{4} < \frac{1}{2}$, so $\frac{1}{4} < \frac{7}{10}$.

Modifying Instruction Some students may need to review comparing fractions with like denominators before comparing those with unlike denominators.

ADDITIONAL EXAMPLE

Example, p. 166

About $\frac{3}{5}$ of the class are bus riders and about $\frac{1}{6}$ are walkers. Are there more bus riders or walkers?

To compare $\frac{3}{5}$ and $\frac{1}{6}$, find the LCM of the denominators, 5 and 6.

5: 5, 10, 15, 20, 25, 30

6: 6, 12, 18, 24, 30

LCM = 30 $\frac{3}{5} = \frac{3 \times 6}{5 \times 6} = \frac{18}{30}$ $\frac{1}{6} = \frac{1 \times 5}{6 \times 5} = \frac{5}{30}$

$\frac{18}{30} > \frac{5}{30}$ so $\frac{3}{5} > \frac{1}{6}$

So, there are more bus riders.

LESSON 8.3

Compare and Order Fractions

Learn how to compare and order fractions.

QUICK REVIEW

1. $\frac{1}{2} = \frac{\blacksquare}{10}$ 5
2. $\frac{2}{3} = \frac{\blacksquare}{9}$ 6
3. $\frac{4}{5} = \frac{\blacksquare}{20}$ 16
4. $\frac{3}{8} = \frac{\blacksquare}{24}$ 9

5. Write the first four multiples of 12. 12, 24, 36, 48

Remember that values increase as you move to the right on a number line. Values decrease as you move left.

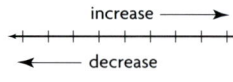

If two fractions have the same denominator, the fraction with the greater numerator is greater. So, $\frac{7}{12} > \frac{5}{12}$ because $7 > 5$.

If fractions do not have common denominators, you can use a number line to compare and order the fractions. The number line shows that $\frac{1}{4} < \frac{3}{8} < \frac{1}{2}$. From least to greatest the order is $\frac{1}{4}, \frac{3}{8}, \frac{1}{2}$.

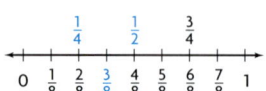

You can also use the least common multiple (LCM) to compare and order fractions.

EXAMPLE

George Washington Carver, one of America's most honored scientists, was born a slave in 1864.

George Washington Carver made over 500 useful agricultural products using peanuts, sweet potatoes, and pecans. About $\frac{7}{10}$ of the products used peanuts and about $\frac{1}{4}$ used sweet potatoes. Did Carver make more products with sweet potatoes or with peanuts?

To compare $\frac{7}{10}$ and $\frac{1}{4}$, find the LCM of the denominators, 10 and 4.

10: 10, 20, 30, 40   *Write multiples of 10.*

4: 4, 8, 12, 16, 20, 24, 28   *Write multiples of 4.*

LCM = 20

$\frac{7}{10} = \frac{7 \times 2}{10 \times 2} = \frac{14}{20}$ *Rewrite the fractions, using the LCM as a common denominator.*

$\frac{1}{4} = \frac{1 \times 5}{4 \times 5} = \frac{5}{20}$

$\frac{14}{20} > \frac{5}{20}$, so $\frac{7}{10} > \frac{1}{4}$.   *Compare $\frac{14}{20}$ and $\frac{5}{20}$.*

So, Carver made more products with peanuts.

Think and Discuss

1. The fractions have the same denominator. So, compare the numerators. $4 < 5$, so $\frac{4}{9} < \frac{5}{9}$.

CHECK FOR UNDERSTANDING

Look back at the lesson to answer each question.

1. **Explain** how to compare $\frac{4}{9}$ and $\frac{5}{9}$.

166 Chapter 8

Guided Practice Compare the fractions. Write <, >, or = for each ●.

2. $\frac{13}{20}$ ● $\frac{9}{20}$ **>**   3. $\frac{1}{4}$ ● $\frac{9}{20}$ **<**   4. $\frac{5}{6}$ ● $\frac{2}{3}$ **>**   5. $\frac{3}{8}$ ● $\frac{6}{16}$ **=**

Use the number line to order the fractions from least to greatest.

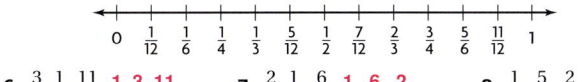

6. $\frac{3}{4}, \frac{1}{3}, \frac{11}{12}$ **$\frac{1}{3}, \frac{3}{4}, \frac{11}{12}$**   7. $\frac{2}{3}, \frac{1}{4}, \frac{6}{12}$ **$\frac{1}{4}, \frac{6}{12}, \frac{2}{3}$**   8. $\frac{1}{3}, \frac{5}{12}, \frac{2}{4}$ **$\frac{1}{3}, \frac{5}{12}, \frac{2}{4}$**

PRACTICE AND PROBLEM SOLVING

Independent Practice Compare the fractions. Write <, >, or = for each ●.

9. $\frac{6}{7}$ ● $\frac{4}{7}$ **>**   10. $\frac{3}{11}$ ● $\frac{4}{11}$ **<**   11. $\frac{4}{12}$ ● $\frac{1}{3}$ **=**   12. $\frac{17}{20}$ ● $\frac{3}{5}$ **>**

13. $\frac{5}{6}$ ● $\frac{15}{18}$ **=**   14. $\frac{7}{9}$ ● $\frac{11}{12}$ **<**   15. $\frac{3}{4}$ ● $\frac{5}{8}$ **>**   16. $\frac{11}{15}$ ● $\frac{2}{3}$ **>**

Use the number line to order the fractions from least to greatest.

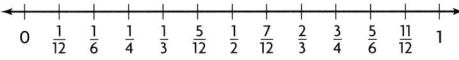

27. Three pieces of each pizza remain. That is $\frac{3}{8}$ of the mushroom and $\frac{3}{12}$ of the cheese pizza. $\frac{3}{12} = \frac{1}{4} = \frac{2}{8} < \frac{3}{8}$, so more of the mushroom pizza is left.

17. $\frac{9}{12}, \frac{1}{2}, \frac{2}{6}$ **$\frac{2}{6}, \frac{1}{2}, \frac{9}{12}$**   18. $\frac{4}{12}, \frac{4}{6}, \frac{7}{12}$ **$\frac{4}{12}, \frac{7}{12}, \frac{4}{6}$**   19. $\frac{7}{12}, \frac{5}{6}, \frac{1}{2}$ **$\frac{1}{2}, \frac{7}{12}, \frac{5}{6}$**

Order the fractions from least to greatest.

20. $\frac{5}{8}, \frac{1}{2}, \frac{3}{4}$ **$\frac{1}{2}, \frac{5}{8}, \frac{3}{4}$**   21. $\frac{2}{5}, \frac{3}{10}, \frac{1}{2}$ **$\frac{3}{10}, \frac{2}{5}, \frac{1}{2}$**   22. $\frac{11}{16}, \frac{3}{4}, \frac{5}{8}$ **$\frac{5}{8}, \frac{11}{16}, \frac{3}{4}$**

23. $\frac{1}{3}, \frac{1}{6}, \frac{1}{2}$ **$\frac{1}{6}, \frac{1}{3}, \frac{1}{2}$**   24. $\frac{2}{3}, \frac{2}{9}, \frac{2}{6}$ **$\frac{2}{9}, \frac{2}{6}, \frac{2}{3}$**   25. $\frac{3}{4}, \frac{1}{12}, \frac{5}{8}$ **$\frac{1}{12}, \frac{5}{8}, \frac{3}{4}$**

Problem Solving Applications

26. During a physical education class, $\frac{1}{3}$ of the students chose to play basketball, $\frac{4}{15}$ chose flag football, and the rest chose tetherball. Which activity was chosen by the most students? **tetherball**

27. ❓ **What's the Error?** For dinner, a mushroom pizza is cut into eighths and a cheese pizza into twelfths. After the meal there are 3 pieces of each left. Pablo tells his mother that the same amount of each pizza is left. What mistake did he make? **See above left.**

28. **REASONING** Find a fraction that has a denominator of 15 and is between $\frac{2}{3}$ and $\frac{4}{5}$. **$\frac{11}{15}$**

MIXED REVIEW AND TEST PREP

Evaluate each expression for $a = 2.3$, $b = 0.7$, and $c = 5.4$. (p. 82)

29. $a - b + c$ **7**   30. $(b \times 5) + a$ **5.8**

31. Find the LCM of 8 and 12. (p. 150) **24**   32. Find the GCF of 16 and 40. (p. 150) **8**

⭐ 33. **TEST PREP** There are 12 cans of soup in 1 case. How many cases should you order if you need 132 cans of soup? (p. 22) **B**
A 10   B 11   C 1,200   D 1,584

EXTRA PRACTICE page H39, Set C

167

3 Practice

⚠️ COMMON ERROR ALERT

When fractions have like numerators, students often choose the fraction with the greater denominator as the greater fraction.

Error	Correction
$\frac{1}{4} > \frac{1}{2}$	$\frac{1}{2} > \frac{1}{4}$

Have students work with fraction circles or fraction bars to reinforce the concept that the fraction with the greater denominator is the lesser fraction. Compare the parts to see that $\frac{1}{4} < \frac{1}{2}$.

Guided Practice

Do Check For Understanding Exercises 1–8 with your students. Identify those having difficulty and use lesson resources to help.

Independent Practice

Note that Exercise 26 is a **multistep or strategy problem**. Assign Exercises 9–28.

MIXED REVIEW AND TEST PREP

Exercises 29–33 provide **cumulative review** (Chapters 1–8).

4 Assess

Summarize the lesson by having students:

DISCUSS Describe a situation where you would want to compare fractions. Possible answer: I ate $\frac{1}{3}$ of my pizza and Jan ate $\frac{2}{5}$ of hers. I want to know who ate more.

✏️ **WRITE** Explain how to compare fractions with unlike denominators. Possible answer: Change the fractions to equivalent fractions with like denominators and then compare the numerators.

Lesson Quiz

Transparency 8.3

Compare the fractions. Write <, >, or = for each ●.

1. $\frac{1}{3}$ ● $\frac{1}{4}$ **>**   2. $\frac{3}{12}$ ● $\frac{7}{12}$ **<**

3. $\frac{2}{7}$ ● $\frac{6}{21}$ **=**

4. Kit used $\frac{5}{8}$ yd of orange string and $\frac{3}{4}$ yd of blue string for her project. Which color did she use more of? **blue**

CHALLENGE 8.3

Let's Compare

For each exercise, choose a fraction from the box at the right.

| $\frac{19}{24}$ | $\frac{7}{24}$ | $\frac{7}{10}$ | $\frac{1}{2}$ |

1. Carl, Philip, and Monica were discussing how they had spent last summer. Carl said that he stayed at his grandmother's house for half the summer. Philip said he stayed at his aunt's house for $\frac{2}{5}$ of the summer. Monica said she was away for a greater fraction of the summer than either of the two boys. If she was away for less than $\frac{3}{4}$ of the summer, what fraction of the summer was she away?
   $\frac{7}{10}$ of the summer

2. When her cat had kittens, Mrs. Banks gave three of them to neighbors while they were still very small. The one she gave to Elise weighed $\frac{2}{5}$ pound. The one that Carmela received weighed $\frac{3}{8}$ pound. The kitten that Denise took home weighed more than Carmela's kitten, but less than Elise's. What fraction of a pound did Denise's kitten weigh?
   $\frac{1}{2}$ lb

3. Each morning, Maria rides her bike $\frac{1}{3}$ mile to school. Angela walks the $\frac{1}{4}$ mile between her home and school. James skateboards from home to school, a distance less than Maria rides, but more than Angela walks. What fraction of a mile does James skateboard to school?
   $\frac{7}{24}$ mi

4. Sandra, Jackson, and Shari all have the same size box of markers. Sandra still has $\frac{5}{6}$ of all the markers that were originally in her box. Jackson has $\frac{3}{4}$ of the original number. Shari has more markers than one of her friends, but less than the other. What fraction of the original number of markers does Shari still have?
   $\frac{19}{24}$ of the markers

PROBLEM SOLVING 8.3

Compare and Order Fractions

Analyze Choose Solve Check

Write the correct answer.

1. Beth has a box of 20 red pencils and a box of 16 blue pencils. If she makes equal-size groups of all red or all blue pencils, what is the greatest number that can be in each group so that no pencils will be left over?
   4 pencils

2. Kim called her brother from her hotel after stopping for the night while on a trip. She told him she had completed $\frac{3}{5}$ of her trip. Had she completed at least half the trip? Explain.
   Yes; $\frac{3}{5} > \frac{1}{2}$.

3. A doughnut shop uses the following formula when selling its doughnuts: $P = \$0.75 \times d$, where d is the number of doughnuts purchased and P is the price the customer pays. What is the greatest number of doughnuts a customer can buy with $5?
   6 doughnuts

4. José read a cake recipe that called for $\frac{3}{4}$ cup flour, $\frac{1}{2}$ cup sugar, and $\frac{2}{5}$ cup milk. He lined up the ingredients in order by the amount, from least to greatest. Which ingredient did José put at the end of the line?
   flour

Choose the letter for the best answer.

5. Evelyn said that she had finished less than half of her homework problems. What fraction of the problems might Evelyn have completed?
   A $\frac{3}{5}$ C $\frac{5}{8}$
   B $\frac{3}{8}$ D $\frac{4}{5}$

6. Four friends are all reading the same book. Gordon has read $\frac{1}{2}$ the book, Nick has read $\frac{2}{5}$, Yvonne has read $\frac{2}{7}$, and Curtis has read $\frac{2}{5}$. Which of them has read the greatest part of the book?
   F Gordon   **H Nick**
   G Yvonne   J Curtis

7. This year's sixth grade in Glenn Middle School has 6 classes. Which is the number of sixth-grade students if there are the same number of students in each class?
   A 184 students   C 170 students
   B 172 students   **D 168 students**

8. Di has 1 red marker, 1 blue marker, and 1 green marker. She plans to make a design having 3 vertical stripes, one of each color. How many different designs can Di make?
   F 3 designs   H 9 designs
   G 6 designs   J 12 designs

9. **Write About It** What are some different ways to compare a fraction to $\frac{1}{2}$?
   Possible answer: Use a number line, find and use the least common denominator, or determine if the denominator of the fraction is greater or less than twice the numerator.

167

LESSON 8.4

ORGANIZER

Objective To convert fractions to decimals

Materials *For each student* 1-cm graph paper, p. TR64, or two 10 × 10 grids (decimal squares), p. TR7; scissors; colored pencils

Lesson Resources E-Lab Recording Sheet • *Equivalent Fractions, Decimals, and Mixed Numbers*

Intervention for Prerequisite Skills Model Fractions (For intervention strategies, see page 159.)

Using the Page

Activity

Some students may confuse tenths with hundredths. Remind students that a 10 × 10 grid column represents one tenth and a square represents one hundredth.

Think and Discuss

What is the same about showing $\frac{6}{10}$ and $\frac{62}{100}$ on a 10 × 10 grid? What is different? *For both you need to shade 6 columns; for $\frac{62}{100}$, you need to shade 2 more squares.*

Practice

Ask students to explain their first step in Exercise 4. *Write $\frac{3}{5}$ as $\frac{6}{10}$.*

Oral Assessment

Write $\frac{1}{4}$ as a decimal. Describe your work. *$\frac{1}{4} = \frac{25}{100} = 0.25$; First I write an equivalent fraction with 100 as the denominator. Then I use the numerator to write the decimal as hundredths.*

USING E-LAB

Students use visual thinking as they play a game that relates fractions and decimals.

The E-Lab Recording Sheets and activities are available on the E-Lab website.

www.harcourtschool.com/elab2002

LESSON 8.4 Explore Fractions and Decimals

MATH LAB

Explore how to convert fractions to decimals.

You need graph paper, scissors, colored pencils.

You can use decimal squares to help you convert fractions to decimals.

Activity

- Cut out a 10 × 10 grid from graph paper. Fold it into 2 equal parts. Then use a colored pencil to shade one of the equal parts. What fraction of the grid is shaded? $\frac{1}{2}$

- How many small squares are in the whole grid? How many of these squares are shaded? What fraction compares these shaded squares to those in the whole grid? What decimal can you write for this fraction? *100 squares; 50 squares; $\frac{50}{100}$; 0.50*

- How many columns are in the whole grid? How many columns are shaded? What fraction compares the shaded columns to the whole grid? What decimal can you write for this fraction? *10 columns; 5 columns; $\frac{5}{10}$; 0.5*

- Are the fractions $\frac{1}{2}$, $\frac{5}{10}$, and $\frac{50}{100}$ equivalent? How do you know? What are two ways to write the fraction $\frac{1}{2}$ as a decimal? **Yes; they all describe the same part of the grid; 0.50, 0.5**

- Cut out another 10 × 10 grid. Fold the grid into 5 equal parts and shade one part. How many rows or columns are shaded? How can you write $\frac{1}{5}$ as a decimal? **2 rows or columns; 0.2 or 0.20**

Think and Discuss

- How can you show tenths in a 10 × 10 grid? **full rows or columns**

- How can you show hundredths in a 10 × 10 grid? **small squares**

- What fractions are easiest to write as decimals? **fractions with denominators of 10 or 100**

- How is the 10 × 10 grid helpful for writing fractions as decimals? **The 10 × 10 grid has 100 squares, so each square is one hundredth of the whole.**

TECHNOLOGY LINK
More Practice: Use E-Lab, *Equivalent Fractions, Decimals, and Mixed Numbers.*
www.harcourtschool.com/elab2002

Practice

Write the fraction as a decimal. Use decimal squares.

1. $\frac{3}{10}$ **0.3**   2. $\frac{60}{100}$ **0.60**   3. $\frac{7}{10}$ **0.7**   4. $\frac{3}{5}$ **0.6**   5. $\frac{90}{100}$ **0.90**
6. $\frac{2}{10}$ **0.2**   7. $\frac{2}{5}$ **0.4**   8. $\frac{6}{10}$ **0.6**   9. $\frac{85}{100}$ **0.85**   10. $\frac{4}{5}$ **0.8**

168 Chapter 8

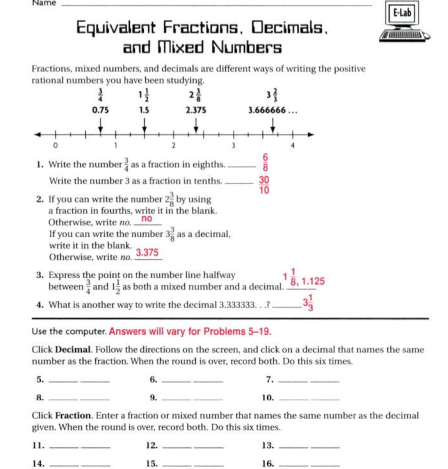

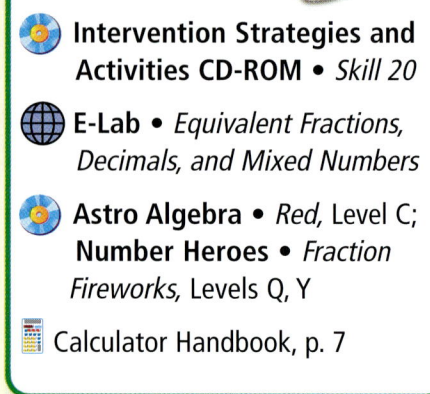

TECHNOLOGY LINK

- Intervention Strategies and Activities CD-ROM • *Skill 20*
- E-Lab • *Equivalent Fractions, Decimals, and Mixed Numbers*
- Astro Algebra • *Red*, Level C; Number Heroes • *Fraction Fireworks*, Levels Q, Y
- Calculator Handbook, p. 7

Fractions, Decimals, and Percents

LESSON **8.5**

LESSON PLANNING

Objective To convert fractions to decimals, decimals to fractions, and fractions to percents

Intervention for Prerequisite Skills
Compare and Order Whole Numbers, Model Percents (For intervention strategies, see page 159.)

Lesson Resources Calculator Handbook, p. 8

NCTM Standards
1. Number and Operations
5. Data Analysis and Probability
6. Problem Solving
8. Communication

Math Background
Consider the following as you help students understand the relationship between fractions, decimals, and percents.
- Use a decimal's place value to write a decimal as a fraction.
- To write a fraction as a decimal, divide the numerator by the denominator.
- To compare fractions and decimals, rewrite the fractions as decimals so that all the numbers are in the same form.
- To write fractions as percents, first rewrite the fractions as decimals, and then convert the decimals to percents.

WARM-UP RESOURCES

 **NUMBER OF THE DAY**

Write the part of the month that has elapsed as 2 equivalent fractions. **Possible answer:** $\frac{12}{30} = \frac{6}{15}$

 **PROBLEM OF THE DAY**

A pizza has 8 slices. Milton ate $\frac{1}{4}$ of the pizza. Earl ate $\frac{1}{2}$ of what was left over. Did Earl eat more or fewer than 4 pieces? Explain. **fewer;** $\frac{1}{4}$ **of 8 = 2; 8 − 2 = 6; Earl ate** $\frac{1}{2}$ **of 6 pieces, or 3 pieces**

Solution Problem of the Day tab, p. PD8

 **DAILY FACTS PRACTICE**

Have students practice multiplication and division facts by completing Set A of *Teacher's Resource Book*, p. TR99.

Vocabulary

terminating decimal a decimal, such as $\frac{1}{2} = 0.5$, for which the division operation results in a remainder of zero

repeating decimal a decimal, such as $\frac{1}{3} = 0.333\ldots$ or $0.\overline{3}$, which shows a pattern of repeating digits

INTERVENTION AND EXTENSION RESOURCES

ALTERNATIVE TEACHING STRATEGY

Have students **use equivalent fractions with denominators of 100 to change fractions to percents.** Display: $\frac{1}{4} = \frac{\square}{100}$

Ask a volunteer to suggest a number to multiply 4 by to equal 100. Then complete the process.

$\frac{1}{4} = \frac{1 \times 25}{4 \times 25} = \frac{25}{100} = 0.25 = 25\%$

Direct students to change these fractions to percents.

$\frac{7}{20} = \frac{\square}{100}$ **35%** $\frac{13}{25} = \frac{\square}{100}$ **52%**

VISUAL
LOGICAL/MATHEMATICAL

ENGLISH LANGUAGE LEARNERS

Reinforce vocabulary by having students **identify numbers as fractions, decimals, or percents.**

- Display the words *decimal, fraction,* and *percent*. Have students copy the words into their math journals leaving a few lines between each word.
- Ask students to list ten examples of a decimal after the word *decimal*.
- Then have them rewrite the decimals as fractions and percents, recording their answers after the appropriate words.

Ask volunteers to share examples with the class.

Check students' work.

VISUAL
VERBAL/LINGUISTIC

MULTISTEP AND STRATEGY PROBLEMS

The following multistep or strategy problem is provided in Lesson 8.5:

Page	Item
171	61

SPECIAL NEEDS

Materials *For each student* two 10 x 10 grids (decimal squares), p. TR7; 2 different colored markers

Have students **model a decimal and a fraction** that they have compared in one of the Exercises 38–43 on page 171. Tell students to model the decimal in one color on one of the grids and to model the fraction in another color on the other grid to show that their answer is correct. Check students' work.

KINESTHETIC
VISUAL/SPATIAL

TECHNOLOGY LINK

 Intervention Strategies and Activities CD-ROM • *Skills 3, 21*

 Astro Algebra • *Red*, Level D

 Calculating Crew • *Nautical Number Line*, Level Q

 Calculator Handbook, pp. 7–8

LESSON 8.5

Fractions, Decimals, and Percents

Learn how to convert fractions to decimals, decimals to fractions, and fractions to percents.

QUICK REVIEW
1. 24 ÷ 4 **6**
2. 144 ÷ 6 **24**
3. 155 ÷ 5 **31**
4. Write 0.73 in words. **seventy-three hundredths**
5. Write 0.026 in words. **twenty-six thousandths**

There are many ways to write numbers. Some ways are as fractions, decimals, and percents.

Sometimes you may have to rewrite a given number in a different form. The easiest conversion is from a decimal to a fraction.

Vocabulary
terminating decimal
repeating decimal

EXAMPLE 1

Write each decimal as a fraction.

A. 0.7 **B.** 0.29

Use the decimal's place value to write each fraction.

$0.7 = \frac{7}{10}$ THINK: "seven tenths"

$0.29 = \frac{29}{100}$ THINK: "twenty-nine hundredths"

Remember that *percent* means "out of one hundred." For example, 25% means "25 out of 100."

To rewrite a fraction as a decimal, use long division or a calculator.

EXAMPLE 2

A newborn koala is about 19 mm long. This is about $\frac{3}{4}$ in. Change $\frac{3}{4}$ to a decimal.

Use long division.

```
  0.75
4)3.00
```
Divide the numerator by the denominator.

Use a calculator.

3 ÷ 4 Enter

So, $\frac{3}{4} = 0.75$.

The decimal 0.75 is an example of a **terminating decimal**. The decimal comes to an end at 5. You know that a decimal terminates if you reach a remainder of zero when you are using long division.

The decimal for the fraction $\frac{4}{11}$ does not terminate. When you divide 4 by 11, you never reach a remainder of zero. This decimal is a **repeating decimal** because it shows a pattern of repeating digits.

To write a repeating decimal, show three dots or draw a bar over the repeating part.

$\frac{4}{11} = 0.363636\ldots$ $\frac{4}{11} = 0.\overline{36}$

```
   0.3636
11)4.0000
   -3 3
     70
    -66
     40
    -33
     70
    -66
      4
```

169

LESSON 8.5 ORGANIZER

Objective To convert fractions to decimals, decimals to fractions, and fractions to percents

Vocabulary terminating decimal, repeating decimal *Review* percent

Lesson Resources Calculator Handbook, pp. 7–8

1 Introduce

QUICK REVIEW provides review of prerequisite skills.

Why Learn This? You can use this skill to convert survey results to percents for graphic displays. *Share the lesson objective with students.*

2 Teach

Guided Instruction

- Discuss the methods of conversion in Examples 1 and 2.

  Why is it simple to convert a decimal to a fraction? The denominator is the same as the decimal place, and the numerator is the number part of the decimal.

  How would you write 0.09 as a fraction? $\frac{9}{100}$

REASONING What kind of fraction results in a decimal containing a whole number? Explain. A fraction greater than 1; the denominator/divisor is less than the numerator/dividend.

- Focus on the concept of terminating and repeating decimals.

  How do you know when to stop dividing when you have a repeating decimal? when you see the pattern of numbers repeat in the quotient

REASONING If $\frac{1}{3} = 0.\overline{3}$, express $\frac{2}{3}$ as a decimal. $0.\overline{6}$

ADDITIONAL EXAMPLES

Example 1, p. 169

Write each decimal as a fraction.

A. 0.3 $\frac{3}{10}$ **B.** 0.41 $\frac{41}{100}$

Example 2, p. 169

Tracy's grade on a quiz was $\frac{7}{9}$. Change $\frac{7}{9}$ to a decimal. $0.7777\ldots$ or $0.\overline{7}$

169

RETEACH 8.5

Fractions, Decimals, and Percents

When you need to change a fraction to a decimal, you can use division. Write the fraction $\frac{3}{5}$ as a decimal.

Step 1 Set up a division problem, dividing the numerator by the denominator.   5)3

Step 2 Place a decimal point after the numerator. Write a zero.   5)3.0

Step 3 Divide as you would with whole numbers.
```
  0.6
5)3.0
  3 0
    0
```

So, written as a decimal, $\frac{3}{5} = 0.6$. Recall that 0.6 is a terminating decimal because it ends after the tenths place.

When you need to change a decimal to a fraction, use place value.

Change 0.364 to a fraction.

Step 1 Identify the place value of the last digit in the decimal number.   0.364 ↑ thousandths

Step 2 Use the place value of the last digit as the denominator. $\frac{364}{1,000}$

So, $0.364 = \frac{364}{1,000}$.

Answer the questions to change the fraction to a decimal.

1. $\frac{1}{4}$
   a. What division problem will you use? **1 ÷ 4**
   b. What is the quotient? **0.25**

2. $\frac{3}{8}$
   a. What division problem will you use? **3 ÷ 8**
   b. What is the quotient? **0.375**

Use place value to write the decimal as a fraction.

3. 0.6 $\frac{6}{10}$ 4. 0.92 $\frac{92}{100}$ 5. 0.48 $\frac{48}{100}$ 6. 0.137 $\frac{137}{1,000}$

PRACTICE 8.5

Fractions, Decimals, and Percents

Write the decimal as a fraction.

1. 0.5 $\frac{5}{10}$ 2. 0.14 $\frac{14}{100}$ 3. 0.06 $\frac{6}{100}$ 4. 0.83 $\frac{83}{100}$
5. 0.62 $\frac{62}{100}$ 6. 0.317 $\frac{317}{1,000}$ 7. 0.805 $\frac{805}{1,000}$ 8. 0.955 $\frac{955}{1,000}$

Write as a decimal. Tell whether the decimal terminates or repeats.

9. $\frac{3}{10}$ **0.3, T**   10. $\frac{6}{9}$ **0.$\overline{6}$, R**   11. $\frac{7}{12}$ **0.583$\overline{3}$, R**   12. $\frac{11}{20}$ **0.55, T**
13. $\frac{7}{30}$ **0.2$\overline{3}$, R**   14. $\frac{9}{10}$ **0.9, T**   15. $\frac{7}{15}$ **0.4$\overline{6}$, R**   16. $\frac{4}{11}$ **0.$\overline{36}$, R**

Compare. Write <, >, or = for each ●.

17. 0.24 ● $\frac{1}{4}$ **<**   18. 0.18 ● $\frac{7}{50}$ **>**   19. $\frac{4}{10}$ ● 0.44 **<**
20. $\frac{1}{5}$ ● 0.19 **>**   21. $\frac{7}{20}$ ● 0.45 **<**   22. $\frac{9}{20}$ ● 0.45 **=**

Write the fraction as a percent.

23. $\frac{3}{5}$ **60%**   24. $\frac{17}{100}$ **17%**   25. $\frac{4}{2}$ **200%**   26. $\frac{1}{500}$ **0.2%**
27. $\frac{9}{25}$ **36%**   28. $\frac{7}{5}$ **140%**   29. $\frac{6}{40}$ **15%**   30. $\frac{17}{20}$ **85%**

Mixed Review

Estimate. **Possible estimates are given.**

31. 56.09 ÷ 7.1 **8**
32. 64.1 − 13.9 **50**
33. 97.6 + 9.8 **10**
34. $1.79 − $0.82 **$1.00**
35. 188.2 × 21.3 **4,000**
36. 602.5 + 102.4 **700**
37. $49.34 × 5 **$250**
38. 711.2 + 798.5 **1,500**

Evaluate the expression.

39. 6 + 4 × 3 **18**   40. 18 − 6 ÷ 2 **14**   41. (10 × 3) ÷ 6 **5**

LESSON 8.5

- Before discussing Example 3, review how to compare decimals.

  How do you compare decimals? Start at the left and compare the digits in each place, one at a time.

- Help students generalize the method for changing decimals to percents.

  What do you do to the decimal point to change any decimal to a percent? move it 2 places to the right

REASONING Challenge students to reverse the process and explain how to write 15% as a fraction. Write 15% as 0.15. Then rewrite the decimal as a fraction, $\frac{15}{100}$. Finally, write the fraction in simplest form, $\frac{3}{20}$.

ADDITIONAL EXAMPLES

Example 3, p. 170

The jar of Spencer's Spices contains $\frac{1}{8}$ lb of cinnamon. The jar of Nature's Best has 0.25 lb of cinnamon. Which container holds more cinnamon? $\frac{1}{8} = 0.125$; $0.25 > 0.125$; Nature's Best

Example 4, p. 170

Stuffed animals are on sale at the zoo's souvenir shop for $\frac{1}{5}$ off. What percent off the original price is the sale price? $\frac{1}{5} = 0.20 = 20\%$

3 Practice

Guided Practice

Do Check for Understanding Exercises 1–17 with your students. Identify those having difficulty and use lesson resources to help.

Additional Answers, Check for Understanding

1. The last digit is in the thousandths place. The numerator is 26 and the denominator is 1,000, so $0.026 = \frac{26}{1000}$.

2. Possible answer: When you use long division to write a fraction as a decimal, you will reach a remainder of zero if the decimal terminates. You will reach a repeating remainder or pattern of remainders if the decimal repeats.

170 Chapter 8

EXAMPLE 3

To compare a fraction and a decimal, you can first rewrite the fraction as a decimal. Then compare the decimals.

A newborn panda weighs about $\frac{1}{4}$ lb. A newborn cocker spaniel weighs about 0.4 lb. Which animal weighs less at birth?

Solve by using long division.

```
   0.25
4)1.00
  − 8
   20
  −20
    0
```
Divide the numerator by the denominator.

Solve by using a calculator.

1 ÷ 4 = 0.25

$0.25 < 0.4$, so $\frac{1}{4} < 0.4$.

So, a newborn panda weighs less than a newborn cocker spaniel.

To write a fraction as a percent, first convert the fraction to a decimal. Then write the decimal as a percent.

EXAMPLE 4

The barrow ground squirrel of Point Barrow, Alaska, is the world's longest-hibernating animal. The squirrel hibernates $\frac{9}{12}$ of the year. What percent of the year does it hibernate?

$\frac{9}{12} = 0.75$ Use long division or a calculator to rewrite the fraction as a decimal.

$0.75 = \frac{75}{100}$ **THINK:** 75 hundredths. Write the decimal as a fraction.

$= 75\%$ **THINK:** Percent means "out of one hundred." So, 75 hundredths is 75 percent.

So, the barrow ground squirrel hibernates 75% of the year.

CHECK FOR UNDERSTANDING

Think and Discuss   Look back at the lesson to answer each question.

1. **Explain** how to use place value to change 0.026 to a fraction. See below.
2. **Compare** a repeating decimal with a terminating decimal. See below.

Guided Practice   Write the decimal as a fraction.

3. 0.7 $\frac{7}{10}$ 4. 0.39 $\frac{39}{100}$ 5. 0.105 $\frac{105}{1,000}$ 6. 0.007 $\frac{7}{1,000}$

Write as a decimal. Tell whether the decimal terminates or repeats.

7. $\frac{1}{4}$ 0.25, T   8. $\frac{7}{20}$ 0.35, T   9. $\frac{2}{3}$ $0.\overline{6}$, R   10. $\frac{8}{11}$ $0.\overline{72}$, R

Compare. Write <, >, or = for each ●.

11. 0.62 ● $\frac{1}{2}$ >   12. $\frac{12}{20}$ ● 0.9   <   13. $\frac{1}{8}$ ● 0.125   =

Write the fraction as a percent.

14. $\frac{7}{10}$ 70%   15. $\frac{1}{5}$ 20%   16. $\frac{1}{4}$ 25%   17. $\frac{40}{100}$ 40%

170   Chapter 8

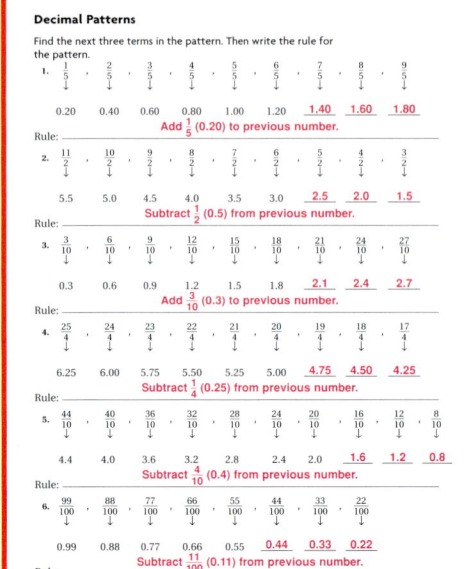

PRACTICE AND PROBLEM SOLVING

Independent Practice

Write the decimal as a fraction.

18. 0.4 $\frac{4}{10}$
19. 0.06 $\frac{6}{100}$
20. 0.35 $\frac{35}{100}$
21. 0.61 $\frac{61}{100}$
22. 0.115 $\frac{115}{1,000}$
23. 0.205 $\frac{205}{1,000}$
24. 0.079 $\frac{79}{1,000}$
25. 0.009 $\frac{9}{1,000}$

Write as a decimal. Tell whether the decimal terminates or repeats.

26. $\frac{1}{5}$ 0.2, T
27. $\frac{1}{6}$ $0.1\overline{6}$, R
28. $\frac{1}{15}$ $0.0\overline{6}$, R
29. $\frac{5}{8}$ 0.625, T
30. $\frac{11}{20}$ 0.55, T
31. $\frac{3}{10}$ 0.3, T
32. $\frac{5}{12}$ $0.41\overline{6}$, R
33. $\frac{1}{9}$ $0.\overline{1}$, R
34. $\frac{11}{12}$ $0.91\overline{6}$, R
35. $\frac{9}{25}$ 0.36, T
36. $\frac{17}{33}$ $0.\overline{51}$, R
37. $\frac{15}{99}$ $0.\overline{15}$, R

TECHNOLOGY LINK
More Practice: Use Mighty Math Calculating Crew, *Nautical Number Line*, Level Q

Compare. Write <, >, or = for each ●.

38. $\frac{1}{10}$ ● 0.04   >
39. 0.15 ● $\frac{3}{20}$ =
40. $\frac{1}{2}$ ● 0.52   <
41. 0.65 ● $\frac{3}{4}$ <
42. $\frac{1}{20}$ ● 0.1   <
43. 0.58 ● $\frac{7}{12}$ <

Write the fraction as a percent.

44. $\frac{9}{10}$ 90%
45. $\frac{3}{4}$ 75%
46. $\frac{1}{2}$ 50%
47. $\frac{6}{100}$ 6%
48. $\frac{3}{5}$ 60%
49. $\frac{25}{50}$ 50%
50. $\frac{3}{2}$ 150%
51. $\frac{1}{200}$ 0.5%

Problem Solving Applications

55. Megan

52. The goal of the East Side Animal Shelter is to have 0.8 of its animals adopted. One week, the shelter found homes for 20 of its 24 animals. Did the shelter reach its goal? Explain.
Yes; $\frac{20}{24} = \frac{5}{6} = 0.8\overline{3} > 0.8$.

Use Data For 53–55, use the table.

53. Write Brian's math test score as a decimal. 0.72

54. Did Juan get a higher score on the math or science test? science

55. Which student got a higher score on the math test than on the science test?

STUDENT	MATH SCORE	SCIENCE SCORE
Brian	$\frac{18}{25}$	0.95
Juan	$\frac{21}{25}$	0.85
Sabina	$\frac{17}{25}$	0.75
Megan	$\frac{23}{25}$	0.90

56. **Write About It** The decimal for $\frac{1}{9}$ is $0.\overline{1}$; for $\frac{2}{9}$, $0.\overline{2}$; and for $\frac{3}{9}$, $0.\overline{3}$. Explain how you could use this information to predict the decimal for $\frac{8}{9}$. Use division to check your method. The numerator repeats after the decimal point, so $\frac{8}{9}$ would be $0.\overline{8}$.

MIXED REVIEW AND TEST PREP

57. Write $4\frac{1}{2}$ as a fraction. (p. 164) $\frac{9}{2}$
58. Write $\frac{36}{5}$ as a mixed number. (p. 164) $7\frac{1}{5}$
59. $79.02 - 2.13$ (p. 66) 76.89
60. $48.541 + 11$ (p. 66) 59.541

★61. **TEST PREP** On his last six math quizzes, Rashard scored 86, 88, 92, 88, 96, and 84. His mean score increased by 1 after his next quiz. What was his seventh score? (p. 109) **A**

A 96   B 95   C 90   D 89

EXTRA PRACTICE page H39, Set D

171

CHAPTER 8

REVIEW/TEST

Purpose To check understanding of concepts, skills, and problem solving presented in Chapter 8

USING THE PAGE

The Chapter 8 Review/Test can be used as a **review** or a **test**.
- Items 1–2 check understanding of concepts and new vocabulary.
- Items 3–29 check skill proficiency.
- Items 30–33 check students' abilities to choose and apply problem solving strategies to real-life problems that involve fractions.

 Suggest that students place the completed Chapter 8 Review/Test in their portfolios.

USING THE ASSESSMENT GUIDE

- Multiple-choice format of Chapter 8 Posttest—See *Assessment Guide*, pp. AG53–54.
- Free-response format of Chapter 8 Posttest—See *Assessment Guide*, pp. AG55–56.

USING STUDENT SELF-ASSESSMENT

The How Did I Do? survey helps students assess what they have learned and how they learned it. This survey is available as a copying master in *Assessment Guide*, p. AGxvii.

172 Chapter 8

CHAPTER 8 REVIEW/TEST

1. **VOCABULARY** When the numerator and denominator of a fraction have no common factors other than 1, the fraction is in __?__ . (p. 161) **simplest form**

2. **VOCABULARY** A number that is made up of a whole number and a fraction is called a __?__ . (p. 164) **mixed number**

Write the fraction in simplest form. (pp. 160–163)

3. $\frac{6}{12}$ **$\frac{1}{2}$**
4. $\frac{12}{16}$ **$\frac{3}{4}$**
5. $\frac{25}{30}$ **$\frac{5}{6}$**

Complete. (pp. 160–163)

6. $\frac{3}{5} = \frac{\blacksquare}{20}$ **12**
7. $\frac{2}{\blacksquare} = \frac{10}{35}$ **7**
8. $\frac{24}{32} = \frac{\blacksquare}{8}$ **6**

Write the fraction as a mixed number or a whole number. (pp. 164–165)

9. $\frac{7}{3}$ **$2\frac{1}{3}$**
10. $\frac{30}{6}$ **5**
11. $\frac{19}{4}$ **$4\frac{3}{4}$**

Write the mixed number as a fraction. (pp. 164–165)

12. $1\frac{5}{6}$ **$\frac{11}{6}$**
13. $3\frac{1}{3}$ **$\frac{10}{3}$**
14. $5\frac{7}{8}$ **$\frac{47}{8}$**

Compare the fractions. Write <, >, or = for each ●. (pp. 166–167)

15. $\frac{7}{8}$ ● $\frac{5}{8}$ **>**
16. $\frac{2}{3}$ ● $\frac{8}{12}$ **=**
17. $\frac{1}{3}$ ● $\frac{1}{2}$ **<**
18. $\frac{1}{2}$ ● $\frac{11}{20}$ **<**
19. $\frac{3}{4}$ ● $\frac{3}{8}$ **>**
20. $\frac{7}{25}$ ● $\frac{1}{5}$ **>**

Write the decimal as a fraction. (pp. 169–171)

21. 0.27 **$\frac{27}{100}$**
22. 0.1 **$\frac{1}{10}$**
23. 0.089 **$\frac{89}{1,000}$**

Write as a decimal. Tell whether the decimal terminates or repeats. (pp. 169–171)

24. $\frac{1}{4}$ **0.25, T**
25. $\frac{5}{6}$ **$0.8\overline{3}$, R**
26. $\frac{7}{20}$ **0.35, T**

Write the fraction as a percent. (pp. 169–171)

27. $\frac{3}{4}$ **75%**
28. $\frac{9}{100}$ **9%**
29. $\frac{11}{25}$ **44%**

30. In the election for class president, Marcus received $\frac{5}{12}$ of the votes, Denise received $\frac{1}{4}$ of the votes, and Alonzo received $\frac{1}{3}$ of the votes. Who won the election? (pp. 166–167) **Marcus**

31. **Use Data** Use the table to find the fraction of the new November films that are rated PG-13. Write your answer in simplest form. What percent is this? (pp. 160–163) **$\frac{2}{5}$; 40%**

32. Of all U.S. car tunnels longer than 1 mile, $\frac{3}{8}$ are in Pennsylvania. Change $\frac{3}{8}$ to a decimal. (pp. 169–171) **0.375**

33. On Library Day, $\frac{13}{20}$ of the students at Pine Street School checked books out of the library. What percent of the students checked out books? (pp. 169–171) **65%**

NOVEMBER FILMS	
Rating	Number
G	3
PG-13	8
PG	5
R	4

172 Chapter 8

CHAPTERS 1–8 STANDARDIZED TEST PREP

TIP! **Understand the problem.**
See item **7**.
To compare fractions and decimals, change them to the same form. Use the form that will be easier to work with. Then find the numbers that are in the proper order.
Also see problem **1**, p. H62.

Choose the best answer.

1. Which group contains fractions that are all equivalent to $\frac{1}{4}$? **C**

   A $\frac{2}{8}, \frac{4}{20}, \frac{11}{44}$ C $\frac{6}{24}, \frac{15}{60}, \frac{50}{200}$

   B $\frac{3}{6}, \frac{20}{80}, \frac{3}{12}$ D $\frac{3}{12}, \frac{25}{100}, \frac{5}{9}$

2. Which is equivalent to 0.36? **G**

   F $\frac{18}{100}$ H $\frac{3}{6}$

   G $\frac{9}{25}$ J Not here

3. Which pair contains numbers that are equivalent? **A**

   A $3\frac{2}{3}, \frac{11}{3}$ C $4\frac{2}{5}, \frac{21}{5}$

   B $\frac{13}{4}, 2\frac{3}{4}$ D $\frac{16}{7}, 3\frac{1}{2}$

4. Eldora measured the length of a remote control sailboat course as $\frac{5}{8}$ of a mile. What is the decimal equivalent of $\frac{5}{8}$? **H**

   F 0.5

   G 0.58

   H 0.625

   J 0.8

5. In which pair are both numbers equivalent to $\frac{3}{4}$? **D**

   A 0.25; 25%   C 80%; $\frac{12}{15}$

   B 30%; $\frac{3}{10}$ D 75%; $\frac{12}{16}$

6. Evan has a bag of fruit. He has 9 apples, 4 oranges, and 3 bananas. What fraction represents the pieces of fruit that are oranges? **G**

   F $\frac{4}{18}$

   G $\frac{1}{4}$

   H $\frac{5}{16}$

   J $\frac{3}{4}$

7. Which numbers are in order from least to greatest? **A**

   A 0.3, $\frac{3}{8}, \frac{2}{5}$

   B $\frac{2}{5}$, 0.3, $\frac{3}{8}$

   C $\frac{3}{8}, \frac{2}{5}$, 0.3

   D $\frac{2}{5}, \frac{3}{8}$, 0.3

8. If 5 packages of hot dogs cost $9.25, what is the cost of 1 package? **H**

   F $0.92

   G $1.15

   H $1.85

   J $2.10

Write What You Know

See below.

9. Explain how the GCF can help you write a fraction in simplest form. Then write $\frac{135}{144}$ in simplest form.

10. Explain how the LCM can help you compare two fractions. Then compare $\frac{5}{6}$ and $\frac{7}{9}$.

173

Write What You Know • Written Response

9. To write a fraction in simplest form, divide the numerator and the denominator by their GCF. The numbers 135 and 144 have a GCF of 9. So, $\frac{135}{144} = \frac{135 \div 9}{144 \div 9} = \frac{15}{16}$.

10. Use the LCM of the denominators to write the fractions with a common denominator. Then compare the numerators. The LCM of 6 and 9 is 18.
    $\frac{5}{6} = \frac{5 \times 3}{6 \times 3} = \frac{15}{18}, \frac{7}{9} = \frac{7 \times 2}{9 \times 2} = \frac{14}{18}$.
    Since $15 > 14$, $\frac{15}{18} > \frac{14}{18}$. So, $\frac{5}{6} > \frac{7}{9}$.

STANDARDIZED TEST PREP •
Chapters 1–8

USING THE PAGE

This page may be used to help students get ready for standardized tests. The test items are written in the same style and arranged in the same format as those on many state assessments. The page is cumulative. It covers math objectives and essential skills that have been taught up to this point in the text. Most of the items represent skills from the current chapter, and the remainder represent skills from earlier chapters.

This page can be assigned at the end of the chapter as classwork or as a homework assignment. You may want to have students use individual recording sheets presented in a multiple-choice (standardized) format. A Test Answer Sheet is available as a blackline master in *Assessment Guide* (p. AGxlii).

You may wish to have students describe how they solved each problem and share their solutions.

ITEM ANALYSIS

Item	Learning Goal	Item	Learning Goal
1	7D	6	1B
2	7C	7	7D
3	7A	8	7B
4	7C	9	5C
5	7D	10	8B

Written Response items for Write What You Know are available as a blackline master in *Performance Assessment*.

SCORING RUBRIC • WRITE WHAT YOU KNOW

2 Demonstrates a complete understanding of the problem and chooses an appropriate strategy to determine the solution

1 Demonstrates a partial understanding of the problem and chooses a strategy that does not lead to a complete and accurate solution

0 Demonstrates little understanding of the problem and shows little evidence of using any strategy to determine a solution

Fraction Concepts 173

CHAPTER 9: Add and Subtract Fractions and Mixed Numbers

CHAPTER PLANNER

PACING OPTIONS
Compacted	6 Days
Expanded	11 Days

Getting Ready for Chapter 9 • Assessing Prior Knowledge and INTERVENTION (See PE and TE page 175.)

LESSON	NCTM STANDARDS	PACING	VOCABULARY*	MATERIALS	RESOURCES AND TECHNOLOGY
9.1 Estimate Sums and Differences pp. 176–179 **Objective** To estimate sums and differences of fractions and mixed numbers	1, 5, 6, 8	2 Days	mixed number		Reteach, Practice, Challenge, Problem Solving 9.1 Worksheets Extra Practice p. H40, Set A Transparency 9.1 Harcourt Math Newsroom Video • *Portable Planetarium*
9.2 Math Lab: Model Addition and Subtraction pp. 180–181 **Objective** To use fraction bars to add and subtract fractions with unlike denominators	1, 8, 10		**unlike fractions** least common multiple (LCM)	For each group fraction bars or fraction strips	E-Lab • *Addition and Subtraction of Unlike Fractions; E-Lab Recording Sheet*
9.3 Add and Subtract Fractions pp. 182–185 **Objective** To add and subtract fractions	1, 6, 7, 8, 9, 10	2 Days (For Lessons 9.2 and 9.3)	**least common denominator (LCD)**		Reteach, Practice, Challenge, Problem Solving 9.3 Worksheets, Extra Practice p. H40, Set B Transparency 9.3 Number Heroes • *Fraction Fireworks* Calculator Handbook, pp. 3–4 Math Jingles™ CD 5–6
9.4 Add and Subtract Mixed Numbers pp. 186–189 **Objective** To add and subtract mixed numbers	1, 5, 6, 8, 9, 10	1 Day	mixed number		Reteach, Practice, Challenge, Problem Solving 9.4 Worksheets Extra Practice p. H40, Set C Transparency 9.4 Calculating Crew • *Nautical Number Line*
9.5 Math Lab: Rename to Subtract pp. 190–191 **Objective** To use fraction bars to rename and subtract mixed numbers	1, 8, 10			For each group fraction bars or fraction strips	E-Lab • *Subtracting Mixed Numbers; E-Lab Recording Sheet*
9.6 Subtract Mixed Numbers pp. 192–193 **Objective** To subtract mixed numbers involving renaming	1, 5, 6, 8	2 Days (For Lessons 9.5 and 9.6)			Reteach, Practice, Challenge, Problem Solving 9.6 Worksheets Extra Practice p. H40, Set D Transparency 9.6 Calculating Crew • *Nautical Number Line*
9.7 Problem Solving Strategy: Draw a Diagram pp. 194–195 **Objective** To use the strategy *draw a diagram* to solve problems	1, 4, 6, 8, 10	1 Day			Reteach, Practice, Challenge, Reading Strategy 9.7 Worksheets Transparency 9.7 Problem Solving Think Along, p. TR1

Ending Chapter 9 • Chapter 9 Review/Test, p. 196 • Standardized Test Prep, p. 197

***Boldfaced** terms are new vocabulary. Other terms are review vocabulary.

174A Chapter 9

CHAPTER AT A GLANCE

Vocabulary Development

The boldfaced words are the new vocabulary terms in the chapter. Have students record the definitions in their Math Journals.

unlike fractions, p. 180

least common denominator (LCD), p. 182

Vocabulary Cards

Have students use the Vocabulary Cards on *Teacher's Resource Book* pp. TR127–128 to make graphic organizers or word puzzles. The cards can also be added to a file of mathematics terms.

Writing Opportunities

PUPIL EDITION
- What's the Question?, p. 185
- What's the Error?, p. 189
- Write About It, p. 193
- Write a Problem, pp. 179, 195

TEACHER'S EDITION
- Write—See the *Assess* section of each TE lesson.
- Writing in Mathematics, pp. 176B, 180, 186B

ASSESSMENT GUIDE
- How Did I Do?, p. AGxvii

NCTM Standards

1. **Number and Operations**
   Lessons 9.1, 9.2, 9.3, 9.4, 9.5, 9.6, 9.7
2. **Algebra**
3. **Geometry**
4. **Measurement**
   Lesson 9.7
5. **Data Analysis and Probability**
   Lessons 9.1, 9.4, 9.6
6. **Problem Solving**
   Lessons 9.1, 9.3, 9.4, 9.6, 9.7
7. **Reasoning and Proof**
   Lesson 9.3
8. **Communication**
   Lessons 9.1, 9.2, 9.3, 9.4, 9.5, 9.6, 9.7
9. **Connections**
   Lessons 9.3, 9.4
10. **Representation**
    Lessons 9.2, 9.3, 9.4, 9.5, 9.7

Family Involvement Activities

These activities provide:
- Letters to the Family
- Math Vocabulary
- Family Game
- Practice (Homework)

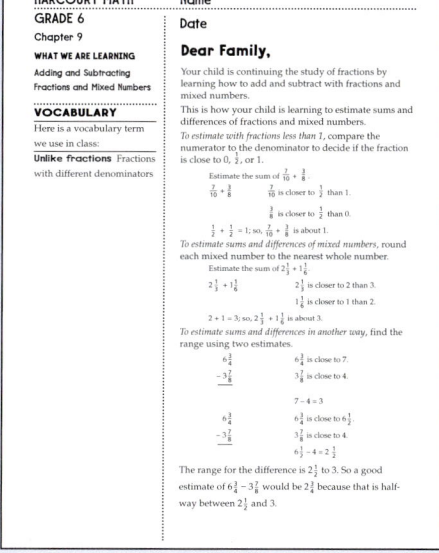

Family Involvement Activities, p. FA33

CHAPTER 9

Add and Subtract Fractions and Mixed Numbers

INTRODUCING THE CHAPTER

Tell students that adding and subtracting fractions and mixed numbers involve combining or finding the difference between these values. Have students focus on the pattern shown from the Navajo rug. Explain that 12 of the 25 squares are dark brown, so $\frac{12}{25}$ of the pattern is dark brown. Ask students what fraction represents all colors in the pattern. $\frac{25}{25}$

USING DATA

To begin the study of this chapter, have students

- Determine what fraction of the pattern is yellow. $\frac{8}{25}$
- Determine what fraction of the squares are neither dark brown nor yellow. $\frac{5}{25}$, or $\frac{1}{5}$

PROBLEM SOLVING PROJECT

Purpose To add and subtract fractions

Grouping pairs or small groups

Materials drawing paper, crayons, rulers

Background An expert Navaho weaver has been known to spend over 370 weaving hours completing both a double saddle blanket ($2\frac{1}{2}$ ft × 5 ft) and a quality rug (3 ft × 5 ft).

Analyze, Choose, Solve, and Check

Have students

- Design a geometric rug pattern by using 15 squares or equilateral triangles.
- Use 3 different colors to decorate the pattern.
- Record the fraction of their pattern that involves each color and the total fraction for each pair of colors.
- Compare and discuss their different patterns and sums. Check students' work.

 Suggest that students display their rug patterns in the classroom before placing them in their portfolios.

CHAPTER 9 Add and Subtract Fractions and Mixed Numbers

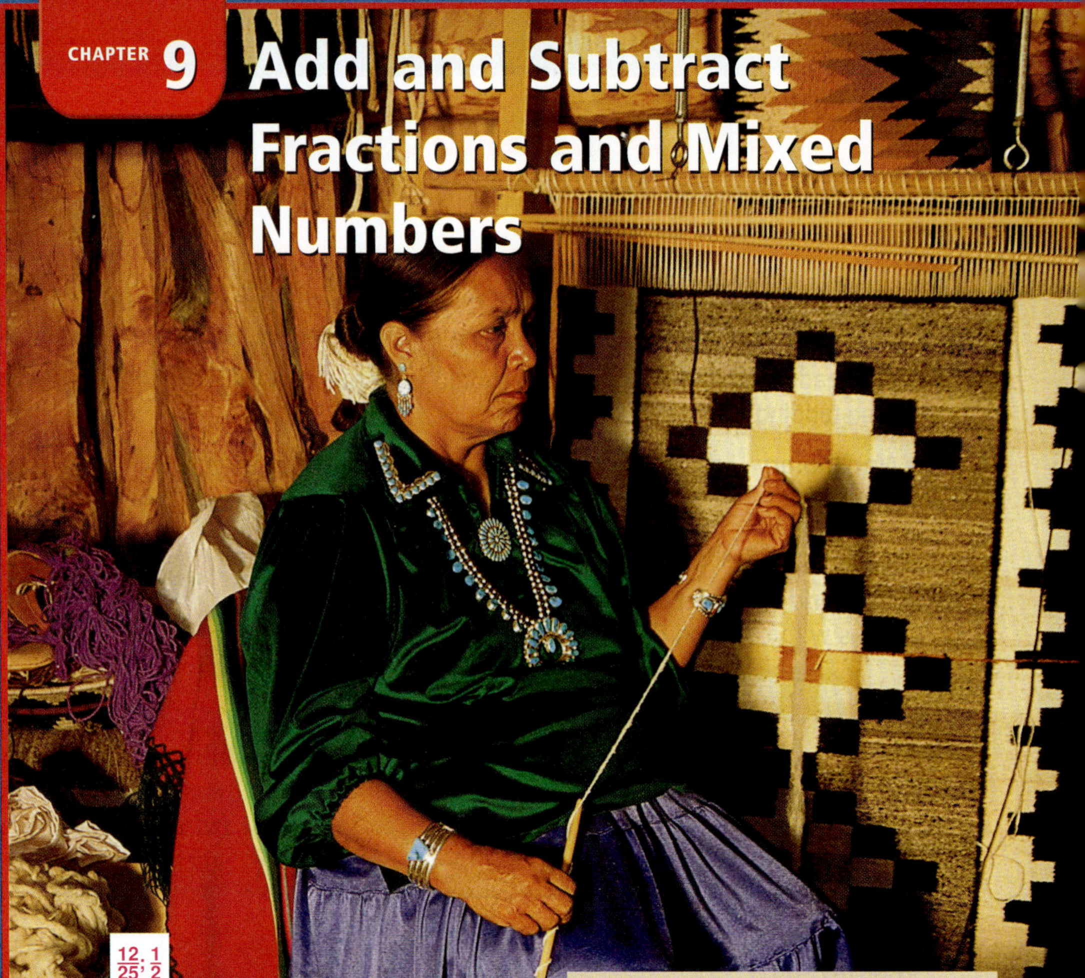

$\frac{12}{25}, \frac{1}{2}$

Traditional Navajo rugs often have geometric patterns woven into them. Completed by hand, a 3 ft x 5 ft rug can require hundreds of hours of work.

PROBLEM SOLVING One of the patterns from the rug in the picture is shown to the right. What fraction of the squares in the picture are dark brown border squares? What benchmark fraction is this closest to?

174 Chapter 9

Why learn math? Explain that the Navajo rug maker employs mathematics when deciding upon the design for a rug. The pattern is based on geometric artistry. The balance of colors and shapes is the artistic employment of fractional parts. Ask: What other workers use fractions and mixed numbers? Possible answer: a seamstress or a tailor

Check What You Know

Use this page to help you review and remember important skills needed for Chapter 9.

✓ Vocabulary

Choose the best term from the box.

> divide
> equivalent
> estimate
> mixed number
> simplest form

1. A number with a whole-number part and a fraction part is called a(n) ___?___. **mixed number**
2. A fraction in which the numerator and denominator have no common factors other than 1 is in ___?___. **simplest form**
3. To write a fraction greater than 1 as a mixed number, ___?___ the numerator by the denominator. **divide**
4. Fractions that name the same number or part are called ___?___ fractions. **equivalent**

✓ Simplify Fractions (For Intervention, see p. H9.)

Write each fraction in simplest form.

5. $\frac{6}{8}$ **$\frac{3}{4}$**
6. $\frac{5}{10}$ **$\frac{1}{2}$**
7. $\frac{4}{12}$ **$\frac{1}{3}$**
8. $\frac{18}{27}$ **$\frac{2}{3}$**
9. $\frac{12}{9}$ **$\frac{4}{3}$, or $1\frac{1}{3}$**
10. $\frac{16}{20}$ **$\frac{4}{5}$**
11. $\frac{12}{8}$ **$\frac{3}{2}$, or $1\frac{1}{2}$**
12. $\frac{9}{15}$ **$\frac{3}{5}$**
13. $\frac{10}{20}$ **$\frac{1}{2}$**
14. $\frac{15}{18}$ **$\frac{5}{6}$**
15. $\frac{26}{39}$ **$\frac{2}{3}$**
16. $\frac{12}{16}$ **$\frac{3}{4}$**
17. $\frac{15}{9}$ **$\frac{5}{3}$, or $1\frac{2}{3}$**
18. $\frac{6}{32}$ **$\frac{3}{16}$**
19. $\frac{17}{51}$ **$\frac{1}{3}$**
20. $\frac{48}{54}$ **$\frac{8}{9}$**
21. $\frac{100}{200}$ **$\frac{1}{2}$**
22. $\frac{25}{10}$ **$\frac{5}{2}$, or $2\frac{1}{2}$**
23. $\frac{80}{64}$ **$\frac{5}{4}$, or $1\frac{1}{4}$**
24. $\frac{84}{96}$ **$\frac{7}{8}$**

✓ Add and Subtract Like Fractions (For Intervention, see p. H10.)

Find the sum or difference. Write each answer in simplest form.

25. $\frac{1}{3} + \frac{1}{3}$ **$\frac{2}{3}$**
26. $\frac{3}{5} + \frac{1}{5}$ **$\frac{4}{5}$**
27. $\frac{3}{8} + \frac{1}{8}$ **$\frac{1}{2}$**
28. $\frac{5}{6} + \frac{1}{6}$ **1**
29. $\frac{3}{4} - \frac{1}{4}$ **$\frac{1}{2}$**
30. $\frac{5}{8} - \frac{3}{8}$ **$\frac{1}{4}$**
31. $\frac{2}{3} - \frac{1}{3}$ **$\frac{1}{3}$**
32. $\frac{4}{5} - \frac{2}{5}$ **$\frac{2}{5}$**
33. $\frac{3}{6} + \frac{2}{6}$ **$\frac{5}{6}$**
34. $\frac{7}{8} - \frac{1}{8}$ **$\frac{3}{4}$**
35. $\frac{3}{4} + \frac{1}{4}$ **1**
36. $\frac{7}{8} - \frac{3}{8}$ **$\frac{1}{2}$**
37. $\frac{4}{9} - \frac{2}{9}$ **$\frac{2}{9}$**
38. $\frac{1}{12} + \frac{7}{12}$ **$\frac{2}{3}$**
39. $\frac{13}{14} - \frac{3}{14}$ **$\frac{5}{7}$**
40. $\frac{1}{15} + \frac{8}{15}$ **$\frac{3}{5}$**

LOOK AHEAD

In Chapter 9 you will
- estimate sums and differences of fractions and mixed numbers
- add and subtract fractions and mixed numbers

LESSON 9.1

Estimate Sums and Differences

LESSON PLANNING

Objective To estimate sums and differences of fractions and mixed numbers

Intervention for Prerequisite Skills
Simplify Fractions, Add and Subtract Like Fractions (For intervention strategies, see page 175.)

NCTM Standards
1. Number and Operations
5. Data Analysis and Probability
6. Problem Solving
8. Communication

Math Background
Students may estimate a sum or difference of fractions, as they may with whole numbers and decimals, to check an answer for reasonableness or to solve a problem in which an estimate is all that is needed.

Experiences with models will help students develop estimating skills for fractions and mixed numbers.

- By comparing fraction bar models, students can picture whether a fraction is closer to 0, $\frac{1}{2}$, or 1.
- Graphing a fraction on a number line can help students determine whether a fraction is closer to 0, $\frac{1}{2}$, or 1.

WARM-UP RESOURCES

 **NUMBER OF THE DAY**

The number of the day is the number that describes today as a day of the month. Write a fraction to describe what part of the current month has passed.
Answer for April 10: $\frac{10}{30} = \frac{1}{3}$

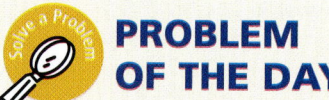 

Mike's and Kay's numbers are both less than 1. The digit in Mike's numerator is the same as the digit in Kay's denominator. Kay's number is $\frac{1}{10}$ greater than Mike's. What are Mike's and Kay's numbers? Mike's number is $\frac{2}{5}$ and Kay's number is $\frac{1}{2}$.

Solution Problem of the Day tab, p. PD9

Have students practice addition and subtraction facts by completing Set B of *Teacher's Resource Book,* p. TR99.

176A Chapter 9

INTERVENTION AND EXTENSION RESOURCES

ALTERNATIVE TEACHING STRATEGY ELL

Materials *For each student* index card

Have students **match exact numbers with estimates.** For half of the class, write sentences with exact numbers, one per card:

- The show lasts 55 min.
- Ginny is 62 in. tall.
- Only 5 eggs are in this carton.

Then write corresponding estimates—*around an hour, approximately 5 ft, about $\frac{1}{2}$ doz*—on cards for the remaining students. Pass out the cards to the class. Have students take turns reading aloud cards with an exact number and then cards with the matching estimates. Check students' work.

See also page 178.

AUDITORY
VERBAL/LINGUISTIC

MULTISTEP AND STRATEGY PROBLEMS

The following multistep and strategy problems are provided in Lesson 9.1:

Page	Item
179	44–45

WRITING IN MATHEMATICS

Have students **complete each statement to include an exact number and an estimate.** Display the following. Possible answers are given.

- Bob watched TV for _____. 35 min, about $\frac{1}{2}$ hour
- Last year Liza grew _____. $2\frac{3}{8}$ in., about $2\frac{1}{2}$ in.
- The recipe calls for _____ of ginger. $2\frac{1}{8}$ tsp., about 2 tsp.
- _____ of the class attended the party. $\frac{2}{3}$, More than $\frac{1}{2}$

LANGUAGE ARTS CONNECTION

Ask students to **research the origins of math terms,** such as *fraction, addition,* and *subtraction,* and to report to the class what they discover. Explain that many English words have Latin origins. The word *fraction,* for example, comes from the Latin word *frangere,* which means "to break." Encourage students to relate this meaning to their understanding of a fraction. Check students' work.

AUDITORY
VERBAL/LINGUISTIC

TECHNOLOGY LINK

 Intervention Strategies and Activities CD-ROM • *Skills 22, 23*

 Harcourt Math Newsroom Video • *Portable Planetarium*

LESSON 9.1

Estimate Sums and Differences

Learn how to estimate sums and differences of fractions and mixed numbers.

QUICK REVIEW

Estimate the sum or difference. *Possible estimates are given.*

1. 823 +116 **900**	2. 364 −232 **130**	3. 736 −381 **300**	4. 589 + 42 **640**	5. 6,755 − 482 **6,300**

You can decide whether a fraction is closest to 0, $\frac{1}{2}$, or 1.

One Way Use a number line.

Look at the number line below. Is $\frac{3}{8}$ closest to 0, $\frac{1}{2}$, or 1?

$\frac{3}{8}$ is closest to $\frac{1}{2}$.

Another Way Compare the numerator to the denominator.

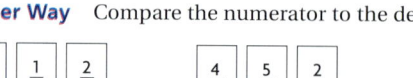

The numerators are much less than half the denominators. So, the fractions are close to 0.	The numerators are about one half the denominators. So, the fractions are close to $\frac{1}{2}$.	The numerators are about the same as the denominators. So, the fractions are close to 1.

Sometimes when you add and subtract fractions, you do not need an exact answer.

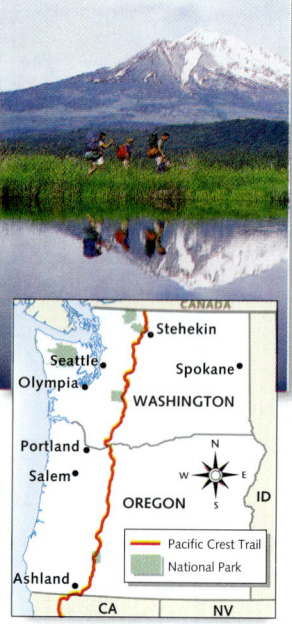

EXAMPLE 1

The Pacific Crest Trail runs from Mexico to Canada, a distance of 1,680 miles. The Tanner family is hiking the 848-mile portion from Ashland, Oregon, to Stehekin, Washington. The family hiked $\frac{3}{10}$ of this portion in June. Then they hiked another $\frac{3}{8}$ in July. About how much of the trail through Oregon and Washington did the Tanners hike in June and July?

Estimate $\frac{3}{10} + \frac{3}{8}$.

$\frac{3}{10} \rightarrow \frac{1}{2}$ $\frac{3}{10}$ is between 0 and $\frac{1}{2}$, but closer to $\frac{1}{2}$.
$+\frac{3}{8} \rightarrow +\frac{1}{2}$ $\frac{3}{8}$ is between 0 and $\frac{1}{2}$, but closer to $\frac{1}{2}$.
$\phantom{+\frac{3}{8} \rightarrow +}1$ The sum is greater than $\frac{1}{2}$, but less than 1.

So, the Tanners hiked more than $\frac{1}{2}$ of the trail, but not all of it.

176 Chapter 9

EXAMPLE 2

Estimate $\frac{7}{8} - \frac{5}{6}$.

$\frac{7}{8} \rightarrow 1$ $\frac{7}{8}$ is between $\frac{1}{2}$ and 1, but closer to 1.

$-\frac{5}{6} \rightarrow -1$ $\frac{5}{6}$ is between $\frac{1}{2}$ and 1, but closer to 1.

$\overline{0}$ Subtract.

The fractions are close to each other and their difference is close to 0.

Remember that a mixed number is a whole number and a fraction combined. $4\frac{2}{5}$ is a mixed number.

Look at the number line below. Is $3\frac{5}{8}$ in. closest to 3 in., $3\frac{1}{2}$ in., or 4 in.?

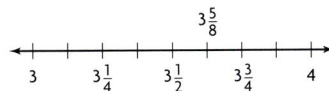

$3\frac{5}{8}$ in. is closest to $3\frac{1}{2}$ in.

To estimate sums and differences of mixed numbers, compare each mixed number to the nearest whole number or the nearest $\frac{1}{2}$.

EXAMPLE 3

Bianca's goal is to jog 15 miles a week. So far this week she has jogged $2\frac{5}{6}$ mi and $3\frac{1}{10}$ mi. About how many more miles must she jog to meet her goal?

Estimate $15 - \left(2\frac{5}{6} + 3\frac{1}{10}\right)$.

$15 - \left(2\frac{5}{6} + 3\frac{1}{10}\right)$ $2\frac{5}{6}$ is close to 3, and $3\frac{1}{10}$ is close to 3.
$\downarrow \downarrow$
$15 - (3 + 3)$ Add.
$15 - 6 = 9$ Subtract.

So, Bianca needs to jog about 9 more miles to meet her goal.

You can find a range to estimate a sum or difference.

EXAMPLE 4

Estimate $5\frac{3}{4} - 4\frac{7}{8}$.

Since $5\frac{3}{4}$ is halfway between $5\frac{1}{2}$ and 6, find two estimates.

$5\frac{3}{4} - 4\frac{7}{8}$	$5\frac{3}{4}$ is close to 6, and $4\frac{7}{8}$ is close to 5.	AND	$5\frac{3}{4} - 4\frac{7}{8}$	$5\frac{3}{4}$ is close to $5\frac{1}{2}$, and $4\frac{7}{8}$ is close to 5.
$\downarrow \downarrow$			$\downarrow \downarrow$	
$6 - 5 = 1$	Subtract.		$5\frac{1}{2} - 5 = \frac{1}{2}$	Subtract.

The range is $\frac{1}{2}$ to 1. A good estimate of $5\frac{3}{4} - 4\frac{7}{8}$ would be $\frac{3}{4}$, halfway between $\frac{1}{2}$ and 1.

LESSON 9.2

ORGANIZER

Objective To use fraction bars to add and subtract fractions with unlike denominators

Vocabulary <mark>unlike fractions</mark> Review least common multiple (LCM)

Materials For each group fraction bars or fraction strips, p. TR18

Lesson Resources E-Lab Recording Sheet • *Addition and Subtraction of Unlike Fractions*

Intervention for Prerequisite Skills Simplify Fractions, Add and Subtract Like Fractions (For intervention strategies, see page 175.)

Using the Pages

Remind students of how to add two like fractions. Point out that the numerator of the answer is the sum of the numerators. Emphasize that they keep the same denominator.

Activity 1

Show students that 12 is the smallest number that is a multiple of both 6 and 4:

multiples of 6: 6, **12**, 18, 24, . . .

multiples of 4: 4, 8, **12**, 16, . . .

Because 12 is a multiple of both 6 and 4, the fractions $\frac{1}{6}$ and $\frac{1}{4}$ can be rewritten with denominators of 12. The fraction bars show that $\frac{1}{6} = \frac{2}{12}$ and $\frac{1}{4} = \frac{3}{12}$.

Think and Discuss

• Discuss adding $\frac{2}{3}$ and $\frac{3}{8}$.

   What must be true of a denominator that could be used to rewrite thirds and eighths? It must be a multiple of both 3 and 8.

Practice

• After students work Exercise 1, ask:

   How did you use the model to find the sum of $\frac{1}{4}$ and $\frac{1}{2}$? I used 2 fourths to fit across 1 half. The sum of 2 fourths plus 1 fourth is 3 fourths.

LESSON 9.2

Explore how to use fraction bars to add and subtract fractions with unlike denominators.

You need fraction bars.

Vocabulary
<mark>unlike fractions</mark>

Remember that the LCM is the least of the common multiples of two or more numbers.
2: 2, 4, **6**, 8, 10, 12, . . .
3: 3, **6**, 9, 12, 15, . . .
The LCM of 2 and 3 is 6.
The LCM can be used to write common denominators of two or more fractions.

Model Addition and Subtraction

QUICK REVIEW
Find the LCM for each set of numbers.
1. 2, 8 **8**   2. 6, 9 **18**
3. 4, 15 **60** 4. 4, 10 **20**
5. 2, 3, 10 **30**

Fractions with the same denominator, such as $\frac{5}{9}$ and $\frac{4}{9}$, are called like fractions. Fractions with different denominators are called <mark>unlike fractions</mark>. You can use fraction bars to rename the denominators before adding.

Activity 1

Find $\frac{1}{6} + \frac{1}{4}$.

• Use fraction bars to show both fractions.

• Which fraction bars fit exactly across $\frac{1}{6}$ and $\frac{1}{4}$? Think about the LCM of 6 and 4. **twelfths**

• What is $\frac{1}{6} + \frac{1}{4}$? **$\frac{5}{12}$**

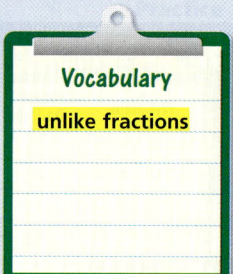

$\frac{1}{6} = \frac{2}{12}$ $\frac{1}{4} = \frac{3}{12}$

Think and Discuss

• Look at the model for $\frac{1}{6} + \frac{1}{4}$. What do you know about $\frac{1}{6}$ and $\frac{2}{12}$? about $\frac{1}{4}$ and $\frac{3}{12}$? **They are equivalent; they are equivalent.**

• How are the denominators of $\frac{1}{6}$, $\frac{1}{4}$, and $\frac{1}{12}$ related? (HINT: Think about common multiples.) **12 is the LCM of 6 and 4.**

Practice

Use fraction bars to find the sum. Draw a diagram of your model.
Check students' diagrams.

1. $\frac{1}{4} + \frac{1}{2}$ **$\frac{3}{4}$**   2. $\frac{1}{2} + \frac{1}{3}$ **$\frac{5}{6}$**   3. $\frac{1}{2} + \frac{2}{5}$ **$\frac{9}{10}$** 4. $\frac{2}{3} + \frac{1}{6}$ **$\frac{5}{6}$**

5. $\frac{1}{3} + \frac{1}{4}$ **$\frac{7}{12}$** 6. $\frac{3}{8} + \frac{1}{4}$ **$\frac{5}{8}$** 7. $\frac{1}{6} + \frac{1}{2}$ **$\frac{2}{3}$** 8. $\frac{3}{8} + \frac{1}{2}$ **$\frac{7}{8}$**

9. $\frac{1}{5} + \frac{1}{2}$ **$\frac{7}{10}$** 10. $\frac{3}{4} + \frac{1}{6}$ **$\frac{11}{12}$** 11. $\frac{1}{3} + \frac{1}{2}$ $\frac{5}{6}$ 12. $\frac{5}{8} + \frac{1}{4}$ **$\frac{7}{8}$**

180 Chapter 9

WRITING IN MATHEMATICS

Have students **write a word problem that can be solved with these fraction bars.**

Possible answer: Jon is swimming across Blue Lake, which is $\frac{5}{6}$ mi long. He has swum $\frac{3}{4}$ mi. How far does he have to go? $\frac{1}{12}$ mi

Intervention and Extension Resources

EARLY FINISHERS

Materials For each group fraction bars or fraction strips, p. TR18

Students **use fraction bars or fraction strips to find the sum of fractions with compatible denominators.** Explain that when the denominator of one fraction is a multiple of the denominator of another fraction, the denominators are *compatible*. Have students solve these exercises.

• $\frac{2}{3} - \frac{1}{6}$ **$\frac{1}{2}$**   • $\frac{4}{5} + \frac{1}{10}$ **$\frac{9}{10}$**   • $\frac{3}{8} + \frac{1}{4}$ **$\frac{5}{8}$**

KINESTHETIC
LOGICAL/MATHEMATICAL

180 Chapter 9

Fraction bars also can be used to subtract unlike fractions.

Activity 2

Find $\frac{1}{2} - \frac{1}{5}$.

- Use fraction bars to show $\frac{1}{2}$ and $\frac{1}{5}$.
- Which fraction bars fit exactly across $\frac{1}{2}$ and $\frac{1}{5}$? Think about the LCM. **tenths**
- Compare $\frac{5}{10}$ and $\frac{2}{10}$. How much more is $\frac{5}{10}$ than $\frac{2}{10}$? **$\frac{3}{10}$ more**
- What is $\frac{5}{10} - \frac{2}{10}$? What is $\frac{1}{2} - \frac{1}{5}$? **$\frac{3}{10}$; $\frac{3}{10}$**

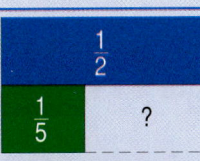
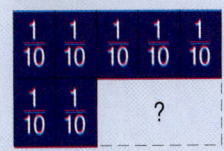

Think and Discuss

- How are the denominators of $\frac{1}{2}$, $\frac{1}{5}$, and $\frac{1}{10}$ related? **The LCM of 2 and 5 is 10.**
- Look at the model of $\frac{3}{4} - \frac{1}{3}$. Which fraction bars will fit exactly across $\frac{3}{4}$ and $\frac{1}{3}$? Explain. **Twelfths; 12 is the LCM of 4 and 3.**
- Which fraction bars will fit exactly across $\frac{1}{2} - \frac{1}{4}$? **fourths, eighths, or twelfths**

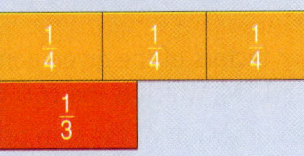

Practice

Use fraction bars to subtract. Draw a diagram of your model. **Check students' diagrams.**

1. $\frac{3}{4} - \frac{1}{3}$ **$\frac{5}{12}$**
2. $\frac{2}{5} - \frac{1}{10}$ **$\frac{3}{10}$**
3. $\frac{1}{3} - \frac{1}{4}$ **$\frac{1}{12}$**
4. $\frac{1}{2} - \frac{1}{3}$ **$\frac{1}{6}$**
5. $\frac{1}{2} - \frac{2}{5}$ **$\frac{1}{10}$**
6. $\frac{1}{2} - \frac{5}{12}$ **$\frac{1}{12}$**
7. $\frac{1}{4} - \frac{1}{6}$ **$\frac{1}{12}$**
8. $\frac{1}{3} - \frac{1}{6}$ **$\frac{1}{6}$**
9. $\frac{1}{4} - \frac{1}{8}$ **$\frac{1}{8}$**
10. $\frac{1}{2} - \frac{1}{4}$ **$\frac{1}{4}$**
11. $\frac{5}{6} - \frac{1}{3}$ **$\frac{1}{2}$**
12. $\frac{7}{8} - \frac{3}{4}$ **$\frac{1}{8}$**

MIXED REVIEW AND TEST PREP

Compare. Write <, >, or = for each . (p. 169)

13. $\frac{3}{8}$ 0.375   **=**
14. 0.15 $\frac{1}{4}$ **<**
15. 0.6 $\frac{11}{20}$ **>**
16. Multiply. 9.25×3.2 (p. 70)   **29.6**

★ 17. **TEST PREP** Of the 80 Moorers at their family reunion, there are 8 fewer females than males. How many males are at the reunion? (p. 22) **D**

   A 32   B 36   C 40   D 44

181

LESSON 9.3

Add and Subtract Fractions

LESSON PLANNING

Objective To add and subtract fractions

Intervention for Prerequisite Skills
Simplify Fractions, Add and Subtract Like Fractions (For intervention strategies, see page 175.)

Lesson Resources Calculator Handbook, pp. 3–4

NCTM Standards
1. Number and Operations
6. Problem Solving
7. Reasoning and Proof
8. Communication
9. Connections
10. Representation

Math Background

These ideas will help students find sums and differences of fractions.

- Multiples or factors of the given denominators can be used to find their LCM, which is the LCD of the given fractions.
- To find the numerator of the sum or difference of two fractions with a common denominator, add or subtract their numerators.
- Use an estimate to decide if an answer is reasonable.

WARM-UP RESOURCES

 **NUMBER OF THE DAY**

We are two common fractions. Our numerators are 1 and the LCM of our denominators is 20. What fractions are we? Possible answer: $\frac{1}{5}$ and $\frac{1}{4}$

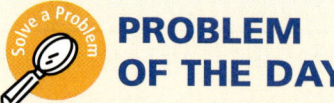 **PROBLEM OF THE DAY**

Kim used $\frac{6}{8}$ of a tank of gas. She bought $\frac{5}{8}$ of a tank and then used $\frac{4}{8}$ of a tank. Kim was almost out of money, so she filled the tank to half full by adding $\frac{2}{8}$ of a tank of gas. What part of a tank of gas did she start with? Kim started with $\frac{7}{8}$ tank of gas.

Solution Problem of the Day tab, p. PD9

 **DAILY FACTS PRACTICE**

Have students practice addition and subtraction facts by completing Set D of *Teacher's Resource Book,* p. TR99.

Vocabulary

least common denominator (LCD)
the least number, other than zero, that is a multiple of two or more denominators

INTERVENTION AND EXTENSION RESOURCES

ALTERNATIVE TEACHING STRATEGY

To **reinforce students' understanding of the least common denominator (LCD),** begin by displaying two fractions, such as $\frac{2}{3}$ and $\frac{5}{6}$.

- Ask four or five volunteers to name possible common denominators. Possible answers: 6, 12, 18, 24, 30
- Ask a volunteer to explain how to choose the LCD. The LCD must be a multiple of 6 because 6 is the greater number; because 6 can also be divided by 3, the LCD is 6.

Repeat with other pairs of fractions.

See also page 184.

AUDITORY
VERBAL/LINGUISTIC

MULTISTEP AND STRATEGY PROBLEMS

The following multistep and strategy problems are provided in Lesson 9.3:

Page	Item
184	44
185	50

SPECIAL NEEDS ELL

Materials *For each group* fraction bars or fraction strips, p. TR18

Ask groups to **use fraction bars or strips to find sums and differences.** Have each group write two fractions using denominators from the fraction bars and then find a sum and a difference: for example, $\frac{5}{8} + \frac{1}{4}$ and $\frac{5}{8} - \frac{1}{4}$. Then have students model the sum and the difference using fractions bars. Finally, have each group write the sum and difference by finding the LCD of the fractions. $\frac{5}{8} + \frac{1}{4} = \frac{7}{8}, \frac{5}{8} - \frac{1}{4} = \frac{3}{8}$

KINESTHETIC
VISUAL/SPATIAL

ADVANCED LEARNERS

Challenge students to **complete the magic square.** Explain to students that the sums of the rows, columns, and diagonals are all the same. The sum in this square is 1.

$\frac{2}{15}$	$\frac{7}{15}$	$\frac{2}{5}$
$\frac{3}{5}$	$\frac{1}{3}$	$\frac{1}{15}$
$\frac{4}{15}$	$\frac{1}{5}$	$\frac{8}{15}$

VISUAL
LOGICAL/MATHEMATICAL

TECHNOLOGY LINK

 Intervention Strategies and Activities CD-ROM • *Skills 22, 23*

 Number Heroes • *Fraction Fireworks,* Levels T, U, W

 Calculating Crew • *Nautical Number Line,* Level M

Calculator Handbook, pp. 3–4

 Math Jingles™ CD 5–6 • *Track 14*

LESSON 9.3 ORGANIZER

Objective To add and subtract fractions

Vocabulary least common denominator (LCD)

Lesson Resources Calculator Handbook, pp. 3–4

1 Introduce

QUICK REVIEW provides review of pre-requisite skills.

Why Learn This? You can use adding and subtracting fractions to solve problems involving time and food preparation. *Share the lesson objective with students.*

2 Teach

Guided Instruction

- Direct students to read Example 1.
  Why is 12 the LCM of 4 and 3? 12 is the least multiple of 4 which is also a multiple of 3.
- As you discuss Example 2, ask:
  Why can we multiply $\frac{1}{2}$ by $\frac{5}{5}$ and $\frac{3}{5}$ by $\frac{2}{2}$? Both $\frac{5}{5}$ and $\frac{2}{2}$ are equal to 1, and you can multiply any number by 1 without changing its value.

Modifying Instruction For Example 2, have students who are having difficulty use fraction bars or strips to make models. Discuss how the models relate to the process shown in the example.

ADDITIONAL EXAMPLES

Example 1, p. 182

Aaron is making chocolate pudding. He combines $\frac{1}{2}$ c of sugar with $\frac{1}{3}$ c of cocoa. What is the total amount of the two ingredients? $\frac{5}{6}$ c

Example 2, p. 182

Find the sum. $\frac{3}{4} + \frac{3}{5}$ $1\frac{7}{20}$

LESSON 9.3 Add and Subtract Fractions

Learn how to add and subtract fractions.

QUICK REVIEW
Solve. Write your answer in simplest form.
1. $\frac{1}{7} + \frac{5}{7}$ $\frac{6}{7}$
2. $\frac{5}{6} - \frac{1}{6}$ $\frac{2}{3}$
3. $\frac{7}{16} - \frac{5}{16}$ $\frac{1}{8}$
4. $\frac{4}{12} + \frac{8}{12}$ 1
5. $\frac{8}{9} - \frac{3}{9}$ $\frac{5}{9}$

Vocabulary
least common denominator (LCD)

You can use a diagram to add and subtract fractions. To help you, think about the LCM of the denominators and about equivalent fractions.

EXAMPLE 1

Kayla is making two recipes. One recipe calls for $\frac{1}{4}$ c raisins, and the other recipe calls for $\frac{1}{3}$ c raisins. How many total cups of raisins does Kayla need?

Complete the diagram to find the sum of $\frac{1}{4}$ and $\frac{1}{3}$.

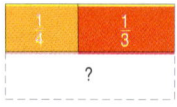

The LCM of 4 and 3 is 12.
Draw twelfths under $\frac{1}{4}$ and $\frac{1}{3}$.
THINK: $\frac{1}{4} = \frac{3}{12}$ and $\frac{1}{3} = \frac{4}{12}$.

So, Kayla needs $\frac{7}{12}$ c of raisins.

Math Idea ▶ To add fractions without using diagrams, you can write equivalent fractions by using the **least common denominator**, or **LCD**. The LCD is the LCM of the denominators.

EXAMPLE 2

Find $\frac{1}{2} + \frac{3}{5}$.

Estimate. Each fraction is close to $\frac{1}{2}$, so the sum is about 1.

$\frac{1}{2} = \frac{1 \times 5}{2 \times 5} = \frac{5}{10}$
$+\frac{3}{5} = +\frac{3 \times 2}{5 \times 2} = +\frac{6}{10}$

The LCM of 2 and 5 is 10, so the LCD of $\frac{1}{2}$ and $\frac{3}{5}$ is 10. Multiply to write equivalent fractions using the LCD.

$\frac{1}{2} = \frac{5}{10}$
$+\frac{3}{5} = +\frac{6}{10}$
$\frac{11}{10}$, or $1\frac{1}{10}$

Add the numerators. Write the sum over the denominator.

Write the answer as a fraction or as a mixed number.

Compare the answer to your estimate. Since $1\frac{1}{10}$ is close to the estimate of 1, the answer is reasonable. So, $\frac{1}{2} + \frac{3}{5} = 1\frac{1}{10}$.

182 Chapter 9

RETEACH 9.3

Add and Subtract Fractions

To add or subtract unlike fractions, first change them to equivalent fractions with the same denominator.
Find the sum. $\frac{1}{4} + \frac{2}{3}$

Step 1: Find the LCM of the denominators. The LCM of 4 and 3 is 12. So, the LCD of $\frac{1}{4}$ and $\frac{2}{3}$ is 12.

Step 2: Multiply to write equivalent fractions, using the LCD.
$\frac{1}{4} = \frac{1 \times 3}{4 \times 3} = \frac{3}{12}$
$\frac{2}{3} = \frac{2 \times 4}{3 \times 4} = \frac{8}{12}$

Step 3: Add the numerators. Write the sum over the denominator.
$\frac{3}{12} + \frac{8}{12} = \frac{11}{12}$ Remember, keep the denominator the same.

Step 4: Write the answer as a fraction or as a mixed number in simplest form. $\frac{11}{12}$ is already in simplest form.

$\frac{1}{4} + \frac{2}{3} = \frac{11}{12}$

Follow the same steps to subtract unlike fractions.

Complete to find each sum. Write the answer in simplest form.

1. Find $\frac{2}{5} + \frac{3}{10}$. The LCM of 5 and 10 is 10.
So, $\frac{2}{5} + \frac{3}{10} = \frac{4}{10} + \frac{3}{10} = \frac{7}{10}$.

2. Find $\frac{7}{8} + \frac{2}{3}$. The LCM of 8 and 3 is 24.
So, $\frac{7}{8} + \frac{2}{3} = \frac{21}{24} + \frac{16}{24} = \frac{37}{24} = 1\frac{13}{24}$.

3. Find $\frac{1}{4} + \frac{5}{6}$. The LCM of 4 and 6 is 12.
So, $\frac{1}{4} + \frac{5}{6} = \frac{3}{12} + \frac{10}{12} = \frac{13}{12} = 1\frac{1}{12}$.

4. Find $\frac{1}{6} + \frac{3}{8}$. The LCM of 6 and 8 is 24.
So, $\frac{1}{6} + \frac{3}{8} = \frac{4}{24} + \frac{9}{24} = \frac{13}{24}$.

Find the sum or difference. Write the answer in simplest form.

5. $\frac{2}{7} + \frac{1}{2}$ $\frac{11}{14}$
6. $\frac{11}{12} - \frac{7}{8}$ $\frac{1}{24}$
7. $\frac{4}{5} + \frac{1}{3}$ $\frac{17}{15}$, or $1\frac{2}{15}$
8. $\frac{3}{4} - \frac{1}{5}$ $\frac{11}{20}$

PRACTICE 9.3

Add and Subtract Fractions

Use the LCD to rewrite the problem by using equivalent fractions.

1. $\frac{3}{8} + \frac{1}{2}$ $\frac{3}{8} + \frac{4}{8}$
2. $\frac{3}{4} - \frac{1}{6}$ $\frac{9}{12} - \frac{2}{12}$
3. $\frac{3}{5} + \frac{4}{5}$ $\frac{10}{15} + \frac{12}{15}$
4. $\frac{8}{9} - \frac{1}{3}$ $\frac{8}{9} - \frac{3}{9}$
5. $\frac{1}{4} + \frac{3}{7}$ $\frac{7}{28} + \frac{12}{28}$

Write the sum or difference in simplest form. Estimate to check.

6. $\frac{1}{2} + \frac{1}{5}$ $\frac{7}{10}$
7. $\frac{6}{7} - \frac{1}{17}$ $\frac{17}{28}$ [sic]
8. $\frac{9}{10} - \frac{3}{5}$ $\frac{3}{10}$
9. $\frac{7}{8} - \frac{1}{2}$ $\frac{3}{8}$
10. $\frac{3}{4} + \frac{5}{8}$ $1\frac{3}{8}$

11. $\frac{4}{5} - \frac{3}{8}$ $\frac{7}{15}$ [sic]
12. $\frac{5}{8} + \frac{1}{10}$ $\frac{29}{40}$
13. $\frac{1}{2} - \frac{1}{6}$ $\frac{1}{3}$
14. $\frac{7}{10} + \frac{1}{4}$ $\frac{19}{20}$
15. $\frac{5}{6} + \frac{1}{3}$ $1\frac{1}{6}$

16. $\frac{11}{12} - \frac{1}{4}$ $\frac{2}{3}$
17. $\frac{3}{10} + \frac{1}{2}$ $\frac{4}{5}$
18. $\frac{3}{4} - \frac{1}{12}$ $\frac{5}{6}$
19. $\frac{5}{7} - \frac{1}{3}$ $\frac{11}{21}$
20. $\frac{4}{5} - \frac{1}{6}$ $\frac{19}{30}$

21. $\frac{3}{4} + \frac{1}{2}$ $1\frac{1}{4}$
22. $\frac{2}{3} - \frac{3}{8}$ $\frac{7}{24}$
23. $\frac{5}{7} + \frac{1}{15}$ $\frac{2}{6}$ [sic]
24. $\frac{13}{14} - \frac{2}{7}$ $\frac{9}{14}$
25. $\frac{3}{5} - \frac{1}{5}$ $\frac{2}{15}$ [sic]

26. $\frac{7}{10} - \frac{2}{5}$ $\frac{3}{10}$
27. $\frac{1}{7} + \frac{2}{7}$ $\frac{9}{14}$
28. $\frac{7}{12} - \frac{1}{4}$ $\frac{1}{3}$
29. $\frac{7}{15} - \frac{2}{5}$ $\frac{1}{15}$
30. $\frac{2}{5} + \frac{1}{3}$ $\frac{11}{15}$

31. $\frac{4}{9} + \frac{1}{2}$ $\frac{17}{18}$
32. $\frac{2}{3} - \frac{2}{7}$ $\frac{8}{21}$
33. $\frac{5}{8} + \frac{1}{3}$ $\frac{23}{24}$
34. $\frac{2}{3} + \frac{1}{9}$ $\frac{7}{9}$ [sic]
35. $\frac{5}{6} - \frac{1}{2}$ $\frac{1}{3}$

Mixed Review

Find the mean, median, and mode.
36. 57, 71, 50, 57, 53, 60 58, 57, 57
37. 21, 25, 29, 18, 31, 27, 24 25, 25, no mode

Find the quotient.
38. $26.98 \div 3.8$ 7.1
39. $1.365 \div 0.07$ 19.5
40. $174.08 \div 27.2$ 6.4

182 Chapter 9

EXAMPLE 3

You can use a similar method to subtract unlike fractions.

Kayla is preparing a pasta dish for a small dinner party. Kayla has $\frac{1}{2}$ c of grated mozzarella cheese. The recipe calls for $\frac{2}{3}$ c of grated mozzarella cheese. How much more mozzarella cheese does Kayla need to grate?

Find $\frac{2}{3} - \frac{1}{2}$.

Estimate. $\frac{2}{3}$ is a little more than $\frac{1}{2}$, so the difference is close to 0.

TECHNOLOGY LINK

More Practice: Use **Mighty Math Number Heroes**, *Fraction Fireworks*, Level W.

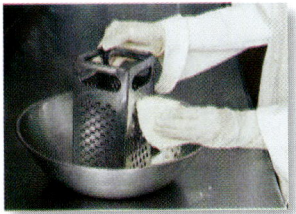

$\frac{2}{3} = \frac{2 \times 2}{3 \times 2} = \frac{4}{6}$ The LCD of $\frac{2}{3}$ and $\frac{1}{2}$ is 6.

$-\frac{1}{2} = -\frac{1 \times 3}{2 \times 3} = -\frac{3}{6}$ Multiply to find the equivalent fractions using the LCD.

$\frac{2}{3} = \frac{4}{6}$
$-\frac{1}{2} = -\frac{3}{6}$
$\phantom{-\frac{1}{2} =} \frac{1}{6}$

Subtract the numerators. Write the difference over the denominator.

Compare the answer to your estimate. Since $\frac{1}{6}$ is close to the estimate of 0, the answer is reasonable.

So, Kayla needs to grate $\frac{1}{6}$ c more cheese.

EXAMPLE 4

A. $\frac{5}{6} = \frac{5 \times 3}{6 \times 3} = \frac{15}{18}$
$-\frac{7}{9} = -\frac{7 \times 2}{9 \times 2} = -\frac{14}{18}$
$\phantom{-\frac{7}{9} =} \frac{1}{18}$

B. $\frac{5}{12} = \frac{5}{12}$
$-\frac{1}{4} = -\frac{1 \times 3}{4 \times 3} = -\frac{3}{12}$
$\phantom{-\frac{1}{4} =} \frac{2}{12} = \frac{1}{6}$

CHECK FOR UNDERSTANDING

Think and Discuss ▶ Look back at the lesson to answer each question.

1. **Tell** how much more cheese Kayla would need to grate if she had $\frac{1}{4}$ c of grated cheese. $\frac{5}{12}$ c more

2. **REASONING Tell** when the LCD of two fractions is equal to the product of the denominators. when the denominators have 1 as their GCF

Guided Practice ▶ Use the LCD to rewrite the problem by using equivalent fractions.

3. $\frac{7}{10} + \frac{1}{5}$ $\frac{7}{10} + \frac{2}{10}$ 4. $\frac{1}{3} + \frac{1}{8}$ $\frac{8}{24} + \frac{3}{24}$ 5. $\frac{4}{5} - \frac{1}{3}$ $\frac{12}{15} - \frac{5}{15}$

183

LESSON 9.3

⚠ COMMON ERROR ALERT

Some students may add both the numerators and the denominators when adding fractions. Have these students use fraction bars to show that the sum of $\frac{1}{2}$ and $\frac{1}{4}$ is $\frac{3}{4}$, not $\frac{3}{8}$.

Error
$$\frac{1}{2} + \frac{1}{4} = \frac{2}{4} + \frac{1}{4} = \frac{3}{8}$$

Correction
$$\frac{1}{2} + \frac{1}{4} = \frac{2}{4} + \frac{1}{4} = \frac{3}{4}$$

Independent Practice

Note that Exercises 44 and 50 are **multistep or strategy problems**. Assign Exercises 14–50.

Multistep or Strategy Problem To solve Exercise 44, students can determine the amount that double the recipe calls for: $2 \times \frac{1}{4}$ cup $= \frac{1}{2}$ cup. Then they subtract that amount from $\frac{7}{8}$ cup. Guide students to conclude they must double the recipe amount before they subtract to find the amount left to drink.

In Exercise 47, students will need to subtract the sum of the given fractions from 1 to answer the question.

In Exercise 49, students should begin by finding the numbers between 21 and 30 that are multiples of 4. Those numbers are 24 and 28. Students can then verify that the GCF of 24 and 28 is 4.

Write the sum or difference in simplest form. Estimate to check.

9. $\frac{17}{12}$, or $1\frac{5}{12}$

6. $\frac{1}{5} + \frac{3}{5} \frac{4}{5}$ 7. $\frac{7}{9} - \frac{4}{9} \frac{1}{3}$ 8. $\frac{7}{9} - \frac{1}{6} \frac{11}{18}$ 9. $\frac{2}{3} + \frac{3}{4}$

10. $\frac{3}{4} - \frac{3}{8} \frac{3}{8}$ 11. $\frac{2}{5} - \frac{1}{3} \frac{1}{15}$ 12. $\frac{2}{5} + \frac{2}{4} \frac{9}{10}$ 13. $\frac{4}{9} + \frac{1}{3} \frac{7}{9}$

PRACTICE AND PROBLEM SOLVING

Independent Practice Use the LCD to rewrite the problem by using equivalent fractions.

14. $\frac{9}{10} - \frac{1}{5} \frac{9}{10} - \frac{2}{10}$ 15. $\frac{6}{7} - \frac{3}{4} \frac{24}{28} - \frac{21}{28}$ 16. $\frac{1}{4} + \frac{5}{8} \frac{2}{8} + \frac{5}{8}$

Write the sum or difference in simplest form. Estimate to check.

17. $\frac{1}{6} + \frac{2}{3} \frac{5}{6}$ 18. $\frac{4}{7} - \frac{1}{7} \frac{3}{7}$ 19. $\frac{1}{2} + \frac{3}{10} \frac{4}{5}$ 20. $\frac{1}{3} - \frac{1}{4} \frac{1}{12}$

21. $\frac{5}{7} - \frac{1}{2} \frac{3}{14}$ 22. $\frac{1}{3} + \frac{2}{3}$ 1   23. $\frac{6}{10} - \frac{4}{10} \frac{1}{5}$ 24. $\frac{3}{8} + \frac{1}{3} \frac{17}{24}$

25. $\frac{11}{8}$, or $1\frac{3}{8}$ 25. $\frac{3}{4} + \frac{5}{8}$ 26. $\frac{7}{12} + \frac{2}{3}$ 27. $1 - \frac{3}{8} \frac{5}{8}$ 28. $\frac{1}{2} - \frac{2}{5} \frac{1}{10}$

26. $\frac{15}{12}$, or $1\frac{1}{4}$ 29. $\frac{4}{5} - \frac{1}{3} \frac{7}{15}$ 30. $\frac{1}{4} + \frac{2}{3} \frac{11}{12}$ 31. $\frac{5}{9} - \frac{1}{3} \frac{2}{9}$ 32. $\frac{3}{8} + \frac{3}{20} \frac{21}{40}$

33. $\frac{6}{8}$, or $\frac{3}{4}$ 33. Find the sum of $\frac{1}{8}$, $\frac{3}{8}$, and $\frac{2}{8}$.   34. Find $\frac{1}{2} + \frac{2}{3} + \frac{1}{6} \frac{4}{3}$, or $1\frac{1}{3}$

35. Find $\frac{3}{4} + 0.5 + 0.75$. **2**   36. Find $0.6 - \frac{3}{8}$. **0.225, or $\frac{9}{40}$**

37. How much longer than $\frac{1}{4}$ mile is $\frac{2}{3}$ mile? $\frac{5}{12}$ mile

Solve each equation mentally. Write the answer in simplest form.

38. $p + \frac{1}{4} = \frac{3}{4}$ $p = \frac{1}{2}$ 39. $r = \frac{5}{12} + \frac{7}{12}$ $r = 1$ 40. $\frac{4}{5} - q = \frac{2}{5}$ $q = \frac{2}{5}$

41. $c = \frac{7}{10} - \frac{1}{10}$ $c = \frac{3}{5}$ 42. $\frac{3}{7} + s = \frac{5}{7}$ $s = \frac{2}{7}$ 43. $m - \frac{1}{6} = \frac{5}{6}$ $m = 1$

Problem Solving Applications **Use Data** For 44–46, use the recipe at the right.

44. Yuji has $\frac{7}{8}$ c of orange juice. How much orange juice does he have left to drink after he doubles the recipe for fruit cups? $\frac{3}{8}$ c

45. How many total teaspoons of $\frac{5}{8}$ tsp vanilla and orange extract does Yuji need to make the fruit cups?

46. Yuji used $\frac{1}{4}$ tsp of orange extract. By how much did he exceed the amount of orange extract in the recipe? $\frac{1}{8}$ tsp

Fruit Cups
2 cups of orange sections
1 cup blueberries
$\frac{1}{4}$ cup orange juice
1 tbsp sugar
1 tsp lemon juice
$\frac{1}{2}$ tsp vanilla extract
$\frac{1}{8}$ tsp orange extract

184 Chapter 9

ALTERNATIVE TEACHING STRATEGY | SCAFFOLDED INSTRUCTION

Purpose Students add and subtract fractions to complete number squares.

Materials *For each group* an addition and a subtraction square; fraction bars or fraction strips, p. TR18

Display the following addition number square.

$\frac{1}{4}$	$\frac{1}{12}$	? $\frac{1}{3}$
$\frac{1}{3}$	$\frac{1}{6}$	? $\frac{1}{2}$
? $\frac{7}{12}$	? $\frac{1}{4}$	? $\frac{5}{6}$

Have students work in groups to copy the addition square and then complete it. Let them use fraction bars if they wish. Discuss any difficulties that students encounter.

Next, display the following subtraction number square. Have students follow the same process as with the addition square.

$\frac{8}{9}$	$\frac{1}{3}$	? $\frac{5}{9}$
$\frac{1}{2}$	$\frac{1}{6}$	? $\frac{1}{3}$
? $\frac{7}{18}$	? $\frac{1}{6}$	? $\frac{2}{9}$

Finally, have groups make number squares for addition and/or subtraction. Tell students that they may write any four fractions to start an addition square. In subtraction squares, the fractions in the top row must be greater than those in the second row. The fractions in the left column must be greater than those in the second column. Have groups find the missing fractions in their squares.

Then ask groups to exchange number squares, solve, and compare solutions.
Check students' work.

47. Each week, Reina spends $\frac{2}{3}$ of her allowance on school lunches and saves $\frac{1}{5}$ of it. What fraction of her allowance is left? Which operation(s) did you use? Why? $\frac{2}{15}$ is left; add to find the total spent and saved; subtract to find how much is left.

48. **What's the Question?** In Mrs. Lucero's class, $\frac{1}{10}$ of the students are wearing blue shirts and $\frac{3}{5}$ of the students are wearing white shirts. The answer is $\frac{3}{10}$ of the class. What part of Mrs. Lucero's class is wearing *neither* blue *nor* white shirts?

49. **REASONING** Joaquin is thinking of two numbers. Each number is between 21 and 30. The GCF of the numbers is 4. What are the numbers? 24 and 28

50. One cup of whole milk contains 166 calories and one cup of skim milk contains 88 calories. How many more calories are there in 4 cups of whole milk than in 1 quart of skim milk? 312 more calories

MIXED REVIEW AND TEST PREP

For 51–52, write in simplest form. (p.160)

51. $\frac{36}{81}$ $\frac{4}{9}$ **52.** $\frac{95}{200}$ $\frac{19}{40}$ **53.** Subtract. 9,285 − 3,153 (p. 20)   6,132

⭐ **54. TEST PREP** A teacher selects 50 students by picking names out of a box without looking. What type of sample is this? (p. 94) **C**

A biased   **B** convenience   **C** random   **D** systematic

⭐ **55. TEST PREP** Eric skated 500 meters in 37.14 seconds. Andre skated the same distance in 37.139 seconds, and Al in 37.12 seconds. What is the correct order of their times in seconds, from fastest to slowest times? (p. 52) **H**

F 37.14, 37.139, 37.12   **H** 37.12, 37.139, 37.14
G 37.12, 37.14, 37.139   **J** 37.139, 37.12, 37.14

PROBLEM SOLVING Thinker's Corner

DOMINO FRACTIONS
Materials: set of dominoes

Dominoes have been played for centuries throughout the world. The Chinese played with them in the 12th century, and a set was even found in a tomb from Ancient Egypt. To play this domino game, think of each of the 15 dominoes shown above as a fraction. If a domino shows 2 circles and 4 circles, the fraction is $\frac{2}{4}$. Play with a partner to combine 5 dominoes at a time to form a fraction sum of $2\frac{1}{2}$. Then break the 15 dominoes into 3 sets of 5, each of which has a sum of $2\frac{1}{2}$. $\frac{3}{4}+\frac{1}{4}+\frac{3}{6}+\frac{1}{2}+\frac{2}{4}$; $\frac{5}{6}+\frac{2}{6}+\frac{1}{3}+\frac{4}{5}+\frac{1}{5}$; $\frac{4}{6}+\frac{1}{6}+\frac{2}{3}+\frac{3}{5}+\frac{2}{5}$

EXTRA PRACTICE page H40, Set B

MIXED REVIEW AND TEST PREP
Exercises 51–55 provide **cumulative review** (Chapters 1–9).

Thinker's Corner

- Have students examine the dominoes pictured. Ask:

  Which domino shows the smallest fraction? (1,6)

  Which domino shows the greatest fraction? (5,6)

  What sets of dominoes show equivalent fractions? (1,2), (2,4), and (3,6); (1,3) and (2,6); (2,3) and (4,6)

REASONING Suppose a domino can have 0 to 6 circles on each side. How many different dominoes are possible? Show your method. There are 28 possibilities. Possible method: Make an organized list. There are 7 dominoes that have 0 as their smallest number of circles, 6 with 1 as smallest, 5 with 2, 4 with 3, 3 with 4, 2 with 5, and 1 with 6.

4 Assess

Summarize the lesson by having students:

DISCUSS Kari wants to subtract $\frac{1}{4}$ from $\frac{7}{8}$. What does she need to do? Rewrite the fractions with an LCD of 8. Then subtract the numerators and write the difference over 8.

WRITE Explain why it is more efficient to use the LCD rather than some other common denominator when adding or subtracting fractions. Using the LCD results in lesser numbers and minimizes the need to simplify answers.

Lesson Quiz

Transparency 9.3

Write the sum or difference in simplest form. Estimate to check.

1. $\frac{1}{3}+\frac{5}{8}$ $\frac{23}{24}$ **2.** $\frac{3}{4}-\frac{1}{3}$ $\frac{5}{12}$
3. $\frac{9}{10}-\frac{2}{5}$ $\frac{1}{2}$ **4.** $\frac{5}{8}+\frac{5}{12}$ $1\frac{1}{24}$
5. $\frac{1}{3}+\frac{5}{6}$ $1\frac{1}{6}$ **6.** $\frac{5}{6}-\frac{5}{8}$ $\frac{5}{24}$

LESSON 9.4

Add and Subtract Mixed Numbers

LESSON PLANNING

Objective To add and subtract mixed numbers

Intervention for Prerequisite Skills

Simplify Fractions, Add and Subtract Like Fractions (For intervention strategies, see page 175.)

NCTM Standards
1. Number and Operations
5. Data Analysis and Probability
6. Problem Solving
8. Communication
9. Connections
10. Representation

Math Background

Because a mixed number is the sum of a whole number and a fraction, two mixed numbers can be added by first adding the fraction parts and then adding the whole numbers. The Associative and Commutative properties justify this procedure. In the lesson, models help students grasp the steps in the algorithm.

To help students master addition and subtraction of mixed numbers, they should follow these steps:

- Write equivalent fractions for the fraction parts, using the LCD of the given fractions.
- Add or subtract the fraction parts.
- Add or subtract the whole numbers next.
- Write the answer in simplest form.

WARM-UP RESOURCES

 **NUMBER OF THE DAY**  Transparency 9.4

Use the number 3, the number of the month, and your age. Write the greatest possible mixed number in simplest form with these three numbers. **Possible answer for 3, 10, and 12: $12\frac{3}{10}$**

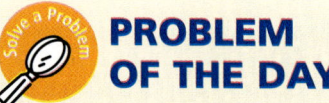 **PROBLEM OF THE DAY**  Transparency 9.4

Complete the Magic Square. The sum is $1\frac{1}{4}$.

$\frac{1}{2}$	$\frac{7}{12}$	$\frac{1}{6}$
$\frac{1}{12}$	$\frac{5}{12}$	$\frac{3}{4}$
$\frac{2}{3}$	$\frac{1}{4}$	$\frac{1}{3}$

Solution Problem of the Day tab, p. PD9

 **DAILY FACTS PRACTICE**

Have students practice addition and subtraction facts by completing Set E of *Teacher's Resource Book*, p. TR99.

INTERVENTION AND EXTENSION RESOURCES

ALTERNATIVE TEACHING STRATEGY ELL

To **reinforce the method for adding fractions and mixed numbers,** present a mixed number, such as $4\frac{1}{4}$. Have pairs of students write and model two different number sentences that have $4\frac{1}{4}$ as the sum.

Possible answer: $2\frac{1}{2} + 1\frac{3}{4} = 4\frac{1}{4}$; $3\frac{3}{4} + \frac{1}{2} = 4\frac{1}{4}$

See also page 188.

AUDITORY
LOGICAL/MATHEMATICAL

MULTISTEP AND STRATEGY PROBLEMS

The following multistep and strategy problems are provided in Lesson 9.4:

Page	Item
189	33, 35

WRITING IN MATHEMATICS

Materials reference books, cookbooks

Have students **apply fractions and mixed numbers** by writing a paragraph about the ingredients in a recipe for a popular rice dish. Tell students that the January harvest festival in South India is called Pongal. A popular rice dish served in India goes by the same name. Challenge students to find this recipe or some other recipe that features rice, copy it into their journals, and then write a paragraph about the recipe that includes the use of fractions and mixed numbers.

Check students' work.

SCIENCE CONNECTION

Have students **apply operations with mixed numbers by using the data below to write two word problems.**

City	Annual Rainfall
Houston, TX	$44\frac{3}{4}$ in.
Phoenix, AZ	$7\frac{1}{10}$ in.
Atlanta, GA	$48\frac{3}{5}$ in.
Chicago, IL	$33\frac{1}{3}$ in.

Check students' work.

VISUAL
VERBAL/LINGUISTIC

TECHNOLOGY LINK

 Intervention Strategies and Activities CD-ROM • *Skills 22, 23*

 Calculating Crew • *Nautical Number Line,* Level N

LESSON 9.4

Independent Practice

Note that Exercises 33 and 35 are **multistep or strategy problems**. Assign Exercises 12–36.

Review the properties of addition and order of operations rules before assigning Exercises 27–29.

Multistep or Strategy Problem To solve Exercise 33, students can add the amounts used to cook. Then they subtract that total from $9\frac{3}{4}$ c. Guide students to conclude that the total amount used to cook must be found before the amount left over can be determined.

Students can solve Exercise 35 by using the strategy *make a table*.

Write the sum or difference in simplest form. Estimate to check.

6. $1\frac{1}{8} + 1\frac{5}{8}$ $2\frac{6}{8}$, or $2\frac{3}{4}$
7. $2\frac{1}{4} + 4\frac{1}{3}$ $6\frac{7}{12}$
8. $5\frac{3}{8} - 1\frac{1}{4}$ $4\frac{1}{8}$
9. $4\frac{1}{3} - 3\frac{1}{6}$ $1\frac{1}{6}$
10. $3\frac{3}{4} + 4\frac{5}{12}$ $8\frac{1}{6}$
11. $6\frac{5}{6} - 5\frac{7}{9}$ $1\frac{1}{18}$

PRACTICE AND PROBLEM SOLVING

Independent Practice

Draw a diagram to find each sum or difference. Write the answer in simplest form. **Check students' diagrams.**

12. $1\frac{5}{12} + 1\frac{1}{4}$ $2\frac{8}{12}$, or $2\frac{2}{3}$
13. $1\frac{1}{3} + 1\frac{1}{6}$ $2\frac{1}{2}$
14. $4\frac{1}{2} - 2\frac{2}{5}$ $2\frac{1}{10}$

Write the sum or difference in simplest form. Estimate to check.

15. $4\frac{1}{2} + 3\frac{4}{5}$ $8\frac{3}{10}$
16. $4\frac{1}{3} - 2\frac{1}{4}$ $2\frac{1}{12}$
17. $5\frac{5}{6} + 4\frac{2}{9}$ $10\frac{1}{18}$
18. $3\frac{1}{4} - 1\frac{1}{6}$ $2\frac{1}{12}$
19. $7\frac{1}{2} - 3\frac{3}{5}$ $4\frac{1}{10}$
20. $3\frac{2}{7} + 8\frac{1}{3}$ $11\frac{13}{21}$
21. $7\frac{3}{4} + 3\frac{2}{5}$ $11\frac{3}{20}$
22. $5\frac{5}{6} - 2\frac{7}{9}$ $3\frac{1}{18}$
23. $4\frac{5}{7} + 3\frac{1}{2}$ $8\frac{3}{14}$

24. How much greater is $5\frac{3}{4}$ than 3? $2\frac{3}{4}$
25. What is the sum of $25\frac{3}{8}$ and $2\frac{3}{4}$? $28\frac{1}{8}$
26. What is the sum of $4\frac{5}{8}$ and 7.8? 12.425, or $12\frac{17}{40}$

TECHNOLOGY LINK
More Practice: Use **Mighty Math Calculating Crew**, *Nautical Number Line*, Level N.

Find the missing number and identify which property of addition you used.

27. $3\frac{7}{8} + \blacksquare = 2\frac{1}{4} + 3\frac{7}{8}$ Commutative, $2\frac{1}{4}$
28. $3\frac{3}{4} + 0 = \blacksquare$ Identity, $3\frac{3}{4}$
29. $\left(\frac{2}{3} + 1\frac{5}{6}\right) + \frac{1}{6} = \frac{2}{3} + \left(\blacksquare + \frac{1}{6}\right)$ Associative, $1\frac{5}{6}$

Problem Solving Applications

Use Data The graph shows the head-and-body lengths of five small mammals. For 30–31, use the graph.

30. How much longer is the harvest mouse than the Kitti's hognosed bat? $1\frac{1}{6}$ in.

31. The masked shrew is $1\frac{2}{3}$ in. long. Is it longer or shorter than the little brown bat? How much? Which operation did you use? Why?

longer; $\frac{1}{6}$ in.; subtraction; to find the difference between the lengths

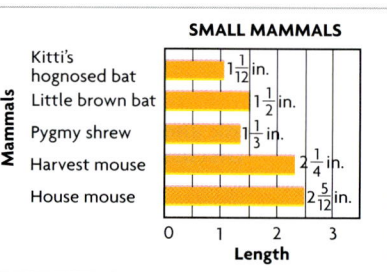

SMALL MAMMALS

Mammals	Length
Kitti's hognosed bat	$1\frac{1}{12}$ in.
Little brown bat	$1\frac{1}{2}$ in.
Pygmy shrew	$1\frac{1}{3}$ in.
Harvest mouse	$2\frac{1}{4}$ in.
House mouse	$2\frac{5}{12}$ in.

188 Chapter 9

ALTERNATIVE TEACHING STRATEGY — SCAFFOLDED INSTRUCTION

Purpose Students will add and subtract mixed numbers by using fractions greater than one.

Give students a mixed-number addition or subtraction problem, such as $3\frac{1}{2} + 1\frac{2}{5}$. Tell them that they can solve a problem like this by changing the mixed numbers to fractions greater than 1. Have students solve the following problem by using both the process shown in the lesson and this method.

Lesson method:
$$3\frac{1}{2} = 3\frac{5}{10}$$
$$+1\frac{2}{5} = +1\frac{4}{10}$$
$$\overline{\phantom{+1\frac{2}{5}}} 4\frac{9}{10}$$

Changing mixed numbers to fractions method:

$$3\frac{1}{2} = \frac{7}{2} = \frac{35}{10}$$
$$+1\frac{2}{5} = +\frac{7}{5} = +\frac{14}{10}$$
$$\phantom{+1\frac{2}{5} ==} \frac{49}{10} = 4\frac{9}{10}$$

Then ask students to find the difference by using both methods.

Method 1
$$3\frac{1}{2} = 3\frac{5}{10}$$
$$-1\frac{2}{5} = -1\frac{4}{10}$$
$$\phantom{-1\frac{2}{5} =} 2\frac{1}{10}$$

Method 2
$$3\frac{1}{2} = \frac{7}{2} = \frac{35}{10}$$
$$-1\frac{2}{5} = -\frac{7}{5} = -\frac{14}{10}$$
$$\phantom{-1\frac{2}{5} ==} \frac{21}{10} = 2\frac{1}{10}$$

Have students compare and contrast the two methods and discuss which one they prefer.

Check students' work.

32. On its way to the shore, a sea turtle traveled $4\frac{1}{4}$ hr the first day. The second day, the turtle traveled $3\frac{1}{2}$ hr. How many hours did the sea turtle travel in the two days? $7\frac{3}{4}$ hr

33. Mrs. Myers used $1\frac{1}{2}$ c of flour to make muffins, $4\frac{1}{4}$ c to make bread, and $\frac{3}{4}$ c to make gravy. If she had $9\frac{3}{4}$ c before she started the meal, how much flour does Mrs. Myers have left? $3\frac{1}{4}$ c

34. **What's the Error?** Izumi added $3\frac{1}{4}$ and $2\frac{2}{3}$ and got $5\frac{3}{12}$. Explain the error. What is the correct sum? See left.

35. Betty, Jim, Manuel, and Rosa won the first four prizes in a design contest. Jim won second prize. Manuel did not win third prize. Rosa won fourth prize. What prize did Betty win? 3rd prize

36. Alexis needs a new blender. She found the same blender on sale at five different stores. The prices are $22.95, $21.85, $22.05, $20.95, and $21.99. Order the prices from least to greatest. $20.95, $21.85, $21.99, $22.05, $22.95

34. Izumi used the LCD, 12, but he forgot to change the numerators before adding; $5\frac{11}{12}$

MIXED REVIEW AND TEST PREP

37. Find the sum of $\frac{2}{3}$ and $\frac{2}{5}$. (p. 182) $\frac{16}{15}$, or $1\frac{1}{15}$

38. Subtract. $425.2 - 51.05$ (p. 66) 374.15

39. Order $\frac{1}{2}$, $\frac{4}{5}$, and $\frac{2}{3}$ from greatest to least. (p. 166) $\frac{4}{5}, \frac{2}{3}, \frac{1}{2}$

40. **TEST PREP** Find the value of $(6 + 4)^2 \div 5$. (p. 44) **D**
    A 4 B 4.4 C 9.2 D 20

41. **TEST PREP** Which is the solution of $x + 3 = 10$? (p. 30) **G**
    F $x = 3$ G $x = 7$ H $x = 10$ J $x = 13$

PROBLEM SOLVING LiNKUP to Reading

Strategy • Choose Relevant Information

Sometimes a word problem contains more information than you need. You must decide which information is relevant, or needed to solve the problem.

The koala of eastern Australia feeds mostly on eucalyptus leaves. It nibbles on about 6 of the 500 species of eucalyptus per day and selects certain trees and leaves over others to find the $1\frac{1}{4}$ pounds of food that it needs. Suppose the koala finds only $1\frac{1}{8}$ pounds of food by the end of the day. How many more pounds of eucalyptus leaves does the koala need? 1. how many more pounds of eucalyptus leaves the koala needs

1. What does the problem ask you to find? 2. needs $1\frac{1}{4}$ pounds of eucalyptus leaves per day; finds $1\frac{1}{8}$ pounds
2. Identify relevant information.
3. What information is not relevant? 3. it nibbles on about six of the 500 species
4. Solve the problem. $1\frac{1}{4} - 1\frac{1}{8} = 1\frac{2}{8} - 1\frac{1}{8} = \frac{1}{8}; \frac{1}{8}$ lb of eucalyptus leaves

EXTRA PRACTICE page H40, Set C

189

MIXED REVIEW AND TEST PREP
Exercises 37–41 provide **cumulative review** (Chapters 1–9).

LiNKUP to READING

• After students read the paragraph about the koala, ask:

How can you decide whether you have too much information? Possible answer: Choose the information you need to solve the problem and see if any is not relevant.

What LCD did you use? 8

REASONING Is it reasonable to assume that a koala eats exactly $1\frac{1}{2}$ lb of food each day? Explain. No; that is probably the average amount a koala eats.

4 Assess

Summarize the lesson by having students:

DISCUSS How could you use diagrams to add $1\frac{1}{2} + 1\frac{1}{3}$? Possible answer: Model $1\frac{1}{2}$ using 1 bar and $\frac{1}{2}$ bar. Model $1\frac{1}{3}$ using 1 bar and $\frac{1}{3}$ bar. Add $\frac{1}{2} + \frac{1}{3}$ by modeling and adding equivalent fractions $\frac{3}{6} + \frac{2}{6} = \frac{5}{6}$. Add the two whole bars for a sum of 2. So, $1\frac{1}{2} + 1\frac{1}{3} = 2\frac{5}{6}$

WRITE Explain how to add two mixed numbers with unlike fractions. Write each fraction part using a common denominator, add the fractions and then the whole numbers, simplify.

Lesson Quiz

Transparency 9.4

Draw a diagram to find each sum or difference. Write the answer in simplest form.

1. $1\frac{1}{8} + 4\frac{3}{4}$ $5\frac{7}{8}$
2. $2\frac{2}{3} - 1\frac{1}{12}$ $1\frac{7}{12}$
3. $5\frac{1}{2} - 2\frac{2}{5}$ $3\frac{1}{10}$
4. $5\frac{2}{3} + 3\frac{5}{6}$ $9\frac{1}{2}$
5. $2\frac{2}{3} - 1\frac{1}{9}$ $1\frac{5}{9}$
6. $1\frac{1}{6} + 2\frac{5}{6}$ 4

READING STRATEGY

K-W-L Chart Before having students read Linkup to Reading, have them read the strategy—*Choose Relevant Information*. Ask them to predict how they will apply this strategy to math. Then have students make a three-column chart headed What I Know, What I Want to Know, and What I Learned. Ask them to fill in the first two columns. Have them fill in the third column as they read through the paragraph and questions.

K-W-L Chart

What I Know	What I Want to Know	What I Learned

LESSON 9.5

ORGANIZER

Objective To use fraction bars to rename and subtract mixed numbers

Materials *For each group* fraction bars or fraction strips, p. TR18

Lesson Resources E-Lab Recording Sheet • *Subtracting Mixed Numbers*

Intervention for Prerequisite Skills Simplify Fractions, Add and Subtract Like Fractions (For intervention strategies, see page 175.)

Using the Pages

Point out to students that in some subtraction problems, the first mixed number must be renamed so that the fraction part is greater than the fraction part of the mixed number being subtracted.

Activity

As students look at the subtraction $2\frac{1}{5} - 1\frac{4}{5}$, ask:

Why can't you do the subtraction as it is given? $\frac{4}{5}$ is greater than $\frac{1}{5}$.

How are the two models in A alike and how are they different? They name the same amount; the first model displays 2 wholes, while the second displays 1 whole plus a second whole divided into fifths.

Modifying Instruction Reinforce the idea that it is possible to model a whole number as a whole or as a fraction. This process is similar to renaming tens and greater place values as necessary when subtracting whole numbers.

Show the similarity between Activity A and regrouping when you subtract $45 - 37$.

```
   3 15
   4 5  = 30 + 10 + 5 = 30 + 15
  -3 7  = 30 +      7 = 30 +  7
                                 8
```

LESSON 9.5 — Rename to Subtract

MATH LAB

Explore how to use fraction bars to subtract mixed numbers.

You need fraction bars.

QUICK REVIEW
Write the difference in simplest form.
1. $\frac{4}{7} - 1\frac{3}{7}$ 2. $\frac{3}{4} - 1\frac{1}{4}\;\frac{1}{2}$ 3. $\frac{7}{10} - \frac{5}{10}\;\frac{1}{5}$
4. $\frac{8}{12} - \frac{4}{12}\;\frac{1}{3}$ 5. $\frac{5}{9} - \frac{2}{9}\;\frac{1}{3}$

Sometimes you need to rename mixed numbers before you can subtract.

Activity

A. Find $2\frac{1}{5} - 1\frac{4}{5}$.

- Use fraction bars to model $2\frac{1}{5}$.

  [bar model] ← $2\frac{1}{5}$

- Here is another way to model $2\frac{1}{5}$.

  [bar model] ← $1\frac{6}{5}$

- From which model can you subtract $1\frac{4}{5}$? **the second one**
- Subtract $1\frac{4}{5}$ from $1\frac{6}{5}$. What is $2\frac{1}{5} - 1\frac{4}{5}$? **$\frac{2}{5}$**

B. Find $2\frac{1}{6} - 1\frac{5}{12}$.

- Use fraction bars to model $2\frac{1}{6}$.

  [bar model] ← $2\frac{1}{6}$

- Since you are subtracting twelfths, think of the LCD for $\frac{1}{6}$ and $\frac{5}{12}$. Change the sixths to twelfths.

  [bar model] ← $2\frac{2}{12}$

- Can you subtract $1\frac{5}{12}$ from either of these models? **no**
- Here is another way to model $2\frac{2}{12}$.

  [bar model] ← $1\frac{14}{12}$

- Subtract $1\frac{5}{12}$ from $1\frac{14}{12}$. What is $2\frac{1}{6} - 1\frac{5}{12}$? **$\frac{9}{12}$, or $\frac{3}{4}$**

190 Chapter 9

SPECIAL NEEDS / ELL

Materials *For each group* three whole bars or strips and four $\frac{1}{4}$ bars or strips, p. TR18

Ask each group to **use fraction bars or strips to model and record two different mixed-number subtraction sentences.** When the groups have finished, have students compare their exercises and solutions. Check students' work.

KINESTHETIC
LOGICAL/MATHEMATICAL

Intervention and Extension Resources

MATH CONNECTION: MEASUREMENT

Materials *For each group* inch rulers, p. TR22; fraction bars or fraction strips, p. TR18

Have groups **measure the lengths of various items and determine the difference** between the longest and shortest measurements. Tell students to measure several books, pencils, and erasers. Then ask groups to model the difference in length between the longest and shortest of each type of item. Check students' work.

KINESTHETIC
VISUAL/SPATIAL

190 Chapter 9

C. Find $2\frac{1}{4} - 1\frac{3}{8}$.

- Use fraction bars to model $2\frac{1}{4}$.

 ← $2\frac{1}{4}$

- Since you are subtracting eighths, think of the LCD for $\frac{1}{4}$ and $\frac{3}{8}$. Change the fourth to eighths.

 ← $2\frac{2}{8}$

TECHNOLOGY LINK
More Practice: Use E-Lab, *Subtraction of Mixed Numbers*.
www.harcourtschool.com/elab2002

- Can you subtract $1\frac{3}{8}$ from either of these models? **no**
- Here is another way to model $2\frac{2}{8}$.

 ← $1\frac{10}{8}$

- Subtract $1\frac{3}{8}$ from $1\frac{10}{8}$. What is $2\frac{1}{4} - 1\frac{3}{8}$? $\frac{7}{8}$

Think and Discuss

- Think about $2\frac{5}{6} - 1\frac{1}{6}$. Do you need to rename before you subtract? Explain.   **No; $\frac{5}{6}$ is greater than $\frac{1}{6}$, and the denominators are the same.**
- Think about $5\frac{2}{5} - 3\frac{4}{5}$. Do you need to rename before you subtract? Explain.   **Yes; $\frac{2}{5}$ is less than $\frac{4}{5}$.**

Practice
Check students' diagrams.
Use fraction bars to subtract. Draw a diagram of your model.

1. $3\frac{1}{3} - 1\frac{2}{3}$ $1\frac{2}{3}$
2. $3\frac{3}{8} - \frac{3}{4}$ $2\frac{5}{8}$
3. $3\frac{1}{9} - 1\frac{4}{9}$ $1\frac{9}{9}$
4. $2\frac{3}{8} - 1\frac{1}{2}$ $\frac{7}{8}$
5. $1\frac{1}{2} - \frac{4}{5}$ $\frac{7}{10}$
6. $3\frac{1}{5} - 1\frac{3}{10}$ $1\frac{9}{10}$
7. $2\frac{2}{3} - 1\frac{3}{4}$ $\frac{11}{12}$
8. $3\frac{1}{12} - 2\frac{5}{6}$ $\frac{1}{4}$

MIXED REVIEW AND TEST PREP

9. Draw a diagram to find the sum of $2\frac{1}{2}$ and $1\frac{1}{4}$. Write the answer in simplest form. (p. 186) $3\frac{3}{4}$
10. Write $\frac{2}{5}$ as a decimal and as a percent. (p. 169) **0.4, 40%**
11. Evaluate $a \times b$ for $a = 4.2$ and $b = 5.1$. (p. 82) **21.42**
12. Write the prime factorization of 245 in exponent form. (p. 148) $7^2 \times 5$
13. **TEST PREP** Tyra earns $5.50 per hour working part-time for a surf shop. Last week she worked 3 hr on Tuesday and twice as long on Saturday. How much did she earn last week? (p. 22) **C**

   A $33.00   B $44.00   C $49.50   D $55.50

191

Remind students that before they model the whole numbers to subtract, they need to be sure that the fractions have been modeled with a common denominator.

Think and Discuss

When discussing these two questions, help students conclude that renaming is necessary whenever the second fraction is greater than the first fraction.

Practice

Before students begin work, ask:
Which exercises require renaming a whole as a fraction? all
Which exercises require renaming the fractions with a common denominator? all but Exercise 1

MIXED REVIEW AND TEST PREP
Exercises 9–13 provide **cumulative review** (Chapters 1–9).

Oral Assessment

Use fraction bars to subtract. Draw a diagram of your model. **Check students' work.**

1. $2\frac{3}{4} - 1\frac{1}{3}$ $1\frac{5}{12}$
2. $3\frac{2}{5} - 2\frac{4}{5}$ $\frac{3}{5}$
3. Did you need to find the LCD to subtract in Exercise 1 or Exercise 3? Why? Yes, in Exercise 3; the denominators are not the same.
4. Did you need to rename a mixed number to subtract in Exercise 5 or Exercise 6? Why? Yes, in both; In Exercise 5, $\frac{1}{2}$ is less than $\frac{4}{5}$; in Exercise 6, $\frac{1}{5}$ is less than $\frac{3}{10}$.

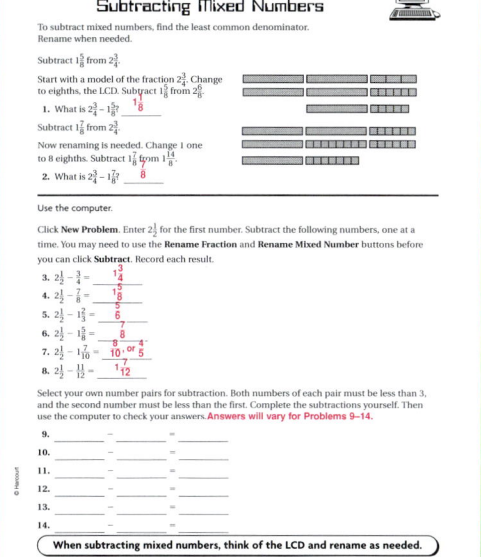

USING E-LAB

Students use visual thinking, reasoning, and a fraction model to subtract mixed numbers.

The E-Lab Recording Sheets and activities are available on the E-Lab website.

www.harcourtschool.com/elab2002

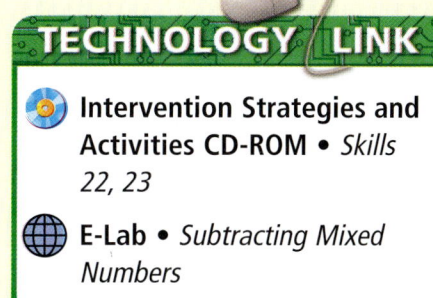

TECHNOLOGY LINK

- Intervention Strategies and Activities CD-ROM • *Skills 22, 23*
- E-Lab • *Subtracting Mixed Numbers*

191

LESSON 9.6 Subtract Mixed Numbers

LESSON PLANNING

Objective To subtract mixed numbers involving renaming

Intervention for Prerequisite Skills
Simplify Fractions, Add and Subtract Like Fractions (For intervention strategies, see page 175.)

NCTM Standards
1. Number and Operations
5. Data Analysis and Probability
6. Problem Solving
8. Communication

Math Background

As with whole numbers and decimals, mixed numbers sometimes must be renamed in order to subtract one from the other.

The following ideas may help students master this algorithm:

- It is easier to compare fractions if they have the same denominator, so the first step is to find the LCD.
- The fraction part of a mixed number must be renamed if it is less than the fraction part of the mixed number to be subtracted from it.
- When renaming a whole, it is helpful to write the whole number as a sum; for example, rewrite 5 as $4 + 1$. The 1 can be renamed as a fraction, for example, $\frac{3}{3}$, $\frac{5}{5}$, or $\frac{8}{8}$.

WARM-UP RESOURCES

 **NUMBER OF THE DAY**

Write your age in years and months as a mixed number.
Possible answer: for 11 years 4 months: $11\frac{1}{3}$

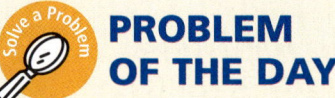 **PROBLEM OF THE DAY**

Melissa rides the bus $1\frac{2}{3}$ mi north and $3\frac{1}{8}$ mi east to get to school. Brandon rides his bike $2\frac{3}{4}$ mi south and $2\frac{1}{6}$ mi west to get to the same school. Who rides farther? Estimate the distances. What can you conclude about the estimates? Brandon rides farther. From the estimates, the distance appears to be the same. To answer the question, you need to find the exact answer.

Solution Problem of the Day tab, p. PD9

 **DAILY FACTS PRACTICE**

Have students practice subtraction facts by completing Set G of *Teacher's Resource Book*, p. TR99.

192A Chapter 9

INTERVENTION AND EXTENSION RESOURCES

ALTERNATIVE TEACHING STRATEGY

To help students **gain proficiency in renaming mixed numbers for subtraction,** have them work with simpler exercises first.

- Tell students to rename $1\frac{2}{3}$, $1\frac{1}{6}$, and $1\frac{3}{4}$ as fractions greater than 1. $\frac{5}{3}, \frac{7}{6}, \frac{7}{4}$
- Have students use their answers to help them rename $5\frac{2}{3}$, $6\frac{1}{6}$, and $3\frac{3}{4}$ as mixed numbers with fractions greater than 1. $4\frac{5}{3}, 5\frac{7}{6}, 2\frac{7}{4}$

VISUAL
LOGICAL/MATHEMATICAL

MULTISTEP AND STRATEGY PROBLEMS

The following multistep or strategy problem is provided in Lesson 9.6:

Page	Item
193	28

SPECIAL NEEDS ELL

Materials *For each group* play money (one-dollar bills, quarters, dimes)

- Discuss the fractional parts of a dollar that a quarter and a dime represent. $\frac{1}{4}, \frac{1}{10}$
- Have students use play money to model these differences: $3\frac{1}{4} - 1\frac{3}{4}$; $5\frac{3}{10} - 2\frac{7}{10}$ 3 dollars, 1 quarter − 1 dollar, 3 quarters; 5 dollars, 3 dimes − 2 dollars, 7 dimes
- Ask what needs to be renamed to find the differences and then solve. 3 dollars, 1 quarter = 2 dollars, 5 quarters; 5 dollars, 3 dimes = 4 dollars, 13 dimes; $1\frac{2}{4} = 1\frac{1}{2}$; $2\frac{6}{10} = 2\frac{3}{5}$
- Repeat for similar subtraction problems.

Check students' work.

KINESTHETIC
LOGICAL/MATHEMATICAL

ADVANCED LEARNERS

Challenge students to **find the next four mixed numbers in the patterns and describe the pattern.** Give students the following:

$1\frac{1}{2}, 3\frac{1}{6}, 4\frac{5}{6}, 6\frac{1}{2}$ $8\frac{1}{6}, 9\frac{5}{6}, 11\frac{1}{2}, 13\frac{1}{6}$; $+ 1\frac{2}{3}$

$2\frac{3}{4}, 4\frac{1}{4}, 3\frac{5}{12}, 4\frac{11}{12}, 4\frac{1}{12}$ $5\frac{7}{12}, 4\frac{3}{4}, 6\frac{1}{4}, 5\frac{5}{12}$; $+ 1\frac{1}{2}, - \frac{5}{6}$

After checking the answers, have each student write a sequence. Then ask pairs of students to exchange and continue each other's sequence. Finally, have pairs compare solutions.

VISUAL
LOGICAL/MATHEMATICAL

TECHNOLOGY LINK

 Intervention Strategies and Activities CD-ROM • *Skills 22, 23*

 Calculating Crew • *Nautical Number Line,* Level N

LESSON 9.6 ORGANIZER

Objective To subtract mixed numbers involving renaming

1 Introduce

QUICK REVIEW provides review of prerequisite skills.

Why Learn This? You can use this skill to find out how much farther you have to travel on a hike to reach a known destination. *Share the lesson objective with students.*

2 Teach

Guided Instruction

- Direct students' attention to the problem about the blizzard.

  How do you determine the greatest and least amounts? The greatest amount is $4\frac{1}{3}$ ft because 4 is the greatest whole number. The least amount is $1\frac{5}{12}$ because $\frac{5}{12}$ is less than $\frac{7}{12}$.

  Modifying Instruction Have students model the exercises with fraction bars or strips.

- Discuss the Example with students.

  Why is it necessary to rename the 5 as a mixed number? so that there is a fraction from which $\frac{4}{5}$ can be subtracted

  How do you decide what denominator to use for the whole number you are renaming? Use the denominator of the fraction in the mixed number.

REASONING How can you check that the answer in the Example is correct? Possible answer: Add $3\frac{1}{5}$ and $1\frac{4}{5}$. The sum should equal 5.

ADDITIONAL EXAMPLE

Example, p. 192

Find the difference. $8 - 3\frac{3}{10}$ $4\frac{7}{10}$

LESSON 9.6

Subtract Mixed Numbers

Learn how to subtract mixed numbers involving renaming.

QUICK REVIEW

Write the number that makes the fraction equivalent to $\frac{1}{2}$.

1. $\frac{\blacksquare}{8}$ 4   2. $\frac{\blacksquare}{6}$ 3   3. $\frac{\blacksquare}{20}$ 10   4. $\frac{\blacksquare}{14}$ 7   5. $\frac{\blacksquare}{100}$ 50

A 1993 blizzard called The Storm of the Century brought record cold, wind, and snow to many cities in the southern and eastern United States.

How much more snow did Mount LeConte receive than Birmingham received?

STORM OF THE CENTURY	
City	Snow (ft)
Birmingham, AL	$1\frac{5}{12}$
Asheville, NC	$1\frac{7}{12}$
Pittsburgh, PA	$2\frac{1}{2}$
Syracuse, NY	$3\frac{7}{12}$
Mount LeConte, TN	$4\frac{1}{3}$

Find $4\frac{1}{3} - 1\frac{5}{12}$.

Estimate. $4\frac{1}{3}$ is close to $4\frac{1}{2}$ and $1\frac{5}{12}$ is close to $1\frac{1}{2}$. So, the difference is about $4\frac{1}{2} - 1\frac{1}{2}$, or 3.

Subtract.

$$\begin{array}{r} 4\frac{1}{3} = 4\frac{4}{12} \\ -1\frac{5}{12} = -1\frac{5}{12} \end{array}$$

The LCD of $\frac{1}{3}$ and $\frac{5}{12}$ is 12.
Write equivalent fractions, using the LCD, 12.

$$\begin{array}{r} 4\frac{1}{3} = 4\frac{4}{12} = 3\frac{16}{12} \\ -1\frac{5}{12} = -1\frac{5}{12} = -1\frac{5}{12} \\ \hline 2\frac{11}{12} \end{array}$$

Since $\frac{5}{12}$ is greater than $\frac{4}{12}$, rename $4\frac{4}{12}$.
$4\frac{4}{12} = 3 + \frac{12}{12} + \frac{4}{12} = 3\frac{16}{12}$.
Subtract the fractions.
Subtract the whole numbers.

The answer is reasonable because it is close to the estimate of 3 ft. So, Mount LeConte received $2\frac{11}{12}$ ft more snow than Birmingham.

EXAMPLE

Jake is running in a 5-km race. So far he has run $1\frac{4}{5}$ km. How far does he have to go?

Find the difference. $5 - 1\frac{4}{5}$

$$\begin{array}{r} 5 = 4\frac{5}{5} \\ -1\frac{4}{5} = -1\frac{4}{5} \\ \hline 3\frac{1}{5} \end{array}$$

Since you are subtracting fifths, rename 5 as $4\frac{5}{5}$.
Subtract the fractions.
Subtract the whole numbers.

So, Jake has $3\frac{1}{5}$ km to go.

192 Chapter 9

RETEACH 9.6

Subtract Mixed Numbers

Mrs. Ruiz buys $4\frac{1}{2}$ lb of apples. She uses $1\frac{2}{3}$ lb to bake apple tarts. How many pounds of apples are left?

Step 1 The LCD of $\frac{1}{2}$ and $\frac{2}{3}$ is 6. Rename the fractions using the LCD.

$$4\frac{1}{2} = 4\frac{3}{6}$$
$$-1\frac{2}{3} = 1\frac{4}{6}$$

Step 2 Since you can't subtract $\frac{4}{6}$ from $\frac{3}{6}$, rename $4\frac{3}{6}$.

Think: $4\frac{3}{6} = 3 + \frac{6}{6} + \frac{3}{6} = 3\frac{9}{6}$

$$4\frac{1}{2} = 4\frac{3}{6} = 3\frac{9}{6}$$
$$-1\frac{2}{3} = 1\frac{4}{6} = 1\frac{4}{6}$$
$$\overline{2\frac{5}{6}}$$

Now, subtract the fractions. Then subtract the whole numbers.

So, $4\frac{1}{2} - 1\frac{2}{3} = 2\frac{5}{6}$

Mrs. Ruiz has $2\frac{5}{6}$ lbs of apples left.

Find the difference. Write the answer in simplest form.

1. $6\frac{2}{5}$
   $-3\frac{7}{10}$
   $\overline{2\frac{7}{10}}$

2. $4\frac{1}{6}$
   $-2\frac{3}{4}$
   $\overline{1\frac{5}{12}}$

3. $9\frac{3}{7}$
   $-5\frac{1}{2}$
   $\overline{3\frac{13}{14}}$

4. $11\frac{1}{3}$
   $-7\frac{2}{9}$
   $\overline{3\frac{4}{9}}$

5. $10\frac{3}{4}$
   $-3\frac{1}{2}$
   $\overline{6\frac{3}{8}}$

6. $6\frac{1}{10}$
   $-4\frac{2}{5}$
   $\overline{1\frac{3}{10}}$

7. $9\frac{1}{8}$
   $-4\frac{3}{4}$
   $\overline{4\frac{3}{8}}$

8. $12\frac{2}{5}$
   $-6\frac{1}{2}$
   $\overline{5\frac{9}{10}}$

9. $8\frac{2}{3}$
   $-5\frac{5}{6}$
   $\overline{2\frac{5}{6}}$

PRACTICE 9.6

Subtract Mixed Numbers

Write the difference in simplest form. Estimate to check.

1. $8\frac{3}{4} - 6\frac{1}{2}$ $2\frac{1}{4}$
2. $4\frac{3}{5} - 2\frac{1}{10}$ $1\frac{1}{2}$
3. $7\frac{1}{4} - 2\frac{2}{3}$ $4\frac{7}{12}$
4. $5\frac{2}{9} - 3\frac{2}{3}$ $1\frac{5}{9}$

5. $3\frac{1}{5} - 2\frac{3}{10}$ $\frac{9}{10}$
6. $5\frac{3}{8} - 4\frac{1}{2}$ $\frac{7}{8}$
7. $6\frac{1}{3} - 2\frac{3}{4}$ $3\frac{7}{12}$
8. $1\frac{7}{9} - 1\frac{2}{3}$ $\frac{1}{9}$

9. $4\frac{2}{3} - 1\frac{1}{2}$ $3\frac{1}{6}$
10. $5\frac{3}{4} - 3\frac{4}{5}$ $2\frac{11}{20}$
11. $3\frac{1}{2} - 1\frac{4}{9}$ $1\frac{1}{9}$
12. $4\frac{5}{8} - 2\frac{1}{4}$ $2\frac{3}{8}$

13. $5\frac{3}{6} - 3\frac{2}{3}$ $1\frac{1}{2}$
14. $4\frac{3}{5} - 2\frac{7}{10}$ $1\frac{9}{10}$
15. $4\frac{1}{8} - 2\frac{3}{4}$ $1\frac{3}{8}$
16. $3\frac{1}{2} - 1\frac{4}{5}$ $1\frac{7}{10}$

17. $5\frac{3}{4} - 2\frac{7}{8}$ $2\frac{7}{8}$
18. $6\frac{1}{4} - 4\frac{2}{17}$ $1\frac{9}{20}$
19. $9\frac{3}{8} - 4\frac{1}{4}$ $5\frac{1}{24}$
20. $5\frac{1}{6} - 1\frac{5}{13}$ $3\frac{13}{24}$

Evaluate each expression for $a = 3\frac{1}{3}$, $b = 2\frac{1}{4}$, $c = 5\frac{1}{6}$.

21. $c - a$ $1\frac{5}{6}$
22. $c - b$ $2\frac{11}{12}$
23. $a - b$ $1\frac{1}{12}$

Mixed Review

Write in exponential form.

24. $5 \times 5 \times 5 \times 5$ 5^4
25. $10 \times 10 \times 10$ 10^3
26. $k \times k \times k \times k \times k$ k^5
27. $w \times w$ w^2

Evaluate each expression.

28. $17.61 - s$ for $s = 12.18$ 5.43
29. $75.6 \div v$ for $v = 6.3$ 12
30. $5f$ for $f = 8.7$ 43.5

CHECK FOR UNDERSTANDING

Think and Discuss — Look back at the lesson to answer each question.

1. Explain how to rename $3\frac{1}{3}$ so you could subtract $1\frac{2}{3}$. $3\frac{1}{3} = 2 + \frac{3}{3} + \frac{1}{3} = 2\frac{4}{3}$

2. **What if** you wanted to find $4\frac{5}{12} - 2\frac{3}{8}$? What equivalent fractions would you write using the LCD? $4\frac{10}{24}$ and $2\frac{9}{24}$

Guided Practice — Write the difference in simplest form. Estimate to check.

3. $4\frac{1}{3} - 2\frac{1}{4}$ $2\frac{1}{12}$
4. $6 - 2\frac{2}{3}$ $3\frac{1}{3}$
5. $7\frac{3}{10} - 3\frac{2}{5}$ $3\frac{9}{10}$
6. $6\frac{1}{5} - 3\frac{7}{10}$ $2\frac{1}{2}$
7. $8\frac{5}{6} - 4\frac{8}{9}$ $3\frac{17}{18}$
8. $16\frac{3}{8} - 7\frac{1}{2}$ $8\frac{7}{8}$

PRACTICE AND PROBLEM SOLVING

Independent Practice — Write the difference in simplest form. Estimate to check.

9. $3\frac{1}{6} - 1\frac{1}{4}$ $1\frac{11}{12}$
10. $5\frac{1}{2} - 3\frac{7}{10}$ $1\frac{4}{5}$
11. $12\frac{1}{9} - 7\frac{1}{3}$ $4\frac{7}{9}$
12. $4\frac{1}{4} - 2\frac{2}{5}$ $1\frac{17}{20}$
13. $8\frac{1}{4} - 5\frac{2}{3}$ $2\frac{7}{12}$
14. $7\frac{5}{9} - 2\frac{5}{6}$ $4\frac{13}{18}$
15. $11\frac{1}{4} - 9\frac{7}{8}$ $1\frac{3}{8}$
16. $5.25 - 2\frac{3}{8}$ $2\frac{7}{8}$, or 2.875
17. $6.2 - 3\frac{1}{2}$ $2\frac{7}{10}$, or 2.7

Evaluate each expression for $j = 5\frac{1}{2}$, $k = 4\frac{3}{5}$, and $m = 2\frac{7}{10}$.

18. $j - k$ $\frac{9}{10}$
19. $j - m$ $2\frac{4}{5}$
20. $k - m$ $1\frac{9}{10}$

21. $\frac{7}{8}$ mi shorter

Problem Solving Applications

21. Whit usually drives $4\frac{7}{8}$ mi on the expressway to work. Sometimes traffic is bad due to weather conditions and he takes another route which is $5\frac{3}{4}$ mi long. How much shorter is his usual route?

22. **Write About It** Why do you write equivalent fractions before you rename? Can you rename before you write equivalent fractions? because it is easier to tell whether renaming is necessary when fractions are equivalent; yes

23. **Number Sense** Prime numbers that differ by 2, such as 3 and 5, or 59 and 61, are called twin primes. Write two other pairs of twin primes between 1 and 50. possible answer: 5, 7; 29, 31

MIXED REVIEW AND TEST PREP

24. Find the sum of $1\frac{1}{3}$ and $2\frac{1}{6}$. (p. 186) $3\frac{1}{2}$

25. Tell which are equivalent numbers. $\frac{11}{3}$, $7\frac{1}{3}$, $6\frac{1}{3}$, $\frac{22}{3}$ (p. 160) $7\frac{1}{3}$, $\frac{22}{3}$

26. Find the median for the data. 30, 36, 39, 38, 36, 33 (p. 106) 36

27. Evaluate $b \div d$, for $b = 2,260$ and $d = 41$. (p. 28) 55 r5

★ 28. **TEST PREP** Alexander buys 3 boxes of computer paper for $10.79 per box, including tax. How much change will he get from $50? (p. 70) C

A $39.21 C $17.63
B $32.37 D $11.00

EXTRA PRACTICE page H40, Set D

193

---

3 Practice

Guided Practice

Do Check for Understanding Exercises 1–8 with your students. Identify those having difficulty and use lesson resources to help.

COMMON ERROR ALERT

Some students may have difficulty writing mixed numbers when renaming one whole. Have them work through the renaming step by first modeling it with fraction bars or strips and recording equivalent values.

Independent Practice

Note that Exercise 28 is a **multistep or strategy problem**. Assign Exercises 9–23.

Algebraic Thinking Review with students how to find prime numbers to answer Exercise 23. They should use factoring and divisibility rules to eliminate composite numbers; for example, all even numbers can be divided by 2, so none, other than 2 itself, can be prime numbers.

MIXED REVIEW AND TEST PREP
Exercises 24–28 provide **cumulative review** (Chapters 1–9).

4 Assess

Summarize the lesson by having students:

DISCUSS Describe the steps in solving Exercise 9. Possible answer: Rename both fractions with the common denominator, 12; rename $3\frac{1}{6}$ as $2\frac{14}{12}$; subtract.

WRITE Describe a way to solve Exercise 4 mentally. Possible answer: Think that $6 - 3 = 3$ and $1 - \frac{2}{3} = \frac{1}{3}$, so the difference is $3\frac{1}{3}$.

Lesson Quiz

Find the difference. Write the answer in simplest form.

1. $2\frac{1}{8} - 1\frac{3}{8}$ $\frac{3}{4}$
2. $2\frac{1}{4} - 1\frac{1}{3}$ $\frac{11}{12}$
3. $3\frac{1}{12} - 1\frac{5}{6}$ $1\frac{1}{4}$
4. $5\frac{1}{2} - 1\frac{7}{10}$ $3\frac{4}{5}$
5. $3\frac{1}{2} - 1\frac{6}{7}$ $1\frac{9}{14}$
6. $12\frac{1}{8} - 9\frac{1}{6}$ $2\frac{23}{24}$

193

LESSON 9.7

Problem Solving Strategy: Draw a Diagram

LESSON PLANNING

Objective To use the strategy *draw a diagram* to solve problems

Intervention for Prerequisite Skills
Simplify Fractions, Add and Subtract Like Fractions (For intervention strategies, see page 175.)

Lesson Resources Problem Solving Think Along, p. TR1

NCTM Standards
1. Number and Operations
4. Measurement
6. Problem Solving
8. Communication
10. Representation

Math Background
By drawing a diagram showing how the data in a problem are related, students often see more clearly how to solve the problem. This is particularly true for problems that involve a physical situation, such as the following:
- distance and direction.
- geometric shapes and perimeter or area.
- patterns, as when students find the number of diagonals in a polygon.

WARM-UP RESOURCES

 **NUMBER OF THE DAY** — Transparency 9.7

Write the present time (hours and minutes) as a mixed number. Estimate what time it will be in $4\frac{5}{6}$ hours.
Possible answer for 10:20 A.M.: $10\frac{1}{3}$; 3:10 P.M.

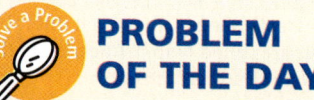 **PROBLEM OF THE DAY** — Transparency 9.7

Write the next 4 numbers. How does each number relate to the one before it?
A. 10, $8\frac{3}{4}$, $7\frac{1}{2}$, $6\frac{1}{4}$ 5, $3\frac{3}{4}$, $2\frac{1}{2}$, $1\frac{1}{4}$; it is $1\frac{1}{4}$ less.
B. 9, $7\frac{7}{8}$, $6\frac{3}{4}$, $5\frac{5}{8}$ $4\frac{1}{2}$, $3\frac{3}{8}$, $2\frac{1}{4}$, $1\frac{1}{8}$; it is $1\frac{1}{8}$ less.
C. $11\frac{7}{10}$, $10\frac{2}{5}$, $9\frac{1}{10}$, $7\frac{4}{5}$ $6\frac{1}{2}$, $5\frac{1}{5}$, $3\frac{9}{10}$, $2\frac{3}{5}$; it is $1\frac{3}{10}$ less.

Solution Problem of the Day tab, p. PD9

 **DAILY FACTS PRACTICE**

Have students practice addition facts by completing Set A of *Teacher's Resource Book*, p. TR100.

194A Chapter 9

INTERVENTION AND EXTENSION RESOURCES

ALTERNATIVE TEACHING STRATEGY

Materials *For each student* ruler (inches), p. TR23

To **review and apply relationships in a rectangle,** have students:

- find and measure the sides of rectangles on this page and other pages in the text.
- state the relationship between opposite sides of a rectangle. They are equal in length.
- write the distances represented by the sides of the rectangle showing the taxi's route on page 194. $8\frac{1}{4}$ mi, $1\frac{1}{3}$ mi, $8\frac{1}{4}$ mi, $1\frac{1}{3}$ mi
- find the distance from location B to where the taxi crosses its own path. $2\frac{2}{3}$ mi
- Add to find the rectangle's perimeter. $19\frac{1}{6}$ mi

KINESTHETIC
LOGICAL/MATHEMATICAL

ENGLISH LANGUAGE LEARNERS

Ask students to **model distances** and **use the strategy draw a diagram.**

- Select a unit, such as a volunteer's shoe length. Have the volunteer pace out several distances in the room. For each, state the length and draw a labeled line on the board to represent the distance.
- Use the sentence frame *How far is it from [Tim's desk] to [the door]?* to ask students about the labeled distances.
- Make up problems like the one on page 194, based on distances in your room. Tell groups to use *draw a diagram* to solve them.

Check students' work.

VISUAL
BODILY/KINESTHETIC

READING STRATEGY

Summarize To use this strategy, students must include the main ideas and all the information they need to solve a problem. As they look at the opening problem on page 194, have them describe:

- what the diagram shows as the given information in the problem. the distances AB, BC, CD, DE, and EF
- how the diagram shows how to find the distances from the point of intersection to C and to E. by showing a rectangle with congruent opposite sides
- how the diagram helps solve the problem. Possible answer: It quickly shows the given data and how they relate to one another.

EARLY FINISHERS

Materials *For each group* a map of a foreign country with distances between major cities labeled

Each group **uses a map to calculate the distance** to a mystery destination. Have each group:

- choose a destination and use the map scale to find the distance from a city to their destination.
- write a description of their destination's location, using the distance and needed directions, such as *north of Athens*.

Tell groups to exchange descriptions and discover each other's destinations. Ask them to draw a diagram that shows directional lines and distances. Check students' work.

VISUAL
LOGICAL/MATHEMATICAL

MULTISTEP AND STRATEGY PROBLEMS

The following multistep and strategy problems are provided in Lesson 9.7:

Page	Item
195	1–11

TECHNOLOGY LINK

Intervention Strategies and Activities CD-ROM • *Skills 22, 23*

LESSON 9.7 ORGANIZER

Objective To use the strategy *draw a diagram* to solve problems

Lesson Resources Problem Solving Think Along, p. TR1

1 Introduce

QUICK REVIEW provides review of prerequisite skills.

Why Learn This? You can use this skill to display data and find solutions to complex word problems. *Share the lesson objective with students.*

2 Teach

Guided Instruction

- Read the problem with students.

  Why are there right angles in the diagram at locations B, C, D, and E? The problem states that the taxi changes directions at each point to a new direction at right angles to the previous one.

  From the point where the taxi crosses its own path, how far is it to location C? to location E? Explain. $1\frac{1}{3}$ mi; $8\frac{1}{4}$ mi; the diagram shows a rectangle, which has opposite sides that are the same length.

- Have volunteers explain steps in the problem solving process.

  What is the advantage of drawing a diagram in this problem? Possible answer: It shows the key points in the taxi's travels, and how they relate to each other geometrically.

  How else could you find a solution in the Solve step? Possible answer: Add the distances from A to B and from B to the point of intersection, plus the lengths of the sides of the rectangle.

LESSON 9.7

PROBLEM SOLVING STRATEGY
Draw a Diagram

Learn how to use the strategy *draw a diagram* to solve problems.

QUICK REVIEW
1. $\frac{3}{4} - \frac{1}{4}$ $\frac{1}{2}$
2. $\frac{4}{5} - \frac{1}{5}$ $\frac{3}{5}$
3. $\frac{2}{3} - \frac{2}{3}$ 0
4. $\frac{1}{5} + \frac{4}{5}$ 1
5. $\frac{9}{10} - \frac{6}{10}$ $\frac{3}{10}$

A taxicab travels $1\frac{1}{3}$ mi west, 4 mi north, $8\frac{1}{4}$ mi east, $1\frac{1}{3}$ mi south, and then 10 mi west. How far has the taxicab traveled when it crosses its own path?

Analyze What are you asked to find?
how far the taxicab has traveled when it crosses its own path
What information is given?
distances and directions in which the taxicab travels
Is there numerical information you will not use? If so, what? No

Choose What strategy will you use?

You can draw a diagram that shows the taxicab's route.

Draw a diagram and label the distances and locations.

Solve How will you solve the problem?

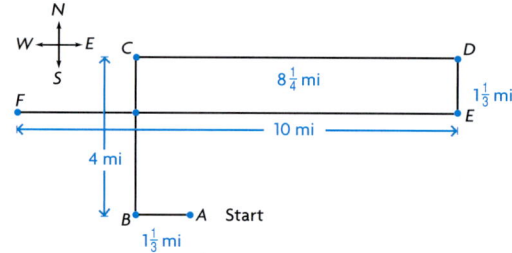

Add all the distances from A to E. Then add the distance from E to the point where the taxicab crosses its own path. This will be the same as the distance from C to D.

$$1\frac{1}{3} + 4 + 8\frac{1}{4} + 1\frac{1}{3} + 8\frac{1}{4}$$

Write equivalent fractions by using the LCD, 12.

$$1\frac{4}{12} + 4 + 8\frac{3}{12} + 1\frac{4}{12} + 8\frac{3}{12} = 22 + \frac{14}{12} = 22 + 1\frac{2}{12} = 23\frac{1}{6}$$

So, the taxicab has traveled $23\frac{1}{6}$ mi when it crosses its own path.

Check How can you check to see if your answer is reasonable?
estimate the sum

194 Chapter 9

RETEACH 9.7

Problem Solving Strategy: Draw a Diagram

Some problems become clearer when you draw a diagram to represent the information you are given.

Ms. Kaminsky is teaching her class a game in which students stand in a circle. Two girls stand together, then 1 boy, then 2 girls, then 1 boy, and so on. Altogether, 7 boys are standing in the circle along with all the girls in the class. If there are 27 students in the class, how many boys are not participating?

Step 1: Think about what you know and what you are asked to find.
- You know that all the girls in the class are in the circle.
- You know that there are 7 boys in the circle.
- You know that for every 2 girls in the circle, there is 1 boy.
- You are asked to find how many boys in the class are not in the circle.

Step 2: Plan a strategy to solve.
- Use the strategy *draw a diagram*.
- Draw what you know from the problem.

Step 3: Solve.
- Carry out the strategy.
  The diagram shows that if 7 boys are in the circle, then there must be 14 girls.
  $7 + 14 = 21$ and $27 - 21 = 6$.

So, 6 boys in the class are not participating.

Use the strategy *draw a diagram* to help you solve.

1. In another game, the arrangement around the circle is reversed. Two boys stand together, then 1 girl, then 2 boys, then 1 girl, and so on. In this version of the game, what is the greatest number of students in Ms. Kaminsky's class that can participate at one time?

   18 students: 12 boys and 6 girls

2. The desks are arranged in rows of 6 from the front of the room to the back. In all, there are 5 rows. Arthur sits in the fourth seat of the first row. Andrea sits 3 rows over and 1 seat back. Charles sits 3 seats in front of Andrea. Juan sits 2 rows over and 2 seats farther back than Charles. Where is Juan compared to Arthur?

   next to Arthur

PRACTICE 9.7

Problem Solving Strategy: Draw a Diagram

Solve by drawing a diagram.

1. In the school art room the students use square tables. Each side of a table is $4\frac{1}{2}$ ft. If some of the tables are placed end-to-end, they form a rectangle with a perimeter of 36 ft. How many tables are used to make the rectangle?

   3 tables

2. The art room is on one side of the hallway with an office, a classroom, and the music room. The art room is between the classroom and the office. The classroom is between the music room and the art room. Which two rooms are on the ends of the hallway?

   office, music room

3. During art class, 2 students can sit at each side of a square table. The students decide to make a large rectangular table by placing 5 square tables end-to-end. How many students will be able to sit at this large table?

   24 students

4. Richard is cutting a hole in a wall to hold an air conditioner. The front of the air conditioner is a rectangle 26 in. wide and 16 in. high. The wall is 72 in. wide. If the air conditioner is centered in the wall, how wide will the wall be on either side of it?

   23 in.

5. Cassandra is training for a charity walk between two towns. The towns are 12 mi apart. On her first day of training, she walks $4\frac{1}{2}$ mi. If she increases her distance by $1\frac{1}{2}$ mi every 3 days, how many days will it take until Cassandra has walked at least 10 mi?

   13 days

6. Marla wants to wrap a present that is in the shape of a cube. She wants to put one piece of ribbon around the top, bottom and two sides. She wants to put a second piece around the top, bottom, and other two sides. The box is $8\frac{1}{2}$ in. on each edge. What is the shortest length of ribbon she needs?

   68 in.

Mixed Review

Write the number in standard form.

7. six hundred and three tenths **600.3**
8. ninety-one hundredths **0.91**
9. ninety and seven hundredths **90.07**
10. eighty and nine tenths **80.9**

Find the GCF for each set of numbers.

11. 10, 15 **5**
12. 16, 40 **8**
13. 18, 45 **9**
14. 20, 28 **4**
15. 24, 56 **8**

PROBLEM SOLVING PRACTICE

Solve by drawing a diagram.

1. A tour bus travels $7\frac{1}{2}$ mi south, $3\frac{1}{3}$ mi east, $4\frac{1}{3}$ mi north, and $11\frac{1}{2}$ mi west. How far has the tour bus traveled when it crosses its own path? **$18\frac{1}{2}$ mi**

2. Tanisha needs a fence that is 33 ft long to separate her two gardens. If she puts one post in the ground every $5\frac{1}{2}$ ft, how many posts will Tanisha need for the fence? **7 posts**

For 3–4, use the information below.

Carla drives south $2\frac{1}{2}$ mi from her home. Next, she drives east $\frac{1}{3}$ mi. Then, she drives south 3 mi.

3. How many total miles does Carla drive from her home? **C**

   A 5 mi   C $5\frac{5}{6}$ mi
   B $5\frac{1}{5}$ mi   D 6 mi

4. In which directions must Carla drive to return home? **H**

   F east and north   H west and north
   G east and south   J west and south

PROBLEM SOLVING STRATEGIES

▶ Draw a Diagram or Picture
  Make a Model
  Predict and Test
  Work Backward
  Make an Organized List
  Find a Pattern
  Make a Table or Graph
  Solve a Simpler Problem
  Write an Equation
  Use Logical Reasoning

MIXED STRATEGY PRACTICE

5. A display in a store has 24 cans in the bottom row, 21 in the second row, and 18 in the third row. If the pattern continues, how many cans are in the fifth row? **12 cans**

6. Tickets to see an Irish dance group cost $52 and $28. Mario bought 7 tickets for his family and paid a total of $316. How many tickets at each price did he buy? **five $52 tickets and two $28 tickets**

7. Ms. Lopez travels north from home to pick up Maria at school. Then she travels $3\frac{1}{2}$ mi east to pick up Marcos and $4\frac{1}{4}$ mi south to pick up Carter. If Ms. Lopez has driven a total of $12\frac{1}{2}$ mi, what is the distance from home to Maria's school? **$4\frac{3}{4}$ mi**

8. Tim has a job in the city. He has to commute $6\frac{1}{2}$ km to work. The bus he takes to work travels about $3\frac{1}{4}$ km in 10 minutes. About how much time does Tim spend on the bus going to and from work? **40 minutes**

9. In a recent contest, Gary scored more points than Catherine, who scored more points than Clara. Christopher scored more points than Clara but fewer than Gary. Who had the most points? **Gary**

10. Use the table below. What is Joshua's total bill if he rented 3 videos and kept them for 6 days? **$20.34**

Video Rental Prices	
1 movie for 1 day	$2.99
Each additional day	$0.99
5 additional days	$3.79

11. ✏️ **Write a problem** that can be solved by using the strategy *draw a diagram*. Explain the steps you would use to solve the problem, and draw the diagram. **Check students' problems and explanations.**

195

3 Practice

Guided Practice

Note that Exercises 1–4 are **multistep or strategy problems**. Do Problem Solving Practice Exercises 1–4 with your students. Identify those having difficulty and use lesson resources to help.

As students draw diagrams for Exercise 1, review the compass directions.

Independent Practice

Note that Exercises 5–11 are **multistep or strategy problems**. Assign Exercises 5–11.

Students may use either the strategy *make a table* or *predict and test* in Exercise 6.

Scaffolded Instruction Use the prompts on Transparency 9 to guide instruction for the multistep or strategy problem in Exercise 10.

4 Assess

Summarize the lesson by having students:

DISCUSS What kinds of problems do you think the strategy *draw a diagram* works well with? **Possible answer: those involving distance or geometric figures**

WRITE Describe the steps you used to solve Exercise 7. **Possible answer: I drew a line segment north from Ms. Lopez's home, then another one east that I marked $3\frac{1}{2}$ mi, then another one south that I marked $4\frac{1}{4}$ mi. Then I subtracted the sum of $3\frac{1}{2}$ mi and $4\frac{1}{4}$ mi from $12\frac{1}{2}$ mi.**

Lesson Quiz

Draw a diagram to solve. **Check students' work.**

1. Kendra rides her bike $4\frac{7}{10}$ mi west from her home to the park. She then rides west to the library. From there, she rides $3\frac{1}{2}$ mi west to Sean's house. If she rides her bike 12 mi in all, how far is it from the park to the library? **$3\frac{4}{5}$ mi**

2. Aaron plans to put braid around the edge of a square tablecloth that measures $6\frac{3}{4}$ ft on each side. How many feet of braid will he need? **27 ft**

CHALLENGE 9.7

Pyramid Patterns in a Diagram

Pascal's Triangle is a diagram that has been used for many years to solve problems. It was named for a French mathematician who may have discovered it.

```
              1              Row 0
            1   1            Row 1
          1   2   1          Row 2
        1   3   3   1        Row 3
      1   4   6   4   1      Row 4
    1   5  10  10   5   1    Row 5
```

Use Pascal's Triangle to answer the following questions.

1. Find each group of numbers in the diagram.

   1 2   3 1   3 3   4 6
   3   4   6   10

   From these groups, can you describe the pattern that is used to create the triangle?
   The bottom number is the sum of the two numbers above it.

2. The next row of the triangle contains these numbers.
   1 _6_ 15 20 _15_ 6 1
   Use the pattern to fill in the row.

3. A section further down the triangle contains the following numbers.
   35   35   21   7   1
   1 _8_   28 _56_ 70 _56_ _28_ 8 1
   Use the pattern to fill in the row.

4. When you add the numbers in the rows, an interesting pattern develops. Add the numbers in rows 0 through 5. Describe what you find.
   The sums of the rows are the powers of 2.

5. What will the sum be for row 6? row 7? **64, 128**

6. Find the diagonals that represent the first 5 counting numbers (1, 2, 3, 4, and 5). If the pattern is extended, in which row will the eighth counting number be found? **row 8**

READING STRATEGY 9.7

Summarize

To **summarize** is to state something in a brief way. Knowing how to summarize information is a useful skill. Sometimes drawing a diagram to display information is a good way to summarize information.

VOCABULARY
summarize

Read the following problem.

Rosie walks dogs to earn money. She leaves home with her own dog, Loki, and picks up a poodle, Dante, $\frac{1}{2}$ mi east of her home. Next she gets Noni, another poodle, who lives $\frac{3}{8}$ mi east of Dante. Another $\frac{1}{4}$ mi east, she picks up a spaniel, Higgins. Rosie then drops off the dogs at their houses in this order: first Loki, then Dante, then Noni, then Higgins. Then Rosie walks home. How far have Rosie and each dog walked?

1. Draw a diagram to summarize the information.
   Check students' diagrams.

2. Solve the problem.
   Rosie walked _6_ mi.   Loki walked _3_ mi.   Dante walked _3_ mi.
   Noni walked _3_ mi.   Higgins walked _3_ mi.

Draw a diagram to summarize the information. Solve the problem.

3. Bill, Samantha, and Tim are doing a science project together. Samantha lives $\frac{7}{10}$ mi west of the school. Tim lives $\frac{2}{5}$ mi west of Samantha. Bill lives $\frac{1}{3}$ mi east of the school. After school, they walk to Bill's house to pick up some equipment. Then they go to Tim's house to work. When they are finished, Bill and Samantha walk home. How far did each student walk after school?

   Check students' diagrams.

   Bill walked $3\frac{1}{6}$ mi.   Samantha walked $2\frac{1}{6}$ mi.   Tim walked $\frac{23}{30}$ mi.

195

CHAPTER 9

REVIEW/TEST

Purpose To check understanding of concepts, skills, and problem solving presented in Chapter 9

USING THE PAGE

The Chapter 9 Review/Test can be used as a **review** or a **test**.

- Items 1–37 check skill proficiency.
- Items 38–40 check students' abilities to choose and apply problem solving strategies to real-life addition and subtraction problems.

 Suggest that students place the completed Chapter 9 Review/Test in their portfolios.

USING THE ASSESSMENT GUIDE

- Multiple-choice format of Chapter 9 Posttest—See *Assessment Guide*, pp. AG57–58.
- Free-response format of Chapter 9 Posttest—See *Assessment Guide*, pp. AG59–60.

USING STUDENT SELF-ASSESSMENT

The How Did I Do? survey helps students assess what they have learned and how they learned it. This survey is available as a copying master in *Assessment Guide*, p. AGxvii.

196 Chapter 9

CHAPTER 9 REVIEW/TEST

Estimate the sum or difference. (pp. 176–179) Possible estimates are given.

1. $\frac{7}{12} + \frac{1}{4}$ **1**
2. $\frac{4}{5} - \frac{2}{7}$ **$\frac{1}{2}$**
3. $\frac{4}{5} + \frac{3}{8}$ **1**
4. $\frac{3}{4} - \frac{1}{3}$ **$\frac{1}{2}$**
5. $4\frac{2}{9} - \frac{1}{7}$ **4**
6. $\frac{9}{20} + 1\frac{4}{5}$ **$2\frac{1}{2}$**
7. $6\frac{1}{4} + 3\frac{2}{9}$ **$9\frac{1}{2}$**
8. $9\frac{4}{5} - 2\frac{7}{9}$ **7**

Write the sum or difference in simplest form. Estimate to check. (pp. 182–185)

9. $\frac{1}{2} + \frac{1}{3}$ **$\frac{5}{6}$**
10. $\frac{3}{4} + \frac{1}{6}$ **$\frac{11}{12}$**
11. $\frac{2}{5} + \frac{2}{4}$ **$\frac{9}{10}$**
12. $\frac{2}{3} + \frac{3}{4}$ **$\frac{17}{12}$, or $1\frac{5}{12}$**
13. $\frac{5}{6} + \frac{2}{3}$ **$\frac{3}{2}$, or $1\frac{1}{2}$**
14. $\frac{1}{3} + \frac{5}{6}$ **$\frac{7}{6}$, or $1\frac{1}{6}$**
15. $\frac{3}{8} + \frac{3}{4}$ **$\frac{9}{8}$, or $1\frac{1}{8}$**
16. $\frac{5}{8} + \frac{1}{6}$ **$\frac{19}{24}$**
17. $\frac{3}{4} - \frac{1}{3}$ **$\frac{5}{12}$**
18. $\frac{7}{8} - \frac{1}{4}$ **$\frac{5}{8}$**
19. $\frac{5}{6} - \frac{2}{9}$ **$\frac{11}{18}$**
20. $\frac{8}{9} - \frac{1}{6}$ **$\frac{13}{18}$**
21. $\frac{7}{8} - \frac{5}{6}$ **$\frac{1}{24}$**
22. $\frac{7}{12} - \frac{5}{12}$ **$\frac{1}{6}$**
23. $\frac{3}{5} - \frac{1}{4}$ **$\frac{7}{20}$**
24. $\frac{4}{5} - \frac{1}{10}$ **$\frac{7}{10}$**

Draw a diagram to find the sum or difference. Write the answer in simplest form. (pp. 186–189) Check students' diagrams.

25. $2\frac{3}{8} + 1\frac{1}{4}$ **$3\frac{5}{8}$**
26. $6\frac{2}{3} - 3\frac{1}{4}$ **$3\frac{5}{12}$**
27. $5\frac{2}{5} - 3\frac{3}{10}$ **$2\frac{1}{10}$**
28. $2\frac{1}{6} + 1\frac{1}{3}$ **$3\frac{1}{2}$**

Write the sum or difference in simplest form. Estimate to check. (pp. 186–189, 192–193)

29. $1\frac{1}{6} + 3\frac{2}{3}$ **$4\frac{5}{6}$**
30. $2\frac{3}{4} + 3\frac{1}{8}$ **$5\frac{7}{8}$**
31. $1\frac{1}{2} + 2\frac{1}{4}$ **$3\frac{3}{4}$**
32. $1\frac{1}{5} + 1\frac{3}{10}$ **$2\frac{1}{2}$**
33. $7\frac{3}{4} - 5\frac{1}{3}$ **$2\frac{5}{12}$**
34. $8\frac{1}{3} - 3\frac{1}{8}$ **$5\frac{5}{24}$**
35. $3\frac{1}{4} - 2\frac{1}{2}$ **$\frac{3}{4}$**
36. $4\frac{1}{2} - 1\frac{2}{3}$ **$2\frac{5}{6}$**
37. $7\frac{1}{5} - 5\frac{4}{9}$ **$1\frac{34}{45}$**

Solve each problem by drawing a diagram. (pp. 194–195)

38. A minibus leaves the garage and travels $9\frac{5}{6}$ mi north to pick up Tanya. Then it travels $3\frac{1}{6}$ mi west to pick up Luis, $4\frac{1}{4}$ mi south to pick up Alissa, and $4\frac{5}{12}$ mi east to the school. How far does the minibus travel before crossing its own path? **$20\frac{5}{12}$ mi**

39. Satoko has a board that is 9 ft long. She needs to cut the board into $2\frac{1}{4}$-ft sections. How many cuts will she have to make? **3 cuts**

40. Del drives $3\frac{1}{4}$ mi north from his home. Next, he drives west $\frac{3}{4}$ mi. Then, he drives north $\frac{1}{2}$ mi. In which directions must Del drive to return home? **east and south**

196 Chapter 9

CHAPTERS 1–9 ★ STANDARDIZED TEST PREP

TIP! Decide on a plan.
See item **2**.
Show the hours practiced as an addition sentence. Find the *least common denominator* before you add.

Also see problem **4**, p. H63.

Choose the best answer.

1. Bryce had $\frac{13}{15}$ gallon of paint. He used $\frac{1}{3}$ gallon for a project. How much paint did he have left? **A**

   A $\frac{8}{15}$ gal C $\frac{1}{3}$ gal
   B $\frac{1}{2}$ gal   D $\frac{1}{5}$ gal

2. Tirzah practiced the piano for $1\frac{2}{3}$ hours on Monday, $\frac{3}{4}$ hour on Wednesday, and $1\frac{1}{2}$ hours on Friday. Which is the total amount of time she spent practicing the piano? **H**

   F $3\frac{2}{5}$ hr   H $3\frac{11}{12}$ hr
   G $3\frac{3}{4}$ hr   J 4 hr

3. Last week Jessie's swimming practice lasted for $\frac{3}{5}$ hour on Monday, $\frac{5}{6}$ hour on Wednesday, and $\frac{9}{10}$ hour on Friday. How many hours did she have swimming practice during the week? **D**

   A $1\frac{1}{3}$ hr   C 2 hr
   B $1\frac{2}{3}$ hr   D $2\frac{1}{3}$ hr

4. What is the value of $n + \frac{1}{2}$ for $n = \frac{5}{6}$? **J**

   F $\frac{1}{3}$ H $1\frac{1}{6}$
   G $\frac{11}{12}$ J $1\frac{1}{3}$

5. In which pair are both numbers equivalent to $\frac{2}{5}$? **C**

   A 0.2; 20%   C 40%; 0.4
   B 25%; $\frac{4}{10}$ D 0.6; $\frac{6}{15}$

6. $14\frac{2}{3} - 9\frac{5}{12}$ **H**

   F $4\frac{1}{3}$ H $5\frac{1}{4}$
   G $4\frac{1}{2}$ J Not here

7. The mean of 5 numbers is 25.6. What is the sum of the numbers? **A**

   A 128   C 26.1
   B 30.6   D Not here

8. Mel hiked $\frac{9}{16}$ mile on Saturday and $\frac{7}{8}$ mile on Sunday. Which is a good estimate for how far Mel hiked in all? **G**

   F about 2 mi
   G about $1\frac{1}{2}$ mi
   H about 1 mi
   J about $\frac{1}{2}$ mi

9. Which is a reasonable estimate of the sum of the fractions $\frac{1}{12}, \frac{4}{9}, \frac{5}{8},$ and $\frac{11}{12}$? **B**

   A 1   C 4
   B 2   D $4\frac{1}{2}$

10. 115,371.9 + 22,671.25 **G**

    F 138,043.015   H 138,043.259
    G 138,043.15   J Not here

See below.

11. Bart passes a sign that read "City Limit $5\frac{1}{2}$ miles." If he drives $2\frac{1}{4}$ miles farther, how far will he be from the city limit? Draw a diagram for the problem and explain how you found your answer.

12. Explain why you must rename to find the difference $1\frac{7}{8} - \frac{9}{10}$. Then find the difference.

197

CHAPTER 10

Multiply and Divide Fractions and Mixed Numbers

INTRODUCING THE CHAPTER

Tell students that as with whole numbers, division of fractions and mixed numbers is the inverse operation of multiplication. Have students estimate the number of cyclists pictured in the photograph. Ask them how many $\frac{1}{2}$ that number is. **about 25 cyclists**

USING DATA

To begin the study of this chapter, have students

- Find the approximate number of kilometers covered during $\frac{1}{2}$ of Stages 9–12. **about 417 km**
- Determine to the nearest hour how long it would take to complete Stages 5–8 at a rate of $17\frac{1}{2}$ km per hour. **about 39 hours**

PROBLEM SOLVING PROJECT

Purpose To use fractions to solve a problem

Grouping pairs or small groups

Background 1 km is equivalent to about $\frac{5}{8}$ mi.

Analyze, Choose, Solve, and Check

Have students

- Convert the distance for each group of stages of the Tour de France to the nearest mile. **Pro.–4: 487 mi, 5–8: 430 mi; 9–12: 521 mi; 13–16: 500 mi; 17–20: 367 mi**
- Use an atlas to map out a bike tour in the United States that is equivalent in distance to the Tour de France.
- Draw a map of the bike tour. Label the distances in miles.

Check students' work.

Suggest that students display the maps in the classroom and then place them in their portfolios.

CHAPTER 10 Multiply and Divide Fractions and Mixed Numbers

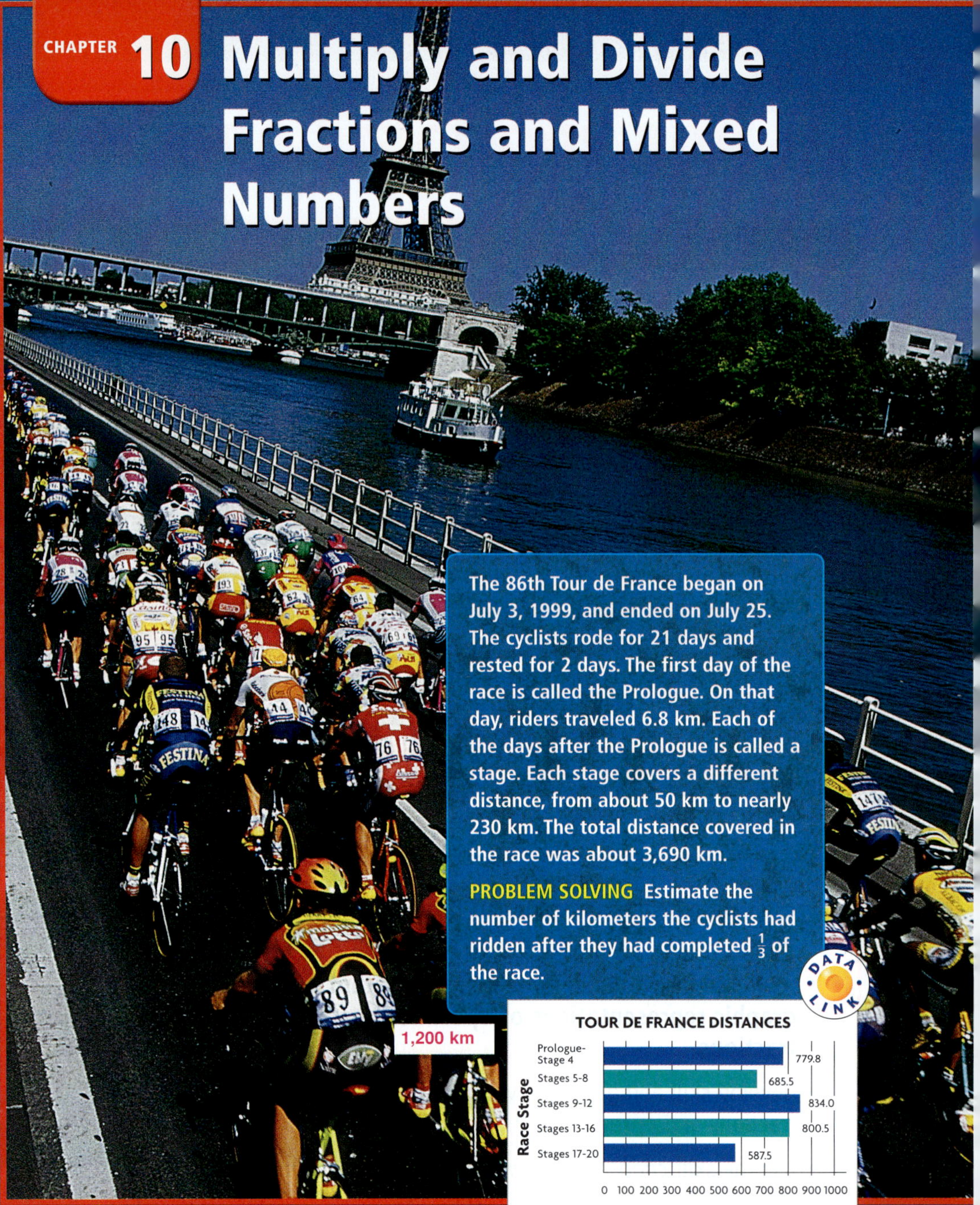

The 86th Tour de France began on July 3, 1999, and ended on July 25. The cyclists rode for 21 days and rested for 2 days. The first day of the race is called the Prologue. On that day, riders traveled 6.8 km. Each of the days after the Prologue is called a stage. Each stage covers a different distance, from about 50 km to nearly 230 km. The total distance covered in the race was about 3,690 km.

PROBLEM SOLVING Estimate the number of kilometers the cyclists had ridden after they had completed $\frac{1}{3}$ of the race.

1,200 km

TOUR DE FRANCE DISTANCES

Race Stage	Kilometers Covered
Prologue–Stage 4	779.8
Stages 5-8	685.5
Stages 9-12	834.0
Stages 13-16	800.5
Stages 17-20	587.5

198 Chapter 10

Why learn math? Explain that cyclists can use fractions to help them calculate their speed as they race. For example, if they decide they want to complete each quarter of a race in a certain amount of time, they can multiply the total distance by $\frac{1}{4}$ to check their time as they complete each quarter. Ask: What other ways do you think cyclists use math? **Possible answer: To determine how many miles per hour they need to travel to complete a race in a specified amount of time**

Check What You Know

Use this page to help you review and remember important skills needed for Chapter 10.

Vocabulary

Choose the best term from the box.

> fraction
> mixed number
> equation

1. An algebraic or numerical sentence that shows two quantities are equal is a(n) __?__. **equation**

2. A number that is made up of a whole number and a fraction is a(n) __?__. **mixed number**

Round Fractions (For Intervention, see p. H10.)

Round each fraction to 0, $\frac{1}{2}$, or 1.

3. $\frac{2}{9}$ **0**
4. $\frac{7}{8}$ **1**
5. $\frac{7}{15}$ **$\frac{1}{2}$**
6. $\frac{5}{8}$ **$\frac{1}{2}$**
7. $\frac{1}{4}$ **$\frac{1}{2}$**
8. $\frac{2}{15}$ **0**
9. $\frac{3}{8}$ **$\frac{1}{2}$**
10. $\frac{5}{6}$ **1**
11. $\frac{1}{3}$ **$\frac{1}{2}$**
12. $\frac{10}{11}$ **1**
13. $\frac{11}{20}$ **$\frac{1}{2}$**
14. $\frac{7}{12}$ **$\frac{1}{2}$**
15. $\frac{1}{6}$ **0**
16. $\frac{4}{7}$ **$\frac{1}{2}$**
17. $\frac{2}{6}$ **$\frac{1}{2}$**

Mental Math and Equations (For Intervention, see p. H11.)

Use mental math to solve.

18. $9.3 + x = 12.5$ **$x = 3.2$**
19. $4c = 128$ **$c = 32$**
20. $x - 160 = 520$ **$x = 680$**
21. $5.11 = 5.28 - x$ **$x = 0.17$**
22. $0.12 = \frac{x}{0.6}$ **$x = 0.072$**
23. $35 - t = 21$ **$t = 14$**
24. $6y = 0.24$ **$y = 0.04$**
25. $r + 3.7 = 6.3$ **$r = 2.6$**

Fractions and Mixed Numbers (For Intervention, see p. H11.)

Write each fraction as a mixed number.

26. $\frac{18}{5}$ **$3\frac{3}{5}$**
27. $\frac{7}{6}$ **$1\frac{1}{6}$**
28. $\frac{16}{15}$ **$1\frac{1}{15}$**
29. $\frac{4}{3}$ **$1\frac{1}{3}$**

Write each mixed number as a fraction.

30. $1\frac{5}{8}$ **$\frac{13}{8}$**
31. $7\frac{1}{2}$ **$\frac{15}{2}$**
32. $3\frac{2}{3}$ **$\frac{11}{3}$**
33. $2\frac{4}{5}$ **$\frac{14}{5}$**

LOOK AHEAD

In Chapter 10 you will
- estimate products and quotients
- multiply and divide fractions and mixed numbers
- use fractions in expressions and equations

Assessing Prior Knowledge

Use the **Check What You Know** page to determine whether your students have mastered the prerequisite skills critical for this chapter.

Intervention

- **Diagnose and Prescribe**
Evaluate your students' performance on this page to determine whether intervention is necessary or if enrichment is appropriate. Options that provide instruction, practice, and a check are listed in the chart below.

CHECK WHAT YOU KNOW RESOURCES

Intervention Card, Copying Master, or CD-ROM
Intervention Strategies and Activities Teaching Transparencies
Intervention Practice Book
Enrichment Book

Were students successful with
CHECK WHAT YOU KNOW?

OPTIONS

NO — INTERVENE

- **ROUND FRACTIONS, 3–17**
  How to Help
  Troubleshooting, Pupil Edition p. H10
  Intervention Strategies and Activities, Skill 24

- **MENTAL MATH AND EQUATIONS, 18–25**
  How to Help
  Troubleshooting, Pupil Edition p. H11
  Intervention Strategies and Activities, Skill 41

- **FRACTIONS AND MIXED NUMBERS, 26–33**
  How to Help
  Troubleshooting, Pupil Edition p. H11
  Intervention Strategies and Activities, Skills 25–26

YES — ENRICH

Check What You Know Enrichment Book, pp. 19–20

LESSON 10.1

Estimate Products and Quotients

LESSON PLANNING

Objective To estimate products and quotients of fractions and mixed numbers

Intervention for Prerequisite Skills

Round Fractions (For intervention strategies, see page 199.)

NCTM Standards

1. Number and Operations
6. Problem Solving
8. Communication

Math Background

Students may use an estimate of fraction or mixed number products and quotients to check an answer's reasonableness. Or, they may estimate a product or quotient because that is all the problem requires.

- Products of mixed numbers can be estimated by rounding. You can round to the nearest whole number or $\frac{1}{2}$ and multiply.
- When working with fractions, products, and quotients, you can get a closer estimate by finding the average of a high estimate and a low estimate.
- Sometimes it is easiest to estimate quotients by using compatible numbers.

WARM-UP RESOURCES

 **NUMBER OF THE DAY**  Transparency 10.1

Write fractions with a denominator of 7 that are close to 0, $\frac{1}{2}$, and 1. **Possible answer: $\frac{1}{7}, \frac{4}{7}, \frac{6}{7}$**

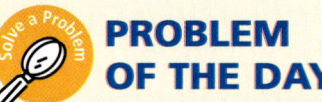 **PROBLEM OF THE DAY**  Transparency 10.1

A is a whole number between 55 and 60. The product of *A* and *B* is between 1,045 and 1,500. Between what two numbers is *B*? **between 19 and 26**

Solution Problem of the Day tab, p. PD10

 **DAILY FACTS PRACTICE**

Have students practice multiplication and division facts by completing Set B of *Teacher's Resource Book*, p. TR100.

INTERVENTION AND EXTENSION RESOURCES

ALTERNATIVE TEACHING STRATEGY

Have students **use number lines to estimate fraction quotients**, such as $8 \div \frac{3}{4}$. Ask them to draw number lines divided into fourths showing 0 through 8. Have them use the number lines to find the number of 1's in 8 and the number of $\frac{1}{2}$'s in 8, since $\frac{3}{4}$ is halfway between 1 and $\frac{1}{2}$.

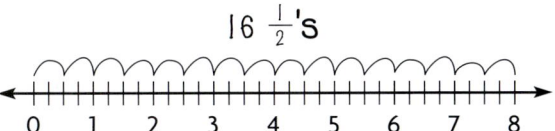

Then, to demonstrate that the average of these first two estimates, 8 and 16, is a reasonable estimate for the quotient, have students use the number line to estimate the number of $\frac{3}{4}$'s in 8. **about 11**

VISUAL
VISUAL/SPATIAL

MULTISTEP AND STRATEGY PROBLEMS

The following multistep and strategy problems are provided in Lesson 10.1:

Page	Item
201	27, 32

ENGLISH LANGUAGE LEARNERS

Review vocabulary with students by discussing and displaying the words *product, quotient, estimate, mixed number,* and *fraction.* Call on volunteers to write an example of each of the terms.

Then combine the words into phrases and ask for volunteers to write examples. For example, *estimate the product of two fractions.* Possible answer: $\frac{7}{8} \times \frac{2}{3}$ is about $\frac{1}{2}$

VISUAL
VERBAL/LINGUISTIC

EARLY FINISHERS

Materials *For each pair* 2 number cubes, labeled 1–6 and 7–12, p. TR75

Have students **practice estimating fraction products.** Working in pairs, each student rolls the cubes and writes a fraction less than 1 from the two numbers. Then students work together to estimate the product of the two fractions they have formed. Possible answer: For $\frac{4}{7} \times \frac{3}{10}$, students might estimate $\frac{1}{4}$.

KINESTHETIC
INTERPERSONAL/SOCIAL

Intervention Strategies and Activities CD-ROM • *Skill 24*

LESSON 10.1 ORGANIZER

Objective To estimate products and quotients of fractions and mixed numbers

1 Introduce

QUICK REVIEW provides review of prerequisite skills.

Why Learn This? You can estimate the number of $\frac{3}{4}$-hr activities you can schedule between 1:00 P.M. and 5:30 P.M. *Share the lesson objective with students.*

2 Teach

Guided Instruction

- Direct students' attention to the opening problem.
- Direct students to understand that 1,000 million is the same as 1 billion.

REASONING Why is an estimate an appropriate answer? *Possible answer: The number $4\frac{2}{5}$ is already an estimate, so it is reasonable to find an estimate rather than an exact answer.*

Modifying Instruction Don't discourage students from reasonable rounding, which may not always be to the nearest whole number. Had the population been $247\frac{7}{10}$ million, rounding to the nearest 10 would make the operation easier.

- Demonstrate the quotient estimate in the Example and ask:

  Why would you not estimate with average estimates if the division were $\frac{5}{6}$? *$\frac{5}{6}$ is halfway between $\frac{2}{3}$ and 1, and $\frac{2}{3}$ is not an easy divisor with which to estimate.*

ADDITIONAL EXAMPLE

Example, p. 200

Mrs. Keenan's students are decorating their classroom with a paper chain for a party. Using construction paper, they are cutting strips $\frac{3}{4}$ in. wide by 9 in. long. Lee has only one sheet of paper, which is 9 in. × 12 in. About how many loops can she make? Estimate. $12 \div \frac{3}{4}$

$12 \div 1 = 12 \qquad 12 \div \frac{1}{2} = 24$

$12 + 24 = 36; \ 36 \div 2 = 18$

So, Lee can make about 18 loops.

200 Chapter 10

LESSON 10.1 — Estimate Products and Quotients

Learn how to estimate products and quotients of fractions and mixed numbers.

QUICK REVIEW
1. 9 × 62 **558**
2. 11 × 45 **495**
3. 4.96 ÷ 10 **0.496**
4. 157 ÷ 12 **13 r1**
5. 8 × 16 **128**

In a landfill, bulldozers spread and compact the garbage into 10-foot layers. Every layer is covered with clean soil.

The landfills in more than half of the states in the United States will soon be full. It is estimated that each person in the United States produces $4\frac{2}{5}$ pounds of garbage a day. If the population of the United States was $249\frac{7}{10}$ million, about how many pounds of garbage would be produced every day?

Remember that when rounding fractions, round to 0, $\frac{1}{2}$, or 1. When rounding mixed numbers, round to the nearest whole number.

One way to estimate the answer is to round the mixed numbers to the nearest whole number.

Estimate. $4\frac{2}{5} \times 249\frac{7}{10}$

$$\begin{array}{ccc} 4\frac{2}{5} & \times & 249\frac{7}{10} \\ \downarrow & & \downarrow \\ 4 & \times & 250 = 1{,}000 \end{array}$$

Round to the nearest whole number.
THINK: $\frac{2}{5}$ rounds to 0, and $\frac{7}{10}$ rounds to 1.
Multiply.

So, about 1,000 million pounds would be produced each day.

You can also estimate by averaging two estimates.

EXAMPLE

Estimate. $8 \div \frac{3}{4}$

Since $\frac{3}{4}$ is halfway between $\frac{1}{2}$ and 1, find the two estimates and then find their average.

Round up.   *Round down.*

$8 \div \frac{3}{4} \to 8 \div 1 = 8 \qquad 8 \div \frac{3}{4} \to 8 \div \frac{1}{2} = 16$

$8 + 16 = 24; \ 24 \div 2 = 12 \qquad$ So, $8 \div \frac{3}{4}$ is about 12.

You can use compatible numbers to estimate a product or quotient.

$23\frac{3}{4} \div 4\frac{1}{2} \to 25 \div 5 = 5$

So, $23\frac{3}{4} \div 4\frac{1}{2}$ is about 5.

200 Chapter 10

RETEACH 10.1

Estimate Products and Quotients

You can often estimate the product or quotient of two mixed numbers by rounding each of them to the nearest whole number. Using a number line may help you round in the appropriate direction.

Use rounding to multiply $2\frac{3}{4} \times 10\frac{1}{8}$

Step 1 Round each mixed number to the nearest whole number.

$\begin{array}{c} 2\frac{3}{4} \times 10\frac{1}{8} \\ \downarrow \quad \downarrow \\ 3 \times 10 \end{array}$

Step 2 Multiply the rounded values.

$3 \times 10 = 30$

The product is about 30.

Use rounding to divide $12\frac{1}{4} \div 5\frac{7}{8}$

Step 1 Round each mixed number to the nearest whole number.

$\begin{array}{c} 12\frac{1}{4} \div 5\frac{7}{8} \\ \downarrow \quad \downarrow \\ 12 \div 6 \end{array}$

Step 2 Divide the rounded values.

$12 \div 6 = 2$

The quotient is about 2.

When a fraction is between 0 and 1, round it to 0, $\frac{1}{2}$, or 1, whichever is closest. Remember that you cannot divide by 0.

Use rounding to multiply $\frac{4}{5} \times 10\frac{1}{8}$

$\begin{array}{c} \frac{4}{5} \times 10\frac{1}{8} \to 1 \times 10 \end{array}$

The product is about 1 × 10, or 10.

Use rounding to divide $12\frac{1}{4} \div \frac{3}{5}$

$12\frac{1}{4} \div \frac{3}{5} \to 12 \div \frac{1}{2}$

The quotient is about $12 \div \frac{1}{2} = 12 \times 2$, or 24.

Complete the estimation of each product or quotient. **Possible answers are given.**

1. Estimate $15\frac{1}{8} \times 1\frac{7}{8}$

   $15\frac{1}{8} \times 1\frac{7}{8} \to \underline{\ 15\ } \times \underline{\ 2\ }$

   $= \underline{\ 30\ }$

2. Estimate $24\frac{5}{6} \div 4\frac{4}{5}$

   $24\frac{5}{6} \div 4\frac{4}{5} \to \underline{\ 25\ } \div \underline{\ 5\ }$

   $= \underline{\ 5\ }$

3. Estimate $\frac{3}{5} \times 48$

   $\frac{3}{5} \times 48 \to \underline{\tfrac{1}{2}} \times \underline{\ 48\ }$

   $= \underline{\ 24\ }$

4. Estimate $59\frac{7}{8} \div 15\frac{1}{8}$

   $59\frac{7}{8} \div 15\frac{1}{8} \to \underline{\ 60\ } \div \underline{\ 15\ }$

   $= \underline{\ 4\ }$

PRACTICE 10.1

Estimate Products and Quotients

Estimate each product or quotient. **Possible answers are given.**

1. $4\frac{1}{4} \times 3\frac{3}{4}$ **16**
2. $20\frac{5}{6} \div 6\frac{3}{4}$ **3**
3. $\frac{3}{4} \times \frac{5}{6}$ **1**
4. $\frac{3}{4} \div \frac{2}{3}$ **1**
5. $45\frac{1}{3} \div 8\frac{3}{8}$ **5**
6. $17\frac{2}{5} \times 1\frac{2}{7}$ **17**
7. $2\frac{3}{5} \times \frac{2}{5}$ **5**
8. $19 \times 6\frac{1}{3}$ **114**
9. $2\frac{3}{4} \times 2\frac{4}{5}$ **9**
10. $36\frac{3}{7} \div 11\frac{3}{8}$ **3**
11. $\frac{7}{9} \times 13\frac{1}{9}$ **13**
12. $\frac{1}{5} \div 20$ **0**
13. $3\frac{3}{4} \div 4\frac{1}{2}$ **1**
14. $42\frac{5}{6} \times 14\frac{4}{9}$ **630**
15. $\frac{1}{10} \times \frac{1}{10}$ **0**
16. $8\frac{1}{3} \times 6\frac{4}{5}$ **56**
17. $12\frac{5}{6} \div 3\frac{2}{5}$ **3**
18. $40\frac{2}{9} \div 7\frac{4}{5}$ **5**
19. $10\frac{5}{6} \times 3\frac{7}{8}$ **44**
20. $18\frac{3}{10} \div 1\frac{6}{7}$ **9**
21. $9\frac{3}{4} \times 17\frac{1}{5}$ **170**

Estimate to compare. Write < or > for each ●.

22. $3\frac{1}{8} \times 5$ ● $12 \div \frac{9}{10}$ **>**
23. $6\frac{1}{2} \div 12$ ● $\frac{5}{8} \div 1\frac{2}{3}$ **>**
24. $5\frac{2}{7} + 1\frac{3}{8}$ ● $2\frac{1}{8} \div 3\frac{7}{8}$ **>**
25. $3\frac{3}{4} \times 1\frac{1}{4}$ ● $31\frac{3}{4} \div 8\frac{1}{4}$ **>**
26. $15\frac{1}{5} \div 4\frac{2}{9}$ ● $1\frac{3}{4} \div 3\frac{4}{5}$ **>**
27. $7\frac{5}{9} \times 1\frac{5}{7}$ ● $36\frac{1}{2} \div 2\frac{7}{8}$ **<**

Mixed Review

Write the fraction as a percent.

28. $\frac{3}{4}$ **75%**
29. $\frac{7}{10}$ **70%**
30. $\frac{1}{20}$ **5%**
31. $\frac{3}{25}$ **12%**
32. $\frac{29}{50}$ **58%**
33. $\frac{13}{10}$ **130%**
34. $\frac{1}{8}$ **12.5%**
35. $\frac{5}{8}$ **62.5%**

CHECK FOR UNDERSTANDING

Think and Discuss — Look back at the lesson to answer each question.

1. **What if** each person in the United States produced $2\frac{1}{5}$ pounds of garbage? About how many pounds of garbage would be produced?

   1. $2 \times 250 = 500$; about 500 million pounds

2. **Tell** what compatible numbers you could use to find $81\frac{3}{5} \div 12\frac{7}{8}$.

   2. Possible answer: $84 \div 12$

Guided Practice — Estimate each product or quotient. *Possible answers are given.*

3. $\frac{7}{8} \times \frac{7}{16}$ $\frac{1}{2}$
4. $10\frac{8}{11} \div 2\frac{1}{5}$ 5
5. $78\frac{3}{7} \div 4\frac{1}{6}$ 20
6. $\frac{3}{5} \times 38$ 19
7. $1\frac{3}{4} \times 35$ 70
8. $21\frac{3}{8} \div 17\frac{1}{3}$ 1
9. $58\frac{3}{4} \times 1\frac{5}{6}$ 118
10. $98\frac{7}{8} \div 23\frac{1}{5}$ 4

PRACTICE AND PROBLEM SOLVING

Independent Practice — Estimate each product or quotient. *Possible answers are given.*

11. $\frac{7}{9} \times \frac{1}{3}$ $\frac{1}{2}$
12. $10\frac{8}{9} \times \frac{5}{6}$ 11
13. $\frac{5}{6} \div \frac{11}{12}$ 1
14. $67\frac{9}{12} \div 2\frac{7}{10}$ 23
15. $24\frac{9}{10} \div 6\frac{2}{3}$ 4
16. $36\frac{5}{8} \div 13\frac{3}{5}$ 4
17. $67\frac{2}{3} \div 23\frac{1}{8}$ 3
18. $97\frac{2}{9} \div 52\frac{5}{8}$ 2
19. $3\frac{11}{12} \times 4\frac{6}{7}$ 20
20. $12\frac{5}{24} \div \frac{8}{12}$ 24
21. $\frac{5}{9} \times \frac{7}{12}$ $\frac{1}{4}$
22. $\frac{2}{5} \div \frac{10}{21}$ 1

Problem Solving Applications — Estimate to compare. Write < or > for each ●.

23. $4\frac{1}{6} \times 3\frac{2}{3}$ ● $7\frac{5}{8} \div 2\frac{1}{3}$ >
24. $7\frac{2}{3} \div 5$ ● $2\frac{4}{8} \times 3\frac{1}{8}$ <

25. Cal runs $5\frac{3}{10}$ miles in $33\frac{4}{5}$ minutes. About how many minutes does it take for Cal to run one mile? **about 7 min**

26. 📖 **Write About It** Explain how you would estimate a quotient of mixed numbers. **Possible answer: Choose two compatible numbers about equal to the given numbers and find their quotient.**

27. Doris picked up used newspapers around her neighborhood for six days. She picked up 0.5 kg the first day, 2 kg the second day, and 3.5 kg the third day. If this pattern continued, how much newspaper did she pick up on the sixth day? **8 kg**

MIXED REVIEW AND TEST PREP

28. Write $\frac{7}{20}$ as a percent. (p. 169) **35%**
29. $5\frac{7}{9} - 2\frac{1}{3}$ (p. 186) $3\frac{4}{9}$
30. Find the mean, median, and mode for the data.
    27, 48, 83, 76, 48, 27 (p. 110) **51.5; 48; 27 and 48**
31. Evaluate the expression $m - 4^2$ for $m = 3^3$. (p. 28) **11**
32. ⭐ **TEST PREP** How much greater is the LCM of 18 and 24 than the LCM of 12 and 8? (p.150) **C**

    A 2   **B** 8   **C** 48   **D** 128

EXTRA PRACTICE page H41, Set A

201

LESSON 10.2

Multiply Fractions

LESSON PLANNING

Objective To multiply fractions

Intervention for Prerequisite Skills

Round Fractions, Fractions and Mixed Numbers (For intervention strategies, see page 199.)

NCTM Standards

1. Number and Operations
6. Problem Solving
7. Reasoning and Proof
8. Communication
10. Representation

Math Background

If you multiply a whole number by a fraction, such as $4 \times \frac{1}{2}$, you can think of it as $\frac{1}{2} + \frac{1}{2} + \frac{1}{2} + \frac{1}{2} = 2$. However, to multiply $\frac{1}{2} \times 4$ or $\frac{1}{2} \times \frac{1}{2}$, you need to think of taking the fractional part of a set or of a fraction.

- Because multiplication is commutative, $\frac{1}{2} \times 4$ is the same as $4 \times \frac{1}{2}$, or 2. So, you can think of half of 4, or one of two equal parts of 4.
- In the same way, you can think of $\frac{1}{2} \times \frac{1}{2}$ as one of two equal parts of $\frac{1}{2}$, or $\frac{1}{4}$.

Both of these problems can be illustrated on a number line or with models, leading to the common algorithm for multiplication of fractions.

WARM-UP RESOURCES

 **NUMBER OF THE DAY**

The number of the day is the total number of days in the current month. If each day this month you save a half-dollar, how much money will you have at the end of the month? Possible answer: for month with 30 days, $15

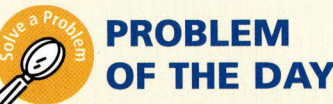 

In a jump-rope marathon, Cara earns $5 for charity for each half hour or fraction of a half hour that she jumps rope. How much money will Cara earn if she jumps rope for 175 min? $30

Solution Problem of the Day tab, p. PD10

 **DAILY FACTS PRACTICE**

Have students practice multiplication facts by completing Set C of *Teacher's Resource Book*, p. TR100.

INTERVENTION AND EXTENSION RESOURCES

ALTERNATIVE TEACHING STRATEGY ELL

Materials *For each pair* 2 number cubes, labeled 1–6 and 4–9, p. TR75

Ask student pairs to **practice multiplying fractions.**

- The pairs roll 2 number cubes twice. The players form 2 fractions.
- The first player estimates whether the product will be equal to, greater than, or less than 1.
- Students check the estimation by finding the product. If the estimation was correct, the first player earns a point.
- Have students take turns estimating the product. The first student to reach 5 points wins. Check students' work.

See also page 204.

VISUAL
LOGICAL/MATHEMATICAL

SPECIAL NEEDS

Materials *For each group* spinner with 8 sections, p. TR72

Have students **practice the multiplication algorithm for fractions.**

- Have the groups of 2 or 3 write a simple fraction in each section of the spinner, such as $\frac{1}{2}$ and $\frac{2}{3}$.
- Each group should use the spinner to form multiplication problems by spinning twice and recording each fraction.
- Then, have the groups find the products.

Check students' work.

VISUAL
LOGICAL/MATHEMATICAL

MULTISTEP AND STRATEGY PROBLEMS

The following multistep and strategy problems are provided in Lesson 10.2:

Page	Item
205	34, 35, 36, 37

WRITING IN MATHEMATICS

Have students **apply their knowledge of the multiplication of fractions** by writing problems.

- Have them work in pairs to choose a topic, such as practicing the piano or making a pizza.
- Then ask each student to write a problem and provide a solution.
- Each pair can share their problems with another pair and talk about the solutions.

Call on volunteers to share their problems with the class. Check students' work.

TECHNOLOGY LINK

 Intervention Strategies and Activities CD-ROM • *Skills 24, 25, 26*

 **Calculating Crew** • *Nautical Number Line,* Level T

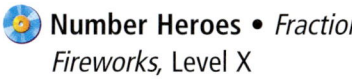 **Number Heroes** • *Fraction Fireworks,* Level X

Calculator Handbook, p. 2

202B

LESSON 10.2 ORGANIZER

Objective To multiply fractions

Lesson Resources Calculator Handbook, p. 2

1 Introduce

QUICK REVIEW provides review of prerequisite skills.

Why Learn This? You can find out how much of each of the ingredients of a recipe you will need if you cut the recipe in half. *Share the lesson objective with students.*

2 Teach

Guided Instruction

- Refer students to the model.

  Why is the paper folded into 4 equal parts? to show fourths

  Does it make sense that the product is in eighths? Explain. Yes; An eighth is half of a fourth.

- Have students explain the relationship between the factors and the product.

  How do you find the denominator of the product? multiply the denominators of the factors

REASONING Why is multiplying fractions usually easier than adding or subtracting fractions? You don't have to find a common denominator in order to multiply.

- After students make the model for $\frac{1}{3} \times \frac{3}{4}$, ask:

  What did you do differently to find $\frac{1}{3} \times \frac{3}{4}$ than to find $\frac{1}{2} \times \frac{3}{4}$? The second fold is in thirds rather than in halves.

202 Chapter 10

LESSON 10.2
Multiply Fractions

Learn how to multiply fractions.

QUICK REVIEW

Write each fraction in simplest form.

1. $\frac{8}{10}$ $\frac{4}{5}$
2. $\frac{21}{28}$ $\frac{3}{4}$
3. $\frac{36}{54}$ $\frac{2}{3}$
4. $\frac{18}{30}$ $\frac{3}{5}$
5. $\frac{12}{8}$ $1\frac{1}{2}$

Carolyn asked $\frac{3}{4}$ of her classmates what time they leave for school in the morning. Of those she asked, $\frac{1}{2}$ leave at 7:00 A.M. What fractional part of the class told her that they leave for school at 7:00 A.M.?

One way to find the fractional part of a fraction is to make a model.

Find $\frac{1}{2}$ of $\frac{3}{4}$, or $\frac{1}{2} \times \frac{3}{4}$.

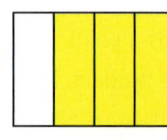

Fold a piece of paper into 4 equal parts. Shade 3 parts to show $\frac{3}{4}$.

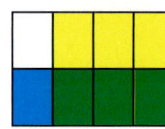

Fold the paper in half. Shade $\frac{1}{2}$ of the paper.

Of the $2 \times 4 = 8$ parts, $1 \times 3 = 3$ are shaded, so $\frac{3}{8}$ of the paper is shaded twice. These parts represent $\frac{1}{2} \times \frac{3}{4}$.

$$\frac{1}{2} \times \frac{3}{4} = \frac{3}{8}$$

So, $\frac{3}{8}$ of the students Carolyn asked leave at 7:00 A.M.

The product of numerators is the numerator of the product, and the product of the denominators is the denominator of the product.

- Compare the numerator and denominator of the product with the numerators and denominators of the factors. What relationship do you see?

You can see this relationship in the solution to the problem.

$$\frac{\text{numerator} \times \text{numerator}}{\text{denominator} \times \text{denominator}} = \frac{\text{numerator}}{\text{denominator}}$$

↑ factor   ↑ factor   ↑ product

- Make a model to find $\frac{1}{3} \times \frac{3}{4}$. Check students' models; $\frac{3}{12}$

202 Chapter 10

You can use this relationship to multiply fractions without making a model.

EXAMPLE 1

Find $\frac{1}{3} \times \frac{3}{7}$. Write the product in simplest form.

$\frac{1}{3} \times \frac{3}{7} = \frac{1 \times 3}{3 \times 7} = \frac{3}{21}$ *Multiply the numerators. Multiply the denominators.*

$= \frac{3 \div 3}{21 \div 3}$ *Divide the numerator and the denominator by the GCF, 3.*

$= \frac{1}{7}$ *Write the product in simplest form.*

So, $\frac{1}{3} \times \frac{3}{7} = \frac{1}{7}$.

Remember that to write a fraction in simplest form, divide the numerator and the denominator by the greatest common factor (GCF).

- Explain why the product, $\frac{1}{7}$, is less than the factor $\frac{3}{7}$. **Possible answer: both factors are less than 1.**

You can also multiply a whole number and a fraction without making a model.

EXAMPLE 2

Ms. Jones's car is being repaired after being in an accident. Her daughter Cindy will have to walk a total of $\frac{9}{10}$ mi to and from school every day for 11 days. How far will Cindy walk in all?

Find $11 \times \frac{9}{10}$.

Estimate. $11 \times 1 = 11$

$11 \times \frac{9}{10} = \frac{11}{1} \times \frac{9}{10}$ *Write the whole number as a fraction.*

$= \frac{11 \times 9}{1 \times 10}$ *Multiply the numerators. Multiply the denominators.*

$= \frac{99}{10}$, or $9\frac{9}{10}$ *Write the answer as a fraction or as a mixed number in simplest form.*

Compare the product to your estimate. $9\frac{9}{10}$ is close to the estimate of 11. The product is reasonable.

So, Cindy will walk $9\frac{9}{10}$ mi.

- **What if** Cindy walked to school for 21 days? How far would Cindy walk in all? **$18\frac{9}{10}$ mi**

203

- Draw students' attention to the Remember. Ask:

  How do you find the greatest common factor of two numbers? Possible answer: Find all the factors of both numbers and determine the greatest one that is a factor of both.

- As students look at Example 1, ask:

  How could you estimate the product? Round $\frac{1}{3}$ to $\frac{1}{2}$ and $\frac{3}{7}$ to $\frac{1}{2}$ and then multiply $\frac{1}{2} \times \frac{1}{2}$, which is $\frac{1}{4}$.

  Why isn't the answer $\frac{3}{21}$? $\frac{3}{21}$ isn't in simplest form.

- As students look at Example 2, ask:

  Why would you want to report the answer as a mixed number? Possible answer: The answer to a word problem is more clearly expressed as a mixed number than as a fraction greater than 1.

ADDITIONAL EXAMPLES

Example 1, p. 203

Find $\frac{4}{5} \times \frac{3}{4}$. Write the product in simplest form. $\frac{3}{5}$

Example 2, p. 203

Nathan is training for a swim meet, which will take place in 2 weeks. He swims $\frac{5}{8}$ mi 6 days a week. How far will Nathan swim before the meet? $\frac{60}{8}$, or $7\frac{1}{2}$ mi

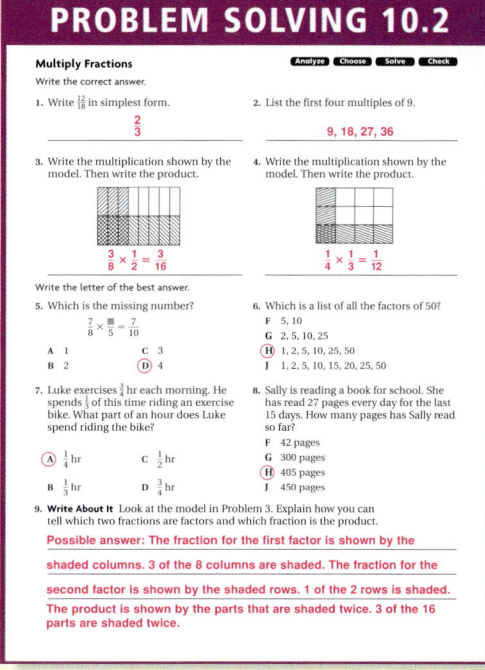

203

LESSON 10.2

- Direct students' attention to the follow-up question to Example 3.

  What is the GCF for the numerators and the denominators? 2

Modifying Instruction Show with Example 3 why you can simplify before you multiply when a numerator and a denominator have a common factor. The Identity Property of 1 allows you to divide the numerator and denominator by the GCF 2:

$$\frac{2 \times 3}{5 \times 4} = \frac{2 \times 3}{5 \times 4} \div 1 = \frac{2 \times 3}{5 \times 4} \div \frac{2}{2}$$

$$= \frac{(2 \div 2) \times 3}{5 \times (4 \div 2)} = \frac{1 \times 3}{5 \times 2} = \frac{3}{10}$$

ADDITIONAL EXAMPLES

Example 3, p. 204

Jeff has $\frac{3}{8}$ of his candy bar supply left over from the candy sale. Kristin helped by selling $\frac{5}{6}$ of his left-over bars. What part of Jeff's candy bar supply was Kristin able to sell? **Kristin sold $\frac{5}{16}$ of Jeff's candy bars.**

Example 4, p. 204

Find $\frac{2}{3} \times \frac{9}{14}$. Use the GCF to simplify the fractions before you multiply.

$$\frac{\overset{1}{\cancel{2}}}{\underset{1}{\cancel{3}}} \times \frac{\overset{3}{\cancel{9}}}{\underset{7}{\cancel{14}}} = \frac{1 \times 3}{1 \times 7} = \frac{3}{7}$$

So, $\frac{2}{3} \times \frac{9}{14} = \frac{3}{7}$.

EXAMPLE 3

When a numerator and a denominator have a common factor, you can simplify before you multiply.

Cheryl has $\frac{3}{4}$ of a box of snacks left from a school party. She gives $\frac{2}{5}$ of the snacks to the people in the school office. What part of the box of snacks does she give the people in the office?

Find $\frac{2}{5} \times \frac{3}{4}$.

Estimate. $\frac{1}{2} \times 1 = \frac{1}{2}$

$\frac{2}{5} \times \frac{3}{4}$ ← The GCF of 2 and 4 is 2.   *Look for a numerator and denominator with common factors. Find the GCF.*

$\frac{\overset{1}{\cancel{2}}}{5} \times \frac{3}{\underset{2}{\cancel{4}}}$ ← $2 \div 2 = 1$
← $4 \div 2 = 2$ *Divide the numerator and denominator by the GCF, 2.*

$\frac{\overset{1}{\cancel{2}}}{5} \times \frac{3}{\underset{2}{\cancel{4}}} = \frac{1 \times 3}{5 \times 2} = \frac{3}{10}$ *Multiply.*

So, Cheryl gives away $\frac{3}{10}$ of the box of snacks.

- What is $\frac{1}{8} \times \frac{6}{7}$? Simplify the fractions before you multiply. $\frac{3}{28}$

EXAMPLE 4

Find $\frac{8}{9} \times \frac{3}{4}$. Use the GCF to simplify the fractions before you multiply.

$\frac{8}{9} \times \frac{3}{4}$ ← The GCF of 8 and 4 is 4.
← The GCF of 3 and 9 is 3.

$\frac{\overset{2}{\cancel{8}}}{\underset{3}{\cancel{9}}} \times \frac{\overset{1}{\cancel{3}}}{\underset{1}{\cancel{4}}} = \frac{2 \times 1}{3 \times 1} = \frac{2}{3}$ *Divide the numerators and denominators by the GCFs, 3 and 4. Multiply.*

So, $\frac{8}{9} \times \frac{3}{4} = \frac{2}{3}$.

CHECK FOR UNDERSTANDING

Think and Discuss Look back at the lesson to answer each question.

1. **Explain** how to make a model to show $\frac{1}{3} \times \frac{3}{8}$. **Divide a rectangle into 8 equal parts and shade 3. Divide the (see left) rectangle into 3 equal parts and shade 1. The 3 parts shaded twice show $\frac{3}{24}$, or $\frac{1}{8}$.**

2. **Tell** how you can rewrite a whole number before you multiply it by a fraction. **write the whole number as a fraction**

Guided Practice Make a model to find the product. **Check students' models.**

3. $\frac{3}{4} \times \frac{1}{2}$ $\frac{3}{8}$
4. $\frac{1}{3} \times \frac{5}{8}$ $\frac{5}{24}$
5. $\frac{2}{5} \times \frac{1}{2}$ $\frac{2}{10}$, or $\frac{1}{5}$
6. $\frac{1}{3} \times \frac{1}{2}$ $\frac{1}{6}$

204 Chapter 10

ALTERNATIVE TEACHING STRATEGY   SCAFFOLDED INSTRUCTION

Purpose Students use fraction strips to investigate multiplication of fractions.

Materials *For each group* 3 pages of fraction strips, p. TR18

Provide each group of 4 with simple multiplication problems such as $\frac{1}{2} \times \frac{3}{4}$.

One student should trace 3 of the 4 fourths on the $\frac{1}{4}$-fraction strip.

Then have another student cut out the traced $\frac{3}{4}$ fraction strip and vertically fold it in half.

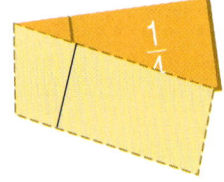

Finally, have another student match the folded section to an equivalent portion of another fraction strip, in this case showing 8 eighths.

The last student should write the problem with the solution: $\frac{1}{2} \times \frac{3}{4} = \frac{3}{8}$.

Students should take turns playing different roles modeling other similar problems.

Check students' work.

Multiply. Write the answer in simplest form.

7. $\frac{3}{4} \times \frac{2}{5}$ **$\frac{3}{10}$**   8. $\frac{2}{5} \times \frac{7}{8}$ **$\frac{7}{20}$**   9. $2 \times \frac{6}{7}$ **$\frac{12}{7}$, or $1\frac{5}{7}$**   10. $\frac{2}{3} \times 16$ **$\frac{32}{3}$, or $10\frac{2}{3}$**

11. $\frac{2}{3} \times 4$ **$\frac{8}{3}$, or $2\frac{2}{3}$**   12. $9 \times \frac{2}{3}$ **6**   13. $\frac{5}{6} \times \frac{2}{3}$ **$\frac{5}{9}$**   14. $\frac{3}{5} \times \frac{5}{6}$ **$\frac{1}{2}$**

PRACTICE AND PROBLEM SOLVING

Independent Practice

Make a model to find the product. **Check students' models.**

15. $\frac{3}{4} \times \frac{1}{4}$ **$\frac{3}{16}$**   16. $\frac{3}{4} \times \frac{2}{3}$ **$\frac{6}{12}$ or $\frac{1}{2}$**   17. $\frac{1}{8} \times \frac{1}{2}$ **$\frac{1}{16}$**   18. $3 \times \frac{1}{4}$ **$\frac{3}{4}$**

Multiply. Write the answer in simplest form.

19. $\frac{1}{3} \times \frac{2}{3}$ **$\frac{2}{9}$**   20. $\frac{3}{4} \times \frac{1}{3}$ **$\frac{1}{4}$**   21. $\frac{1}{5} \times \frac{2}{3}$ **$\frac{2}{15}$**   22. $\frac{1}{4} \times \frac{2}{7}$ **$\frac{1}{14}$**

23. $\frac{4}{5} \times \frac{7}{8}$ **$\frac{7}{10}$**   24. $\frac{2}{9} \times \frac{3}{4}$ **$\frac{1}{6}$**   25. $\frac{1}{8} \times \frac{4}{5}$ **$\frac{1}{10}$**   26. $\frac{5}{9} \times \frac{3}{10}$ **$\frac{1}{6}$**

27. $\frac{4}{9} \times \frac{3}{5}$ **$\frac{4}{15}$**   28. $\frac{2}{3} \times 21$ **14**   29. $24 \times \frac{1}{12}$ **2**   30. $\frac{1}{8} \times 16$ **$\frac{2}{1}$, or 2**

Compare. Write <, >, or = for ●.

31. $\frac{2}{9} \times \frac{3}{10}$ ● $\frac{2}{9}$ **<**   32. $\frac{5}{6} \times 5$ ● $\frac{5}{6}$ **>**   33. $8 \times \frac{1}{9}$ ● 8 **<**

Problem Solving Applications

34. Sandra takes $\frac{1}{2}$ hr to walk to school. She spends $\frac{1}{2}$ of that time walking down her street. What part of an hour does Sandra spend walking down her street? How many minutes is this? **$\frac{1}{4}$ hr; 15 min**

35. There are 144 registered voters in Booker County. In the last election, $\frac{1}{4}$ of them did not vote. How many voters did vote? **108 voters**

36.  **What's the Question?** Natalie runs $\frac{2}{3}$ the distance her brother runs in one week. Her brother runs 15 miles one week. The answer is 10 miles. **Possible question: How many miles does Natalie run?**

37. **REASONING** Scott chose a number, added 2, multiplied the sum by 4, and divided the product by 8. The final number was 4. What number had Scott chosen? **6**

MIXED REVIEW AND TEST PREP

38. $3\frac{2}{3} \times 2\frac{1}{7}$ (p. 206) **$7\frac{6}{7}$**   39. Solve. $5b = 60.45$ (p. 82) **$b = 12.09$**   40. 267.45×2.8 (p. 70) **748.86**

★ 41. **TEST PREP** The manager of a grocery store asks customers in the store on a given day to fill out a card naming their favorite cookie. Which type of sample is this? (p. 98) **A**

    A convenience   **B** biased   **C** random   **D** systematic

★ 42. **TEST PREP** Which shows the GCF for 130 and 75? (p. 150) **H**

    F 3   **G** 4   **H** 5   **J** 6

EXTRA PRACTICE page H41, Set B

---

3 Practice

COMMON ERROR ALERT

When multiplying a whole number by a fraction, some students may multiply the whole number by the denominator as well as by the numerator. Encourage students always to rewrite the whole number as a fraction before multiplying.

Error	Correction
$7 \times \frac{4}{5} = \frac{7 \times 4}{7 \times 5} = \frac{28}{35}$	$\frac{7}{1} \times \frac{4}{5} = \frac{28}{5} = 5\frac{3}{5}$

Guided Practice

Do Check for Understanding Exercises 1–14 with your students. Identify those having difficulty and use lesson resources to help.

Independent Practice

Note that Exercises 34, 35, 36, and 37 are **multistep or strategy problems**. Assign Exercises 15–37.

Scaffolded Instruction Use the prompts on Transparency 10 to guide instruction for the multistep or strategy problem in Exercise 37.

MIXED REVIEW AND TEST PREP

Exercises 38–42 provide **cumulative review** (Chapters 1–10).

4 Assess

Summarize the lesson by having students:

DISCUSS Why is the product the same whether you multiply and then simplify or simplify and then multiply? **Possible answer: In either case you divide by the same common factors.**

WRITE Why does $\frac{3}{4} \times \frac{8}{5} = \frac{3}{1} \times \frac{2}{5}$? **The denominator/numerator pair of 4 and 8 are divided by the common factor of 4, resulting in the simplified form.**

Lesson Quiz

Multiply. Write the answer in simplest form.

1. $14 \times \frac{5}{7}$ **10**   2. $\frac{7}{12} \times \frac{3}{14}$ **$\frac{1}{8}$**

3. $\frac{6}{7} \times \frac{7}{8}$ **$\frac{3}{4}$**   4. $\frac{3}{4} \times \frac{1}{6}$ **$\frac{1}{8}$**

LESSON 10.3

Multiply Mixed Numbers

LESSON PLANNING

Objective To multiply mixed numbers

Intervention for Prerequisite Skills

Fractions and Mixed Numbers (For intervention strategies, see page 199.)

NCTM Standards

1. Number and Operations
6. Problem Solving
8. Communication

Math Background

The rule for multiplying mixed numbers is the same as that for multiplying fractions, so students must first write each mixed number as a fraction and then apply the rule for multiplying fractions.

Consider the following as you help students understand how to multiply mixed numbers:

- The estimate is important because it provides a check of how reasonable your answer is. If you make a mistake writing a mixed number as a fraction or simplifying the answer, the check may alert you to the error.
- Before you multiply numerators and denominators, apply the GCF process so that the answer will be in simplest form.
- The solution is not complete until the answer is written in simplest form or as a whole or mixed number.

WARM-UP RESOURCES

 **NUMBER OF THE DAY**  Transparency 10.3

Write your age in years and months as a mixed number. Then express it as a fraction greater than one. **Possible answer for 11 yr, 9 mo: $11\frac{3}{4}, \frac{47}{4}$**

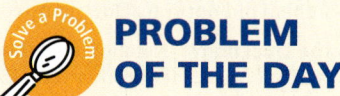 **PROBLEM OF THE DAY**  Transparency 10.3

It took André 1 min to fill his aquarium $\frac{1}{3}$ full. How long will it take him to fill the aquarium $\frac{3}{4}$ full? **$2\frac{1}{4}$ min**

Solution Problem of the Day tab, p. PD10

 **DAILY FACTS PRACTICE**

Have students practice multiplication facts by completing Set D of *Teacher's Resource Book,* p. TR100.

INTERVENTION AND EXTENSION RESOURCES

ALTERNATIVE TEACHING STRATEGY — ELL

Materials *For each pair* eight $\frac{1}{2}$-fraction strips, p. TR18

Have students **model multiplication of mixed numbers.** To illustrate $2\frac{1}{2} \times 1\frac{3}{5}$, or $\frac{5}{2} \times \frac{8}{5}$, have students use fraction strips to model $\frac{5}{2}$. Then, ask them to model $\frac{1}{5}$ of $\frac{5}{2}$, which is $\frac{1}{2}$. Finally, have them model 8 of the $\frac{1}{2}$ portions, which is $8 \times \frac{1}{2}$, or 4.

Students can also use fraction strips to model the Example on page 206. Have them compare their modeling to the use of the Distributive Property and the renaming of the mixed number in the example. Check students' work.

VISUAL
BODILY/KINESTHETIC

CAREER CONNECTION

Present this information and have students **solve the problem below by multiplying by a mixed number.**

A political campaign manager arranges for the press conferences, coffees, dinners, and other campaign appearances the candidate makes. He or she may also help plan the advertising that will be done and be responsible for being sure the necessary local, state, and federal reports are filed.

A campaign manager is paid $28 per hour. She works for $50\frac{3}{4}$ hr one week. How much does she make that week? $1,421

AUDITORY
VERBAL/LINGUISTIC

MULTISTEP AND STRATEGY PROBLEMS

The following multistep or strategy problem is provided in Lesson 10.3:

Page	Item
207	26

ADVANCED LEARNERS

Challenge students to discover the error in the **use of the Distributive Property.** Tell students that Mai Ling used the Distributive Property to multiply $6\frac{1}{2} \times 5\frac{1}{2}$ and got $30\frac{1}{4}$ as an answer. Challenge them to show whether she is correct. No; $(6 + \frac{1}{2}) \times (5 + \frac{1}{2}) = 6 \times (5 + \frac{1}{2}) + \frac{1}{2} \times (5 + \frac{1}{2}) = (6 \times 5) + (6 \times \frac{1}{2}) + (\frac{1}{2} \times 5) + (\frac{1}{2} \times \frac{1}{2}) = 30 + 3 + 2\frac{1}{2} + \frac{1}{4} = 35\frac{3}{4}$

AUDITORY
LOGICAL/MATHEMATICAL

TECHNOLOGY LINK
Intervention Strategies and Activities CD-ROM •
Skills 25, 26

LESSON 10.3 ORGANIZER

Objective To multiply mixed numbers

1 Introduce

QUICK REVIEW provides review of prerequisite skills.

Why Learn This? You will be able to find the perimeter of squares and the area of rectangles whose sides are given as mixed numbers. *Share the lesson objective with students.*

2 Teach

Guided Instruction

- Have students consider the opening problem.
  How is multiplying two mixed numbers like multiplying two fractions? *Once the mixed numbers are written as fractions, the process is the same. Divide by the GCFs as needed, multiply the numerators, and multiply the denominators.*

- Guide students to compare the relative values of products and their factors when multiplying fractions and mixed numbers.
  Describe the size of the factors in relation to the product in each of the following: $\frac{1}{2} \times \frac{3}{4} = \frac{3}{8}$, $3 \times \frac{1}{2} = 1\frac{1}{2}$, $1\frac{1}{3} \times 2\frac{1}{4} = 3$. *The product of two fractions is less than either factor. The product of a whole number and a fraction is less than the whole number and greater than the fraction. The product of two mixed numbers is greater than either factor.*

Algebraic Thinking Point out to students that it is possible to use the Distributive Property to multiply the two mixed numbers from the opening problem.

$2\frac{1}{2} \times 3\frac{1}{5} = (2 \times 3) + (2 \times \frac{1}{5}) + (\frac{1}{2} \times 3) + (\frac{1}{2} \times \frac{1}{5})$
$\phantom{2\frac{1}{2} \times 3\frac{1}{5}} = 6 + \frac{2}{5} + \frac{3}{2} + \frac{1}{10}$
$\phantom{2\frac{1}{2} \times 3\frac{1}{5}} = 8$

ADDITIONAL EXAMPLE

Example, p. 206

Multiply. $4 \times 2\frac{7}{9}$

$4 \times 2\frac{7}{9} = (4 \times 2) + (4 \times \frac{7}{9})$
$\phantom{4 \times 2\frac{7}{9}} = 8 + \frac{28}{9} = 8 + 3\frac{1}{9} = 11\frac{1}{9}$
So, $4 \times 2\frac{7}{9} = 11\frac{1}{9}$

206 Chapter 10

LESSON 10.3
Multiply Mixed Numbers

Learn how to multiply mixed numbers.

QUICK REVIEW
Write the missing numerator.
1. $1\frac{2}{3} = \frac{\blacksquare}{3}$ 5
2. $6\frac{3}{5} = \frac{\blacksquare}{5}$ 33
3. $7\frac{3}{7} = \frac{\blacksquare}{7}$ 52
4. $9\frac{3}{8} = \frac{\blacksquare}{8}$ 75
5. $3\frac{2}{5} = \frac{\blacksquare}{5}$ 17

Remember that you can write a mixed number as a fraction.

$2\frac{3}{4} = \frac{(2 \times 4) + 3}{4}$
$\phantom{2\frac{3}{4}} = \frac{11}{4}$

Ann and Sheri are training for a bicycle race. On one day, Ann rides $3\frac{1}{5}$ mi. Sheri rides $2\frac{1}{2}$ times as far as Ann. How many miles does Sheri ride?

Find $2\frac{1}{2} \times 3\frac{1}{5}$. Estimate. $3 \times 3 = 9$

$2\frac{1}{2} \times 3\frac{1}{5} = \frac{5}{2} \times \frac{16}{5}$ *Write the mixed numbers as fractions.*

$\phantom{2\frac{1}{2} \times 3\frac{1}{5}} = \frac{\cancel{5}^1}{\cancel{2}_1} \times \frac{\cancel{16}^8}{\cancel{5}_1}$ *Simplify the fractions. Multiply.*

$\phantom{2\frac{1}{2} \times 3\frac{1}{5}} = \frac{8}{1}$, or 8   *Write the answer in simplest form or as a whole or mixed number.*

So, Sheri rides 8 mi. The answer is reasonable since it is close to the estimate of 9 mi.

You can use the Distributive Property to multiply a whole number by a mixed number.

EXAMPLE

Multiply. $5 \times 2\frac{3}{8}$

$5 \times 2\frac{3}{8} = 5 \times (2 + \frac{3}{8})$

$\phantom{5 \times 2\frac{3}{8}} = (5 \times 2) + (5 \times \frac{3}{8})$ *Use the Distributive Property.*

$\phantom{5 \times 2\frac{3}{8}} = (5 \times 2) + (\frac{5}{1} \times \frac{3}{8})$ *Write the whole number as a fraction. Find 5×2 and $\frac{5}{1} \times \frac{3}{8}$.*

$\phantom{5 \times 2\frac{3}{8}} = 10 + \frac{15}{8}$

$\phantom{5 \times 2\frac{3}{8}} = 10 + 1\frac{7}{8} = 11\frac{7}{8}$ *Write the fraction as a mixed number. Find the sum.*

So, $5 \times 2\frac{3}{8} = 11\frac{7}{8}$.

206 Chapter 10

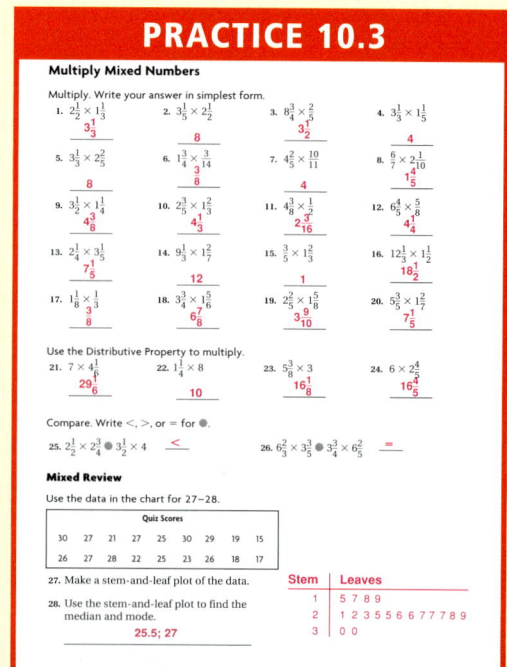

RETEACH 10.3
Multiply Mixed Numbers

Barbara bought $2\frac{1}{4}$ dozen donuts. Her family ate $\frac{2}{3}$ of them. How many dozen did her family eat?

Step 1 What is $\frac{2}{3}$ of $2\frac{1}{4}$?
Find $\frac{2}{3} \times 2\frac{1}{4}$.

Step 2 Write the mixed number as a fraction.

Step 3 Use the GCF to simplify.
The GCF of 2 and 4 is 2.
The GCF of 3 and 9 is 3.

Step 4 Multiply.

So, Barbara's family ate $1\frac{1}{2}$ dozen.

PRACTICE 10.3
Multiply Mixed Numbers

Multiply. Write your answer in simplest form.

CHECK FOR UNDERSTANDING

Think and Discuss Look back at the lesson to answer each question.

1. $3 \times 4\frac{2}{3} = (3 \times 4) +$
$(3 \times \frac{2}{3}); 12 + 2 = 14.$
1. **Show** how to use the Distributive Property to find $3 \times 4\frac{2}{3}$.

2. **Discuss** whether the product of two mixed numbers is greater than or less than the factors. Give two examples.
greater than; Possible example: $4\frac{1}{2} \times 6\frac{1}{3} = 28\frac{1}{2}, 3\frac{2}{3} \times 5\frac{3}{5} = 20\frac{8}{15}$

Guided Practice Multiply. Write your answer in simplest form.

3. $\frac{3}{4} \times 1\frac{1}{2}$ $1\frac{1}{8}$
4. $\frac{1}{2} \times 2\frac{1}{3}$ $1\frac{1}{6}$
5. $1\frac{1}{2} \times 1\frac{1}{2}$ $2\frac{1}{4}$
6. $1\frac{2}{5} \times 2\frac{1}{4}$ $3\frac{3}{20}$

Use the Distributive Property to multiply.

7. $6\frac{1}{8} \times 3$ $18\frac{3}{8}$
8. $3 \times 9\frac{4}{5}$ $29\frac{2}{5}$
9. $1\frac{1}{8} \times 2$ $2\frac{1}{4}$
10. $6 \times 4\frac{1}{4}$ $25\frac{1}{2}$

PRACTICE AND PROBLEM SOLVING

Independent Practice Multiply. Write your answer in simplest form.

11. $4\frac{2}{3} \times 1\frac{3}{4}$ $8\frac{1}{6}$
12. $1\frac{3}{8} \times 4\frac{2}{3}$ $6\frac{5}{12}$
13. $5\frac{1}{2} \times 6$ 33
14. $2 \times 3\frac{1}{7}$ $6\frac{2}{7}$

15. $4\frac{1}{6} \times 3\frac{3}{5}$ 15
16. $1\frac{3}{4} \times 3$ $5\frac{1}{4}$
17. $10\frac{1}{5} \times 8\frac{1}{3}$ 85
18. $5 \times 1\frac{5}{6}$ $9\frac{1}{6}$

Use the Distributive Property to multiply.

19. $3 \times 2\frac{2}{5}$ $7\frac{1}{5}$
20. $4 \times 8\frac{5}{6}$ $35\frac{1}{3}$
21. $3\frac{3}{4} \times 6$ $22\frac{1}{2}$
22. $1\frac{1}{2} \times 12$ 18

Compare. Write <, >, or = for each ●.

23. $3\frac{1}{3} \times 2\frac{1}{7}$ ● $3\frac{1}{4} \times 5$ <
24. $7 \times 7\frac{3}{7}$ ● $6\frac{3}{4} \times 4\frac{4}{5}$ >

Problem Solving Applications

25. Mr. Jackson rides his bicycle $1\frac{2}{3}$ mi every day. His wife rides $1\frac{1}{4}$ times as far as he does. How many miles does Mrs. Jackson ride her bicycle? $2\frac{1}{12}$ mi

26. John works part time for $6.50 an hour. He works $3\frac{1}{2}$ hr each on Monday, Tuesday, and Thursday afternoons. How much does he earn those three days? $68.25

27. **Write About It** Without multiplying, tell whether the product $\frac{2}{3} \times \frac{3}{4}$ is a fraction, a whole number, or a mixed number.
Since both factors are fractions less than 1, the product is a fraction.

MIXED REVIEW AND TEST PREP

28. $\frac{4}{9} \times \frac{2}{3}$ (p. 202) $\frac{8}{27}$
29. $8\frac{3}{4} - 2\frac{2}{5}$ (p. 186) $6\frac{7}{20}$
30. Simplify. $\frac{27}{45}$ (p. 160) $\frac{3}{5}$
31. $414,089 - 62,036$ (p. 20) $352,053$

★ 32. **TEST PREP** Which is the mean of the data? (p. 110)
90, 94, 65, 90, 84, 94, 85 **B**

A 29   B 86   C 90   D 94

EXTRA PRACTICE page H41, Set C

207

3 Practice

Guided Practice

Do Check for Understanding Exercises 1–10 with your students. Identify those having difficulty and use lesson resources to help.

Independent Practice

Note that Exercise 26 is a **multistep or strategy problem**. Assign Exercises 11–27.

Encourage students to estimate each product to check for reasonableness after they multiply.

MIXED REVIEW AND TEST PREP
Exercises 28–32 provide **cumulative review** (Chapters 1–10).

4 Assess

Summarize the lesson by having students:

DISCUSS What steps did you follow to find the answer for Exercise 12? Write the mixed numbers as fractions, $\frac{11}{8}$ and $\frac{14}{3}$. Divide by the GCF, multiply the numerators, and multiply the denominators: $\frac{11}{4} \times \frac{7}{3} = \frac{77}{12}$. Write as a mixed number: $6\frac{5}{12}$.

WRITE Explain how to use the Distributive Property to multiply a whole number and a mixed number.
Possible answer: First multiply the whole number times the whole number part of the mixed number. Then write the whole number as a fraction and multiply that times the fraction part of the mixed number. Find the sum of the two products.

Lesson Quiz

Transparency 10.3

Multiply. Write your answer in simplest form.

1. $3 \times \frac{3}{8}$ $1\frac{1}{8}$
2. $4\frac{1}{3} \times 2\frac{2}{5}$ $10\frac{2}{5}$
3. $5\frac{5}{6} \times 2\frac{1}{4}$ $13\frac{1}{8}$
4. $6 \times 3\frac{1}{3}$ 20

CHALLENGE 10.3

Fraction Analogies

PROBLEM SOLVING 10.3

Multiply Mixed Numbers

LESSON 10.4

ORGANIZER

Objective To model division of fractions

Vocabulary reciprocal

Materials *For each student* fraction circles, pp. TR19–20

Lesson Resources E-Lab Recording Sheet • *Exploring Division of Fractions*

Intervention for Prerequisite Skills Fractions and Mixed Numbers (For intervention strategies, see page 199.)

Using the Pages

Encourage students to discuss times they have used fractions or division of fractions, such as with crafts and cooking.

Activity 1

Students expect the quotient to be less than the dividend. This thinking must be revised with the division of fractions. The modeling process shows that the quotient is greater than the dividend when dividing a whole number by a fraction.

REASONING How many $\frac{1}{3}$'s of a circle do you need to make $\frac{1}{6}$? only $\frac{1}{2}$ of a $\frac{1}{3}$ section

Think and Discuss

As they discuss the meaning of $2 \div \frac{1}{6} = 12$, you may want to have students model the division. Then ask: **How can you use the inverse to write the division sentence as a multiplication sentence?** $12 \times \frac{1}{6} = 2$

Practice

Even though students may know the answers to the first two exercises without modeling, encourage them to model the divisions. This ability will help them solve more complex problems.

LESSON 10.4 Division of Fractions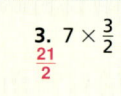

Explore how to model division of fractions.

You need fraction circles.

QUICK REVIEW
1. $3 \times \frac{1}{3}$ 1
2. $\frac{1}{4} \times 4$ 1
3. $7 \times \frac{3}{2}$ $\frac{21}{2}$
4. $2 \times \frac{8}{12}$ $1\frac{1}{3}$
5. $\frac{2}{5} \times \frac{5}{2}$ 1

Vocabulary
reciprocal

TECHNOLOGY LINK
More Practice: Use E-Lab, *Exploring Division of Fractions.*
www.harcourtschool.com/elab2002

Using models will help you understand division of fractions.

Activity 1

A. Use fraction circles to find $4 \div \frac{1}{3}$, or the number of thirds in 4 wholes.

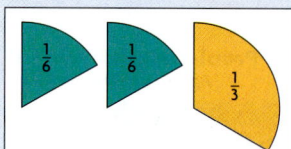

• Trace 4 whole circles on your paper.
• Model $4 \div \frac{1}{3}$ by tracing $\frac{1}{3}$-circle pieces on the 4 circles.

One whole equals three thirds.

• How many thirds are in 4 wholes? What is $4 \div \frac{1}{3}$? **12; 12**

B. Use fraction circles to find $\frac{1}{3} \div \frac{1}{6}$, or the number of sixths in $\frac{1}{3}$.

• Place as many $\frac{1}{6}$ pieces as you can on the $\frac{1}{3}$ piece.
• How many sixths are in $\frac{1}{3}$? What is $\frac{1}{3} \div \frac{1}{6}$? **2; 2**

Think and Discuss

• If $4 \div \frac{1}{3} = 12$, explain what $8 \div \frac{1}{3}$ must be. **24**
• If $2 \div \frac{1}{6} = 12$, explain what $8 \div \frac{1}{6}$ must be. **48**

Practice

Use fraction circles to model each problem. Draw a diagram of your model. **Check students' models.**

1. $3 \div \frac{1}{3}$ **9**
2. $4 \div \frac{1}{4}$ **16**
3. $\frac{1}{2} \div \frac{1}{8}$ **4**
4. $\frac{3}{4} \div \frac{1}{8}$ **6**

208 Chapter 10

ALTERNATIVE TEACHING STRATEGY

Materials *For each group* play money— 5 one-dollar bills, 10 half-dollars, and 20 quarters

Have students **use play money to model division with fractions.** Have students identify what part of a dollar each coin is. Then present problems such as $3 \div \frac{1}{2}$ and $4 \div \frac{1}{4}$. Have volunteers model the divisions with the play money, first showing the amounts with dollars and then using the coins to represent the fractions.

KINESTHETIC
LOGICAL/MATHEMATICAL

Intervention and Extension Resources

WRITING IN MATHEMATICS

Have students **describe applications of fractions** by writing a description of how a person in one of various occupations would use fractions in his or her work.

Occupations: stockbroker, chef, tailor, dietitian, fitness trainer, musician

Check students' work.

VISUAL
VERBAL/LINGUISTIC

Two numbers are **reciprocals** if their product is 1.

$$\frac{1}{2} \times 2 = 1 \qquad \frac{3}{4} \times \frac{4}{3} = 1 \qquad 6 \times \frac{1}{6} = 1$$
$$\uparrow \quad \uparrow \qquad\qquad \uparrow \quad \uparrow \qquad\qquad \uparrow \quad \uparrow$$
$$\text{reciprocals} \qquad \text{reciprocals} \qquad \text{reciprocals}$$

By using inverse operations, you can write related number sentences.

$$1 \div \frac{1}{2} = 2 \qquad 1 \div \frac{3}{4} = \frac{4}{3} \qquad 1 \div 6 = \frac{1}{6}$$
$$1 \div 2 = \frac{1}{2} \qquad 1 \div \frac{4}{3} = \frac{3}{4} \qquad 1 \div \frac{1}{6} = 6$$

You can use reciprocals and inverse operations when you divide.

Activity 2

• Study these problems.

Find $6 \div \frac{1}{2}$.
$1 \div \frac{1}{2} = 2$ *Think of the reciprocal of $\frac{1}{2}$.*
Since $1 \div \frac{1}{2} = 1 \times 2$, $6 \div \frac{1}{2} = 6 \times 2$.
So, $6 \div \frac{1}{2} = 6 \times 2 = 12$.

Find $\frac{3}{4} \div \frac{1}{3}$.
$1 \div \frac{1}{3} = 3$ *Think of the reciprocal of $\frac{1}{3}$.*
Since $1 \div \frac{1}{3} = 1 \times 3$, $\frac{3}{4} \div \frac{1}{3} = \frac{3}{4} \times 3$.
So, $\frac{3}{4} \div \frac{1}{3} = \frac{3}{4} \times 3 = \frac{9}{4}$, or $2\frac{1}{4}$.

Think and Discuss

- When you divide 1 by a number, what is the quotient? *the reciprocal of the number*
- If $1 \div \frac{1}{3} = 3$, then what is $2 \div \frac{1}{3}$? What is $5 \div \frac{1}{3}$? *$2 \times 3 = 6$; $5 \times 3 = 15$*
- If $1 \div \frac{2}{5} = \frac{5}{2}$, then what is $2 \div \frac{2}{5}$? What is $\frac{1}{2} \div \frac{2}{5}$? *$2 \times \frac{5}{2} = 5$; $\frac{1}{2} \times \frac{5}{2} = \frac{5}{4}$, or $1\frac{1}{4}$*

Practice

Find the value of n.

1. If $1 \div \frac{1}{2} = 2$, then $2 \div \frac{1}{2} = n$. *n = 4*
2. If $1 \div \frac{4}{5} = \frac{5}{4}$, then $4 \div \frac{4}{5} = n$. *n = 5*
3. If $1 \div \frac{2}{3} = \frac{3}{2}$, then $\frac{1}{2} \div \frac{2}{3} = n$. *n = $\frac{3}{4}$*
4. If $1 \div \frac{3}{5} = \frac{5}{3}$, then $\frac{3}{4} \div \frac{3}{5} = n$. *n = $\frac{5}{4}$, or $1\frac{1}{4}$*

Find the quotient.

5. $6 \div \frac{3}{4}$ *8*
6. $4 \div \frac{1}{2}$ *8*
7. $\frac{1}{2} \div \frac{1}{3}$ *$\frac{3}{2}$, or $1\frac{1}{2}$*
8. $\frac{2}{3} \div \frac{1}{8}$ *$\frac{16}{3}$, or $5\frac{1}{3}$*

MIXED REVIEW AND TEST PREP

9. $2\frac{4}{5} \times 3\frac{3}{8}$ (p. 206) *$9\frac{9}{20}$*
10. Complete. $\frac{3}{8} = \frac{\blacksquare}{24}$ (p. 160) *9*
11. Compare 606.64 and 606.074. Use <, >, or =. (p. 52) *606.64 > 606.074*
12. $46.08 - 19.204$ (p. 66) *26.876*
★13. **TEST PREP** How many times greater is the GCF of 15 and 18 than the GCF of 25 and 33? (p. 150) *B*

   A 2   B 3   C 5   D 6

209

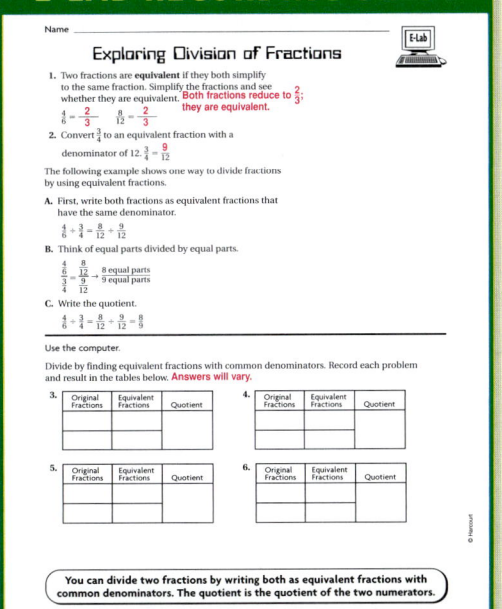

USING E-LAB

Students use a model as they develop a new algorithm for division of fractions.

E-Lab Recording Sheets and activities are available on the E-Lab website.

www.harcourtschool.com/elab2002

TECHNOLOGY LINK

- **Intervention Strategies and Activities CD-ROM** • *Skills 25, 26*
- **E-Lab** • *Exploring Division of Fractions*
- **Math Jingles™ CD 5–6** • *Track 15*

LESSON 10.5

Divide Fractions and Mixed Numbers

LESSON PLANNING

Objective To divide with fractions and mixed numbers

Intervention for Prerequisite Skills

Fractions and Mixed Numbers (For intervention strategies, see page 199.)

NCTM Standards
1. Number and Operations
2. Algebra
6. Problem Solving
8. Communication
9. Connections

Math Background

The first step in dividing one fraction by another is to rewrite the problem as a multiplication problem in which the divisor is replaced by its reciprocal. As with the rule for multiplication of fractions and of mixed numbers, the rule for division of fractions applies to dividing with mixed numbers.

Consider the following as you help students master the division algorithm for fractions.

- Before applying the division algorithm, both the dividend and the divisor must be written in fraction form.
- The reciprocal of the divisor is formed by exchanging the numerator with the denominator.
- An estimate is especially helpful because it can alert you to a possible error in rewriting the divisor.

WARM-UP RESOURCES

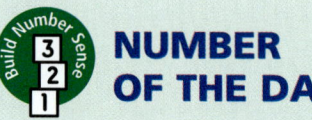 

Take your age and multiply it by $2\frac{1}{2}$. **Possible answer:** $12 \times 2\frac{1}{2} = 30$

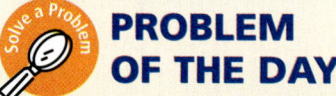 

A hawk flies $\frac{1}{3}$ mi in 30 sec. How far can the hawk fly in 1 min? How fast does it fly in miles per hour? **$\frac{2}{3}$ mi; 40 mph**

Solution Problem of the Day tab, p. PD10

Have students practice division facts by completing Set F of *Teacher's Resource Book*, p. TR100.

INTERVENTION AND EXTENSION RESOURCES

ALTERNATIVE TEACHING STRATEGY

Materials *For each pair* 6 index cards

Have students work in pairs to **practice the division algorithm for fractions.** Have each pair choose 3 exercises from Exercises 26–37, page 212, and copy the exercises one to a card. On the remaining 3 cards, have them write the multiplication step showing the divisor's reciprocal.

Have pairs trade their sets of cards. Ask students to work together to match the divisions with the multiplications and then solve the problems. Check students' work.

See also page 212.

VISUAL
LOGICAL/MATHEMATICAL

ENGLISH LANGUAGE LEARNERS ELL

Review with students the steps for **dividing one mixed number by another.** Display the steps below and have students illustrate each step with an example.

- Write the mixed numbers as fractions.
- Use the reciprocal of the divisor to write the problem as a multiplication problem.
- Simplify.
- Multiply.
- Write the answer in simplest form.

Ask volunteers to present their examples to the class.

Check students' work.

AUDITORY
VERBAL/LINGUISTIC

MULTISTEP AND STRATEGY PROBLEMS

The following multistep and strategy problems are provided in Lesson 10.5:

Page	Item
213	53

ALGEBRA CONNECTION

Challenge students to **solve mixed number and fraction riddles** using mental math. Call on volunteers to explain their answers.

- What number times $3\frac{1}{2}$ is $10\frac{1}{2}$? **3**
- $\frac{3}{4}$ divided by what number is $\frac{15}{16}$? **$\frac{4}{5}$**
- What number times $\frac{7}{10}$ is 1? **$\frac{10}{7}$, or $1\frac{3}{7}$**
- $3\frac{1}{3}$ divided by what number is 3? **$1\frac{1}{9}$**

AUDITORY
LOGICAL/MATHEMATICAL

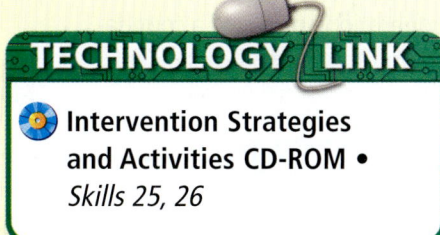

Intervention Strategies and Activities CD-ROM •
Skills 25, 26

LESSON 10.5 ORGANIZER

Objective To divide with fractions and mixed numbers

1 Introduce

QUICK REVIEW provides review of prerequisite skills.

Why Learn This? You can use division of fractions to find out how many half-cup servings are in a container of juice. *Share the lesson objective with students.*

2 Teach

Guided Instruction

- Direct students' attention to the Math Idea. Ask:

  How does the reciprocal relate to this method? You multiply the reciprocal of the divisor by the dividend.

  Review with students the meaning of reciprocal.

  What is the product of a number and its reciprocal? 1

- Work through Example 1. Then display for students a follow-up problem, $\frac{1}{8} \div \frac{4}{5}$.

  For which fraction do you need to write the reciprocal? $\frac{4}{5}$

Algebraic Thinking Using the reciprocal of the divisor to write a multiplication problem provides the opportunity to teach students about inverse operations as they apply to dividing with fractions. Understanding the relationship between inverse operations provides a foundation for understanding the process of solving equations.

ADDITIONAL EXAMPLE

Example 1, p. 210

Find $\frac{7}{8} \div \frac{1}{3}$.

$\frac{7}{8} \div \frac{1}{3} = \frac{7}{8} \times \frac{3}{1}$

$= \frac{21}{8}$, or $2\frac{5}{8}$

So, $\frac{7}{8} \div \frac{1}{3} = 2\frac{5}{8}$.

LESSON 10.5

Divide Fractions and Mixed Numbers

Learn how to divide with fractions and mixed numbers.

QUICK REVIEW

Write each mixed number as a fraction.

1. $2\frac{1}{3}$ $\frac{7}{3}$ 2. $1\frac{3}{8}$ $\frac{11}{8}$ 3. $4\frac{1}{2}$ $\frac{9}{2}$

4. $3\frac{9}{10}$ $\frac{39}{10}$ 5. $6\frac{2}{3}$ $\frac{20}{3}$

Stacey is bringing juice to be served after the school council meeting. Stacey said she would bring enough juice for everyone at the meeting to have a glass. Each glass holds $\frac{1}{5}$ liter. How many glasses can be served if Stacey brings 8 liters of juice?

Find $8 \div \frac{1}{5}$.

Math Idea ▶ When you divide by a fraction, you can use multiplication to find the quotient.

Rewrite the division problem as a multiplication problem by using the reciprocal.

$8 \div \frac{1}{5} = \frac{8}{1} \div \frac{1}{5}$ *Write the whole number as a fraction.*

$= \frac{8}{1} \times \frac{5}{1}$ *Use the reciprocal of the divisor to write a multiplication problem.*

$= \frac{8}{1} \times \frac{5}{1} = \frac{40}{1}$, or 40   *Multiply.*

So, 8 liters can fill 40 glasses.

You can also use the reciprocal of the divisor to divide fractions and mixed numbers.

EXAMPLE 1

Find $\frac{2}{3} \div \frac{4}{7}$.

$\frac{2}{3} \div \frac{4}{7} = \frac{2}{3} \times \frac{7}{4}$ *Use the reciprocal of the divisor to write a multiplication problem.*

$= \frac{\overset{1}{2}}{3} \times \frac{7}{\underset{2}{4}}$ *Divide the numerator and denominator by the GCF, 2.*

$= \frac{7}{6}$, or $1\frac{1}{6}$ *Multiply.*

So, $\frac{2}{3} \div \frac{4}{7} = 1\frac{1}{6}$.

210 Chapter 10

RETEACH 10.5

Divide Fractions and Mixed Numbers

PRACTICE 10.5

Divide Fractions and Mixed Numbers

EXAMPLE 2

Each member of the school council will write his or her name on a strip of paper at the meeting. Jeremy has pieces of paper $5\frac{1}{4}$ in. long. Each strip of paper should be $1\frac{3}{4}$ in. How many strips can Jeremy cut from the length of one piece of paper?

Find $5\frac{1}{4} \div 1\frac{3}{4}$.

Estimate. $5 \div 2 = 2\frac{1}{2}$

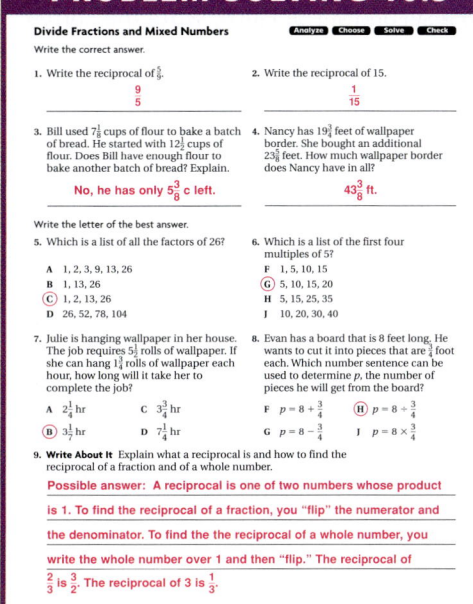

$5\frac{1}{4} \div 1\frac{3}{4} = \frac{21}{4} \div \frac{7}{4}$ *Write the mixed numbers as fractions.*

$= \frac{21}{4} \times \frac{4}{7}$ *Use the reciprocal of the divisor to write a multiplication problem.*

$= \frac{\overset{3}{\cancel{21}}}{\underset{1}{\cancel{4}}} \times \frac{\overset{1}{\cancel{4}}}{\underset{1}{\cancel{7}}}$ *Simplify and multiply.*

$= \frac{3}{1}$, or 3

Compare the product to your estimate. 3 is close to the estimate of $2\frac{1}{2}$. The product is reasonable.

So, Jeremy can cut 3 strips from each piece of paper.

• What is $2\frac{3}{4} \div 1\frac{2}{5}$? **$1\frac{27}{28}$**

Sometimes you can use mental math to divide whole numbers and fractions.

EXAMPLE 3

Use mental math to solve.

A $9 \div \frac{1}{2}$ **THINK:** $9 \times 2 = 18$.   *Dividing by $\frac{1}{2}$ is the same as multiplying by 2.*
So, $9 \div \frac{1}{2} = 18$.   *There are 18 halves in 9.*

B $13 \div \frac{1}{3}$ **THINK:** $13 \times 3 = 39$.   *Dividing by $\frac{1}{3}$ is the same as multiplying by 3.*
So, $13 \div \frac{1}{3} = 39$.   *There are 39 thirds in 13.*

C $20 \div \frac{2}{5}$ **THINK:** $20 \times 5 = 100$.   *Dividing by $\frac{2}{5}$ is the same as multiplying by $\frac{5}{2}$.*
$100 \div 2 = 50$.
So, $20 \div \frac{2}{5} = 50$.

• Use mental math to find $15 \div \frac{1}{6}$. **90**

211

• Have students consider the word problem for Example 2.

Which mixed number is the dividend and which is the divisor? $5\frac{1}{4}$ is the dividend and $1\frac{3}{4}$ is the divisor.

In the follow-up problem, how would you rewrite the mixed numbers? $2\frac{3}{4}$ as $\frac{11}{4}$, and $1\frac{2}{5}$ as $\frac{7}{5}$

• Encourage students to explain in their own words the mental math in Example 3.

How could you describe dividing by a unit fraction in terms of multiplying? Possible answer: Multiply by the denominator of the unit fraction.

REASONING **How could you use mental math to divide 9 by $\frac{3}{2}$?** Think $9 \times 2 = 18$. Then $18 \div 3 = 6$.

ADDITIONAL EXAMPLES

Example 2, p. 211

The recipe for Thirst Ender drink requires $1\frac{1}{2}$ cups of sugar for each gallon of water. If you have only 9 cups of sugar, how many gallons of Thirst Ender can you make? **6 gallons**

Example 3, p. 211

Use mental math to solve.

A. $12 \div \frac{1}{5}$ Think: $12 \times 5 = 60$.
So, $12 \div \frac{1}{5} = 60$.

B. $7 \div \frac{1}{8}$ Think: $7 \times 8 = 56$.
So, $7 \div \frac{1}{8} = 56$.

C. $12 \div \frac{2}{3}$ Think: $12 \times 3 = 36$.
$36 \div 2 = 18$.
So, $12 \div \frac{2}{3} = 18$.

211

LESSON 10.5

3 Practice

⚠ COMMON ERROR ALERT

When dividing two fractions, students may find the reciprocal for the dividend instead of the divisor.

Have students copy the division and then circle the divisor to help them remember for which fraction they need to write the reciprocal.

Error	Correction
$\frac{3}{4} \div \frac{1}{5} = \frac{4}{3} \times \frac{1}{5}$	$\frac{3}{4} \div \textcircled{$\frac{1}{5}$} = \frac{3}{4} \times \frac{5}{1}$

Guided Practice

Do Check for Understanding Exercises 1–15 with your students. Identify those having difficulty and use lesson resources to help.

Independent Practice

Note that Exercise 53 is a **multistep or strategy problem**. Assign Exercises 16–57.

Multistep or Strategy Problem To solve Exercise 53, students should add the weights of the 3 packages of ground turkey; $2 + 7 + 3 = 12$. They then divide the sum by $\frac{1}{4}$; $12 \div \frac{1}{4} = 48$.

For Exercises 46–51, encourage students to recopy the division, making the substitution before writing the expression as a multiplication expression and evaluating the expression.

CHECK FOR UNDERSTANDING

Think and Discuss — Look back at the lesson to answer each question.

1. **Tell** what the reciprocal of a number is. Give an example. See left.
2. **Give** an example of a fraction or mixed-number division problem where the quotient is greater than the dividend. See left.

Guided Practice

1. Possible answer: a number that when multiplied by the given number gives a product of 1; $\frac{4}{3}$ is the reciprocal of $\frac{3}{4}$.

2. Possible answer: $3\frac{1}{4} \div \frac{3}{4} = 4\frac{1}{3}$ or $\frac{3}{4} \div \frac{1}{3} = 2\frac{1}{4}$

Write the reciprocal of the number.

3. $\frac{2}{3}$ $\frac{3}{2}$ 4. $\frac{3}{4}$ $\frac{4}{3}$ 5. 7 $\frac{1}{7}$ 6. $2\frac{3}{8}$ $\frac{8}{19}$ 7. $4\frac{1}{3}$ $\frac{3}{13}$

Find the quotient. Write the answer in simplest form.

8. $\frac{1}{3} \div \frac{1}{2}$ $\frac{2}{3}$ 9. $\frac{1}{5} \div \frac{1}{4}$ $\frac{4}{5}$ 10. $\frac{1}{4} \div \frac{1}{2}$ $\frac{1}{2}$ 11. $\frac{1}{8} \div 3$ $\frac{1}{24}$

12. $4 \div \frac{2}{3}$ 6 13. $3\frac{3}{5} \div 1\frac{1}{5}$ 3 14. $2\frac{3}{5} \div 4$ $\frac{13}{20}$ 15. $3\frac{1}{4} \div 2\frac{2}{3}$ $1\frac{7}{32}$

PRACTICE AND PROBLEM SOLVING

Independent Practice

Write the reciprocal of the number.

16. $\frac{5}{8}$ $\frac{8}{5}$ 17. 10 $\frac{1}{10}$ 18. $\frac{1}{6}$ 6 19. $\frac{2}{9}$ $\frac{9}{2}$ 20. $3\frac{1}{2}$ $\frac{2}{7}$

21. $\frac{15}{7}$ $\frac{7}{15}$ 22. $\frac{1}{5}$ 5 23. 9 $\frac{1}{9}$ 24. $1\frac{5}{6}$ $\frac{6}{11}$ 25. $2\frac{3}{5}$ $\frac{5}{13}$

Find the quotient. Write the answer in simplest form.

26. $\frac{3}{8} \div \frac{1}{2}$ $\frac{3}{4}$ 27. $\frac{2}{3} \div \frac{4}{7}$ $1\frac{1}{6}$ 28. $\frac{7}{8} \div \frac{1}{3}$ $2\frac{5}{8}$ 29. $8 \div \frac{6}{7}$ $9\frac{1}{3}$

30. $12 \div \frac{3}{5}$ 20 31. $\frac{3}{4} \div \frac{1}{3}$ $2\frac{1}{4}$ 32. $\frac{4}{9} \div \frac{3}{5}$ $\frac{20}{27}$ 33. $4 \div \frac{4}{5}$ 5

34. $3\frac{2}{5} \div 1\frac{1}{5}$ $2\frac{5}{6}$ 35. $3\frac{4}{5} \div \frac{3}{4}$ $5\frac{1}{15}$ 36. $4\frac{1}{2} \div \frac{1}{4}$ 18 37. $4\frac{1}{5} \div 2\frac{3}{5}$ $1\frac{8}{13}$

Use mental math to find each quotient.

38. $10 \div \frac{1}{2}$ 20 39. $12 \div \frac{2}{3}$ 18 40. $6 \div \frac{1}{4}$ 24 41. $8 \div \frac{1}{4}$ 32

42. $3 \div \frac{1}{6}$ 18 43. $9 \div \frac{1}{9}$ 81 44. $7 \div \frac{1}{2}$ 14 45. $11 \div \frac{1}{4}$ 44

🅐**ALGEBRA** Evaluate the expression.

46. $4 \div a$ for $a = \frac{2}{3}$ 6 47. $b \div 2\frac{1}{3}$ for $b = 5\frac{1}{2}$ $2\frac{5}{14}$

48. $1\frac{4}{5} \div a$ for $a = 2\frac{3}{5}$ $\frac{9}{13}$ 49. $c \div 7\frac{4}{5}$ for $c = 2\frac{1}{6}$ $\frac{5}{18}$

50. $b \div 7$ for $b = \frac{1}{5}$ $\frac{1}{35}$ 51. $3\frac{1}{5} \div b$ for $b = 4$ $\frac{4}{5}$

212 Chapter 10

ALTERNATIVE TEACHING STRATEGY SCAFFOLDED INSTRUCTION

Purpose Students use an activity to reinforce the concept of division of fractions.

Materials large cards or sheets of paper

Write a number of division problems, such as $\frac{1}{2} \div \frac{2}{3}$, $\frac{2}{5} \div 3\frac{1}{2}$, and so on, on large cards, one to a card.

$\frac{1}{2} \div \frac{2}{3}$	$\frac{2}{5} \div 3\frac{1}{2}$

On a second set of cards, write the corresponding multiplication problems, one to a card.

$\frac{1}{2} \times \frac{3}{2}$	$\frac{2}{5} \times \frac{2}{7}$

Hold up the first division problem and ask students what they would do to solve. Encourage them to give answers describing the process rather than solving the division.

Then mix up the division and multiplication cards and distribute them one to a student. Ask students with the division problems to stand around the room and hold up their cards. Then have students with multiplication cards find the corresponding division cards. Have each pair show their cards to the class, and ask for volunteers to simplify and solve. Check students' work.

Problem Solving ▶ Applications

56. She wrote the reciprocal of the dividend instead of the divisor; $1\frac{1}{2}$

52. The members of the school council were told they could divide the 15 acres of land next to the school into $1\frac{1}{2}$-acre sections to be used for new building projects for the school. How many $1\frac{1}{2}$-acre sections will there be? **10 sections**

53. Katie bought 2-lb, 7-lb, and 3-lb packages of ground turkey. How many $\frac{1}{4}$-lb turkey burgers can she make? **48 burgers**

54. In a $\frac{1}{4}$-mi relay, each runner on a team runs $\frac{1}{16}$ mi. How many runners are in the relay? **4 runners**

55. Gerald has $8\frac{3}{4}$ yd of fabric. This is 7 times the amount he needs to make one costume for the school play. How much fabric does he need for each costume? $1\frac{1}{4}$ **yd**

56. ❓ **What's the Error?** Jamie worked the problem below. What mistake did Jamie make? What is the correct answer in simplest form?

$$3\frac{1}{3} \div 2\frac{2}{9} = \frac{10}{3} \div \frac{20}{9} = \frac{3}{10} \times \frac{20}{9} = \frac{2}{3}$$

57. Jacki wants to buy a blouse for $12.95, a T-shirt for $15.95, a book for $10.50, and two pens for $3.75 each. What is the total cost? **$46.90**

MIXED REVIEW AND TEST PREP

58. $3\frac{1}{3} \times 2\frac{2}{5}$ (p. 206) **8**

59. $10\frac{1}{2} + 2\frac{7}{8}$ (p. 186) $13\frac{3}{8}$

60. Evaluate $5.2x$ for $x = 3.41$. (p. 82) **17.732**

⭐ 61. **TEST PREP** Which is the sum for 234,607 + 84,395? (p. 20) **D**

A 218,002   B 218,992   C 318,902   D 319,002

⭐ 62. **TEST PREP** Which shows $4\frac{3}{8}$ as a fraction? (p. 172) **H**

F $\frac{11}{8}$ G $\frac{15}{8}$ H $\frac{35}{8}$ J $\frac{39}{8}$

PROBLEM SOLVING LINKUP to Careers

Architecture Horace King was born in South Carolina in 1807. As an adult, he built more than 100 covered bridges in Georgia and nearby states. A bridge builder must first plan on paper a structure that will support heavy loads and withstand forces of nature. The mathematics that a bridge builder uses must be exact and correct at all times.

- The panels used on a covered bridge are $\frac{3}{4}$ ft wide. How many panels are needed to cover a side of a bridge that is 90 ft long? **120 panels**

EXTRA PRACTICE page H41, Set D

213

READING STRATEGY

K-W-L Chart Before having students read the Link Up, have them look at the picture and the title. Ask them to predict what the Link Up will be about. Then have students make a three-column chart headed What I Know, What I Want to Know, and What I Learned. Ask them to fill in the first two columns. Have them fill in the third column as they read through the paragraph.

K-W-L Chart

What I Know	What I Want to Know	What I Learned

MIXED REVIEW AND TEST PREP

Exercises 58–62 provide **cumulative review** (Chapters 1–10).

LINKUP to CAREERS

- *Draw students' attention to the Link Up. Have them tell what they know about the work a building engineer does.*

  Why is it important that a bridge builder know how to work with fractions? Possible answer: Measurements involve fractions. Since the pieces of a bridge must fit together exactly, a misunderstanding of fractions could lead to a bridge that falls down.

REASONING Suppose a bridge builder wanted to make the surface of a bridge 6 ft wide. Wood planks are $3\frac{1}{2}$ in. wide. Explain whether 6 ft seems like a good width for a bridge or why you would suggest that the builder use a different width. Possible answer: Use a different width. Six feet = 72 in. and 72 in. can not be evenly divided by $3\frac{1}{2}$ in. Use 20 planks across to get a width of 70 in. or 21 planks across to get a width of $73\frac{1}{2}$ in.

4 Assess

Summarize the lesson by having students:

DISCUSS What division expression would you use to solve Exercise 53 and what would be the divisor's reciprocal? $12 \div \frac{1}{4}; \frac{4}{1}$, or 4

WRITE How is a reciprocal used in division of fractions? To divide by a fraction, you multiply by the reciprocal of the divisor.

Lesson Quiz

Transparency 10.5

Write the reciprocal of the number.

1. $\frac{1}{5}$ **5**
2. $\frac{3}{7}$ $\frac{7}{3}$
3. $6\frac{1}{3}$ $\frac{3}{19}$

Find the quotient. Write the answer in simplest form.

4. $\frac{3}{7} \div \frac{3}{8}$ $\frac{8}{7}$, or $1\frac{1}{7}$
5. $3\frac{1}{7} \div \frac{2}{7}$ **11**
6. $9 \div \frac{1}{5}$ **45**

213

LESSON 10.6

Problem Solving Skill: Choose the Operation

LESSON PLANNING

Objective To *choose the operation* to solve a problem

Intervention for Prerequisite Skills
Fractions and Mixed Numbers (For intervention strategies, see page 199.)

Lesson Resources Problem Solving Think Along, p. TR1

NCTM Standards
1. Number and Operations
6. Problem Solving
8. Communication

Math Background
Choosing the correct operation is a very important skill for students. The following ideas will help them understand this process.

- Not all problems involving several numbers indicate addition. Some numbers may be extraneous to the solution.
- Joining groups of equal sizes implies multiplication, while joining different-sized groups requires addition.
- Modeling the problem may help students determine the operation needed.
- Using whole numbers in place of fractions or decimals may make it easier to select the needed operation.

WARM-UP RESOURCES

 **NUMBER OF THE DAY**

Write a number sentence dividing the number of the day of the month by $\frac{1}{4}$. **Possible answer:** $5 \div \frac{1}{4} = 20$

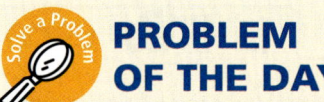 

Maria has $2\frac{1}{2}$ as many trading cards as Roberto, who has $\frac{1}{4}$ as many as Julia. Karen and Tim each have 25, which is 15 fewer than Julia has. Who has the most cards? **Julia**

Solution Problem of the Day tab, p. PD10

 **DAILY FACTS PRACTICE**

Have students practice multiplication and division facts by completing Set G of *Teacher's Resource Book*, p. TR100.

INTERVENTION AND EXTENSION RESOURCES

ALTERNATIVE TEACHING STRATEGY

To help students **analyze word problems for choosing the operation,** call on a volunteer to read Problem A on page 214. Have students discuss the concepts that indicate addition and multiplication. joining groups of different sizes and joining equal-sized groups

Call on volunteers to change the problem so a different operation would be needed. For example, "How much farther did José drive to Jim's house than to play practice?"

Repeat the activity for Problems B, C, and D.

Check students' work.

VISUAL
VERBAL/LINGUISTIC

ENGLISH LANGUAGE LEARNERS

Materials *For each group* poster board

Have students work in small groups to make a poster to **illustrate one of the four operations.** Review the descriptions given on page 214. Have each group choose an operation and make a poster that includes a description of the operation, a word problem that uses the operation, and an illustration. Encourage students to share their posters and display them for the class.

Check students' work.

VISUAL
VISUAL/SPATIAL

READING STRATEGY

Multiple-Meaning Words Some words or phrases can have more than one meaning, depending on the surrounding words or sentences. Have students analyze the use of the words *how much* in Problems C and D on page 214 and in Exercises 1 and 2 on page 215. Point out that those words themselves do not indicate an operation. In fact, they suggest different operations, based on the context. Have students make up two problems using the phrase *how much* to indicate two different operations. Check students' work.

EARLY FINISHERS

Materials *For each group* 4 index cards

Have students **practice choosing the operation** needed to solve word problems.

- Give each group of students 4 index cards.
- Have them write one operation on each card: add, subtract, multiply, and divide.
- Ask students to take turns drawing a card and then finding a word problem in their textbook whose solution requires that operation.
- Another student verifies that the operation is correct. Check students' work.

VISUAL
LOGICAL/MATHEMATICAL

MULTISTEP AND STRATEGY PROBLEMS

The following multistep and strategy problems are provided in Lesson 10.6:

Page	Item
215	3b, 4b, 5, 6

TECHNOLOGY LINK

Intervention Strategies and Activities CD-ROM •
Skills 25, 26

LESSON 10.6 ORGANIZER

Objective To *choose the operation* to solve a problem

Lesson Resources Problem Solving Think Along, p. TR1

1 Introduce

QUICK REVIEW provides review of pre-requisite skills.

Why Learn This? This skill will be useful in solving a wide variety of real-life problems, such as finding the total number of hours the library is open during a week. *Share the lesson objective with students.*

2 Teach

Guided Instruction

- Direct students' attention to the four operations outlined on this page.

  Describe a problem situation where you would add. Possible answer: I know how many people are in our class and how many are in another class. I would add to find how many in all.

- Ask similar questions about the other three operations. Then discuss the examples.

  When would you combine addition and multiplication to solve a problem? Possible answer: in a 2-step problem when you want to join a variety of equal-sized groups

---

LESSON 10.6

PROBLEM SOLVING SKILL
Choose the Operation

Learn how to choose the operation needed to solve a problem.

QUICK REVIEW

1. $\frac{5}{6}$ plus $\frac{3}{8}$ $1\frac{5}{24}$
2. $\frac{2}{3}$ divided by $\frac{1}{2}$ $1\frac{1}{3}$
3. $\frac{5}{6}$ minus $\frac{1}{2}$ $\frac{1}{3}$
4. $\frac{5}{7} \times \frac{2}{3}$ $\frac{10}{21}$
5. $\frac{1}{2} \div \frac{3}{4}$ $\frac{2}{3}$

Use the chart to help you decide how the numbers in a problem are related.

Then choose the operation needed to solve each problem.

ADD	• Joining groups of different sizes
SUBTRACT	• Taking away or comparing groups
MULTIPLY	• Joining equal-sized groups
DIVIDE	• Separating into equal-sized groups • Finding out how many in each group

Read each problem and decide how you would solve it.

A In one week, José drove $3\frac{1}{4}$ mi to play practice, $3\frac{1}{4}$ mi to the hardware store to buy supplies, and $7\frac{4}{5}$ mi to Jim's house to work on the costumes for the play. How many miles did he drive in all?	**B** Carla bought 6 yd of terry cloth to make collars for some of the costumes for the play. Each collar takes $\frac{2}{3}$ yd of fabric. How many collars can Carla make?
C Leanne makes props for the school play. She uses $4\frac{1}{8}$ lb of clay to make each vase and $2\frac{3}{4}$ lb to make each bowl. How much more clay does Leanne use to make one vase than one bowl?	**D** Sean bought $12\frac{2}{3}$ yd of pine lumber for $6 per yd to build some scenery for the school play. How much money did he spend?

A add, or multiply and add
B divide
C subtraction
D multiply

- What operation would you use to solve each problem?
- Which problems could you solve by using a combination of operations? **Problem A; add or multiply and add**
- How did the chart above help you decide which operation to use for each problem? **Answers will vary.**

RETEACH 10.6

Problem Solving Skill: Choose the Operation

It is often helpful to think about the kind of answer you need to solve a problem before you decide which operation or operations to use. Here are some different problem types and the operations used to solve them.

- **Add** to combine two or more like measures, such as length or weight.

  Lolly put $3\frac{1}{2}$ lb of sugar into a can containing 6 lb of sugar. The total amount of sugar in the can is $3\frac{1}{2} + 6$, or $9\frac{1}{2}$ lb.

- **Subtract** to take away an amount or to compare like measures.

  Perry spilled $2\frac{1}{3}$ lb of flour out of a 10-lb bag. The flour remaining in the bag weighs $10 - 2\frac{1}{3}$, or $7\frac{2}{3}$ lb. There is $7\frac{2}{3} - 2\frac{1}{3}$, or $5\frac{1}{3}$ lb more flour in the bag than was spilled.

- **Multiply** to combine a number of equal measures or to calculate a new type of measure.

  A square $3\frac{1}{2}$ yd on each side has a perimeter of $4 \times 3\frac{1}{2}$, or 14 yd. The square also has an area of $3\frac{1}{2} \times 3\frac{1}{2}$, or $12\frac{1}{4}$ yd².

- **Divide** to determine how many parts of equal size are in a measure or to determine the size of several equal parts.

  A $9\frac{1}{2}$ in. long board is cut into 4 equal lengths. Each piece has a length of $9\frac{1}{2} \div 4$, or $2\frac{3}{8}$ in. If a 6 ft board is cut in $1\frac{1}{2}$ ft lengths, there will be $6 \div 1\frac{1}{2}$, or 4 pieces.

Name the operation you would use to solve the problem. Then solve it.

1. Orange juice comes in 1-gal (128-oz) containers. William uses $12\frac{4}{5}$ oz of orange juice to make his favorite fruit smoothie. How many smoothies should he be able to make from one container of orange juice?

   divide: $128 \div 12\frac{4}{5}$;
   about 10 smoothies

2. William spills about $3\frac{1}{2}$ oz of liquid each time he makes a smoothie. He makes an average of 44 smoothies each day. How many ounces of liquid would he spill on a typical day?

   multiply: $3\frac{1}{2} \times 44$;
   about 154 oz

PRACTICE 10.6

Problem Solving Skill: Choose the Operation

Solve. Name the operations used.

1. Marie practiced piano a total of $17\frac{1}{2}$ hr last week. If she practiced the same amount of time each day, how long did she practice daily?

   $2\frac{1}{2}$ hr, division

2. Sylvan withdrew $\frac{2}{5}$ of the amount in his savings account, and spent $\frac{7}{10}$ of that money. What fraction of his total savings does he still have?

   $\frac{18}{25}$, multiplication and subtraction

3. Ike practices guitar $2\frac{1}{2}$ hr per day, but Jenn only practices $\frac{3}{4}$ hr. How much longer does Ike practice?

   $1\frac{3}{4}$ hr, subtraction

4. A painter is going to paint a wall that measures $2\frac{2}{3}$ yd by $4\frac{1}{2}$ yd. What is the area of the wall?

   12 yd², multiplication

5. José gives each of his 15 patio plants $\frac{3}{4}$ qt of water daily in warm weather. How much water does José use on his plants on a warm day?

   $11\frac{1}{4}$ qt, multiplication

6. José waters each of his 15 patio plants with $\frac{1}{2}$ qt water daily in cool weather. How much water can José expect to use on his patio plants during a cool week?

   52.5 qt, multiplication

7. Marisol rode her scooter $1\frac{1}{2}$ mi to Athena's home, then $\frac{3}{4}$ mi to Ariel's home, then $1\frac{1}{4}$ mi back to her home. How far did Marisol ride?

   $3\frac{1}{2}$ mi, addition

8. Bill can polish a car in $2\frac{3}{4}$ hr. Lara and Danny can do the same job working together in $1\frac{1}{2}$ hr. How much faster than Bill can Lara and Danny do the job when working together?

   $1\frac{1}{4}$ hr faster, subtraction

Mixed Review

Write each fraction in simplest form.

9. $\frac{5}{10}$ $\frac{1}{2}$
10. $\frac{20}{50}$ $\frac{2}{5}$
11. $\frac{15}{25}$ $\frac{3}{5}$
12. $\frac{22}{32}$ $\frac{11}{16}$
13. $\frac{21}{24}$ $\frac{7}{8}$

PROBLEM SOLVING PRACTICE

Solve. Name the operations used.

1. It takes Karen $5\frac{1}{2}$ minutes to walk to the community center. It takes Jackie $7\frac{1}{3}$ minutes. How much longer does it take Jackie? **$1\frac{5}{6}$ min; subtraction**

2. Rob makes candles that weigh $1\frac{3}{16}$ lb each. How much do 24 of them weigh? **$28\frac{1}{2}$ lb; multiplication**

3. Katie decorates travel bags. Each bag requires $1\frac{1}{3}$ yd of trim around the top and 1 yd of trim on the handle.
   a. Which operations could you use to find how much trim is needed for a number of bags? **B**
      A subtraction and division
      B addition and multiplication
      C division and multiplication
      D addition and division
   b. How much trim would it take to make 4 bags? **H**
      F $8\frac{1}{3}$ yd
      G 8 yd
      H $9\frac{1}{3}$ yd
      J 9 yd

MIXED APPLICATIONS

4. Marcie has 18 oz of dough left in a container. She needs $3\frac{3}{5}$ oz of dough to make one ornament.
   a. Write the expression you would use to find the number of ornaments she can make. Solve the problem. **$18 \div 3\frac{3}{5}$; 5 ornaments**
   b. How many ounces of dough are left in the container if she makes 4 ornaments? **$3\frac{3}{5}$ oz**

Use Data For 5–8, use the map.

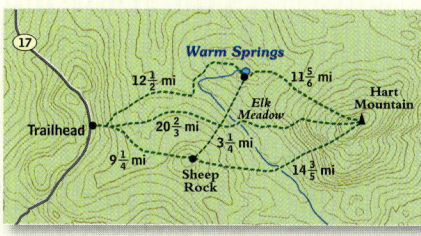

5. Darrin hiked the shortest route from the trailhead to Hart Mountain in $7\frac{3}{4}$ hours. What is the average number of miles he hiked in 1 hour? **$2\frac{2}{3}$ mi**

6. How much shorter was the trail that Darrin took than the trail through Warm Springs? **$3\frac{2}{3}$ mi**

7. Elk Meadow is halfway along the trail from the trailhead to Hart Mountain. How far is it from the trailhead to Elk Meadow? **$10\frac{1}{3}$ mi**

8. Sharlene left the trailhead at 8:30 A.M. for a daylong hike. She returned to the trailhead $7\frac{1}{2}$ hours after she left. What time did she return? **4 P.M.**

9. **Write About It** Explain how you decide what operation to use when solving a problem.
   Possible answer: I read the problem and see how the numbers are related.

215

3 Practice

Guided Practice

Do Problem Solving Practice Exercises 1–3 with your students. Identify those having difficulty and use lesson resources to help.

Independent Practice

Note that Exercises 3b, 4b, 5, and 6 are **multi-step or strategy problems**. Assign Exercises 4–9.

You may want to have students use fraction circles to model the solution to Exercise 4.

4 Assess

Summarize the lesson by having students:

DISCUSS What did you do to solve Exercise 5? Possible answer: I divided the shortest distance by $7\frac{3}{4}$.

WRITE Describe two ways to solve Exercise 3b. What property do these two ways illustrate? Add $1\frac{1}{3}$ and 1 and then multiply by 4, or multiply $1\frac{1}{3}$ by 4 and 1 by 4 and then add; the Distributive Property.

Lesson Quiz

Transparency 10.6

Solve.

1. A recipe for cookies calls for $3\frac{3}{4}$ cups of flour. How much flour would you need for half a recipe? **$1\frac{7}{8}$ cups**

2. Kayla uses $3\frac{2}{3}$ yd of plaid material and $1\frac{1}{2}$ yd of solid-color material for each outfit she makes. How many yards does it take for 3 outfits? **$15\frac{1}{2}$ yd**

CHALLENGE 10.6

ABCD Methods

Ann, Badri, Cristina, and Devon measured a redwood deck to determine its area. They made the following sketch to record their measurement.

(Sketch with dimensions: $3\frac{1}{2}$, $4\frac{1}{2}$, $4\frac{1}{4}$, $3\frac{1}{8}$)

Here's how each of them calculated the area of the deck. Perform each calculation.

1. **Ann:** $4\frac{1}{2} \times 3\frac{1}{2} + 3\frac{1}{8} \times 3\frac{1}{2} + 4\frac{1}{4} \times 3\frac{1}{8} =$ _____

2. **Badri:** $3\frac{1}{2} \times (4\frac{1}{2} + 3\frac{1}{8}) + 4\frac{1}{4} \times 3\frac{1}{8} =$ _____

3. **Cristina:** $4\frac{1}{2} \times 3\frac{1}{2} + 3\frac{1}{8} \times (3\frac{1}{2} + 4\frac{1}{4}) =$ _____

4. **Devon:** $(4\frac{1}{2} + 3\frac{1}{8}) \times (3\frac{1}{2} + 4\frac{1}{4}) - (4\frac{1}{2} \times 4\frac{1}{4}) =$ _____

5. Who used the correct method? Explain.
   The area is $39\frac{31}{32}$ yd². All four of the methods are correct. Each person chose to do their computation in a different way. Notice that Devon chose to visualize the deck as a large rectangle with a smaller rectangle removed.

READING STRATEGY 10.6

Multiple-Meaning Words   Analyze Choose Solve Check

Some problems contain words that have more than one meaning. The words may have the same spelling and different pronunciations or the same sound but different meanings. You can use information given in the problem to determine which meaning of the word is being used. Read the following problem.

VOCABULARY multiple-meaning

The Continental Divide, or Great Divide, is the watershed of North America. This means that it is the high point of land that separates the waters that flow east from those that flow west. The chart below shows precipitation information for the Continental Divide. How much greater is the annual precipitation at the highest elevation than at the lowest elevation?

Elevation	4,000–7,000 ft	7,000–11,000 ft	11,000–14,000 ft
Annual Precipitation	11 in.	20 in.	40 in.

1. Which word has both a mathematical meaning and an everyday meaning? **divide**
2. What operation is needed to solve the problem? **subtraction**
3. Solve the problem. **29 in.**

Read each problem carefully. Then solve.

Mr. Winston is building an addition onto his house. The area of the addition is 150 square feet. The contractor is charging him $200 per square foot. How much will the addition cost?

4. Which word has both a mathematical meaning and an everyday meaning? **addition**
5. What operation is needed to solve the problem? **multiplication**
6. Solve the problem. **$30,000**

Mr. Winston's house is in a suburban area. The original house had an area of 1,750 square feet. When construction is complete, what will be the area of the new house?

7. Which word has both a mathematical meaning and an everyday meaning? **area**
8. What operation is needed to solve the problem? **addition**
9. Solve the problem. **1,900 sq ft**

LESSON 10.7

Algebra: Fraction Expressions and Equations

LESSON PLANNING

Objective To evaluate expressions with fractions and to solve equations with fractions by using substitution and mental math

Intervention for Prerequisite Skills
Fractions and Mixed Numbers (For intervention strategies, see page 199.)

NCTM Standards
1. Number and Operations
2. Algebra
6. Problem Solving

Math Background

The skills students have learned in evaluating expressions and solving equations with whole numbers and decimals can be used in working with expressions and equations with fractions and mixed numbers.

Consider the following as you help students understand the process of working with fraction expressions and equations:

- Once the variable has been replaced with a number, use the rules for adding, subtracting, multiplying, or dividing fractions and mixed numbers.
- Rewrite a division equation as a multiplication equation before solving.
- To find solutions for addition or subtraction equations, rewrite the fractions with common denominators.

WARM-UP RESOURCES

 **NUMBER OF THE DAY**

Take the number that represents the month of the year. Use it as the numerator in a fraction with a denominator of 8. Multiply the fraction times 4. What is the product? Possible answer: $\frac{3}{8} \times 4 = 1\frac{1}{2}$

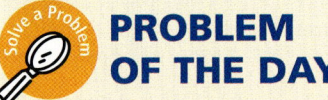 

Ming Li ran 90 ft from first base to second base. Each stride was about $3\frac{1}{2}$ ft long. If she takes about 2 strides per second, about how long did it take her to get to second base? Possible answer: about 13 sec

Solution Problem of the Day tab, p. PD10

 **DAILY FACTS PRACTICE**

Have students practice multiplication and division facts by completing Set A of *Teacher's Resource Book*, p. TR101.

216A Chapter 10

INTERVENTION AND EXTENSION RESOURCES

ALTERNATIVE TEACHING STRATEGY — ELL

Materials *For each group* 1 index card per student

Have smaller groups of students **work with algebraic expressions**. Have each group member choose a different fraction to write on an index card. Have the groups exchange cards. Give each group an algebraic expression such as $c \div \frac{2}{3} + \frac{3}{4}$. Have the group members work together to evaluate the expression for each of the fractions on the cards. Check students' work.

VISUAL
LOGICAL/MATHEMATICAL

MULTISTEP AND STRATEGY PROBLEMS

The following multistep or strategy problem is provided in Lesson 10.7:

Page	Item
217	24

WRITING IN MATHEMATICS

Explain how you would **solve an equation** as in Exercise 24 on page 217. Possible answer: The product of $3\frac{1}{2}$ and the number of days equals the total weight of the insects, $17\frac{1}{2}$. So, for b days, $3\frac{1}{2} \times b = 17\frac{1}{2}$. To solve, make a table and substitute values for b such as 3, 4, and 5 until one works. Then check the answer by substituting it in the equation.

ADVANCED LEARNERS

Challenge students to **solve equations with mixed numbers and fractions**.

- Ask students to write number sentences involving fractions or mixed numbers. For example, $14\frac{5}{9} - 5\frac{1}{3} = 9\frac{2}{9}$ or $1\frac{1}{2} \div \frac{3}{8} = 4$.
- Have a volunteer present his or her number sentence as an algebraic equation. For example, "If I subtract $5\frac{1}{3}$ from a number, I get $9\frac{2}{9}$."
- The other students then write the equation and solve.
  $n - 5\frac{1}{3} = 9\frac{2}{9}$; $n = 14\frac{5}{9}$

VISUAL
VERBAL/LINGUISTIC

TECHNOLOGY LINK

Intervention Strategies and Activities CD-ROM •
Skills 25, 26, 41

Math Jingles™ CD 5–6 •
Tracks 8 and 9

LESSON 10.7 ORGANIZER

Objective To evaluate expressions with fractions and to solve equations with fractions by using substitution and mental math

1 Introduce

QUICK REVIEW provides review of prerequisite skills.

Why Learn This? Being able to evaluate expressions will help you check the solutions to equations that have fractions and mixed numbers. Share the lesson objective with students.

2 Teach

Guided Instruction

- As you discuss the opening problem and writing an expression, guide students in determining the correct expression.

  If you wanted to find the weight of 2 brochures, how would you do it? Multiply 2 times $\frac{3}{4}$.

  How can you use your answer to help find an expression? To find the weight of b brochures, multiply b by $\frac{3}{4}$.

Modifying Instruction For students having difficulty substituting values for variables in expressions or equations, display the expression and write the value on a piece of paper. Place the paper over the variable.

ADDITIONAL EXAMPLE

Example, p. 216

Solve the equation $w \times \frac{1}{3} = 4$ by using mental math.

$w \times \frac{1}{3} = 4$ Remember that multiplying by $\frac{1}{3}$ is like dividing by 3. What number divided by 3 is 4?

$w = 12$

$12 \times \frac{1}{3} = 4$ Check your answer. Replace w with 12.

$\frac{12}{1} \times \frac{1}{3} = \frac{12}{3}$, or 4

So, $w = 12$.

216 Chapter 10

LESSON 10.7

ALGEBRA
Fraction Expressions and Equations

Learn how to evaluate expressions and solve equations with fractions.

QUICK REVIEW

Solve.
1. $6h = 24$ 2. $\frac{g}{2} = 16$ 3. $3m = 2.1$
    $h = 4$ $g = 32$ $m = 0.7$
4. $r + 9.10 = 10.08$ 5. $x - 1.7 = 1.7$
    $r = 0.98$ $x = 3.4$

Venus's-flytrap is found in the coastal regions of North and South Carolina.

Ann will be mailing some brochures about the different types of plants that eat insects. The price to mail them depends on the total weight. Each brochure weighs $\frac{3}{4}$ ounce. How much do the brochures weigh in all?

You can write and evaluate an expression. Choose a variable to represent the number of brochures.

Let b = the number of brochures.

$b \times \frac{3}{4}$ or $\frac{3}{4}b$ *Write the expression.*

The number of brochures Ann mails changes every week. What was the total weight of the 23 brochures she mailed last week?

$b \times \frac{3}{4}$ or $\frac{3}{4}b$ *Write the expression.*

$23 \times \frac{3}{4}$ *Replace b with 23.*

$\frac{69}{4}$, or $17\frac{1}{4}$ *Multiply.*

So, the total weight is $17\frac{1}{4}$ oz.

You have solved equations with whole numbers and decimals by using mental math. You can use mental math to solve some equations with fractions.

EXAMPLE

Solve the equation $n \div \frac{1}{4} = 8$ by using mental math.

$n \div \frac{1}{4} = 8$ *Remember, when dividing by $\frac{1}{4}$, you multiply by the reciprocal, 4.*

$n = 2$ *What number times 4 equals 8?*

$2 \div \frac{1}{4} = 8$ *Check your answer. Replace n with 2.*

$\frac{2}{1} \times \frac{4}{1} = \frac{8}{1}$, or 8   So, $n = 2$.

- Use mental math to solve $x + \frac{2}{3} = \frac{5}{6}$. $x = \frac{1}{6}$

216 Chapter 10

RETEACH 10.7

Algebra: Fraction Expressions and Equations

Algebraic expressions can have fractions in them. Also, fractional values can replace a variable in the expression.

Let's explore the fraction expressions $x + \frac{2}{5}$ and $\frac{3}{4}y$ for several values of the variables x and y.

x	$x + \frac{2}{5}$
$\frac{1}{5}$	$\frac{1}{5} + \frac{2}{5} = \frac{3}{5}$
1	$1 + \frac{2}{5} = 1\frac{2}{5}$
10	$10 + \frac{2}{5} = 10\frac{2}{5}$

y	$\frac{3}{4}y$
$\frac{1}{2}$	$\frac{3}{4} \times \frac{1}{2} = \frac{3}{8}$
2	$\frac{3}{4} \times 2 = 1\frac{1}{2}$
10	$\frac{3}{4} \times 10 = 7\frac{1}{2}$

Notice that the expression $x + \frac{2}{5}$ adds $\frac{2}{5}$ to any value of x, and that the expression $\frac{3}{4}y$ multiplies any value of y by $\frac{3}{4}$.

When solving an equation involving fractions and fractional expressions, often you can use your number sense to decide what value for the variable makes the equation true.

- Solve the equation $x + \frac{2}{5} = \frac{3}{5}$.
  Ask yourself: *What number can I add to $\frac{2}{5}$ to get $\frac{3}{5}$?*
  You can add $\frac{1}{5}$ to $\frac{2}{5}$ to get $\frac{3}{5}$. So, $x = \frac{1}{5}$.

- Solve the equation $\frac{3}{4}y = \frac{30}{4}$.
  Ask yourself: *What can I multiply $\frac{3}{4}$ by to get $\frac{30}{4}$?*
  You can multiply $\frac{3}{4}$ by 10 to get $\frac{30}{4}$. So, $y = 10$.

Evaluate the expression.

1. $x + \frac{1}{10}$ for $x = \frac{1}{2}$ $\frac{3}{5}$ 2. $\frac{1}{9}y$ for $y = \frac{4}{9}$ $\frac{4}{18}$ 3. $\frac{1}{3} - z$ for $z = \frac{1}{4}$ $\frac{1}{12}$

Use mental math to solve the equation.

4. $y + \frac{1}{2} = \frac{5}{2}$ $y = 2$ 5. $\frac{1}{10}m = 2$ $m = 20$ 6. $\frac{1}{3}t = 1$ $t = 3$

7. $\frac{1}{4}x = 2$ $x = 8$ 8. $n + \frac{3}{8} = \frac{1}{2}$ $n = \frac{1}{8}$ 9. $w - \frac{3}{10} = \frac{1}{5}$ $w = \frac{1}{2}$

10. $\frac{5}{6} - x = \frac{3}{4}$ $x = \frac{1}{12}$ 11. $t - \frac{1}{3} = 1$ $t = \frac{4}{3}$ 12. $a + \frac{1}{3} = 7$ $a = 6\frac{2}{3}$

PRACTICE 10.7

Algebra: Fraction Expressions and Equations

Evaluate the expression.

1. $2\frac{1}{4} + x$ for $x = 2\frac{1}{2}$ $4\frac{6}{8}$ 2. $2\frac{1}{4} + x$ for $x = \frac{1}{2}$ $2\frac{3}{4}$ 3. $2\frac{1}{4} + x$ for $x = \frac{3}{8}$ $2\frac{5}{8}$

4. $y - 2\frac{3}{5}$ for $y = 5\frac{4}{5}$ $3\frac{1}{5}$ 5. $y - 2\frac{3}{5}$ for $y = 4\frac{7}{10}$ $2\frac{1}{10}$ 6. $y - 2\frac{3}{5}$ for $y = 6$ $3\frac{2}{5}$

7. $\frac{3}{5}s$ for $s = 2$ $1\frac{1}{5}$ 8. $\frac{3}{5}s$ for $s = \frac{1}{3}$ $\frac{1}{5}$ 9. $\frac{3}{5}s$ for $s = 1\frac{2}{3}$ 1

10. $6\frac{7}{9}p$ for $p = \frac{1}{2}$ $3\frac{1}{7}$ 11. $6\frac{7}{9}p$ for $p = \frac{7}{3}$ $14\frac{2}{3}$ 12. $6\frac{7}{9}p$ for $p = 2\frac{5}{8}$ $16\frac{1}{2}$

13. $x \div 1\frac{1}{3}$ for $x = 4\frac{1}{2}$ 3   14. $x \div 3\frac{1}{4}$ for $x = \frac{1}{4}$ $\frac{1}{13}$ 15. $x \div 2\frac{1}{3}$ for $x = 2\frac{1}{3}$ 1

Use mental math to solve the equation.

16. $x + 5\frac{3}{8} = 5\frac{1}{2}$ $x = \frac{1}{8}$ 17. $\frac{1}{2}y = \frac{1}{12}$ $y = \frac{1}{6}$ 18. $z - 8\frac{1}{5} = 12\frac{2}{5}$ $z = 20\frac{3}{5}$ 19. $w + \frac{9}{20} = \frac{5}{9}$ $w = \frac{1}{4}$

20. $\frac{4}{5}n = 3$ $n = 3\frac{3}{4}$ 21. $c + 4\frac{1}{3} = 7\frac{5}{6}$ $c = 3\frac{1}{2}$ 22. $m - 6\frac{1}{2} = 5\frac{7}{8}$ $m = 12\frac{3}{8}$ 23. $\frac{3}{4}d = 9\frac{3}{4}$ $d = 13$

Mixed Review

Add or subtract. Write the answer in simplest form.

24. $\frac{7}{8} - \frac{3}{4}$ $\frac{1}{8}$ 25. $\frac{5}{12} - \frac{1}{12}$ $\frac{1}{3}$ 26. $\frac{1}{6} + \frac{1}{2}$ $\frac{2}{3}$ 27. $\frac{2}{5} + \frac{9}{20}$ $\frac{17}{20}$

28. $\frac{4}{5} - \frac{3}{7}$ $\frac{13}{35}$ 29. $\frac{4}{9} + \frac{3}{10}$ $\frac{67}{90}$ 30. $\frac{7}{10} - \frac{1}{6}$ $\frac{8}{15}$ 31. $\frac{5}{4} + \frac{8}{15}$ $\frac{19}{20}$

32. $4\frac{1}{2} + 2\frac{1}{4} + 1\frac{1}{8}$ $7\frac{7}{8}$ 33. $4\frac{3}{8} - 2\frac{1}{4}$ $\frac{15}{8}$

CHECK FOR UNDERSTANDING

Think and Discuss

Look back at the lesson to answer each question.

1. **What if** Ann mails 32 brochures? Show how you would find the total weight of the brochures. $b \times \frac{3}{4}$; $32 \times \frac{3}{4}$; 24 oz

2. **Explain** how you would solve the equation $d \div \frac{1}{3} = 9$.
Think: What number times 3 equals 9? $d = 3$

Guided Practice

Evaluate the expression.

3. $x - 1\frac{1}{5}$ for $1x = 3\frac{2}{5}$ $2\frac{1}{5}$
4. $\frac{1}{4}y$ for $y = \frac{1}{2}$ $\frac{1}{8}$
5. $t + 2\frac{1}{4}$ for $t = 3\frac{3}{8}$ $5\frac{5}{8}$

Use mental math to solve the equation.

6. $x - 4\frac{1}{4} = 2\frac{1}{2}$ $x = 6\frac{3}{4}$
7. $\frac{2}{3}y = \frac{1}{4}$ $y = \frac{3}{8}$
8. $2\frac{5}{8} = t + 1\frac{1}{4}$ $t = 1\frac{3}{8}$

PRACTICE AND PROBLEM SOLVING

Independent Practice

Evaluate the expression.

9. $c + 2\frac{2}{3}$ for $c = 5\frac{1}{6}$ $7\frac{5}{6}$
10. $4\frac{5}{8} m$ for $m = 3\frac{1}{3}$ $15\frac{5}{12}$
11. $b \div 2\frac{2}{5}$ for $b = 3\frac{1}{8}$ $1\frac{1}{4}$
12. $4\frac{5}{8} + a$ for $a = 4\frac{4}{8}$ $9\frac{1}{8}$
13. $b \times \frac{1}{5}$ for $b = \frac{1}{4}$ $\frac{1}{20}$
14. $\frac{5}{8} \div c$ for $c = 5$ $\frac{1}{8}$
15. $\frac{21}{24} - a$ for $a = \frac{1}{2}\frac{3}{8}$
16. $c \div 1\frac{1}{4}$ for $c = 1\frac{2}{3}$ $1\frac{1}{3}$
17. $\frac{1}{3} + \frac{1}{2} + d$ for $d = \frac{1}{6}$ 1

Use mental math to solve the equation.

18. $x - 3\frac{1}{2} = 2\frac{1}{4}$ $x = 5\frac{3}{4}$
19. $\frac{1}{5}x = \frac{2}{5}$ $x = 2$
20. $14\frac{2}{3} + x = 28$ $x = 13\frac{1}{3}$
21. $8x = 2$ $x = \frac{1}{4}$
22. $\frac{2}{3}t = 2$ $t = 3$
23. $3\frac{1}{3} + x = 4\frac{2}{3}$ $x = 1\frac{1}{3}$

Problem Solving Applications

24. Ann's brochures tell about the eating habits of pitcher plants, sundew plants, and Venus's-flytraps. Suppose one type of plant traps about $3\frac{1}{2}$ oz of insects each day. Choose an operation. Then write and solve an equation to find how many days the plant takes to trap a total of $17\frac{1}{2}$ oz of insects. See above left.

24. Possible answer: multiplication; let b equal the number of days; $3\frac{1}{2} b = 17\frac{1}{2}$; $b = 5$ days

Sundew Plant

25. **What's the Error?** Robin evaluated $\frac{3}{4}h$ for $h = \frac{3}{8}$ in this way:
$\frac{3}{4} \times \frac{3}{8} = \frac{6}{8} \times \frac{3}{8} = \frac{9}{8}$. Explain her mistake. Robin did not multiply the numerators or multiply the denominators to get $\frac{9}{32}$.

MIXED REVIEW AND TEST PREP

26. $4\frac{5}{8} \div 2\frac{1}{6}$ (p. 210) $2\frac{7}{52}$
27. Order from greatest to least. $\frac{3}{4}, \frac{3}{8}, \frac{3}{5}$ (p. 166) $\frac{3}{4}, \frac{3}{5}, \frac{3}{8}$
28. 25.95×13.3 (p. 70) 345.135
29. Solve. $1.5 = \frac{a}{3}$ (p. 82) $a = 4.5$

⭐ 30. **TEST PREP** Which decimal is 10 times as great as 34.62? (p. 70) **D**

A 0.34   B 3.4   C 3.46   D 346.2

EXTRA PRACTICE page H41, Set E

217

3 Practice

Guided Practice

Do Check for Understanding Exercises 1–8 with your students. Identify those having difficulty and use lesson resources to help.

Independent Practice

Note that Exercise 24 is a **multistep or strategy problem**. Assign Exercises 9–25.

As students evaluate the expressions in Exercises 9–17, encourage them to think about how the value of the expressions will relate to the number being substituted.

MIXED REVIEW AND TEST PREP

Exercises 26–30 provide **cumulative review** (Chapters 1–10).

4 Assess

Summarize the lesson by having students:

DISCUSS What if you decided 32 was the correct answer to the Example on page 216? What would you do to check it?
Possible answer: Check by substituting 32 into the original equation. When you multiply by 4 and see that it is wrong, rethink your solution.

WRITE Explain how you solve an equation such as the one in Exercise 18. Possible answer: I think of number families and remember that the sum of the difference and the number subtracted equals the number you subtracted from. I also remember the check procedure for subtraction.

Lesson Quiz

Evaluate the expression.

1. $\frac{3}{4}x$ for $x = 5\frac{1}{3}$ 4
2. $2\frac{3}{5} + b$ for $b = 4\frac{7}{15}$ $7\frac{1}{15}$

Use mental math to solve the equation.

3. $n + \frac{3}{4} = 1\frac{1}{2}$ $n = \frac{3}{4}$
4. $a \div 3 = \frac{1}{5}$ $a = \frac{3}{5}$

217

CHAPTER 10

REVIEW/TEST

Purpose To check understanding of concepts, skills, and problem solving presented in Chapter 10

USING THE PAGE

The Chapter 10 Review/Test can be used as a **review** or a **test**.

- Item 1 checks understanding of concepts and new vocabulary.
- Items 2–35 check skill proficiency.
- Items 36–40 check students' abilities to choose and apply problem solving strategies to real-life problems involving multiplication and division of fractions.

 Suggest that students place the completed Chapter 10 Review/Test in their portfolios.

USING THE ASSESSMENT GUIDE

- Multiple-choice format of Chapter 10 Posttest—See *Assessment Guide*, pp. AG61–62.
- Free-response format of Chapter 10 Posttest—See *Assessment Guide*, pp. AG63–64.

USING STUDENT SELF-ASSESSMENT

The How Did I Do? survey helps students assess what they have learned and how they learned it. This survey is available as a copying master in *Assessment Guide*, p. AGxvii.

218 Chapter 10

CHAPTER 10 REVIEW/TEST

1. **VOCABULARY** Two numbers are __?__ if their product is 1. (pp. 208–209) **reciprocals**

Estimate each product or quotient. (pp. 200–201) **Possible answers are given.**

2. $\frac{2}{9} \times \frac{1}{6}$ **0**
3. $\frac{7}{8} \div \frac{11}{12}$ **1**
4. $\frac{7}{15} \div \frac{8}{9}$ **$\frac{1}{2}$**
5. $2\frac{4}{5} \times 3\frac{3}{10}$ **9**
6. $4\frac{2}{15} \times 5\frac{4}{7}$ **24**
7. $31\frac{3}{8} \div 4\frac{1}{2}$ **8**

Multiply. Write the answer in simplest form. (pp. 202–207)

8. $\frac{1}{6} \times \frac{3}{5}$ **$\frac{1}{10}$**
9. $\frac{2}{3} \times \frac{4}{7}$ **$\frac{8}{21}$**
10. $16 \times \frac{5}{12}$ **$\frac{20}{3}$, or $6\frac{2}{3}$**
11. $\frac{3}{8} \times 10$ **$\frac{15}{4}$, or $3\frac{3}{4}$**
12. $1\frac{1}{2} \times \frac{3}{4}$ **$\frac{9}{8}$, or $1\frac{1}{8}$**
13. $4\frac{1}{2} \times 2\frac{1}{3}$ **$\frac{21}{2}$, or $10\frac{1}{2}$**
14. $1\frac{1}{2} \times \frac{2}{3}$ **1**
15. $2\frac{1}{3} \times 3\frac{1}{7}$ **$\frac{22}{3}$, or $7\frac{1}{3}$**

Find the quotient. Write it in simplest form. (pp. 210–213)

16. $\frac{6}{7} \div \frac{3}{5}$ **$\frac{10}{7}$, or $1\frac{3}{7}$**
17. $\frac{3}{4} \div \frac{1}{3}$ **$\frac{9}{4}$, or $2\frac{1}{4}$**
18. $3\frac{1}{3} \div 2\frac{4}{5}$ **$1\frac{4}{21}$**
19. $9\frac{1}{2} \div 1\frac{3}{8}$ **$6\frac{10}{11}$**
20. $8 \div \frac{6}{7}$ **$\frac{28}{3}$, or $9\frac{1}{3}$**
21. $\frac{4}{5} \div 4$ **$\frac{1}{5}$**
22. $2\frac{3}{5} \div 4\frac{1}{5}$ **$\frac{13}{21}$**
23. $\frac{5}{8} \div 10$ **$\frac{1}{16}$**

Evaluate the expression. (pp. 216–217)

24. $x - 3\frac{1}{3}$ for $x = 6\frac{1}{5}$ **$2\frac{13}{15}$**
25. $\frac{3}{4} \div r$ for $r = 4$ **$\frac{3}{16}$**
26. $25\frac{1}{8} + m$ for $m = 6\frac{2}{3}$ **$31\frac{19}{24}$**
27. $x + 3\frac{1}{10}$ for $x = 1\frac{1}{2}$ **$4\frac{3}{5}$**
28. $\frac{7}{8} a$ for $a = \frac{2}{5}$ **$\frac{7}{20}$**
29. $\frac{3}{4} b$ for $b = 1\frac{1}{6}$ **$\frac{7}{8}$**

Use mental math to solve the equation. (pp. 216–217)

30. $x - 6\frac{1}{9} = 3\frac{2}{3}$ **$x = 9\frac{7}{9}$**
31. $x \div \frac{1}{3} = 12$ **$x = 4$**
32. $12\frac{3}{4} + x = 15\frac{7}{8}$ **$x = 3\frac{1}{8}$**
33. $1\frac{1}{8} b = 4\frac{1}{20}$ **$b = 3\frac{3}{5}$**
34. $c - \frac{1}{5} = \frac{1}{2}$ **$c = \frac{7}{10}$**
35. $\frac{3}{7} \div a = \frac{1}{28}$ **$a = 12$**

Solve.

36. Mike rides his bicycle $6\frac{1}{2}$ min to school. Cami rides her bicycle $1\frac{1}{2}$ times as long. How long does it take Cami to ride to school? (pp. 206–207) **$9\frac{3}{4}$ min**

37. Eric practiced $\frac{3}{4}$ hr on Monday and $\frac{2}{5}$ hr on Saturday. How much longer did Eric practice on Monday? (pp. 214–215) **$\frac{7}{20}$ hr**

38. A race course is $\frac{3}{4}$ mi long. Racers want to run three equal sprints. How far apart should the markers be for the sprints? (pp. 210–213) **$\frac{1}{4}$ mi**

39. Sara has $\frac{5}{6}$ yd of ribbon and Mark has $\frac{3}{4}$ yd of ribbon. How much ribbon do Sara and Mark have altogether? (pp. 214–215) **$\frac{19}{12}$ yd, or $1\frac{7}{12}$ yd**

40. The sum of Kiesha's and Brent's heights is $127\frac{1}{4}$ in. Kiesha's height is $64\frac{1}{2}$ in. Write an equation to find Brent's height. (pp. 216–217) **$64\frac{1}{2} + a = 127\frac{1}{4}$; $62\frac{3}{4}$ in.**

218 Chapter 10

CHAPTERS 1–10 ★ STANDARDIZED TEST PREP

 **Understand the problem.**
See item **2**.
You know that this week's playing time is $\frac{2}{3}$ of last week's playing time. Use a *variable* to write any relationship like this one as an equation.
Also see problem **1**, p. H62.

Choose the best answer.

1. Carla rode her bicycle $1\frac{2}{5}$ miles on Monday, $2\frac{1}{4}$ miles on Wednesday, and $\frac{4}{5}$ mile on Friday. Which is a reasonable estimate of the total distance Carla rode her bicycle? **B**

   A 3 mi **B** $4\frac{1}{2}$ mi **C** $5\frac{1}{2}$ mi **D** $6\frac{1}{2}$ mi

2. Vince is trying to decrease the amount of time he spends playing video games. This week he spent $5\frac{2}{3}$ hours playing video games. That was $\frac{2}{3}$ of the amount of time he played video games last week. How long did Vince play video games last week? **J**

   F $4\frac{1}{2}$ hr **G** $6\frac{1}{3}$ hr **H** 7 hr **J** $8\frac{1}{2}$ hr

3. Eva ordered $2\frac{1}{2}$ pounds of potato salad, $2\frac{3}{4}$ pounds of fruit salad, and $1\frac{5}{8}$ pounds of cole slaw. Which operation would be best to use to find out how much more fruit salad she ordered than cole slaw? **C**

   A addition **C** subtraction
   B multiplication **D** division

4. Which expression can be used to find $\frac{3}{5} \div \frac{1}{3}$? **H**

   F $\frac{3}{5} + \frac{1}{3}$ **H** $\frac{3}{5} \times \frac{3}{1}$
   G $\frac{3}{5} - \frac{1}{3}$ **J** $\frac{5}{3} \times \frac{1}{3}$

5. Which is a reasonable estimate for $21\frac{4}{5} \div 3$? **D**

   A 3 **B** 4 **C** 5 **D** 7

6. Solve.
   $$b - 4\frac{1}{2} = 2\frac{2}{3}$$ **J**

   F $b = 1\frac{5}{6}$ **H** $b = 6\frac{5}{6}$
   G $b = 6\frac{3}{5}$ **J** $b = 7\frac{1}{6}$

7. Cal has 4 bags of apples with masses of 2.5 kilograms, 2.05 kilograms, 3.12 kilograms, and 2.18 kilograms. Which shows the masses in order from lightest to heaviest? **A**

   A 2.05, 2.18, 2.5, 3.12
   B 2.5, 2.18, 3.12, 2.05
   C 3.12, 2.5, 2.18, 2.05
   D 2.18, 2.5, 2.05, 3.12

8. The sum of the prime factors of 12 is 7. What is the sum of the prime factors of 72? **F**

   F 12 **H** 42
   G 27 **J** Not here

Write What You Know

See below.

9. Textbooks are $1\frac{1}{2}$ in. thick. They are packed in boxes with 2 stacks of books in each box. A box is 15 in. high. How many textbooks can one box hold? Explain the method you used to solve the problem.

10. Explain how the Distributive Property can be used to find the product of a whole number and a mixed number. Use what you wrote to find the product $8 \times 6\frac{1}{2}$.

219

Write What You Know • Written Response

9. A box can hold 20 textbooks. Find the number of textbooks in a stack. Since a stack is 15 in. high and a textbook is $1\frac{1}{2}$ in. thick, a stack can hold $15 \div 1\frac{1}{2}$ textbooks. And $15 \div 1\frac{1}{2} = \frac{15}{1} \div \frac{3}{2} = \frac{15}{1} \times \frac{2}{3} = 10$. Since there are 2 stacks, there are $2 \times 10 = 20$ textbooks in a box.

10. A mixed number can be written as the sum of the whole number and the fraction. To find $8 \times 6\frac{1}{2}$, rewrite the problem as $8 \times (6 + \frac{1}{2})$.
    Then $8 \times (6 + \frac{1}{2}) = (8 \times 6) + (8 \times \frac{1}{2}) = 48 + 4 = 52$.

STANDARDIZED TEST PREP • Chapters 1–10

USING THE PAGE

This page may be used to help students get ready for standardized tests. The test items are written in the same style and arranged in the same format as those on many state assessments. The page is cumulative. It covers math objectives and essential skills that have been taught up to this point in the text. Most of the items represent skills from the current chapter, and the remainder represent skills from earlier chapters.

This page can be assigned at the end of the chapter as classwork or as a homework assignment. You may want to have students use individual recording sheets presented in a multiple-choice (standardized) format. A Test Answer Sheet is available as a blackline master in *Assessment Guide* (p. AGxlii).

You may wish to have students describe how they solved each problem and share their solutions.

ITEM ANALYSIS

Item	Learning Goal	Item	Learning Goal
1	9A	6	10C
2	10B	7	3A
3	10D	8	7B
4	10B	9	9C
5	10A	10	10C

Written Response items for Write What You Know are available as a blackline master in *Performance Assessment*.

SCORING RUBRIC • WRITE WHAT YOU KNOW

2 Demonstrates a complete understanding of the problem and chooses an appropriate strategy to determine the solution

1 Demonstrates a partial understanding of the problem and chooses a strategy that does not lead to a complete and accurate solution

0 Demonstrates little understanding of the problem and shows little evidence of using any strategy to determine a solution

Multiply and Divide Fractions and Mixed Numbers 219

UNIT 3

MATH DETECTIVE
On a Roll

Purpose To use deductive reasoning to solve problems involving fractions

USING THE PAGE

- Direct students' attention to the Reasoning section. Have students read through each of the Fraction Mystery clues before they complete the mystery fractions. Ask them to solve Fraction Mystery 1.

- **How did you pick the numbers to write the pair of equivalent fractions for Fraction Mystery 1?** Answers will vary.

- Have students read the clue for Fraction Mystery 2.

    When is a fraction greater than 1? when the numerator is greater than the denominator

- After students solve Fraction Mysteries 3 and 4, have them explain their thinking.

    How did you find a fraction to solve Fraction Mystery 3? Answers will vary.

    Were you able to solve both Fraction Mysteries 3 and 4 using the numbers you rolled? Explain. Answers will vary.

    If you were not able to complete Fraction Mysteries 3 and/or 4, could you complete them using numbers that you used in Fraction Mysteries 1 and 2? Explain. Answers will vary.

    Think It Over! After students complete the Write About It, have them compare the numbers they rolled and their solutions to the Fraction Mystery numbers.

PROBLEM SOLVING

MATH DETECTIVE

On a Roll

REASONING Roll a number cube ten times. Record the numbers that you roll. Use your knowledge of fractions to attempt to solve the problems below. You may use each of your ten numbers exactly once. Once you use a number, cross it out so that you don't use it again. Depending on which numbers you roll, you may or may not be able to solve all four mysteries.
Check students' solutions.

Fraction Mystery 1
Write a pair of equivalent fractions.

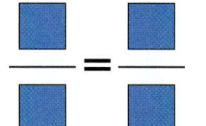

Fraction Mystery 2
Write a fraction with a value greater than 1.

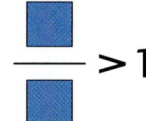

Fraction Mystery 3
Write a fraction with a value greater than $\frac{1}{2}$ and less than 1.

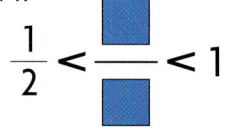

Fraction Mystery 4
Write a fraction that is not in simplest form.

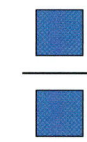

Think It Over!

- **Write About It** Salvador rolled all 2's and 6's. Explain why he couldn't solve all four mysteries.
Using just 2's and 6's, you can't write a fraction with a value greater than $\frac{1}{2}$ and less than 1 (Mystery 3).

- **Stretch Your Thinking** Write a set of ten numbers with which you could solve only two of the mysteries.
Possible answer: 1, 1, 1, 1, 1, 3, 3, 3, 3, 3. Neither Mystery 3 nor Mystery 4 can be solved using these numbers.

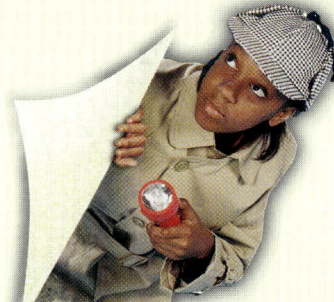

Intervention and Extension Resources

SPECIAL NEEDS

MATERIALS *For each pair* index cards

Have students **use index cards to model fractions.** Working in pairs, students should record each number they roll on a separate index card. Then have them use the index cards to make and stack fractions.

Have pairs use the index cards to model several different fractions. Then allow them to use their cards more than once to find each Fraction Mystery number. Finally, ask pairs to find each Fraction Mystery number using each index card only once.
Check students' work.

KINESTHETIC; VISUAL/SPATIAL

Challenge: Mixed Numbers and Time

Learn how to write times as mixed numbers, and how to add and subtract times.

The ticket agent at the airport told Mario that his flight from Portland, OR, to St. Paul, MN, would take "about $4\frac{1}{2}$ hours." The agent expressed the time as a mixed number.

To write a time as a mixed number, use the fact that there are 60 minutes in 1 hour.

EXAMPLE 1

Mario's flight from Portland to St. Paul took 4 hours 24 minutes. Write the time as a mixed number.

$24 \text{ min} = 24 \times \frac{1}{60} \text{ hr} = \frac{24}{60} \text{ hr}$ *Think: 60 min = 1 hr. So, 1 min = $\frac{1}{60}$ hr.*

$= \frac{2}{5} \text{ hr}$ *Write the fraction in simplest form.*

$4 \text{ hr } 24 \text{ min} = 4\frac{2}{5} \text{ hr}$ *Write the time.*

So, Mario's flight took $4\frac{2}{5}$ hr.

To add or subtract times, use the same methods you use to add or subtract mixed numbers.

EXAMPLE 2

Solve.

A. 2 hr 18 min
    + 3 hr 52 min
    ─────────────
      5 hr 70 min

$= 5 \text{ hr} + (60 + 10) \text{ min}$
$= 5 \text{ hr} + 1 \text{ hr} + \frac{10}{60} \text{ hr}$
$= 6 \text{ hr} + \frac{1}{6} \text{ hr}$
$= 6\frac{1}{6} \text{ hr}$

B. 8 hr 14 min → 7 hr (60 + 14) min
    −3 hr 20 min → −3 hr 20 min
    ──────────────────────────────
                     7 hr 74 min
                    −3 hr 20 min
                    ─────────────
                     4 hr 54 min

$= 4 \text{ hr} + \frac{54}{60} \text{ hr}$
$= 4\frac{9}{10} \text{ hr}$

TALK ABOUT IT

- Tell how you would write the time 7 hours 35 minutes as a mixed number. *Write 35 minutes as $\frac{35}{60}$, simplify the fraction, and then add it to 7 to get $7\frac{7}{12}$ hr.*

TRY IT

Write as a mixed number.

1. 1 hr 30 min $1\frac{1}{2}$ hr
2. 4 hr 55 min $4\frac{11}{12}$ hr
3. 1 hr 19 min $1\frac{19}{60}$ hr
4. 28 min $\frac{7}{15}$ hr
5. 45 min $\frac{3}{4}$ hr
6. 3 hr 42 min $3\frac{7}{10}$ hr

Solve. Write as a mixed number.

7. 5 hr 12 min + 8 hr 18 min
   $13\frac{1}{2}$ hr

8. 6 hr 10 min − 4 hr 20 min
   $1\frac{5}{6}$ hr

Chapters 7–10 **221**

Intervention and Extension Resources

EARLY FINISHERS

Have students **extend their understanding of mixed numbers and time**. Ask them to make a list of common fractional times with which they are familiar, including $\frac{1}{2}, \frac{1}{3}, \frac{1}{4}, \frac{1}{5}$, and $\frac{1}{6}$ of an hour, and multiples of the fractional times. Then have them write the number of minutes that correspond to each fractional unit of time. *Check students' work. Answers should include the following (and appropriate multiples): $\frac{1}{2}$ hr = 30 min, $\frac{1}{3}$ hr = 20 min, $\frac{1}{4}$ hr = 15 min, $\frac{1}{5}$ hr = 12 min, $\frac{1}{6}$ hr = 10 min.*

VISUAL; LOGICAL/MATHEMATICAL

CHALLENGE

Mixed Numbers and Time

Objective To extend the concepts and skills of Chapters 7–10

USING THE PAGE

- Have a volunteer read the introduction. Then direct students' attention to Example 1.

  What fraction that relates the number of minutes in an hour is equivalent to 1 hour? *$\frac{60 \text{ min}}{60 \text{ min}} = 1$ hr*

  If Mario's flight took 4 hours 12 minutes, how would you write the time as a mixed number? *$4\frac{1}{5}$ hr*

- After students have read Example 2, ask:

  Why do you rename 70 minutes in the first example? *70 minutes is greater than 1 hour, and you want to write the answer in simplest form.*

  Explain how to rename to subtract in the second example. *14 is less than 20, so you cannot subtract the minutes. Since 1 hr = 60 min, you must change 8 hr 14 min to 7 hr 74 min.*

- Have students complete the Talk About It and then extend their thinking.

  Reasoning If you wanted to add 15 minutes to $7\frac{7}{12}$ hr, explain how you would find the sum as a mixed number. *Possible answer: 15 min = $\frac{1}{4}$ hr, so add $7\frac{7}{12} + \frac{1}{4} = 7\frac{7}{12} + \frac{3}{12} = 7\frac{10}{12} = 7\frac{5}{6}$ hr*

Try It Before assigning Try It Exercises 1–6, remind students that when they write time as a mixed number, the minutes are expressed as a fraction of an hour. So, the units of time for each of Exercises 1–6 will be expressed as hours.

Have students write Exercises 7–8 in vertical form to solve. Remind students to write these answers as mixed numbers.

Fraction Concepts and Operations **221**

UNIT 3 CHAPTERS 7–10

STUDY GUIDE AND REVIEW

Purpose To help students review concepts and skills presented in Chapters 7–10

USING THE PAGES

✓ Assessment Checkpoint

The Study Guide and Review includes content from Chapters 7–10.

Chapter 7
- 7.1 Divisibility
- 7.2 Prime Factorization
- 7.3 Least Common Multiple and Greatest Common Factor
- 7.4 Problem Solving Strategy: *Make an Organized List*

Chapter 8
- 8.1 Equivalent Fractions and Simplest Form
- 8.2 Mixed Numbers and Fractions
- 8.3 Compare and Order Fractions
- 8.4 Math Lab: Explore Fractions and Decimals
- 8.5 Fractions, Decimals, and Percents

Chapter 9
- 9.1 Estimate Sums and Differences
- 9.2 Math Lab: Model Addition and Subtraction
- 9.3 Add and Subtract Fractions
- 9.4 Add and Subtract Mixed Numbers
- 9.5 Math Lab: Rename to Subtract
- 9.6 Subtract Mixed Numbers
- 9.7 Problem Solving Strategy: *Draw a Diagram*

Chapter 10
- 10.1 Estimate Products and Quotients
- 10.2 Multiply Fractions
- 10.3 Multiply Mixed Numbers
- 10.4 Math Lab: Divide Fractions
- 10.5 Divide Fractions and Mixed Numbers
- 10.6 Problem Solving Skill: *Choose the Operation*
- 10.7 Algebra: Fraction Expressions and Equations

The blue page numbers in parentheses provided with each group of exercises indicate the pages on which the concept or skill was presented. The red number given with each group of exercises identifies the Learning Goal for the concept or skill.

UNIT 3 | Study Guide and Review

VOCABULARY

1. The largest of the common factors of two numbers is the __?__. (p. 151) **greatest common factor**
2. To add unlike fractions, you can write equivalent fractions by using the __?__. (p. 182) **least common denominator**

EXAMPLES

Chapter 7

• Write the prime factorization of a number. (pp. 148–149) **7B**

Find the prime factorization of 150.
150
15×10
$3 \times 5 \times 2 \times 5$
$2 \times 3 \times 5^2$

• Find the GCF and LCM of two or more numbers. (pp. 150–153) **7C**

Find the LCM of 9 and 15.

9: 9, 18, 27, 36, 45 *Find multiples of 9*
15: 15, 30, 45 *and 15.*
LCM = 45 *Find the LCM.*

Chapter 8

• Write fractions in simplest form. (pp. 160–163) **8A**

Write $\frac{18}{30}$ in simplest form.

$\frac{18}{30} = \frac{18 \div 6}{30 \div 6} = \frac{3}{5}$ *Divide by the GCF.*

• Write mixed numbers as fractions and fractions as mixed numbers. (pp. 164–165) **8B**

Write $3\frac{5}{6}$ as a fraction.

$3\frac{5}{6} = \frac{(3 \times 6)}{6} + \frac{5}{6} = \frac{23}{6}$

• Convert among fractions, decimals, and percents. (pp. 169–171) **8C**

Write $\frac{1}{8}$ as a percent.

$\frac{1}{8} = 1 \div 8 = 0.125$ *Change $\frac{1}{8}$ to a decimal.*
$0.125 = 12.5\%$
So, $\frac{1}{8} = 12.5\%$.

EXERCISES

Write the prime factorization of each number in exponent form.

3. 52 $2^2 \times 13$
4. 125 5^3
5. 180 $2^2 \times 3^2 \times 5$
6. 27 3^3
7. 100 $2^2 \times 5^2$
8. 72 $2^3 \times 3^2$

Find the GCF for each set of numbers.

9. 15, 30 **15**
10. 9, 12 **3**
11. 8, 16, 12 **4**

Find the LCM for each set of numbers.

12. 6, 10 **30**
13. 6, 8 **24**
14. 5, 9, 12 **180**

Write the fraction in simplest form.

15. $\frac{8}{12}$ $\frac{2}{3}$
16. $\frac{15}{20}$ $\frac{3}{4}$
17. $\frac{40}{50}$ $\frac{4}{5}$
18. $\frac{36}{48}$ $\frac{3}{4}$
19. $\frac{18}{21}$ $\frac{6}{7}$
20. $\frac{90}{40}$ $\frac{9}{4}$

Write the mixed number as a fraction.

21. $4\frac{2}{3}$ $\frac{14}{3}$
22. $2\frac{8}{9}$ $\frac{26}{9}$
23. $8\frac{1}{2}$ $\frac{17}{2}$

Write the fraction as a mixed number.

24. $\frac{17}{2}$ $8\frac{1}{2}$
25. $\frac{40}{7}$ $5\frac{5}{7}$
26. $\frac{33}{4}$ $8\frac{1}{4}$

Write the decimal as a fraction.

27. 0.75 $\frac{3}{4}$
28. 0.6 $\frac{6}{10}$, or $\frac{3}{5}$
29. 0.43 $\frac{43}{100}$

Write the fraction as a percent.

30. $\frac{5}{8}$ 62.5%
31. $\frac{2}{5}$ 40%
32. $\frac{5}{2}$ 250%

Chapter 9

- **Add and subtract fractions and mixed numbers.** (pp. 186–193) **9A**

Subtract. $4\frac{1}{4} - 2\frac{2}{3}$

$$4\frac{1}{4} = 4\frac{3}{12} = 3\frac{15}{12}$$
$$-2\frac{2}{3} = -2\frac{8}{12} = -2\frac{8}{12}$$
$$\overline{\phantom{-2\frac{2}{3}}} \quad \overline{\phantom{-2\frac{8}{12}}} \quad 1\frac{7}{12}$$

Write equivalent fractions. Rename as needed. Subtract fractions. Subtract whole numbers.

Add or subtract. Write the answer in simplest form.

33. $\frac{3}{4} + \frac{1}{3}$ $\frac{13}{12}$, or $1\frac{1}{12}$
34. $\frac{3}{5} + \frac{1}{2}$ $\frac{11}{10}$, or $1\frac{1}{10}$
35. $\frac{9}{10} - \frac{3}{5}$ $\frac{3}{10}$
36. $\frac{5}{6} - \frac{1}{4}$ $\frac{7}{12}$
37. $2\frac{3}{8} + 3\frac{3}{4}$ $6\frac{1}{8}$
38. $4\frac{3}{5} + 2\frac{1}{3}$ $6\frac{14}{15}$
39. $4\frac{1}{3} - 1\frac{5}{6}$ $2\frac{1}{2}$
40. $5\frac{1}{8} - 3\frac{2}{3}$ $1\frac{11}{24}$

Chapter 10

- **Multiply and divide fractions and mixed numbers.** (pp. 202–213) **10B**

Divide. $3\frac{3}{4} \div 4\frac{1}{2}$

$$3\frac{3}{4} \div 4\frac{1}{2} = \frac{15}{4} \div \frac{9}{2}$$

Write mixed numbers as fractions.

$$= \frac{15}{4} \times \frac{2}{9}$$

Multiply by the reciprocal.

$$= \frac{5}{6}$$

Multiply or divide. Write the answer in simplest form.

41. $\frac{1}{3} \times \frac{2}{5}$ $\frac{2}{15}$
42. $\frac{5}{8} \times \frac{7}{10}$ $\frac{7}{16}$
43. $\frac{5}{9} \div \frac{1}{3}$ $\frac{5}{3}$, or $1\frac{2}{3}$
44. $9 \div \frac{4}{5}$ $\frac{45}{4}$, or $11\frac{1}{4}$
45. $3\frac{3}{4} \times 3\frac{1}{3}$ $12\frac{1}{2}$
46. $3\frac{1}{8} \times 1\frac{1}{5}$ $3\frac{3}{4}$
47. $1\frac{2}{3} \div 3\frac{5}{6}$ $\frac{10}{23}$
48. $1\frac{7}{9} \div 2\frac{2}{5}$ $\frac{20}{27}$

- **Evaluate an algebraic expression using fractions.** (pp. 216–217) **10C**

Evaluate $f + 5\frac{1}{8}$ for $f = 2\frac{1}{3}$.

$f + 5\frac{1}{8}$

$2\frac{1}{3} + 5\frac{1}{8}$ Replace f with $2\frac{1}{3}$.

$2\frac{8}{24} + 5\frac{3}{24} = 7\frac{11}{24}$ Add.

Evaluate each expression.

49. $d + 2\frac{2}{3}$ for $d = 2\frac{5}{6}$ $5\frac{1}{2}$
50. $k \times \frac{3}{4}$ for $k = 1\frac{3}{5}$ $1\frac{1}{5}$
51. $m - 3\frac{4}{9}$ for $m = 6\frac{5}{18}$ $2\frac{5}{6}$
52. $w \div 1\frac{1}{3}$ for $w = 2\frac{1}{2}$ $1\frac{7}{8}$

PROBLEM SOLVING APPLICATIONS

53. Katelynn has two sheets of paper $8\frac{1}{2}$ in. by 11 in. She cuts out a 3-in. square from the center top edge of one of the sheets. Which has the greater perimeter, the sheet with the square cut out or the uncut sheet? How much greater? (pp. 194–195) **cut sheet; 6 in. greater** **9C**

54. Juan goes to the fruit market every 6 days. Anita goes to the market every 4 days. They meet at the market on August 31. When will they meet at the market again? (pp. 154–155) **September 12** **7D**

Chapters 7–10 **223**

Assessment Checkpoint

Portfolio Suggestions The portfolio represents the growth, talents, achievements, and reflections of the mathematics learner. Students might spend a short time selecting work samples for their portfolios and completing A Guide to My Math Portfolio from *Assessment Guide*, page AGxix.

You may want to have students respond to the following questions:

- What new understanding of math have I developed in the past several weeks?
- What growth in understanding or skills can I see in my work?
- What can I do to improve my understanding of math ideas?
- What would I like to learn more about?

For information about how to organize, share, and evaluate portfolios, see *Assessment Guide*, page AGxviii.

Use the item analysis in the **Intervention** chart to diagnose students' errors. You may wish to reinforce content or remediate misunderstandings by using the text pages or lesson resources.

STUDY GUIDE AND REVIEW INTERVENTION

How to Help Options

Learning Goal	Items	Text Pages	Reteach and Practice Resources
7B See page 144C for Chapter 7 learning goals	3–8	148–149	Worksheets for Lesson 7.2
7C See page 144C for Chapter 7 learning goals	9–14	150–153	Worksheets for Lesson 7.3
7D See page 144C for Chapter 7 learning goals	54	154–155	Worksheets for Lesson 7.4
8A See page 144C for Chapter 8 learning goals	15–20	160–163	Worksheets for Lesson 8.1
8B See page 144C for Chapter 8 learning goals	21–26	164–165	Worksheets for Lesson 8.2
8C See page 144C for Chapter 8 learning goals	27–32	169–171	Worksheets for Lesson 8.5
9A See page 144C for Chapter 9 learning goals	33–40	186–189, 190, 191, 192–193	Worksheets for Lessons 9.4, 9.5, 9.6
9C See page 144C for Chapter 9 learning goals	53	194–195	Worksheets for Lesson 9.7
10B See page 144C for Chapter 10 learning goals	41–48	202–205, 206–207, 208–209, 210–215	Worksheets for Lessons 10.2, 10.3, 10.4, 10.5
10C See page 144C for Chapter 10 learning goals	49–52	20–21, 22–25	Worksheets for Lesson 10.2

Fraction Concepts and Operations 223

UNIT 3

PERFORMANCE ASSESSMENT

Purpose To provide performance assessment tasks for Chapters 7–10

USING THE PAGE
- Have students work individually or in pairs as an alternative to formal assessment.
- Use the performance indicators and work samples below to evaluate Tasks A–B.

See *Performance Assessment* for
- a complete scoring rubric, p. PA20, for this unit.
- additional student work samples for this unit.
- copying masters for this unit.

 You may suggest that students place completed Performance Assessment tasks in their portfolios.

Performance Indicators
Task A

A student with a Level 3 paper should

___ Find the product of two mixed numbers using the standard algorithm.

___ Show understanding of how to use the Distributive Property to find the product of two mixed numbers.

___ Exhibit understanding of the Distributive Property by explaining how it is being used incorrectly in a specific example.

Task B

A student with a Level 3 paper should

___ Use fractions to describe parts of a class, hour, day, or week.

___ Determine the decimal equivalent of a fraction or the fractional equivalent of a decimal.

___ Show work and explain how the answers were determined.

Performance Assessment

TASK A • Fraction Fun   See additional answers p.225

Margo finds the product of $3\frac{1}{2}$ and $8\frac{2}{3}$ by writing each as a fraction and multiplying numerators and denominators. Chen said he can find the product by using the Distributive Property. Here's how he starts:

$$3\frac{1}{2} \times 8\frac{2}{3} = (3 + \tfrac{1}{2}) \times (8 + \tfrac{2}{3})$$
$$= 3 \times (8 + \tfrac{2}{3}) + \tfrac{1}{2} \times (8 + \tfrac{2}{3})$$

a. Finish Chen's solution to find the answer.

b. Use Margo's method to find $3\frac{1}{2} \times 8\frac{2}{3}$. Do you get the same answer as with Chen's method?

c. Tom said he thinks $(3 + \tfrac{1}{2}) \times (8 + \tfrac{2}{3}) = 3 \times 8 + \tfrac{1}{2} \times \tfrac{2}{3}$. Is he correct? Explain.

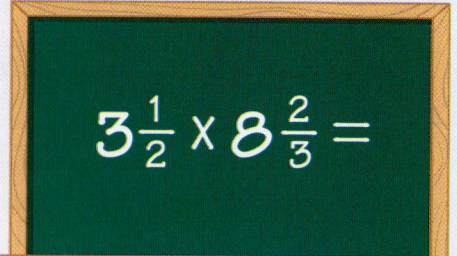

TASK B • Pen Pal   See additional answers p.225

Assume that you have a pen pal and are going to start writing. This is your first letter. Write a short paragraph describing yourself and your activities. Include the following in the paragraph:

a. How you divide your time during a typical week

b. How your school day is divided by classes and other activities

c. What you like to do after school

As you write your letter, use fractions or decimals to name parts of classes, hours, days, or weeks. Each time you use a fraction, write the equivalent decimal in parentheses next to it. Do the reverse each time you use a decimal.

Work Samples for Task A and Task B

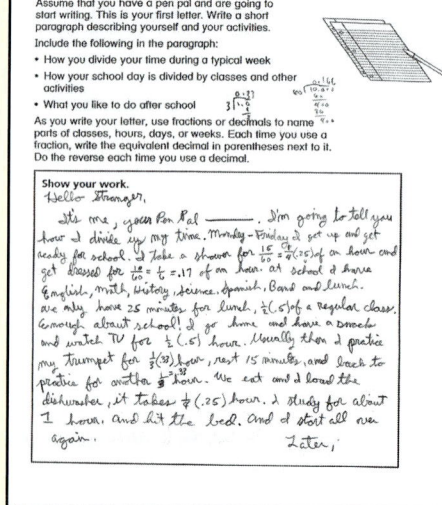

Level 3 This student shows good understanding of the task and has used correct procedures. Computations are correct. All tasks are completed. The last sentence is unclear.

Level 3 The student effectively integrates fractions and decimals into the letter. School activities and nonschool activities are discussed. Accurate equivalents are given.

Mighty Math Astro Algebra
Fractions, Decimals, and Percents

Click 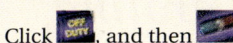, and then .

- Choose Red Level D. Click . Then click and review the topics on translating among decimals, fractions, and percents.
- Close the AstroNet. Complete the mission by sorting the fractions, decimals, and percents.

Practice and Problem Solving

1. Click . Choose Red Level I. Click 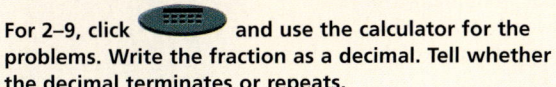 and then . Complete the mission by sorting the fractions, decimals and percents.

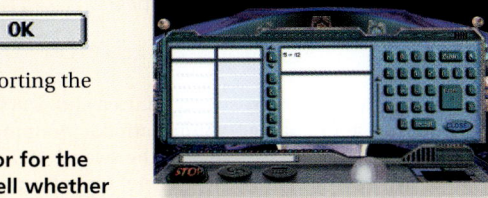

For 2–9, click and use the calculator for the problems. Write the fraction as a decimal. Tell whether the decimal terminates or repeats.

2. $\frac{5}{12}$ 0.41$\overline{6}$, R
3. $\frac{7}{50}$ 0.14, T
4. $\frac{15}{32}$ 0.46875, T
5. $\frac{5}{11}$ 0.$\overline{45}$, R

Write as a fraction.

6. 0.78 $\frac{39}{50}$
7. 0.045 $\frac{9}{200}$
8. 68% $\frac{17}{25}$
9. 85% $\frac{17}{20}$

10. In Ms. Dehmel's class, $\frac{4}{5}$ of the students ride the bus to school. In Mr. Eaton's class, 78% of the students ride the bus. Which class has a greater percent of students who ride the bus?
Ms. Dehmel's; $\frac{4}{5}$ = 80%; 80% > 78%

11. **REASONING** Erika needs to score 91% on her math test to get an "A" for the semester. She gets a score of $\frac{23}{25}$. Does Erika get an "A"? Explain.
Yes; $\frac{23}{25}$ = 92%; 92% > 91%

Multimedia Math Glossary www.harcourtschool.com/mathglossary

12. **Vocabulary** Look up *prime factorization* in the Multimedia Math Glossary. use a factor tree to write the prime factorization of 256. 2^8

Chapters 7–10 **225**

Additional Answers for Performance Task A, PE p. 224

Students should show the following to finish Chen's solution:

$3 \times (8 + \frac{2}{3}) + \frac{1}{2} \times (8 + \frac{2}{3}) = 3 \times 8 + 3 \times \frac{2}{3} + \frac{1}{2} \times 8 + \frac{1}{2} \times \frac{2}{3} = 24 + 2 + 4 + \frac{1}{3} = 30\frac{1}{3}$

Margo's solution: $3\frac{1}{2} \times 8\frac{2}{3} = \frac{7}{2} \times \frac{26}{3} = \frac{7}{1} \times \frac{13}{3} = \frac{91}{3} = 30\frac{1}{3}$

Students should see that they get the same solution. The mistake that Tom is making is that 3 must be multiplied by both 8 and $\frac{2}{3}$, as must $\frac{1}{2}$.

Additional Answers for Performance Task B, PE p. 224

As students write their letters to their imaginary pen pals, they should think of things they do in school or at home every day and estimate the part of a class period, hour, day, or week that they spend on that activity. As they write about these activities, they should add details about their lives as well.

Check their letters for imagination in describing their lives, how well they integrate the use of fractions and decimals, and how accurately they write the decimal equivalents of fractions and fraction equivalents of decimals.

TECHNOLOGY LINKUP

Objective To write fractions as decimals; to identify terminating and repeating decimals; to write decimals and percents as fractions

USING THE PAGE

Remind students that to change a fraction to a decimal, they should divide the numerator by the denominator. To change a decimal to a fraction, they should write a fraction with a denominator of 10, 100, or 1,000, then simplify the fraction.

Using Mighty Math Astro Algebra

You may need to demonstrate for students how to use the Astro Algebra calculator to rewrite fractions, decimals, and percents in other forms.

Practice and Problem Solving

Before students use the Astro Algebra calculator, you may want to have them complete some exercises using paper and pencil, using the calculator to check their answers.

Multimedia Math Glossary

Prime factorization, and all other vocabulary words in this unit, can be found in the Harcourt Multimedia Math Glossary.
www.harcourtschool.com/mathglossary

Fraction Concepts and Operations 225

UNIT 3

PROBLEM SOLVING
On Location in South Carolina

Purpose To provide additional practice for concepts and skills in Chapters 7–10

USING THE PAGE
Trees

- Have students refer to the table comparing tree heights.

How can you use decimals to compare the heights of the trees listed in the table to the height of the water tupelo tree? Use division to write each fraction as a decimal. Round each decimal to two decimal places. black locust: 27 ÷ 38 = 0.71; sassafras: 25 ÷ 57 = 0.44; southern catalpa: 109 ÷ 114 = 0.96

Is it easier to compare a tree's height to the height of the water tupelo tree using fractions or decimals? Explain. Possible answer: decimals, since you can compare them by comparing their digits. To compare fractions, you must first rewrite them with common denominators.

How tall is the sassafras tree? 50 feet

- After Exercise 6, ask this question.

A quaking aspen tree at the Clemson University Arboretum is shorter than the striped maple tree at Paris Mountain State Park. Its height is a factor of the striped maple's height. What is the tallest the quaking aspen could be? 14 feet

Extension Discuss students' solutions to Exercise 6. Then challenge students to make up a similar problem using a diagram. They can compare the height, length, width, depth, weight, mass, or any single dimension of objects or figures. Check students' work.

PROBLEM SOLVING ON LOCATION

In South Carolina

The palmetto tree is South Carolina's state tree.

Trees

South Carolina is a state known for its beautiful trees. In fact, the state takes its nickname, the Palmetto State, from a tree. The palmetto tree, a familiar sight in South Carolina, appears on the state flag and the state seal.

1. South Carolina's tallest sweetgum tree and water tupelo tree are found at Congaree Swamp National Monument. The water tupelo is 114 feet tall and the sweetgum is 159 feet tall. Express the water tupelo's height as a fraction of the sweetgum's height, in simplest form. $\frac{38}{53}$

2. The diameter of the sweetgum tree is shown in the diagram. Write the diameter as a fraction. $\frac{132}{25}$

   $5\frac{7}{25}$ ft

3. A butternut tree in Oconee County, South Carolina, is 71 feet tall. Classify the number 71 as prime or composite. prime

4. A live oak tree on John's Island, near Charleston, South Carolina, is believed to be about 1,400 years old. Write the prime factorization of 1,400. $7 \times 5^2 \times 2^3$

Use Data For 5, use the table.

5. The table shows how the heights of three tall South Carolina trees compared with the height of the water tupelo tree at Congaree Swamp. Order the three trees from tallest to least tall. Southern Catalpa, Black Locust, Sassafras

Heights Compared to Water Tupelo Tree	
Black Locust	$\frac{27}{38}$
Sassafras	$\frac{25}{57}$
Southern Catalpa	$\frac{109}{114}$

Table Rock State Park has the most challenging hiking trails in South Carolina.

State Parks

Each year, thousands of visitors enjoy swimming, boating, fishing, and other outdoor activities in South Carolina's state parks. Many of the parks have hiking trails that allow close-up views of palmettos and others of the state's many varieties of trees.

Use Data For 1–9, use the table.

1. Which trails are the same length?
   Oconee Bell and Sandhills
3. Suppose you hike both of the listed trails at Table Rock. How many miles will you hike in all? $5\frac{2}{5}$ mi
5. How much longer is the Sulphur Springs Trail than the Brissey Ridge Trail? $1\frac{7}{10}$ mi

Mr. Ristorcelli hiked the Pinnacle Mountain Trail every summer for the past 15 years.

6. How far has he hiked altogether? **51 mi**
7. How many times would he have to hike the Yemasee Trail in order to cover the same distance? **153 times**
8. Kendra hiked four trails for a total of $8\frac{7}{10}$ miles. Three of the trails she hiked were Oconee Bell, Sandhills, and Pinnacle Mountain. What was the fourth trail?
   Brissey Ridge
9. One weekend Dean hiked the Sandhills Trail. The next weekend he hiked a trail that was a little more than twice as long. Which trail did he hike the second weekend? **Pinnacle Mountain**

2. Which park has the longest trail listed?
   Paris Mountain
4. Ari hiked the Sandhills Trail at an average rate of $\frac{3}{4}$ mile per hour. How long did the hike take? **2 hr**

South Carolina State Park Trails

Park	Trail	Approx. Length (mi)
Devil's Fork	Oconee Bell	$1\frac{1}{2}$
Lake Warren	Yemasee	$\frac{1}{3}$
Paris Mountain	Sulphur Springs	4
Paris Mountain	Brissey Ridge	$2\frac{3}{10}$
Sesquicentennial	Sandhills	$\frac{3}{2}$
Table Rock	Carrick Creek	2
Table Rock	Pinnacle Mountain	$3\frac{2}{5}$

USING THE PAGE

State Parks

- Direct students' attention to the table.

Suppose you hike both trails listed at Paris Mountain. How many miles will you hike in all? $6\frac{3}{10}$ miles

Eric hiked $\frac{3}{4}$ mile along the Pinnacle Mountain Trail and then realized that he forgot his camera. He returned to the trailhead to get the camera and then hiked the entire trail. How many miles did Eric hike? $4\frac{9}{10}$ miles

Frances hiked the Brissey Ridge Trail 12 times in 1 year. How many miles did she hike in all? $27\frac{3}{5}$ miles

Suppose you want to hike three trails so that your total hiking distance is as close as possible to 6 miles. Which trails should you hike? What will your total hiking distance be? Possible answers: Yemasee, Sulphur Spring, Sandhill, $5\frac{5}{6}$ miles; Yemasee, Brissey Ridge, Pinnacle Mountain, $6\frac{1}{30}$ miles. The latter choice is closest to 6 miles, but some students may presume that the answer should be less than 6 miles.

Extension Have students order the trail lengths from least to greatest. Then have them use their data to write a problem comparing different trails. Ask them to exchange problems with another student and solve. Check students' work.

Teaching Notes

Additional Ideas:

Good Questions to Ask:

Additional Resources:

Notes for Next Time:

UNIT 4 Algebra: Integers

UNIT AT A GLANCE

CHAPTER 11
Algebra: Number Relationships.........226

Lesson 1 — Understand Integers
Lesson 2 — Rational Numbers
Lesson 3 — Compare and Order Rational Numbers
Lesson 4 — Problem Solving Strategy: *Use Logical Reasoning*

CHAPTER 12
Algebra: Operations with Integers....240

Lesson 1 — Math Lab: Model Addition of Integers
Lesson 2 — Add Integers
Lesson 3 — Math Lab: Model Subtraction of Integers
Lesson 4 — Subtract Integers
Lesson 5 — Multiply and Divide Integers
Lesson 6 — Explore Operations with Rational Numbers

UNIT 4 Algebra: Integers
Assessment Options

Assessing Prior Knowledge
Determine whether students have the required prerequisite concepts and skills.
Check What You Know PE pp. 227, 241

Test Preparation
Provide review and practice for chapter and standardized tests.
Standardized Test Prep PE pp. 239, 261
Mixed Review and Test Prep
See the last page of each PE skill lesson.
Study Guide and Review, PE pp. 264–265

Formal Assessment
Assess students' mastery of chapter concepts and skills.
Chapter Review/Test
PE pp. 238, 260
Pretest and Posttest Options
  Chapter Test, Form A
    pp. AG73–74, 77–78
  Chapter Test, Form B
    pp. AG75–76, 79–80
Unit 4 Test • Chapters 11–12, pp. AG81–88

Daily Assessment
Obtain daily feedback on students' understanding of concepts.
Quick Review
  See the first page of each PE lesson.
Mixed Review and Test Prep
  See the last page of each PE skill lesson.
Number of the Day
  See the first page of each TE skill lesson.
Problem of the Day
  See the first page of each TE skill lesson.
Lesson Quiz
  See the *Assess* section of each TE skill lesson.

Performance Assessment
Assess students' understanding of concepts applied to real-world situations.
Performance Assessment (Tasks A–B)
  PE, p. 266; pp. PA30–31

Student Self-Assessment
Have students evaluate their own work.
How Did I Do?, p. AGxvii
A Guide to My Math Portfolio, p. AGxix
Math Journal
  See *Write* in the *Assess* section of each TE skill lesson and TE pages 230B, 250B, 256B.

 Harcourt Electronic Test System Math Practice and Assessment
Make and grade chapter tests electronically.
This software includes:
- multiple-choice items
- free-response items
- customizable tests
- the means to make your own tests

Portfolio
Portfolio opportunities appear throughout the Pupil and Teacher's Editions.
Suggested work samples:
Problem Solving Project TE pp. 226, 240
Write About It, PE pp. 235
Chapter Review/Test PE pp. 238, 260

KEY   **AG** Assessment Guide   **TE** Teacher's Edition   **PA** Performance Assessment   **PE** Pupil Edition

Correlation to STANDARDIZED TESTS

LEARNING GOAL	TAUGHT IN LESSONS	CAT	CTBS/ TERRA NOVA	ITBS	MAT	SAT	YOUR STATE TEST
11A To identify and write integers, opposites, and absolute values	11.1	•		•	•	•	
11B To identify and represent relationships among sets of numbers by using a variety of methods, including number lines	11.2	•	•		•	•	
11C To compare and order rational numbers	11.3	•	•	•	•	•	
11D To solve problems by using an appropriate strategy such as *use logical reasoning*	11.4	•	•	•	•	•	
12A To write sums and differences of integers by using a variety of methods, including models and number lines	12.1, 12.2, 12.3, 12.4	•	•		•	•	
12B To write products and quotients of integers	12.5	•			•	•	
12C To write sums, differences, products, and quotients of rational numbers and combinations of these operations	12.6	•	•		•		

UNIT 4 Algebra: Integers
Technology Links

🌐 The Harcourt Learning Site

Visit The Harcourt Learning Site for related links, activities, and resources:

- Integers activity *(Use with Chapter 12.)*
- *Multimedia Math Glossary*
- E-Lab interactive learning experiences
- professional development and instructional resources

www.harcourtschool.com

Harcourt Math Newsroom Videos

These videos bring exciting news events to your classroom from the leaders in news broadcasting. For each unit, there is a **Harcourt Math Newsroom Video** that helps students see the relevance of math concepts to their lives. You may wish to use the data and concepts shown in the video for real-life problem solving or class projects.

See **Technology Linkup** in the Pupil Edition, page 267.

TECHNOLOGY CORRELATION

Intervention Strategies and Activities This CD-ROM helps you assess students' knowledge of prerequisite concepts and skills.

Mighty Math CD-ROM Series includes *Astro Algebra*. It provides levels of difficulty that increase from A up to Z.

E-Lab is a collection of electronic learning activities.

The chart below correlates technology activities to specific lessons.

LESSON	ACTIVITY/LEVEL	SKILL
11.2	**Astro Algebra** • *Red*, Level L	Classify rational numbers
11.3	**Astro Algebra** • *Red*, Level K **Harcourt Math Newsroom Video** • *Slow Down Light*	Compare and order rational numbers
12.1	**Astro Algebra** • *Red*, Level A	Add integers
12.2	**Astro Algebra** • *Red*, Level A **Calculator Handbook**, *p. 11*	Add integers
12.3	**E-Lab** • *Modeling Subtraction of Integers* **Astro Algebra** • *Red*, Level F	Model subtraction of integers
12.4	**Astro Algebra** • *Red*, Level F **Calculator Handbook**, *p. 12*	Subtract integers
12.5	**Astro Algebra** • *Red*, Levels G, M	Multiply and divide integers

For the Student

 Intervention Strategies and Activities

Review and practice the prerequisite skills for Chapters 11–12.

E-LAB These interactive learning experiences reinforce and extend the skills taught in Chapters 11–12.

- Skill development
- Practice

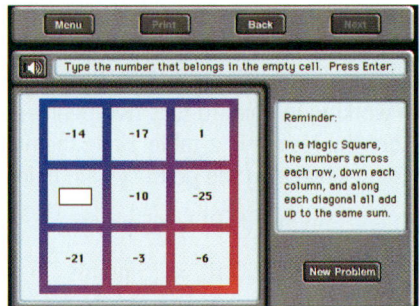

 Mighty Math

The learning activities in this comprehensive math software series complement, enrich, and enhance the Pupil Edition lessons.

Astro Algebra • *Red*

For the Teacher

 Teacher Support Software

- **Intervention Strategies and Activities** Provide instruction, practice, and a check of the prerequisite skills for each chapter.
- **Electronic Lesson Planner** Quickly prepare daily and weekly lessons for all subject areas.
- **Harcourt Electronic Test System Math Practice and Assessment** Edit and customize Chapter Tests or construct unique tests from large item banks.

For the Parent

 The Harcourt Learning Site

- Encourage parents to visit The Harcourt Learning Site to help them reinforce mathematics vocabulary, concepts, and skills with their children.
- Have them click on *Math* for vocabulary, activities, real-life connections, and homework tips for Chapters 11–12.

www.harcourtschool.com

Internet

Teachers can find rational number activities and resources.

Students can learn more about integers and reinforce the critical concepts and skills for Chapters 11–12.

Parents can use The Harcourt Learning Site's resources to help their children with the vocabulary, concepts, and skills needed for Chapters 11–12.

Visit The Harcourt Learning Site
www.harcourtschool.com

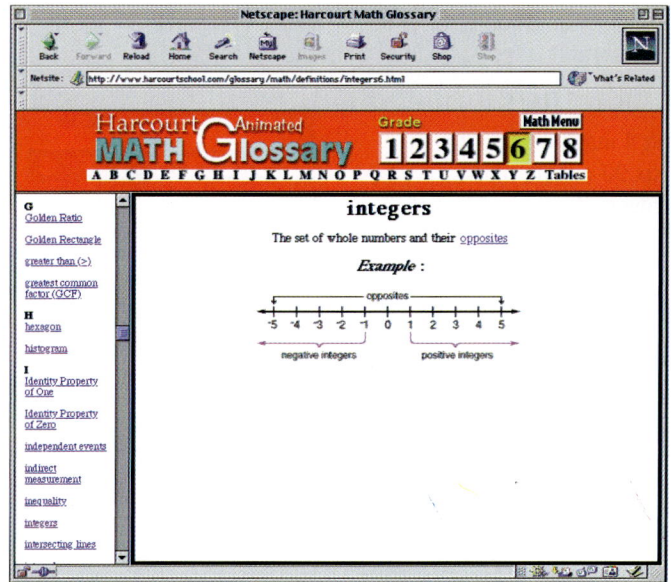

Algebra: Integers 226E

UNIT 4 Algebra: Integers

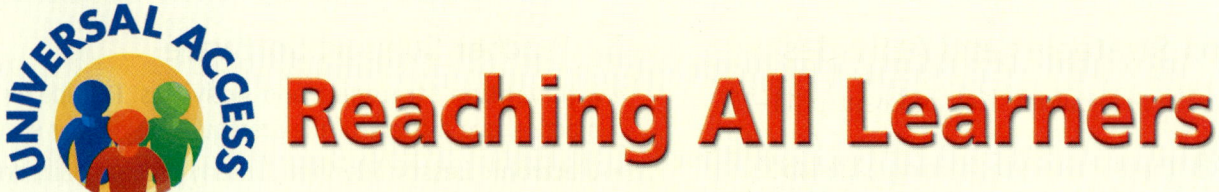

Reaching All Learners

ADVANCED LEARNERS

Challenge students to **apply their knowledge of multiplication of integers**. Have them determine if there is a rule for finding the sign of the product of more than two integers. Suggest that students make and complete a table, such as the following, to develop a pattern and then use the pattern to decide on the rule.

PRODUCT OF INTEGERS		
Integers	Number of Negative Integers	Sign of Product
$^-2 \times 4$	1	$-$
$^-2 \times {^-2}$	2	$+$
$^-2 \times {^-2} \times {^-2}$	3	$-$

Students should conclude that for an even number of negative integers the product is positive, and that for an odd number of negative integers the product is negative. *Use with Lesson 12.5.*

VISUAL; LOGICAL/MATHEMATICAL

SPECIAL NEEDS

MATERIALS *For each pair* 2 pennies

Reinforce the concept of addition and multiplication of integers. Ask students working in pairs to toss two pennies at the same time and record the results in a table. Have them keep a running total of their scores by using the point system below.

$$(H,T) = {^+5} \text{ points} \quad (H,H) = {^-5} \text{ points}$$
$$(T,T) = {^-5} \text{ points} \quad (T,H) = {^+5} \text{ points}$$

Then have students calculate their final scores by multiplying the number of tosses with a positive value by 5; multiplying the number of tosses with a negative value by $^-5$; adding the two products to the running total to determine the final score. *Use with Lessons 12.1–12.2 and 12.5.*

VISUAL
BODILY/KINESTHETIC

BLOCK SCHEDULING

INTERDISCIPLINARY COURSES
- Science — Connect integers to the Celsius and Fahrenheit temperature scales.
- Geography — Use positive and negative integers to compare the altitudes of various mountains and seafloor features.
- Physical Science — Explore the charged particles called electrons and protons.
- Physical Education — Use absolute value to describe how runners' times differ from the mean. If the mean is 11 sec, then 14 sec and 8 sec each differ from the mean by 3.

COMPLETE UNIT

Unit 4 may be presented in
- six 90-minute blocks.
- seven 75-minute blocks.

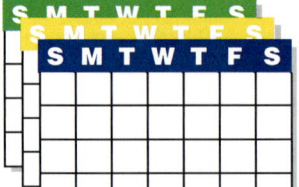

INTERDISCIPLINARY SUGGESTIONS

PURPOSE To connect *Algebra: Integers* to other subjects with these activities

CHAPTER 11 — Physical Education

Students research the procedure and ratings for the sit-and-reach flexibility test. For example, Good: 2 to 4 in.; Average: 0 to 2 in.; Fair: $^-2$ to 0 in.; and so on.

CHAPTER 12 — Earth Science

In the winter, people feel colder when the wind is blowing. Students write a summary explaining any patterns they see in the wind-chill table on page TE250B.

VISUAL; VERBAL/LINGUISTIC

ACTIVITIES AND GAMES FOR HOME OR SCHOOL

You may choose to use this activity and game in the classroom or send it home for students to do with family members.

ACTIVITY • *Hop the Number Line*

PURPOSE To add integers
Use after Lesson 12.2.

VARIATIONS Have students subtract instead of add the integers.

GAME • *80 Feet Down*

PURPOSE To multiply and divide integers
Use with Lesson 12.5.

KINESTHETIC; VISUAL/SPATIAL

ENGLISH LANGUAGE LEARNERS

VOCABULARY PREVIEW Have students make a math dictionary for the vocabulary in the unit. The dictionary will provide students with an easy review tool for tests and will help them build toward independence when answering word problems. Have students make a three-column chart and label the columns. As students learn about each new term, they can fill in their dictionary. *Use with Lessons 11.1–12.1.*

Term	Example	Explanation

Negative Integer Relate the terms *negative integer* and *positive integer* to right and left of zero. *Use with Lesson 11.1.*

Absolute Value The absolute value of a number is its distance from zero. Have students work with a number line and ask questions about positive and negative integers and their distance from zero. *Use with Lesson 11.1.*

VISUAL; LOGICAL/MATHEMATICAL

LITERATURE CONNECTIONS

These books provide students with additional ways to explore integers and other rational numbers.

Snow Bound by Harry Mazer (Bantam, 1975) tells of two teen runaways who find themselves trapped in a car in a snowstorm and who must work together to survive.

- The average minimum temperature in upstate New York is $^{-}8°C$. Have students find the current temperature in °C and tell how many degrees warmer it is compared to that average. *Use with Lesson 12.4.*

The Facts on File Children's Atlas by David and Jill Wright (Facts on File, Inc., 1993) includes photos, maps, comparative facts, figures, puzzles, and activity boxes that enhance the study of geography.

- Use the fact box on Asia to find the locations of and the difference between the highest and lowest points in that continent. *Use with Lessons 12.3–12.4.*

Algebra: Integers **226G**

CHAPTER 11 Algebra: Number Relationships

CHAPTER PLANNER

PACING OPTIONS
Compacted	4 Days
Expanded	7 Days

Getting Ready for Chapter 11 • Assessing Prior Knowledge and INTERVENTION (See PE and TE page 227.)

LESSON	NCTM STANDARDS	PACING	VOCABULARY*	MATERIALS	RESOURCES AND TECHNOLOGY
11.1 Understand Integers pp. 228–229 **Objective** To identify integers and to find absolute value	1, 2, 6, 7, 8	1 Day	integers, opposites, positive integers, negative integers, absolute value		Reteach, Practice, Challenge, Problem Solving 11.1 Worksheets; Extra Practice p. H42, Set A; Transparency 11.1; Math Jingles™ CD 5–6
11.2 Rational Numbers pp. 230–233 **Objective** To classify sets of numbers and to find another rational number between two rational numbers	1, 2, 6, 8	2 Days	ratio, rational number, Venn diagram		Reteach, Practice, Challenge, Problem Solving 11.2 Worksheets; Extra Practice p. H42, Set B; Transparency 11.2; Astro Algebra • Red
11.3 Compare and Order Rational Numbers pp. 234–235 **Objective** To compare and order rational numbers	1, 2, 6, 8	1 Day			Reteach, Practice, Challenge, Problem Solving 11.3 Worksheets; Extra Practice p. H42, Set C; Transparency 11.3; Astro Algebra • Red; Harcourt Math Newsroom Video • *Slow Down Light*
11.4 Problem Solving Strategy: *Use Logical Reasoning* pp. 236–237 **Objective** To solve problems by using the strategy *use logical reasoning*	1, 6, 7	1 Day			Reteach, Practice, Challenge, Reading Strategy 11.4 Worksheets; Transparency 11.4; Problem Solving Think Along, p. TR1

Ending Chapter 11 • **Chapter 11 Review/Test**, p. 238 • **Standardized Test Prep**, p. 239

***Boldfaced** terms are new vocabulary. Other terms are review vocabulary.

CHAPTER AT A GLANCE

Vocabulary Development

The boldfaced words are the new vocabulary terms in the chapter. Have students record the definitions in their Math Journals.

integers, p. 228
opposites, p. 228
positive integers, p. 228
negative integers, p. 228
absolute value, p. 228
ratio, p. 230
rational number, p. 230
Venn diagram, p. 230

Vocabulary Cards
Have students use the Vocabulary Cards on *Teacher's Resource Book* pp. TR129–132 to make graphic organizers or word puzzles. The cards can also be added to a file of mathematics terms.

NCTM Standards

1. **Number and Operations**
   Lessons 11.1, 11.2, 11.3, 11.4
2. **Algebra**
   Lessons 11.1, 11.2, 11.3
3. **Geometry**
4. **Measurement**
5. **Data Analysis and Probability**
6. **Problem Solving**
   Lessons 11.1, 11.2, 11.3, 11.4
7. **Reasoning and Proof**
   Lessons 11.1, 11.4
8. **Communication**
   Lessons 11.1, 11.2, 11.3
9. **Connections**
10. **Representation**

Writing Opportunities

PUPIL EDITION
- Write a Problem, p. 229
- What's the Error?, p. 233
- Write About It, p. 235
- What's the Question?, p. 237

TEACHER'S EDITION
- Write—See the *Assess* section of each TE lesson.
- Writing in Mathematics, p. 230B

ASSESSMENT GUIDE
- How Did I Do?, p. AGxvii

Family Involvement Activities

These activities provide:
- Letter to the Family
- Math Vocabulary
- Family Game
- Practice (Homework)

Family Involvement Activities, p. FA41

226I

CHAPTER 11 Algebra: Number Relationships

MATHEMATICS ACROSS THE GRADES

SKILLS TRACE ACROSS THE GRADES

GRADE 5
Represent and order integers on a number line; find opposites and absolute values of integers

GRADE 6
Identify and write integers, opposites, and absolute values; represent relationships among sets of numbers using a variety of methods; compare and order rational numbers

GRADE 7
Compare and order integers and rational numbers; represent relationships among sets of numbers

SKILLS TRACE FOR GRADE 6

LESSON	FIRST INTRODUCED	TAUGHT AND PRACTICED	TESTED	REVIEWED
11.1	Grade 4	PE pp. 228–229, H42, p. RW51, p. PW51, p. PS51	PE p. 238, pp. AG73–76	PE pp. 238, 239, 268–269
11.2	Grade 6	PE pp. 230–233, H42, p. RW52, p. PW52, p. PS52	PE p. 238, pp. AG73–76	PE pp. 238, 239, 268–269
11.3	Grade 6	PE pp. 234–235, H42, p. RW53, p. PW53, p. PS53,	PE p. 238, pp. AG73–76	PE pp. 238, 239, 268–269
11.4	Grade 4	PE pp. 236–237, p. RW54, p. PW54, p. PS54	PE p. 238, pp. AG73–76	PE pp. 238, 239, 268–269

KEY PE Pupil Edition PS Problem Solving Workbook RW Reteach Workbook
 PW Practice Workbook AG Assessment Guide

Looking Back  Prerequisite Skills

To be ready for Chapter 11, students should have the following understandings and skills:

- **Vocabulary**—*positive numbers, negative numbers, whole numbers*
- **Locate Points on a Number Line**—locate integers on a number line
- **Sets of Numbers**—classify numbers as counting, whole, even, odd
- **Compare Fractions**—compare fractions as $<$, $>$, or $=$
- **Temperature**—read temperature on a thermometer

Check What You Know

Use page 227 to determine students' knowledge of prerequisite concepts and skills.

Intervention

Help students prepare for the chapter by using the intervention resources described on TE page 227.

Looking at Chapter 11  Essential Skills

Students will

- develop skill identifying and writing integers, opposites, and absolute values.
- **understand the relationships among the sets of rational numbers, integers, and whole numbers.**
- develop skill comparing and ordering rational numbers.
- use the strategy *use logical reasoning* to solve problems.

EXAMPLE

Find a rational number between $^-5.5$ and $^-5.0$.

Venn Diagram	Number Line
Rational Numbers / Integers / Whole Numbers	Think of a number line to find a number between the two numbers. $^-5.5$ $^-5.4$ $^-5.3$ $^-5.2$ $^-5.1$ $^-5.0$ So $^-5.4$, $^-5.3$, and $^-5.2$ are some of the numbers between $^-5.5$ and $^-5.0$.

Looking Ahead  Applications

Students will apply what they learn in Chapter 11 to the following new concepts:

- Operations with Integers (Chapter 12)
- Addition and Subtraction Equations (Chapter 14)
- Expressions with Squares and Square Roots (Chapter 13)
- Multiplication and Division Equations (Chapter 15)

Algebra: Number Relationships   226K

CHAPTER 11

Algebra: Number Relationships

INTRODUCING THE CHAPTER

Tell students that integers are used to describe real-life relationships. Have students focus on the photograph. Ask them what kind of integer they would use to describe how many feet below the summit the climbers are. **a negative integer**

USING DATA

To begin the study of this chapter, have students

- Name an integer that describes the altitude that is 500 ft below the base camp. **⁻500**
- Name an integer that describes any altitude between Camp 2 and Camp 3. **Possible answer: ⁺10,000**
- Make a graph to show the camp and altitude data in the table. **Check students' work.**

PROBLEM SOLVING PROJECT

Purpose To use integers to solve a problem

Grouping pairs or small groups

Background Mt. Everest in Nepal is the highest mountain in the world at 29,028 ft above sea level.

Analyze, Choose, Solve, and Check

Have students

- Use an almanac or other reference to find the five highest mountains in North America and the five highest mountains in South America.
- Record the altitudes of the mountains on a graph to the nearest hundred feet.
- Order the mountains from highest to lowest.

Check students' work.

 Suggest that students place the graphs and ordered data in their portfolios.

CHAPTER 11 Algebra: Number Relationships

656 ft

The elevation of Mt. McKinley is 20,320 feet. It is the highest mountain in North America and has one of Earth's steepest vertical rises. With the high altitude, unpredictable weather, and steep, icy slopes, this mountain is a challenge to climb.

PROBLEM SOLVING Climbers are flown into base camp, where they start their climb at 7,200 ft. If a climb takes 20 days, what is the average altitude gained per day?

Mt. McKINLEY

Camp	Altitude	Possible Time
Base Camp	7,200 ft	0 day
Camp 2	9,500 ft	1st day
Camp 3	11,000 ft	3rd day
Camp 4	14,200 ft	10th day
Camp 5	16,200 ft	12th day
High Camp	17,200 ft	15th day
Summit	20,320 ft	20th day

226 Chapter 11

Why learn math? Explain that mountain climbers can use integers to describe the distance they have climbed or have left to climb. For example, 250 ft from the top of the mountain can be described as ⁻250 ft since it is 250 ft below the summit. Ask: If you were a mountain climber, how would you use an integer to describe where you are if you climbed 2,500 ft from the base of the mountain? **⁺2,500 ft**

Check What You Know

Use this page to help you review and remember important skills needed for Chapter 11.

✓ Vocabulary

Choose the best term from the box.

> positive numbers
> negative numbers
> whole numbers

1. Numbers to the left of zero on the number line are ___?___. **negative numbers**
2. Numbers greater than zero are ___?___. **positive numbers**

✓ Locate Points on a Number Line (For Intervention, see p. H13.)

Copy each number line. Graph the numbers on the number line. **Check students' answers.**

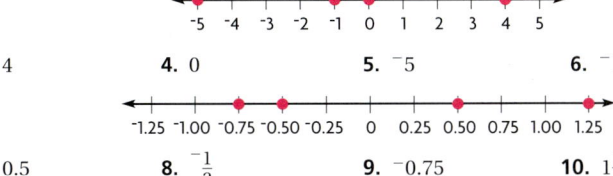

3. 4   4. 0   5. ⁻5   6. ⁻1

7. 0.5   8. $\frac{-1}{2}$ 9. ⁻0.75   10. $1\frac{1}{4}$

✓ Sets of Numbers (For Intervention, see p. H12.)

Give four examples from each set of numbers. **Possible answers are given.**

11. whole numbers
    0, 1, 2, 3, 4
12. counting numbers
    1, 2, 3, 4
13. odd numbers
    1, 3, 5, 7

✓ Compare Fractions (For Intervention, see p. H12.)

Compare. Write <, >, or = for each ●.

14. $9\frac{1}{2}$ ● $9\frac{3}{4}$ **<**
15. $\frac{2}{3}$ ● $\frac{3}{2}$ **<**
16. $\frac{3}{4}$ ● $\frac{3}{5}$ **>**
17. $\frac{16}{2}$ ● $\frac{24}{3}$ **=**
18. $2\frac{3}{5}$ ● $2\frac{1}{4}$ **>**
19. $\frac{12}{2}$ ● $2\frac{1}{12}$ **>**

✓ Temperature (For Intervention, see p. H12.)

Tell the temperature shown by each letter on the thermometer.

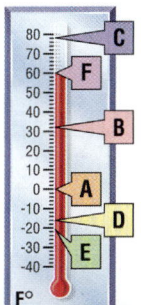

20. A **0°F**
21. B **32°F**
22. C **78°F**
23. D **⁻16°F**
24. E **⁻20°F**
25. F **60°F**

LOOK AHEAD

In Chapter 11 you will
- identify integers and rational numbers
- compare and order rational numbers
- use number lines

LESSON 11.1

Understand Integers

LESSON PLANNING

Objective To identify integers and to find absolute value

Intervention for Prerequisite Skills
Locate Points on a Number Line, Temperature (For intervention strategies, see page 227.)

NCTM Standards
1. Number and Operations
2. Algebra
6. Problem Solving
7. Reasoning and Proof
8. Communication

Vocabulary

integers all whole numbers and their opposites

opposites pairs of integers that are the same distance from 0

positive integers integers greater than 0

negative integers integers less than 0

absolute value the distance of an integer from 0 on the number line

Math Background

Integers include the set of whole numbers and their opposites. Integers allow us to find solutions to equations such as $x + 4 = 2$, and answers to subtraction problems such as $4 - 9$.

These ideas will help students understand integers.

- Every integer has an opposite. The opposite of a positive integer is negative, the opposite of a negative integer is positive, and the opposite of zero is itself, zero.

- The absolute value of an integer is the integer's distance from zero on the number line. Because distance is always non-negative, the absolute value of an integer is always non-negative.

WARM-UP RESOURCES

 **NUMBER OF THE DAY**

Two consecutive whole numbers have a product of 110. What is the greater number? 11

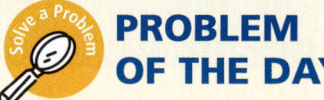 **PROBLEM OF THE DAY**

One side of Jessica's square array is 2 tiles longer than a side of Dave's square array. Together they use a total of 100 tiles. How many tiles are on each side of Dave's array? 6 tiles on a side

Solution Problem of the Day tab, p. PD11

 **DAILY FACTS PRACTICE**

Have students practice addition facts by completing Set B of *Teacher's Resource Book*, p. TR101.

INTERVENTION AND EXTENSION RESOURCES

ALTERNATIVE TEACHING STRATEGY

Materials red and blue chalk or markers

Use a thermometer to **provide examples of applications of negative numbers.** Make a large thermometer showing temperatures from $^-10°F$ to $20°F$ without the signs. Have students identify temperatures above zero and below zero. Ask volunteers to insert the positive signs in red and the negative signs in blue on the thermometer. Point out that positive temperatures are warmer than negative temperatures, so positive numbers are greater than negative numbers. Use the thermometer to illustrate the concept of absolute value and have students identify the absolute values of several positive and negative integers.

Check students' work.

VISUAL
BODILY/KINESTHETIC

ENGLISH LANGUAGE LEARNERS

Reinforce vocabulary associated with integers. Draw a large number line from $^-10$ to $^+10$. Display these vocabulary words above the number line: *integers, opposites, positive integers, negative integers*.

Have volunteers find the number that matches a description that you give. For example,

- negative 2 $^-2$
- positive 8 $^+8$
- the opposite of 6 $^-6$
- a positive integer Possible answer: $^+5$

Then, reverse the process by asking students to practice describing the integer you point out on the number line. Check students' work.

VISUAL
BODILY/KINESTHETIC

MULTISTEP AND STRATEGY PROBLEMS

The following multistep and strategy problems are provided in Lesson 11.1:

Page	Item
229	27, 30

VOCABULARY STRATEGY

Have students **illustrate the integer vocabulary.** Make a bulletin board display with separate areas labeled: *integers, positive integers, negative integers, opposites,* and *absolute value*.

Group students and assign each group an area. Have them make and display examples to illustrate the concept. Discuss their completed bulletin board and refer to it as needed.

VISUAL
VERBAL/LINGUISTIC

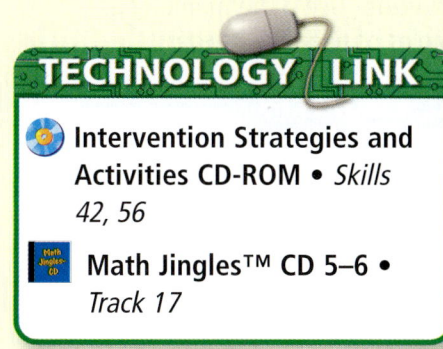

Intervention Strategies and Activities CD-ROM • *Skills 42, 56*

Math Jingles™ CD 5–6 • *Track 17*

LESSON 11.1 ORGANIZER

Objective To identify integers and to find absolute value

Vocabulary integers, opposites, positive integers, negative integers, absolute value

1 Introduce

QUICK REVIEW provides review of prerequisite skills.

Why Learn This? You can use this skill to describe positive or negative values, such as bank deposits and withdrawals. *Share the lesson objective with students.*

2 Teach

Guided Instruction

- After introducing the lesson vocabulary, draw students' attention to the number line.

REASONING Are there any integers between $^+1$ and $^+2$? Explain. No; integers are positive whole numbers and their opposites.

- As students look at the number line showing absolute value, ask:

Why do both 3 and $^-3$ have the same absolute value? They are the same distance from 0.

REASONING Relate 0 to the absolute value of integers. Absolute value is defined as the distance from zero on a number line.

Modifying Instruction Use a thermometer as you discuss examples of negative integers.

ADDITIONAL EXAMPLES

Example 1, p. 228

Name an integer to represent the situation:

A. a loss of 9 yards in a football game $^-9$
B. a temperature of 3° above zero $^+3$
C. a bank withdrawal of $40 $^-40$

Example 2, p. 228

Use the number line to find absolute value.
A. $|^-2|$ 2 B. $|^+3|$ 3
C. $|^+1|$ 1 D. $|^+4|$ 4

228 Chapter 11

LESSON 11.1

Understand Integers

Learn how to identify integers and find absolute value.

QUICK REVIEW
Order from greatest to least.
1. 1, 0, 3   2. 21, 17, 19   3. 46, 64, 55
4. 201, 199, 200   5. 800, 808, 880

New Orleans is located along the Mississippi River.

Vocabulary
integers
opposites
positive integers
negative integers
absolute value

Mount Sunflower in Kansas is 4,039 ft above sea level. New Orleans is 8 ft below sea level. Sea level equals 0 ft. You can use the integers $^+4,039$ and $^-8$ to represent these elevations.

Integers include all whole numbers and their **opposites**. Each integer has an opposite that is the same distance from 0 but on the opposite side of 0. The opposite of positive 8 ($^+8$) is negative 8 ($^-8$). The opposite of 0 is 0.

Integers greater than 0 are **positive integers**. Integers less than 0 are **negative integers**. The integer 0 is neither positive nor negative.

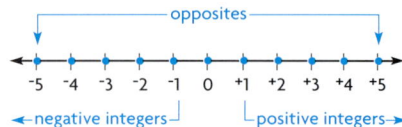

Negative integers are written with a negative sign, $^-$.

Positive integers can be written with or without a positive sign, $^+$.

EXAMPLE 1

Name an integer to represent the situation.
A. a gain of 12 yd   B. 30° below zero   C. a deposit of $100
   $^+12$ $^-30$ $^+100$

The **absolute value** of an integer is its distance from 0. Look at $^+3$ and $^-3$. They are both 3 units from 0.

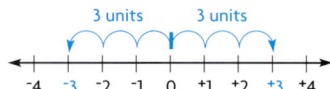

Write: $|^-3| = 3$ Read: The absolute value of negative three is three.
Write: $|^+3| = 3$ Read: The absolute value of positive three is three.

EXAMPLE 2

Use the number line to find each absolute value.
A. $|^-4|$ B. $|^+2|$ C. $|^-1|$ D. $|^+3|$
   4   2   1   3

228 Chapter 11

RETEACH 11.1

Understand Integers

A diver started out at the bottom of the ocean, 250 feet below sea level. He came to the surface and then climbed a hill 300 feet above sea level. How can you represent these numbers?
Integers are numbers that can show opposite directions.
250 feet below sea level is $^-250$ ft.   300 feet above sea level is $^+300$ ft.

Every integer, except zero, has an opposite.

The absolute value of a number is used to show how far a number is from 0.
Find $|^-2|$. This is read as the absolute value of negative 2.

So, $|^-2| = 2$.

Write the opposite integer.

1. $^-8$	2. $^+7$	3. $^+11$	4. $^-9$	5. $^-14$	6. $^+17$
$^+8$	$^-7$	$^-11$	$^+9$	$^+14$	$^-17$
7. $^-3$	8. $^+12$	9. $^+23$	10. $^-30$	11. $^+33$	12. $^-50$
$^+3$	$^-12$	$^-23$	$^+30$	$^-33$	$^+50$

Find the absolute value.

13. $|^-6|$	14. $|^+6|$	15. $|^-8|$	16. $|^-21|$	17. $|^+13|$	18. $|^-26|$
6	6	8	21	13	26
19. $|^-45|$	20. $|^+56|$	21. $|^-77|$	22. $|^+92|$	23. $|^-345|$	24. $|^+880|$
45	56	77	92	345	880

PRACTICE 11.1

Understand Integers

Vocabulary

Complete.
1. ___Integers___ include all whole numbers and their opposites.
2. The ___absolute value___ of an integer is its distance from 0.

Write an integer to represent each situation.
3. earning 7 dollars   4. digging a hole 2 feet deep
   $^+7$ $^-2$
5. taking 10 steps backward   6. climbing up a mountain 20 feet
   $^-10$ $^+20$

Find the absolute value.

7. $|^-3|$	8. $|^+3|$	9. $|^-2|$	10. $|^-6|$	11. $|^+9|$	12. $|^-15|$
3	3	2	6	9	15
13. $|^-32|$	14. $|^+32|$	15. $|^-47|$	16. $|^+78|$	17. $|^-180|$	18. $|^+574|$
32	32	47	78	180	574

Write the opposite integer.

19. $^-5$	20. $^+13$	21. $^+21$	22. $^-19$	23. $^-25$	24. $^+37$
$^+5$	$^-13$	$^-21$	$^+19$	$^+25$	$^-37$

Mixed Review

Multiply. Write the answer in simplest form.

25. $\frac{1}{5} \times \frac{6}{7}$ 26. $\frac{4}{9} \times \frac{3}{5}$ 27. $\frac{4}{5} \times 30$

$\frac{6}{35}$ $\frac{4}{15}$ 24

28. $2\frac{7}{10} \times \frac{2}{3}$ 29. $3\frac{3}{4} \times 2\frac{2}{5}$ 30. $1\frac{1}{2} \times 3\frac{1}{3}$

$\frac{9}{5}$, or $1\frac{4}{5}$ 9   5

CHECK FOR UNDERSTANDING

Think and Discuss — Look back at the lesson to answer each question.

1. **Give an example** of a scale in real life in which zero is used together with other integers. Possible answer: temperature
2. **Tell** what the absolute value of an integer is. It is its distance from zero on a number line.

Guided Practice — Write an integer to represent each situation.

3. 350 ft below sea level −350
4. an increase of 78 points +78
5. 14 degrees below zero −14

Write the opposite integer.

6. −289 +289
7. −25 +25
8. +315 −315
9. +742 −742
10. +993 −993

PRACTICE AND PROBLEM SOLVING

Independent Practice — Write an integer to represent each situation.

11. a $5.00 decline in value −5
12. an increased attendance of 477 +477
13. a gain of 12,000 ft in altitude +12,000
14. a decrease of 50 points −50

Write the opposite integer.

15. −2 +2
16. −14 +14
17. −31 +31
18. +88 −88
19. +207 −207

Find the absolute value.

20. |+390| 390
21. |−28| 28
22. |−727| 727
23. |+660| 660
24. |+795| 795

Problem Solving Applications

25. The elevation of the Dead Sea is about 1,310 ft below sea level. Write the elevation using an integer. −1,310 ft
26. **REASONING** What values can n have if $|n| = 5$? +5 or −5
27. Chris has a stack of coins containing 5 quarters, 5 dimes, and 15 nickels. What fraction of the number of coins are quarters? $\frac{1}{5}$
28. **Write a problem** about two campers who camped at altitudes of 9,470 ft and 7,200 ft. Use positive and negative integers to describe the change in elevation. Check students' problems.

The Dead Sea

MIXED REVIEW AND TEST PREP

29. Multiply. $\frac{4}{5} \times \frac{3}{4}$ (p. 202) $\frac{12}{20}$, or $\frac{3}{5}$
30. Lisa works for $6.50 an hour, 3 hours a day, 4 days a week. How many weeks will it take her to earn more than $300? (p. 76) 4 weeks
31. Find the value of the expression $9 \times (12 − 4) \div 2^3 + 4$ (p. 44) 13
32. Write the prime factorization of 72. (p. 148) $3 \times 3 \times 2 \times 2 \times 2$, or $3^2 \times 2^3$
33. **TEST PREP** Six soccer teams compete for the regional play-offs. Each team plays each of the other teams only once. How many games do they play? (p. 154) C

   A 6   B 12   C 15   D 30

EXTRA PRACTICE page H42, Set A

229

---

3 Practice

COMMON ERROR ALERT

Students may read numbers such as +3 and −2 as "plus 3" and "minus 2." Urge them to say "positive 3" and "negative 2." Explain that the words *plus* and *minus* are used for addition or subtraction of numbers.

Guided Practice

Do Check for Understanding Exercises 1–10 with your students. Identify those having difficulty and use lesson resources to help.

Independent Practice

Note that Exercises 27 and 30 are **multistep or strategy problems**. Assign Exercises 11–28.

Algebraic Thinking As students approach Exercise 28, remind them that some equations can have more than one solution.

Scaffolded Instruction Use the prompts on Transparency 11 to guide instruction for the multistep or strategy problem in Exercise 30.

Transparency 11

MIXED REVIEW AND TEST PREP

Exercises 29–33 provide **cumulative review** (Chapters 1–11).

4 Assess

Summarize the lesson by having students:

DISCUSS When you write an integer, what does the + or − sign tell you about where the integer is? Possible answer: The + sign indicates that the integer is to the right of 0, and the − sign indicates that the integer is to the left of 0.

WRITE What are integers? Integers are all whole numbers and their opposites.

Lesson Quiz

Transparency 11.1

Write the opposite integer.

1. +60 −60
2. −100 +100
3. +75 −75
4. −97 +97

Find the absolute value.

5. |−13| 13
6. |+27| 27
7. |−350| 350
8. |+105| 105

229

LESSON 11.2

Rational Numbers

LESSON PLANNING

Objective To classify sets of numbers and to find another rational number between two rational numbers

Intervention for Prerequisite Skills
Locate Points on a Number Line, Sets of Numbers, Compare Fractions (For intervention strategies, see page 227.)

NCTM Standards
1. Number and Operations
2. Algebra
6. Problem Solving
8. Communication

Vocabulary
ratio a comparison of two numbers

rational number any number that can be written as a ratio $\frac{a}{b}$ where a and b are integers and $b \neq 0$

Venn diagram a diagram that shows the relationships between sets

Math Background
The set of rational numbers includes fractions and mixed numbers as well as integers and decimals. Any number that can be written in the form $\frac{a}{b}$, where a and b are integers and $b \neq 0$, is a rational number.

To help students understand rational numbers, stress these points.

- A mixed number can be written in the form $\frac{a}{b}$ by multiplying the whole number by the denominator of the fraction, adding the numerator, and writing the result over the denominator.

- Any terminating decimal can be written in the form $\frac{a}{b}$ by using place value: $4.5 = 4\frac{5}{10} = \frac{45}{10}$.

- You can find a rational number between any two given rational numbers by finding their mean.

WARM-UP RESOURCES

 **NUMBER OF THE DAY**

Write the opposite of the number representing the day of the month. **Possible answer for April 4: ⁻4**

 **PROBLEM OF THE DAY**

I am a palindrome number, and the sum of my 7 digits is 25. My tens digit is 4 times as great as my ones digit and 4 more than my hundreds digit. My thousands digit and my thousandths digit are each 7 greater than my ones digit. What number am I? **8,041.408**

Solution Problem of the Day tab, p. PD11

 **DAILY FACTS PRACTICE**

Have students practice subtraction facts by completing Set C of *Teacher's Resource Book*, p. TR101.

INTERVENTION AND EXTENSION RESOURCES

ALTERNATIVE TEACHING STRATEGY ELL

Materials *For each student* ruler, p. TR22; tracing paper

Reinforce the concept of rational numbers. Have students draw a number line from $^-5$ to $^+5$ on tracing paper, marking equal increments. After discussing positive rational numbers, ask students to identify 10 positive rational numbers. Have them write the numbers in the form $\frac{a}{b}$ and graph them on the number line.

Then have students fold the number line at the zero point and mark the opposites of all the rational numbers they have graphed so far.

Discuss the negative rational numbers as opposites, and have students write them in the form $\frac{a}{b}$. Finally, have them graph another negative rational number between two pairs of adjacent numbers.

Check students' work.

See also page 232.

VISUAL
BODILY/KINESTHETIC

MULTISTEP AND STRATEGY PROBLEMS

The following multistep and strategy problems are provided in Lesson 11.2:

Page	Item
233	39, 40

WRITING IN MATHEMATICS

Have students copy the Venn diagram showing the **relationship between whole numbers, integers, and rational numbers.** Then, write a paragraph describing how to determine into which box to place a number such as 2.5.

Possible answer: First check to see if the number is a whole number or an integer. 2.5 is neither. Then, to decide if it is a rational number, see if the number can be written as a ratio of two whole numbers. Because $2.5 = 2\frac{5}{10} = \frac{25}{10}$, it is a rational number. If the number could not be written as a whole number, integer, or rational number, it would not be placed in any of the boxes.

GEOMETRY CONNECTION

Materials reference materials

Enhance students' knowledge of a commonly-used irrational number. This decimal used in geometry is the approximate value of π, a number that never terminates or repeats.

3.14159265358979...

The fact that it never terminates or repeats was proven in 1767 by Johann Lambert. Have students research pi. Possible answer: In 1989, Yasumasa Kanada and Yoshiaki Tamura proved once again when they computed π to 1,073,740,000 places that it never terminates or repeats.

AUDITORY
LOGICAL/MATHEMATICAL

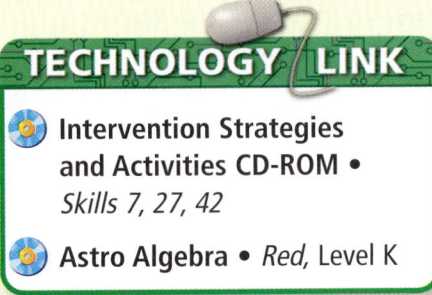

TECHNOLOGY LINK

- Intervention Strategies and Activities CD-ROM • *Skills 7, 27, 42*
- Astro Algebra • *Red,* Level K

LESSON 11.2 ORGANIZER

Objective To classify sets of numbers and to find another rational number between two rational numbers

Vocabulary ratio, rational number, Venn diagram

1 Introduce

QUICK REVIEW provides review of prerequisite skills.

Why Learn This? This skill is necessary in algebra to identify and understand how different types of numbers are related. *Share the lesson objective with students.*

2 Teach

Guided Instruction

- As you discuss the sample rational numbers and Example 1, ask:

  How can you write any whole number as a ratio of integers? Possible answer: Write it as a fraction that has the whole number in the numerator and 1 in the denominator.

  How can you write a mixed number as a ratio? Write it as a fraction greater than one.

- Before discussing A–D in Example 2, draw another Venn diagram showing how cats, mammals, and animals are related. Ask:

  Are all cats mammals? yes

  Are all animals mammals? no

Modifying Instruction Have students verbalize the relationships between sets of numbers: *The set of whole numbers is contained in the set of integers. Every integer is also a rational number.*

ADDITIONAL EXAMPLE

Example 1, p. 230

Write each rational number as a ratio $\frac{a}{b}$.

A. $1\frac{5}{8}$ B. 0.1   C. 65   D. $^-1.5$

$1\frac{5}{8} = \frac{13}{8}$ $0.1 = \frac{1}{10}$ $65 = \frac{65}{1}$ $^-1.5 = \frac{-3}{2}$

See also page 231.

LESSON 11.2 Rational Numbers

Learn how to classify rational numbers and find another rational number between two rational numbers.

QUICK REVIEW

Write as a decimal and as a fraction.
1. eight tenths   2. fifty-four hundredths   3. three tenths
4. nineteen hundredths   5. forty thousandths

1. $0.8, \frac{8}{10}$ 2. $0.54, \frac{54}{100}$ 3. $0.3, \frac{3}{10}$ 4. $0.19, \frac{19}{100}$ 5. $0.040, \frac{40}{1,000}$

A **ratio** is a comparison of two numbers, a and b, written as a fraction $\frac{a}{b}$. A **rational number** is any number that can be written as a ratio $\frac{a}{b}$, where a and b are integers and $b \neq 0$. The numbers below are all rational numbers since they can be expressed as a ratio $\frac{a}{b}$.

$$3\frac{2}{5} \qquad 0.6 \qquad 42 \qquad ^-2.5$$

Vocabulary
ratio
rational number
Venn diagram

EXAMPLE 1

Write each rational number as a ratio $\frac{a}{b}$.

A. $3\frac{2}{5}$ B. 0.6   C. 42   D. $^-2.5$

$3\frac{2}{5} = \frac{17}{5}$ $0.6 = \frac{6}{10}$ $42 = \frac{42}{1}$ $^-2.5 = \frac{-5}{2}$

The **Venn diagram** shows how the sets of rational numbers, integers, and whole numbers are related.

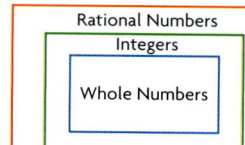

EXAMPLE 2

Use the Venn diagram to determine in which set or sets each number belongs.

A. 80 — The number 80 belongs in the sets of whole numbers, integers, and rational numbers.

B. $^-2$ — The number $^-2$ belongs in the sets of integers and rational numbers but not in the set of whole numbers.

C. $6\frac{1}{2}$ — The number $6\frac{1}{2}$ belongs in the set of rational numbers but not in the set of integers or the set of whole numbers.

D. 7.09 — The number 7.09 belongs in the set of rational numbers but not in the set of integers or the set of whole numbers.

- Name two integers that are not also whole numbers. Possible answer: $^-7, ^-4$

230 Chapter 11

RETEACH 11.2

Rational Numbers

Between any two rational numbers, you can always find other rational numbers.

Fractions or Mixed Numbers

Find a rational number between $3\frac{3}{4}$ and $3\frac{7}{8}$.

- Use a common denominator to write equivalent fractions. 8 is a common denominator of 4 and 8. $\frac{3}{4} = \frac{6}{8}$ $\frac{7}{8}$

- There are no eighths between $\frac{6}{8}$ and $\frac{7}{8}$. Use a greater common denominator. Try 16. $\frac{3}{4} = \frac{12}{16}$ $\frac{7}{8} = \frac{14}{16}$ $3\frac{3}{4} = 3\frac{12}{16}$ $3\frac{7}{8} = 3\frac{14}{16}$

So, $3\frac{13}{16}$ is between $3\frac{3}{4}$ and $3\frac{7}{8}$. $3\frac{13}{16}$ is between $3\frac{12}{16}$ and $3\frac{14}{16}$

Decimals

Find a rational number between 1.3 and 1.4.

- Add a zero to each number.   1.3 = 1.30   1.4 = 1.40
- Find a number between the two numbers.   1.31, 1.32, 1.33, 1.34, 1.35, 1.36, 1.37, 1.38, and 1.39 are between 1.30 and 1.40.

So, 1.31, 1.33, 1.36, and 1.38 are some of the numbers between 1.3 and 1.4.

Find a rational number between the two given numbers. Possible answers are given.

1. $\frac{1}{2}$ and $\frac{3}{4}$ — $\frac{5}{8}$
2. $\frac{1}{8}$ and $\frac{1}{4}$ — $\frac{3}{16}$
3. $1\frac{1}{3}$ and $1\frac{2}{3}$ — $1\frac{1}{2}$
4. $^-3\frac{1}{2}$ and $^-3\frac{3}{4}$ — $^-3\frac{5}{8}$
5. 3.6 and 3.7 — 3.61
6. 1.2 and 1.3 — 1.22
7. $^-5.9$ and $^-6$ — $^-5.95$
8. 2.11 and 2.12 — 2.111
9. $5\frac{1}{4}$ and $5\frac{3}{4}$ — $5\frac{1}{2}$
10. $^-2.75$ and $^-2.76$ — $^-2.757$
11. $\frac{7}{10}$ and 0.8 — $\frac{3}{4}$, or 0.75
12. $1\frac{1}{10}$ and 1.2 — $\frac{23}{20}$ or 1.15

PRACTICE 11.2

Rational Numbers

Use the number line to find a rational number between the two given numbers. Possible answers are given.

1. 2 and $2\frac{1}{2}$ — $2\frac{1}{4}$
2. $2\frac{1}{2}$ and 3 — $2\frac{3}{4}$
3. 3 and $3\frac{1}{2}$ — $3\frac{1}{4}$
4. $3\frac{1}{2}$ and 4 — $3\frac{3}{4}$

Find a rational number between the two given numbers. Possible answers are given.

5. $\frac{3}{8}$ and $\frac{4}{4}$ — $\frac{7}{24}$
6. $\frac{3}{8}$ and $\frac{2}{5}$ — $\frac{1}{2}$
7. $1\frac{7}{8}$ and $1\frac{3}{5}$ — $1\frac{15}{16}$
8. $^-3$ and $^-3\frac{1}{2}$ — $^-3\frac{1}{4}$
9. 3.1 and 3.2 — 3.15
10. $^-1.7$ and $^-1.8$ — $^-1.72$
11. $^-5.6$ and $^-5.7$ — $^-5.68$
12. 3.04 and 3.05 — 3.041

Write each rational number in the form $\frac{a}{b}$. Possible answers are given.

13. $3\frac{1}{2}$ — $\frac{7}{2}$
14. 0.3 — $\frac{3}{10}$
15. 0.45 — $\frac{45}{100}$ or $\frac{9}{20}$
16. 11.2 — $\frac{112}{10}$ or $\frac{56}{5}$
17. $2\frac{1}{4}$ — $\frac{9}{4}$
18. 3.15 — $\frac{315}{100}$ or $\frac{63}{20}$
19. 15 — $\frac{15}{1}$
20. 27 — $\frac{27}{1}$
21. $3\frac{1}{5}$ — $\frac{16}{5}$
22. 0.59 — $\frac{59}{100}$
23. 370 — $\frac{370}{1}$
24. $4\frac{1}{7}$ — $\frac{29}{7}$

Use the Venn diagram at the right to determine in which set or sets the number belongs.

25. 1.8 — R
26. $5\frac{2}{3}$ — R
27. 48 — all

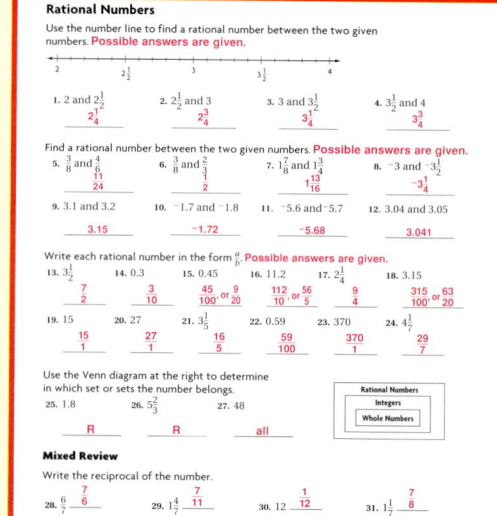

Mixed Review

Write the reciprocal of the number.

28. $\frac{6}{7}$ — $\frac{7}{6}$
29. $1\frac{4}{7}$ — $\frac{7}{11}$
30. 12 — $\frac{1}{12}$
31. $1\frac{1}{7}$ — $\frac{7}{8}$

Find the quotient. Write the answer in simplest form.

32. $\frac{2}{5} \div \frac{1}{3}$ — $1\frac{1}{5}$
33. $6 \div \frac{8}{9}$ — $6\frac{3}{4}$
34. $3\frac{3}{5} \div 1\frac{4}{5}$ — $1\frac{7}{8}$

Christopher is training to run in a 5-km road race. Yesterday he ran $4\frac{1}{4}$ km, and he plans to run $4\frac{1}{2}$ km tomorrow. What distance could he run today if he wants to run between $4\frac{1}{4}$ km and $4\frac{1}{2}$ km?

Think of the distances of Christopher's training runs as rational numbers.

One Way You can use a number line to find numbers between two rational numbers.

EXAMPLE 3

Find a distance between $4\frac{1}{4}$ km and $4\frac{1}{2}$ km, using the number line.

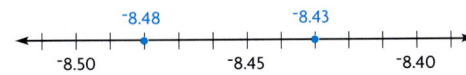

Notice that there is a mark between $4\frac{1}{4}$ and $4\frac{1}{2}$. That could be a distance Christopher could run.

So, Christopher could run $4\frac{3}{8}$ km today.

Another Way You can use a common denominator to find a number between two rational numbers.

EXAMPLE 4

Find a number between $4\frac{1}{4}$ and $4\frac{1}{2}$.

$4\frac{1}{4} = 4\frac{2}{8}$ $4\frac{1}{2} = 4\frac{4}{8}$ *Use a common denominator to write equivalent fractions.*

$4\frac{3}{8}$ is between $4\frac{2}{8}$ and $4\frac{4}{8}$.   *Find a rational number between the two numbers.*

So, $4\frac{3}{8}$ is between $4\frac{1}{4}$ and $4\frac{1}{2}$.

- What are some common denominators you could use to find other rational numbers between $4\frac{1}{4}$ and $4\frac{1}{2}$? **Possible answers: 16, 24, 32**

You can also find a number between two rational numbers in decimal form.

EXAMPLE 5

Find a rational number between ⁻8.4 and ⁻8.5.

⁻8.4 = ⁻8.40   *Add a zero to each decimal.*

⁻8.5 = ⁻8.50

Remember that you can add any number of zeros to the right of a decimal without changing its value.

Use a number line to find a number between the two decimals.

So, ⁻8.43, ⁻8.45, and ⁻8.48 are some of the numbers between ⁻8.4 and ⁻8.5.

231

- When discussing Example 3, ask:

  If the distances were marked in sixteenths, what other distances would be between $4\frac{1}{4}$ km and $4\frac{1}{2}$ km? $4\frac{5}{16}$ km, $4\frac{6}{16}$ km, $4\frac{7}{16}$ km

- After discussing Examples 4 and 5, demonstrate two other possibilities for finding a number between two rational numbers.

  Could you use the mean to find a rational number between the two fractions? Explain. Yes; The mean is the average of the two numbers, and would be located in between them, similar to the balance point on the line plot discussed in Chapter 6.

Modifying Instruction Point out that another simple way to find a fraction between two other fractions, such as $\frac{1}{2}$ and $\frac{3}{4}$, is to add the numerators and add the denominators: $\frac{1+3}{2+4} = \frac{4}{6} = \frac{2}{3}$. Therefore, $\frac{2}{3}$ lies between $\frac{1}{2}$ and $\frac{3}{4}$.

ADDITIONAL EXAMPLES

Example 2, p. 230

Use the Venn diagram on page 230 to determine in which set or sets each number belongs.

A. 0.06 rational numbers

B. ⁻7 integers, rational numbers

C. $5\frac{1}{2}$ rational numbers

D. 11 whole numbers, integers, rational numbers

Example 3, p. 231

Find a distance between $4\frac{1}{2}$ km and $4\frac{7}{8}$ km, using the number line on page 231. Possible answers: $4\frac{5}{8}$ or $4\frac{6}{8}$

Example 4, p. 231

Find a rational number between $3\frac{1}{2}$ and $3\frac{3}{4}$. Possible answer: $3\frac{5}{8}$

Example 5, p. 231

Find a rational number between ⁻4.5 and ⁻4.6. Possible answer: ⁻4.52

231

LESSON 11.2

3 Practice

Guided Practice

Do Check for Understanding Exercises 1–11 with your students. Identify those having difficulty and use lesson resources to help.

COMMON ERROR ALERT

Students may think that integers are not rational numbers. Remind them that integers such as 5 and ⁻7 can be written in ratio form $\frac{a}{b}$.

$$5 = \frac{5}{1}; \quad ^-7 = \frac{^-7}{1}$$

Independent Practice

Note that Exercises 39 and 40 are **multistep or strategy problems**. Assign Exercises 12–35.

Before beginning Exercises 20–25, remind students that more than one answer is possible. Another rational number can always be found between any two rational numbers.

For Exercises 20–25, you may want to help students rewrite one or both of the numbers so that they are in like form and, therefore, easier to compare.

CHECK FOR UNDERSTANDING

Think and Discuss

Look back at the lesson to answer each question.

1. All integers can be written in the form $\frac{a}{1}$; possible example: $^-3 = \frac{^-3}{1}$

1. **Tell** why every integer is a rational number. Give an example to support your answer.

2. **Tell** what numbers would be between $4\frac{1}{4}$ and $4\frac{1}{2}$ if you divided a number line into sixteenths. $4\frac{5}{16}, 4\frac{6}{16}, 4\frac{7}{16}$

Guided Practice

Write each rational number in the form $\frac{a}{b}$. Possible answers are given.

3. $^-0.37$ $\frac{^-37}{100}$ 4. $2\frac{4}{5}$ $\frac{14}{5}$ 5. 0.889 $\frac{889}{1,000}$ 6. 7.31 $\frac{731}{100}$ 7. $^-7\frac{1}{3}$ $\frac{^-22}{3}$

Use the number line to find a rational number between the two given numbers. Possible answers are given.

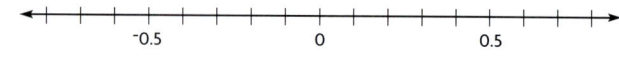

8. 1 and $1\frac{1}{2}$ $1\frac{1}{4}$ 9. $\frac{^-3}{4}$ and $\frac{^-1}{4}$ $\frac{^-1}{2}$ 10. $\frac{1}{2}$ and $1\frac{3}{4}$ 11. $\frac{^-1}{2}$ and $\frac{1}{2}$ 0

PRACTICE AND PROBLEM SOLVING

Independent Practice

Write each rational number in the form $\frac{a}{b}$. Possible answers are given.

12. $9\frac{2}{3}$ $\frac{29}{3}$ 13. $^-0.71$ $\frac{^-71}{100}$ 14. 80.4 $\frac{804}{10}$ 15. $^-2\frac{5}{8}$ $\frac{^-21}{8}$ 16. 3.18 $\frac{318}{100}$

Use the number line to find a rational number between the two given numbers. Possible answers are given.

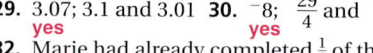

17. $^-0.2$ and $^-0.4$ $^-0.3$ 18. 0 and 0.2 0.1   19. $^-0.6$ and $^-0.8$ $^-0.7$

Find a rational number between the two given numbers. Possible answers are given.

20. $\frac{3}{4}$ and $\frac{1}{2}$ $\frac{5}{8}$ 21. $^-7$ and $\frac{^-15}{2}$ $^-7\frac{1}{4}$ 22. 104.1 and $103\frac{7}{8}$ 104

23. 16.1 and 16.01 16.05   24. $3\frac{5}{8}$ and $\frac{27}{8}$ $\frac{28}{8}$ 25. $\frac{^-3}{4}$ and $\frac{^-3}{8}$ $\frac{^-5}{8}$

Tell if the first rational number is between the second and third rational numbers. Write *yes* or *no*.

26. $\frac{1}{3}; \frac{1}{2}$ and $\frac{3}{4}$ no   27. 0.97; 0.85 and 0.99 yes   28. 3.29; 3.20 and 3.25 no

29. 3.07; 3.1 and 3.01 yes   30. $^-8; \frac{^-29}{4}$ and $\frac{^-33}{4}$ yes   31. $\frac{^-1}{8}; \frac{^-1}{16}$ and $\frac{^-1}{4}$ yes

Problem Solving Applications

32. Marie had already completed $\frac{1}{2}$ of the running race but had not yet reached the point $\frac{3}{4}$ of the way through the race. Could she have completed $\frac{5}{8}$ of the race? Explain. Yes; $\frac{5}{8}$ is between $\frac{1}{2}$ and $\frac{3}{4}$.

232 Chapter 11

ALTERNATIVE TEACHING STRATEGY | SCAFFOLDED INSTRUCTION

Purpose Students play a game to reinforce understanding of rational numbers.

Materials 11 index cards labeled from ⁻1.5 through 1.5 in increments of tenths

Before playing the game, guide students to name rational numbers between tenths. Remind them that ⁻1.5 is the same as ⁻1.50 and ⁻1.4 is the same as ⁻1.40. Guide them to realize that they can use hundredths.

Ask: Name several rational numbers between ⁻1.50 and ⁻1.40. Possible answers: ⁻1.45, ⁻1.43, ⁻1.46

Display the following number line:

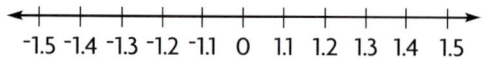

Play the following game with the class:

1. Ask two students to each pick a card and show them to the class.

2. Have players take turns picking cards. The first player to pick a card that shows a number between the first two numbers stands up and directs one of the card holders to take his or her seat.

3. Repeat Step 2 until two players have two consecutive tenths.

4. Challenge the rest of the class to name a rational number that comes between the two numbers shown.

Check students' work.

232 Chapter 11

33. Susan follows the directions of a treasure map and walks 45 steps north, then $\frac{24}{3}$ steps east, $\frac{75}{5}$ steps south, and 22 steps west. How many steps does Susan walk? **90 steps**

34. Is it easier to find a rational number between $\frac{1}{2}$ and $\frac{3}{4}$ or between 0.50 and 0.75? Explain your reasoning.
See below.

35. ❓ **What's the Error?** Jeff says that every whole number is an integer and that every integer is a whole number. Explain his error.
Not all integers are whole numbers; ⁻8 is not a whole number.

MIXED REVIEW AND TEST PREP

36. Find the absolute value. $|-88|$ (p. 228) **88**
37. Write the decimal and fraction equivalents of 34%. (p. 169) **0.34; $\frac{34}{100}$**
38. $\frac{2}{5} + \frac{4}{10} + \frac{4}{5}$ (p. 182) **$1\frac{3}{5}$**

34. Possible answer: Between 0.50 and 0.75; you don't have to find a common denominator.

⭐ 39. **TEST PREP** What is the difference between the range and the mean of the number set 30, 8, 13, 20, and 24? (p. 106) **D**

   A 0   B 1   C 2   D 3

⭐ 40. **TEST PREP** Which is the GCF of 54 and the number that is the LCM of 3, 5, 9, and 15? (p. 150) **G**

   F 6   G 9   H 18   J 45

PROBLEM SOLVING — LiNKUP to Reading

Strategy • Use Graphic Aids Graphic aids such as Venn diagrams, charts, and tables provide specific or important information in a visual form rather than in text. Sometimes, the information needed to solve a problem may be provided only in a graphic aid.

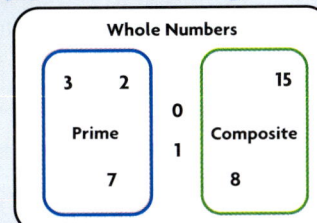

Look at the Venn diagram to the right. It shows the relationships among whole, prime, and composite numbers.

- Is a composite number always, sometimes, or never a whole number? **Always; the diagram shows that all composite numbers are also within the set of whole numbers.**
- Is a whole number always, sometimes, or never a prime number? **Sometimes; the diagram shows that some numbers within the set of whole numbers are not also prime.**

Use the Venn diagram to solve the following problems.

1. How would you describe the numbers 7 and 15? Are they whole numbers? Are they prime or composite? **yes; 7 is prime, 15 is composite**

2. Is 1 a whole number? Is it prime or composite? Explain. **yes; neither; it falls within the whole number category, but outside both prime and composite number groups**

EXTRA PRACTICE page H42, Set B

233

READING STRATEGY

K-W-L Chart Before having students read the Link Up to Reading, have them look at the title and Venn diagram. Ask them to predict what the Link Up will be about. Then have students make a three-column chart headed What I Know, What I Want to Know, and What I Learned. Ask them to fill in the first two columns. Have them fill in the third column as they read the paragraph.

K-W-L Chart

What I Know	What I Want to Know	What I Learned

MIXED REVIEW AND TEST PREP
Exercises 36–40 provide **cumulative review** (Chapters 1–11).

Multistep or Strategy Problem To solve Exercise 39, students can find the range and the mean of the data set. Then they can find the difference between the two.

LiNKUP to READING

- Review with students the definitions of prime numbers and composite numbers. After students have studied the Venn diagram, ask:

  Why is the 0 outside the prime and composite sets? It is neither prime nor composite.

  Where would the integers be in this diagram? outside the whole numbers

REASONING What is the only even prime number? **2**

What is the greatest odd prime number? Since there is an infinite number of prime numbers, the greatest odd prime cannot be determined.

4 Assess

Summarize the lesson by having students:

DISCUSS If you have a positive rational number and a negative rational number, is there always a rational number between them? Explain. Yes; The number 0 is between any positive rational number and any negative rational number.

 **WRITE** How would you find a rational number between ⁻1 and $-\frac{3}{4}$? Possible answer: Write ⁻1 as the fraction $-\frac{4}{4}$. Next use the common denominator 8 to write equivalent fractions. Then find a rational number between the two numbers.

Lesson Quiz

Write each rational number in the form $\frac{a}{b}$. Possible answers are given.

1. $20\frac{4}{5}$ $\frac{104}{5}$
2. ⁻13.5 $-\frac{135}{10}$ or $-\frac{27}{2}$

Find a rational number between the two given numbers. Possible answers are given.

3. ⁻4 and $-4\frac{1}{2}$ $-4\frac{1}{4}$
4. 5.7 and 5.8   5.73

233

LESSON 11.3

Compare and Order Rational Numbers

LESSON PLANNING

Objective To compare and order rational numbers

Intervention for Prerequisite Skills
Locate Points on a Number Line, Compare Fractions, Temperature (For intervention strategies, see page 227.)

NCTM Standards
1. Number and Operations
2. Algebra
6. Problem Solving
8. Communication

Math Background
These concepts will help students understand how to compare two or more rational numbers.
- A number line can be used to compare integers. If one integer is to the left of another on the number line, it is less than the integer to the right.
- As negative integers increase in absolute value, they decrease in numeric value.
- Any negative rational number is less than every positive one.
- To compare a decimal and a fraction, write both as decimals, or write both as fractions or mixed numbers with a common denominator.

WARM-UP RESOURCES

 **NUMBER OF THE DAY**

Graph the number associated with the month of the year on a number line. Move 10 units to the left. What number did you find? Possible answer for 2: 10 units to the left of 2 is $^-8$.

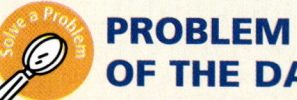 **PROBLEM OF THE DAY**

Leon's number is less than Bev's.

Shannon has a negative number.

Leon's and Tony's numbers were the least and the greatest.

Who has each of these numbers:

$^-2\frac{1}{5}$, $\frac{19}{18}$, 2.05, $^-2.25$?

Leon $^-2.25$; Shannon $^-2\frac{1}{5}$; Bev $\frac{19}{18}$; Tony 2.05

Solution Problem of the Day tab, p. PD11

 **DAILY FACTS PRACTICE**

Have students practice multiplication facts by completing Set D of *Teacher's Resource Book*, p. TR101.

INTERVENTION AND EXTENSION RESOURCES

ALTERNATIVE TEACHING STRATEGY

By relating the concepts of "greater than" and "less than" to numbers on a thermometer, many students will find it easier to **compare rational numbers.** Draw the number line vertically instead of horizontally.

Refer to Examples 1 and 2 on PE page 234. Call on volunteers to show the numbers you are comparing on the vertical number line. Then write the comparison statements and relate them to the numbers on the horizontal number line.

VISUAL
LOGICAL/MATHEMATICAL

MULTISTEP AND STRATEGY PROBLEMS

The following multistep or strategy problem is provided in Lesson 11.3:

Page	Item
235	27

ENGLISH LANGUAGE LEARNERS

Materials *For each student* 2 index cards

To **reinforce the concept of ordering rational numbers,** have each student write a rational number of their choosing on one index card and then write the opposite number on the other index card.

Have two volunteers each display one of their number cards. Then have a third volunteer write either $<$, $>$, or $=$ between the two numbers and explain his or her reasoning to the class. Check students' work.

VISUAL
VERBAL/LINGUISTIC

WRITING IN MATHEMATICS

Materials *For each group* 15 index cards

Have students **practice comparing rational numbers.** Give each group of 3 students 15 index cards. Have students write a different rational number on each card, including both positive and negative rational numbers.

The groups shuffle their cards and place them face down in a pile. Each student turns over one card, and the student with the greatest number takes all the cards that are face up. They repeat the activity, but this time the student with the least number takes all the cards that are face up. Play continues, alternating greatest and least taking all, until one student has all the cards. Check students' work.

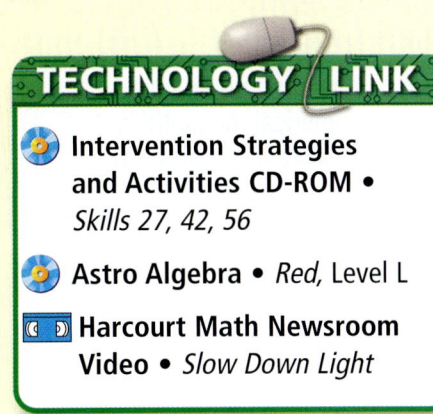

TECHNOLOGY LINK

- Intervention Strategies and Activities CD-ROM • *Skills 27, 42, 56*
- Astro Algebra • *Red,* Level L
- Harcourt Math Newsroom Video • *Slow Down Light*

LESSON 11.3 ORGANIZER

Objective To compare and order rational numbers

1 Introduce

QUICK REVIEW provides review of pre-requisite skills.

Why Learn This? Comparing and ordering numbers helps you decide which product has a higher rating or is more costly, and helps you organize information from a survey. *Share the lesson objective with students.*

2 Teach

Guided Instruction

- As you discuss Example 1, display a number line to illustrate the examples. Ask:

  How do you decide which number is less? The number farther to the left on the number line is less.

- When students are reading Example 2, ask:

  Why is 8.3 rewritten as 8.30? so the numbers will have the same number of decimal places and will be easier to compare

REASONING If you compare two negative numbers and the absolute value of the first number is greater than the absolute value of the second number, which number is greater? Explain. The second number; the number with the greater absolute value is farther to the left on the number line and, therefore, is less than the second number.

Algebraic Thinking Understanding number relationships shown by numerical inequalities will help students as they solve inequalities in algebra.

ADDITIONAL EXAMPLES

Example 1, p. 234

Compare the integers. Use $<$ and $>$. Think about their positions on a number line.

A. 1 and $^-2$ $\ ^-2 < 1$ or $1 > \ ^-2$

B. $^-4$ and $^-7$ $\ ^-4 > \ ^-7$ or $^-7 < \ ^-4$

Example 2, p. 234

Order $6\frac{1}{2}$, $^-6\frac{1}{3}$, and 6.4 from least to greatest.
$^-6\frac{1}{3}$, 6.4, $6\frac{1}{2}$

LESSON 11.3

Compare and Order Rational Numbers

Learn how to compare and order rational numbers.

QUICK REVIEW
Order the numbers from greatest to least.
1. 12, 14, 10   14; 12; 10
2. $^-10$, $^-21$, $^-20$ $^-10$; $^-20$; $^-21$
3. 3.10, 3.05, 3.15   3.15; 3.10; 3.05
4. $\frac{1}{4}$, $\frac{1}{2}$, $\frac{3}{4}$ $\frac{3}{4}$; $\frac{1}{2}$; $\frac{1}{4}$
5. $\frac{3}{5}$, $\frac{3}{6}$, $\frac{3}{4}$ $\frac{3}{4}$; $\frac{3}{5}$; $\frac{3}{6}$

TECHNOLOGY LINK
To learn more about comparing rational numbers, watch the *Harcourt Math Newsroom Video* Slow Down Light.

Temperature commonly is measured on a scale in units of degrees. The scale contains both negative and positive numbers, like a number line. For example, the temperature in Death Valley has reached a high of 132°F, and the record low in Alaska is $^-80$°F.

You can use a number line to compare integers. On a number line, each number is greater than any number to its left and less than any number to its right. The number line below shows that $^-80° < 132°$ and $132° > \ ^-80°$.

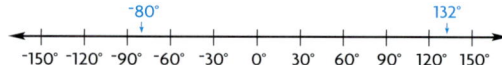

EXAMPLE 1

Compare the integers. Use $<$ and $>$. Think about their positions on a number line.

A. 2 and $^-3$
$^-3$ is to the left of 2 on the number line.
So, $^-3 < 2$, or $2 > \ ^-3$.

B. $^-2$ and $^-4$
$^-2$ is to the right of $^-4$ on the number line.
So, $^-2 > \ ^-4$, or $^-4 < \ ^-2$.

It is easier to compare and order rational numbers when they are all expressed as decimals or as fractions with a common denominator.

EXAMPLE 2

Order $8\frac{1}{4}$, $^-8\frac{1}{2}$, and 8.3 from least to greatest.

Since $^-8\frac{1}{2}$ is the only negative number, it is the least number.

Compare $8\frac{1}{4}$ and 8.3.

$8\frac{1}{4} = 8.25$ and $8.3 = 8.30$ *Write the numbers as decimals with the same number of decimal places.*

$8.25 < 8.30$, or $8\frac{1}{4} < 8.3$ *Compare by looking at the place values.*

$^-8\frac{1}{2} < 8\frac{1}{4} < 8.3$ *Order the three numbers.*

So, from least to greatest the numbers are $^-8\frac{1}{2}$, $8\frac{1}{4}$, and 8.3.

234 Chapter 11

RETEACH 11.3

Compare and Order Rational Numbers

You can compare rational numbers in decimal or fraction form.

Using Decimals

Compare $\frac{7}{25}$ and 0.35.

- Write the number that is not a decimal in decimal form. $\frac{7}{25} = 25\overline{)7.00}$ 0.28
- Compare decimals using place value. $0.28 < 0.35$

So, $\frac{7}{25} < 0.35$.

Using Fractions

Compare $\frac{1}{4}$ and 0.3.

- Write the number that is not a fraction in fraction form. $0.3 = \frac{3}{10}$
- Rewrite the fractions with the same denominator. Use 20 as a common denominator. $\frac{1}{4} = \frac{5}{20}$ $\frac{3}{10} = \frac{6}{20}$
- Compare the two fractions. $\frac{5}{20} < \frac{6}{20}$

So, $\frac{1}{4} < 0.3$.

Compare. Write $<$ or $>$.

1. $\frac{2}{5}$ __>__ 0.2
2. 0.65 __<__ $\frac{2}{3}$
3. 8.9 __>__ $8\frac{4}{5}$
4. $^-4\frac{1}{8}$ __>__ $^-4.3$

Compare the rational numbers and order them from least to greatest.

5. 4.2, 2.4, $\frac{8}{2}, \frac{2}{5}$ $\frac{2}{5} < 2.4 < \frac{8}{2} < 4.2$
6. $\frac{7}{5}, \frac{2}{3}, 0.2, 0.8$ $0.2 < \frac{2}{3} < 0.8 < \frac{7}{5}$
7. 0.1, 0.6, 0.9, 0, $\frac{1}{9}$ $0 < 0.1 < \frac{1}{9} < 0.6 < 0.9$
8. $\frac{1}{3}, \frac{1}{5}, \frac{1}{8}, 0.1$ $0.1 < \frac{1}{8} < \frac{1}{5} < \frac{1}{3}$
9. $^-1.4, ^-1.5, 1.2, \frac{6}{4}$ $^-1.5 < \ ^-1.4 < 1.2 < \frac{6}{4}$
10. 2.1, $^-3.8, \frac{10}{2}, \ ^-\frac{3}{4}$ $^-3.8 < \ ^-\frac{3}{4} < 2.1 < \frac{10}{2}$
11. 5.8, 4.9, 5.7, $2\frac{3}{5}, 0.58$ $0.58 < 2\frac{3}{5} < 4.9 < 5.7 < 5.8$
12. $^-5.6, \ ^-4.62, 2.34, \ ^-4.68$ $^-5.6 < \ ^-4.68 < \ ^-4.62 < 2.34$

PRACTICE 11.3

Compare and Order Rational Numbers

Compare. Write $<$ or $>$ for $\bigcirc$.

1. 0.25 $\bigcirc$ 0.4   **<**
2. $\frac{3}{8}$ $\bigcirc$ 0.2   **>**
3. $^-2\frac{1}{5}$ $\bigcirc$ $^-2.3$ **>**
4. $^-\frac{5}{8}$ $\bigcirc$ $^-\frac{3}{10}$ **<**
5. 5 $\bigcirc$ $^-2$ **>**
6. $^-\frac{7}{10}$ $\bigcirc$ $\frac{4}{5}$ **<**
7. $^-2.6$ $\bigcirc$ $^-2.62$ **>**
8. $\frac{3}{4}$ $\bigcirc$ $\frac{5}{6}$ **<**
9. $3.8 + 2.2 \bigcirc 2\frac{1}{6} + 3\frac{4}{5}$ **>**
10. $3\frac{1}{2} \times 2 \bigcirc 4\frac{1}{5} + 2.8$ **<**
11. $7\frac{1}{4} + 3\frac{1}{2} \bigcirc 1\frac{5}{8} \times 6$ **<**

Order the rational numbers from least to greatest.

12. 2.9, $^-1.7, \frac{3}{4}, 2.9$ $^-1.7; \frac{3}{4}; 2.9; \frac{9}{3}$
13. $^-\frac{1}{5}, \frac{1}{9}, \frac{1}{10}, ^-0.1$ $^-\frac{1}{5}; \ ^-0.1; \frac{1}{10}; \frac{1}{9}$
14. 0, 0.8, $^-1.4, \ ^-0.6, \frac{3}{5}$ $^-1.4; \ ^-0.6; 0; \frac{3}{5}; 0.8$
15. 8.7, $^-9.2, \ ^-7.3, 6.2, 6\frac{1}{2}, 8\frac{7}{8}$ $^-9.2; \ ^-7.3; 6.2; 6\frac{1}{2}; 8.7; 8\frac{7}{8}$
16. $4\frac{1}{4}, 4\frac{3}{5}, 4.9, 4.08, 0.49$ $0.49; 4.08; 4\frac{1}{4}; 4\frac{3}{5}; 4.9$

Order the rational numbers from greatest to least.

17. 7.3, 6, $\frac{7}{8}$, 2 $7.3; 6; 2; \frac{7}{8}$
18. 2.4, $^-1.4, \ ^-3, 4.7, 3.8$ $4.7; 3.8; 2.4; \ ^-1.4; \ ^-3$
19. $\frac{7}{10}, 0.5, \ ^-0.6, 0.42$ $0.5; 0.42; \frac{7}{10}; \ ^-0.6$

Mixed Review

Find the LCM of each set of numbers.
20. 4, 10   **20**
21. 7, 12   **84**
22. 8, 18, 24   **72**
23. 5, 15, 20   **60**

Find the GCF of each set of numbers.
24. 12, 20   **4**
25. 16, 42   **2**
26. 15, 50, 75   **5**
27. 36, 54, 72   **18**

Find a pair of numbers for each set of conditions. **Possible answers are given.**
28. The LCM is 30. The GCF is 2.   **6 and 10**
29. The LCM is 36. The GCF is 6.   **12 and 18**

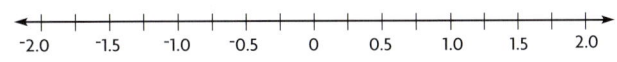

CHECK FOR UNDERSTANDING

Think and Discuss Look back at the lesson to answer each question.

1. **Describe** how you would compare 2.62 and $2\frac{3}{5}$.
   Possible answer: convert $2\frac{3}{5}$ to a decimal and compare place values.
2. **Give an example** of a rational number that is greater than $^-2.5$ and one that is less than $^-2.5$. Possible answer: $^-1.4$ and $^-3.2$

Guided Practice Compare. Write <, >, or = for each ●.

[number line from -2.0 to 2.0]

3. $^-1.5$ ● $^-0.5$ 4. 0.5 ● $^-1.0$ 5. $1\frac{1}{4}$ ● 1.5 6. $^-2$ ● $^-1\frac{1}{2}$
   <   >   <   <

PRACTICE AND PROBLEM SOLVING

Independent Practice Compare. Write <, >, or = for each ●.

7. $\frac{1}{2}$ ● $\frac{3}{4}$ 8. $\frac{1}{4}$ ● $^-\frac{1}{2}$ 9. 0.5 ● $\frac{3}{8}$ 10. $^-\frac{1}{4}$ ● 0.25
   <   >   >   <

11. 1.25 ● 1.75 12. $^-\frac{1}{4}$ ● $^-\frac{1}{3}$ 13. 2 ● $^-3$ 14. $\frac{4}{5}$ ● 0.9
    <   <   >   <

17. 0.4; 0.46; 0.6

18. $^-\frac{3}{4}$; $^-\frac{1}{2}$; $^-\frac{3}{8}$; $^-\frac{1}{8}$

19. $^-\frac{1}{4}$; $^-0.2$; 0; $\frac{1}{4}$

15. $3.2 + 4.4$ ● $4\frac{3}{4} + 2\frac{3}{4}$ 16. $2\frac{3}{4} \times 4$ ● $3\frac{1}{4} + 8.5$
    >   <

Order the rational numbers from least to greatest.

17. $0.6, 0.4, 0.46$ 18. $^-\frac{1}{8}, ^-\frac{1}{2}, ^-\frac{3}{8}, ^-\frac{3}{4}$ 19. $^-0.2, \frac{1}{4}, 0, ^-\frac{1}{4}$

Problem Solving Applications

20. Lynda's times for running a mile are $5\frac{1}{2}$ min, 5.48 min, 5.51 min, and $5\frac{2}{5}$ min. What is the longest she has taken to run a mile? **5.51 min.**

21. The mean temperatures for three days were $^-3°$, $^-5°$, and $1°$. Order the temperatures from highest to lowest. $1°, ^-3°, ^-5°$

22. Possible answer: The negative integer is least. Compare the positive numbers by converting the fraction to a decimal and compairing it with the decimal.

22. ✏️ **Write About It** Explain how you would order three numbers that include a positive fraction and decimal and a negative integer.

MIXED REVIEW AND TEST PREP

23. Write an integer to represent 450 ft below sea level. (p. 228) $^-450$

24. Round 2.0955 to the nearest thousandth. (p. 58) **2.096**

25. Write $3\frac{3}{7}$ as a fraction. (p. 164) $\frac{24}{7}$

26. $\frac{7}{8} - \frac{1}{2}$ (p. 182) $\frac{3}{8}$

⭐ 27. **TEST PREP** The swim team practiced for 8 hours the first week, $1\frac{1}{4}$ times as long the second week, 12 hours the third week, and $\frac{24}{4}$ hours the fourth week. How many hours total did the swim team practice? (p. 206) **C**

   A 26   B 28   **C 36**   D 38

EXTRA PRACTICE page H42, Set C

235

---

3 Practice

Guided Practice

Do Check for Understanding Exercises 1–6 with your students. Identify those having difficulty and use lesson resources to help.

⚠️ **COMMON ERROR ALERT**

When comparing two negative numbers, students may think that the number with the greater absolute value is the greater number. Remind students that the number that is farther to the right on the number line is always greater.

Independent Practice

Note that Exercise 27 is a **multistep or strategy problem**. Assign Exercises 7–22.

Before students begin Exercises 7–16, remind them that positive numbers are always greater than negative numbers.

MIXED REVIEW AND TEST PREP

Exercises 23–27 provide **cumulative review** (Chapters 1–11).

4 Assess

Summarize the lesson by having students:

DISCUSS How does the number line show that all negative numbers are less than all positive numbers? All negative numbers are to the left of all positive numbers and zero.

📖✏️ **WRITE** What are three ways you might compare two positive rational numbers? Possible answer: Show them on a number line. Write them as fractions with a common denominator. Compare them in decimal form.

Lesson Quiz

Transparency 11.3

Compare. Write <, >, or = for each ●.

1. 3.25 ● 3.19 >
2. $^-1.5$ ● $^-\frac{1}{4}$ <
3. $\frac{1}{3}$ ● $^-\frac{2}{3}$ >

Order the rational numbers from least to greatest.

4. $0.1, 0.4, ^-0.2, ^-0.1$ $^-0.2, ^-0.1, 0.1, 0.4$
5. $^-2, 0.2, \frac{1}{4}, ^-\frac{1}{2}$ $^-2, ^-\frac{1}{2}, 0.2, \frac{1}{4}$

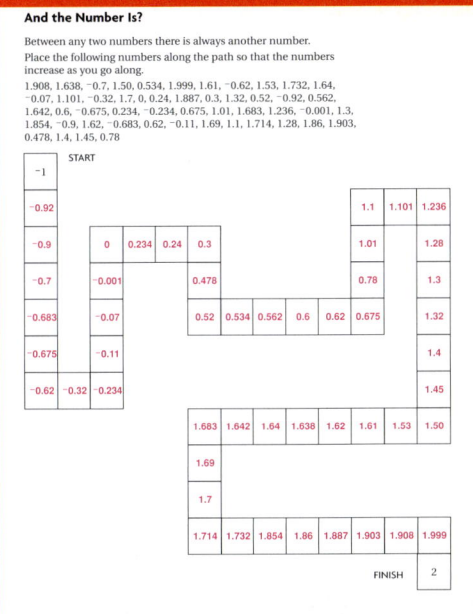

CHALLENGE 11.3

And the Number Is?

Between any two numbers there is always another number.
Place the following numbers along the path so that the numbers increase as you go along.

1.908, 1.638, $^-0.7$, 1.50, 0.534, 1.999, 1.61, $^-0.62$, 1.53, 1.732, 1.64, $^-0.07$, 1.101, $^-0.32$, 1.7, 0, 0.24, 1.887, 0.3, 1.32, 0.52, $^-0.92$, 0.562, 1.642, 0.6, $^-0.675$, 0.234, $^-0.234$, 0.675, 1.01, 1.683, 1.236, $^-0.001$, 1.3, 1.854, $^-0.9$, 1.62, $^-0.683$, 0.62, $^-0.11$, 1.69, 1.1, 1.714, 1.28, 1.86, 1.903, 0.478, 1.4, 1.45, 0.78

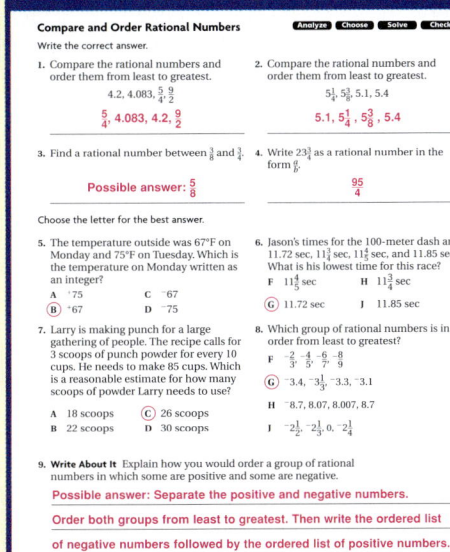

PROBLEM SOLVING 11.3

Compare and Order Rational Numbers Analyze · Choose · Solve · Check

Write the correct answer.

1. Compare the rational numbers and order them from least to greatest.
   $4.2, 4.083, \frac{5}{4}, \frac{9}{2}$
   $\frac{5}{4}, 4.083, 4.2, \frac{9}{2}$

2. Compare the rational numbers and order them from least to greatest.
   $5\frac{1}{4}, 5\frac{3}{8}, 5.1, 5.4$
   $5.1, 5\frac{1}{4}, 5\frac{3}{8}, 5.4$

3. Find a rational number between $\frac{1}{8}$ and $\frac{3}{4}$.
   Possible answer: $\frac{5}{8}$

4. Write $23\frac{3}{4}$ as a rational number in the form $\frac{a}{b}$.
   $\frac{95}{4}$

Choose the letter for the best answer.

5. The temperature outside was 67°F on Monday and 75°F on Tuesday. Which is the temperature on Monday written as an integer?
   A $^+75$ C $^-67$
   B $^+67$ D $^-75$

6. Jason's times for the 100-meter dash are 11.72 sec, $11\frac{3}{4}$ sec, $11\frac{4}{5}$ sec, and 11.85 sec. What is his lowest time for this race?
   F $11\frac{4}{5}$ sec   H $11\frac{3}{4}$ sec
   G 11.72 sec   J 11.85 sec

7. Larry is making punch for a large gathering of people. The recipe calls for 3 scoops of punch powder for every 10 cups. He needs to make 85 cups. Which is a reasonable estimate for how many scoops of powder Larry needs to use?
   A 18 scoops   **C 26 scoops**
   B 22 scoops   D 30 scoops

8. Which group of rational numbers is in order from least to greatest?
   F $^-\frac{2}{3}, ^-\frac{4}{5}, ^-\frac{6}{7}, ^-\frac{8}{9}$
   G $^-3.4, ^-3\frac{1}{3}, ^-3.3, ^-3.1$
   H $^-8.7, 8.07, 8.007, 8.7$
   J $^-2\frac{1}{2}, ^-2\frac{1}{3}, 0, ^-2\frac{1}{4}$

9. **Write About It** Explain how you would order a group of rational numbers in which some are positive and some are negative.
   Possible answer: Separate the positive and negative numbers. Order both groups from least to greatest. Then write the ordered list of negative numbers followed by the ordered list of positive numbers.

235

LESSON 11.4

Problem Solving Strategy: *Use Logical Reasoning*

LESSON PLANNING

Objective To solve problems by using the strategy *use logical reasoning*

Intervention for Prerequisite Skills
Compare Fractions (For intervention strategies, see page 227.)

Lesson Resources Problem Solving Think Along, p. TR1

NCTM Standards
1. Number and Operations
6. Problem Solving
7. Reasoning and Proof

Math Background
Students are often faced with real-life problems that can be solved with the strategy *use logical reasoning*. It is often helpful to use the strategy *make a table* when solving a problem using logical reasoning.

The following ideas will help students understand this strategy.

- When you make a table like the one on page 236, each column can contain only one "yes" because each amount of money is associated with only one person.
- Each row can contain only one "yes" because each person has only one amount of money. So, once a *yes* is entered in the table, a *no* can be written in each of the other boxes in that row and column.
- It is not necessary to take the clues in order. Sometimes later clues give pieces of information that allow the student to complete one or more boxes.

WARM-UP RESOURCES

 **NUMBER OF THE DAY**

Write the numbers associated with the month of the year, the day of the month, and their opposites. Place them in order. **Possible answer for February 15:** ⁻15, ⁻2, 2, 15

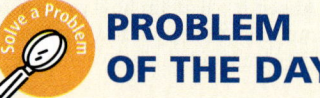 **PROBLEM OF THE DAY**

Cara is twice as old as Lee. In 5 years she will be $1\frac{1}{2}$ times as old as he is. How old is Lee? **5 years old**

Solution Problem of the Day tab, p. PD11

 **DAILY FACTS PRACTICE**

Have students practice division facts by completing Set E of *Teacher's Resource Book*, p. TR101.

236A Chapter 11

INTERVENTION AND EXTENSION RESOURCES

ALTERNATIVE TEACHING STRATEGY ELL

Materials *For each student* 8 index cards

Ask students to **use logical reasoning** to solve the problem on page 236. Have students write the four amounts and the four names given in the problem on cards, one to a card. Begin by asking students to put the numbers in order from least to greatest.

Ask students which clue can be used first. Once they have chosen the clue "Leticia has twice as much money as Shelley," have them try each pair of numbers to see which fit. Once they have chosen $5 and $10, have them match the names and the amounts. Continue working with the remaining names and amounts to solve the problem.

VISUAL
LOGICAL/MATHEMATICAL

SPECIAL NEEDS

To help students see how to **use other strategies to solve problems involving logical reasoning**, suggest they work in groups to act out Exercise 5 on page 237.

Once students have lined up in a possible order to solve, have other students check the clues to see if the order shown fits the clues given. Encourage them to keep track of the orders they have tried so they do not repeat them.

When students have completed their activity, ask them to verbalize the correct order. *Arlene, Kathy, and Helene*

KINESTHETIC
INTERPERSONAL/SOCIAL

READING STRATEGY

Analyze Information Have students use the reading strategy *analyze information* to help them understand the problem on page 236.

- Direct students to focus on the phrase "but not necessarily in that order." Have students explain what the phrase tells them. *that Stephen does not necessarily have $7.75 just because both are listed first*
- Then ask how they will use the clue that Leticia has twice as much money as Shelley. *Look for two amounts such that one is twice the other.*

VOCABULARY STRATEGY

Challenge students to work in groups of 2 or 3 to **write logical reasoning problems** similar to Exercises 1–4. Suggest that they start with data such as four people's names and information about them, such as favorite subject, number of siblings, and so on. Also suggest that two of the numbers be related in some way that is easy to describe. Then, have them write clues from the data. Remind students to check the solutions to their own problems to be sure they work before exchanging with others to solve. *Check students' work.*

AUDITORY
LOGICAL/MATHEMATICAL

MULTISTEP AND STRATEGY PROBLEMS

The following multistep and strategy problems are provided in Lesson 11.4:

Page	Item
237	1–10

Intervention Strategies and Activities CD-ROM • *Skill 27*

LESSON 11.4 ORGANIZER

Objective To solve problems by using the strategy *use logical reasoning*

Lesson Resources Problem Solving Think Along, p. TR1

1 Introduce

QUICK REVIEW provides review of prerequisite skills.

Why Learn This? In the future, this strategy will help you solve more challenging word problems. *Share the lesson objective with students.*

2 Teach

Guided Instruction

- After discussing the problem, and the steps "analyze" and "choose", direct students' attention to the table. Ask:

  How does the table help solve the problem? Possible answer: It helps you organize your information.

  Why can you write *no* in all the other spaces in a row and in a column once you write a *yes* in that row or column? Possible answer: A money amount can belong to only one person and each person can have only one amount.

- Help students analyze the reasoning process.

REASONING **Which clue did you use first? Explain.** Possible answer: Leticia has twice as much money as Shelley; the other clue has 2 possibilities and this clue provides 2 definite answers.

LESSON 11.4

PROBLEM SOLVING STRATEGY
Use Logical Reasoning

Learn how to solve problems by using logical reasoning.

QUICK REVIEW

Compare. Use <, >, or = for each ●.

1. $\frac{3}{8}$ ● $\frac{7}{8}$ <
2. 3.5 ● 3.8   <
3. 2 ● -2   >
4. $8\frac{1}{2}$ ● $8\frac{4}{8}$ =
5. Order $7\frac{1}{2}$, 7.75, $\frac{21}{3}$ from least to greatest. $\frac{21}{3}$, $7\frac{1}{2}$, 7.75

Stephen, Leticia, Shelley, and Ed emptied their banks. They found $7.75, $4.35, $5.00, and $10.00, but not necessarily in that order. Leticia has twice as much money as Shelley. Stephen has an amount between Shelley's and Leticia's. Who has $4.35?

Analyze What are you asked to find? who has $4.35

What information are you given? names of people, amounts of money, plus information about how amounts are related

Choose What strategy will you use?

You can *use logical reasoning*.

Solve How will you solve the problem?

Take the clues one at a time. Use a table to help.

	$4.35	$5.00	$7.50	$10.00
Stephen	no	no	yes	no
Leticia	no	no	no	yes
Shelley	no	yes	no	no
Ed	yes	no	no	no

Only one box in each row and column can have a "yes."

Leticia has twice as much money as Shelley. So, Leticia must have $10.00 and Shelley $5.00. Fill in "yes" in those boxes, and fill in "no" in the rest of the boxes in those rows and columns.

Stephen has an amount between Shelley's and Leticia's. $7.50 is between $5.00 and $10.00, so Stephen must have $7.50. Fill in the rest of the boxes with "yes" and "no."

So, Ed has $4.35.

Check How can you check your answer? Go back to the problem and check that the numbers fit the clues.

Give the clue that Leticia has $2\frac{1}{2}$ times as much money as Shelley.

What if the amounts were $7.50, $4.35, $5.00, and $12.50? How could you change the clues in the problem? See left.

236   Chapter 11

RETEACH 11.4

Problem Solving Strategy: Use Logical Reasoning
When a problem presents a lot of information, logical reasoning can often be used in combination with organizing the information in a table.

Chris, Keiko, Rosa, and Jamal are officers in their school's student government. One of them is president, one is vice president, one is secretary, and one is treasurer. Rosa is the president. Chris is not the treasurer. Keiko is the vice president. What office does Jamal hold?

Step 1: Think about what you know and what you are asked to find.
You know: the names of the officers and the offices;
Rosa is the president, Chris is not the treasurer, and Keiko is the vice president.
You are asked to find the office held by Jamal.

Step 2: Plan a strategy to solve.
- Use the strategy *use logical reasoning*.
- Organize the information in a table, using one at a time.

Step 3: Solve.
- Since Rosa is president, put a Y for *yes* in the box where the Rosa column and the President row cross. Put an N for *no* in the other boxes in this row and column, since Rosa cannot hold any other office and nobody else can be president.
- Since Chris is not the treasurer, put an N where the Chris column and the Treasurer row cross.
- Since Keiko is the vice president, put a Y where the Keiko column and the Vice President row cross. Put Ns in the other boxes in this row and column.
- The only office left for Chris is secretary. Put a Y where the Chris column and the Secretary row cross.
- Therefore, Jamal must be the treasurer.

	Chris	Keiko	Rosa	Jamal
President	N	N	Y	N
Vice President	N	Y	N	N
Secretary	Y	N	N	N
Treasurer	N	N	N	Y

Solve the problem using logical reasoning.

1. The Raiders, Rangers, Cougars, and Lions are in one division of a baseball league. Currently, the Raiders are ahead of the Rangers and Cougars. The Raiders and Cougars are behind the Lions. The Cougars are in last place. Who is in first place? **Lions**

2. José, Cody, and Sid are students. One of them is in sixth grade, one is in seventh, and one is in eighth. The seventh-grader and José walk to school together. Sid plays ball with the eighth-grader. Cody is in sixth grade. Who is in seventh grade? **Sid**

PRACTICE 11.4

Problem Solving Strategy: Use Logical Reasoning
Solve the problems by using logical reasoning.

1. Tamara, Alex, Elena, and Fred entered their dogs in the county dog show. The dogs were a terrier, a setter, a golden retriever, and a Great Dane. Neither girl owned the Great Dane. Neither boy entered a setter. Tamara owns a golden retriever. What breed of dog did Elena enter in the show? **setter**

2. Bobby, Ken, Sam, and Ayesha each participate in one sport at school. They play softball, football, basketball, and soccer. Ayesha plays first base. Ken does not play football. If Sam plays soccer, what sport does Bobby participate in? **football**

3. Adel, James, Erica, and An were comparing how far they live from school. An lives only $\frac{1}{3}$ as far as Adel. James lives twice as far as Erica and 4 times as far as An. If Adel lives 9 blocks from school, how far away does Erica live? **Erica: 6 blocks**

4. Ahmed looked over his math homework problems. He saw that $\frac{1}{2}$ of the problems were about fractions, $\frac{1}{3}$ were about decimals, and the rest were about geometry. If there were 4 geometry problems, how many problems did he have in all? **24 homework problems**

5. Robert, Stanley, and Keith are brothers. Robert is 4 years younger than Stanley. Keith is 3 years older than Robert. Robert is 9 years older than his cousin Richard. If Richard is 11, how old is each brother? **Robert: 20; Keith: 23; Stanley 24**

6. Adam, Carin, Dana, and Juanita are lined up for a photograph. As the photographer looks at them, Juanita is to the right of Carin. Adam is on one end. Dana is between Carin and Adam. Give their order from left to right. **Adam, Dana, Carin, Juanita**

Mixed Review
Determine whether each number is divisible by 2, 3, 4, 5, 6, 8, 9, or 10.

7. 125 **5**
8. 336 **2, 3, 4, 6, 8**
9. 1,010 **2, 5, 10**
10. 249 **3**
11. 9,072 **2, 3, 4, 6, 8, 9**

Multiply. Write the answer in simplest form.

12. $\frac{1}{2} \times \frac{2}{5}$ **$\frac{1}{5}$**
13. $\frac{3}{8} \times \frac{2}{3}$ **$\frac{1}{4}$**
14. $\frac{5}{6} \times \frac{1}{4}$ **$\frac{5}{24}$**
15. $\frac{3}{4} \times \frac{5}{6}$ **$\frac{5}{8}$**

236   Chapter 11

PROBLEM SOLVING PRACTICE

Solve the problems by using logical reasoning.

1. Arthur, Victoria, and Jeffrey are in sixth, seventh, and eighth grades, although not necessarily in that order. Victoria is not in eighth grade. The sixth grader is in chorus with Arthur and in band with Victoria. Which student is in each grade? **Jeffrey, sixth grade; Victoria, seventh grade; Arthur, eighth grade**

2. Use the following information to tell which numbers in the box at the right are A, B, C, D, and E.

-3.5	$2\frac{1}{2}$
	0.43
4.3	-0.43

   - A is greater than D and less than C.
   - A and D are opposites.
   - E is the greatest number.

   A = 0.43; B = -3.5; C = $2\frac{1}{2}$; D = -0.43; E = 4.3

For 3–4, use this information.

Amir, Katherine, Patrick, and Lee are comparing their stamp collections. Lee has twice as many stamps as Amir. Patrick has 5 fewer stamps than Katherine, who has 5 fewer stamps than Lee. Their collections consist of 15, 20, 25, and 30 stamps.

3. If Amir has 15 stamps, how many stamps does Patrick have? **C**

   A 30   C 20
   B 25   D 15

4. The four have a total of 90 stamps. How many stamps does Katherine have? **G**

   F 30   H 20
   G 25   J 15

MIXED STRATEGY PRACTICE

5. Kathy, Arlene, and Helene are in line for concert tickets, though not necessarily in that order. Helene is behind Kathy. Kathy is not first in line. Tell the order of the three in line. **Arlene, Kathy, Helene**

6. Mel sold 25 beanbag toys at a fair. He sold some for $8 and some for $5. He made $170. How many toys did he sell for $8? **15**

7. A guidebook costs $3.75 in Canadian dollars in a Vancouver, B.C., bookstore. It also has a price of $3.50 in United States dollars. If the exchange rate is $1.12 Canadian dollars for each U.S. dollar, would you rather pay in Canadian or U.S. dollars? **$3.50 × $1.12 = $3.92, so it is cheaper in Canadian dollars.**

8. Shelters are located at regular intervals along a wilderness trail. The distance from the first shelter to the fourth shelter is 6 mi. What is the distance from the seventh shelter to the thirteenth shelter? **12 mi.**

9. Sandy has to be at the airport for a 5:20 P.M. flight. She wants to arrive 1 hr 15 min early. If it takes 50 min to drive to the airport, when should she leave for the airport? **3:15 P.M.**

10. **What's the Question?** Anne bought 28 fish. She bought three times as many goldfish as angelfish. The answer is 7. **How many angelfish did Anne buy?**

PROBLEM SOLVING STRATEGIES

- Draw a Diagram or Picture
- Make a Model
- Predict and Test
- Work Backward
- Make an Organized List
- Find a Pattern
- Make a Table or Graph
- Solve a Simpler Problem
- Write an Equation
- ▶ **Use Logical Reasoning**

3 Practice

Guided Practice

Note that Exercises 1–4 are **multistep or strategy problems**. Do Problem Solving Practice Exercises 1–4 with your students. Identify those having difficulty and use lesson resources to help.

Before assigning Exercise 2, call on a volunteer to define a number's opposite.

Independent Practice

Note that Exercises 5–10 are **multistep or strategy problems**. Assign Exercises 5–10.

In Exercise 8, students may find it helpful to draw a diagram.

4 Assess

Summarize the lesson by having students:

DISCUSS Which clue did you use first to solve Exercise 1? Possible answer: "The sixth grader is in chorus with Arthur and in band with Victoria."

WRITE Should you always start solving problems involving logical reasoning with the first clue? Explain. No; start with the clue that clearly eliminates one case or allows you to write a *yes* in one box.

Lesson Quiz

1. The numbers 1.4, -3.5, 5, -1.2, and 1.6 are labeled A, B, C, D, and E, though not necessarily in that order. The sum of A and C is 3, and C is greater than A. D is between A and B. Tell which number goes with each letter. A, 1.4; B, -3.5; C, 1.6; D, -1.2; E, 5

2. Kelly, Maria, Rob, and Kevin participated in a school walkathon to raise money for the library. They walked 1.5 mi, 3 mi, 7 mi, and 10 mi. Kevin walked twice as far as Kelly, and Maria walked 4 more miles than Kevin. Who walked the farthest? Rob

---

CHALLENGE 11.4
Logically Speaking

1. Danielle, Effie, and Hannah are in the school orchestra. Each girl plays one instrument: the oboe, the French horn, or the clarinet. Hannah does not play the oboe, nor does she play the clarinet. Effie has never taken clarinet lessons. Which instrument does each girl play?

   Danielle: clarinet; Effie: oboe; Hannah: French horn

2. Austin and Koby each completed a reading assignment. Austin read more pages than Koby. Neither of the boys read more than 40 pages. The difference between the lengths of the reading assignments is 16 pages. The product of the lengths of the reading assignments is 720. How many pages did each boy read?

   Austin: 36 pages; Koby: 20 pages

3. Sabrina, Tom, Carlos, and Fran like to go fishing. On one day they each caught a fish: a salmon, a tuna, a catfish, and a flounder. No one caught a fish that begins with the same letter as his or her name. Neither boy caught a salmon. Neither girl caught a tuna. One of the boys caught a flounder. What kind of fish did each one catch?

   Sabrina: catfish; Tom: flounder
   Carlos: tuna; Fran: salmon

4. Shandra is trying to guess Helen's secret number. Helen gives Shandra the following clues:
   - The number is between 1 and 10
   - If you divide the number by 2, the result is greater than 3.
   - If you triple the number, the result is greater than 24.
   What is Helen's secret number?

   9

5. There are 12 soft drink bottles in Martha's refrigerator. The bottles are either cherry soda or root beer. If Martha takes a cherry soda and replaces it with a new bottle of root beer, there will be an equal number of each flavor. However, if she drinks a root beer and replaces it with a new bottle of cherry soda, there will be twice as many cherry sodas as root beers. How many bottles of each flavor are in the refrigerator right now?

   7 cherry, 5 root beer

6. Lori, Lisa, and Lauren are sisters. When they add their ages together, they get a sum of 19. Three years from now, Lori will be twice as old as Lisa, and Lisa will be twice as old as Lauren. How old is each girl now?

   Lori: 13, Lisa: 5, Lauren: 1

READING STRATEGY 11.4
Analyze Information

The information in a problem can offer clues about how to solve it. Analyze, or look carefully at, the problem. Underline or record details that help you understand the problem.

VOCABULARY analyze

Read the following problem.

Abigail, Bart, Carlotta, and Donald each play a different sport, soccer, basketball, ice hockey, or lacrosse, but not necessarily in that order. Abigail plays a sport that uses a round ball. Carlotta needs a stick to play her sport. Donald can't play his sport outside in the summer. Bart's sport isn't played on grass. Which sport does each play?

1. Analyze the problem. Underline or record details that will help you solve the problem. Which sport does each clue suggest?

   Check students' connections of each clue to the different sports.

2. Solve the problem.

   Abigail—soccer; Bart—basketball; Carlotta—lacrosse; Donald—ice hockey

Analyze the problem. Underline or record details that help you reach an understanding. Then solve.

3. Ari, Latanya, Mary, and Jed each make a different dinner course, soup, salad, main course, or dessert, but not necessarily in that order. Mary is the only one whose recipe doesn't require vegetables. Latanya is the only one who doesn't need to use a stove. Jed's course is the only one that requires a spoon. What did they each prepare?

   Ari: main course
   Latanya: salad
   Mary: dessert
   Jed: soup

4. Fred, Georgia, Hal, and Inez all participated in the Geo-Bee. Their scores were 92%, 75%, 100%, and 83%, but not necessarily in that order. Georgia's score was $\frac{3}{4}$ of Hal's score. Inez's score was 9 points less than Fred's score. What score did each receive?

   Georgia; 100%;
   Fred; 92%;
   Inez; 83%;
   Hal; 75%

CHAPTER 11

REVIEW/TEST

Purpose To check understanding of concepts, skills, and problem solving presented in Chapter 11

USING THE PAGE

The Chapter 11 Review/Test can be used as a **review** or a **test**.

- Items 1–2 check understanding of concepts and new vocabulary.
- Items 3–38 check skill proficiency.
- Items 39–40 check students' abilities to choose and apply problem solving strategies to real-life problems involving integers.

 Suggest that students place the completed Chapter 11 Review/Test in their portfolios.

USING THE ASSESSMENT GUIDE

- Multiple-choice format of Chapter 11 Posttest—See *Assessment Guide*, pp. AG73–74.
- Free-response format of Chapter 11 Posttest—See *Assessment Guide*, pp. AG75–76.

USING STUDENT SELF-ASSESSMENT

The How Did I Do? survey helps students assess what they have learned and how they learned it. This survey is available as a copying master in *Assessment Guide*, p. AGxvii.

CHAPTER 11 REVIEW/TEST

1. **VOCABULARY** Positive whole numbers, their opposites, and 0 make up the set of __?__. (p. 228) **integers**

2. **VOCABULARY** A number that can be written as a ratio $\frac{a}{b}$, where a and b are integers and $b \neq 0$, is a(n) __?__. (p. 230) **rational number**

Write an integer to represent each situation. (pp. 228–229)

3. an increase of 15 points **$^+15$**
4. 6 degrees below zero **$^-6°$**
5. a loss of 20 pounds **$^-20$**

Write the opposite integer. (pp. 228–229)

6. $^-32$ **32**
7. 12 **$^-12$**
8. $^-15$ **15**
9. $^-289$ **289**
10. 0 **0**

Find the absolute value. (pp. 228–229)

11. $|^-12|$ **12**
12. $|^-4|$ **4**
13. $|^+17|$ **17**
14. $|^+8|$ **8**
15. $|^-347|$ **347**

Write each rational number in the form $\frac{a}{b}$. (pp. 230–233) **Possible answers are given.**

16. $2\frac{2}{1}$ **$\frac{2}{1}$**
17. $^-0.89$ **$\frac{^-89}{100}$**
18. $3\frac{2}{3}$ **$\frac{11}{3}$**
19. 5.4 **$\frac{54}{10}$**
20. 14 **$\frac{14}{1}$**
21. 0.334 **$\frac{334}{1,000}$**

Find a rational number between the two given numbers. (pp. 230–233) **Possible answers are given.**

22. $\frac{1}{4}$ and $\frac{2}{3}$ **$\frac{5}{12}$**
23. 1.3 and 1.32 **1.31**
24. $\frac{2}{5}$ and $\frac{3}{4}$ **$\frac{9}{20}$**
25. $^-4.3$ and $^-4.4$ **$^-4.36$**
26. 3.4 and 3.52 **3.45**
27. $2\frac{1}{5}$ and $2\frac{1}{2}$ **$2\frac{1}{4}$**
28. $3\frac{5}{8}$ and $3\frac{9}{10}$ **$3\frac{4}{5}$**
29. 0.9 and 0.94 **0.92**

Compare. Write <, >, or = for each ●. (pp. 234–235)

30. $7\frac{5}{8}$ ● $7\frac{10}{16}$ **=**
31. 1.01 ● 1.10 **<**
32. $^-3.4$ ● $^-4.3$ **>**
33. $\frac{^-25}{3}$ ● $\frac{^-17}{2}$ **<**
34. $4\frac{3}{4}$ ● 4.77 **<**

Order the rational numbers from greatest to least. (pp. 234–235)

35. 3.7, 3.2, $3\frac{5}{8}$ **3.7; $3\frac{5}{8}$; 3.2**
36. $^-8, ^-3, \frac{^-77}{11}$ **$^-3; \frac{^-77}{11}; ^-8$**
37. $^-1.2, 1.2, 0.12$ **1.2; 0.12; $^-1.2$**
38. $\frac{5}{7}, \frac{9}{14}, \frac{11}{8}$ **$\frac{11}{8}, \frac{5}{7}, \frac{9}{14}$**

39. Elizabeth, Doria, Emily, and Claudia each earned money doing odd jobs. They earned $4.50, $6.50, $8.00, and $9.00. Doria earned twice as much as Elizabeth. Emily earned $1.50 more than Claudia. Claudia earned $2.00 more than Elizabeth. How much did each person earn? (pp. 236–237) **Elizabeth: $4.50, Doria: $9.00, Emily: $8.00, Claudia: $6.50**

40. Louisa, Chris, and Vicki each have one pet: a dog, a cat, and an iguana. Louisa does not have a cat and Chris does not have a dog. Louisa's pet is a reptile. What pet does each girl have? (pp. 236–237) **Louisa: iguana, Vicki: dog, Chris: cat**

238 Chapter 11

CHAPTERS 1–11 ★ STANDARDIZED TEST PREP

Look for important words.
See item **7**.
Important words are *least to greatest*. Write the numbers as decimals and order them from least to greatest.
Also see problem **2**, p. H62.

Choose the best answer.

1. Which of the following can be used to represent the depth of a cave that goes 37 feet below ground level? **A**

   A ⁻37   C ⁺3.7
   B ⁻3.7   D ⁺37

2. Which points on the number line show opposite integers? **G**

   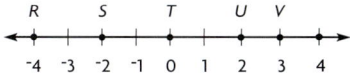

   F R and S   H U and V
   G S and U   J T and R

3. With respect to sea level, the average depth of the Atlantic Ocean is ⁻11,730 feet, the Pacific Ocean is ⁻12,925 feet, and the Gulf of Mexico is ⁻5,297 feet. Which shows these values in order from least to greatest? **C**

   A ⁻11,730, ⁻12,925, ⁻5,297
   B ⁻11,730, ⁻5,297, ⁻12,925
   C ⁻12,925, ⁻11,730, ⁻5,297
   D ⁻12,925, ⁻5,297, ⁻11,730

4. Last week, the dance band practiced for $5\frac{1}{3}$ hours. The band usually practices for $8\frac{3}{4}$ hours per week. How many fewer hours did the band practice last week? **H**

   F $2\frac{3}{4}$ hr   H $3\frac{5}{12}$ hr
   G $3\frac{1}{3}$ hr   J $4\frac{1}{5}$ hr

5. Earl read that $\frac{3}{8}$ of his town's budget is spent on road repairs. Which is that amount expressed as a decimal? **A**

   A 0.375   B 0.4   C 0.67   D 0.875

6. Which symbol makes this number sentence true? **F**

   ⁻3 ● 0

   F <   G >   H =   J ≥

7. Which shows the numbers in order from least to greatest? **D**

   $1.7, \frac{1}{2}, ⁻0.3, \frac{⁻3}{5}$

   A $\frac{1}{2}, 1.7, \frac{⁻3}{5}, ⁻0.3$
   B $\frac{⁻3}{5}, ⁻0.3, 1.7, \frac{1}{2}$
   C $⁻0.3, \frac{1}{2}, \frac{⁻3}{5}, 1.7$
   D Not here

8. Vicki has 8 mysteries, 6 biographies, and 10 adventure books. What part of her book collection is biographies? **H**

   F $\frac{1}{8}$ G $\frac{1}{6}$ H $\frac{1}{4}$ J $\frac{1}{3}$

Write What You Know   See below.

9. Four baseball players are standing in line. Owen is standing between Paul and Jim. Casey is in front of Paul. Jim is last. Find how they are arranged in line. Explain how you found your answer.

10. Explain what it means for a number to be rational. Then show that both 1.02 and $3\frac{1}{3}$ are rational.

Write What You Know • Written Response

9. Jim is last, so Paul is in front of Jim. Since Owen is between Paul and Jim, their order is Paul, Owen, Jim. Since Casey is in front of Paul, the order of the four players is Casey, Paul, Owen, Jim.

10. A rational number is a number that can be written as a ratio $\frac{a}{b}$, where a and b are integers and $b \neq 0$. The decimal number 1.02 can be, rewritten as $1 + 0.02 = 1 + \frac{2}{100} = \frac{100}{100} + \frac{2}{100} = \frac{102}{100}$. So, 1.02 is rational. The mixed number $3\frac{1}{3}$ can be rewritten as $3 + \frac{1}{3} = \frac{9}{3} + \frac{1}{3} = \frac{10}{3}$. So, $3\frac{1}{3}$ is rational.

STANDARDIZED TEST PREP •
Chapters 1–11

USING THE PAGE

This page may be used to help students get ready for standardized tests. The test items are written in the same style and arranged in the same format as those on many state assessments. The page is cumulative. It covers math objectives and essential skills that have been taught up to this point in the text. Most of the items represent skills from the current chapter, and the remainder represent skills from earlier chapters.

This page can be assigned at the end of the chapter as classwork or as a homework assignment. You may want to have students use individual recording sheets presented in a multiple-choice (standardized) format. A Test Answer Sheet is available as a blackline master in *Assessment Guide* (p. AGxlii).

You may wish to have students describe how they solved each problem and share their solutions.

ITEM ANALYSIS

Item	Learning Goal	Item	Learning Goal
1	11A	6	11C
2	11A	7	8A
3	9B	8	11C
4	8C	9	11C
5	11C	10	11C

Written Response items for Write What You Know are available as a blackline master in *Performance Assessment*.

SCORING RUBRIC • WRITE WHAT YOU KNOW

2 Demonstrates a complete understanding of the problem and chooses an appropriate strategy to determine the solution

1 Demonstrates a partial understanding of the problem and chooses a strategy that does not lead to a complete and accurate solution

0 Demonstrates little understanding of the problem and shows little evidence of using any strategy to determine a solution

Algebra: Number Relationships

CHAPTER 12 Algebra: Operations with Integers

CHAPTER PLANNER

PACING OPTIONS
- Compacted: 4 Days
- Expanded: 7 Days

Getting Ready for Chapter 12 • Assessing Prior Knowledge and INTERVENTION (See PE and TE page 241.)

LESSON	NCTM STANDARDS	PACING	VOCABULARY*	MATERIALS	RESOURCES AND TECHNOLOGY
12.1 Math Lab: Model Addition of Integers pp. 242–243 **Objective** To use two-color counters to add integers	1, 2, 8, 10		**additive inverse**	For each pair 15 two-color counters	Astro Algebra • Red
12.2 Add Integers pp. 244–247 **Objective** To use a number line to add integers	1, 2, 6, 8	2 Days (For Lessons 12.1 and 12.2)	absolute value		Reteach, Practice, Challenge, Problem Solving 12.2 Worksheets Extra Practice p. H43, Set A Transparency 12.2 Astro Algebra • Red Calculator Handbook, p. 11 Math Jingles™ CD 5–6
12.3 Math Lab: Model Subtraction of Integers pp. 248–249 **Objective** To use two-color counters to subtract integers	1, 2, 8, 10			For each group 25 two-color counters	E-Lab • *Modeling Subtraction of Integers*; E-Lab Recording Sheet Astro Algebra • Red
12.4 Subtract Integers pp. 250–251 **Objective** To use a number line to subtract integers	1, 6, 8	1 Day (For lessons 12.3 and 12.4)			Reteach, Practice, Challenge, Problem Solving 12.4 Worksheets Extra Practice p. H43, Set B Transparency 12.4 Astro Algebra • Red Calculator Handbook, p. 12 Math Jingles™ CD 5–6
12.5 Multiply and Divide Integers pp. 252–255 **Objective** To multiply and divide integers	1, 2, 6, 8	1 Day		For each group 12 two-color counters	Reteach, Practice, Challenge, Problem Solving 12.5 Worksheets Extra Practice p. H43, Set C Transparency 12.5 Astro Algebra • Red
12.6 Explore Operations with Rational Numbers pp. 256–259 **Objective** To add, subtract, multiply, and divide with rational numbers	1, 2, 6, 7, 8, 9, 10	1 Day			Reteach, Practice, Challenge, Problem Solving 12.6 Worksheets Extra Practice p. H43, Set D Transparency 12.6

Ending Chapter 12 • Chapter 12 Review/Test, p. 260 **• Standardized Test Prep,** p. 261
Ending Unit 4 • Math Detective, p. 262; **Challenge,** p. 263; **Study Guide and Review,** pp. 264–265; **Performance Assessment,** p. 266; **Technology,** p. 267; **Problem Solving On Location,** pp. 267A–267B

***Boldfaced** terms are new vocabulary. Other terms are review vocabulary.

CHAPTER AT A GLANCE

Vocabulary Development

The boldfaced word is the new vocabulary term in the chapter. Have students record the definition in their Math Journals.

additive inverse, p. 243

additive inverse

Vocabulary Cards
Have students use the Vocabulary Cards on *Teacher's Resource* pp. TR131–132 to make graphic organizers or word puzzles. The cards can also be added to a file of mathematics terms.

NCTM Standards

1. **Number and Operations**
   Lessons 12.1, 12.2, 12.3, 12.4, 12.5, 12.6
2. **Algebra**
   Lessons 12.1, 12.2, 12.3, 12.4, 12.5, 12.6
3. **Geometry**
4. **Measurement**
5. **Data Analysis and Probability**
6. **Problem Solving**
   Lessons 12.2, 12.4, 12.5, 12.6
7. **Reasoning and Proof**
   Lesson 12.6
8. **Communication**
   Lessons 12.1, 12.2, 12.3, 12.4, 12.5, 12.6
9. **Connections**
   Lesson 12.6
10. **Representation**
    Lessons 12.1, 12.3, 12.6

Writing Opportunities

PUPIL EDITION	TEACHER'S EDITION	ASSESSMENT GUIDE
• What's the Error?, p. 247 • What's the Question?, p. 251	• Write—See the *Assess* section of each TE lesson. • Writing in Mathematics, p. 250B	• How Did I Do?, p. AGxvii

Family Involvement Activities

These activities provide:
- Letter to the Family
- Math Vocabulary
- Family Game
- Practice (Homework)

Family Involvement Activities, p. FA45

CHAPTER 12 Algebra: Operations with Integers

MATHEMATICS ACROSS THE GRADES

SKILLS TRACE ACROSS THE GRADES

GRADE 5	GRADE 6	GRADE 7
Add and subtract positive and negative integers	Write sums, differences, products, and quotients of integers and rational numbers	Add, subtract, multiply, and divide rational numbers; evaluate expressions with rational numbers

SKILLS TRACE FOR GRADE 6

LESSON	FIRST INTRODUCED	TAUGHT AND PRACTICED	TESTED	REVIEWED
12.1	Grade 5	PE pp. 242–243	PE p. 260, pp. AG77–80	PE pp. 260, 261, 264–265
12.2	Grade 5	PE pp. 244–247, H43, p. RW55, p. PW55, p. PS55	PE p. 260, pp. AG77–80	PE pp. 260, 261, 264–265
12.3	Grade 5	PE pp. 248–249	PE p. 260, pp. AG77–80	PE pp. 260, 261, 264–265
12.4	Grade 5	PE pp. 250–251, H43, p. RW56, p. PW56, p. PS56	PE p. 260, pp. AG77–80	PE pp. 260, 261, 264–265
12.5	Grade 6	PE pp. 252–255, H43, p. RW57, p. PW57, p. PS57	PE p. 260, pp. AG77–80	PE pp. 260, 261, 264–265
12.6	Grade 6	PE pp. 256–259, H43, p. RW58, p. PW58, p. PS58	PE p. 260, pp. AG77–80	PE pp. 260, 261, 264–265

KEY
PE Pupil Edition
PW Practice Workbook
PS Problem Solving Workbook
AG Assessment Guide
RW Reteach Workbook

Looking Back  Prerequisite Skills

To be ready for Chapter 12, students should have the following understandings and skills:

- **Vocabulary**—*positive numbers, integers*
- **Understand Integers**—use positive and negative numbers to represent situations
- **Number Lines**—write integers on a number line
- **Multiplication and Division Facts**—find products and quotients using mental math; solve for *n* in basic facts number sentences

Check What You Know
Use page 241 to determine students' knowledge of prerequisite concepts and skills.

Intervention
Help students prepare for the chapter by using the intervention resources described on TE page 241.

Looking at Chapter 12  Essential Skills

Students will

- **develop understanding of integer addition and subtraction by using two-color counters to add and subtract integers.**
- use number lines to add and subtract integers.
- use absolute values to add and subtract integers.
- understand the concepts of integer multiplication and division.
- explore addition, subtraction, multiplication, and division of rational numbers.

EXAMPLE

Use counters to find the sum.

$$^-2 + {}^+5$$

So, $^-2 + {}^+5 = {}^+3$.

Looking Ahead  Applications

Students will apply what they learn in Chapter 12 to the following new concepts:

- Evaluate Expressions (Chapter 13)
- Solve Multiplication and Division Equations (Chapter 15)
- Solve Addition and Subtraction Equations (Chapter 14)

Operations with Integers   240D

CHAPTER 12

Operations with Integers

INTRODUCING THE CHAPTER

Remind students that integers show how quantities or measures compare to zero. Integers, like other numbers, can be added and subtracted. After students read the paragraph, explain that the difference between two negative numbers is the same as the difference between two positive numbers, but the sign may be different. Ask students what the estimated difference is between 193 and 129. **about 60** Have them find the opposite of 60. **⁻60**

USING DATA

To begin the study of this chapter, have students

- Draw and label a number line to show the temperature data for the Earth.
- Draw and label a number line to show the temperature data for Mars.
- Make a double bar graph of all the data in the table.

Check students' work.

PROBLEM SOLVING PROJECT

Purpose To compare data expressed as integers

Background Mars is the next planet in the solar system after the Earth. NASA's *Pathfinder* and *Sojourner* missions provided valuable scientific data, including temperature, about Mars.

Analyze, Choose, Solve, and Check

Have students

- Analyze the data given in the table and make comparisons.
- Write a paragraph that compares temperature extremes on Earth to temperature extremes on Mars.
- Support the comparison with data from the table and explain how the data was interpreted. **Check students' work.**

 Suggest that students place their paragraphs in their portfolios.

CHAPTER 12 Algebra: Operations with Integers

NASA's Pathfinder and Sojourner space probes were built to withstand extreme temperatures as they traveled from Earth to Mars.

PROBLEM SOLVING The south pole on Mars is made not of frozen water but of frozen carbon dioxide, which has a temperature of ⁻193°F. The coldest day recorded on Earth was ⁻129°F, in Antarctica. Which temperature is lower? By about how much?

Temperature	Earth	Mars
Lowest Recorded	⁻129°F	⁻220°F
Highest Recorded	136°F	68°F
Average	57°F	⁻81°F

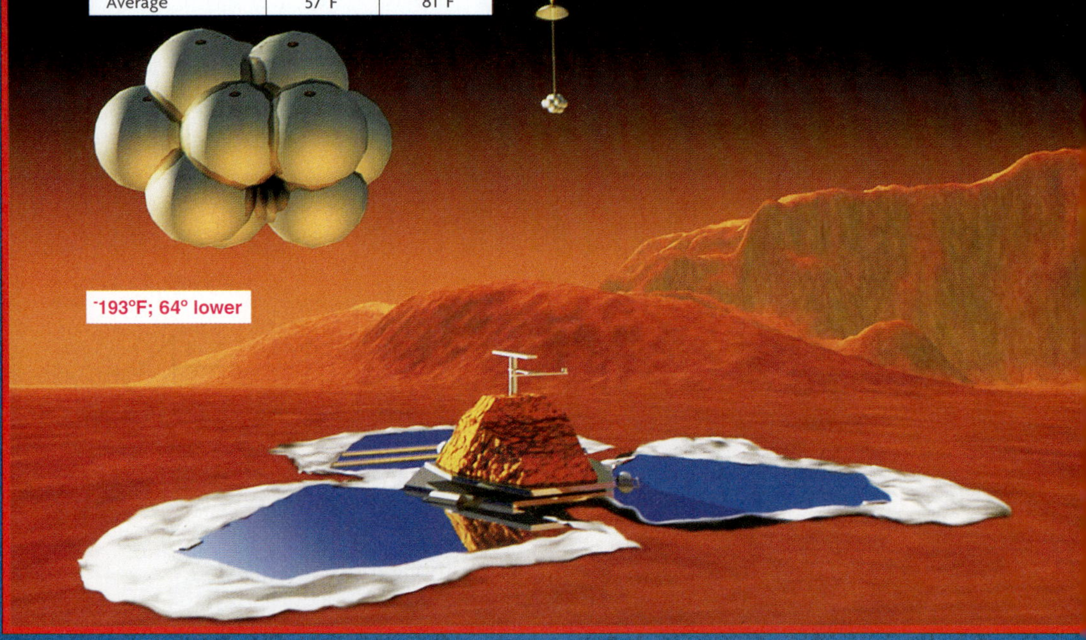

⁻193°F; 64° lower

Why learn math? Explain that scientists and engineers who design equipment for space exploration must understand integers and be comfortable computing with them. Ask: How would addition and subtraction of integers be used in a career? **Possible answer: A bank teller uses addition and subtraction of integers to balance accounts.**

TECHNOLOGY LINK
To find out more about integers, visit The Harcourt Learning Site.
www.harcourtschool.com

Check What You Know

Use this page to help you review and remember important skills needed for Chapter 12.

Vocabulary

Choose the best term from the box.

> positive numbers
> integers

1. The set of whole numbers and their opposites is the set of __?__. **integers**

Understand Integers (For Intervention, see p. H12.)

Write a positive or negative integer to represent each situation.

2. 14° below zero **−14**
3. 62 degrees above zero **+62**
4. 10 ft above sea level **+10**
5. 13 m below sea level **−13**
6. bottom of a well, 50 ft below the surface **−50**
7. a gain of 12 yards **+12**
8. a bank deposit of $280 **+280**
9. 3 ft below ground level **−3**

Number Lines (For Intervention, see p. H13.)

Name the integer that corresponds to the point.

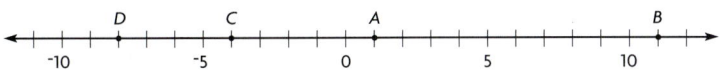

10. A **+1**
11. B **+11**
12. C **−4**
13. D **−8**

Multiplication and Division Facts (For Intervention, see p. H13.)

Find the product or the quotient.

14. 4×6 **24**
15. 9×7 **63**
16. $81 \div 9$ **9**
17. $80 \div 8$ **10**
18. $24 \div 3$ **8**
19. 4×1 **4**
20. $120 \div 12$ **10**
21. 8×7 **56**
22. 12×6 **72**
23. $108 \div 9$ **12**
24. 3×11 **33**
25. $45 \div 9$ **5**
26. 7×12 **84**
27. $144 \div 12$ **12**

Solve for n.

28. $2 \times n = 14$ **n = 7**
29. $n \times 5 = 15$ **n = 3**
30. $n \div 6 = 9$ **n = 54**
31. $88 \div n = 11$ **n = 8**
32. $42 = n \times 6$ **n = 7**
33. $11 = n \div 12$ **n = 132**

LOOK AHEAD

In Chapter 12 you will
- add and substract integers
- multiply and divide integers
- explore operations with rational numbers

241

LESSON 12.1

ORGANIZER

Objective To use two-color counters to add integers

Vocabulary additive inverse

Materials *For each pair* 15 two-color counters

Intervention for Prerequisite Skills
Understanding Integers (For intervention strategies, see page 241.)

Using the Pages

Introduce students to the counters used to model positive and negative integers. Show the yellow side of a counter and call on volunteers to describe what it represents. Repeat with the red side.

Activity 1

Ask students to explain their models and the process they used to find Lin's and Nina's scores.

Think and Discuss

Have students use their counters to show that changing the order of the addends for Lin's and Nina's scores does not change the sum. Ask:

What property of addition have you illustrated? the Commutative Property

Practice

Ask students to predict which sums in Exercises 1–4 will be positive and which will be negative. The sums in 1 and 4 are positive; the sums in 2 and 3 are negative.

242 Chapter 12

LESSON 12.1 Model Addition of Integers

Explore how to use two-color counters to add integers.

You need two-color counters.

QUICK REVIEW
1. $40 + 50$ **90**   2. $120 - 32$ **88**
3. $19 + 21$ **40**   4. $118 - 70$ **48**
5. $120 + 130$ **250**

Vocabulary
additive inverse

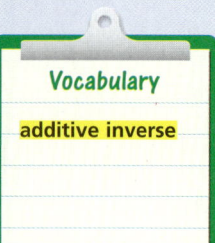

Lin and Nina are playing a board game. To keep track of points, they are using yellow counters to represent positive points, or points gained, and red counters to represent negative points, or points lost.

Activity 1

• Lin earned 6 points during the first round. She earned 3 points during the second round. Use yellow counters to model Lin's total score.

First-round points   Second-round points
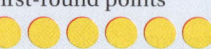

$6 + 3 = 9 \leftarrow$ total score

• Nina lost 2 points in the first round. Then she lost 5 points in the second round. Use red counters to model Nina's total score.

First-round points   Second-round points
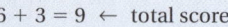

$^{-}2 + ^{-}5 = ^{-}7 \leftarrow$ total score

Think and Discuss

• How is adding the scores like adding whole numbers? How is it different? **You add just as you do with whole numbers. The difference is that Nina's scores are both negative numbers.**
• Would changing the order when adding Lin's or Nina's points change their scores? Why or why not? **No; Changing the order does not change the sum.**
• How would you model $2 + 7$? $^{-}2 + ^{-}7$? **by showing 2 and 7 yellow counters; by showing 2 and 7 red counters**

Practice

Use counters to find the sum.

1. $4 + 9$ 2. $^{-}3 + ^{-}7$ 3. $^{-}6 + ^{-}4$ 4. $5 + 8$
   $^{+}13$ $^{-}10$ $^{-}10$ $^{+}13$

242 Chapter 12

SPECIAL NEEDS ELL

Materials *For each group* 30 two-color counters

Provide students with additional practice in **adding integers** by using counters to solve the word problem below. Let red counters represent the borrowed money and let yellow counters represent the money paid back.

• Peter borrowed $15 from Sarah. A week later, he paid back $5. How much does Peter still owe Sarah? $^{-}15 + ^{+}5 = ^{-}10$; Peter still owes Sarah $10.

KINESTHETIC
INTERPERSONAL/SOCIAL

Intervention and Extension Resources

Activity 2

The **additive inverse** of an integer is its opposite. 1 and ⁻1 are the additive inverses of each other. When you add an integer and its additive inverse, the sum is always 0. You can model this using counters.

- Model the sum of 1 and its additive inverse, ⁻1.

 = 0

- Model the sum of 5 and its additive inverse, ⁻5.

 = 0

- During a game, Carmen gained 8 points and then lost 5 points. To find her total score, Carmen paired points gained with points lost. Use yellow and red counters to model Carmen's total score. Remember that pairs of red and yellow counters equal 0.

 8 + ⁻5 = 3

- Robert gained 3 points and then lost 7 points. Use yellow and red counters to model Robert's total score.

 3 + ⁻7 = ⁻4

Think and Discuss

- Why is Carmen's score positive? **There are more yellow counters than red counters.**
- Why is Robert's score negative? **There are more red counters than yellow counters.**

Practice

Use counters to find each sum.

1. 4 + ⁻6 **⁻2**
2. ⁻2 + 6 **4**
3. 7 + ⁻7 **0**
4. ⁻3 + 8 **5**
5. 5 + 2 **7**
6. 3 + 1 **4**
7. ⁻4 + ⁻5 **⁻9**
8. ⁻3 + ⁻8 **⁻11**

MIXED REVIEW AND TEST PREP

Order the rational numbers from least to greatest. (p. 234)

9. ⁻6.4, ⁻6.2, ⁻6.8
   ⁻6.8, ⁻6.4, ⁻6.2
10. $\frac{1}{3}, \frac{2}{5}, \frac{1}{4}, \frac{3}{5}$ **$\frac{1}{4}, \frac{1}{3}, \frac{2}{5}, \frac{3}{5}$**
11. $-3\frac{4}{7}, -3.6, -3\frac{1}{2}$
    $-3.6, -3\frac{4}{7}, -3\frac{1}{2}$
12. Evaluate the expression $c + \frac{1}{2}$ for $c = \frac{1}{3}$. (p. 216) **$\frac{5}{6}$**
13. **TEST PREP** Which shows the difference $6\frac{5}{6} - 5\frac{3}{4}$? (p. 186) **B**

    A $\frac{1}{9}$ B $1\frac{1}{12}$ C $1\frac{1}{2}$ D $1\frac{3}{4}$

243

Activity 2

Direct students' attention to Activity 2. Ask:

How do you model a gain of 8 points? a loss of 5 points? with 8 yellow counters; with 5 red counters

How do you know from the problem that Robert's total score will be negative? Robert lost more points than he gained.

Think and Discuss

Ask:

How do the exercises in Activity 1 differ from those in Activity 2? The exercises in Activity 1 involve adding integers with the same sign; those in Activity 2 involve adding negative integers to positive integers.

Practice

Although some students may be able to solve the exercises in their heads, modeling the exercises will help them better understand the concept of adding integers. This understanding will be important in the next lesson when students add greater integers.

MIXED REVIEW AND TEST PREP

Exercises 9–13 provide **cumulative review** (Chapters 1–12).

Oral Assessment

What is the sign of the sum of two positive integers? positive

What is the sign of the sum of two negative integers? negative

What is the sign of the sum of a positive and a negative integer? the sign of the addend with the greater absolute value

What integers could you add to equal 0? Give an example. any integer and its opposite; ⁺2 + ⁻2 = 0

EARLY FINISHERS

Materials For each group 10 two-color counters

Have students **use integers**. Explain that in golf, *par* is the standard number of strokes to hit a ball into a specific hole. Have groups use counters to find Bruce's scores.

BRUCE'S SCORES			
Hole	Par	Above or Below Par	Bruce's Score
1	4	⁻1	? **⁺3**
2	3	⁻2	? **⁺1**
3	5	⁺1	? **⁺6**
4	5	⁺2	? **⁺7**

KINESTHETIC
BODILY/KINESTHETIC

TECHNOLOGY LINK

- Intervention Strategies and Activities CD-ROM • Skill 9
- Astro Algebra • Red, Level A

243

LESSON 12.2
Algebra: Add Integers
LESSON PLANNING

Objective To use a number line to add integers

Intervention for Prerequisite Skills
Understand Integers, Locate Points on a Number Line (For intervention strategies, see page 241.)

Lesson Resources Calculator Handbook, p. 13

NCTM Standards
1. Number and Operations
2. Algebra
6. Problem Solving
8. Communication

Math Background

Adding two positive integers or two negative integers is much like adding whole numbers: the two numbers are added and then the sign of the numbers is attached. When adding a positive number and a negative number, students must first actually find the difference of the absolute values of the two numbers.

The following points may help students understand the process of adding integers:

- The sum of two positive integers is positive.
- The sum of two negative integers is negative.
- The sum of a positive and a negative number may be positive or negative. The answer is the difference of the two absolute values, and the sign is that of the number with the greater absolute value.

WARM-UP RESOURCES

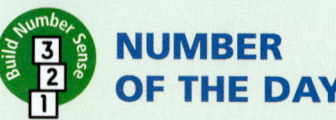 

NUMBER OF THE DAY

How many minutes after the hour is it right now? Find its opposite. Possible answer for 23 minutes past the hour: ⁻23

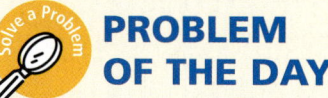 

PROBLEM OF THE DAY

On a number line, the distance from 0 to a negative integer is four times as great as the distance to a positive integer. The sum of their absolute values is 35. What are the numbers? ⁻28, 7

Solution Problem of the Day tab, p. PD12

DAILY FACTS PRACTICE

Have students practice addition facts by completing Set G of *Teacher's Resource Book,* p. TR101.

INTERVENTION AND EXTENSION RESOURCES

ALTERNATIVE TEACHING STRATEGY

Reinforce the concept of **adding integers.** Have students complete each statement with *always, sometimes,* or *never.*

- The sum of a positive integer and a negative integer is _____ positive. sometimes
- A negative integer added to a negative integer _____ equals a positive integer. never
- The sum of two positive integers is _____ positive. always
- The sum of two negative integers is _____ negative. always

See also page 246.

AUDITORY
VERBAL/LINGUISTIC

MULTISTEP AND STRATEGY PROBLEMS

The following multistep and strategy problems are provided in Lesson 12.2:

Page	Item
247	38, 39

ENGLISH LANGUAGE LEARNERS

Materials *For each group* 20 two-color counters

Have students **model addition of integers** by solving this riddle:

- When you add <u>me</u> to <u>6</u> you get <u>3</u>. What am I? ⁻3

Read the riddle aloud several times and model it with counters. Then have students work in groups to solve the riddles shown below.

- When you add <u>me</u> to <u>⁻5</u>, you get <u>0</u>. What am I? ⁺5
- When you add <u>me</u> to <u>⁻9</u>, you get <u>⁺2</u>. What am I? ⁺11

Ask students to make a riddle to share with the class. Have them read the riddle aloud, then model the answer with counters. Check students' work.

AUDITORY, KINESTHETIC
VERBAL/LINGUISTIC

TECHNOLOGY • *CALCULATOR*

Have students use this activity to **explore integers on their calculators.**

- Have students enter 7 and press the +/− key. Ask: What number is displayed? ⁻7
- Ask them to press the +/− key again. Ask: Now what number is displayed? 7
- Ask students to describe the keys they would press to find the sum of ⁻7 and ⁻2. Ask: What is the sum? 7, +/−, +, 2, +/−, =; ⁻9
- Request that students repeat the above process to find the sum of ⁻8 and ⁺5. 8, +/−, +, 5, =; ⁻3

VISUAL
BODILY/KINESTHETIC

TECHNOLOGY LINK

 Intervention Strategies and Activities CD-ROM • *Skills 9, 42*

 Astro Algebra • *Red,* Level A

 Calculator Handbook, p. 11

 Math Jingles™ CD 5–6 • *Track 17*

244B

LESSON 12.2 ORGANIZER

Objective To use a number line to add integers

Vocabulary Review absolute value

Lesson Resources Calculator Handbook, p. 11

1 Introduce

QUICK REVIEW provides review of prerequisite skills.

Why Learn This? You can use addition of integers to determine net gain or loss of yardage in a football game and other types of games. Share the lesson objective with students.

2 Teach

Guided Instruction

- Direct students' attention to the first number line. Ask:

  What number does each arrow represent? $^-3$ and $^-2$

  What operation is shown by placing the tail of the arrow representing $^-2$ at the head of the one representing $^-3$? addition

- Have students consider the second number line. Ask:

  Why do the arrows point in opposite directions? One arrow represents a positive number and the other represents a negative number.

- Direct students' attention to Example 1.

  Can you place the tail of the arrow representing $^-6$ at 0? Explain. No; to show addition, you have to place the tail of the second arrow for $^-6$ at the same point on the number line as the head of the first arrow for 4.

REASONING **Can you draw the arrow for $^-6$ first and then the arrow for 4 and still get the same answer? Explain.** Yes; addition is commutative.

ADDITIONAL EXAMPLE

Example 1, p. 244

Use a number line to find the sum $^-3 + 5$. 2

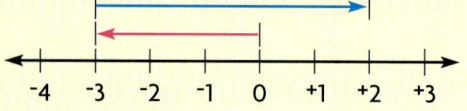

244 Chapter 12

LESSON 12.2 Add Integers

Learn how to use a number line to add integers.

QUICK REVIEW
1. 25 + 42 67
2. 240 − 60 180
3. 13 + 17 30
4. 112 + 35 147
5. 250 − 220 30

Jeb and Raul made up a game using a number line. Play starts at 0. A spinner is used to show positive moves and negative moves.

Jeb's first spin was $^-3$, and his second spin was $^-2$. What is Jeb's position on the number line?

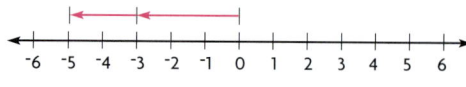

$^-3 + ^-2 = ^-5$ ← Jeb is at $^-5$.

Remember that you can write a positive number without the $^+$ sign.

$^+7 = 7$

Raul's first spin was 4, and his second spin was $^-9$. Where is Raul on the number line?

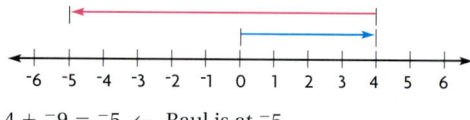

$4 + ^-9 = ^-5$ ← Raul is at $^-5$.

You can use a number line to find the sum of two integers.

EXAMPLE 1

Use a number line to find the sum $4 + ^-6$.

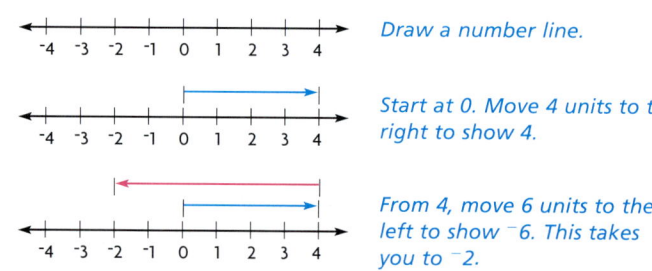

Draw a number line.

Start at 0. Move 4 units to the right to show 4.

From 4, move 6 units to the left to show $^-6$. This takes you to $^-2$.

So, $4 + ^-6 = ^-2$.

When integers with the same sign are added, the arrows point in the same direction. When integers with different signs are added, the arrows point in different directions.

- When integers are added on a number line, when do the arrows point in the same direction and when do the arrows point in different directions? See at left.

244 Chapter 12

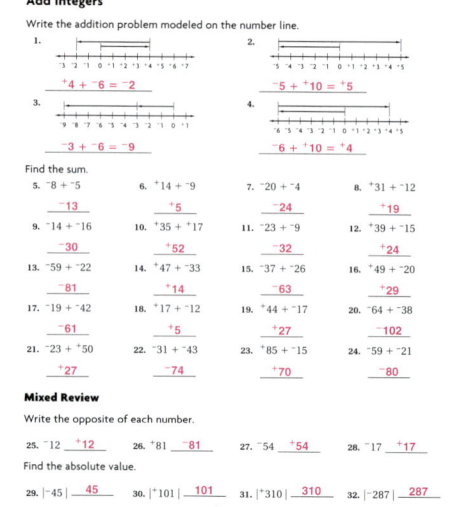

Remember that the absolute value of an integer is its distance from 0 on the number line.

When adding integers, you can use their absolute values to find the sum.

Adding with the Same Sign
When adding integers with like signs, add the absolute values of the integers. Use the sign of the addends for the result.

EXAMPLE 2

Find the sum $^-7 + {}^-2$.

$^-7 + {}^-2$

$|^-7| + |^-2| = 7 + 2$ *Add the absolute values of the integers.*

$\qquad\qquad\quad = 9$

So, $^-7 + {}^-2 = {}^-9$.   *Use the sign of the original addends.*

Adding with Different Signs
When adding integers with unlike signs, subtract the lesser absolute value from the greater absolute value. Use the sign of the addend with the greater absolute value for the result.

EXAMPLE 3

A. Find the sum $^-8 + 3$.

$^-8 + 3$

Subtract the lesser absolute value from the greater absolute value.

$|^-8| - |3| = 8 - 3$

$\qquad\qquad = 5$

Use the sign of the addend with the greater absolute value.

$|^-8| > |3|$ *The sum is negative.*

So, $^-8 + 3 = {}^-5$.

B. Find the sum $^-5 + 9$.

$^-5 + 9$

Subtract the lesser absolute value from the greater absolute value.

$|9| - |^-5| = 9 - 5$

$\qquad\qquad = 4$

Use the sign of the addend with the greater absolute value.

$|9| > |^-5|$ *The sum is positive.*

So, $^-5 + 9 = 4$.

• Find the sum $9 + {}^-12$. **$^-3$**

EXAMPLE 4

On the first play of a football game, the Cobras gained 21 yards. On the second play, they lost 9 yards. Find the total number of yards gained or lost by the Cobras on the first two plays.

$21 + {}^-9$ *Use 21 for yards gained and $^-9$ for yards lost.*

$|21| - |^-9| = 21 - 9 = 12$ *Subtract the lesser absolute value from the greater absolute value.*

$|21| > |^-9| \rightarrow 21 + {}^-9 = 12$ *Use the sign of the addend with the greater absolute value.*

So, the Cobras gained a total of 12 yards on the first two plays.

245

Modifying Instruction To reinforce the rules for adding integers, use a number line to model the Examples on page 245 and the Additional Examples below.

• Review the concept of absolute value. Then ask students to think about the rule for adding two integers with the same sign and direct their attention to Example 2.

Why is the answer negative in Example 2? The addends are both negative.

• Ask students to consider the rule for adding two integers with opposite signs as in Example 3.

What will be the sign of the sum in A? Why? Negative; $^-8$ has the greater absolute value.

What will be the sign of the sum in B? Why? Positive; 9 has the greater absolute value.

ADDITIONAL EXAMPLES

Example 2, p. 245

Find the sum $^-5 + {}^-4$. **$^-9$**

Example 3, p. 245

A. Find the sum $3 + {}^-7$. **$^-4$**

B. Find the sum $12 + {}^-4$. **8**

Example 4, p. 245

On the last two plays of the half, the Ramchargers gained 5 yards and lost 8 yards. Find the total number of yards gained or lost by the Ramchargers on the last two plays of the half. **lost 3 yards**

CHALLENGE 12.2

Sum It Up

The integer at the top of each rectangle is the sum of four addends contained in the rectangle. Shade the boxes containing the addends you use to get the sum. You will use one addend in each row.

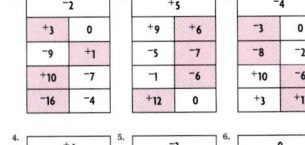

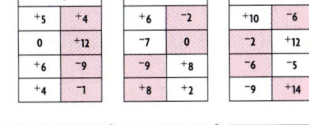

PROBLEM SOLVING 12.2

Add Integers

Analyze Choose Solve Check

Write the correct answer.

1. Find the missing number in the pattern.

$3 + 2 = 5$
$3 + 1 = 4$
$3 + 0 = 3$
$3 + {}^-1 = 2$
$3 + {}^-2 = 1$
$3 + {}^-3 = \blacksquare$

0

2. Find the missing number in the pattern.

$^-3 + 1 = {}^-2$
$^-3 + 0 = {}^-3$
$^-3 + {}^-1 = {}^-4$
$^-3 + {}^-2 = {}^-5$
$^-3 + {}^-3 = {}^-6$
$^-3 + {}^-4 = \blacksquare$

$^-7$

3. Carmen wants to share her money with her cousin Jasmine. Together they have $48. If Carmen gives Jasmine $3, they will each have the same amount of money. How much money does each girl have now?

Carmen has $27, Jasmine has $21.

4. Five students were waiting in line to return books at the library. There were 3 students ahead of John. There were 3 students behind Leila. Carla was first in line. Paul was last. What number in line was Sara?

third

Choose the letter for the best answer.

5. On three consecutive plays, a football team lost 2 yards, gained 5 yards, and gained 7 yards. Which expression could be used to find the total yards gained by the team on these three plays?

A $^-2 + {}^-5 + {}^-7$ C $^-2 - {}^-5 - {}^+7$
B $^-2 + {}^-5 + {}^+7$ **(D) $^-2 + {}^+5 + {}^+7$**

6. By 10:00 A.M., the temperature had risen 7°C from a morning low temperature of $^-15$°C. What was the temperature at 10:00 A.M.?

F $^-8$°C   H 7°C
G $^-7$°C   J 8°C

7. Kirk is 131.9 centimeters tall and Thad is 162.3 centimeters tall. Which is the best estimate of how much taller Thad is than Kirk?

A 10 cm   **(C) 30 cm**
B 20 cm   D 40 cm

8. Patty had 25.8 meters of wire to install lights in her backyard. She used only 19.4 meters. How much wire was left?

F 7.4 m   H 5.4 m
(G) 6.4 m   J 3.4 m

9. **Write About It** Explain why $8 + {}^-3 = {}^-3 + 8$.

Addition is commutative. Changing the order of the addends does not change the sum.

245

LESSON **12.2**

3 Practice

Guided Practice

Do Check for Understanding Exercises 1–13 with your students. Identify those having difficulty and use lesson resources to help.

⚠ COMMON ERROR ALERT

Students may assign the wrong sign to the sum of two integers with different signs. To avoid this mistake, have students begin each exercise by circling the integer farther from zero. Remind students that this integer has the greater absolute value and the sign of this integer will be the sign used in the sum.

Independent Practice

Note that Exercises 38 and 39 are **multistep or strategy problems**. Assign Exercises 14–41.

Multistep or Strategy Problem To solve Exercise 39, students can first add 15 and ⁻9, and then add ⁻8. Guide students to conclude that a loss is represented by a negative number.

Algebraic Thinking For students to picture the value of *x* in each exercise, have them use a number line. For example, in Exercise 32 have students locate ⁻5 on a number line and then determine what number is needed to get to ⁻7.

CHECK FOR UNDERSTANDING

Think and Discuss ▶ Look back at the lesson to answer each question.

1. **Explain** how you determine the sign of the sum of two integers with the same sign. Use the sign of the addends.
2. **Explain** how you determine if the sum of two integers with different signs is positive or negative. Use the sign of the addend with the greater absolute value.
3. **Tell** how you know the Cobras had a gain of 12 yards instead of a loss of 12 yards in Example 4. The sum was 12, which indicates a gain of 12 yards, not a loss of 12 yards.

Guided Practice ▶ Write the addition problem modeled on the number line.

4.  ⁻2 + ⁻4 = ⁻6

5. 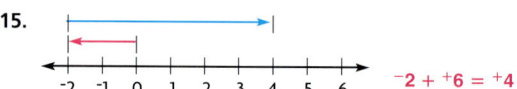 5 + ⁻9 = ⁻4

Find the sum.

6. ⁻9 + 6 **⁻3**   7. ⁻3 + ⁻4 **⁻7**   8. ⁻8 + 2 **⁻6**   9. 5 + ⁻7 **⁻2**
10. ⁻3 + 7 **4**   11. ⁻8 + ⁻2 **⁻10**   12. 11 + ⁻5 **6**   13. 6 + ⁻6 **0**

PRACTICE AND PROBLEM SOLVING

Independent Practice ▶ Write the addition problem modeled on the number line.

14. ⁺2 + ⁺4 = ⁺6

15. ⁻2 + ⁺6 = ⁺4

Find the sum.

16. ⁻5 + 8 **3**   17. 2 + ⁻3 **⁻1**   18. 7 + 2 **9**   19. ⁻1 + 4 **3**
20. 8 + 7 **15**   21. ⁻12 + ⁻8 **⁻20**   22. ⁻15 + ⁻10 **⁻25**   23. ⁻17 + 25 **8**
24. ⁻2 + 5 **3**   25. ⁻12 + ⁻16 **⁻28**   26. ⁻17 + 5 **⁻12**   27. 25 + ⁻37 **⁻12**
28. 24 + 12 **36**   29. 30 + ⁻41 **⁻11**   30. |16| + |⁻9| **25**   31. |⁻64| + |36| **100**

ALGEBRA Use mental math to find the value of *x*.

32. ⁻5 + *x* = ⁻7   *x* = ⁻2
33. *x* + ⁻6 = ⁻13   *x* = ⁻7
34. *x* + ⁻10 = ⁻4   *x* = 6
35. ⁻8 + *x* = ⁻3   *x* = 5

246 Chapter 12

ALTERNATIVE TEACHING STRATEGY — SCAFFOLDED INSTRUCTION

Purpose Students use counters to add integers.

Materials *For each group* 20 two-color counters

Give each group 20 counters. Remind students that the yellow side indicates a positive number and the red side indicates a negative number. Ask each group to use their counters to show ⁻3 and ⁻2.

⁻3

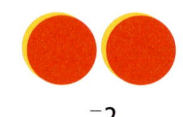
⁻2

Then ask them to use the counters to find the sum ⁻3 + ⁻2. Call on a volunteer to explain the addition process.

⁻3 + ⁻2 = ⁻5

Next have students use their counters to show 4 and ⁻7. Ask them what happens when they make a pair of one counter of each color. The sum is zero. Have them use the counters to find the sum 4 + ⁻7. Ask a volunteer to explain the addition process.

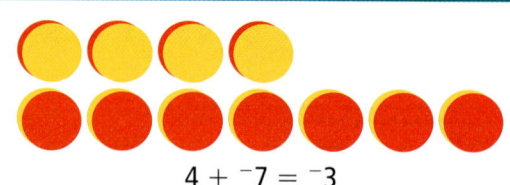
4 + ⁻7 = ⁻3

Finally, have groups use their counters to find sums for other pairs of integers. Call on volunteers to demonstrate the addition using counters, and record the number sentences. Check students' work.

Problem Solving ▶ Applications

36. In the morning the temperature was ⁻6°F. By noon it had risen 11°F. What was the temperature at noon? **5°F**

37. In the evening the temperature was ⁻9°F. By midnight it had dropped 3°F. What was the temperature at midnight? **⁻12°F**

38. Tina played a game in which she owned stock worth $33. During the game, the stock increased $11 in value and then decreased $15 in value. Write an addition sentence to find the new value of the stock. **33 + 11 + ⁻15 = 29; $29**

39. On the first three plays of a football game, the Wildcats gained 15 yards, lost 9 yards, and lost 8 yards. Find the total number of yards gained or lost by the Wildcats on the first three plays.
The Wildcats lost 2 yards.

40. **What's the Error?** Ken says that ⁻6 + 2 = 8. What is his error? What is the correct sum? **He ignored the fact that 6 is negative; ⁻6 + 2 = ⁻4.**

41. **Geometry** A rectangle with an area of 180 cm² has a length of 15 cm. Find the width. **12 cm**

MIXED REVIEW AND TEST PREP

42. Is ⁻1.5 less than, greater than, or equal to ⁻1$\frac{3}{6}$? (p. 234) **equal to**

43. Is 8$\frac{3}{5}$ less than, greater than, or equal to 8.5? (p. 234) **greater than**

44. Is ⁻3.4 less than, greater than, or equal to ⁻3$\frac{1}{5}$? (p. 234) **less than**

45. **TEST PREP** Find the quotient $4\frac{2}{3} \div 3\frac{1}{2}$. (p. 210) **C**

46. **TEST PREP** Find the sum $5\frac{3}{4} + 6\frac{1}{3}$. (p. 186) **H**

A $\frac{3}{4}$ B $1\frac{1}{4}$ C $1\frac{1}{3}$ D $1\frac{1}{2}$

F $11\frac{1}{12}$ G $11\frac{4}{7}$ H $12\frac{1}{12}$ J $12\frac{11}{12}$

PROBLEM SOLVING **Thinker's Corner**

Math Fun • Opposites Distract Remember that every number has an opposite that is the same distance from zero but is on the opposite side on the number line. Use a number line to solve these riddles.

1. I am the opposite of a number between $2\frac{2}{3}$ and $3\frac{5}{6}$. **Possible answer: ⁻3$\frac{3}{4}$**

2. I am the first integer that is less than the opposite of a number between $2\frac{1}{2}$ and $2\frac{3}{4}$. **⁻3**

3. I am the opposite of a number between ⁻4.3 and ⁻4$\frac{3}{8}$. **Possible answer: 4.37**

4. We are between ⁻2.4 and ⁻2.6. If you add our opposites, you get 5. **Possible answer: ⁻2.43 and ⁻2.57**

5. I am the opposite of the integer between the sum ⁻5 + 2 and the sum ⁻15 + 10. **4**

6. I am the second integer that is less than the sum ⁻22 + ⁻17. **⁻41**

7. I am the opposite of the even integer between the sum 35 + ⁻14 and the sum 12 + 6. **⁻20**

8. I am the integer that is 4 times the sum 42 + ⁻35. **28**

EXTRA PRACTICE page H43, Set A

247

MIXED REVIEW AND TEST PREP
Exercises 42–46 provide **cumulative review** (Chapters 1–12).

 Thinker's Corner

Have students work in pairs to solve these riddles. Point out that some riddles have more than one possible answer.

• As a follow-up to Exercise 4, ask:

REASONING What are two other possible answers? Possible answers: ⁻2.41 and ⁻2.59; ⁻2.44 and ⁻2.56; ⁻2.401 and ⁻2.599

4 Assess

Summarize the lesson by having students:

DISCUSS Describe what the arrows on a number line will look like to show the sum ⁻4 + ⁻6. They will both point in the negative direction. One will start at 0 and end at ⁻4 and the other will start at ⁻4 and end at ⁻10.

 **WRITE** Describe how you would add two integers with opposite signs. Find the absolute value of each number and subtract the lesser from the greater. Then determine the sign of the sum by choosing the sign of the number with the greater absolute value.

Lesson Quiz

Find the sum.

1. ⁻11 + 3 **⁻8**   2. ⁻5 + ⁻2 **⁻7**
3. 9 + ⁻16 **⁻7**   4. ⁻21 + 29 **8**
5. ⁻85 + 85 **0**   6. ⁻33 + ⁻13 **⁻46**

LESSON 12.3

ORGANIZER

Objective To use two-color counters to subtract integers

Materials *For each group* 25 two-color counters

Lesson Resources E-Lab Recording Sheet • *Modeling Subtraction of Integers*

Intervention for Prerequisite Skills Understanding Integers (For intervention strategies, see page 241.)

Using the Pages

Remind students that there is more than one way to show the same value. For example, they have used two-color counters and number lines to model the addition of integers. Briefly review using counters to model integers.

How can you model 4? ⁻6? 0? with four yellow counters; with six red counters; with an equal number of red and yellow counters

Activity 1

Refer to the expression ⁻7 − 5.

Why are 5 yellow counters needed? to subtract 5

Why do we need to add both red and yellow counters to the model? Adding the same number of both colors does not change the value of the number you are subtracting from.

Refer to the expression 7 − ⁻2.

Describe the steps you will use to model 7 − ⁻2. First, model 7 with 7 yellow counters. Then, add two pairs of red and yellow counters to the model. Finally, remove the 2 red counters and write the answer 9.

LESSON 12.3 — Model Subtraction of Integers

Explore how to use two-color counters to subtract integers.

You need two-color counters.

QUICK REVIEW
1. 17 − 12 **5**
2. 42 + 20 **62**
3. 224 − 19 **205**
4. 132 − 18 **114**
5. 89 + 19 **108**

You can use red and yellow counters to subtract integers. Subtracting integers is similar to subtracting whole numbers.

Activity 1

- Find ⁻9 − ⁻4. First, make a row of 9 red counters.

- Then, take away 4 of them.

 → ⁻9 − ⁻4

- How many counters are left? What is ⁻9 − ⁻4? **5; ⁻5**

Using red and yellow counters, model ⁻7 − 5.

- First, make a row of 7 red counters.

- Recall that a red counter paired with a yellow counter equals 0. Adding a red counter paired with a yellow counter does not change the value of ⁻7. Show another way to model ⁻7 that includes 5 yellow counters.

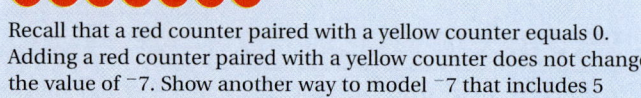

- Use your model to find ⁻7 − 5. Take away 5 yellow counters.

- What does your model show now? What is ⁻7 − 5? **12 red counters; ⁻12**

Now model 7 − ⁻2.

- Model 7. Put down pairs of yellow and red counters until you can take away ⁻2. What is 7 − ⁻2? **9**

248 Chapter 12

SPECIAL NEEDS / ELL

Materials *For each pair* 2 blank number cubes, p. TR75; 25 two-color counters

Ask each pair to **model subtraction of integers** by labeling the faces of one cube with positive integers 1–6 and the faces of the other with negative integers ⁻1 − ⁻6. Have each pair roll their cubes, write a subtraction problem using the numbers on the cubes, and model the subtraction with counters. Tell them to find and record the answer. Have pairs repeat the process to form 10 subtraction problems, modeling each with counters. Check students' work.

KINESTHETIC, VISUAL
INTERPERSONAL/SOCIAL

Intervention and Extension Resources

Activity 2

Addition and subtraction of integers are related.

- Copy the model for ⁻7 below.

- Use the model to find ⁻7 − ⁻3. What is ⁻7 − ⁻3? **⁻4**

- Model ⁻7 again. Then add three yellow counters to find ⁻7 + 3. What is ⁻7 + 3? **⁻4**

TECHNOLOGY LINK
More Practice: Use E-Lab, *Modeling Subtraction of Integers.*
www.harcourtschool.com/elab2002

Think and Discuss

- The models above show that ⁻7 − ⁻3 = ⁻7 + 3. How are ⁻3 and 3 related? How are subtraction and addition related? **They are opposites; they are opposite, or inverse, operations.**
- How are the two models different? **One has only red counters, and the other has red and yellow counters.**
  You can write a subtraction problem as an addition problem by adding the opposite of the number you are subtracting.

  $$6 - {}^-2 = 6 + 2 \leftarrow \text{Add the opposite of } {}^-2.$$

- How can you write ⁻6 − 2 as an addition problem? **by changing subtraction to addition and using ⁻2 instead of 2**

Practice

Use counters to find the difference.

1. ⁻7 − 4 **⁻11**
2. ⁻8 − ⁻5 **⁻3**
3. ⁻13 − 9 **⁻22**
4. ⁻9 − 4 **⁻13**

Complete the addition problem.

5. ⁻6 − ⁻2 = ⁻6 + ■   **2**
6. ⁻9 − ⁻3 = ⁻9 + ■ **3**
7. ⁻7 − 5 = ⁻7 + ■ **⁻5**
8. ⁻19 − 12 = ⁻19 + ■ **⁻12**

MIXED REVIEW AND TEST PREP

Find the sum. (p. 244)

9. ⁻3 + 9 **6**
10. 2 + ⁻7 **⁻5**
11. ⁻1 + ⁻7 **⁻8**

12. Write an integer to represent 217 m below sea level. (p. 228) **⁻217**

★ 13. **TEST PREP** Three friends shared a pizza that had 8 slices. There are $1\frac{1}{2}$ slices left. Alana had $2\frac{1}{2}$ slices, and Jill had $1\frac{1}{2}$ slices. How many slices did George have? (p. 182) **D**

   A $5\frac{1}{2}$ **B** 5   **C** 4   **D** $2\frac{1}{2}$

Activity 2

- Have students explore the relationship between the addition and subtraction of integers.

  Why do you remove 3 red counters for ⁻7 − ⁻3? You need to subtract, or take away, ⁻3.

  Why do you take away 3 red counters for ⁻7 + 3? They are paired with yellow counters and the pairs equal zero.

Think and Discuss

When discussing these questions, have students outline the steps they must follow to rewrite a subtraction problem as an addition problem. Change the subtraction sign to addition; write the opposite of the number being subtracted.

Practice

As students work through Exercises 5–8, have them model each exercise as an addition problem and a subtraction problem to check that addition and subtraction are opposite operations.

MIXED REVIEW AND TEST PREP

Note that Exercise 13 is a **multistep or strategy problem**. Exercises 9–13 provide **cumulative review** (Chapters 1–12).

Oral Assessment

How can you use addition to subtract integers? Possible answer: Change the number being subtracted to its opposite and then add the integers.

Explain how to solve Exercise 4. Possible answer: Make one row of 9 red counters and one row of 4 yellow counters and 4 red counters. Then take away the 4 yellow counters and count the remaining red counters.

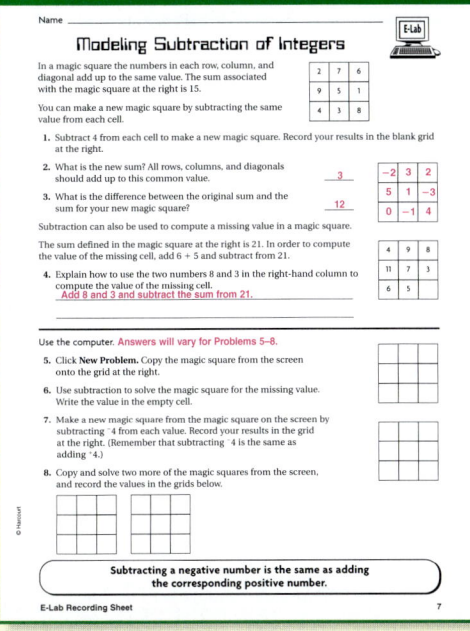

USING E-LAB

Students use addition and subtraction of integers to solve magic squares and to make new magic squares from old ones.

The E-Lab Recording Sheets and activities are available on the E-Lab website.

www.harcourtschool.com/elab2002

TECHNOLOGY LINK

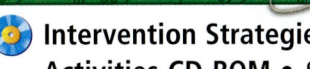 Intervention Strategies and Activities CD-ROM • *Skill 9*

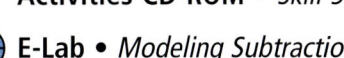 E-Lab • *Modeling Subtraction of Integers*

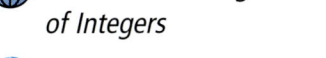 Astro Algebra • *Red,* Level F

LESSON 12.4

Algebra: Subtract Integers

LESSON PLANNING

Objective To use a number line to subtract integers

Intervention for Prerequisite Skills
Understand Integers, Locate Points on a Number Line (For intervention strategies, see page 241.)

Lesson Resources Calculator Handbook, p. 12

NCTM Standards
1. Number and Operations
2. Algebra
6. Problem Solving
8. Communication

Math Background
Subtraction of integers is defined in terms of addition. To find the difference of two integers, add the opposite of the one being subtracted. This procedure will help students understand how to subtract integers.

- First, rewrite the subtraction problem as an addition problem. Make sure to write the opposite of the number being subtracted.
- Then use the rules of adding integers.

WARM-UP RESOURCES

 **NUMBER OF THE DAY**

Write the day of the month. Add the opposite of that number to 30. What is the sum? *Possible answer:* $30 + {}^-21 = 9$

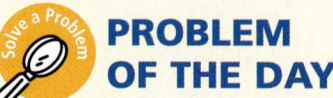 **PROBLEM OF THE DAY**

Write the missing numbers. Each of the numbers in the first four rows is the sum of the two numbers below it.

$$\begin{array}{cccc} & & {}^-4 & \\ & {}^-3 & {}^-1 & \\ & {}^-1 & {}^-2 & 1 \\ {}^-4 & 3 & {}^-5 & 6 \\ {}^-14 & 10 & {}^-7 & 2 & 4 \end{array}$$

Solution Problem of the Day tab, p. PD12

 **DAILY FACTS PRACTICE**

Have students practice subtraction facts by completing Set B of *Teacher's Resource Book,* p. TR102.

250A Chapter 12

INTERVENTION AND EXTENSION RESOURCES

ALTERNATIVE TEACHING STRATEGY — ELL

Reinforce the concept of using a number line to subtract integers. Draw a large number line on the floor and display the following: $14 - 6$.

Place one student at 6 and another at 14. Have a third student walk from the number being subtracted, 6, to 14. Have students describe the path and record the solution. **8 steps in the positive direction**

Repeat the activity for $^-2 - 5$. Place students at 5 and $^-2$. Ask a third student to start at 5, walk to $^-2$, and describe the path. **7 steps in the negative direction**

KINESTHETIC
BODILY/KINESTHETIC

MULTISTEP AND STRATEGY PROBLEMS

The following multistep or strategy problem is provided in Lesson 12.4:

Page	Item
251	29

TECHNOLOGY LINK

- Intervention Strategies and Activities CD-ROM • *Skills 9, 42*
- Astro Algebra • *Red*, Level F
- The Harcourt Learning Site
- Calculator Handbook, p. 12
- Math Jingles™ CD 5–6 • *Track 17*

WRITING IN MATHEMATICS

Students can use graphic aids to **practice integer subtraction.** Display the graph below that shows the high and low temperatures for one week at a weather station in northern Minnesota.

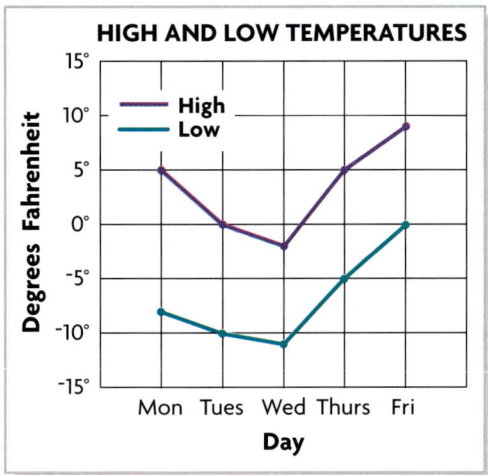

Have students write subtraction problems by using the graph. **Check students' work.**

VISUAL
VISUAL/SPATIAL

ADVANCED LEARNERS

Challenge students to **apply their knowledge of subtracting integers** by using the chart below. The chart shows the windchill factor. Explain that the windchill factor is how cold it feels at a given temperature and wind speed.

WINDCHILL TEMPERATURE					
Wind Speed (in mph)					
	0	5	10	15	20
10°	10°	6°	$^-9°$	$^-18°$	$^-24°$
5°	5°	0°	$^-15°$	$^-25°$	$^-31°$
0°	0°	$^-5°$	$^-22°$	$^-25°$	$^-31°$
$^-5°$	$^-5°$	$^-10°$	$^-27°$	$^-38°$	$^-46°$
$^-10°$	$^-10°$	$^-15°$	$^-34°$	$^-45°$	$^-53°$

Have students write subtraction problems based on information in the table. **Check students' work.**

VISUAL
VISUAL/SPATIAL

LESSON 12.4 ORGANIZER

Objective To use a number line to subtract integers

Lesson Resources Calculator Handbook, p. 12

1 Introduce

QUICK REVIEW provides review of prerequisite skills.

Why Learn This? You can find how much temperature changes over time by subtracting integers. *Share the lesson objective with students.*

2 Teach

Guided Instruction

- Direct students' attention to page 250 and the description of a temperature change on Mars.

REASONING If the temperature had been 1°F instead of ⁻1°F, how would you write the subtraction and addition problem? $8 - 1 = 8 + {}^-1$

Modifying Instruction Use counters to model the temperature change on Mars to help students understand the subtraction process.

- Ask students to think about the first step in the Example while one volunteer uses counters to represent $^-9 - {}^-22$ and another models $^-9 + 22$.

  Are $^-9 - {}^-22$ and $^-9 + 22$ equal? Explain. Yes; they both equal 13.

  Does the same kind of reasoning apply for $8 - 4$? Explain. Yes; $8 - 4 = 8 + {}^-4 = 4$.

ADDITIONAL EXAMPLE

Example, p. 250

When the mountain climbers started their expedition, the temperature was ⁻8°F. By late afternoon the temperature was ⁻13°F. What was the range of temperatures? 5°F

250 Chapter 12

LESSON 12.4

Subtract Integers

Learn how to use a number line to subtract integers.

QUICK REVIEW
1. 29 − 24 **5**
2. 14 + 8 **22**
3. 217 − 12 **205**
4. 97 + 17 **114**
5. 365 − 295 **70**

During the summer of 1997, NASA landed the Mars Pathfinder on the planet Mars. On July 9, Pathfinder reported a temperature of ⁻1°F. On July 10, Pathfinder reported a temperature of 8°F. Find the range of temperatures reported by Pathfinder from July 9 to July 10.

Pathfinder endured temperatures as low as ⁻89°F.

To find the range of temperatures, you need to find the difference of 8 and ⁻1, or $8 - {}^-1$. You can find the difference of two integers by adding the opposite of the integer you are subtracting. You can then use the rules for addition of integers.

The opposite of ⁻1 is 1. So, $8 - {}^-1$ becomes $8 + 1$.

$8 - {}^-1 = 8 + 1 = 9$

So, the range of temperatures was 9°F.

EXAMPLE

During an experiment, a scientist recorded a high temperature of ⁻9°C and a low temperature of ⁻22°C. What was the range of temperatures during the experiment?

$^-9 - {}^-22 = {}^-9 + 22$ — Write the subtraction problem as an addition problem. Use the rules for addition of integers.

$^-9 + 22$

$|22| - |{}^-9| = 22 - 9$ — Subtract the lesser absolute value from the greater absolute value.

$= 13$

$|22| > |{}^-9| \rightarrow {}^-9 - {}^-22 = 13$ — Use the sign of the addend with the greater absolute value.

So, the range of temperatures during the experiment was 13°C.

- During the afternoon, the temperature fell from 7°F to ⁻5°F. What was the range of temperatures? **12°F**

250 Chapter 12

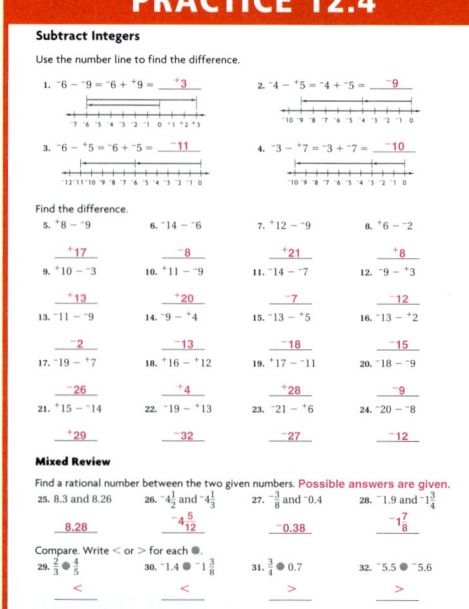

CHECK FOR UNDERSTANDING

Think and Discuss — Look back at the lesson to answer the question.
1. **Tell** how you would write the subtraction problem as an addition problem if the temperature at the end of the experiment in the example was ⁻40°C. ⁻9 + 40

Guided Practice — Rewrite the subtraction problem as an addition problem.

2. 7 − 10
   7 + ⁻10
3. 3 − ⁻6
   3 + 6
4. ⁻1 − ⁻8
   ⁻1 + 8
5. ⁻4 − 6
   ⁻4 + ⁻6

Find the difference.

6. 4 − 8
   ⁻4
7. ⁻7 − ⁻2
   ⁻5
8. 4 − ⁻5
   9
9. 1 − 8
   ⁻7

PRACTICE AND PROBLEM SOLVING

Independent Practice — Rewrite the subtraction problem as an addition problem.

10. 12 − 15
    12 + ⁻15
11. 8 − ⁻11
    8 + 11
12. ⁻6 − ⁻13
    ⁻6 + 13
13. ⁻9 − 11
    ⁻9 + ⁻11

Find the difference.

14. 6 − 11   ⁻5
15. ⁻9 − ⁻5   ⁻4
16. 2 − ⁻1   3
17. 3 − 5   ⁻2
18. 7 − 11   ⁻4
19. ⁻5 − ⁻5   0
20. 8 − ⁻3   11
21. 4 − 9   ⁻5
22. 31 − 37   ⁻6
23. ⁻35 − ⁻39 4
24. |⁻43| − |12| 31
25. |⁻27| − |⁻32| ⁻5

Evaluate.

26. ⁻3 − ⁻5 + ⁻8   ⁻6
27. 6 − ⁻4 + ⁻5   5
28. 8 − ⁻6 − 10   4

Problem Solving Applications

29. On Friday morning, the temperature in Anchorage, Alaska, was ⁻12°F. By that evening, the temperature had fallen 7°. The temperature on Saturday morning was 5° higher than the temperature on Friday evening. What was the temperature on Saturday morning? ⁻14°F

30. The water level of a river was 3 ft above normal. After an unusually dry season, the water level is 6 ft below normal. Find the range of the water levels of the river. 9 ft

31. **? What's the Question?** The temperature at noon was 15°F. By midnight the temperature was ⁻3°F. The answer is 18°F.
    Possible question: What was the range of temperatures?

MIXED REVIEW AND TEST PREP

32. Find the sum ⁻4 + 9. (p. 244) 5
33. Write the opposite integer for 213. (p. 228) ⁻213
34. Find the quotient $5\frac{3}{4} \div 2\frac{1}{4}$. (p. 210) $2\frac{5}{9}$
35. Find the prime factorization of 84. (p. 148) 2 × 2 × 3 × 7
★ 36. **TEST PREP** Solve x + 9 = 21 using mental math. (p. 30) C
    A x = 9   B x = 11   C x = 12   D x = 30

EXTRA PRACTICE page H43, Set B

251

---

3 Practice

Guided Practice

Do Check for Understanding Exercises 1–9 with your students. Identify those having difficulty and use lesson resources to help.

COMMON ERROR ALERT

Students often remember to write the opposite of the number being subtracted but forget to change the problem to addition.

7 − 10 = 7 − ⁻10

Remind students to write the opposite of the number being subtracted *and* change the subtraction to addition.

7 − 10 = 7 + ⁻10

Independent Practice

Note that Exercise 29 is a **multistep or strategy problem**. Assign Exercises 10–31.

MIXED REVIEW AND TEST PREP

Exercises 32–36 provide **cumulative review** (Chapters 1–12).

4 Assess

Summarize the lesson by having students:

DISCUSS Explain how to find the opposite of an integer. Possible answer: Write the absolute value and change the sign.

WRITE Explain how you solved Exercise 30. Possible answer: Write the subtraction problem 3 − ⁻6. Rewrite the problem as 3 + 6. Add. The sum is 9.

Lesson Quiz

Transparency 12.4

Find the difference.

1. 6 − 14   ⁻8
2. ⁻9 − ⁻11   2
3. ⁻5 − 10   ⁻15
4. 3 − ⁻24   27
5. 31 − ⁻17   48
6. ⁻19 − 5   ⁻24

CHALLENGE 12.4

Create the Problem

Create a word problem that can be solved with each subtraction problem below. Then trade problems with a classmate, and solve each other's problems. Check students' problems.

1. ⁻12 − ⁺7 = ⁻19
2. ⁺15 − ⁻9 = ⁺24
3. ⁻25 − ⁻17 = ⁻8
4. ⁻32 − ⁺14 = ⁻46
5. ⁻55 − ⁻23 = ⁻32
6. ⁺78 − ⁻19 = ⁺97

PROBLEM SOLVING 12.4

Subtract Integers

Write the correct answer.

1. Find the missing number in the pattern.
   4 − 2 = 2
   4 − 1 = 3
   4 − 0 = 4
   4 − ⁻1 = 5
   4 − ⁻2 = 6
   4 − ⁻3 = ■
   7

2. Find the missing number in the pattern.
   ⁻2 − 1 = ⁻3
   ⁻2 − 0 = ⁻2
   ⁻2 − ⁻1 = ⁻1
   ⁻2 − ⁻2 = 0
   ⁻2 − ⁻3 = 1
   ⁻2 − ⁻4 = ■
   2

3. Five years ago, Sean was three times as old as his brother. Today Sean is twice as old as his brother. How many years older than his brother is Sean? How old is Sean now?
   10 years older, 20 years old

4. For one week of work, Roberto earned $615. He worked 25 hours at his regular pay of $15 per hour. He also worked overtime hours, for which he was paid $20 per hour. How many overtime hours did Roberto work?
   12 hours overtime

Choose the letter for the best answer.

5. Which addition problem is equivalent to the subtraction problem ⁻8 − ⁻17?
   A ⁺8 + ⁻17
   B ⁻8 + ⁻17
   C ⁺8 + ⁺17
   D ⁻8 + ⁺17

6. Four hours ago, the temperature outside was +6°F. Since then the temperature has dropped 13°F. What is the temperature outside now?
   F +19°F   **H** ⁻7°F
   G +7°F   J ⁻19°F

7. John collected n gadgets. Frank gave him 18 more gadgets. John now has 51 gadgets. Which equation could be used to find the number of gadgets John had before Frank gave him more?
   A n + 18 = 51   C n + 51 = 18
   B n − 51 = 18   D n − 18 = 51

8. Maria keeps old records stored in special boxes. Each box can hold 45 old records. If she has 16 boxes full of old records, how many records does she have?
   F 690 records   H 710 records
   G 700 records   **J** 720 records

9. **Write About It** Explain why 3 − 2 ≠ 2 − 3.
   Subtraction is not commutative. You will not get the same difference if you change the order of the numbers in a subtraction problem.

251

LESSON 12.5 Multiply and Divide Integers

LESSON PLANNING

Objective To multiply and divide integers

Intervention for Prerequisite Skills

Understand Integers, Number Lines, Multiplication and Division Facts (For intervention strategies, see page 241.)

NCTM Standards
1. Number and Operations
2. Algebra
6. Problem Solving
8. Communication

Math Background

The following ideas will help students understand multiplication and division of negative integers.

- You can find multiples of a number by adding the number repeatedly.
- By establishing a pattern, it can be shown that the product of a negative number and a positive number is negative, and that the product of two negative numbers is positive.
- Division is the inverse of multiplication.
- The rules for determining the sign when multiplying integers are the same for dividing integers.

WARM-UP RESOURCES

 **NUMBER OF THE DAY**

The number of the day is the opposite of the day of the month. Write 2 addition number sentences using integers with the number of the day as the sum. **Possible answer for January 4: opposite = ⁻4; ⁻6 + 2 = ⁻4; ⁻2 + ⁻2 = ⁻4**

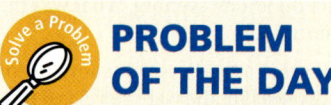 

We are two integers. Our sum is ⁻15 and our difference is 3. Who are we? **⁻6 and ⁻9**

Solution Problem of the Day tab, p. PD12

Have students practice multiplication facts by completing Set C of *Teacher's Resource Book,* p. TR102.

252A Chapter 12

INTERVENTION AND EXTENSION RESOURCES

ALTERNATIVE TEACHING STRATEGY — ELL

Materials *For each group* two-color counters

Have students **use counters to show the product of two negative integers,** such as $^-2 \times {^-2}$. Remind students that $^-2$ is the opposite of 2, and $^-2 \times {^-2}$ can be thought of as removing two groups of $^-2$. Have students display 2 sets of 2 zero pairs and remove 2 sets of $^-2$.

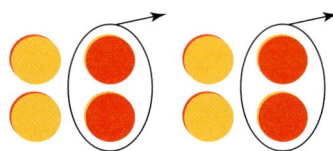

Ask: What is the product of $^-2 \times {^-2}$? 4

Ask students to model other products in a similar manner. **Check students' work.**

VISUAL
BODILY/KINESTHETIC

MULTISTEP AND STRATEGY PROBLEMS

The following multistep and strategy problems are provided in Lesson 12.5:

Page	Item
255	33, 40

SPECIAL NEEDS

Reinforce the concept of multiplying integers by discussing each of the scenarios below. Then have students write the appropriate number sentence to illustrate it. Point out that the word *not* can sometimes be translated as a negative sign.

- Finding $2 on the floor three days in a row is the same as finding $6. $3 \times 2 = 6$
- Paying $2 for lunch on 3 separate days is the same as paying out $6. $3 \times {^-2} = {^-6}$
- Not having to pay $2 for each of 3 videos is the same as getting $6. $^-3 \times {^-2} = 6$

AUDITORY
VERBAL/LINGUISTIC

EARLY FINISHERS

Materials *For each pair* 2 number cubes without numbers, p. TR75

Have students **practice dividing integers.** Give each pair 2 number cubes. Have them write 2, $^-9$, 6, $^-4$, 12, and $^-36$ on the faces of one cube and $^-2$, 9, $^-6$, 18, $^-4$, and 36 on the faces of the other.

Each partner rolls one of the cubes, divides 36 by the number shown, and records the number sentence. Have students repeat the activity 3 times. Then ask them to exchange cubes and repeat 3 more times. Finally have each pair list all the different number sentences they formed. **Answers will vary.**

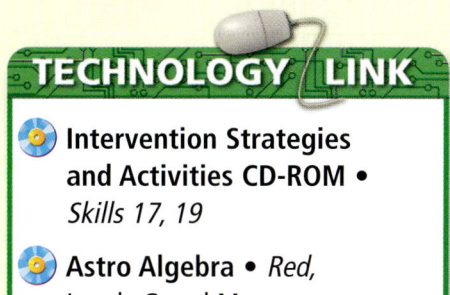

TECHNOLOGY LINK

- **Intervention Strategies and Activities CD-ROM** • *Skills 17, 19*
- **Astro Algebra** • *Red, Levels G and M*

LESSON 12.5 ORGANIZER

Objective To multiply and divide integers

Materials *For each group* 12 two-color counters

1 Introduce

QUICK REVIEW provides review of prerequisite skills.

Why Learn This? You can determine the average high and low temperatures for your city. Share the lesson objective with students.

2 Teach

Guided Instruction

- Check students' understanding of the models.

  How is a yellow counter different from a red counter? A yellow counter represents $^+1$. A red counter represents $^-1$.

  How could you use counters to model the product 3 × 5? the product $^-4$ × 6? What are the products? Make 3 groups of 5 yellow counters; make 6 groups of 4 red counters; 15; $^-24$.

- Direct students' attention to the submarine information at the bottom of page 252.

  Why is a number line a better model for this situation than counters? Possible answer: it would take too many counters to model the problem.

252 Chapter 12

LESSON 12.5

Multiply and Divide Integers

Learn how to multiply and divide integers.

QUICK REVIEW
1. 80 × 6   2. 4 × 25   3. 50 × 6   4. 3,600 ÷ 4   5. 540 ÷ 60
   480   100   300   900   9

Use red and yellow counters to model multiplication of integers. A red counter represents $^-1$ and a yellow counter represents $^+1$.

Activity 1

You need: two-color counters

- Use yellow counters to model the product 2 × 3.

  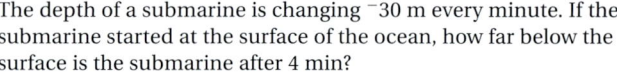 ← 2 groups of $^+3$

  2 × 3 = 6

- Use red counters to model the product 2 × $^-3$.

  ← 2 groups of $^-3$

  2 × $^-3$ = $^-6$

- Use red counters to model the product $^-2$ × 3. Using the Commutative Property, you can write $^-2$ × 3 as 3 × $^-2$.

   ← 3 groups of $^-2$

  $^-2$ × 3 = 3 × $^-2$ = $^-6$

- How could you model the product 3 × 4? **You could use 3 groups of 4 yellow counters.**
- How could you model the product 3 × $^-4$? **You could use 3 groups of 4 red counters.**
- What do you notice about the product of two positive integers? of a positive integer and a negative integer? **The product is positive. The product is negative.**

The depth of a submarine is changing $^-30$ m every minute. If the submarine started at the surface of the ocean, how far below the surface is the submarine after 4 min?

Use a number line to find the product 4 × $^-30$.

The number line shows that the depth of the submarine changed $^-120$ m. So, the submarine is 120 m below the surface.

252 Chapter 12

RETEACH 12.5

Multiply and Divide Integers

You can use patterns to multiply integers. Look at the patterns.

3 × 2 = 6	$^-2$ × 3 = $^-6$
3 × 1 = 3	$^-2$ × 2 = $^-4$
3 × 0 = 0	$^-2$ × 0 = 0
3 × $^-1$ = $^-3$	$^-2$ × $^-2$ = 4
3 × $^-2$ = $^-6$	$^-2$ × $^-3$ = 6

When you multiply a positive integer by a negative integer, the product is a negative integer:

$8 × ^-2 = ^-16$ and $^-8 × 2 = ^-16$

When you multiply two negative integers, the product is a positive integer.

$^-6 × ^-5 = 30$

You can use related multiplication problems to divide integers.

3 × 2 = 6	6 ÷ 3 = 2
$^-3$ × 2 = $^-6$	$^-6$ ÷ 3 = $^-2$
3 × $^-2$ = $^-6$	$^-6$ ÷ 3 = $^-2$
$^-3$ × $^-2$ = 6	6 ÷ $^-3$ = $^-2$

When you divide two negative integers, the quotient is positive.

$^-12 ÷ ^-2 = 6$

When you divide a positive integer by a negative integer, the quotient is negative.

$12 ÷ ^-2 = ^-6$

When you divide a negative integer by a positive integer, the quotient is negative.

$^-12 ÷ 2 = ^-6$

Find the product.

1. $^-8 × 7$ = $^-56$
2. $10 × ^-5$ = $^-50$
3. $^-11 × ^-2$ = 22
4. $9 × ^-3$ = $^-27$
5. $^-100 × 2$ = $^-200$
6. $6 × 9$ = 54
7. $^-80 × ^-10$ = 800
8. $5 × ^-25$ = $^-125$
9. $^-20 × 7$ = $^-140$
10. $63 × 9$ = 567
11. $^-8 × ^-17$ = 136
12. $11 × ^-25$ = $^-275$

Find the quotient.

13. $^-32 ÷ 8$ = $^-4$
14. $18 ÷ ^-3$ = $^-6$
15. $^-24 ÷ ^-6$ = 4
16. $48 ÷ ^-8$ = $^-6$
17. $^-100 ÷ 25$ = $^-4$
18. $63 ÷ 9$ = 7
19. $^-60 ÷ ^-3$ = 20
20. $50 ÷ ^-25$ = $^-2$

PRACTICE 12.5

Multiply and Divide Integers

Find the product or quotient.

1. $^-3 × 7$ = $^-21$
2. $8 × ^-3$ = $^-24$
3. $^-14 ÷ ^-2$ = 7
4. $24 ÷ ^-3$ = $^-8$
5. $^-150 ÷ 25$ = $^-6$
6. $36 ÷ 9$ = 4
7. $^-80 ÷ ^-4$ = 20
8. $75 ÷ ^-25$ = $^-3$
9. $^-130 ÷ ^-5$ = 26
10. $^-4 × 6$ = $^-24$
11. $9 × ^-6$ = $^-54$
12. $^-6 × ^-7$ = 42
13. $^-12 × ^-2$ = 24
14. $90 ÷ ^-5$ = $^-18$
15. $160 ÷ 16$ = 10
16. $^-88 ÷ 11$ = $^-8$
17. $42 ÷ 3$ = 14
18. $^-70 ÷ ^-7$ = 10
19. $^-4 × 25$ = $^-100$
20. $^-35 × ^-2$ = 70
21. $^-5 × 12$ = $^-60$
22. $^-14 × ^-7$ = 98
23. $^-200 ÷ ^-40$ = 5
24. $11 × ^-11$ = $^-121$

ALGEBRA Use mental math to find the value of y.

25. $y × ^-4 = ^-16$ $y = 4$
26. $y ÷ ^-8 = 5$ $y = ^-40$
27. $^-6 × y = 60$ $y = ^-10$
28. $^-21 ÷ y = ^-7$ $y = 3$
29. $y × ^-12 = 12$ $y = ^-1$
30. $y ÷ ^-3 = ^-9$ $y = 27$

Mixed Review

Find the sum or difference.

31. $^-2 + ^-13$ = 15
32. $^-16 - ^-2$ = 14
33. $^-3 - 24$ = 27
34. $^-5 + 10$ = 5

Write the mixed number as a fraction.

35. $3\frac{2}{5}$ = $\frac{17}{5}$
36. $5\frac{2}{9}$ = $\frac{47}{9}$
37. $1\frac{8}{11}$ = $\frac{19}{11}$
38. $9\frac{3}{8}$ = $\frac{75}{8}$
39. $4\frac{1}{4}$ = $\frac{17}{4}$

You can use patterns to find rules to multiply integers.

EXAMPLE 1

Complete the pattern.

$4 \times 3 = 12$
$4 \times 2 = 8$
$4 \times 1 = 4$
$4 \times 0 = 0$
$4 \times {}^-1 = \blacksquare$
$4 \times {}^-2 = \blacksquare$
$4 \times {}^-3 = \blacksquare$

Study the pattern. As the second factor decreases by 1, the product decreases by 4. Use this to complete the pattern.

$4 \times 3 = 12$
$4 \times 2 = 8$
$4 \times 1 = 4$
$4 \times 0 = 0$
$4 \times {}^-1 = {}^-4$
$4 \times {}^-2 = {}^-8$
$4 \times {}^-3 = {}^-12$

So, the missing products are $^-4$, $^-8$, and $^-12$.

- What is the sign of the product of a positive integer and a negative integer? **The sign is negative.**

EXAMPLE 2

Complete the pattern.

$^-4 \times 3 = {}^-12$
$^-4 \times 2 = {}^-8$
$^-4 \times 1 = {}^-4$
$^-4 \times 0 = 0$
$^-4 \times {}^-1 = \blacksquare$
$^-4 \times {}^-2 = \blacksquare$
$^-4 \times {}^-3 = \blacksquare$

Study the pattern. As the second factor decreases by 1, the product increases by 4. Use this to complete the pattern.

$^-4 \times 3 = {}^-12$
$^-4 \times 2 = {}^-8$
$^-4 \times 1 = {}^-4$
$^-4 \times 0 = 0$
$^-4 \times {}^-1 = 4$
$^-4 \times {}^-2 = 8$
$^-4 \times {}^-3 = 12$

So, the missing products are 4, 8, and 12.

- What is the sign of the product of two positive integers? two negative integers? **The sign is positive.**

Examples 1 and 2 lead to the rules below.

> The product of two integers with like signs is positive.
> The product of two integers with unlike signs is negative.

Multiplication and division are inverse operations. To solve a division problem, think about the related multiplication problem.

$42 \div 7 = \blacksquare \rightarrow 7 \times 6 = 42$, so $42 \div 7 = 6$.

Use related multiplication problems to determine the sign of the quotient when dividing integers.

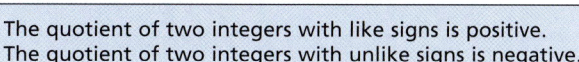

$8 \times 3 = 24$, so $24 \div 8 = 3$. $^-8 \times {}^-3 = 24$, so $24 \div {}^-8 = {}^-3$.
$^-8 \times 3 = {}^-24$, so $^-24 \div {}^-8 = 3$. $8 \times {}^-3 = {}^-24$, so $^-24 \div 8 = {}^-3$.

TECHNOLOGY LINK
More Practice: Use **Mighty Math Astro Algebra,** Red, Levels G and M.

In the problems above, look at the sign of the quotient of two positive integers and the sign of the quotient of two negative integers. Then look at the sign of the quotient of a positive integer and a negative integer. The rules below apply when dividing integers.

> The quotient of two integers with like signs is positive.
> The quotient of two integers with unlike signs is negative.

253

- Have students look at the pattern in Example 1.

  How would you find the pattern? by looking at how the factors and the products change from number sentence to number sentence

  REASONING Describe how the pattern changes. The second factor of each number sentence is 1 less than the preceding one and each product is 4 less than the previous one.

- Before discussing integer division, remind students that addition and subtraction are *inverse operations*. Ask:

  What other operations are inverse operations? multiplication and division

  REASONING Describe two ways you could write $5 \times 3 = 15$ as a division sentence. $15 \div 3 = 5$, $15 \div 5 = 3$

Modifying Instruction To prepare for the lesson on operations with rational numbers, you may want to display the rules for integer multiplication and division, along with examples, after completing this lesson.

ADDITIONAL EXAMPLES

Example 1, p. 253   **Example 2, p. 253**

Complete the pattern.   Complete the pattern.

$3 \times 3 = 9$ $^-3 \times 3 = {}^-9$
$3 \times 2 = 6$ $^-3 \times 2 = {}^-6$
$3 \times 1 = 3$ $^-3 \times 1 = {}^-3$
$3 \times 0 = 0$ $^-3 \times 0 = 0$
$3 \times {}^-1 = {}^-3$ $^-3 \times {}^-1 = 3$
$3 \times {}^-2 = {}^-6$ $^-3 \times {}^-2 = 6$
$3 \times {}^-3 = {}^-9$ $^-3 \times {}^-3 = 9$

CHALLENGE 12.5

Different Names for Numbers

Write an expression for each integer by using the integers $^+4$, $^+4$, $^-4$, and $^-4$ and any of the four operations. You can use an integer more than once in an expression. An example for Exercise 1 is done for you.
Possible answers are given.

1. $^-2$ $(^-4 \div {}^+4) + (^-4 \div {}^+4)$
2. 0 $(^+4 + {}^-4) + (^-4 + {}^+4)$
3. $^-1$ $(^-4 - {}^+4) \div (^+4 - {}^-4)$
4. 8 $(^+4 - {}^-4) + (^-4 + {}^+4)$
5. 24 $(^-4 \times {}^-4) + (^+4 + {}^+4)$
6. 15 $(4 \times 4) + (^+4 \div {}^-4)$
7. $^-64$ $(^-4 + {}^-4) \times (^+4 + {}^+4)$
8. 2 $(^-4 \times {}^-4) \div (^+4 + {}^+4)$
9. 256 $(^-4 \times {}^+4) \times (^-4 \times {}^+4)$
10. 16 $(^-4 - {}^-4) + (^+4 \times {}^+4)$
11. 1 $(^+4 \times {}^-4) \div (^+4 \times {}^-4)$
12. $^-16$ $(^+4 \times {}^-4) + (^-4 + {}^+4)$

PROBLEM SOLVING 12.5

Multiply and Divide Integers   Analyze Choose Solve Check

Write the correct answer.

1. During the last three months, Miller Aviation showed its losses as $^-\$54.09$ per share. What is the average loss per month for each share of stock?

   $^-\$18.03$ per month

2. Fred needs $\frac{2}{3}$ cups of milk for a recipe that makes 8 biscuits. He wants to make 2 dozen biscuits. How much milk does he need?

   2 cups of milk

3. Hilary needed to cut the following lengths of ribbon for an art project: $\frac{3}{4}$ ft of blue, 8 in. of red, 0.5 ft of yellow, and 12.5 in. of green ribbon. List the ribbon colors in order from least to greatest length.

   yellow, red, blue, green

4. For a science experiment, Kelly must measure the amount of water that evaporates from a cylinder. Each week the amount of water decreases by 2 cm. If this rate continues, what will be the change in the water level after 8 weeks?

   $^-16$ cm

Choose the letter for the best answer.

5. Yesterday the temperature dropped by 3° each half hour from 3 A.M. until 6 A.M. What was the temperature change during that time?
   A $^-18°$ C 12°
   B $^-12°$ D 18°

6. During a Tigers football game, the total loss for the last 4 plays was $^-24$ yd. Which integer describes the average loss per play?
   F 96 yd   (H) $^-6$ yd
   G $^-96$ yd   J 6 yd

7. Jan charged the following items on her credit card last month: $49.97 for new CDs, $97.99 for a new portable CD player, and $147.50 on food. What is the best estimate of her credit card bill?
   A $200
   (B) $300
   C $400
   D $500

8. Grace scored 8, 7, 8, 8, 10, 6, 7, 8, 7, 6, and 9 on her last 11 math quizzes. What term describes her most frequent score, 8?
   F mean
   G median
   (H) mode
   J range

9. **Write About It** Explain how you decided whether the answer to Exercise 4 was positive or negative.

   Possible answer: Since the amount of water decreases each week, the total change must be negative.

253

LESSON 12.5

ADDITIONAL EXAMPLES

Example 3, p. 254

Find the quotient.

A. ⁻96 ÷ ⁻8 12   B. 48 ÷ ⁻12 ⁻4

Example 4, p. 254

Find the average of ⁻8, 6, ⁻12, and ⁻18. ⁻8

3 Practice

Guided Practice

Do Check for Understanding Exercises 1–10 with your students. Identify those having difficulty and use lesson resources to help.

COMMON ERROR ALERT

Students may show the product or quotient of two negative integers as negative.

⁻4 × ⁻3 = ⁻12   ⁻35 ÷ ⁻7 = ⁻5

Direct them to solve in two steps. Have them first write the sign and then the product or quotient.

Independent Practice

Note that Exercises 33 and 40 are **multistep or strategy problems**. Assign Exercises 11–40.

Algebraic Thinking For Exercises 27–32, have students begin by deciding whether the variable will be positive or negative.

EXAMPLE 3 Find the quotient.

A. ⁻84 ÷ ⁻7
   ⁻84 ÷ ⁻7 = 12   *Divide as with whole numbers. The quotient is positive since the integers have like signs.*

B. ⁻55 ÷ 11
   ⁻55 ÷ 11 = ⁻5   *Divide as with whole numbers. The quotient is negative since the integers have unlike signs.*

• Will the quotient ⁻72 ÷ 8 be positive or negative? Will the quotient ⁻72 ÷ ⁻8 be positive or negative? **negative, positive**

EXAMPLE 4 The low temperatures for five days in Fairbanks, Alaska, were ⁻3°F, ⁻8°F, 2°F, 3°F, and ⁻4°F. Find the average low temperature for the five days.

Find the sum. Divide the sum by 5.

⁻3 + ⁻8 + 2 + 3 + ⁻4 = ⁻10
⁻10 ÷ 5 = ⁻2

So, the average low temperature was ⁻2°F.

CHECK FOR UNDERSTANDING

Think and Discuss Look back at the lesson to answer each question.

1. **Find** the product 5 × ⁻7. Find the product ⁻5 × ⁻7. **⁻35; 35**
2. **Tell** how the rules for multiplying two integers compare with the rules for dividing two integers. **The rules are the same.**

Guided Practice Find the product or quotient.

3. ⁻9 × 6 **⁻54**   4. ⁻3 × ⁻4 **12**   5. ⁻8 × 2 **⁻16**   6. 8 × ⁻7 **⁻56**
7. ⁻21 ÷ 7 **⁻3**   8. ⁻16 ÷ ⁻2 **8**   9. 120 ÷ ⁻5 **⁻24**   10. ⁻132 ÷ 6 **⁻22**

PRACTICE AND PROBLEM SOLVING

Independent Practice Find the product or quotient.

11. ⁻5 × 8 **⁻40**   12. 2 × ⁻3 **⁻6**   13. 7 × 2 **14**   14. ⁻9 × 4 **⁻36**
15. 84 ÷ 7 **12**   16. ⁻72 ÷ ⁻8 **9**   17. ⁻63 ÷ ⁻7 **9**   18. ⁻70 ÷ 7 **⁻10**
19. 24 × 12 **288**   20. 30 × ⁻12 **⁻360**   21. ⁻75 × 4 **⁻300**   22. ⁻62 × ⁻20 **1,240**
23. 432 ÷ 12 **36**   24. 255 ÷ ⁻15 **⁻17**   25. ⁻960 ÷ ⁻12 **80**   26. ⁻4,978 ÷ 19 **⁻262**

254 Chapter 12

ALTERNATIVE TEACHING STRATEGY   SCAFFOLDED INSTRUCTION

Purpose Students will use two-color counters to model integer division.

Materials *For each pair* 24 two-color counters

Have each pair arrange 12 yellow counters in 3 equal groups. Then ask them to write the number sentence they have modeled.
12 ÷ 3 = 4

Ask students to write a multiplication sentence for the number sentence they have modeled.
4 × 3 = 12 or 3 × 4 = 12

Next, ask students to divide 12 red counters into 2 equal groups. Again, ask students to write the number sentence they have modeled.
⁻12 ÷ 2 = ⁻6

Ask students to write a multiplication sentence for the number sentence they have modeled.
⁻6 × 2 = ⁻12 or 2 × ⁻6 = ⁻12

Finally, have students model and solve similar problems, such as 14 ÷ 2 and ⁻18 ÷ 2. **7, ⁻9**

254 Chapter 12

 **ALGEBRA** Use mental math to find the value of y.

27. $y \times {}^-6 = {}^-18$
    $y = 3$
28. $y \times {}^-10 = 40$
    $y = {}^-4$
29. ${}^-8 \times y = 32$
    $y = {}^-4$
30. $y \div {}^-6 = {}^-3$
    $y = 18$
31. $y \div {}^-7 = 5$
    $y = {}^-35$
32. ${}^-18 \div y = 2$
    $y = {}^-9$

Problem Solving Applications

33. Drought conditions caused the water level in a lake to change by ${}^-3$ in. per month in May and June and by ${}^-4$ in. per month in July and August. Write the change in water level over these four months as a negative number. ${}^-14$ in.

34. The low temperatures for the four days of a winter festival were ${}^-8°F$, ${}^-6°F$, ${}^-9°F$, and ${}^-1°F$. Find the average low temperature for the four days. ${}^-6°F$

35. Use the Associative Property to help you find the product.
    ${}^-3 \times {}^-2 \times {}^-5$. ${}^-30$

36. During the first 6 months of business, a sporting goods store showed its losses as ${}^-\$3,054$. How would it show the average monthly loss? ${}^-\$509$

37. The depth of a submarine changed ${}^-630$ ft over a period of 9 minutes. If the submarine descended at a constant rate, what was the change of depth per minute? ${}^-70$ ft per minute

38.  **What's the Error?** Beth says that ${}^-5 \times {}^-2 = {}^-10$. What is her error? What is the correct product? **The product of integers with like signs is positive; ${}^-5 \times {}^-2 = 10$**

39. **What's the Question?** In a division problem, the divisor is ${}^-250$ and the dividend is 10. The answer is negative. **What's the sign of the quotient?**

40. **NUMBER SENSE** Joe picked a number, added 5, multiplied the sum by 3, subtracted 10, and doubled the result. His final result was 28. What number had Joe picked? 3

MIXED REVIEW AND TEST PREP

41. $16 - {}^-6$ (p. 250) **22**
42. ${}^-32 + {}^-42$ (p. 244) **${}^-74$**
43. $5\frac{2}{3} \times 6\frac{3}{4}$ (p. 206) **$38\frac{1}{4}$**

44. **TEST PREP** Evaluate $2k$ for $k = 9.6$. (p. 82) **A**
    A 19.2 B 11.6 C 7.6 D 4.8

45. **TEST PREP** Write the fraction $\frac{5}{8}$ as a percent. (p. 169) **H**
    F 0.625% G 6.25% H 62.5% J 625%

EXTRA PRACTICE page H43, Set C

255

LESSON **12.6**

Explore Operations with Rational Numbers

LESSON PLANNING

Objective To add, subtract, multiply, and divide with rational numbers

Intervention for Prerequisite Skills
Understanding Integers, Number Lines, Multiplication and Division Facts (For intervention strategies, see page 241.)

NCTM Standards
1. Number and Operations
2. Algebra
6. Problem Solving
7. Reasoning and Proof
8. Communication
9. Connections
10. Representation

Math Background
The following ideas will help students understand operations with rational numbers.

- A rational number is a number that can be written in the form $\frac{a}{b}$, where a and b are integers and $b \neq 0$.
- The set of rational numbers includes positive and negative fractions and mixed numbers as well as integers and decimals.
- To operate on rational numbers, use the order of operations, along with the rules for adding, subtracting, multiplying, and dividing integers.
- The sign of the product or quotient of two rational numbers with the same sign is positive. The sign of the product or quotient of two rational numbers with unlike signs is negative.

WARM-UP RESOURCES

 **NUMBER OF THE DAY**

Begin with the number of the month of the year. Find its opposite. Then write a division sentence with that number as the quotient. Possible answer for 11: ⁻22 ÷ 2 = ⁻11

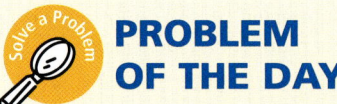 **PROBLEM OF THE DAY**

Write the missing integers.

⁻2 × ■3 × 4 = ⁻24
× ■5 × ⁻4 × ■⁻2 = ■40
⁻10 × ■⁻12 × ⁻8 = ⁻960

Solution Problem of the Day tab, p. PD12

 **DAILY FACTS PRACTICE**

Have students practice multiplication and division facts by completing Set F of *Teacher's Resource Book*, p. TR102.

INTERVENTION AND EXTENSION RESOURCES

ALTERNATIVE TEACHING STRATEGY — ELL

Materials *For each student* grid paper

Have students **use a number line to operate on rational numbers.** Ask students to use grid paper to draw four number lines, each running from ⁻2 to 2 and marked in tenths. Then have them use a number line to solve each of the following, and have them explain their methods.

⁻1.5 + 0.6   ⁻0.7 − 1.2   ⁻0.2 × 3   ⁻1.6 ÷ 0.4

⁻1.5 + 0.6: Start at ⁻1.5 and move right 0.6; ⁻1.5 + 0.6 = ⁻0.9; ⁻0.7 − 1.2: Rewrite as ⁻0.7 + ⁻1.2; Start at ⁻0.7 and move left 1.2; ⁻0.7 + ⁻1.2 = ⁻1.9; ⁻0.2 × 3: Start at 0 and move left 0.2 units three times; ⁻0.2 × 3 = ⁻0.6; ⁻1.6 ÷ 4: Divide the portion of a number line between 0 and ⁻1.6 into four equal parts; ⁻1.6 ÷ 4 = ⁻0.4.

VISUAL
VERBAL/LINGUISTIC

SPECIAL NEEDS

Materials *For each pair* number cube

To **practice operating with rational numbers,** each student of a pair writes a positive or negative fraction. Students should designate one fraction the *first* fraction and the other the *second* fraction. One student rolls a number cube, interpreting the roll as follows:

1: add   2: subtract   3: multiply   4: divide

5: choose an operation   6: roll again

Students start with the first fraction, perform the indicated operation with the second fraction, and compare their answers. Then they repeat the game, starting with a positive or negative decimal.

KINESTHETIC
VISUAL/SPATIAL

MULTISTEP AND STRATEGY PROBLEMS

The following multistep and strategy problems are provided in Lesson 12.6:

Page	Item
259	52, 53, 54

WRITING IN MATHEMATICS

At the conclusion of the lesson, ask students to summarize what they have learned by comparing and contrasting the rules for operating on rational numbers with the rules for operating on integers. Possible answer: Use the same rules to find the sign of a sum, difference, product, or quotient of two rational numbers as you use to find the sum, difference, product, or quotient of two integers. To operate on rational numbers that are not integers, however, you must also use the rules for operating on fractions, mixed numbers, and decimals.

TECHNOLOGY LINK

 Intervention Strategies and Activities CD-ROM •
Skills 17, 19

LESSON 12.6 ORGANIZER

Objective To add, subtract, multiply, and divide with rational numbers

1 Introduce

QUICK REVIEW provides review of prerequisite skills.

Why Learn This? You can use this skill to determine stock prices after several changes. Share the lesson objective with students.

2 Teach

Guided Instruction

- Review with students the rules for adding, subtracting, multiplying, and dividing integers. After looking at Example 1 and Example 2, ask:

  Is the method for operating with rational numbers different from the method for operating with integers? No, the methods are the same.

- After looking at Example 2, ask:

  Why is the sign of the product positive? Both factors are negative.

  What does this tell you about the method for finding the sign of a product of two rational numbers? Use the same method you use to find the sign of the product of two integers.

ADDITIONAL EXAMPLES

Example 1, p. 256

Low tide on July 16 was 4.3 ft higher than the $^-0.8$ ft low tide on July 9. How high was the low tide on July 16? 3.5 ft

Example 2, p. 256

Find the product $2\frac{1}{3} \times {}^-\frac{3}{4}$. $^-1\frac{3}{4}$

LESSON 12.6

Explore Operations with Rational Numbers

Learn how to add, subtract, multiply, and divide with rational numbers.

QUICK REVIEW

1. $2 \times {}^-5$ **2.** $^-7 \times {}^-3$
   $^-10$ 21
3. $^-4 \times 9$ **4.** $18 \div {}^-6$
   $^-36$ $^-3$
5. $^-24 \div {}^-8$
   3

LOW TIDES

DATE	A.M.	Feet	P.M.	Feet
Dec. 7	6:41	2.6	7:35	$^-0.4$
Dec. 8	7:19	3.1	8:09	$^-0.4$
Dec. 9	7:56	3.3	8:43	$^-0.3$
Dec. 10	8:32	3.4	9:15	$^-0.1$
Dec. 11	9:10	3.5	9:47	0.1
Dec. 12	9:51	3.6	10:22	0.3
Dec. 13	10:37	3.6	11:01	0.6

When you add, subtract, multiply, and divide rational numbers, use the same rules for determining signs as you use with integers.

EXAMPLE 1

Ocean tides are caused by the gravitational pull of the moon and the sun. On most days there are two high tides and two low tides. The table above shows the heights of the low tides, in feet, for one week in Astoria, Oregon. What is the range of the heights of the low tides?

The highest low tide measures 3.6 ft. The lowest measures $^-0.4$ ft.

$3.6 - {}^-0.4 = 3.6 + 0.4$ *Write the subtraction problem as an addition problem.*

$\phantom{3.6 - {}^-0.4} = |3.6| + |0.4|$ *Add the absolute values of the integers.*

$\phantom{3.6 - {}^-0.4} = 3.6 + 0.4$

$\phantom{3.6 - {}^-0.4} = 4.0$

So, the range of the heights of the low tides is 4.0 ft.

• Low tide on December 1 was 2.5 ft higher than on December 9. How high was the low tide on December 1? **2.2 ft**

EXAMPLE 2

Find the product. $^-3\frac{1}{2} \times {}^-4$

$^-3\frac{1}{2} \times {}^-4 = \frac{^-7}{2} \times \frac{^-4}{1}$ *Write the mixed number as a fraction.*

$\phantom{^-3\frac{1}{2} \times {}^-4} = \frac{28}{2}$ *Multiply. The fractions have like signs, so the product is positive.*

$\phantom{^-3\frac{1}{2} \times {}^-4} = 14$ *Simplify.*

So, $^-3\frac{1}{2} \times {}^-4 = 14$.

EXAMPLE 3

As the tide went out, the water level dropped $15\frac{1}{2}$ in. in $2\frac{1}{2}$ hours. What was the average change in the water level per hour? Estimate to check that your answer is reasonable.

Find $^-15\frac{1}{2} \div 2\frac{1}{2}$.

Estimate. $^-16 \div 2 = ^-8$

$^-15\frac{1}{2} \div 2\frac{1}{2} = \frac{^-31}{2} \times \frac{2}{5}$ *Multiply by the reciprocal of the divisor. The numbers have unlike signs, so the product is negative.*

$= \frac{^-31}{5}$ *Simplify.*

$= ^-6\frac{1}{5}$ *Write the quotient as a mixed number.*

$^-6\frac{1}{5}$ is close to the estimate of $^-8$. The quotient is reasonable.

So, the average change in the water level was $^-6\frac{1}{5}$ in. per hour.

EXAMPLE 4

Remember the order of operations.
1. Operate inside parentheses.
2. Clear exponents.
3. Multiply and divide from left to right.
4. Add and subtract from left to right.

At 4 A.M. the air temperature was $^-5.2°F$. For the next 3 hours, the temperature fell 1.5° per hour. Then the sun came up and the temperature rose 4.6°. What was the final temperature?

Write an expression to find the change in temperature.

$^-5.2 + (3 \times ^-1.5) + 4.6$

$^-5.2 + ^-4.5 + 4.6$ *Operate inside parentheses.*

$^-9.7 + 4.6$ *Add from left to right.*

$^-5.1$

So, the final temperature was $^-5.1°F$.

CHECK FOR UNDERSTANDING

Think and Discuss   Look back at the lesson to answer each question.
For 1–2, see Additional Answers, p. 259.

1. **Compare and contrast** multiplication of a positive decimal and a negative decimal with multiplication of a positive fraction and a negative fraction.

2. **Tell** if the order of operations process changes depending upon whether you are operating on positive or negative numbers. Give an example to illustrate your answer.

257

- Review the order of operations rules with students. After looking at Example 4, ask:

REASONING Compare using the order of operations with rational numbers to using the order of operations with whole numbers. Possible answer: The order of operations rules are the same. But because rational numbers can be negative, you also have to use the rules for determining the sign of an answer.

ADDITIONAL EXAMPLES

Example 3, p. 257

Following the flood, the water level on Main Street dropped $4\frac{1}{8}$ in. in $2\frac{1}{4}$ hours. What was the average change in the water level per hour? $^-1\frac{5}{6}$ in. per hour

Example 4, p. 257

At noon, the air temperature was $^-12.8°$. For the next 4 hours, the temperature rose 3.4° per hour. Then the temperature fell 2.9° per hour for 3 hours. What was the final temperature? $^-7.9°$

3 Practice

Guided Practice

Do Check for Understanding Exercises 1–18 with your students. Identify those having difficulty and use lesson resources to help.

CHALLENGE 12.6

Mental Properties

You can use properties of addition and multiplication to mentally solve equations with rational numbers.

Properties of Addition	Properties of Multiplication
Commutative: $3.5 + ^-7.4 = ^-7.4 + 3.5$	**Commutative:** $\frac{1}{4} \times \frac{3}{8} = \frac{3}{8} \times \frac{1}{4}$
Associative: $\frac{1}{2} + (\frac{1}{4} + \frac{2}{3}) = (\frac{1}{2} + \frac{1}{4}) + \frac{2}{3}$	**Associative:** $1.2 \times (4 \times ^-3.1) = (1.2 \times 4) \times ^-3.1$
Additive Inverse: $\frac{6}{9} + \frac{^-6}{9} = 0$	**Distributive:** $^-2 \times (5.2 + ^-7.2) = (^-2 \times 5.2) + (^-2 \times ^-7.2)$
Identity Property of Zero: $^-24.5 + 0 = ^-24.5$	**Property of Zero:** $^-6.9 \times 0 = 0$
	Identity Property of One: $\frac{5}{6} \times 1 = \frac{5}{6}$

Write the property you can use to solve each equation. Then use mental math to solve for n.

1. $\frac{1}{2} \times (\frac{2}{3} + \frac{4}{5}) = (\frac{1}{2} \times \frac{2}{3}) + (n \times \frac{4}{5})$ — Distributive Property; $n = \frac{1}{2}$
2. $n + 5\frac{3}{4} = 5\frac{3}{4} + ^-1\frac{7}{8}$ — Commutative Property of Addition; $n = ^-1\frac{7}{8}$
3. $^-4 \times (0.8 + 2.5) = n$ — Distributive Property; $n = ^-13.2$
4. $\frac{^-5}{12} \times \frac{9}{10} = \frac{9}{10} \times n$ — Commutative Property of Multiplication; $n = \frac{^-5}{12}$
5. $n \times ^-4\frac{1}{12} = ^-4\frac{1}{12}$ — Identity Property of One; $n = 1$
6. $^-7.8 + n = 0$ — Additive Inverse Property; $n = 7.8$
7. $^-0.5 \times (^-2.2 + 4) = n$ — Distributive Property; $n = ^-0.9$
8. $n + ^-11.7 = ^-11.7$ — Identity Property of Zero; $n = 0$
9. $n + 1\frac{5}{6} = 0$ — Additive Inverse Property; $n = ^-1\frac{5}{6}$
10. $6.2 + (3.8 + ^-11.9) = n$ — Associative Property of Addition; $n = ^-1.9$
11. $^-9\frac{11}{12} \times n = 0$ — Property of Zero; $n = 0$
12. $n \times 1 = ^-30.7$ — Identity Property of One; $n = ^-30.7$

PROBLEM SOLVING 12.6

Explore Operations with Rational Numbers   Analyze Choose Solve Check

Write the correct answer.

1. The high temperature for a winter day in Ossining was 8.5°F. The low temperature that day was $^-2.8°F$. What is the difference between the high and low temperatures?
   5.7°F

2. At 3 A.M. the temperature was 3.5°F. It changed $^-0.5°$ each half hour until 5:30 A.M. What was the temperature at 5:30 A.M.?
   1°F

3. Keisha works 15 hours each week and earns $8.50 per hour. She has saved $20 toward a bicycle that costs $167.50. How many more weeks does she need to work in order to buy the bicycle?
   2 weeks

4. Jan runs 0.75 mi in 7.5 min. Rita runs $\frac{7}{8}$ mi in 450 sec. Who runs faster? Explain.
   Rita, because they both run for the same amount of time, but $\frac{7}{8}$ mi is greater than 0.75 mi.

Choose the letter for the best answer.

5. Tom deposits $100 into his bank account every month. A $2.50 service fee is deducted each month. Which expression represents the total deposits and service charges for one year?
   A ($2.50 − $100) × 12
   B ($100 + $^-2.50) × 12
   C $2.50 + $100 × 12
   D (12 − $2.50) + $100

6. In a game of shuffleboard, Pauline had the following scores: $8\frac{1}{2}$, $^-5\frac{1}{4}$, 12, $7\frac{1}{4}$, $^-14$, $6\frac{1}{2}$, $^-9$, and $^-11\frac{1}{2}$. What was her total score for that game?
   F $4\frac{1}{2}$ H $^-1\frac{1}{2}$
   G 2   **J $^-5\frac{1}{2}$**

7. In his last four basketball games, Jason scored 21 points, 17 points, 19 points, and 15 points. Jason wants to have an average of 20 points per game. How many points must he score in the next game to achieve this average?
   A 27   C 29
   B 28   D 30

8. Mr. Smith started a diet when he weighed 252.5 lb. He lost 3.5 lb each month for the first 3 months. He went off the diet for 2 months and gained 4 lb each month. Last month he lost 4 lb. How much did Mr. Smith weigh after 6 months?
   F 241 lb   H 246 lb
   G 243 lb   J 248 lb

9. **Write About It** Explain how you arrived at your answer to Exercise 5.
   Possible answer: The service fee is deducted so it is negative. The sum of the deposits and the service fee for each month is multiplied by 12 to find the total for 1 year.

LESSON 12.6

Independent Practice

Note that Exercises 52, 53, and 54 are multistep or strategy problems. Assign Exercises 19–55.

For Exercises 19–36, encourage students to decide on the sign of the answer and to jot it down before they perform their calculations.

For Exercises 37–44, caution students not to try to carry out too many operations at once. They can use order of operations rules to decide which operation they are going to use at any one time. Then they should decide on the sign the operation will yield. Next they should carry out the operation, giving them a simpler expression to work with. Then they can start the process over again, looking at the simplified expression to decide which operation they will perform next.

Multistep or Strategy Problem To solve Exercise 53, students can subtract $3\frac{1}{4}$ from $5\frac{1}{2}$ and then divide that amount by 2. This is the amount Ellen gave to Ruben. Guide students to subtract before they divide.

Guided Practice Find the sum or difference. Estimate to check.

3. $-3\frac{4}{5} + 4\frac{1}{2}$ **1**
4. $7\frac{3}{4} - 8\frac{1}{2}$ **$-\frac{3}{4}$**
5. $2.5 + -3.2$ **-0.7**
6. $2.4 - -1.5$ **3.9**
7. $9\frac{2}{5} - -3\frac{4}{5}$ **$13\frac{1}{5}$**
8. $\frac{1}{2} + -3\frac{1}{2}$ **-3**
9. $-3.9 - 4.1$ **-8.0**
10. $-10 + 8\frac{2}{3}$ **$-1\frac{1}{3}$**

Find the product or quotient. Estimate to check.

11. -4.5×2.4 **-10.8**
12. $-6 \div -1\frac{1}{2}$ **4**
13. $3.8 \div 0.2$ **19**
14. $\frac{4}{5} \times -2\frac{1}{2}$ **-2**
15. $4.9 \div 1.4$ **3.5**
16. $-\frac{1}{2} \times -6$ **3**
17. $1\frac{1}{4} \times -1\frac{3}{5}$ **-2**
18. $-4\frac{1}{2} \div \frac{1}{2}$ **-9**

PRACTICE AND PROBLEM SOLVING

Independent Practice Find the sum or difference. Estimate to check.

19. $-2.5 - -1.5$ **-1**
20. $-1.5 + 5.8$ **4.3**
21. $-5\frac{1}{2} - 3\frac{3}{4}$ **$-9\frac{1}{4}$**
22. $6.5 + 0.9$ **7.4**
23. $3\frac{1}{2} + -7$ **$-3\frac{1}{2}$**
24. $10\frac{1}{2} - -12\frac{1}{2}$ **23**
25. $-27.6 - -43.2$ **15.6**
26. $-15\frac{1}{2} + -22$ **$-37\frac{1}{2}$**
27. $-34.6 + -52.8$ **-87.4**

Find the product or quotient. Estimate to check.

28. $\frac{3}{5} \times \frac{5}{6}$ **$\frac{1}{2}$**
29. $6.8 \div -0.4$ **-17**
30. -1.4×-20 **28**
31. $-7\frac{1}{2} \div -\frac{5}{8}$ **12**
32. 0.6×-2.5 **-1.5**
33. $-\frac{2}{3} \div -3$ **$\frac{2}{9}$**
34. $8.8 \div 0.2$ **44**
35. $-8 \div \frac{1}{6}$ **-48**
36. $16\frac{2}{3} \times \frac{2}{5}$ **$6\frac{2}{3}$**

Evaluate the expression.

37. $6.4 + 17.1 \div -3$ **0.7**
38. $\left(-\frac{1}{2}\right)^2 - \left(-3\frac{1}{2} + 2\right)$ **$1\frac{3}{4}$**
39. $(3.6 - 5.4) \div (-2.5 + 3.7)$ **-1.5**
40. $-7 - -5 - (-3 \times -1)$ **-5**
41. $8 \times -1.5 \div \left(-\frac{2}{3} + \frac{1}{6}\right)$ **24**
42. $\frac{2}{3} + \frac{8}{9} \div -\frac{2}{3} + 1$ **$\frac{1}{3}$**
43. $4.4 \div (3.8 - 6) \div -0.5$ **4**
44. $\left(-\frac{1}{2} + \frac{1}{3}\right)^2 \times -72$ **-2**

ALGEBRA Evaluate the expression for $x = -2.4$.

45. $x - 8.1$ **-10.5**
46. $-3.7 + x$ **-6.1**
47. $x - -6.8$ **4.4**
48. $10.7 - x$ **13.1**
49. $x + -22.6$ **-25**
50. $12 + (x - 1.2)$ **8.4**

258 Chapter 12

ALTERNATIVE TEACHING STRATEGY SCAFFOLDED INSTRUCTION

Purpose Students use a four-step process to operate on rational numbers.

Materials For each student calculator

Explain that students can use an organized method to add, subtract, multiply, and divide rational numbers, and to check their answers. The method involves four steps.

Write Write the problem.

Sign Decide on the sign of the answer and jot it down.

Operate Carry out the operation.

Check Use your calculator to check the answer.

Demonstrate the method by finding the sum $-6.5 + 2.3$. Have students divide a sheet of paper into two columns.

Write $-6.5 + 2.3$

Sign $|-6.5| > |+2.3|$, so the sum is negative.

Operate Subtract the lesser absolute value from the greater absolute value.
$6.5 - 2.3 = 4.2$

The sum is negative, so $-6.5 + 2.3 = -4.2$

Check (-)6.5 (+) 2.3 (=)
$-6.5 + 2.3 = -4.2$

Have students use the method to find the sum, difference, product, and quotient of such rational number pairs as $(-3.6, 0.9)$ and $\left(\frac{4}{5}, -\frac{1}{2}\right)$.
$-2.7, -4.5, -3.24, -4; \frac{3}{10}, 1\frac{3}{10}, -\frac{2}{5}, -1\frac{3}{5}$

Problem Solving ▶ Applications

51. Science The daytime temperature on the moon can reach 265.9°F. At night the temperature can fall to ⁻291.5°F. Find the difference between the daytime and nighttime temperatures.   **557.4°F**

52. In a football game, Sean gained $4\frac{1}{2}$ yd on one play, lost 8 yd on the next play, and gained $6\frac{1}{2}$ yd on the third play. What was his total gain?   **3 yd**

53. Ellen had $5\frac{1}{2}$ lb of flour. She used $3\frac{1}{4}$ lb and gave half of what was left to Ruben. How much flour did she give to Ruben?   **$1\frac{1}{8}$ lb**

54. During a storm, rain fell at a rate of 0.9 in. per hour from 9 P.M. until 1:30 A.M. From 1:30 A.M. until 4 A.M., the rate of rainfall was 0.6 in. per hour. What was the total amount of rainfall?   **5.55 in.**

55. REASONING For what values of a will $⁻2 \times (a - 3)$ be positive? Explain.   **for $a < 3$; for the product to be positive, $(a - 3)$ must be negative, which means a must be less than 3**

MIXED REVIEW AND TEST PREP

56. Find the sum. ⁻15 + ⁺9   (p. 244)   **⁻6**

57. Write the percent for 0.085.   (p. 60)   **8.5%**

58. Evaluate the expression $n + \frac{5}{6}$ for $n = \frac{1}{2}$.   (p. 216)   **$1\frac{1}{3}$**

★ **59. TEST PREP** Mr. Tilson works 40 hours per week. One year he worked 50 weeks and earned $24,000. How much did he earn per hour?   (p. 22)   **D**

A $10.00   **B** $11.00   **C** $11.50   **D** $12.00

★ **60. TEST PREP** Ari's test scores were 88, 79, 74, 79, and 90. What is the median of his scores?   (p. 106)   **H**

F 16   **G** 74   **H** 79   **J** 82

PROBLEM SOLVING LiNKUP to Careers

Veterinarian A veterinarian is a doctor who treats animals. Even though a veterinarian spends most of his or her time tending to patients, there is also some mathematics to be done. For example, after a dog undergoes surgery, it may need medication. The amount to give depends on the weight of the dog. Too much medicine could harm the dog. Figuring out the right amount of medicine requires mathematics.

- A certain medication is given at a rate of 0.2 mg per pound daily. Georgia's dog Barkum is undergoing surgery. Barkum weighs $30\frac{1}{2}$ lb. If needed, how much medicine should Barkum receive daily?   **6.10 mg**

EXTRA PRACTICE page H43, Set D

259

MIXED REVIEW AND TEST PREP

Exercises 56–60 provide **cumulative review** (Chapters 1–12).

LiNKUP to CAREERS

- Direct students' attention to the Link Up. Have them tell what they know about the work that a veterinarian does.

What are some other ways that a veterinarian uses math on the job? Possible answer: Scheduling appointments, keeping track of bills, weighing patients, writing prescriptions.

REASONING A veterinarian is writing a prescription for a bottle of medicine for a cat. The dose is 0.4 ml per pound of weight, twice a day for 1 week. If the cat weighs $7\frac{1}{2}$ lb, what should the veterinarian write down for the amount of medicine to be put in the bottle?   42 ml

4 Assess

Summarize the lesson by having students:

DISCUSS What number would you get if you removed the parentheses from Exercise 39?   9.46

WRITE Outline the steps you would use to solve Exercise 38. First, square $⁻\frac{1}{2}$. Then add $⁻3\frac{1}{2} + 2$, and subtract the sum from $(⁻\frac{1}{2})^2$.

Lesson Quiz

Transparency 12.6

Evaluate the expression.

1. $⁻12.7 - ⁻2.6$ ⁻10.1
2. $12 \times ⁻4.25$ ⁻51
3. $⁻3 \div \frac{2}{3}$ $⁻4\frac{1}{2}$
4. $⁻8\frac{1}{2} + 9\frac{3}{4}$ $1\frac{1}{4}$
5. $8.5 - 3 \div ⁻0.8$ 12.25

ADDITIONAL ANSWERS

Lesson 12.6, p. 257

1. Possible answer: The methods for multiplying decimals and fractions are different but the method for finding the sign of the product is the same for both.

2. Possible example: $4 + 10 \div 2 = 4 + 5 = 9$; $⁻4 + ⁻10 \div ⁻2 = ⁻4 + 5 = 1$

CHAPTER 12 REVIEW/TEST

1. **VOCABULARY** The opposite of an integer is its ___?___ . (p. 243) **additive inverse**

Write the addition problem modeled on each number line. (pp. 244–247)

2. $4 + {}^-5 = {}^-1$

3. $4 + {}^-6 = {}^-2$

Find the sum. (pp. 244–247, 256–259)

4. $7 + {}^-6$ **1**
5. ${}^-5 + {}^-3$ **${}^-8$**
6. ${}^-7 + 4$ **${}^-3$**
7. $4 + {}^-1$ **3**
8. ${}^-37 + 24$ **${}^-13$**
9. $17 + {}^-19$ **${}^-2$**
10. ${}^-17 + {}^-41$ **${}^-58$**
11. $21 + 17$ **38**
12. ${}^-7.2 + {}^-4$ **${}^-11.2$**
13. $6\frac{1}{2} + {}^-2\frac{1}{4}$ **$4\frac{1}{4}$**
14. ${}^-2\frac{4}{5} + 1\frac{1}{5}$ **${}^-1\frac{3}{5}$**
15. ${}^-9.7 + {}^-1.3$ **${}^-11$**

Find the difference. (pp. 250–251, 256–259)

16. $7 - 11$ **${}^-4$**
17. ${}^-2 - 8$ **${}^-10$**
18. ${}^-3 - {}^-5$ **2**
19. $4 - {}^-4$ **8**
20. ${}^-1 - 4$ **${}^-5$**
21. ${}^-6 - {}^-8$ **2**
22. $51 - {}^-23$ **74**
23. ${}^-41 - 18$ **${}^-59$**
24. ${}^-8\frac{5}{6} - {}^-2\frac{1}{3}$ **${}^-6\frac{1}{2}$**
25. $5\frac{1}{4} + {}^-2\frac{1}{8}$ **$3\frac{1}{8}$**
26. ${}^-16.2 + {}^-5.9$ **22.1**
27. $36.4 + {}^-25.1$ **11.3**

Find the product or quotient. (pp. 252–255, 256–259)

28. ${}^-9 \times {}^-5$ **45**
29. ${}^-12 \times 5$ **${}^-60$**
30. ${}^-36 \div {}^-6$ **6**
31. $144 \div {}^-12$ **${}^-12$**
32. $\frac{1}{3} \times {}^-\frac{1}{8}$ **${}^-\frac{1}{24}$**
33. ${}^-4\frac{1}{2} \div \frac{1}{2}$ **${}^-9$**
34. ${}^-9.6 \div 0.4$ **${}^-24$**
35. ${}^-2.1 \times {}^-20$ **42**

Solve.

36. A submarine started one leg of its voyage at ${}^-300$ ft. At the end of that leg of the voyage, it was at ${}^-1,250$ ft. What was the difference between the two depths? (pp. 250–251) **950 ft.**

37. Tasha needed to find the average temperature during three days in the winter. Her first step was to add the temperatures together. If the temperatures were ${}^-6°$, ${}^-8°$, and $11°$, what was the sum? (pp. 244–247) **${}^-3°$**

38. At midnight the temperature was ${}^-15°$F. The temperature continued to fall until 7:00 A.M., when it was ${}^-22°$F. How many degrees had the temperature fallen during the seven hours? (pp. 250–251) **7°**

39. During the last glacial age, sea level in one body of water changed by an average of ${}^-3$ feet every 200 years. The glacial age lasted 10,000 years. What was the total change in sea level? (pp. 252–255) **${}^-150$ ft.**

40. During the first nine months of the year, a worldwide club showed its total loss of members as ${}^-2,718$. What integer shows the average loss of members per month? (pp. 252–255) **${}^-302$**

260   Chapter 12

CHAPTERS 1–12 ★ STANDARDIZED TEST PREP

Get the information you need.
See item **3**.
Think about how you can write the subtraction of integers as addition. Then find the expression that has the same solution as the given subtraction problem.
Also see problem **3**, p. H63.

Choose the best answer.

1. $^-54 \div {^-9}$? **C**
   - A $^-486$
   - B $^-6$
   - C 6
   - D Not here

2. What is the value of $29 + k$ for $k = {^-26}$? **H**
   - F $^-55$
   - G $^-3$
   - H 3
   - J 55

3. Which problem has the same solution as $^-3 - {^-5}$? **C**
   - A $^-3 - 5$
   - B $^-3 + {^-5}$
   - C $^-3 + 5$
   - D $3 + 5$

4. $^-1.8 \times 30$ **F**
   - F $^-54$
   - G 0.06
   - H 54
   - J Not here

5. Use mental math to find the value of x. **D**
   $$x + {^-9} = 5$$
   - A $x = {^-14}$
   - B $x = {^-4}$
   - C $x = 4$
   - D $x = 14$

6. The graph shows the number of sixth-grade students at a middle school who are in each club listed. How many more students are in the Science Club than in the Chess Club? **G**

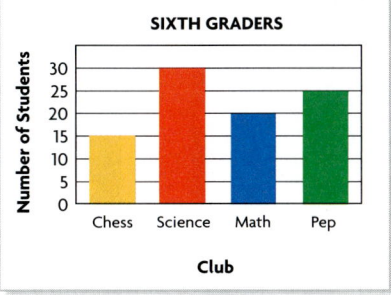

   - F 2
   - G 15
   - H 20
   - J 30

7. How is $2 \times 2 \times 2 \times 2$ written in exponent form with a base of 2? **A**
   - A 2^4
   - B 2^5
   - C 4^2
   - D 2^{10}

8. Ellen got a score of 50 on her history exam. The mean score on that exam was 45. Which statement must be true? **F**
   - F Ellen's score is above the average score.
   - G Ellen got the top score.
   - H Ellen's score is higher than most of the other students' scores.
   - J Ellen's score is below the average score.

9. $0.053 - 0.019$ **C**
   - A 3.4
   - B 0.34
   - C 0.034
   - D Not here

Write What You Know

See below.

10. Explain how any integer subtraction problem can be rewritten as an integer addition problem. Use your method to find $^-4 - {^+8}$.

11. Estimate the quotient $9\frac{1}{6} \div 1\frac{7}{8}$ using mental math and explain the method you used. Then find the quotient and explain how you found your answer.

261

Write What You Know • Written Response

10. To convert a subtraction problem to an addition problem, change the subtraction sign to an addition sign and write the opposite of the integer being subtracted. Then use the rules for addition. $^-4 - {^+8} = {^-4} + {^-8} = {^-12}$.

11. $9\frac{1}{6}$ is close to 10 and $1\frac{7}{8}$ is close to 2. So, to estimate the quotient, divide 10 by 2 to get 5. To find the exact answer, write the mixed numbers as fractions: $9\frac{1}{6} = \frac{55}{6}$ and $1\frac{7}{8} = \frac{15}{8}$. Then use the reciprocal of the divisor to write a multiplication problem: $\frac{55}{6} \div \frac{15}{8} = \frac{55}{6} \times \frac{8}{15} = \frac{44}{9}$, or $4\frac{8}{9}$.

STANDARDIZED TEST PREP •
Chapters 1–12

USING THE PAGE

This page may be used to help students get ready for standardized tests. The test items are written in the same style and arranged in the same format as those on many state assessments. The page is cumulative. It covers math objectives and essential skills that have been taught up to this point in the text. Most of the items represent skills from the current chapter, and the remainder represent skills from earlier chapters.

This page can be assigned at the end of the chapter as classwork or as a homework assignment. You may want to have students use individual recording sheets presented in a multiple-choice (standardized) format. A Test Answer Sheet is available as a blackline master in *Assessment Guide* (p. AGxlii).

You may wish to have students describe how they solved each problem and share their solutions.

ITEM ANALYSIS

Item	Learning Goal	Item	Learning Goal
1	12B	7	2B
2	12A	8	5C
3	12A	9	4A
4	12B	10	12A
5	12A	11	10A
6	6A		

Written response items for the Write What You Know are available as a blackline master in *Performance Assessment*.

SCORING RUBRIC • WRITE WHAT YOU KNOW

2 Demonstrates a complete understanding of the problem and chooses an appropriate strategy to determine the solution

1 Demonstrates a partial understanding of the problem and chooses a strategy that does not lead to a complete and accurate solution

0 Demonstrates little understanding of the problem and shows little evidence of using any strategy to determine a solution

Operations with Integers **261**

UNIT 4

MATH DETECTIVE
Play by the Rules

Purpose To use deductive reasoning to solve problems involving operations of integers

USING THE PAGE

- Direct students' attention to the Reasoning section and Function Machines 1 and 2.

  If you know the output for Function Machine 1, how would you use the rule to find the input? Possible answer: Use the inverse operation and subtract ⁻6 from the output, or add 6 to the output.

  Can you write an equivalent rule for Function Machine 1? for Function Machine 2? Subtract 6; add ⁻4.

- Have students look at the rule and the input values for Function Machine 3.

  Do you think all of the outputs for Function Machine 3 will be positive, negative, or a combination of the two? Explain. A combination; there are both positive and negative inputs and adding 1 will not change the output as much as multiplying by ⁻3.

- Ask students to find and discuss the rule for Function Machine 4.

  Reasoning **If you know only one input and output for a function machine, can you find a unique rule for the function machine? Explain.** No, Possible answer: For example, for Function Machine 4 if you know only the output for ⁻4 is 2, then the rule could be to divide by ⁻2 or it could be to add 6.

  Think It Over! After students complete the Write About It, have them compare their methods for finding the rule. Guide them to understand that both forms of the rule are equivalent.

262 Unit 4 • Chapters 11–12

MATH DETECTIVE
PROBLEM SOLVING

Play by the Rules

REASONING Each function machine is missing something. For Machines 1, 2, and 3, an input or an output value is missing in each step. For Machine 4, the rule is missing. Use your knowledge of integer operations to find what is missing for each machine.

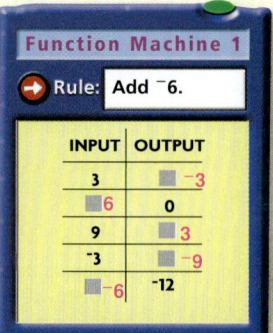

Function Machine 1
Rule: Add ⁻6.

INPUT	OUTPUT
3	⁻3
6	0
9	3
⁻3	⁻9
⁻6	⁻12

Function Machine 2
Rule: Subtract 4.

INPUT	OUTPUT
5	1
11	7
1	⁻3
⁻2	⁻6
4	0

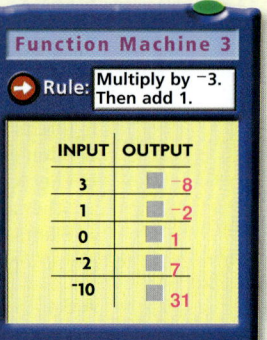

Function Machine 3
Rule: Multiply by ⁻3. Then add 1.

INPUT	OUTPUT
3	⁻8
1	⁻2
0	1
⁻2	7
⁻10	31

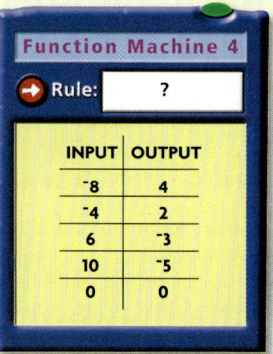

Function Machine 4
Rule: ?

INPUT	OUTPUT
⁻8	4
⁻4	2
6	⁻3
10	⁻5
0	0

Divide by ⁻2 or multiply by $-\frac{1}{2}$.

Think It Over!

- **Write About It** Explain how you found a rule for Function Machine 4. Check students' answers.
- **Stretch Your Thinking** Find a rule for this function machine. Multiply by 2; then add 3.

INPUT	OUTPUT
0	3
1	5
2	7
3	9

Intervention and Extension Resources

ADVANCED LEARNERS

Challenge students to **find the output using each rule** below for inputs 12 and ⁻9.

Rule 1: Divide by ⁻3. Then add ⁻6. ⁻10; ⁻3

Rule 2: Multiply by $-\frac{2}{3}$. Then subtract ⁻4. ⁻4; 10

Rule 3: Divide by $-\frac{1}{2}$. Then add 8. ⁻16; 26

Rule 4: Multiply by $1\frac{1}{3}$. Then add ⁻12. 4; ⁻24

Next ask students to make their own function table by using a rule involving two integer operations. Have them complete the table by finding the output for 5 different inputs. Then challenge them to exchange completed tables without the rule and find the rule for the table. Check students' work.

VISUAL
LOGICAL/MATHEMATICAL

Challenge
Negative Exponents

Learn how to work with negative exponents and how to write small numbers by using scientific notation.

Activity

Copy and complete the pattern. Use a calculator as needed.

$10^3 = 1{,}000$

$10^2 = 100$

$10^1 = 10$

$10^0 = \blacksquare\ 1$

$10^{-1} = \blacksquare\ 0.1$

$10^{-2} = \blacksquare\ 0.01$

$10^{-3} = \blacksquare\ 0.001$

Notice that as the value of the exponent decreases, each number is 0.1, or $\frac{1}{10}$, as great as the previous number.

- For powers of 10, how is the negative exponent related to the number of decimal places? **The negative exponent tells the number of places to the right of the decimal point.**

Negative exponents are used to write very small numbers in scientific notation.

$$0.004 = 4 \times 0.001 = 4 \times 10^{-3}$$

Replace 0.001 with 10^{-3}.

Look at the relationship between 0.004 and 4×10^{-3}. To write 0.004 as 4, move the decimal point 3 places to the right, multiplying by 1,000. Use $^-3$ as the exponent of 10 to show the corresponding division by 1,000.

TECHNOLOGY LINK

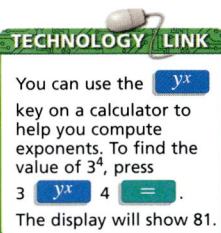

You can use the y^x key on a calculator to help you compute exponents. To find the value of 3^4, press 3 y^x 4 =. The display will show 81.

EXAMPLE

Write 0.0000245 in scientific notation.

0.0000245
5 places

2.45×10^{-5}

Count the number of places the decimal point must be moved to the right to form a number that is at least 1 but less than 10.

Write the number. Since the decimal point moved 5 places to the right, the exponent of 10 is $^-5$.

TALK ABOUT IT

- When writing a very small number in scientific notation, how can you tell what the exponent should be? **Since the decimal is moved to the right, the exponent is the negative of the number of places moved.**

TRY IT

Write using scientific notation.

1. 0.002 2×10^{-3}
2. 0.00034 3.4×10^{-4}
3. 0.06 6×10^{-2}
4. 0.005365 5.365×10^{-3}
5. 0.0084 8.4×10^{-3}
6. 0.0000794 7.94×10^{-5}
7. 0.0000008 8×10^{-7}
8. 0.000202 2.02×10^{-4}

Chapters 11–12 **263**

CHALLENGE
Negative Exponents

Objective To extend the concepts and skills of Chapters 11–12

USING THE PAGE

- Have students complete the Activity and extend the pattern.

  What is 0.0001 as a power of 10? 10^{-4}
  What is 10^{-6}? 0.000001

- Work through the discussion of negative exponents and the Example.

  Compare how to write a very small number in scientific notation to writing a very large number in scientific notation. How are they similar? different? In both cases, the number is a factor that is greater than or equal to 1 but less than 10 and multiplied by a power of 10. For large numbers, the power of 10 has a positive exponent, and for small numbers the power of 10 has a negative exponent.

- Have students complete the Talk About It.

  How can you check that you have written a number in scientific notation correctly? Possible answer: Convert the number back into standard form.

Try It Before assigning Try It Exercises 1–8, suggest that students copy each number and draw arrows in the standard form of the number, as shown in the Example, to help them write the needed exponent.

Intervention and Extension Resources

SCIENCE CONNECTION

Reinforce students' understanding of scientific notation. Tell students the following:
Numerical constants used in science can be very large or very small numbers. Scientific notation makes it easier to record some of these constants. For example, the mass of an electron is 9.110×10^{-31} kg, the mass of a neutron is 1.675×10^{-27} kg, and the mass of a proton is 1.673×10^{-27} kg.

Have students order the masses of the particles from greatest to least. Ask them to explain their thinking. neutron, proton, electron; Compare the exponents first. $^-31$ is less than $^-27$, so its mass is the least. Then compare numbers that have the exponent $^-27$. 1.675 is greater than 1.673. So, a neutron's mass is the greatest.

Finally have students work in pairs to record the mass of each of the particles in standard form. Check students' work.

VISUAL; LOGICAL/MATHEMATICAL

Algebra: Integers **263**

UNIT 4 CHAPTERS 11–12

STUDY GUIDE AND REVIEW

Purpose To help students review concepts and skills presented in Chapters 11–12

USING THE PAGES

✓ Assessment Checkpoint

The Study Guide and Review includes content from Chapters 11–12.

Chapter 11
11.1 Understand Integers
11.2 Rational Numbers
11.3 Compare and Order Rational Numbers
11.4 Problem Solving Strategy: *Use Logical Reasoning*

Chapter 12
12.1 Math Lab: Model Addition of Integers
12.2 Add Integers
12.3 Math Lab: Model Subtraction of Integers
12.4 Subtract Integers
12.5 Multiply and Divide Integers
12.6 Explore Operations with Rational Numbers

The blue page numbers in parentheses provided with each group of exercises indicate the pages on which the concept or skill was presented. The red number given with each group of exercises identifies the Learning Goal for the concept or skill.

UNIT 4 Study Guide and Review

VOCABULARY

1. The distance an integer is from zero is its __?__. (p. 228) **absolute value**
2. All whole numbers and their opposites are the set of __?__. (p. 228) **integers**
3. A comparison of two numbers, a and b, written as a fraction $\frac{a}{b}$ is a __?__. (p. 230) **ratio**

EXAMPLES | EXERCISES

Chapter 11

- **Write the absolute value of an integer.** (pp. 228–229) **11A**

  $|{-5}| = 5$ *Write the distance -5 is from 0 on the number line.*

  Write the absolute value.

  4. $|{-8}|$ **8**   5. $|{+6}|$ **6**   6. $|{-28}|$ **28**
  7. $|{-73}|$ **73**   8. $|{+49}|$ **49**   9. $|0|$ **0**

- **Classify numbers as whole numbers, integers, and rational numbers.** (pp. 230–233) **11B**

  -5 is an integer and a rational number.
  $\frac{3}{5}$ is a rational number.

  Name the sets to which each number belongs.

  10. $3\frac{5}{9}$ **rational**   11. 0.35 **rational**
  12. -42 **rational, integer**   13. $\frac{6}{3}$ **whole, integer, rational**

- **Find a rational number between two rational numbers.** (pp. 230–233) **11B**

  Find a rational number between -8.5 and -8.45.
  -8.50 -8.45 *Write each decimal using hundredths.*
  -8.46 is between -8.5 and -8.45. Some other numbers are -8.47, -8.48, and -8.482.

  Find a rational number between the two given numbers. Possible answers are given.

  14. $\frac{1}{8}$ and $\frac{1}{2}$ $\frac{1}{4}$ 15. -2.3 and -2.4 -2.35
  16. 18.01 and 18.02 **18.015**   17. $\frac{-1}{8}$ and $\frac{1}{10}$ $\frac{-1}{10}$
  18. $\frac{1}{10}$ and 0.2 **0.15**   19. $-1\frac{1}{2}$ and $-1\frac{3}{4}$ $-1\frac{5}{8}$

- **Compare and order rational numbers.** (pp. 234–235) **11C**

  Compare. Write <, >, or = for the ●.
  -5.8 ● $-5\frac{3}{4}$
  $-5\frac{3}{4} = -5.75$ *Write the fraction as a decimal.*
  $-5.80 < -5.75$ *Compare the decimals.*
  So, $-5.8 < -5\frac{3}{4}$.

  Compare. Write <, > or = for each ●.

  20. $\frac{2}{5}$ ● 0.38 **>**   21. $3\frac{3}{4}$ ● 3.89 **<**
  22. -0.25 ● $\frac{-1}{4}$ **=**   23. $4\frac{3}{10}$ ● 4.03 **>**

  Order from least to greatest.

  24. 1.55, $\frac{2}{3}$, -3, 4.2, -1.8 $-3, -1.8, \frac{2}{3}, 1.55, 4.2$
  25. $\frac{-3}{4}$, 0, 6.5, -8, -3.6 $-8, -3.6, \frac{-3}{4}, 0, 6.5$

264 Unit 4

Chapter 12

- **Add and subtract integers.** (pp. 242–251) **12A**

  Subtract. $6 - {}^-4$
  $6 - {}^-4 = 6 + 4$ *Write as an addition sentence.*
  $\phantom{6 - {}^-4} = 10$

  Add. $15 + {}^-3$
  $|15| - |{}^-3| = 15 - 3$ *Subtract the absolute values.*
  $\phantom{|15| - |{}^-3|} = 12$ *The sum is positive.*

Find the sum or difference.

26. ${}^-3 + 9$ **6**
27. ${}^-8 + {}^-9$ **${}^-17$**
28. $12 + {}^-5$ **7**
29. ${}^-4 - 10$ **${}^-14$**
30. ${}^-9 - {}^-13$ **4**
31. $15 - {}^-6$ **21**

- **Multiply and divide integers.** (pp. 252–255) **12B**

  ${}^-4 \times 7 = {}^-28$ *The product or quotient of a positive and a negative integer is negative.*

  ${}^-72 \div {}^-9 = 8$ *The product or quotient of two positive or two negative integers is positive.*

Find the product or quotient.

32. ${}^-16 \cdot 3$ **${}^-48$**
33. ${}^-12 \cdot {}^-12$ **144**
34. $15 \cdot {}^-6$ **${}^-90$**
35. ${}^-14 \cdot {}^-9$ **126**
36. ${}^-54 \div 9$ **${}^-6$**
37. $124 \div {}^-4$ **${}^-31$**
38. ${}^-75 \div {}^-15$ **5**
39. ${}^-119 \div 7$ **${}^-17$**

- **Operate with rational numbers.** (pp. 256–259) **12C**

  Find the product.

  $2\frac{2}{3} \times {}^-1\frac{1}{3} = \frac{8}{3} \times \frac{{}^-4}{3}$ *Write mixed numbers as fractions.*

  $\phantom{2\frac{2}{3} \times {}^-1\frac{1}{3}} = \frac{{}^-32}{9}$, or ${}^-3\frac{5}{9}$ *Multiply. The fractions have unlike signs so the product is negative.*

Add, subtract, multiply, or divide.

40. $36.7 + {}^-51.4$ **${}^-14.7$**
41. ${}^-87.3 - 49.1$ **${}^-136.4$**
42. $3\frac{3}{4} \times \frac{{}^-1}{3}$ **${}^-1\frac{1}{4}$**
43. ${}^-3\frac{1}{8} \div {}^-5$ **$\frac{5}{8}$**

PROBLEM SOLVING APPLICATIONS

44. Four students spent money on books. The amounts they spent were $2.50, $3.25, $5.00, and $5.75. Ari spent $3.25 more than Nina. Nina spent half as much as Jamal. Kenny spent $1.75 less than Jamal. How much did each person spend? (pp.236–237) **Jamal: $5.00, Kenny: $3.25, Nina: $2.50, Ari: $5.75 11D**

45. Natalie, Aaron, Toral, and Casey are each on a different sports team. Toral is not on the baseball team. Casey's team plays on the ice. Aaron's team practices in water. The four teams are soccer, hockey, swimming, and baseball. Who plays on each team? (pp. 236–237) **Natalie: baseball, Aaron: swimming, Toral: soccer, Casey: hockey 11D**

46. Rachel is the owner of a restaurant. Her profits and losses for the past four weeks are ${}^-$380, $420, ${}^-$145, and $620. How much was her profit or loss for the past four weeks? (pp.244–247) **12A Her profit was $515.**

47. Frances claimed that the average of the low temperatures last week was greater than ${}^-4°C$. Patty says that the average was less than ${}^-4°C$. The low temperatures last week were ${}^-6°C$, ${}^-8°C, {}^-3°C, {}^-4°C, 0°C, {}^-9°C$, and ${}^-5°C$. Who is correct? Explain. (pp.252–255)

47. Patty is correct since the average low temperature was ${}^-5°C$. 12B

Chapters 11–12 **265**

✓ Assessment Checkpoint

 **Portfolio Suggestions** The portfolio represents the growth, talents, achievements, and reflections of the mathematics learner. Students might spend a short time selecting work samples for their portfolios and completing A Guide to My Math Portfolio from *Assessment Guide*, page AGxix.

You may want to have students respond to the following questions:

- What new understanding of math have I developed in the past several weeks?
- What growth in understanding or skills can I see in my work?
- What can I do to improve my understanding of math ideas?
- What would I like to learn more about?

For information about how to organize, share, and evaluate portfolios, see *Assessment Guide*, page AGxviii.

Use the item analysis in the **Intervention** chart to diagnose students' errors. You may wish to reinforce content or remediate misunderstandings by using the text pages or lesson resources.

STUDY GUIDE AND REVIEW INTERVENTION

How to Help Options

Learning Goal	Items	Text Pages	Reteach and Practice Resources
11A See page 226C for Chapter 11 learning goals	4–9	228–229	Worksheets for Lesson 11.1
11B See page 226C for Chapter 11 learning goals	10–13	230–233	Worksheets for Lesson 11.2
11C See page 226C for Chapter 11 learning goals	14–19, 20–25	230–233, 234–235	Worksheets for Lessons 11.2, 11.3
11D See page 226C for Chapter 11 learning goals	44–45	236–237	Worksheets for Lesson 11.4
12A See page 226C for Chapter 12 learning goals	26–31, 46	242–251	Worksheets for Lessons 12.1, 12.2, 12.3, 12.4
12B See page 226C for Chapter 12 learning goals	32–39, 47	252–255	Worksheets for Lesson 12.5
12C See page 226C for Chapter 12 learning goals	40–43	256–259	Worksheets for Lesson 12.6

Algebra: Integers **265**

UNIT 4

PERFORMANCE ASSESSMENT

Purpose To provide performance assessment tasks for Chapters 11–12

USING THE PAGE
- Have students work individually or in pairs as an alternative to formal assessment.
- Use the performance indicators and work samples below to evaluate Tasks A–B.

See *Performance Assessment* for
- a complete scoring rubric, p. PAx, for this unit.
- additional student work samples for this unit.
- copying masters for this unit.

 You may suggest that students place completed Performance Assessment tasks in their portfolios.

Additional Answers for Task B
d. Possible answer: The low temperature on Thursday is 4 times the high temperature. The high temperature on Thursday is 12° greater than the low temperature.

Performance Indicators

Task A
A student with a Level 3 paper
____ Exhibits understanding of operations with integers.
____ Exhibits understanding of the use of a simple four-function calculator with memory keys.
____ Shows work and explains how the answers were determined.

Task B
A student with a Level 3 paper
____ Interprets a graph showing integer values by writing statements about the graph.
____ Determines the median value of a set of integers.
____ Graphs integers on a number line.

266 Unit 4 • Chapters 11–12

Performance Assessment

TASK A • Calculate This!
Materials: simple four-function calculator

Suppose the +/− key on a simple calculator does not work. However, it does have memory +, memory −, and memory recall keys.

a. Explain how to use the calculator to find 7 − ⁻8. What is the answer? 8 [M−] 7 [−] [MR] [=] 15

b. Explain how to use the calculator to find ⁻9 + 7. What is the answer? 9 [M−] 7 [+] [MR] [=] ⁻2

c. Explain how to find the product or quotient of a positive and a negative number and of two negative numbers by using the calculator. With just one negative number, use the [M−] and [MR] keys. With 2 negative numbers, enter their absolute values.

TASK B • Brrrr!
Materials: Local newspaper weather report

The graph shows the high and low temperatures in Barrow, Alaska, for one week.

a. Find the median high temperature. 1°F
b. Find the median low temperature. ⁻8°F
c. Graph the low temperatures on a number line. Check students' number lines.
d. Write two statements comparing the high and low temperatures on Thursday. See left.

Find a similar graph in your local newspaper for where you live. Repeat the activity using that graph. Check students' work.

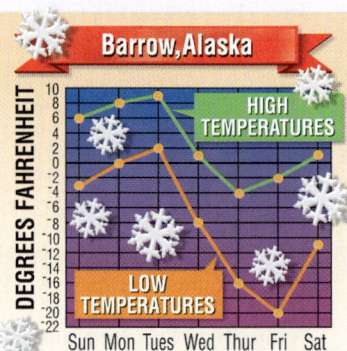

266 Unit 4

Work Samples for Task A and Task B

Level 3 The student shows good understanding of using a calculator. The work is correct and complete.

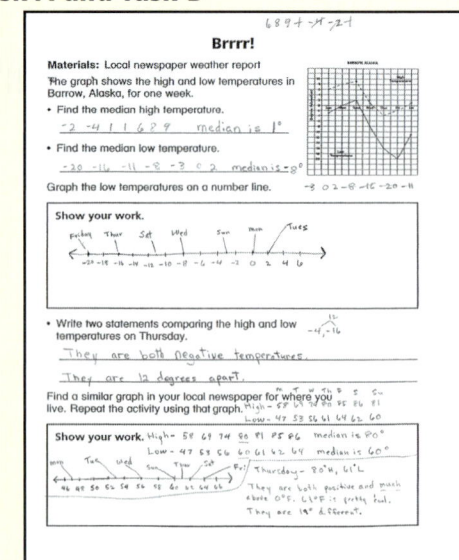

Level 3 The student demonstrates a good understanding of the task. All answers are complete and accurate. Work demonstrates understanding of negative numbers.

Mighty Math Calculating Crew
Operations with Integers

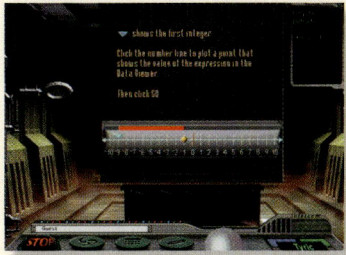

Click [OFF DUTY]. Then click [icon].

- Choose Red Level A. Click [OK]. Then click [ON DUTY]. Complete the mission by finding the value of each addition expression.

- Click [icon]. Choose Red Level G. Click [OK] and then [ON DUTY]. Find the value of each multiplication expression.

Practice and Problem Solving

1. Click [icon]. Choose Red Level F. Click [OK] and then [ON DUTY]. Find the value of each addition or subtraction expression.

2. Click [icon]. Choose Red Level M. Click [OK] and then [ON DUTY]. Find the value of each division expression.

Click [calculator]. Use the calculator to find the value of the expressions.

3. $24 + {}^-18$
   6
4. $34 + {}^-50$
   ⁻16
5. $^-7 + {}^-21$
   ⁻28
6. $14 - 51$
   ⁻37
7. $^-17 - 5$
   ⁻22
8. $^-22 - {}^-32$
   10
9. 34×6
   204
10. $18 \times {}^-12$
    ⁻216
11. $^-21 \times 7$
    ⁻147
12. $168 \div 12$
    14
13. $^-128 \div 4$
    ⁻32
14. $^-144 \div {}^-9$
    16

Write and evaluate the expression.

15. The amount of dog food in the bag changes by ⁻8 oz a day. What is the total change after 2 weeks? **14 × ⁻8; ⁻112 oz**

16. The high temperature was 12°F. The low temperature was ⁻5°F. What was the range of the temperatures? **12 − ⁻5; 17°F**

Multimedia Math Glossary   www.harcourtschool.com/mathglossary

17. **Vocabulary** Locate *additive inverse* in the Multimedia Math Glossary. Use a model to represent the sum of 4 and its additive inverse.

Chapters 11–12 **267**

TECHNOLOGY LINKUP

Objective To use Mighty Math software to find the values of expressions involving positive and negative integers

USING THE PAGE

Review with students the rules for evaluating expressions involving positive and negative integers.

Using Mighty Math Astro Algebra

Students who are still conceptualizing the idea of positive and negative integers may need to begin with Level A, Add Integers on a Number Line. Then they can move on to Levels F, G, and M as they are ready.

Practice and Problem Solving

Be sure students are using the correct operations. You may wish to have them state whether the expression will have a positive or negative value before they use the calculator to solve.

Multimedia Math Glossary

Frequency table, and all other vocabulary words in this unit, can be found in the Harcourt Multimedia Math Glossary.
www.harcourtschool.com/mathglossary

Algebra: Integers **267**

UNIT 4

PROBLEM SOLVING
On Location in Kentucky

Purpose To provide additional practice for concepts and skills in Chapters 11–12

USING THE PAGE
Temperatures

- After Exercise 1, have students extend their thinking.

  Suppose the temperature rose 0.5°F the first hour, 1°F the second hour, and 1.5°F the third hour after the record low was recorded in Shelbyville. If this pattern continued for two more hours, what would the temperature be at the end of the 5-hour period? ⁻29.5°F

- After Exercise 3, refer students again to the record temperature table.

  Is the difference between the record high and record low temperatures greater in February or March? February

- After Exercise 7, have students examine the wind-chill table again.

  Describe any patterns you see in the first two rows of the table. Possible answer: A 5-mph wind makes the temperature feel about 3 degrees cooler; a 10 mph wind makes the temperature feel about 15 degrees cooler.

Extension Have students discuss ways that they could present the data in the wind-chill table using a graph. Then have groups of students prepare graphs. Possible answer: Use a multiple line graph to show the data, one line for each wind speed. Check students' graphs.

PROBLEM SOLVING ON LOCATION
In Kentucky

Temperatures

Kentucky has experienced some extreme weather in the past 100 years. In 1930, the temperature hit 114°F in Greensburg. More recently, Shelbyville residents braved temperatures of ⁻37°F on January 19, 1994, following a blizzard that closed state highways for two days.

1. Find the difference between Kentucky's record high temperature and record low temperature. **151°F**

Use Data For 2–3, use the table.

2. By how many degrees did Kentucky's record high February temperature exceed its record low February temperature? **118°F**

3. Find the difference between Kentucky's record high March temperature and record high February temperature. **8°F**

4. Here's a useful rule for estimating temperature change: temperature decreases by about 3.6°F for every 1,000 feet that you climb in elevation. Suppose you hiked 2,500 feet up Black Mountain. By about how many degrees does the temperature decrease as you hike? **about 9°F**

Monthly Record Kentucky Temperatures		
	February	March
Low	⁻32°F	⁻14°F
High	86°F	94°F

Because of the "wind-chill factor," air temperatures feel colder when the wind is blowing. The stronger the wind, the colder it feels. The table below shows how much colder it feels.

Use Data For 5–7, use the table.

5. Suppose the wind is blowing at 10 mph. How many degrees colder will the wind make it feel if the air temperature is 30°F? **14°F**

6. Suppose the wind is blowing at 20 mph. How many degrees colder will the wind make it feel if the air temperature is 25°F? **28°F**

Wind-Chill Factor	Air Temperature (°F)			
Wind speed	35	30	25	20
5 mph	32	27	22	16
10 mph	22	16	10	3
15 mph	16	9	2	⁻5
20 mph	12	4	⁻3	⁻10

7. Describe any pattern you see in the row for 20 mph. **Possible answer: When there is a 20 mph wind, for each 5° drop in temperature, it feels like the temperature has dropped by 7–8°.**

Elevations

Being a hilly and mountainous state, all of Kentucky lies above sea level. Its highest point is Black Mountain in Harlan County, 4,139 feet above sea level. Its lowest point is along the Mississippi River in Fulton County, 257 feet above sea level. The lowest point in the United States is Death Valley, California, where the elevation is 282 feet below sea level ($^-282$). Closer to Kentucky, New Orleans, Louisiana, has an elevation of 8 feet below sea level ($^-8$).

Use Data For 1–6, use the diagram.

1. What is the difference in elevation between the lowest point in Kentucky and the highest point in Kentucky? **3,882 ft**

2. What is the difference in elevation between Death Valley and the top of Black Mountain? **4,421 ft**

3. What is the difference in elevation between Death Valley and the lowest point in Kentucky? **539 ft**

4. About how many times the elevation of the Mississippi River in Fulton County is the elevation of Black Mountain? **about 16 times**

5. About how many times the elevation of New Orleans is the elevation of Death Valley? **about 35 times**

6. **REASONING** Evaluate this statement: "Death Valley is about as far below sea level as the Mississippi River is above sea level in Fulton County, Kentucky." **The statement is fair. The absolute values of the two elevations, 257 and 282, are reasonably close to each other.**

Black Mountain, Kentucky's highest point

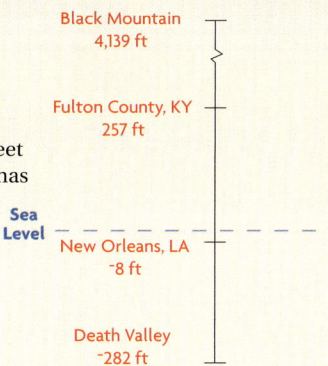

Black Mountain 4,139 ft
Fulton County, KY 257 ft
Sea Level
New Orleans, LA $^-8$ ft
Death Valley $^-282$ ft

USING THE PAGE

Elevations

• Direct students' attention to the elevation diagram.

Suppose the Mississippi River is 85 feet deep near Fulton County. What is the elevation of the riverbed at that point? 172 ft

At its deepest point, the Mississippi River is about 200 feet deep. This occurs near New Orleans. Estimate the elevation of the Mississippi riverbed at this point. $^-208$ ft

Write and solve an equation to find the difference in elevation between New Orleans and Death Valley. Possible equation: $d = {^-8} - {^-282}$; $d = 274$ ft

Extension Remind students that temperature decreases by about 3.6°F for every 1,000 ft in elevation that you climb. Have them suppose that they start a 6,000-ft mountain hike at 8 A.M. where the temperature is 60°F. They will climb 1,500 feet per hour. Direct students to make a table that shows the temperature each hour during the climb up the mountain. Temperature decreases by about 5.4°F each hour. 8 A.M.: 60°F; 9 A.M.: 54.6°F; 10 A.M.: 49.2°F; 11 A.M.: 43.8°F; 12 noon: 38.4°F. Check students' work.

Student Handbook

Troubleshooting H2

Prerequisite Skills Review Do you have the math skills needed to start a new chapter? Use this list of skills to review and remember your skills from last year.

Skill	Page
Properties	H2
Represent Decimals	H2
Write and Read Decimals	H3
Compare and Order Whole Numbers	H3
Round Whole Numbers and Decimals	H4
Whole-Number Operations	H4
Remainders	H5
Read a Table	H5
Mean, Median, Mode, and Range	H6
Read Bar Graphs	H6
Read Stem-and-Leaf Plots	H7
Prime and Composite Numbers	H7
Factors and Multiples	H8
Model Fractions	H8
Model Percents	H9
Simplify Fractions	H9
Add and Subtract Like Fractions	H10
Round Fractions	H10
Mental Math and Equations	H11
Fractions and Mixed Numbers	H11
Compare Fractions	H12
Understand Integers	H12
Number Lines	H13
Multiplication and Division Facts	H13
Classify Lines	H14
Patterns	H14
Exponents	H15
Order of Operations	H15
Add and Subtract	H16
Inverse Operations	H16

Skill	Page
Words for Operations	H17
Words and Equations	H17
Evaluate Expressions	H18
Classify Angles	H18
Name Angles	H19
Identify Polygons	H19
Identify Solid Figures	H20
Faces, Edges, and Vertices	H20
Write Equivalent Fractions	H21
Solve Multiplication Equations	H21
Congruent and Similar Figures	H22
Write Fractions as Decimals	H22
Multiply with Fractions and Decimals	H23
Fractions, Decimals, and Percents	H23
Certain, Impossible, Likely, Unlikely	H24
Analyze Data	H24
Customary Units	H25
Metric Units	H25
Solve Proportions	H26
Perimeter	H26
Change Units	H27
Find the Square and Cube of a Number	H27
Areas of Squares, Rectangles, and Triangles	H28
Areas of Circles	H28
Compare Numbers	H29
Function Tables	H29
Slides, Flips, and Turns	H30
Line Symmetry	H30
Measure Angles	H31
Ordered Pairs	H31

Extra Practice H32

When learning new skills, it may help you to practice a little more. On these pages you'll find extra practice exercises so you can be sure of your new skills!

Sharpen Your Test-Taking Skills H62

Before a test, sharpen your test-taking skills by reviewing these pages. Here you can find tips, such as how to get ready for a test, how to understand the directions, and how to keep track of time.

Skills Review H66

Review addition, subtraction, multiplication, and division of whole numbers and decimals to improve your skills.

Glossary H69

This glossary will help you speak and write the language of mathematics. Use the glossary to check the definitions of important terms.

Selected Answers H79

Check your answers to part of an assignment to see how well you're doing. If you see that you need help, you can review the lesson before completing the assignment.

Index H94

Use the index when you want to review a topic. It lists the page numbers where the topic is taught.

Table of Measures .... Back Cover

All the important measures used in this book are in these tables. If you've forgotten exactly how many feet are in a mile, these tables will help you.

Student Handbook **H1**

Troubleshooting

Properties
The charts list the basic properties of addition and multiplication.

Addition

PROPERTY	EXAMPLE WITH NUMBERS	EXAMPLE WITH VARIABLES
Commutative	$3 + 7 = 7 + 3$	$a + b = b + a$
Associative	$(4 + 5) + 2 = 4 + (5 + 2)$	$(a + b) + c = a + (b + c)$
Identity Property of Zero	$9 + 0 = 9$ and $0 + 9 = 9$	$a + 0 = a$ and $0 + a = a$

Multiplication

PROPERTY	EXAMPLE WITH NUMBERS	EXAMPLE WITH VARIABLES
Commutative	$8 \times 6 = 6 \times 8$	$a \times b = b \times a$
Associative	$(2 \times 9) \times 5 = 2 \times (9 \times 5)$	$(a \times b) \times c = a \times (b \times c)$
Identity Property of One	$6 \times 1 = 6$ and $1 \times 6 = 6$	$a \times 1 = a$ and $1 \times a = a$
Property of Zero	$7 \times 0 = 0$ and $0 \times 7 = 0$	$a \times 0 = 0$ and $0 \times a = 0$
Distributive	$3 \times (5 + 7) = (3 \times 5) + (3 \times 7)$	$a \times (b + c) = (a \times b) + (a \times c)$

Practice
Name the property shown.
1. $0 \times 12 = 0$ — Multiplication Property of Zero
2. $(8 \times 21) \times 5 = 8 \times (21 \times 5)$ — Associative of Multiplication
3. $72 \times 5 = 5 \times 72$ — Commutative of Multiplication
4. $4 \times (3 + 2) = (4 \times 3) + (4 \times 2)$ — Distributive
5. $9 + 0 = 9$ — Identity Property of Zero
6. $k + 7 = 7 + k$ — Commutative of Addition
7. $(m + 4) + h = m + (4 + h)$ — Associative of Addition
8. $c \times (2 + 3) = (c \times 2) + (c \times 3)$ — Distributive
9. $1 \times x = x$ — Identity Property of One

Represent Decimals
You can use decimal squares to represent decimals. A decimal square is divided into 100 parts. Each part represents 1 hundredth of the whole, or 0.01. Count the number of shaded parts. Write this number as hundredths.

24 out of 100 small squares are shaded.
0.24 of the whole is shaded.

Practice
Write the decimal that is represented.
1. 0.30
2. 0.42
3. 0.07
4. 0.80

H2 Prerequisite Skills Review

Write and Read Decimals
A place-value chart can help you write and read numbers. Each three-digit group, such as ones or millions, is called a **period**.

Read: "thirty million, one hundred twenty-eight thousand, five hundred ninety-seven and forty-six thousandths."

MILLIONS			THOUSANDS			ONES					
Hundreds	Tens	Ones	Hundreds	Tens	Ones	Hundreds	Tens	Ones	Tenths	Hundredths	Thousandths
	3	0	1	2	8	5	9	7	0	4	6

Example
Name the place value of the digit, 8, in the chart.
To write the value of the 8 in the chart, multiply the digit times the value of the place-value position.
Think: $8 \times 1{,}000$, or 8,000.

Practice
Name the place value of the digit 4.
1. 327,489,223.78 — hundred thousands
2. 198,238,042.08 — tens
3. 149,678,935.91 — ten millions
4. 728,035.84 — hundredths

Give the value of the blue digit.
5. 538.92 — 8
6. 82,901,733.006 — 80,000,000
7. 29.35 — 0.05
8. 125,674.173 — 0.003

Compare and Order Whole Numbers
You can compare and order numbers by comparing the digits in each place-value position.

Example
Write <, >, or = to compare the numbers. 14,675 ● 14,228

Step 1 Compare the ten thousands.
14,675
↓ same number of
14,228 ten thousands

Step 2 Compare the thousands.
14,675
↓ same number of
14,228 thousands

Step 3 Compare the hundreds.
14,675
↓ 6 > 2
14,228 So, 14,675 > 14,228.

Practice
Write <, >, or = to compare the numbers.
1. 3,919 ● 3,991 <
2. 188,937 ● 189,066 <
3. 70,001 ● 70,001 =

Order the numbers from greatest to least.
4. 9,774; 9,718; 9,762 — 9,774; 9,762; 9,718
5. 82,056; 82,856; 81,978 — 82,856; 82,056; 81,978
6. 23,091; 23,910; 23,109 — 23,910; 23,109; 23,091

Student Handbook H3

Troubleshooting

Round Whole Numbers and Decimals
Follow these steps to round a number to a given place.

Example
Round 168,279 to the nearest ten thousand.

Step 1 Find the digit in the place being rounded, the ten thousands place.
16**8**,279

Step 2 Look at the next digit to the right. If it is 5 or greater, round up. Otherwise, round down.
16**8**,279 8 > 5, so round up.

Step 3 Change to zero each digit to the right of the place being rounded.
168,279 → 170,000

Round to the nearest whole number.
1. 7.97 — 8
2. 15.58 — 16

Round to the nearest thousand.
3. 3,378 — 3,000
4. 7,607 — 8,000

Round to the nearest ten thousand.
5. 530,410 — 530,000
6. 12,677 — 10,000

Round to the nearest tenth.
7. 16.53 — 16.5
8. 2.96 — 3.0

Whole-Number Operations
When adding, subtracting, multiplying, and dividing whole numbers, align digits that have the same place value.

Examples
A. Find the sum. $247 + 1{,}496 + 89$

```
  22    Think: 7 + 6 + 9 = 22
  247   Regroup 22 ones as
1,496   2 tens 2 ones.
+  89   Add the other columns
1,832   in a similar way.
```

B. Find the difference. $31 - 12$

```
  2 11  Regroup when necessary.
  3̸1̸
- 1 2
  1 9
```

C. Find the product. 27×86

```
   27    Multiply by the 6 ones.
 × 86    Multiply by the 8 tens.
  162    Add the products.
+2 160
 2,322
```

D. Find the quotient. $146 \div 9$

```
   16 r2   Divide the 14 tens.
9)146      Multiply and subtract.
 - 9       Bring down the 6 ones.
   56      Divide the 56 ones.
 - 54      Multiply and subtract.
    2      Write the remainder.
```

Practice
Add, subtract, multiply, or divide.
1. $79 + 56 + 99$ — 234
2. $345 - 26$ — 319
3. 67×76 — 5,092
4. $376 \div 15$ — 25 r1
5. $700 - 388$ — 312
6. 19×203 — 3,857
7. $155 + 9 + 4{,}823$ — 4,987
8. $524 \div 31$ — 16 r28

H4 Prerequisite Skills Review

Remainders
When dividing whole numbers, you can write a remainder as a whole number, as a decimal, or as a fraction.

Example
Divide: $15 \div 4$. Write the remainder as a decimal and as a fraction.

Step 1 Write the dividend as a two-place decimal.
4)15.00

Step 2 Divide.
  3.75
4)15.00

Step 3 Write the decimal as a fraction in simplest form.
$0.75 = \frac{75}{100} = \frac{75 \div 25}{100 \div 25} = \frac{3}{4}$

$15 \div 4 = 3\frac{3}{4}$

Practice
Divide. Write the remainder as a decimal and as a fraction.
1. 4)30 — 7.5; $7\frac{1}{2}$
2. 8)50 — 6.25; $6\frac{1}{4}$
3. 10)3,315 — 331.5; $331\frac{1}{2}$
4. 12)2,442 — 203.5; $203\frac{1}{2}$
5. 6)21 — 3.5; $3\frac{1}{2}$
6. 8)94 — 11.75; $11\frac{3}{4}$
7. 20)74 — 3.7; $3\frac{7}{10}$
8. 25)210 — 8.4; $8\frac{2}{5}$

Read a Table
Use the title of a table to understand what the data represent. Use the labels to understand what the items represent.

Example
How far did Angie jog on Tuesday?

DAILY JOGGING RECORD (IN MILES)

Day	Sun	Mon	Tue	Wed	Thu	Fri	Sat
Milo	4	3	5	5	0	7	6
Angie	6	3	4	0	5	8	0

The number in the row marked "Angie" and the column marked "Tue" is 4. So, Angie jogged 4 miles on Tuesday.

Practice
Use the data in the table above to answer the questions.
1. How far did Milo jog on Friday? 7 mi
2. Which person jogged 5 miles on Wednesday? Milo
3. On which day did Angie jog 6 miles? Sun
4. On which day did Milo and Angie jog the same distance? Mon
5. How far did Angie jog on the two days that she jogged the same distance? 0 mi
6. Who ran the greater total distance during the week? Milo

Student Handbook H5

H2–H5 Troubleshooting

Troubleshooting

Mean, Median, Mode, and Range

Example

Find the mean, median, mode, and range for this set of data.
6, 19, 6, 9, 11, 15

Mean Find the sum of the data items. Divide the sum by the number of items.	6 + 19 + 6 + 9 + 11 + 15 = 66 mean = 66 ÷ 6 = 11
Median Arrange the items from least to greatest. The median is the middle value. If there are two middle values, the median is the average of the two values.	6, 6, **9, 11**, 15, 19 9 and 11 are the middle numbers. median = (9 + 11) ÷ 2 = 10
Mode Arrange the items from least to greatest. The mode is the value or values that repeat most often. If no value repeats, there is no mode.	**6, 6**, 9, 11, 15, 19 mode = 6
Range Arrange the items from least to greatest. The range is the difference of the greatest and least values.	**6**, 6, 9, 11, 15, **19** range = 19 − 6 = 13

Practice

Find the mean, median, mode, and range for each set of data.

1. 2, 1, 3, 5, 9 *4; 3; none; 8*
2. 6, 52, 41, 21, 35 *31; 35; none; 46*
3. 11, 15, 6, 11, 22 *13; 11; 11; 16*
4. 9, 5, 2, 5, 6, 1, 14 *6; 5; 5; 13*

Read Bar Graphs

A **bar graph** uses bars of different lengths to show and compare data.

Example

How many books did Anita read last month?

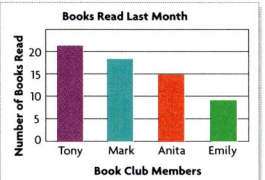

Find the bar representing Anita.
Look at the top of the bar. Read the scale at the left for the number the bar represents.
Anita read 15 books last month.

Practice

Use the bar graph to answer the questions.

1. How many books did Mark read? *18 books*
2. Who read 22 books? *Tony*
3. Who read the least number of books? *Emily*
4. How many more books did Tony read than Anita? *7 books*

H6 Prerequisite Skills Review

Read Stem-and-Leaf Plots

In a **stem-and-leaf plot**, data are organized by "stems," or the tens digits, and "leaves," or the ones digits. The stems and leaves are listed in order.

Example

Each entry in the stem-and-leaf plot gives the number of cards that were stacked in a house of cards when it collapsed. How many times were there from 40 to 49 cards when the collapse occurred?

Stem	Leaves
1	2 5 7 8 9
2	0 0 1 3 5 8
3	1 2 4 7
4	2 8
5	1 6

The numbers from 40 to 49 have the stem of 4, because the tens digit is 4. There are two such numbers in the table:
Stem 4, leaf 2 shows **42**.
Stem 4, leaf 8 shows **48**.
So, twice there were from 40 to 49 cards. Once there were 42 cards, and once there were 48.

Practice

Use the stem-and-leaf plot to answer the questions.

1. How many times were there from 10 to 19 cards when the collapse occurred? *6*
2. What was the greatest number of cards stacked? the least number? *56; 12*
3. How many houses of cards were built in this competition? *20*
4. What are the mode, median, and range of the data in the table? *mode: 18 and 20; median: 24; range: 44*

Prime and Composite Numbers

A **prime number** is a whole number that has exactly two factors, itself and 1. A **composite number** has more than two factors.

Examples

Decide whether the number is *prime* or *composite*.

A. 13 13 has exactly two factors, 13 and 1. These are the only whole numbers that divide 13 evenly. So, 13 is prime.

B. 12 12 has the factors 1, 2, 3, 4, 6, and 12. Each of these whole numbers divides 12 evenly. Since 12 has more than two factors, it is composite.

Practice

Decide whether the number is *prime* or *composite*.

1. 19 *prime*
2. 6 *composite*
3. 10 *composite*
4. 2 *prime*
5. 15 *composite*
6. 28 *composite*
7. 31 *prime*
8. 99 *composite*
9. 71 *prime*
10. 456 *composite*

Student Handbook H7

Troubleshooting

Factors and Multiples

Multiples of a number are the products that result when the number is multiplied by 0, 1, 2, 3, 4, and so on.
27 is a multiple of 9 because 27 = 9 × 3.

Factors are numbers that divide a whole number evenly.
8 is a factor of 32 because 8 divides 32 evenly: 32 ÷ 8 = 4.

Examples

A. List the next three multiples of 7: 7, 14, 21.

7, 14, 21, **28, 35, 42**
↑ ↑ ↑ ↑ ↑ ↑
7 × 1 7 × 2 7 × 3 7 × 4 7 × 5 7 × 6

So, the next three multiples are 28, 35, and 42.

B. Find all the factors of 20.

Think: 1 × 20 = 20, so 1 and 20 are factors of 20. 2 × 10 = 20, so 2 and 10 are factors of 20. 4 × 5 = 20, so 4 and 5 are factors of 20.

So, the factors of 20 are 1, 2, 4, 5, 10, and 20.

Practice

List the next three multiples of the number.

1. 5: 5, 10, 15 *20, 25, 30*
2. 9: 9, 18, 27 *36, 45, 54*
3. 6: 12, 18, 24 *30, 36, 42*
4. 20: 40, 60, 80 *100, 120, 140*

Find all of the factors of the number.

5. 13 *1, 13*
6. 12 *1, 2, 3, 4, 6, 12*
7. 30 *1, 2, 3, 5, 6, 10, 15, 30*
8. 24 *1, 2, 3, 4, 6, 8, 12, 24*
9. 15 *1, 3, 5, 15*
10. 22 *1, 2, 11, 22*
11. 28 *1, 2, 4, 7, 14, 28*
12. 36 *1, 2, 3, 4, 6, 9, 12, 18, 36*

Model Fractions

A **fraction** is a number that names a part of a whole or a part of a group.

Example

Write the fraction represented by the shaded part.

The group has 7 squares. The part that is shaded is 4 squares.
So, the fraction shaded is $\frac{4}{7}$.

Practice

Write the fraction represented by the shaded part.

1. 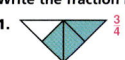 $\frac{3}{4}$
2.  $\frac{5}{9}$
3. $\frac{3}{8}$
4. $\frac{5}{6}$

H8 Prerequisite Skills Review

Model Percents

Percent (%) means "per hundred." You can use a hundred square to model percents.

Example

Write the percent for the shaded part of the square.

There are 100 parts, and 75 of the parts are shaded.
So, 75% of the parts are shaded.

Practice

Write the percent for the shaded part of the square.

1. 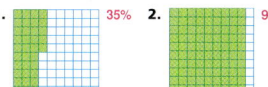 *35%*
2. 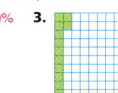 *90%*
3. 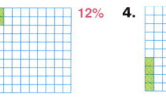 *12%*
4. 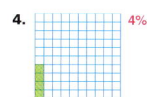 *4%*

Simplify Fractions

A fraction is in **simplest form** when the numerator and denominator have no common factors other than 1.

$\frac{10}{15}$ is not in simplest form, because 10 and 15 have the common factor 5.

$\frac{7}{12}$ is in simplest form, because 7 and 12 have no common factors other than 1.

You can write a fraction in simplest form by dividing the numerator and denominator by the greatest common factor (GCF).

Example

Write the fraction in simplest form. $\frac{16}{28}$

Step 1

Find the greatest common factor of 16 and 28.

factors of 16: 1, 2, **4**, 8, 16
factors of 28: 1, 2, **4**, 7, 14, 28

The *greatest* common factor is 4.

Step 2

Divide the numerator and the denominator by the GCF.

$\frac{16 \div 4}{28 \div 4} = \frac{4}{7}$ Since the GCF of 4 and 7 is 1, $\frac{4}{7}$ is in simplest form.

Practice

Write the fraction in simplest form.

1. $\frac{9}{15}$ *$\frac{3}{5}$*
2. $\frac{8}{12}$ *$\frac{2}{3}$*
3. $\frac{5}{20}$ *$\frac{1}{4}$*
4. $\frac{18}{6}$ *$\frac{3}{1}$, or 3*
5. $\frac{24}{32}$ *$\frac{3}{4}$*
6. $\frac{24}{30}$ *$\frac{4}{5}$*

Student Handbook H9

Troubleshooting

Add and Subtract Like Fractions

Like fractions are fractions that have the same denominator. To add or subtract like fractions, add or subtract the numerators. Keep the same denominator.

Example

Find the sum. $\frac{4}{9} + \frac{2}{9}$

Step 1
Add the numerators. Write the sum over the like denominator.
$\frac{4}{9} + \frac{2}{9} = \frac{4+2}{9} = \frac{6}{9}$

Step 2
If necessary, rewrite the fraction in simplest form.
$\frac{6}{9} = \frac{6 \div 3}{9 \div 3} = \frac{2}{3}$

Practice

Find the sum or difference. Write the answer in simplest form.

1. $\frac{1}{7} + \frac{4}{7}$ $\frac{5}{7}$
2. $\frac{4}{5} - \frac{2}{5}$ $\frac{2}{5}$
3. $\frac{3}{8} + \frac{3}{8}$ $\frac{3}{4}$
4. $\frac{3}{4} + \frac{1}{4}$ 1
5. $\frac{7}{10} - \frac{5}{10}$ $\frac{1}{5}$
6. $\frac{8}{12} - \frac{4}{12}$ $\frac{1}{3}$
7. $\frac{15}{11} - \frac{9}{11}$ $\frac{6}{11}$
8. $\frac{5}{9} + \frac{8}{9}$ $\frac{13}{9}$, or $1\frac{4}{9}$
9. $\frac{2}{10} + \frac{6}{10}$ $\frac{4}{5}$
10. $\frac{12}{8} - \frac{4}{8}$ 1
11. $\frac{3}{6} + \frac{1}{6}$ $\frac{2}{3}$
12. $\frac{5}{8} + \frac{1}{8}$ $\frac{3}{4}$
13. $\frac{7}{16} + \frac{5}{16}$ $\frac{3}{4}$
14. $\frac{17}{20} - \frac{9}{20}$ $\frac{2}{5}$
15. $\frac{11}{25} + \frac{4}{25}$ $\frac{3}{5}$

Round Fractions

To round a fraction to 0, $\frac{1}{2}$, or 1, compare the numerator with the denominator.

Examples

Round each fraction to 0, $\frac{1}{2}$, or 1.

A. $\frac{2}{11}$
Compared to 11, 2 is close to 0.
Round the fraction to 0.

B. $\frac{7}{15}$
7 is about half of 15.
Round the fraction to $\frac{1}{2}$.

C. $\frac{13}{16}$
13 is close to 16.
Round the fraction to 1.

Practice

Round each fraction to 0, $\frac{1}{2}$, or 1.

1. $\frac{3}{4}$ 1
2. $\frac{1}{5}$ 0
3. $\frac{2}{7}$ 0
4. $\frac{6}{13}$ $\frac{1}{2}$
5. $\frac{5}{12}$ $\frac{1}{2}$
6. $\frac{14}{15}$ 1
7. $\frac{4}{18}$ 0
8. $\frac{3}{7}$ $\frac{1}{2}$
9. $\frac{4}{5}$ 1
10. $\frac{12}{20}$ $\frac{1}{2}$
11. $\frac{11}{24}$ $\frac{1}{2}$
12. $\frac{17}{20}$ 1
13. $\frac{11}{18}$ $\frac{1}{2}$
14. $\frac{3}{25}$ 0
15. $\frac{15}{16}$ 1

H10 Prerequisite Skills Review

Mental Math and Equations

You can use mental math to solve equations. Try using a related equation to find the value of the variable.

Examples

Use mental math to solve the equation.

A. $k + 6 = 10$
Think: When 6 is added to k, the sum is 10. That means that $k = 10 - 6$.
So, $k = 4$. Check: $4 + 6 = 10$ ✔

B. $m - 4 = 11$
Think: When 4 is subtracted from m, the difference is 11. That means that $m = 11 + 4$.
So, $m = 15$. Check: $15 - 4 = 11$ ✔

C. $c \div 5 = 7$
Think: When c is divided by 5, the quotient is 7. That means that $c = 7 \times 5$.
So, $c = 35$. Check: $35 \div 5 = 7$ ✔

D. $h \times 4 = 36$
Think: When h is multiplied by 4, the product is 36. That means that $h = 36 \div 4$.
So, $h = 9$. Check: $9 \times 4 = 36$ ✔

Practice

Use mental math to solve the equation.

1. $g - 5 = 6$ 11
2. $r \times 2 = 12$ 6
3. $d \div 3 = 6$ 18
4. $x + 9 = 17$ 8
5. $v \div 6 = 5$ 30
6. $y - 6 = 9$ 15
7. $a + 3 = 15$ 12
8. $n \times 3 = 21$ 7

Fractions and Mixed Numbers

A **mixed number** is made up of a whole number and a fraction. You can write a mixed number as a fraction greater than 1, and a fraction greater than 1 as a mixed number.

Examples

A. Write $\frac{14}{3}$ as a mixed number.

$3\overline{)14}$ → 4 r2 Divide the numerator by the denominator.

$\frac{14}{3} = 4\frac{2}{3}$ Write the remainder as the numerator of a fraction.

B. Write $3\frac{4}{5}$ as a fraction.

$3 \times 5 = 15$ Multiply the denominator by the whole number.

$15 + 4 = 19$ Add the numerator.

$3\frac{4}{5} = \frac{19}{5}$ Write the sum over the denominator.

Practice

Write the fraction as a mixed number.

1. $\frac{13}{7}$ $1\frac{6}{7}$
2. $\frac{18}{5}$ $3\frac{3}{5}$
3. $\frac{10}{3}$ $3\frac{1}{3}$
4. $\frac{22}{9}$ $2\frac{4}{9}$
5. $\frac{17}{6}$ $2\frac{5}{6}$

Write the mixed number as a fraction.

6. $2\frac{4}{5}$ $\frac{14}{5}$
7. $1\frac{3}{8}$ $\frac{11}{8}$
8. $3\frac{3}{4}$ $\frac{15}{4}$
9. $2\frac{22}{25}$ $\frac{72}{25}$
10. $2\frac{11}{16}$ $\frac{43}{16}$

Student Handbook **H11**

Troubleshooting

Compare Fractions

To compare fractions with unlike denominators, rename the fractions so that they have like denominators. Then compare the numerators.

Examples

Compare. Use <, >, or = for ●.

A. $\frac{5}{8}$ ● $\frac{7}{8}$
The denominators are like. Compare the numerators.
$5 < 7$, so $\frac{5}{8} < \frac{7}{8}$.

B. $\frac{7}{10}$ ● $\frac{3}{5}$
The denominators are unlike. Rename one or both fractions.
$\frac{3 \times 2}{5 \times 2} = \frac{6}{10}$
$7 > 6$, so $\frac{7}{10} > \frac{3}{5}$.

C. $2\frac{5}{12}$ ● $2\frac{1}{3}$
The whole numbers are equal. Rename one or both fractions.
$\frac{1 \times 4}{3 \times 4} = \frac{4}{12}$
$5 > 4$, so $2\frac{5}{12} > 2\frac{1}{3}$.

Practice

Compare. Use <, >, or = for ●.

1. $\frac{4}{5}$ ● $\frac{3}{5}$ >
2. $\frac{11}{15}$ ● $\frac{13}{15}$ <
3. $1\frac{4}{6}$ ● $1\frac{2}{3}$ =
4. $2\frac{5}{8}$ ● $2\frac{1}{2}$ >
5. $\frac{2}{3}$ ● $\frac{5}{12}$ >
6. $\frac{7}{8}$ ● $\frac{3}{4}$ >
7. $\frac{5}{15}$ ● $\frac{2}{6}$ =
8. $\frac{8}{9}$ ● $\frac{2}{3}$ >
9. $\frac{13}{16}$ ● $\frac{7}{8}$ <
10. $\frac{5}{9}$ ● $\frac{14}{27}$ >

Understand Integers

The **integers** are the whole numbers and their opposites. Integers greater than 0 are **positive integers** and are found to the right of 0 on a number line. Integers less than 0 are **negative integers** and are found to the left of 0.

$\leftarrow$ -6 -5 -4 -3 -2 -1 0 1 2 3 4 5 6 $\rightarrow$

Example

Name the temperatures indicated by point A and point B.
The temperature at point A is $^-35°$. The temperature at point B is $^+68°$.

Practice

Give four examples of each set of numbers. *Possible answers are given.*

1. whole numbers 0, 1, 2, 3
2. negative integers $^-1, ^-2, ^-3, ^-4$

Write a positive or negative integer for each situation.

3. thermometer point C $0°F$
4. thermometer point D $^+105°F$
5. thermometer point E $^-7°F$
6. thermometer point F $^+32°F$
7. 35 ft above sea level $^+35$
8. a loss of $5 $^-5$

H12 Prerequisite Skills Review

Number Lines

You can use a number line to graph numbers and to compare and order numbers. If two numbers are graphed on a number line, the number to the right is greater.

Example

Graph $^-3$, 2, and $^-1.5$ on the number line. Then order the numbers from least to greatest.

Graph each number by placing a dot on the number line.

$\leftarrow$ -5 -4 -3 -2 -1 0 1 2 3 4 5 $\rightarrow$

Notice that $^-1.5$ appears halfway between $^-1$ and $^-2$. Since values increase as you move to the right on a number line, the order of the numbers from least to greatest is $^-3, ^-1.5, 2$.

Practice

Name the number graphed by each point.

A E   C B   D
$\leftarrow$ -5 -4 -3 -2 -1 0 1 2 3 4 5 $\rightarrow$

1. A $^-5$
2. B 0
3. C $^-1$
4. D 3.5
5. E $^-3.5$

Graph the numbers on a number line. Then order the numbers from least to greatest. *Check students' number lines.*

6. 4, 2, $^-1$ $^-1$, 2, 4
7. $^-1, ^-3, 0$ $^-3, ^-1, 0$
8. 1.5, 2, $^-1.5$ $^-1.5, 1.5, 2$
9. 3, $^-3, ^-4$ $^-4, ^-3, 3$
10. $^-3.5, ^-3, ^-5$ $^-5, ^-3.5, ^-3$

Multiplication and Division Facts

To multiply a number by a power of 10, move the decimal point one place to the right for each zero. To divide a number by a power of 10, move the decimal point one place to the left for each zero.

If you forget a multiplication fact that has 11 or 12 as a factor, think of 11 as 10 + 1, and think of 12 as 10 + 2.

Example

Find the product. 12×8

$12 \times 8 = (10 + 2) \times 8 = (10 \times 8) + (2 \times 8) = 80 + 16 = 96$

Practice

Find the product or quotient.

1. 56×10 560
2. $780 \div 10$ 78
3. 4.3×100 430
4. $5,280 \div 1,000$ 5.28
5. $2.61 \times 1,000$ 2,610
6. 0.48×100 48
7. $124 \div 10$ 12.4
8. $5.77 \div 100$ 0.0577
9. 11×10 110
10. 12×9 108

Student Handbook **H13**

Troubleshooting

Classify Lines
Parallel lines are lines in a plane that are always the same distance apart. **Intersecting lines** cross at exactly one point. **Perpendicular lines** intersect to form 90° angles, or right angles.

Examples
Classify the lines.

A.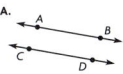

Line AB and line CD are in the same plane and are always the same distance apart. The lines are *parallel*.
$\overleftrightarrow{AB} \parallel \overleftrightarrow{CD}$ (∥ means "is parallel to.")

B.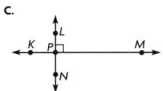

Line EH and line GF cross at exactly one point, J. The lines are *intersecting*.

C.

Lines KM and LN intersect at P to form right angles. The lines are *perpendicular* and *intersecting*.
$\overleftrightarrow{KM} \perp \overleftrightarrow{LN}$ (⊥ means "is perpendicular to.")

Practice
Classify the lines.

1. 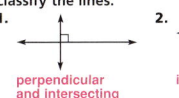 perpendicular and intersecting
2.  intersecting
3. 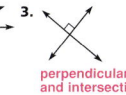 perpendicular and intersecting
4. 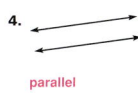 parallel

Patterns
Look for a rule for the pattern. Then use it to extend the pattern.

Example
Find the next three possible numbers in the pattern. 1, 2, 4, 8

Step 1 Find a rule.
Each number in the series is 2 times the number before it.
$1 \times 2 = 2; 2 \times 2 = 4; 4 \times 2 = 8$
The rule is "multiply by 2."

Step 2 Use the rule.
Multiply each number by 2 to find the next number:
$8 \times 2 = 16, 16 \times 2 = 32, 32 \times 2 = 64$
So, the next three numbers are 16, 32, and 64.

Practice
Find the next three possible numbers in the pattern. Write the rule.

1. 10, 14, 18, 22, 26 30, 34, 38; add 4
2. 729, 243, 81, 27 9, 3, 1; divide by 3
3. 1, 4, 16 64, 256, 1,024; multiply by 4
4. 63, 55, 47, 39 31, 23, 15; subtract 8
5. 49, 37, 25, 13 1, −11, −23; subtract 12
6. $2\frac{1}{2}, 1\frac{3}{4}, 1, \frac{1}{4}$ $-\frac{1}{2}, -1\frac{1}{4}, -2$; subtract $\frac{3}{4}$

H14 Prerequisite Skills Review

Exponents
You can use exponents to express powers of numbers. An **exponent** tells how many times the **base** is used as a factor.

$6 \times 6 \times 6 \times 6 = 6^4$

The exponent 4 shows that the base 6 is used 4 times as a factor.

Examples
Find the value of 5^3.
A. $5^3 = 5 \times 5 \times 5 = 125$

Write 49 by using an exponent.
B. $49 = 7 \times 7 = 7^2$

Practice
Find the value.

1. 2^4 16 **2.** 4^3 64 **3.** 5^2 25 **4.** 10^3 1,000 **5.** 3^4 81

Write by using an exponent.

6. 9 3^2 **7.** 32 2^5 **8.** 36 6^2 **9.** 100 10^2 **10.** 27 3^3

Order of Operations
When more than one operation is used in an expression, follow these rules to evaluate the expression.

RULES FOR ORDER OF OPERATIONS
1. First, do the operations **in parentheses**.   3. Next, **multiply and divide** from left to right.
2. Next, evaluate **exponents**.   4. Finally, **add and subtract** from left to right.

Examples
Evaluate each expression.

A. $3^2 + 5 \times 2$
1. There are no parentheses.
2. Evaluate exponents. $3^2 + 5 \times 2 = 9 + 5 \times 2$
3. Multiply. $9 + 5 \times 2 = 9 + 10$
4. Add. $9 + 10 = 19$

B. $(21 - 6) \div 3$
1. Evaluate parentheses. $(21 - 6) \div 3 = 15 \div 3$
2. There are no exponents.
3. Divide. $15 \div 3 = 5$
4. There is no addition or subtraction.

Practice
Evaluate each expression.

1. $4 + 6 \times 9$ 58 **2.** $(5 + 4) \times 3$ 27 **3.** $25 - 12 \times 3$ −11 **4.** $2^2 + 3^2$ 13
5. $(2 + 3)^2$ 25 **6.** $\frac{10 - 4}{2} \times 4^2$ 48 **7.** $5^2 \div (1^5 + 4)$ 5 **8.** $2 + 3 \times 3 \times 3$ 29

Student Handbook H15

Troubleshooting

Add and Subtract
To add or subtract unlike fractions, write equivalent fractions with common denominators. Then add or subtract the like fractions.

To add or subtract decimals, align the decimal points. Write equivalent decimals. Then add or subtract as you would with whole numbers.

Examples
Find the sum. $\frac{1}{6} + \frac{1}{2}$

A. $\frac{1}{6} + \frac{1}{2}$ Write equivalent fractions with a denominator of 6.
$\frac{1 \times 3}{2 \times 3} = \frac{3}{6}$
$\frac{1}{6} + \frac{3}{6} = \frac{4}{6}$, or $\frac{2}{3}$ Add the numerators and simplify.

Find the difference. $34.7 - 3.651$

B.   34.700   Align the decimal points.
  − 3.651   Write an equivalent decimal.
  31.049   Subtract as for whole numbers.

Practice
Find the sum or difference.

1. $3.12 + 2.7$ 5.82 **2.** $1.197 - 1.09$ 0.107 **3.** $6 - 3.4$ 2.6 **4.** $2.17 + 141.8$ 143.97
5. $\frac{3}{4} + \frac{1}{6}$ $\frac{11}{12}$ **6.** $\frac{5}{8} - \frac{3}{16}$ $\frac{7}{16}$ **7.** $\frac{3}{5} + \frac{5}{4}$ $\frac{17}{20}$ **8.** $\frac{1}{3} + \frac{5}{12}$ $\frac{3}{4}$

Inverse Operations
Addition and subtraction are **inverse** operations. This means that you can check a sum or difference by using the inverse operation. You can also check a product or quotient by using the inverse operation.

Examples
Use the inverse operation to check the answer.

A. 12 27
   + 15 − 15
   27 12

B. 322 142
   − 180 + 180
   142 322

C. 12 12
   × 9 9)108
   108

D. 4 14
   14)56 × 4
   56

Practice
Use the inverse operation to check the answer. Possible answers are given.

1. $23 + 15 = 38$ 38 − 15 = 23
2. $57 - 31 = 26$ 26 + 31 = 57
3. $8 \times 7 = 56$ 56 ÷ 7 = 8
4. $144 \div 6 = 24$ 6 × 24 = 144
5. $14 \times 3 = 42$ 42 ÷ 3 = 14
6. $366 - 218 = 148$ 148 + 218 = 366
7. $586 + 255 = 841$ 841 − 255 = 586
8. $250 \div 5 = 50$ 5 × 50 = 250

H16 Prerequisite Skills Review

Words for Operations
Many different words and phrases can be used in numerical and algebraic expressions to represent the operations of addition, subtraction, multiplication, and division.

Examples
Write the operation described by the phrase.

A. the sum of 6 and 7
To find a sum means to add. The operation is addition.

B. the product of k and 15
To find a product means to multiply. The operation is multiplication.

Practice
Write the operation described by the phrase.

1. k greater than 75 addition
2. 42 less than m subtraction
3. the sum of h and 6 addition
4. the quotient of 11 and b division
5. c times w multiplication
6. the difference of a and 45 subtraction
7. 9 increased by 7 addition
8. the product of 12 and p multiplication
9. 2 reduced by m subtraction

Words and Equations
You can write equations to represent some sentences. Use a variable to represent what is unknown in the sentence. Use operation signs or other mathematical signs to represent words or phrases in the sentence.

Example
Write an equation for this sentence: A number increased by 7 is 12.

Represent "a number" by a variable such as x.
Represent "increased by" with a plus sign.
Represent "is" with an equal sign.

So, the equation is $x + 7 = 12$.

Practice
Write an equation for the sentence. *Variables may vary.*

1. The product of a number and 7 is 49. $y \times 7 = 49$
2. 8 less than a number is 50. $m - 8 = 50$
3. 12 and a number have a quotient of 2. $\frac{12}{b} = 2$
4. 13 times a number is 91. $13k = 91$
5. 7 more than a number is 12. $n + 7 = 12$
6. 16 divided by a number is $\frac{4}{5}$. $\frac{16}{x} = \frac{4}{5}$
7. A number reduced by 13 is 9. $m - 13 = 9$
8. 45 greater than a number is 26. $y + 45 = 26$

Student Handbook H17

Troubleshooting

Evaluate Expressions

An expression that includes one or more variables is called an **algebraic expression**. To evaluate algebraic expressions, replace the variables with the given numbers. Then evaluate as you would with a numerical expression. Follow the rules for the order of operations.

Example

Evaluate $3 + m \times 2$ for $m = 6$.
$3 + m \times 2 = 3 + 6 \times 2$ Replace m with 6.
$ = 3 + 12$ Multiply before adding.
$ = 15$ Add.

Practice

Evaluate the expressions for the given value of the variables.

1. $4m$ for $m = 9$ 36
2. $\frac{3}{4}p$ for $p = 8$ 6
3. $\frac{z}{24}$ for $z = 96$ 4
4. $t - 5$ for $t = 19$ 14
5. $6h$ for $h = 9$ 54
6. $5 + c$ for $c = -13$ -8
7. $72 \div k$ for $k = 6$ 12
8. $p - 14$ for $p = 27$ 13
9. $-8f$ for $f = -7$ 56

Classify Angles

You can classify angles by their sizes.

| An *acute* angle has a measure greater than 0° and less than 90°. | A *right* angle forms a square corner. It measures 90°. | An *obtuse* angle has a measure greater than 90° and less than 180°. | A *straight* angle has a measure of 180°. |

Practice

Classify each angle by stating whether it is *acute, obtuse, right,* or *straight*.

1. K obtuse
2. P straight
3. Z acute
4. N right
5.  V acute
6.  M obtuse
7.  Q right
8. 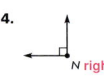 T acute

H18 Prerequisite Skills Review

Name Angles

You can name an angle by using one letter, three letters, or a number.

Examples

Name the angle formed by the blue rays.

Use the vertex letter to name the angle. $\angle G$

There are two angles with vertices at N. So, use three letters. $\angle ANV$ or $\angle VNA$

Use a number. $\angle 5$

Practice

Name the angle formed by the blue rays.

1. 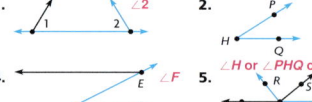 $\angle 2$
2.   $\angle H$ or $\angle PHQ$ or $\angle QHP$
3. 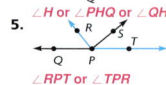 $\angle N$
4. $\angle F$
5. 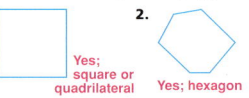 $\angle RPT$ or $\angle TPR$
6. 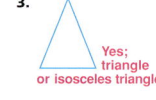 $\angle 2$

Identify Polygons

A **polygon** is a closed plane figure formed by three or more line segments. A polygon is classified by its number of sides.

Examples

State whether the figure is a polygon or not. If it is, classify it.

A. The figure is not closed, so it is not a polygon.

B. The figure is a 5-sided polygon called a pentagon.

Practice

State whether the figure is a polygon or not. If it is, classify it.

1. Yes; square or quadrilateral
2. Yes; hexagon
3. Yes; triangle or isosceles triangle
4. not a polygon

Student Handbook **H19**

Troubleshooting

Identify Solid Figures

A **polyhedron** is a solid figure with flat faces that are polygons. A **prism** is a polyhedron with two congruent, parallel **bases**. Its **lateral faces** are rectangles. A **pyramid** has a polygon for its base and triangles for its lateral faces.

A **cone** has a circular base. A **cylinder** has two congruent parallel circular bases. Cones and cylinders have curved lateral surfaces.

Practice

Identify each solid figure.

1. cube or square prism
2. triangular pyramid
3. cylinder
4. hexagonal prism
5. cone
6. pentagonal prism
7. hexagonal pyramid
8. triangular prism

Faces, Edges, and Vertices

Each polygon that forms a solid figure is a **face** of the figure. A line segment where two faces meet is an **edge**. A point where three or more edges meet is a **vertex**.

Example

Tell the number of faces, vertices, and edges of the cube.
The cube has 6 faces, 8 vertices, and 12 edges.

Practice

Give the number of faces, edges, and vertices of each solid figure.

1. 8; 18; 12
2. 7; 12; 7
3. 6; 10; 6
4. 8; 12; 6

H20 Prerequisite Skills Review

Write Equivalent Fractions

To write a fraction equivalent to a given fraction, multiply or divide the numerator and denominator by the same number.

Examples

Complete the number sentence to find an equivalent fraction.

A. $\frac{18}{24} = \frac{3}{\blacksquare}$

Think: To change 18 to 3, divide the numerator of $\frac{18}{24}$ by 6. Then divide the denominator by 6.

$\frac{18}{24} = \frac{18 \div 6}{24 \div 6} = \frac{3}{4}$

So, the equivalent fraction is $\frac{3}{4}$.

B. $\frac{3}{5} = \frac{\blacksquare}{20}$

Think: To change 5 to 20, multiply the denominator of $\frac{3}{5}$ by 4. Then multiply the numerator by 4.

$\frac{3}{5} = \frac{3 \times 4}{5 \times 4} = \frac{12}{20}$

So, the equivalent fraction is $\frac{12}{20}$.

Practice

Complete each number sentence to find an equivalent fraction.

1. $\frac{1}{2} = \frac{\blacksquare}{8}$ 4
2. $\frac{12}{18} = \frac{2}{\blacksquare}$ 3
3. $\frac{3}{4} = \frac{15}{\blacksquare}$ 20
4. $\frac{\blacksquare}{6} = \frac{25}{30}$ 5
5. $\frac{8}{24} = \frac{\blacksquare}{12}$ 4
6. $\frac{\blacksquare}{15} = \frac{30}{45}$ 10
7. $\frac{3}{5} = \frac{48}{\blacksquare}$ 80
8. $\frac{\blacksquare}{40} = \frac{7}{8}$ 35
9. $\frac{27}{36} = \frac{\blacksquare}{12}$ 9
10. $\frac{\blacksquare}{16} = \frac{8}{32}$ 4
11. $\frac{12}{\blacksquare} = \frac{6}{10}$ 20
12. $\frac{\blacksquare}{100} = \frac{11}{20}$ 55

Solve Multiplication Equations

To solve a multiplication equation, use an inverse operation. Divide both sides of the equation by the number that multiplies the variable.

Example

Solve and check. $8n = 40$

$8n = 40$
$\frac{8n}{8} = \frac{40}{8}$ Divide both sides by 8.
$n = 5$ Simplify.

Check: $8n \stackrel{?}{=} 40$
$8 \times 5 \stackrel{?}{=} 40$
$40 = 40$ ✔

Practice

Solve and check.

1. $6k = 18$ 3
2. $3h = 27$ 9
3. $5e = 75$ 15
4. $10d = 700$ 70
5. $120 = 8m$ 15
6. $24y = 108$ 4.5
7. $40 = 160b\frac{1}{4}$
8. $-1p = 12$ -12
9. $15n = 135$ 9
10. $4k = 30$ $7\frac{1}{2}$
11. $1.2\,t = 96$ 80
12. $-5b = 35$ -7
13. $198 = 18s$ 11
14. $9 = \frac{1}{2}p$ 18
15. $-7f = -98$ 14
16. $-84 = 14q$ -6

Student Handbook **H21**

Troubleshooting

Congruent and Similar Figures

Two figures are **similar** if they have the same shape. One may be an enlargement or a reduction of the other. Two figures are **congruent** if they have the same size and shape.

Examples

Tell if the figures in each pair appear to be congruent, similar, both, or neither.

A. The figures are neither congruent or similar.

B. The figures are similar.

C. Both; the figures are congruent and similar.

Practice

Tell if the figures in each pair appear to be congruent, similar, both, or neither.

1. similar
2. neither
3. both

Write Fractions as Decimals

To write a fraction as a decimal, divide the numerator by the denominator. Or, write an equivalent fraction with a denominator of 10, 100, or 1,000. Then rewrite the fraction as a decimal.

Example

Write $\frac{3}{4}$ as a decimal.

One Way

$4\overline{)3.00}$ Divide 3 by 4. So, $\frac{3}{4} = 0.75$.
-28
20
-20
0

Another Way

$\frac{3}{4} = \frac{3 \times 25}{4 \times 25} = \frac{75}{100}$ Write an equivalent fraction with a denominator of 100.

$\frac{75}{100} = 0.75$ Write the equivalent fraction as a decimal.

So, $\frac{3}{4} = 0.75$.

Practice

Write the fraction as a decimal.

1. $\frac{1}{2}$ 0.5
2. $\frac{7}{10}$ 0.7
3. $\frac{1}{4}$ 0.25
4. $\frac{3}{8}$ 0.375
5. $\frac{7}{20}$ 0.35
6. $\frac{2}{5}$ 0.4
7. $\frac{17}{25}$ 0.68
8. $\frac{43}{50}$ 0.86
9. $\frac{48}{250}$ 0.192
10. $\frac{19}{50}$ 0.38

Multiply with Fractions and Decimals

When multiplying decimals, the number of decimal places in the product is the total of decimal places in the two factors. When multiplying fractions, write the product of the numerators over the product of the denominators.

Examples

Find the product.

A.
2.57 2 decimal places
$\times 1.2$ 1 decimal place
514
$+ 2570$
3.084 $2 + 1 = 3$ decimal places

B. $\frac{3}{4} \times \frac{8}{9} = \frac{3 \times 8}{4 \times 9} = \frac{24}{36}$ Multiply numerators. Multiply denominators.

$\frac{24 \div 12}{36 \div 12} = \frac{2}{3}$ Simplify.

Practice

Find the product. Write the answer in simplest form. 3. 16.845; 4. 18.4275; 5. 395.5

1. 7.2×0.9 6.48
2. 8.4×1.1 9.24
3. 112.3×0.15
4. 17.55×1.05
5. 15.82×25
6. $\frac{1}{3} \times \frac{2}{3}$ $\frac{2}{9}$
7. $\frac{5}{6} \times \frac{3}{4}$ $\frac{5}{8}$
8. $\frac{2}{3} \times \frac{5}{12}$ $\frac{5}{18}$
9. $\frac{5}{8} \times \frac{3}{5}$ $\frac{3}{8}$
10. $\frac{1}{2} \times \frac{1}{4}$ $\frac{1}{8}$

Fractions, Decimals, and Percents

You can write a percent as a decimal and a decimal as a percent. You can write a percent as a fraction and a fraction as a percent.

Examples

A. Write 42% as a decimal.
$42\% = 42 \div 100$ Divide by 100.
So, $42\% = 0.42$.

B. Write 0.781 as a percent.
$0.781 = 0.781 \times 100$ Multiply by 100.
So, $0.781 = 78.1\%$.

C. Write 24% as a fraction.
$24\% = \frac{24}{100} = \frac{6}{25}$ Write the percent over 100. Simplify.
So, $24\% = \frac{6}{25}$.

D. Write $\frac{11}{25}$ as a percent.
$\frac{11}{25} = 11 \div 25 = 0.44$ Divide numerator by denominator.
So, $\frac{11}{25} = 44\%$. Write the percent.

Practice

Write each decimal and each fraction as a percent.
Write each percent as a decimal and as a fraction.

1. $\frac{2}{5}$ 40%
2. 10% 0.1, $\frac{1}{10}$
3. 0.34 34%
4. 25% 0.25, $\frac{1}{4}$
5. $\frac{1}{2}$ 50%
6. 0.81 81%
7. $\frac{7}{25}$ 28%
8. 60% 0.60, $\frac{3}{5}$
9. $\frac{18}{50}$ 36%
10. 0.02 2%

H22 Prerequisite Skills Review

Student Handbook H23

Troubleshooting

Certain, Impossible, Likely, Unlikely

An event is **certain** if it is sure to happen. It is **impossible** if it can never happen. It is **likely** if there is a strong chance that it will happen. It is **unlikely** if there is a strong chance that it will not happen.

Examples

Tell if the event is certain, impossible, likely, or unlikely.

A. $2 + 2$ will equal 5 next Tuesday.
$2 + 2$ can never equal 5.
So, the event is *impossible*.

B. You toss a penny 10 times and get at least 1 head.
The probability of getting a head is $\frac{1}{2}$ each time you toss a coin.
So, the event is *likely*.

Practice

Tell if the event is certain, impossible, likely, or unlikely.

1. Rain will fall at least once this year. likely
2. There will be no June next year. impossible
3. You toss a quarter and get either a head or a tail. certain
4. A person who does not know you will guess your phone number. unlikely

Analyze Data

A **line graph** shows changes over time. How many people attended Game 2?

The dot on the vertical "Game 2" line is beside 140. So, 140 people attended the game.

A **circle graph** shows data as parts of a whole. How many people bought peanuts at the fair?

The section of the circle labeled "Peanuts" shows 16. So, 16 people bought peanuts.

Practice

For 1–2, use the line graph.

1. How many people attended Game 4? 150
2. What was the total attendance at all the games? 740

For 3–4, use the circle graph.

3. What was the most popular snack? popcorn
4. Popcorn and one other snack were half the sales. What was the other snack? candied apples

H24 Prerequisite Skills Review

Customary Units

The table shows equivalents in the customary system of measurement. To change from a larger unit to a smaller one, multiply. To change from a smaller unit to a larger one, divide.

UNITS OF LENGTH	UNITS OF CAPACITY	UNITS OF WEIGHT
12 inches (in.) = 1 foot (ft)	2 cups (c) = 1 pint (pt)	16 ounces (oz) = 1 pound (lb)
3 feet = 1 yard (yd)	2 pints = 1 quart (qt)	2,000 pounds = 1 ton (T)
5,280 feet = 1 mile (mi)	4 quarts = 1 gallon (gal)	
1,760 yards = 1 mile		

Examples

Change to the given unit.

A. 24 in. = ■ ft **Think:** An inch is smaller than a foot. To change from inches to feet, divide by 12.
$24 \div 12 = 2$, so 24 in. = 2 ft.

B. 4 gal = ■ qt **Think:** A gallon is larger than a quart. To change from gallons to quarts, multiply by 4.
$4 \times 4 = 16$, so 4 gal = 16 qt.

Practice

Change to the given unit.

1. 2 yd = ■ ft 6
2. 4 c = ■ pt 2
3. 4 lb = ■ oz 64
4. 2.5 gal = ■ qt 10

Metric Units

The table shows how prefixes used with the basic units **meter (m)**, **liter (L)**, and **gram (g)** are related in the metric system. To change units, multiply by 10 for each place you move to the right. Divide by 10 for each place you move to the left.

kilo (k)	hecto (h)	deka (da)	BASIC UNIT	deci (d)	centi (c)	milli (m)

Examples

Change to the given unit.

A. 8 m = ■ cm **Think:** From meter to centimeter, the move is 2 places to the right. Multiply by 10×10, or 100.
$8 \times 100 = 800$, so 8 m = 800 cm.

B. 4,000 mg = ■ g **Think:** From milligrams to grams, the move is 3 places to the left. Divide by $10 \times 10 \times 10$, or 1,000.
$4,000 \div 1,000 = 4$, so 4,000 mg = 4 g.

Practice

Change to the given unit.

1. 2 m = ■ dm 20
2. 6 L = ■ mL 6,000
3. 3,000 g = ■ kg 3
4. 5,000 mg = ■ cg 500

Student Handbook H25

Troubleshooting

Solve Proportions

A **proportion** is a number sentence which states that two ratios are equal. You can solve a proportion by using the fact that in a proportion, the **cross products** are equal.

cross products: $2 \times 6 = 3 \times 4$

Example

Solve for n. $\frac{2}{3} = \frac{n}{12}$

$2 \times 12 = 3 \times n$ Use the cross products to write an equation.
$24 = 3 \times n$ Simplify.
$\frac{24}{3} = \frac{3 \times n}{3}$ Divide both sides of the equation by 3.
$8 = n$ Write the solution.

Practice

Solve for n.

1. $\frac{n}{3} = \frac{6}{9}$ $n = 2$
2. $\frac{4}{n} = \frac{5}{10}$ $n = 8$
3. $\frac{8}{10} = \frac{n}{15}$ $n = 12$
4. $\frac{20}{15} = \frac{12}{n}$ $n = 9$
5. $\frac{n}{6} = \frac{25}{30}$ $n = 5$
6. $\frac{6}{5} = \frac{9}{15}$ $n = 3$
7. $\frac{8}{n} = \frac{32}{16}$ $n = 4$
8. $\frac{48}{36} = \frac{n}{9}$ $n = 12$

Perimeter

Perimeter is the distance around a figure. To find the perimeter of a polygon, find the sum of the lengths of the sides.

Example

The opposite sides of a rectangle are congruent, so the sides of the figure that are not labeled measure 4 cm and 2 cm.
So, the perimeter is 12 cm. $P = 4\text{ cm} + 2\text{ cm} + 4\text{ cm} + 2\text{ cm}$

Practice

Find the perimeter of the figure.

1. 12 cm
2. 44 cm
3. 15.5 cm
4. 36 cm
5. 40 in.
6. 26 in.

H26 Prerequisite Skills Review

Change Units

When you change from one unit to another, you must decide whether there are more or fewer of the new unit. If there are more units, multiply. If there are fewer units, divide.

Examples

A rug is 4 yd long. How many feet long is this?

There will be *more* feet in the measurement than there are yards. Since 1 yd = 3 ft, multiply to find the answer.

$3 \times 4 = 12$. So, the rug is 12 ft long.

Practice

Change the measurement to the given unit.

1. a 3-ft table, to in. **36 in.**
2. a 5-gal tank, to qt **20 qt**
3. a 15-ft room, to yd **5 yd**
4. a 1,000-m race, to cm **10,000 cm**

Find the Square and Cube of a Number

The **square** of a number is the product of the number used twice as a factor. The **cube** of a number is the product of the number used three times as a factor.

Examples

A. Find the square of 5.

$5 \times 5 = 25$ Find the product, using 5 as a factor twice.

So, 5 squared is 25, or $5^2 = 25$.

B. Find the cube of 4.

$4 \times 4 \times 4 = 64$ Find the product, using 4 as a factor three times.

So, 4 cubed is 64, or $4^3 = 64$.

Practice

Find the square of each number.

1. 7 **49**
2. 14 **196**
3. 25 **625**
4. 80 **6,400**
5. 5.6 **31.36**

Find the cube of each number.

6. 2 **8**
7. 3 **27**
8. 5 **125**
9. 10 **1,000**
10. 0 **0**
11. 6 **216**
12. 11 **1,331**
13. $\frac{1}{2}$ **$\frac{1}{8}$**
14. $\frac{3}{5}$ **$\frac{27}{125}$**
15. 0.3 **0.027**

Student Handbook **H27**

Troubleshooting

Areas of Squares, Rectangles, and Triangles

The area of a rectangle or square is the product of the length and the width. The area of a triangle is *half* the product of the base and the height.

$A = l \times w$
$A = \frac{1}{2} \times b \times h$

Examples

Find the area of the figure.

A. 6 ft, 5 ft
B. 9 in., 9 in.
C. 8 cm, 12 cm

$A = 6 \times 5 = 30$, or 30 ft² | $A = 9 \times 9 = 9^2 = 81$, or 81 in.² | $A = \frac{1}{2} \times 12 \times 8 = 48$, or 48 cm²

Practice

Find the area of the figure.

1. 13 yd — **169 yd²**
2. 12 ft, 7 ft — **42 ft²**
3. 15 cm, 10.4 cm — **156 cm²**
4. $14\frac{1}{2}$ in., 4 in. — **29 in.²**

Areas of Circles

The area of a circle is the product of π and the square of the radius. Use 3.14 for the value of π.

$A = \pi r^2$

Examples

Find the area of the circle.

A. radius = 3 cm

$A \approx 3.14 \times 3^2$
$\approx 3.14 \times 9$
≈ 28.26 cm²

B. The radius is half of the diameter, or 8 in.

$A \approx 3.14 \times 8^2$
$\approx 3.14 \times 64$
≈ 200.96 in.²

Practice

Find the area of the circle. Use 3.14 for π.

1. 10 cm — **314 cm²**
2. 12 in. — **452.16 in.²**
3. 18 cm — **254.34 cm²**
4. 42 in. — **1,384.74 in.²**

H28 Prerequisite Skills Review

Compare Numbers

| To compare integers, use a number line. The number to the right on the line is greater. | To compare decimals, align the decimal points. Then compare the digits from left to right. | To compare fractions, rename unlike fractions, then compare the numerators. |

Examples

Compare. Write < or > for each ●.

A. 2 ● ⁻3

2 is to the right of ⁻3, so 2 > ⁻3.

B. 6.25 ● 6.179

6.25
6.179

$0.2 > 0.1$, so $6.25 > 6.179$.

C. $\frac{1}{2}$ ● $\frac{2}{3}$

$\frac{1 \times 3}{2 \times 3} = \frac{3}{6}$ $\frac{2 \times 2}{3 \times 2} = \frac{4}{6}$

$3 < 4$, so $\frac{1}{2} < \frac{2}{3}$.

Practice

Compare. Write < or > for each ●.

1. 485 ● 579 **<**
2. 3.03 ● 3.3 **<**
3. $\frac{1}{2}$ ● $\frac{1}{3}$ **>**
4. ⁻11 ● ⁻3 **<**
5. 5 ● ⁻4 **>**
6. 14.97 ● 14.9 **>**
7. 523.6 ● 532.8 **<**
8. 0.007 ● 0.01 **<**
9. ⁻0.5 ● ⁻0.4 **<**
10. $\frac{3}{4}$ ● $\frac{5}{8}$ **>**

Function Tables

To find an output value in a function table, replace the variable with an input value. Then evaluate the algebraic expression.

Example

Complete the function table.

k	$k + 4$
7	
9	
11	
16	

Replace k in $k + 4$ with 7: $k + 4 = 7 + 4 = 11$.
Replace k in $k + 4$ with 9: $k + 4 = 9 + 4 = 13$.
Replace k in $k + 4$ with 11: $k + 4 = 11 + 4 = 15$.
Replace k in $k + 4$ with 16: $k + 4 = 16 + 4 = 20$.

Input / Output

k	$k + 4$
7	11
9	13
11	15
16	20

Practice

Copy and complete the function table.

1.

x	$x + 7$
9	16
13	20
18	25
24	31

2.

b	$b \times 16$
2	32
5	80
8	128
13	208

3.

m	$m \div 6$
192	32
150	25
90	15
51	8.5

4.

r	$r - 7.4$
25.8	18.4
19.2	11.8
13.05	5.65
9.1	1.7

Student Handbook **H29**

Troubleshooting

Slides, Flips, and Turns

There are three ways that you can **transform** a figure. You can **slide,** or **translate,** the figure along a straight line. You can **flip,** or **reflect,** the figure over a line. Or you can **turn,** or **rotate,** the figure around a point.

Examples

Identify the transformation as a *slide, flip,* or *turn.*

A.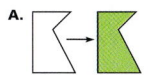
The figure has been translated along a straight line to the right. This is a *slide.*

B.
The figure has been reflected across a vertical line. This is a *flip.*

C.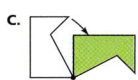
The figure has been rotated clockwise around a point. This is a *turn.*

Practice

Identify the transformation as a *slide, flip,* or *turn.*

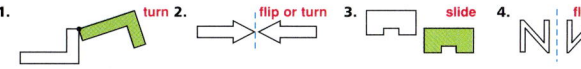

1. turn   2. flip or turn   3. slide   4. flip

Line Symmetry

A figure has **line symmetry** if it can be folded or reflected so that the two parts of the figure match, or are congruent. A figure can have more than one line of symmetry.

Examples

Is the dashed line a line of symmetry? Write *yes* or *no.*

A. 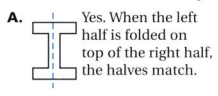 Yes. When the left half is folded on top of the right half, the halves match.

B. 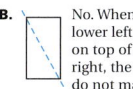 No. When the lower left is folded on top of the upper right, the halves do not match.

Practice

Is the dashed line a line of symmetry? Write *yes* or *no.*

1.  No
2. 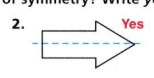 Yes
3. Yes
4. 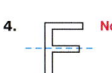 No

Measure Angles

To measure ∠*ABC* with a protractor, place the center of the protractor at *B*.

Align the 0° mark on the protractor with one ray of the angle.

Read the angle measure where the other ray passes through the scale on the protractor.

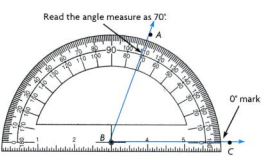
Read the angle measure as 70°.

Practice

Use a protractor to measure the angle.

1. 60°   2. 160°   3. 100°   4. 25°

Ordered Pairs

You can use two numbers called an **ordered pair** to locate a point on a grid. The first number tells you how far to move horizontally from (0,0). The second number tells you how far to move vertically from (0,0).

Example

Write the ordered pair for point *A*.

To reach point *A* from (0,0), go
⁺3 units right (horizontally) and
⁻2 units down (vertically).

So, the ordered pair is (3,⁻2).

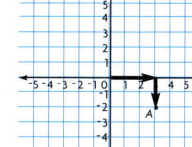

Practice

Write the ordered pair for each point.

1. point $A(^-3,3)$
2. point $C(^-2,0)$
3. point $D(^-4,^-2)$
4. point $E(^-2,^-4)$
5. point $F(0,^-4)$
6. point $G(2,^-2)$
7. point $I(4,2)$
8. point $J(2,5)$
9. point $B(^-1,2)$
10. point $H(4,^-2)$

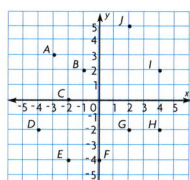

CHAPTER 1 Extra Practice

Set A (pp. 16–19)

Estimate. *Possible estimates are given.*

1. 589 + 342 **900**
2. 865 − 383 **500**
3. 8,926 + 1,674 **10,600**
4. 8,960 9,043 + 8,756 **27,000**
5. 3,406 − 792 **26,000**
6. 4,496 ÷ 51 **90**
7. 47 × 98 **5,000**
8. 321 × 43 **12,000**
9. 3,552 ÷ 58 **60**
10. 6,251 ÷ 12 **500**

Tell whether the estimate is an overestimate or underestimate.

11. 848 + 692 ≈ 1,550 **overestimate**
12. 398 × 66 ≈ 28,000 **overestimate**
13. 317 − 67 ≈ 230 **underestimate**

Set B (pp. 20–21)

Find the sum or difference.

1. 6,343 − 1,145 **5,198**
2. 19,826 − 3,517 **16,309**
3. 22,912 + 84,718 **107,630**
4. 532,047 + 36,603 **568,650**
5. 6,052 + 2,791 + 3,047 + 654 **12,544**
6. 7,380 − 1,890 − 408 − 792 **4,290**

Set C (pp. 22–25)

Multiply or divide, writing any remainder as a fraction.

1. 83 × 62 **5,146**
2. $16\overline{)869}$ **$54\frac{5}{16}$**
3. $45\overline{)1,675}$ **$37\frac{2}{9}$**
4. 458 × 194 **88,852**

Set D (pp. 28–29)

Write a numerical or algebraic expression for the word expression.

1. nine dollars less than twelve dollars **12 − 9**
2. thirteen more than a number, x **$x + 13$**
3. twenty-two dogs times two bowls **22 × 2**
4. fifty-two divided by two **52 ÷ 2**

Evaluate each expression.

5. $8 \times a$, for $a = 35$ **280**
6. $p − 38$, for $p = 97$ **59**
7. $175 ÷ k$, for $k = 5$ **35**
8. $96 + m$, for $m = 48$ **144**

Set E (pp. 30–31)

Determine which of the given values is the solution of the equation.

1. $5w = 70$; $w = 12, 13,$ or 14 **14**
2. $16 − a = 7$; $a = 8, 9,$ or 10 **9**
3. $48 ÷ x = 8$; $x = 6, 7,$ or 8 **6**

Solve each equation by using mental math.

4. $23 + c = 28$ **$c = 5$**
5. $k ÷ 9 = 5$ **$k = 45$**
6. $q − 25 = 25$ **$q = 50$**
7. $c \times 6 = 426$ **$c = 71$**

H32 Extra Practice

CHAPTER 2 Extra Practice

Set A (pp. 36–39)

Use mental math to find the value.

1. $26 + (4 + 18)$ **48**
2. $6 \times 4 \times 5$ **120**
3. $28 + 9 + 41$ **78**
4. $83 − 36$ **47**
5. $4 + 19 + 26$ **49**
6. 38×6 **228**
7. $47 − 29$ **18**
8. $7 + 22 + 13$ **42**
9. 51×8 **408**
10. $(6 \times 8) \times 5$ **240**
11. $19 + 31$ **50**
12. $92 − 29$ **63**
13. $65 + 17$ **82**
14. $10,000 ÷ 25$ **400**
15. $3 \times 10 \times 7$ **210**
16. $51 − 29$ **22**
17. $4 \times 8 \times 2$ **64**
18. $19 + 52 + 21$ **92**
19. Ben has three cousins under the age of 13. The product of their ages is 336. How old are they? **8, 7, and 6 years old**

Set B (pp. 40–41)

Write the equal factors. Then find the value.

1. 10^4 **10 × 10 × 10 × 10; 10,000**
2. 5^6 **5 × 5 × 5 × 5 × 5 × 5; 15,625**
3. 2^3 **2 × 2 × 2; 8**
4. 3^2 **3 × 3; 9**
5. 7^4 **7 × 7 × 7 × 7; 2,401**
6. 1^8 **1 × 1 × 1 × 1 × 1 × 1 × 1 × 1; 1**
7. 27^2 **27 × 27; 729**
8. 15^1 **15; 15**
9. 12^3 **12 × 12 × 12; 1,728**
10. 9^4 **9 × 9 × 9 × 9; 6,561**

Write in exponent form.

11. $3 \times 3 \times 3 \times 3$ **3^4**
12. 10×10 **10^2**
13. $5 \times 5 \times 5 \times 5 \times 5 \times 5$ **5^6**
14. $12 \times 12 \times 12$ **12^3**
15. $1 \times 1 \times 1 \times 1 \times 1$ **1^4**
16. $21 \times 21 \times 21 \times 21 \times 21$ **21^5**
17. Harold was hired to take a survey of 625 people. He wants to write this number in exponent form. If the base is 5, what is the exponent? How did you arrive at your answer? **4; explanations will vary**

Set C (pp. 44–45)

Give the correct order of operations.

1. $5 \times 7 + 3 ÷ 2$ **multiply, divide, add**
2. $(6 + 15) ÷ 3 \times 1 − 6$ **parentheses, divide, multiply, subtract**
3. $12^2 \times 10 − 10^3 − 100$ **exponents, multiply, subtract**
4. $10^2 + 5^2 ÷ 5^2 − 5$ **exponents, divide, add, subtract**

Evaluate the expression.

5. $7 + (2 \times 2)^4 − 9 \times 9$ **182**
6. $90 \times 5 − 4 \times (18 ÷ 6)$ **438**
7. $3^2 \times (4 + 5)^2 − 36$ **693**
8. $15^2 ÷ (4^2 + 9) + 8^1$ **17**
9. $(8^2 + 3^3) \times (9 − 5)^3$ **5,824**
10. $10^3 ÷ (10^2 ÷ 10^1) + 10^2$ **200**
11. Scott bought 5 pounds of nails that cost $2.75 per pound and 3 pounds of screws that cost $3.50 per pound. How much did Scott spend? **$24.25**

Student Handbook H33

CHAPTER 3 Extra Practice

Set A (pp. 52–55)

Write the value of the blue digit.

1. 6.12053 **2 hundredths**
2. 0.0231 **3 thousandths**
3. 8.7 **7 tenths**
4. 0.849 **4 hundredths**

Write the number in expanded form.

5. 0.00309 **0.003 + 0.00009**
6. 5.015 **5 + 0.01 + 0.005**
7. 3.032 **3 + 0.03 + 0.002**
8. 20.0518 **20 + 0.05 + 0.001 + 0.0008**
9. 200.05 **200 + 0.05**
10. 5.16 **5 + 0.1 + 0.06**

Compare the numbers. Write <, >, or = for ●.

11. 5.099 ● 5.999 **<**
12. 226.5 ● 226.4 **>**
13. 251.36 ● 241.36 **>**
14. 18.3 ● 18.30 **=**
15. 4.18 ● 4.28 **<**
16. 49.089 ● 49.098 **<**

Write the numbers in order from least to greatest.

17. 82.16, 82, 82.15 **82, 82.15, 82.16**
18. 141.14, 114.41, 141.41 **114.41, 141.14, 141.41**
19. 5.09, 5.49, 5.23 **5.09, 5.23, 5.49**

Set B (pp. 58–59)

Estimate. *Possible estimates are given.*

1. 3.7 + 3.15 + 2.98 **10**
2. 62.8 × 6 **378**
3. 109.7 − 53.622 **56**
4. 788.3 × 92 **72,000**
5. 5.92 + 3.15 + 4.07 **13**
6. 21.513 × 9.8 **220**
7. 5.816 + 3.215 + 1.6 **11**
8. 465.09 − 73.46 **400**
9. 728 ÷ 8.1 **90**
10. 8.1 − 2.456 **6**
11. 20.8 ÷ 7 **3**
12. 123.95 ÷ 61 **2**

Set C (pp. 60–61)

Write the decimal and percent for the shaded part.

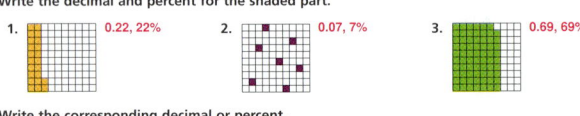

1. **0.22, 22%**
2. **0.07, 7%**
3. **0.69, 69%**

Write the corresponding decimal or percent.

4. 60% **0.6**
5. 0.9 **90%**
6. 39% **0.39**
7. 0.04 **4%**
8. 0.46 **46%**
9. 18% **0.18**
10. 0.41 **41%**
11. 0.38 **38%**
12. 7% **0.07**
13. 90% **0.9**

H34 Extra Practice

CHAPTER 4 Extra Practice

Set A (pp. 66–69)

Add or subtract. Estimate to check.

1. 12.8 − 4.1 **8.7**
2. $21.85 + $17.48 **$39.33**
3. 17.3 − 16.5 **0.8**
4. 8.36 + 5.216 + 0.09 **13.666**
5. 8 + 7.317 + 3.06 **18.377**
6. 5.08 − 2.261 **2.819**

Copy the problem. Place the decimal point correctly in the answer.

7. 13.601 − 10.311 = 329 **3.29**
8. 18.56 + 3.12 = 2168 **21.68**
9. 5 − 3.021 = 1979 **1.979**

Set B (pp. 70–73)

Tell the number of decimal places there will be in the product.

1. 21.3 × 18.4 **2**
2. 7.03 × 7.05 **4**
3. 9.2 × 2.13 **3**

Copy the problem. Place the decimal point in the product.

4. 360.05 × 12.6 = 4536630 **4,536.630**
5. 762 × 3.285 = 2503170 **2,503.170**
6. 10.11 × 1.02 = 103122 **10.3122**

Multiply. Estimate to check.

7. 37.5 × 10.26 **384.750**
8. 6.42 × 9.1 **58.422**
9. 0.05 × 2.9 **0.145**
10. 19.3 × 2.41 **46.513**
11. 2.39 × 7.6 **18.164**
12. 2 × 6.005 **12.010**

Set C (pp. 76–79)

Rewrite the problem so that the divisor is a whole number.

1. 16.92 ÷ 0.12 **1692 ÷ 12**
2. 661.44 ÷ 31.2 **6614.4 ÷ 312**
3. 20.2 ÷ 0.53 **2020 ÷ 53**

Copy the problem. Place the decimal point in the quotient.

4. 4.48 ÷ 2.8 = 16 **1.6**
5. 28.68 ÷ 1.2 = 239 **23.9**
6. 9.87 ÷ 2 = 4935 **4.935**

Divide. Estimate to check.

7. 9.72 ÷ 1.2 **8.1**
8. 25.0224 ÷ 3.12 **8.02**
9. 20.801 ÷ 6.1 **3.41**
10. 80.4 ÷ 4.8 **16.75**

Set D (pp. 82–83)

Evaluate each expression.

1. $8m$ for $m = 4.2$ **33.6**
2. $6.3 + 5.04 + k$ for $k = 8.4$ **19.74**
3. $(3.4 − c) + 63$ for $c = 2.47$ **63.93**

Solve each equation by using mental math.

4. $p + 2.8 = 7.9$ **$p = 5.1$**
5. $\frac{k}{3} = 5.6$ **$k = 16.8$**
6. $a − 17.1 = 9.5$ **$a = 26.6$**
7. $6s = 7.2$ **$s = 1.2$**
8. $x + 22.6 = 30.8$ **$x = 8.2$**
9. $15n = 4.5$ **$n = 3$**

Student Handbook H35

CHAPTER 5 Extra Practice

Set A (pp. 94–97)

Determine the type of sample. Write *convenience*, *random*, or *systematic*.

1. The teacher randomly selected a student and then surveyed every third student on his attendance list. **systematic**
2. The teacher asked students in English class about their favorite books. **convenience**

Set B (pp. 98–99)

A survey is to be conducted about the favorite foods of middle-school students. Tell whether the sampling method is *biased* or *unbiased*.

1. Randomly survey 10 people who get off the school bus one morning. **biased**
2. Randomly survey 1 of every 10 students in sixth, seventh, and eighth grades. **unbiased**

Set C (pp. 102–105)

Use the table at the right.

1. Copy and complete the table. **Check students' tables.**
2. How large was the sample size? **35**

Favorite Color	Tally	Frequency	Cumulative Frequency
Blue	☓☓	■ 11	■ 11
Green	☓☓☓	■ 9	■ 20
Red	☓☓☓	■ 8	■ 28
Yellow	☓	■ 5	■ 33
Orange	☓	■ 2	■ 35

Set D (pp. 106–108)

Find the mean, median, and mode. **Mean, median and mode are listed in order.**

1. 4, 5, 7, 8, 8 **6.4; 7; 8**
2. 6.1, 8.1, 7.4, 7.2 **7.2; 7.3; none**
3. 12, 14, 18, 12, 22 **15.6; 14; 12**
4. 1, 1, 7, 9, 3, 4, 2, 3 **3.75; 3; 1 and 3**

Set E (pp. 109–111)

Use the data in the table.

NUMBER OF HOURS SPENT ON HOMEWORK										
Week	1	2	3	4	5	6	7	8	9	10
Hours	8	9	10	9	7	9	8	0	7	2

1. Find the mean, median, and mode with the outliers and then without the outliers. **8.25, 8.5, 9; 6.8, 8, 9**
2. How does including the outliers affect the mean? the median? the mode? **mean and median decrease; mode doesn't change**

Set F (pp. 112–115)

Write *yes* or *no* to tell whether the conclusion is valid. Explain.

1. The first ten females to enter a diner say their favorite drink is orange juice. You conclude that orange juice is the favorite drink of all the customers. **No. The sample is biased.**
2. Two hundred students are randomly selected and asked what their favorite kind of music is. Seventy percent say pop music. You conclude that most students prefer pop music. **Yes. The sample and the population are unbiased.**

H36 Extra Practice

CHAPTER 6 Extra Practice

Set A (pp. 120–123)

1. Make a multiple-bar graph using the data in the table below. **See Additional Answers, p. H61A.**

CLUB MEETING ATTENDANCE				
	Sept	Oct	Nov	Dec
Male	42	31	65	61
Female	56	47	51	60

2. Make a multiple-line graph using the data in the table below. **See Additional Answers, p. H61A.**

AVERAGE RAINFALL (IN INCHES)				
	Jan	Feb	Mar	Apr
City A	4	5	3	5
City B	8	12	4	7

Set B (pp. 124–125)

1. Make a line graph using the data in the table at the right. **See Additional Answers, p. H61A.**
2. If the trend continues, how many miles will be traveled on the sixth day? **about 265 mi**

DISTANCE TRAVELED					
Day	1	2	3	4	5
Miles	45	85	129	175	220

Set C (pp. 126–128)

1. Make a stem-and-leaf plot of the data below. Find the mode and median. **See Additional Answers, p. H61A.**

14 22 37 41 13 18 22 29 33
36 25 21 30 48 44 19 17 28

2. Use the data in the chart below to make a histogram. **See Additional Answers, p. H61A.**

NUMBER OF HOURS OF MONTHLY EXERCISE					
0–2	3–5	6–8	9–11	12–15	16–18
12	15	19	22	21	6

Set D (pp. 130–131)

For 1–3, use the box-and-whisker graph. The graph shows the number of tickets sold.

1. What was the greatest number of tickets sold? **95**
2. What is the median? **92**
3. What are the lower and upper quartiles? **86 and 94**

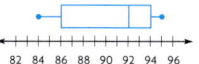

Set E (pp. 132–135)

For 1–3, use the bar graph at the right.

1. About how many times as high is the bar for Central than the bar for Lee? **about 4 times as high**
2. Has Central won four times as many championships as Lee? Explain. **No. Central has won 60 championships while Lee has won 30 championships.**
3. How can you change the graph so it is not misleading? **Start the scale at zero and have equal intervals.**

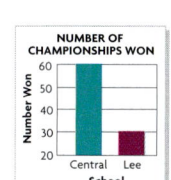

Student Handbook H37

CHAPTER 7 Extra Practice

Set A (pp. 146–147)

Tell whether each number is divisible by 2, 3, 4, 5, 6, 8, 9, or 10.

1. 80 **2, 4, 5, 8, 10**
2. 99 **3, 9**
3. 105 **3, 5**
4. 126 **2, 3, 6, 9**
5. 234 **2, 3, 6, 9**
6. 370 **2, 5, 10**
7. 591 **3**
8. 1,620 **2, 3, 4, 5, 6, 9, 10**
9. 3,048 **2, 3, 4, 6, 8**
10. 8,020 **2, 4, 5, 10**

Set B (pp. 148–149)

Use division or a factor tree to find the prime factorization.

1. 18 $2 \times 3 \times 3$
2. 40 $2 \times 2 \times 2 \times 5$
3. 36 $2 \times 2 \times 3 \times 3$
4. 27 $3 \times 3 \times 3$
5. 100 $2 \times 2 \times 5 \times 5$
6. 150 $2 \times 3 \times 5 \times 5$
7. 280 $2 \times 2 \times 2 \times 5 \times 7$
8. 12 $2 \times 2 \times 3$
9. 108 $2 \times 2 \times 3 \times 3 \times 3$
10. 420 $2 \times 2 \times 3 \times 5 \times 7$

Write the prime factorization of each number in exponent form.

11. 28 $2^2 \times 7$
12. 420 $2^2 \times 3 \times 5 \times 7$
13. 48 $2^4 \times 3$
14. 81 3^4
15. 72 $2^3 \times 3^2$
16. 80 $2^4 \times 5$
17. 144 $2^4 \times 3^2$
18. 121 11^2
19. 168 $2^3 \times 3 \times 7$
20. 1,475 $5^2 \times 59$

Solve for n to complete the prime factorization.

21. $2 \times 5 \times n = 50$ $n = 5$
22. $2 \times n \times 5 = 30$ $n = 3$
23. $2 \times 3 \times n = 18$ $n = 3$
24. $n \times 3 \times 3 = 27$ $n = 3$
25. $2 \times 2 \times 2 \times n = 16$ $n = 2$
26. $n \times 2 \times 3 = 42$ $n = 7$
27. $2 \times 3 \times 3 \times n = 54$ $n = 3$
28. $3 \times n \times 5 = 75$ $n = 5$
29. $n \times n \times 3 = 12$ $n = 2$

Set C (pp. 150–153)

List the first five multiples of each number.

1. 4 **4, 8, 12, 16, 20**
2. 9 **9, 18, 27, 36, 45**
3. 14 **14, 28, 42, 56, 70**
4. 5 **5, 10, 15, 20, 25**
5. 30 **30, 60, 90, 120, 150**
6. 8 **8, 16, 24, 32, 40**
7. 12 **12, 24, 36, 48, 60**
8. 21 **21, 42, 63, 84, 105**

Find the LCM of each set of numbers.

9. 45, 54 **270**
10. 10, 35 **70**
11. 12, 20 **60**
12. 18, 27 **54**
13. 150, 60 **300**
14. 54, 18 **108**
15. 55, 33, 44 **660**
16. 35, 25, 49 **1,225**

17. Jean needs eggs and muffins for the class breakfast. Eggs come in cartons of 12 and muffins come in packages of 8. What is the least number of eggs and muffins she can buy to have an equal number of each? **24 of each**

Find the GCF of each set of numbers.

18. 12, 108 **12**
19. 148, 84 **4**
20. 132, 108 **12**
21. 75, 105 **15**
22. 45, 108 **9**
23. 32, 128 **32**
24. 252, 336 **84**
25. 56, 280, 400 **8**

H38 Extra Practice

CHAPTER 8 Extra Practice

Set A (pp. 160–163)

Write the fraction in simplest form.

1. $\frac{24}{80}$ $\frac{3}{10}$
2. $\frac{14}{63}$ $\frac{2}{9}$
3. $\frac{24}{56}$ $\frac{3}{7}$
4. $\frac{15}{60}$ $\frac{1}{4}$
5. $\frac{16}{40}$ $\frac{2}{5}$
6. $\frac{50}{90}$ $\frac{5}{9}$

Complete.

7. $\frac{9}{12} = \frac{3}{\blacksquare}$ **4**
8. $\frac{7}{42} = \frac{1}{\blacksquare}$ **6**
9. $\frac{25}{\blacksquare} = \frac{5}{10}$ **50**
10. $\frac{3}{7} = \frac{21}{\blacksquare}$ **49**
11. $\frac{1}{12} = \frac{\blacksquare}{144}$ **12**
12. $\frac{9}{20} = \frac{\blacksquare}{60}$ **27**
13. $\frac{5}{\blacksquare} = \frac{15}{27}$ **9**
14. $\frac{12}{\blacksquare} = \frac{3}{8}$ **32**

Set B (pp. 164–165)

Write the fraction as a mixed number or a whole number.

1. $\frac{7}{3}$ $2\frac{1}{3}$
2. $\frac{9}{2}$ $4\frac{1}{2}$
3. $\frac{36}{5}$ $7\frac{1}{5}$
4. $\frac{72}{8}$ 9
5. $\frac{10}{2}$ 5
6. $\frac{13}{4}$ $3\frac{1}{4}$
7. $\frac{36}{6}$ 6
8. $\frac{12}{9}$ $1\frac{1}{3}$

Write the mixed number as a fraction.

9. $4\frac{1}{4}$ $\frac{17}{4}$
10. $5\frac{2}{3}$ $\frac{17}{3}$
11. $2\frac{1}{8}$ $\frac{17}{8}$
12. $7\frac{1}{9}$ $\frac{64}{9}$
13. $3\frac{4}{7}$ $\frac{25}{7}$
14. $1\frac{6}{11}$ $\frac{17}{11}$
15. $4\frac{3}{8}$ $\frac{35}{8}$
16. $15\frac{3}{4}$ $\frac{63}{4}$

Set C (pp. 166–167)

Compare the fractions. Write <, >, or = for each ●.

1. $\frac{5}{9}$ ● $\frac{1}{2}$ **>**
2. $\frac{3}{8}$ ● $\frac{1}{5}$ **>**
3. $\frac{7}{25}$ ● $\frac{3}{5}$ **<**
4. $\frac{3}{5}$ ● $\frac{12}{20}$ **=**

Order the fractions from least to greatest.

5. $\frac{1}{3}, \frac{1}{6}, \frac{1}{10}$ $\frac{1}{10}, \frac{1}{6}, \frac{1}{3}$
6. $\frac{4}{9}, \frac{2}{3}, \frac{3}{8}, \frac{4}{9}$ $\frac{3}{8}, \frac{4}{9}, \frac{2}{3}$
7. $\frac{1}{9}, \frac{3}{4}, \frac{5}{12}, \frac{1}{2}$ $\frac{1}{9}, \frac{5}{12}, \frac{1}{2}, \frac{3}{4}$

Set D (pp. 169–171)

Write the decimal as a fraction.

1. 0.9 $\frac{9}{10}$
2. 0.081 $\frac{81}{1,000}$
3. 0.29 $\frac{29}{100}$

Write as a decimal. Tell whether the decimal terminates or repeats.

4. $\frac{3}{8}$ **0.375, T**
5. $\frac{5}{16}$ **0.3125, T**
6. $\frac{1}{3}$ **0.3, R**
7. $\frac{7}{9}$ **0.7, R**

Compare. Write <, >, or = for each ●.

8. $\frac{3}{10}$ ● 0.03 **>**
9. $\frac{2}{3}$ ● 0.7 **<**
10. 0.79 ● $\frac{5}{8}$ **>**
11. 0.15 ● $\frac{3}{20}$ **=**

Write each fraction as a percent.

12. $\frac{6}{25}$ **24%**
13. $\frac{7}{10}$ **70%**
14. $\frac{9}{20}$ **45%**
15. $\frac{4}{5}$ **80%**

Student Handbook H39

CHAPTER 9 Extra Practice

Set A (pp. 176–179)

Estimate the sum or difference. **Possible estimates are given.**

1. $\frac{4}{7} - \frac{3}{6}$ **0**
2. $3\frac{7}{9} + 1\frac{10}{11}$ **6**
3. $5\frac{4}{5} - 3\frac{1}{2}$ **$2\frac{1}{2}$**
4. $\frac{8}{9} + \frac{3}{5}$ **$1\frac{1}{2}$**
5. $\frac{4}{9} + \frac{2}{3}$ **1**
6. $\frac{8}{9} - \frac{5}{7}$ **0**
7. $\frac{4}{9} - \frac{1}{6}$ **$\frac{1}{2}$**
8. $3\frac{10}{13} + 4\frac{1}{9}$ **8**
9. $\frac{3}{7} + 1\frac{1}{9}$ **$\frac{1}{2}$**
10. $5\frac{3}{4} - 1\frac{1}{8}$ **5**
11. $\frac{5}{9} - \frac{3}{7}$ **0**
12. $8\frac{7}{8} + 5\frac{1}{2}$ **$14\frac{1}{2}$**

13. Shawna practiced the tuba for $\frac{5}{8}$ hr on Monday, $\frac{1}{3}$ hr on Wednesday, and $\frac{1}{4}$ hr on Friday. About how many hours did Shawna practice last week? **Possible answer: about 2 hours**

Set B (pp. 182–185)

Write the sum or difference in simplest form. Estimate to check.

1. $\frac{3}{4} + \frac{1}{2}$ **$\frac{5}{4}$, or $1\frac{1}{4}$**
2. $\frac{4}{7} - \frac{3}{21}$ **$\frac{5}{7}$**
3. $\frac{1}{6} + \frac{5}{18}$ **$\frac{4}{9}$**
4. $\frac{9}{10} - \frac{2}{5}$ **$\frac{1}{2}$**
5. $\frac{5}{6} + \frac{1}{4}$ **$\frac{13}{12}$, or $1\frac{1}{12}$**
6. $\frac{6}{20} + \frac{3}{10}$ **$\frac{9}{20}$** (unclear)
7. $\frac{5}{9} - \frac{1}{18}$ **$\frac{1}{2}$**
8. $\frac{7}{8} - \frac{1}{16}$ **$\frac{13}{16}$**
9. $\frac{3}{5} - \frac{1}{9}$ **$\frac{22}{45}$**
10. $\frac{5}{20} + \frac{11}{20}$ **$\frac{4}{5}$**
11. $\frac{11}{12} - \frac{1}{6}$ **$\frac{3}{4}$**
12. $\frac{1}{10} + \frac{3}{8}$ **$\frac{19}{40}$**

13. Rico read $\frac{1}{5}$ of a book on Saturday and $\frac{1}{3}$ of the same book on Sunday. What portion of the book does he have left to read? **$\frac{7}{15}$ of the book**

Set C (pp. 186–189) Check students' diagrams.

Draw a diagram to find each sum or difference. Write the answer in simplest form.

1. $2\frac{3}{8} + 1\frac{1}{4}$ **$3\frac{5}{8}$**
2. $2\frac{1}{3} - 1\frac{1}{6}$ **$1\frac{1}{6}$**
3. $1\frac{1}{5} + 1\frac{7}{10}$ **$2\frac{9}{10}$**
4. $3\frac{1}{3} - 1\frac{3}{4}$ **$1\frac{7}{12}$**

Write the sum or difference in simplest form. Estimate to check.

5. $4\frac{1}{12} - 2\frac{1}{6}$ **$2\frac{1}{4}$**
6. $5\frac{2}{5} - 3\frac{3}{10}$ **$2\frac{1}{10}$**
7. $2\frac{4}{9} + 1\frac{2}{5}$ **$3\frac{23}{45}$**
8. $4\frac{2}{3} + 6\frac{1}{6}$ **$10\frac{5}{6}$**

9. Stefan bought $1\frac{1}{2}$ lb of potato salad, $2\frac{1}{4}$ lb of coleslaw, and $4\frac{3}{4}$ lb of chicken at a deli. What is the total weight of his purchase? **$8\frac{1}{2}$ lb**

Set D (pp. 192–193)

Write the difference in simplest form. Estimate to check.

1. $4\frac{5}{9} - 3\frac{5}{6}$ **$\frac{13}{18}$**
2. $5\frac{1}{3} - 4\frac{1}{9}$ **$1\frac{2}{9}$**
3. $6\frac{1}{4} - 3\frac{2}{3}$ **$2\frac{7}{12}$**
4. $6\frac{1}{3} - 4\frac{7}{9}$ **$1\frac{5}{9}$**
5. $4 - 1\frac{3}{7}$ **$2\frac{4}{7}$**
6. $5\frac{3}{10} - \frac{2}{5}$ **$4\frac{9}{10}$**
7. $7\frac{1}{6} - \frac{1}{2}$ **$6\frac{2}{3}$**
8. $4\frac{1}{4} - 1\frac{3}{8}$ **$2\frac{5}{8}$**

9. Mr. Norman had $51\frac{1}{3}$ ft of plastic pipe. He installed $25\frac{3}{4}$ ft in the bathroom. How much pipe does he have left? **$25\frac{7}{12}$ ft**

CHAPTER 10 Extra Practice

Set A (pp. 200–201)

Estimate each product or quotient. **Possible answers are given.**

1. $\frac{8}{9} \times \frac{3}{5}$ **$\frac{1}{2}$**
2. $\frac{1}{6} \times \frac{4}{9}$ **0**
3. $5\frac{1}{2} \div \frac{5}{6}$ **2** (unclear)
4. $\frac{7}{9} \div \frac{3}{4}$ **1**
5. $3\frac{1}{3} \times 5\frac{1}{5}$ **15**
6. $8\frac{3}{5} \div 2\frac{2}{3}$ **4**
7. $9\frac{7}{8} \times 10\frac{1}{2}$ **100**
8. $18\frac{3}{4} \div 5\frac{1}{8}$ **3**

9. Liz is making cookies to give to some new neighbors. The recipe calls for $1\frac{3}{4}$ cups of flour. About how much flour does Liz need if she makes $2\frac{1}{2}$ times the recipe? **about 4 cups**

Set B (pp. 202–205)

Multiply. Write the answer in simplest form.

1. $\frac{2}{4} \times \frac{3}{4}$ **$\frac{3}{8}$**
2. $\frac{2}{5} \times \frac{1}{4}$ **$\frac{1}{10}$**
3. $\frac{3}{8} \times \frac{3}{5}$ **$\frac{9}{40}$**
4. $3 \times \frac{7}{8}$ **$2\frac{5}{8}$**
5. $\frac{2}{7} \times 8$ **$2\frac{2}{7}$**
6. $\frac{5}{6} \times \frac{3}{8}$ **$\frac{5}{16}$**
7. $\frac{5}{9} \times \frac{1}{5}$ **$\frac{1}{9}$**
8. $\frac{1}{9} \times \frac{3}{4}$ **$\frac{1}{12}$**

9. To make a dye for art class, Jan needs $\frac{1}{4}$ tsp of red coloring and $\frac{1}{2}$ that amount of green coloring. How much green coloring does Jan need to make dye? **$\frac{1}{8}$ tsp**

Set C (pp. 206–207)

Multiply. Write the answer in simplest form.

1. $2\frac{2}{3} \times 2\frac{1}{4}$ **6**
2. $2\frac{1}{2} \times \frac{1}{5}$ **$\frac{1}{2}$**
3. $\frac{1}{5} \times 2\frac{2}{8}$ **$\frac{9}{20}$**
4. $3\frac{1}{6} \times 5\frac{1}{4}$ **$16\frac{5}{8}$**
5. $2\frac{3}{4} \times \frac{5}{6}$ **$2\frac{7}{24}$**
6. $2\frac{2}{7} \times \frac{5}{5}$ **$\frac{32}{35}$**
7. $3\frac{2}{3} \times 2\frac{3}{5}$ **$9\frac{8}{15}$**
8. $6\frac{8}{9} \times 2\frac{3}{7}$ **$16\frac{46}{63}$**

Set D (pp. 210–213)

Find the quotient. Write the answer in simplest form.

1. $\frac{2}{9} \div \frac{8}{18}$ **$\frac{1}{2}$**
2. $\frac{3}{8} \div 2\frac{1}{4}$ **$\frac{1}{6}$**
3. $2 \div 2\frac{2}{3}$ **$\frac{3}{4}$**
4. $1\frac{11}{4} \div 1\frac{5}{6}$ **$1\frac{1}{22}$**
5. $\frac{3}{4} \div \frac{1}{8}$ **6**
6. $\frac{5}{6} \div 1\frac{2}{3}$ **$\frac{1}{2}$**
7. $3\frac{3}{4} \div 2\frac{7}{12}$ **$1\frac{14}{31}$**
8. $\frac{5}{7} \div \frac{10}{11}$ **$\frac{11}{14}$**

Set E (pp. 216–217)

Evaluate the expression.

1. $3r$ for $r = \frac{1}{2}$ **$1\frac{1}{2}$**
2. $\frac{5}{6} + d$ for $d = \frac{3}{10}$ **$1\frac{2}{15}$**
3. $3\frac{3}{5} \div a$ for $a = \frac{1}{2}$ **$7\frac{1}{5}$**
4. $y + 3\frac{5}{8}$ for $y = 1\frac{1}{4}$ **$4\frac{7}{8}$**
5. $2\frac{5}{6}k$ for $k = 1\frac{1}{3}$ **$3\frac{7}{9}$**
6. $a \div 4\frac{1}{2}$ for $a = 3\frac{3}{5}$ **$\frac{4}{5}$**

Solve the equation.

7. $c + \frac{3}{4} = 1\frac{1}{4}$ **$c = \frac{1}{2}$**
8. $\frac{5}{9}s = \frac{5}{27}$ **$s = \frac{1}{3}$**
9. $x - 6\frac{1}{2} = 12\frac{1}{4}$ **$x = 18\frac{3}{8}$**

CHAPTER 11 Extra Practice

Set A (pp. 228–229)

Write an integer to represent each situation.

1. a temperature increase of 5° **$^+5$**
2. the wind speed decreases by 12 mph **$^-12$**
3. depositing $510 into a savings account **$^+510$**
4. a temperature of 7° below zero **$^-7$**

Write the opposite integer.

5. $^-3$ **$^+3$**
6. $^-12$ **$^+12$**
7. $^+360$ **$^-360$**
8. $^-1$ **$^+1$**
9. $^+160$ **$^-160$**
10. $^-3,047$ **$^+3,047$**
11. $^-1,119$ **$^+1,119$**
12. $^-942$ **$^+942$**

Set B (pp. 230–233)

Write each rational number in the form $\frac{a}{b}$.

1. $3\frac{1}{6}$ **$\frac{19}{6}$**
2. 0.5 **$\frac{5}{10}$**
3. 0.27 **$\frac{27}{100}$**
4. 13.4 **$\frac{134}{10}$**
5. $2\frac{2}{5}$ **$\frac{12}{5}$**
6. 3.18 **$\frac{318}{100}$**
7. 10.02 **$\frac{1,002}{100}$**
8. 300 **$\frac{300}{1}$**
9. 0.36 **$\frac{36}{100}$**
10. $5\frac{1}{3}$ **$\frac{16}{3}$**
11. 312 **$\frac{312}{1}$**
12. $4\frac{1}{4}$ **$\frac{17}{4}$**

Find a rational number between the two given numbers. **Possible answers are given.**

13. $\frac{3}{8}$ and $\frac{5}{6}$ **$\frac{5}{12}$**
14. $\frac{1}{8}$ and $\frac{1}{4}$ **$\frac{3}{16}$**
15. $\frac{5}{9}$ and $\frac{11}{15}$ **$\frac{5}{8}$**
16. 1.8 and 1.9 **1.85**
17. $^-1.5$ and $^-1.3$ **$^-1.4$**
18. 4.23 and 4.235 **4.231**
19. 3.8 and 3.82 **3.81**
20. $^-5$ and $^-4.9$ **$^-4.94$**
21. $^-12\frac{1}{4}$ and $^-12\frac{1}{3}$ **$^-12\frac{3}{10}$**

Set C (pp. 234–235)

Compare. Write <, >, or = for each ●.

1. 0.4 ● 0.38 **>**
2. $\frac{2}{7}$ ● 0.25 **>**
3. $^-0.6$ ● $^-\frac{2}{5}$ **<**
4. $\frac{7}{9}$ ● 0.8 **<**
5. 0.28 ● $\frac{2}{7}$ **<**
6. $\frac{3}{13}$ ● 0.13 **>**
7. $^-\frac{4}{9}$ ● $^-3.1$ **>**
8. $^-3$ ● 0.31 **<**
9. $2\frac{1}{5}$ ● $2\frac{4}{13}$ **<**
10. 0.87 ● 0.868 **>**
11. $^-\frac{5}{8}$ ● $\frac{5}{8}$ **<**
12. $2\frac{1}{16}$ ● $2\frac{1}{10}$ **<**

Compare the rational numbers and order them from least to greatest.

13. $\frac{1}{3}, \frac{2}{5}, 0.38, \frac{1}{2}$ **0.38, $\frac{1}{3}, \frac{2}{5}, \frac{1}{2}$**
14. $\frac{2}{7}, \frac{1}{3}, \frac{1}{4}, 0.26$ **$\frac{1}{4}$, 0.26, $\frac{2}{7}, \frac{1}{3}$**
15. 0.92, $\frac{9}{8}, \frac{8}{9}$, 0.924 **$\frac{8}{9}$, 0.92, 0.924, $\frac{9}{8}$**
16. 0.23, $\frac{2}{9}, \frac{1}{4}, \frac{1}{5}$ **$\frac{1}{5}, \frac{2}{9}$, 0.23, $\frac{1}{4}$**
17. $^-\frac{2}{5}, ^-\frac{1}{3}, ^-\frac{1}{2}, ^-\frac{3}{5}$ **$^-\frac{3}{5}, ^-\frac{1}{2}, ^-\frac{2}{5}, ^-\frac{1}{3}$**
18. $\frac{1}{10}, ^-3, ^-0.3$ **$^-3, ^-0.3, \frac{1}{10}$**
19. $^-7, 6, 1, ^-5$ **$^-7, ^-5, 1, 6$**
20. $^-0.1, 0.01, 10, ^-10$ **$^-10, ^-0.1, 0.01, 10$**
21. $\frac{1}{2}, \frac{1}{6}, \frac{1}{9}, 0.1$ **0.1, $\frac{1}{9}, \frac{1}{6}, \frac{1}{2}$**

CHAPTER 12 Extra Practice

Set A (pp. 244–247)

Find the sum.

1. $^-3 + 9$ **6**
2. $1 + ^-5$ **$^-4$**
3. $5 + 4$ **9**
4. $^-2 + 8$ **6**
5. $^-9 + 17$ **8**
6. $^-14 + ^-8$ **$^-22$**
7. $^-19 + 8$ **$^-11$**
8. $22 + ^-36$ **$^-14$**
9. $31 + 19$ **50**
10. $32 + ^-45$ **$^-13$**
11. $63 + ^-47$ **16**
12. $^-71 + 32$ **$^-39$**

13. At 9:00 A.M., the temperature outside was $^-9°C$. By 3:00 P.M., the temperature had risen 5°C. What was the temperature at 3:00 P.M.? **$^-4°C$**

Set B (pp. 250–251)

Find the difference.

1. $2 - 9$ **$^-7$**
2. $^-8 - ^-3$ **$^-5$**
3. $4 - ^-2$ **6**
4. $4 - 9$ **$^-5$**
5. $8 - 19$ **$^-11$**
6. $14 - ^-2$ **16**
7. $^-12 - ^-12$ **0**
8. $3 - 7$ **$^-4$**
9. $23 - 32$ **$^-9$**
10. $14 - 29$ **$^-15$**
11. $^-48 - 17$ **$^-65$**
12. $^-24 - ^-39$ **15**

13. The water level in the city water tower was 8 ft above normal. Three weeks later, the level was 4 ft below normal. Find the range of the water levels. **12 ft**

Set C (pp. 252–255)

Find the product or quotient.

1. $^-7 \times 6$ **$^-42$**
2. $^-6 \times ^-8$ **48**
3. $^-20 \times 4$ **$^-80$**
4. $50 \times ^-8$ **$^-400$**
5. 32×10 **320**
6. $40 \times ^-9$ **$^-360$**
7. $^-27 \div 9$ **$^-3$**
8. $^-64 \div ^-8$ **8**
9. $^-140 \div 10$ **$^-14$**
10. $225 \div ^-25$ **$^-9$**
11. $360 \div 15$ **24**
12. $216 \div ^-12$ **$^-18$**

13. The depth of a diver changed $^-5$ ft every 30 sec. How much of a depth change did the diver have after 2 min? **$^-20$ ft**

Set D (pp. 256–259)

Find the sum or difference.

1. $3.8 + 0.7$ **4.5**
2. $8\frac{1}{2} - 17\frac{1}{2}$ **$^-26$**
3. $^-4.5 - ^-2.5$ **$^-2$**
4. $^-8.5 + 1.8$ **$^-6.7$**
5. $^-7\frac{1}{2} - 4\frac{3}{4}$ **$^-12\frac{1}{4}$**
6. $2\frac{1}{2} + ^-8$ **$^-5\frac{1}{2}$**
7. $^-4.9 - 4.1$ **$^-9.0$**
8. $^-17\frac{1}{2} + ^-34$ **$^-51\frac{1}{2}$**

Find the product or quotient.

9. $^-8 \times ^-2\frac{1}{2}$ **3.2** (unclear)
10. $^-2.6 \times ^-20$ **52**
11. $7.7 \div 0.1$ **77**
12. $\frac{4}{7} \times \frac{7}{8}$ **$\frac{1}{2}$**
13. $0.5 \times ^-2.6$ **$^-1.3$**
14. $85 \div ^-17$ **$^-5$**
15. $3\frac{1}{4} \times ^-1\frac{3}{5}$ **$^-5\frac{1}{5}$**
16. $\frac{-3}{4} \div ^-4$ **$\frac{3}{16}$**

CHAPTER 13 Extra Practice

Set A (pp. 270–271)

Write an algebraic expression for the word expression.
1. 9 less than y $y - 9$
2. 13 more than a number, x $x + 13$
3. $\frac{3}{4}$ increased by y $\frac{3}{4} + y$
4. 7 more than the quotient of 52 and k $\frac{52}{k} + 7$
5. Fred has 32 more than twice the number of baseball cards that José has. Write an algebraic expression for the number of baseball cards Fred has. $32 + 2b$

Write a word expression for each. Possible word expressions are given.
6. $9^2 + a$ the sum of 9 squared and a
7. $7x + 12$ 12 added to 7 times a number x
8. $24 - \frac{1}{5}c$ 24 decreased by $\frac{1}{5}c$
9. $y + \frac{1}{2}x$ y increased by $\frac{1}{2}x$

Set B (pp. 272–275)

Evaluate the algebraic expression for the given value of the variable.
1. $x - 6$ for $x = {}^-12$ $^-18$
2. $y + 13$ for $y = {}^-25$ $^-12$
3. $42 - k$ for $k = {}^-30$ 72
4. $^-38 + p$ for $p = 22$ $^-16$
5. $a^2 - 25$ for $a = 3$ $^-16$
6. $x^3 - 19$ for $x = 2$ $^-11$
7. $3(a + b)^2 - c$ for $a = {}^-3$, $b = 5$, and $c = {}^-8$ 20
8. $6xyz$ for $x = {}^-4$, $y = {}^-3$, and $z = {}^-6$ $^-432$
9. $3pq + r$ for $p = 2$, $q = {}^-8$, and $r = {}^-15$ $^-63$

Evaluate the algebraic expression for $x = 2, 3, 4,$ and 5.
10. $7x + 12$ $26, 33, 40, 47$
11. $^-4x + 10$ $2, ^-2, ^-6, ^-10$
12. $3x - 26$ $^-20, ^-17, ^-14, ^-11$
13. $^-9x - 4$ $^-22, ^-31, ^-40, ^-49$
14. $\frac{60}{x} - 14$ $16, 6, 1, ^-2$
15. $\frac{^-60}{x} - 14$ $^-44, ^-34, ^-29, ^-26$

Simplify the expression. Then evaluate the expression for the given value of the variable.
16. $5x + 3x + 12$ for $x = {}^-4$ $8x + 12; ^-20$
17. $2y + 7y - 25$ for $y = 2$ $9y - 25; ^-7$
18. $42z - 30z - 20$ for $z = {}^-2$ $12z - 20; ^-44$

Set C (pp. 278–279)

Evaluate the expression.
1. $\sqrt{100} - 2 + 6$ 14
2. $3 \times (9^2 - 41)$ 120
3. $4 \times \sqrt{25} - 3 \times 2$ 14
4. $\sqrt{36} \cdot \sqrt{36}$ 36
5. $530 \times (\sqrt{4} - 2)^2$ 0
6. $8 + \sqrt{49} - 3^2$ 6

Evaluate the expression for $a = 16$, $b = 3$, and $c = 6$.
7. $^-a + (b + c)^2$ 65
8. $\frac{c}{b} - \sqrt{a}$ $^-2$
9. $\sqrt{a + b + c}$ 5

H44 Extra Practice

CHAPTER 14 Extra Practice

Set A (pp. 284–285)

Write an equation for the word sentence. Choice of variable may vary.
1. $\frac{2}{3}$ of a number is 12. $\frac{2}{3}n = 12$
2. 1.6 more than a number is 5. $1.6 + n = 5$
3. $1\frac{1}{2}$ less than a number is 6. $n - 1\frac{1}{2} = 6$
4. A number divided by $^-6$ is 30. $\frac{n}{^-6} = 30$
5. 9 times a number is 53. $9n = 53$
6. 1.5 increased by a number is 4.2. $1.5 + n = 4.2$
7. $3\frac{1}{2}$ decreased by a number is $\frac{1}{2}$. $3\frac{1}{2} - n = \frac{1}{2}$
8. The quotient of a number and 3.3 is 99. $\frac{n}{3.3} = 99$
9. The sum of 11.2 and a number is 65.07. $11.2 + n = 65.07$
10. 64 divided by a number is 3.2. $\frac{64}{n} = 3.2$
11. Leroy sold 250 boxes of apples during a fund-raiser. This was 5 times as many apples as Mary sold. Write an equation that represents the situation. $5m = 250$

Set B (pp. 287–289)

Solve and check.
1. $x + 6 = 15$ $x = 9$
2. $15 = a + 2$ $a = 13$
3. $11 + k = 25$ $k = 14$
4. $z + 2.7 = 19.6$ $z = 16.9$
5. $5.7 = b + 8.6$ $b = 2.9$
6. $24.8 = 17.2 + c$ $c = 7.6$
7. $y + 8\frac{2}{3} = 16$ $y = 7\frac{1}{3}$
8. $13\frac{1}{4} = 4\frac{4}{5} + s$ $s = 8\frac{9}{20}$
9. $13\frac{3}{8} = t + 7\frac{5}{6}$ $t = 5\frac{13}{24}$
10. $13.2 = x + 7.12$ $x = 6.08$
11. $t + 7\frac{1}{3} = 10\frac{1}{12}$ $t = 2\frac{3}{4}$
12. $4.06 + r = 13.56$ $r = 9.5$
13. A carpenter cut a 72-in. board into two pieces. One of the pieces is 24 in. long. How long is the second piece? 48 in.

Set C (pp. 290–291)

Solve and check.
1. $x - 7 = 11$ $x = 18$
2. $31 = a - 7$ $a = 38$
3. $k - 10 = 42$ $k = 52$
4. $z - 3.9 = 15.8$ $z = 19.7$
5. $6.5 = b - 21.3$ $b = 27.8$
6. $56.7 = c - 19.8$ $c = 76.5$
7. $y - 5\frac{5}{6} = 13$ $y = 18\frac{5}{6}$
8. $22\frac{1}{3} = s - 4\frac{3}{4}$ $s = 27\frac{1}{12}$
9. $8\frac{3}{5} = t - 4\frac{7}{7} = t = 13\frac{11}{35}$
10. $a - 27 = 18$ $a = 45$
11. $60.3 = b - 8.07$ $b = 68.37$
12. $4\frac{7}{9} = x + 1\frac{2}{3}$ $x = 6\frac{4}{9}$
13. Reggie withdrew $175 from his checking account so he could go shopping for the new school year. His new balance was $234. How much was in the account before the withdrawal? $409

Student Handbook H45

CHAPTER 15 Extra Practice

Set A (pp. 297–299)

Solve and check.
1. $3x = 12$ $x = 4$
2. $7k = 56$ $k = 8$
3. $\frac{p}{16} = {}^-3$ $p = {}^-48$
4. $\frac{a}{14} = 2$ $a = 28$
5. $27 = {}^-3x$ $x = {}^-9$
6. $8.8 = 2.2n$ $n = 4$
7. $4x = 24$ $x = 6$
8. $8x = {}^-32$ $s = {}^-4$
9. $9 = \frac{p}{3}$ $p = 27$
10. $56 = {}^-7p$ $p = {}^-8$
11. $21 = \frac{s}{3}$ $s = 63$
12. $45 = 9n$ $n = 5$
13. $180 = 3d$ $d = 60$
14. $46.2 = \frac{a}{3}$ $a = 138.6$
15. $1,486 = \frac{a}{2}$ $a = 2,972$
16. $\frac{c}{1.2} = 5.6$ $c = 6.72$
17. $^-12a = {}^-216$ $a = 18$
18. $3.8m = 57$ $m = 15$

For 19–20, write and solve an equation to answer the question. Variables will vary.
19. Celia divided her baseball cards equally among 4 friends. Each friend got 23 baseball cards. How many baseball cards did Celia have originally? $\frac{b}{4} = 23$; $b = 92$; 92 baseball cards
20. Julie earns $6.75 per hour at her job. She wants to save $324.00. How many hours does she have to work to earn $324.00? $6.75h = 324$; $h = 48$; 48 hours

Set B (pp. 300–303)

Use the formula $d = rt$ to complete.
1. $d = \blacksquare$ mi $r = 35$ mi per hr $t = 4$ hr 140 mi.
2. $d = 1,600$ km $r = \blacksquare$ km per min $t = 400$ min 4 km per min
3. $d = 2,100$ km $r = 70$ km per sec $t = \blacksquare$ sec 30 sec
4. $d = \blacksquare$ ft $r = 90.7$ ft per sec $t = 31$ sec 2,811.7 ft
5. $d = 567$ mi $r = \blacksquare$ mi per hr $t = 17.5$ hr 32.4 mi per hr
6. $d = 4,850$ m $r = 250$ m per sec $t = \blacksquare$ sec 19.4 sec

Convert the temperature to degrees Fahrenheit. Write your answer as a decimal.
7. $40°C$ $104°F$
8. $2.3°C$ $36.1°F$
9. $14°C$ $57.2°F$
10. $20°C$ $68°F$

Convert the temperature to degrees Celsius. Write your answer as a decimal and round to the nearest tenth of a degree.
11. $42°F$ $5.6°C$
12. $47°F$ $8.3°C$
13. $79°F$ $26.1°C$
14. $100°F$ $37.8°C$
15. The Concorde jet has a cruising speed of 1,354 mi per hr. Suppose the Concorde maintained this speed for $3\frac{1}{2}$ hr. How far would the Concorde travel? 4,739 mi
16. The air conditioner is on only when the room temperature is greater than 75°F. Room temperature is 22°C. Is the air conditioner on? Explain. No. 22°C is equal to 71.6°F, which is less than 75°F.

H46 Extra Practice

CHAPTER 16 Extra Practice

Set A (pp. 318–319)

For 1–4, use the figure at the right. Tell how many of each you can name. Then name them.
1. points 4; R, S, T, W
2. line segments 6; $\overline{RS}, \overline{ST}, \overline{TW}, \overline{RT}, \overline{SW}, \overline{RW}$
3. rays 6; $\overrightarrow{RS}, \overrightarrow{ST}, \overrightarrow{TW}, \overrightarrow{WT}, \overrightarrow{TS}, \overrightarrow{SR}$
4. lines 1; Possible name: $\overleftrightarrow{RW}$

Name the geometric figure.
5. $\overleftrightarrow{DH}$ or $\overleftrightarrow{HD}$
6. possible answer: plane GSZ
7. $\overline{CK}$ or $\overline{KC}$
8. point F

Set B (pp. 322–325)

Tell if the angles are vertical, adjacent, complementary, supplementary, or none of these.
1. $\angle CFD$ and $\angle DFE$ adjacent
2. $\angle AFB$ and $\angle CFD$ vertical
3. $\angle EFA$ and $\angle CFD$ none of these
4. AFB and BFC supplementary and adjacent

Find the unknown angle measure. The type of angle pair is given.
5. complementary 70°
6. complementary 45°
7. supplementary 145°
8. supplementary 105°

Find the measure of each angle.
9. $\angle QSP$ 60°
10. $\angle LSR$ 90°
11. $\angle QSN$ 150°
12. $\angle RSP$ 90°
13. $\angle MSN$ 30°
14. $\angle RSN$ 180°

Set C (pp. 326–327)

Use the figure.
1. Name all the lines that are parallel to $\overleftrightarrow{BE}$. $\overleftrightarrow{AG}$
2. Name a line that is perpendicular to and intersects $\overleftrightarrow{BE}$. $\overleftrightarrow{AF}$
3. Name all the lines that are perpendicular to $\overleftrightarrow{AD}$. $\overleftrightarrow{BE}$
4. Name a line that is parallel to $\overleftrightarrow{AC}$. $\overleftrightarrow{GE}$
5. Name all the lines that intersect $\overleftrightarrow{GE}$. $\overleftrightarrow{AG}, \overleftrightarrow{BE}, \overleftrightarrow{AD}$

Student Handbook H47

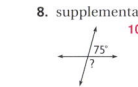

Student Handbook H44–H47

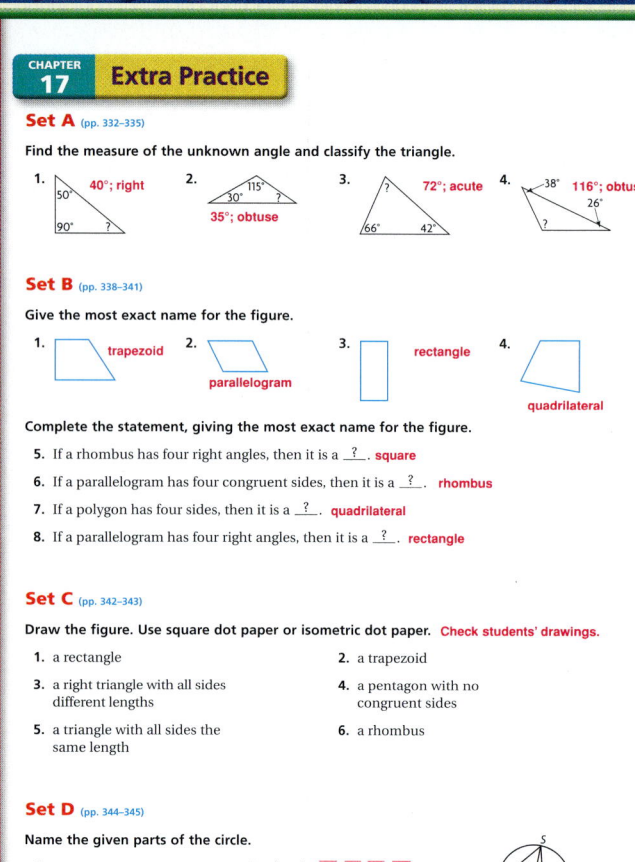
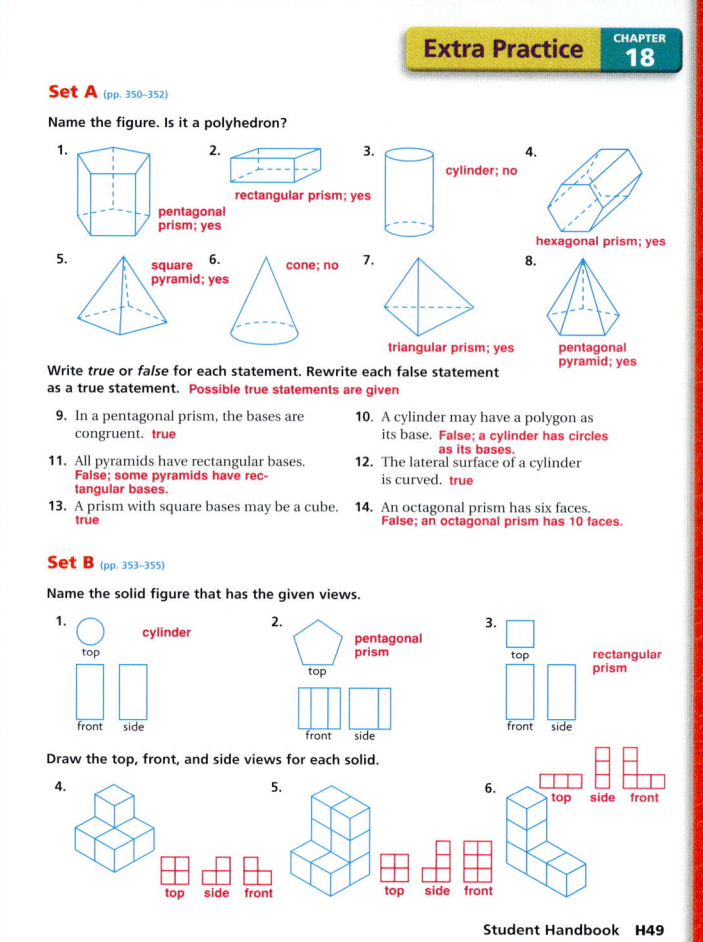
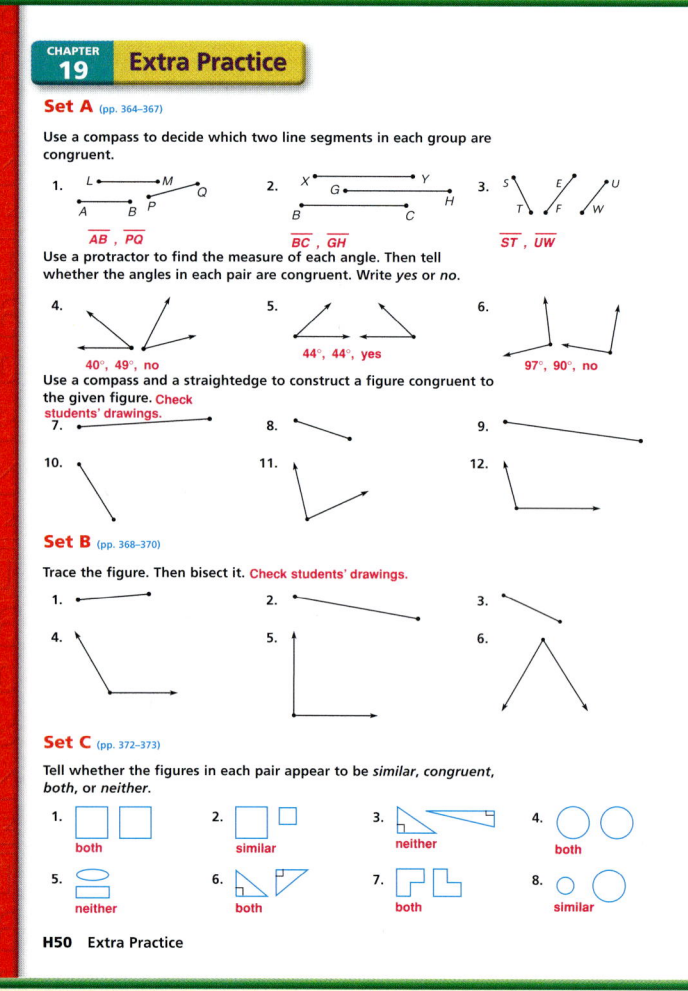
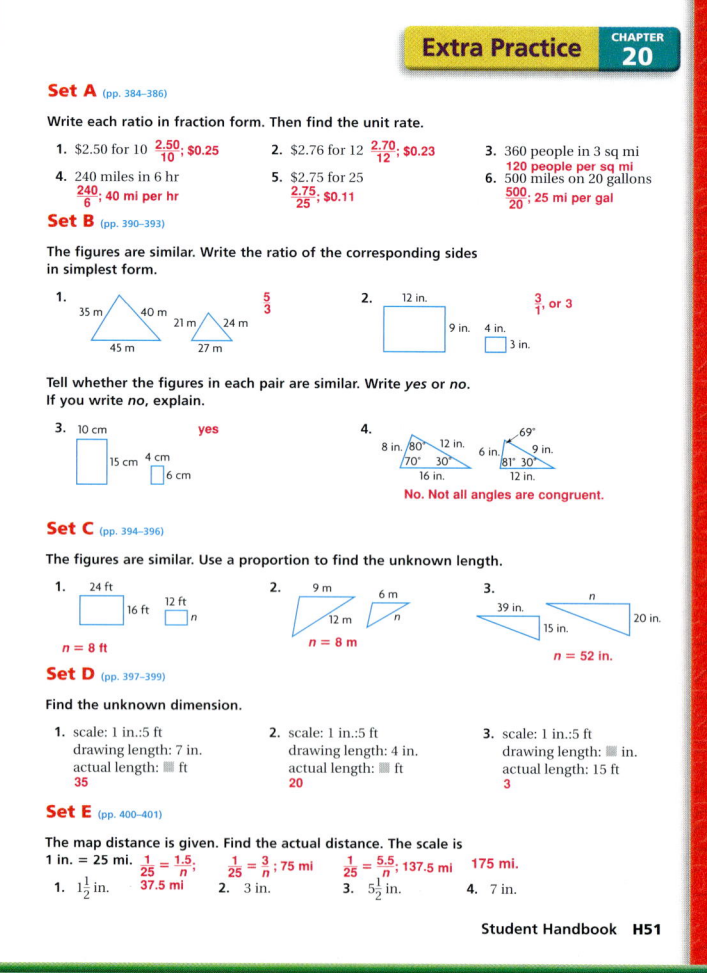

CHAPTER 21 Extra Practice

Set A (pp. 406–407)
Write the percent that is shaded.

1. 50% 2. 73% 3. 75% 4. 12.5%

Set B (pp. 408–411)
Write as a percent.

1. 0.3 **30%**
2. 0.09 **9%**
3. 0.43 **43%**
4. $\frac{7}{20}$ **35%**
5. $\frac{3}{8}$ **$37\frac{1}{2}$%**

Set C (pp. 412–415)
Find the percent.

1. 20% of 8 **1.6**
2. 30% of 90 **27**
3. 45% of 75 **33.75**
4. 50% of 58 **29**
5. Of all the cookies Wendy baked, 40% were chocolate chip. If she baked 200 cookies, how many were chocolate chip? **80**

Set D (pp. 418–421)
Find the sale price.

1. regular price: $31.00; 25% off **$23.25**
2. regular price: $65.00; 50% off **$32.50**
3. regular price: $42.00; 75% off **$10.50**

Find the regular price.

4. sale price: $45.00; 25% off **$60**
5. sale price: $24.95; 50% off **$49.90**
6. sale price: $14.40; 20% off **$18.00**

Set E (pp. 422–423)
Find the simple interest.

	Principal	Rate	Interest for 1 Year	Interest for 5 Years
1.	$65,000.00	4%	$2,600.00	$13,000.00
2.	$735.00	7%	$51.45	$257.25
3.	$1,300.00	3.9%	$50.70	$253.50
4.	$2,250.00	2.3%	$51.75	$258.75

5. Nancy put $2,500 in a savings account for 3 years at a simple interest rate of 8%. How much interest did she earn? **$600**

H52 Extra Practice

CHAPTER 22 Extra Practice

Set A (pp. 428–431)
For 1–5, use the spinner at the right. Find each probability. Write each answer as a fraction, a decimal, and a percent.

1. P(Dee) $\frac{1}{8}$, 0.125, 12.5%
2. P(Miles or Lili) $\frac{1}{4}$, 0.25, 25%
3. P(not Cara) $\frac{7}{8}$, 0.875, 87.5%
4. P(Marta) 0, 0.00, 0%
5. P(Hugo, Miwa, or Chen) $\frac{3}{8}$, 0.375, 37.5%

A number cube is labeled 5, 10, 25, 50, 100, and 2,000. Find each probability. Write each answer as a fraction.

6. P(25) $\frac{1}{6}$
7. P(5 or 25) $\frac{1}{3}$
8. P(a number ending in zero) $\frac{2}{3}$
9. P(1,000) 0
10. P(1,000 or 2,000) $\frac{1}{6}$
11. P(not 500) 1

Cards showing pictures of team mascots are placed in a hat. There are 3 lions, 5 bears, 4 cheetahs, and 8 tigers. You choose one card without looking. Find each probability.

12. P(bear) $\frac{1}{4}$
13. P(tiger or lion) $\frac{11}{20}$
14. P(member of cat family) $\frac{15}{20}$, or $\frac{3}{4}$

A bag contains some buttons: 8 blue, 12 brown, 10 red, 4 green, and 6 yellow. You choose one button without looking. Find each probability.

15. P(yellow) $\frac{3}{20}$
16. P(black) 0
17. P(not brown) $\frac{7}{10}$

Set B (pp. 436–437)
A spinner is divided into 5 equal sections. Each section is labeled with one of the letters A, E, I, O, and U. Anna spins the pointer 100 times and records her results in the table below.

Letter	A	E	I	O	U
Times Landed On	15	5	30	40	10

1. P(A) $\frac{3}{20}$
2. P(E) $\frac{1}{20}$
3. P(I) $\frac{3}{10}$
4. P(O) $\frac{2}{5}$
5. P(U) $\frac{1}{10}$
6. P(A or U) $\frac{1}{4}$
7. P(E, I or U) $\frac{9}{20}$
8. P(not A) $\frac{17}{20}$

9. Based on her experimental results, how many times can Anna expect the pointer to land on O in the next 20 spins? **8 times**
10. Based on her experimental results, how many times can Anna expect the pointer to land on E if she spins 2,000 times? **100 times**

Student Handbook H53

CHAPTER 23 Extra Practice

Set A (pp. 444–446)
Draw a tree diagram or make a table to find the number of possible outcomes for each situation. **Check students' diagrams.**

1. choosing vanilla, chocolate, or strawberry yogurt, with cherry or chocolate sauce, and sprinkles or nuts **12 outcomes**
2. tossing a penny and spinning the pointer on the spinner **16 choices**

Use the Fundamental Counting Principle to find the number of outcomes for each situation.

3. a choice of pancakes, french toast, or waffles, and juice, milk, or tea **9 outcomes**
4. a choice of 6 salads and 10 dressings **60 outcomes**
5. rolling 2 number cubes labeled A to F **36 outcomes**

Set B (pp. 447–449)
Write independent or dependent to describe the events.

1. You have a bag of 6 red marbles and 4 green marbles. You draw one marble, record the color, place the marble back in the bag, and draw again. **independent**
2. Ana draws one name from a box to select the winner of a movie pass and then draws another name from the same box for the winner of a CD. **dependent**

A box contains five cards labeled C, L, A, S, S. Without looking in the box, you select a card, replace it, and then select another. For 3–8, find the probability of each event. Then find the probability assuming the first card is not replaced.

3. P(C, L) $\frac{1}{25}$; $\frac{1}{20}$
4. P(L, C) $\frac{1}{25}$; $\frac{1}{20}$
5. P(A, S) $\frac{2}{25}$; $\frac{1}{10}$
6. P(S, L) $\frac{2}{25}$; $\frac{1}{10}$
7. P(C, L or S) $\frac{3}{25}$; $\frac{3}{20}$
8. P(C or L, S) $\frac{4}{25}$; $\frac{1}{5}$

Set C (pp. 450–451)
Seventy-five students from Park Middle School were randomly surveyed about favorite pizza toppings. The results are shown in the table.

FAVORITE PIZZA TOPPINGS	
Topping	Number of Students
Pepperoni	30
Black Olives	15
Sausage	12
Mushrooms	10
Other	8

1. Suppose there are 225 students who attend Park Middle School. Predict the number of students who prefer black olives as a pizza topping. **about 45 students**
2. Suppose there are 314 students at Park Middle School. Predict the number of students who prefer sausage as a pizza topping. **about 50 students**

H54 Extra Practice

CHAPTER 24 Extra Practice

Set A (pp. 462–463)
Use a proportion to convert to the given unit.

1. 15 ft = m yd **5**
2. 12 c = r pt **6**
3. 8 gal = h qt **32**
4. 120 oz = x lb **$7\frac{1}{2}$**
5. 6 days = b hr **144**
6. 12 in. = p yd **$\frac{1}{3}$**

Set B (pp. 464–465)
Use a proportion to convert to the given unit.

1. 0.22 m = c km **0.00022**
2. 450 mm = h cm **45**
3. 0.0030 kL = p L **3**
4. 1,800 g = ■ kg **1.8**
5. 10,000 cm = ■ m **100**
6. 0.35 L = ■ mL **350**
7. Paul buys 2 L of orange juice. He drinks 250 mL with breakfast. How many milliliters are left? **1,750 mL**
8. Tiffany runs 5,000 m. Ashley runs 3.5 km. Who runs farther and by how many meters? **Tiffany; 1,500 m**

Set C (pp. 466–467)
Use a proportion to convert to the given unit. Use the table on page 456. Round to the nearest hundredth if necessary.

1. 9 in. ≈ ■ cm **22.86**
2. 11 yd ≈ ■ m **10.01**
3. 3 mi ≈ ■ km **4.83**
4. 10 L ≈ ■ gal **2.64**
5. 55 cm ≈ ■ ft **1.80**
6. 3.5 kg ≈ ■ lb **7.78**
7. Eve weighs 51 kg. Beth weighs 116 lb. Who weighs more and by about how many pounds? **Beth; about 4 lb**
8. Zack has a board that is 6 ft long. He wants to cut a length that is 150 cm. About how many centimeters will be left? **about 33 cm**

Set D (pp. 468–471)
Measure the line segment to the given length.

1. nearest half inch; nearest inch **$2\frac{1}{2}$ in.; 2 in.**

2. nearest centimeter; nearest millimeter **3 cm; 33 mm**

Tell which measurement is more precise.

3. 7 fl oz or 1 cup **7 fl oz**
4. 2 qt or 9 c **9 c**
5. 1,245 m or 1 km **1,245 m**

Name an appropriate customary or metric unit of measure for each item.

6. weight or mass of a bag of sugar **lb or kg**
7. length of a shoelace **in. or cm**
8. water in a fishbowl **qt or liter**

Student Handbook H55

CHAPTER 25 **Extra Practice**

Set A (pp. 479–481)

Find the perimeter.

1. 18 in., 9 in. **54 in.**
2. 3.3 cm, 3.3 cm, 3.3 cm, 3.3 cm, 3.3 cm, 3.3 cm **19.8 cm**
3. 490 cm, 6.5 m, 9.8 m **21.2 m**

The perimeter is given. Find the unknown length.

4. 18 cm, 41 cm, 32 cm, 52 cm, 52 cm, x; $P = 230$ cm **x = 35 cm**
5. 15 m, 10 m, 9 m, 8.2 m, y; $P = 54.5$ m **y = 12.3 m**
6. $13\frac{5}{6}$ ft, $13\frac{2}{6}$ ft, $18\frac{1}{2}$ ft, g; $P = 54$ ft **g = 18 ft**
7. Tanya wants to put a string of lights around a rectangular window that is 40 in. wide and 48 in. high. How long will the string of lights need to be to go around the window one time? **176 in.**
8. Linda is building a raised flowerbed $10\frac{1}{2}$ ft long and 4 ft wide. How many feet of lumber will she need to go around the flowerbed? **29 ft**

Set B (pp. 484–487)

Find the circumference. Use 3.14 or $\frac{22}{7}$ for π. Round to the nearest whole number.

1. 7 in. **44 in.**
2. 12 ft **38 ft**
3. 49 cm **308 cm**
4. $4\frac{1}{2}$ in. **28 in.**
5. 2.7 cm **8 cm**
6. 154 ft **968 ft**
7. $r = 16$ ft **100 ft**
8. $d = 23.5$ cm **74 cm**
9. $d = 28$ ft **88 ft**
10. $r = 25.9$ m **163 m**
11. $d = 82.1$ mm **258 mm**
12. $d = 3.21$ m **10 m**
13. A blue circular rug in Dominique's room has a diameter of $6\frac{1}{2}$ ft. To the nearest foot, what is the circumference of the rug? **20 ft**
14. A child's hat has a radius of 8 cm. To the nearest tenth of a centimeter, what is the circumference of the hat? **50.2 cm**

Extra Practice CHAPTER 26

Set A (pp. 494–497)

Estimate the area. Each square on the grid represents 1 cm^2.

1. **about 23 cm^2**
2. **about 7 cm^2**
3. **about 11 cm^2**

Find the area.

4. 24 cm, 32 cm **384 cm^2**
5. 9 ft, 15 ft **67.5 ft^2**
6. 5 in., 8 in. **20 in.^{2}**
7. 9.5 m, 5.2 m **49.4 m^2**

Set B (pp. 498–500)

Find the area of each figure.

1. 12 yd, 7 yd **84 yd^2**
2. 6 ft, 4 ft, 10 ft **32 ft^2**
3. 5.3 m, 8.6 m **45.58 m^2**
4. 18 cm, 12 cm, 23 cm **246 cm^2**
5. A wall plaque is shaped like a trapezoid with a height of 6 in. and bases that measure 3.5 in. and 9.5 in. Find the area. **39 in.^{2}**

Set C (pp. 502–503)

Find the area of each circle to the nearest whole number.

1. 9 yd **254 yd^2**
2. 12 mm **113 mm^2**
3. 7 ft **154 ft^2**
4. 4.2 in. **55 in^2**
5. $d = 24$ cm **452 cm^2**
6. $r = 9.2$ mm **266 mm^2**
7. $d = 8$ yd **50 yd^2**
8. $d = 7\frac{1}{2}$ ft **44 ft^2**

Set D (pp. 504–507)

Find the surface area.

1. 2 in., 8 in., 5 in. **132 in.^{2}**
2. 4 m, 2 m, 2 m **20 m^2**
3. 7 m, 4 m, 3 m **122 m^2**
4. 4 in., 5 in., 5 in., 9 in., 6 in. **168 in.^{2}**

CHAPTER 27 **Extra Practice**

Set A (pp. 512–515)

Find the volume.

1. 3 in., 4 in., 6 in. **72 in.^{3}**
2. 10.9 m, 5 m, 2.5 m **136.25 m^3**
3. 4.9 m, 10.2 m, 12.5 m **312.375 m^3**

Find the unknown length.

4. 6 cm, 14 cm, x; $V = 420$ cm^3 **x = 5 cm**
5. 12 cm, 5.5 cm, x; $V = 247.5$ cm^3 **x = 7.5 cm**
6. 17 in., 6 in., x; $V = 153$ in.^{3} **x = 3 in.**

Set B (pp. 518–519)

Find the volume.

1. 17 ft, 8 ft, 8 ft **362.7 ft^3**
2. 9.5 m, 7 m, 5 m **110.8 m^3**
3. 79 in., 41 in., 42 in. **45,346 in.^{3}**
4. Find the volume of a rectangular pyramid with a length of 18 cm, a width of 16 cm, and a height of 12 cm. **1,152 cm^3**
5. Find the volume of a rectangular pyramid with a base of 300 ft^2 and a height of 50 ft. **5,000 ft^3**

Set C (pp. 521–523)

Find the volume. Round to the nearest whole number.

1. 17 ft, 5 ft **about 4,537 ft^3**
2. 6.2 mm, 4.5 mm **about 136 mm^3**
3. 7 cm, 18 cm **about 2,769 cm^3**

Find the volume of the inside cylinder to the nearest whole number.

4. 6.7 cm, 4 cm, 1.5 cm **about 131 cm^3**
5. 3 ft, 9 ft, 13 ft **about 1,470 ft^3**
6. 5 mm, 2 mm, 9 mm **about 254 mm^3**

Extra Practice CHAPTER 28

Set A (pp. 537–538)

Write a rule for each sequence.

1. 2, 6, 18, 54, . . . **multiply by 3**
2. $\frac{1}{10}, \frac{1}{5}, \frac{3}{10}, \frac{2}{5}, \ldots$ **add $\frac{1}{10}$ and simplify**
3. $^-35, ^-20, ^-5, 10, \ldots$ **add 15**
4. 440, 44, 4.4, 0.44, . . . **divide by 10**

Find the next three possible terms in each sequence.

5. $\frac{2}{5}, \frac{4}{5}, 1\frac{1}{5}, \ldots$ **$1\frac{3}{5}, 2, 2\frac{2}{5}$**
6. 2.3, 3.9, 5.5, . . . **7.1, 8.7, 10.3**
7. 7, 17, 37, 67, . . . **107, 157, 217**
8. $27, ^-9, 3, ^-1, \ldots$ **$\frac{1}{3}, ^-\frac{1}{9}, \frac{1}{27}$**
9. Vittorio practices 6 weeks for the skateboard championship. The first 4 weeks he practices $9\frac{1}{2}$ hr, $11\frac{1}{4}$ hr, 13 hr, and $14\frac{3}{4}$ hr. Following this pattern, how many hours will Vittorio practice the sixth week? **$18\frac{1}{4}$ hr**

Set B (pp. 539–542)

Write an equation to represent the function.

1. **$y = x + 3$**

x	0	1	2	3	4
y	3	4	5	6	7

2. **$y = x - 4$**

x	10	9	8	7	6
y	6	5	4	3	2

3. **$y = 8x$**

x	0	1	2	3	4
y	0	8	16	24	32

4. **$y = x \div 3$**

x	30	27	24	21	18
y	10	9	8	7	6

Set C (pp. 543–545)

Draw the next two figures in the pattern.

1. 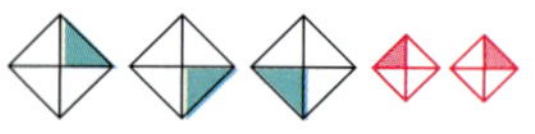
2. 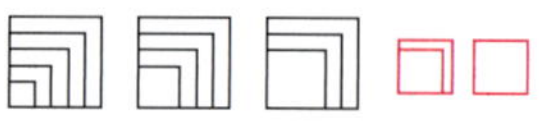

Draw the next two solids in the pattern.

3. 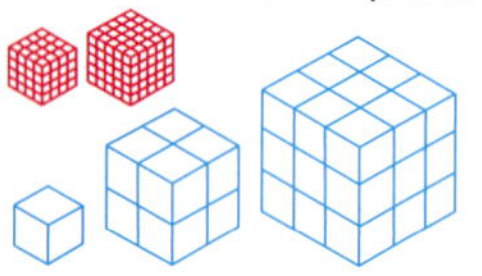
4. 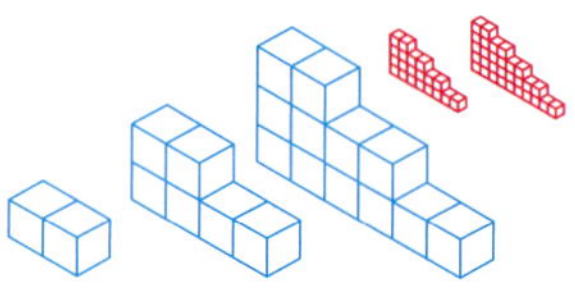

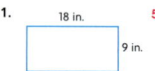

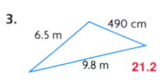

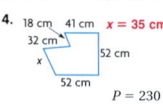

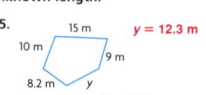

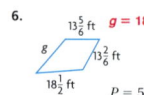

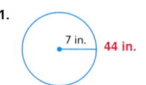

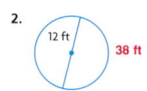

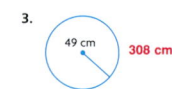

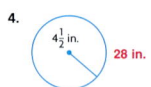

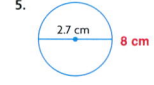

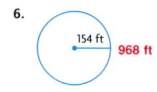

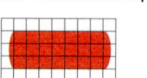

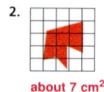

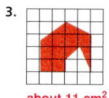

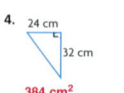

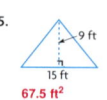

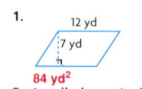

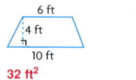

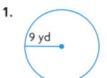

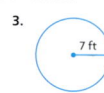

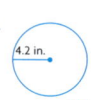

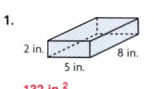

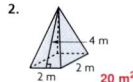

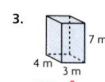

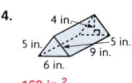

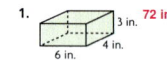

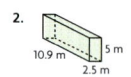

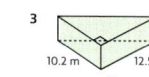

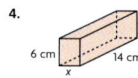

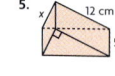

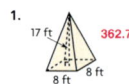

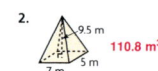

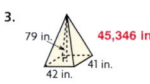

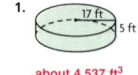

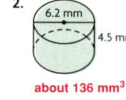

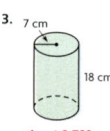

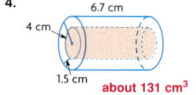

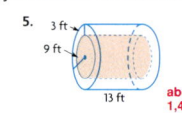

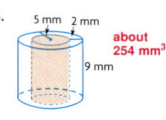

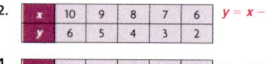

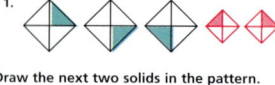

CHAPTER 29 Extra Practice

Set A (pp. 550–552)

Tell which type or types of transformation the second figure is of the first figure. Write *translation*, *rotation*, or *reflection*.

1. translation
2. rotation
3. rotation
4. reflection or rotation

Set B (pp. 553–555)

Trace and cut out several of each shape. Tell whether the shape can be used repeatedly to form a tessellation. Write *yes* or *no*.

1. yes
2. yes
3. yes
4. no

Set C (pp. 558–559)

Tell how many ways you can place the solid figure on the outline.

1. 12 ways
2. 8 ways
3. 2 ways

Set D (pp. 560–563)

Trace the figure. Draw the lines of symmetry.

1.
2.
3.
4.

Tell whether each figure has rotational symmetry, and, if so, identify the symmetry as a fraction of a turn and in degrees.

5. no
6. yes; $\frac{1}{2}$; 180°
7. yes; $\frac{1}{9}$; 40°
8. no

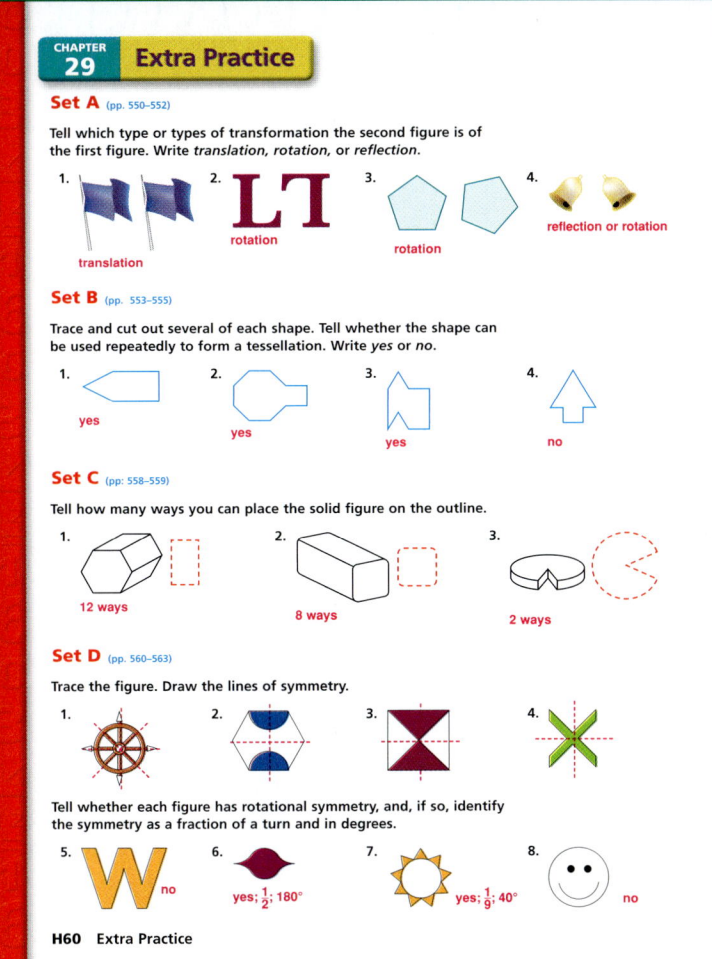

H60 Extra Practice

CHAPTER 30 Extra Practice

Set A (pp. 568–569)

Graph the solutions of the inequality. See Additional Answers, p. H61A.

1. $x < 4$
2. $x > 9$
3. $x \geq 3$
4. $x < {}^-5$

Solve the inequality and graph the solutions on a number line. See Additional Answers, p. H61A.

5. $7a \leq 14$ $a \leq 2$
6. $n + 5 > 8$ $n > 3$
7. $m - 2 < 0$ $m < 2$
8. $5c \geq 5$ $c \geq 1$

Set B (pp. 570–573)

Write the ordered pair for each point on the coordinate plane.

1. point A (2,6)
2. point B (${}^-5$,${}^-3$)
3. point C (2,${}^-2$)
4. point D (${}^-4$,7)
5. point E (0,7)
6. point F (${}^-4$,2)
7. point G (3,0)
8. point H (${}^-2$,${}^-5$)
9. point I (6,${}^-4$)
10. point J (6,4)

Set C (pp. 574–575)

Complete the table. Then write an equation relating y to x.

1.
x	1	2	3	4	5
y	3	4	5	▪	▪
6; 7; $y = x + 2$

2.
x	1	2	3	4	5
y	6	12	18	▪	▪
24; 30; $y = 6x$

3.
x	8	9	10	11	12
y	1	2	3	▪	▪
4; 5; $y = x - 7$

4.
x	8	4	0	${}^-4$	${}^-8$
y	${}^-2$	${}^-1$	0	▪	▪
1; 2; $y = x \div {}^-4$

Set D (pp. 580–583)

Copy the figure onto a coordinate plane. Transform the figure according to the directions given. Name the new coordinates.
See Additional Answers, p. H61A.

1. 2 units to the right
A'(2,2), B'(4,2), C'(2,0), D'(4,0)

2. rotate 90° clockwise about (${}^-2$,0)
E'(0,0), F'(0,${}^-4$), G'(${}^-2$,0), H'(${}^-2$,${}^-4$)

3. reflect across the y-axis
I'(3,${}^-2$), J'(0,${}^-2$), K'(0,2)

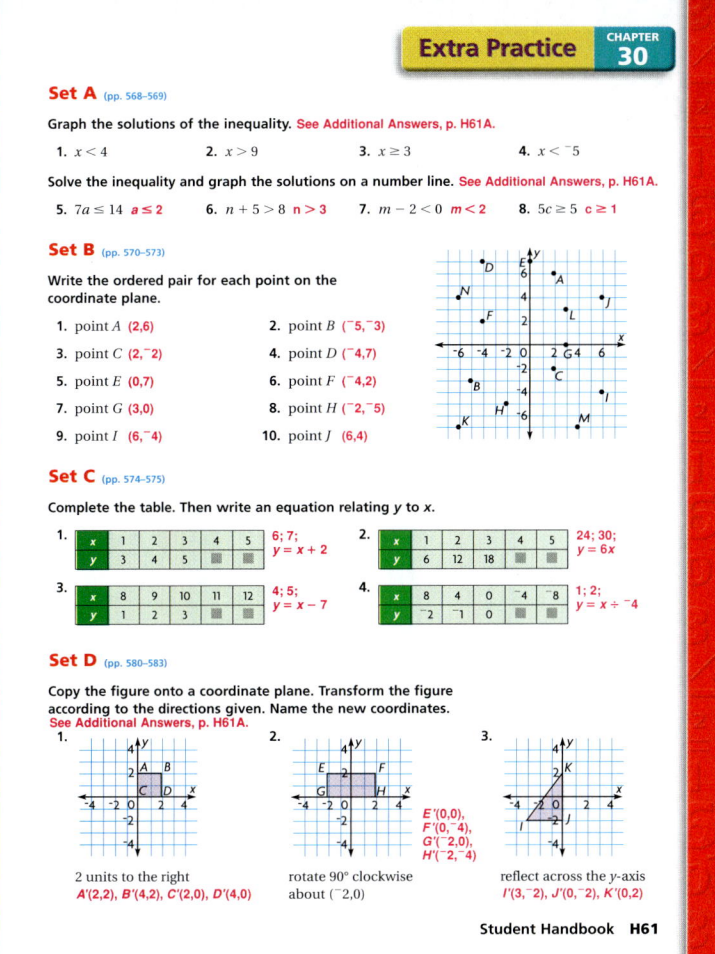

Student Handbook H61

Additional Answers

Chapter 6 Extra Practice, page H37

1.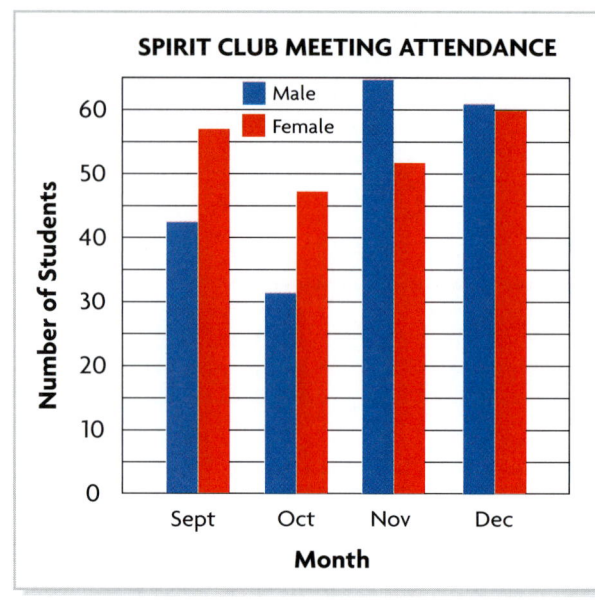

2.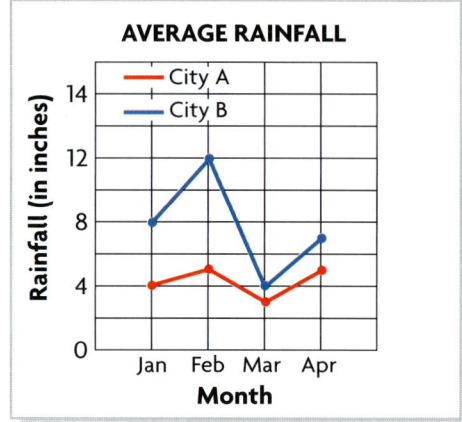

3.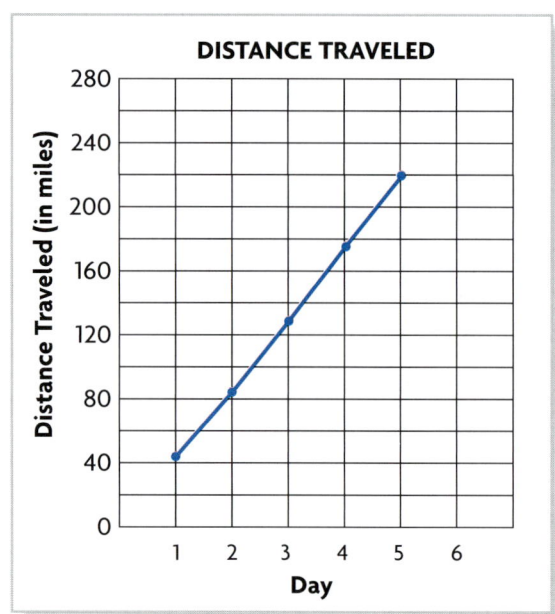

4.

Stems	Leaves
1	3 4 7 8 9
2	1 2 2 5 8 9
3	0 3 6 7
4	1 4 8

mode: 22; median: 26.5

5.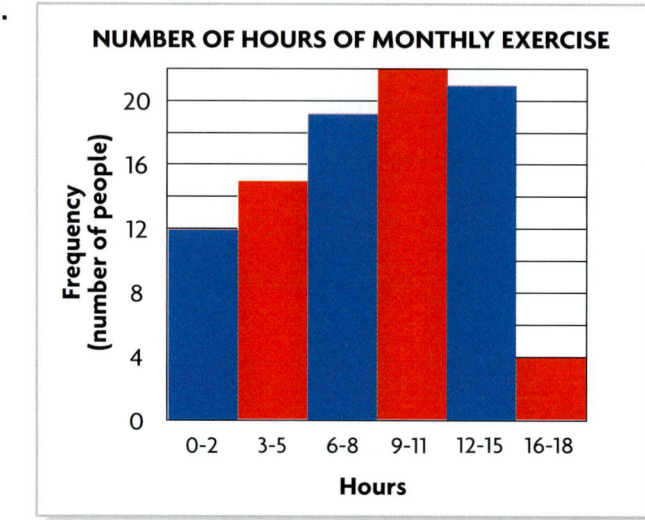

10.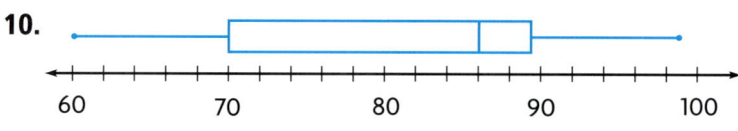

H61A Student Handbook

Additional Answers

Chapter 30 Extra Practice, Set A, page H61

1.
2.
3.
4.
5.
6.
7.
8.

Chapter 30 Extra Practice, Set D, page H61

1.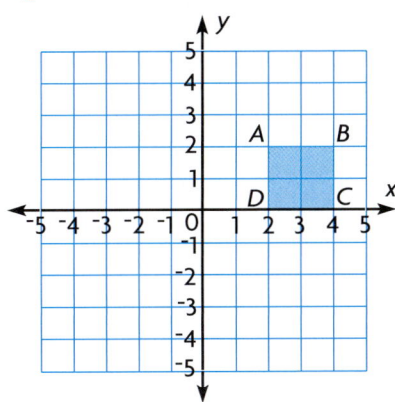

2.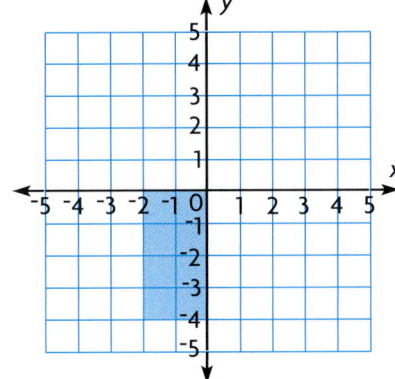

3.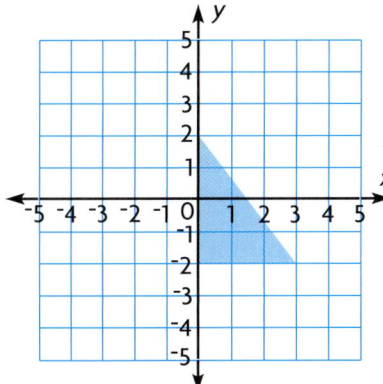

Sharpen Your Test-Taking Skills

TIPS FOR TAKING MATH TESTS

Being a good test-taker is like being a good problem solver. When you answer test questions, you are solving problems. Remember to ANALYZE, CHOOSE, SOLVE, and CHECK.

- Analyze
- Choose
- Solve
- Check

Analyze

Read the problem.
- Look for math terms and recall their meanings.
- Reread the problem and think about the question.
- Use the details in the problem and the question.

1. The difference between two numbers is 37. Their sum is 215. What are the numbers?

   A 37 and 178   C 89 and 126
   B 83 and 120   D 107 and 108

   TIP! Understand the problem. The problem requires you to find two numbers for which the difference and sum are given. Reread the problem to compare the details to the answer choices. You can use estimation instead of calculating the sum and difference of each pair of numbers. The answer is **C**.

- Each word is important. Missing a word or reading it incorrectly could cause you to get the wrong answer.
- Pay attention to words that are in **bold** type, all CAPITAL letters, or *italics* and words like *round*, *best*, or *least to greatest*.

2. Florinda took $\frac{1}{2}$ hour to complete a race. Doris took $\frac{3}{8}$ hour and Violet took $\frac{2}{3}$ hour to complete the same race. Which lists the three runners from fastest to slowest?

   F Florinda, Doris, Violet
   G Doris, Violet, Florinda
   H Violet, Florinda, Doris
   J Doris, Florinda, Violet

   TIP! Look for important words. The words *fastest to slowest* are important. The fastest runner takes the least amount of time. Think about where each fraction would be placed on a number line between 0 and 1. Then put the times in order from least to greatest. The answer is **J**.

Choose

Think about how you can solve the problem.
- See if you can solve the problem with the information given.
- Pictures, charts, tables, and graphs may have the information you need.
- You may need to recall information not given.
- Sometimes the answer choices have information to help solve the problem.

3. Jeremy wants to make a graph to see the trend of the profit in his lawn-mowing business over the past 10 months. What type of graph would best show this?

   A circle graph   C line graph
   B histogram   D stem-and-leaf plot

   TIP! Get the information you need. The answer choices give four different types of graphs or plots. Think about each one and the kind of data that is appropriate for it. The problem states that Jeremy wants to see the trend in his profit over the past 10 months, and line graphs show trends. The answer is **C**.

- You may need to write a number sentence and solve it to answer the question.
- Some problems have two steps or more.
- In some problems you need to look at relationships instead of computing an answer.
- If the path to the solution isn't clear, choose a problem solving strategy and use it to solve the problem.

4. A square tile with 4-inch sides is rotated along the line below. If the tile stopped so that the letter P appears in an upright position, which of these distances could it have been rotated?

   F 40 in.   H 48 in.
   G 44 in.   J 52 in.

   TIP! Decide on a plan. From the distances given, you must find the one that could allow for complete rotations of the square tile. Since each side of the square is 4 inches long, it moves 16 inches in one complete rotation. If you *use logical reasoning*, you see that only one of the choices is a multiple of 16. The answer is **H**.

Solve

Follow your plan, working logically and carefully.
- Estimate your answer. Compare it to the answer choices.
- Use reasoning to find the most likely choices.
- Make sure you solved all the steps needed to answer the problem.
- If your answer does not match any of the answer choices, check the numbers you used. Then check your computation.

5. Glen and Doug painted a fence. It took Glen three times as long to paint one side of the fence as it took Doug to paint the other side. If *h* represents the number of hours Doug painted, which expression shows the hours Glen painted?

   A $3 \div h$ C $3 + h$
   B $3 - h$ D $3 \times h$

   TIP! Eliminate choices. It is important to understand that *h* represents the hours it took Doug to paint one side of the fence. Since Glen painted three times as long as Doug, you can eliminate answers **A** and **B** because it does not make sense to divide by or subtract the hours Doug painted. Answer **C** means three more than *h*, so it can be eliminated. The answer is **D**.

- If your answer still does not match one of the choices, look for another form of the number, such as a decimal instead of a fraction.
- If answer choices are given as pictures, look at each one by itself while you cover the other three.
- If you do not see your answer and the answer choices include Not here, make sure your work is correct and then mark Not here.
- Read answer choices that are statements and relate them to the information in the problem one by one.

6. Tanya and her class helped plant 120 tulip and daffodil bulbs in front of the school. If 30 of the bulbs were tulips, what percent of the bulbs were daffodils?

   F 90%   H 25%
   G $66\frac{2}{3}$%   J Not here

   TIP! Choose the answer. Since 30 of the bulbs were tulips, 90 were daffodils. The question asks what percent are daffodils, so you need to find what percent of 120 is 90 (90 ÷ 120). If your answer doesn't match one of the answer choices, check your computation. If you know your work is correct, mark the letter for Not here. The answer is **J**.

Check

Take time to catch your mistakes.
- Be sure you answered the question asked.
- Check for important words you might have missed.
- Be sure you used all the information you needed.
- Check your computation by using a different method.

7. The temperature was 6°F today. A forecaster predicted that the temperature would drop by 5° each day for the next four days. Which sequence could be used to find the predicted temperatures?

   A 6°F, 11°F, 16°F, 21°F, 26°F
   B 6°F, 1°F, 6°F, 11°F, 16°F
   C 6°F, 1°F, ⁻6°F, ⁻11°F, ⁻16°F
   D 6°F, 1°F, ⁻4°F, ⁻9°F, ⁻14°F

   TIP! Check your work. You need to find the sequence that shows a drop of 5° in temperature for the next four days. Draw a thermometer or a number line to check your computation. The answer is **D**.

Don't Forget!

Before the test...
- Listen to the teacher's directions and read the instructions.
- Write down the ending time if the test is timed.
- Know where and how to mark your answers.
- Know whether you should write on the test page or use scratch paper.
- Ask any questions you have before the test begins.

During the test...
- Work quickly but carefully. If you are unsure how to answer a question, leave it blank and return to it later.
- If you cannot finish on time, look over the questions that are left. Answer the easiest ones first. Then go back to answer the others.
- Fill in each answer space carefully. Erase completely if you change an answer. Erase any stray marks.
- Check that the answer number matches the question number, especially if you skip a question.

Skills Review — Adding and Subtracting

1. 70 + 8 = **78**
2. 43 + 5 = **48**
3. 58 + 7 = **65**
4. 67 + 18 = **85**
5. 24 + 36 = **60**
6. 264 + 58 = **322**
7. 836 + 54 = **890**
8. 1,641 + 385 = **2,026**
9. 3,231 + 578 = **3,809**
10. 4,605 + 2,493 = **7,098**
11. 5,619 + 2,537 = **8,156**
12. 84,520 + 3,864 = **88,384**
13. 274,051 + 40,318 = **314,369**
14. 68 − 5 = **63**
15. 74 − 43 = **31**
16. 50 − 38 = **12**
17. 149 − 58 = **91**
18. 394 − 145 = **249**
19. 560 − 217 = **343**
20. 750 − 192 = **558**
21. 1,460 − 316 = **1,144**
22. 4,960 − 681 = **4,279**
23. 8,000 − 562 = **7,438**
24. 35,120 − 6,299 = **28,821**
25. 36,009 − 19,148 = **16,861**
26. 0.4 + 0.9 = **1.3**
27. 2.6 + 8.1 = **10.7**
28. 0.31 + 2.04 = **2.35**
29. 5.92 + 3.15 = **9.07**
30. 0.218 + 2.143 = **2.361**
31. 2.14 + 6.08 = **8.22**
32. 7.04 + 0.13 = **7.17**
33. 0.126 + 0.408 = **0.534**
34. 8.360 + 5.216 = **13.576**
35. 5.816 + 3.215 = **9.031**
36. 17.31 + 12.06 = **29.37**
37. 8.26 + 26.15 = **34.41**
38. 31.18 + 125.50 = **156.68**
39. 11.4 − 6.2 = **5.2**
40. 12.8 − 4.1 = **8.7**
41. 17.3 − 16.5 = **0.8**
42. 21.8 − 17.4 = **4.4**
43. 13.2 − 8.17 = **5.03**
44. 35.51 − 23.17 = **12.34**
45. 8.026 − 7.317 = **0.709**
46. 19.408 − 1.582 = **17.826**
47. 5.848 − 5.261 = **0.587**
48. 13.601 − 10.311 = **3.290**
49. 5 − 3.21 = **1.79**
50. 630.71 − 527.21 = **103.5**

Skills Review — Multiplying

1. 20 × 4 = **80**
2. 30 × 2 = **60**
3. 30 × 3 = **90**
4. 24 × 2 = **48**
5. 31 × 3 = **93**
6. 46 × 2 = **92**
7. 37 × 4 = **148**
8. 74 × 6 = **444**
9. 63 × 8 = **504**
10. 58 × 9 = **522**
11. 61 × 20 = **1,220**
12. 84 × 40 = **3,360**
13. 76 × 30 = **2,280**
14. 68 × 50 = **3,400**
15. 92 × 60 = **5,520**
16. 26 × 14 = **364**
17. 93 × 15 = **1,395**
18. 50 × 38 = **1,900**
19. 81 × 54 = **4,374**
20. 79 × 43 = **3,397**
21. 680 × 35 = **23,800**
22. 710 × 52 = **36,920**
23. 825 × 173 = **142,725**
24. 522 × 286 = **149,292**
25. 463 × 836 = **387,068**
26. 0.5 × 0.9 = **0.45**
27. 3.6 × 7.1 = **25.56**
28. 1.3 × 2.5 = **3.25**
29. 5.9 × 3.1 = **18.29**
30. 4.2 × 8.7 = **36.54**
31. 3.14 × 6.8 = **21.352**
32. 7.24 × 0.8 = **5.792**
33. 9.65 × 0.4 = **3.86**
34. 7.25 × 1.8 = **13.05**
35. 9.97 × 0.9 = **8.973**
36. 17.31 × 2.06 = **35.6586**
37. 8.26 × 0.87 = **7.1862**
38. 31.82 × 15.5 = **493.21**
39. 15.4 × 6.27 = **96.558**
40. 12.8 × 4.16 = **53.248**
41. 17.3 × 16.5 = **285.45**
42. 165.4 × 7.4 = **1,223.96**
43. 139.2 × 8.17 = **1,137.264**
44. 935.5 × 3.25 = **3,040.375**
45. 987.2 × 7.8 = **7,700.16**
46. 246.28 × 25.5 = **6,280.14**
47. 950.8 × 52.4 = **49,821.92**
48. 825.2 × 9.63 = **7,946.676**
49. 440.7 × 425.2 = **187,385.64**
50. 2,145.2 × 527.5 = **1,131,593**

Skills Review — Dividing

1. 8)72 = **9**
2. 27)108 = **4**
3. 6)138 = **23**
4. 4)816 = **204**
5. 5)525 = **105**
6. 6)624 = **104**
7. 7)763 = **109**
8. 4)836 = **209**
9. 5)205 = **41**
10. 2)164 = **82**
11. 7)642 = **91 r5**
12. 9)700 = **77 r7**
13. 2)785 = **392 r1**
14. 5)277 = **55 r2**
15. 3)343 = **114 r1**
16. 90 ÷ 30 = **3**
17. 40 ÷ 20 = **2**
18. 160 ÷ 40 = **4**
19. 360 ÷ 40 = **9**
20. 540 ÷ 90 = **6**
21. 630 ÷ 70 = **9**
22. 3,000 ÷ 60 = **50**
23. 100 ÷ 20 = **5**
24. 560 ÷ 70 = **8**
25. 3,500 ÷ 50 = **70**
26. 630 ÷ 58 = **10 r50**
27. 4,801 ÷ 37 = **129 r28**
28. 100 ÷ 21 = **4 r16**
29. 560 ÷ 82 = **6 r68**
30. 1,875 ÷ 19 = **98 r13**
31. 900 ÷ 300 = **3**
32. 480 ÷ 240 = **2**
33. 840 ÷ 105 = **8**
34. 1,500 ÷ 300 = **5**
35. 9,800 ÷ 800 = **12 r200**
36. 1.2 ÷ 4 = **0.3**
37. 0.12 ÷ 4 = **0.03**
38. 3.5 ÷ 5 = **0.7**
39. 6.4 ÷ 8 = **0.8**
40. 0.18 ÷ 9 = **0.02**
41. 3.69 ÷ 3 = **1.23**
42. 83.7 ÷ 9 = **9.3**
43. 44.8 ÷ 4 = **11.2**
44. 56.8 ÷ 8 = **7.1**
45. 19.75 ÷ 5 = **3.95**
46. 2.24 ÷ 4 = **0.56**
47. 4.48 ÷ 2.8 = **1.6**
48. 3.78 ÷ 3 = **1.26**
49. 12.1 ÷ 1.1 = **11**
50. 229.6 ÷ 8.2 = **28**
51. 0.38)13.3 = **35**
52. 0.55)2.42 = **4.4**
53. 2.48)1.3392 = **0.54**
54. 6.41)135.892 = **21.2**
55. 15)10.8 = **0.72**
56. 9)43.65 = **4.85**
57. 18.2)378.56 = **20.8**
58. 49.3)201.144 = **4.08**
59. 186.24 ÷ 29.1 = **6.4**
60. 378.56 ÷ 18.2 = **20.8**

Glossary

Pronunciation Key

a	add, map	f	fit, half	n	nice, tin	p	pit, stop	yo͞o	fuse, few
ā	ace, rate	g	go, log	ng	ring, song	r	run, poor	v	vain, eve
â(r)	care, air	h	hope, hate	o	odd, hot	s	see, pass	w	win, away
ä	palm, father	i	it, give	ō	open, so	sh	sure, rush	y	yet, yearn
b	bat, rub	ī	ice, write	ô	order, jaw	t	talk, sit	z	zest, muse
ch	check, catch	j	joy, ledge	oi	oil, boy	th	thin, both	zh	vision, pleasure
d	dog, rod	k	cool, take	ou	pout, now	t͟h	this, bathe		
e	end, pet	l	look, rule	o͝o	took, full	u	up, done		
ē	equal, tree	m	move, seem	o͞o	pool, food	û(r)	burn, term		

ə the schwa, an unstressed vowel representing the sound spelled *a* in above, *e* in sicken, *i* in possible, *o* in melon, *u* in circus

Other symbols:
- • separates words into syllables
- ′ indicates stress on a syllable

A

absolute value [ab′sə•lo͞ot val′yo͞o] The distance of an integer from zero (p. 228)

acute angle [ə•kyo͞ot′ an′gəl] an angle whose measure is greater than 0° and less than 90° (p. 320)

acute triangle [ə•kyo͞ot′ trī′an•gəl] A triangle with all angles less than 90° (p. 332)
Example:

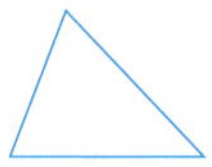

Addition Property of Equality [ə•dish′ən prä′pər•tē əv i•kwol′ə•tē] The property that states that if you add the same number to both sides of an equation, the sides remain equal (p. 290)

additive inverse [ad′ə•tiv in′vûrs] The opposite of a given number (p. 243)

adjacent angles [ə•jā′sənt an′gəlz] Angles that are side by side and have a common vertex and ray (p. 322)
Example:

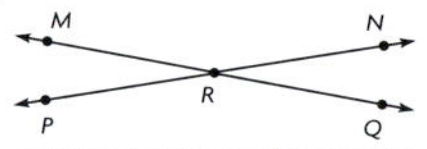
M
N
R
P
Q
∠*MRN* and ∠*NRQ* are adjacent angles.

algebraic expression [al•jə•brā′ik ik•spre′shən] An expression that includes at least one variable (p. 28)
Examples: $x + 5$, $3a - 4$

algebraic operating system [al•jə•brā′ik ä′pə•rā•ting sis′təm] A way for calculators to follow the order of operations when evaluating expressions (p. 43)

angle [an′gəl] A figure formed by two rays with a common endpoint (p. 320)
Example:

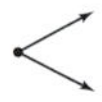

arc [ärk] A part of a circle, named by its endpoints (p. 344)
Example:

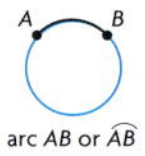
A
B
arc *AB* or $\overparen{AB}$

area [âr′ē•ə] The number of square units needed to cover a given surface (p. 494)

Associative Property [ə•sō′shē•ā•tiv prä′pər•tē] The property that states that the way addends are grouped or factors are grouped does not change the sum or the product (p. 36)
Examples: $12 + (5 + 9) = (12 + 5) + 9$
$(9 \times 8) \times 3 = 9 \times (8 \times 3)$

Student Handbook **H69**

axes [ak′sēz] The horizontal number line (*x*-axis) and the vertical number line (*y*-axis) on the coordinate plane (p. 570)

B

bar graph [bär′graf] A graph that displays countable data with horizontal or vertical bars (p. 120)

base [bās] A number used as a repeated factor (p. 40)
Example: $8^3 = 8 \times 8 \times 8$; 8 is the base.

base [bās] A side of a polygon or a face of a solid figure by which the figure is measured or named (pp. 350, 495)
Examples:

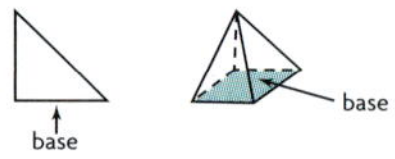
base
base

biased question [bī′əst kwes′chən] A question that leads to a specific response or excludes a certain group (p. 98)

biased sample [bī′əst sam′pəl] A sample is biased if individuals or groups from the population are not represented in the sample. (p. 98)

bisect [bī•sekt′] To divide into two congruent parts (p. 368)

box-and-whisker graph [bäks•ənd•hwis′kər graf] A graph that shows how far apart and how evenly data are distributed (p. 129)

C

Celsius [səl′sē•əs] A metric scale for measuring temperature (p. 301)

certain [sûr′tən] Sure to happen (p. 429)

chord [kôrd] A line segment with its endpoints on a circle (p. 344)
Example:

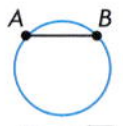
A
B
chord: $\overline{AB}$

circle [sûr′kəl] A closed plane figure with all points of the figure the same distance from the center (p. 344)
Example:

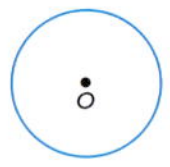
O

circle graph [sûr′kəl graf] A graph that lets you compare parts to the whole and to other parts (p. 122)
Example:

FAVORITE HOBBIES

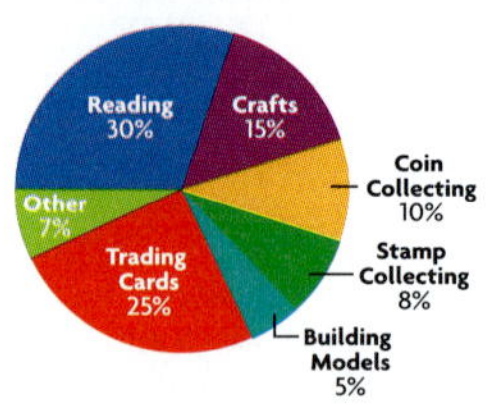
Reading 30%
Crafts 15%
Coin Collecting 10%
Stamp Collecting 8%
Building Models 5%
Trading Cards 25%
Other 7%

circumference [sûr•kum′fər•əns] The distance around a circle (p. 484)

clustering [klus′tər•ing] A method used to estimate a sum when all addends are about the same (p. 16)

Commutative Property [kə•myo͞o′tə•tiv prä′pər•tē] The property that states that if the order of addends or factors is changed, the sum or product stays the same (p. 36)
Examples: $6 + 5 + 7 = 5 + 6 + 7$
$8 \times 9 \times 3 = 3 \times 8 \times 9$

compensation [kom•pən•sā′shən] A mental math strategy for some addition and subtraction problems (p.37)

complementary angles [kom•plə•men′tər•ē an′gəlz] Two angles whose measures have a sum of 90° (p. 323)
Example:

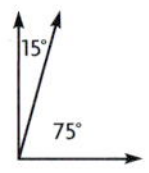
15°
75°

H70 Glossary

Multimedia Math Glossary **www.harcourtschool.com/mathglossary**

composite number [käm•pä′zət num′bər] A whole number greater than 1 that has more than two whole-number factors (p. 148)

compound event [käm′pound i•vent′] An event made of two or more simple events (p. 444)

congruent [kən•gro͞o′ənt] Having the same size and shape (p. 390)

convenience sample [kən•vēn′yənts sam′pəl] Sampling the most available subjects in the population to obtain quick results (p. 95)

coordinate plane [kō•ôr′də•nit plān] A plane formed by two perpendicular number lines called axes; every point on the plane can be named by an ordered pair of numbers. (p. 570)

corresponding angles [kôr•ə•spän′ding an′gəlz] Angles that are in the same position in different plane figures (p. 390)
Example:

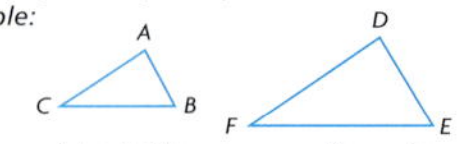
A
C
B
D
F
E
∠*A* and ∠*D* are corresponding angles.

corresponding sides [kôr•ə•spän′ding sīdz] Sides that are in the same position in different plane figures (p. 390)
Example:

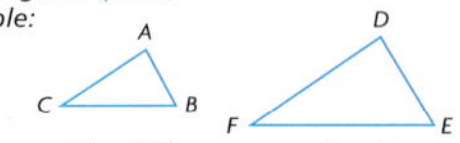
A
C
B
D
F
E
$\overline{CA}$ and $\overline{FD}$ are corresponding sides.

cube [kyo͞ob] A rectangular solid with six congruent faces (p. 350)
Example:

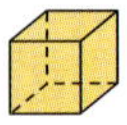

cumulative frequency [kyo͞o′myə•lə•tiv frē′kwən•sē] A running total of the number of subjects surveyed (p. 103)

D

decimal [de′sə•məl] A number with one or more digits to the right of the decimal point (p. 52)

denominator [di•nä′mə•nā•tər] The part of a fraction that tells how many equal parts are in the whole (p. 159)
Example: $\frac{3}{4}$ ←denominator

dependent events [di•pen′dənt i•vənts′] Events for which the outcome of the second event depends on the outcome of the first event (p. 448)

diameter [dī•am′ə•tər] A line segment that passes through the center of a circle and has its endpoints on the circle (p. 344)
Example:

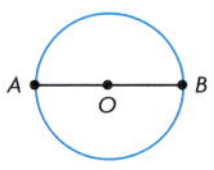
A
O
B
diameter: $\overline{AB}$

dimension [di•men′shən] The length, width, or height of a figure (p. 511)

discount [dis′kount] An amount that is subtracted from the regular price of an item (p. 418)

Distributive Property of Multiplication [di•strib′yə•tiv prä′pər•tē əv mul•tə•plə•kā′shən] The property that states that multiplying a sum by a number is the same as multiplying each addend by the number and then adding the products (p. 36)
Example: $14 \times 21 = 14 \times (20 + 1) =$
$(14 \times 20) + (14 \times 1)$

dividend [di′və•dend] The number that is to be divided in a division problem
Example: In $56 \div 8$, 56 is the dividend.

Division Property of Equality [di•vi′zhən prä′pər•tē əv i•kwol′ə•tē] The property that states that if you divide both sides of an equation by the same nonzero number, the sides remain equal (p. 297)

divisor [di•vī′zər] The number that divides the dividend
Example: In $45 \div 9$, 9 is the divisor.

E

equally likely [ē′kwə•lē lī′klē] Having the same chance of occurring (p. 428)

Student Handbook **H71**

equation [i•kwā′zhən] A statement that shows that two quantities are equal (p. 30)

equilateral triangle [ē•kwə•la′tə•rəl trī′an•gəl] A triangle with three congruent sides (p. 332)
Example:

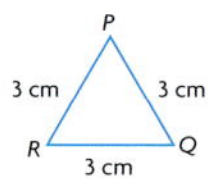
P
3 cm
3 cm
R
Q
3 cm

equivalent fractions [ē•kwiv′ə•lənt frak′shənz] Fractions that name the same amount or part (p. 160)

equivalent ratios [ē•kwiv′ə•lənt rā′shē•ōz] Ratios that name the same comparisons (p. 384)

estimate [es′tə•mit] An answer that is close to the exact answer and that is found by rounding, by clustering, or by using compatible numbers (p. 16)

evaluate [i•val′yo͞o•āt] Find the value of a numerical or algebraic expression (p. 28)

event [i•vent′] A set of outcomes (p. 427)

experimental probability [ik•sper•ə•men′təl prä•bə•bil′ə•tē] The ratio of the number of times an event occurs to the total number of trials or times the activity is performed (p. 436)

exponent [ik•spō′nənt] A number that tells how many times a base is used as a factor (p. 40)
Example: $2^3 = 2 \times 2 \times 2 = 8$;
3 is the exponent.

F

face [fās] One of the polygons of a solid figure (p. 493)

Example:

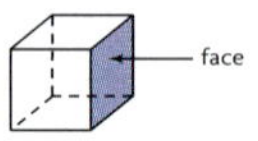
face

factor [fak′tər] A number that is multiplied by another number to find a product

Fahrenheit [fâr′ən•hīt] A customary scale for measuring temperature (p. 301)

formula [fôr′myə•lə] A rule that is expressed with symbols (p. 300)
Example: $A = lw$

fractal [frak′təl] A figure with repeating patterns containing shapes that are like the whole but of different sizes throughout (p. 543)

frequency table [frē′kwən•sē tā′bəl] A table representing totals for individual categories or groups (p. 103)

function [funk′shən] A relationship between two quantities in which one quantity depends on the other (p. 539)

Fundamental Counting Principle [fun•də•men′təl koun′ting prin′sə•pəl] If one event has *m* possible outcomes and a second independent event has *n* possible outcomes, then there are $m \times n$ total possible outcomes. (p. 444)

G

greatest common factor (GCF) [grā′təst kä′mən fak′tər] The greatest factor that two or more numbers have in common (p. 151)

H

height [hīt] A measure of a polygon or solid figure, taken as the length of a perpendicular from the base of the figure (p. 495)
Example:

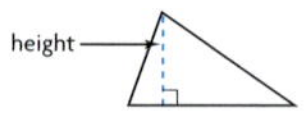
height

hexagon [heks′ə•gon] A six-sided polygon

histogram [his′tə•gram] A bar graph that shows the number of times data occur in certain ranges or intervals (p. 127)

hypotenuse [hī•pot′ə•n(y)o͞os′] In a right triangle, the side opposite the right angle (p. 488
Example:

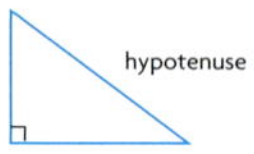
hypotenuse

I

Identity Property of Zero [ī•den′tə•tē prä′pər•tē əv zir′ō] The property that states that the sum of zero and any number is that number (p. H2)
Example: $25 + 0 = 25$

H72 Glossary

H69–H72 Glossary

Multimedia Math Glossary — www.harcourtschool.com/mathglossary

Identity Property of One [ī•den′tə•tē prä′pər•tē əv wun] The property that states that the product of any number and 1 is that number (p. H2)
Example: $12 \times 1 = 12$

impossible [im•pos′ə•bəl] Never able to happen (p. 429)

independent events [in′di•pen′dənt i•vents′] Events for which the outcome of the second event does not depend on the outcome of the first event (p. 447)

indirect measurement [in•di•rekt′ mezh′ər•mənt] The technique of using similar figures and proportions to find a measure (p. 394)

inequality [in′i•kwäl′ə•tē] An algebraic or numerical sentence that contains the symbol <, >, ≤, ≥, or ≠ (p. 568)
Example: $x + 3 > 5$

integers [in′ti•jərz] The set of whole numbers and their opposites (p. 228)

isosceles triangle [ī•sä′sə•lēz trī′an•gəl] A triangle with exactly two congruent sides (p. 331)
Example:

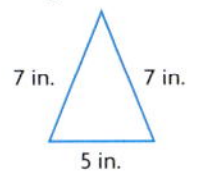

L

lateral faces [lat′ər•əl fās′əz] The faces in a prism or pyramid that are not bases (p. 350)

least common denominator (LCD) [lēst kä′mən di•nä′mə•nāt•ər] The least common multiple of two or more denominators (p. 182)

least common multiple (LCM) [lēst kä′mən mul′tə•pəl] The smallest number, other than zero, that is a common multiple of two or more numbers (p. 150)

leg [leg] In a right triangle, either of the two sides that form the right angle (p. 488)
Example:

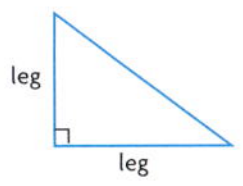

like terms [līk tûrmz] Expressions that have the same variable with the same exponent (p. 273)

line [līn] A straight path that extends without end in opposite directions (p. 318)
Example:

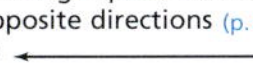

line graph [līn graf] A graph that uses a line to show how data change over time (p. 121)

line of symmetry [līn əv si′mə•trē] A line that divides a figure into two congruent parts (p. 560)

line plot [līn plät] A graph that shows frequency of data along a number line (p. 121)
Example:
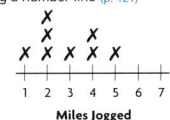
Miles Jogged

line segment [līn seg′mənt] A part of a line with two endpoints (p. 318)
Example:

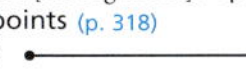

line symmetry [līn si′mə•trē] A figure has line symmetry if a line can separate the figure into two congruent parts. (p. 560)

lower extreme [lō′ər ik•strēm′] The least number in a set of data (p. 129)

lower quartile [lō′ər kwôr′til] The median of the lower half of a set of data (p. 129)

M

mean [mēn] The sum of a group of numbers divided by the number of addends (p. 106)

median [mē′dē•ən] The middle value in a group of numbers arranged in order (p. 106)

midpoint [mid′point] The point that divides a line segment into two congruent line segments (p. 368)

mixed number [mikst num′bər] A number represented by a whole number and a fraction (p. 164)

mode [mōd] The number or item that occurs most often in a set of data (p. 106)

multiple-bar graph [mul′tə•pəl bär′graf] A bar graph that represents two or more sets of data (p. 120)

multiple-line graph [mul′tə•pəl līn′graf] A line graph that represents two or more sets of data (p. 121)

multiple [mul′tə•pəl] The product of a given whole number and another whole number (p. 150)

Multiplication Property of Equality [mul′tə•plə•kā′shən prä′pər•tē əv i•kwol′ə•tē] The property that states that if you multiply both sides of an equation by the same number, the sides remain equal (p. 298)

N

negative integers [ne′gə•tiv in′ti•jərz] Integers to the left of zero on the number line (p. 228)

net [net] An arrangement of two-dimensional figures that folds to form a polyhedron (p. 356)
Example:

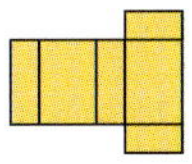

numerator [noō′mə•rā•tər] The part of a fraction that tells how many parts are being used (p. 159)
Example: $\frac{3}{4}$ ← numerator

numerical expression [noō•mâr′i•kəl ik•spre′shən] A mathematical phrase that uses only numbers and operation symbols (p. 28)

O

obtuse angle [äb•toōs′ an′gəl] An angle whose measure is greater than 90° and less than 180° (p. 320)
Example:
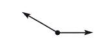

obtuse triangle [äb•toōs′ trī′an•gəl] A triangle with one angle greater than 90° (p. 332)
Example:
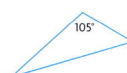

opposites [ä′pə•zəts] Two numbers that are an equal distance from zero on the number line (p. 228)

order of operations [ôr′dər əv ä•pə•rā′shənz] The process for evaluating expressions: first perform the operations in parentheses, clear the exponents, perform all multiplication and division, and then perform all addition and subtraction (p. 42)

ordered pair [ôr′dərd pâr] A pair of numbers that can be used to locate a point on the coordinate plane (p. 570)
Examples: (0,2), (3,4), (⁻4,5)

origin [ôr′ə•jən] The point where the x-axis and the y-axis in the coordinate plane intersect, (0,0) (p. 570)

outcome [out′kəm] A possible result of a probability experiment (p. 428)

outlier [out′lī•ər] A data value that stands out from others in a set; outliers can significantly affect measures of central tendency. (p. 110)

overestimate [ō•vər•es′tə•mət] An estimate that is greater than the exact answer (p. 17)

P

parallel lines [pâr′ə•lel līnz] Lines in a plane that are always the same distance apart (p. 326)
Example:
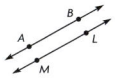

percent (%) [pər•sent′] The ratio of a number to 100; percent means "per hundred." (p. 60)

perimeter [pə•ri′mə•tər] The distance around a figure (p. 477)

Multimedia Math Glossary — www.harcourtschool.com/mathglossary

perpendicular lines [pər•pen•dik′yə•lər līnz] Two lines that intersect to form right, or 90°, angles (p. 326)
Example:
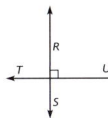

pi (π) [pī] The ratio of the circumference of a circle to its diameter; $\pi \approx 3.14$ or $\frac{22}{7}$ (p. 484)

plane [plān] A flat surface that extends without end in all directions (p. 318)

point [point] An exact location in space, usually represented by a dot (p. 318)

point of rotation [point əv rō•tā′shən] The central point around which a figure is rotated (p. 561)

polygon [pä′lē•gän] A closed plane figure formed by three or more line segments (p. 331)

polyhedron [pä•lē•hē′drən] A solid figure with flat faces that are polygons (p. 350)
Example:

Hexagonal Prism

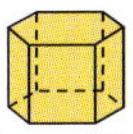

population [pä•pyə•lā′shən] The entire group of objects or individuals considered for a survey (p. 94)

positive integers [pä′zə•tiv in′ti•jərz] Integers to the right of zero on the number line (p. 228)

prime factorization [prīm fak•tə•rī•zā′shən] A number written as the product of all of its prime factors (p. 148)
Example: $24 = 2^3 \times 3$

prime number [prīm num′bər] A whole number greater than 1 whose only factors are 1 and itself (p. 148)

principal [prin′sə•pəl] The amount of money borrowed or saved (p. 422)

prism [priz′əm] A solid figure that has two congruent, polygon-shaped bases, and other faces that are all rectangles (p. 350)
Example:

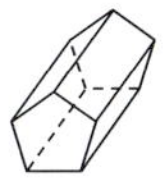

probability [prä•bə•bil′ə•tē] See theoretical probability and experimental probability

product [prä′dəkt] The answer in a multiplication problem (p. 15)

Property of Zero [prä′pər•tē əv zē′rō] The property that states that the product of any number and zero is zero (p. 35)

proportion [prə•pôr′shən] An equation that shows that two ratios are equal (p. 387)
Example: $\frac{1}{3} = \frac{3}{9}$

pyramid [pir′ə•mid] A solid figure with a polygon base and triangular sides that all meet at a common vertex (p. 351)
Example:

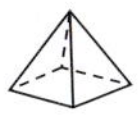

Pythagorean Theorem [pə•thag•ə•rē′ənthē′ə•rem] In any right triangle, if a and b are the lengths of the legs and c is the length of the hypotenuse, then $a^2 + b^2 = c^2$ (p. 488)
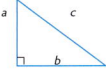

Q

quadrants [kwäd′rənts] The four regions of the coordinate plane (p. 570)

quadrilateral [kwä•dri•lat′ə•rəl] A polygon with four sides and four angles (p. 331)

quotient [kwō′shənt] The number, not including the remainder, that results from dividing (p. 23)

R

radius [rā′dē•əs] A line segment with one endpoint at the center of a circle and the other endpoint on the circle (p. 344)
Example:
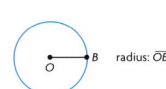
radius: $\overline{OB}$

random sample [ran′dəm sam′pəl] A sample in which each subject in the overall population has an equal chance of being selected (p. 95)

range [rānj] The difference between the greatest and least numbers in a group (p. 103)

rate [rāt] A ratio that compares two quantities having different units of measure (p. 385)

ratio [rā′shē•ō] A comparison of two numbers, a and b, written as a fraction $\frac{a}{b}$ (p. 230)

rational number [ra′shə•nəl num′bər] Any number that can be written as a ratio $\frac{a}{b}$, where a and b are integers and $b \neq 0$ (p. 230)

ray [rā] A part of a line with a single endpoint (p. 318)
Example:

ray: $\overrightarrow{JK}$

reciprocal [ri•sip′rə•kəl] Two numbers are reciprocals of each other if their product equals 1. (p. 209)

reflection [ri•flek′shən] A movement of a figure by flipping it over a line (p. 550)

regular polygon [reg′yə•lər pä′lē•gän] A polygon in which all sides are congruent and all angles are congruent (p. 336)
Example:

repeating decimal [ri•pēt′ing de′sə•məl] A decimal that doesn't end, because it shows a repeating pattern of digits after the decimal point (p. 169)

right angle [rīt an′gəl] An angle that has a measure of 90° (p. 320)
Example:

right triangle [rīt trī′an•gəl] A triangle with one right angle (p. 332)
Example:

rotation [rō•tā′shən] A movement of a figure by turning it around a fixed point (p. 550)

rotational symmetry [rō•tā′shən•əl si′mə•trē] The property of a figure that can be rotated less than 360° around a central point and still be congruent to the original figure (p. 561)

S

sales tax [sālz taks] A percent of the cost of an item, added onto the item's cost (p. 420)

sample [sam′pəl] A part of a population (p. 94)

sample space [sam′pəl spās] The set of all possible outcomes (p. 428)

scale [skāl] A ratio between two sets of measurements (p. 398)

scale drawing [skāl drô′ing] A drawing that shows a real object smaller than (a reduction) or larger than (an enlargement) the real object (p. 397)

scalene [skā′lēn] A triangle with no congruent sides (p. 331)

scatterplot [skat′ər•plät] A graph with points plotted to show a relationship between two variables (p. 139)

Multimedia Math Glossary www.harcourtschool.com/mathglossary

sector [sek′tər] A region enclosed by two radii and the arc joining their endpoints (p. 344)
Example:

sequence [sē′kwəns] An ordered set of numbers (p. 536)

similar figures [si′mə-lər fig′yərz] Figures with the same shape but not necessarily the same size (p. 372)

simple interest [sim′pəl in′trəst] A fixed percent of the principal, paid yearly (p. 422)

simplest form [sim′pləst fôrm] The form in which the numerator and denominator of a fraction have no common factors other than 1 (p. 161)

solution [sə-lōō′shən] A value that, when substituted for a variable in an equation, makes the equation true (p. 30)

square [skwâr] The product of a number and itself; a number with the exponent 2 (p. 276)

square [skwâr] A rectangle with four congruent sides (p. 511)

square root [skwâr rōōt] One of two equal factors of a number (p. 277)

stem-and-leaf plot [stem ənd lēf plät] A type of graph that shows groups of data arranged by place value (p. 126)

straight angle [strāt an′gəl] An angle whose measure is 180° (p. 320)
Example:

Subtraction Property of Equality [sub-trak′shən prä′pər-tē əv i-kwol′ə-tē] The property that states that if you subtract the same number from both sides of an equation, the sides remain equal (p. 287)

sum [sum] The answer to an addition problem (p. 15)

supplementary angles [sup-lə-men′tə-rē an′gəlz] Two angles whose measures have a sum of 180° (p. 323)
Example:

surface area [sûr′fəs âr′ē-ə] The sum of the areas of the faces of a solid figure (p. 504)

survey [sûr′vā] A method of gathering information about a group (p. 94)

systematic sample [sis-tə-ma′tik sam′pəl] A sampling method in which one subject is selected at random and subsequent subjects are selected according to a pattern (p. 95)

T

term [tûrm] Each number in a sequence (p. 536)

terms [tûrmz] The parts of an expression that are separated by an addition or subtraction sign (p. 273)

terminating decimal [tûr′mə-nāt-ing de′sə-məl] A decimal that ends, having a finite number of digits after the decimal point (p. 169)

tessellation [tes-ə-lā′shən] A repeating arrangement of shapes that completely covers a plane, with no gaps and no overlaps (p. 553)

theoretical probability [thē-ə-re′ti-kəl prä-bə-bil′ə-tē] A comparison of the number of favorable outcomes to the number of possible equally likely outcomes (p. 428)

transformation [trans-fər-mā′shən] A rigid transformation is a movement that does not change the size or shape of a figure (p. 550)

translation [trans-lā′shən] A movement of a figure along a straight line (p. 550)

tree diagram [trē dī′ə-gram] A diagram that shows all possible outcomes of an event (p. 444)

triangular number [trī-an′gyə-lər num′bər] A number that can be represented by a triangular array (p. 536)

U

unbiased sample [un-bī′əst sam′pəl] A sample is unbiased if every individual in the population has an equal chance of being selected. (p. 98)

underestimate [un-dər-es′tə-māt] An estimate that is less than the exact answer (p. 17)

unit rate [yōō′nət rāt] A rate that has 1 unit as its second term (p. 385)
Example: $1.45 per pound

unlike fractions [un′līk frak′shənz] Fractions with different denominators (p. 180)

upper extreme [up′ər ik-strēm′] The greatest number in a set of data (p. 129)

upper quartile [up′ər kwôr′til] The median of the upper half of a set of data (p. 129)

V

variable [vâr′ē-ə-bəl] A letter or symbol that stands for one or more numbers (p. 28)

Venn diagram [ven dī′ə-gram] A diagram that shows relationships among sets of things (p. 230)

vertex [vûr′teks] The point where two or more rays meet; the point of intersection of two sides of a polygon; the point of intersection of three or more edges of a solid figure; the top point of a cone (pp. 320, 351)
Examples:

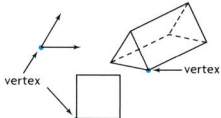

vertical angles [vûr′ti-kəl an′gəlz] A pair of opposite congruent angles formed where two lines intersect (p. 322)
Example:

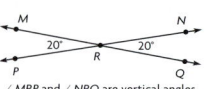

∠MRP and ∠NRQ are vertical angles.

volume [väl′yəm] The number of cubic units needed to occupy a given space (p. 512)

W

whole number [hōl num′bər] One of the numbers 0, 1, 2, 3, 4, ... The set of whole numbers goes on without end.

X

x-axis [eks-ak′səs] The horizontal number line on a coordinate plane (p. 570)

x-coordinate [eks-kō-ôr′də-nət] The first number in an ordered pair; it tells the distance to move right or left from (0,0).

Y

y-axis [wī-ak′səs] The vertical number line on a coordinate plane (p. 570)

y-coordinate [wī-kō-ôr′də-nət] The second number in an ordered pair; it tells the distance to move up or down from (0,0).

Selected Answers

Chapter 1
Page 15
1. product **3.** difference **5.** sum
7. hundreds **9.** hundred thousands
11. hundred millions **13.** ten millions
15. 90,000 **17.** 300,000,000
19. 20,000,000,000 **21.** 2,000 **23.** 29,000
25. 35,000 **27.** 6,000 **29.** 135,000
31. 60,000 **33.** 50,000 **35.** 10,000
37. 640,000 **39.** 90,000

Pages 18–19
1. underestimate since both addends are rounded down **3.** 1,500 **5.** 12,000 **7.** 600
9. 24,000 **11.** 360 **13.** 1,500 **15.** 80
17. 70 **19.** 6,000 **21.** 18,000 **23.** 350
25. 14,000 **27.** 16,000 **29.** 4,000
31. 150,000 **33.** 90 **35.** 12 **37.** 3,200,000
39. over; 400 + 700 **41.** over; 100 × 20
43. over; 300 × 30 **45.** > **47.** < **49.** <
51. about 4,500,000 books **53.** 9,936 people
55. 235 and 345 **57.** 36 in.; 54 in.^{2}

Page 21
3. 1,439 **5.** 26,208 **7.** 36,374 **9.** 1,111,384
11. 46,395,619 **13.** 75,125 **17.** 11,306
19. 1, 2, 3, 4, 6, 8, 12, 24

Pages 24–25
1. There were 102 full packages. The 18 newspapers left over were not enough to make a full package. **3.** 792,456 **5.** 3,553,275 **7.** 54 **9.** 38,480 **11.** 850,038
13. 56 r4 **15.** 961,184 **17.** 3,496,458
19. 1,745,677 **21.** 26,522 **23.** 14
25. 20,900 **27.** 26 r2 **29.** 89 **31.** 1,312,800
33. 1,855,998 **35.** 84 **37.** $367\frac{9}{20}$ **39.** $846\frac{1}{4}$
41. 415 **43.** $159 **45.** 10 **47.** 64

Page 27
1. 13 apples, 27 oranges **3.** B **5.** April 26, May 2, and May 8 **7.** 48 cards **9.** 121 stamps **11.** 5 min

Page 29
1. An algebraic expression has one or more variables. **3.** 125 − 46 **5.** $y \div 15$, or $\frac{y}{15}$
7. 46 **9.** 25 × 20 **11.** 76 − k **13.** 465
15. 9,000 **17.** 140 **19.** 3 **21.** $n + 12$
23. Deidre **25.** 215 **27.** 7,643

Page 31
1. No, because 4 + 3 equals 7; $x = 6$
3. $f = 21$ **5.** $x = 4$ **7.** $k = 4$ **9.** $x = 5$
11. $k = 18$ **13.** $m = 9$ **15.** $x = 20$
17. $k = 29$ **19.** $v = 20$ **21.** $d = 9$
23. $p = 6$ **25.** 25 students **27.** 41
29. 4,990

Chapter 2
Page 35
1. 27 **3.** 64 **5.** 10,000 **7.** 512 **9.** 343
11. 256 **13.** Commutative of Addition
15. Property of Zero **17.** Associative of Addition **19.** Distributive **21.** Identity of Multiplication **23.** Distributive **25.** $a = 5$ **27.** $r = 3$ **29.** $t = 8$ **31.** 20 **33.** 32
35. 8 **37.** 32 **39.** 18

Pages 38–39
3. 204 **5.** 157 **7.** 56 **9.** 145 **11.** 55
13. 142 **15.** 93 **17.** 43 **19.** 168 **21.** 495
23. 185 **25.** 108 **27.** 212 **29.** 72
31. 244 **33.** 85 **35.** 45 **37.** 1,500
39. 1,028 **41.** 160 **43.** 64 **45.** 52
49. 72 CDs **51.** 18 CDs each **53.** 68 more signatures **57.** 33,672

Student Handbook H79

Page 41

1. 7 **3.** $2 \times 2 \times 2$; 8 **5.** $3 \times 3 \times 3 \times 3$; 81
7. $1 \times 1 \times 1 \times 1$; 1 **9.** $7 \times 7 \times 7$; 343
11. $5 \times 5 \times 5$; 125 **13.** 34×34; 1,156
15. $10 \times 10 \times 10 \times 10 \times 10 \times 10 \times 10 \times 10$; 100,000,000 **17.** $2 \times 2 \times 2 \times 2 \times 2 \times 2 \times 2 \times 2 \times 2 \times 2$; 1,024 **19.** 1 **21.** 25; 25
23. 12^3 **25.** 4^4 **27.** n^2 **29.** 8^2 **31.** 10^4
33. 1940 **35.** How many more games does Scott have than Aaron? **37.** 46,000
39. 745 r12

Pages 42–43

1. parentheses, exponent, multiplication, addition; 154 **3.** multiplication, division, addition; 16 **1.** No; 17 is the correct value, not 9 **3.** $15 \div 3 + 12$ **5.** 29 **7.** 216
9. 31

Pages 44–45

1. $(420 - 100) \div 40 = 8$ **3.** 25 **5.** 28
7. 25 **9.** 68 **11.** 100 **13.** 1,218
15. 504 **17.** 60 **19.** 77 **21.** 11
23. $\$20 - (\$3.25 + 3 \times \$1.29) = \$12.88$
25. 18,447 people **27.** 204 **29.** 32 r3

Page 47

1. Aquarium Tour, Underwater Acrobats, Animal Acts, Whale Acts **3.** H **5.** 3 books, 4 magazines **7.** $4,700 **9.** 24 nails

Chapter 3

Page 51

1. 0.2 **3.** 0.09 **5.** 836.23 **7.** 93,450.38
9. 306,007.06 **11.** 7 **13.** 1 tenth **15.** <
17. < **19.** > **21.** > **23.** > **25.** >
27. 1 **29.** 42 **31.** 72 **33.** 26.4 **35.** 8.6
37. 14.5 **39.** 55.6

Pages 54–55

1. 2.0769 **3.** 0.007 **5.** 0.06 **7.** $0.001 + 0.00003$ **9.** $300 + 40 + 2 + 0.04 + 0.006$
11. = **13.** 1.351, 1.361, 1.363 **15.** 0.0007
17. 0.00006 **19.** $0.03 + 0.006 + 0.0002$
21. $2 + 0.4 + 0.05 + 0.006$ **23.** < **25.** =
27. > **29.** < **31.** 0.405, 1.05, 1.125, 1.25, 1.45 **33.** 8.91, 9.082, 9.285, 9.82, 9.85
35. 125.4, 125.35, 125.33, 125.3 **37.** 41.01, $14\frac{1}{10}$, $14\frac{3}{100}$, 14.01 **39.** $94,563,020; ninety-four million, five hundred sixty-three thousand, twenty **43.** 228 **45.** 29

Page 57

1. C **3.** air conditioner **5.** calculator, $10.50; pen, $1.50; notebook, $2.00 **7.** 60 cards **9.** Blaster

Page 59

3. 40 **5.** 90 **7.** 210 **9.** 32 **11.** 9
13. 50 **15.** 140 **17.** 1,800 **19.** 27,000
21. $17,000 **23.** 36 **25.** > **27.** Possible answer: about 120 lb **31.** 2.235, 2.325, 2.523, 2.532 **33.** 1,121 r18

Page 61

1. 18% means 18 per hundred, so 18 are red.
3. 0.06, 6% **5.** 0.7 or 0.70 **7.** 3% **9.** 0.5, 0.50 **11.** 0.11, 11% **13.** 0.62 **15.** 0.28
17. 0.53 **19.** 85% **21.** 40% **23.** 0.79, 79%
25. Write 0.6 as 0.60 to show hundredths. Then write 60 hundredths as 60%. **27.** 172
29. 82 r26

Chapter 4

Page 65

1. 47 **3.** 44 **5.** 10 **7.** 222 **9.** 252
11. 1,444 **13.** 4,000 **15.** 3,000 **17.** 24
19. 47 **21.** 14 **23.** 29 **25.** 43 **27.** 840
29. 19,405 **31.** 8.75 **33.** 12.5 **35.** 160.8
37. $9\frac{5}{8}$ **39.** $39\frac{2}{7}$

H80 Selected Answers

Pages 68–69
3. $0.63   **5.** 18.585   **7.** 20.923   **9.** 37.31
11. 124.94   **13.** 87.80   **15.** 1,438.66
17. 0.6   **19.** 287.673   **21.** 488.55
23. 529.484   **25.** 257.14   **27.** 9.42   **29.** yes
31. no   **33.** 1.25   **35.** 12.3   **37.** 2.216
39. 0.090   **41.** 210; more than   **43.** 441 lb
45. 0.46

Pages 72–73
3. 0.35   **5.** 2   **7.** 3   **9.** 2.87   **11.** 1.218
13. 25.84   **15.** 0.24   **17.** 3   **19.** 6
21. 194.208   **23.** 0.410   **25.** 13.8   **27.** 36.5
29. 4.92   **31.** 0.08   **33.** 0.441   **35.** 5.5080
37. 26.8467   **39.** 0.39501   **41.** 0.8934
43. $7.31   **47.** 65.65   **49.** 299 r14

Page 74
1. 1.01   **3.** 0.45

Page 75
1. 6   **3.** 3   **5.** 4   **7.** 3   **9.** 8   **11.** 0.57
13. 89 r21

Pages 78–79
1. multiply by 100   **3.** 96 ÷ 16
5. 482.4 ÷ 24   **7.** 14.86   **9.** 1.97   **11.** 4.4
13. 819 ÷ 9   **15.** 239 ÷ 5   **17.** 2,335.8 ÷ 102
19. 6.1   **21.** 6.9   **23.** 9.91   **25.** 0.21
27. 21.2   **29.** 8.2   **31.** 15.05   **33.** 220
35. 2.36   **37.** 3.03   **39.** 5 weeks
41. 30 pages   **43.** greater than

Page 81
1. C   **3.** 4 magnets   **5.** 8 hr 5 min
7. 65 min, or 1 hr 5 min   **9.** 108 pieces

Page 83
1. $a \times 6.45$; 12×6.45; $77.40   **3.** 11.7
5. 5.08   **7.** $m = 19$ **9.** 4.6   **11.** 14   **13.** 3.9
15. $r = 7.7$ **17.** $a = 8.2$ **19.** $p = 17.6$
21. $n \div 6$; 4.8 mi   **23.** 4 mi   **25.** 37
27. 3.08, 3.508, 3.58, 3.85

Chapter 5
Page 93
1. median   **3.** range   **5.** 13 more girls
7. 29 girls   **9.** 8   **11.** 8.475   **13.** 28; 32
15. 11; 10.8   **17.** 6

Pages 96–97
5. population, because there are not too many students and they are easily available
7. systematic   **9.** convenience
11. population; There are not too many basketball team members.   **13.** systematic
15. convenience   **21.** 9.3   **23.** 2

Pages 98–99
1. Do you enjoy movies that are longer than two hours?   **3.** biased; excludes males
5. unbiased; all members have an equal chance   **7.** unbiased; all shoppers have an equal chance   **9.** biased; excludes people who shop in other stores   **11.** biased
13. Yes. It leads others to agree with the opinion.   **15.** 4   **17.** 9,000

Page 101
1. D   **3.** 12 members   **5.** 1 3-point, 7 2-point; 3 3-point, 4 2-point; 5 3-point, 1 2-point   **7.** $42.50   **9.** 37 birds
11. Saturday

Pages 104–105
1. 1–10, 11–20; Since $2 \times 10 = 20$, you would have 2 intervals that include 10 numbers
5. 16   **13.** 9   **15.** No. The data are categories and not numerical.   **19.** unbiased

Pages 107–108
1. The mean would become 21.2. The mean or median would be most useful to describe the data.   **3.** 7 hours   **5.** 24; 23; 22
7. 15.375 points   **9.** 10 and 12   **11.** 336.8; 360; no mode   **15.** Tim should divide by 4, not 3; 6   **17.** 71 people   **19.** 90%

Pages 110–111

1. The median and mode would not change. The mean would not increase as much.
3. 16; 14; 12   **5a.** Abe: mean 7.5, median 6, mode 7; Bart: mean 7.5, median 8.5, mode 0;
5b. Abe: mean 9.3, median 7, mode 7 and 20; Bart: mean 6.4, median 5, mode 0;
5c. Abe's: all increase; Bart's: mean/median decrease; mode doesn't change   **7a.** 75
7b. No. With a perfect score of 100, the mean would be 65.7.   **9.** 5.75; 5.5; 5   **11.** 9.5; 10; 10 and 12

Pages 113–115

1. No. The sample would be a convenience sample and not a random sample.
3. Yes. The sample is random and representative; the question is unbiased.
5. No. The sample is not representative.
7. No. The sample is not representative of the population.   **9.** Yes. An unbiased question was asked of a random sample from the correct population.   **11.** The interest in crafts decreased.   **13.** The interest in trading cards increased.   **15.** A random sample was chosen from the correct population and the question was unbiased.   **17.** The sample is biased since it only includes middle school athletes.   **19.** 4.25, 4, 4 and 6   **21.** 7, 7, no mode

Chapter 6

Page 119

1. Saturn V, Ariane IV, Titan-Centaur, Titan IIIC   **3.** about 55 m   **7.** about 15,000 sq mi
9. 92   **11.** 22   **13.** 3 or more   **15.** 85

Pages 122–123

1. so you can tell the difference between the sets of data   **3.** line graph   **5.** bar graph
7. circle graph   **11.** Each month, more comedies were rented.   **13.** Comedy video rentals will increase and action video rentals will decrease.   **17.** No. The sample is not representative of the student population.
19. 0.56

Page 125

1. Extend the graph until it intersects with 24 mi and find the corresponding time.
3. 5 hr   **5.** 6 hr   **7.** 8 hr   **9.** about 45 mi
11. 5.6; 5; 4   **13.** 6.5; 7; 9

Pages 127–128

5. bar graph   **7.** bar graph   **9b.** 24.5 in.; 28 in.   **13.** random   **15.** 73%

Pages 130–131

1. A box-and-whisker graph shows the distribution of data, not all of the data.
3. 11; 22; 11   **5.** 58; 67; 9   **7.** 85; 104; 19
9. 72 points; 97 points   **13.** $4.07   **15.** 0.02 or two hundredths

Pages 133–135

3. Yes. The question is biased and could lead people to choose the Mustangs as the best baseball team.   **5.** No. Angel Falls is about 3,200 ft high while Tugela Falls is about 3,000 ft high.   **7.** Yes. The question is biased and could lead people to choose oranges.   **9.** No. Brand B costs $30 and Brand A costs $20.
13. Lin looked only at the graph and did not look at the scales.   **15.** 43 and 55

Chapter 7

Page 145

1. prime number   **3.** yes   **5.** no   **7.** yes
9. yes   **11.** yes   **13.** no   **15.** yes   **17.** no
19. yes   **21.** no   **23.** 16, 20, 24   **25.** 48, 60, 72   **27.** 20, 25, 30   **29.** 6, 12, 18, 24, 30
31. 30, 60, 90, 120, 150   **33.** 9, 18, 27, 36, 45
35. 1, 2, 4, 8   **37.** 1, 11   **39.** 1, 2, 3, 6, 9, 18, 27, 54

Page 147

3. 2, 4, 8   **5.** 2, 4, 8   **7.** 3   **9.** 2, 3, 4, 6, 9
11. 2, 4   **13.** 3, 5   **15.** 2, 3, 4, 6, 8   **17.** 3, 5
19. 2, 3, 6, 9   **21.** 3   **23.** T   **25.** T
29. $28.88   **31.** 5%   **33.** 16,302

H82   Selected Answers

Page 149

1. $2^2 \times 3 \times 13$ 3. $2 \times 2 \times 3$ 5. $2 \times 2 \times 2 \times 2$ 7. 3×7 9. 2×127 11. $2 \times 2 \times 2 \times 2 \times 2 \times 2$ 13. $2 \times 2 \times 19$ 15. $2 \times 3 \times 3; 2 \times 3^2$ 17. $7 \times 7; 7^2$ 19. $2 \times 2 \times 7 \times 19; 2^2 \times 7 \times 19$ 21. 2×373 23. $n = 2$ 25. $n = 5$ 27. $c = 2$ or 3 31. 60%   33. 13

Pages 152–153

3. 3, 6, 9, 12, 15   5. 11, 22, 33, 44, 55   7. 21   9. 18   11. 3   13. 3   15. 4, 8, 12, 16, 20   17. 16, 32, 48, 64, 80   19. 10, 20, 30, 40, 50   21. 14, 28, 42, 56, 70   23. 24   25. 60   27. 120   29. 108   31. 2   33. 3   35. 1   37. 8   39. 9 and 12   41. a. 60   b. 4 packages of cereal samples, 3 packages of pamphlets   43. She found the GCF instead of LCM. The LCM is 30.   45. 124 r14

Page 155

1. October 15   3. H   5. 9 in.   7. 18 blocks

Chapter 8

Page 159

1. >   3. denominator   5. >   7. <   9. >   11. 62,000; 61,600; 61,060   13. $\frac{5}{6}$ 15. $\frac{1}{3}$ 17. 20%   19. 75%

Pages 162–163

1. Also multiply the denominator by 5; $\frac{10}{15}$ 3. 12   5. 4   7. 1   9. 16   11. 1, 2, 4   13. 1, 2   15. $\frac{1}{8}$ 17. $\frac{1}{6}$ 19. $\frac{11}{4}$ 21. $\frac{3}{10}$ 23. 5   25. 4   27. 6   29. 8   31. 12   33. 7   35. 1   37. 1, 3   39. 1, 5   41. 1   43. $\frac{1}{6}$ 45. $\frac{1}{8}$ 47. $\frac{5}{9}$ 49. $\frac{1}{5}$ 51. $\frac{2}{3}$ 53. $\frac{10}{7}$ 55. $\frac{1}{4}$ 57. $\frac{1}{5}$ 59. $\frac{1}{2}$ 61. $\frac{2}{3}$ 63. $\frac{1}{4}$ 65. What fractions of the muffins are bran?   67. 24   69. 0.035

Page 165

1. It has a whole number part and a fraction part.   3. $3\frac{1}{2}$ 5. $3\frac{2}{3}$ 7. $\frac{5}{4}$ 9. $\frac{8}{3}$ 11. $\frac{37}{7}$ 13. $4\frac{1}{2}$ 15. $5\frac{3}{4}$ 17. $5\frac{1}{6}$ 19. $12\frac{6}{7}$ 21. $16\frac{2}{3}$ 23. $3\frac{2}{3}$ 25. $\frac{13}{2}$ 27. $\frac{19}{10}$ 29. $\frac{37}{4}$ 31. $\frac{53}{11}$ 33. $\frac{93}{5}$ 35. no, only a fraction whose numerator is greater than its denominator   37. $2^2 \times 3^2$ 39. 23×1

Pages 166–167

1. The denominators are the same. $4 < 5$, so $\frac{4}{9} < \frac{5}{9}$ 3. <   5. =   7. $\frac{1}{4}, \frac{6}{12}, \frac{2}{3}$ 9. >   11. =   13. =   15. >   17. $\frac{2}{6}, \frac{1}{2}, \frac{9}{12}$ 19. $\frac{1}{2}, \frac{7}{12}, \frac{5}{6}$ 21. $\frac{3}{10}, \frac{2}{5}, \frac{1}{2}$ 23. $\frac{1}{6}, \frac{1}{3}, \frac{1}{2}$ 25. $\frac{1}{12}, \frac{5}{8}, \frac{3}{12}$ 27. 3 pieces of each pizza remain. That is $\frac{3}{8}$ of the mushroom and $\frac{3}{12}$ cheese pizza. $\frac{3}{12} = \frac{1}{4} = \frac{2}{8} < \frac{3}{8}$, so more of the mushroom pizza is left.   29. 7   31. 24

Page 168

1. 0.3   3. 0.7   5. 0.90   7. 0.4   9. 0.85

Pages 170–171

3. $\frac{7}{10}$ 5. $\frac{105}{1,000}$ 7. 0.25; T   9. $0.\overline{6}$; R   11. >   13. =   15. 20%   17. 40%   19. $\frac{6}{100}$ 21. $\frac{61}{100}$ 23. $\frac{205}{1,000}$ 25. $\frac{9}{1,000}$ 27. $0.1\overline{6}$; R   29. 0.625; T   31. 0.3; T   33. 0.1; R   35. 0.36; T   37. $0.\overline{15}$; R   39. =   41. <   43. <   45. 75%   47. 6%   49. 50%   51. 0.5%   53. 0.72   55. Megan   57. $\frac{9}{2}$ 59. 76.89

Chapter 9

Page 175

1. mixed number   3. divide   5. $\frac{3}{4}$ 7. $\frac{1}{3}$ 9. $\frac{4}{3}$, or $1\frac{1}{3}$ 11. $\frac{3}{2}$, or $1\frac{1}{2}$ 13. $\frac{1}{2}$ 15. $\frac{2}{3}$ 17. $\frac{5}{3}$, or $1\frac{2}{3}$ 19. $\frac{1}{3}$ 21. $\frac{1}{2}$ 23. $\frac{5}{4}$, or $1\frac{1}{4}$ 25. $\frac{2}{3}$ 27. $\frac{1}{2}$ 29. $\frac{1}{2}$ 31. $\frac{1}{3}$ 33. $\frac{5}{6}$ 35. 1   37. $\frac{2}{9}$ 39. $\frac{5}{7}$

Pages 178–179

3. close to 1   5. between 0 and $\frac{1}{2}$ 7. $1\frac{1}{2}$ 9. $\frac{1}{2}$ 11. $7\frac{1}{2}$ 13. 2   15. close to 1   17. close to 0   19. 2   21. $\frac{1}{2}$ 23. 17   25. 3   27. 0   29. $4\frac{1}{2}$ 31. 9 to $9\frac{1}{2}$; $9\frac{1}{4}$ 33. $4\frac{1}{2}$ to 5; $4\frac{3}{4}$ 35. $5\frac{1}{2}$ to 6; $5\frac{3}{4}$ 37. about 7   39. about 6 feet   41. about 26   43. about 30   47. about $1\frac{1}{2}$ yd   49. March 15, 24; April 2, 11, 20, 29   51. 45%

Student Handbook   H83

Page 180

1. $\frac{3}{4}$ 3. $\frac{9}{10}$ 5. $\frac{7}{12}$ 7. $\frac{2}{3}$ 9. $\frac{7}{10}$ 11. $\frac{1}{2}$

Page 181

1. $\frac{5}{12}$ 3. $\frac{1}{12}$ 5. $\frac{1}{10}$ 7. $\frac{1}{12}$ 9. $\frac{1}{8}$ 11. $\frac{1}{2}$
13. = 15. >

Pages 183–185

1. $\frac{5}{12}$ c more 3. $\frac{7}{10} + \frac{2}{10}$ 5. $\frac{12}{15} - \frac{5}{15}$ 7. $\frac{1}{3}$
9. $1\frac{5}{12}$ 11. $\frac{1}{15}$ 13. $\frac{7}{9}$ 15. $\frac{24}{28} - \frac{21}{28}$ 17. $\frac{5}{6}$
19. $\frac{4}{5}$ 21. $\frac{3}{14}$ 23. $\frac{1}{5}$ 25. $1\frac{3}{8}$ 27. $\frac{5}{8}$ 29. $\frac{7}{15}$
31. $\frac{2}{9}$ 33. $\frac{6}{8}$, or $\frac{3}{4}$ 35. 2 37. $\frac{5}{12}$ mile
39. $r = 1$ 41. $c = \frac{3}{5}$ 43. $m = 1$ 45. $\frac{5}{8}$ tsp
47. $\frac{2}{15}$ is left; add to find the total spent and saved; subtract to find how much is left.
49. 24 and 28 51. $\frac{4}{9}$ 53. 6,132

Pages 187–189

3. $3\frac{4}{5}$ 5. $1\frac{3}{10}$ 7. $6\frac{7}{12}$ 9. $1\frac{1}{6}$ 11. $1\frac{1}{18}$
13. $2\frac{1}{2}$ 15. $8\frac{3}{10}$ 17. $10\frac{1}{18}$ 19. $4\frac{1}{10}$ 21. $11\frac{3}{20}$
23. $8\frac{3}{14}$ 25. $28\frac{1}{8}$ 27. Commutative, $2\frac{1}{4}$
29. Associative, $1\frac{5}{6}$ 31. longer; $\frac{1}{6}$ in.; subtraction; to find the difference in length
33. $3\frac{1}{4}$ c 35. 3rd prize 37. $\frac{16}{15}$, or $1\frac{1}{15}$
39. $\frac{4}{5}, \frac{2}{3}, \frac{1}{2}$

Page 191

1. $1\frac{2}{3}$ 3. $1\frac{4}{9}$ 5. $\frac{7}{10}$ 7. $\frac{11}{12}$ 9. $3\frac{3}{4}$ 11. 21.42

Page 193

1. $3\frac{1}{3} = 2\frac{4}{3}$ 3. $2\frac{1}{12}$ 5. $3\frac{9}{10}$ 7. $3\frac{17}{18}$ 9. $1\frac{11}{12}$
11. $4\frac{7}{9}$ 13. $2\frac{7}{12}$ 15. $1\frac{3}{8}$ 17. $2\frac{7}{10}$, or 2.7
19. $2\frac{4}{5}$ 21. $\frac{7}{8}$ mi shorter 25. $7\frac{1}{3}, \frac{22}{3}$
27. 55 r5

Page 195

1. $18\frac{1}{2}$ mi 3. C 5. 12 cans 7. $4\frac{3}{4}$ mi
9. Gary

Chapter 10

Page 199

1. equation 3. 0 5. $\frac{1}{2}$ 7. $\frac{1}{2}$ 9. $\frac{1}{2}$ 11. $\frac{1}{2}$
13. $\frac{1}{2}$ 15. 0 17. $\frac{1}{2}$ 19. $c = 32$ 21. $x = 0.17$ 23. $t = 14$ 25. $r = 2.6$ 27. $1\frac{1}{6}$

29. $1\frac{1}{3}$ 31. $\frac{15}{2}$ 33. $\frac{14}{5}$

Page 201

1. about 500 million pounds 3. $\frac{1}{2}$ 5. 20
7. 70 9. 118 11. $\frac{1}{2}$ 13. 1 15. 4 17. 3
19. 20 21. $\frac{1}{4}$ 23. > 25. about 7 min
27. 8 kg 29. $3\frac{4}{9}$ 31. 11

Pages 204–205

3. $\frac{3}{8}$ 5. $\frac{2}{10}$, or $\frac{1}{5}$ 7. $\frac{3}{10}$ 9. $\frac{12}{7}$, or $1\frac{5}{7}$ 11. $\frac{8}{3}$, or $2\frac{2}{3}$ 13. $\frac{5}{9}$ 15. $\frac{3}{16}$ 17. $\frac{1}{16}$ 19. $\frac{2}{9}$ 21. $\frac{2}{15}$
23. $\frac{7}{10}$ 25. $\frac{1}{10}$ 27. $\frac{4}{15}$ 29. 2 31. <
33. < 35. 108 voters 37. 6 39. $b = 12.09$

Page 207

1. $3 \times 4\frac{2}{3} = (3 \times 4) + (3 \times \frac{2}{3}); 12 + 2 = 14$.
3. $1\frac{1}{8}$ 5. $2\frac{1}{4}$ 7. $18\frac{3}{8}$ 9. $2\frac{1}{4}$ 11. $8\frac{1}{6}$
13. 33 15. 15 17. 85 19. $7\frac{1}{5}$ 21. $22\frac{1}{2}$
23. < 25. $2\frac{1}{12}$ mi 29. $6\frac{7}{20}$ 31. 352,053

Page 208

1. 9 3. 4

Page 209

1. $n = 4$ 3. $n = \frac{3}{4}$ 5. 8 7. $\frac{3}{2}$, or $1\frac{1}{2}$ 9. $9\frac{9}{20}$
11. $606.64 > 606.074$

Pages 212–213

3. $\frac{3}{2}$ 5. $\frac{1}{7}$ 7. $\frac{3}{13}$ 9. $\frac{4}{5}$ 11. $\frac{1}{24}$ 13. 3
15. $1\frac{7}{32}$ 17. $\frac{1}{10}$ 19. $\frac{9}{2}$ 21. $\frac{7}{15}$ 23. $\frac{1}{9}$
25. $\frac{5}{13}$ 27. $1\frac{1}{6}$ 29. $9\frac{1}{3}$ 31. $2\frac{1}{4}$ 33. 5
35. $5\frac{1}{15}$ 37. $1\frac{8}{13}$ 39. 18 41. 32 43. 81
45. 44 47. $2\frac{5}{14}$ 49. $\frac{5}{18}$ 51. $\frac{4}{5}$ 53. 48 burgers 55. $1\frac{1}{4}$ yd 57. $46.90 59. $13\frac{3}{8}$

Page 215

1. $1\frac{5}{6}$ min; subtraction 3a. B 3b. H
5. $2\frac{2}{3}$ mi 7. $10\frac{1}{3}$ mi

Page 217

1. $b \times \frac{3}{4}; 32 \times \frac{3}{4}$; 24 oz 3. $2\frac{1}{5}$ 5. $5\frac{5}{8}$
7. $y = \frac{3}{8}$ 9. $7\frac{5}{6}$ 11. $1\frac{1}{4}$ 13. $\frac{1}{20}$ 15. $\frac{3}{8}$
17. 1 19. $x = 2$ 21. $x = \frac{1}{4}$ 23. $x = 1\frac{1}{3}$
27. $\frac{3}{4}, \frac{3}{5}, \frac{3}{8}$ 29. $a = 4.5$

Chapter 11

Page 227

1. negative numbers   11. 1, 2, 3, 4
13. 1, 3, 5, 7   15. <   17. =   19. >
21. 32°F   23. ⁻16°F   25. 60°F

Page 229

3. ⁻350   5. ⁻14   7. ⁺25   9. ⁻742   11. ⁻5
13. ⁺12,000   15. ⁺2   17. ⁺31   19. ⁻207
21. 28   23. 660   25. ⁻1,310 ft   27. $\frac{1}{5}$
29. $\frac{12}{20}$, or $\frac{3}{5}$ 31. 13

Pages 232–233

1. All integers can be written in the form $\frac{a}{1}$
3. $\frac{-37}{100}$ 5. $\frac{889}{1,000}$ 7. $\frac{-22}{3}$ 9. $\frac{-1}{2}$ 11. 0
13. $\frac{-71}{100}$ 15. $\frac{-21}{8}$ 17. ⁻0.3   19. ⁻0.7
21. $-7\frac{1}{4}$ 23. 16.05   25. $\frac{-5}{8}$ 27. yes
29. yes   31. yes   33. 90 steps   35. Not all integers are whole numbers; ⁻8 is not a whole number.   37. 0.34; $\frac{34}{100}$

Page 235

3. <   5. <   7. <   9. >   11. <   13. >
15. >   17. 0.4, 0.46, 0.6   19. $-\frac{1}{4}$, ⁻0.2, 0, $\frac{1}{4}$
21. 1°, ⁻3°, ⁻5°   23. ⁻450   25. $\frac{24}{7}$

Page 237

1. Jeffrey, sixth grade; Victoria, seventh grade; Arthur, eighth grade   3. C   5. Arlene, Kathy, Helene   7. Canadian Dollars   9. 3:15 P.M.

Chapter 12

Page 241

1. integers   3. ⁺62   5. ⁻13   7. ⁺12   9. ⁻3
11. ⁺11   13. ⁻8   15. 63   17. 10   19. 4
21. 56   23. 12   25. 5   27. 12
29. $n=3$ 31. $n=8$ 33. $n=132$

Page 242

1. ⁺13   3. ⁻10

Page 243

1. ⁻2   3. 0   5. ⁺7   7. ⁻9   9. ⁻6.8, ⁻6.4, ⁻6.2   11. ⁻3.6, $-3\frac{4}{7}$, $-3\frac{1}{2}$

Pages 246–247

1. Use sign of the addends.   3. The sum was 12, which indicates a gain of 12 yards, not a loss of 12 yards   5. 5 + ⁻9 = ⁻4   7. ⁻7
9. ⁻2   11. ⁻10   13. 0   15. ⁻2 + 6 = 4
17. ⁻1   19. 3   21. ⁻20   23. 8   25. ⁻28
27. ⁻12   29. ⁻11   31. 100   33. $x = -7$
35. $x = 5$ 37. ⁻12°F   39. The Wildcats lost 2 yards.   41. 12 cm   43. greater than

Page 249

1. ⁻11   3. ⁻22   5. 2   7. ⁻5   9. 6   11. ⁻8

Page 251

1. ⁻9 + 40   3. 3 + 6   5. ⁻4 + ⁻6   7. ⁻5
9. ⁻7   11. 8 + 11   13. ⁻9 + ⁻11   15. ⁻4
17. ⁻2   19. 0   21. ⁻5   23. ⁻4   25. ⁻5
27. 5   29. ⁻14°F   31. What was the range of temperatures?   33. ⁻213   35. 2 × 2 × 3 × 7

Pages 254–255

1. ⁻35; 35   3. ⁻54   5. ⁻16   7. ⁻3   9. ⁻24
11. ⁻40   13. 14   15. 12   17. 9   19. 288
21. ⁻300   23. 36   25. 80   27. $y=3$
29. $y=-4$ 31. $y=-35$ 33. ⁻14 in.   35. ⁻30
37. ⁻70 ft per minute   39. What's the sign of the quotient?   41. 22   43. $38\frac{1}{4}$

Pages 258–259

3. 1   5. ⁻0.7   7. $13\frac{1}{5}$ 9. ⁻8.0   11. ⁻10.8
13. 19   15. 3.5   17. ⁻2   19. ⁻1   21. $-9\frac{1}{4}$
23. $-3\frac{1}{2}$ 25. 15.6   27. ⁻87.4   29. ⁻17
31. ⁻12   33. $\frac{2}{9}$ 35. ⁻48   37. 0.7   39. ⁻1.5
41. 24   43. 4   45. ⁻10.5   47. 4.4   49. ⁻25
51. 557.4°F   53. $1\frac{1}{8}$ lb   57. 8.5%

Chapter 13

Page 269

1. exponent   3. algebraic expression   5. 32
7. 125   9. 16   11. 81   13. 49   15. 216
17. 25   19. 7   21. 2   23. 23   25. 20
27. 4   29. 21   31. 1, 2, 4, 8, 16   33. 1, 3, 11, 33   35. 1, 17   37. 1, 2, 3, 6, 7, 14, 21, 42
39. 1, 2, 4, 5, 8, 10, 20, 40

Student Handbook   **H85**

Page 271
3. $y \div 1.5$ **5.** 54 less than 19 times x
7. $3c + 2.9$ **9.** $h \times 4j \times k$ **11.** $6.3p - 5m$
19. $14.25 + 3.5h$ **21.** $2 + 2x$ **25.** $^-36$
27. 6 and 8

Pages 274–275
1. You will have to perform fewer computations when evaluating the expression if you simplify it first.
3. $^-16, ^-11, ^-6, ^-1$ **5.** $^-2, 6, 14, 22$ **7.** $1, ^-2, ^-3, ^-2$ **9.** $7x - 8; ^-29$ **11.** $^-16, ^-20, ^-24, ^-28$ **13.** $20\frac{1}{2}, 20, 19\frac{1}{2}, 19$ **15.** $^-5, 1, 3, 4$
17. $12x - 41; ^-89$ **19.** $356 + 6a; 236$
21. $^-2$ **23.** Associative; 67 **25.** $x = 2$
27. $x = ^-1$ **31.** $^-5$

Page 276
1. 4 **3.** 49

Page 277
1. 8 **3.** 14 **5.** 22 **7.** 81; 9 **9.** $^-49$
11. $>$

Page 279
1. Operate inside the parentheses, evaluate $\sqrt{36}$, then multiply; 12 **3.** $^-33$ **5.** 24
7. 150 **9.** $^-24$ **11.** $^-10$ **13.** 77 **15.** $>$
17. 14 ft **19.** $^-2, ^-6, ^-10, ^-14$ **21.** $2, ^-4, ^-6, ^-7$

Chapter 14
Page 283
1. equation **3.** 37 **5.** 23 **7.** 11.37
9. 2.7 **11.** $1\frac{7}{12}$ **13.** $\frac{1}{2}$ **15.** $7\frac{11}{12}$ **17.** $2\frac{9}{28}$
19. $43 - 19 = 24$ **21.** $16 + 3 = 19$
23. $46 + 196 = 242$ **25.** $125 \div 5 = 25$
27. $2 \cdot 6 = 12$ **29.** $21 \cdot 12 = 252$
31. $25 \cdot 16 = 400$ **33.** addition
35. subtraction **37.** multiplication
39. subtraction

Page 285
1. any quantity that you do not know
3. $x + 7 = 20$ **5.** $5 \cdot m = 35$; or $5m = 35$
7. $14 = n + 12$ **9.** $n \div 2\frac{3}{4} = \frac{5}{6}$
11. $72,000 = 9e$ **15.** $8x - 15; ^-31$
17. $25 + 9z; ^-47$

Page 286
1. $x = 4$ **3.** $x = 7$ **5.** $x = 3$ **7.** $x = 1$

Pages 288–289
3. $x = 7$ **5.** $c = 2.7$ **7.** $m = 15\frac{1}{4}$ **9.** $x = 8$
11. $k = ^-41$ **13.** $b = 2.9$ **15.** $y = 8\frac{3}{4}$
17. $t = ^-19$ **19.** $x =$ unknown length; $x + 12 + 10 = 29; x = 7$, or 7 cm **21.** $11 = n + 8$
23. 13

Page 291
1. You add the number that is being subtracted from the variable. **3.** $b = ^-6$
5. $y = 23.2$ **7.** $d = 34\frac{7}{15}$ **9.** $a = 33$
11. $z = 17.0$ **13.** $c = 23.3$ **15.** $s = 22\frac{11}{12}$
17. $m = 11.1$ **19.** $f = \frac{19}{24}$ **21.** $s =$ original amount in savings; $s - (110 + 90 + 40) = 527$; $s = 767$ **23.** $61 - 12 - 13 - 14 + 5 - 7$
25. $b = 4.4$ **27.** 9, 7, 5, 3

Chapter 15
Page 295
1. Celsius **3.** $18 = n + 6$ **5.** $\frac{n}{2} = \frac{2}{3}$
7. $2x = 47$ **9.** 14 **11.** $^-73$ **13.** $^-140$
15. 9 **17.** 25 **19.** 99 **21.** 27.3 **23.** $3\frac{4}{5}$
25. $y = 5$ **27.** $a = 9$ **29.** $c = 30$
31. $q = 56$ **33.** $y = 0.04$

Page 296
1. $c = 4$ **3.** $b = 2$

Page 299
1. You use inverse operations. **3.** $x = ^-7$
5. $y = 7.2$ **7.** $k = 6$ **9.** $a = 18$ **11.** $p = 16$
13. $n = ^-5$ **15.** $a = 16.48$ **17.** $w = 2.73$
19. $a = 10$ **21.** $\frac{m}{3} = 14; m = 42$; 42 marbles
23. $0.45 = 0.9w; w = 0.5$ cm **25.** $x = 22$
27. $z = ^-12$

Pages 302–303
1. The rate of speed is in feet per minute.
5. 20 **7.** 50°F **9.** 86°F **11.** 41°F
13. 13.9°C **15.** 37.8°C **17.** 36.8 **19.** 2.5
21. 98.6°F **23.** 203°F **25.** 0°C **27.** 8.3°C
29. 34.4°C **31.** 87.5 mi per hr **33.** Yes. The shuttle would be traveling 17,550 mi per hr.
35. Earth **39.** $y = {^-}6$

Page 305
1. $4x + 1 = 5$; $x = 1$ **3.** $y = 2$ **5.** Add 1 to both sides, then divide by 2; $x = 3$. **7.** 3
9. 374

Page 307
1. 18 mi **3.** C **5.** 4:20 P.M. **7.** 8 dimes, 4 nickels **9.** 6 students

Chapter 16
Page 317
1. angle **3.** line **5.** obtuse **7.** acute
9. straight **11.** acute **13.** ∠MNP or ∠PNM
15. ∠3 **17.** ∠a **19.** ∠PBF or ∠FBP

Page 319
1. plane, point, line, line segment, or ray
3. point R **5.** point C **7.** plane DHY
9. $\overleftrightarrow{XY}$ **11.** $\overrightarrow{PQ}, \overrightarrow{QR}, \overrightarrow{PR}$ **13.** $\overrightarrow{PQ}, \overrightarrow{QP}, \overrightarrow{RQ}, \overrightarrow{QR}, \overrightarrow{RP}, \overrightarrow{PR}$ **15.** a point **17.** a point
23. $n = 96$ **25.** ${^-}2$

Page 321
1. 135°; obtuse **3.** 73°; acute **9.** 45°
11. 90° **13.** $b = 8$ **15.** $\frac{2}{15}$

Pages 324–325
3. ∠AED and ∠BEC, ∠AEB and ∠DEC
5. ∠AED and ∠BEC **7.** 51° **9.** 33°
11. ∠DOC **13.** ∠AOB, ∠DOC **15.** 18°
17. 131° **19.** 90° **21.** supplementary, adjacent **23.** none of these **25.** ∠1, ∠2; ∠2, ∠3; ∠3, ∠4; ∠4, ∠5; ∠5, ∠1; They are side by side and have a common vertex.
27. ∠1, ∠5; ∠4, ∠5; Together they form a straight line. **29.** Their measures are equal.
31. $\overrightarrow{PQ}$ or $\overrightarrow{QP}$ **33.** ${^-}6$

Pages 326–327
1. perpendicular and intersecting
3. intersecting / perpendicular
5. intersecting **7.** intersecting
9. intersecting **11.** $\overleftrightarrow{AC}, \overleftrightarrow{BD}, \overleftrightarrow{CG}, \overleftrightarrow{DH}$
13. $\overleftrightarrow{BD}, \overleftrightarrow{FH}, \overleftrightarrow{CD}, \overleftrightarrow{GH}$ **17.** Perpendicular lines always intersect but intersecting lines are not necessarily perpendicular.
19. 59.2 mi per hr **21.** $\frac{5}{12}$

Chapter 17
Page 331
1. isosceles **3.** pentagon **5.** hexagon
7. quadrilateral **9.** triangle **15.** add 4; 20, 24, 28 **17.** subtract 6; 13, 7, 1 **19.** multiply by $\frac{1}{2}$; $\frac{1}{32}, \frac{1}{64}, \frac{1}{128}$

Pages 333–335
3. 49°; acute **5.** 51°; obtuse **7.** 61°; acute
9. 20°; obtuse **11.** 40°; acute **13.** 45°
15. 75° **17.** 30° **19.** 115° **21.** 120°
23. 55° **25.** 90° **27.** 46° **29.** 18°
31. 153° **33.** perpendicular **35.** 21

Page 337
1. 16, 22, and 29 dimes **3.** C
5. Silvia: blue; Rhoda: green; David: brown
7. about $4 billion **9.** 24 mi per gal

Pages 340–341
3. trapezoid **5.** square **7.** rectangle or square **9.** square **11.** rectangle
13. quadrilateral **15.** rhombus **17.** square
19. quadrilateral **21.** parallelogram, trapezoid **23.** 11 in. × 15 in. **27.** 84°
29. 0.08

Pages 430–431

1. 1, 2, 3, 4, 5, 6 **3.** $\frac{1}{4}$, 0.25, 25% **5.** $\frac{1}{4}$, 0.25, 25% **7.** $\frac{3}{4}$, 0.75, 75% **9.** $\frac{1}{4}$, 0.25, 25% **11.** $\frac{0}{8}$, 0, 0% **13.** $\frac{1}{2}$, 0.50, 50% **15.** $\frac{1}{6}$ **17.** $\frac{1}{3}$ **19.** $\frac{1}{2}$ **21.** $\frac{1}{3}$ **23.** 1 **25.** = **27.** > **29.** < **31.** 0.6 **33.** $\frac{5}{12}$ **35.** 77% **37.** the probability the event will not occur **41.** 9 blue; 15 are not blue **43.** ⁻7.75, $7\frac{1}{4}$, $7\frac{3}{8}$, 7.5

Page 433

1. *too much*; $\frac{4}{7}$ **3.** *too little*; need the 1970 land speed record **5.** D **7.** 4 hr **9.** 8,528 steps

Page 435

1. $\frac{5}{8}$ **3.** 89

Page 437

3. $\frac{1}{5}$ **5.** $\frac{1}{10}$ **7.** $\frac{2}{15}$ **9.** $\frac{1}{4}$ **11.** $\frac{2}{5}$ **13.** $\frac{2}{5}$ **15.** 15 times **17.** number of times 4 lands ÷ total number of rolls **19.** $\frac{4}{7}$ **21.** 9.75

Chapter 23

Page 441

1. sample space **3.** outcome **5.** $\frac{1}{3}$ **7.** $\frac{3}{16}$ **9.** $\frac{2}{5}$ **11.** $\frac{1}{8}$ **13.** $\frac{1}{25}$ **15.** about 26 in. **17.** 100 students **19.** 16 students

Page 443

1. 12 choices **3.** B **5.** 1,870,737 **7.** 4 groups of 10; 3 groups of 8

Pages 445–446

1. 24 **3.** 9 outcomes **5.** 36 outcomes **7.** 16 outcomes **9.** 18 outcomes **11.** 216 outcomes **13.** yes; 12 · 4 · 8 = 384, and 384 > 365 **15.** $\frac{1}{5}$ **17.** 163°

Pages 448–449

1. 16%; 36%; yes, because there are more odd numbers on the spinner than even numbers, and the probability of two odd numbers is greater. **3.** dependent **5.** $\frac{1}{36}$ **7.** 0 **9.** $\frac{1}{20}$ **11.** $\frac{1}{5}$ **13.** independent **15.** $\frac{1}{36}$; $\frac{1}{30}$ **17.** $\frac{1}{6}$, $\frac{1}{5}$ **19.** $\frac{1}{9}$; $\frac{2}{15}$ **21.** $\frac{1}{27}$; 0 **23.** $\frac{1}{108}$; 0 **27.** 15 outcomes **29.** 18 outcomes

Page 451

1. Write and solve the proportion $\frac{5}{75} = \frac{n}{210}$ **3.** $\frac{7}{20}$, 0.35, or 35% **5.** about 102 sixth graders **7.** about 600 cars **9.** $\frac{1}{3}$ **11.** $\frac{1}{2}$

Chapter 24

Page 461

1. feet **3.** multiply **5.** 3 **7.** 56 **9.** 4 **11.** 24 **13.** 12 **15.** 5 **17.** 1,000 **19.** 10 **21.** 1,000 **23.** 2 **25.** 4,000 **27.** 9 **29.** $n = 45$ **31.** $n = 9$ **33.** $n = 10$

Page 463

1. Use the ratio $\frac{4 \text{ qt}}{1 \text{ gal}}$ on one side of the proportion and the number of quarts over x gallons on the other side. **3.** 4 **5.** 128 **7.** 12 **9.** $\frac{1}{2}$ **11.** $1\frac{1}{2}$ **13.** 60 **15.** $2\frac{1}{4}$ **17.** $3\frac{1}{4}$ **19.** = **21.** $4\frac{1}{2}$ yd **25.** $1\frac{13}{55}$ **27.** 2.361

Page 465

1. Use the ratio $\frac{10 \text{ dm}}{1 \text{ m}}$ on one side of the proportion and the number of decimeters over x meters on the other side. **3.** 0.005 **5.** 9,000 **7.** 200,000 **9.** 0.5 **11.** 1.2 **13.** 440,000 **15.** 18,000 **17.** 425 **19.** > **21.** = **25.** $\frac{1}{3}$ **27.** 40% **29.** ⁻9

Page 467

1. greater; it takes 1.61 km to make 1 mi. **3.** 25.4 **5.** 427.7 **7.** 10 **9.** 76.2 **11.** 4.75 **13.** 26.37 **15.** < **17.** > **19.** 1.38 in. **21.** 18°C **23.** ⁻10 **25.** $\frac{7}{6}$

H90 Selected Answers

Pages 470–471
1. gram; because the gram is a smaller unit than the kilogram   **3.** 2 cm; 23 mm   **5.** 85 in.   **7.** 65 oz   **9.** meter, yard, or foot   **11.** gram or ounce   **13.** $1\frac{1}{2}$ in.; $1\frac{3}{4}$ in.   **15.** $\frac{3}{4}$ in.; $\frac{7}{8}$ in.   **17.** 8 ft   **19.** 9 oz   **21.** millimeter or part of an inch   **23.** =   **25.** =   **27.** To the nearest millimeter because a millimeter is a tenth of a centimeter, which is smaller than a half centimeter.   **31.** 17.055 L   **33.** ⁻96

Page 473
1. estimate; no   **3.** estimate; no   **5.** C   **7.** $9\frac{3}{20}$ in.   **9.** $2\frac{3}{8}$ in. below   **11.** 23 years old

Chapter 25
Page 477
1. perimeter   **3.** 32 ft   **5.** 20 cm   **7.** 104 in.   **9.** 4   **11.** 36   **13.** 4   **15.** 3,000   **17.** 5,000   **19.** 0.00009   **21.** 88   **23.** 2,625   **25.** 526.5   **27.** 60   **29.** 73.6

Page 481
3. 13.57 m   **5.** 9 ft or 108 in.   **7.** 43 m   **9.** $x = 13.5$ cm   **13.** 5.40   **15.** $a = 1,264$

Page 483
1. 66 in.   **3.** B   **7.** 42.4%   **9.** 1 in.

Pages 486–487
1. They both use π; one uses the diameter and the other uses 2 times the radius, which is equal to the diameter.   **3.** 16 m   **5.** 12 cm   **7.** 283 yd   **9.** 27 in.   **11.** 33 cm   **13.** 20 in.   **15.** 9 yd   **17.** 220 cm   **19.** 316 in.   **21.** 22 ft   **23.** 16 cm   **25.** 4.5 ft   **27.** 9.3 cm   **29.** The circumference is twice as long.   **31.** 88 yd   **33.** 149 m

Page 489
1. 25   **3.** 41   **5.** no   **7.** yes   **9.** 109.9 mm   **11.** 314 in.

Chapter 26
Page 493
1. square   **3.** faces   **5.** 144   **7.** 400   **9.** 81   **11.** 5.76   **13.** 2,401   **15.** 16,641   **17.** 25   **19.** 192   **21.** 16   **23.** 5 faces   **25.** 4 faces

Pages 496–497
1. The area of the triangle is $\frac{1}{2}$ the area of the rectangle.   **3.** about 20 m²   **5.** 117 in.²   **7.** 45.5 in.²   **9.** about 6 m²   **11.** 558.25 mm²   **13.** 0.2 m²   **15.** 22.5 ft²   **17.** 273 yd²   **19.** 12   **21.** 60%

Pages 499–500
1. $A = \frac{1}{2}h(b_1 + b_2) = \frac{1}{2} \times 4 \times (4.2 + 6.5) = 21.4$; 21.4 m²   **3.** 10 ft²   **5.** 102.3 m²   **7.** 136.95 m²   **9.** 40 ft²   **11.** 1,008 cm²   **15.** 24 m²   **17.** 0.003

Page 501
1. 3 m²   **3.** 154 m²

Page 503
3. 28 cm²   **5.** 1,809 ft²   **7.** 50 yd²   **9.** 7 m²   **11.** 50 mm²   **13.** 3,419 in.²   **15.** 77 cm²   **17.** 804 ft²   **19.** 100 ft²   **21.** 3

Pages 506–507
1. Find the area of each pentagonal face and the area of the five rectangular faces and add.   **3.** 184 ft²   **5.** 336 cm²   **7.** 340 m²   **9.** 73.5 m²   **11.** 108 cm²   **13.** 69.36 cm²   **15a.** $3 \times 6 \times 12$; 252 m²   **15b.** $3 \times 6 \times 6$; 144 ft²   **15c.** $10 \times 5 \times 15$; 550 in.²   **15d.** $2 \times 8 \times 32$; 672 cm²   **17.** 2 cans   **19.** 5 in.   **21.** 113 ft²   **23.** 0.25

Chapter 27
Page 511
1. dimensions   **3.** height   **5.** 64   **7.** 27   **9.** 216   **11.** 0.008   **13.** 1,728   **15.** 20 ft²   **17.** 21 m²   **19.** $58\frac{7}{12}$ ft²   **21.** 52.36 m²   **23.** 141 cm²   **25.** 346 cm²

Pages 514–515
1. Find 26 × 3 × 18, or 1,404 cubes.
3. 24 in.^{3} **5.** 48 cm^3 **7.** 324 ft^3
9. 294 ft^3 **11.** 144 in.^{3} **13.** $x = 10$ m
15. 189 ft^3 **17.** 648 ft^3 **19.** 45.63

Page 517
1. twice the volume of the original container
3. B   **7.** 10 shirts, 0 shorts; 0 shirts, 8 shorts; 5 shirts, 4 shorts   **9.** Mike, Sharon, Jasmine, Hugh

Page 519
1. Both include the area of the base times the height; prism: $V = bh$; pyramid: $V = \frac{1}{3}Bh$
3. 48 m^3 **5.** 224 cm^3 **7.** 24 cm^3, or 24,000 mm^3 **9.** 4,500 ft^3 **11.** 24 yd^3, or 648 ft^3
15. 110   **17.** 0.79

Page 520
1. about 75 cm^3

Pages 522–523
1. 6,029 in.^{3} **3.** πr^2 represents the area of the base and h represents the height.
5. about 346 in.^{3} **7.** about 942 in.^{3}
9. about 1,409 cm^3 **11.** about 311 cm^3
13. about 2,374 cm^3 **15.** about 7,436 ft^3
17. The volume is about eight times as large.
21. about 4 ft^3 **23.** $\frac{1}{2} \div \frac{1}{10}$

Chapter 28
Page 533
1. equation   **3.** <   **5.** <   **7.** >   **9.** >
11. <   **13.** >   **15.** <   **17.** >   **19.** 10; 21
21. 25; 25.5   **23.** 15   **25.** ⁻5   **27.** 63
29. 29   **31.** $\frac{3}{14}$ **33.** 3.1

Page 535
1. the eighth day   **3.** A   **5.** 7 videos
7. 660 ft   **9.** 40 ft^2

Pages 537–538
1. Multiply by 4; 2,048; 8,192; 32,768   **3.** Add 15 to each successive term   **5.** 43, 54, 65
7. Divide each term by 3.   **9.** Add 0.89 to each term.   **11.** 335, 485, 665   **13.** 81, 76, 70
15. 9, 12.7, 16.4, 20.1, . . .   **17.** add $54; $462
21. 754 ft^3 **23.** 264

Pages 541–542
3. $b = a - 6$ **5.** $d = c + 1.1$ **7.** $l = 4w$
9. $g = k \div 2$; 27   **11.** $d = c \div ⁻4$; ⁻8
15. in $y^2 = x$, if $x = 1$, $y = 1$ or $y = ⁻1$
17. $t = 0.75n$; $37.50   **19.** $209   **21.** $2\frac{5}{104}$

Page 545
9. $5^0, 5^1, 5^2, 5^3, \ldots$ **11.** 16 prisms; 25 prisms
13. $y = ⁻9, ⁻3, 3, 9, 15$ **15.** 5×13

Chapter 29
Page 549
1. congruent   **3.** slide   **5.** slide   **7.** turn
17. 25°   **19.** 135°

Pages 551–552
1. The figure does not change in size or shape.   **3.** rotation or reflection
5. reflection   **7.** rotation, reflection
9. rotation or translation   **11.** rotation
13. translation, reflection, reflection
19. horizontal reflection   **25.** $\frac{5}{8}$

Pages 554–555
3. yes   **5.** no   **11.** yes   **13.** no
19. 288 sq yd.   **21.** rotation   **23.** $\frac{11}{12}$

Page 557
1. triangle and square; square and octagon
3. C   **5.** 63°   **7.** a   **9.** Gerta, Bernie, Nash, Lisa

H92   Selected Answers

Page 559

1. Yes, the figure's size and shape do not change.   **3.** 8 ways   **5.** 6 ways   **7.** 8 ways   **9.** 12 ways   **11.** rotate it 180°   **13.** reflection   **15.** 24 cm

Pages 562–563

1. Line symmetry means that the figure can be folded into two congruent halves that are mirror images. Rotational symmetry means that the figure matches itself when rotated less than 360°.   **11.** yes; $\frac{1}{4}$; 90°   **13.** yes; $\frac{1}{2}$; 180°   **27.** yes; $\frac{1}{5}$; 72°   **29.** no   **31.** 8 pieces; yes   **33.** To have rotational symmetry, it must match up with a rotation of less than 360°   **35.** 125.6 in.^{3}

Chapter 30

Page 567

1. ordered pair   **3.** perpendicular   **5.** =   **7.** >   **9.** <   **11.** >   **13.** (⁻2,2)   **15.** (0,0)   **17.** (2,1)   **19.** (0,3)   **21.** (⁻2,⁻2)

Page 569

1. Yes.   **19.** $p > {}^-11$ **21.** $a < 20{,}320$ **25.** 37.5%   **27.** $\frac{3}{5}$

Pages 572–573

1. (5,4) is above the x-axis. (⁻5,⁻4) is below the x-axis.   **3.** (2,4)   **5.** (0,0)   **7.** (5,⁻5)   **9.** (0,⁻5)   **11.** (2,⁻3)   **13.** A, H   **15.** B   **17.** (2,6)   **19.** (2,⁻2)   **21.** (0,7)   **23.** (3,0)   **25.** (6,⁻4)   **27.** (⁻7,⁻6)   **29.** A, L, J   **31.** B, H, K   **41.** triangle; 28 sq units   **43.** It is 0; it is 0.   **45.** $y \geq {}^-7$ **47.** 157 cm

Page 575

1. a table, a graph of ordered pairs.   **7.** (45,15), (48,16), (51,17), (54,18); $w = \frac{1}{3}c$ **9.** Quadrant I   **11.** 40 cm^2

Page 577

1. C   **3.** $y = 9x$ **5.** 1,333 mi   **7.** 11:00 A.M.   **9.** about 800

Pages 582–583

3. $A'(0,3)$, $B'(2,3)$, $C'(2,1)$, $D'(0,1)$ **5.** $A'(0,0)$, $B'(⁻6,0)$, $C'(⁻4,2)$, $D'(⁻2,2)$ **7.** $E'(0,⁻1)$, $F'(3,⁻1)$, $G'(3,⁻4)$, $H'(0,⁻4)$ **9.** $E'(1,⁻3)$, $F'(4,⁻3)$, $G'(4,⁻1)$, $H'(1,⁻1)$ **11.** No.   **13.** $y = 7x$ **15.** $\frac{1}{4}$

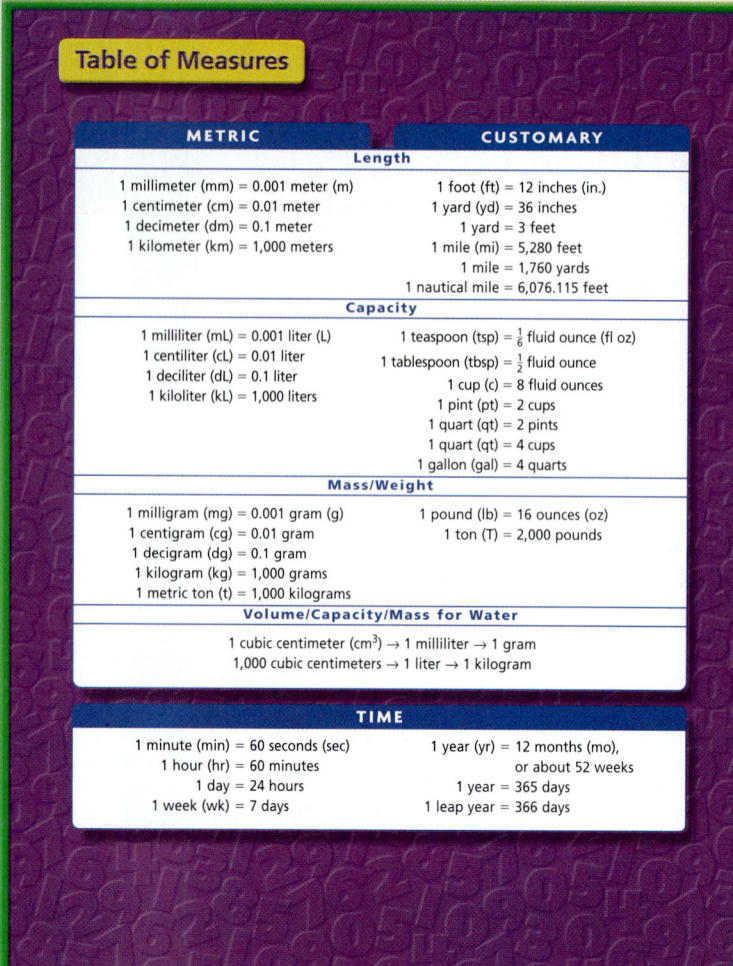
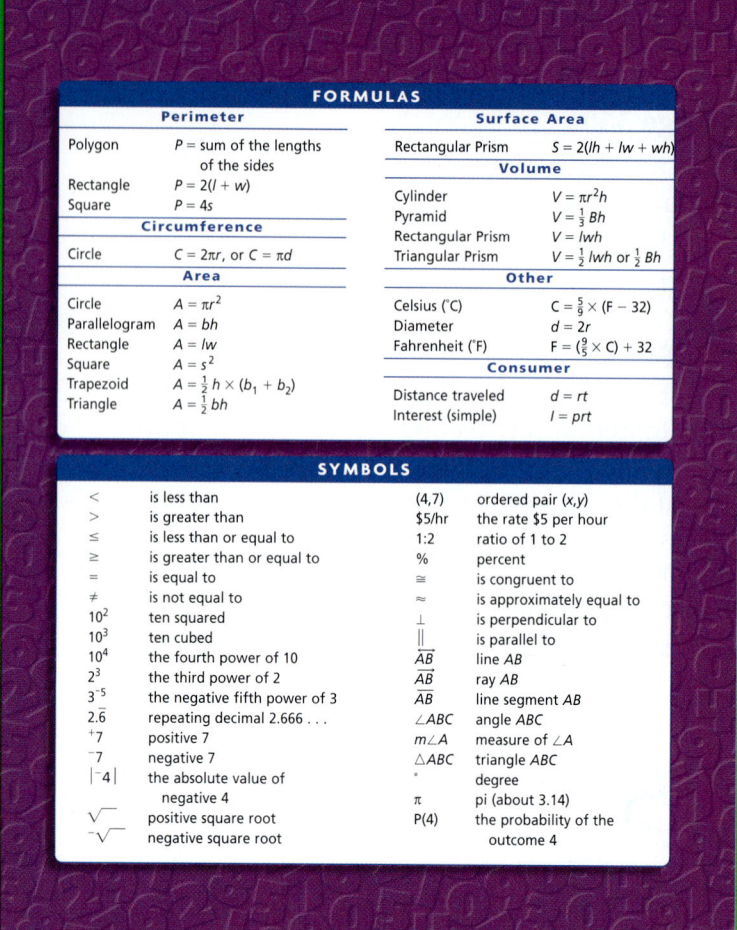

Table of Measures

HARCOURT Math

PROBLEM OF THE DAY

This section provides complete solutions for the **Problem of the Day** in each lesson plan. The problems include all types—one-step, multi-step, applied, process, nonroutine, open-ended, and puzzle problems—and provide options for students to develop their ability to use logical reasoning to choose and apply problem-solving strategies to varied and interesting situations.

The **Problem of the Day** for a lesson is also available on the Daily Transparency. Each Daily Transparency includes the **Problem of the Day**, the **Quick Review**, and the **Mixed Review and Test Prep** in Grades 1 and 2. In Grades 3–6, each Daily Transparency includes the **Number of the Day**, the **Problem of the Day**, and the **Lesson Quiz**.

Problem of the Day

Chapter 1 Answer Key

Lesson 1.1

Problem

Find the three-digit number that rounds to 440 and includes a digit that is the quotient of $24 \div 3$. Is there more than one possible answer? Explain.

Solution

Strategy: Use Logical Reasoning
The numbers that round to 440 are 435, 436, 437, 438, 439, 440, 441, 442, 443, 444. One of the digits is the quotient of $24 \div 3 = 8$. The only one of those numbers that rounds to 440 and has 8 as a digit is 438.

438; No. The numbers that round to 440 are 435–444. $24 \div 3 = 8$. Only one of those numbers has 8 as a digit.

Lesson 1.2

Problem

Presidents George Washington, John Adams, and Thomas Jefferson lived a total of 240 years. Adams lived the longest, 23 years longer than Washington. Jefferson lived 16 years longer than Washington. How old was each president when he died?

Solution

Strategy: Predict and Test
Note that the average age is 80.
Begin by guessing, for example, that Washington is 70, Adams is 93, and Jefferson is 86. The sum of their ages is then 249, which is 9 too large. Subtract 3 from each age to get the solution.

Washington, 67; Adams, 90; Jefferson, 83

Lesson 1.3

Problem

Find the product. Compare the product with the first factor. Write a rule for multiplying a 2-digit number by 11 and a rule for multiplying greater numbers by 11.

Solution

Strategy: Find a Pattern
1. $13 \times 11 = 143$
2. $72 \times 11 = 792$
3. $326 \times 11 = 3,586$
4. $6,045 \times 11 = 66,495$

A shortcut for multiplying a 2-digit factor by 11:
$34 \times 11 =$
1. Moving from right to left, write the ones digit of the other factor. 4
2. Write the sum of the ones and tens digits. 74
3. Write the tens digit. 374

For multiplying greater factors by 11:
1. Write the ones digit of the other factor.
2. Write the sum of the ones and tens digits, the sum of the tens and hundreds digits, etc., as necessary.
3. Write the digit in the greatest place-value position. Some students may try problems in which the sum of 2 digits is greater than 9. The pattern remains the same, but the 1 that is "carried" is added to the next pair of digits or to the first digit.

Lesson 1.4

Problem

At noon on Monday, Latisha sets her watch to the correct time. If her watch loses one minute each hour, and she does not correct it, what time will her watch show at noon on Wednesday? At noon on what day will her watch show 10:00?

Solution

Strategy: Make a Table/Write a Number Sentence
$24 \times 1 = 24$ minutes are lost each day.
Subtract 24 minutes each day.
Results:

Monday noon:	12:00
Tuesday noon:	11:36
Wednesday noon:	11:12
Thursday noon:	10:48
Friday noon:	10:24
Saturday noon:	10:00

At noon on Wednesday her watch will show 11:12. At noon on Saturday her watch will show 10:00.

Chapter 1 Answer Key

Lesson 1.5

Problem

In a number game, when Tina says *three*, Jay says *ten*. When Tina says *five*, Jay says *sixteen*. When Tina says *nine*, Jay says *twenty-eight*. When Tina says *eight*, what does Jay say? If Jay says *one*, what number has Tina said?

Solution

Strategy: Make a Table
If Tina said the counting numbers, the numbers would have a difference of 1. Enter the given numbers and look for a pattern.

TINA	0	1	2	3	4	5	6	7	8	9
JAY	1	4	7	10	13	16	19	22	25	28

When Tina says eight, Jay says *twenty-five*.
If Jay says *one*, Tina must have said *zero*.
For discussion: Students note that when Tina says *zero*, Jay said *one*, so he must have added 1. However in the other numbers he didn't just add 1. What else did he do that made the number share a difference of 3? He multiplied by 3 and added 1. $3x + 1$.

Lesson 1.6

Problem

Martin saves n dollars each week. Kara saves twice as much as Martin. In 15 weeks their combined savings total $450. How much does Martin save each week? How can you use mental math to solve it?

Solution

Strategy: Use Logical Reasoning
THINK: $450 is saved in 15 weeks and the same amount is saved each week.
$450 \div 15 = 30$.
Martin and Kara save a total of $30 each week.
Kara saves twice as much as Martin, so $10 + 20 = 30$.
Martin saves $10 each week.

PD1B Problem of the Day Solutions

Chapter 2 Answer Key

Lesson 2.1

Problem
Replace the ■ with the digits 0–9 to make correct number sentences. Use each digit only once.

■ × ■ = 18
■ × ■ = 24
■ × ■ = 0
■ × ■ = 28
■ × ■ = 6

Solution
Strategy: Use Logical Reasoning
Students can write the digits 0–9, and then cross them off as they use them. Students look for a product that can have only one correct set of factors, i.e. 28. $4 \times 7 = 28$ or $7 \times 4 = 28$ (Digits may be in any order.)
Since 4 is already used, only $3 \times 8 = 24$.
Since 3 is used, only $2 \times 9 = 18$.
Since 2 and 3 are used, only $6 \times 1 = 6$.
The only digits left are 5 and 0, so $5 \times 0 = 0$.

$2 \times 9 = 18$
$3 \times 8 = 24$
$5 \times 0 = 0$
$4 \times 7 = 28$
$1 \times 6 = 6$

Lesson 2.2

Problem
Replace the letters *a*, *b*, and *c* with the numbers 3, 4, and 5 to make a true sentence.
$2^a + 2^a = b^c$

Solution
Strategy: Predict and Test
$2^5 + 2^5 = 4^3$
$(2 \times 2 \times 2 \times 2 \times 2) + (2 \times 2 \times 2 \times 2 \times 2) = (4 \times 4 \times 4)$
$32 + 32 = 64$

Lesson 2.4

Problem
Complete the expression using the numbers 3, 4, and 5 so that it equals 19.
____ + ____ × ____

Solution
Strategy: Predict and Test
Substitute one of the numbers for each of the blanks and evaluate the expression:
$3 + 4 \times 5 = 3 + 20 = 23$, too big
$5 + 3 \times 4 = 5 + 12 = 17$, too small
$4 + 3 \times 5 = 4 + 15 = 19$

$4 + 3 \times 5$

Lesson 2.5

Problem
Look at the following figure. Start at point *A*. Write the sequence that allows you to go around the entire figure, covering each segment only once. Is there only one way? Can you do the same thing if you start at *B*?

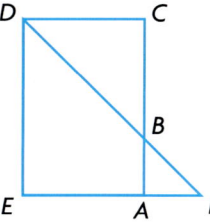

Solution
Strategy: Predict and Test
One possible answer is: Go from *A* to *F*, from *F* to *D*, from *D* to *E*, from *E* to *A*, from *A* to *C*, and from *C* to *D*. There are several others that work. Students can find a route by tracing the figure with a finger or pencil. If you start at *B*, there is no sequence of steps that allows you to trace all the segments only once.

Chapter 3 Answer Key

Lesson 3.1

Problem
The money that Mrs. Frey deposited in her bank account was in $10 bills. The sum of the digits in the amount she deposited was 18. If she had deposited $10 more, the sum of the digits would have been 1. How much did Mrs. Frey deposit?

Solution
Strategy: Use Logical Reasoning

Since the deposit was in $10 bills the amount must be a multiple of 10.

$9 + 9 + 0 = 18$

$\$990 + \$10 = \$1,000$

$1 + 0 + 0 + 0 = 1$

$990

Lesson 3.2

Problem
Jake knows that Uranus is farther from Earth than Saturn but not as far as Neptune. Match each planet with its distance from Earth.

0.744 billion mi

2.7 billion mi

1.6 billion mi

Solution
Strategy: Make an Organized List

Students can compare the three decimals and list them in order from least to greatest since all the exponent factors are the same.

0.744   Saturn

1.6   Uranus

2.7   Neptune

Once they have them in order, they can attach Saturn to the smallest number, Uranus to the next larger one, and Neptune to the largest one.

Lesson 3.3

Problem
Randall, Kira, and Sean have 100 baseball cards in all. Randall has twice as many as Kira and 10 more than Sean. How many cards does each person have?

Solution
Strategy: Predict and Test

Students can make a table and *predict and test* to find the number of cards each person has.

RANDALL	KIRA	SEAN	TOTAL
50	25	40	115
46	23	36	105
44	22	34	100

Randall has 44 cards, Kira has 22 cards, and Sean has 34 cards.

Lesson 3.4

Problem
An estimate of the sum of two decimals is 27 and an estimate of the product is 140. Give two decimals that satisfy these requirements.

Solution
Strategy: Predict and Test

Students can determine two whole numbers that satisfy the requirements, for example, 20 and 7. Then, they can select any decimals that round to 20 and to 7.

Possible answer: 6.85 and 19.61

Chapter 4 Answer Key

Lesson 4.1

Problem

Replace each [♥] with a different digit from 0–9 to make a true number sentence.

[♥].[♥] [♥] [♥] + [♥] [♥].[♥] [♥] + [♥].[♥] = 22.815

Solution

Strategy: Predict and Test

Possible answer:

$0.725 + 13.69 + 8.4 = 22.815$

Lesson 4.2

Problem

Rhonda is making an input/output table. When she sees 5 she writes 3.0. When she sees 10 she writes 6.0. When she sees 4, she writes 2.4. What will she write when she sees 12? What did she see if she wrote 4.2?

Solution

Strategy: Find a Pattern

$5 \times \underline{} = 3.0$

$10 \times \underline{} = 6.0$

$5 \times 0.6 = 3.0$

$10 \times 0.6 = 6.0$

Students can determine that each number Rhonda sees is multiplied by 0.6 to get the number she writes. They can determine this number by dividing 3.0 by 5 or by choosing a sensible estimate of the number and testing to see if they are correct.

Then, to find what she writes for 12:

$12 \times 0.6 = 7.2$.

To find what she saw when she wrote 4.2, divide: $4.2 \div 0.6 = 7$. 7.2; 7

Lesson 4.4

Problem

The sum of two decimal numbers is 9.3. Their difference is 4.3, and their product is 17.00. What are the numbers?

Solution

Strategy: Use Logical Reasoning

There is a 3 in the tenths place for the sum *and* the difference, so one of the digits in the tenths place must be a zero or a 5. If one tenths digit is 0, the other tenths digit is 3; if one tenths digit is 5, the other tenths digit is 8. 17.00 is divisible by 6.5 and 6.8, but not by 6.3. So the digits in the tenths place are 5 and 8.

The digit 1 may not be great enough for the ones place.

THINK: One number is 4.3 greater than the other.

Try $2.8 + 4.3 = 7.1$ (Not correct. There should be a 5 in the tenths place.)

Try $2.5 + 4.3 = 6.8$ (One number has a 5 and the other has an 8 in the tenths place.)

Check: $2.5 + 6.8 = 9.3$

$6.8 - 2.5 = 4.3$

$2.5 \times 6.8 = 17.00$

The two numbers are 2.5 and 6.8.

Some students may need the strategy *Predict and Test*.

Lesson 4.5

Problem

Marge had five 32-oz bottles of milk. Monday she drank 6 oz from the first bottle, Tuesday she drank 12 oz from the second bottle, Wednesday she drank 18 oz from the third bottle, and so on. Each day she divided the remaining milk among her five cats so that each got a whole number of ounces. She saved the remainder. On the sixth day, she drank what was left. How much did she drink on Day 6?

Solution

Strategy: Find a Pattern

In order to solve, students must first find the pattern in the amounts she drinks each day: 6, 12, 18, 24, 30. Then they can find the amount of milk she had to divide among the cats: 26, 20, 14, 8, and 2 oz.

Finally, students need to recognize that the amount she had left to drink each day is the remainder in the division problem in which the dividend is the amount left each day and the divisor is 5. The remainders are 1, 0, 4, 3, and 2, so she has 10 oz to drink on the 6th day.

10 oz

Chapter 4 Answer Key

Lesson 4.6

Problem

Lashonda and Mark each have the same number of coins. Lashonda has $8.25 in quarters. Mark has all dimes. How much more money does Lashonda have than Mark?

Solution

Strategy: Write a Number Sentence

Lashonda: $8.25 ÷ 0.25 = 33 quarters

Mark: 33 dimes × $0.10 = $3.30

$8.25 − $3.30 = $4.95

Lashonda has $4.95 more than Mark.

Chapter 5 Answer Key

Lesson 5.1

Problem
Dana's survey showed that 3 out of 8 students preferred pepperoni pizza and 1 out of 8 students preferred cheese pizza. How many more of the 72 students surveyed by Dana liked pepperoni pizza than liked cheese pizza?

Solution
Strategy: Write a Number Sentence

1⁄8 of 72 = 9 students

3⁄8 of 72 = 27 students

27 − 9 = 18 more students liked pepperoni pizza.

Lesson 5.2

Problem
In a survey of students about a field trip to a nearby factory, 12 students were undecided, 5 times that many were in favor of the field trip, and half as many were against it as were in favor of it. How many students participated in the survey?

Solution
Strategy: Use Logical Reasoning

There are 12 students who are undecided.

$5 \times 12 = 60$, so there are 60 students in favor of the field trip.

Half of 60 is 30, so there are 30 students against the field trip.

$12 + 60 + 30 = 102$

102 students

Lesson 5.3

Problem
It takes 4 yd of material and 3 yd of trim to make 2 pumpkin decorations. Rhonda needs to make 10 decorations. How many yards of material and trim will she need?

Solution
Strategy: Make an Organized List

Students may find it helpful to organize information by making a list or table showing the number of yards of material and trim needed for 2, 4, 6, 8, and 10 decorations.

NUMBER OF DECORATIONS	MATERIAL	TRIM
2	4 yd	3 yd
4	8 yd	6 yd
6	12 yd	9 yd
8	16 yd	12 yd
10	20 yd	15 yd

20 yards of material and 15 yards of trim

Lesson 5.4

Problem
Anne's line plot of ages of students has a range of 4. Each age has twice as many x's as the previous one. If the last age has 16 x's, how many students did Anne include in her data?

Solution
Strategy: Use Logical Reasoning

If does not matter what the ages were, but there are 4 sets of ages. So work backward, dividing by 2 each time. Find the sum.

$16 + 8 + 4 + 2 + 1 = 31$ students

Lesson 5.5

Problem
A set of 9 numbers has a mean of 7. When one more number is added to the set, the mean is doubled. What number was added to the data set?

Solution
Strategy: Work Backward

The new mean is 14 and the new set has 10 numbers. So, the sum of the numbers is 10×14, or 140. The sum of the original nine numbers was 9×7 or 63. The difference between the two sums is 77, so 77 must have been added to the original set of numbers.

77

Chapter 5 Answer Key

Lesson 5.6

Problem

Rita has 1 brother and 3 sisters. If the mean age of all the children is 5 years, what will their mean age be 7 years from now?

Solution

Strategy: Use Logical Reasoning

There are 5 children in all: Rita, 1 brother, and 3 sisters. Since the mean of their current ages is 5, the total of their current ages is 5 × 5 years = 25 years.

Seven years from now each child will be 7 years older. Then the total of their ages will be 25 + 5 × 7 = 25 + 35 = 60. To find the mean in 7 years, use the number sentence 60 ÷ 5 = 12. So in 7 years the mean of their ages will be 12 years.

12 years

Lesson 5.7

Problem

The mean of these numbers is 14. The greatest number is 21 more than the least. The mode is 18. What are the missing numbers?

3 6 9 ◆ ◆ ◆ ◆

Solution

Strategy: Use Logical Reasoning

3 6 9 ◆ ◆ ◆ 24

(3 + 21 = 24)

There are 7 numbers with a mean of 14, so

7 × 14 = 98, the total. The mode is 18. Two or three of the missing numbers must be 18. Try three 18's and find the total.

3 6 9 18 18 18 24

(Sum is 96, which is 2 less than 98.)

So try two 18's and 20 (18 + 2)

3 6 9 18 18 20 24

(Yes, the sum is 98.)

PD5B Problem of the Day Solutions

Chapter 6 Answer Key

Lesson 6.1

Problem
Maurie is the second-youngest of 4 teenagers, all 2 years apart in age. His mother is 3 times as old as he is and 24 years younger than her father. How old is Maurie's grandfather?

Solution
Strategy: Use Logical Reasoning

The teenagers must be 13, 15, 17, and 19 years old.
Maurie must be 15 years old.
His mother is 45 years old. (3×15)
His grandfather is 69 years old. ($45 + 24$)

Lesson 6.2

Problem
A line graph shows that the temperature at 6 A.M. was 4° warmer than at 4 A.M. In the hours between midnight and 4 A.M., the temperature had fallen an average of 3° per hour. If the temperature at 6 A.M. was 10°F, what was the temperature at midnight?

Solution
Strategy: Make a Table or a Graph

Students may reproduce the line graph to help solve. They should begin by graphing the temperature at 6 A.M., which was 10°F. Then, using the clue that this temperature was 4° warmer than at 4 A.M., they can graph 6° at 4 A.M. Because the temperature fell an average of 3° per hour for 4 hours, the temperature at midnight was $6° + 12° = 18°F$. 18°F at midnight

Lesson 6.3

Problem
What is the least number that can be divided evenly by each of the numbers 1 through 12?

Solution
Strategy: Use Logical Reasoning

Numbers 1, 2, 3, 4, and 6 are factors of 12, so if the number is divisible by 12 it is divisible by all these also. Remember, if a number is divisible by 5 and 2, it is divisible by 10. Multiply the odd numbers $5 \times 7 \times 9 \times 11 = 3,465$.

3,465 is not divisible by 8 or 12, so multiply it by 8.
$3,465 \times 8 = 27,720 \quad 27,720 \div 12 = 2,310$
27,720 is divisible by each of the numbers 1 through 12.

Lesson 6.5

Problem
Jason made a stem-and-leaf plot of the ages of all the adults at a family reunion. His father's age was the median age. Jason is 1 year younger than $\frac{1}{3}$ his father's age. How old is Jason?

Ages of the Adults

Stems	Leaves
2	2 6
3	0 3 7
4	2 3 3
5	5 8
6	8

Solution
Strategy: Use Logical Reasoning

Use the stem-and-leaf plot to find the median age of 42. So, Jason's father is 42 years old.
One-third of 42 is $\frac{1}{3} \times 42 = 14$.
Jason is 1 year younger than 14: $14 - 1 = 13$.
13 years old

Lesson 6.6

Problem
A circle graph shows that half of the 120 ancestors of the students surveyed came to the U.S. from Europe or South America, a quarter came from Africa, and the rest from Asia and Australia. Five times as many came from Europe as from South America. How many of the students' ancestors came from South America?

Solution
Strategy: Predict and Test

$120 \div 2 = 60$ ancestors from Europe or South America
Find 2 addends for 60, one of which must be 5 times the other. The smaller addend represents the South American ancestors.
$20 + 40 = 60$ ($40 = 2 \times 20$) Incorrect
$15 + 45 = 60$ ($45 = 3 \times 15$) Incorrect
$10 + 50 = 60$ ($50 = 5 \times 10$) Correct
10 ancestors

Problem of the Day Solutions **PD6**

Chapter 7 Answer Key

Lesson 7.1

Problem
Corrine is having a party. She has 45 different party favors and wants to give each guest the same number of favors. How many guests could she invite and how many favors would they get?

Solution
Strategy: Make an Organized List

Once students recognize this as a factoring problem, they can list the factors of 45:

1, 3, 5, 9, 15, and 45.

Then, to answer the question, they match pairs of factors. If Corrine invites 1 person, that person gets 45 favors, because $1 \times 45 = 45$. If she invites 3 people, each gets 15, because $3 \times 15 = 45$, and so on.

1 guest would get 45 favors; 3 guests would get 15 favors each; 5 guests would get 9 favors; 9 guests would get 5 favors; 15 guests would get 3 favors; 45 guests would get 1 favor.

Lesson 7.2

Problem
The sum of the ages of Mr. and Mrs. Olsen and their two children is 108. Their ages are sets of twin prime numbers. What are their ages? NOTE: Twin primes are two prime numbers whose difference is 2.

Solution
Strategy: Predict and Test

Students can list prime numbers and circle twin primes. They predict and test to see which sets total 108. 3, 5, 7, 11, 13, 17, 19, 29, 31, 41, 43, 59, 61, ...

Ages: $11 + 13 + 41 + 43 = 108$

Lesson 7.3

Problem
Find the least values for n and m such that the value of the first expression is twice that of the second expression.

Hint: n and m are both less than 6.

$114 \times n \qquad 95 \times m$

Solution
Strategy: Use Logical Reasoning

Students can use prime factorization to find values for n and m.

$114 \times n \qquad 95 \times m$

Prime factorization for 114: $2 \times 3 \times 19$

Prime factorization for 95: 5×19

Both expressions have the common factor of 19. If 114 is multiplied by 5 and 95 is multiplied by 3, the first expression will have a value twice that of the second expression.

$114 \times 5 = 570$ and $95 \times 3 = 285$;

$570 = 2 \times 285$

$n = 5$ and $m = 3$

Lesson 7.4

Problem
Jon is making birdhouses. He is cutting 1- by 6-in. boards into pieces that are 11 in. long to make the sides of the birdhouses. Which board length would produce the least waste—4 ft, 6 ft, 8 ft, or 10 ft? Explain.

Solution
Strategy: Make an Organized List

Students should first list the multiples of 11: 11, 22, 33, 44, 55, 66, 77, 88, 99, 110, and 121.

Then they can write the lengths of the board in inches: 4 ft = 48 in., 6 ft = 72 in., 8 ft = 96 in., and 10 ft = 120 in.

By comparing the multiples to the lengths, they can see that the 4-ft board provides the least waste.

4 ft long

Chapter 8 Answer Key

Lesson 8.1

Problem
John has 3 coins, 2 of which are the same. Ellen has 1 fewer coin than John, and Anna has 2 more coins than John. Each girl has only 1 kind of coin. Who has coins that could equal the value of a half-dollar?

Solution
Strategy: Use Logical Reasoning

The value of a half-dollar is 50 cents. Ellen has 1 fewer coin than John, so she has 2 coins. Two quarters have a value of 50 cents. So Ellen's coins could equal 50 cents.

Anna has 2 more coins than John, so she has 5 coins. 5 dimes have a value of 50 cents. So Anna's coins could equal 50 cents.

John has 3 coins and 2 of them are the same. There is no combination of 2 like coins and 1 other coin that would equal 50 cents. Ellen and Anna are the ones whose coins could equal the value of a half-dollar.

Lesson 8.2

Problem
At the car show, there are 20 vehicles on display. Some are motorcycles and some are cars. All 56 wheels on the vehicles need to be polished. What fraction of the vehicles are motorcycles?

Solution
Strategy: Predict and Test

Students need to find 2 numbers, a and b, which meet these conditions:

$a + b = 20$ and $2a + 4b = 56$.

Prediction 1	$a + b$	$2a + 4b$
10 motorcycles		
10 cars	$10 + 10 = 20$	$20 + 40 = 60$
Prediction 2		
12 motorcycles		
8 cars	$12 + 8 = 20$	$24 + 32 = 56$

There are 12 motorcycles and 8 cars. The fraction of vehicles that are motorcycles is $\frac{12}{20}$ or $\frac{3}{5}$.

Lesson 8.3

Problem
From 4:00 to 5:30, Jacob, Lisa, and Chelsea took turns playing the same computer game. Jacob played for $\frac{1}{2}$ hour and Lisa played for $\frac{3}{4}$ hour. For how many minutes did Chelsea play the game?

Solution
Strategy: Work Backward

Jacob, Lisa, and Chelsea played the computer game a total of 90 minutes. The amount of time left after Jacob and Lisa played is the number of minutes that Chelsea played.

Jacob played for $\frac{1}{2}$ hour or 30 minutes.

$90 - 30 = 60$ minutes

Lisa played for $\frac{3}{4}$ hour or 45 minutes.

$60 - 45 = 15$ minutes.

Chelsea played the computer game for 15 minutes.

Lesson 8.5

Problem
A pizza has 8 slices. Milton ate $\frac{1}{4}$ of the pizza. Earl ate $\frac{1}{2}$ of what was left over. Did Earl eat more or fewer than 4 pieces? Explain.

Solution
Strategy: Use Logical Reasoning

$\frac{1}{4}$ of 8 pieces = 2 pieces, so Milton ate 2 pieces of pizza.

$8 - 2 = 6$

$\frac{1}{2}$ of 6 pieces = 3 pieces, so Earl ate 3 pieces.

$3 < 4$

Earl ate fewer than 4 pieces.

Chapter 9 Answer Key

Lesson 9.1

Problem
Mike's and Kay's numbers are both less than 1. The digit in Mike's numerator is the same as the digit in Kay's denominator. Kay's number is $\frac{1}{10}$ greater than Mike's. What are Mike's and Kay's numbers?

Solution
Strategy: Predict and Test

Mike's number is less than 1 and $\frac{1}{10}$ less than Kay's.
Predict: $\frac{1}{2}$ for Kay's number.
$\frac{1}{2} = \frac{5}{10}$ $\frac{5}{10} - \frac{1}{10} = \frac{4}{10} = \frac{2}{5}$
If Kay's number is $\frac{1}{2}$, then Mike's number is $\frac{2}{5}$.
Test: The 2 in Mike's numerator is the same digit as the 2 in Kay's denominator.

Mike's number is $\frac{2}{5}$ and Kay's is $\frac{1}{2}$.

Lesson 9.3

Problem
Kim used $\frac{6}{8}$ of a tank of gas. She bought $\frac{5}{8}$ of a tank and then used $\frac{4}{8}$ of a tank. Kim was almost out of money, so she filled the tank to half full by adding $\frac{2}{8}$ of a tank of gas. What part of a tank of gas did she start with?

Solution
Strategy: Work Backward

$\frac{1}{2} = \frac{4}{8}$, so at the end she had $\frac{4}{8}$ tank of gas.
$\frac{4}{8} - \frac{2}{8} + \frac{4}{8} - \frac{5}{8} + \frac{6}{8} = \frac{7}{8}$. amount Kim started with

Kim started with $\frac{7}{8}$ of a tank of gas.

Lesson 9.4

Problem
Complete the Magic Square. The sum is $1\frac{1}{4}$.

$\frac{1}{2}$	?	$\frac{1}{6}$
?	$\frac{5}{12}$	?
?	?	?

Solution
Strategy: Write an Equation
Sample equations:
For the first row:
$\frac{1}{2} + \frac{1}{6} + n = 1\frac{1}{4}$; $n = \frac{7}{12}$.

For the second column:
$\frac{7}{12} + \frac{5}{12} + n = 1\frac{1}{4}$; $n = \frac{1}{4}$
For the diagonal:
$\frac{1}{2} + \frac{5}{12} + n = 1\frac{1}{4}$; $n = \frac{1}{3}$
If students need more help, show the answer this way.

$\frac{6}{12}$	$\frac{7}{12}$	$\frac{2}{12}$
$\frac{1}{12}$	$\frac{5}{12}$	$\frac{9}{12}$
$\frac{8}{12}$	$\frac{3}{12}$	$\frac{4}{12}$

$\frac{1}{2}$	$\frac{7}{12}$	$\frac{1}{6}$
$\frac{1}{12}$	$\frac{5}{12}$	$\frac{3}{4}$
$\frac{2}{3}$	$\frac{1}{4}$	$\frac{1}{3}$

Lesson 9.6

Problem
Melissa rides the bus $1\frac{2}{3}$ mi north and $3\frac{1}{8}$ mi east to get to school. Brandon rides his bike $2\frac{3}{4}$ mi south and $2\frac{1}{6}$ mi west to get to the same school. Who rides farther? Estimate the distances. What can you conclude?

Solution
Strategy: Write an Equation
Estimate of Melissa's distance: $2 + 3 = 5$
Estimate of Brandon's distance: $3 + 2 = 5$
Actual Distance
Melissa: $1\frac{2}{3} + 3\frac{1}{8} = 4\frac{19}{24}$
Brandon: $2\frac{3}{4} + 2\frac{1}{6} = 4\frac{11}{12}, 4\frac{22}{24}$

Brandon rides farther. From the estimates the distances appear to be the same. To answer the question you need to find the exact answer.

Lesson 9.7

Problem
Write the next 4 numbers. How does each number relate to the one before it?

A. 10, $8\frac{3}{4}$, $7\frac{1}{2}$, $6\frac{1}{4}$

B. 9, $7\frac{7}{8}$, $6\frac{3}{4}$, $5\frac{5}{8}$

C. $11\frac{7}{10}$, $10\frac{2}{5}$, $9\frac{1}{10}$, $7\frac{4}{5}$

Solution
Strategy: Use a Pattern

For each sequence find the difference between the first and second terms. Check to make sure it is the same between the second and third terms and the third and fourth terms.

A. 10, $8\frac{3}{4}$, $7\frac{1}{2}$, $6\frac{1}{4}$, 5, $3\frac{3}{4}$, $2\frac{1}{2}$, $1\frac{1}{4}$; it is $1\frac{1}{4}$ less

B. 9, $7\frac{7}{8}$, $6\frac{3}{4}$, $5\frac{5}{8}$, $4\frac{1}{2}$, $3\frac{3}{8}$, $2\frac{1}{4}$, $1\frac{1}{8}$; it is $1\frac{1}{8}$ less

C. $11\frac{7}{10}$, $10\frac{2}{5}$, $9\frac{1}{10}$, $7\frac{4}{5}$, $6\frac{1}{2}$, $5\frac{1}{5}$, $3\frac{9}{10}$, $2\frac{3}{5}$; it is $1\frac{3}{10}$ less

Chapter 10 Answer Key

Lesson 10.1

Problem
A is a whole number between 55 and 60. The product of A and B is between 1,045 and 1,500. Between what two numbers is B?

Solution
Strategy: Work Backward

It is given that the product of A and B is between 1,045 and 1,500. Since A is a number between 55 and 60, then the lower range for $A \times B = 1,045$ and the upper range for $A \times B = 1,500$.

$1,045 \div 55 = 19$ and $1,500 \div 60 = 25$.

So B must be between 19 and 25.

between 19 and 25

Lesson 10.2

Problem
In a jump-rope marathon, Cara earns $5 for charity for each half hour or fraction of a half hour that she jumps rope. How much money will Cara earn if she jumps rope for 175 min?

Solution
Strategy: Predict and Test

$\frac{1}{2}$ hour = 30 minutes

$175 \div 30 = 5$ r25 or 6 half hours

$6 \times \$5 = \$30$

Lesson 10.3

Problem
It took André 1 min to fill his aquarium $\frac{1}{3}$ full. How long will it take him to fill the aquarium $\frac{3}{4}$ full?

Solution
Strategy: Draw a Picture

It takes André 1 minute to fill the aquarium one-third full. Shade one third of the picture. To fill it $\frac{3}{4}$ full, divide the aquarium into twelfths. Now each third is divided into fourths. Each section takes $\frac{1}{4}$ minute or 15 seconds to fill. $\frac{3}{4} = \frac{9}{12}$ so if he fills the aquarium $\frac{3}{4}$ full, he fills 9 sections.

4 sections = 1 min

1 section = 15 sec

9 sections = 2 min and 15 sec or $2\frac{1}{4}$ min

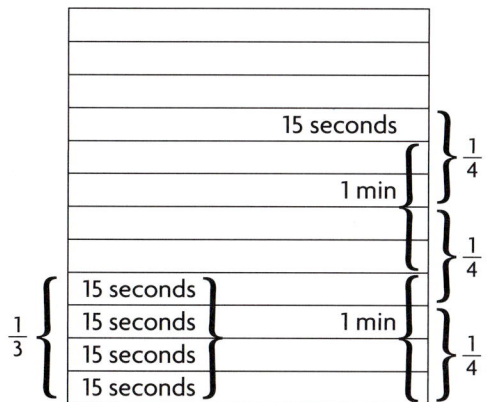

Lesson 10.5

Problem
A hawk flies $\frac{1}{3}$ mi in 30 sec. How far can the hawk fly in 1 min? How fast does it fly in miles per hour?

Solution
Strategy: Write a Number Sentence

A hawk flies $\frac{1}{3}$ mi in 30 sec, so it flies $2 \times \frac{1}{3}$ mi in 2×30 sec or $\frac{2}{3}$ mi in 60 sec, or 1 min.

If the hawk flies $\frac{2}{3}$ mi in 1 min, it flies $60 \times \frac{2}{3}$ mi in 60×1 min, or 40 mi in 60 min, or 1 hr.

So the hawk's speed is 40 miles per hour.

Lesson 10.6

Problem
Maria has $2\frac{1}{2}$ times as many trading cards as Roberto, who has $\frac{1}{4}$ as many as Julia. Karen and Tim each have 25, which is 15 fewer than Julia. Who has the most cards?

Solution
Strategy: Use Logical Reasoning

Students can start with Karen and Tim, who each have 25. That is 15 fewer than Julia, so she has 40. Roberto has $\frac{1}{4}$ as many as Julia, so he has 10; and Maria has $2\frac{1}{2}$ times as many as Roberto, so she has 25. Hence, Julia has the most.

Julia

Chapter 10 Answer Key

Lesson 10.7

Problem
Ming Li ran 90 ft from first base to second base. Each stride was about $3\frac{1}{2}$ ft long. If she takes about 2 strides per second, about how long did it take her to get to second base?

Solution
Strategy: Use Logical Reasoning

Ming Li does 2 strides per sec, so $2 \times 3\frac{1}{2} = 7$. Ming Li can cover 7 ft per sec. It is 90 ft to second base, so estimate $10 \times 7 = 70$. In 10 sec she covers 70 ft. In 3 more seconds, she covers 3×7 or 21 more feet. $70 + 21 = 91$ ft

Possible answer: about 13 sec

Chapter 11 Answer Key

Lesson 11.1

Problem
One side of Jessica's square array is 2 tiles longer than a side of Dave's square array. Together they use a total of 100 tiles. How many tiles are on each side of Dave's array?

Solution
Strategy: Predict and Test

Dave's array	$5 \times 5 = 25$	$6 \times 6 = 36$
Jessica's array	$7 \times 7 = 49$	$8 \times 8 = 64$
	74	100

There are 6 tiles on each side of Dave's array.

Lesson 11.2

Problem
I am a palindrome number, and the sum of my 7 digits is 25. My tens digit is 4 times as great as my ones digit and 4 more than my hundreds digit. My thousands digit and my thousandths digit are each 7 greater than my ones digit. What number am I?

Solution
Strategy: Predict and Test, Use Logical Reasoning

The ones digit must be less than 3 or the tens digit will be greater than the greatest digit (9). If the ones digit is 0, the tens digit will also be 0, but 0 cannot be 4 more than the hundreds digit. You are the number 8,041.408.

Lesson 11.3

Problem
Leon's number is less than Bev's.
Shannon has a negative number.
Leon's and Tony's numbers are the least and the greatest.
Who has each of these numbers:
$-2\frac{1}{5}$, $\frac{19}{18}$, 2.05, -2.25?

Solution
Strategy: Use Logical Reasoning

For help, students should write all numbers as fractions.
$-2\frac{1}{5}$, $\frac{19}{18}$, $2\frac{1}{20}$, $-2\frac{1}{4}$

Ordered from least to greatest:

-2.25,	$-2\frac{1}{5}$,	$\frac{19}{18}$,	2.05
Leon	Shannon	Bev	Tony
($<$ Bev's)	(negative)	(Boys are least and greatest)	

Leon, -2.25; Shannon $-2\frac{1}{5}$; Bev $\frac{19}{18}$; Tony, 2.05

Lesson 11.4

Problem
Cara is twice as old as Lee. In 5 years she will be $1\frac{1}{2}$ times as old as he is. How old is Lee?

Solution
Strategy: Predict and Test

In five years Cara's age will be $1\frac{1}{2}$ times Lee's age. Students may realize that Lee's present age must be an odd number. In order to multiply $1\frac{1}{2}$ by Lee's age in 5 years and get an integer, his present age would need to be an odd number.

Try 3 and 6 for their present ages.

In 5 years, the ages would be 8 and 11, which does not work.

Try 5 and 10 for the present ages. In 5 years, they would be 10 and 15, and 15 is $1\frac{1}{2}$ times 10.

5 years old

Chapter 12 Answer Key

Lesson 12.2

Problem
On a number line, the distance from 0 to a negative integer is four times as great as the distance to a positive integer. The sum of their absolute values is 35. What are the numbers?

Solution
Strategy: Predict and Test

Students may use the predict and test strategy, for example, trying 5 and ⁻20, 6 and ⁻24, and 7 and ⁻28. As an alternative, students may employ logical reasoning and divide 35 into 5 parts. The numbers are ⁻28 and 7.

Lesson 12.4

Problem
Write the missing numbers.

Each of the numbers in the first 4 rows is the sum of the two numbers below it.

```
                    ?
                ⁻3      ?
            ?       ⁻2      1
        ⁻4      ?       ?       6
   ⁻14     10     ⁻7       ?       4
```

Solution
Strategy: Write a Number Sentence

```
                    ⁻4
                ⁻3      ⁻1
            ⁻1      ⁻2      1
        ⁻4      3       ⁻5      6
   ⁻14     10     ⁻7       2       4
```

Lesson 12.5

Problem
We are two integers. Our sum is ⁻15 and our difference is 3. Who are we?

Solution
Strategy: Predict and Test; Make an organized List

Since the difference of the two integers is positive and the sum is negative, the two integers are most likely negative. So make a list of two negative integers that have a sum of ⁻15. Then check to see if their difference is 3.

Sum	Difference
⁻1 + ⁻14 = ⁻15	⁻1 + ⁻14 = 13
⁻2 + ⁻13 = ⁻15	⁻2 + ⁻13 = 11
⁻3 + ⁻12 = ⁻15	⁻3 + ⁻12 = 9
⁻4 + ⁻11 = ⁻15	⁻4 + ⁻11 = 7
⁻5 + ⁻10 = ⁻15	⁻5 + ⁻10 = 5
⁻6 + ⁻9 = ⁻15	⁻6 + ⁻9 = 3

The two integers are ⁻6 and ⁻9.

Lesson 12.6

Problem
Write the missing integers.

```
  ⁻2  ×   ?   ×   4   =  ⁻24
  ×?  ×  ⁻4   ×   ?   =   ?
  ⁻10 ×   ?   ×  ⁻8   =  ⁻960
```

Solution
Strategy: Write a Number Sentence

```
  ⁻2  ×   3   ×   4   =  ⁻24
  ×5  ×  ⁻4   ×  ⁻2   =   40
  ⁻10 ×  ⁻12  ×  ⁻8   =  ⁻960
```

HARCOURT Math

SCOPE & SEQUENCE AND CORRELATIONS

This section contains the following:

▶ **Scope & Sequence**
The scope and sequence shows the development of all strands of math across the grades—Kindergarten through Grade 6. In addition there is a detailed scope and sequence specific to the grade level.

▶ **Correlation to Standardized Tests**
The standardized test correlations will assist you as you prepare students for the following standardized tests:
- **CAT**–California Achievement Test
- **CTBS/Terra Nova**–Comprehensive Test of Basic Skills
- **ITBS**–Iowa Test of Basic Skills
- **MAT**–Metropolitan Achievement Test
- **SAT**–Stanford Achievement Test

▶ **Manipulatives Chart**
This chart is a correlation of the Pupil Edition and Teacher's Edition with the manipulative kits designed to accompany the program.

Scope and Sequence

NUMBER SENSE

NUMBER AND QUANTITATIVE REASONING	K	1	2	3	4	5	6
WHOLE NUMBERS							
Meaning of numbers	●	●	●	●	●	●	●
Read and write numbers							
to 30	●	▲					
to 100		●	●	▲	▲		
to 1,000			●	●	▲	▲	▲
in the ten thousands				●	●	▲	▲
in the millions					●	●	▲
in the billions						●	▲
Count	●	●	●	●			
Place value							
tens and ones		●	●	●			
to 100		●	●	▲	▲		
to 1,000			●	●	▲		
to 10,000				●	▲	▲	▲
in the millions					●	●	▲
in the billions						●	▲
Expanded form		●	●	●	●	●	▲
Compare and order							
to 10 (with objects)	●	▲					
to 100 (using symbols)		●	●	▲	▲	▲	
to 1,000 (using symbols)				●	●	●	▲
to 10,000 (using symbols)					●	●	●	▲
in the millions (using symbols)					●	●	▲
in the billions (using symbols)						●	▲
Make reasonable estimates		●	●	●	●	●	●
Rounding							
to nearest ten, hundred, or thousand				●	●	●	●
to nearest ten thousand through nearest million					●	●	●
Even/odd			●	●	▲	▲	▲
Ordinal numbers	●	●	●				
Multiples						●	●	●
Divisibility						●	●
Prime and composite					●	●
Least common multiple						●	●
Common factors						●	●
Greatest common factor						●	●
Powers and exponents						●	●
Factor whole numbers					●	●	●
Prime factors					●	●	●
Prime factorization						●	●
Square numbers and square roots					●	●	●

● Teach   ▲ Reinforce and Maintain

NUMBER AND QUANTITATIVE REASONING

WHOLE NUMBERS
Make reasonable estimates 17–19

Rounding
  to nearest ten, hundred, or thousand 16
  to nearest ten thousand through nearest million 16

Multiples 16–19, 150, 154–155
Divisibility 146–147, H38
Prime and composite 148, 233
Least common multiple 150–153, 180–182, H38
Common factors 151
Greatest common factor 151–153, 204, H38
Powers and exponents 40–41, 276–279, H33
  Challenge • Scientific notation 87
  Challenge • Negative exponents 263
Factor whole numbers 151–153
Prime factors 148–149, 233
Prime factorization 148–149, 150–153, H38
Square numbers and square roots
  squares 276–279, H45
  square roots 277–279, H45

Type printed in red indicates that a topic is being introduced for the first time.

MEASUREMENT AND GEOMETRY

SPATIAL SENSE	K	1	2	3	4	5	6
VISUAL THINKING							
Patterns		●	●	●	●	●	●
tessellations				●	●	●	●
nets				●	●	●	●
Congruence			●	●	●	●	●
Symmetry			●	●	●	●	●
line (bilateral)			●	●	●	●	●
point (rotational)				●	●	●	●
Similarity					●	●	●
Transformations							
translations, reflections, and rotations				●	●	●	●
dilations							●
Representing							
building, drawing 3-D figures					●	●	●
different views						●	●
Perspective							●
Networks							●
COORDINATE GEOMETRY							
COORDINATE PLANE							
Ordered pairs				●	●	●	●
Graph points and figures					●	●	●
Graph linear relationships					●	●	●
Graph equations					●	●	●
Relations and functions					●	●	●
Identify functions							
linear functions					●	●	●
nonlinear functions							●
Translations, reflections, rotations					●	●	●

● Teach   ▲ Reinforce and Maintain

SPATIAL SENSE
VISUAL THINKING
Patterns 543–545, H59
    tessellations 553–555, 556–557, H60
    nets 356–357, 512, 518
Congruence 322, 390–393
Symmetry
    line (bilateral) 560–563, H60
    point (rotational) 560–563
Similarity 390–393, 394–396, 397–399, H51
Transformations
    translations, reflections, and rotations 550–552, 558–559, 580–583, H60, H61
    Challenge • Dilations (Stretching) 587
Representing
    building, drawing 3–D figures 353–355
    different views 353–355
    Challenge • Perspective 377
    Challenge • Networks 527

COORDINATE GEOMETRY
COORDINATE PLANE
Ordered pairs 570–573, 574–575, H61
Graph points and figures 570–573, 574–575, 580–583, 587, H61
Graph linear relationships 574–575
Graph equations 574–575, 578–579
Relations and functions 574–575, 578–579, H61
Identify functions
    linear functions 578–579
    nonlinear functions 578–579
Translations, reflections, rotations 580–583, H61

Type printed in red indicates that a topic is being introduced for the first time.

STATISTICS, DATA ANALYSIS, AND PROBABILITY

STATISTICS AND DATA ANALYSIS	K	1	2	3	4	5	6
COLLECTING DATA							
Use systematic way to record			●	●	●	▲	▲
Pose question/collect data	●	●	●	●	●	▲	▲
Formulate question				●	●	▲	▲
Analyze question					●	●	●
Conduct survey			●	●	●	●	●
Sampling							●
determine when appropriate							●
bias/errors							●
select in different ways (convenience, random, systematic)							●
determine most representative							●
ORGANIZING DATA							
Sort objects/data and describe categories	●	●	●	●			
Tally table/chart	●	●	●	●	●	●	▲
Frequency table/chart				●	●	●	▲
cumulative frequency					●	●	▲
Organized list				●	●	●	●
Stem-and-leaf plot					●	●	▲
Line plot				●	●	●	▲
DISPLAYING DATA							
Objects/pictures	●	●					
Picture graph	●	●	●				
Pictograph			●	●	▲	▲	
Bar graph	●	●	●	●	●	●	▲
Line graph				●	●	●	▲
identify ordered pairs				●	●	●	▲
write ordered pairs				●	●	●	▲
graph ordered pairs					●	●	●
Circle graph					●	●	●
Histogram						●	●
Box-and-whisker graph							●
Scatterplot							●
Represent same data in different ways			●	●	●	●	●	●
Choose an appropriate graph					●	●	●

● Teach   ▲ Reinforce and Maintain

STATISTICS AND DATA ANALYSIS

COLLECTING DATA
Analyze question 94–97
Conduct survey 94–97
Sampling 94–97
   determine when appropriate 94–97
   bias 98–99, H36
   select in different ways (convenience, random, systematic) 95–97, H36
   determine most representative 94–97

ORGANIZING DATA
Organized list 154–155, 442–443

DISPLAYING DATA
Line graph 121–123, H37
   graph ordered pairs 570–573
   multiple-line graphs 121–123
Circle graph 122–123, 416–417
   with degrees 416–417
   with percents 122–123
Histogram 127–128, H37
Box-and-whisker graphs 129–131, H37
   Challenge • Scatterplot 139
Represent same data in different ways 120–123, 132–135
Choose an appropriate graph 120–123, 126–128

Type printed in red indicates that a topic is being introduced for the first time.

STATISTICS, DATA ANALYSIS, AND PROBABILITY

STATISTICS AND DATA ANALYSIS	K	1	2	3	4	5	6
ANALYZING DATA							
Ask/answer questions about data	●	●	●	●	●	▲	▲
Interpret one-variable graphs	●	●	●	●	●	●	▲
Interpret two-variable graphs					●	●	●
Interpret tables	●	●	●	●	●	●	▲
Compare data		●	●	●	●	●	▲
Compare data sets of different sizes						●	▲
Compare/choose appropriate representations					●	●	●
Identify misleading graphs							●
Choose scale					●	●	●
Identify outliers					●	●	●
Find range			●	●	●	●	▲
Measures of central tendency							
find mean (average)					●	●	▲
find median					●	●	▲
find mode			●	●	●	●	▲
compare/analyze measures						●	●
determine effects on measures of adding data						●	●
determine effects of outliers							●
Relate to conclusions the way data is displayed						●	●
Evaluate conclusions based on data					●	●	●
Make predictions				●	●	●	●
Misleading graphs							●

● Teach ▲ Reinforce and Maintain

STATISTICS AND DATA ANALYSIS

ANALYZING DATA

Interpret two-variable graphs 578–579

Compare/choose appropriate representations 94–97, 120–123, 126–128, 132–135

Identify misleading graphs 132–135, H37

Choose scale 120

Identify outliers 110–111, H36

Measures of central tendency
   compare/analyze measures 106–108, H36
   determine effects on measures of adding data 109–111, H36
   determine effects of outliers 110–111, H36

Relate to conclusions the way data is displayed 112–115, H36

Evaluate conclusions based on data 112–115, H36

Make predictions 124–125

Misleading graphs 132–135, H37

Type printed in red indicates that a topic is being introduced for the first time.

STATISTICS, DATA ANALYSIS, AND PROBABILITY

PATTERNS	K	1	2	3	4	5	6
GEOMETRIC PATTERNS							
Identify and describe	●	●	●	●	●	●	●
Extend	●	●	●	●	●	●	●
Generate	●	●	●	●	●	●	●
COLOR/NUMERIC/RHYTHMIC PATTERNS							
Write a rule				●	●	●	●
Identify and describe	●	●	●	●	●	●	●
Extend	●	●	●	●	●	●	●
Generate	●	●	●	●	●	●	●
LINEAR NUMBER							
Identify and describe		●	●	●	●	●	●
Extend		●	●	●	●	●	●
Generate				●	●	●	●

● Teach ▲ Reinforce and Maintain

PATTERNS

GEOMETRIC PATTERNS
Identify and describe 543–545, H59
Extend 543–545, H59
Generate 543–545, H59

COLOR/NUMERIC/RHYTHMIC PATTERNS
Write a rule 536–538, H59
Identify and describe 536–538
Extend 536–538, H59
Generate 536–538

LINEAR NUMBER
Identify and describe 539–542, H59
Extend 539–542
Generate 539–542

Type printed in red indicates that a topic is being introduced for the first time.

Standardized Test Correlations

LEARNING GOALS FOR GRADE 6		CAT	CTBS/TERRA NOVA	ITBS	MAT	SAT
Chapter 11	**Number Relationships**					
11A	To identify and write integers, opposites, and absolute values	CAT		ITBS	MAT	SAT
11B	To identify and represent relationships among sets of numbers by using a variety of methods including number lines	CAT	CTBS/Terra Nova		MAT	SAT
11C	To compare and order rational numbers	CAT	CTBS/Terra Nova	ITBS	MAT	SAT
11D	To solve problems by using an appropriate problem solving strategy such as *use logical reasoning*	CAT	CTBS/Terra Nova	ITBS	MAT	SAT
Chapter 12	**Operations with Integers**					
12A	To write sums and differences of integers by using a variety of methods including models and number lines	CAT	CTBS/Terra Nova		MAT	SAT
12B	To write products and quotients of integers	CAT			MAT	SAT
12C	To write sums, differences, products, and quotients of rational numbers and combinations of these operations	CAT	CTBS/Terra Nova		MAT	
Chapter 13	**Expressions**					
13A	To write and evaluate algebraic expressions	CAT			MAT	
13B	To evaluate expressions with squares and square roots				MAT	
Chapter 14	**Addition and Subtraction Equation**					
14A	To write verbal sentences as equations					
14B	To use models to solve one-step equations	CAT	CTBS/Terra Nova	ITBS	MAT	SAT
14C	To solve addition and subtraction equations	CAT	CTBS/Terra Nova	ITBS	MAT	SAT
Chapter 15	**Multiplication and Division Equations**					
15A	To solve multiplication and division equations, and to use models to solve multiplication equations	CAT	CTBS/Terra Nova	ITBS	MAT	SAT
15B	To solve real-world problems by using formulas	CAT	CTBS/Terra Nova	ITBS		SAT
15C	To use models to solve two-step equations		CTBS/Terra Nova			SAT
15D	To solve problems by using an appropriate strategy such as *work backward*	CAT	CTBS/Terra Nova	ITBS	MAT	SAT
Chapter 16	**Geometric Figures**					
16A	To identify, classify, and draw points, rays, lines, and planes	CAT	CTBS/Terra Nova	ITBS	MAT	SAT
16B	To identify, classify, measure, and draw angles	CAT	CTBS/Terra Nova	ITBS	MAT	SAT
16C	To recognize the relationships among angles		CTBS/Terra Nova	ITBS	MAT	SAT
Chapter 17	**Plane Figures**					
17A	To identify, classify, and draw triangles, quadrilaterals, and other two-dimensional figures	CAT	CTBS/Terra Nova	ITBS	MAT	SAT
17B	To identify and measure parts of a circle	CAT		ITBS	MAT	SAT
17C	To solve problems by using appropriate strategies such as *find a pattern*	CAT	CTBS/Terra Nova	ITBS	MAT	SAT
Chapter 18	**Solid Figures**					
18A	To identify, classify, and draw solid figures		CTBS/Terra Nova	ITBS	MAT	SAT
18B	To identify solid figures from different points of view					
18C	To identify nets and patterns for solid figures					
18D	To solve problems by using an appropriate strategy such as *solve a simpler problem*	CAT	CTBS/Terra Nova	ITBS	MAT	SAT
Chapter 19	**Congruence and Similarity**					
19A	To identify and construct congruent line segments and angles					
19B	To bisect line segments and angles					
19C	To identify and analyze congruent and similar figures	CAT	CTBS/Terra Nova	ITBS	MAT	SAT
19D	To construct parallel lines					
Chapter 20	**Ratio and Proportion**					
20A	To write ratios, rates, unit rates, and proportions	CAT		ITBS	MAT	SAT
20B	To use ratios and proportions to solve problems involving similar figures, scale drawings, and maps	CAT	CTBS/Terra Nova	ITBS	MAT	SAT
20C	To solve problems by using an appropriate strategy such as *write an equation*			ITBS		SAT
Chapter 21	**Percent and Change**					
21A	To write ratios as percents		CTBS/Terra Nova	ITBS	MAT	SAT
21B	To write equivalent forms of percents, decimals, and fraction	CAT	CTBS/Terra Nova	ITBS	MAT	SAT
21C	To solve real-life and application percent problems such as those involving tips, discounts, sales tax, and simple interest and to estimate and find the percent of a number	CAT		ITBS	MAT	SAT
21D	To make circle graphs using percents	CAT		ITBS		
Chapter 22	**Probability of Simple Events**					
22A	To calculate the likelihood of an event, to find the theoretical probabilities of simple events, and to express probabilities as fractions, decimals, and percents	CAT		ITBS	MAT	SAT
22B	To make predictions based on experimental probabilities		CTBS/Terra Nova	ITBS	MAT	SAT
22C	To solve problems by using an appropriate skill, such as *too much or too little information*	CAT	CTBS/Terra Nova	ITBS	MAT	SAT

Standardized Test Correlations

LEARNING GOALS FOR GRADE 6		CAT	CTBS/ TERRA NOVA	ITBS	MAT	SAT
Chapter 23	**Probability of Compound Events**					
23A	To identify and find the probabilities of compound, independent, and dependent events by using a variety of methods					
23B	To find probabilities, and to make predictions by using sample data	CAT	CTBS/Terra Nova	ITBS	MAT	SAT
23C	To solve problems using an appropriate strategy such as *make an organized list*					
Chapter 24	**Units of Measure**					
24A	To convert between customary measures of length, weight, and capacity and to convert between metric measures of length, mass, and capacity	CAT	CTBS/Terra Nova			SAT
24B	To estimate and write conversions between units in customary and metric systems	CAT	CTBS/Terra Nova			SAT
24C	To measure to a given degree of precision using appropriate units and tools		CTBS/Terra Nova			
24D	To solve problems using an appropriate skill such as *estimate or find exact answer*	CAT	CTBS/Terra Nova	ITBS	MAT	SAT
Chapter 25	**Length and Perimeter**					
25A	To estimate, measure, and calculate perimeters of plane figures	CAT	CTBS/Terra Nova	ITBS	MAT	SAT
25B	To find the circumference of a circle					
25C	To solve problems by using an appropriate strategy such as *draw a diagram*		CTBS/Terra Nova			
Chapter 26	**Area**					
26A	To estimate and write the area of polygons	CAT	CTBS/Terra Nova	ITBS		SAT
26B	To estimate and write the area of a circle					
26C	To write the surface area of prisms and pyramids					
Chapter 27	**Volume**					
27A	To estimate and write the volume of triangular and rectangular prisms		CTBS/Terra Nova			
27B	To estimate and write the volume of triangular and rectangular pyramids		CTBS/Terra Nova			
27C	To estimate and write the volume of cylinders					
27D	To solve problems by using an appropriate strategy such as *make a model*		CTBS/Terra Nova			
Chapter 28	**Patterns**					
28A	To identify, extend, and make number patterns in function tables and sequences and to write a rule to define a pattern	CAT	CTBS/Terra Nova	ITBS	MAT	SAT
28B	To identify and extend geometric patterns, and to write a rule to define a pattern	CAT	CTBS/Terra Nova	ITBS	MAT	SAT
28C	To solve problems by using an appropriate strategy, such as *find a pattern*					
Chapter 29	**Geometry and Motion**					
29A	To identify, analyze, and draw transformations of plane and solid figures		CTBS/Terra Nova	ITBS		
29B	To identify, analyze, and build or make tessellations					
29C	To identify and analyze line and rotational symmetry in geometric figures		CTBS/Terra Nova	ITBS	MAT	SAT
29D	To solve problems by using an appropriate strategy, such as *make a model*		CTBS/Terra Nova			
Chapter 30	**Graph Relationships**					
30A	To write, solve, and graph algebraic inequalities on a number line					
30B	To identify, locate, and graph points, relations, and transformations on a coordinate plane, and to write a rule for relations by using tables and graphs	CAT	CTBS/Terra Nova	ITBS	MAT	SAT
30C	To identify linear and nonlinear relationships					
30D	To solve problems by using an appropriate skill such as *make generalizations*	CAT	CTBS/Terra Nova	ITBS	MAT	SAT

Manipulatives Chart

MANIPULATIVES	KIT SOURCES	PUPIL EDITION PAGES	TEACHER EDITION PAGES
Algebra Tiles	Core Manipulative Kit Teacher Modeling Kit Build-A-Kit® Module N	286, 296, 304, 305	
Base-Ten Units (1 cm)	Core Manipulative Kit Teacher Modeling Kit Build-A-Kit® Module E	354, 512–513, 516	20B, 40B, 146B
Square Tiles	Core Manipulative Kit Teacher Modeling Kit Build-A-Kit® Module M	276–277, 578, 579	40B, 276–277, 392, 578–579
Equabeam™ Balance	Teacher Modeling Kit		
Fraction Circles	Core Manipulative Kit Teacher Modeling Kit Build-A-Kit® Module J	208	208
Fraction Bars	Core Manipulative Kit Teacher Modeling Kit Build-A-Kit® Module I	160–161, 180–182, 186–187, 190–191	160, 180–181, 182B, 182, 184, 190–191
Fraction Tower	Core Manipulative Kit Teacher Modeling Kit		
Geoboards (11 x11)	Core Manipulative Kit Teacher Modeling Kit Build-A-Kit® Module Q		
Pattern Blocks	Core Manipulative Kit Build-A-Kit® Module L	553	553, 560B, 562
Spinners	Core Manipulative Kit Teacher Modeling Kit Build-A-Kit® Module K	55, 108, 434	38, 58B, 202B, 430, 434, 436B
Two-Color Counters	Core Manipulative Kit Teacher Modeling Kit Build-A-Kit® Module C	242–243, 248–249, 252, 387, 412	290B, 387, 412, 534B, 536B

HARCOURT Math

REVIEW OF RESEARCH

As the content and pedagogy of *Harcourt Math* was developed, the primary goal of the authors, advisors, editors, and reviewers was to ensure the accuracy of the mathematical content and the validity of the pedagogical approach. Research about effective ways to develop children's mathematical competencies, to intervene to help those children whose performance levels were below expectation, and to provide teachers with suggested instructional strategies was consulted and used as the basis for developing the philosophy of the program and the organizational structure of the chapters and lessons in the program.

In the following section, best practices as documented by research are defined; supporting studies are cited and described; and the ways in which these best practices are implemented in the program are shown. Research summaries for the following best practices are included:

Research Reviews	Page
Intervention	RS1
The Value of Visual Images in the Learning Process	RS5
Explicit Instruction: Delivering Instruction That Is Clear and Direct	RS10
The Use of Manipulatives, or Concrete Materials	RS14
The Use of Practice or Review to Improve Performance and Retention	RS18
Lesson Closure	RS22

Research Articles

What Research Says About...
Intervention

Overview

Intervention in this research review is defined as "the set of strategies that a teacher uses to accommodate students' diverse skill levels, interests, and learning preferences and to maximize learning for all students." Intervention is closely related to an approach to teaching known as differentiated instruction. Intervention includes specific accommodations made for individual students, but this report explores intervention as a system and presents the overarching principles that support such a system.

Though built on long-standing beliefs and practices, intervention and differentiated instruction are relatively new concepts, and at this time no specific educational research has been done on the effectiveness of intervention or differentiated instruction as a holistic, systematic approach. However, research in the areas of neurology, psychology, education, and other fields supports various components of the approach. These principles form the basis of this research review.

Research Findings

Intervention is based on a set of principles that, when put into practice, yield increased learning for all students. These principles are supported by decades of research in such areas as neurobiology, psychology, anthropology, and education. The principles discussed in this review are the following:

- Assessment is a tool for instruction.
- Students at all levels learn best when they face a moderate challenge.
- The human brain is designed to seek meaning and recognize patterns, and it has a limited focus.
- Varied classroom activities accommodate and motivate various learning profiles.

Proponents of intervention emphasize that in order for these principles to produce positive results, the following two structural components must be established:

- For intervention methods adopted in individual classrooms to be successful, a school must have the resources and attitude to support them (Kame'enui and Simmons, 1998).
- Schools and teachers must reexamine the purpose of assessment and use it as a building block for an intervention or differentiated instruction program.

According to Tomlinson (1999), assessment is essential to differentiated instruction in that it provides each student with an entry point for instruction—and from that entry point teachers should lead each student on an individualized course.

Assessment is a tool for instruction.

In an article that draws specifically from examples in the area of elementary mathematics, Beattie and Algozzine (1982) outlined the following steps that a teacher can follow in incorporating into instruction the information learned from diagnostic tests.

1. Determine from diagnostic test results not just students' scores but also the types of items for which correct or incorrect responses were provided.

2. Form tentative conclusions about the nature of each student's abilities.
3. Administer informal, teacher-created tests that target the identified area of weakness to determine the incorrect processes in which a student may be engaging.
4. Analyze content and process, and then begin remediation.

The authors warn, however, that a few experiences with a task will not yield complete understanding and that mere repetition will also not be of much benefit. The authors suggest varying the activities that foster the targeted skill area and providing four examples.

The integration of assessment and instruction can be achieved by two methods presented by Valencia and Wixson (1991): *alternative methods* and *scaffolding approaches*. The alternative-methods approach involves having a student try out several distinct alternatives to solving a problem. The scaffolding approach involves modifying different levels of support, from the least assistance to increasing assistance, as a student engages in a task. An effective system of evaluation to accompany the alternative-methods approach is to compare a student's performance, motivation, or knowledge during and after the administration of the different interventions. To accompany the scaffolding approaches, evaluation should focus on a comparison of the student's learning, performance, motivation, or knowledge at various levels of support for the purpose of finding an optimal level—one that presents appropriate challenge without frustration. The authors also note that interviewing students following the intervention can be a source of valuable information, and they provide sample questions that can be included in such an interview.

Students at all levels learn best when they face a moderate challenge.

Csikszentmihalyi, Rathunde, and Whalen (1993) studied the relationship between challenge and skill during moments in a classroom when students were engaged in academic work. They concluded that "only when challenges and skills were felt to be high and working in tandem did all the varied components of well-being—cognitive, emotional, and motivation—come together for the students. Concentration was far above its normal classroom level, and self-esteem, potency, and involvement also reached their highest levels" (p. 186). The engagement of skills without challenge maintained high esteem, though esteem was lower than in the first scenario. Challenge without skill maintained attention, but esteem dropped significantly. The worst profile was when both challenge and skill were absent—but the study also showed that this situation, which occurred when students were reading, watching films, or listening to lectures, accounted for 29 percent of all classroom experiences. The authors also found that teenagers were willing to accept challenges and overcome obstacles when the problems were interesting and the necessary skills were within the individual's reach.

The human brain is designed to seek meaning and recognize patterns, and it has a limited focus.

According to Howard (2000) in a survey of cognitive research, two effective ways to organize information and focus attention are the use of *advanced organizers* and *chunking*. In a summary of research findings by Walter Kintsch (1994), Howard states that advanced organizers presented before students interact with a text increase learning. The type of learning that occurs is dependent upon the arrangement of the organizers. When organizers are arranged in the same way as the target text, students score higher on recall of information. When organizers are arranged differently from the target text, comprehension scores are raised. A different arrangement, Howard notes, forces more participation from the learner, yielding a deeper understanding. Howard also points to research done by G. A. Miller in 1956 that demonstrated that seven pieces of new and unassociated pieces of information is the maximum that most individuals can handle at a time. Howard recommends that information be presented in "chunks" of no more than seven items, which students must master before another chunk is introduced.

Varied classroom activities accommodate and motivate various learning profiles.

Dunn, Beaudry, and Klavas (1989) reviewed a number of correlational studies and experimental research studies on instructional environments, as well as perceptual, sociological, time-of-day, and mobility preferences, and concluded that students' achievement increases when teaching methods match their learning styles. For example, eight studies throughout the late 1970s and 1980s revealed that when students were taught with *resources that initially matched* their preferred modalities, the students' scores increased; when those same students were reinforced with *resources that mismatched* their modalities, their scores increased even more.

Recommended Instructional Practices

The following best practices are derived from the research literature:

- Provide opportunities for enrichment for all students.
- Provide graphic or pictorial organizers for texts, notes, and concepts, and encourage students to develop their own.
- Organize curriculum around concepts and ideas, and for each concept identify essential questions, not facts.
- Regularly relate concepts and skills to students' own lives.
- Present no more than seven units of information to students at a time, and allow for that set to be recalled and reinforced before moving on.
- Identify the essence, or most important aspect, of a concept or body of information, and focus students' attention on it.
- Vary the ways in which students interact with concepts and information and demonstrate skills and knowledge.

How the Research Is Implemented in *Harcourt Math*

Intervention is the cornerstone on which *Harcourt Math* was developed. The assessment program is designed to provide teachers with diagnostic instruments of various types to determine each child's mathematical strengths and weaknesses. These assessments are linked to program-specific materials that provide review, reteaching, and remediation so that the teacher at each grade level has the necessary materials to meet the wide range of learning abilities found in every classroom.

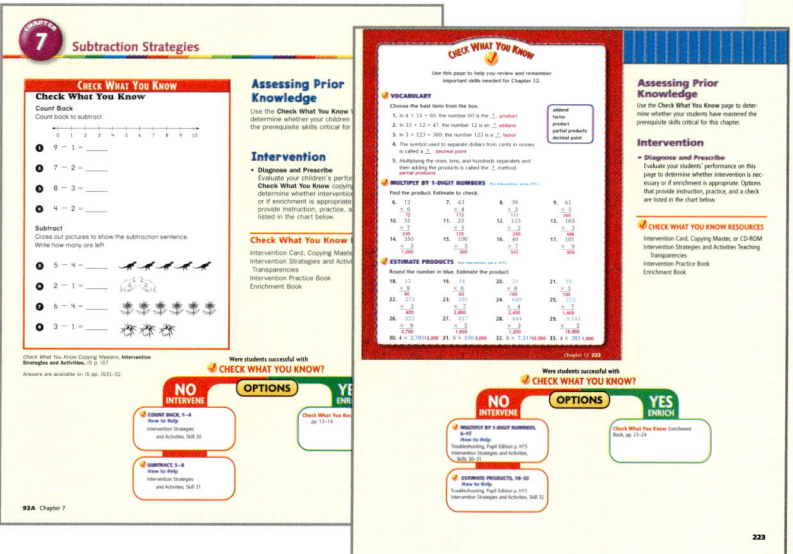

The Check What You Know assessment is a tool for instruction. It helps teachers define each student's strengths and weaknesses on prerequisite skills for each chapter and identify the student's entry point for instruction in the chapter.

Chapters in the Pupil Edition are organized around a key mathematical idea. Before students begin work in each chapter, there is an assessment tool called Check What You Know that helps students review and recall prerequisite skills critical for success in the chapter. This diagnostic instrument provides the teacher with information about students' areas of strength and weakness. For each of these prerequisite skills, there is an intervention strategy in the *Intervention Strategies and Activities Kit*, which may be used to help the student before beginning instruction.

In addition, the *Teacher's Edition* contains suggestions throughout each lesson to vary classroom instruction to match students' preferred learning modalities, learning styles, and interests. These activities provide a balanced program designed to develop students' conceptual understanding, skill proficiency, and problem solving abilities.

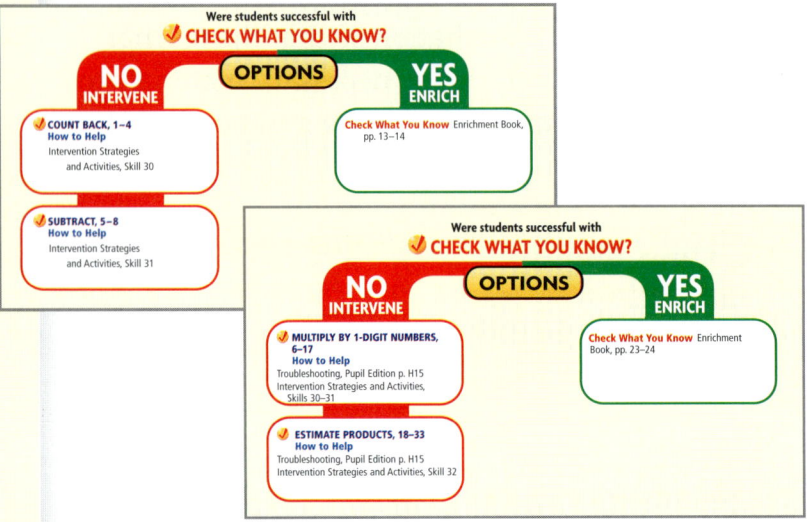

The Options provide varied activities that foster development of the skills on which students need assistance and that can help students make continuous progress toward achieving grade level objectives.

Through the use of *Harcourt Math*, teachers can help students move along a continuum of continuous progress based on careful diagnosis of strengths and weaknesses. The goal of the program is prevention rather than remediation.

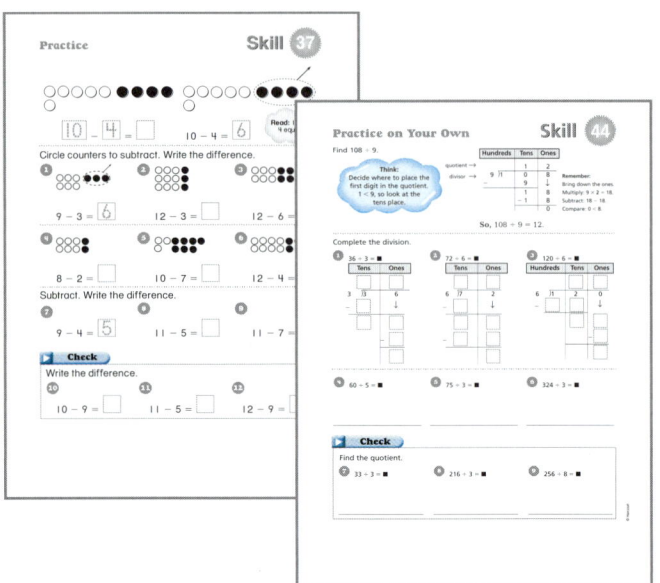

These activities provide visual, verbal, and symbolic representation of key concepts and skills targeted as essential prerequisite skills for grade level work. Notice that the level of scaffolding support is modified as students work through the problems. The independent Practice provides a Check section at the end to assess whether the student has been successful.

References

Beattie, J., and B. Algozzine. (1982). Testing for teaching. *Arithmetic Teacher*, 30, 47–51.

Csikszentmihalyi, M., K. Rathunde, and S. Whalen. (1993). *Talented Teenagers: The Roots of Success and Failure*. New York: Cambridge University Press.

Dunn, R., J. Beaudry, and A. Klavas. (1989). Survey of research on learning styles. *Educational Leadership*, 46, (6), 50–58.

Howard, P. (2000). *The Owner's Manual for the Brain*. Austin, TX: Bard Press.

Kame'enui, E. J., and D.C. Simmons. (1998). Beyond effective practice to schools as host environments: Building and sustaining a schoolwide intervention model in beginning reading. *Oregon School Study Council Bulletin*, 41, (3), 3–24.

Kintsch, W. (1994). Text Comprehension, Memory, and Learning. *American Psychologist*, 49, (4), 294–303.

Tomlinson, C. A. (1999). *The Differentiated Classroom: Responding to the Needs of All Learners*. Alexandria, VA: Association for Supervision and Curriculum Development.

Valencia, S. W., and K. K. Wixson. (1991). Diagnostic teaching. *Reading Teacher*, 44, 420–423.

What Research Says About...

The Value of Visual Images in the Learning Process

Overview

Visual learning has been defined as "the acquisition and construction of knowledge as a result of interaction with visual phenomena." While the use of visual images abounds in many content areas, in mathematics we most readily turn to the use of charts, graphs, pictorial representations, and concrete models as visual aids to the learning process.

Research Findings

Through several studies of older learners, we know something about learning from visual models. Barsalou (1992) describes visual modeling as a process in which specific visual attributes and relationships are analogous to the actual physical parts and relationships in the situations they represent. Visuals help learners isolate and identify important material, recall prior knowledge, provide interaction with content, and enhance information acquisition (Dwyer et al, 1987). Winn (1987) also summarizes the advantages of visuals, noting that they make the abstract more concrete. It is the concreteness that stimulates the imaginal process.

Despite the data provided by these researchers, particularly by the prolific Dwyer, we cannot generalize all of their findings to younger students of mathematics yet. There is a need for further research. Instead, we must rely at this point on reports and naturalistic observations from elementary mathematics classrooms.

Using visuals enhances concept development and skill development and increases students' interest in a topic.

Korithoski and Korithoski (1993) gave fifth-grade students hands-on activities to help them understand the concept of arithmetic mean and gain experience in using mathematical models. The majority of students in this study came from classrooms in which use of manipulatives and hands-on learning had not been common practice. For most of the students, the concept of division was related more to an algorithmic process than to an activity in which physical objects were actually being distributed. The sequence of instruction consisted of the following steps:

1. Finding the mean by using models and redistributing them to show the mean
2. Using a number-line model representing candy pieces in a container along with movement along a number line with no remainders
3. Using a number-line model with all pieces not represented
4. Using only movement along a number line with no remainders
5. Using movement along a number-line model only with remainders
6. Calculating the mean with the algorithm with no model while making intuitive connections with the models used previously

The authors observed that many unexpected benefits occurred by looking at these concepts from a different perspective. The fifth graders enjoyed the experience, since determining the arithmetic mean was anchored in real time and space with objects that already had meaning to the students. The activity also involved the class in meaningful discourse, which led to valuable insights as to how the students reasoned and thought through problems. Finally, the authors noted that the students were active in their learning process, too busy developing an understanding of the concepts to notice the passage of time.

Visuals provide students clearer mental images and help them develop a deeper and more permanent understanding.

Englert and Sinicrope (1994) used graphic area models of multiplication to provide students with visual, concrete representations to teach two-digit multiplication. Fourth-grade teacher Englert observed that as her students made the transition from multiplication by a single-digit number to that by a two-digit number, the students became confused. Rather than using strategies that were based on their understanding of place value, the students tended to create nonsensical algorithms.

Englert and Sinicrope felt that Englert's students needed a concrete representation of the process before they could move on to the abstract. Englert cites in her report (Kennedy, 1986, 6), which advocated the use of manipulatives: "Students who see and manipulate a variety of objects have clearer mental images and can represent abstract ideas more completely." Yet Englert also considered the warning by Heddens (1986, 14): "Simply using manipulatives... is not sufficient; teachers must guide children to develop skills in thinking."

Since Schultz (1991) recommends using an area model, Englert used this semi-concrete model to enable students to develop the mental images necessary to reason abstractly. The fourth graders began by reviewing graphic area models, with the drawings showing the partial products. Next, they enhanced their models by using colors to represent the partial products. The third step was to draw the rectangular area on graph paper to represent the problem and then go on to partition this rectangle through use of color to represent place values. Students began to see patterns and solve problems of increasing difficulty and to make more observations. Although the time spent in developing the multiplication algorithm using the visual approach took longer than a more traditional approach, Englert felt that less time was needed for review and reteaching. It also provided "meaningfulness," something Brownell (1986) stressed as important to the learning of mathematics. Because multiplication was meaningful, the students had a deeper and more permanent understanding.

Skemp (1971) said using models and manipulatives is necessary for children to abstract concepts into appropriate mathematical structures, enabling them to learn more advanced mathematics. Therefore, Englert and Sinicrope believe the extra time invested in using this model for whole-number multiplication has both immediate and long-term benefits.

Bennett and Nelson (1994) worked with sixth graders who were learning to understand applications of percents, using a 10 x 10 grid to visualize the "parts per hundred" aspect of percent. This visual model offered a means of representing given information as well as suggesting different approaches to finding solutions. Once students have a visual image of percents and recognize that 100 percent is represented by one whole square, or unit square, and that 1 percent is represented by one small square, or one-hundredth of a unit square, then 10 x 10 grids can be shaded to illustrate percents accordingly.

Exploring their desire to have the school cafeteria serve pizza more often, students in Julia Mason's fourth-grade class carried out experiments to determine probabilities by using the visual tools of spinners and charts. The fourth graders had the opportunity to formulate this problem from an issue they found compelling, one that involved probability, prediction, modeling, and simulation. On the basis of the data they gathered, the students decided that their "lunch wheel spinners" gave them more favorable results. Mason also noted that the activity produced a number of valid and creative student approaches to the problem posed and enabled her to identify students who still had serious misconceptions about probability.

Hershkowitz and Markovits (1992) developed a 36-unit mathematics curriculum for third graders that introduced students to basic visual concepts and applied visual abilities and visual thinking to learning tasks such as ratio and proportion and numerical intuition. Begun as a continuing study of 25 pairs of matched classes when these children were preschoolers (1985–87), the study noted that students' counting strategies dropped dramatically, whereas the use of three estimation strategies increased, with the most dramatic increase occurring in global estimation. Though this study only provides naturalistic observations of students' success, its longitudinal aspect provides insight.

In conclusion, we once again call for further study of visual learning by young mathematics students. As we pursue instructional practices using these visual tools, we also need to consider Clark's view (1983, 1994). He argues that it is not the medium that causes gains in learning but the instructional strategy embedded in the media presentation.

Recommended Instructional Practices

The following best practices are derived from the research literature:

- Visuals should be used when students need to learn concrete concepts or when they must identify spatial relationships.

- Visuals are very useful in learning tasks which involve memory. The information received from visuals appears to remain longer in memory.
- Visuals should be incorporated to give a real-life meaningfulness to student learning.
- Concrete materials can be an effective aid to students' thinking and to successful teaching, but effectiveness is contingent on teachers' continual efforts to emphasize understanding, not just doing.

How the Research Is Implemented in *Harcourt Math*

Visuals are used throughout *Harcourt Math* to facilitate students' understanding of concepts and the mathematics that underlies each of the mathematical procedures. Questions that encourage students to reason about the visuals are included in the Pupil Edition, and questions in the Guided Instruction in the *Teacher's Edition* provide the scaffolding essential to derive meaning from visuals.

Visuals are also used to help students appreciate the usefulness of mathematics in everyday activities. Engaging students' interest in mathematics is a critical variable in assuring their engagement in daily lessons and in motivating them to put forth the effort in mathematics necessary for successful progress.

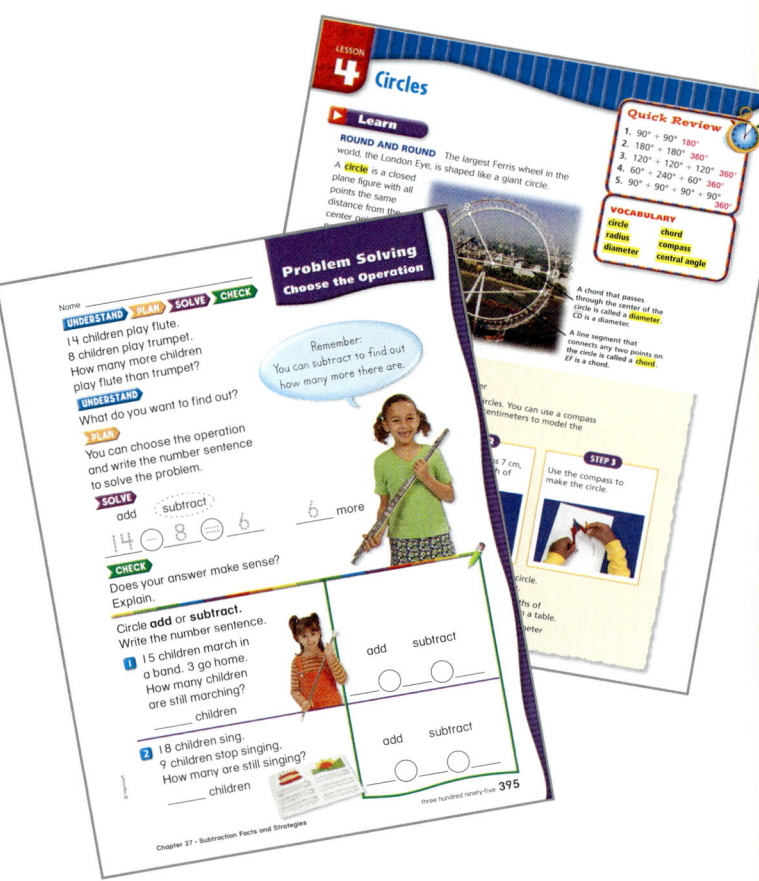

The use of real-life visuals to illustrate concepts and skills makes mathematics more meaningful and helps students see the usefulness of mathematics in everyday experiences.

The use of visuals provides a clear mental image and enhances understanding of a concept. Students' interest levels are higher when they are involved in using visuals as learning tools.

In this media age, children live in a visual world. However, many of these visuals pass rapidly before children's eyes, and little learning results. Teaching children to read visuals and to link visuals and text is critical to their academic success. Visuals in mathematics are powerful teaching tools that children need to learn to "read" and use as they develop their mathematical abilities

References

Barsalou, Lawrence (1992). *Cognitive Psychology: An Overview for Cognitive Scientists*. Hillsdale, NJ: Lawrence Erlbaum Associates.

Bennett, A. B., Jr., and T. L. Nelson (1994). A conceptual model for solving percent problems. *Mathematics Teaching in the Middle School*, 1, (1), 20–25.

Brownell, W. A. (1986). *AT* classic: The revolution in Arithmetic. *Arithmetic Teacher*, 34, (2), 38–42.

Clark, R. E. (1994b). Media and method. *Educational Technological Research and Development*, 42, (3), 7–10.

_____. (1994a). Media will never influence learning. *Educational Technological Research and Development*, 42, (2), 21–29.

_____. (1983). Reconsidering research on learning from media. *Review of Educational Research*, 53, 445–459.

Dwyer, F. M., ed. (1987). *Enhancing Visualized Instruction—Recommendations for Practitioners*. State College, PA: Learning Services.

_____. (1978). *Strategies for Improving Visual Learning*. State College, PA: Learning Services.

_____. (1972). *A Guide for Improving Visualized Instruction*. State College, PA: Learning Services.

Englert, G. R., and R. Sinicrope (1994). Making connections with two-digit multiplication. *Arithmetic Teacher*, 41, (8), 446–448.

Heddens, J. W. (1986). Bridging the gap between the concrete and the abstract. *Arithmetic Teacher*, 33, (6), 14–17.

Hershkowitz, R., and Z. Markovits (1992). Conquer mathematics concepts by developing visual thinking. *Arithmetic Teacher*, 39, (9), 38–41.

Kennedy, L. M. (1986). A rationale. *Arithmetic Teacher*, 33, (6), 6–7.

Korithoski, T., and P. Korithoski (1993). Mean or meaningless. *Arithmetic Teacher*, 41, (4), 194–197.

Schultz, J. E. (1991). Area models—Spanning the mathematics of grades 3–9. *Arithmetic Teacher*, 39, (2), 42–46.

Skemp, R. S. (1971). *The Psychology of Learning Mathematics*. Harmondsworth, England: Penguin Books.

Winn, W. (1987). Charts, graphs, and diagrams in educational materials. In D. M. Willows and H. A. Houghton (Eds.), *The Psychology of Illustration: Basic Research* (Vol. 1). New York: Springer-Verlag.

What Research Says About...

Explicit Instruction:
Delivering Instruction That Is Clear and Direct

Overview

Explicit instruction is a procedure for making specific skills or strategies known to a learner. The teacher is charged with presenting material in such a way as to focus the student's attention on something specific to be learned (Wilkinson, 1999). Several terms are used in the literature for this concept. These terms include direct instruction, direct teaching, explicit teaching, explicit instruction, and direct explanation. In addition, recent research literature focuses on using alternatives in instruction including explicit instruction, self-directed learning, exploratory learning, discovery, and constructivism (Rieber, 1991; Guskey and Passaro, 1992).

Bangert-Downs and Bankert (1990, 8) reviewed 250 articles on teaching critical thinking as a general process and as a specific tool in several content area domains. Twenty studies were suitable for a meta-analysis; each of the 20 compared a group that received explicit instruction with a group that did not. Bangert-Downs and Bankert state, "It is most striking that the studies reviewed in this meta-analysis so consistently produced findings favorable to explicit instruction in critical thinking". These studies included students at the elementary, secondary, and post-secondary levels. Bangert-Downs and Bankert stress the importance of teaching young children to assess the "trustworthiness" of a statement. They suggest that children in younger grades benefit greatly from explicit instruction more than older students who have been making "trustworthiness" judgments for a longer time.

Lester (1983), in his review of trends and issues in mathematical problem solving research, stated that the role of the teacher is often overlooked in research studies. He recommends that research be directed toward understanding how the teacher affects problem solving behavior. He suggests including the teacher as a variable in research studies rather than ignoring or factoring out the teacher.

Research Findings
Explicit instruction is an effective strategy for teaching basic math skills and procedures.

Past and current instructional innovations use explicit teaching. Peterson, Swing, Braverman, and Buss (1981) taught a short unit on probability to fifth- and sixth-grade students. The teaching method followed a direct instructional model. After viewing themselves on videotape, students were interviewed about their thought processes during the lesson. The results showed that students who used specific cognitive strategies such as relating what was being taught to prior knowledge did better on an achievement test. Higher-ability students reported specific understandings and/or cognitive strategies, whereas lower-ability students gave vague reasons for what they did or did not understand.

Din (1998) developed individualized programs for students aged 7–16 who were referred for help in math by their parents. After three weeks of instruction, all 19 students showed significant improvement in basic math skills. The program provided direct instruction in numeration concepts, computational procedures, multiplication tables, and application.

Explicit instruction in interpreting the language used in word problems helps students select the correct algorithm to solve.

Stein (1998) examined understanding of word problems in mathematics. She found that after explicit instruction designed to help students interpret the language used in word problems, students were better able to select the correct algorithm to solve them. Stein stressed that students need to attend to the specific language used in a problem. Students used a math story chart based on Polya's four-step model. This chart helped them interpret language used in mathematical word problems (Polya, 1957). Polya's first step, Tell (the information given in the problem), formed the foundation for explicit teaching of problem solving.

Rudnitsky, Etheredge, Freeman, and Gilbert (1995) explicitly taught 401 students in Grades 3 and 4 a structured sequence of problem solving steps. These steps, (1) understand the problem, (2) make a plan, (3) use the plan, and (4) check the answer, are similar to Polya's four-step model. One group of students also wrote their own story problems. Results showed significantly higher problem solving scores for students in the structured-sequence-plus-writing group. This group also performed significantly better on a retention test ten weeks later.

Explicit instruction in writing about mathematics increases students' conceptual understanding, procedural knowledge, and mathematical communication.

Several studies examined writing as part of learning mathematics. In research with 540 fifth-grade students, Niemi (1996) found that students who received explicit instruction showed higher levels of principled understanding of fractions. Students were told that they were going to do a TV demonstration and had to write an explanation of everything a fifth-grade student should know about fractions. They were given a set of guiding questions to help them with their explanations. The students who had explicit instruction on fraction principles expressed more principles in their explanations. Even with a very short time period, $7\frac{1}{2}$ days, students' understanding was sensitive to cognitive changes.

Jurdah and Abu Zein (1998) compared two groups of upper-elementary students. One group wrote in math journals for 7–10 minutes at the end of math class three times a week for 12 weeks. Students in the journal-writing group had significantly higher posttest scores at the level of conceptual understanding, procedural knowledge, and mathematical communication. An important result was that teachers could intercede, provide supportive responses, and plan an explicit lesson based on students' perceived weaknesses.

In conclusion, explicit teaching appears to have favorable instructional consequences. Regardless of whether it is called explicit instruction, direct instruction, direct teaching, explicit teaching, or direct explanation, students benefit from instruction that is direct, explicit, and clearly stated.

Recommended Instructional Practices

The following best practices are derived from the research literature:

- Procedures based on Polya's four-step model are effective: (1) understand the problem, (2) devise a plan, (3) carry out the plan, (4) look back.
- It is useful to use a chart to show the series of steps to follow to solve a problem or complete a mathematical procedure.
- Writing about mathematics helps students understand both the "how" and the "why" of a mathematical procedure.

- Writing in mathematics is useful to improve mathematical communication, an area of instruction that is sometimes neglected.
- It is useful to observe learners as they work through problems and then to explicitly reteach the steps as needed.
- Short, intensive periods of explicit strategy instruction can have lasting effects.

How the Research Is Implemented in *Harcourt Math*

The structure of *Harcourt Math* lends itself to explicit teaching. Throughout the program, mathematical skills and procedures are clearly presented with models, explanations, and questions that can be used by the teacher to focus students' attention on key concepts and procedures. The *Teacher's Edition* facilitates explicit teaching through the use of guided instruction.

The development of problem solving skills and strategies is explicitly done through a four-step process modeled on Polya's work. Not only are these four steps developed in the Pupil Edition, but a *Problem Solving Think Along*, a format that is integral to the program, provides questions that help the students apply Polya's four-step process and record their thinking as they work through the steps.

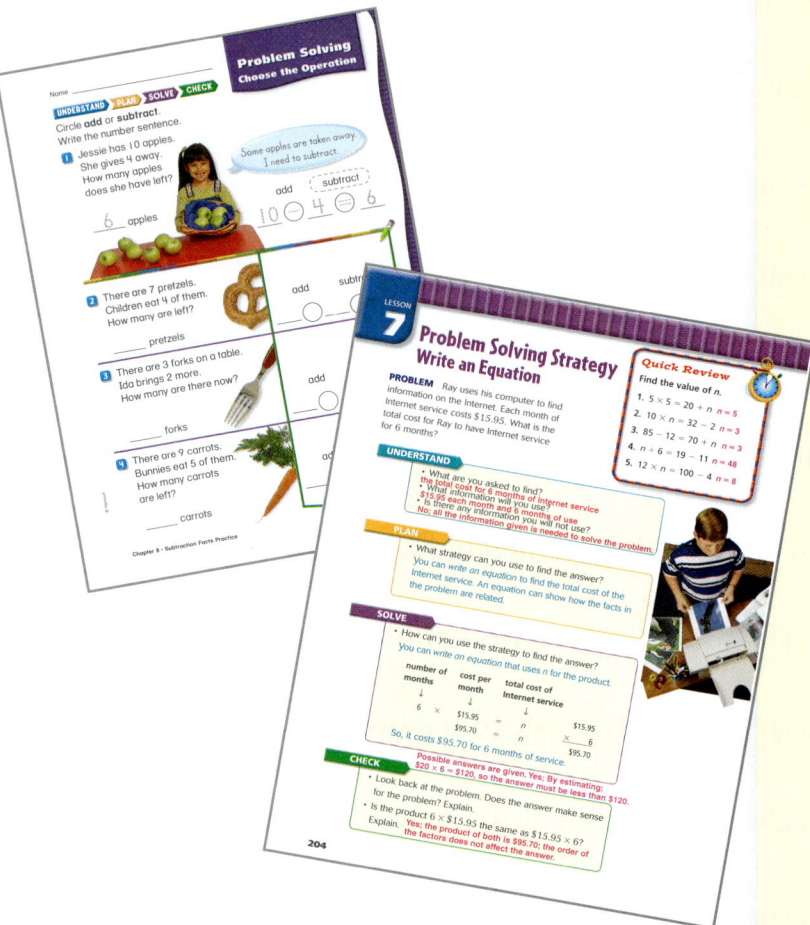

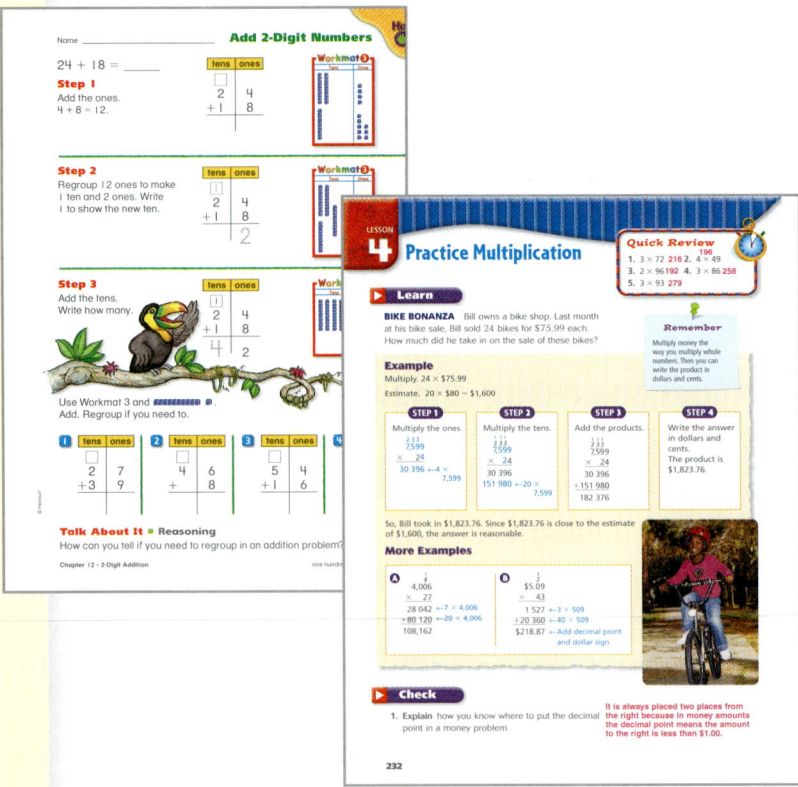

Explicit step-by-step instruction that includes hands-on experiences, visual models, verbal descriptions of the process, and symbolic notation helps students both understand and apply algorithmic procedures.

Explicit instruction in interpreting the language of word problems, in planning and solving the problem, and in checking the solution for reasonableness by using the four-step process—Understand, Plan, Solve, and Check—helps student select the correct operation and equation to solve the problem.

Writing is a part of students' daily work. Daily exercise sets require that students write their own problems, explanations of how they solved a given problem, justifications for their solutions, and conclusions and generalizations drawn from the mathematics they are learning. The use of a journal is encouraged, and teachers are given concrete suggestions for the types of entries that might be included.

The goal of the pedagogical approach in *Harcourt Math* is to provide a balance of suggested instructional strategies with explicit instruction forming the core of the suggested strategies so that teachers can be assured that they can guide students' mathematical development efficiently and effectively.

References

Bangert-Downs, R. L., and E. Bankert (1990). *Meta-analysis of effects of explicit instruction for critical thinking*. Paper presented at the annual meeting of the American Educational Research Association, Boston, April 16–20, 1990. 12 pp. (ERIC Document Reproduction Service No. ED 328 614.)

Din, F. S. (1998). *Direct instruction in remedial math instructions*. Paper presented at the National Conference on Creating the High Quality School, Arlington, VA, March 25–28, 1998. 14 pp. (ERIC Document Reproduction Service No. ED 417 955.)

Guskey, T. R., and P. D. Passaro (1992). *How mastery learning can address our nation's science education needs*. Paper presented at the annual meeting of the American Educational Research Association, San Francisco, April 10–14, 1992. 35 pp. (ERIC Document Reproduction Service No. ED 370 760.)

Jurdah, M., and R. Abu Zein (1998). The effect of journal writing on achievement in and attitudes toward mathematics. *School Science and Mathematics*, 98, (8), 412–419.

Lester, F. K., Jr. (1983). Trends and issues in mathematical problem-solving research. In R. Lesh and M. Landau (Eds.), *Acquisition of Mathematics Concepts and Processes* (pp. 229–257). New York: Academic Press.

Niemi, D. (1996). A fraction is not a piece of pie: Assessing exceptional performance and deep understanding in elementary school mathematics. *Gifted Child Quarterly*, 40, 70–80.

Peterson, P. L., S. R. Swing, M. T. Braverman, and R. Buss (1981). *Students' aptitudes and their reports of cognitive processes during direct instruction*. Paper presented at the annual meeting of the American Educational Research Association, Los Angeles, April 13–17, 1981. 45 pp. (ERIC Document Reproduction Service No. ED 211 234.)

Polya, G. (1957). *How to Solve It: A New Aspect of Mathematical Method*. Princeton, NJ: Princeton University Press.

Rieber, L. P. (1991). Computer-based microworlds: A bridge between constructivism and direct instruction. In M. R. Simonson and C. Hargrave (Eds.), *Proceedings of the Association for Educational Communications and Technology*, Orlando, FL, February 13–17, 1991. 18 pp. (ERIC Document Reproduction Service No. ED 335 007.)

Rudnitsky, A. M., S. Etheredge, J. M. Freeman, and T. Gilbert (1995). Learning to solve addition and subtraction word problems through a structure-plus-writing approach. *Journal for Research in Mathematics Education*, 26, 467–486.

Stein, M. G. (1998). *Strategic learning: The implications of language in successful math problem-solving*. 40 pp. (ERIC Document Reproduction Service No. ED 416 501.)

Wilkinson, L. (1999). An introduction to the explicit teaching of reading. In J. Hancock (Ed.), *The Explicit Teaching of Reading* (pp. 1–12). Newark, DE: International Reading Association.

What Research Says About...

The Use of Manipulatives, or Concrete Materials

Overview

In the mathematics classroom, manipulatives include such objects as counters, base-ten blocks, fraction rods, three-dimensional geometric models, geoboards, tangrams, and spinners. (See, for example, National Council of Teachers of Mathematics (1989), pp. 17 and 67–68.) Most research supports the view that such materials are effective in assisting students to develop new concepts by working from the concrete to the more abstract. When used appropriately by teachers, these materials, therefore, tend to lead to students having better understanding, achievement, and attitude.

Research Findings

The use of manipulatives increases students' conceptual understanding and achievement in mathematics.

There are a number of studies that show that the use of manipulatives (or concrete materials) can increase students' understanding of mathematical concepts and, hence, their achievement in mathematics overall. These studies range across all grade levels, from kindergarten to college, and across many mathematical topics, including basic computation, geometry, and algebra. They also compare the use of manipulatives with traditional textbook presentations, with the use of diagrams, and with computer presentations. Most of these studies demonstrate that using manipulatives has a positive effect on students' achievement, although a few studies show little effect. (See, for example, Chester, Davis, and Reglin (1991), Hiebert and Wearne (1992), and Peck and Connell (1991) for positive results, and Thompson's (1992) discussion of the inconsistencies among similar studies.)

Earlier work is summarized by Sowell (1989) in a meta-analysis combining the results of 60 studies that compare the effects of using manipulatives with the effects of using more abstract instruction. Sowell concludes "that mathematics achievement is increased through the long-term use of concrete instructional materials and that students' attitudes toward mathematics are improved when they have instruction with concrete materials provided by teachers knowledgeable about their use. Instruction with pictures and diagrams does not appear to differ in effectiveness from instruction with symbols."

Two main points emerge from these conclusions:
- Long-term use of manipulatives is important for success.
- Teachers need to have clear purposes for using manipulatives in a particular situation.

A discussion of more recent studies that address these two issues follows.

Several researchers have surveyed teachers about the nature and extent of the use of manipulatives in their classrooms. Gilbert and Bush (1988) asked teachers of Grades 1–3 about 11 different materials and discovered that overall use was only moderate and decreased from Grade 1 to Grade 3. In a follow-up study, after the publication of the NCTM standards (National Council of Teachers of Mathematics, 1989) that recommended the use of manipulatives, Hatfield (1994) found that little had changed. In her survey of K–6 teachers, she found limited use of manipulatives and diminishing use through the grades. Even the use of fraction bars, which are of more relevance in Grades 4–6, declined in these higher grades. Such results were confirmed more recently in a survey of Grades 1–6 by Marlow and Inman (1997), who found that, while counters were commonly used, other manipulatives such as fraction rods, geoboards, and geometric models were not often used by students in the classrooms of the teachers surveyed. Marlow and Inman also asked about barriers to the use of manipulatives and found that the main reasons for not using manipulatives were low parent expectations, lack of materials, discipline problems, and lack of preparation time.

Effective use of manipulatives is dependent upon teachers' guidance of students' interaction between the manipulatives and the concept or skill represented by the manipulatives.

The second point noted above in connection with Sowell's meta-analysis may also help explain those studies in which results were not positive about the use of manipulatives: the context in which particular manipulatives are used is very important and varies from class to class. Meira (1998) suggests that the link between manipulatives and learning is mediated by students' prior knowledge, culture, and so forth, and that therefore context is vital in determining the interaction between the concrete materials and students' skills acquisition. Linked to this is the finding of Schram, Felman-Nemser, and Ball (1990) that the teachers in their study did not discriminate among concrete materials and assumed students' seeing or touching such materials automatically produced understanding. These authors suggest that an important role for textbooks lies in developing teachers' understanding of the appropriate use of manipulatives.

Manipulatives should be used frequently and over the long term to ensure gains in achievement.

The research literature suggests very strongly that students who use manipulatives in their mathematics classrooms frequently and over the long term will gain in understanding, achievement, and attitude. Therefore, teachers should incorporate the use of concrete materials much more into their instruction, but not indiscriminately. Careful thought is needed to make the manipulatives appropriate to the teaching purpose and to the students—and with the first of these, teachers can be assisted by well-written textbooks. However, teachers will always need to monitor the effectiveness of particular manipulatives and switch to others when the need arises, either to apply to the whole class or to adapt to individual students' learning styles. In addition, change is needed to alter the prevailing attitude among teachers that manipulatives are for young students only. Students of all ages need concrete support with new abstract ideas (e.g., in algebra or coordinate geometry), and teachers at all grade levels should provide suitable manipulative materials to introduce any new concept that will benefit from their use.

Recommended Instructional Practices

The following best practices are derived from the research literature:

- Manipulatives should be used frequently and over the long term to help students understand concepts and procedures.
- Teachers should monitor their students' use of manipulatives to determine whether one manipulative is more effective than another for individual students.
- The link between the use of manipulatives and learning is facilitated by the teacher's use of questions that help students see the relationship of the manipulatives to the concept that is being modeled.

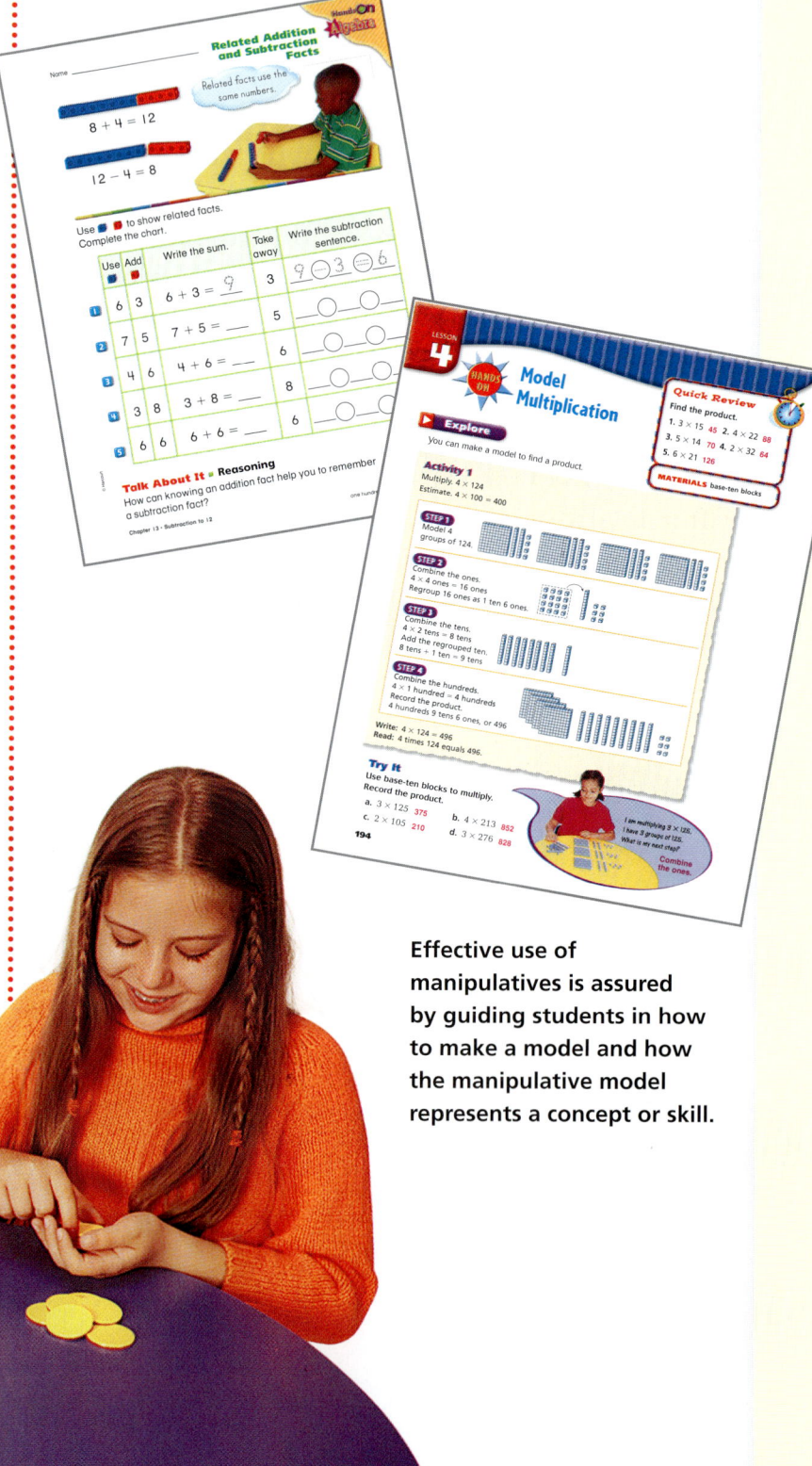

Effective use of manipulatives is assured by guiding students in how to make a model and how the manipulative model represents a concept or skill.

How the Research Is Implemented in *Harcourt Math*

Manipulatives are used throughout the grade levels to facilitate students' development of conceptual understanding of key concepts in each of the mathematical strands. The use of a wide variety of manipulatives is suggested. Learning experiences with manipulatives are accompanied by questions in the Pupil Edition that help students interact with the models to better understand concepts and procedures. Questions provided in the *Teacher's Edition* help teachers facilitate the interaction between the use of manipulatives and the concepts modeled.

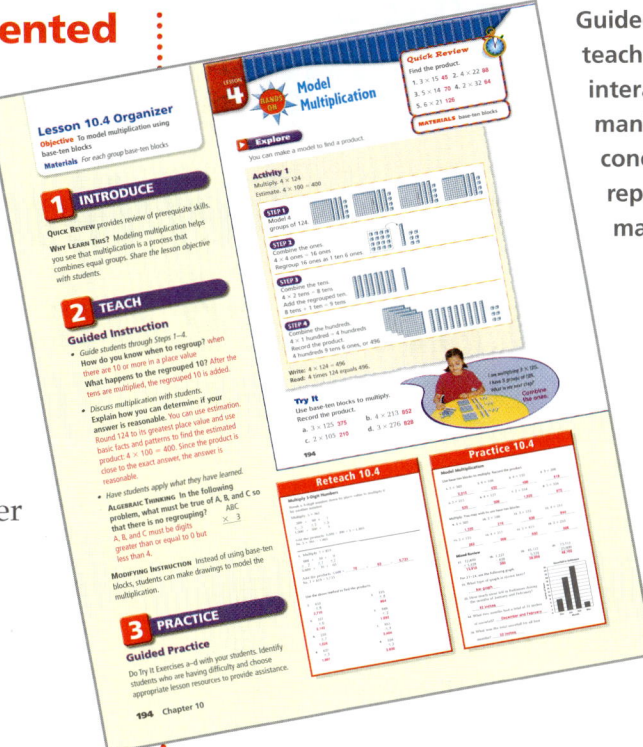

Guided Instruction helps teachers guide students' interaction between the manipulatives and the concept or skill represented by the manipulatives.

References

Chester, J., J. Davis, and G. Reglin (1991). *Math manipulatives use and math achievement of third-grade students.* (ERIC Document Reproduction Service No. ED 339 591.)

Gilbert, R. K., and W. S. Bush (1988). Familiarity, availability, and use of manipulative devices in mathematics at the primary level. *School Science and Mathematics*, 88, (6), 459–469.

Hatfield, M. M. (1994). Use of manipulative devices: Elementary school cooperating teachers self-report. *School Science and Mathematics*, 94, (6), 303–309.

Hiebert, J., and D. Wearne (1992). Links between teaching and learning place value with understanding in first grade. *Journal for Research in Mathematics Education*, 23, (2), 98–122.

Marlow, L., and D. Inman (1997). *Status report on teaching in the elementary school: Math, science, and social studies.* Paper presented at a meeting of the Eastern Educational Research Association, Hilton Head, SC, February 1997.

Meira, L. (1998). Making sense of instructional devices: The emergence of transparency in mathematical activity. *Journal for Research in Mathematics Education*, 29, (2), 121–142.

National Council of Teachers of Mathematics (1989). *Curriculum and Evaluation Standards for School Mathematics.* Reston, VA: author.

Peck, D. M., and M. L. Connell (1991). Using physical materials to develop mathematical intuition in fraction part-whole situations. *Focus on Learning Problems in Mathematics*, 13, (4), 3–12.

Schram, P., S. Felman-Nemser, and D. L. Ball (1990). *Thinking about teaching subtraction with regrouping: A comparison of beginning and experienced teachers' responses to textbooks.* Research Report 89–5, East Lansing, MI: National Center for Research on Teacher Education.

Sowell, E. J. (1989). Effects of manipulative materials in mathematics instruction. *Journal for Research in Mathematics Education*, 20, (6), 498–505.

Thompson, P. W. (1992). Notations, conventions, and constraints: Contributions to effective uses of concrete materials in elementary mathematics. *Journal for Research in Mathematics Education*, 23, (2), 123–147.

What Research Says About...
The Use of Practice or Review to Improve Performance and Retention

Overview

Studies have demonstrated that in many areas of learning, there is a considerable amount of forgetting over time. In mathematics instruction, this is of special concern because of the building-block nature of math instruction. As Geary (1994) discusses, education in mathematics must focus on both procedural skills and conceptual knowledge, and one of the primary ways to develop foundational skills is through practice.

Research has demonstrated that practice is important for reinforcing students' knowledge and for preparing students to move on to new topics and types of problems. Although both word problems and drill-type problems have been demonstrated to lead to higher performance, research addresses the greater effectiveness of higher-level word problems and worked examples to review, rather than large amounts of frequent review of lower-level drill-type problems.

Research Findings

There are numerous studies that show that review helps students retain knowledge and improve performance. Much of the early research was done in the field of cognitive psychology, but more recent research demonstrates that the results of these studies can be generalized and applied specifically to the instruction of mathematics topics.

In mathematics instruction, review is especially important for the development of automaticity with basic math facts. Once facts have been learned to the point that they can be used automatically, students are able to focus on higher-level problem solving and allocate their attention to the other components of task performance. Studies also show that the type of review is important. Although drill-type review does help retention, it helps more for retention of lower-level concepts, while more thought-provoking questions help higher-level retention and generalizability.

Practice/Review is important for retention and automaticity.

In their reviews of research in mathematics, both Geary (1994) and Suydam and Dessart (1980) conclude that practice is essential for mastering skills and developing the automaticity that will allow these skills to be used routinely in other practices.

A number of research studies have addressed the importance of practice for skill retention. Bahrick and Hall (1991) studied 1,726 individuals to examine the life span retention of content acquired in math courses. The study found that talent and achievement had some impact, but the primary variable for success was practice that had occurred over time. Individuals who had learned the math content over a short period without practice over time showed declines in mathematics performance.

Ausubel (1966) studied the specific timing of reviews to determine whether early review (after one day) was more or less effective than delayed review (after seven days). He found that both produced results and that one did not seem more effective than the other. As a result, he concluded that both serve a purpose—the early review to consolidate the knowledge and the delayed review to allow students to relearn any forgotten information.

Although the inclusion of regular early and delayed review is only one variable in the experimental design used by Good and Grouws in their 1979 study, this study supports the idea that review at regular intervals is significant for the retention of mathematical knowledge. This study of Grade 4 math students determined that those receiving an experimental instructional program that included regular reviews—daily, weekly, and monthly—performed significantly higher on math achievement than did a control group not receiving the experimental instruction.

Goldman, Mertz, and Pellegrino (1989) concluded that repeated exposure to basic addition facts over a 12-week practice period increased the speed and accuracy of responses among Grade 3 and 4 students, thereby indicating the increased likelihood of direct retrieval as a result of practice.

In a series of experiments conducted primarily with Grade 8 students, Cooper and Sweller (1987) found that practice with algebraic word problems enhanced schema acquisition and automation, allowing students to solve both similar test problems and related transfer problems. The development of automation, which led to the ability to transfer skills to new types of problems, took more time than the development of schema acquisition, essentially background knowledge in one type of problem. Cooper and Sweller theorize that familiarity with the basic operations necessary for solving problems allowed the students to use greater cognitive capacity to deal with those parts of the problem that are unfamiliar.

Thought-provoking questions are important types of review questions.

Burns (1960) studied whether thought-provoking review questions would help retention of learning when spaced throughout arithmetic instruction in Grade 6. Interested in evaluating both the achievement and the attitudes of students involved, Burns determined that the inclusion of thought-provoking questions, rather than repetitive drill, encouraged discussion, improved performance, and proved more interesting to students.

In a 1980 study, Lee randomly assigned 60 seventh-grade students to experimental groups receiving different types of review questions and review passages. Groups received either word-type review questions, which required a thorough understanding of concepts and the application of these concepts to new situations; calculation-type review questions, which required comprehension of a narrower range of concepts and rules; or no review questions. Review questions of both types were found to facilitate retention. The group given word-type review questions received the highest score on a posttest of math aptitude, followed by those given calculation-type review questions, followed by the no-review group.

Although Cooper and Sweller's 1987 study did not specifically address using worked problems (sample problems that show the steps of the process already worked out) versus student-worked practice problems, they concluded by comparing the results of their 1987 study with previous research studies on practice that the use of worked examples may help speed up the process of automation as a result of practice.

Recommended Instructional Practices

The following best practices are derived from the research literature:
- Practice and regular review should be incorporated into daily lesson planning to encourage automaticity with mathematical skills.
- Practice should involve the use of a variety of problems and a mixture of procedures.

- Depending on the instructional goals, effective review can include basic drill-type items, worked problems, or more thought-provoking word-type review problems.
- Including worked problems and models enhance the development of automaticity.

How the Research Is Implemented in *Harcourt Math*

In Grades 1 and 2, the Daily Routine in each lesson helps children review prior-taught skills, and spaced reviews throughout the student text provide ongoing review of critical skills. In Grades 3 through 6, every lesson begins with a Quick Review and ends with a Mixed Review. These reviews of a lesson's prerequisite skills and of prior-taught skills cover a wide range of key mathematical topics and represent a mixture of procedures and skills.

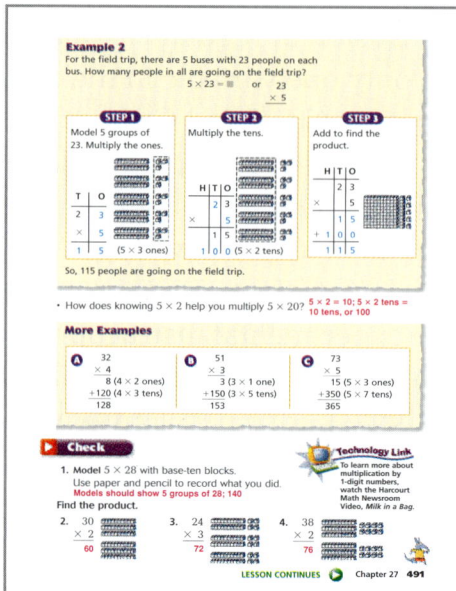

Step-by-step procedures shown with models and worked examples provide thought-provoking examples that deepen students' understanding of a procedure and make practice more effective in building toward automaticity.

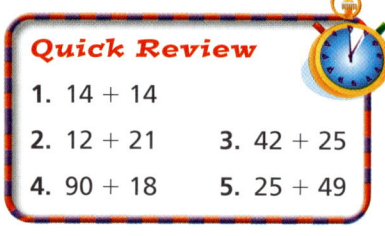

Regular reviews in every lesson help students develop automaticity with skills in all strands of mathematics.

Models that show the step-by-step application of algorithmic procedures, and worked examples that show the variety of possible types of problems within a procedure are included throughout the program. These models are often accompanied by pictures of manipulatives to allow students to deepen their conceptual understanding of a procedure as they apply the steps in the procedure. At the end of a group of chapters that cover a major mathematical topic, a Study Guide and Review shows worked examples of the types of problems on which students should develop automaticity with page references to help students review if necessary.

In Grades 3 through 6, practice exercises for every lesson include a variety of problems. Basic drill-type items are included in lessons emphasizing basic fact acquisition and procedural fluency. Thought-provoking word problems that apply the procedure, skill, or concept taught in the lesson and that review prior-taught skills are included in every lesson. The emphasis in many of these problems is on logical reasoning; solving multistep problems; and explaining, justifying, and proving solutions. Exercises that require writing questions given a possible solution, writing problems with given conditions, and analyzing errors develop students' conceptual and procedural knowledge of mathematics.

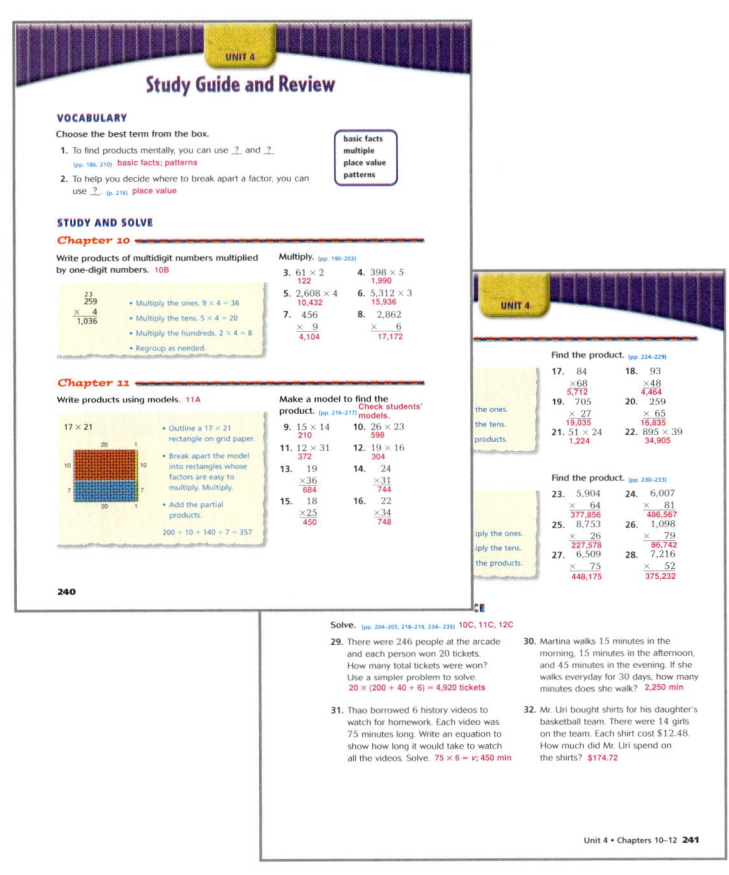

Review over a chunk of instruction on a concept and related skills helps students make their learning more permanent. Providing worked examples with page references to lessons in which that skill or concept is taught ensures that students can find help that will allow them to be successful with the practice.

Review of word problems involves choosing and using strategies taught throughout the program, solving logical reasoning and multistep problems, and applying skills and concepts learned earlier.

References

Ausubel, D. P. (1966). Early versus delayed review in meaningful learning. *Psychology in the Schools*, 3, 195–198.

Bahrick, H. P., and L. K. Hall (1991). Lifetime maintenance of high school mathematics content. *Journal of Experimental Psychology: General*, 120, 22–33.

Burns, P. C. (1960). Intensive review as a procedure in teaching arithmetic. *Elementary School Journal*, 60, 205–211.

Cooper, G., and J. Sweller (1987). Effects of schema acquisition and rule automation on mathematical problem-solving transfer. *Journal of Educational Psychology*, 79, 347–362.

Geary, D. C. (1994). *Children's Mathematical Development: Research and Practical Applications*. Washington, D.C.: American Psychological Association.

Goldman, S. R., D. L. Mertz, and J. W. Pellegrino (1989). Individual differences in extended practice functions and solution strategies for basic addition facts. *Journal of Educational Psychology*, 81, 481–496.

Good, T. L., and D. A. Grouws (1979). The Missouri Mathematics Effectiveness Project: An experimental study in fourth-grade classrooms. *Journal of Educational Psychology*, 71, 355–362.

Lee, H. (1980). The effects of review questions and review passages on transfer skills. *Journal of Educational Research*, 73, 330–335.

Suydam, M. N., and D. J. Dessart (1980). Skill learning. In Shumway, R. (Ed.), *Research in Mathematics Education*. Reston, VA: National Council of Teachers of Mathematics.

What Research Says About...
Lesson Closure

Overview

Most educational experts agree that effective lesson closure is essential to good teaching. In analyzing lesson plan models used over the past 150 years, Kelly (1997) found that closure was included in more than half of the lesson types. Closure gives students an opportunity to think and talk about what they have learned and a chance to internalize the skills and content taught.

Closure is generally considered one of the last steps in a lesson, but its exact placement in the instructional sequence varies from model to model. In some instructional planning models, closure is paired with an initial, introductory step called a set. Set activities prepare students for the upcoming instruction, and closure summarizes the lesson. Romberg and Wilson (1973) recommend a variation of this model with lesson parts called an advance organizer and a post organizer. They explain that when mathematics teachers finish a presentation, "closing with a concise summarization is usually desirable."

Others contend that closure can occur at any time throughout a lesson when a teacher wants to clarify key points and check whether students are understanding. So, it can be done in intervals instead of as a final conclusion to the lesson.

A review of the literature also reveals two contrasting styles of closure. Some experts (Hunter, 1991) recommend an explicit, teacher-directed approach to closure that is concerned primarily with restating or summarizing what has been taught. Others advocate a more reflective, student-centered approach that engages students in a variety of activities such as repeating important concepts, summarizing critical points, questioning procedures used in the lesson, giving assignments, projecting future activities, and acknowledging student contributions to the lesson (Bailey, 1980).

Research Findings

Researchers who have surveyed existing instructional practices have found that teachers often do not use closure effectively. Wolf and Supon (1994) noted that teachers often provide all the pertinent information to convey a concept but then quickly conclude by assigning homework or telling students to get out the necessary materials for the next subject. Likewise, in a case study of math instruction, Welch (1978) found that a lesson frequently consisted of a review of the previous lesson, some explanation of new materials, the assignment of new problems, and teacher coaching, without any closure.

Is closure effective? Does closure enhance students' learning? Because of the complexity of the teaching-learning process, it is difficult to isolate a single instructional variable such as closure and assess its effect on students' learning. Consequently, we must rely on the results of correlational studies and learning theory to deduce the value of closure.

A variety of closure strategies should be a regular and consistent part of lesson planning.

Rosenshine and Meister (1994) analyzed the results of 16 studies and found that closure activities appeared to help students perform better on measures used in the studies. These closure strategies included summarization, summary writing, and question generation.

Drawing on schema theory, Romberg (1992) recommends closure as a way to help students make connections between newly learned concepts and previously acquired knowledge. He calls for having students learn *how* as opposed to learning *what* in mathematics. Welch also calls for an emphasis on learning *how* as opposed to learning *what*.

Paul, Binker, and Weil (1995) contend that one of the main goals of math instruction is to get students to think independently as problem solvers and to use their problem solving abilities to make sense of the world. Teachers can promote such thinking, they suggest, through closure activities that include questioning *how* and *why* a math concept or skill is useful and by having students pose problems to other students as extended practice.

Recommended Instructional Practices

The following best practices are derived from the research literature:

- Make closure a regular and consistent part of lesson planning.
- Use a variety of closure strategies or activities. Some may be more explicit, teacher-directed activities such as repeating important concepts, summarizing critical points, and making connections between the new lesson and previous lessons; others may be more reflective, learner-centered activities such as encouraging students to question procedures used in the lesson, having students explain how and why the concept may be applied to real-life situations, and engaging students in discussions to validate their own thinking.
- Offer students opportunities to demonstrate their knowledge in a variety of ways through oral, written, and graphic summarization.
- Use closure, when appropriate, throughout a lesson to clarify key points and check student understanding.
- Use closure as a decision-making tool for deciding which students have fully grasped the concept of the lesson and can move on and which students could profit from additional review or reteaching.

How the Research Is Implemented in *Harcourt Math*

Closure is a regular and consistent part of every lesson plan. Step 4 of the lesson plan provides a variety of closure activities—oral discussion; skill check; and written explanations, such as generalizations or restatements of the big idea of the lesson, or problem solving activities that require reasoning and justification.

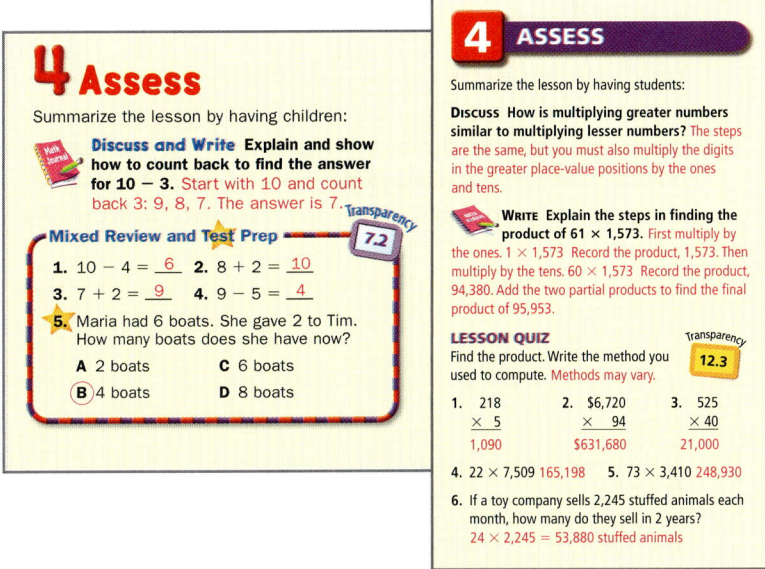

Closure is a part of every lesson. Students are actively involved in summarizing the content of the lesson through oral discussion and through writing. An integral part of closure for every lesson is a brief Lesson Quiz or Review that helps teachers evaluate students' progress and make decisions about future instruction.

In addition, closure is developed throughout the lessons through the use of reasoning questions, Math Idea statements, and a short section called Check that guides students in applying the lesson skills and concepts before moving to independent practice.

Students participate in many different kinds of activities that help them develop their understanding of the connections among mathematical ideas and that encourage them to use different ways to demonstrate their understanding and knowledge of the mathematics they are learning.

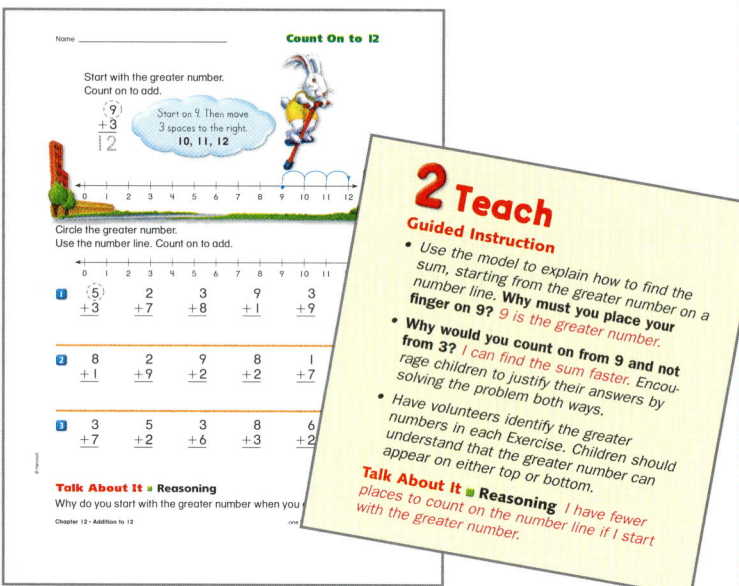

The Guided Instruction in the *Teacher's Edition* and the Talk About It question in the Pupil Edition provide closure within the lesson. Closure helps students prepare for the independent practice and provides diagnostic information to the teacher about whether students are ready for independent practice.

The wide variety of closure activities and their placement throughout the lesson and at the end of every lesson provide important ongoing information about students' progress and provide solid data that teachers can use to make decisions about individual students' needs and effective ways to plan future lessons based on overall class achievement and needs.

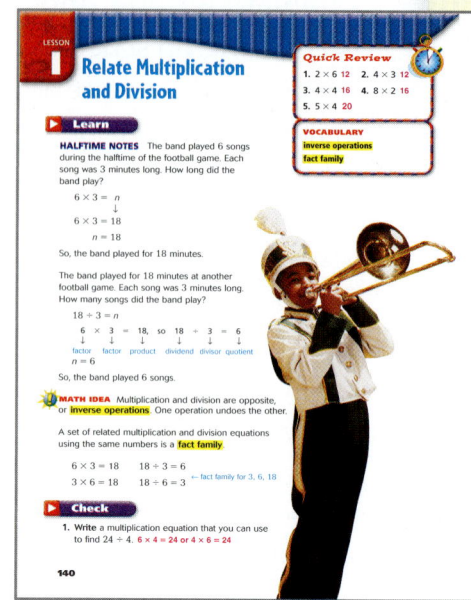

In Grades 3-6, certain lessons in a sequence of lessons on a related topic provide opportunities within the lesson to summarize and bring closure to math ideas. Every lesson includes a Check set of exercises and problems that provide closure to the instructional part of the lesson and a bridge to the independent practice section of the lesson.

References

Bailey, G. D. (April 1980). Set and closure revisited. *NASSKP Bulletin*, 64, (435), 103–110.

Hunter, M. (October 1991). Forum: Generic lesson design—The case for. *The Science Teacher*, 58, (7), 26–28.

Kelly, K. B. (1997). *Evolution/role of lesson plans in instructional planning*. Paper presented at the Eighth Annual Reading/Literacy Conference, Bakersfield, CA. (ERIC Document Reproduction Services No. ED 414 261.)

Paul, R., A. J. A. Binker, and D. Weil (1995). *Critical Thinking: Handbook K–3: A Guide for Remodelling Lesson Plans in Language Arts, Social Studies, and Science*. Santa Rosa, CA: Foundation for Critical Thinking.

Romberg, T. A. (1992). Mathematics learning and teaching: What we have learned in ten years. In C. Collins and J. Mangieri (Eds.), *Teaching Thinking: An Agenda for the Twenty-First Century* (pp. 49–64). Hillsdale, NJ: Lawrence Erlbaum Associates.

Romberg, T. A., and J. W. Wilson (March 1973). The effect of an advanced organizer, cognitive set, and post organizer on the learning and retention of written materials. *Journal for Research in Mathematics Education*, 4, (2), 68–76.

Rosenshine, B., and C. Meister (Winter 1994). Reciprocal teaching: A review of the research. *Review of Educational Research*, 64, (4), 479–530.

Welch, W. (1978). Science education in Urbanville: A case study. In R. Stake and J. Easley (Eds.), *Case Studies in Science Education* (pp. 515–533). Urbana: University of Illinois.

Wolf, P., and V. Supon (1994). *Winning through student participation in lesson closure*. (ERIC Document Reproduction Services No. ED 368 694.)

HARCOURT Math

PROFESSIONAL HANDBOOK

▶ **Section 1**
Articles by respected mathematicians describe the conceptual foundations of key math topics.

Marvelous Decimals	PH1
Doing Decimal Arithmetic: Addition	PH5
Doing Decimal Arithmetic: Multiplication	PH9
Estimation and Approximation	PH13
Estimation and Arithmetic	PH17
A "Blind Spot" in the Order of Operations	PH21
Models for Fractions	PH24
Geometry and Measurement	PH26

▶ **Section 2**
Articles by the authors of *Harcourt Math* describe the philosophy and research base that guided the development of the program and provide examples of implementation of the philosophy.

Conceptual Understanding—The Power of Models and Visuals	PH29
Effective Practice: Memorizing the Number Facts	PH33
Effective Practice: Building Computational and Procedural Efficiency	PH35
Developing Algebraic Thinking: Preparation for Algebra Begins in Kindergarten	PH38
Problem Solving: *The* Reason to Teach Mathematics	PH41
Instructional Strategies: Best Practices Defined by Research	PH43
Assessment—An Integral Part of the Learning Process	PH48
Technology—Pathways to Teaching and Learning	PH52
Linking Home and School: Making Mathematics Part of the Family's Daily Routine	PH55
Mathematics for All Doesn't Mean Mathematics for Some!	PH57
Making Math Memorable: Brain-Compatible Learning	PH60

Professional Handbook

Marvelous Decimals

by Roger Howe

It is hard to see clearly how wonderful something as familiar as our decimal number system really is, but pause a moment to contemplate its marvels.

Of the many virtues of the decimal system, consider the following six:

1. **Efficiency** It represents numbers with astounding (in a sense, perfect) compactness and efficiency.

2. **Sophistication** It is highly sophisticated. It uses all the operations of basic algebra (addition, multiplication, and exponentiation [raising to powers]) merely to represent numbers.

3. **Ease of calculation** It makes calculations fast and simple to perform. More technically, it supports efficient, general, easily implemented algorithms for calculation.

4. **Ease of comparison** It is compatible with our ideas of ordering and magnitude. We can easily tell which of two decimal numbers is larger.

5. **Ease of approximation** It deals smoothly with errors and approximation.

6. **Scale independence** It scales easily, allowing representation of arbitrarily small numbers as well as arbitrarily large ones by essentially the same scheme. The arithmetic operations are scale-independent.

That the decimal system does so much so well and so easily makes it a marvel for users, but it presents a challenge for educators. All the structure built into decimals, and all the benefits they confer, are hard to appreciate without both extended practice and extended thought. It takes a long time for a student to unpack the intellectual treasure chest we know as the decimal place-value number system, and it takes careful guidance from teachers to strike the right balance between helping students master the mechanics of decimal computation and providing them with insight into the various computational recipes that have been developed. However, done right, attention to concepts can provide a solid foundation for computation. It can make the sometimes seemingly arcane procedures of arithmetic appear as pragmatic solutions to essential problems.

History of the Decimal Place-Value System

Let's look briefly at the history of the development of the decimal number system and its possible effects on mathematics and science. The decimal system took a long time to develop. After several thousand years of precursors, this way of writing numbers was invented in India in about A.D. 500. It was adopted and developed by Islamic scholars and finally introduced into Europe in the late Middle Ages. An early pioneer and promoter of decimal notation in Europe was Leonardo of Pisa, also known as Fibonacci. Before Fibonacci, Roman numerals were used to keep records and calculation was done with a counting board. Use of decimals gradually displaced Roman numerals and became standard by the fifteenth century. Although the effects of a change of practice such as the transition from Roman numerals to decimals are diffuse and subtle to gauge, some writers attribute to decimals both a significant role in the European commercial boom of the late Middle Ages and Renaissance and a great stimulus to science and mathematics.

This collection of articles surveys a few highlights of the features of the decimal system listed above. This article discusses the first two items—efficiency and sophistication. The following articles discuss how the sophisticated structure built into decimal numbers promotes efficient arithmetic and how it allows for estimation.

The Efficiency of the Decimal Place-Value System

Writing Numbers Let's look first at writing numbers. The earliest representations of numbers were simple tallies: one mark for one object of some sort. So to indicate 10, it would take 10 tally marks: I I I I I I I I I I. Our decimal system represents up to 9 objects with just one symbol, but that does not begin to reveal its efficiency. As we use more digits, the system gets increasingly efficient. Each additional digit increases our descriptive reach *tenfold*. One digit can tell us the number of people in a typical nuclear family. With two digits, we can name the number of people in an extended family or in a typical school classroom. With three digits, we can specify the number of students in a moderate-sized school. Four digits allow us to name the populations of a large public high school, a moderate-sized college, or a small town. Five digits can number the largest campuses of state universities or large towns and small cities. With six digits, we can count the populations of sizable cities (San Francisco, Cleveland) and some states (Delaware, Wyoming). Seven digits allow us to name the populations of New York City and Paris, most states, and many countries. Using eight digits, we can number the populations of the largest cities (Tokyo, Mexico City, São Paulo), states (California, New York), and most countries (England, Canada, Argentina, Nigeria). Nine digits count the people in large countries like the United States, Japan, or Indonesia—all but China and maybe India. With ten digits, the same number as tallies to count our fingers, we can record the entire human population of the world. With ten digits, we can also write an arbitrary U.S. telephone number. This efficient representation lets us dial anyone in the country in a matter of seconds. (Of course, we will probably get the person's answering machine.)

The principle behind this efficiency is the same principle that makes language work. Start with a small set of basic symbols and create an arbitrarily large vocabulary by making lists formed from the basic symbols—that is, by putting several basic symbols in sequential order. Thus, spoken words are sequences (in time) of basic sounds (called phonemes by linguists). In English and other languages that use an alphabet, the same system governs writing. Each written word is a sequence (on the page) of letters that encode spoken sounds. So, the decimal system is merely the application to numbers of a general technique for presenting information. We can call it the *alphabetic principle* or *digitization*. The decimal system, however, embodies digital representation at its ultimate. It represents every (positive whole) number, with no exceptions and in exactly one way for each—no misses, no repetitions. This contrasts with the situation for language. There are many sequences of letters and many sequences of syllables that do not make words, but rather only nonsense. Additionally, there is more than one way to pronounce or spell some words.

The Sophistication of the Decimal Number System

Very Round Numbers Admirable as it is, the representational efficiency of our decimal notation hardly touches on the sophistication of the system as a way of representing numbers specifically, rather than some other indefinitely large collection, such as words or telephone numbers. In assigning telephone numbers, we are not really using the numbers as numbers—they are just labels. It doesn't make any sense to add, multiply, or round off telephone numbers. We could just as well use any collection of symbols to write telephone numbers. There still is a (redundant) labeling of the telephone buttons also by letters, which these days is exploited by firms with 800 numbers to provide customers with a catchy mnemonic for remembering how to call them—for example, 1-800-CALL-XYZ. Numbers are convenient for labeling things, but they are much more. They have a rich structure—arithmetic operations, ordering, and magnitude—that makes them much more valuable than a simple list. The decimal system is admirably compatible with all these structures. This compatibility comes out of the sophisticated way the decimal system uses algebraic operations and principles for the mere task of writing numbers. The basis of the system is numbers of a very special type—a single digit followed by zeros. We will call such numbers "very round." Thus, 7,000; 30; 600,000; 200; 5; and 800,000,000,000 are very round numbers.

Every positive whole number is a sum of very round numbers—one for each decimal place. Indeed, the digits in a given number tell us how to express it as a sum of very round numbers. The following examples show 742 and 3,805 expressed in this way. When we teach children to express numbers in this way, we call it *expanded form*.

$$742 = 700 + 40 + 2$$
$$3,805 = 3,000 + 800 + 5$$

Notice that in 3,805 the zero in the tens place indicates that no very round number with only one zero is needed. This is the place-value principle in action. We record the absence of the multiple of 10 in order to signal that the 8 is representing 800, not 80. Very round numbers are the purest expression of the place-value principle. They are the elements from which all numbers are formed. Thus the zero, which might seem wasteful since it stands for nothing, is the key to making decimal place-value representation work. We will call the very round numbers used to form a general decimal whole number the *very round components* of the number. Thus, the very round components of 742 are 700, 40, and 2.

Structure of Very Round Numbers The use of addition in writing numbers is supplemented by the use of multiplication in forming the very round numbers. Each very round number is the product of a digit times a power of 10 as shown in the following examples.

$$700 = 7 \times 100$$
$$40 = 4 \times 10$$
$$3,000 = 3 \times 1,000$$

Of course, this is also true for the single digits such as $2 = 2 \times 1$, etc., but we don't usually bother to make this explicit.

Each very round number with digit 1 is a power of 10—the product of 10 multiplied by itself repeatedly. The number of multiplications by 10 is just the number of zeros as shown in the following examples.

$1 = 10^0$

$10 = 10^1$

$100 = 10 \times 10 = 10^2$

$1,000 = 10 \times 10 \times 10 = 10^3$

Polynomials in the Variable "10" If we take into account the multiplicative structure of the very round numbers, in addition to the additive structure of non-round numbers, we see that every decimal number is composed from a very limited collection of basic components. In fact, every decimal number is made by combining just the digits and 10 by means of the basic operations of algebra—addition, multiplication, and exponentiation (raising to powers). Here is how this looks for our examples of 742 and 3,805.

$742 = 700 + 40 + 2$
$= (7 \times 100) + (4 \times 10) + (2 \times 1)$
$= 7 \times 10^2 + 4 \times 10^1 + 2 \times 10^0$

$3,805 = 3,000 + 800 + 5$
$= (3 \times 1,000) + (8 \times 100) + (5 \times 1)$
$= 3 \times 10^3 + 8 \times 10^2 + 5 \times 10^0$

Thus, simply to record numbers, our decimal system implicitly uses all the operations of basic algebra. Furthermore, the expressions we use to write decimal numbers are of a kind familiar from algebra. A quantity that is formed by taking some number, raising it to various powers, and multiplying by some coefficients and summing is referred to as a *polynomial*. Thus, we see that decimal notation is a shorthand device for expressing numbers as "polynomials in 10." This way of describing our familiar decimal notation may seem to be making something simple complicated, but the similarity is not superficial. There are strong parallels between decimal numbers and polynomials, not only in the way of writing them but also in calculations and other properties. The standard procedures for addition, multiplication, and so forth, exploit the structure just elaborated. Furthermore, the "Laws of Algebra" are heavily involved in justifying and comparing algorithms for computation. Understanding and appreciation of the decimal system is greatly enhanced by recognition of this pervasive role of algebra.

Doing Decimal Arithmetic: Addition

by Roger Howe

The article "Marvelous Decimals" (pp. PH1–PH4), describes the algebraic sophistication that our standard decimal notation brings to the writing of numbers. Decimal numbers implicitly treat ordinary whole numbers as "polynomials in 10." What do we get from this algebraic sophistication? We get remarkable power, not only to record but also to manipulate numbers, to perform the usual operations of arithmetic. We get *ease of calculation in the form of efficient, easily implemented algorithms.* Addition and multiplication are essentially combinations of single-digit operations (plus keeping track of decimal places). More exactly, we add or multiply general numbers by appropriate combinations of additions or multiplications of very round numbers, and these essentially amount to calculations with single-digit numbers. Let's see what this means for addition.

Adding Very Round Numbers

Remember that a "very round number" is a number with a single digit followed by zeros: 9; 40; 100; 20,000; 600,000,000, and so forth. The general principle behind addition of decimal numbers is that every addition can be done by an appropriate combination of additions of very round numbers. So, let's examine the sum of two very round numbers.

Let's call the number of digits in a very round number the *length of the number*. Adding very round numbers is interesting only when both numbers have the same length. To add very round numbers of different lengths, all we do is to put the leading digit of each number in its place in the sum. For example, $7{,}000 + 40 = 7{,}040$. This procedure holds for the sums of many very round numbers, as long as they all have different lengths. For example, $3{,}000 + 800 + 5 = 3{,}805$. This is, of course, the basic principle that is used to express a general whole number as a sum of its very round components. Students refer to this expression of a number as *expanded form*. However, when we add two very round numbers of the same length, such as $3{,}000 + 2{,}000 = 5{,}000$; or $40 + 40 = 80$; or $700 + 600 = 1{,}300$, it is not a simple matter of recording digits in appropriate places. Instead, we see that we essentially have to do a single-digit addition and also keep track of the number of zeros in the very round summands. Treating the above examples more formally, we see these processes taking place:

$$3{,}000 + 2{,}000 = (3 \times 1{,}000) + (2 \times 1{,}000) = (3 + 2) \times 1{,}000 = 5 \times 1{,}000 = 5{,}000$$
$$40 + 40 = (4 \times 10) + (4 \times 10) = (4 + 4) \times 10 = 8 \times 10 = 80$$
$$700 + 600 = (7 \times 100) + (6 \times 100) = (7 + 6) \times 100 = 13 \times 100 = 1{,}300$$

The crucial fact we need to know to find the sum is a single-digit addition—an "addition fact." Another way of thinking about factoring out the power of 10 is to treat the power of 10 as a unit. Thus, for $3{,}000 + 2{,}000 = 5{,}000$, we can say, "Three thousands plus two thousands makes five thousands," in analogy with "Three apples plus two apples makes five apples." If the sum of the digits of the numbers we are adding is less than 10, the sum will again be very round and will be the same length as the two original numbers. However, if the sum of the digits is 10 or larger, the sum of the very round numbers will be longer than the original numbers and may not be very round. If we have to do further calculations with it, we should decompose it into a sum of its very round components. This would lead us to write:

$$700 + 600 = 1{,}300 = 1{,}000 + 300$$

We can now continue the computation, using the 1,000 and the 300 individually, as called for. This, of course, is the source of "carrying," or "regrouping."

Adding Any Whole Numbers

To add two general numbers, we decompose each of them into their very round components and add components with the same length. Thus, to compute $26 + 53$, we write:

$$26 + 53 = (20 + 6) + (50 + 3) =$$
$$(20 + 50) + (6 + 3) = 70 + 9 = 79$$

This example is very simple. It has only two-digit addends and requires no carrying or regrouping. Before looking at more complicated examples, the following points should be made. First, although we have written everything on a line, with liberal use of parentheses, what we have done, in essence, is the same as the standard procedure for this addition. Standard procedure tells us to line the numbers up one under the other and add each column:

$$\begin{array}{r} 26 \\ +53 \\ \hline 79 \end{array}$$

We see that the process of alignment and column-wise operation forces us to add the 6 and 3 and the 20 and 50 (represented only by the digits, with the zeros being implicit in the location of the digit in the tens column). Thus, the standard procedure is a way to achieve automatically what we have done explicitly. Although we will not do it in each case, all the illustrations that follow also translate in similar fashion into the standard format.

Second, we should take note of all the rearrangement we have done to get from $(20 + 6) + (50 + 3)$ to $(20 + 50) + (6 + 3)$. These rearrangements are, of course, recognized as being legitimate by anyone experienced with arithmetic. More formally, they are justifiable by means of some of the Laws of Algebra. Specifically, this rewriting used the Commutative and Associative Laws for addition several times. Also, as discussed above, in the addition $20 + 50 = 70$, we are implicitly invoking the Distributive Law, which connects addition and multiplication. The Distributive Law is the law that allows us to factor out the 10 in the process.

$$20 + 50 = (2 \times 10) + (5 \times 10) =$$
$$(2 + 5) \times 10 = 7 \times 10 = 70$$

Since the purpose of this article is not to give a formal treatment of arithmetic, but only to discuss some key ideas, we usually do not mention these Laws of Algebra again. However, they are implicitly involved in virtually all arithmetic calculations, and a full understanding of decimal arithmetic does involve fluency with these Laws.

As previewed on page PH6, sometimes the sum of two digits is 10 or larger, and results in a sum that and has one more digit than the original numbers. This results in "carrying," or "regrouping." We must decompose the result into its very round components and combine the longer one with the original components of that length. Thus,

$$76 + 53 = (70 + 6) + (50 + 3) =$$
$$(70 + 50) + (6 + 3) =$$
$$120 + 9 = 100 + 20 + 9 = 129$$

This is quite a transparent process if the carrying only affects the largest power of 10 as above. If the overflow occurs in a smaller place, the overflow from one sum must be combined with the sum for the next higher place-value position.

$$26 + 57 = (20 + 6) + (50 + 7) =$$
$$(20 + 50) + (6 + 7) = 70 + 13 =$$
$$70 + (10 + 3) = (70 + 10) + 3 =$$
$$80 + 3 = 83$$

In adding multi-digit numbers, the need to carry, or regroup, may affect several places, and there may be a cascade effect, whereby an overflow at one place affects several larger places, as in

$$146 + 57 = (100 + 40 + 6) + (50 + 7) =$$
$$100 + (40 + 50) + (6 + 7) =$$
$$100 + 90 + 13 = 100 + 90 + (10 + 3) =$$
$$100 + (90 + 10) + 3 = 100 + 100 + 3 =$$
$$200 + 3 = 203$$

Despite these complications, the principle is clear. All additions can be done by suitable combinations of additions of very round numbers.

Ease of Calculation

It is also important that this "suitable combination" involves only a fairly small number of additions. To add any two numbers under 1,000 (three-digit numbers), we need at most three basic additions and perhaps three more to allow for carrying, or regrouping (six at most). To add two numbers under 10,000, it is only slightly worse—a maximum of eight one-digit additions. Compare this with the effort of counting the objects in a set made by joining two sets, each with several thousand objects! Just as it eases the labor of writing numbers, decimal notation reduces the effort needed for addition. Furthermore, we should remember that all the complications are inherent already in one-digit addition—$7 + 6 = 13$ is greater than 10, and we can't help that. Our decimal system accommodates this fact of life as gracefully as one could hope.

Extra Efficiency of the Standard Algorithm

The standard format for doing the last addition is:

$$\begin{array}{r} \scriptstyle 1\ 1 \\ 146 \\ +\ \ 57 \\ \hline 203 \end{array}$$

The ones above the 4 and the 1 represent the regroupings. They remind us to convert 10 ones to a ten and 10 tens to a hundred and to then add these newly created larger units to the ones already there.

The standard algorithm is much more compact and involves much less writing than the computation above. The reason for the compactness is twofold.

- First, rather than explicitly performing the space-consuming decomposition of each number into very round numbers, the standard procedure forces this by lining up the numbers according to their place values and prescribing column addition.

- Second, the standard procedure goes from right to left, adding the ones and regrouping a 10, if necessary, *before* adding the tens, with this process being repeated similarly for each place-value position.

This short-circuits the space-consuming recombination and recalculation steps in our first version. **Thus, the standard procedure is designed for compactness and efficiency of exact calculation.**

Flexibility for Mental Math and Estimation

However, the principle that our computation emphasizes—that multi-digit addition is a combination of additions of very round numbers of the same length—is worth understanding. For one thing, it allows us to think flexibly about addition. It shows us that the operations needed to compute the sum may be done in many orders—thanks to the Commutative and Associative Laws. The standard procedure picks one order—the order that keeps rewriting to a minimum. However, other orders are possible and may sometimes be useful. In particular, with mental math, it often seems more natural to start by adding the places farthest to the left, since these are the largest parts of the numbers being added. Thus, to add 146 and 57 mentally, many people would start by adding 140 + 50 to get 190, and then add the 6 + 7 = 13 to the result, getting 190 + 13 = 203. The point is that the 190 is much larger than the 13, so with just that part of the addition completed, you already know "most of" the answer. If you did not need to know the exact answer, but just approximately how large the sum is, you could say that it is "about 190" or "about 200" (since it is clearly more than 190, and 200 is a very round number). By contrast, the standard procedure starts by computing the ones place, which is the smallest part of the answer, and doesn't find the main part of the result until the end of the calculation. The same ideas let us quickly estimate the size of a multi-digit sum, so that we can check an answer on a calculator for reasonableness. This is discussed more thoroughly in "Estimation and Arithmetic" (pp. PH17–PH20).

Doing Decimal Arithmetic: Multiplication

by Roger Howe

In the article "Doing Decimal Arithmetic—Addition" (pp. PH5–PH8), we saw how decimal notation provides a framework for efficient addition. The story with multiplication is similar, but in some ways more remarkable. Everything again depends on combining operations on very round numbers.

Multiplying Very Round Numbers

Recall that very round numbers are numbers consisting of a single digit followed by some zeros. Examples are 2,000 and 4 and 90. They may also be described as a single-digit number times a power of 10. Multiplying very round numbers amounts to multiplying the digits, plus keeping track of the powers of 10. More precisely, to multiply two very round numbers, we multiply their digits, then on the right we append as many zeros as are in both factors together. Here are some examples of multiplications of very round numbers:

$$20 \times 40 = (2 \times 10) \times (4 \times 10) = (2 \times 4) \times (10 \times 10) = 8 \times 100 = 800;$$
$$30 \times 3{,}000 = (3 \times 10) \times (3 \times 1{,}000) = (3 \times 3) \times (10 \times 1{,}000) = 9 \times 10{,}000 = 90{,}000;$$
$$600 \times 7 = (6 \times 100) \times 7 = (6 \times 7) \times 100 = 42 \times 100 = 4{,}200 = 4{,}000 + 200$$

As the last example shows, when the product of the two digits is 10 or more, the product of the very round numbers may not be very round, and in any case will have one more digit than when the product is less than 10. This is not an essential problem, but it is one source of regrouping in the standard procedures for multiplication.

One case that might cause confusion is when one of the digits is 5 and the other is even. When this happens, the product of the digits will be a multiple of 10 and so the whole product will be very round, and it will seem to have an extra zero. But this case is computed in the same way as all the others. The "extra" zero is contributed by the product of the digits. Thus, in the examples below, the extra zero in the second product results because $5 \times 4 = 20$:

$$50 \times 300 = (5 \times 10) \times (3 \times 100) = (5 \times 3) \times (10 \times 100) = 15 \times 1{,}000 = 15{,}000, \text{ but}$$
$$50 \times 400 = (5 \times 10) \times (4 \times 100) = (5 \times 4) \times (10 \times 100) = 20 \times 1{,}000 = 20{,}000$$

Multiplying Any Whole Numbers

Once we know how to multiply very round numbers, the Distributive Law tells us what to do to multiply general multi-digit numbers. To multiply two in the real world—for example, in weather. Future weather can be very sensitive to small changes in current conditions. This sensitivity may put a serious limit on how good our weather predictions can possibly be. Thus, the issue of approximation in general is complex.

We can discuss only a few general principles and simple situations. Two quite different types of accuracy are used frequently in discussing approximation. It is important to be able to distinguish between them and to tell whether one or the other is appropriate. One is *absolute accuracy*, in which errors are compared to a fixed number and amounts smaller than the fixed number are ignored. This is seen often in financial statements, which often give figures stated in units of a thousand dollars. The other kind is *relative accuracy*, in which errors are compared to the number being approximated. Relative accuracy is what is proposed above as the appropriate kind of accuracy to be considered in a discussion of government waste.

In rounding numbers, absolute accuracy refers to the number of digits that we neglect. Relative accuracy refers to the number of digits we retain. The retained digits are called *significant digits*. For example, if we round 4,286,419 to 4,286,000, we have absolute accuracy to the thousands and relative accuracy to four significant digits. The type of accuracy that is the more relevant depends on the context.

We should note a possible ambiguity in rounding. The usual convention in dealing with rounded numbers is that the non-zero digits are accurate (except that the last digit may be off by one) and that the digits represented by zeros could have been more or less anything. However, it might happen that the last significant digit is a zero. Then the number appears to be less accurate than it is. Thus, 4,280,419 rounded to the thousands is 4,280,000, and under usual conventions, this looks as if it is rounded to ten thousands. Sometimes context can tell us that some zeros at the end of a rounded number are in fact significant, but without further information, one assumes that they represent deleted or unknown digits.

Equipped with the language of significant digits, we again ask, "How accurate is accurate enough?" When measuring physical quantities, each additional significant digit is hard won. In many situations, one or two significant digits are enough to be useful. The usefulness of percents is based on the fact that, frequently, when discussing amounts that are part of some whole, anything less than one one-hundredth of the total is small enough to be negligible. Specifying percentages is akin to having two significant figures of accuracy. To have more than four significant digits takes painstaking work, and frequently even four-digit significance is not realistic. Consider again the example of the radius of the Earth. Since this is approximately 4,000 miles, four significant figures would specify it to at least the nearest mile. However, the difference in altitude between the deepest sea trenches and the highest mountains is over 10 miles. In this situation, and many others involving physical quantities, four significant figures of accuracy are simply not attainable.

Estimation and Arithmetic

by Roger Howe

In the article "Estimation and Approximation" (pp. PH13–PH16), **the issues of approximation and, in particular, the technique of rounding are discussed. The present article will discuss how rounding interacts with arithmetic—specifically, with the basic operations of addition and multiplication of positive whole numbers.**

ROUNDING SIMPLIFIES ARITHMETIC

Since all decimal arithmetic is based on single-digit arithmetic, rounding simplifies the mechanics of the arithmetic operations. In the addition of round numbers, the zeros can just be carried along. For example:

$$87 + 37 = 124 \text{ and } 87{,}000 + 37{,}000 = 124{,}000$$

This means that when we round off numbers, it becomes less work to add them. It is easier to add 87,000 and 37,000 than to add 87,266 and 37,495.

Zeros on the end of rounded numbers are also easy to deal with in multiplication. They can just be deleted and then reinserted after the multiplication is done. The total number of added zeros in the product is the sum of the numbers of zeros ending the two factors. For example:

$$200 \times 34 = (2 \times 100) \times 34 = (2 \times 34) \times 100 = 68 \times 100 = 6{,}800$$
$$20 \times 340 = (2 \times 10) \times (34 \times 10) = (2 \times 34) \times (10 \times 10) = 68 \times 100 = 6{,}800$$
$$2 \times 3{,}400 = 2 \times (34 \times 100) = (2 \times 34) \times 100 = 68 \times 100 = 6{,}800$$

So, rounding simplifies multiplication. It is easier to compute 20×340 than to compute 23×345. As with multiplication of very round numbers, zeros created by the product of the non-round parts of the numbers may seem to be "extra zeros." They make it seem that there are more zeros than normal. But they do not change the process. One simply has to keep in mind that the "extra" zeros come from the multiplication of the retained digits. In the multiplication $20 \times 350 = (2 \times 10) \times (35 \times 10) = (2 \times 35) \times (10 \times 10) = 70 \times 100 = 7{,}000$, the final result has three zeros. One zero came from the product 2×35, and the other two zeros, one from the 20 and one from the 350, were appended to the end of the 70.

Accuracy of Rounded Arithmetic

Although rounding simplifies arithmetic, it introduces a new issue. If we add or multiply numbers that are only approximations of actual quantities, we must ask how well the result approximates the actual sum or product. Since a full discussion of this question gets rather involved, only a few basic observations are given here.

Recall the two types of accuracy—absolute and relative—discussed in "Estimation and Approximation." Although both apply to the same rounding process, they are quite different ways of thinking about error. *Absolute error* refers to the number of decimal places ignored in rounding, while *relative error* refers to the number of decimal places that are retained. Suppose we round 49,248 down to 49,200. To describe the absolute accuracy of this, we would say we have rounded (down) to the hundreds place. To describe the relative accuracy, we would say that we have retained three significant digits.

In doing addition, absolute accuracy is the relevant consideration. The main point is that the absolute accuracy of a sum can be no better than the worst absolute accuracy of the addends. For example, consider the sum $4{,}300 + 280 = 4{,}580$. This sum appears to be rounded to the tens place. However, since 4,300 is standing for any number between 4,300 and 4,399 (since we are talking about rounding down), and 280 means a number between 280 and 289, all we know about the true sum is that it is between 4,580 and 4,688. This cannot actually be represented by a single rounded number according to standard conventions, but we frequently fudge it and write it as 4,600. If we decide to represent it by a single decimal number, 4,600 is the best choice. Certainly the third digit of accuracy that is implied by the formal sum is unwarranted.

This principle is the source of humor in the story of a family visiting a natural history museum. After looking at the skeleton of a huge dinosaur, they approached the guard in the room to ask how old it was. He answered, "70 million and 12 years."

They replied, "Oh, that's very old! How do you know?"

He answered, "Well, when I started working here, they told me that it was 70 million years old, and I've been here 12 years, so now it must be 70 million and 12."

In a case in which two numbers of substantially different sizes are being added, the need to round both to the same absolute accuracy can result in seemingly paradoxical situations that involve ignoring fairly large numbers. For example, in adding the rounded numbers $40{,}000 + 2{,}800$, where 40,000 has been rounded to the nearest 10,000, the 2,800 must also be rounded to that place, which means that it will round to 0! It can be very difficult to round 42,800 back to 40,000, completely writing off the 2,800. However, retaining the extra digits in the formally correct sum creates a false impression of accuracy.

For products, the parallel principle is that *relative* accuracy of a product can be no better than the *least* relative accuracy of the factors. In standard arithmetic, $80 \times 74 = 5{,}920$. However, if 80 is standing for some number between 80 and 89, then even if the 74 is completely accurate, the number 5,920 is standing for some number between 80×74 and $89 \times 74 = 6{,}586$. We see from the large range of possibilities in the hundreds and smaller places that it makes no sense to state a result with more than one significant digit. Even the first digit may not be completely certain, but if we want to represent the product by a single number, 6,000 is as good as we can do. The extra digits of implied accuracy in the value 5,920 are quite misleading. When using a calculator, we need to make extra effort to keep in mind that most of the digits in the rapidly produced result of a multiplication may be meaningless.

If nothing further is to be done with a number, retaining meaningless digits may do little harm—it produces some unnecessary mental clutter, but there is no reason to think that the four-digit number 5,920 is a worse approximation of the actual number it represents than 6,000 is. The real problem with retaining meaningless digits is that they may produce a false sense of accuracy, leading us to use them in calculations whose results are meaningless and misleading. This is particularly likely to happen when we do subtraction. Suppose we have two numbers, 5,920 and 6,160, neither of which is accurate to more than one significant digit but for which we have retained three digits. If we now subtract them, we get 240, but since the two numbers involved are, in fact, only accurate to the nearest 1,000, this number is totally meaningless—none of its digits represent any reality. If we now use this number in further computations, whatever we produce will be nonsense. Calculators are quite willing to produce this kind of nonsense at the push of a button. It is up to us to know when to round off.

BALLPARK ESTIMATES USING ROUNDING

We can use the fact that rounding simplifies arithmetic to estimate a sum or a product by doing the same operation on rounded numbers. This gives us the opportunity to check that a calculation done on a calculator is roughly correct. Such checks can guard against gross errors when we punch in the numbers to be operated on. We can simplify the arithmetic as much as we want, depending on how accurate we want our check to be. The fewer digits we retain, the easier our check computation will be. Unfortunately, it will also be less accurate. The simplest thing to check would be when we retain only the largest very round component of a number. As we have noted already in "Doing Decimal Arithmetic: Addition" and "Doing Decimal Arithmetic: Multiplication," the sum or the product of the largest very round components represents the largest single contribution to the sum. However, due to regrouping, this single very round contribution may not determine even the leading digit in the actual sum. This problem is worse for multiplication than for addition. Consider the product for

$$14 \times 65 = (10 + 4) \times (60 + 5)$$
$$= (10 \times 60) + (4 \times 60) + (10 \times 5) + (4 \times 5)$$
$$= 600 + 240 + 50 + 20$$
$$= 910$$

We see that, of the four products of very round components, the largest is $10 \times 60 = 600$. However, the next two terms, $4 \times 60 = 240$ and $10 \times 5 = 50$, raise the amount from 600 to near 900, and the last term, $4 \times 5 = 20$, actually puts the total over 900. Thus, although the leading product gives a substantial chunk of the final product, the other terms contribute enough so that it is a matter of judgment as to whether the 600 is a "good approximation" of the actual product. Such judgments might be challenging to young students just learning multiplication.

One way to eliminate the need for such judgments is to sandwich the actual result between two products of rounded approximations to the original factors. This would create a "ballpark" in which the exact answer should lie. This strategy relies on the following basic properties of addition and multiplication:

- When adding two numbers, if either number increases, the sum also increases.
- When multiplying two positive numbers, if either number increases, the product also increases.

These facts are referred to as *monotonicity* of addition and multiplication, respectively. The monotonicity properties suggest the following strategy for getting "ballparks" in which a sum or a product of two numbers should lie. This strategy involves rounding up as well as rounding down. Rounding up is slightly more complicated than rounding down but is still quite simple. To round down, drop the last (meaning rightmost) digits of a number and replace them with zeros. In rounding up, do the same but also add a 1 to the last non-rounded digit of the number. This will always produce a number larger than the original. Thus, 47,283 rounded down to the thousands is 47,000. Rounded up to the thousands, it is 48,000. Effectively, we have written 47,283 = 47,000 + 283, and replaced 283 by 1,000 to do the rounding up. Rounded down to the hundreds, 47,283 is 47,200; rounded up it is 47,300. Occasionally, rounding up will create a regrouping situation, and one or more digits could turn over, creating zeros that actually represent significant digits. Thus, 49,936 rounded up to the thousands is 49,000 + 1,000 = 50,000. Rounded up to the hundreds, it is 49,900 + 100 = 50,000—the same as rounding up to the thousands. In our application, this will not cause problems.

Here is the "ballpark estimation strategy" for sums or products of two positive whole decimal numbers.

1. Round the numbers down (to any desired accuracy).

2. Round the numbers up (to any desired accuracy).

3. Do the same operation (that is, add or multiply) on the rounded numbers (on the two rounded up numbers and on the two rounded down numbers) as specified for the original numbers.

4. The actual sum or product should lie between the two answers calculated in Step 3 above.

Of course, the simplest arithmetic will result if the rounding leaves only one significant digit. Here are some examples. For the product $14 \times 65 = 910$, the rounded down product is $10 \times 60 = 600$ and the rounded up product is $20 \times 70 = 1,400$. We have $600 < 910 < 1,400$, showing that the product is in the right ballpark. Of course, in this case, 1,400 is more than twice as large as 600, so the ballpark is rather large! But this is to some extent unavoidable. When the leading digit is a 1, the longest very round component says less about the number than when the leading digit is greater than 1.

Here is another example, using larger leading digits. Consider the product $43 \times 826 = 35,518$. Round up and down to one significant digit. Rounding down gives $40 \times 800 = 32,000$. Rounding up gives $50 \times 900 = 45,000$. We do indeed have $32,000 < 35,518 < 45,000$, as we must if we had done the actual multiplication correctly. If we round 826 to two significant digits, the ballpark rounded down is $40 \times 820 = 32,800$ and rounded up is $50 \times 830 = 41,500$. So, we have $32,800 < 35,518 < 41,500$. If we round 826 to one significant digit and leave 43 at two significant digits, the ballpark is $43 \times 800 = 34,400$ and $43 \times 900 = 38,700$. If we retain two significant digits in both factors, the ballpark is $43 \times 820 = 35,260$ and $43 \times 830 = 35,690$. With one-digit accuracy in both factors, the ballpark includes numbers with a leading digit of either 3 or 4. It did not completely determine even the first digit, although it limited the possibilities to 3 or 4. With two-digit accuracy in both factors, we also have two-digit accuracy in the product. Thus, we can narrow the ballpark as much as we want at the price of doing more work. The true answer must always lie in the ballpark.

A "Blind Spot"
IN THE ORDER OF OPERATIONS

by Liping Ma

Liping Ma is a mathematics educator and researcher. She was an elementary school teacher in China before moving to the United States in 1988. She later attended Michigan State University and Stanford University as a doctoral student. In addition to her experiences as a mathematics teacher, she is the mother of two children who attend schools in the United States. Her many experiences with the educational systems of both countries have given her a unique perspective. In the following essay, she shares her concern about a "blind spot" she has noticed in the way many American educators address the order of operations in expressions or equations.

Many teachers in the United States are fond of using the mnemonic phrase "**P**lease **E**xcuse **M**y **D**ear **A**unt **S**ally." This mnemonic is intended to be a helpful reminder of the order in which operations should be addressed in expressions or equations. "Always do **p**arentheses first, **e**xponents second, then **m**ultiplication, then **d**ivision, then **a**ddition, and then **s**ubtraction." However, using this mnemonic can cause confusion and can lead to an incorrect answer. Consider, for example, this problem.

$$5 - 3 + 2 = \blacksquare$$

If we follow the mnemonic, it means that we should add $3 + 2$ first and then subtract that answer from 5. Then our result would be 0, which is incorrect. The correct answer is 4. Do $5 - 3$ first, then add 2. To help clarify the rationale for the priority of the operations, I've provided the following hypothetical discussion between two teachers.

Teacher A: This morning, one of my students asked me why we use "Please Excuse My Dear Aunt Sally" to determine the order of operations. I wasn't really sure of the answer. Is the priority order of operations really so arbitrary?

Teacher B: I don't think it is arbitrary. In fact, I think using that mnemonic can sometimes lead to errors. First, let's think about the four basic operations: addition, subtraction, multiplication, and division. If I asked you to put the four operations into two groups, what would you do?

Professional Handbook **PH21**

Teacher A: I would put addition and subtraction in one group, since I know they are closely related: $2 + 3 = 5$ gives you $5 - 3 = 2$ or $5 - 2 = 3$. I would put multiplication and division in another group. They are also closely related: $2 \times 3 = 6$ gives you $6 \div 3 = 2$ or $6 \div 2 = 3$. Also, children learn addition and subtraction first at about the same time, and then they learn multiplication and division.

Teacher B: There is a name for the relationship of the operations in each group—*inverse operation*. Addition and subtraction are a pair of inverse operations, and multiplication and division are another pair of inverse operations. And the operation of squaring a number, for example $2^2 = 4$, and the operation of finding a square root, for example $\sqrt{4} = 2$ forms yet another pair of inverse operations.

Teacher A: Inverse operations? So we put operations that are inverses of each other in the same group? Interesting!

Teacher B: Have you ever noticed that subtracting a number is the same as adding its negative?[1] For example, $5 - 3 = 5 + (^-3)$. On the other hand, dividing a number is equivalent to multiplying by its reciprocal. For example, $10 \div 2 = 10 \times \frac{1}{2}$.

Teacher A: I see! In this sense, we can say that addition and subtraction are essentially the same operation, and multiplication and division are essentially the same operation. That is why there is no priority between addition and subtraction and between multiplication and division! But I still don't understand why some operations have priority over others.

Teacher B: Well, you just mentioned that addition and subtraction are closely related, and so are multiplication and division. How about addition and multiplication? Do you see any connections between them?

Teacher A: Yes. Multiplication is like adding the same number many times. But I think multiplication is much more powerful than addition. For example, to know how much 7 times 5 is, we do $5 + 5 + 5 + 5 + 5 + 5 + 5$ with addition and 7×5 with multiplication. The latter is much more efficient—not to mention what would happen if we were using big numbers.

Teacher B: Now let me ask you to solve a word problem. Suppose that your school has 45 fourth graders and 32 fifth graders, and each fourth grader has 4 books and each fifth grader has 7 books. How many books do all the fourth and fifth graders have?

[1] This is why in an equation with only addition and subtraction, the position of a number can be changed if the number is given the appropriate sign.

Teacher A: This is a three-step problem. I first do $45 \times 4 = 180$, get the number of fourth graders' books, then do $32 \times 7 = 224$, get the number of fifth graders' books, and then do $180 + 224 = 404$ to get all the books for the two grades.

Teacher B: But can you put the three steps in one equation?

Teacher A: I haven't thought about it.

Teacher B: Putting them together you get $45 \times 4 + 32 \times 7 = 404$ books. Doesn't this look more efficient and concise than the three separate equations?

Teacher A: Now I see. Because multiplication has priority over addition, we are able to deal with two chunks of computation in one equation. Otherwise, we can only deal with individual operations. So, this equation has two layers— one layer deals with multiplication and the other deals with addition.

Teacher B: I like the word "chunk." The problems that mathematics deals with may be much more complicated than the word problem I just gave you. But with the priority system, an equation can have many more layers. So it can deal with chunks, sub-chunks, even several layers of sub-chunks.

Teacher A: Oh, I remember what I learned in high school. In addition to parentheses ( ), there are also brackets [ ] and braces { }. They all have priority over all operations, though there is no fixed order in which they are handled. When parentheses, brackets, and braces are in an equation, there can be as many as six "layers" in the equation.

Teacher B: Yes. Of course, in elementary school our students learn a brief version of this priority system with only three layers.

Teacher A: This short version is: do parentheses first, multiplication *and* division second, and addition *and* subtraction last. I believe that once students learn this version well and feel comfortable with it, they will be ready to face the whole operation system when they learn more advanced mathematics.

Teacher B: I think your brief version makes lots of sense. From this version, young students will learn an important concept of mathematics operations.

Teacher A: Actually, it is from your explanation! Thank you so much!

Models for FRACTIONS

The study of fractions is the deepest and most interesting part of elementary school arithmetic. It builds on all the work that has happened earlier and sets the stage for algebra. A teacher should have several models handy to help students understand the basic concepts. With practice one gets better at seeing which model will help a given student the most in a certain situation.

BY TOM ROBY

The "slices of pie" model (or "pizza" model, though pizzas come in different shapes) is almost a cliché—but for good reason. It's easier to visualize $\frac{1}{3}$ of a circle accurately and distinguish it from $\frac{1}{4}$ of a circle than it is to grasp the same fractions in a bar model. At some level this seems to be hardwired into students' minds, but it also is trained by the clock reading that students begin as soon as they get to school. The "pie" shape also works well for adding fractions since it is easy to visualize and conceptualize pieces of pie being subdivided into smaller pieces to form common denominators.

Another important model for fractions is the number line. It helps students see how the new numbers they are learning relate to familiar ones. It gives a geometric feeling to addition in that adding $\frac{1}{3}$ to $\frac{1}{2}$ can be seen as starting $\frac{1}{3}$ to the right of zero and proceeding another $\frac{1}{2}$ unit to the right. The number line also becomes essential later in the mathematical sequence when working with positive and negative fractions. The slice of pie can still be used, but only if one can convince students to think of "pie demerits," or some way that a slice of pie can count negatively!

For a concrete practical model, it's hard to beat money. Students can draw on their direct experience. It is probably worth asking students to think about why we call the twenty-five cent piece a "quarter." For a student having trouble with some concept, asking a similar question in a monetary context can pay real dividends. For example, if adding fractions is a stumbling block, give students the problem $\frac{1}{2} + \frac{1}{4}$. Ask them to tell you what they would get if they added a half-dollar to a quarter. When they answer "seventy-five cents," respond with "Good, now what's that in quarters?" "How can we see it, thinking only in quarters?" "Can I trade my half-dollar for something equivalent?" The disadvantage of money is that it only models well those fractions corresponding to the kinds of money we have—halves, fourths, tenths, twentieths, and hundredths. Just try modeling $\frac{1}{3} + \frac{1}{7}$ using money! This illustrates why it is critical for students to move beyond this model to more general ones as soon as they are able.

For a hands-on practical model, use strips of paper. Students can draw lines to divide a strip into three equal segments and then shade the first two segments. Now fold the paper along the $\frac{1}{3}$ line so that the shaded portion is folded on top of itself. Fold that in half again. When unfolded, the paper will have fold marks at $\frac{1}{6}$, $\frac{2}{6}$ (or $\frac{1}{3}$), and $\frac{3}{6}$ (or $\frac{1}{2}$). This helps students see the equivalent fractions. An extra folding in half will produce twelfths, and so on.

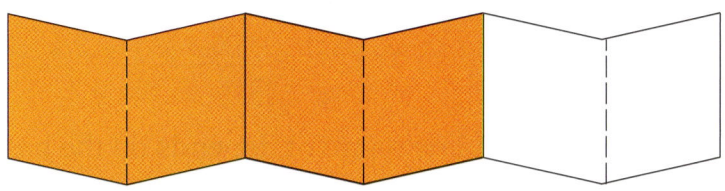

Paper or geometric models are excellent for helping students understand the reason underlying the rule for multiplication of fractions:

$$\text{product} = \frac{\text{product of numerators}}{\text{product of denominators}}$$

For example, to model $\frac{3}{4} \times \frac{2}{3}$, divide a (not too thin) strip of paper into thirds, starting from the left edge. Fold it in half, this time top to bottom, so that the fold runs through the middle of the strip. One more top-to-bottom fold will place fold marks at heights $\frac{1}{4}$, $\frac{1}{2}$, and $\frac{3}{4}$. Now unfold the paper and shade the small rectangles that fall within the lower three horizontal strips and the leftmost two vertical strips. You should get something like this:

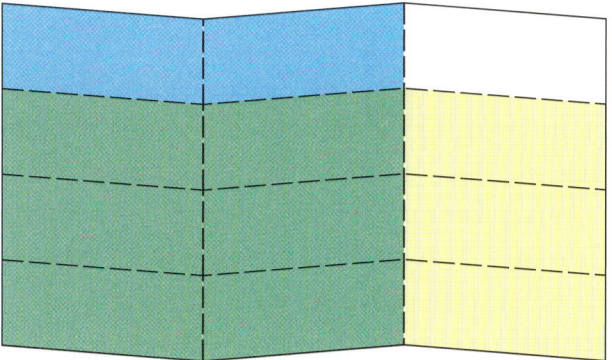

In this example, $\frac{3}{4}$ represents the fraction along the left side, and $\frac{2}{3}$ represents the fraction running along the bottom. So, the product of the numerators represents the number of rectangles that get shaded, and the product of the denominators represents the total number of small rectangles.

$$\frac{3}{4} \times \frac{2}{3} = ?$$

$$\frac{3 \times 2 \text{ number of shaded rectangles}}{4 \times 3 \text{ total number of small rectangles}}$$

We divide the number of shaded rectangles by the total number of rectangles to get the product.

$$\frac{3 \times 2}{4 \times 3} = \frac{6}{12}, \text{ or } \frac{1}{2}$$

Actual paper isn't necessary to show this model to students—a carefully drawn picture will do. This is faster when reviewing the concept, but the paper folding drives the concept home when students are learning it for the first time. It's important that students work with strips of paper on their own to model a couple of multiplication problems after they have seen it presented. Then they will never have any trouble with the rule for multiplying fractions.

Visual models are helpful even for whole numbers and their operations. Because fractions are harder to conceptualize, few students can be successful in using them without the benefits of visual models of several types.

GEOMETRY and Measurement

by David G. Wright

Geometry is visual and beautiful. It is a strand in mathematics in which precise definitions give meaning to shapes, measurements, and even addition and multiplication of numbers. In this article we look at understanding measurement of length, area, and volume and developing spatial sense.

Linear Measurement—Measuring One Dimension

The number of units needed to measure a line segment depends on the basic unit chosen. The basic unit can be an inch, a foot, a yard, a centimeter, a meter, or a kilometer. It can even be the length of a paper clip. Whatever the basic unit of length, students should know that a length of 3 means that when 3 of the basic units are abutted, the total length is the length of the object being measured. The 3 units cover the whole length, but they do not overlap. As simple as this idea is, it must be completely understood by students before any other discussion of measurement of length, area, or volume takes place.

What do we mean when we say that an object has a length of $3\frac{2}{7}$? If one of our unit lengths is divided into seven equal pieces, the length of any one of them is $\frac{1}{7}$. The length of two of them abutted has a length of $\frac{2}{7}$. Thus, a length of $3\frac{2}{7}$ means that if three unit lengths and two $\frac{1}{7}$ lengths are abutted, the resulting length is the same as that of the object being measured. So, we see that linear measurement gives a geometric understanding to fractions. It also gives a geometric meaning to the addition of fractions. The sum of $\frac{4}{3}$ and $\frac{7}{2}$ is simply the length of two line segments of lengths $\frac{4}{3}$ and $\frac{7}{2}$ when they are abutted. Linear measure is also used to measure the perimeter of a polygon and the circumference of a circle.

Area Measurement—Measuring Two Dimensions

Now let us consider measurement of area—a measurement of two dimensions. The unit of measurement is now a square with a side length of 1. We call such a square a *unit square*. Thus, measuring the area of a shape is the same as saying how many unit squares are needed to exactly fill the shape so that the squares abut but do not overlap. This is best understood in the case of a rectangle in which the lengths of the sides are counting numbers. For instance, a rectangle with sides of lengths 2 and 3 can be covered with no overlaps by two rows of 3 copies of the unit square with a side length of 1. Thus, the area of this rectangle is 2×3, or 6 square units.

In order to teach geometry to young children, teachers must develop competence in the following areas:

- **Developing spatial sense, including an understanding of one, two, and three dimensions,**
- **Understanding basic shapes and their properties,**
- **Communicating geometric ideas,**
- **Understanding length, area, and volume.**

CBMS Mathematical Education of Teachers Project, Draft Report March, 2000

In general, the area of a rectangle is the product of the lengths of the sides, or the dimensions of length and width. This works even for fractional lengths and gives meaning to the multiplication of fractions. For instance, consider a rectangle with sides of lengths $\frac{2}{7}$ and $\frac{3}{5}$. What could the area of this rectangle be? The first thing that should be noticed is that it must be less than 1 square unit because it fits inside a unit square. So, the answer should be a fraction of a unit square. But what fraction is it? Consider a unit square and divide the vertical side into 7 equal pieces (the denominator of one dimension—$\frac{2}{7}$) and the horizontal side into 5 equal pieces (the denominator of the second dimension—$\frac{3}{5}$). Using the divisions on the vertical side, slice the square into 7 equal rectangles. Now, using the divisions on the vertical side, slice each of the 7 equal rectangles into 5 equal pieces. Thus, we have $5 \times 7 = 35$ equal rectangles that do not overlap filling the unit square. So, the area of any one of these rectangles must be $\frac{1}{35}$. Furthermore, a rectangle with sides of lengths $\frac{2}{7}$ and $\frac{3}{5}$ is filled up with 6 of these rectangles (the product 2×3), each of which has an area of $\frac{1}{35}$. So, we can see that the area of the rectangle is $\frac{6}{35}$, or the product $\frac{2}{7} \times \frac{3}{5}$. This example gives a geometric meaning to the product of two fractions.

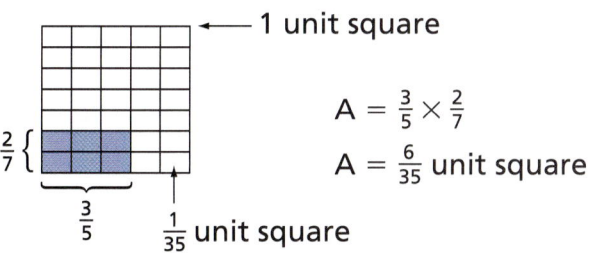

Area is also used to measure triangles. A triangle cannot be filled up by unit squares without some cutting. A simpler way to get the area of a right triangle is to find the area of two copies. These two copies can fit together to form a rectangle. The area of each right triangle is half the area of the rectangle.

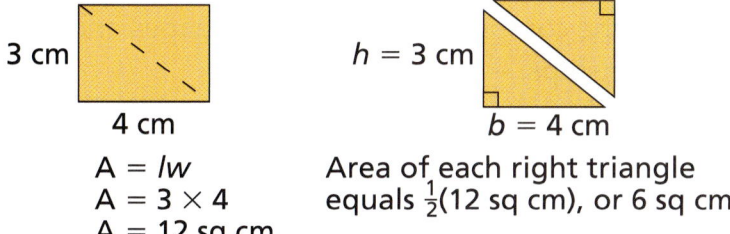

In general, knowing how to find the area of right triangles enables you to find the area of any triangle by looking at the sum or difference of areas of right triangles. The area of the following triangle BDF placed inside a rectangular grid can be found by computing the area of the rectangle and then subtracting the area of the right triangles.

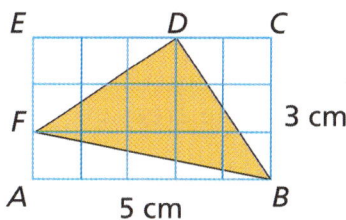

Area of $\triangle BDF$ equals area of rectangle $ABCE$ minus the sum of the areas of $\triangle ABF$, $\triangle BCD$, and $\triangle DEF$.

Area also measures other plane shapes, like parallelograms, trapezoids, and other polygons. These areas can be computed by breaking the shape, or an even larger shape, into triangles or rectangles that do not overlap and then computing the area of each piece.

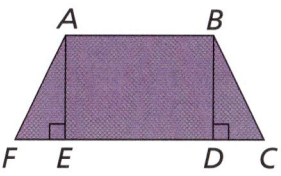

Area of trapezoid $ABCF$ equals the sum of the areas of $\triangle AEF$, $\triangle BDC$, and rectangle $ABDE$.

We also find area of a circle (technically inside the circle). The area of a circle cannot be computed as above. A precise description about how area is computed requires the ideas of calculus, but the following picture shows that a circle can be cut to almost form a rectangle whose height is the radius and whose base is half the circumference. The area of this "rectangle" is the area of the circle. Since the circumference is $2 \times \pi \times$ radius, we can see that the formula for the area of the circle is πr^2.

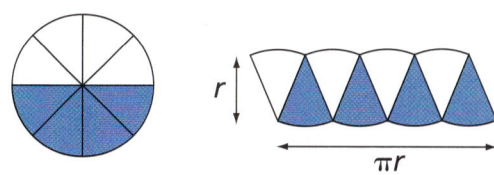

Area also measures the surface of a solid. This is most easily done for a solid like a square pyramid, where the faces are triangles and the base is a square. Finding the surface area of a solid requires finding the area of each of the faces of the solid and finding the sum of the areas.

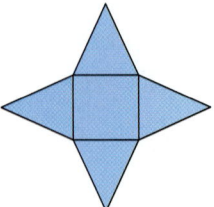 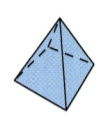

Surface area of a square pyramid equals area of square base plus the sum of the areas of the four triangular faces.

Volume Measurement— Measuring Three Dimensions

The volume of a solid is computed by finding how many unit cubes (where each edge is length 1) are needed to fill the solid with no gaps or overlaps. This is most easily done for a rectangular prism (box) where each of the edges has a natural number for a length. For example, a box whose edge lengths are 2, 3, and 4 has two layers, each of which contains $3 \times 4 = 12$ cubes. So the total volume is $2 \times 3 \times 4 = 24$ cubic units.

We measure volume for other solid objects, such as cones, pyramids, and spheres (filled in). If the previous examples are well understood, the idea of volume should not be too difficult. However, except for the rectangular prism, the volume formulas are a bit harder to explain.

Developing Spatial Sense— Understanding One, Two, and Three Dimensions

Spatial sense is "the ability to form a mental image of an object or set of objects, to recognize the structure of such objects, and to decompose those objects into component parts and recombine them in correct relation to one another" (CBMS Mathematical Education of Teachers Project, Draft Report, March 2000). Much of what went on in our discussion of understanding length, area, and volume also required the use of spatial sense. Here is an excellent problem that will help develop spatial sense.

> A number of white cubes with an edge length of 1 are fitted together to build a cube with an edge length of 3. The outside of this cube is painted red. How many unit cubes have been painted on
> at least one face?
> exactly one face?
> exactly two faces?
> exactly three faces?

Repeat the problem for cubes with edge lengths of 4, 5, or some other number greater than 3.

There are various strategies for solving the first question. One is to carefully count the painted unit cubes in some order so as to find the total and not have any repetitions. A simpler strategy is to notice that the volume of the cube is 27. When the painted cubes are taken away, there is only one cube remaining. So, the number of painted cubes is 26.

Basic two- and three-dimensional shapes have mathematical definitions. A rectangle, for instance, is a quadrilateral (four-sided polygon) with right angles. From this definition, it follows that a square is also a rectangle. It is important for students to know that a square is a special kind of rectangle. A square is a rectangle in which all the sides have equal length. A rectangle may or may not be a square depending on whether the sides have equal length. A triangle is a polygon with 3 sides. Since the definition says nothing about equal sides, students need to be exposed to a variety of triangles so that they do not get the idea that all triangles must be equilateral. Through such experience students can learn to communicate geometric ideas and to form mental images of shapes so that they understand the component parts of the shapes and their relationships.

CONCEPTUAL UNDERSTANDING
The Power of Models and Visuals

by Evan Maletsky

Conceptual understanding is the foundation upon which mathematical thinking is built. For students to be good critical thinkers and effective problem solvers in mathematics, they need both understanding and skill in dealing with the concepts they have been taught. How do we build a solid understanding of mathematical concepts in our classrooms?

Models—Powerful Tools for Building Understanding

Many abstract concepts have their beginnings in concrete, hands-on experiences. This is especially true when it comes to the learning process that takes place in the mathematics classroom. Models offer powerful tools for developing solid understanding of mathematical concepts.

Sometimes, the best models for the classroom are made from the simplest of things. A 2-in. × 8-in. strip of paper may not look like much to work with at first glance. But it may, in fact, be just the needed visual model to give meaning to an otherwise abstract concept.

Have students fold the strip in half and in half again, open it up, and look at the unfolded strip. What mathematical ideas would you want your students to see when they look at this unfolded strip?

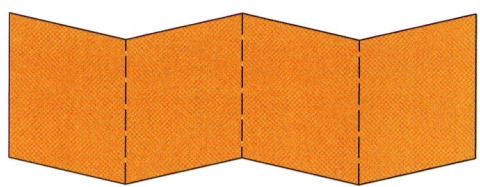

Compare the areas of the newly formed squares with that of the original rectangle. Each of the 2-in. squares has $\frac{1}{4}$ the area of the original rectangle, but it does not have $\frac{1}{4}$ of the perimeter. Many students confuse area with perimeter, and many think that the terms mean the same thing. Understanding the one-dimensional property of perimeter and comparing it with the two-dimensional property of area is fundamental to an understanding of the concepts of both perimeter and area. This model can serve to reinforce these key concepts.

Of course, there is much more that students need to see in this folded strip of paper. For example, everyone sees the 4 squares, but not everyone sees the other 6 rectangles that are there. To help students see all 10 rectangles, label the four square parts, in order, as A, B, C, and D. Then use the letters to name the different rectangles.

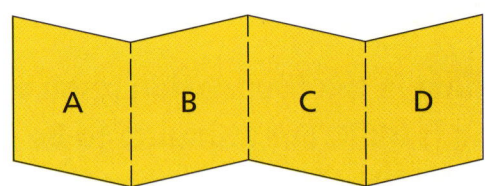

Rectangles: **ABCD, ABC, BCD**
AB, BC, CD, A, B, C, and D

Identify 1 rectangle made from all four squares, 2 rectangles from three squares, 3 rectangles from two squares, and 4 rectangles from one square. One quickly sees the sum of the number of rectangles, $1 + 2 + 3 + 4 = 10$. If the strip were folded into five parts, there would be $1 + 2 + 3 + 4 + 5 = 15$ rectangles. As shown in the table, the number of rectangles shows a nice pattern involving the triangular numbers.

Number of creases	0	1	2	3	4	5	6
Number of parts	1	2	3	4	5	6	7
Number of rectangles	1	3	6	10	15	21	28

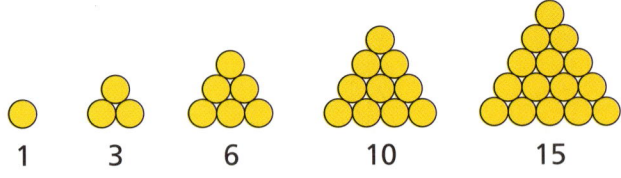

In general, fold the strip into n connected parts and use the formula $\frac{n(n+1)}{2}$ to find the total number of rectangles. This number is a triangular number.

This same two-dimensional flat strip can easily be folded into models that are three-dimensional when studying the concept of volume. Use 1-inch wooden cubes as units of volume. Fill the models of a 2-in. cube and a 1-in. × 3-in. × 2-in. rectangular prism. Use these volumes to estimate that of a triangular prism.

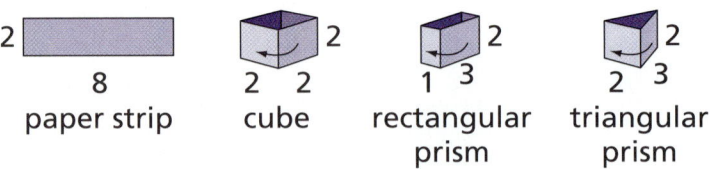

This activity sets in motion the development of the relationship of linear dimensions to volume. With different folding patterns, prisms of other shapes can easily be constructed and explored by your students. This same strip of paper, folded in different ways, creates models of prisms with different volumes. Fold it into eight equal 1-in. parts to form a prism with a regular octagon as a base. Its volume will be greater than that of the cube. For the maximum volume, don't fold it at all. Just curl it around to form a cylinder.

Models help bring in the dynamics of geometry. Only by seeing the action and change that comes from forming these different solids will students really begin to understand fully the concept of volume.

Visuals—A Way of Seeing Mathematics

Many arithmetic and algebraic concepts are abstractions in the eyes of the students because they don't see any reality in these concepts. Here is where models can offer visual support to numerical ideas.

Consider for a moment the concept of percent. Many students struggle with percent. They've been taught the computational algorithms, but what do they see? Even many of those who get the correct numerical results to percent problems have little if anything to say about what those numbers mean.

Think again about our folded strip. If the whole strip represents 100 percent, then each of the four squares is a visual model for 25 percent. The squares can be combined with the eye to show 50 percent and 75 percent as well.

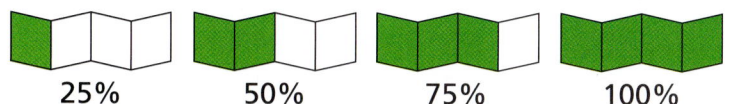

25% 50% 75% 100%

Of course, if a small square represents the whole, or 100 percent, then, from the very same model, we can see 100 percent, 200 percent, 300 percent, and 400 percent. But then, we can assign any value at all to the squares or to the strip. Suppose we call the whole strip the number 12. Then, through areas, we quickly see the numbers 3, 6, and 9 as well.

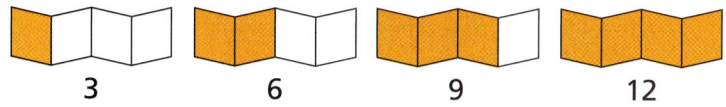

3 6 9 12

Maybe your students see even more. The same number of squares are shaded to show both 9 and 75 percent. Does this mean that 75 percent of 12 is 9? Clearly, it does.

On one of the past National Assessment of Educational Progress tests, eighth graders were asked to find 75 percent of 12. Sadly, less than half answered correctly. One wonders what it was that they saw, if anything, in trying to do this problem.

Our students need to develop good number sense. That comes from more than just practice in computation. It requires a vast and varied set of experiences that include concrete models and visual images of number relationships. This is especially important when it comes to estimation.

As an example, consider again our paper strip. Have each student fold it at some random point of their choice. Then assign some number to the area of the original strips and have the students estimate the corresponding number for the areas of the two parts. Watch how they work. Note if they do any computation. Ask them to write down the process they use. See if they use rounding or compatible numbers or just wild guesses. The results may well be very revealing.

If the original rectangle were assigned an area of 492, how would you go about estimating the areas of the two parts shown below?

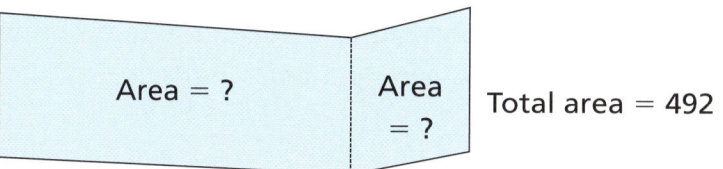

Concrete to Pictorial to Abstract

Models and visuals play a critical part in the concrete and pictorial stages of learning. They are essential to building the conceptual understanding that is needed, ultimately, to do abstract thinking. In *Harcourt Math* the approach to building mathematical concepts is to provide learning experiences based on reasoning before presenting practice. Throughout the years, mathematics has been the model for logic and reasoning. The foundation for success in this arena is built on the understanding of the basic concepts developed in the elementary school years.

Think again about our folded strip. This time, number the squares 1, 2, 3, and 4 on one side and 5, 6, 7, and 8 on the other side, with the 8 behind the 1. Now tear the strip apart into four squares, numbered front and back. Think of the numbers as digits, and ask some arrangement questions to reinforce the concept of place value.

- How many two-digit numbers can be formed with the digit 2 in the tens place?

- Do you see why the number 27 cannot be formed?

- What are the 24 different three-digit numbers possible with 2 in the tens place?
- What are the 48 different four-digit numbers possible with 2 in the tens place?

Adapt the example to any grade level you desire. Simplify the problem by using only the digits 1, 2, 3, and 4 on one side. At a more challenging level, make the problem a cooperative learning activity in which students use the digits 1 through 8 and count all possible arrangements with 1, 2, 3, or 4 digits.

one-	two-	three-	four-digit choices
$8 +$	$8 \times 6 +$	$8 \times 6 \times 4 +$	$8 \times 6 \times 4 \times 2$

$= 8 + 48 + 192 + 384 = 632$

Moving from a simple modeling of two-digit numbers and the place-value concept at the concrete stage, this activity can quickly take us into the visual and abstract stages of mathematical reasoning.

At every level, a mathematics program must offer a rich blend of concept development, skill-oriented activities, and problem-solving opportunities. *Harcourt Math* captures these ideas in a new and refreshing way. It is the best way to reach and teach our students mathematics.

Mathematics and the mathematical experience must tickle the senses as well as sharpen and stretch the mind. It is through handling, seeing, and thinking experiences that students can sense the excitement, appreciate the beauty, and share in the creativity of the subject.

References

National Council of Teachers of Mathematics (1989). *Curriculum and Evaluation Standards for School Mathematics.* Reston, Virginia.

National Council of Teachers of Mathematics (2000). *Principles and Standards for School Mathematics.* Reston, Virginia.

National Council of Teachers of Mathematics (1991). *Professional Standards for Teaching Mathematics.* Reston, Virginia.

Sobel, M., and E. Maletsky (1999). *Teaching Mathematics: A Sourcebook of Aids, Activities, and Strategies.* Needham Heights, Massachusetts: Allyn and Bacon.

Effective Practice: Memorizing the Number Facts

by Grace M. Burton

Almost everyone agrees that children need to memorize the number facts. Indeed, when teachers and parents talk, one of the most frequently asked questions about school mathematics is, "How can I get the children to stop counting on their fingers?" After nearly a century of educational research, some strategies for assuring memorization have been developed. You may wish to consider them as you plan your mathematics program. To encourage the memorization of number facts:

- **Build understanding first.** All of us memorize more easily when we understand what is to be memorized. This necessary understanding can be developed when children have many chances to model the facts with real objects and to illustrate the meaning of the facts with pictures.

$3 \times 5 = \underline{15}$

- **Make clear that the goal is memorization.** When children are told to "learn their facts," it may not be obvious to them that adults mean they should immediately produce a sum or product when given two numbers. While there is a time in early instruction when "figuring out" is important, direct recall is the final goal, and children should be aware of this.

- **Have children thoroughly explore the Commutative and the Identity Properties for addition and multiplication.** When children truly believe that 3×6 and 6×3 (or $3 + 6$ and $6 + 3$) have the same answer, their memorization task is cut in half. When they accept that the sum of any number plus 0 is the number they started with (and that any number times 1 is the number they started with), there are 19 fewer addition or multiplication facts to be memorized.

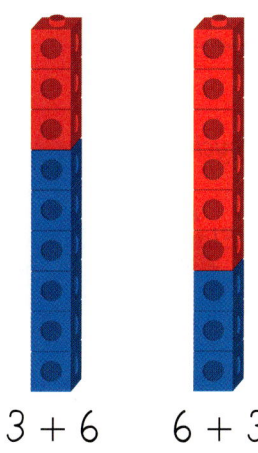

$3 + 6 \quad 6 + 3$

$3 \times 6 = \underline{18}$

$6 \times 3 = \underline{18}$

Professional Handbook **PH33**

Test results show (Kieran) that students are much more successful solving equations when they have had experiences with number models with missing numbers in various formats, such as

___ = 8 + 2   4 + ___ = 10   ___ + 5 = 10 + 2

3 + 10 = ___ + 7   ■ + 5 = 9 + 2

In *Harcourt Math*, the development of algebraic thinking is a focus at every grade level. Algebra is a way of thinking, and the goal of the program is to develop children's abilities to think algebraically about the concepts and skills of mathematics and to apply algebraic thinking to solving problems of all types. Helping students understand mathematics from a perspective of identifying and describing patterns in both number and shape, making and testing generalizations, understanding the concept of equivalence as they memorize basic facts, developing proficiency with the algorithms, relating geometric concepts to algebraic concepts, and exploring the idea of function are some of the many ideas that are threads throughout the program.

The algebraic thinking introduced in the program beginning in kindergarten prepares the way for later formal instruction in algebra. By providing opportunities for our students to build confidence and competence, we help students see that algebra is something they can understand—not a hurdle to clear, but just another step along the way to mathematical literacy.

References

Carroll, Bill. *University of Chicago School Mathematics Project.*

Kenney, Patricia, and Silver, Ed. Eds. (1997) *Results from the 6th Mathematics Assessment of the National Assessment of Educational Progress.* Reston, Va.: NCTM.

Kieran, C. (1992) "The learning and teaching of school algebra." *Handbook of Research on Mathematics Teaching and Learning.* New York: Macmillan.

Steen, L. A. (1992) "Does everybody need to study algebra?" *Basic Education 37.*

———— (1988) "The science of patterns." *Science* 240.

Problem Solving:
The Reason to Teach Mathematics

by Jan Scheer

Gerry and I have been friends since before his son, Jeremy, started school. Jeremy always had problems in math, and Gerry frequently called to ask for my advice, which was usually ignored. (You're never a hero in your own family or with your closest friends.) The trouble persisted, and Jeremy was sent to an "after-school program" where memorization was the primary learning approach.

Jeremy was at my home recently. Knowing that his class of 30 students would be taking a big bus trip, I inquired as to the number of buses that would be needed if a bus holds 20 students. Without taking a breath, Jeremy responded by saying, "That's easy! One and a half." While I stood dumbfounded, his dad marveled at how much Jeremy had learned at his "after-school program." I composed myself and asked Jeremy if he would take the half of the bus with the driver or the half without! Jeremy informed me that I was confusing him and left in a huff.

This incident really disturbed me. Although Jeremy could, indeed, divide 30 by 20 and get an answer of $1\frac{1}{2}$, he had no understanding of the problem. Understanding and application would indicate that in this case $\frac{30}{20}$ should proffer an answer of 2, not $1\frac{1}{2}$.

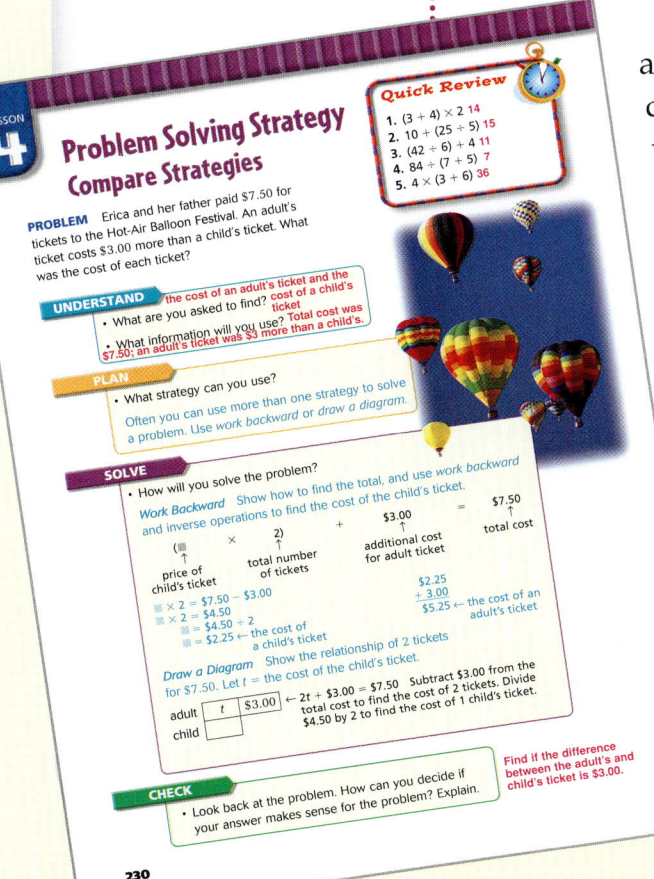

In real life, when do we solve problems? I have never observed anyone in any place take out a pencil and paper to solve a long division problem. I have never seen anyone in a supermarket whip out base-ten blocks to figure which can of soup is the better buy. Nor have I seen anyone use fraction circles to determine what stock to purchase. But every day I see people who are using their mathematical understanding, experiences, and skills to deal with problem situations. They **understand** what the problem is; they **make a plan;** they **solve** the problem; and, finally, they **check** to be sure that their answer is reasonable. If Jeremy had followed this plan, he would have known that his answer was not reasonable.

In *Harcourt Math,* problem solving is part of every lesson. Many lessons present a problem as the vehicle for teaching a skill so that children can see the connection between the skill they are learning and its application. There are lessons devoted specifically to teaching children to use a problem solving strategy and helping them understand

Professional Handbook

the types of problems for which that strategy is a good choice. Children practice problem solving in every lesson—problems that require skills taught earlier in their mathematics experiences, that require two or more steps to solve, that require reading and analyzing data, that require logical reasoning, and that apply the skill presently being taught.

The program helps children develop their abilities to think through a problem by focusing on questions that guide their understanding of the problem, their development of a plan as to which strategy to try or what approach to take, their solution of the problem, and their reflection on whether or not their answer is reasonable in the context of the problem. Reminders to use this process—**Understand, Plan, Solve,** and **Check**—are in every chapter. A Problem Solving Think Along provides a recording device for students as they address each of the above processes. This encourages the "stop and think" approach to solving problems rather than just random attempts. The teaching support that is part of *Harcourt Math* includes a Problem of the Day for every lesson—a rich problem that requires application of this thinking process. In addition, exercises in the practice sets for lessons require students to write their own word problems; write a question from given information; and explain, prove, and/or justify their solutions.

Harcourt Math has a strong focus on problem solving so that as children are becoming proficient with basic facts and the algorithmic procedures, they are seeing the usefulness of mathematics in solving problems that relate to their everyday experience. Children need to be taught that sometimes there can be more than one right answer; often, there can be more than one way to get the answer. They need to be able to use a variety of problem solving strategies. In short, we always need to be teaching children to **think.** *Problem solving is not one reason to teach mathematics; it is the only reason to teach mathematics.*

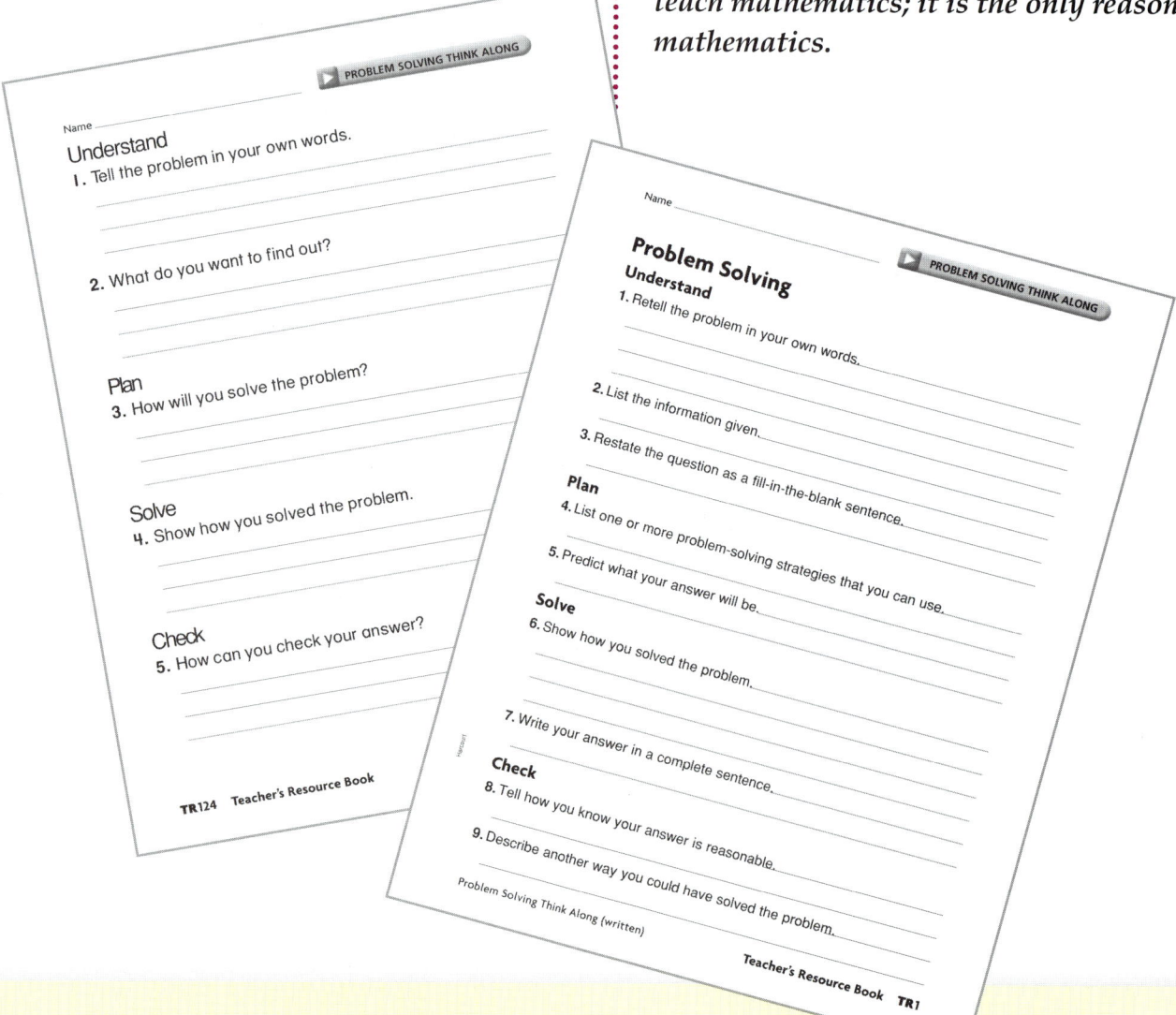

Instructional STRATEGIES:
BEST PRACTICES DEFINED BY RESEARCH

by Joyce McLeod

Instructional strategies used in the classroom usually mirror what teachers think about how children learn. So, choosing instructional strategies from among the best practices defined by research is based upon assumptions about learning and should reflect what is known about learning. The following research-based statements answer the question "What is learning?"

- Learning is goal–oriented.
- Learning is linking new information to prior knowledge.
- Learning is organizing information.
- Learning is acquiring a repertoire of cognitive and metacognitive structures.
- Learning occurs in phases yet is nonlinear.
- Learning is influenced by development (Jones, et al., 1987).

Best practices reflect these statements about learning. As we have learned more about how the brain functions as a learning organ and how learning modalities play a critical role in learning, our repertoire of best practices has been refined to reflect these understandings about learners.

Choosing appropriate instructional strategies in mathematics is a critical factor in ensuring that students make continuous progress. The hierarchical nature of the mathematics curriculum makes it imperative that teachers diagnose and intervene quickly to prevent severe deficits in children's mathematical development. In order to provide a balanced program, best practices for developing conceptual understanding, skill and procedural fluency, and reasoning and problem solving abilities must be aligned with the content. In addition, strategies that focus on effective learning modalities and memory strategies to ensure automaticity with the basic facts must be a part of the teacher's repertoire. Among the best practices for teaching mathematics, the following stand out:

- practice
- explicit instruction
- questioning strategies
- use of visuals
- reading and vocabulary development strategies
- use of manipulatives
- intervention

Let's look briefly at each of these best practices and how they are implemented in *Harcourt Math*.

Practice

Practice is important for reinforcing students' knowledge and for preparing students to move on to new topics and new types of problems. Review helps students retain knowledge and improve performance. In a review of research, Geary (1994) concludes that practice is essential for mastering skills and developing the automaticity that will allow these skills to be used routinely in other situations. Bahrick and Hall (1991) studied the life span retention of content acquired in math courses. Their study found that talent and achievement had some impact, but the primary variable for success was practice that had occurred over time. Individuals who had learned the math content over a short period without practice over time showed declines in performance.

Daily Routine

In *Harcourt Math*, practice and review are designed to reflect these research findings. In Grades 1 and 2, the Daily Routine in each lesson helps children review prior-taught skills, and spaced reviews throughout the student text provide ongoing review of critical skills. In Grades 3 through 6, every lesson begins with a Quick Review and ends with a Mixed Review and Test Prep. These reviews of lesson prerequisite skills and of prior-taught skills cover a wide range of key mathematical topics and represent a mixture of procedures and skills.

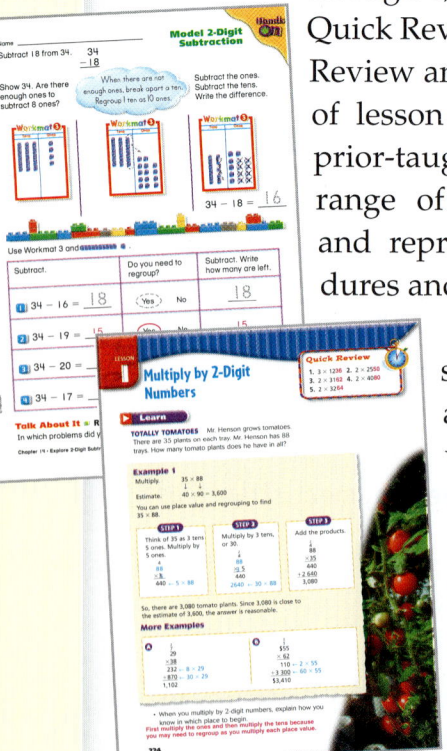

Models that show the step-by-step application of algorithmic procedures and worked examples showing the variety of possible types of problems within a procedure are included throughout the program. These models are often accompanied by pictures of manipulatives to allow students to deepen their conceptual understanding of a procedure as they apply the steps in the procedure. At the end of a group of chapters that cover a major mathematical topic, a Study Guide and Review shows worked examples of the types of problems in which students should develop automaticity, with page references to help students review if necessary.

In Grades 3 through 6, practice exercises for every lesson include a variety of problems. Basic drill-type items are included in lessons emphasizing basic fact acquisition and procedural fluency. Thought-provoking word problems that apply the procedure, skill, or concept taught in the lesson and that review prior-taught skills are included in every lesson. The emphasis in many of these problems is using logical reasoning; solving multistep problems; and explaining, justifying, and proving solutions. Exercises that require writing questions given a possible solution, writing problems with given conditions, and analyzing errors develop students' conceptual and procedural knowledge of mathematics.

Explicit Instruction

Explicit instruction is one of many terms that describe a teaching practice focused on making specific skills or strategies known to a learner. Explicit instruction involves explaining, demonstrating, and/or modeling mathematical concepts and procedures to students.

Liping Ma, in her book *Knowing and Teaching Elementary Mathematics*, emphasizes that in order for a teacher to instruct in mathematics, he or she must have these four properties of understanding:

- **Basic ideas** are the "simple but powerful basic concepts and principles of mathematics" that should be revisited and reinforced.

PH44 Professional Handbook

- **Connectedness** is the understanding of the connections between mathematical ideas that prevents students' learning from being fragmented and helps them see that math is a unified body of knowledge.
- **Multiple perspectives** are the different facets of an idea, the various approaches to the solution of a problem, and the explanations that students make of the different facets.
- **Longitudinal coherence** is the understanding of the whole mathematics curriculum that helps teachers understand what students have studied previously and what they are going to learn later (Ma, p. 122).

In *Harcourt Math*, explicit instruction is facilitated by clear development in every lesson. In Grades 3–6, the instructional part of the student's lesson consists of vocabulary development, models and/or examples, and questions that help students connect new learning to previous learning, consider different approaches to thinking about a basic idea in mathematics, or choose different solution methods for a given problem. In Grades 1–2, the students' lessons show models and reinforce vocabulary. In the *Teacher's Edition* for all grade levels, explicit instruction is developed through Guided Instruction, which includes questions that help students connect the lesson topic to previously learned material, facilitate conceptual understanding and efficient skill development, and help students avoid common errors.

Questioning Strategies

All learning begins with questions. Questions cause interactions, and the quality of those interactions is determined by the character of the question. Questions are fundamental to teaching because they provide information, help students become more actively involved in a lesson, and guide students toward the highest levels of learning in which they apply what they have learned in a variety of ways.

Good questions focus students' attention on concepts, generalizations, laws, and principles and help them think critically and see relationships. In any one lesson, four or five good, open-ended questions challenge students to analyze, apply, react to, or reflect on content. Providing students time to answer (usually about a three- to four-second lapse following a question) results in more comprehensive, higher-quality answers (Rowe 1974).

In *Harcourt Math*, questions that guide students' thinking and help them analyze concepts and skills are included in the Pupil Edition. These questions are designed to help students develop strategies for solving problems and for memorizing basic facts, procedural processes, and key mathematical definitions. We know that memorizing something that is not understood or that has little meaning to an individual is virtually an impossible task. Think, for example, about your Social Security number. Memorizing and retaining that sequence of numbers in your long-term memory if it had no relevance to you would be impossible. You would have to memorize that sequence again and again if you were asked to repeat it at random or to use it in some isolated example. It would only become a part of your long-term memory and easily accessed if it had relevance for you and was connected to other experiences in your memory. So, good questions form the basis for helping students make connections and store information in their long-term memories so that they can easily access it when needed. The basic facts form a large part of the memories that we need to make permanent in students' long-term memories, and since memorizing what we do not understand is almost impossible, questioning for understanding forms the basis for memory.

Use of Visuals

Research clearly shows that the use of visuals enhances learning. Visuals help learners:

- isolate and identify important material.

Professional Handbook PH45

- recall prior knowledge.
- provide interaction with content.
- enhance information acquisition (Dwyer, 1994).

Picture viewing is more exploratory than reading. Fixation durations are generally longer in picture viewing. Visuals serve as aids to memory because information received from visuals appears to remain longer in memory. Visuals also provide motivation when they give a real-life meaning to the mathematics. However, the most important aspect to remember about using visuals is that it is not the visual itself that causes gains in learning, but rather the instructional strategy in which the visual is embedded.

In *Harcourt Math*, visuals and accompanying questions and teaching suggestions are included throughout the program. Visuals are used to help teachers provide critical scaffolds to understanding and, therefore, memory. They are also used to help students appreciate the usefulness of mathematics in everyday activities. Our students live in a visual world, but many of these visuals pass rapidly before their eyes with very little learning occurring. So, teaching children to read visuals and to link visuals and text is a critical part of the focus in the program. The adage "Before you can see to learn, you must first learn to see," is taken seriously by the authors of this program.

Reading and Vocabulary Development Strategies

Reading in mathematics is critical for students' success. Mathematics is a language and a way of thinking. Therefore, teaching students to read mathematics for understanding and developing the unique vocabulary of mathematics are essential for success. Students' reading abilities in mathematics can be helped by relating their personal knowledge and experience to the information in the text, relating one part of the text to another, providing the lesson objective to students at the beginning of the lesson, and discussing the meaning of important new words (Ornstein, 1995).

In *Harcourt Math*, development of students' content reading skills forms a focus in the presentation of each lesson. Some of the helps to content-area reading include providing instructional objectives to focus students' thinking, providing a key question to guide the lesson development, identifying and defining key mathematical terms used in the lesson, and providing prompts to help students remember prior-taught material necessary for understanding the lesson. Review exercises throughout the program reinforce the content-area reading developed around each topic.

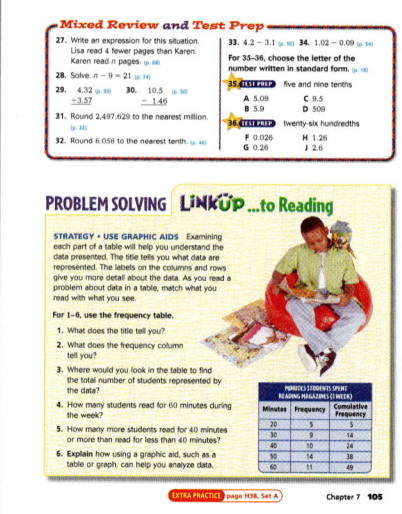

Use of Manipulatives

Research supports the use of manipulatives to increase students' conceptual understanding and achievement in mathematics. Two main points that emerge from the research are:

- Long-term use of manipulatives is important for success.
- Teachers need to have clear purposes for using manipulatives in a particular situation (Sowell, 1989).

In *Harcourt Math*, manipulatives and the pictorial representations of manipulatives are used throughout the grade levels. Concepts are introduced through manipulative activities accompanied by questions that help students see the link between the concrete material or a pictorial representation and the mathematical idea being modeled. These structured manipulative activities demonstrate the purpose for using the manipulative and provide support to teachers in making the most efficient and effective use of manipulatives.

Basic-fact strategies and the procedures for each of the computational algorithms are developed through the use of manipulatives. These activities support memorization of basic facts and procedures because they provide concrete visuals that children can use to link related facts and procedures and, therefore, make the task of memorization easier. In the beginning, memorization of basic facts is best facilitated by use of basic-fact strategies supported by concrete experiences. As the goal of automaticity with the basic facts is reached, the basic-fact strategies are no longer used in the process of memorization but remain in the students' memories to act as helps for solving word problems. The same premise holds true as children develop automaticity with the basic algorithmic procedures.

Intervention

Intervention refers to a given set of strategies used by a teacher to accommodate the diverse skill levels, interests, and learning preferences of students. In order to meet the learning needs of all students, attention must be given to tailoring instruction so that the needs of auditory, visual, and kinesthetic learners are met; students' personal interests are considered; and instruction begins where students are. Carol Tomlinson, in her book *The Differentiated Classroom,* makes the following recommendations for differentiating instruction in the classroom:

- Build instruction around the essential concepts, principles, and skills of a subject.
- Attend to individual differences.
- Use assessment as today's means of understanding how to modify tomorrow's instruction.
- Use assessment data to modify the content students are to learn, the processes through which content is to be taught, and the product by which the students demonstrate what they have learned.

In *Harcourt Math,* intervention is woven through the program as the means by which teachers differentiate instruction. The variety of assessment instruments provides effective tools that help teachers diagnose students' strengths and weaknesses and allow them to make informed decisions about the classroom curriculum. In Grades 3–6, a Student Handbook in the back of the *Pupil Edition* provides intervention activities and exercises. *Intervention Strategies and Activities* links intervention activities to the grade-level content and, therefore, allows teachers to present grade-level content differentiated to meet the needs of all learners.

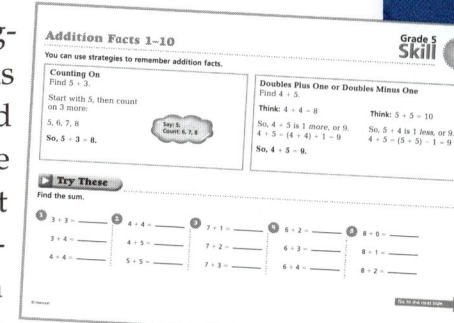

The instructional strategies used in the classroom are an important determinant of students' achievement. The instructional strategies developed in *Harcourt Math* will help you make instructional decisions based on the goals of your classroom curriculum as mandated by your school or district, by the needs of each of your students, and by preparation for tests that your students are required to take.

References

Bahrick, H. P., & Hall, L. K. (1991). "Lifetime maintenance of high school mathematics content." *Journal of Experimental Psychology: General,* 120: 22–23.

Chuska, Kenneth. (1995). *Improving Classroom Questions.* Bloomington, Indiana: Phi Delta Kappa Educational Foundation.

Dwyer, Francis M., with Morre, David M. (1994). *Visual Literacy: A spectrum of visual learning.* Englewood Cliffs, New Jersey: Education Technology Publications.

Geary, D. C. (1994). *Children's Mathematical Development: Research and Practical Applications.* American Psychological Association. Washington, D.C.

Jones, B., Palincsar, A., Ogle, D., & Carr, E. (1987). *Strategic Teaching and Learning: Cognitive Instruction in the Content Areas.* Alexandria, Virginia: Association for Supervision and Curriculum Development.

Ma, Liping. (1999). *Knowing and Teaching Elementary Mathematics.* Mahwah, New Jersey: Lawrence Erlbaum Associates, Publisher.

Ornstein, Allan. (1995). *Strategies for Effective Teaching* (2nd Ed). Chicago: Brown & Benchmark Publishers.

Rowe, Mary Budd. "Wait-time and rewards as instruction variables: Their influence on language, logic, and fate control: Part I: Wait-time." *Journal of Research in Science Teachings* 11, no. 2: 81–84.

Sowell, E. J. (1989). "Effects of manipulative materials in mathematics instruction." *Journal for Research in Mathematics Education,* 29 (2), 121–142.

Tomlinson, Carol. (1999). *The Differentiated Classroom: Responding to the Needs of All Learners.* Alexandria, Virginia. Association for Supervision and Curriculum Development.

ASSESSMENT
An Integral Part of the Learning Process

by Lynda Luckie

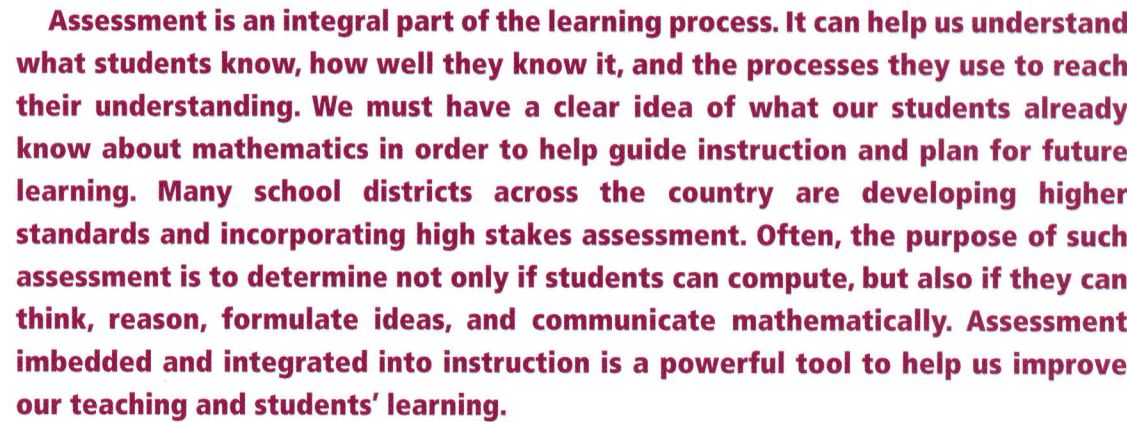

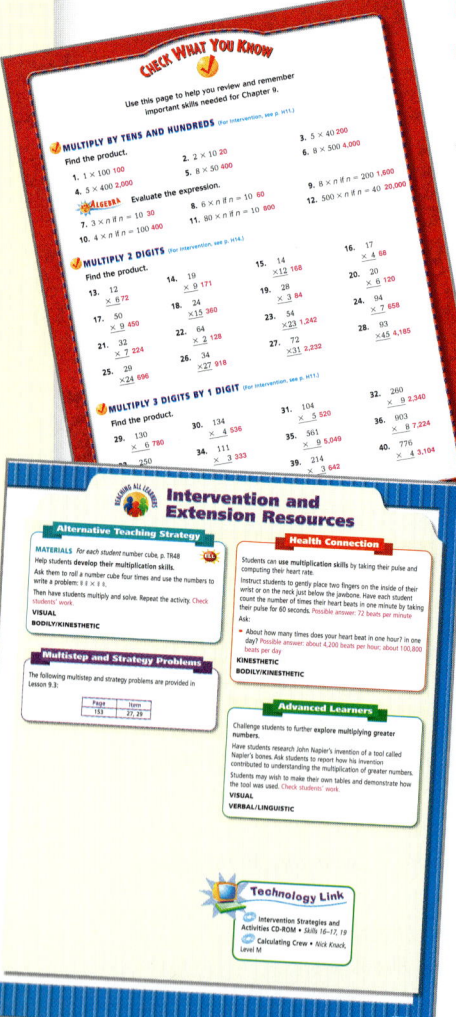

Assessment is an integral part of the learning process. It can help us understand what students know, how well they know it, and the processes they use to reach their understanding. We must have a clear idea of what our students already know about mathematics in order to help guide instruction and plan for future learning. Many school districts across the country are developing higher standards and incorporating high stakes assessment. Often, the purpose of such assessment is to determine not only if students can compute, but also if they can think, reason, formulate ideas, and communicate mathematically. Assessment imbedded and integrated into instruction is a powerful tool to help us improve our teaching and students' learning.

Entry-Level assessment is included in *Harcourt Math* to determine whether students enter a grade level with weaknesses and gaps in their skill proficiency and understanding or whether students show strengths and are, therefore, capable of thriving in a challenging curriculum. This entry-level assessment, designed to determine where students are in their skill development, conceptual understanding, and ability to solve problems, is provided at the beginning of each year. In addition, the *Check What You Know* at the beginning of each chapter provides an entry-level assessment for the content in that chapter. This assessment is linked to diagnostic and prescriptive information to assist the teacher in intervening to prevent students from experiencing failure with new material.

Progress monitoring is a part of every lesson. The third step of the lesson plan in *Harcourt Math* provides activities to check students' conceptual understanding and skill development, providing immediate diagnostic information to the teacher and immediate feedback to the student. In addition, end-of-chapter and end-of-unit progress monitoring provides students, their parents, and their teachers information as to whether students' learning has been connected with related mathematical ideas, can be used to solve problems, and, for skills, is progressing toward the level of automaticity.

Summative evaluation occurs at the end of each chapter, at the end of each unit, and at the end of the year. There are a variety of types of assessment instruments—multiple-choice tests, free-response tests, performance tasks, portfolio suggestions, and writing prompts—that may be used for summative evaluation. The assessment program in *Harcourt Math* is designed to develop a complete portrait of each student's mathematical development—conceptual development, skill proficiency, and problem solving ability. The underlying philosophy of the assessment program is described below.

Varied assessment tools help students value mathematics.

While a good assessment program helps teachers develop a solid plan for instruction and learning, it should also help students learn to value mathematics. Students' definitions and understanding of assessment dictate their perceptions of what we value. For instance, if assessment consists solely of computation-driven tests, then students will believe that is all we value. On the other hand, if we want students to value development of problem solving skills, reasoning, and mathematical communication, then we must find ways to make these connections more explicit for students (Davinoy, Bliem, and Mayfield). If we value strategies and thought processes, we must look at not only the "what" but also the "how."

Just as a balanced diet includes more than one food group, so does a balanced assessment program use more than one format. Students should be given opportunities in multiple formats to demonstrate what they know and can do. While the importance of computational skills should not be minimized, it is almost impossible to accurately assess students' understanding of the big ideas of mathematics without a variety of assessment tools. Indeed, some elements of mathematics learning can only be measured in ways other than multiple-choice tests (Stenmark, Mathematics Assessment NCTM). One of the most powerful arguments for implementing various assessment techniques is the value it has as a diagnostic tool for improved instruction.

Varied assessment tools prepare students for standardized tests.

Some useful assessment models include performance assessments, portfolios, math journals, individual or small-group projects, formal tests that include multiple-choice and free-response items, and self-assessment instruments. Good teaching also includes some informal student assessment such as observations and student interviews. Anecdotal record sheets and checklists can be very useful for this kind of documentation. Included in a balanced assessment program is a good preparation for local or national standardized testing. Exposure to and familiarity with the format and types of problems can be the key to building students' fluency and self-confidence in their abilities to take tests and perform well. Then the high-stakes tests will not be something for students to dread, but rather another opportunity to show what they know and can do.

Varied assessment tools provide useful diagnostic information to teachers about students' strengths and weaknesses.

Portfolios can be one of the best ways to present a clear, visible, continuous, and comprehensive picture of students' progress. It can be a way to showcase students' work. You may want to include several different types of assessment in portfolios, such as performance assessment tasks, paper-and-pencil tests, projects, and students' writing about mathematics. A suggestion for a writing piece is for students to record how they view themselves as mathematicians, both at the beginning and toward the end of the school year. You may see a huge difference in their perceptions over time. Many teachers find that students particularly enjoy selecting their best work to include in their portfolios. Moreover, a student portfolio can provide an invaluable communication tool as you conference with parents about their child's progress.

Performance Assessment consists of presenting a mathematical task or project for students to work on and then making a determination of what they know and can do. This can be accomplished by using any combination of observations, student interviews, and rubrics that can be developed by teachers and/or by students. Performance assessments often incorporate a wide variety of mathematical skills and give evidence of a rich mathematics curriculum. As students work on these tasks, they become engaged in the activities and quickly learn that math is more than just learning how to complete an algorithm. They also learn that there is often more than one right answer and more than one way to solve a problem.

Math Journals provide invaluable insight into a student's progress. They can reveal understanding as well as attitudes and perceptions. Entries in a journal can be as simple and unstructured as responding to a prompt or can become more complex as students are asked to explain a mathematical process or justify an answer. As in their portfolios, students might record their changing view of themseves as mathematicians. Students should be encouraged to use pictures and diagrams to help explain their thinking. Many teachers use a math journal as a personal communication tool by responding in a student's journal and developing an ongoing conversation about mathematics. One caution to be considered is that students need to know that a journal entry is a viable part of the classroom mathematics program. If journals are never reviewed and never receive affirmation, then they will likely become just another chore in the minds of students.

Standardized and Free-Response Format Tests provide one way to diagnose your students' abilities to apply the skills and procedures of mathematics. These tests can be used before instruction on a topic is begun, to determine whether

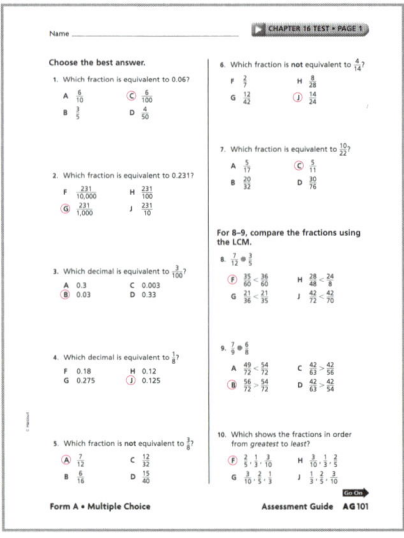

or not each student has the prerequisite skills for the new content. They can be used at the end of a sequence of lessons on a topic to determine whether an appropriate skill level has been attained. These tests can also assess students' abilities to apply skills and concepts in problem solving situations. Cumulative Review tests are a valuable way to measure remembering and retention over time of both conceptual and procedural knowledge and the ability to apply that knowledge in problem solving situations.

Self-Assessment tools help you understand the attitudes and perceptions your students have about mathematics, their assessment of their own abilities to do mathematics, and their preferred ways of working. For many students, cooperative learning groups are an effective way for students to develop their problem solving abilities and their abilities to

communicate about mathematics they are learning and to practice in order to develop skill and procedural fluency. Other students may work better individually for some of these types of activities. Self-assessment tools give you valuable information not only about the mathematics students are learning but also about the ways in which they prefer to learn.

References

Stenmark, J. K. (1989). *Mathematics Assessment: Myths, Models, Good Questions, and Practical Suggestions.* Berkley, California: University of California.

Davinroy, K. H., Bliem, C. L., & Mayfield, V. (1995) "How Does My Teacher Know What I Know?: Third Graders' Perceptions of Math, Reading, and Assessement." Boulder, Colorado: CRESST (National Center for Research on Evaluation, Standards, and Student Testing), University of Colorado.

The purpose of assessment should be to help students see the real-life application of mathematics.

As you develop your assessment program, keep in mind the big ideas. For example, what you want to assess should help you decide on the assessment tool. Whichever form you choose should embrace an "authentic assessment" mentality. In other words, assessment should help foster making a transition from what students learn in school to the mathematics they will need to know in "real life." There is also a very real place in your program for daily and periodic review of what students are learning. Curriculum should spiral and revisit ideas within the context of new material. For example, finding the area of given parcels of land can be imbedded in a problem solving task that involves representing data in a graph about national parks.

Building a Comprehensive Assessment Program in Your Classroom

If you are already using multiple forms of assessment, then you already realize the benefits. If not, one of the purposes of this article is to encourage you to begin to make some changes. The decision to make changes comes slowly and with deliberate effort. Some suggestions include:

- Take small steps.
- Try one new thing at a time.
- Allow yourself to make mistakes and adjustments.
- Collaborate with fellow teachers, parents, and students.
- Persevere until it becomes a natural, integrated component of your teaching.

Harcourt Math provides a varied toolbox of assessment instruments designed to provide ongoing diagnostic information about your students' strength and weaknesses. Students can only make progress in mathematics when teachers have varied sources of information about the abilities of their students. Your choice of assessment instruments should be linked to the kinds of information that you need to make your mathematics classroom a rich experience for your students; to prepare students for the standardized assessment required by your state, district, or school; and to provide both the student and his or her family members with a complete profile of the student's mathematical competency. The choices are many and varied and will allow you to continue to modify your assessment program to meet your needs and your students' needs and to document to parents, school administrators, and the community the richness of your classroom mathematics program.

Keep in mind that assessment should not be something we "do" after we "teach." Often we can assess students' learning in the process of teaching. Teachers do this every day. We watch students develop new skills and strategies as they become problem solvers. We know that we can imbed skill practice within tasks that are interesting and exciting for our students. We watch as students learn from each other and construct their own understanding on the foundation of what they already know, enabling them to transfer that knowledge to new situations. Based on these ongoing and informal observations that are supported in the daily lessons in *Harcourt Math*, teachers can make good choices as to which of the many and varied assessment options included with the program are right for their classroom. When teachers allow students to experience success in a variety of assessment formats, they confirm their belief that all students can learn, and they empower students to view themselves as mathematicians.

Technology
Pathways to Teaching and Learning

by Howard Johnson

Increasingly today, children are using computers for what we have long considered traditional activities of childhood. They use this technology to play, learn, communicate, and develop relationships. Historically, education has centered around a model of learning that focuses on presenting information that students are to learn. Then, through repetition and practice, it was assumed that facts, concepts, and skills were stored and integrated to form knowledge structures. Testing could then measure certain outcomes and behaviors to assess the degree to which the material had been learned.

Recently, however, it has become clear that learner-centered education improves students' motivation to learn. This shift changes the teacher's role to that of creating and structuring the learning environment so that interaction with the teacher, with other students, and with all types of instructional materials and technology are practiced. Teachers are even more critical and valued in this learning setting because they diagnose students' strengths and weaknesses and use all of the instructional tools they have to reach each individual student. Indeed, as pointed out by Banchoff, "The future potential of the Internet . . . will change the way we do mathematics, the way we write about it, and the way we present it in our classrooms at all levels. It is a very exciting time to be a teacher."

Technology as a Tool for Learning

The benefits students receive from using technology are determined by how their teachers connect its use to the classroom curriculum. In every case, however, these tools provide an effective alternative means of representing mathematical concepts, skills, and ideas. Technology can be used to help students engage in problem solving experiences with real data—problems that students themselves may encounter in the world, problems that can be approached with a variety of strategies, and problems that may have more than one solution. Jensen and Williams report that using technology has a positive effect on "both problem solving achievement and attitudes toward the activity of problem solving." The computer can be an invaluable tool to help the problem solver focus on generating ideas, trying out various approaches, and checking hunches. Computers put students in the position of being able to test their conjectures since the machine can produce an endless stream of output that students can use as data for testing and revising their conjectures.

Careful planning is essential in using technology in classrooms. Campbell and Stewart offer the following guidelines for teachers to consider:

- Use technology to enhance your curriculum goals, not for its own sake.
- Use technology to supplement manipulative activities, not to replace them.
- Use software that is developmentally appropriate.
- Use interactive software that capitalizes on the potential of the computer.
- Encourage students to question and help each other.
- Help students determine when technology is appropriate and when other approaches are more appropriate or efficient.

Technology as a Tool for Teaching

In *Harcourt Math,* technology is linked directly to the lessons in the program and is designed to reinforce and extend content, to provide practice on basic skills, to provide activities that help students see the relevance of mathematics to their daily lives, and to diagnose and intervene to help those students who are not ready for grade-level instruction. The following technology components are for student use:

Harcourt Learning Site (www.harcourtschool.com), which provides the following resources:

- a multimedia glossary
- activities for each grade level, designed to present interesting situations and to provide practice in problem solving and computation

- *News Breaks* that provide articles to help students see the relevance of mathematics to solving everyday problems
- *Video Updates* that provide timely information and link mathematics to world events
- *E-Lab,* a series of activities that support the hands-on lessons in the textbook by providing a slightly different perspective on the concept being developed

Mighty Math, a series of six CD-ROMs that provide support for concept development, computation practice, and problem solving experiences:

- *Carnival Countdown* and *Zoo Zillions*— Kindergarten through Grade 3
- *Calculating Crew* and *Number Heroes*— Grades 3–6
- *Astro Algebra* and *Cosmic Geometry*— Grade 6

Intervention Strategies and Activities is a series of six CD-ROMs (one for each grade level, 1–6) that provide instruction, practice, and a short assessment for critical skills. Students work on skills and concepts from prior-taught topics that they had difficulty with, as shown by their performance on the Check What You Know inventory for every chapter in the program. Students are provided immediate feedback that focuses not only on the wrong answer but also on the error made. Each activity includes the following:

- *Show Me* section, which teaches the skill using oral and visual prompts
- *Try These* section, which provides scaffolded support in doing the skill
- *Practice* section, to reinforce the skill
- *Check* section, which is a short assessment

Technology as a Planning and Management Tool

To assist teachers in planning lessons and in developing assessment instruments tailored to their individual classroom requirements, the following technology components are provided:

Lesson Planner CD-ROM, which provides templates that you can use to customize your lesson plans so they match your local curriculum.

Harcourt Electronic Test Bank: Math Practice and Assessment, which provides features that allow you to:

- construct alternative tests and practice sets by using the item bank
- assign the Chapter Review/Test found in the *Pupil Edition* to the entire class or to individuals
- have students take the test on the computer with all multiple-choice answers scored electronically
- weight answers and weight the test score in relation to the overall grade
- present items one at a time on the screen to help those students with special needs

Intervention Strategies and Activities, described above, has the following features to help teachers manage the intervention program:

- assign the Check What You Know diagnostic inventory for each chapter
- set up specific activities as class work or as individual assignments
- view and/or print class or individual reports showing scores on activities
- view and/or print skill reports for the complete list of skills for the entire grade level

There is no longer a question as to whether or not technology should be used in the learning and teaching of mathematics. Teachers will have the time-saving convenience of technology to help with lesson planning, meeting the learning needs of each student, and developing assessment instruments aligned with the classroom curriculum. Students will have these powerful tools to enhance mathematics learning and prepare them for this century with the skills and knowledge they will need to be competitive in their work world and in their everyday lives.

References

Banchoff, Thomas F. (2000). The Mathematician as a Child and Children as Mathematicians. *Teaching Children Mathematics,* Vol. 6, Number 6, 350–356.

Campbell, P., & E. Stewart, (1993). Calculators and Computers. In R. J. Jensen (ed.), *Research Ideas for the Classroom: Early Childhood Mathematics* (pp. 251–268). New York, NY: Macmillan.

Jensen, R., & B. Williams. (1993). Technology: Implications for Middle Grades Mathematics. In D. T. Owens (ed.), *Research Ideas for the Classroom: Middle Grades Mathematics* (pp. 225–243). New York, NY: Macmillan.

Linking Home and School
Making Mathematics Part of the Family's Daily Routine

by Vicki Newman

School-home communication is one of the key components of an effective mathematics program. Parents have a better understanding of their child's mathematics program and become more supportive of a standards-based curriculum when the communication between school and home is ongoing. With a clear picture of what their child is learning at school and how they can provide support at home, families play a vital role in their child's mathematics education.

Children benefit when parents and teachers work together to emphasize the importance of mathematics in their children's daily lives. If children see a reason for studying math at school, they learn to value mathematics and develop a positive attitude and interest. The link between school and home is essential in ensuring a successful mathematics program. The following suggestions may help you foster the link between school and home and provide valuable information to parents.

- Identify grade-level mathematics content standards to provide parents with an outline of essential concepts and skills their child should know.

- Explain the goals of a balanced mathematics program through parent newsletters, parent conferences, or school bulletin boards. Have children share classroom projects and daily work that reflect how their mathematics program provides a balance between conceptual competence, computational and procedural competence, and mathematical reasoning.

- Help parents recognize their role and their child's role in learning mathematics. Suggest ways parents can support mathematics learning at school and extend learning experiences at home. Identify strategies parents can use at home to help their children with homework.

- Share information with parents regarding their child's progress in mathematics. Include samples of daily work and share the variety of assessment tools you frequently use to identify their child's progress.

- Encourage parents and the community to become involved in mathematics education by letting them know that their comments, concerns, and suggestions are valued.

The school-home link in *Harcourt Math* will help you implement these suggestions as you share the following materials with parents:

- **Family Involvement Activities** in kindergarten through Grade 2 outline the mathematics taught in each chapter, provide a list of the math vocabulary (with definitions) that will be introduced and reviewed, and give examples of the types of questions and problems children will be working on in the classroom. One important component of the school-home link is helping family members understand the thinking their child must do in order to be successful with the skills, concepts, and problem solving activities presented in the chapter. The Family Involvement materials include problems that children and their parents can solve together and a Family Fun activity that helps children see that mathematics can be fun and that solving math problems is a real-life skill. In addition, at the bottom of each of the lesson pages in the Pupil Edition, there is a Home Activity that describes the mathematics the child is learning and provides a suggestion for a way parents can help. All of these school-home links provide practical information that will help you involve parents as partners in supporting their child's progress in meeting mathematics standards.

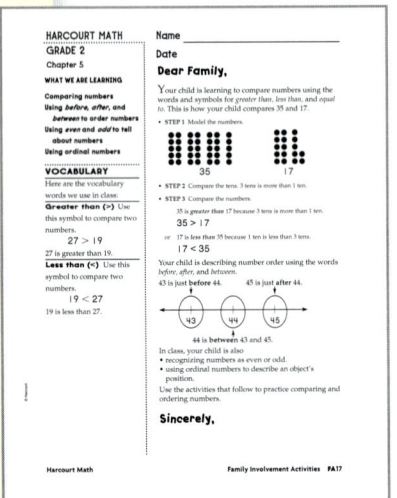 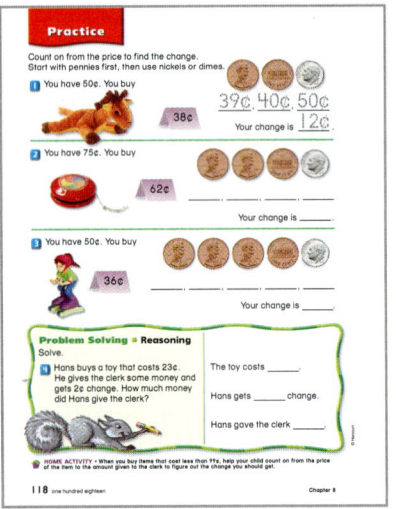

■ **Family Involvement Activities** for Grades 3 through 6 also include a description of what the child will be learning, a list of vocabulary terms and definitions, and suggested ways for parents to help, including activities that provide more practice and a game that helps families have fun while helping their child with the mathematics they are learning at school.

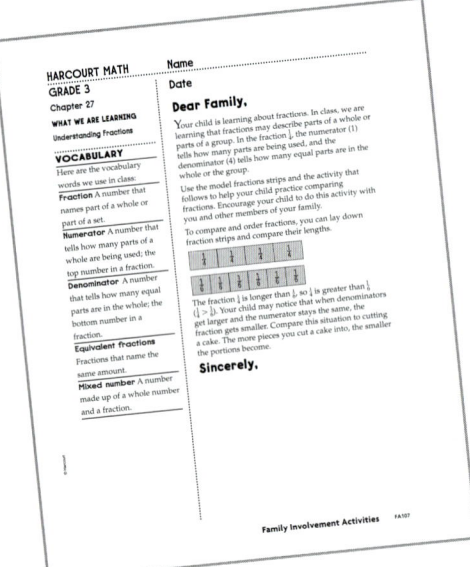

Making Mathematics Part of the Family's Daily Routines

Parents can work with teachers to help their children see the relevance of mathematics in their everyday lives. When parents are filling out a mail-order form, figuring out how many days until a special family event, or calculating the mileage on a family trip, they can involve their children in these mathematical moments. "Parents' attitudes toward mathematics have an impact on how mathematics will be viewed by their children. The child whose parents show enthusiasm for mathematics in the home will be more likely to develop enthusiasm" (Brosnan et al., 329).

As children solve word problems in contexts that are meaningful, they become more confident in applying their math skills. When parents listen to their children explain how they solved a problem, they become more aware of the math concepts and skills their child is mastering. By helping their child apply classroom lessons in their home environment, parents become advocates of their child's education and develop a partnership with teachers that links school and home. The materials in *Harcourt Math* are designed to help you build that critical school-home link.

References

Brosnan, Patricia; Diamantis, Maria; and Hartog, Martin D. Feb. 1998. "Doing Mathematics with Your Child," *Teaching Children Mathematics,* Vol. 4, No. 6, pp. 326–330.

California Department of Education. 1999. *Mathematics Framework for California Schools, Kindergarten Though Grade Twelve.*

Ensign, Jacque. Feb. 1998. "Parents, Portfolios, and Personal Mathematics," *Teaching Children Mathematics,* Vol. 4 No. 6, pp. 331–337.

Follmer, Robin; Ford, Marilyn Sue; and Litz, Kathleen K. Feb. 1998. "School-Family Partnerships: Parents, Children, and Teachers Benefit!" *Teaching Children Mathematics,* Vol. 4, No. 6, pp. 310–312.

Moldavan, Carla C. Feb. 2000. "A Parent's Portfolio: Observing the Power of Matt, the Mathematician," *Teaching Children Mathematics,* Vol. 6, No. 6, pp. 372–374.

Mathematics for All Doesn't Mean Mathematics for Some!

by Jennie M. Bennett

The National Council of Teachers of Mathematics has a public service announcement that says "Do math and you can do anything." What does the reality for "doing math" mean for the youth of today? Today's mathematics opens doors to many opportunities, including entry into tomorrow's jobs—jobs that have not yet been defined. What does this mean for elementary classroom teachers?

Elementary teachers lay the mathematical foundation for students. This foundation must include developing students' mathematical thinking and infusing the elementary curriculum with algebraic concepts. Teachers must encourage students to pursue algebra and other higher mathematics courses at the secondary level and help students see that success in these higher-level courses will prepare them to participate fully in our ever-changing technological society. Students are now expected to achieve the higher standards set by the National Council of Teachers of Mathematics and by their local districts. High expectations are being placed on teachers to ensure that **all** students achieve these higher standards.

Teachers are being challenged to help every child achieve regardless of the child's academic ability. As teachers face this challenge, they must develop strategies to ensure that

- the needs of culturally different students are met;
- gender equity problems are identified and addressed;
- the needs of the mathematically gifted students are met; and
- students with special needs are included in the classroom curriculum.

All children have a thirst for knowledge. A thirst to learn mathematics is an opportunity for every child to participate fully in the classroom program and ultimately in the work force. For children of color, this opportunity is especially important.

The population of the United States is changing to reflect a more diverse society. In 1976, the nonwhite population of United States schools was 24 percent. By 1984, the population of nonwhite students in United States schools had increased to 29 percent. In the new millennium, the projected percent of nonwhite students will increase between 30 percent and 40 percent (Tiedt and Tiedt, 1999). What do these statistics mean for teachers? The ever-changing cultural makeup of America means that teachers must fully address the challenges of equity and excellence. Both equity and excellence must be a part of a guide for students as well (Tomlinson, 1999).

Strategies for Addressing Equity and Excellence in the Classroom

When students arrive in the classroom, they do not leave their cultural backgrounds and experiences at the door. Mathematical ideas have natural cultural links for many students. Building on these links will help teachers include all children. In *Harcourt Math*, cultural contributions and experiences are embedded so that students can see the mathematical ideas and achievements of people from their own background and the backgrounds of their classmates. As defined by James Banks (Banks and Banks, 1993), this level of cultural integration is the lowest level. However, this level still remains important for students and teachers.

Encouraging culturally different students to participate in mathematics includes using the following instructional strategies:

- asking probing questions;
- allowing adequate wait time;
- engaging students in meaningful problem solving situations;
- asking them to explain their thinking;
- working in cooperative learning groups; and
- giving students positive feedback that helps them make progress.

Research states that multicultural education infused in classroom instruction impacts and supports individual esteem, teaches empathy for others, and provides equity for all students (Tiedt and Tiedt, 1999).

The issue of gender equity in the mathematics classroom continues to be deeply rooted in culture and tradition. Teachers must not let the stereotype that "girls cannot do much math" perpetuate in their classrooms from year to year. Girls can do math and should be presented the same rigorous activities that are presented to the boys in the class. They should be encouraged to participate in classroom activities and to freely communicate their mathematical thinking to their classmates. Teachers must be aware of and reduce factors in the classroom that do not equally enhance both boys' and girls' mathematical abilities.

Similarly, students with physical and mental challenges must have equal access to a high-quality mathematics instructional program. Barriers to their success in the mathematics classroom must be lowered. *Harcourt Math* offers inclusion opportunities so that lessons can be modified to allow these students to interact with other students and experience success in mathematics. The use of manipulatives addresses various learning styles and is an effective link to enable students with physical and mental challenges to succeed in the classroom.

The instruction of gifted and talented students also presents challenges. Many teachers are faced with including gifted students in the classroom instructional program, and they are challenged to meet the needs of these students and ensure that their potential is fully developed. The National Council of Teachers of Mathematics has broadened the definition of gifted and talented students to include "mathematically promising" students. Students showing mathematical promise come from all cultures and across all economic strata and have the potential to become problem solvers of the future (Bennett, 1999). Teachers need to make a special effort to challenge, encourage, and support promising students. *Harcourt Math* provides opportunities for students to develop their full capacities for learning mathematics.

The goal of *Harcourt Math* is to provide an instructional program that can be tailored to the instructional needs of every student. Some of the significant opportunities provided to students in the program are:

- providing problem solving activities at various levels of difficulty;
- asking students to make conjectures and justify their thinking;
- asking students to write about the mathematics they are learning and to write their own problems;
- having students use manipulatives to develop their conceptual understanding and model their thinking;
- asking students probing questions that require critical thinking and logical reasoning;
- having students use technology to develop both their conceptual understanding and skill proficiency; and
- having students work with their peers to share their thinking and build their knowledge.

Teachers play a pivotal role in providing quality learning experiences. It is through the expertise of the teacher that students learn mathematics and develop their full potential (NCTM, 2000). When teachers use *Harcourt Math*, they have opportunities to provide rich experiences to challenge each student. Some of the elements that teachers can use to form a consistent approach to developing their classroom instructional program are

- asking students at the end of each lesson to reflect on what they learned;
- asking students to explain and justify their answers to problem solving situations;
- enhancing students' styles of learning by engaging them in cooperative group work;
- using a variety of questioning strategies to build students' mathematical reasoning abilities;
- integrating culturally relevant connections;
- having students show multiple representations of concepts and problems;
- infusing algebraic thinking and introducing algebraic concepts in the early grades;
- engaging students in active, stimulating, and challenging mathematics; and
- communicating with parents about how to help their child at home.

Linking teachers, students, and parents to the mathematics developed in *Harcourt Math* ensures that all students, not just some of them, will be challenged to develop their mathematical abilities to their fullest potential.

References

Banks, James, and Cherry Banks. (1993). *Multicultural Education: Issues and Perspectives,* Second Edition. Boston: Allyn and Bacon.

Bennett, Jennie. (1999). "The Missing Link: Connecting Parents of Mathematically Promising Students to Schools". *Developing Mathematically Promising Students.* Reston, VA.: National Council of Teachers of Mathematics.

National Council of Teachers of Mathematics. (2000). *Principles and Standards for School Mathematics.* Reston, VA.: National Council of Teachers of Mathematics.

National Council of Teachers of Mathematics. (2000). *Changing the Faces of Mathematics: Perspectives on African Americans.* Reston, VA.: National Council of Teachers of Mathematics.

Tiedt, Pamela, and Iris Tiedt. (1999). *Multicultural Teaching: A Handbook of Activities, Information, and Resources,* Fifth Edition. Boston: Allyn and Bacon.

Tomlinson, Carol. (1999). *The Differentiated Classroom: Responding to the Needs of All Learners.* Alexandria, VA.: Association for Supervision and Curriculum Development.

Making Math Memorable: Brain-Compatible Learning

by Marilee Sprenger

It seems that the brain has a mind of its own. When the brain receives any information, including mathematical information, it sorts the information, discards information it doesn't believe is important, and places information that is meaningful in different pathways for later access. We used to believe that the brain took in all information and held it for later retrieval, but it is now believed that up to 99 percent of sensory input to the brain is simply dropped (Wolf, 2000).

How does the brain know what to keep and what to dispose of? Learning appears to take place under the following two conditions:

1. If the brain can connect new material to previously stored information, there is a stronger possibility for encoding. With prior knowledge, the brain has already set in place a network of information or ideas that help the new learning make sense. It is through this meaning-making that new memories can be stored for later access.
2. The other avenue for learning is to stimulate the brain with a novel situation or experience. When the brain experiences a new situation, novelty causes fresh stimulation that sparks attention. This novelty has an emotional component to it, and emotion drives attention and learning (Sylwester, 2000).

Brain-Compatible Learning in *Harcourt Math*

The Harcourt system for math instruction is an excellent approach to brain-compatible learning because the approach is consistent in leading the brain through the learning process. Consistency is a key to brain-compatible learning. The brain requires certain rituals, or parameters, that signal the brain that all is well. It thus feels safe to take in new, unanticipated experiences and new information for careful examination. Without this feeling of security, the brain may quickly initiate a stress response whenever a novel experience presents itself, preventing meaningful learning. When this need for security is met, information can be used at higher brain levels for complex thinking (Jensen, 1998).

The sequence of instruction in *Harcourt Math* helps create the ritual, and the individual components of each lesson allow new connections to be made. Each lesson begins by accessing prior knowledge, an activity that helps the child retrieve memories that will help him or her relate to the new material. Recall that one way we learn is to connect new information to previously stored information.

Reviewing skills is the next step in the lesson sequence. Optimal learning requires a state of "flow" in the brain (Jensen, 2000). This is a pattern of activity that combines challenge, skills, and desire. Students must feel that their skills are equal to the task at hand to allow them to work without stress. Reminding students of their previously learned and mastered skills aids them in attaining this state. The introduction of a lesson also includes an explanation of *why* the information is important to learn. It is through this understanding that students may be able to relate their learning to real-life situations.

The introduction provides a framework for learning the new material. *Harcourt Math* focuses on concrete, hands-on learning experiences. Research on brain development supports concrete learning as a hallmark of effective teaching. It is through concrete experiences that the brain can take steps toward higher-level thinking and abstraction (Sprenger, 1999). Modifications for meeting the needs of each individual learner are suggested at each step in the learning sequence. Since each brain is unique, provision for responding to that uniqueness is crucial. A variety of teaching strategies, as well as many opportunities for enrichment, pave the way for meeting the needs of each child. Activities are aimed at accommodating the different learning styles of students. Since many individuals learn better with visual representations, *Harcourt Math* emphasizes the use of pictures, graphs, and tables. For the kinesthetic, hands-on learner, activities using a variety of manipulatives are suggested. This exploration may be the simulation of a concept or skill that helps the brain that has no prior knowledge. If the meaning cannot be connected to stored memories, this method may create the understanding necessary for new learning and memory and allow the student options for taking the information into his or her brain in the most accessible format. Questioning strategies that follow the guided instruction allow the brain to connect more meaning to the learning.

Guided practice, a key step in the learning sequence in *Harcourt Math*, is necessary for complete learning and memory. Short-term memory is the first step in the process of retaining semantic learning (facts). Sometimes short-term memory is called immediate memory, and initially the area of the brain involved in this procedure holds information for only seconds. Immediate memory can be extended through a rehearsal process called working memory. Without this rote or elaborate rehearsal, memories are not stored and are easily forgotten (Squire and Kandel, 1999). It is only through this procedure that the information to be stored for the long term can be catalogued by a small structure in the brain called the hippocampus. Because of its size and its processing speed, large amounts of information cannot be stored quickly (Jensen, 2000). The process of encoding for long-term storage takes place over time. With each successive practice session, more of the learning can be placed in long-term storage. Independent practice allows for a second memory pathway to be utilized. This is procedural memory, which is sometimes called "muscle memory." Repetition of the process of problem solving allows the brain to store this information in another area of the brain. Storage in more memory pathways leads to better learning and easier retrieval.

Assessment is an ongoing process in *Harcourt Math*. Recent research emphasizes the importance of providing feedback to students on a regular basis. Feedback affects both attention and learning. It can also affect stress levels. In one study, students who were given no feedback had higher stress levels than those receiving negative feedback. It is more stressful for the brain to be uninformed than to be told that it needs improvement (Goleman, 1998). Multiple assessment methods provided in *Harcourt Math* create different types of feedback for both teacher and learner. By using multiple assessments, different memory pathways can be accessed separately or together. Students can discover that information has been stored in one way and that practice must be done for other storage processes. Self-assessment is also an excellent tool for feedback and greater understanding of the learning process. Results of self-assessment are metacognition and identification of new learning goals (Marzano, 2000).

Powerful Learning

The brain utilizes different areas for processing math. For instance, brain-imaging studies indicate that subtraction and multiplication activate separate brain regions (Talan, 1999). Fundamental knowledge that is stored through practice, such as the multiplication tables, provide one type of memory. Practical applications of a skill stimulate another memory pathway. With frequent feedback, these applications enhance both memory and recall. Greater effort is required for the application of skills, but understanding how learning relates in real-life situations stimulates cognition (Kessler, Learning Brain, 2000). Computer-assisted activities included in *Harcourt Math* aid in both drill and practice and practical applications.

Just as multiple memory pathways aid in retention and retrieval of memories, integrating disciplines also enhances memory. Students find more relevance in what they are learning through interdisciplinary and cross-disciplinary methods. They will see more easily how learning relates to real life and have more links to the learning (Jensen, 1998). In *Harcourt Math*, students are shown connections to other disciplines. Through science, social studies, health, physical education, or literature, students find many associations to activate their learning and memory.

As a classroom teacher, I often had to scramble to create lessons for advanced students, slower students, or just students who happened to retrieve information quickly and finished work early. Since my workload seemed to always be increasing, I hoped for a textbook series that would fill in these blanks for me. This series provides these resources. I cannot stress enough the importance of having activities and challenges for all students. The brain makes connections through active engagement, and the emotional brain responds well to the knowledge that its needs are being met.

Meeting needs of all involved is a strong point in *Harcourt Math*. I am not only referring to students. The series also provides information for teachers and parents that enhance each child's learning experience. The *Skills Trace Across the Grades* provides the teacher necessary information to not only stay on track, but also to really know when students have the skills, understanding, and concepts to move forward. It's not just the usual "This is the end of the chapter; it's time to move on!"

Parents are also included in the program. Through *Family Involvement Activities*, parents and caregivers understand what the student is learning and can become involved in the process. This gives the entire program a cooperative effort and can give the student a sense of belonging, which is a basic brain instinct. The link between home and school is vital as parents and the home are the primary sources of comfort and happiness for most children (Kessler, 2000). Providing this connection strengthens the math program and the learning.

Back to the Basics of Learning

There are some basic principles of brain-based learning that should be examined by every educator. By understanding these principles and applying them, the classroom becomes an arena for rich experiences where standards are woven into large themes that students can relate to in many ways.

The brain learns through patterns. It looks for meaning by organizing and categorizing information. The brain can perceive and generate its own patterns. If information fits into an established pattern, the brain can apply the new learning and easily remember it. This principle is particularly relevant to math. It is necessary for students to be able to recognize patterns in problem solving. Since students are always seeking meaning through patterning, we must present material to help them identify and use patterns.

Emotions are critical to patterning. Any activity that creates an emotional tie causes the brain to release chemicals that enhance memory. Emotions cannot be separated from cognition; therefore, the emotional climate of the classroom must be monitored at all times. It is important that math be more than a verbal subject. Students need activities that invite emotion and, therefore, meaning. Each year teachers face students who may have internalized a pattern about math that is negative based on an earlier experience. It is particularly important that the teacher use as many resources as possible to change this thought pattern. It may be helpful to use *Technology Linkups* with these students. Or perhaps the *Math Detective* activities in the Pupil Editions will help students develop a positive feeling about math.

Learning is social. The brain learns better when it works in cooperation with other brains. For this reason, it is important to provide opportunities for group work. Dyads, triads, and small groups may provide the stimulation and meaning necessary for the learning to be made permanent. It may be through the experiences of others that students find the connection they need. The need for socialization may date back in history to times when people needed to be together for survival. The idea of safety in numbers may now be applied to math class!

Harcourt Math provides many opportunities for students to work together to solve problems. This group work provides two distinct opportunities to enhance learning. It gives students the chance to review concepts and skills, hear explanations from their peers rather than always from the teacher, and do some peer teaching. The students also have the opportunity to work on their social and emotional skills. This important arena needs to be cultivated and will result in stronger cognitive abilities (Kessler, 2000).

Learning involves conscious and unconscious processes. In memory terms this principle refers to the two most fundamental kinds of memory—implicit and explicit. *Implicit* memory occurs without conscious attention. That is to say that the memory is formed without much effort. *Explicit* memory requires conscious attention and is formed through the rehearsal process (Sprenger, 1999). In *Harcourt Math* we may see implicit learning through the *Problem of the Day* or *Number of the Day* activities in every lesson plan in the *Teacher's Editions*. As students enjoy problem solving together, they are conceptualizing the problem solving process.

Brains perceive both parts and wholes simultaneously. The left hemisphere of the brain handles details as one of its functions. The right hemisphere examines the big picture. The two hemispheres work together for complete understanding. It is through the integration of parts and wholes that students make connections. For instance, if students learn about graphs in math class, they might create a graph in geography that represents the amounts of rainfall in a given area over time.

Learning involves mind, body, and movement. Current research shows that some of our memories are stored in our bodies (Pert, 1997). Consequently, the more we involve the entire physiology, the better the learning and memory will be. Researchers have found that muscle tension varies a great deal during cognitive tasks. It appears that every mental task is also somewhat physical (Jensen, 2000). Whenever there is physical movement, heart and respiratory rates increase. This may also increase attention. So, when teaching a graphing lesson, rather than simply having students make a graph from given information, have students do a physical activity that can then be graphed. An activity such as this might involve mind, body, and movement.

High stress or threat can impair learning and even kill brain cells. This information has a great impact on teaching practice. Handling a classroom by threat is unfortunately not an uncommon practice. Threats and stress cause the brain to operate at a survival level where no higher-level thinking can take place. Students learn best when they are taught through ritual, novelty, and challenge. The brain also requires choice and feedback. If these practices were the norm in every classroom, there would be little need for threat or stress because students would be too busy learning to do anything inappropriate. Good curriculum and textbooks that provide these five necessities can lower stress for both teachers and students.

Brain Research—Not Just Another Fad

When I made my first brain-research-based presentation many years ago, the staff development director came up to me and said these words, "Brain research is here to stay!" We all know that in education there are some trends that come and go. Although we are learning new things about the brain on a daily basis, we have good information now to allow us to make better decisions about instructional strategies and classroom management. Teachers must make hundreds of decisions every day. Making them based on brain-based theories and applications that enhance learning and memory and provide for individual differences makes good sense. It also creates classrooms that make the mathematics experience a memorable one.

References

Goleman, Daniel. (1998). *Working with Emotional Intelligence*. New York, NY: Bantam.

Jensen, Eric. (2000). *Brain-Based Learning*. San Diego, CA: The Brain Store.

_____. (2000). *Learning with the Body in Mind*. San Diego, CA: The Brain Store.

_____. (1998). *Teaching with the Brain in Mind*. Alexandria, VA: A.S.C.D.

Kessler, Rachel. (2000). *Brain-Based Techniques Improve Test Scores*. San Diego, CA. The Learning Brain Newsletter.

_____. (2000). *The Soul of Education*. Alexandria, VA: A.S.C.D.

Marzano, Robert J. (2000). *Transfroming Classroom Grading*. Alexandria, VA: A.S.C.D.

Pert, Candace. (1997). *Molecules of Emotion*. New York: Scribner.

Sprenger, Marilee. (1999). *Learning and Memory: The Brain in Action*. Alexandria, VA: A.S.C.D.

Squire, Larry, and Eric Kandel. (1999). *Memory: From Mind to Molecules*. New York: Scientific American Library.

Sylwester, Robert. (2000) *A Biological Brain in a Cultural Classroom*. Thousand Oaks, CA: Corwin Press.

Talan, Jamie. (1999). "Simple division: Brain splits up math work." Seattle Times. Available online at: http://seattletimes.nwsource.com/news/nation-world/html98/math_19990507.html

Wolf, Pat. (2000). "Chalk Talk: From Discovery to the Blackboard." The Brain Connection to Education Spring Conference 2000. (audiocassette)

HARCOURT Math

BIBLIOGRAPHY AND INDEX

The following bibliography contains references to:

▶ **Fiction and nonfiction books for students**
▶ **Technology resources**
▶ **Professional books and magazines**

These materials will assist you in creating an interesting learning environment. The references to literature are provided to help you work through your media center to acquire literature selections that you can use with *Harcourt Math*. The math activities developed to correlate to these books will help you build a math curriculum to meet the needs of all students.

The index contains information for both the Pupil Edition and Teacher's Edition. The italicized entries are found in the Teacher's Edition.

Bibliography

Books for Students

The Alaska Purchase. Cohen, Daniel. Millbrook, 1996.

The Amazing Book of Shapes. Sharman, Lydia. Dorling Kindersley, 1994.

American History Math: 50 Problem-Solving Activities That Link Math to Key Events in U.S. History. Glasthal, Jacqueline B. Scholastic, 1996.

"Hockey" by Scott Blaine in *American Sports Poems.* Knudson, R.R., and May Swenson, eds. Orchard Books, 1995.

Arithmetricks: 50 Easy Ways to Add, Subtract, Multiply, and Divide Without a Calculator. Julius, Edward H. John Wiley & Sons, 1995.

Bats, Bugs, and Biodiversity: Adventures in the Amazonian Rain Forest. Goodman, Susan E. Atheneum Books, 1995.

"Mother and Daughter" in *Baseball in April and Other Stories.* Soto, Gary. Harcourt, 1990.

The Circuit. Jimenez, Francisco. University of New Mexico Press, 1997.

"Up the Slide" in *The Complete Short Stories of Jack London.* London, Jack. Stanford University Press, 1993.

The Diary of Anne Frank. Goodrich, Frances, and Albert Hackett. Heinemann, 1995.

Exploring the Night Sky. Dickinson, Terrence. Camden House, 1987

Exploring the Titanic. Ballard, Robert D. Scholastic, Inc., 1988.

Face on the Milk Carton. Cooney, Caroline B. Laurel Leaf, 1994.

The Facts On File Children's Atlas. Wright, David & Jill. Facts on File, Inc., 1993.

Force & Motion: Eyewitness Science. Lafferty, Peter. Dorling Kindersley, 1992.

From the Mixed-Up Files of Mrs. Basil E. Frankweiler. Konigsburg, E.L. Atheneum, 1987.

Funny and Fabulous Fraction Stories. Greenberg, Dan. Scholastic, 1996.

G Is for Googol. Schwartz, David M. Tricycle Press, 1998.

The Gilded Cat. Dexter, Catherine. William Morrow & Company, 1992.

Go Figure! The Numbers You Need for Everyday Life. Hopkins, Nigel J., John W. Mayne, and John R. Hudson. Gale Research Inc., 1992.

"Raymond's Run" in *Gorilla, My Love.* Bambara, Toni Cade. Random House, 1992.

Gulliver's Travels. Swift, Jonathan. Running Press Book Publishers, 1992.

I Have A Dream: The Life and Words of Martin Luther King, Jr. Haskins, James. Millbrook Press, 1992.

Incredible Comparisons. Ash, Russell. Dorling Kindersley, 1996.

Island of the Blue Dolphins. O'Dell, Scott. Houghton Mifflin, 1990.

Janice Van Cleave's Math for Every Kid. Van Cleave, Janice. John Wiley & Sons, 1991.

Jigsaw Jackson. Birchman, David F. Lothrop, Lee & Shepard, 1996.

King of the Wind. Henry, Marguerite. Macmillan, Inc., 1991.

The Librarian Who Measured the Earth. Lasky, Kathryn. Little, Brown and Company, 1994.

Life by the Numbers. Devlin, Keith. John Wiley & Sons, 1998.

Little Farm in the Ozarks. MacBride, Roger Lea. HarperCollins, 1994.

"The School Play" in *Local News.* Soto, Gary. Harcourt, 1993.

The Maltese Cat. Kipling, Rudyard. Creative Education, Inc., 1991.

The Math Chef. D'Amico, Joan, and Karen Eich Drummond, R. D. John Wiley & Sons, 1997.

Math Mini Mysteries. Markle, Sandra. Atheneum, 1993.

Melisande. Nesbit, E. Harcourt, 1989.

"Tranquility Base" in *Men from Earth.* Aldrin, Buzz. Bantam Books, 1991.

Mountains. Simon, Seymour. Morrow Junior Books, 1994.

Nearer Nature. Arnosky, Jim. Midaya Press, 1996.

A New Way of Life (formerly titled ***The Vietnamese in America***). Rutledge, Paul. Lerner Publications Company, 1991.

Not for a Billion, Gazillion Dollars. Danzinger, Paula. Delacorte Press, 1992.

Nothing But the Truth. Avi. Avon, 1991.

Number the Stars. Lowry, Lois. Houghton Mifflin, 1989.

The Phantom Tollbooth. Juster, Norton. Random House, 1989.

Pyramid. Macaulay, David. Houghton Mifflin, 1975.

The Red Pony. Steinbeck, John. Viking Penguin, 1992.

Real-Life Math Investigations. Lee, Martin, and Marcia Miller. Scholastic, 1997.

Robinson Crusoe. Defoe, Daniel. Running Press Book Publishers, 1990.

Secrets of the Shopping Mall. Peck, Richard. Dell, 1991.

The Secrets of Vesuvius Bisel, Sara C. Madison Press, 1990.

She Flew No Flags. Manley, Joan B. Houghton Mifflin Company, 1995.

Skinnybones. Park, Barbara. Bullseye Books, 1995.

Snow Bound. Mazer, Harry. Dell, 1990.

The Star Fisher. Yep, Laurence. William Morrow & Company, 1991.

Sticks. Bauer, Joan. Delacorte Press, 1996.

Summer Ice: Life along the Antarctic Peninsula. McMillan, Bruce. Houghton Mifflin, 1995.

"Dog of Pompeii" by Louis Untermeyer in ***Teach Your Children Well: A Parent's Guide to the Stories, Poems, Fables, and Tales That Instill Traditional Values.*** Allison, Christine, ed. Delacorte Press, 1993.

The Tarantula in My Purse. George, Jean Craighead. HarperCollins, 1996.

"Emergency Landing" from ***Visiting Mrs. Nabakov and Other Excursions.*** Williams, Ralph. Vintage Books, 1995.

Volcano: The Eruption and Healing of Mount St. Helens. Lauber, Patricia. Bradbury Press, 1986.

The Walrus and the Carpenter. Carroll, Lewis. Penguin Books USA Inc., 1992.

What Hearts. Brooks, Bruce. HarperCollins, 1992.

What Is a Wall, After All? Allen, Judy. Candlewick Press, 1993.

Where Am I? The Story of Maps and Navigation. Smith, A. G. Stoddart Kids, 1997.

The Whipping Boy. Fleischman, Sid. William Morrow & Company, 1986.

The Widow and the Parrot. Woolf, Virginia. Harcourt, 1992.

Math Software

TITLE	PUBLISHER	DESCRIPTION	SYSTEM REQUIREMENTS
JumpStart Adventures 4th Grade: Haunted Island (Windows/Macintosh Hybrid CD-ROM)	Knowledge Adventure	Students apply lessons in math, and other curriculum areas to rescue 13 lost friends before they turn into freakish fiends forever! **Math Topics** • Equations • Division • Multiplication • Addition • Subtraction • Decimals • Fractions • Units of Measure	**PC:** 486 or Pentium or higher. Windows 3.1 or 95/98. 256 color SVGA monitor, 8 MB RAM, 9 MB hard disk space, 2X CD-ROM drive, sound card and speakers **Macintosh:** 68040 processor; System 7.1 or higher, 256 color monitor 640 × 480, 8 MB RAM, 9 MB hard disk space, 2X CD-ROM drive.
JumpStart Adventures 5th Grade: Jo Hammet, Kid Detective (Windows/Macintosh Hybrid CD-ROM)	Knowledge Adventure	While visiting the museum on a field trip, Jo Hammet uncovers a sinister plot to destroy all of the city's factories! Her mission (with student help) is to find clues leading to the capture of the mad genius Dr. X. **Math Topics** • Fractions • Decimals • Equations • Division • Multiplication • Geometry • Ratios	**PC:** 486 or Pentium or higher. Windows 3.1 or 95/98. 256 color SVGA monitor, 16 MB RAM, 15 MB hard disk space, 2X CD-ROM drive, sound card and speakers. **Macintosh:** 68040 processor; System 7.1 or higher, 256 color monitor 640 × 480, 8 MB RAM, 15 MB hard disk space, 2X CD-ROM drive.
MathKeys (Windows/Macintosh Hybrid CD-ROM)	The Learning Company	Students calculate probability by operating spinners, coin flippers, and other playful machines. They can see data in graph, chart, number, and word sentence forms. **Math Topics** • Whole Numbers • Probability • Geometry • Measurement • Fractions	**PC:** 486; Windows 95 or higher; 4 MB RAM; hard disk; 256 color graphics; Windows Compatible sound card; mouse **Macintosh:** 68020 processor; Mac OS 7.0 or higher; 4 MB RAM; hard disk; 256 colors; mouse
Mighty Math Calculating Crew (Windows/Macintosh Hybrid CD-ROM)	Edmark	This program teaches students the concepts, facts, and thinking skills necessary to build math confidence and develop a strong, lasting understanding of math. **Math Topics** • Multiplication • Division • Decimals • Number Lines • 3-D Solids • Money Transactions	**PC:** 486DX or better, 8 MB RAM, and 5 MB hard disk, 2X CD-ROM drive, Windows 3.1 or 95. 256 Color SVGA monitor **Macintosh:** 68030/25 MHz or better. 8 MB RAM, 2X CD-ROM drive, 14" or larger color monitor, System 7.0.1 or higher
Mighty Math Number Heroes (Windows/Macintosh Hybrid CD-ROM)	Edmark	The combined programs include over 250 reading and math activities, hundreds of animations, and five original songs. Each activity offers hints and includes three levels of difficulty. **Math Topics** • Multiplication • Division • Fractions • 2-D Geometry • Probability	**PC:** 486DX or better, 8 MB RAM, and 5 MB hard disk, 2X CD-ROM drive, Windows 3.1 or 95. 256 Color SVGA monitor **Macintosh:** 68030/25 MHz or better. 8 MB RAM, 2X CD-ROM drive, 14" or larger color monitor, System 7.0.1 or higher
The ClueFinders 5th Grade Adventure (Windows/Macintosh Hybrid CD-ROM)	Mattel	A sudden tsunami has shipwrecked the gang on an uncharted volcanic island, and Owen and Leslie have disappeared! Students collect the mysterious Cryp Tiles to solve the mystery before the volcano blows! **Math Topics** • Multiplication • Division • Geometry	**PC:** Pentium or better; Windows 95/98; 16 MB RAM; 25 MB hard disk space; VGA/SVGA monitor; 256 colors; Supports mouse & sound card **Macintosh:** Power Mac; Hard Drive System 7.1 and up, 16 MB RAM, 25 MB hard disk space; 4X CD-ROM drive; 256 colors
The ClueFinders 6th Grade Adventures (Windows/Macintosh Hybrid CD-ROM)	Mattel	The 9 interactive puzzles, customized workbooks, and 50 printable activities help students learn important skills in math in addition to a full curriculum of subjects. **Math Topics** • Decimals • Percents • Fractions • Statistics • Estimations • Ratios	**PC:** Pentium or better; Windows 95/98; 16 MB RAM; 25 MB hard disk space; VGA/SVGA monitor; 256 colors; Supports mouse & sound card **Macintosh:** Power Mac; Hard Drive System 7.1 and up, 16 MB RAM, 25 MB hard disk space; 4X CD-ROM drive; 256 colors

Math Software

TITLE	PUBLISHER	DESCRIPTION	SYSTEM REQUIREMENTS
The ClueFinders 6th Grade Math Adventures (Windows/Macintosh Hybrid CD-ROM)	Mattel	Students travel to a small village high in the Himalayas to solve the mystery of the disappearing treasures. The software continually adjusts the program to match students' growing abilities. **Math Topics** • Multiplication • Division • Geometry • Problem Solving	**PC:** 486 processor or better; Windows 95/98; 16 MB RAM; 25 MB hard disk space; VGA/SVGA monitor; 256 colors; Supports mouse & sound card **Macintosh:** Power Mac; Hard Drive System 7.1 and up, 16 MB RAM, 25 MB hard disk space; 4X CD-ROM drive; 256 colors
Carmen Sandiego Math Detective (Windows/Macintosh Hybrid CD-ROM)	Broderbund	The program includes three levels of over 250 activities with thousands of math problems, as well as over 400 word problems and customizable problem sets. **Math Topics** • Numeration • Geometry • Measurement • Problem Solving	**PC:** 586 processor & up; Windows 3.1 & up; 20 MB hard drive space, 8 MB RAM; VGA/SVGA monitor; 2X CD-ROM drive; sound card **Macintosh:** 68040 processor; System 7.1 & up, 6 MB RAM, 20MB hard disk space, 2X CD-ROM drive, 256 colors
Community Construction Kit (Windows/Macintosh Hybrid CD-ROM)	Tom Snyder Productions	An opportunity for students to develop and create their own communities by designing historically accurate buildings and landscaping. **Math Topics** • Geometry • Map Skills • 3-D Concepts	**PC:** 386 processor & up; Windows 3.1 & up; 14 MB hard disk space, 4 MB RAM; VGA/SVGA monitor; 2X CD-ROM, sound card **Macintosh:** 68020 processor and up, System 7.0 & up, 4 MB RAM, 8 MB hard disk space, 2X CD-ROM drive, 256 colors
CornerStone Mathematics – Level B (Windows/Macintosh Hybrid CD-ROM)	Skillsbank	The process includes: Warm-up, Review, Quiz, Take Five, and Worksheet to demonstrate, guide, test, reinforce, and provide extra practice of concepts. **Math Topics** • Number Concepts • Estimation • Whole Number Computation • Decimals • Fractions • Percents • Data and Graphs	**PC:** 486 processor & up; Windows 3.1 & up; 10 MB hard disk space, 8 MB RAM; VGA/SVGA monitor; 2X CD-ROM drive; sound card **Macintosh:** 68030 processor, System 7.0 & up, 8 MB RAM, 5 MB hard disk space, 2X CD-ROM drive, 256 colors
The Cruncher (Windows/Macintosh Hybrid CD-ROM)	Knowledge Adventure	Program features include a full-featured spreadsheet, step-by-step animated tutorials, ten real-world projects and templates, colorful charts, and graphs. **Math Topics** • Spreadsheets • Graphs • Statistics • Surveys • Investments	**PC:** 386; Windows 3.1 or 95; 4 MB RAM; 14 MB hard disk space; 2X CD-ROM drive; SVGA; sound card **Macintosh:** 68020/16 MHz; System 7.0; 4 MB RAM; 8 MB hard disk space; 2X CD-ROM drive; 256 colors
Fraction Operations (Windows/Macintosh Hybrid CD-ROM)	Tenth Planet	Fraction Operations combines hands-on techniques with multimedia technology. Math concepts are presented in a variety of ways to accommodate a range of learning styles and ability levels. **Math Topics** • Common Denominators • Addition • Equivalent Fractions • Subtraction • Dividing Fractions • Multiplication	**PC:** 386 processor & up; Windows; 14 MB hard disk space, 4 MB RAM; VGA/SVGA monitor; 2X CD-ROM drive **Macintosh:** 68020 processor; System 7.0 & up, 4 MB RAM, 8 MB hard disk space, 2X CD-ROM drive, 256 colors
Geometer's Sketchpad (Windows/Macintosh Hybrid CD-ROM)	Key Curriculum Press	The software includes specific lessons for investigations, explorations, demonstrations, and constructions. Geometric figures designed by the student can be manipulated, transformed, and distorted while preserving geometric relationships. **Math Topics** • Geometry • Visualization • Analysis • Informal Deduction	**PC:** 386/25 MHz; DOS 3.1 or later; 1 MB RAM; 7 MB hard drive space; VGA. **Macintosh:** MacPlus or better, with 1 MB RAM or more, running System Software Version 6.0 or better
The Graph Club (Windows/Macintosh Hybrid CD-ROM)	Tom Snyder Productions	The Graph Club is an easy-to-use graphing tool that can be used for self-directed exploration, lessons, and presentations, or creative class projects. **Math Topics** • Gather, Sort, and Classify • Interpret Tables and Graphs • Analyze Data and Graph	**PC:** Windows 3.1: CPU 386SX / 16 MHz; 8 MB RAM; Monitor: 13" VGA 640 × 480 256 Colors **Macintosh:** LC II or better: System 7.1; 16 MHz; 4 MB RAM; Monitor: 13" 640 × 480 resolution, 256 Colors

Math Software

TITLE	PUBLISHER	DESCRIPTION	SYSTEM REQUIREMENTS
Interactive Math Journey (Windows/Macintosh Hybrid CD-ROM)	The Learning Company	Children take a trip through five math lands to learn key concepts. **Math Topics** • Patterns and Shapes • Addition • Measurement • Subtraction • Fractions • Multiplication	**PC:** 486DX/66 MHz; Windows; 256 color SVGA monitor; 8 MB RAM; sound card, 8 MB hard disk space; 2X CD-ROM drive **Macintosh:** 68030 processor; System 7.01; 256 Color Monitor; 8 MB RAM; 8 MB hard disk space; 2X CD-ROM drive, 13" display
Reader Rabbit's Math Ages 6–9 (Windows/Macintosh Hybrid CD-ROM)	The Learning Company	Students help Reader Rabbit escape from a pirate-infested island by outwitting pirates by using their outstanding math abilities. **Math Topics** • Addition • Subtraction • Geometry • Place Value • Problem Solving	**PC:** 486DX/66 MHz; DOS 5; 256 color SVGA monitor; 8 MB RAM; sound card, 20 MB hard disk space; 4X CD-ROM drive **Macintosh:** Power PC, System 7.1; 256 Color Monitor; 8 MB RAM; 20 MB hard disk space; 4X CD-ROM drive, 13" display
Logical Journey of the Zoombinis (Windows/Macintosh Hybrid CD-ROM)	Broderbund	A map displays students' progress on the journey and provides access to options for turning on/off background music, dialog, sound effects, and transition screens. The randomly generated puzzle solutions automatically adjust to a player's skill level. **Math Topics** • Relationships • Patterns • Grouping • Sorting • Matching • Graphing • Deductive Reasoning	**PC:** 386DX/66 MHz; Windows; 256 color SVGA monitor; 4 MB RAM; sound card, 14 MB hard disk space; 2X CD-ROM drive **Macintosh:** 68020 processor; System 7.0; 256 Color Monitor; 4 MB RAM; 8 MB hard disk space; 2X CD-ROM drive, 12" display
The Lost Mind of Dr. Brain (Windows/Macintosh Hybrid CD-ROM)	Knowledge Adventure	The program's story line is based on an experiment gone awry in which Dr. Brain has transferred too much of his brain to Rathbone, his lab rat. The student's mission is to repair Dr. Brain's brain and restore Rathbone to his ordinary cheese-loving state. **Math Topics** • Problem Solving • Data Analysis • Logical and Deductive Reasoning • Visual/Spatial Relationships	**PC:** 386/33 MHz; Windows 3.1 or 95; 8 MB RAM; 14 MB hard disk space; 2X CD-ROM drive; SVGA; sound card **Macintosh:** 68020/16 MHz; System 7.0; 4 MB RAM; 8 MB hard disk space; 2X CD-ROM drive; 256 colors
Math Blaster Ages 9–12 (Windows/Macintosh Hybrid CD-ROM)	Knowledge Adventure	Student progress can be charted using the math skills screen, which keeps track of mastered subjects and levels. The program automatically adjusts to individual student levels. Online help and math tips provide explanations upon request. **Math Topics** • Addition • Subtraction • Multiplication • Division • Multi-Digit Numbers • Fractions • Decimals • Percents	**PC:** 386/33 MHz; Windows 3.1 or 95; 4 MB RAM; 14 MB hard disk space; 2X CD-ROM drive; SVGA; sound card **Macintosh:** 68020/16 MHz; System 7.0; 4 MB RAM; 8 MB hard disk space; 2X CD-ROM drive; 256 colors
Math Blaster Pre-Algebra (Windows/Macintosh Hybrid CD-ROM)	Knowledge Adventure	Earth's inhabitants are being "zapped" of their mathematical abilities by the magnetic brain invented by Dr. Dabble, the mad scientist. Students solve word problems involving pre-algebra and logical-thinking skills as they attempt to locate the disembodied brain in Dr. Dabble's mansion. **Math Topics** • Decimals • Factors • Integers • Fractions • Prime Numbers • Multiples • Percents	**PC:** 386/33 MHz; Windows 3.1 or 95; 4 MB RAM; 14 MB hard disk space; 2X CD-ROM drive; SVGA; sound card **Macintosh:** 68020/16 MHz; System 7.0; 4 MB RAM; 8 MB hard disk space; 2X CD-ROM drive; 256 colors
Math for the Real World (Windows or Macintosh CD-ROM)	Knowledge Adventure/ Kaplan	Students solve practical real-world problems as they travel the country with an up-and-coming rock band. **Math Topics** • Time • Charts • Logic • Fractions • Money	**PC:** 386/33 MHz; Windows 3.1 or 95; 4 MB RAM; 14 MB hard disk space; sound card; 2X CD-ROM drive **Macintosh:** 68020 processor; System 7.0 or later; 4 MB RAM; 8 MB hard disk space; 256 colors; 2X CD-ROM drive

Math Software

TITLE	PUBLISHER	DESCRIPTION	SYSTEM REQUIREMENTS
Math Workshop™ Deluxe (Windows/Macintosh Hybrid CD-ROM)	Broderbund	Interface divides activities into beginning (downstairs) and more advanced (upstairs). Custom play option allows teachers to create their own "problem sets." **Math Topics** • Computation • Estimation • Logical Reasoning • Fractions • Spatial Visualization	**PC:** 486/33 MHz; 256 color SVGA monitor; 8 MB RAM; Soundcard, 4MB hard disk space; CD-ROM drive **Macintosh:** Power PC; 256 Color Monitor; 12 MB RAM; 25 MB hard disk space; CD-ROM drive
Snootz Math Trek (Windows/Macintosh Hybrid CD-ROM)	Theatrix Interactive	"Snootian Translator" allows students to share secret messages with friends and aliens. The program provides several levels of difficulty for activities. **Math Topics** • Sequencing • Map Reading • Shapes • Symbols • Cause and Effect	**PC:** 386/33MHz; Windows 3.1 or 95; 4 MB RAM; 14 MB hard disk space; 2X CD-ROM drive; SVGA; sound card **Macintosh:** 68020/16 MHz; System 7.0; 4 MB RAM; 8 MB hard disk space; 2X CD-ROM drive; 256 colors
Thinkin' Things Collection 2 (Windows/Macintosh Hybrid CD-ROM)	Edmark	Students are led to use logical reasoning and creative imagination as they perform various tasks in this engaging program. **Math Topics** • Critical Thinking • Spatial Awareness • Problem Solving • Perspective	**PC:** 386/33 MHz; Windows 3.1 or 95; 4 MB RAM; 14 MB hard disk space; 2X CD-ROM drive; SVGA; sound card **Macintosh:** 68020/16 MHz; System 7.0; 4 MB RAM; 8 MB hard disk space; 2X CD-ROM drive; 256 colors
Thinkin' Things Collection 3 (Windows/Macintosh Hybrid CD-ROM)	Edmark	This program allows students to use logical reasoning to solve various activities. They are encouraged to be creative in their solutions. **Math Topics** • Logical Reasoning • Analyze and Synthesize Information	**PC:** 386/33 MHz; Windows 3.1 or 95; 4 MB RAM; 14 MB hard disk space; 2X CD-ROM drive; SVGA; sound card **Macintosh:** 68020/16 MHz; System 7.0; 4 MB RAM; 8 MB hard disk space; 2X CD-ROM drive; 256 colors
Widget Workshop (Windows/Macintosh Hybrid CD-ROM)	Maxis	A hands-on laboratory that allows students to produce experiments and inventions while exploring principles of science, mathematics, logic, computer science, and physics. **Math Topics** • Logic • Puzzles • Math Functions • Logic Gates • Random Numbers	**PC:** 386/33 MHz; Windows 3.1 or 95; 4 MB RAM; 14 MB hard disk space; 2X CD-ROM drive; SVGA; sound card **Macintosh:** 68020/16 MHz; System 7.0; 4 MB RAM; 8 MB hard disk space; 2X CD-ROM drive; 256 colors
Mathville Starway (Windows/Macintosh Hybrid CD-ROM)	Ingenuity Works	Ten skill-testing activities from all areas of Grades 5–7 math curriculum will keep students intrigued and interested in learning. The software supports National Council of Teachers of Mathematics' standards. **Math Topics** • Composite Numbers • Fractions • Number Lines • Percents • Angles • Coordinates	**PC:** 486 or higher, Windows 3.x or Windows 9x, SVGA, 640 × 480, 256 color monitor, 8 MB RAM CD-ROM drive, sound card **Macintosh:** System 7 or higher, 640 × 480 color monitor, 8 MB RAM, CD-ROM drive
Mathville Waterway (Windows or Macintosh CD-ROM)	Ingenuity Works	The program contains a virtual water world that overflows with TWELVE entertaining and skill-testing math activities. It is designed to stretch student abilities by challenging them to apply skills to a variety of new problem-solving situations. **Math Topics** • Decimals • Fractions • Transformations • Coordinates • Number Patterns • Metric Units • Probability and Data • Logic • Problem Solving	**PC:** 486 or higher, Windows 3.x or Windows 9x, SVGA, 640 × 480, 256 color monitor, 8 MB RAM CD-ROM drive, sound card **Macintosh:** System 7 or higher, 640 × 480 color monitor, 8 MB RAM, CD-ROM drive

Books for Teachers

Baratta-Lorton, M. *Mathematics Their Way: An Activity-Centered Mathematics Program for Early Childhood Education.* Addison-Wesley, 1995.

Baratta-Lorton, R. *Mathematics: A Way of Thinking.* Addison-Wesley, 1977.

Benson, D. *The Moment of Proof: Mathematical Epiphanies.* Oxford University Press, 1999.

Berk, L., and A. Winsler. *Scaffolding Children's Learning: Vygotsky and Early Childhood Education.* National Association for the Education of Young Children, 1995.

Bloomer, A., and P. Carlson. *Activity Math: Using Manipulatives in the Classroom.* Addison-Wesley, 1993.

Bresser, R., and C. Holtzman. *Developing Number Sense—Grades 3–6.* Math Solutions Publications, 1999.

Bright, G., and J. Harvey. *Basic Math Games.* Dale Seymour Publications, 1987.

Brodie, J. P. *Constructing Ideas About Large Numbers.* Creative Publications, 1995.

Burk, D., A. Snider, and P. Symonds. *Box It or Bag It Mathematics: Teachers' Resource Guide, First–Second.* The Math Learning Center, 1988.

Burns, M. *About Teaching Mathematics.* Math Solutions Publications, 1993.

Burns, M. *About Teaching Mathematics: A K–8 Resource.* Math Solutions Publications, 1992.

Burns, M. *A Collection of Math Lessons from Grades 6–8.* Math Solutions Publications, 1990.

Burns, M. *Math and Literature (K–3).* Math Solutions Publications, 1992.

Burns, M. *Math By All Means: Division, Grades 3 and 4.* Math Solutions Publications, 1994.

Burns, M. *Math By All Means: Multiplication, Grade 3.* Math Solutions Publications, 1994.

Burns, M. *Math By All Means: Probability, Grades 3 and 4.* Math Solutions Publications, 1994.

Burns, M. *MATH: Facing an American Phobia.* Math Solutions Publications, 1998.

Burns, M. *Probability, Grades 2–3.* Math Solutions Publications, 1994.

Burns, M. *This Book Is About Time.* Yolla Bolly Press, 1978.

Burton, G. *Towards a Good Beginning: Teaching Early Childhood Mathematics.* Addison-Wesley, 1985.

Burton, G., D. Clements, et al. *Addenda Series, Grades K–6.* NCTM, 1991–1992.

Burton, G., et al. *Addenda Series, Grades K–6, Number Sense and Operations.* NCTM, 1993.

Butterworth, B. *The Mathematical Brain.* Macmillan, 1999.

Caine, R. and G. *Unleashing the Power of Perceptual Change: The Potential of Brain-Based Teaching.* ASCD, 1997.

Carpenter, T., E. Fennema, M. Franke, L. Levi, and S. Empson. *Children's Mathematics—Cognitively Guided Instruction.* Heinemann, 1999.

Cathcart. W., Y. Pothier, J. Vance, and N. Bezuk. *Learning Mathematics in Elementary and Middle Schools.* Merrill, 2000.

Childs, L., and L. Choate. *Nimble with Numbers.* Dale Seymour Publications, 1999.

Clapham, C. *Concise Dictionary of Mathematics.* Oxford University Press, 1996.

Coates, G., and J. Stenmark. *Family Math for Young Children.* Lawrence Hall of Science, 1997.

Coburn, T., et al. *Addenda Series, Grades K–6, Patterns,* NCTM, 1993.

Cohen, J. "The First 'R': Reflective Capacities." *Educational Leadership,* Vol. 57, ASCD, September 1999.

Cooney, M., ed. *Celebrating Women in Mathematics and Science.* NCTM, 1996.

Copley, J., ed. *Mathematics in the Early Years.* NCTM, 1999.

Cowan, T., and J. Maguire. *Timelines of African-American History: 500 Years of Black Achievement.* Berkley Publishing Group, 1994.

Crawford, M. and M. Witte. "Strategies for Mathematics: Teaching in Context." *Educational Leadership,* Vol. 57, ASCD, November 1999.

Curcio, F. "Developing Number Sense in the Middle Grades," *Addenda Series.* NCTM, 1991.

Curcio, F., and N. Bezuk, et al. *Addenda Series, Grades 5–8, Understanding Rational Numbers and Proportions.* NCTM, 1994.

Danielson, C., and L. Abrutyn. *An Introduction to Using Portfolios in the Classroom.* ASCD, 1997.

Del Grande, J., and L. Morrow, *Addenda Series, Grades K–6, Geometry and Spatial Sense,* NCTM, 1993.

Drake, S. *Planning Integrated Curriculum.* ASCD, 1993.

Eby, J., and E. Kujawa. *Reflective Planning, Teaching and Evaluation: K–12.* Merrill, 1994.

Elliott, P., ed. *Communication in Mathematics, K–12 and Beyond (1996 Yearbook).* NCTM, 1996.

Fennelland, F., and D. Williams. "Decimal Dash" in *The Arithmetic Teacher.* NCTM, 1986.

Ferrini-Mundy, J., K. Graham, L. Johnson, and G. Mills, eds. *Making Change in Mathematics Education: Learning from the Field.* NCTM, 1998.

Flournoy, V., et al. *The Patchwork Quilt.* Scholastic, 1996.

Forte, I., and S. Schurr. *Interdisciplinary Units and Projects for Thematic Instruction.* Incentive Publications Inc., 1994.

Franco, B., et al. "Geometry Concentration" in *Understanding Geometry.* Great Source Education Group, 1998.

Franco, B., et al. *Understanding Geometry.* Great Source Education Group, 1998.

Fuson, K. C., and Y. Kwon. "Korean Children's Understanding of Multidigit Addition and Subtraction," *Child Development,* Vol. 63, 491–506, 1992.

Garland, T. *Fibonacci Fun: Fascinating Activities with Intriguing Numbers.* Dale Seymour Publications, 1998.

Geary, D. C. *Children's Mathematical Development: Research and Practical Applications.* American Psychological Association, Washington, D.C., 1994.

Geary, D. C. "Reflections of Evolution and Culture in Children's Cognition: Implications for Mathematics Development and Mathematics Instruction," *American Psychologist,* Vol. 50, 24–27, 1995.

Geary, D. C., C. C. Bow-Tomas, and Y. Yao. "Counting Knowledge and Skill in Cognitive Addition: A Comparison of Normal and Mathematically Disabled Children," *Journal of Experimental Child Psychology,* Vol. 54, 372–91, 1992.

Geary, D. C., et al. "A Biocultural Model of Academic Development," in *Global Prospects for Education: Development, Culture, and Schooling.* Edited by S. G. Paris and H. M. Wellman, Washington, D.C.: American Psychological Association, 1998.

Geary, D. C., et al. "Development of Arithmetical Competencies in Chinese and American Children: Influence of Age, Language, and Schooling," *Child Development,* Vol. 67, 2022–44, 1996.

Geary, D. C., and K. F. Widamin. "Numerical Cognition: On the Convergence of Componential and Psychometric Models," *Intelligence,* Vol. 16, 47–80, 1992.

Geddes, D., et al. *Addenda Series, Grades 5–8, Geometry in the Middle Grades.* NCTM, 1992.

Geddes, D., et al. *Addenda Series, Grades 5–8, Measurement in the Middle Grades.* NCTM, 1994.

Gelfand, I., and A. Shen. *Algebra.* Birkhauser, 1993.

Glassman, B., ed. *Macmillan Visual Almanac.* Blackbirch Press, 1996.

Glatzer, D., and J. Glatzer. *Math Connections.* Dale Seymour Publications, 1989.

Goldsmith, L., and J. Mark. "What Is Standards-Based Mathematics Curriculum?" *Educational Leadership*, Vol. 57, ASCD, November 1999.

Greenes, C., and G. Immerzeel. *Problem Solving Focus: Time and Money.* Dale Seymour Publications, 1993.

Grouws, D., ed. *Handbook of Research on Mathematics Teaching and Learning.* Macmillan, 1992.

Han, S. T., and B. Ford. *The Master Revealed—A Journey with Tangrams.* Cuisenaire.

Heaton, R. *Teaching Mathematics to the New Standards: Relearning the Dance.* Teachers College Press, 2000.

Henderson, J. *Reflective Teaching: Becoming an Inquiring Educator.* Macmillan, 1992.

Hiebert, J., T. Carpenter, E. Fennema, K. Fuson, D. Wearne, H. Murray, A. Olivier, and P. Humam. *Making Sense: Teaching and Learning Mathematics with Understanding.* Heinemann, 1997.

Hoffman, P. *The Man Who Loved Only Numbers: The Story of Paul Erdos and the Search for Mathematical Truth.* Hyperion, 1998.

House, P., and A. Coxford. *Connecting Mathematics Across the Curriculum.* NCTM, 1995.

Hynes, M. E., ed. *Mission Mathematics: K–6.* NCTM, 1997.

Irvin, J., ed. *What Current Research Says to the Middle Level Practitioner.* National Middle School Association, 1997.

Jacobs, H. *Interdisciplinary Curriculum: Design and Implementation.* ASCD, 1989.

Jurgens, H., E. Maletsky, H. O. Peitgen, T. Perciante, D. Saupe, and L. Yunker. *Fractals for the Classroom: Strategic Activities, Vols. 1 & 2.* NCTM. Copublished with Springer-Verlag, 1991-1992.

Kamii, C., and L. Housman. *Young Children Reinvent Arithmetic: Implications of Piaget's Theory.* Teachers College Press, 1999.

Kaplan, J. *Basic Decimals.* Educational Design, Inc., 1996.

Kaplan, J. *Basic Fractions.* Educational Design, Inc., 1996.

Kaplan, J. *Strategies for Solving Math Word Problems.* Educational Design, Inc., 1996.

Kenney, P., and E. Silver. *Results from the Sixth Mathematics Assessment of the National Assessment of Educational Progress.* NCTM, 1997.

Krause, M. *Multicultural Mathematics Materials.* NCTM, 1993.

Lamancusa, J. *Kid Cash: Creative Money-Making Ideas.* TAB Books, 1993.

Lee, M., and M. Miller. *Great Graphing.* Scholastic Professional Books, 1993.

Leutzinger, L., ed. *Mathematics in the Middle.* NCTM, Copublished with the National Middle School Association, 1998.

Levia, M., et al. "Oh How We've Changed!" in *Addenda Series: Fourth Grade.* NCTM, 1992.

Lindquist, M., et al. *Making Sense of Data. Addenda Series, Grades K–6.* NCTM, 1992.

Ma, Liping. *Knowing and Teaching Elementary Mathematics.* Lawrence Erlbaum Associates, 1999.

Madfes, T., Project Director. *Learning from Assessment: Tools for Examining Assessment through Standards.* (Includes PBS Mathline Video). NCTM, 1999.

Maletsky, E. *Teaching with Student Math Notes.* NCTM, 1993.

Mamchur, C. *A Teacher's Guide to Cognitive Type Theory and Learning Style.* ASCD, 1996.

The Math Learning Center. "Fractions on a Geoboard," in *Opening Eyes to Mathematics, Volume 3.* 1995.

McIntosh, A., B. Reys, R. Reys, and J. Hope. *Number SENSE: Simple Effective Number Sense Experiences, Grades 4–6.* Dale Seymour Publications, 1997.

Means, B., C. Chelener, and M. Knapp. *Teaching Advanced Skills to At-Risk Students.* Jossey-Bass Inc., 1991.

Mendlesohn, E. *Teaching Primary Math with Music.* Dale Seymour Publications, 1990.

Merrill, W. *A Calculator Tutorial.* Dale Seymour Publications, 1996.

Miller, D., and A. McKinnon. *The Beginning School Mathematics Project.* ASCD, 1995.

Miller, E. *Read It! Draw It! Solve It! Problem Solving for Primary Grades.* Dale Seymour Publications, 1997.

Morrison, P., and P. Morrison. *Powers of Ten.* W. H. Freeman and Morrison Company, 1982.

Morrow, L., ed. *The Teaching and Learning of Algorithms in School Mathematics (1998 Yearbook).* NCTM, 1998.

Moses, B. ***Algebraic Thinking, Grades K–12: Readings from NCTM's School-Based Journals and Other Publications.*** NCTM, 1999.

Myren, C. ***Posing Open-Ended Questions in the Primary Classroom.*** Teaching Resource Center, 1997.

Newman, V. ***Math Journals, Grades K–5.*** Teaching Resource Center, 1994.

Newman, V. ***Numbercises—A Fitness Program: Strategies for Addition and Subtraction.*** Teaching Resource Center, 1998.

Norton-Wolf, S. ***Base-Ten Block Activities.*** Learning Resources, 1990.

O'Connor, V., and M. Hynes, ***Mission Mathematics: 5–8.*** NCTM, 1997.

Ohanian, S. ***Garbage, Pizza, Patchwork Quilts, and Math Magic.*** W. H. Freeman and Company, 1992.

Olson, A. ***Mathematics Through Paper Folding.*** NCTM, 1975.

Pappas, T. ***Fractals, Googols and Other Mathematical Tales.*** Wild World Publishing/Tetra, 1993.

Pappas, T. ***The Magic of Mathematics—Discovering the Spell of Mathematics.*** Wild World Publishing/Tetra, 1994.

Parker, M., ed. ***She Does Math!—Real-Life Problems from Women on the Job.*** The Mathematical Association of America, 1995.

Perrone, V., ed. ***Expanding Student Assessment.*** ASCD, 1991.

Phillips, E., et al. ***Addenda Series, Grades 5–8, Patterns and Functions.*** NCTM, 1991.

Phillips, L. M., ed. ***Mathematics: Teacher Resource Handbook.*** Kraus International Publications, 1993.

Piccirilli, R. ***Mental Math: Computation Activities for Anytime.*** Scholastic Professional Books, 1996.

Pohl, V. ***How to Enrich Geometry Using String Designs.*** NCTM, 1986.

Pollard, J. ***Building Toothpick Bridges.*** Dale Seymour Publications, 1985.

Project AIMS. ***AIMS Activities.*** AIMS Educational Foundation, 1988–1995.

Reys, B., et al. ***Addenda Series, Grades 5–8.*** NCTM, 1991.

Rich, D. ***MegaSkills.*** Houghton Mifflin, 1992.

Richardson, K. ***Developing Number Concepts: Book 1, Counting, Comparing and Patterns.*** Dale Seymour Publications, 1999.

Richardson, K. ***Developing Number Concepts: Book 2, Addition and Subtraction.*** Dale Seymour Publications, 1999.

Richardson, K. ***Developing Number Concepts: Book 3, Place Value, Multiplication, and Division.*** Dale Seymour Publications, 1998.

Ringenberg, L. ***A Portrait of 2.*** NCTM, 1995.

Rommel, Carol A. ***Integrating Beginning Math & Literature.*** Incentive Publications, Inc., 1991.

Satariano, P. ***Storytime, Mathtime: Math Explorations in Children's Literature.*** Dale Seymour Publications, 1997.

Schechter, B. *My Brain Is Open: The Mathematical Journeys of Paul Erdos.* Simon & Schuster, 1998.

Scheidt, T. *Fantasy Baseball.* Giant Step Press, 1994.

Schifter, D., and C. Fosnot. *Reconstructing Mathematics Education: Stories of Teachers Meeting the Challenge of Reform.* Teachers College Press, 1993.

Schoenfeld, A. "When Good Teaching Leads to Bad Results: The Disasters of Well-Taught Mathematics Courses," *Educational Psychologist,* Vol. 23, 145–66, 1998.

Schullman, D., and E. Rebeka. *Growing Mathematical Ideas in Kindergarten.* Math Solutions Publications, 1999.

Schultz, K., et al. *Mathematics for Every Young Child.* Merrill, 1990.

Seymour, D. *Getting Smarter Every Day.* Prentice Hall, 1999.

Seymour, D. *Probability Model Masters.* Dale Seymour Publications, 1990.

Sheffield, L. *Developing Mathematically Promising Students.* NCTM, 1999.

Silverman, R., W. Welty, and S. Lyon. *Case Studies for Teacher Problem Solving.* McGraw-Hill, Inc., 1992.

Singer, Margie, et al. *Between Never and Always*. Dale Seymour Publications, 1997.

Skinner, P. *It All Adds Up!* Math Solutions Publications (Adapted by permission of Addison-Wesley Longman, Australia), 1999.

Skinner, P. *What's Your Problem? Posing and Solving Mathematical Problems, K–2.* Heinemann, 1990.

Slavin, R. E., N. L. Karweit, and B. A. Wasik, eds. *Preventing Early School Failure: Research, Policy, and Practice.* Allyn and Bacon, 1994.

Sobel, M., and E. Maletsky. *Teaching Mathematics: A Sourcebook of Aids, Activities, and Strategies.* Allyn & Bacon, 1998.Sonnabend, T. *Mathematics for Elementary Teachers—An Interactive Approach.* Saunders College Publishing, Harcourt Brace College Publishers, 1993.

Steen, L., ed. *On the Shoulders of Giants—New Approaches to Numeracy.* National Research Council, 1990.

Steen, L., ed. *Why Numbers Count—Quantitative Literacy for Tomorrow's America.* NCTM, 1997.

Stenmark, J., V. Thompson, and R. Cossey. *Family Math.* University of California, 1986.

Stenmark, J., ed. *Mathematics Assessment: Myths, Models, Good Questions, and Practical Suggestions.* NCTM, 1991.

Sternberg, R., and W. Williams. *How to Develop Student Creativity.* ASCD, 1996.

Stevenson, F. *Exploratory Problems in Mathematics.* NCTM, 1992.

Stewart, K., and K. Walker. *20 Thinking Questions for Base-Ten Blocks, Grades 3–6.* Creative Publications, 1995.

Stiff, L., and F. Curcio, eds. *Developing Mathematical Reasoning in Grades K–12 (1999 Yearbook).* NCTM, 1999.

Sylvester, R. *A Celebration of Neurons—An Educator's Guide to the Human Brain.* ASCD, 1995.

A Teacher's Guide to Performance-Based Learning and Assessment. Educators in Connecticut's Pomperaug Regional School District 15. ASCD, 1996.

Thiessen, D., and M. Mathias. ***The Wonderful World of Mathematics: A Critically Annotated List of Children's Books in Mathematics.*** NCTM, 1992.

Thornton, C., and N. Bley, eds. ***Windows of Opportunity: Mathematics for Students with Special Needs.*** NCTM, 1994.

Threewit, F. ***Estimation Destinations.*** Cuisenaire, 1994.

Tomlinson, Carol Ann. ***How to Differentiate Instruction in Mixed-Ability Classrooms.*** ASCD, 1995.

Trafton, P., and D. Thiesen. ***Learning Through Problems: Number Sense and Computational Strategies/A Resource for Teachers.*** Heinemann, 1999.

Van Cleave, J. "Graphing," in Van Cleave's ***Math for Every Kid: Easy Activities That Make Learning Math Fun.*** Wiley, 1991.

Van Cleave, J. ***Math for Every Kid.*** Wiley, 1991.

Van de Walle, J. ***Elementary and Middle School Mathematics: Teaching Developmentally, Third Edition.*** Dale Seymour Publications, 1997.

Walter, M. ***Boxes, Squares, and Other Things.*** NCTM, 1995.

Webb, N., and T. Romberg. ***Reforming Mathematics Education in America's Cities: The Urban Mathematics Collaborative Project.*** Teachers College Press, 1994.

Welchman-Tischler, R. ***How to Use Children's Literature to Teach Mathematics.*** NCTM, 1992.

Wu, H. "The 1997 Mathematics Standards War in California," in ***What Is at Stake at the K–12 Standards Wars.*** Edited by S. Stotsky. New York: Peter Lang Publishers, 1999.

Zaslavsky, C. ***Fear of Math—How to Get Over It and Get On with Your Life.*** Rutgers University Press, 1994.

Zaslavsky, C. ***Multicultural Math: Hands-On Math Activities from Around the World.*** Scholastic Professional Books, 1994.

Zawojewski, J., et al. ***Addenda Series, Grades 5–8, Dealing with Data and Chance.*** NCTM, 1991.

Zemelman, S., H. Daniels, and A. Hyde. ***Best Practice: New Standards for Teaching and Learning in America's Schools.*** Heinemann, 1998.

Index

Absolute value, 228–229
Activities
> Home or School Activities, 14G, 92G, 144G, 226G, 268G, 316G, 382G, 460G, 532G. See also Family Involvement Activities
> Literature Connections, 14G, 86, 92G, 144G, 226G, 268G, 316G, 382G, 460G, 532G

Acute angles, 320–321, 332, H18
Acute triangles, 332–335, H49
Addends. See Addition
Addition
> Associative Property, 36–39, H2, H33
> Commutative Property, 36–39, H2, H33
> of decimals, 66–69, H35
> equations, 286–289, H46
> estimating sums, 16–19, 58–59, 176–179, H32, H34, H40
> of fractions, 180, 182–185, H10, H16, H40
> Identity Property of Zero in, H2
> of integers, 242–247, H14, H43
> mental math and, 36–39, H33
> missing addends, 30–31
> of mixed numbers, 186–189, H40
> models of, 180–181, 186–188, 242–244, 246
> properties of, 290–291, H2
> Property of Equality, 290–291
> of rational numbers, 256–259
> subtraction as an inverse operation, H16
> of whole numbers, 20–21, H4, H32

Additive inverse, 243
Adjacent angles, 322–325, H48
Advanced Learners, 14F, 20B, 22B, 26B, 44B, 70B, 76B, 80B, 92F, 94B, 106B, 109B, 126B, 130B, 144F, 146B, 164B, 182B, 192B, 206B, 216B, 226F, 250B, 262, 268F, 272B, 290B, 306B, 316F, 322B, 332B, 344B, 353B, 358B, 372B, 382F, 394B, 418B, 422B, 436B, 442B, 460F, 462B, 479B, 489, 502B, 516B, 521B, 532F, 534B, 553B, 560B, 574B, 579, 580B

Algebra
> absolute value, 228–229
> coordinate plane, graphing ordered pairs, 570–575, 578–579, H61
> equations
>> models of, 286, 296, 304
>> solving, 30–31, 82–83, 163, 216–217, 286–291, 296–299, H11, H21, H46–H47
>> two-step, 304–305
>> writing, 82–83, 284–285, 388–389, 540–542, 574–577, H46
> in exercise sets, 24, 45, 69, 108, 212, 246, 255, 258
> exponents, 40–41, 263, 276–279, H15, H33
> expressions
>> describing simple geometric patterns with, 336–337
>> evaluating for up to three variables, 28–29, 69, 212, 216–217, 246, 272–275, 278–279, H15, H18, H32
>> simplifying, 273–275, H45
>> writing for up to three variables, 28–29, 82–83, 270–271, H45
> finding a rule, 574–577, H14, H61
> formulas, using, 124–125, 300–303, 422–423, 479–481, 484–487, 495–507, 513–515, 518–523, H47
> functions
>> finding, 539–542, 574–577, H59, H61
>> using to graph ordered pairs, 574–575, H61
> inequalities
>> evaluating, 24
>> graphing, 568–569, H61
>> solving, 24, 568–569, H61
>> writing, 569
> input/output patterns, 539, H29
> inverse operations, H16
> in Math Detective, 86, 138, 220, 262, 310, 376, 454, 526, 586
> missing numbers, 108, 480, 515
> order of operations
>> on calculator, 43
>> by paper and pencil, 42–45, 257–259, 272, H15
> parentheses, 42–45, 257–259, 272, H15
> patterns and functions, 539–542, 574–577, H59, H61
> properties
>> Associative, H2
>> Commutative, H2
>> Distributive, 206–207, 274–275, H2
>> of Equality, 287–291
>> Identity Property of Zero in Multiplication, H2
> proportions, 387–401, 462–467, H26, H51
> rates, 385–386, H51
> recognizing a linear pattern, 578–579, H61
> in Thinker's Corner, 39, 163, 289, 542
> variables, 28–29

Algebraic equations. See Equations
Algebraic expressions
> evaluating for up to three variables, 28–29, 69, 212, 216–217, 246, 272–275, 278–279, H18, H32
> writing for up to three variables, 28–29, 82–83, 270–271, H45

Algebraic inequalities, 24, 568–569, H61
Algebraic operating system (AOS), 43
Algebraic Thinking, 26, 29, 31, 37, 41, 52, 79, 82, 120, 151, 193, 206, 210, 229, 234, 246, 254, 270–271, 284, 299, 319, 323, 332, 365, 370, 391, 395, 400, 413, 419, 422, 430, 449, 462, 470, 480, 495, 505, 513, 516, 518, 540, 560, 574 See also Algebra and Professional Handbook, Angela Giglio Andrews

Alternative Teaching Strategy, 16B, 18, 20B, 22B, 24, 26B, 28B, 30B, 36B, 38, 40B, 44B, 46B, 52B, 54, 56B, 58B, 60B, 66B, 68, 70B, 72, 76B, 78, 80B, 82B, 87, 94B, 96, 98B, 100B, 102B, 104, 106B, 109B, 112B, 114, 120B, 122, 124B, 126B, 130B, 132B, 134, 146B, 148B, 150B, 152, 154B, 160B, 162, 164B, 166B, 169B, 176B, 178, 182B, 184, 186B, 188, 192B, 194B, 200B, 202B, 204, 206B, 208, 210B, 212, 214B, 216B, 228B, 230B, 232, 234B, 236B, 244B, 246, 250B, 252B, 254, 256B, 258, 270B, 272B, 274, 278B, 284B, 287B, 290B, 297B, 300B, 302, 304, 306B, 318B, 322B,

324, 326B, 332B, 334, 336B, 338B, 340, 342B, 344B, 350B, 353B, 358B, 364B, 366, 368B, 372B, 384B, 388B, 390B, 392, 394B, 397B, 400B, 406B, 408B, 410, 412B, 414, 418B, 420, 422B, 428B, 430, 432B, 436B, 442B, 444B, 447B, 450B, 462B, 464B, 466B, 468B, 470, 472B, 479B, 482B, 484B, 486, 494B, 496, 498B, 502B, 504B, 506, 512B, 514, 516B, 518B, 521B, 527, 534B, 536B, 539B, 541, 543B, 550B, 553B, 556B, 558B, 560B, 562, 568B, 570B, 572, 574B, 576B, 580B, 582, 587 See also Professional Handbook, Joyce McLeod

Analyzing data, 100–115, 120–135, H5, H24

Angles
in a circle, 416–417
acute, 320–321, 332, H18
adjacent, 322–325, H48
bisecting, 369–370, H50
in circle graphs, 416–417
classifying, 320–325, H48
complementary, 322–325, 333, H48
congruent, 322, 365–367, 391, H50
corresponding, 391–393
degrees in, 320–321, H18, H31
measuring, 320–321, H31
obtuse, 320–321, 332, H18
in polygons, 338–341, H49
in quadrilaterals, 338–341, H49
right, 320–321, 332, H18
solving to find an unknown angle, 322–325, 332–335, H48, H49
straight, 320–321, 332, H18
supplementary, 322–325, 333, H48
in triangles, 332–335, H49
vertical, 322–325, 333, H48

Answer Key, H83–H97

Answers to Selected Exercises, H83–H97

Applications, *14K, 34D, 50D, 64D, 92K, 118D, 144K, 158D, 174D, 198D, 226K, 240D, 268K, 282D, 294D, 316K, 330D, 348D, 362D, 382K, 404D, 426D, 440D, 460K, 476D, 492D, 510D, 532K, 548D, 566D See also* Problem solving

Architecture, 316, 348, 382, 492

Arcs, 344–345, H49

Area
of circles, 501–503, H28, H57
of combined shapes, 495–497, H57
estimating, 494, 496
of irregular figures, 494, 496
of parallelograms, 498–500, H57
of rectangles, 494–497, H28, H57
square units, 494
of squares, 495–497, H28
surface, of a solid, 504–507, H57
of trapezoids, 499–500, H57
of triangles, 495–500, H28, H57

Art Connection, *270B, 356, 484B, 518B*

Assessing Prior Knowledge, *14B, 15, 35, 51, 65, 92B, 93, 119, 144B, 145, 159, 175, 199, 226B, 227, 241, 268B, 269, 283, 295, 316B,*

317, 331, 349, 363, 382B, 383, 405, 427, 441, 460B, 461, 477, 493, 511, 532B, 533, 549, 567

Assessment
Chapter Review/Test, 32, 48, 62, 84, 116, 136, 156, 172, 196, 218, 238, 260, 280, 292, 308, 328, 346, 360, 374, 402, 424, 438, 452, 474, 490, 508, 524, 546, 564, 584, *T16–T17*
Cumulative Review. *See* Mixed Applications, Mixed Review and Test Prep, Mixed Strategy Practice, *and* Standardized Test Prep
Daily, 14B, 92B, 144B, 226B, 268B, 316B, 382B, 460B, 532B, T16–T17
Formal, 14B, 92B, 144B, 226B, 268B, 316B, 382B, 460B, 532B, T16–T17
Harcourt Electronic Test System, 14B, 92B, 144B, 226B, 268B, 316B, 382B, 460B, 532B, T16–T17
Lesson Quiz. See last page of each lesson
Math Practice and Assessment, 14B, 92B, 144B, 226B, 268B, 316B, 382B, 460B, 532B, T16–T17
Mixed Review and Test Prep, 19, 21, 25, 29, 31, 39, 41, 43, 45, 55, 59, 61, 69, 73, 75, 79, 83, 97, 99, 105, 108, 111, 115, 123, 125, 128, 131, 135, 147, 149, 153, 163, 165, 167, 171, 179, 181, 185, 189, 191, 193, 201, 205, 207, 209, 213, 217, 229, 233, 235, 243, 247, 249, 251, 255, 259, 271, 275, 277, 279, 285, 289, 291, 299, 303, 305, 319, 321, 325, 327, 335, 341, 343, 345, 352, 355, 357, 367, 370, 373, 386, 393, 396, 399, 401, 407, 411, 415, 417, 421, 423, 431, 435, 437, 446, 449, 463, 465, 467, 471, 487, 497, 500, 503, 507, 515, 519, 523, 538, 542, 545, 552, 555, 559, 563, 569, 573, 575, 579, 583
Oral, 43, 75, 129, 168, 181, 191, 219, 243, 249, 277, 286, 296, 305, 321, 357, 371, 387, 417, 435, 478, 489, 501, 520, 579
Performance, 14B, 92B, 144B, 226B, 268B, 316B, 382B, 460B, 532B, T16–T17
Standardized Test Prep, 33, 49, 63, 85, 117, 137, 157, 173, 197, 219, 239, 261, 281, 293, 309, 329, 347, 361, 375, 403, 425, 439, 453, 475, 491, 509, 525, 547, 565, 585
Student, 14B, 92B, 144B, 226B, 268B, 316B, 382B, 460B, 532B, T16–T17
Study Guide and Review, 88–89, 140–141, 222–223, 264–265, 312–313, 378–379, 456–457, 528–529, 588–589
Test Preparation, 14B, 92B, 144B, 226B, 268B, 316B, 382B, 460B, 532B, T16–T17
See also Professional Handbook, Lynda Luckie

Assessment Options, 14B, 92B, 144B, 226B, 268B, 316B, 382B, 460B, 532B

Associative Property
of Addition, 36–39, H2, H33
of Multiplication, 36–39, H2, H33

Astro Algebra. *See Mighty Math Astro Algebra*

Average. *See* Mean, Median, *and* Mode

Average speed, 300–303, 385

Axes, 570

Bar graphs, 120–123, H6, H37

Bases
of plane figures, 495–496, 498–500
of powers, 40–41, H15
of solid figures, 350–355, 358–359, H50

Basic Facts. *See Daily Facts Practice*

Biased sample, 98–99, H36

Bibliographies
Computer, BI3–BI6
Professional, BI7–BI13
Student, BI1–BI2

Binary number system, 542

Bisects
angles, 369–370, H50
line segments, 368–370, H50

Block Scheduling, 14F, 92F, 144F, 226F, 268F, 316F, 382F, 460F, 532F

Box-and-whisker graphs, 129–131, H37

Calculating Crew. *See Mighty Math Calculating Crew*

Calculator Handbook, 40, 70, 76, 87, 126, 129, 139, 169, 182, 202, 244, 250, 416, 422, 436, 444, 479, 484, 494, 498, 502, 512, 536, 574

Calculators, 43, 77, 163, 169, 170, 276–277, 413, 435, 484, 485, 591

Capacity
customary, 462–463, H25, H55
metric, 464–465, H25, H55

Career Connection, 16B, 28B, 70B, 112B, 130B, 154B, 166B, 206B, 412B, 504B, 550B, 570B

Careers, 153, 213, 393

Celsius temperature, 301–303

Center, of circle, 344

Centimeters, 464–465, H25, H55

Challenge
Explore Perspective, 377
Explore Scatterplots, 139
Mixed Numbers and Time, 221
Negative Exponents, 263
Networks, 527
Percent of Increase and Decrease, 455
Reflexive, Symmetric, and Transitive Properties, 311
Scientific Notation, 87
Stretching Figures, 587

Chapter at a Glance, 14I, 34B, 50B, 64B, 92I, 118B, 144I, 158B, 174B, 198B, 226I, 240B, 268I, 282B, 294B, 316I, 330B, 348B, 362B, 382I, 404B, 426B, 440B, 460I, 476B, 492B, 510B, 532I, 548B, 566B

Chapter Planner, 14H, 34A, 50A, 64A, 92H, 118A, 144H, 158A, 174A, 198A, 226H, 240A, 268H, 282A, 294A, 316H, 330A, 348A, 362A, 382H, 404A, 426A, 440A, 460H, 476A, 492A, 510A, 532H, 548A, 566A

Chapter Project. *See Problem Solving Projects*

Chapter Review/Test, 32, 48, 62, 84, 116, 136, 156, 172, 196, 218, 238, 260, 280, 292, 308, 328, 346, 360, 374, 402, 424, 438, 452, 474, 490, 508, 524, 546, 564, 584

Check What You Know, 15, 35, 51, 65, 93, 119, 145, 159, 175, 199, 227, 241, 269, 283, 295, 317, 331, 349, 363, 383, 405, 427, 441, 461, 477, 493, 511, 533, 549, 567

Choose the Operation, 214–215

Chords, 344–345, H49

Circle graphs, 122–123, 416–417, H52

Circles
arc, 344–345, H49
area of, 501–503, H28, H57
as base of cylinder, 351–352
center of, 344
chord, 344–345, H49
circumference of, 484–487, H56
degrees in, 416–417
diameter of, 344–345, H49
fractions of, 160, 208
radius of, 344–345, H49
sector of, 344–345, H49
See also Geometry

Circumference, 484–487, H56

Classroom organization. *See Organizer at beginning of each lesson*

Clustering (in estimation), 16–19, 58–59

Combinations, 446

Combined shapes, area of, 495–497, H57

Common Error Alert, 18, 21, 24, 38, 45, 54, 59, 61, 68, 78, 83, 96, 99, 107, 123, 125, 147, 149, 153, 165, 167, 171, 184, 193, 205, 212, 229, 232, 235, 246, 251, 254, 274, 279, 289, 291, 299, 302, 319, 324, 334, 345, 352, 355, 366, 373, 386, 392, 399, 401, 411, 415, 421, 423, 430, 437, 445, 451, 463, 465, 481, 483, 486, 496, 503, 517, 519, 538, 541, 551, 562, 572, 582

Common factors, 151

Common multiples, 150

Communication
Opportunities to communicate mathematical thinking are contained throughout the book. Some examples: 16, 20, 22, 28, 30, 36, 40, 42, 44, 46, 52, 58, 60, 66, 70, 74, 76, 80, 82, 94, 98, 102, 106, 109, 120, 124, 126, 129, 130, 132, 146, 148, 150, 160, 164, 166, 168, 169, 176, 180, 182, 186, 190, 192, 194, 200, 202, 206, 208, 210, 214, 228, 230, 234, 242, 244, 248, 250, 252, 256, 270, 272, 276, 278, 284, 286, 287, 290, 296, 297, 300, 304, 318, 320, 322, 326, 332, 336, 338, 342, 344, 350, 353, 356, 358, 364, 368, 371, 372, 384, 387, 388, 390, 394, 397, 400, 406, 408, 412, 416, 418, 422, 432, 434, 436, 442, 444, 447, 450, 462, 464, 466, 468, 478, 479, 484, 488, 494, 498, 501, 502, 504, 512, 516, 518, 520, 521, 536, 539, 543, 550, 553, 556, 558, 560, 568, 570, 574, 578, 580

See also Math Labs, Writing in Mathematics

Commutative Property
of Addition, 36–39
of Multiplication, 36–39

Compare
    decimals, 53–55, 231–232, H29
    fractions, 166–167, 231–232, H12, H29
    integers, 228–229, H29, H42
    mixed numbers, 234–235
    on the number line, 166–167, 231–232, 234–235, H13, H29
    percents, 406–407
    rational numbers, 234–235
    whole numbers, H3

Compass, 371, 416–417, 484
Compatible numbers, 17–19
Compensation, 37–39
Complementary angles, 322–325, H48
Composite numbers, 148, H7
Compound events, 444–446, H54
Computer, 353 *See also* Cross-curricular connections E-Lab activities Mighty Math Software *and* Technology
Computer software, BI3–BI6 *See* Technology Links
Cones, 350–352, H50
Congruent
    angles, 322, 365–367, 391, H50
    figures, 322, 372–373, 390–393, H22, H50
    line segments, 364–367, H50

Connections. *See* Cross-curricular connections
Constructing
    congruent angles, 365–367, H50
    congruent line segments, 364–367, H50
    parallel lines, 371, H14, H50

Consumer, 52, 294, 404, 416, 418, 422
Consumer Connection, 568B
Convenience samples, 95–97, H36
Coordinate grid. *See* Coordinate plane
Coordinate plane
    graphing on, 570–575, 578–579, H61
    quadrants of, 570–573, H61
    transformations on, 580–583, H61

Corresponding
    angles, 391–393, H51
    sides, 391–396, H51

Cosmic Geometry. *See* Mighty Math Cosmic Geometry
Counting principle. *See* Fundamental Counting Principle
Critical thinking. *See* Reasoning
Cross products, 388–389, H26, H51
Cross-curricular connections
    Architecture, 316, 348, 382, 492
    Art, *270B*, 325, 356, *484B*, *518B*
    Career, *16B, 28B, 70B, 112B, 130B, 154B, 166B, 206B, 412B, 504B, 550B, 570B*
    Computer, 353
    Consumer, 52, 294, 404, 416, 418, 422, *568B*
    Fine Arts, *310, 318B, 466B, 472B*
    Geography, 92, 133, 176, 226, 228, 284, 415, 566
    Health, 50, 174
    History, 154, 166, 440
    Language Arts, *176B, 342B, 390B, 558B*
    Literature. *See* Literature Connections
    Math, 42, *150B*, 190, *210B, 230B*, 276, 526, *536B*, 586
    Music, *576B*
    Physical Education, *139, 166B, 278B, 450B*
    Reading, 25, 135, 189, 233, 275, 341, 415, 515, 563
    Science, 14, 20, *22B*, 34, 36, *44B*, 66, 69, *82B*, 91A, 91B, 102, 107, 118, 120, 130, 143A, 143B, 158, 161, 164, 170, 186, *186B*, 187, 192, 200, 216, 234, 240, 250, *263, 284B*, 297, *300B*, 332, 384, *384B*, 387–388, *397B, 428B, 444B, 468B, 534B*, 548, 553, 560, 573, *574B*, 580
    Social Studies, *26B*, 30, 60, *60B, 66B, 98B*, 103, *124B*, 255A, 255B, *297B*, 300, 315A, 315B, *338B, 350B*, 381A, 394, *400B*, 410, 416, 459A, 476, 484, 487, *504B*, 510, 581
    Sports, 16, 198, 282, 426, 460, 532
    Technology, 64, 330, 344
    Visual Arts, 498B, 550B

Cubes, 350–352, H50
Cubic units, 512
Cumulative frequencies, 103–105
Cumulative Review. *See* Mixed Applications, Mixed Review and Test Prep, Mixed Strategy Practice, *and* Standardized Test Prep
Cups, 462–463, H25, H55
Curriculum connections. *See* Cross-curricular connections
Customary system
    capacity, 462–463, H25, H55
    changing units, 462–463, H25, H27, H55
    Fahrenheit temperature, 301–303
    length, 462–463, H55
    metric units compared to, 466–467, H55
    weight, 462–463, H25, H55
    See also Table of Measures *on back cover*

Cylinders, 350–352, H50

Daily Facts Practice, 16A, 20A, 22A, 26A, 28A, 30A, 36A, 40A, 44A, 46A, 52A, 56A, 58A, 60A, 66A, 70A, 76A, 80A, 82A, 94A, 98A, 100A, 102A, 106A, 109A, 112A, 120A, 124A, 126A, 130A, 132A, 146A, 148A, 150A, 154A, 160A, 164A, 169A, 176A, 182A, 186A, 192A, 194A, 200A, 202A, 206A, 210A, 214A, 216A, 228A, 230A, 234A, 236A, 244A, 250A, 252A, 256A, 258A, 260A, 262A, 270A, 272A, 278A, 284A, 287A, 290A, 297A, 300A, 306A, 318A, 322A, 326A, 332A, 336A, 338A, 342A, 344A, 350A, 353A, 358A, 364A, 368A, 384A, 388A, 390A, 394A, 397A, 400A, 406A, 408A, 412A, 418A, 422A, 428A, 432A, 436A, 442A, 444A, 447A, 450A, 462A, 464A, 466A, 468A, 472A, 479A, 484A, 494A, 498A, 502A, 504A, 512A, 516A, 518A, 521A, 534A, 536A, 539A, 543A, 550A, 553A, 556A, 558A, 560A, 568A, 570A, 574A, 576A, 580A *See also* Professional Handbook, Grace M. Burton

Data
   analyzing, 100–115, 120–135, H5, H24
   bar graphs, 120–123, H6, H37
   box-and-whisker graphs, 129–131, H37
   circle graphs, 122–123, 416–417
   collecting, 94–99, H36
   comparing measures of central tendency, 106–111, H36
   conclusions based on, 112–115, H36
   evaluating validity of claims, 112–115
   frequency tables, 103–105, H36
   histograms, 127–128, H37
   interpreting, 120–135
   line graphs, 121–125, H37
   line plots, 102–105
   making tables, 56–57, 100–101
   measures of central tendency
      mean, 106–111, H6, H36
      median, 106–111, H6, H36
      mode, 106–111, H6, H36
   misleading graphs, 132–135, H37
   organized lists, 154–155, 442–443
   organizing, 56–57, 100–105, 120–123, 126–131
   outcomes, 428
   outliers, 110–111, H36
   predictions from, 124–125
   range, 103, H6
   representing, 120–123, 126–131, H37
   samples
      biased, 98–99, H36
      convenience, 95–97, H36
      random, 95–97, H36
      responses to survey, 94–97
      systematic, 95–97, H36
      unbiased, 98–99, H36
      when to use, 94–97
   scatterplots, 139
   stem-and-leaf plots, 126–128, H7, H37
   surveys, 94–99
   tally tables, 100–101
   tree diagrams, 444–446, H54
   use data, 19, 21, 38, 41, 47, 55, 57, 59, 61, 73, 91A, 91B, 114, 129, 143A, 143B, 163, 171, 179, 184, 188, 225A, 225B, 267A, 267B, 291, 303, 315A, 315B, 337, 345, 381B, 433, 443, 451, 459A, 459B, 471, 473, 483, 507, 517, 531A, 531B, 542, 557, 569
   Venn diagrams, 230, 233
Data ToolKit, 91, 92E, 100B, 102, 102B, 120B, 126B, 130B, 315, 382E, 406B, 460E, 466B, 512B, 531
Decimal points, placing, 52, 66–73, 76–79
Decimals
   adding, 66–69, H35
   calculators and, 77
   comparing, 53–55, H29
   dividing, 74–79, H35
   equivalent, 52–55, H34
   estimating, 58–59, 67–69
   to fractions, 169–171
   fractions to, H22
   metric measures and, 464–465
   models of, 70–75, H2
   money and, 66
   multiplying, 70–73, H23, H35
   ordering on a number line, 53–55, 231–232
   patterns in, 77
   to percents, 60–61, 408–411, H23
   percents to, 60–61, 408–411, H23
   place value in, 52–55, H2
   placing the decimal point, 52, 66–73, 76–79
   and powers of ten, 87, 263
   repeating, 169–171
   rounding, 58, H4
   subtracting, 66–69, H35
   terminating, 169–171
   in words, 52–55, H3
   See also Professional Handbook, Roger Howe
Decimeters, 464–465, H25, H55
Degrees
   in angles, 320–321, H18, H31
   in a circle, 416–417
   measuring with a protractor, 320–321
   in a quadrilateral, 336
   on a thermometer, 301–303
   in a triangle, 332–335, H49
Denominators
   least common, 182–187
   in like fractions, 180
   in unlike fractions, 180
Dependent events, 448–449
Diameters, 344–345, H49
Differences. *See* Subtraction
Digits, place value of, H3
Discounts, 418–421
Discuss. See last page of each lesson
Disjoint events, 444–446
Distance formula, 124–125, 300–303
Distributive Property, 36–39, 206–207, 274–275
Dividends. *See* Division
Divisibility rules, 146–147, H38
Division
   with calculators, 77
   connecting to multiplication, 210–213
   of decimals, 74–79, H35
   divisibility rules, 146–147
   equations, 298–299, H47
   and equivalent fractions, 161
   estimation in, 200–201
   to find unit rates, 385–386, H51

of fractions, 208–213, H41
of integers, 253–255, H43
interpreting remainders, 80–81
mental math and, 36–39, 211–212, H13
of mixed numbers, 211–213
models of, 74–75, 208–209
multiplication as an inverse operation, H16
patterns in, 77, H13
of powers of ten, H13
Property of Equality, 297–299
of rational numbers, 257–259
remainders, 23–24, H5
of whole numbers, 23–25, 37–39, H4

Divisor. *See* Division
Double-bar graphs, 120–123
Double-line graphs, 121–123
Draw a Diagram strategy, 2, 194–195
Draw a Picture strategy. *See* Draw a Diagram strategy

Early Finishers, *20B, 28B, 46B, 56B, 58B, 66B, 76B, 100B, 102B, 132B, 150B, 160B, 180, 194B, 200B, 214B, 221, 243, 252B, 278B, 304, 326B, 336B, 357, 364B, 377, 388B, 406B, 408B, 412B, 432B, 447B, 464B, 472B, 478, 494B, 512B, 536B, 539B, 556B, 570B*
Edges, 358–359, H20, H50
E-Lab activities, 43, 75, 129, 143, 162, 168, 181, 191, 208, 249, 277, 286, 296, 305, 353, 369, 387, 417, 435, 459, 484, 501, 520, 534, 544, 581
See also Technology Links
Endpoints, 318, 320
English Language Learners, *14G, 30B, 40B, 56B, 60B, 80B, 82B, 92G, 98B, 100B, 102B, 120B, 144G, 148B, 160B, 169B, 194B, 200B, 210B, 214B, 226G, 228B, 234B, 244B, 260B, 268G, 270B, 284B, 297B, 316G, 321, 326B, 338B, 353B, 368B, 382G, 384B, 390B, 400B, 408B, 434, 447B, 454, 460G, 468B, 479B, 484B, 494B, 521B, 532G, 543B, 558B, 580B*
Equations
addition, 286–289, H46
decimals in, 82–83
division, 298–299
fractions in, 216–217
integers in, 284–291, H46
models of, 286, 296, 304
multiplication, 296–299, H47
proportions, 387–389, 394–401, 462–467, H26, H51
solving, 30–31, 82–83, 216–217, 286–291, 296–299, H11, H21, H46–H47
subtraction, 286, 290–291, H46
two-step, 304–305
writing, 82–83, 284–285, 388–389, 540–542, 574–577, H46, H59
Equilateral triangles, 332
Equivalent decimals, 52–55, H34
Equivalent fractions, 160–163, H21, H39

Equivalent ratios, 384–386, H51
Error analysis. *See* What's the Error?
Errors, finding. *See* Common Error Alert
Essential Skills, *14K, 34D, 50D, 64D, 92K, 118D, 144K, 158D, 174D, 198D, 226K, 240D, 268K, 282D, 294D, 316K, 330D, 348D, 362D, 382K, 404D, 426D, 440D, 460K, 476D, 492D, 510D, 532K, 548D, 566D*
Estimate or Find Exact Answer, 472–473
Estimation
of area, 494, 496
of circumference, 484
by clustering, 16–19, 58–59
compatible numbers, 17–19, 58–59
of decimals, 58–59, 67–69
of differences, 16–19, 177–178
of fractions, 176–179, 200–201
multiples of ten, 16–19
overestimate, 17–19
of percent, 412–415
of perimeter, 478
of products, 17–19, 200–201
of quotients, 17–19, 200–201
reasonableness, 66–69
and rounding, 16–19, 58–59
of sums, 16–17, 58–59, 176–179, H32, H34, H40
underestimate, 17–19
of unknown quantities, 124–125
of volume, 512
when to estimate, 472–473
See also Professional Handbook, Roger Howe
Events
compound, 444–446
dependent, 448–449
impossible versus certain, 429
independent, 447–449, H54
Expanded form, 52–55
Experimental probability, 434–437, H53
Exponents, 40–41, 263, 276–279, H15, H33
negative, 263
Expressions
algebraic, 28–29, 69, 82–83, 212, 216–217, 246, 270–275, 278–279, H18, H32
evaluating, 82–83, 216–217, 272–275, 278–279, H15, H18
numerical, 28–29, 278–279, H15
writing, 28–29, 82–83, 216–217, 270–271, H45
Extra Practice, H32–H61

Faces, 350–355, 358–359, H20, H50
Fact Practice. *See* Daily Facts Practice on the first page of every TE lesson
Factor trees, 148–149

Index **BI19**

Factors
  common, 151
  finding prime factors of a number, 148–149, H38
  greatest common, 151–153
  missing, 30–31
  of numbers, H8
  prime, 148–149
Fahrenheit temperature, 301–303
Family Involvement Activities, 14I, 34B, 50B, 64B, 92I, 118B, 144I, 158B, 174B, 198B, 226I, 240B, 268I, 282B, 294B, 316I, 330B, 348B, 362B, 382I, 404B, 426B, 440B, 460I, 476B, 492B, 510B, 532I, 548B, 566B
Family Support, T18 *See also* Professional Handbook, Vicki Newman
Feet, 462–463, H25, H55
Fibonacci sequence, 538
Find a Pattern strategy, 7, 336–337, 534–535
Finding a rule. *See* Algebra *and* Functions
Fine Arts Connection, 310, 318B, 466B, 472B
Flips, H30
  See also Reflections
Fluid ounces, 462–463
Focus on Problem Solving, 1–13
Formulas
  for area
    of circles, 501–503
    of parallelograms, 498–500
    of rectangles, 494–497
    of squares, 495–497, H28
    of trapezoids, 499–500
    of triangles, 495–497
  Celsius to Fahrenheit, 301–303
  for circumference of circles, 484–487
  distance, 124–125, 300–303
  Fahrenheit to Celsius, 301–303
  generating, 495, 498–499, 501
  for perimeter of polygons, 479
  for simple interest, 422–423
  for surface area of solids, 504–507
  for volume
    of cylinders, 521–523
    of rectangular prisms, 513–515
    of triangular prisms, 514–515
  See also Table of Measures *on back cover*
Fractals, 544
Fractions
  adding, 180, 182–185, H10, H16, H40
  of circles, 160, 208
  comparing, 166–167, H12, H29
  decimals and, 168–171, H22, H23
  dividing, 208–213, H41
  in equations, 216–217
  equivalent, 160–163, H21, H39
  estimating, 176–179
  finding common denominators, 180–185, H40
  greatest common factor (GCF), 203–205
  improper, 164
  least common denominator (LCD), 182–185
  least common multiple (LCM), 150–153, 180–182
  like, 180, H10
  in measurements, 462–463
  in mixed numbers, 164–165, H11
  models of, 160–161, 180, 202, 208–209, H8
  multiplying, 202–205, H23, H41
  negative, 230–235, H42
  on number lines, 166–167, 176–178
  ordering, 166–167, 231–232
  as parts of a group, H8
  as parts of a whole, H8
  percent and, 170–171, 406–411, H23, H39
  probability and, 428–431, 436–437
  as ratios, 384–386, H51
  rounding, 176–178, 200, H10
  simplest form of, 161–163, H9
  subtracting, 181, 183–185, H10, H16, H40
  See also Professional Handbook, Tom Roby
Frequency tables, 103–105
Function tables. *See* Functions
Functions, 262, 539–542, 574–577, H59, H61
Fundamental Counting Principle, 444–446, H54

Gallons, 462–463, H25, H55
Geography, 92, 133, 176, 226, 228, 284, 415, 566
Geometric patterns, 543–545, H59
Geometric probability, 431
Geometry
  angles
    acute, 320–321, 332, H18
    adjacent, 322–325, H48
    bisecting, 369–370, H50
    in a circle, 416–417
    in circle graphs, 416–417
    classifying, 320–325, H48
    complementary, 322–325, 333, H48
    congruent, 333, 365–367, 391, H50
    corresponding, 391–393
    degrees in, 320–321, H18, H31
    drawing, 320–321
    measuring, 320–321, H31
    obtuse, 320–321, 332, H18
    in polygons, 338–341, H49
    in quadrilaterals, 338–341, H49
    right, 320–321, 332, H18, H48

solving to find an unknown, 322–325, 332–335, H48, H49
straight, 320–321, 332, H18, H48
supplementary, 322–325, 333, H48
in triangles, 332–335, H49
vertical, 322–325, 333, H48
area
  of circles, 501–503, H28, H57
  of parallelograms, 498–500, H57
  of rectangles, 494–497, H28, H57
  of squares, 499–500, H28
  of trapezoids, 495–497, H57
  of triangles, 495–497, H28, H57
circles
  arc, 344–345, H49
  area, 501–503, H28, H57
  chord, 344–345, H49
  circumference, 484–487, H56
  diameter, 344–345, H49
  radius, 344–345, H49
  sector, 344–345, H49
classifying
  angles, 320–325, H18, H48
  figures as similar, 390–393, H51
  kinds of symmetry, 560–563
  lines, 326–327
  polygons, 332–335, 338–341, H19, H49
  polyhedra, 350–352, H20, H50
  prisms and pyramids, 350–352, H50
  quadrilaterals, 338–341, H49
  tessellating figures, 553–555, H60
  triangles, 332–335, H49
congruent angles, 322, 365–367, 391, H50
congruent figures, 322, 372–373, 390–393, H22, H50
congruent line segments, 364–367, H50
constructing
  congruent angles, 365–367, H50
  congruent line segments, 364–367, H50
  parallel lines, 371, H14, H50
coordinate plane, representing points on, 570–575, H61
degrees
  in an angle, 320–321, H18, H31
  in a circle, 416–417, H52
  in a quadrilateral, 336
  in a triangle, 332–335, H49
flips, H30 *See also* Reflections
irregular figures, 478, 494, 496
lines
  angle relationships and, 322–325
  intersecting, 326–327, H48
  parallel, 326–327, 371, H48
  perpendicular, 326–327, H48
  ray, defined, 318
  segments, 318, 364–367
motion. *See* Transformations

networks, 527
perimeter, 478, H26, H56
pi, 79, 484–487
plane, defined, 318–319
plane figures
  circles, 344–345, H49
  drawing, 342–343, H49
  hexagons, 336
  octagons, 336
  pentagons, 336
  polygons, 332–341, H19
  quadrilaterals, 338–341, H49
  triangles, 332–335, H49
point, defined, 318–319
Pythagorean Theorem, 488–489
quadrilaterals
  parallelograms, 338–341, H49
  rectangles, 338–341, H49
  rhombuses, 338–341, H49
  squares, 338–341, H49
  trapezoids, 338–341, H49
similar figures, 372–373, 390–393, H22, H50, H51
slides, H30 *See also* Translations
solid figures
  cones, 350–352, H50
  constructing, from a pattern, 356–357, H50
  cubes, 350–352, H50
  cylinders, 350–352, H50
  drawing, 353–355
  edges of, 358–359, H20, H50
  faces of, 350–355, 358–359, H20, H50
  prisms, 350–352, H50
  pyramids, 350–352, H50
  vertices of, 358–359, H20, H50
surface area of solids, 504–507, H57
symmetry
  line, 560–563, H30, H60
  rotational, 560–563, H60
tessellations, 553–557, H60
three-dimensional figures. *See* Solid figures
transformations
  reflections (flips), 550–552, 558–559, 580–583, H30, H61
  rotations (turns), 550–552, 558–559, 580–583, H30, H61
  stretching, 587
  tessellations and, 553–557, H60
  translations (slides), 550–552, 558–559, 580–583, H30, H61
triangles
  acute, 332–335, H49
  equilateral, 332
  isosceles, 332, 342
  obtuse, 332–335, H49
  right, 332–335, 488–489, H49, H50
  scalene, 332
  similar, 395

Index **B121**

turns, H30 *See also* Rotations
two-dimensional figures. *See* Plane figures
volume
  of cylinders, 520–523, H58
  of prisms, 512–515, H58
  of pyramids, 518–519, H58
  See also Professional Handbook, David G. Wright
Geometry Connection, 230B, 276, 536B
Gifted and Talented. *See* Advanced Learners
Glossary, H73–H82
Grams, 464–465, H25, H55
Graphs
  analyzing, 120–135
  bar, 120–123, H6, H37
  box-and-whisker, 129–131, H37
  choosing, 120–123, 126–128
  circle, 122–123, 416–417, H52
  comparing, 120–123, 132–135
  coordinate plane, 570–575, 578–579, H61
  of equations, 578–579
  of functions, 574–575, H61
  histograms, 127–128, H37
  inequalities, 568–569, H61
  interpreting, 120–135
  intervals of, 127–128
  key of, 120
  labels on, 416–417
  line, 121–125, H37
  line plots, 102–105
  making, 102–105, 120–123, 126–128, 129, H36–H37
  misleading, 132–135, H37
  multiple-bar, 120–123, H37
  multiple-line, 121–123, H37
  ordered pairs, 570–575, 578–579, H61
  predictions from, 124–125
  as problem solving strategy, 8
  scatterplots, 139
  stem-and-leaf plots, 126–128, H7, H37
  transformations, 580–583
  tree diagram, 444–446, H54
"Greater than" symbol, 568–569, H3
Greatest common factor (GCF), 151–153, 161–163, 203–205
Guess and Check strategy. *See* Predict and Test strategy
Guided Instruction. *See* the Teacher's Edition for each lesson
Guided Practice. *See* the Teacher's Edition for each lesson

Harcourt Math Newsroom Videos, 18, 234, 273, 338, 390, 484, 539
Health, 50, 174

Height
  of parallelograms, 498–500
  of prisms, 512–515
  of pyramids, 505–506, 518–519
  of trapezoids, 498–500
  of triangles, 495–497
Hexagonal prisms, 350–352
Hexagonal pyramids, 350–352
Hexagons, 336
Histograms, 127–128, H37
History, 154, 166, 440
Home-School Connection. *See* Family Involvement Activities
Hour, 462–463
How to Help Options. *See* Intervention
Hypotenuse, 488–489

Identity Property
  of Addition, H2
  of Multiplication, H2
Images
  reflection, 550–552, H60
  translation, 550–552, H60
Impossible events, 429, H24
Improbable events. *See* Unlikely outcomes
Improper fractions, 164
Inches, 462–463, H25, H55
Inclusion. *See* Special Needs
Independent events, 447–449, H54
Independent Practice. *See* the Teacher's Edition for each lesson
Indirect measurement, 394–396, H51
Inequalities, 24, 568–569, H61
Input/output tables, 539–542, H29
Integer equations, 284–291, H46
Integers
  adding, 242–247, H14, H43
  comparing and ordering, 228–229, H29
  concept of, 228–229, H12
  dividing, 253–255, H44
  in equations, 284–291, H46
  in expressions, 257
  models of, 242–244, 246, 248–249, 252
  multiplying, 252–255, H43
  negative, 228–229, 263
  on a number line, 228–229, 244, 246, 252, H12, H29
  opposites, 228–229
  subtracting, 248–251, H14, H43
Interdisciplinary Suggestions, 14F, 92F, 144F, 226F, 268F, 316F, 382F, 460F, 532F

Interest, simple, 422–423, H52

Internet Activities, 14E, 92E, 144E, 226E, 268E, 316E, 382E, 460E, 532E *See also* E-Lab Activities *and* Technology Links

Interpret the remainder, 80–81

Intersecting lines, 326–327, H48

Intervals, 127–128

Intervention, 15, 35, 51, 65, 89, 93, 119, 141, 145, 159, 175, 199, 223, 227, 241, 265, 269, 283, 295, 313, 317, 331, 349, 363, 379, 383, 405, 427, 441, 457, 461, 477, 493, 511, 529, 533, 549, 567, 589, T15

Intervention and Extension Resources, 16B, 20B, 22B, 26B, 28B, 30B, 36B, 40B, 42, 44B, 46B, 52B, 56B, 58B, 60B, 66B, 70B, 74, 76B, 80B, 82B, 86–87, 94B, 98B, 100B, 102B, 106B, 109B, 112B, 120B, 124B, 126B, 130B, 132B, 138–139, 146B, 148B, 150B, 154B, 160B, 164B, 166B, 169B, 176B, 180, 182B, 186B, 190, 192B, 194B, 200B, 202B, 206B, 208, 210B, 214B, 216B, 220–221, 228B, 230B, 234B, 236B, 242–243, 244B, 248, 250B, 252B, 256B, 262–263, 270B, 272B, 276, 278B, 284B, 287B, 290B, 297B, 300B, 304, 306B, 310–311, 318B, 320–321, 322B, 326B, 332B, 336B, 338B, 342B, 344B, 350B, 353B, 356–357, 358B, 364B, 368B, 372B, 376–377, 384B, 388B, 390B, 394B, 397B, 400B, 406B, 408B, 412B, 416, 418B, 422B, 428B, 432B, 434, 436B, 442B, 444B, 447B, 450B, 454–455, 462B, 464B, 466B, 468B, 472B, 478, 479B, 482B, 484B, 488, 489, 494B, 498B, 502B, 504B, 512B, 516B, 518B, 521B, 526–527, 534B, 536B, 539B, 543B, 550B, 553B, 556B, 558B, 560B, 568B, 570B, 574B, 576B, 578–579, 580B, 586–587 *See also* Reaching All Learners *and* Technology Links

Intervention for Prerequisite Skills, 16A, 20A, 22A, 26A, 28A, 30A, 36A, 40A, 44A, 46A, 52A, 56A, 58A, 60A, 66A, 70A, 74, 76A, 80A, 82A, 94A, 98A, 100A, 102A, 106A, 109A, 112A, 120A, 124A, 126A, 129, 130A, 132A, 146A, 148A, 150A, 154A, 160A, 164A, 166A, 169A, 176A, 182A, 186A, 192A, 194A, 200A, 202A, 206A, 210A, 214A, 216A, 228A, 230A, 234A, 236A, 242, 244A, 248, 250A, 252A, 256A, 258A, 260A, 262A, 270A, 272A, 276, 278A, 284A, 286, 287A, 290A, 296, 297A, 300A, 304, 306A, 318A, 320, 322A, 326A, 332A, 336A, 338A, 342A, 344A, 350A, 353A, 356, 358A, 364A, 371, 372A, 384A, 387, 388A, 390A, 394A, 397A, 400A, 406A, 408A, 412A, 416, 418A, 422A, 428A, 432A, 434, 436A, 442A, 444A, 447A, 450A, 462A, 464A, 466A, 468A, 472A, 478, 479A, 484A, 494A, 498A, 501, 502A, 504A, 512A, 516A, 518A, 520, 521A, 534A, 536A, 539A, 543A, 550A, 553A, 556A, 558A, 560A, 568A, 570A, 574A, 576A, 578, 580A

Intervention Strategies and Activities, 14E, 16B, 20B, 22B, 26B, 28B, 30B, 36B, 40B, 43, 44B, 46B, 52B, 56B, 58B, 60B, 66B, 70B, 75, 76B, 80B, 82B, 92E, 94B, 98B, 100B, 102B, 106B, 109B, 112B, 120B, 124B, 126B, 129, 130B, 132B, 144E, 146B, 148B, 150B, 154B, 160B, 164B, 166B, 168, 169B, 176B, 181, 182B, 186B, 191, 192B, 194B, 200B, 202B, 206B, 209, 210B, 214B, 216B, 226E, 228B, 230B, 234B, 236B, 243, 244B, 249, 250B, 252B, 256B, 268E, 270B, 272B, 277, 278B, 284B, 286, 287B, 290B, 296, 297B, 300B, 305, 306B, 316E, 318B, 321, 322B, 326B, 332B, 336B, 338B, 342B, 344B, 350B, 353B, 357, 358B, 364B, 368B, 371, 372B, 382E, 384B, 387, 388B, 390B, 394B, 397B, 400B, 406B, 408B, 412B, 417, 418B, 422B, 428B, 432B, 435, 436B, 442B, 444B, 447B, 450B, 460E, 462B, 464B, 466B, 468B, 472B, 478, 479B, 482B, 484B, 494B, 498B, 501, 502B, 504B, 512B, 516B, 518B, 520, 521B, 532E, 534B, 536B, 539B, 543B, 550B, 553B, 556B, 558B, 560B, 568B, 570B, 574B, 576B, 579, 580B

Inverse operations, 306, H16

Irregular figures
    area of, 494, 496
    estimate perimeter of, 478

Isosceles triangles, 332, 342

Journal. *See* Write About It *and* Write a Problem

Key, of graph, 120

Kilograms, 464–465, H25, H55

Kiloliters, 464–465, H25, H55

Kilometers, 464–465, H25, H55

Labels, on graphs, 416–417

Language Arts Connection, 176B, 342B, 390B, 558B

Language Support. *See* English Language Learners

Lateral faces, 350–352

Learning Goals, 14C, 92C, 144C, 226C, 268C, 316C, 382C, 460C, 532C

Least common denominator (LCD), 182–187

Least common multiple (LCM), 150–153, 166–167, 180–182

Leg, 488–489

Length
    customary system, 462–463, H25
    measuring, 468–471, H55
    metric system, 464–465, H25, H55

"Less than" symbol, 568–569, H3

Lesson Planning, 16A, 20A, 22A, 26A, 28A, 30A, 36A, 40A, 44A, 52A, 56A, 58A, 60A, 66A, 70A, 76A, 80A, 82A, 94A, 98A, 100A, 102A, 106A, 109A, 112A, 120A, 124A, 126A, 130A, 132A, 146A, 148A, 150A, 154A, 160A, 164A, 169A, 176A, 182A, 186A, 192A, 194A, 200A, 202A, 206A, 210A, 214A, 216A, 228A, 230A, 234A, 236A, 244A, 250A, 252A, 256A, 258A, 260A, 262A, 270A, 272A, 278A, 284A, 287A, 290A, 297A, 300A, 306A, 318A, 322A, 326A, 332A, 336A, 338A, 342A, 344A, 350A, 353A, 358A, 364A, 368A, 372A, 384A, 388A, 390A, 394A, 397A, 400A, 406A, 408A, 412A, 418A, 422A, 428A, 432A, 436A, 442A, 444A, 447A, 450A, 462A, 464A, 466A, 468A, 472A, 479A, 484A, 494A, 498A, 502A, 504A, 512A, 516A, 518A, 521A, 534A, 536A, 539A, 543A, 550A, 553A, 556A, 558A, 560A, 568A, 570A, 574A, 576A, 580A

Lesson Quiz. *See* last page of each lesson

Lesson Resources, 42, 46A, 46, 56A, 56, 74, 80A, 80, 100A, 100, 129, 154A, 160A, 160, 168, 180, 190, 194A, 194, 208, 214A, 214, 236A, 236, 276, 286, 296, 304, 306A, 306, 336A, 336, 353A, 353, 358A, 358, 371,

387, 388A, 388, 397A, 397, 416, 432A, 432, 434, 442A, 442, 472A, 472, 479A, 484A, 501, 516A, 516, 520, 534A, 534, 539A, 539, 543A, 543, 556A, 556, 576A, 576, 580A, 580

Like fractions, 180, H10

Likely events, 428–431, H24

Line graphs, 121–125, H37

Line plots, 102–105

Line segments
- bisecting, 368–370, H50
- congruent, 364–367, H50
- defined, 318–319

Line symmetry, 560–563, H30, H60

Linear relationships, 578–579

Lines
- angle relationships and, 322–325
- as axes, 570
- intersecting, 326–327, H48
- parallel, 326–327, 371, H48, H50
- perpendicular, 326–327, H48
- of symmetry, 560–563

Linkup
- to Art, 325
- to Careers, 153, 213, 303, 393
- to Reading, 25, 135, 189, 233, 275, 341, 415, 515, 563
- to Science, 69, 573
- to Social Studies, 108, 487
- *See also* Problem solving

List, organized, 154–155, 442–443

Literature Connection, 86
- *The Amazing Book of Shapes,* by Lydia Sharman, 316G
- *The Circuit,* by Francisco Jimenez, 382G
- *Exploring the Titanic,* by Robert D. Ballard, 92G
- *The Facts on File Children's Atlas,* by David and Jill Wright, 226G
- *From the Mixed-Up Files of Mrs. Basil E. Frankweiler,* by E. L. Konigsberg, 14G
- *Funny and Fabulous Fraction Stories,* by Dan Greenberg, 144G
- *Incredible Comparisons,* by Russell Ash, 460G
- *Island of the Blue Dolphins,* by Scott O'Dell, 144G
- *Janice Van Cleave's Math for Every Kid,* by Janice Van Cleave, 532G
- *Life by the Numbers,* by Keith Devlin, 382G
- *Melisande,* by E. Nesbit, 268G
- *Not for a Billion, Gazillion Dollars,* by Paula Danziger, 14G
- *Number the Stars,* by Lois Lowry, 268G
- *The Phantom Tollbooth,* by Norton Juster, 316G
- *The Secrets of Vesuvius,* by Sara C. Bisel, 532G
- *Snow Bound,* by Harry Mazer, 226G
- *What Hearts,* by Bruce Brooks, 92G
- *The Whipping Boy,* by Sid Fleischman, 460G

Liters, 464–465, H25, H55

Logical Reasoning strategy. *See* Use Logical Reasoning strategy

Lower quartile, 129–131, H37

Make a Graph strategy, 8

Make a Model strategy, 3, 516–517, 556–557

Make a Table strategy, 8, 56–57, 100–101

Make an Organized List strategy, 6, 154–155, 442–443

Make Generalizations, 576–577

Manipulatives and visual aids
- algebra tiles, 286, 296, 304
- centimeter cubes, 354, 512–513
- compass, 371, 416–417, 484, 501
- decimal squares, 70–72, 74–75
- dot paper, 342–343, 350
- fraction bars, 160–163, 180–181, 190–191
- fraction circles, 208
- geometric solids, 353
- graph paper, 354, 390, 397, 488, 498, 520, 578–583
- grid paper, 356–357
- hundred chart, 146
- nets, 356–357, 512, 518
- number lines, 53–55, 166–167, 176–178, 228–229, 231–232, 234–235, 244, 246, 252, 568–569, H12, H29
- pattern blocks, 553–555
- protractors, 320, 332, 416–417
- rulers, 356–357, 367, 397, 478–479, 484
- spinners, 434
- square tiles, 276, 578–579
- two-color counters, 242–243, 252, 387, 412

Manipulatives Chart, SC32
- *See also* Professional Handbook, Evan Maletsky

Maps, 400–401, H51

Mass, 464–465, H25, H55

Math Background, 16A, 20A, 22A, 26A, 28A, 30A, 36A, 40A, 44A, 46A, 52A, 56A, 58A, 60A, 66A, 70A, 76A, 80A, 82A, 94A, 98A, 100A, 102A, 106A, 109A, 112A, 120A, 124A, 126A, 130A, 132A, 146A, 148A, 150A, 154A, 160A, 164A, 169A, 176A, 182A, 186A, 192A, 194A, 200A, 202A, 206A, 210A, 214A, 216A, 228A, 230A, 234A, 236A, 244A, 250A, 252A, 256A, 258A, 260A, 262A, 270A, 272A, 278A, 284A, 287A, 290A, 297A, 300A, 306A, 318A, 322A, 326A, 332A, 336A, 338A, 342A, 344A, 350A, 353A, 358A, 364A, 368A, 384A, 388A, 390A, 394A, 397A, 400A, 406A, 408A, 412A, 418A, 422A, 428A, 432A, 436A, 442A, 444A, 447A, 450A, 462A, 464A, 466A, 468A, 472A, 479A, 484A, 494A, 498A, 502A, 504A, 512A, 516A, 518A, 521A, 534A, 536A, 539A, 543A, 550A, 553A, 556A, 558A, 560A, 568A, 570A, 574A, 576A, 580A

Math Connection, 526, 586
- Algebra, 42, 150B, 210B
- Geometry, 230B, 276, 536B
- Measurement, 190

Math Detective, 86, 138, 220, 262, 310, 376, 454, 526, 586

Math Jingles™ CD 5–6, 22B, 28B, 40B, 44B, 75, 80B, 82B, 106B, 146B, 182B, 209, 216B, 228B, 244B, 250B, 270B, 282B, 284B, 332B, 350B, 372B, 408B, 428B, 464B, 536B, 550B, 558B, 570B, 580B

Math Labs
- Algebra: Explore Proportions, 387

Algebra: Model Addition of Integers, 242–243
Algebra: Model Multiplication of Integers, 252
Algebra: Model Subtraction of Integers, 248–249
Angles, 320–321
Area of a Circle, 501
Construct Circle Graphs, 416–417
Construct Parallel Lines, 371
Division of Fractions, 208–209
Estimate Perimeter, 478
Explore Box-and-Whisker Graphs, 129
Explore Division of Decimals, 74–75
Explore Fractions and Decimals, 168
Explore Linear and Nonlinear Relationships, 578–579
Explore Order of Operations, 42–43
Model Addition and Subtraction, 180–181
Model and Solve One-Step Equations, 286
Model Multiplication Equations, 296
Models of Solid Figures, 356–357
Rename to Subtract, 190–191
Simulations, 434–435
Squares and Square Roots, 276–277
The Pythagorean Theorem, 488–489
Two-Step Equations, 304–305
Volume of a Cylinder, 520

Mathematics Across the Grades, 14J, 34C, 50C, 64C, 92J, 118C, 144J, 158C, 174C, 198C, 226J, 240C, 268J, 282C, 294C, 316J, 330C, 348C, 362C, 382J, 404C, 426C, 440C, 460C, 476C, 492C, 510C, 532J, 548C, 566C

Mean, 106–111, H6, H36

Measurement
of angles, 320–321, H31
area, 494–503, H57
area/perimeter relationship, 497
capacity, 462–465, H25, H55
Celsius degrees, 301–303
changing units, 462–467, 497, H25, H27
choosing a reasonable unit, 468–471, H55
choosing an appropriate measurement tool, 468–471, H55
comparing, 462–467, H55
converting between customary and metric, 466–467, H55
customary system
cups, 462–463, H25, H55
feet, 462–463, H25, H55
gallons, 462–463, H25, H55
inches, 462–463, H25, H55
miles, H25
ounces, 462–463, H25, H55
pints, 462–463, H25, H55
pounds, 462–463, H25, H55
quarts, 462–463, H25, H55
tons, 462–463, H25
yards, 462–463, H25, H55
decimals and, 462–467, H25, H55
of distance, 462–465, H55
estimating
area, 494, 496
perimeter, 478
volume of solids, 512
Fahrenheit degrees, 301–303
fluid measure, 462–465, 468–471, H55
fractions in, 462–463
length, 462–465, 468–471, H55
liquid volume, 462–465, H55
mass, 464–465, 468–471, H25, H55
metric system
centimeters, 464–465, H25, H55
decimeters, 464–465, H25
grams, 464–465, H25, H55
kilograms, 464–465, H25, H55
kiloliters, 464–465, H25, H55
kilometers, 464–465, H25, H55
liters, 464–465, H25, H55
meters, 464–465, H25, H55
milliliters, 464–465, H25, H55
one-dimensional figures. *See* Lines, Line segments, *and* Rays
perimeter of a polygon, 478–481, H26, H56
Pythagorean Theorem, 488–489
square units, 494
surface area of solids, 504–507
Table of Measures, back cover
temperature, 301–303
three-dimensional figures. *See* Solid figures
two-dimensional figures. *See* Plane figures
unit conversions, 301–303, 462–467
volume of solids, 512–515, 518–523, H58
weight, 462–463, H25, H55

Measures of central tendency, 106–111, H6, H36
effects of additional data, 109–111
effects of outliers, 110–111
mean, 106–111, H6, H36
median, 106–111, H6, H36
mode, 106–111, H6, H36

Median, 106–111, H6, H36

Mental math
addition, 36–39, H33
division, 37–39, 211–212, H33
multiplication, 36–39, H33
to solve equations, 30–31, 82–83, 303, H11, H35
subtraction, 37–39

Meters, 464–465, H25, H55

Metric system
capacity
kiloliters, 464–465, H25, H55
liters, 464–465, H25, H55
milliliters, 464–465, H25, H55
Celsius temperature, 301–303
changing units in, 464–465, H25, H55
customary units compared to, 466–467, H55
decimals and, 464–465, H25, H55

Number lines
   absolute value of numbers on, 228
   comparing on, 166–167, 231–232, 234–235, H13, H29
   decimals on, 53–55, 231–232
   fractions on, 166–167, 176–178
   inequalities on, 568–569
   integers on, 228–229, 244, 246, 252, H12, H29
   multiplying on, 252
   ordering on, 231–232, 234–235
   rational numbers on, 231–232, H13
   rounding on, 176

Number of the Day, 16A, 20A, 22A, 26A, 28A, 30A, 36A, 40A, 44A, 46A, 52A, 56A, 58A, 60A, 66A, 70A, 76A, 80A, 82A, 94A, 98A, 100A, 102A, 106A, 109A, 112A, 120A, 124A, 126A, 130A, 132A, 146A, 148A, 150A, 154A, 160A, 164A, 169A, 176A, 182A, 186A, 192A, 194A, 200A, 202A, 206A, 210A, 214A, 216A, 228A, 230A, 234A, 236A, 244A, 250A, 252A, 256A, 258A, 260A, 262A, 270A, 272A, 278A, 284A, 287A, 290A, 297A, 300A, 306A, 318A, 322A, 326A, 332A, 336A, 338A, 342A, 344A, 350A, 353A, 358A, 364A, 368A, 384A, 388A, 390A, 394A, 397A, 400A, 406A, 408A, 412A, 418A, 422A, 428A, 432A, 436A, 442A, 444A, 447A, 450A, 462A, 464A, 466A, 468A, 472A, 479A, 484A, 494A, 498A, 502A, 504A, 512A, 516A, 518A, 521A, 534A, 536A, 539A, 543A, 550A, 553A, 556A, 558A, 560A, 568A, 570A, 574A, 576A, 580A

Number patterns, 77, 253, 536–542, H14, H59

Number sense
   The use of number sense to solve problems is a focal point at Grade 6 and is found throughout this book. Some examples: 193, 255, 285
   See Number of the Day

Number sentences. *See* Equations

Number theory
   divisibility, 146–147, H38
   least common multiple, 150–153
   prime factorization, 148–150

Numbers
   absolute value, 228–229
   adding, 20–21, 66–69, 180–189, 242–247
   binary, 542
   comparing, 53–55, 166–167, 228–229, 231–232, 406–407, H3, H12, H29, H42
   compatible, 17–19
   composite, 148, H7
   cubes of, H27
   decimal, 52–55, 58–61, 66–79, H2, H22–H23
   dividing, 23–25, 37–39, 74–79, 208–213, 253–255
   expanded form of, 52–55
   factors of, H8
   fractions, 176–185, 200–205, 208–213, H8, H22–H29
   integers, 228–229, H12
   millions, H3
   mixed, 164–165, 186–195, 200–201, 206–207
   multiples of, H8
   multiplying, 22–25, 70–73, 202–207, 252–255
   negative, 228–235
      as exponents, 263
   opposites, 228–229
   ordered pairs of, 570–573
   ordering, 52–55
   percents of, 406–407, 412–415, H9
   pi, 484–487
   place value in, 52–55, H3
   positive, 228–229
   prime, 148, H7
   rational, 230–233
   reading and writing, 52–55, 228–229, H3
   reciprocal, 209
   rounding, 16, 176–178, 186, H4, H10
   square of, 276, 278–279, H27, H45
   square roots of, 276–277, 278–279, H45
   standard form of, 52–55
   subtracting with, 20–21, 37–39, 66–69, 181, 183–185, 187–193, 248–251
   triangular, 536
   word form, 52–55

Numerical expressions
   evaluating, 28–29
   writing, 28–29

Numerical patterns, 77, 253, 536–542, H14, H59

Obtuse angles, 320–321, 332, H18
Obtuse triangles, 332–335, H49
Octagons, 336
Operations
   choose the, 214–215
   inverse, 306, H16
   order of, 42–45, 257–259, 272, H15
   words for, H17

Opposites, 228–229

Order
   decimals, 53–55, 231–232
   fractions, 166–167, 231–232, H12, H29
   integers, 228–229, H42
   mixed numbers, 234–235
   on number lines, 166–167, 231–232, 234–235
   rational numbers, 234–235
   whole numbers, H3

Order of operations, 42–45, 257–259, 272, H15 *See also Professional Handbook, Liping Ma*

Ordered pairs, 570–575
   equations from, 574–575
   graphing, 570–575
   identifying on a graph, 570–573
   writing, 570–573, H31

Organized lists, 154–155, 442–443

Origin, 570
Ounces, 462–463, H55
Outcomes, 428–431, 436–437, 444–449
Outliers, 110–111, H36
Overestimates, 17–19

Pacing. See Chapter Planner
Parallel lines, 326–327, 371
Parallelograms, 338–341, H49
Parentheses, 42–45, 257–259, 272, H15
Patterns
    in division, 77, H13
    extending, 536–537, H14, H59
    finding, 536–537, H59
    fractals, 544
    functions and, 539–542, H59
    of geometric figures, 543–545, H59
    input/output, 539, H29
    with multiples, 253
    in multiplication, 253–255
    number, 77, 253, 536–542, H14, H59
    in polygons, 336–337
    sequences, 536–538, 586, H59
    for solid figures, 356–357
    tessellations in, 553–557, H60
    triangular numbers, 536
    See also Find a Pattern strategy *and* Problem solving strategies
Pentagonal prisms, 350–352
Pentagonal pyramids, 350–352
Pentagons, 336
Percents
    and circle graphs, 416–417, H52
    to decimals, 408–415, H23
    decimals to, 408–411, H52
    of decrease, 455
    discounts, 418–421, H52
    estimate, 412–415
    to fractions, 170–171, 408–415, H23, H52
    fractions to, 170–171, 408–411, H23, H52
    of increase, 455
    models of, 406–407, H9
    of numbers, 412–415
    for parts of a whole, 416–417
    and proportions, 415
    to ratios, 408–411, H23
    ratios to, 406–407
    sales tax, 418–421
    simple interest, 422–423
    tips, 412–415

Performance Assessment, 90, 142, 224, 266, 314, 380, 458, 530, 590
Perimeter, 478–481, H26, H56
Perpendicular lines, 326–327
Perspective, 377
Physical Education Connections, 139, 166B, 278B, 450B
Pi, 79, 484–487
Pints, 462–463, H25, H55
Place value
    to hundredths, 52–55, H2
    to millions, 52–55, H3
    models of, H2–H3
    periods, H3
    powers of ten and, 40
    stem-and-leaf plots and, 126
    to ten-thousandths, 52–55
    to thousandths, 52–55, H3
Plane figures
    area of, 494–503, H57
    congruent, 390–393
    drawing, 342–343, H49
    perimeter of, 478–481, H56
    similar, 390–396, H51
    symmetry in, 560–563
    transforming, 550–552, 580–583
    See also Circles, Geometry, *and* Polygons
Planes, 318–319, H48
Plots
    line, 102–105
    stem-and-leaf, 126–128, H7, H37
Points, 318–319
Points, with ordered pairs, 570–573
Polygons
    classifying, 332–335, 338–341, H19, H49
    hexagons, 336
    octagons, 336
    parallelograms, 338–341, H49
    pentagons, 336
    quadrilaterals, 336, 338–341, H49
    rectangles, 338–341, H49
    regular, 336
    rhombuses, 338–341, H49
    squares, 338–341, H49
    trapezoids, 338–341, H49
    triangles, 332–335, H49
    See also Geometry
Polyhedron, 350–352, H50
Populations, 94–97
Portfolio, 14B, 92B, 144B, 226B, 268B, 316B, 382B, 460B, 532B
Portfolio Suggestions, 14, 32, 34, 48, 50, 62, 64, 84, 89, 92, 118, 136, 141, 144, 156, 158, 172, 174, 196, 198, 218, 223, 226, 238, 240, 260, 268, 280, 282, 292, 294, 308, 313, 316, 328, 330, 346, 348, 360,

362, 374, 379, 382, 402, 404, 424, 426, 438, 440, 452, 457, 460, 474, 476, 490, 492, 508, 510, 524, 529, 532, 546, 548, 564, 566, 584, 589

Positive integers, 228–229

Possible outcomes, 428, 444–445, H54

Pounds, 462–463, H55

Powers of ten, 40, 87, 263, H13 See also Exponents

Precision, 468–471, H55

Predict and Test strategy, 4, 26–27

Predictions
   about populations, 450–451, H54
   of certain events, 436–437, H53
   of outcomes, 436–437

Prefixes, in metric units, 464–465, H25, H55

Prerequisite Skills, *14K, 34D, 50D, 64D, 92K, 118D, 144K, 158D, 174D, 198D, 226K, 240D, 254D, 268K, 282D, 294D, 316K, 330D, 348D, 382K, 404D, 426D, 440D, 460K, 476D, 492D, 510D, 532K, 548D, 566D*

Prime factorization, 148–150

Prime numbers, 148, H7

Prisms
   bases of, 350–352, H50
   classifying, 350–352, H50
   heights of, 512–515
   models with nets, 356–357, 512
   surface area of, 504–507, H57
   volume of, 512–515, H58

Probability
   calculating for
      one event followed by another independent event, 447–449, H54
      two disjoint events, 444–446
   certain events, 429, H24
   compound events, 444–445, H54
   dependent events, 448–449, H54
   equally likely events, 428–431
   estimating future events, 450–451, H54
   experimental, 434–437, H53
   Fundamental Counting Principle, 444–446, H54
   geometric, 431
   impossible events, 429, H24
   independent events, 447–449, H54
   likely outcomes, 429, H24
   possible outcomes, 428, 444–445
   predictions, 436–437, 450–451, H53–H54
   random, 445
   sample space, 428, 444–445
   simulations, 434–435
   that an event will not occur, 429–431, 447–449
   theoretical, 428–431, 447–451, H53
   tree diagrams, 444–446, H54
   unlikely events, 429, H24
   verify reasonableness, 430, 448

Problem of the Day, *16A, 20A, 22A, 26A, 28A, 30A, 36A, 40A, 44A, 46A, 52A, 56A, 58A, 60A, 66A, 70A, 76A, 80A, 82A, 94A, 98A, 100A, 102A, 106A, 109A, 112A, 120A, 124A, 126A, 130A, 132A, 146A, 148A, 150A, 154A, 160A, 164A, 169A, 176A, 182A, 186A, 192A, 194A, 200A, 202A, 206A, 210A, 214A, 216A, 228A, 230A, 234A, 236A, 244A, 250A, 252A, 256A, 258A, 260A, 262A, 270A, 272A, 278A, 284A, 287A, 290A, 297A, 300A, 306A, 318A, 322A, 326A, 332A, 336A, 338A, 342A, 344A, 350A, 353A, 358A, 364A, 368A, 384A, 388A, 390A, 394A, 397A, 400A, 406A, 408A, 412A, 418A, 422A, 428A, 432A, 436A, 442A, 444A, 447A, 450A, 462A, 464A, 466A, 468A, 472A, 479A, 484A, 494A, 498A, 502A, 504A, 512A, 516A, 518A, 521A, 534A, 536A, 539A, 543A, 550A, 553A, 556A, 558A, 560A, 568A, 570A, 574A, 576A, 580A*

Problem of the Day Solutions, *PD1–PD30*

Problem solving. See Problem solving applications, Problem solving Linkup, Problem solving on Location, Problem solving skills, Problem solving strategies, Problem solving Thinkers' Corner, *Professional Handbook, Jan Scheer;* Scaffolded Instruction

Problem solving applications
   art, 396
   astronomy, 399
   geography, 415
   geometry, 247, 289, 573
   graphs, 303
   measurement, 569
   number sense, 193, 255, 285
   reasoning, 29, 99, 111, 123, 149, 205, 259, 279, 286, 302, 325, 335, 342, 345, 352, 386, 396, 415, 421, 423, 431, 465, 487, 497, 500, 507, 515, 541, 552, 563
   science, 285, 303, 437
   space, 303
   technology, 163
   use data, 19, 21, 38, 41, 47, 55, 57, 59, 61, 73, 114, 163, 171, 179, 184, 188, 291, 303, 337, 345, 433, 443, 451, 471, 473, 483, 507, 517, 542, 557, 569, 577

Problem Solving Projects, *14, 34, 50, 64, 92, 118, 144, 158, 174, 198, 226, 240, 268, 282, 294, 316, 330, 348, 362, 382, 404, 426, 440, 460, 476, 492, 510, 532, 548, 566*

Problem Solving on Location, 91A–91B, 143A–143B, 225A–225B, 267A–267B, 315A–315B, 381A–381B, 459A–459B, 531A–531B, 591A–591B

Problem solving skills
   choose the operation, 214–215
   compare strategies, 12
   estimate or find exact answer, 472–473
   interpret the remainder, 80–81
   make generalizations, 576–577
   multistep problems, 13
   sequence and prioritize information, 46–47
   too much or too little information, 432–433

Problem solving strategies
   draw a diagram/picture, 2, 194–195, 482–483
   find a pattern, 7, 336–337, 534–535
   make a model, 3, 516–517, 556–557
   make a table or graph, 8, 56–57, 100–101
   make an organized list, 6, 154–155, 442–443
   predict and test, 4, 26–27

solve a simpler problem, 9, 358–359
use logical reasoning, 11, 236–237
work backward, 5, 306–307
write an equation, 10, 388–389

Products. *See* Multiplication

Professional Handbook, *PH1–PH64*

Andrews, Angela Giglio, "Developing Algebraic Thinking," PH38–PH40
Bennett, Jennie M., "Mathematics for All Doesn't Mean Mathematics for Some!," PH57–PH59
Burton, Grace M.
  "Effective Practice: Building Computational and Procedural Efficiency," PH35–PH37
  "Effective Practice: Memorizing the Number Facts," PH33–PH34
Howe, Roger
  "Doing Decimal Arithmetic: Addition," PH5–PH8
  "Doing Decimal Arithmetic: Multiplication," PH9–PH12
  "Estimation and Approximation," PH13–PH16
  "Estimation and Arithmetic," PH17–PH20
  "Marvelous Decimals," PH1–PH4
Johnson, Howard, "Technology—Pathways to Teaching and Learning," PH52–PH54
Luckie, Lynda, "Assessment: An Integral Part of the Learning Process," PH48–PH51
Ma, Liping, "A Blind Spot in the Order of Operations," PH21–PH23
Maletsky, Evan, "Conceptual Understanding: The Power of Models and Visuals," PH29–PH32
McLeod, Joyce, "Instructional Strategies: Best Practices Defined by Research," PH43–PH47
Newman, Vicki, "Linking Home and School," PH55–PH56
Roby, Tom, "Models for Fractions," PH24–PH25
Scheer, Jan, "Problem Solving: The Reason to Teach Mathematics," PH41–PH42
Sprenger, Marilee, "Making Math Memorable: Brain-Compatible Learning," PH60–PH64
Wright, David G., "Geometry and Measurement," PH26–PH28

Properties
Associative, 36–39, H2, H33
Commutative, 36–39, 274–275, H2
Distributive, 36–39, 206–207, 274–275, H2
of Equality, 287–291, 297–299
Identity, of One, H2
Identity, of Zero, H2
Reflexive, 311
Symmetric, 311
Transitive, 311
of Zero in Multiplication, H2

Proportional reasoning
area, 497
conversions, 301–303, 462–465
geometry, 394–396
measurement, 462–465
percent, 412–415, 418–423
perimeter, 478
predictions, 450–451

probability, 450–451
proportion, 394–401, H51
ratio, 384–386
samples, 94–97
volume, 516–517

Proportions
and indirect measurement, 394–396, H51
making predictions, 450–451
models of, 387
and percents, 412–415, 418–423
and scale drawings, 397–399, H51
and similar figures, 394–396, H51
solving, 388–401, H26, H51
use tables and rules, 386, 462–467
writing, 387–389, 462–467, H51

Protractor, 320–321, 416–417, H31

Pyramids
hexagonal, 350–352
models with nets, 356–357
pentagonal, 350–352
rectangular, 350–352
square, 350–352
surface area of, 505–506, H57
triangular, 350–352
volume of, 518–519, H58

Pythagorean Theorem, 488–489

Quadrants, 570–573, H61

Quadrilaterals
angles of, 338–341, H49
degrees in, 336
parallelograms, 338–341, H49
rectangles, 338–341, H49
rhombuses, 338–341, H49
sides of, 338–341, H49
squares, 338–341, H49
trapezoids, 338–341, H49

Quartile, 129–131, H37

Quarts, 462–463, H25, H55

Quick Review, 16, 20, 22, 26, 28, 30, 36, 40, 42, 44, 46, 52, 56, 58, 60, 66, 70, 74, 76, 80, 82, 94, 98, 100, 102, 106, 120, 124, 126, 130, 132, 146, 148, 150, 154, 160, 164, 166, 169, 176, 180, 182, 186, 190, 192, 194, 200, 202, 206, 208, 210, 214, 216, 228, 230, 234, 236, 242, 244, 248, 250, 252, 256, 270, 272, 276, 278, 284, 287, 290, 297, 300, 304, 306, 318, 320, 322, 326, 332, 336, 338, 342, 344, 350, 353, 356, 358, 364, 368, 372, 384, 388, 390, 394, 397, 400, 406, 408, 412, 416, 418, 422, 428, 432, 434, 436, 442, 444, 447, 450, 462, 464, 466, 468, 472, 479, 482, 484, 488, 494, 498, 502, 504, 512, 516, 518, 521, 534, 536, 539, 543, 550, 553, 556, 558, 560, 568, 570, 574, 576, 578, 580

Quotients. *See* Division

Radius, 344–345, H49
Random numbers, 435
Random sample, 95–97, H36
Range, 103, H6
Rates, 385–386, H51
  unit rate, 385–386
  use tables, graphs, and rules, 124–125
Rational numbers, 230–233, 234–235, 256–259, H13, H43
Ratios, 384–386
  equivalent, 384–386, H51
  making predictions, 450–451
  percents, 406–407
  rates, 385–386, H51
  and rational numbers, 230–233
  and similar figures, 390–393, H51
  unit rates, 385–386, H51
  writing, 384–386, H51
Rays, 318–319, H48
Reaching All Learners, T14
  Advanced Learners, 14F, 20B, 22B, 26B, 44B, 70B, 76B, 80B, 92F, 94B, 106B, 109B, 126B, 130B, 144F, 146B, 164B, 182B, 192B, 206B, 216B, 226F, 250B, 262, 268F, 272B, 290B, 306B, 316F, 322B, 332B, 344B, 353B, 358B, 372B, 382F, 394B, 418B, 422B, 436B, 442B, 460F, 462B, 479B, 489, 502B, 516B, 521B, 532F, 534B, 553B, 560B, 574B, 579, 580B
  Alternative Teaching Strategy, 16B, 18, 20B, 22B, 24, 26B, 28B, 30B, 36B, 38, 40B, 44B, 46B, 52B, 54, 56B, 58B, 60B, 66B, 68, 70B, 72, 76B, 78, 80B, 82B, 87, 94B, 96, 98B, 100B, 102B, 104, 106B, 109B, 112B, 114, 120B, 122, 124B, 126B, 130B, 132B, 134, 146B, 148B, 150B, 152, 154B, 160B, 162, 164B, 166B, 169B, 176B, 178, 182B, 184, 186B, 188, 192B, 194B, 200B, 202B, 204, 206B, 208, 210B, 212, 214B, 216B, 228B, 230B, 232, 234B, 236B, 244B, 246, 250B, 252B, 254, 256B, 258, 270B, 272B, 274, 278B, 284B, 287B, 290B, 297B, 300B, 302, 304, 306B, 318B, 322B, 324, 326B, 332B, 334, 336B, 338B, 340, 342B, 344B, 350B, 353B, 358B, 364B, 366, 368B, 372B, 384B, 388B, 390B, 392, 394B, 397B, 400B, 406B, 408B, 410, 412B, 414, 418B, 420, 422B, 428B, 430, 432B, 436B, 442B, 444B, 447B, 450B, 462B, 464B, 466B, 468B, 470, 472B, 479B, 482B, 484B, 486, 494B, 496, 498B, 502B, 504B, 506, 512B, 514, 516B, 518B, 521B, 527, 534B, 536B, 539B, 541, 543B, 550B, 553B, 556B, 558B, 560B, 562, 568B, 570B, 572, 574B, 576B, 580B, 582, 587
  Art Connection, 270B, 356, 484B, 518B
  Block Scheduling, 14F, 92F, 144F, 226F, 268F, 316F, 382F, 460F, 532F
  Career Connection, 16B, 28B, 70B, 112B, 130B, 154B, 166B, 206B, 412B, 504B, 550B, 570B
  Early Finishers, 20B, 28B, 46B, 56B, 58B, 66B, 76B, 100B, 102B, 132B, 150B, 160B, 180, 194B, 200B, 214B, 221, 243, 252B, 278B, 304, 326B, 336B, 357, 364B, 377, 388B, 406B, 408B, 412B, 432B, 447B, 464B, 472B, 478, 494B, 512B, 536B, 539B, 556B, 570B
  English Language Learners, 14G, 30B, 40B, 56B, 60B, 80B, 82B, 92G, 98B, 100B, 102B, 120B, 144G, 148B, 160B, 169B, 194B, 200B, 210B, 214B, 226G, 228B, 234B, 244B, 268G, 270B, 284B, 297B, 316G, 321, 326B, 338B, 353B, 368B, 382G, 384B, 390B, 400B, 408B, 434, 447B, 454, 460G, 468B, 479B, 484B, 494B, 521B, 532G, 543B, 558B, 580B
  Fine Arts Connection, 310, 318B, 466B, 472B, 482B
  Health Connection, 50, 174
  History Connection, 154, 166, 440
  Interdisciplinary Suggestions, 14F, 92F, 144F, 226F, 268F, 316F, 382F, 460F, 532F
  Language Arts Connection, 176B, 342B, 390B, 558B
  Learning Styles. See Modalities
  Literature Connections, 14G, 86, 92G, 144G, 226G, 268G, 316G, 382G, 460G, 532G
  Math Connection, 526, 586
    Algebra, 42, 150B, 210B, 488
    Geometry, 230B, 276, 536B
    Measurement, 190
  Physical Education Connection, 139, 166B, 278B, 450B
  Reading Strategy, 26B, 46B, 80B, 100B, 214B, 236B, 306B, 336B, 358B, 388B
  Science Connection, 22B, 44B, 82B, 186B, 263, 284B, 300B, 384B, 397B, 428B, 444B, 468B, 534B, 574B
  Social Studies Connection, 26B, 60B, 66B, 98B, 124B, 297B, 338B, 350B, 368B, 400B, 416, 504B
  Special Needs, 14F, 30B, 36B, 52B, 74, 92F, 109B, 120B, 144F, 146B, 154B, 169B, 182B, 190, 192B, 202B, 220, 226F, 236B, 242, 248, 252, 256B, 258B, 262B, 268F, 276, 290B, 306B, 316F, 320, 322B, 342B, 356, 358B, 364B, 382B, 394B, 397B, 406B, 436B, 442B, 460F, 464B, 488, 502B, 512B, 532F, 539B, 553B, 578
  Technology, 46B, 52B
  Vocabulary Strategy, 36B, 94B, 228B, 320, 336B, 344B, 376, 428B, 444B, 472B, 518B, 560B, 568B
  Writing in Mathematics, 16B, 40B, 58B, 74, 106B, 112B, 124B, 126B, 132B, 148B, 164B, 176B, 180, 186B, 202B, 208, 216B, 230B, 250B, 256B, 262B, 272B, 287B, 300B, 318B, 332B, 350B, 372B, 388B, 418B, 422B, 432B, 450B, 462B, 472B, 498B, 516B, 543B, 556B, 576B, 578
Reading, 25, 135, 189, 275, 341, 415, 515
  See also Reading Strategies
Reading Strategies
  activate prior knowledge, 516B
  analyze information, 236B, 515
  cause and effect, 534B
  choose relevant information, 189
  classify and categorize, 135
  compare, 26B
  follow directions, 388B
  form mental image, s 556B
  K-W-L chart, 69, 79, 153, 189, 213, 233, 275, 335, 367, 415, 487, 497, 515, 542, 573
  make generalizations, 576B
  make inferences, 336B
  multiple-meaning words, 214B

paraphrasing, 358B
sequence, 46B, 415
use context, 25, *80B*
use graphic aids, 56B, 100B, 233, 275

Reasoning
applying strategies from a simpler problem to solve a more complex problem, 9, 358–359, 526
breaking a problem into simpler parts, 13, 335
checking validity of results, 194, 279, 310
choosing a problem solving strategy, 27, 57, 101, 155, 195, 237, 307, 337, 359, 389, 443, 483, 517, 535, 557
estimating
knowing when to estimate, 472–473
using estimation to verify reasonableness of an answer, 20–25, 66–73, 182–189, 192–193, 203, 206, 211
evaluating reasonableness of solutions, 20–25, 66–73, 182–189, 192–194, 203, 206, 211
generalizing beyond a particular problem to other situations, 149, 342, 487, 577
identifying missing information, 56, 432–433, 442, 482, 556
identifying relationships among numbers, 79, 86, 147, 149, 167, 183, 185, 220, 229, 423, 478
in Math Detective, 86, 138, 220, 310, 376, 454, 526
mathematical conjectures, 146, 253, 332, 335, 484, 498
observing patterns, 7, 123, 262, 296, 336–337, 358, 534–535
recognizing relevant and irrelevant information, 26–27, 100–101, 189, 194–195, 306–307, 358–359, 388–389, 432–433, 442–443, 482–483, 516–517, 534–535, 556–557
sequencing and prioritizing information, 29, 46–47, 138, 205, 431
in Thinker's Corner, 79, 247, 303, 325, 335, 431, 471, 507
Use Logical Reasoning strategy, 11, 236–237
Opportunities to explain reasoning are contained in the Guided Instruction and Guided Practice portions of the Teacher's Edition for many lessons. 17, 20, 23, 26, 27, 28, 30, 36, 39, 40, 44, 55, 56, 58, 60, 69, 72, 75, 76, 79, 80, 82, 86–87, 94–95, 98, 100, 103–104, 106–110, 121, 124, 132, 135, 138, 146, 153–154, 161, 163, 169–170, 176, 178, 183, 185–186, 189, 192, 194, 200, 202, 208, 211, 213, 220–221, 228, 233–234, 236, 244, 247, 250, 253, 257, 259, 262, 270, 272, 278, 284, 287–290, 298, 300–301, 303, 306, 310–311, 318, 322–323, 325–326, 332, 335–336, 339, 341–342, 344, 350–351, 353, 357–358, 367, 368, 370, 376, 384–385, 388, 390–395, 397–398, 400, 408–409, 411–413, 415–419, 422, 431, 432B, 442B, 446, 454–455, 471, 472B, 478, 484, 487, 494–495, 497–498, 502, 505, 512–513, 515, 522, 526–527, 534, 536–537, 542, 551, 553–554, 560, 563, 570–571, 573–574, 576, 578, 580, 586–587

Reasoning and Proof. *See Reasoning*
Reciprocals, 209
Rectangles
area of, 495–497, H28, H57
as faces of solid figure, 350–352
perimeter of, 479–481, H56
Rectangular prisms
classifying, 350–352, H50

cubes, 350–352, H50
surface area of, 504, 506–507, H57
volume of, 512–515, H58
Rectangular pyramids, 350–352, H50
Reflections, 550–552, 558–559, 580–583, H30, H61
Reflexive Property, 311
Regrouping, or renaming
in addition, 186–189
with mixed numbers, 186–191
in subtraction, 190–191
Regular polygons, 336
Remainders, 23–24, 80–81, H5
Repeating decimals, 169–171
Representation
Opportunities to create and use representation are contained throughout the book. Some examples: 56, 60, 70, 74, 100, 102, 106, 120, 126, 132, 154, 160, 180, 182, 186, 190, 194, 202, 244, 248, 256, 276, 286, 300, 304, 344, 364, 387, 388, 390, 394, 397, 400, 442, 444, 464, 482, 501, 516, 550, 553, 556, 558, 560, 578
Research Articles, RS1–RS24
Review. *See Mixed Applications, Mixed Review and Test Prep, Mixed Strategy Practice, and Standardized Test Prep*
Rhombuses, 338–341, H49
Right angles, 320–321, 332, H18
Right triangles, 332–335, H49
Pythagorean Theorem, 488–489
Rotational symmetry, 560–563
Rotations, 550–552, 558–559, 580–583
Rounding
decimals, 58, H4
fractions, 176–178, 200, H10
mixed numbers, 186
whole numbers, 16, H4

Sales tax, 418–421
Sample space, 428–431, 442–443, 444–446
Samples
of a population, 94–99
biased, 98–99, H36
characteristics of, 94–99
convenience, 95–97, H36
different ways of selecting, 94–97
limitations of, 94–99
random, 95–97, H36
responses to survey, 94–97
systematic, 95–97, H36
unbiased, 98–99, H36
when to use, 94–97

Index **BI33**

Scaffolded Instruction
    Alternative Teaching Strategy, 18, 38, 54, 68, 72, 78, 96, 104, 114, 122, 134, 152, 162, 178, 184, 188, 204, 212, 232, 246, 254, 258, 274, 302, 324, 334, 340, 366, 392, 414, 420, 430, 470, 486, 496, 506, 514, 541, 562, 572, 582
    Problem solving, 21, 39, 57, 81, 96, 123, 155, 162, 195, 205, 229, 255, 279, 285, 302, 325, 345, 359, 370, 389, 421, 433, 446, 463, 483, 497, 519, 535, 557, 577

Scale, 398–401, H51

Scale drawings, 397–399, H51

Scalene triangles, 332

Scatterplots, 139

Science, 14, 20, 34, 36, 66, 69, 91A, 91B, 102, 107, 118, 120, 130, 143A, 143B, 158, 161, 164, 170, 186, 187, 192, 200, 216, 234, 240, 250, 259, 267A, 267B, 297, 332, 381A, 384, 387–388, 459A, 531A, 531B, 548, 553, 560, 573, 591A

Science Connection, 22B, 44B, 82B, 186B, 263, 284B, 300B, 384B, 397B, 428B, 444B, 468B, 534B, 574B

Scientific notation, 87, 263

Scope and Sequence, SC1–SC28

Sequence and prioritize information, 46–47

Sequences, 536–538, 586, H59

Shapes. *See* Geometry

Sides. *See* Polygons *and* Polyhedron

Similar figures
    corresponding parts, 391–396, H51
    identifying, 390–393, H22, H51
    indirect measurement, 394–396, H51
    proportions, 394–396, H51
    ratios, 390–393, H51

Simple interest, 422–423, H52

Simplest form of fractions, 161–163, H9

Simulations, 434–435

Skills Review, H66–H68

Skills Trace Across the Grades, 14J, 34C, 50C, 64C, 92J, 118C, 144J, 158C, 174C, 198C, 226J, 240C, 268J, 282C, 294C, 316J, 330C, 348C, 362C, 382J, 404C, 426C, 440C, 460J, 476C, 492C, 510C, 532J, 548C, 566C

Skills Trace for Grade 6, 14J, 34C, 50C, 64C, 92J, 118C, 144J, 158C, 174C, 198C, 226J, 240C, 268J, 282C, 294C, 316J, 330C, 348C, 362C, 382J, 404C, 426C, 440C, 460J, 476C, 492C, 510C, 532J, 548C, 566C

Slides, H30
    See also Translations

Social Studies, 30, 60, 103, 225A, 225B, 315A, 315B, 300, 381A, 394, 410, 459A, 476, 484, 487, 510

Social Studies Connection, 26B, 60B, 66B, 98B, 124B, 297B, 338B, 350B, 368B, 400B, 416, 504B

Solid figures
    cones, 350–352, H50
    cubes, 350–352, H50
    cylinders, 350–352, H50
    drawing, 353–354
    edges of, 358–359, H20
    faces of, 350–355, 358–359, H20, H50
    prisms, 350–352, H50
    pyramids, 350–352, H50
    surface area of, 504–507, H57
    transformations of, 558–559
    views of, 353–355
    volume of, 512–515, 518–519, H58

Solve a Simpler Problem strategy, 358–359

Solving equations, 30–31, 82–83, 216–217, 286–291, 296–299, H11, H21, H46–H47

Special Needs, 14F, 30B, 36B, 52B, 74, 92F, 109B, 120B, 144F, 146B, 154B, 169B, 182B, 190, 192B, 202B, 220, 226F, 236B, 242, 248, 252B, 256B, 268F, 276, 290B, 306B, 316F, 320, 322B, 342B, 356, 358B, 382F, 394B, 397B, 406B, 436B, 442B, 460F, 464B, 488, 502B, 512B, 532F, 539B, 553B, 578

Spiral Review. *See* Mixed Review and Test Prep

Sports, 16, 198, 282, 426, 460, 532

Square pyramids, 350–352

Square roots, 277–279, H45

Square unit, 494

Squares
    area of, 495–497, H28
    of numbers, 276, 278–279, H27, H45

Standard form, 52–55

Standardized Test Correlation. *See* Correlation to Standardized Tests

Standardized Test Prep, 33, 49, 63, 85, 117, 137, 157, 173, 197, 219, 239, 261, 281, 293, 309, 329, 347, 361, 375, 403, 425, 439, 453, 475, 491, 509, 525, 547, 565, 585

Statistics
    analyzing data, 100–115, 120–135, H5, H24
    averages. *See* Measures of central tendency
    bar graph, 120–123, H6, H37
    circle graph, 122–123, 416–417
    collecting data, 94–99, H36
    frequency tables, 103–105, H56
    histogram, 127–128, H37
    line graph, 121–125, H37
    measures of central tendency
        mean, 106–111, H6, H36
        median, 106–111, H6, H36
        mode, 106–111, H6, H36
    organizing data, 56–57, 100–105, 120–123, 126–131
    outliers, 110–111, H36
    range, 103, H6
    samples
        biased, 98–99, H36
        convenience, 95–97, H36
        of a population, 94–99
        random, 95–97, H36
        systematic, 95–97, H36
        unbiased, 98–99, H36

stem-and-leaf plots, 126–128, H7, H37
surveys, 94–99
tally tables, 100–101
See also Data
Stem-and-leaf plots, 126–128, H7, H37
Straight angles, 320–321, H18
Stretching, 587
Student Handbook, H1–H110
Study Guide and Review, 88–89, 140–141, 222–223, 264–265, 312–313, 378–379, 456–457, 528–529, 588–589
Subtraction
addition as an inverse operation, H16
of decimals, 66–69, H35
equations, 286, 290–291, H46
estimation and, 16–19, 177–179
of fractions, 181–185, H10, H16, H40
of integers, 248–251, H14, H43
mental math and, 37–39
of mixed numbers, 187–193, H40
models of, 180–181, 186–188, 248–249
Property of Equality, 287–288
of rational numbers, 256–259, H43
of whole numbers, 20–21, 37–39, H4
Summarize. *See the Discuss and Write questions in the Assess section of each TE lesson*
Sums. *See* Addition
Supplementary angles, 322–325
Surface area, 504–507, H57
Surveys, 94–99
Symbols
angle, 320
degree, 320
is approximately equal to, 17
less than and greater than, 568–569, H3
line, 318
line segment, 318
ray, 318
See also Table of Measures *on back cover*
Symmetric Property, 311
Symmetry
line, 560–563, H30, H60
mirror images, 552
rotational, 560–563
verifying, H30
Systematic samples, 95–97, H36

Table of Measures, back cover
Tables and charts
analyzing data from, 100–101, H5

frequency, 102–105
function, 574–575
input/output, 539–542, H29
making, 100–101
organizing data in, 100–101
tally tables, 100–101
Tally tables, 100–101
Technology
Calculator Handbook, 40B, 70B, 76B, 87B, 126B, 129B, 139B, 169B, 182B, 244B, 250B, 416B, 422B, 436B, 444B, 479B, 484B, 494B, 498B, 502B, 512B, 536B, 574B, 578B
calculators, 43, 77, 163, 169, 170, *244B,* 276–277, 413, 435, *466B,* 484, 485, 591
cross-curricular connections, 64, 330, 344
Data ToolKit, 91, 102, 315, 531
E-Lab activities, 43, 75, 129, 162, 168, 181, 191, 208, 249, 277, 286, 296, 305, 353, 369, 387, 417, 435, 459, 484, 501, 520, 534, 544, 581, *T19*
Harcourt Math Newsroom Videos, 18, 234, 273, 338, 391, 484, 539, *T19*
Math Jingles™ CD 5–6, 22B, 28B, 40B, 44B, 75, 80B, 82B, 106B, 146B, 182B, 209, 216B, 228B, 244B, 250B, 270B, 282B, 284B, 332B, 350B, 372B, 408B, 428B, 464B, 536B, 550B, 558B, 570B, 580B
Mighty Math software, 24, 53, 171, 183, 188, 225, 253, 267, 298, 333, 334, 345, 381, 448, 554, *T19*
Technology Links, 24, 43, 53, 91, 102, 122, 129, 162, 168, 171, 181, 183, 188, 191, 208, 225, 234, 249, 253, 267, 273, 277, 286, 296, 298, 305, 315, 323, 334, 338, 345, 369, 381, 387, 435, 448, 459, 484, 501, 520, 534, 539, 541, 544, 554, 581, *T19*
See also Professional Handbook, Howard Johnson
Technology Correlations, 14D, 92D, 144D, 226D, 268D, 316D, 382D, 460D, 532D
Technology Links
Astro Algebra, 14E, 28B, 30B, 40B, 43, 44B, 144E, 168, 169B, 226E, 230B, 234B, 243, 244B, 249, 250B, 252B, 268E, 270B, 272B, 278B, 284B, 287B, 290B, 296, 297B, 305, 382E, 384B, 387, 406B, 412B, 418B, 532E, 568B, 570B
Calculating Crew, 14E, 16B, 22B, 52B, 66B, 70B, 144E, 164B, 169B, 182B, 186B, 192B, 202B, 316E, 350B, 353B, 357
Cosmic Geometry, 316E, 321, 322B, 326B, 332B, 338B, 344B, 372B, 382E, 390B, 397B, 460E, 479B, 484B, 494B, 498B, 502B, 504B, 521B, 532E, 550B, 553B
Data ToolKit, 92E, 100B, 102B, 120B, 126B, 130B, 382E, 406B, 460E, 466B, 512B
E-Lab, 14E, 43, 75, 92E, 129, 144E, 160B, 168, 181, 191, 209, 226E, 249, 268E, 277, 286, 296, 305, 316E, 353B, 382E, 387, 397B, 417, 435, 460E, 501, 520, 532E, 539B, 543B, 580B
Harcourt Learning Site, 46B, 132B, 154B, 250B, 306B, 358B, 450B, 504B, 543B
Harcourt Math Newsroom Video, 16B, 94B, 176B, 272B, 338B, 397B, 484B, 539B
Intervention Strategies and Activities, 14E, 16B, 20B, 22B, 26B, 28B, 30B, 36B, 40B, 43, 44B, 46B, 52B, 56B, 58B, 60B, 66B, 70B, 75, 76B, 80B, 82B, 92E, 94B, 98B, 100B, 102B, 106B, 109B, 112B, 120B,

124B, 126B, 129, 130B, 132B, 144E, 146B, 148B, 150B, 154B, 160B, 164B, 166B, 168, 169B, 176B, 181, 182B, 186B, 191, 192B, 194B, 200B, 202B, 206B, 209, 210B, 214B, 216B, 226E, 228B, 230B, 234B, 236B, 243, 244B, 249, 250B, 252B, 256B, 268E, 270B, 272B, 277, 278B, 284B, 286, 287B, 290B, 296, 297B, 300B, 305, 306B, 316E, 318B, 321, 322B, 326B, 332B, 336B, 338B, 342B, 344B, 350B, 353B, 357, 358B, 364B, 368B, 371, 372B, 382E, 384B, 387, 388B, 390B, 394B, 397B, 400B, 406B, 408B, 412B, 417, 418B, 422B, 428B, 432B, 435, 436B, 442B, 444B, 447B, 450B, 460E, 462B, 464B, 466B, 468B, 472B, 478, 479B, 482B, 484B, 494B, 498B, 501, 502B, 504B, 512B, 516B, 518B, 520, 521B, 532E, 534B, 536B, 539B, 543B, 550B, 553B, 556B, 558B, 560B, 568B, 570B, 574B, 576B, 579, 580B

Number Heroes, 14E, 20B, 22B, 76B, 144E, 160B, 168, 182B, 202B, 316E, 318B, 332B, 338B, 382E, 428B, 444B, 447B, 450B, 494B, 532E, 543B, 553B, 560B

See also Technology

Temperatures
    Celsius, 301–303
    Fahrenheit, 301–303

Terminating decimals, 169–171, H39

Terms, 273–275

Tessellations, 553–557, H60

Test Prep. *See* Mixed Review and Test Prep

Test-taking skills. *See* Tips for Taking Math Tests

Theoretical probability
    of compound events, 447–451, H54
    of simple events, 428–431, H53

Thinker's Corner
    Algebra, 163, 497, 542
    Domino Fractions, 135
    Equation Relation, 289
    Explore Combinations, 446
    Figure It Out, 335
    Geometric Geography, 367
    Geometric Probability, 431
    Math Fun, 39, 247
    Math Match, 411
    Reasoning, 79, 507
    Spin a Decimal, 55
    What's the Unit, 471
    See also Problem Solving

Three-dimensional figures. *See* Solid figures

Time, 124–125, 221, 300–303, 462–463

Tips, 412–415

Tips for Taking Math Tests, 33, 49, 63, 85, 117, 137, 157, 173, 197, 219, 239, 261, 281, 293, 309, 329, 347, 361, 375, 403, 425, 439, 453, 475, 491, 509, 525, 547, 565, 585, H62–H65

Tons, 462–463, H25

Too Much or Too Little Information, 432–433

Transformations
    reflections (flips), 550–552, 558–559, 580–583, H30, H61
    rotations (turns), 550–552, 558–559, 580–583, H30, H61
    on the coordinate plane, 580–583, H61
    stretching, 587
    tessellations and, 553–557, H60
    translations (slides), 550–552, 558–559, 580–583, H30, H61

Transitive Property, 311

Translations, 550–552, 558–559, 580–583, H30, H61

Trapezoids, 338–341, H49
    area of, 499–500
    height of, 499–500

Tree diagram, 444–446, H54

Triangles
    acute, 332–335, H49
    angles in, 332–335, H49
    area of, 495–497, H28, H57
    degrees in, 332–335, H49
    equilateral, 332
    as faces of solid figure, 350–357
    height of, 495–497
    isosceles, 332, 342
    obtuse, 332–335, H49
    right, 332–335, 448–449, H49
    scalene, 332

Triangular numbers, 536

Triangular prisms, 350–352

Triangular pyramids, 350–352

Troubleshooting, H2–H31

Turns, H30 *See* Rotations

Two-dimensional figures. *See* Plane figures

Two-step equations, 304–305

Underestimates, 17–19

Unit at a Glance, *14A, 92A, 144A, 226A, 268A, 316A, 382A, 460A, 532A*

Unit prices. *See* Unit rates

Unit rates, 385–386, H51

Units, converting, 462–467

Universal Access
    Advanced Learners, 14F, 20B, 22B, 26B, 44B, 70B, 76B, 80B, 92F, 94B, 106B, 109B, 126B, 130B, 144F, 146B, 164B, 182B, 192B, 206B, 216B, 226B, 250B, 268F, 272B, 290B, 306B, 316F, 322B, 332B, 344B, 353B, 358B, 372B, 382F, 394B, 416, 418B, 422B, 436B, 442B, 460F, 462B, 479B, 489, 502B, 516B, 521B, 532F, 534B, 553B, 560B, 574B, 579, 580B
    Alternative Teaching Strategy, 16B, 18, 20B, 22B, 24, 26B, 28B, 30B, 36B, 38, 40B, 44B, 46B, 52B, 54, 56B, 58B, 60B, 66B, 68, 70B, 72, 76B, 78, 80B, 82B, 87, 94B, 96, 98B, 100B, 102B, 104, 106B, 109B, 112B, 114, 120B, 122, 124B, 126B, 130B, 132B, 134, 146B, 148B, 150B, 152, 154B, 160B, 162, 164B, 166B, 169B, 176B, 178, 182B,

184, 186B, 188, 192B, 194B, 200B, 202B, 204, 206B, 208, 210B, 212, 214B, 216B, 228B, 230B, 232, 234B, 236B, 244B, 246, 250B, 252B, 254, 256B, 258, 270B, 272B, 274, 278B, 284B, 287B, 290B, 297B, 300B, 302, 304, 306B, 318B, 322B, 324, 326B, 332B, 334, 336B, 338B, 340, 342B, 344B, 350B, 353B, 358B, 364B, 366, 368B, 372B, 384B, 388B, 390B, 392, 394B, 397B, 400B, 406B, 408B, 410, 412B, 414, 418B, 420, 422B, 428B, 430, 432B, 436B, 442B, 444B, 447B, 450B, 462B, 464B, 466B, 468B, 470, 472B, 479B, 482B, 484B, 486, 494B, 496, 498B, 502B, 504B, 506, 512B, 514, 516B, 518B, 521B, 534B, 536B, 539B, 541, 543B, 550B, 553B, 556B, 558B, 560B, 562, 568B, 570B, 572, 574B, 576B, 580B, 582, 587

Art Connection, 270B, 356, 484B, 518B
Block Scheduling, 14F, 92F, 144F, 226F, 268F, 316F, 382F, 460F, 532F
Career Connection, 16B, 28B, 70B, 112B, 130B, 154B, 166B, 206B, 412B, 504B, 550B, 570B
Early Finishers, 20B, 28B, 46B, 56B, 58B, 66B, 76B, 100B, 102B, 132B, 150B, 160B, 180, 194B, 200B, 214B, 221, 243, 252B, 278B, 304, 326B, 336B, 357, 364B, 377, 388B, 406B, 408B, 412B, 432B, 447B, 464B, 472B, 478, 494B, 512B, 536B, 539B, 556B, 570B
English Language Learners, 14G, 30B, 40B, 56B, 60B, 80B, 82B, 92G, 98B, 100B, 102B, 120B, 144G, 148B, 160B, 169B, 194B, 200B, 210B, 214B, 226G, 228B, 234B, 244B, 268G, 270B, 284B, 297B, 316G, 321, 326B, 338B, 353B, 368B, 382G, 384B, 390B, 400B, 408B, 434, 447B, 454, 460G, 468B, 479B, 484B, 494B, 521B, 532G, 543B, 558B, 580B
Fine Arts Connection, 310, 318B, 466B, 472B, 482B
Health Connection, 50, 174
History Connection, 154, 166, 440
Interdisciplinary Suggestions, 14F, 92F, 144F, 226F, 268F, 316F, 382F, 460F, 532F
Language Arts Connection, 176B, 342B, 390B, 558B
Literature Connection, 14G, 86, 92G, 144G, 226G, 268G, 316G, 382G, 460G, 532G
Math Connection, 526, 586
    Algebra, 42, 150B, 210B, 488
    Geometry, 230B, 276, 536B
    Measurement, 190
Physical Education Connection, 139, 166B, 278B, 450B
Reading Strategy, 26B, 46B, 80B, 100B, 214B, 236B, 306B, 336B, 358B, 388B
Science Connection, 22B, 44B, 82B, 186B, 263, 284B, 300B, 368B, 384B, 397B, 428B, 444B, 468B, 534B, 574B
Social Studies Connection, 26B, 60B, 66B, 98B, 124B, 297B, 338B, 350B, 368B, 400B, 416, 504B
Special Needs, 14F, 30B, 36B, 52B, 74, 92F, 109B, 120B, 144F, 146B, 154B, 169B, 182B, 190, 192B, 202B, 220, 226F, 236B, 242, 248, 252B, 256B, 268F, 276, 290B, 306B, 316F, 320, 322B, 342B, 356, 358B, 364B, 382F, 394B, 397B, 406B, 436B, 442B, 460F, 464B, 488, 502B, 512B, 532F, 539B, 553B, 578
Technology, 46B, 52B
Vocabulary Strategy, 36B, 94B, 228B, 320, 336B, 344B, 376, 428B, 444B, 472B, 518B, 560B, 568B
Writing in Mathematics, 16B, 40B, 58B, 74, 106B, 112B, 124B, 126B, 132B, 148B, 164B, 176B, 180, 186B, 202B, 208, 216B, 230B, 250B, 256B, 272B, 287B, 300B, 318B, 332B, 350B, 372B, 388B, 418B, 422B, 432B, 450B, 462B, 472B, 498B, 516B, 543B, 556B, 576B, 578

Unlike fractions, 180–181
Unlikely outcomes, 429, H24
Upper quartile, 129–131, H37
Use Logical Reasoning strategy, 11, 236–237
Using Data, 14, 34, 50, 64, 92, 118, 144, 158, 174, 198, 226, 240, 268, 282, 294, 316, 330, 348, 362, 382, 404, 426, 440, 460, 476, 510, 532, 548, 566 See also Data

Variables, 28–29, 270–275, 284–291, 296–305
Venn diagrams, 230, 233, 372
Vertical angles, 322–325, 333, H48
Vertices, 358–359, H20
Visual Arts Connection, 498B, 550B
Vocabulary, 16A, 16, 28A, 28, 30A, 30, 36A, 36, 40A, 40, 42, 60A, 60, 94A, 94, 98A, 98, 102A, 102, 106, 109A, 109, 120A, 120, 126A, 126, 129, 146, 148A, 148, 150A, 150, 160A, 160, 164A, 164, 169A, 169, 176, 182A, 182, 186, 208, 228A, 228, 230A, 230, 242, 244, 272A, 272, 276, 284, 287A, 287, 290A, 290, 297A, 297, 318A, 318, 320, 322A, 322, 326A, 326, 332A, 332, 336A, 336, 342, 344A, 344, 350A, 350, 356, 364A, 364, 368A, 368, 372A, 372, 384A, 384, 387, 388, 390A, 390, 394A, 394, 397A, 397, 408, 418A, 418, 422A, 422, 428A, 428, 436A, 436, 444A, 444, 447A, 447, 450, 464, 494A, 494, 504A, 504, 512A, 512, 536A, 536, 539A, 539, 543A, 543, 550A, 550, 553A, 553, 558, 560A, 560, 568A, 568, 570A, 570, 580
Vocabulary Cards, 14I, 34B, 50B, 92I, 118B, 144I, 158B, 174B, 198B, 226I, 240B, 268I, 282B, 294B, 316I, 330B, 348B, 362B, 382I, 404B, 426B, 440B, 476B, 492B, 510B, 532B, 548B, 566B
Vocabulary Development, 14I, 34B, 50B, 64B, 92I, 118B, 144I, 158B, 174B, 198B, 226I, 240B, 254B, 268I, 282B, 294B, 316I, 330B, 348B, 362B, 382I, 404B, 426B, 440B, 460I, 476B, 492B, 510B, 532B, 548B, 566B
Vocabulary Strategy, 36B, 94B, 228B, 320, 336B, 344B, 376, 428B, 444B, 472B, 518B, 560B, 568B
Volume
    and changing dimensions, 516–517
    cubic units, 512
    of cylinders, 520–523, H58
    estimating, 512
    of prisms, 512–515, H58
    of pyramids, 518–519, H58

Warm-Up Resources. See Number of the Day, Problem of the Day, and Daily Facts Practice
Weight, 462–463, H25, H55

Index **BI37**

What If?, 26, 38, 56, 80, 100, 107, 113, 193, 201, 203, 236, 271, 289, 291, 306, 388, 442, 445, 482, 513, 514, 515, 516, 521, 582

What's the Error?, 25, 45, 55, 79, 83, 108, 115, 123, 153, 165, 167, 189, 213, 217, 233, 247, 255, 275, 291, 299, 327, 341, 343, 389, 396, 423, 463, 487, 523, 535, 563, 569

What's the Question?, 27, 31, 41, 57, 81, 105, 131, 155, 163, 185, 205, 237, 251, 255, 279, 289, 307, 325, 335, 370, 386, 411, 431, 449, 465, 483, 507, 517, 577, 583

Whole numbers
comparing, H3
ordering, H3
place value of, H3
rounding, 16, H4
See also Addition, Division, Multiplication, *and* Subtraction

Why Learn Math?, 14, 34, 50, 64, 92, 118, 144, 158, 174, 198, 226, 240, 268, 282, 294, 316, 330, 348, 382, 404, 426, 440, 460, 476, 510, 532, 548, 566

Why Learn This?, 16, 20, 22, 26, 28, 30, 36, 40, 44, 46, 52, 56, 58, 60, 66, 70, 76, 80, 82, 94, 98, 100, 102, 106, 109, 112, 120, 124, 126, 130, 132, 146, 148, 150, 154, 160, 164, 166, 169, 176, 182, 186, 192, 200, 202, 206, 210, 214, 216, 228, 230, 234, 236, 244, 250, 270, 272, 278, 284, 287, 290, 297, 300, 306, 318, 322, 326, 332, 336, 338, 342, 344, 350, 353, 358, 384, 388, 390, 394, 397, 400, 406, 408, 412, 418, 422, 428, 432, 436, 442, 444, 447, 450, 462, 464, 466, 468, 472, 479, 484, 494, 498, 502, 504, 512, 516, 518, 521, 534, 536, 539, 543, 550, 553, 556, 558, 560, 568, 570, 574, 576, 580

Word expressions, translating, 270–271, 284–285, H17, H46

Word form, 52–55

Work Backward strategy, 306–307

World Wide Web. *See* E-Lab activities

Write. See last page of each lesson

Write a Problem, 21, 59, 73, 101, 125, 163, 179, 195, 229, 271, 285, 303, 337, 407, 433, 443, 451, 467, 473, 481, 523, 538, 545, 575

Write About It, 19, 29, 39, 47, 61, 69, 86, 97, 99, 128, 138, 149, 165, 171, 193, 201, 215, 220, 235, 310, 319, 327, 355, 367, 373, 376, 393, 399, 401, 437, 446, 454, 471, 519, 526, 555, 559, 573, 586

Write an Equation strategy, 10, 388–389

Write What You Know, 33, 49, 63, 85, 117, 137, 157, 173, 197, 219, 239, 261, 281, 293, 309, 329, 347, 361, 375, 403, 425, 439, 453, 475, 491, 509, 525, 547, 565, 585

Writing equations, 82–83, 284–285, 388–389, 540–542, 574–577, H46, H59

Writing in math. *See* What's the Question?, Write About It, Write a Problem, *and* Write What You Know

Writing in Mathematics, 16B, 40B, 58B, 74, 106B, 112B, 124B, 126B, 132B, 148B, 164B, 176B, 180, 186B, 202B, 208, 216B, 230B, 250B, 272B, 287B, 300B, 318B, 332B, 350B, 372B, 388B, 418B, 422B, 432B, 450B, 462B, 472B, 498B, 516B, 543B, 556B, 576B, 578

Writing Opportunities, 14I, 34B, 50B, 64B, 92I, 118B, 144I, 158B, 174B, 198B, 226I, 240B, 254B, 256B, 268I, 282B, 294B, 316I, 330B, 348B, 362B, 382I, 404B, 426B, 440B, 460I, 476B, 492B, 510B, 532I, 548B, 566B

X-axis, 570
X-coordinate, 570

Yards, 462–463, H25, H55
Y-axis, 570
Y-coordinate, 570
Year, 462–463

Zero properties

HARCOURT Math

NCTM Standards Correlations

This section contains the following:

▶ **Correlations to the NCTM Standards and Expectations**

These correlations demonstrate how *Harcourt Math* supports the NCTM Content Standards. The chart indicates the lessons and chapters that correlate to each Standard for the appropriate grade span.

Correlations to NCTM Standards

STANDARD	GRADES 6–8 EXPECTATIONS	CORRELATION TO NCTM STANDARDS
Instructional programs from prekindergarten through grade 12 should enable all students to—	*In grades 6–8 all students should—*	
1. NUMBERS AND OPERATIONS		
Understand numbers, ways of representing numbers, relationships among numbers, and number systems	• work flexibly with fractions, decimals, and percents to solve problems; • compare and order fractions, decimals, and percents efficiently and find their approximate locations on a number line; • develop meaning for percents greater than 100 and less than 1; • understand and use ratios and proportions to represent quantitative relationships; • develop an understanding of large numbers and recognize and appropriately use exponential, scientific, and calculator notation; • use factors, multiples, prime factorization, and relatively prime numbers to solve problems; • develop meaning for integers and represent and compare quantities with them.	Lessons 1.1, 1.2, 1.3, 1.4, 1.5, 1.6, 2.1, 2.2, 2.3, 2.4, 2.5, 3.1, 3.2, 3.3, 3.4, 4.1, 4.2, 4.3, 4.4, 4.5, 4.6, 7.1, 7.2, 7.3, 7.4, 8.1, 8.2, 8.3, 8.4, 8.5, 9.1, 9.2, 9.3, 9.4, 9.5, 9.6, 9.7, 10.1, 10.2, 10.3, 10.4, 10.5, 10.6, 10.7, 11.1, 11.2, 11.3, 11.4, 12.1, 12.2, 12.3, 12.4, 12.5, 12.6, 13.1, 13.2, 13.3, 13.4, 14.1, 14.2, 14.3, 14.4, 15.1, 15.2, 15.3, 15.4, 15.5, 17.1, 17.2, 17.3, 17.4, 17.5, 20.1, 20.2, 20.3, 20.4, 20.5, 20.6, 20.7, 21.1, 21.2, 21.3, 21.4, 21.5, 21.6, 22.1, 22.2, 22.3, 22.4, 23.3, 23.4, 24.1, 24.2, 24.3, 24.4, 24.5, 25.1, 25.2, 25.4, 25.5, 26.1, 26.2, 26.3, 26.4, 26.5 27.1, 27.2, 27.3, 27.5, 28.1, 28.2, 28.3, 30.1
Understand meanings of operations and how they relate to one another	• understand the meaning and effects of arithmetic operations with fractions, decimals, and integers; • use the associative and commutative properties of addition and multiplication and the distributive property of multiplication over addition to simplify computations with integers, fractions, and decimals; • understand and use the inverse relationships of addition and subtraction, multiplication and division, and squaring and finding square roots to simplify computations and solve problems.	
Compute fluently and make reasonable estimates	• select appropriate methods and tools for computing with fractions and decimals from among mental computation, estimation, calculators or computers, and paper and pencil, depending on the situation, and apply the selected methods; • develop and analyze algorithms for computing with fractions, decimals, and integers and develop fluency in their use; • develop and use strategies to estimate the results of rational-number computations and judge the reasonableness of the results; • develop, analyze, and explain methods for solving problems involving proportions, such as scaling and finding equivalent ratios.	

STANDARD	GRADES 6–8 EXPECTATIONS	CORRELATION TO NCTM STANDARDS
Instructional programs from prekindergarten through grade 12 should enable all students to—	In grades 6–8 all students should—	
2. ALGEBRA		
Understand patterns, relations, and functions	• represent, analyze, and generalize a variety of patterns with tables, graphs, words, and when possible, symbolic rules; • relate and compare different forms of representation for a relationship; • identify functions as linear or nonlinear and contrast their properties from tables, graphs, or equations.	Lessons 1.3, 1.5, 1.6, 2.2, 2.3, 2.4, 4.1, 4.6, 5.2, 6.2, 10.4, 10.5, 10.7, 11.1, 11.2, 11.3, 12.1, 12.2, 12.3, 12.4, 12.5, 12.6, 13.1, 13.2, 13.4, 14.1, 14.2, 14.3, 14.4, 15.1, 15.2, 15.3, 15.4, 15.5, 17.1, 18.4, 20.1, 20.2, 20.3, 20.4, 20.5, 20.6, 20.7, 21.1, 21.3, 21.4, 21.5, 23.4, 24.1, 24.2, 24.3, 25.2, 25.3, 25.4, 25.5, 27.1, 27.3, 27.5, 26.1, 26.2, 26.3, 26.4, 26.5, 28.1, 28.2, 28.3, 30.1, 30.2, 30.3, 30.4, 30.5, 30.6
Represent and analyze mathematical situations and structures using algebraic symbols	• develop an initial conceptual understanding of different uses of variables; • explore relationships between symbolic expressions and graphs of lines, paying particular attention to the meaning of intercept and slope; • use symbolic algebra to represent situations and to solve problems, especially those that involve linear relationships; • recognize and generate equivalent forms for simple algebraic expressions and solve linear equations.	
Use mathematical models to represent and understand quantitative relationships	• model and solve contextualized problems using various representations, such as graphs, tables, and equations.	
Analyze change in various contexts	• use graphs to analyze the nature of changes in quantities in linear relationships.	

STANDARD	GRADES 6–8 EXPECTATIONS	CORRELATION TO NCTM STANDARDS
Instructional programs from prekindergarten through grade 12 should enable all students to—	*In grades 6–8 all students should—*	
3. GEOMETRY		
Analyze characteristics and properties of two- and three-dimensional geometric shapes and develop mathematical arguments about geometric relationships	• precisely describe, classify, and understand relationships among types of two- and three-dimensional objects using their defining properties; • understand relationships among the angles, side lengths, perimeters, areas, and volumes of similar objects; • create and critique inductive and deductive arguments concerning geometric ideas and relationships, such as congruence, similarity, and the Pythagorean relationship.	Lessons 16.1, 16.2, 16.3, 16.4, 17.1, 17.2, 17.3, 17.4, 17.5, 18.1, 18.2, 18.3, 18.4, 19.1, 19.2, 19.3, 19.4, 20.4, 20.5, 20.6, 25.2, 25.4, 25.5, 26.1, 26.2, 26.3, 26.4, 26.5, 27.1, 27.3, 28.2, 28.4, 29.1, 29.2, 29.3, 29.4, 29.5, 30.2, 30.6
Specify locations and describe spatial relationships using coordinate geometry and other representational systems	• use coordinate geometry to represent and examine the properties of geometric shapes; • use coordinate geometry to examine special geometric shapes, such as regular polygons or those with pairs of parallel or perpendicular sides.	
Apply transformations and use symmetry to analyze mathematical situations	• describe sizes, positions, and orientations of shapes under informal transformations such as flips, turns, slides, and scaling; • examine the congruence, similarity, and line or rotational symmetry of objects using transformations.	
Use visualization, spatial reasoning, and geometric modeling to solve problems	• draw geometric objects with specified properties, such as side lengths or angle measures; • use two-dimensional representations of three-dimensional objects to visualize and solve problems such as those involving surface area and volume; • use visual tools such as networks to represent and solve problems; • use geometric models to represent and explain numerical and algebraic relationships; • recognize and apply geometric ideas and relationships in areas outside the mathematics classroom, such as art, science, and everyday life.	

Correlations to NCTM Standards

STANDARD	GRADES 6–8 EXPECTATIONS	CORRELATION TO NCTM STANDARDS
Instructional programs from prekindergarten through grade 12 should enable all students to—	*In grades 6–8 all students should—*	
4. MEASUREMENT		
Understand measurable attributes of objects and the units, systems, and processes of measurement	• understand both metric and customary systems of measurement; • understand relationships among units and convert from one unit to another within the same system; • understand, select, and use units of appropriate size and type to measure angles, perimeter, area, surface area, and volume.	Lessons 9.7, 15.3, 16.2, 16.3, 17.5, 19.1, 19.2, 20.4, 20.5, 20.6, 20.7, 21.6, 24.1, 24.2, 24.3, 24.4, 24.5, 25.1, 25.2, 25.4, 25.5, 26.1, 26.2, 26.3, 26.4, 26.5, 27.1, 27.3, 27.4, 27.5, 29.3, 30.1
Apply appropriate techniques, tools, and formulas to determine measurements	• use common benchmarks to select appropriate methods for estimating measurements; • select and apply techniques and tools to accurately find length, area, volume, and angle measures to appropriate levels of precision; • develop and use formulas to determine the circumference of circles and the area of triangles, parallelograms, trapezoids, and circles and develop strategies to find the area of more-complex shapes; • develop strategies to determine the surface area and volume of selected prisms, pyramids, and cylinders; • solve problems involving scale factors, using ratio and proportion; • solve simple problems involving rates and derived measurements for such attributes as velocity and density.	

STANDARD	GRADES 6–8 EXPECTATIONS	CORRELATION TO NCTM STANDARDS
Instructional programs from prekindergarten through grade 12 should enable all students to—	*In grades 6–8 all students should—*	
5. DATA ANALYSIS AND PROBABILITY		
Formulate questions that can be addressed with data and collect, organize, and display relevant data to answer them	• formulate questions, design studies, and collect data about a characteristic shared by two populations or different characteristics within one population; • select, create, and use appropriate graphical representations of data, including histograms, box plots, and scatterplots.	Lessons 1.1, 1.2, 1.3, 5.1, 5.2, 5.3, 5.4, 5.5, 5.6, 6.1, 6.2, 6.3, 6.4, 6.5, 6.6, 8.1, 8.2, 8.5, 9.1, 9.4, 9.6, 14.4, 15.3, 18.4, 20.3, 21.4, 22.1, 22.2, 22.3, 22.4, 23.2, 23.3, 23.4, 24.4, 24.5, 29.3 30.5
Select and use appropriate statistical methods to analyze data	• find, use, and interpret measures of center and spread, including mean and interquartile range; • discuss and understand the correspondence between data sets and their graphical representations, especially histograms, stem-and-leaf plots, box plots, and scatterplots.	
Develop and evaluate inferences and predictions that are based on data	• use observations about differences between two or more samples to make conjectures about the populations from which the samples were taken; • make conjectures about possible relationships between two characteristics of a sample on the basis of scatterplots of the data and approximate lines of fit; • use conjectures to formulate new questions and plan new studies to answer them.	
Understand and apply basic concepts of probability	• understand and use appropriate terminology to describe complementary and mutually exclusive events; • use proportionality and basic understanding of probability to make and test conjectures about the results of experiments and simulations; • compute probabilities for simple compound events, using such methods as organized lists, tree diagrams, and area models.	

STANDARD		CORRELATION TO NCTM STANDARDS
6. PROBLEM SOLVING		
Instructional programs from prekindergarten through grade 12 should enable all students to—	• Build new mathematical knowledge through problem solving • Solve problems that arise in mathematics and in other contexts • Apply and adapt a variety of appropriate strategies to solve problems • Monitor and reflect on the process of mathematical problem solving	Lessons 1.1, 1.2, 1.3, 1.4, 1.5, 1.6, 2.1, 2.2, 2.4, 2.5, 3.1, 3.2, 3.3, 3.4, 4.1, 4.2, 4.4, 4.5, 4.6, 5.1, 5.2, 5.3, 5.4, 5.5, 5.6, 6.1, 6.2, 6.3, 6.5, 6.6, 7.1, 7.2, 7.3, 7.4, 8.1, 8.2, 8.3, 8.5, 9.1, 9.3, 9.4, 9.6, 9.7, 10.1, 10.2, 10.3, 10.5, 10.6, 10.7, 11.1, 11.2, 11.3, 11.4, 12.2, 12.4, 12.5, 12.6, 13.1, 13.2, 13.4, 14.1, 14.3, 14.4, 15.2, 15.3, 15.5, 16.1, 16.3, 16.4, 17.1, 17.2, 17.3, 17.4, 17.5, 18.1, 18.2, 18.4, 19.1, 19.2, 19.4, 20.1, 20.3, 20.4, 20.5, 20.6, 20.7, 21.1, 21.2, 21.3, 21.5, 21.6, 22.1, 22.2, 22.4, 23.1, 23.2, 23.3, 23.4, 24.1, 24.2, 24.3, 24.4, 24.5, 25.2, 25.3, 25.4, 26.1, 26.2, 26.4, 26.5, 27.1, 27.2, 27.3, 27.5, 28.1, 28.2, 28.3, 28.4, 29.1, 29.2, 29.3, 29.4, 29.5, 30.1, 30.2, 30.3, 30.4, 30.6
7. REASONING AND PROOF		
Instructional programs from prekindergarten through grade 12 should enable all students to—	• Recognize reasoning and proof as fundamental aspects of mathematics • Make and investigate mathematical conjectures • Develop and evaluate mathematical arguments and proofs • Select and use various types of reasoning and methods of proof	Lessons 1.4, 1.5, 3.2, 5.2, 5.4, 5.6, 6.1, 6.6, 7.1, 7.2, 8.3, 9.3, 10.2, 11.1, 11.4, 12.6, 13.4, 14.3, 14.4, 15.1, 16.3, 17.1, 17.4, 17.5, 18.1, 19.4, 20.1, 20.4, 20.5, 20.7, 21.3, 21.5, 21.6, 22.1, 23.3, 24.2, 24.4, 24.5, 25.1, 25.4, 25.5, 26.1, 26.2, 26.5, 27.1, 28.2, 28.3, 29.1, 29.2, 29.3, 29.4, 29.5, 30.4

STANDARD		CORRELATION TO NCTM STANDARDS
8. COMMUNICATION		
Instructional programs from prekindergarten through grade 12 should enable all students to—	• Organize and consolidate their mathematical thinking through communication • Communicate their mathematical thinking coherently and clearly to peers, teachers, and others • Analyze and evaluate the mathematical thinking and strategies of others • Use the language of mathematics to express mathematical ideas precisely	Lessons 1.1, 1.2, 1.3, 1.5, 1.6, 2.1, 2.2, 2.3, 2.4, 2.5, 3.1, 3.3, 3.4, 4.1, 4.2, 4.3, 4.4, 4.5, 4.6, 5.1, 5.2, 5.4, 5.5, 5.6, 6.1, 6.2, 6.3, 6.4, 6.5, 6.6, 7.1, 7.2, 7.3, 8.1, 8.2, 8.3, 8.4, 8.5, 9.1, 9.2, 9.3, 9.4, 9.5, 9.6, 9.7, 10.1, 10.2, 10.3, 10.4, 10.5, 10.6, 11.1, 11.2, 11.3, 12.1, 12.2, 12.3, 12.4, 12.5, 12.6, 13.1, 13.2, 13.3, 13.4, 14.1, 14.2, 14.3, 14.4, 15.1, 15.2, 15.3, 15.4, 16.1, 16.2, 16.3, 16.4, 17.1, 17.2, 17.3, 17.4, 17.5, 18.1, 18.2, 18.3, 18.4, 19.1, 19.2, 19.3, 19.4, 20.1, 20.2, 20.4, 20.5, 20.6, 20.7, 21.1, 21.2, 21.3, 21.4, 21.5, 21.6, 22.1, 22.3, 22.4, 23.1, 23.2, 23.3, 23.4, 24.1, 24.2, 24.3, 24.4, 25.1, 25.2, 25.4, 25.5, 26.1, 26.2, 26.3, 26.4, 26.5, 27.1, 27.3, 27.4, 27.5, 28.2, 28.3, 28.4, 29.1, 29.2, 29.3, 29.4, 29.5, 30.1, 30.2, 30.3, 30.5, 30.6
9. CONNECTIONS		
Instructional programs from prekindergarten through grade 12 should enable all students to—	• Recognize and use connections among mathematical ideas • Understand how mathematical ideas interconnect and build on one another to produce a coherent whole • Recognize and apply mathematics in contexts outside of mathematics	Lessons 1.3, 2.5, 4.1, 5.1, 5.4, 5.5, 5.6, 7.3, 8.2, 9.3, 9.4, 10.5, 12.6, 13.2, 14.1, 15.3, 16.4, 17.3, 19.1, 19.3, 20.4, 20.5, 20.6, 21.3, 22.4 25.4, 27.1, 29.5, 30.2
10. REPRESENTATION		
Instructional programs from prekindergarten through grade 12 should enable all students to—	• Create and use representations to organize, record, and communicate mathematical ideas • Select, apply, and translate among mathematical representations to solve problems • Use representations to model and interpret physical, social, and mathematical phenomena	Lessons 3.2, 3.4, 4.2, 4.3, 5.3, 5.4, 5.5, 6.1, 6.3, 6.6, 7.4, 8.1, 9.2, 9.3, 9.4, 9.5, 9.7, 10.2, 12.2, 12.3, 12.6, 13.3, 14.2, 15.3, 15.4, 17.5, 19.1, 20.2, 20.3, 20.4, 20.5, 20.6, 20.7, 23.1, 23.2, 24.2, 25.3, 26.2, 26.3, 27.2, 29.1, 29.2, 29.3, 29.4, 29.5, 30.5

Teaching Notes

Additional Ideas:

Good Questions to Ask:

Additional Resources:

Notes for Next Time:

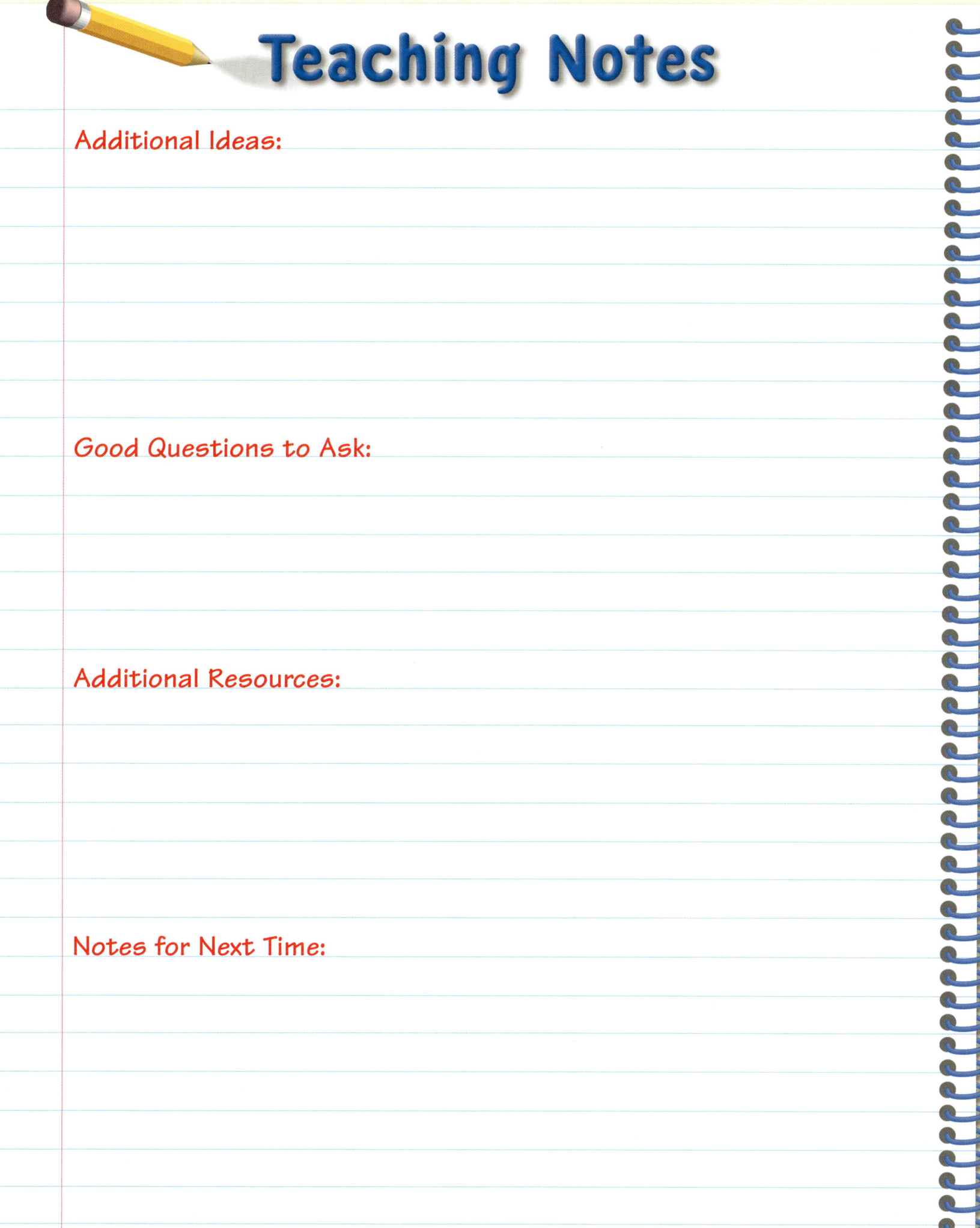

Teaching Notes

Additional Ideas:

Good Questions to Ask:

Additional Resources:

Notes for Next Time:

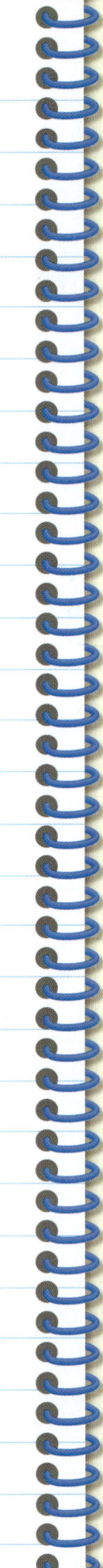